Studies in Natural Products Chemistry

Volume 29
Bioactive Natural Products (Part J)

Studies in Natural Products Chemistry
edited by Atta-ur-Rahman

FOREWORD

Volume 29 of "Studies in natural Products Chemistry" contains a number of articles in frontier areas written by leading experts. The volume contains synthetic approaches, structural studies as well as structure-activity relationships to a number of classes of bioactive compounds.

The article on ecdysteriod structure-activity relationships by Dinan describes recent work on insect steroid hormones with the objective of preparing more potent compounds with lesser side effects. Another article on potentially cancer chemopreventive and anti-inflammatory terpenoids from natural sources by Akihisa, Yusukawa and Tokuda presents studies on natural substances which can be involved in cancer prevention. Suzuki and coworkers have presented their studies on total synthesis of biologically instriguing drimane-type sesquiterpenoids via intramolecular Diel-Alder approaches. Another article by Banerjee and coworkers describes the synthesis of bioactive diterpenes. Microbial sources have, in the past, provided a large number of important biologically active compounds many of which have found their way into medicine. The article by Ishibashi on search for bioactive natural products from unexploited microbial resources presents recent studies in this area. The review by Cali and coworkers on polyhydroxy-p-terphenyls and related p-terphenyloquinones from fungi presents recent work in this field.

Rezanka and Spizek have described work carried out on halogen-containing antibiotics from streptomycetes. Antoher interesting article by Baudoin and Gueritte reviews studies on natural bridged bialys with axial chirality and antimitotic properties. The review by Toyooka and Nemoto describes synthetic studies on biologically active alkaloids starting from lactams.

Carbasugars are an important class of compounds exhibiting a variety of different biological activities. The review by Rassu and coworkers presents the work carried out on chemical synthesis of carbasugars and analogues. Anti-carbohydrate antibodies are produced by immunization of animals with carbohydrate containing antigens. The article by Pazur on biosynthesis and properties of anti-carbohydrate antibodies should prove to be of interest to scientists working in this field.

Plant protease inhibitors have potential for pharmaceutical development especially in relation to Alzheimer's disease, angiogenesis, cancer, inflammatory disease and viral and protozoal infections. The article by Polya in this area presents recent studies in this field. Other articles of interest include those on natural products library for cycloozygenase and lipoxygenase dual inhibitors by Jia, chemistry and biology of lapachol and related natural products by Ravels and coworkers, and chemistry and pharmacology of the genus *Dorstenia* by Ngadjui and Abegaz.

I hope that this volume will be received with the same enthusiasm and interest as the previous volumes of this series.

I would like to express my thanks to Dr. Zareen Amtul for her assistance in the preparation of the index. I am also grateful to Mr. Waseem Ahmad for typing and to Mr. Mahmood Alam for secretarial assistance.

Atta-ur-Rahman
Ph.D. (Cantab.), Sc.D. (Cantab.)
Federal Minister / Chairman
Higher Education Commission
Government of Pakistan

July, 2003

PREFACE

All organisms rely on nature to live. It is no surprise that the official record of herbs, a source for remedial effects, started in China as early as around 2800 BC, two of the most famous being *Panax ginseng* and *Ginkgo biloba*. Herbal knowledge was also preserved in documents by the Egyptians and Indians, while the Assyrians documented 1000 medicinal plants around 1900–400 BC. In contrast, in Africa, Central and South America, where the rain forests are abundant in organisms of interest to our lives due to the rich biodiversity, folk treatments were conveyed from one generation to the next via shamans without written records. Thus studies of natural products provide us with one of the earliest areas in chemistry.

Natural products chemistry began with isolation, followed by purification, painstaking structure determination and synthesis. It underwent dramatic changes with the advent of spectroscopy around the early 1930s and subsequent advancements in isolation techniques as well as synthetic methodologies. Natural products chemistry is now moving towards the interface of dynamic biosciences, including biology, medicine and pharmacology, and is becoming increasingly multidisciplinary. Natural products chemistry is entering a new era now that it is becoming possible to investigate how bioactive compounds interact with their receptors on a molecular structural and time-resolved basis.

The area covered by natural products chemistry is boundless as it deals with nature itself. There is literally no limit to the topics to be dealt with. This volume 29 continues the tradition of supplying us with superb review articles written by experts. The articles in this volume deal with the screening, isolation, structure, synthesis, biosynthesis, and pharmacology of plant and microbial natural products that exhibit antimitotic, cancer chemotherapeutic, enzyme inhibitory, anti-inflammatory, antibiotic and molting hormone activities. The compound types also cover a huge range of natural products, i.e., polyketides, terpenoids, sugars, alkaloids, proteins, and enzymes.

Professor Atta-ur-Rahman once again has succeeded in compiling a precious volume carrying review articles on a wide variety of subjects that we might have missed otherwise. We are most grateful for his efforts in identifying the authors and topics and presenting them in review form to the chemical community.

Professor Koji Nakanishi
Department of Chemistry
Columbia University
3000 Broadway
New York, NY 10027
USA

CONTENTS

CONTRIBUTORS

Berhanu M. Abegaz

Department of Chemistry, Faculty of Science, University of Botswana, Private Bag, 0022, Gaborone, Botswana

Toshihiro Akihisa

College of Science and Technology, Nihon University, 1-8 Kanda Surugadai, Chiyoda-ku, Tokyo 101-8308, Japan

Luciana Auzzas

CNR, Istituto di Chimica Biomolecolare – Sezione di Sassari, I-07100 Sassari, Italy

Ajoy K. Banerjee

Centro de Química, Instituto Venezolano de Investigaciones Científicas (IVIC), Apartado 21827, Caracas 1020-A, Venezuela

Lucia Battistini

Università degli Studi di Parma, Dipartimento Farmaceutico, I-43100 Parma, Italy

Olivier Baudoin

Institut de Chimie des Substances Naturelles, CNRS, Avenue de la Terrasse, 91198 Gif-sur-Yvette Cedex, France

Valeria Calì

Dipartimento di Scienze Chimiche, Viale A. Doria 6, I-95125 Catania, Italy

Claudio Curti

Università degli Studi di Parma, Dipartimento Farmaceutico, I-43100 Parma, Italy

Laurence Dinan

Department of Biological Sciences, University of Exeter, Hatherly Laboratories, Prince of Wales Road, Exeter, Devon, EX4 4PS, UK.

Ana Estévez-Braun

Instituto Universitario de Bio-Orgánica "Antonio González", Instituto Canario de Investigación del Cáncer. Avda. Astrofisico Francisco Sánchez N°2, 38206, Universidad de La Laguna, La Laguna, Tenerife, Spain

Françoise Guéritte

Institut de Chimie des Substances Naturelles, CNRS, Avenue de la Terrasse, 91198 Gif-sur-Yvette Cedex, France

Masami Ishibashi

Graduate School of Pharmaceutical Sciences, Chiba University, 1-33 Yayoi-cho, Inage-ku, Chiba 263-8522, Japan

Qi Jia — Department of Natural Products Chemistry, Unigen Pharmaceuticals, Inc. 100 Technology Drive, Broomfield, Colorado 80021, USA

Manuel S. Laya — Centro de Química, Instituto Venezolano de Investigaciones Científicas (IVIC), Apartado 21827, Caracas 1020-A, Venezuela

Hideo Nemoto — Faculty of Pharmaceutical Sciences, Toyama Medical and Pharmaceutical University, Sugitani 2630, Toyama 930-0194, Japan

Bonaventure T. Ngadjui — Department of Organic Chemistry, Faculty of Science, University of Yaounde I BP 812 Yaounde, Cameroon

John H. Pazur — Department of Biochemistry and Molecular Biology, The Pennsylvania State University, University Park, PA 16802, USA

Elisa Pérez-Sacau — Instituto Universitario de Bio-Orgánica "Antonio González", Instituto Canario de Investigación del Cáncer. Avda. Astrofisico Francisco Sánchez N°2, 38206, Universidad de La Laguna, La Laguna, Tenerife, Spain

Luigi Pinna — Università degli Studi di Sassari, Dipartimento di Chimica, I-07100 Sassari, Italy

Gideon M. Polya — Department of Biochemistry, La Trobe University, Bundoora, Melbourne, Victoria 3086, Australia

PO.S. Poon NG — Centro de Química, Instituto Venezolano de Investigaciones Científicas (IVIC), Apartado 21827, Caracas 1020-A, Venezuela

Gloria Rassu — CNR, Istituto di Chimica Biomolecolare – Sezione di Sassari, I-07100 Sassari, Italy

Ángel G. Ravelo — Instituto Universitario de Bio-Orgánica "Antonio González", Instituto Canario de Investigación del Cáncer. Avda. Astrofisico Francisco Sánchez N°2, 38206, Universidad de La Laguna, La Laguna, Tenerife, Spain

T. Řezanka — Institute of Microbiology, Academy of Sciences of the Czech Republic, Vídeňská, 1083, 142 20, Prague 4, Czech Republic

Carmela Spatafora — Dipartimento di Scienze Chimiche, Viale A. Doria 6, I-95125 Catania, Italy

J. Spížek	Institute of Microbiology, Academy of Sciences of the Czech Republic, Vídeňská, 1083, 142 20, Prague 4, Czech Republic
Yoshikazu Suzuki	Department of Applied Chemistry, Keio University, Hiyoshi, Kohoku-ku, Yokohama 223-8522, Japan
Kin-Ichi Tadano	Department of Applied Chemistry, Keio University, Hiyoshi, Kohoku-ku, Yokohama 223-8522, Japan
Ken-Ichi Takao	Department of Applied Chemistry, Keio University, Hiyoshi, Kohoku-ku, Yokohama 223-8522, Japan
Harukuni Tokuda	Department of Biochemistry, Kyoto Prefectural University of Medicine, Hirokoji-agaru, Kawaharacho-dori, Kamigyo-ku, Kyoto 602-0841, Japan
Naoki Toyooka	Faculty of Pharmaceutical Sciences, Toyama Medical and Pharmaceutical University, Sugitani 2630, Toyama 930-0194, Japan
Corrado Tringali	Dipartimento di Scienze Chimiche, Viale A. Doria 6, I-95125 Catania, Italy
Ken Yasukawa	College of Pharmacy, Nihon University, 7-7-1, Narashinodai, Funabashi-shi, Chiba 274-8555, Japan

Bioactive Natural Products

Atta-ur-Rahman (Ed.) *Studies in Natural Products Chemistry, Vol. 29*
© 2003 Elsevier B.V. All rights reserved.

ECDYSTEROID STRUCTURE-ACTIVITY RELATIONSHIPS

LAURENCE DINAN

Department of Biological Sciences, University of Exeter, Hatherly Laboratories, Prince of Wales Road, Exeter, Devon, EX4 4PS, U.K (Tel: +44-1392-264605; Fax: +44-1392-263700; E-mail: L.N.Dinan@ex.ac.uk)

ABSTRACT:
The purposes of this review are i) to provide a compilation of the biological activity data for insect steroid hormones (ecdysteroids) and ii) to summarise the findings concerning their structure-activity relationships (SAR), the contexts within which they are performed and their applications. All ecdysteroid SAR studies have been considerably helped by the availability of ecdysteroid analogues (phytoecdysteroids) in certain plant species, where they occur in much higher concentrations and with greater structural variety than in arthropods. Initial SAR studies, using a variety of bioassay systems were of necessity fragmentary and empirical. However, they led to some very important conclusions about the structural features of ecdysteroids associated with biological activity, which largely still stand today. However, a number of technical advances have revolutionised what is possible in ecdysteroid QSAR since the beginning of the 1990s. Firstly, with the cloning and characterisation of ecdysteroid receptor genes from a number of arthropod species, our understanding of the mode of action of ecdysteroids and their role in signal transduction has greatly increased. Secondly, even minor phytoecdysteroids can be isolated, unambiguously identified and thoroughly purified in amounts suitable for quantiative bioassay. Thirdly, new bioassays have been developed which permit the rapid comparison of the relative potencies of large numbers of analogues. Lastly, the development of powerful computer-based software permits complex QSAR analysis, molecular modelling, activity prediction and analogue design. Recent developments include homology models for the ligand-binding domains of several insect ecdysteroid receptor proteins, Comparative Molecular Field Analysis and 4D-QSAR models for ecdysteroid binding to intracellular receptors. These approaches are generating experimentally testable predictions which are informing synthetic strategies for ecdysteroid analogues. The applications of ecdysteroid QSAR to agriculture and medicine are discussed, especially the design of novel pest control agents and the application of ecdysteroid receptors in gene switch systems for the control of transfected genes in mammalian or plant cells.

INTRODUCTION

Ecdysteroids are the steroid hormones of arthropods and probably of other invertebrate phyla too (see [1] for reviews). In insects, which are the most extensively studied class of the arthropods, ecdysteroids have been implicated in the regulation of moulting, metamorphosis, aspects of reproduction, embryological development and diapause [1]. In Crustacea, ecdysteroids regulate moulting and have been implicated in aspects of reproduction. Ecdysteroids have been detected in almost all invertebrate groups, but a hormonal role has not been identified for them in most of these phyla [2]. In view of the large number of species in the Insecta alone, the ecdysteroids are the most widespread family of steroid hormones in the animal kingdom. Further, analogues (phytoecdysteroids) have been detected in certain plant species, where they generally occur in much higher concentration and greater diversity than ecdysteroids in invertebrates (zooecdysteroids). However, the most common phytoecdysteroid is 20-hydroxyecdysone (Figure (**1**): structure **1**), which is generally accepted to be the endogenous ecdysteroid in insects. The availability of ecdysteroids from plant sources has greatly facilitated endocrinological studies on insects and other invertebrates and has also provided a wonderful natural resource for ecdysteroid structure-activity relationships [3].

The purpose of this review is to summarise the effects, modes of action, structural diversity and biological activity of ecdysteroids and to show how ecdysteroid QSAR and molecular modelling are being applied in agriculture and medicine.

DEFINITION OF AN ECDYSTEROID

The term ecdysteroid was originally defined as "all compounds structurally related to ecdysone" [4]. However, Lafont and Horn [5] differentiated between true ecdysteroids and ecdysteroid-related compounds. True ecdysteroids possess a steroid nucleus with an A/B-*cis*-ring junction and a 14α-hydroxy-7-en-6-one chromophore, irrespective of their biological activity, while ecdysteroid-related compounds do not fulfil all these criteria.

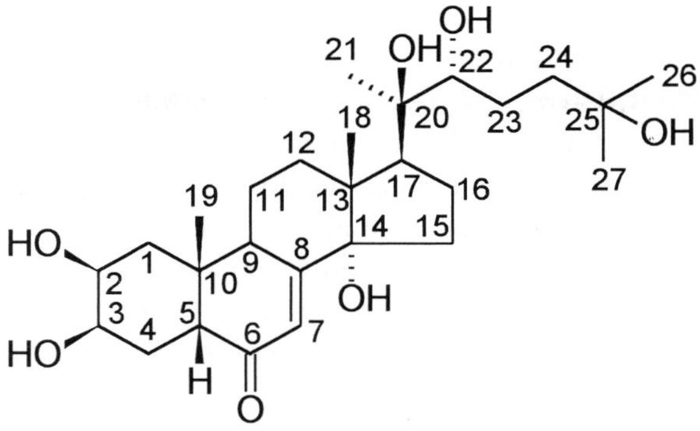

1: 20-hydroxyecdysone (20E)

Figure 1: Structure of 20-hydroxyecdysone showing the conventional steroid numbering.

ROLES OF ECDYSTEROIDS IN ARTHROPODS

A detailed summary of the roles of ecdysteroids during arthropod development and reproduction is beyond the scope of this review. This aspect has been extensively reviewed elsewhere [1,2,6-9].

Insects have been most extensively studied, where ecdysteroids regulate moulting and metamorphosis and have been implicated in the regulation of reproduction, embryonic development and diapause. In larval insects the synthesis of ecdysteroids (E {2-1} and 3DE {2-9}) by the prothoracic gland (although other sites of ecdysteroid production are also possible [10]) is stimulated by prothoracicotropic hormone. These ecdysteroids circulate in the haemolymph and are converted to 20E (**1-1**) in the peripheral tissues (especially the fat-body). Each moult is preceded by a peak of ecdysteroid in the haemolymph, which induces apolysis and subsequent events of moulting, some of which are induced by the declining titre after the peak has been passed. During the last larval instar there is also an earlier, smaller peak of ecdysteroids, which commits tissues to adult (hemimetabolous insects) or pupal (holometabolous insects) differentiation. The biosynthesis and further metabolism of ecdysteroids, even in insects, is not yet fully elucidated. Ecdysteroids are not the only hormonal regulators of moulting and metamorphosis, as they act in concert with juvenile hormones, ecdysis-triggering hormone, eclosion hormone and bursicon.

Ecdysteroid tritres rise again during the adult stage. This has been more extensively studied in female insects, where high titres correlate with ovarian development, but ecdysteroids also appear to be involved in stimulation of spermatogenesis in male adults [11]. The titres, identities and functions of ecdysteroids in adult female insects have been most extensively studied in Orthopteran insects (*Locusta migratoria* [12], *Schistocerca gregaria* [13], *Acheta domesticus* [14], *Gryllus bimaculatus* [15]), but certain lepidopteran insects have also been studied (*Bombyx mori* [16], *Manduca sexta* [17]). The role of ecdysteroids during ovarian maturation in dipteran insects appears to possess differences to the orpthopteran pattern [18].

Ecdysteroid titres also fluctuate during embryogenesis, and evidence exists to correlate ecdysteroid peaks with the formation of various cuticles during embryogenesis [19].

The hormonal regulation of diapause is complex, depending on the stage in the life-cycle when it occurs and the insect species. The reader is referred to a review devoted to this aspect [20].

In decapod crustaceans, the production of ecdysteroids by the Y-organs is inhibited by moult-inhibiting hormone from the eye-stalks [21,22]. The ecdysteroids released are E (**2-1**), 25dE (**2-2**) and 3DE (**2-9**), which are converted to 20E (**1-1**) and poA (**2-7**). Although elevated levels of ecdysteroids are associated with ovarian maturation and eggs of decapod crustaceans, their functions have not yet been elucidated.

Some information exists on the hormonal roles of ecdysteroids in other arthropod classes (arachnids [23], myriapods [24]) and on the presence of ecdysteroids in other invertebrate phyla (helminths and annelids [25]), but further studies are required to elucidate their roles.

ROLES OF ECDYSTEROIDS IN PLANTS

Two main hypotheses have been put forward to account for the occurrence of ecdysteroids in the plant world. The first is that PEs have a hormonal role within the plant, but there is very little hard evidence in support of this hypothesis (reviewed in [26]). Alternatively, PEs participate in the defence of plants against non-adapted phytophagous invertebrates. Deterrent effects of 20E on non-adapted insect species are

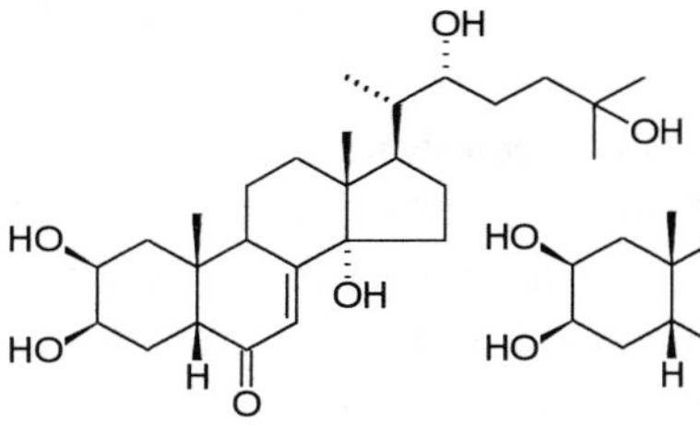

1: ecdysone (E)

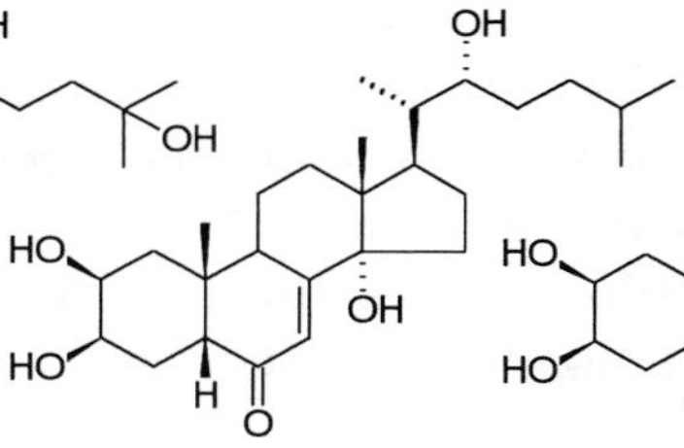

2: 25-deoxyecdysone (25dE)

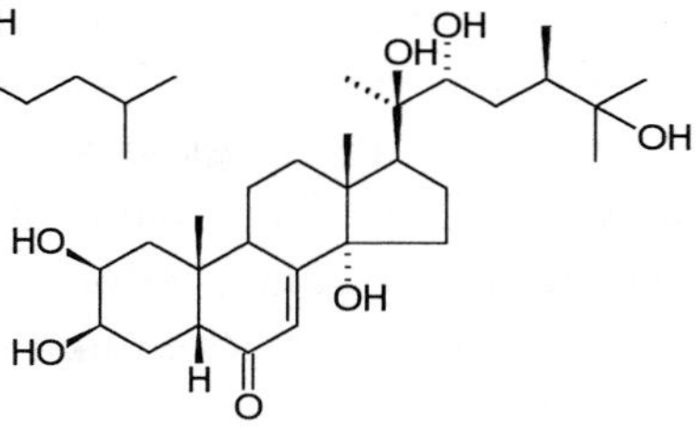

3: makisterone A (MaA)

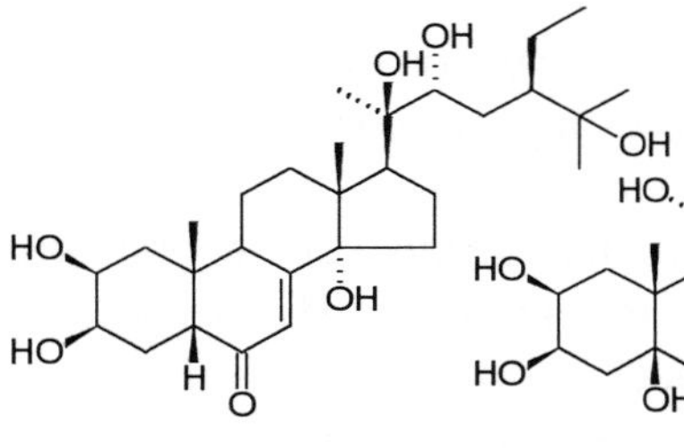

4: makisterone C (MaC)

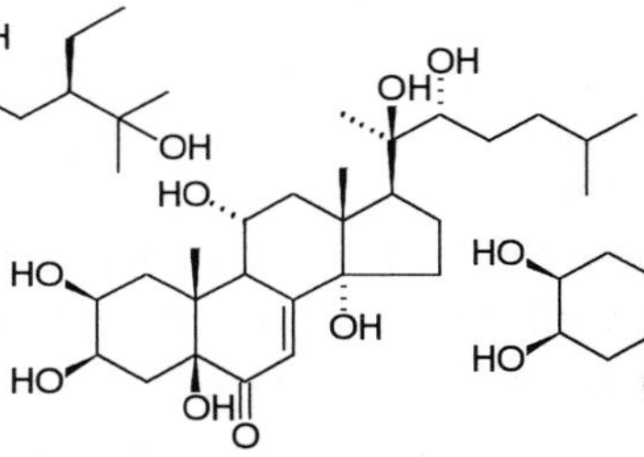

5: muristerone A (MuA)

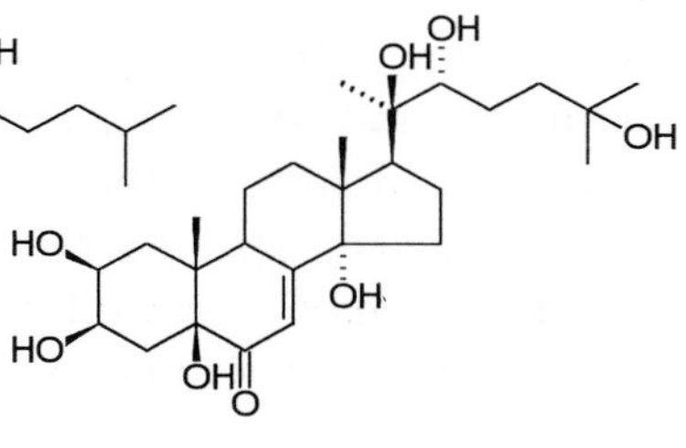

6: polypodine B (PolB)

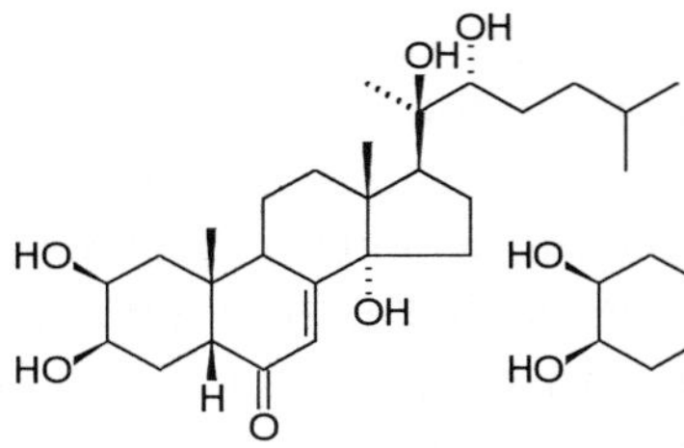

7: ponasterone A (PoA)

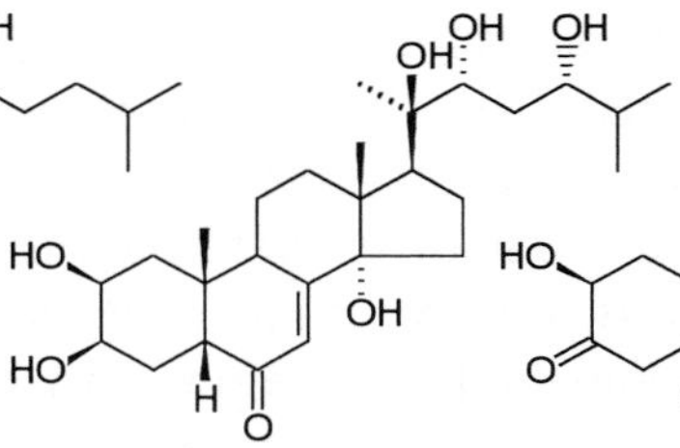

8: pterosterone

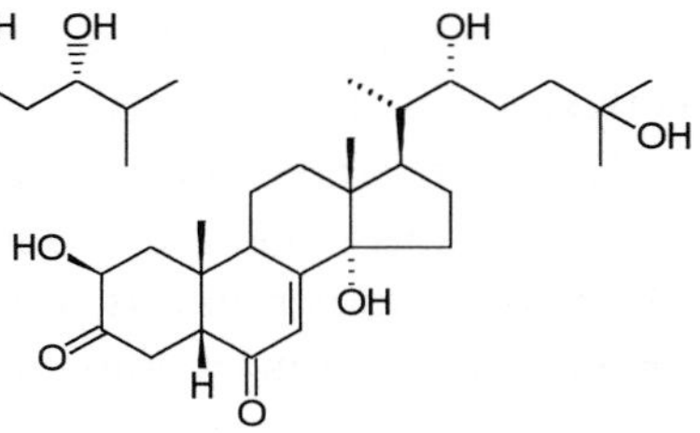

9: 3-dehydroecdysone (3DE)

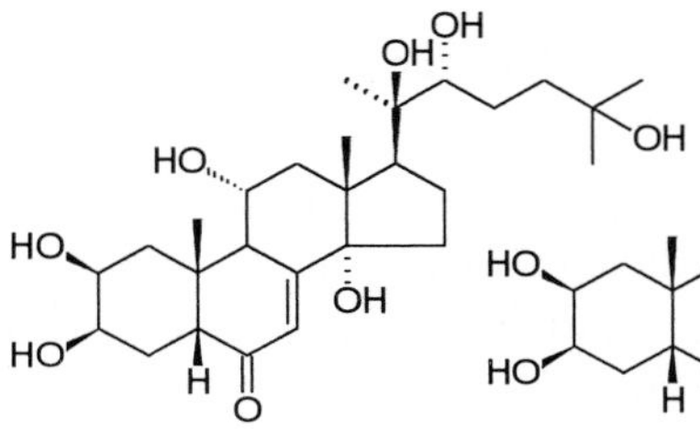

10: turkesterone

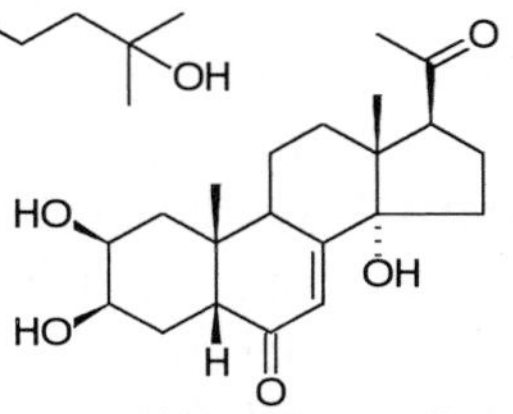

11: poststerone (Pos)

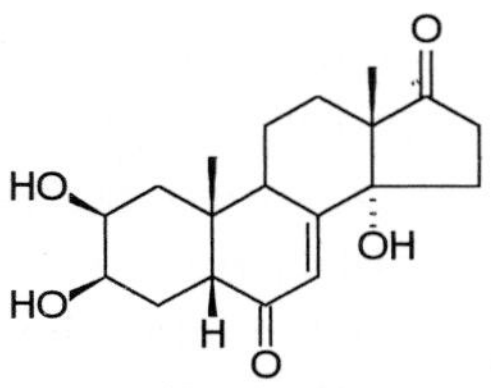

12: rubrosterone

Figure 2: Structures of some common ecdysteroids.

observed at 2 – 25 ppm [27, and references cited therein], which is well within the concentration range found in ecdysteroid-containing plants. Almost all PEs demonstrate quantifiable ecdysteroid receptor agonist activity [28]. They have also been shown to affect insect growth and reproduction when added to natural or artificial diets, which would otherwise support optimum growth and reproduction [28-32]. The activity of a PE depends on its affinity for, and the effective concentration at, the target site (normally assumed to be the intracellular ecdysteroid receptor complex), and, if it is ingested, its potency depends also on the amount ingested, the ability to cross the gut wall and the consequences of the metabolism. PEs may have allelochemical properties as hormonal disruptors and toxins for insects and other invertebrates [30,33-37]. It has also been demonstrated that ecdysteroids possess antifeedant activity against certain species of insects and that some species possess taste receptors which respond to ecdysteroids [38-40]. However, there are several highly polyphagous insect species which develop normally on diets containing very high concentrations (400 ppm) of 20E [41-47]. Hence, PEs can be considered as a chemical defence against certain, but not all, herbivorous insects [5].

20E is generally accompanied in plants by one or two other major ecdysteroids (e.g. polB), together accounting for 90-97% of the total ecdysteroid, and many other minor ecdysteroids. The relative contribution of the various major and minor ecdysteroids to a potential defensive role of PE is a moot point at present. It has been suggested that the minor PEs provide evolutionary flexibility to respond to predation if and when the major compounds are no longer effective; mutation results in plants with altered ecdysteroid profiles and new major ecdysteroids which are effective in deterring the predators [48,49].

Recently, it has been demonstrated that ecdysteroid accumulation in spinach is inducible by mechanical [50] or insect [51] damage to roots. Evidence was further provided for the involvement of jasmonates in the induction of *de novo* ecdysteroid synthesis. Also, short- and long-term labelling of 20E from [2-^{14}C]mevalonic acid in spinach demonstrate that ecdysteroids are metabolically stable in this species, which fits much better with a role in plant defence, rather than a phytohormonal role [52]. Most recently, it has been demonstrated that root predation by the fungus gnat *Bradysia impatiens* results in elevated ecdysteroid levels in spinach and a significant reduction in larval establishment of *B. impatiens* [53].

Definitive assessment of the contribution of PEs to the deterrence of invertebrate predators must await the comparison of the susceptiblity of genetic lines of the same plant species which differ in their ecdysteroid levels and/or profiles [26].

BIOLOGICAL EFFECTS OF ECDYSTEROIDS ON VERTEBRATES

This aspect has been thoroughly reviewed elsewhere [54-56] The widespread occurrence of ecdysteroids in invertebrates and plants suggests that human consumption of these steroidal compounds may not be insignificant. Spinach, for example, contains *ca.* 50 µg/g f.w. [57]. Although no thorough toxicological study appears to have been performed for any ecdysteroid, there is evidence that the acute toxicity of these compounds to mammals is extremely low [58,59]. There is a growing interest in the effects of (phyto)ecdysteroids on vertebrates, especially mammals (and *Homo sapiens* in particular) firstly because of the largely beneficial pharmacological (including anabolic) effects which have been reported and, secondly, because of the development of ecdysteroid-regulated gene switch systems. However, it should be stated at the outset that our knowledge in this area is limited and fragmentary. Very few of the studies concerning pharmacological effects are extensive or fully convincing and very little is known about ecdysteroid metabolism in vertebrates. It appears that ecdysteroids are rapidly cleared from mammals following at least partial metabolism. At least circumstantial evidence has been provided that ecdysteroids can i) stimulate growth [60], ii) enhance physical performance [61], iii) affect cellular proliferation and/or differentiation [62], iv) enhance protein synthesis [63], v) reduce hypoglycaemia [64], vi) enhance the conversion of cholesterol to bile acids (antiatherosclerotic effect [65]). Thus, ecdysteroid-containing preparations have been used as tonics or medicines with roborant, anabolic, antidepressive and adaptogenic effects and to improve nerve function, stimulate hepatic functions, improve heart and lung functions, improve renal function, improve wound healing and affect the immune system. Even a brief survey of the internet can reveal dozens of companies retailing ecdysteroid-containing preparations for sportsmen and body-builders, purporting to have roborant and anabolic effects with far fewer side-effects than other anabolic steroids [56]. However, it must be emphasised that thorough clinical trials are required to substantiate and confirm the pharmacological effects of ecdysteroids. The generally positive pharmacological effects may be compatible with ecdysteroids being used as an elicitor in gene switching systems, but far more extensive studies are required on the pharmacokinetics, metabolism and mode(s) of action of ecdysteroids in mammalian systems.

STRUCTURAL DIVERSITY OF NATURAL ECDYSTEROIDS

Zooecdysteroids

E (**2-1**) and 20E (**1-1**) were the first ecdysteroids to be isolated and identified from pupae of *Bombyx mori* [66,67], with 20E also being isolated from the crayfish *Jasus lalandii* [68]. Since then, approximately 50 ecdysteroid analogues have been identified from arthropods. The structural diversity of these compounds is summarised in Figure (**3**). Zooecdysteroids can be categorised according to their biochemical role: precursors (e.g. 2dE), active molecules (20E), further metabolites and excretory compounds (e.g. 20,26-dihydroxyecdysone, 20-hydroxyecdysonoic acid) and ovarian/embryonic storage compounds (e.g. ecdysteroid phosphates), although this classification is not absolute, as some ecdysteroids (e.g. 3DE {**2-9**}) are regarded as both prohormones and further metabolites. The majority of known zooecdysteroids are C_{27} compounds, but certain insects generate C_{28} (e.g. maA {**2-3**}) or C_{29} (e.g. maC {**2-4**}) compounds, either because they cannot dealkylate the corresponding phytosterols, or because their ecdysteroid receptors can apparently accommodate C_{27}, C_{28} and C_{29} ecdysteroids.

Phytoecdysteroids

The first phytoecdysteroids were identified in the mid-1960s by four research groups almost simultaneously: from the leaves of *Podocarpus nakaii* [69], from the bark of *Podocarpus elatus* [70], from the roots of *Achyranthes fauriei* [71], from the pinnae of Pteridium aquilinum [72] and from the rhizomes of *Polypodium vulgare* [73]. The high concentrations present in these plants prompted the screening [74] of many more plant species and the rapid isolation and identification of many ecdysteroid analogues. Approximately 250 analogues have been identified to date, of which some (20E {**1-1**}, E {**2-1**}, maA {**2-3**), poA {**2-7**} etc) are also found amongst the zooecdysteroids. The structural diversity of phytoecdysteroids has been reviewed recently [3] and is summarised in Figure (**2**). As many permutations and combinations of structural variations occur, it is to be expected that many other phytoecdysteroids have yet to be identified. Novel modifications (e.g. C-21 hydroxylation [75,76]) are still being detected, so probably over 1000 possible analogues could exist in nature. Side-chain lactones are relatively common amongst phytoecdysteroids, but have not been found in the zooecdysteroids (although ecdysonoic acid and 20-hydroxyecdysonoic acid are found). Carboxyl groups at C-24, C-26 or C-29 form internal esters with a hydroxyl group at C-20, C-22, C-25, C-28 or C-29, to form a variety of 5- or 6-membered lactone ring analogues (e.g. ajugalactone {**4-1**}, carthamosterone {**4-3**}, cyasterone {**4-4**}, sidisterone {**4-6**}). A furan ring may be generated by cyclisation of the side-chain of a C_{27} ecdysteroid through dehydration of C-22 and C-25 hydroxyls, to form analogues such as ajugasterone D (**4-2**) and stachysterone D (**4-5**). C_{18} (rubrosterone {**2-12**}), C_{19} (poststerone {**2-11**}) and C_{24} (sidisterone) ecdysteroids are regarded as arising through side-chain cleavage and are regarded as inactivation metabolites as they generally possess very low biological activity in insect assays. Ecdysteroids have also been isolated from certain fungal species, but it is currently not clear if fungi have the capacity to synthesise these mycoecdysteroids themselves, or whether they just modify ecdysteroids which they have taken up from the substratum. A compilation of all known ecdysteroid analogues, together with their physicochemical and biological data (where available), is available in book form [77] and on the World-Wide Web [78].

12

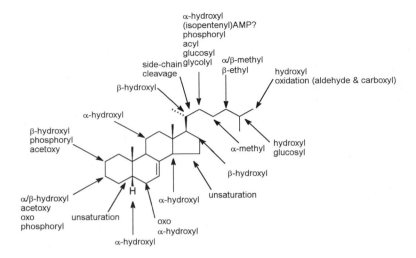

ZOOECDYSTEROIDS

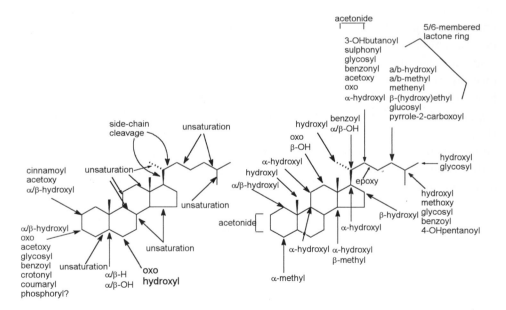

PHYTOECDYSTEROIDS

Figure 3: Structural diversity of zoo- and phytoecdysteroids. The diagrams represent the known structural variations. "?" indicates not conclusively proven

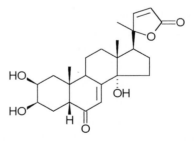

1: ajugalactone

2: ajugasterone D

3: carthamosterone

4: cyasterone

5: stachysterone D
(shidasterone)

6: sidisterone

Figure 4: Examples of phytoecdysteroids with restricted side-chain conformations

MODE(S) OF ACTION OF ECDYSTEROIDS IN INVERTEBRATES

Intracellular receptors

The mode of action of ecdysteroids in arthropod systems is summarised in Figure (**5**). Ecdysteroid receptors are members of the nuclear receptor superfamily [79], which are characterised by a domain structure. The N-terminal A/B-domain is highly variable and is associated with transcriptional activation. The C-domain is highly conserved and is involved in binding the receptor complex to specific response elements in the DNA. The D-domain is variable and represents a hinge region between the DNA-binding domain and the ligand-binding domain (E-domain). The E-domain is not only responsible for ligand binding, but also has been implicated in receptor dimerisation and interactions with other transcriptional activators. There may also be a C-terminal F-domain, which, if present, is highly variable between even closely related nuclear receptors [80]. Nuclear receptors regulate gene expression as dimers, either as homodimers or as heterodimers with another member of the nuclear receptor superfamily. One of the most promiscuous heterodimeric partners for vertebrate nuclear receptors is RXR, of which the equivalent in insects is Ultraspiracle (USP [81]). In the case of ecdysteroid receptors, only the EcR:USP (or EcR:RXR) [82] complex is able to bind the ecdysteroid ligand with high affinity and the presence of ecdysteroid promotes complex formation. The ecdysteroid binds to the EcR protein. No definitive ligand for USP has been identified, but it has been suggested that juvenile hormones (or methyl farnesoate in Crustacea) may bind to this receptor component and modify the transactivation capacity of the complex [83].

The most extensively studied ecdysteroid receptor system in arthropods is that of *Drosophila melanogaster*, where three isoforms (A, B1 and B2) of EcR occur [84,85]. These isoforms arise through alternative promoter usage and differential splicing, resulting in different A/B-domains, but they all possess common DNA- and ligand-binding domains. The EcR isoforms show tissue- and stage-specificity. Although there is only one form of USP in *D. melanogaster*, two or more isoforms have been found in other arthropods. USP isoforms also show tissue- and stage-specificity [86].

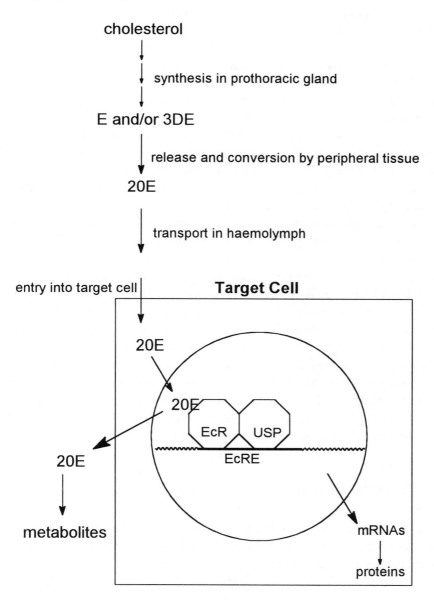

Figure 5: Cartoon summarising the mode of action of ecdysteroids.

EcR and USP gene homologues have now been characterised from a variety of arthropod species: *Aedes aegypti* [86,87], *Amblyomma americana* [88], *Bombyx mori* (silkworm) [89], *Chironomus tentans* (midge) [90,91], *Choristoneura fumiferana* [92,93], *Heliothis virescens* [94], *Locusta migratoria* (migratory locust) [95,96], *Lucilia cuprina* (Austalian green blowfly) [97,98], *Manduca sexta* (tobacco hornworm) [99,100], *Ostrinia nubilalis* (European corn-borer) [101], *Sarcophaga crassipalpis* [102], *Tenebrio molitor* (yellow mealworm) [103,104], *Uca pugilator* (fiddler crab) [105].

The biochemical characterisation of ecdysteroid receptor complexes lags well behind that of vertebrate steroid hormone receptors and has been in a period of quiescence for the past decade, as emphasis has been placed on the characterisation and expression of the genes. The generally accepted ligand for ecdysteroid receptors in arthropods is 20E, but this does not preclude the other ecdysteroids being significant at particular stages of development or in certain tissues [106]. In fact, ecdysteroid receptor complexes recognise a wide range of ecdysteroid structural analogues and sophisticated structure-activity and molecular modelling studies are now beginning to be performed [28,107-109]. Owing to the importance of ecdysteroid receptors in the regulation of arthropod development, they are seen as an appropriate target for the development of new pest control agents. In the context of this review, the identification of bisacylhydrazines as non-steroidal ecdysteroid agonists [110-112] is particularly worth mentioning as, in addition to several of these molecules being commercialised as insecticides, other analogues appear appropriate as gene switching elicitors. Antagonists for ecdysteroid receptors are also being identified [113].

Non-genomic actions of ecdysteroids

In addition to the actions mediated by intracellular receptors, steroid hormones also demonstrate certain rapid (occurring within seconds) actions, which appear to be mediated by membrane effects/receptors; this is true for vertebrate [114] and invertebrate [115] steroids. One presumes that the SAR studies for ecdysteroids which have been performed to date are predominantly determined by interaction of ecdysteroids with nuclear receptors, but one should bear in mind that in certain assay systems the measured activity may be modulated by interaction with nongenomic sites.

Taste receptors

It has been recognised for some time that certain insect species have the ability to perceive ecdysteroids [38,116,117]. This is probably a general phenomenon for many phytophagous insect species, which would encounter phytoecdysteroids in their diets [26]. Only recently have more extensive electrophysiological studies on ecdysteroid taste receptor cells been performed [40,118], but very little is currently known about their specificity [119]. This will clearly be an important area for future research, if ecdysteroids (or their analogues) are to be developed further as crop protection agents.

CHEMICAL SYNTHESIS OF ECDYSTEROID ANALOGUES

Ecdysteroid chemistry has been reviewed previously [120-122], so I shall not cover synthetic procedures here, but rather indicate what the main motivations are in this area and highlight where current and future effort will be.

Generation of rare natural ecdysteroids

Phytoecdysteroid-containing plants generally contain a few (1-3) major ecdysteroids, which make up ca. 95% of the total ecdysteroid, and a plethora of minor ecdysteroids, accounting for the rest. The major phytoecdysteroids are predominantly 20E and polB [3]. This means that the majority of the structural diversity is associated with the minor ecdysteroids. In addition to raising the question about the biological significance of the minor ecdysteroids, it means that most ecdysteroids are isolated only in small amounts, which, while sufficient for thorough identification and bioassay, are not adequate for physiological experimentation or *in vivo* testing. Thus, it is desirable to be able also to generate the most significant of the minor ecdysteroids chemically. The total synthesis of ecdysteroids has been achieved [123-129], but the overall yields were very low, so generation of analogues is essentially by partial synthesis from readily available ecdysteroids (mainly 20E). The cost of ecdysteroids was for a long time inhibitory to ecdysteroid chemistry, but the interest in incorporating ecdysteroids into anabolic preparations in recent years has resulted in 20E at least becoming commercially available in much larger amounts at much reduced prices (US$50 - 100/g). Owing to the relatively late elucidation of the structure of the first ecdysteroid, ecdysteroids did not benefit directly from the great era in steroid chemistry, so the literature on ecdysteroid chemistry is somewhat limited. Further, the large number of hydroxyl groups characteristically present in ecdysteroid molecules mean that extensive protection/deprotection strategies need to be employed in order to generate the desired analogue in reasonable yield. This said, there have been a small group of ecdysteroid chemists located around the world who have developed effective procedures for the generation of ecdysteroid analogues. Some recent examples are: calonysterone and 9,20-dihydroxyecdysone [130], 2-dehydro-3-*epi*-20-hydroxyecdysone [131], 25-deoxyecdysone [132], 2-deoxy-3-dehydroecdysone [133], 2-deoxy-20-hydroxyecdysone [134], 3-deoxyecdysteroids [135], 24-*epi*-abutasterone [136], 3-*epi*-2-deoxy-20-hydroxyecdysone [137], 20E glycosides [138], shidasterone [139] and 22-*O*-alkyl ether and acyl ester derivatives of 20E [140].

Synthesis of non-natural analogues

The structural variety of natural zoo-, phyto- and mycoecdysteroids is wide [78], but it does not encompass some structural features which would be desirable for certain applications or for certain SAR studies. There are a number of areas in which novel ecdysteroids would be very helpful:

a) *Analogues for QSAR and molecular modelling*

For example, the generation of fluorinated ecdysteroids would be informative in determining which groups in the molecule are involved in H-bonding and how (as donors or acceptors). Thus, 25-fluoroponasterone A (**6-1**) [141] and 2β,3β,20*R*,22*R*-tetrahydroxy-25-fluoro-5β-cholest-8,14-dien-6-one (**6-2**) [142] have been generated, but the overall experience has been that fluorinated ecdysteroids are difficult to generate. Also, it might be desirable to synthesise specific target molecules to test predictions arising from molecular models. For example, molecular modelling of ecdysteroid activity in the B_{II} bioassay [28,108] resulted in the prediction that there would be steric space around C-11 of the steroid. This has been tested by synthesising a series of 11α-acyl esters of turkesterone ([143], **6-3** and Table {**5**}).

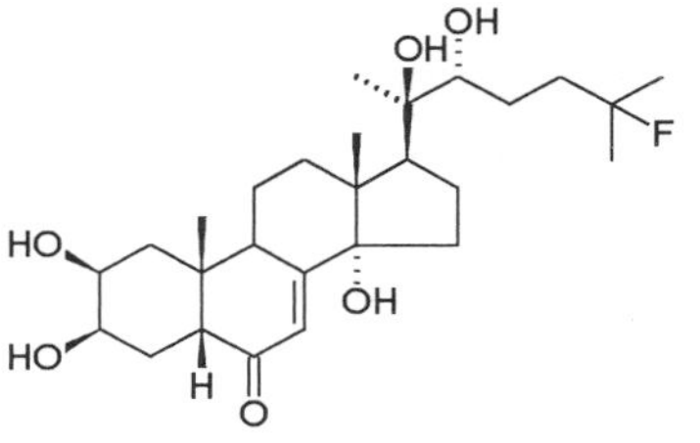

1: 25-fluoroponasterone A

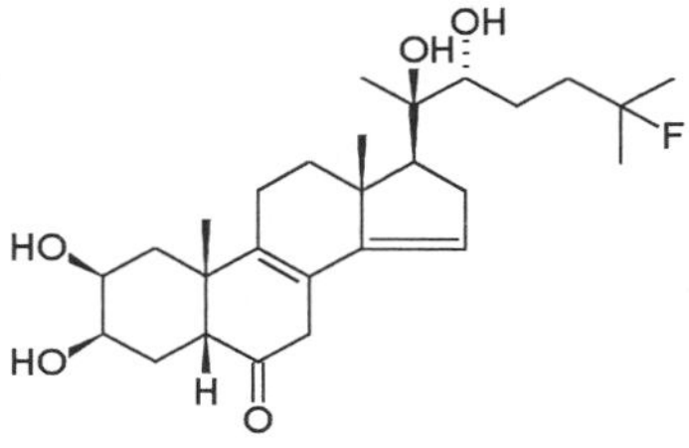

2: 2β,3β,20*R*,22*R*-tetrahydroxy-25-fluoro-5β-cholest-8,14-dien-6-one

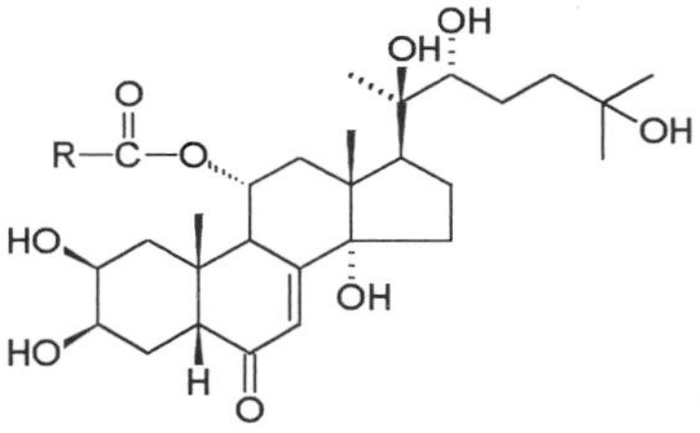

3: 11α-acyl turkesterone
(R = -CH$_3$ to -(CH$_2$)$_{18}$CH$_3$)

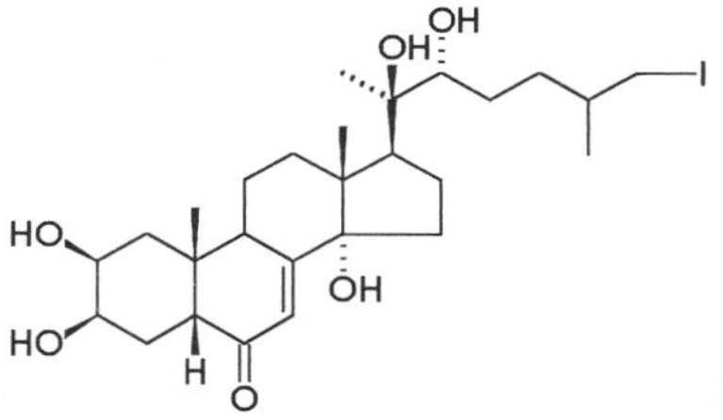

4: 26-iodoponasterone A

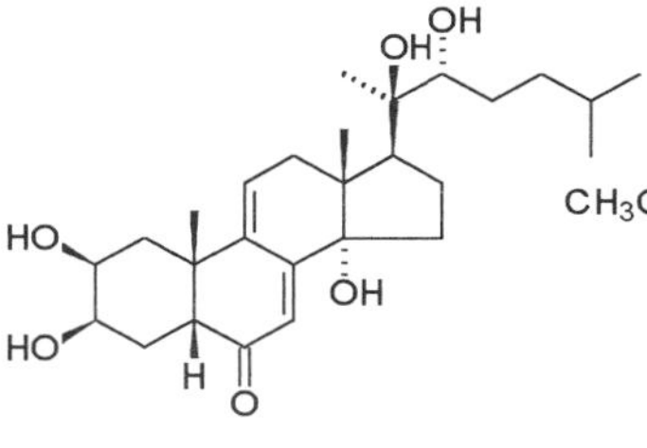

5: dacryhainansterone
(5-deoxykaladasterone)

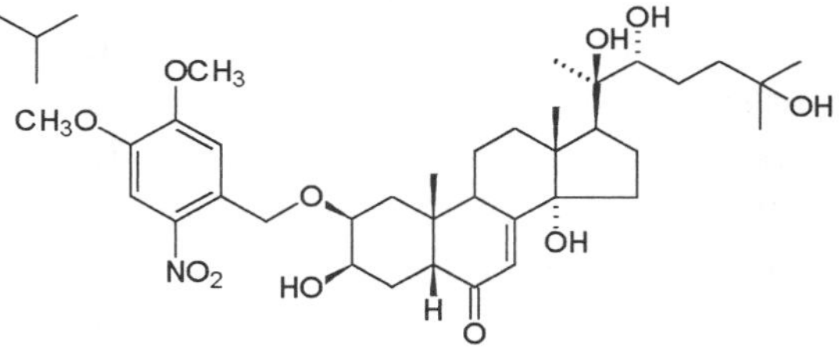

6: 2-(4,5-dimethoxy-2-nitrobenzyl)-20-hydroxyecdysone

Figure 6: Important synthetic ecdysteroid analogues.

b) *Radiolabelled or fluorescently-tagged ecdysteroids for receptor assays*:
currently [24,25,26,27-^{3}H]poA is used for most cell-free ecdysteroid
receptor assays, as it has a high affinity for the receptor (Kd ~ 10^{-9}M) and
is available at high specific activity (>120Ci/mmol [144]). However, it is
very expensive to purchase commercially, requires regular purification
and may not have a high enough detectability for some applications.
Thus, high affinity ligands with greater detectability are being sought.
These might be radiolabelled, non-natural ecdysteroids retaining high
receptor affinity (e.g. 26-iodoponasterone A {6-4} [145]) or could be
high-affinity fluorescently-tagged ecdysteroids, suitable for new types of
receptor assay.

c) *Photochemical transformations*

Since the ecdysteroids generally possess a characterisitic UV absorbance,
one way of generating novel chemical analogues is to irradiate the
compounds with UV light, thereby generating radicals. Such chemistry is
rather unpredicatable, but generates analogues which cannot readily
prepared by other means. Thus, this approach has been used for the
generation of a range of dimeric ecdysteroids (7-1 - 7-3), which retain
surprisingly high biological activity ([146,147] and Table {5}).

d) *Affinity labelling ecdysteroids*

One reason for the slow progress in the biochemical characterisation of
ecdysteroid receptor proteins is the non-covalent interaction between the
receptor and the ligand, which dissociates during purification. One
solution to this would be to generate a covalent link between the ligand
and the receptor by identifying a high affinty ligand which was either
already chemically reactive or which could be activated (e.g. by UV
light). Molecular modelling identified ecdysteroid 7,9(11)-dien-6-ones
(e.g. dacryhainansterone {6-5}) as high affinity ligands with the potential
for photoactivation. Initial evidence indicates that such compounds would
be effective photoaffinity labels for ecdysteroid receptors [148]. Using
molecular models to predict high affinity ligands has the potential to save
considerable synthetic effort.

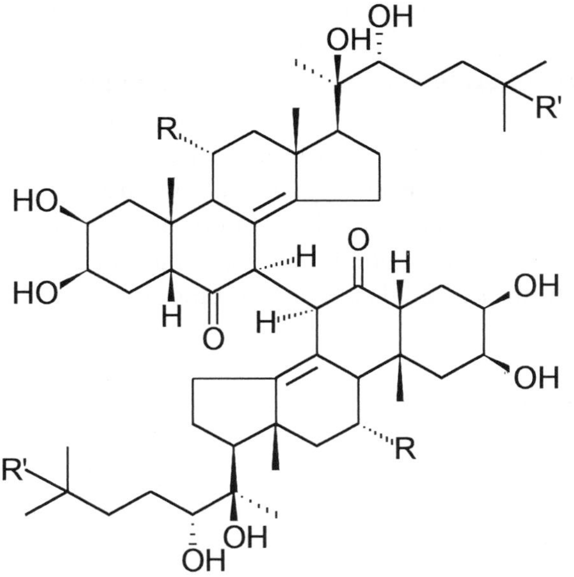

	R	**R'**
1: ponasterone A dimer:	-H	-H
2: ajugasterone C dimer:	-OH	-H
3: 20-hydroxyecdysone dimer:	-OH	-OH

Figure 7: Dimeric ecdysteroids.

e) Analogues for gene-switching systems

Lin et al. [149] have recently reported the synthesis of 2-(4,5-dimethoxy-2-nitrobenzyl)-20-hydroxyecdysone (**6-6**), which itself has a very low affinity for ecdysteroid receptors expressed in mammalian 293T cells, but can be photolysed at 348 nm to release 20E, which then activates the receptor-regulated reporter gene. With this system, it will be possible not only to control the temporal activation of transfected genes, but also their spatial activation. The authors recognise that the effectiveness of this system can be considerably improved in future by derivatising an ecdysteroid (poA or muA) which is more active in ecdysteroid-responsive mammalian cells.

NON-STEROIDAL ECDYSTEROID RECEPTOR AGONISTS

Ecdysteroids suffer a number of disadvantages for applied purposes because they are chemically complex, environmentally and metabolically unstable and far too costly to isolate or synthesise in large amounts [150]. Non-steroidal ecdysteroid analogues have greater potential for application in these respects. Further if such compounds are demonstrated to interact with the ecdysteroid-binding site of the EcR/USP complex, then analogues within these classes of compounds can be used to extend QSAR studies.

Bisacylhydrazines

The most significant class of non-steroidal ecdysteroid agonists identified to date are the bisacylhydrazines [110,111]. These compounds (Fig. {**8**}) are not only proving to be effective experimental tools of the investigation of ecdysteroid-regulated processes, but several members of this class have been/are being developed as insecticidal compounds. The first compound of this class to be identified was RH-5849 (**8-1**). The BAHs are weak agonists of the ecdysteroid receptor, but are metabolised only slowly in insects, allowing them to reach adequate concentrations to induce premature and incomplete moulting, thereby killing the insect. Further, the BAHs are Order-selective, for example, tebufenozide (RH-5992 {**8-2**} [151,152]) and methoxyfenozide (RH-2485 {**8-4**} [153]) are much more effective against Lepidoptera than against Diptera and halofenozide (RH-0345 {**8-3**} [112]) shows selectivity for Coleoptera and Lepidoptera. Additionally, BAHs demonstrate very low mammalian toxicity. Owing to the straightforward chemical synthesis of BAH analogues, it has been possible to conduct QSAR studies on insect systems *in vivo* and *in vitro*. Although it is clear that BAHs interact with the ligand-binding site of the EcR protein, it has not yet been resolved whether ecdysteroids and BAHs interact with this site in the same way; several different 'overlap' hypotheses have been proposed [154-157], but none is currently fully convincing. QSAR studies for BAHs are beginning to be reported, using a variety of bioassay systems which have been used for ecdysteroid QSAR: *Chilo suppressalis* larvae [158-160], *C. suppressalis* integument *in vitro* [161-163], B_{II} bioassay [164,165]. BAHs are also being developed as elicitors for modified ecdysteroid receptors used in gene switch systems (e.g. RG-102240, GS-E {**8-5**}).

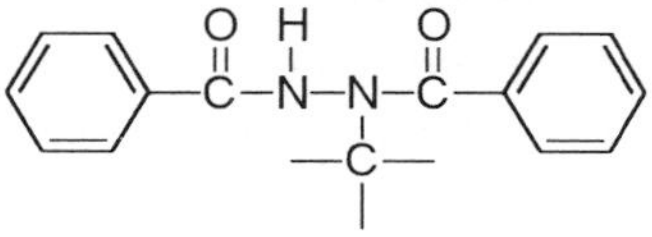

1: RH-5849

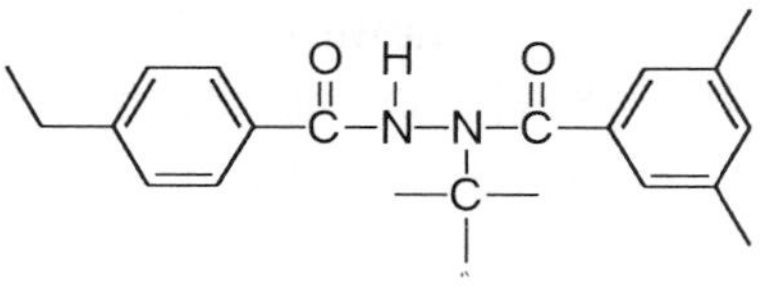

2: RH-5992
(tebufenozide)

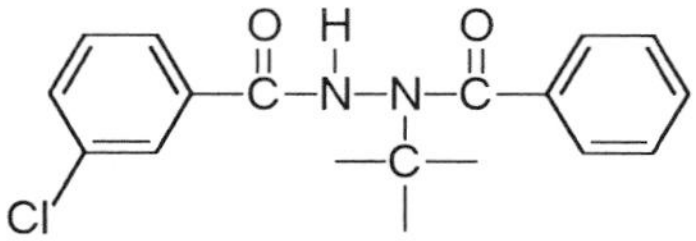

3: RH-0345
(halofenozide)

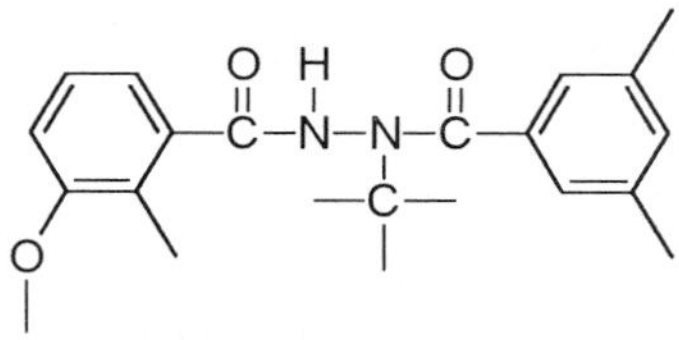

4: RH-2485
(methoxyfenozide)

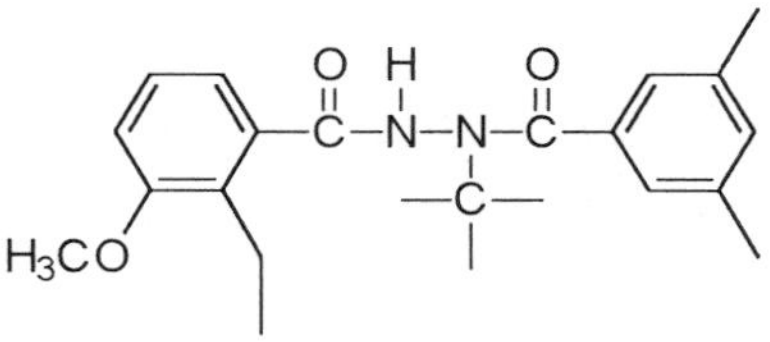

5: RG-102240
(GS-E)

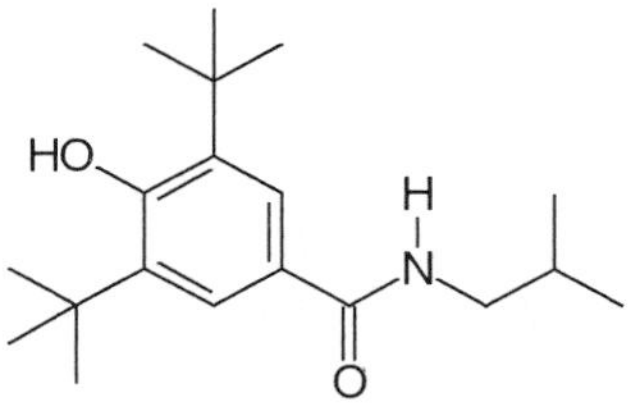

6: DTBHIB

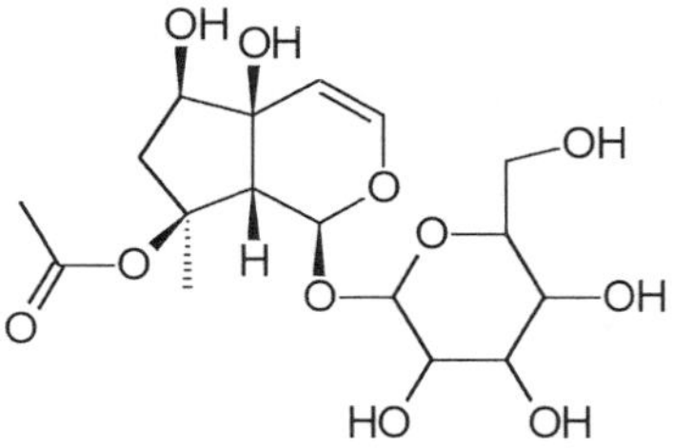

7: 8-O-acetylharpagide

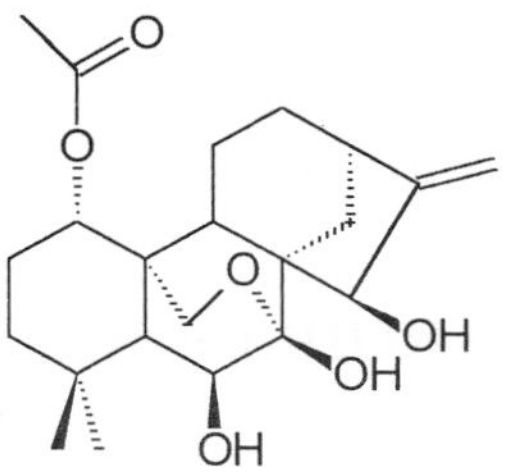

8: maocrystal E

Figure 8: Non-steroidal ecdysteroid agonists.

DTBHIB

The synthetic compound 3,5-di-*tert*-4-hydroxy-*N*-isobutylbenzamide (DTBHIB {8-6}) has been reported to be a weak ecdysteroid agonist, by inducing ecdysteroid-specific changes in the *Drosophila melanogaster* K_c cell line ($EC_{50} = 3 \times 10^{-6}M$) and by being able to displace [^{3}H]poA from receptors *in vitro* ($EC_{50} = 6 \times 10^{-6}M$). However, the isopropyl analogue showed no activity in either test [166].

8-*O*-Acetylharpagide

Elbrecht et al. [167] reported that the iridoid glucoside 8-*O*-acetylharpagide (8-7), isolated from *Ajuga reptans*, possessed ecdysteroid agonist activity. Unfortunately, *A. reptans* is also a rich source of phytoecdysteroids and it is now clear that 8-*O*-acetylharpagide, *per se*, possesses no agonistic activity [168].

Tetrahydroquinolines

Researchers at FMC Corp. (Princeton, NJ, U.S.A.) reported that 4-phenylamino-1,2,3,4-tetrahydroquinolines also act as ecdysteroid agonists [169].

Maocrystal E

As part of an extensive survey of natural products to detect those showing ecdysteroid agonist or antagonist activities in the B_{II} bioassay [170], it was found that the diterpene maocrystal E (8-8) showed agonist activity ($EC_{50} = 5 \times 10^{-6}M$).

ECDYSTEROID BIOASSAY SYSTEMS

The various bioassay systems used for ecdysteroid structure-activity studies have been reviewed previously [120,150,171]. Only those systems which have been used extensively and those which are recent are described below.

Whole animal bioassays

Calliphora bioassay

This was the original ecdysteroid activity assay, developed by Becker and Plagge [172] and was used to monitor the original isolation of ecdysone from *Bombyx mori* pupae [66]. Prepupae of *Calliphora erythrocephala* (= *C. vicina*) or *C. stygia* [173] are ligated below the ring gland and those not pupariating in the posterior portion within 24 h are used for bioassay, with the test solution being injected into the posterior portion, with the abdomens being scored for cuticle darkening (sclerotisation) 24 h later. The biological activities of a range of ecdysteroids are summarised in Table (**1**).

Musca bioassay

A similar, but more sensitive, bioassay using larvae of *Musca domestica* was developed [174-177]. This bioassay is still finding application today [140,178]. Table (**1**) summarises the biological activities of ecdysteroids in this bioassay.

Chilo suppressalis dipping assay

Ecdysteroids are generally regarded as not being able to penetrate insect cuticle readily [179]. However, the rice stem borer, *C. suppressalis*, has a thin cuticle which allows ecdysteroids to penetrate when ligated last instar larvae are dipped into a methanolic solution of the test compound for 5 s [180]. The proportion of pupated abdomina are assessed after 48 h. This assay is less sensitive than the *Calliphora* bioassay, but is easier to perform. Its main application has been in screening plant extracts for the presence of phytoecdysteroids [74].

Table 1: Biological activities of ecdysteroids and related compounds in *in vivo* assays. Data taken from Bergamasco and Horn [171], unless otherwise indicated. The number in brackets after the name of the ecdysteroid refers to the number allocated to it in [171], where the structures for each of these compounds is given.

Ecdysteroid	**Activity***
Calliphora bioassay	
ecdysone (1)	100
20-hydroxyecdysone (2)	100
ponasterone A (3)	100
3-dehydroecdysone (4)	10
(25*R/S*)inokosterone (5)	100
(25*R/S*)20,26-dihydroxyecdysone (7)	10
22-deoxyecdysone (8)	5
22,25-dideoxyecdysone (9)	10
2,22,25-trideoxyecdysone (11)	33
2β,3β,20R,22R,25-pentahydroxy-5β-cholest-8,14-dien-6-one (13)	20
14-deoxyecdysone (14)	7
3β-hydroxy-5β-cholest-7-en-6-one (15)	1-2
2-deoxyecdysone (17)	100
2-deoxy-20-hydroxyecdysone (18)	100
2,25-dideoxyecdysone (19)	50
25-deoxyecdysone (20)	7
14α-hydroxy-5β-cholest-7-en-6-one (21)	15
3-*epi*-2-deoxy-20-hydroxyecdysone (24)	33
2α,3α,14α,20R,22R-pentahydroxy-5β-cholest-7-en-6-one (25)	7-10
3β-methoxy-14α-hydroxy-5β-cholest-7-en-6-one (26)	20
3β-ethoxy-14α-hydroxy-5β-cholest-7-en-6-one (27)	4
20-hydroxyecdysone 3-*O*-cinnamate (30)	55
polypodine B (33)	400
3-oxo-14α-hydroxy-5β-cholest-7-en-6-one (34)	20
3α-methyl-3β,14α-dihydroxy-5β-cholest-7-en-6-one (35)	33
5β-hydroxypterosterone (37)	~100
20-hydroxyecdysone 25-acetate [viticosterone E] (40)	7
22-deoxy-20-hydroxyecdysone [taxisterone] (41)	1-2
22-dehydro-25-deoxyecdysone (42)	10
20-(2',2'-dimethyl-(22R)-furanyl)ecdysone (44)	20
20-(2',2'-dimethyl-(22S)-furanyl)ecdysone (45)	10
makisterone A (46)	100
makisterone C (47)	100
3β-hydroxy-5β-cholest-8,14-dien-6-one (54)	inactive
3β-hydroxy-5β-cholest-7,14-dien-6-one (55)	inactive
7,8-dihydroecdysone (56)	inactive
25-deoxy-7,8-dihydroecdysone (57)	inactive
3β-acetoxy-14α-hydroxy-5β-cholest-7-en-6-one (59)	inactive

2β,14α,-dihydroxy-5α-cholest-7-en-6-one (63)   inactive
14α-hydroxy-3α,5α-cyclocholest-7-en-6-one (64)   inactive
2,3-seco-ecdysone (65)   inactive
3α,14α-dihydroxy-5α-cholest-7-en-6-one (66)   inactive
3β-methyl-3α,14α-dihydroxy-5β-cholest-7-en-6-one (67)   inactive
3α-ethyl-3β,14α-dihydroxy-5β-cholest-7-en-6-one (68)   inactive
3β-ethyl-3α,14α-dihydroxy-5β-cholest-7-en-6-one (69)   inactive
5β-methyl-3β,14α-dihydroxycholest-7-en-6-one (70)   inactive
3β,5α,14α-trihydroxycholest-7-en-6-one (71)   inactive
poststerone (74)   inactive
6-oxo-3α,14α-dihydroxy-5β-chol-7-enoic acid methyl ester (75)   inactive
24-*nor*-3β,14α,25-trihydroxy-5β-cholest-7-en-6-one (76)   inactive
22-*epi*-ecdysone (77)   inactive
22-*epi*-20-hydroxyecdysone (78)   inactive
24*R*-methyl-2β,14α-dihydroxy-5β-cholest-7,22-dien-6-one (80)   inactive
(14α-H)3β-hydroxycholest-4,7-dien-6-one (81)   inactive
3β,14α-dihydroxycholest-4,7-dien-6-one (82)   inactive
3β-hydroxy-5β-cholest-6,8(14)-diene (83)   inactive
3-oxocholest-4,6,8(14)-triene (84)   inactive
3β-hydroxy-5β-cholest-7ene (85)   inactive
3β-hydroxy-5α,8-epidioxycholest-6-ene (86)   inactive
3β,7ξ-dihydroxycholest-5-ene (87)   inactive
(14α-H)cholest-4,7-dien-3,6-dione (88)   inactive
14α-hydroxycholest-4,7-dien-3,6-dione (89)   inactive
6α-hydroxycholest-4,7-dien-3-one (90)   inactive
6β-hydroxycholest-4,7-dien-3-one (91)   inactive
cholest-4,7-dien-3-one (92)   inactive
6,24-dioxo-3β,14α-dihydroxy-5β-chol-7-en-24-dimethylamine (93)   inactive

Musca bioassay
20-hydroxyecdysone (2)   100
(25*R/S*)26-hydroxyecdysone (6)   7-10
(25*R/S*)20,26-dihydroxyecdysone (7)   7-10
22,25-dideoxyecdysone (9)   17
2,3-bis(trimethylsilyl) 22,25-dideoxyecdysone (10)   17
22,25-dideoxy-6(α/β)-hydroxyecdysone (12)   ~1
3-*epi*-20-hydroxyecdysone (22)   7-10
3-*epi*-ecdysone (23)   10-20
3β-methoxy-2β,14α-dihydroxy-5β-cholest-7-en-6-one (28)   17
20-hydroxyecdysone 2-*O*-cinnamate (31)   10
polypodine B 2-*O*-cinnamate (32)   10
capitasterone (43)   100
makisterone C (47)   0
cyasterone (48)   100
2β,3β,14α-tris(trimethylsilyloxy)-5β-cholest-7-en-6-one (58)   inactive

3β-methoxy-2β,14α-dihydroxy-5β-cholest-7-en-6-one (60)	inactive
2β,3β,14α-trihydroxy-5β-cholest-7en-6-one 2,3-acetonide (61)	inactive
2β,3β,14α-trihydroxy-5β-cholest-7en-6-one 3-mesylate (62)	inactive
27-nor-2β,3β,14α-trihydroxy-5β-cholest-7-en-6-one (73)	inactive
20-hydroxyecdysone	100[a]
pinnatasterone	2[a]
canescensterone	84[a]
20-hydroxyecdysone 22-acetate	76[a]
22-O-methyl-20-hydroxyecdysone	64[a]
22-O-ethyl-20-hydroxyecdysone	67[a]
22-O-palmitoyl-20-hydroxyecdysone	52[a]
20-hydroxyecdysone 22-O-(pyrrole-2-carboxylate)	211[a]
20-hydroxyecdysone 22-O-(furan-2-carboxylate)	30[a]
20-hydroxyecdysone 22-O-(thiophene-2-carboxylate)	11[a]
20-hydroxyecdysone 22-O-glycolate	516[a]
20-hydroxyecdysone 22-O-acetylglycolate	276[a]
20-hydroxyecdysone 22-O-chloroacetate	57[a]

Sarcophaga bioassay

ponasterone A 2β-glucoside (29)	++
sengosterone (36)	+
pterosterone (38)	++
stachysterone C (39)	++
capitasterone (43)	120
cyasterone (48)	100
amarasterone A (51)	++
amarasterone B (52)	++
ajugasterone C (53)	++
rubrosterone (72)	inactive

Chilo dipping test

2β,3β,20R,22R,25-pentahydroxy-14β-methyl-5β-cholest-7,12-dien-6-one (16)	++
ajugasterone B (49)	++
makisterone D (50)	++
shidasterone (stachysterone D) (79)	inactive

* activities are expressed either as a percentage relative to the activity of 20-hydroxyecdysone (100%) or as active (+) or highly active (++)

a: data taken from [140]

Other whole insect assays

Sláma et al. [31] have compared the potencies of 15 ecdysteroids in 3 bioassays based on the induction of pup(ari)ation in ligated last-instar larvae of *Dermestes vulpinus* (Coleoptera), *Galleria mellonella* (Lepidoptera) and *Sacrophaga bullata* (Diptera). Their data are summarised in Table (**2**). The most striking finding was the relative biological activity of turkesterone (11α,20-dihydroxyecdysone, **2-10**), which was 10-fold more potent than 20E in the *Sarcophaga* and *Dermestes* assays, but only poorly active in the *Galleria* assay. Further, the authors demonstrated that several ecdysteroid conjugates (silenosides A, C and E) are inactive when injected into the haemolymph of these species. 20E 2,3-acetonide was also inactive, while 20E 20,22-acetonide was weakly active in the *Sarcophaga* assay (ED_{50} = 31.2 μg/g). The *Sarcophaga* and *Galleria* assays were also used to demonstrate that a range of conjugates of 20E (fatty acyl esters, benzoates and glucosides) were predominantly inactive, regardless of whether they were topically applied or injected; only 2-oleoyl 20E and 20E 25-benzoate were active on injection, but not when applied topically [181].

The activities of 4 ecdysteroids (20E {**1-1**}, E {**2-1**}, polB {**2-6**} and poststerone {**2-11**}) have been compared after spraying on to the food (potato leaves) of the Colorado beetle, *Leptinotarsa decemlineata* [182]. Surprisingly, poststerone gave a high level of toxicity.

Taken together these data indicate that different insect species are susceptible to ecdysteroid analogues to differing extents. This may go some way to explaining why phytoecdysteroid-containing plant species possess different ecdysteroid profiles.

Tissue bioassays

***Chilo suppressalis* cultured integument assay**

The stimulation of the incorporation of [^{14}C]*N*-acetylglucosamine into the cultured integument of *C. suppressalis* larvae by ecdysteroids has been used to compare the potencies of several analogues [160]: poA (EC_{50} = 3 x 10^{-8}M), 20E (1.8 x 10^{-7}M), Cya (5.2 x 10^{-7}M), Ino (6.7 x 10^{-7}M), MaA (1.9 x 10^{-6}M) and E (>5.4 x 10^{-6}M).

Table 2: Potencies of ecdysteroids in *in vivo* assays based on the induction of pup(ari)ation in ligated last-instar larvae of *Dermestes vulpinus*, *Galleria mellonella* and *Sarcophaga bullata*. Data are taken from Sláma et al. [31] and are expressed as ED$_{50}$ values (μg/g). n.d. : not determined.

Ecdysteroid	D. vulpinus	EC$_{50}$ value G. mellonella	S. bullata
ecdysone	42.0	7.8	4.2
2-deoxyecdysone	n.d.	15.6	31.3
20-hydroxyecdysone	42.0	7.8	2.6
2-deoxy-20-hydroxyecdysone	n.d.	12.5	26.0
20E 20-*O*-benzoate	n.d.	15.6	2.6
20E 22-*O*-benzoate	n.d.	7.8	2.6
20E 2,25-dibenzoate	n.d.	31.2	104.0
polypodine B	29.0	7.8	7.8
ponasterone A	17.0	12.5	7.3
cyasterone	21.0	3.8	10.4
turkesterone	4.2	62.5	0.3
24(28)-dehydromakisterone A	n.d.	12.5	10.4
integristerone A	n.d.	15.6	15.6
nusilsterone	n.d.	12.5	52.0
rapisterone D	n.d.	7.8	7.8

Imaginal disc bioassays

The evagination of imaginal discs of Dipteran and Lepidopteran insects is a well-characterised ecdysteroid-dependent system, which has been used by several authors to develop bioassays [183-187]. However, they have been used to only a limited extent to compare the biological activities of ecdysteroids (Table {**3**}).

Insect cell-based bioassays

K$_c$ cell bioassay

The ecdysteroid-responsive Kc permanent cell line was established from disrupted embryos of *D. melanogaster* [188]. Consequently, it is not known what *in vivo* cell type (if any) they represent. However, the response (aggregation of cells, cessation of division and the formation of axon-like projections) is characteristic and can form the basis of a somewhat time-consuming bioassay [189], which has been used to compare the biological activities of 60 steroids, most of which were inactive. The biological activities of a number of ecdysteroids in this Kc-H cell bioassay are summarised in Table (**4**).

Table 3: Imaginal disc bioassays, giving the concentration of ecdysteroid required for 50% evagination (EC_{50}) and the IC_{50} or K_d value determined from a competion assay for specifically bound [^{3}H]ponasterone A.

Ecdysteroid	EC_{50}	IC_{50}/K_d
Calliphora vicina[a]		
ponasterone A	$1.4 \times 10^{-9}M$	$1.0 \times 10^{-9}M$
muristerone A	$1.5 \times 10^{-9}M$	$1.3 \times 10^{-9}M$
20-hydroxyecdysone	$7.9 \times 10^{-8}M$	$3.7 \times 10^{-8}M$
makisterone A	$1.2 \times 10^{-7}M$	$1.5 \times 10^{-8}M$
inokosterone	$1.5 \times 10^{-7}M$	$1.5 \times 10^{-8}M$
Drosophila melanogaster[b]		
ponasterone A	$1.0 \times 10^{-9}M$	$2.0 \times 10^{-9}M$
20-hydroxyecdysone	$4.2 \times 10^{-8}M$	$5.2 \times 10^{-8}M$
ecdysone	$1.5 \times 10^{-5}M$	$7.2 \times 10^{-5}M$
Drosophila melanogaster[c]		
20-hydroxyecdysone	$6.0 \times 10^{-8}M$	
cyasterone	$5.5 \times 10^{-8}M$	
inokosterone	$4.5 \times 10^{-7}M$	
ecdysone	$2.2 \times 10^{-5}M$	
rubrosterone	$7.9 \times 10^{-3}M$	

a: data taken from Terentiou et al. [187] (K_d values presented)
b: data taken from Yund [254] (IC_{50} values presented)
c: data taken from Chihara et al. [184]

The inhibition of uptake of [^{3}H]poA into Kc cells has also been used as a measure of the affinities of several ecdysteroids [190]. More recently, a similar approach has been used for ecdysteroids and BAHs for the ecdysteroid receptors present in these cells [191]: poA ($IC_{50} = 1.3 \times 10^{-9}$ M), 20E ($4.6 \times 10^{-8}M$), Cya ($6.2 \times 10^{-8}M$), Ino ($9.1 \times 10^{-8}M$), MaA ($1.1 \times 10^{-7}M$) and E ($2.6 \times 10^{-6}M$).

Kc cells have been transfected by electroporation with a reporter gene consisting of the 23-bp EcRE from the hsp27 gene ligated to the firefly luciferase gene. The cells are then cultured with or without ecdysteroids for 24 h, before being lysed to measure luciferase activity [192]. 20E and E induce luciferase activity up to 90-fold, with EC_{50}

values of 2×10^{-7}M and 2×10^{-5}M, respectively. Although faster than cellular assays based on morphological responses (e.g. Kc-H assay, B_{II} bioassay), this assay is technically much more complicated and has not been generally adopted.

Table 4: Activities of ecdysteroids in the Kc-H cell assay

Ecdysteroid	Relative activity	EC_{50}
14-deoxymuristerone A[b]	80	1.8×10^{-10}M
ponasterone A[a]	8	1.8×10^{-9}M
muristerone A[a]	7	2.0×10^{-9}M
muristerone A[b]	10	1.4×10^{-9}M
ponasterone C[a]	5	2.8×10^{-9}M
20-hydroxyecdysone[a]	1	1.4×10^{-8}M
makisterone A[a]	0.6	2.3×10^{-8}M
cyasterone[a]	0.5	2.8×10^{-8}M
viticosterone E[a]	0.20	7.0×10^{-8}M
inokosterone[a]	0.11	1.3×10^{-7}M
podecdysone A[a]	0.11	1.3×10^{-7}M
$2\beta,3\beta,5\beta,14a$-tetrahydroxycholest-7-en-6-one[a]	0.03	4.7×10^{-7}M
2-deoxy-20-hyroxyecdysone[a]	0.03	4.7×10^{-7}M
ecdysone[a]	0.006	2.3×10^{-6}M
$3\beta,14\alpha$-dihydroxy-5β-cholest-7-en-6-one[a]	inactive	
$3\beta,5\beta,14\alpha$-trihydroxycholest-7-en-6-one[a]	inactive	
$2\beta,3\beta$-dihydroxy-5β-cholest-7-en-6-one[a]	inactive	
$2\beta,3\beta$-dihydroxycholesta-7,14-dien-6-one[a]	inactive	
$2\beta,3\beta$-dihydroxy-5β-cholest-6-one[a]	inactive	
$2\beta,3\beta,5\beta$-trihydroxycholest-6-one[a]	inactive	
$2\beta,3\beta,25$-trihydroxy-5β-cholest-6-one[a]	inactive	
3β-hydroxy-5β-cholest-6-one[a]	inactive	

a: Data taken from Cherbas et al. [189]; compounds tested up to a maximal concentration of 2×10^{-6}M.
b: Data taken from Cherbas et al. [255]

Chironomus tentans epithelial cell line

This cell line does not synthesise ecdysteroids and metabolises exogenous ecdysteroids only slowly [193]. Acetylcholinesterase activity is specifically induced by ecdysteroids and this has been used as the basis of a bioassay to compare a limited range of ecdysteroids: 20E ($EC_{50} = 7.5 \times 10^{-7}$M), turkesterone ($8.0 \times 10^{-7}$M), MuA ($6 \times 10^{-8}$M).

Drosophila melanogaster imaginal disc cell lines

The effect of a few ecdysteroids and the BAHs, RH5849 and RH5992, on cell density of the cell lines C18+ and C18R were determined [194]. All compounds had similar effects, but C18R was 10-100-fold less sensitive to the ecdysteroid analogues. It was not reported if these cell lines metabolise ecdysteroids. The activities with the C18+ cells were: 20E (EC_{50} = 6 x 10^{-8}M), Ino (4 x 10^{-7}M), MaA (5 x 10^{-8}M) and MuA (6 x 10^{-9} M).

Sf-9 cell line

The ecdysteroid receptor-containing *Spodoptera frugiperda* Sf-9 cell line has been used to compare the activities of ecdysteroids and BAHs in a cell uptake competition assay using [^{3}H]poA [195]. The activities of the ecdysteroids tested were: PoA (IC_{50} = 8.9 x 10^{-9}M), 20E (1.7 x 10^{-7}M), Cya (2.7 x 10^{-7}M), Ino (5.6 x 10^{-7}M), MaA (3.9 x 10^{-7}M) and E (2.3 x 10^{-6} M).

B_{II} cell bioassay

The B_{II} cell bioassay was developed in the early 1990s [196,197]. It is based on the ecdysteroid-responsive *Drosophila melanogaster l(2)mbn* tumorous haemocyte cell line [198]. On the addition of 20E (or other ecdysteroids), the cells undergo clumping and an induction of phagocytosis. This dramatic response forms the basis of the microplate-based bioassay, since the extent of the response can be measured turbidometrically (with an automated microplate reader) and the specificity of the response can be assessed by examination of the microplate wells *in situ* with an inverted microscope. Further, the bioassay can be used to detect not only ecdysteroid-like activities (agonists), but also ecdysteroid-inhibiting activities (antagonists). B_{II} cells do not metabolise E, 20E or poA [144]. The bioassay has proved simple, sensitive, fast and remarkably robust, being able to cope with relatively crude plant extracts [113]. The bioassay can also be used quantitatively to compare the potencies of pure compounds by determining EC_{50} values (Table {5}). Agonistic compounds almost certainly interact with the ecdysteroid binding site of the receptor. However, the mode of action of antagonistic substances is not initially so clear-cut. Prevention of binding of ecdysteroid to the ligand-binding domain of the receptor protein is just one possibility, but any compound which prevents the induction of the ecdysteroid response (e.g. from preventing the entry of the steroid into the cell to preventing the synthesis of the induced proteins) would also appear as an antagonist. Thus, a number of subsequent assays are required to pin down the point of action of antagonists. For example, gel-shift assays can be used to determine if the EcR/USP complex interacts normally with EcREs or radioligand displacement assay can be used to determine if the antagonist interacts with the ecdysteroid-binding site.

Table 5: Biological activities of ecdysteroids in the *Drosophila melanogaster* B_{II} bioassay. Activities are presented as the concentration required to half the maximal response (EC_{50}). Where compounds show no activity at $10^{-4}M$, they are described as "inactive". Where the solubility is lower than $10^{-4}M$, the activity is given as >[maximal solubility].

Ecdysteroid	EC_{50}
abutasterone	$1.4 \times 10^{-7} M^{ah}$
ajugalactone	$1.6 \times 10^{-7} M^{cdh}$
ajugasterone B	$3.3 \times 10^{-7} M^{m}$
ajugasterone C	$3.0 \times 10^{-8} M^{cdfh}$, $8.0 \times 10^{-8} M^{bcdg}$
	$6.2 \times 10^{-8} M^{j}$
ajugasterone C dimer	$8.2 \times 10^{-7} M^{m}$
amarasterone A	$5.6 \times 10^{-7} M^{m}$
atrotosterone	$2.5 \times 10^{-8} M^{b}$
atrotosterone B	$6.5 \times 10^{-6} M^{b}$
atrotosterone C	$3.0 \times 10^{-6} M^{bcd}$
bombycosterol	$>2.7 \times 10^{-5} M^{c}$
calonysterone	inactive^{c}
canescensterone	$5.3 \times 10^{-10} M^{m}$
castasterone	$>10^{-3} M^{c}$
cheilanthone B	$1.3 \times 10^{-7} M^{m}$
cyasterone	$5.3 \times 10^{-8} M^{cd}$, $1.2 \times 10^{-8} M^{h}$
cyasterone 3-acetate	$4.3 \times 10^{-7} M^{cd}$
dacryhainansterone	$5.2 \times 10^{-9} M^{g}$
2-dansyl-20-hydroxyecdysone	$2.2 \times 10^{-6} M^{m}$
24(28)-dehydroamarasterone B	$5.2 \times 10^{-7} M^{cd}$
3-dehydro-2-deoxyecdysone (silenosterone)	$4.6 \times 10^{-5} M^{m}$
3-dehydro-2-deoxyecdysone 22-acetate	$4.0 \times 10^{-5} M^{m}$
3-dehydroecdysone	$6.0 \times 10^{-6} M^{cd}$
22-dehydroecdysone	$4.5 \times 10^{-8} M^{cd}$
2-dehydro-3-*epi*-20-hydroxyecdysone	$4.0 \times 10^{-7} M^{dik}$
4-dehydro-20-hydroxyecdysone	$2.8 \times 10^{-7} M^{cd}$
22-dehydro-12-hydroxycyasterone	$1.3 \times 10^{-6} M^{cd}$
22-dehydro-12-hydroxynorsengosterone	$1.3 \times 10^{-5} M^{cd}$

22-dehydro-12-hydroxysengosterone	$2.7 \times 10^{-6} M^{cd}$
22-dehydro-20-hydroxyecdysone	$1.7 \times 10^{-7} M^{cd}$
22-dehydro-20-*iso*-ecdysone	$3.0 \times 10^{-6} M^{cd}$
24(28)-dehydromakisterone A	$4.0 \times 10^{-9} M^{cd}$
24(25)-dehydroprecyasterone	$7.3 \times 10^{-8} M^{m}$
14-dehydroshidasterone	$1.5 \times 10^{-5} M^{m}$
14-deoxy-14,18-cyclo-20-hydroxyecdysone	$6.3 \times 10^{-7} M^{m}$
2-deoxy-20,26-dihydroxyecdysone	$5.3 \times 10^{-6} M^{m}$
3-deoxy-1β,20-dihydroxyecdysone	$1.0 \times 10^{-7} M^{m}$
2-deoxyecdysone	$5.0 \times 10^{-5} M^{cdh}$, $2.0 \times 10^{-5} M^{m}$
2-deoxyecdysone 22-glucoside	$2.1 \times 10^{-5} M^{m}$
22-deoxyecdysone	ca. $10^{-5} M^{cd}$
25-deoxyecdysone	$1.0 \times 10^{-8} M^{cd}$
2-deoxyecdysone 22-acetate	$1.4 \times 10^{-6} M^{acd}$
2-deoxyecdysone 3.22-diacetate	$>5 \times 10^{-5} M^{m}$
2-deoxyecdysone 25-glucoside	$7.3 \times 10^{-6} M^{m}$
(20*R*)22-deoxy-20,21-dihydroxyecdysone	$2.0 \times 10^{-7} M^{dei}$
(14α-H) 14-deoxy-25-hydroxydacryhainansterone	$2.2 \times 10^{-7} M^{g}$
(14β-H)14-deoxy-25-hydroxydacryhainansterone	$1.3 \times 10^{-6} M^{g}$
2-deoxy-20-hydroxyecdysone	$6.6 \times 10^{-7} M^{acd}$, $1.3 \times 10^{-6} M^{m}$
2-deoxy-21-hydroxyecdysone	$4.3 \times 10^{-6} M^{d}$
(5α-H)2-deoxy-21-hydroxyecdysone	$9.5 \times 10^{-5} M^{d}$
2-deoxy-20-hydroxyecdysone 3-acetate	$4.3 \times 10^{-6} M^{d}$
2-deoxy-20-hydroxyecdysone 22-acetate	$2.4 \times 10^{-5} M^{d}$
2-deoxy-20-hydroxyecdysone 22-benzoate	$6.3 \times 10^{-6} M^{cd}$
2-deoxy-20-hydroxyecdysone 3,22-diacetate	inactive^{d}
14-deoxy(14α-H)-20-hydroxyecdysone	$3.0 \times 10^{-8} M^{cd}$, $2.2 \times 10^{-8} M^{f}$
14-deoxy(14β-H)-20-hydroxyecdysone	$8.3 \times 10^{-7} M^{cdf}$
22-deoxy-20-hydroxyecdysone (taxisterone)	$1.4 \times 10^{-8} M^{cd}$, $9.5 \times 10^{-8} M^{j}$
(5α-H)2-deoxyintegristerone A	$5.0 \times 10^{-6} M^{m}$
5-deoxykaladasterone	$5.2 \times 10^{-10} M^{acdh}$
25-deoxypolypodine B	$1.0 \times 10^{-8} M^{cd}$
2-deoxypolypodine B 3-glucoside	$1.6 \times 10^{-5} M^{m}$
24,25-didehydrodacryhaninansterone	$2.2 \times 10^{-8} M^{m}$
25,26-didehydrodacryhainansterone	$2.7 \times 10^{-8} M^{m}$
25,26-didehydroponasterone A	$1.5 \times 10^{-7} M^{cd}$
22,25-dideoxyecdysone	ca. $10^{-6} M^{cd}$
2,22-dideoxy-20-hydroxyecdysone	ca. $5 \times 10^{-5} M^{cd}$
22,23-diepi-geradiasterone	$4.0 \times 10^{-5} M^{cd}$
dihydropoststerone	$7.4 \times 10^{-4} M$ & $1.2 \times 10^{-4} M^{m}$
dihyropoststerone tribenzoate	inactive^{m}
(5α-H)dihydrorubrosterone	$5.6 \times 10^{-6} M^{d}$
(5β-H)dihydrorubrosterone	$>10^{-5} M^{d}$
dihydrorubrosterone 17-acetate	$>10^{-4} M^{cdi}$
9,20-dihydroxyecdysone	$1.6 \times 10^{-5} M^{cd}$
(22*S*)20-(2,2'-dimethylfuranyl)ecdysone	$7.3 \times 10^{-6} M^{cd}$

(22*R*)20-(2,2'-dimethylfuranyl)ecdysone	$1.0 \times 10^{-5} M^{cd}$
ecdysone	$2.0 \times 10^{-7} M^{ab}$, $1.1 \times 10^{-6} M^{cdh}$
	$1.0 \times 10^{-6} M^{m}$
ecdysone 2,3-acetonide	$> 3.0 \times 10^{-5} M^{c}$
ecdysone 6-carboxymethyloxime	$5.0 \times 10^{-7} M^{a}$
ecdysone 22-hemisuccinate	$7.5 \times 10^{-6} M^{a}$
24-*epi*-abutasterone	$1.2 \times 10^{-6} M^{di}$
24-*epi*-castasterone	$> 10^{-4} M^{c}$
22-*epi*-ecdysone	$3.5 \times 10^{-6} M^{cd}$, $5.3 \times 10^{-6} M^{cdh}$
3-*epi*-20-hydroxyecdysone (coronatasterone)	$1.3 \times 10^{-7} M^{d}$, $1.6 \times 10^{-7} M^{j}$
1-*epi*-integristerone A	$2.5 \times 10^{-7} M^{cd}$
3-*epi*-22-*iso*-ecdysone	$4.4 \times 10^{-6} M^{cd}$
14-*epi*-20-hydroxyecdysone	$1.7 \times 10^{-4} M^{f}$
24-*epi*-makisterone A	$2.2 \times 10^{-7} M^{bcd}$
25-fluoropolypodine B	$4.7 \times 10^{-8} M^{m}$
25-fluoroponasterone A	$5.1 \times 10^{-9} M^{cdh}$
geradiasterone	$4.0 \times 10^{-7} M^{cd}$
28-homobrassinolide	$> 10^{-4} M^{c}$
14α-hydroperoxy-20-hydroxyecdysone	$1.6 \times 10^{-8} M^{f}$
5β-hydroxyabutasterone	$2.3 \times 10^{-8} M^{cdi}$
25-hydroxyatrotosterone	$1.0 \times 10^{-5} M^{b}$
25-hydroxyatrotosterone B	$5.0 \times 10^{-5} M^{b}$
24-hydroxycyasterone	$2.0 \times 10^{-6} M^{m}$
25-hydroxydacryhainansterone	$2.6 \times 10^{-7} M^{g}$
5β-hydroxy-25,26-didehydroponasterone A	$6.3 \times 10^{-8} M^{cd}$
20-hydroxyecdysone	$7.6 \times 10^{-9} M^{a}$, $7.5 \times 10^{-9} M^{bcdfgh}$
(5α-H)20-hydroxyecdysone	$3.3 \times 10^{-6} M^{cd}$
20-hydroxyecdysone 2-acetate	$4.0 \times 10^{-7} M^{cd}$
20-hydroxyecdysone 3-acetate	$4.7 \times 10^{-7} M^{cd}$
20-hydroxyecdysone 22-acetate	$4.0 \times 10^{-6} M$, $2.0 \times 10^{-7} M^{d}$
20-hydroxyecdysone 22-benzoate	$4.3 \times 10^{-9} M^{d}$
20-hydroxyecdysone 6-carboxymethyloxime	inactive[a]
20-hydroxyecdysone dimer	$2.1 \times 10^{-7} M^{f}$
20-hydroxyecdysone 2-β-D-glucopyranoside	$2.0 \times 10^{-5} M^{bcd}$
20-hydroxyecdysone 3-β-D-glucopyranoside	$1.3 \times 10^{-5} M^{bcd}$
20-hydroxyecdysone 22-β-D-glucopyranoside	$4.7 \times 10^{-5} M^{bcd}$
20-hydroxyecdysone 25-β-D-glucopyranoside	$8.5 \times 10^{-6} M^{bcd}$
20-hydroxyecdysone 2-hemisuccinate	$8.3 \times 10^{-7} M^{a}$
20-hydroxyecdysone 22-hemisuccinate	inactive[a]
20-hydroxyecdysone 2,3,22-triacetate	$> 1.6 \times 10^{-5} M^{c}$
20-hydroxyecdysone 3β-D-xylopyranoside	$1.6 \times 10^{-6} M^{cdi}$
6β-hydroxy-20-hydroxyecdysone	$1.7 \times 10^{-7} M^{cd}$
6α-hydroxy-20-hydroxyecdysone	$2.0 \times 10^{-6} Mcd$
26-hydroxypolypodine B	$4.8 \times 10^{-7} M^{a}$
5β-hydroxystachysterone C	$3.5 \times 10^{-8} M^{cd}$
(25*R/S*)inokosterone	$1.1 \times 10^{-7} M^{ah}$

(25R)-inokosterone	1.5x10^{-7}M^{cd}
(25S)-inokosterone	2.7x10^{-7}M^{cd}
inokosterone 26-hemisuccinate	3.8x10^{-6}M^{a}
integristerone A	1.8x10^{-7}M, 8.3x10^{-7}M,
	2.0x10^{-7}M^{m}
20-iso-ecdysone	1.0x10^{-4}M^{cd}
20-iso-22-epi-ecdysone	1.0x10^{-4}M^{cd}
iso-homobrassinolide	>10^{-5}M^{c}
isostachysterone C ($\Delta^{25(26)}$)	9.2 x 10^{-9}M^{m}
kaladasterone	3.4x10^{-7}M^{g}
ketodiol	>6.0x10^{-6}M^{c}
5α-ketodiol	>2.3x10^{-5}M^{c}
limnantheoside C	1.3x10^{-6}M^{l}
makisterone A	1.3x10^{-8}M^{abcdh}
makisterone C	2.0x10^{-7}M^{cdi}
malacosterone	9.0x10^{-6}M^{bcd}
24-methylecdysone (20-deoxymakisterone A)	2.1x10^{-6}M^{m}
14-methyl-12-en-15,20-dihydroxyecdysone	2.3x10^{-6}M^{m}
14-methyl-12-en-shidasterone	4.0x10^{-6}M^{m}
muristerone A	2.2x10^{-8}M^{acdfgh}
29-norcyasterone	1.2x10^{-8}M^{cd}
29-norsengosterone	1.3x10^{-7}M^{cd}
paxillosterone	4.2x10^{-7}M^{b}
paxillosterone 20,22-p-hydroxybenzylidene acetal	3.0x10^{-7}M^{b}
pinnatasterone	4.0x10^{-7}M^{m}
podecdysone B	1.2x10^{-5}M^{fg}
polypodine B	1.0x10^{-9}M^{bcdh}
polyporusterone B	2.1x10^{-9}M^{m}
ponasterone A	3.1x10^{-10}M^{abcdh}, 2.0x10^{-10}M^{m}
ponasterone A 6-carboxymethyloxime	7.5x10^{-6}M^{a}
ponasterone A dimer	4.6x 10^{-8}M^{m}
ponasterone A 2-hemisuccinate	3.1x10^{-7}M^{a}
ponasterone A 22-hemisuccinate	5.0x10^{-7}M^{a}
ponasterone A 3β-D-xylopyranoside	1.5x10^{-5}M^{cd}
ponasterone A 3β-D-xylopyranoside	1.5x10^{-5}M^{i}
poststerone	ca. 2x10^{-5}M^{cd}, 2.0x10^{-5}M^{h}
poststerone 20-dansylhydrazine	6.0x10^{-6}M^{m}
punisterone	8.3x10^{-7}M^{cdi}
rapisterone B	2.3x10^{-7}M^{m}
rapisterone C	3.9x10^{-7}M^{m}
rapisterone D	1.0x10^{-9}M^{cd}
rubrosterone	>10^{-4}M^{cdi}
sengosterone	9.0x10^{-8}M^{cd}
shidasterone	1.5x10^{-6}M, 4.0x10^{-6}M^{cd}
sidisterone	4.3x10^{-6}M^{m}
silenoside A (20-hydroxyecdysone 22-galactoside)	4.1x10^{-5}M^{m}

silenoside C (integristerone A 22-glactoside) 1.0×10^{-4}M[m]

silenoside D (20-hydroxyecdysone 3-galactoside) 3.0×10^{-5}M[m]

silenoside E (2-deoxyecdysone 3-galactoside)   inactive[m]

stachysterone B 8.2×10^{-8}M[g]

stachysterone C 1.4×10^{-8}M[cd]

2,14,22,25-tetradeoxy-5α-ecdysone $>2.5\times10^{-5}$M[c]

2α,3α,22S,25-tetrahydroxy-5α-cholestan-6-one $>10^{-4}$M[c]

2β,3β,20R,22R,-tetrahydroxy-25-fluoro-5β-cholest-8,14-dien-6-one 7.2×10^{-9}M[cd]

2β,3β,6α-trihydroxy-5β-cholestane $>2.4\times10^{-5}$M[c]

2β,3β,6β-trihydroxy-5β-cholestane $>2.4\times10^{-5}$M[c]

turkesterone 1.3×10^{-6}M[acdgh], 3.0×10^{-7}M[cd] 8.0×10^{-7}M[m]

turkesterone 2-acetate 5.2×10^{-6}M[h]

turkesterone 11α-acetate 4.0×10^{-6}M[h]

turkesterone 11α-arachidate 2.2×10^{-5}M[h]

turkesterone 11α-butanoate 4.0×10^{-5}M[h]

turkesterone 11α-decanoate 1.1×10^{-6}M[h]

turkesterone 2,11α-diacetate 2.0×10^{-4}M[h]

turkesterone 2,22-diacetate 1.0×10^{-3}M[h]

turkesterone 11α-hexanoate 2.2×10^{-5}M[h]

turkesterone 11α-laurate 3.4×10^{-6}M[h]

turkesterone 11α-myristate 1.3×10^{-6}M[h]

turkesterone 11α-propionate 8.3×10^{-6}M[h]

viticosterone E 1.0×10^{-7}M[cd]

a: taken from [197]

b:taken from [216]

c: taken from [28]

d: taken from [108]

e: taken from [76]

f: taken from [147]

g: taken from [148]

h: taken from [143]

i: taken from [164]

j: taken from [256]

k: taken from [257]

l: taken from [258]

m: [unpublished]

Transgenic ecdysteroid-responsive mammalian cells

There is considerable interest developing in the use of (modified) ecdysteroid receptors as components in gene switching systems in transformed mammalian or plant cells [56]. Figure (9) provides a generalised diagram of how such a system functions. Some of these systems have been commercialised as tools for the examination of the effects of temporally- and spatially-regulated gene expression (Invitrogen system [http://www.invitrogen.com] and RHeoGene system [http://rheogene.com]). These systems show low basal activity of the transformed gene, high inducibility and ligand-dependent induction. However, gene expression systems in mammalian and plants cells possess markedly different ecdysteroid specificities to ecdysteroid receptors in insect systems. Both the affinity and specificity seem to be affected. Thus, the generally accepted endogenous hormone in insects, 20E (1-1), is inactive in transgenic systems. Two phytoecdysteroids, muA (2-5) and poA (2-7), are normally used to activate the transgenic systems, but even these are required at at least 100-fold higher concentrations than in insect systems; e.g. EC_{50} values for MuA and PoA in the *D. melanogaster* B_{II} bioassay are 2.2 x 10^{-8}M and 3.1 x 10^{-10}M, respectively [28], while concentrations of 1-10μM are required to induce transgenic expression. The basis of this altered affinity/specificity is not clear and it could derive from: i) altered metabolism, ii) use of RXR, rather than USP, iii) altered transportation into cells, iv) fusion of the EcR ligand-binding domain to the GR DNA-binding domain and/or VP16 to form VgEcR.

A limited investigation of the ecdysteroid specificity of VgEcR/RXR in CV-1 cells has been performed [199]. MuA, PoA and 14dMuA were almost equivalently active (EC_{50} = ca. 5 x 10^{-7}M), with ponasterone C being moderately active (EC_{50} = 2 x 10^{-5}M), PolB (2-6) being weakly active and 20E (1-1), Ino, MaA (2-3), E (2-1), 2dE, 20E22Ac and 2d20E being inactive or only very weakly active at 10^{-4}M. This study also showed that the presence of a natural (9-*cis*-retinoic acid) or synthetic (LG268 or LG1069) RXR ligand, while inactive in itself, potentiated the activity of ponA by 3- to 5-fold. More extensive QSAR studies of ligand potency under these conditions could contribute to the identification of more potent ligands for use with transfected cells.

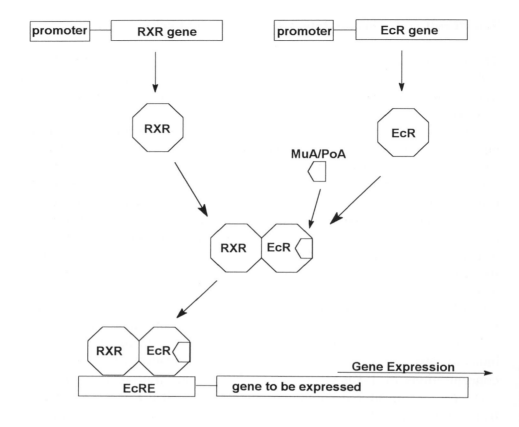

Figure 9: Generalised scheme for an ecdysteroid-regulated gene switch system; RXR = retinoid X receptor, EcR = ecdysteroid receptor, EcRE = ecdysteroid response element, MuA = muristerone A, PoA = ponasterone A.

Cell-free assays

Radioligand displacement assays

Cell-free receptor assays have not been extensively used to investigate ligand specificity of ecdysteroid receptors. This is largely a consequence of the need to extract the receptor from a suitable biological source in adequate amounts to carry out assays with several analogues and the lability of the receptor. Two radiolabelled ecdysteroids ([^{3}H]PoA and [^{125}I]26-iodoPoA) have been employed as ligands in the competition assays, using adsorption to either charcoal or nitrocellulose membranes, respectively, for separation of bound radioligand from free. In most cases, the assessment was limited to just 3 ecdysteroids (E, 20E and PoA): *D. melanogaster* Kc cell extract [200], *Calliphora erythrocephala* larval extracts [201], *Galleria mellonella* [202] *Chironomus tentans* larval extract [203]. However, one more extensive comparison of ecdysteroids has been conducted with receptor-containing extract from B_{II} cells (Table {**6**} [143]). With the advent of cloned ecdysteroid receptor genes (EcR and USP) from various insect species, it is now possible to translate these *in vitro* or bacterially express the proteins to provide an easier strategy to provide larger amounts of the receptor proteins. This approach has been used to compare the potencies of a series of turkesterone 11α-acyl esters in the cellular B_{II} bioassay with their affinities for dmEcR/dmUSP complexes [143]. It can be expected that this approach will be more widely applied in the future. For example, Makka et al. [204] have very recently determined the relative binding affinities of the ecdysteroids present in *Bombyx mori* eggs for *in vitro* expressed BmEcR-B1/BmUSP receptor complexes: PoA (IC50 = 8.9×10^{-8}M), 20E (9.5×10^{-7}M), 2d20E (5.6×10^{-5}M), 22d20E (8.1×10^{-5}M), 20E22Ac (1.1×10^{-4}M), E (1.8×10^{-4} M), 2dE (5.1×10^{-3}M) and E22P (5.9×10^{-3}M).

Gel-shift assays

A further possibility for examining the influence of ecdysteroid structure on ecdysteroid receptor function is gel mobility assays involving the co-incubation of the ecdysteroid with EcR/USP-containing preparation and an appropriate EcRE (usually the hsp27 EcRE) in a labelled form. After equilibration, the reaction mixtures are separated by PAGE, which retards the migration of EcR/USP/EcRE complexes relative to unbound EcRE. Ecdysteroid promotes high-affinity interaction between EcR/USP and EcRE and thus by conducting a series of incubations with ecdysteroids at differing concentrations, one can, at least in a semi-quantitative manner, determine the relationship between ecdysteroid structure and the activation of the receptor complex. This approach is also valuable to characterise the mode of action of ecdysteroid antagonists [205], as failure to activate the receptor complex in spite of interaction with the LBD is an important type of antagonism.

Table 6: K_d values for a range of ecdysteroids and two bisacylhydrazines in cell-free extracts of B_{II} cells. Competition assays were performed with [^{3}H]ponasterone A (0.2 nM), followed by charcoal assay, to determine IC_{50} values, which were then converted to K_d values. Data taken from [143].

Ecdysteroid	K_d value
ponasterone A	7.2×10^{-10}M
ajugasterone C	1.3×10^{-9}M
muristerone A	2.2×10^{-9}M
25-fluoro-20-hydroxyecdysone	5.2×10^{-9}M
polypodine B	7.7×10^{-9}M
5-deoxykaladasterone	1.0×10^{-8}M
abutasterone	1.1×10^{-8}M
cyasterone	1.7×10^{-8}M
20-hydroxyecdysone	2.5×10^{-8}M
inokosterone	3.5×10^{-8}M
makisterone A	6.4×10^{-8}M
turkesterone	1.4×10^{-7}M
RH5992	2.1×10^{-7}M
ajugalactone	3.1×10^{-7}M
RH5849	1.5×10^{-7}M
ecdysone	1.7×10^{-6}M
2-deoxyecdysone	3.8×10^{-6}M
22-*epi*-ecdysone	6.7×10^{-6}M
poststerone	1.5×10^{-5}M

Taste receptors

From the observation that the common shore crab (*Carcinus maenas*; Crustacea, Decapoda) rejects ecdysteroid-containing food, Tomaschko et al. [206] developed an *in vivo* oesophagus dilation assay, capable of comparing ecdysteroid analogues: 20E (EC_{50} = 1.8 x 10^{-4}M) and 20E22Ac (4.6 x 10^{-4}M).

Electrophysiological experiments to determine the relative abilities of contact chemoreceptors to perceive E, 20E and PoA in larvae of *Mamestra brassicae* [40], *Bombyx mori*, *Spodoptera littoralis* and *Ostrinia nubilalis* [118]. In *M. brassicae* a specific cell is present in the lateral sensilla of the maxilla which responds to 20E and PoA, but not to E. In *B. mori*, the medial sensilla were responsive to 20E and PoA, but not E. This finding confirmed previous observations [39]. Both lateral and medial sensilla were responsive to E, 20E and PoA in *O. nubilalis*, with a threshold of perception of 1 μM. Lateral and medial sensilla of *S. littoralis* larvae perceived 20E (with a threshold of 10 μM), but not E or PoA. The detected concentrations are compatible with those found in phytoecdysteroid-containing plants. These findings indicate the perception of PEs is common amongst phytophagous insect species.

A more extensive study of the perception of ecdysteroid intracellular receptor agonists and antagonists [113] by medial sensilla of larvae of *O. nubilalis* has shown that the compounds could be categorised into 4 groups [119]: Group 1 (inactive: RH5849, RH5992, cucurbitacin B, 2,3-dihydro-3β-methoxywithacnistine, 2,3-dihydro-3β-hydroxy-withacnistine and 2,3-dihydro-3β-methoxyiochromolide), Group 2 (weakly active: 5-*epi*-20-hydroxyecdysone, 9,20-dihydroxyecdysone, canescensterone, ecdysone 22-myristate), Group 3 (active: E, 20E, Ino, MaA and pinnatasterone), Group 4 (very active: PoA {2-7}, polB {2-6}, Cya {4-4} and pterosterone {2-8}). Thus, it appears that antagonists (cucurbitacins and withanolides) and non-natural agonists (BAHs) are not perceived, while ecdysteroids differ considerably in their perception. The activity sequence is not the same as for intracellular ecdysteroid receptors, and this may go some way to explain why ecdysteroid-accumulating plant species possess different ecdysteroid profiles, as, depending on the specific plant-insect interaction, the ecdysteroids may be targeting the intracellular or taste receptors.

CONCLUSIONS FROM EARLY EMPIRICAL QSAR STUDIES

The early literature on ecdysteroid structure-activity studies has been reviewed [120,150,171]. The various bioassays employed vary in their complexity of performance and interpretation. A number of factors can affect the potency of a test compound to a greater or lesser extent depending on the assay system: penetration, metabolism, excretion rate, sequestration and target site activity. *In vivo* assays will give a measure of all of these acting in concert, but not allow the individual contributions to be identified. On the other hand, cell-free receptor assays probably give a direct measure of target activity, but may bear little resemblance to the *in vivo* situation.

Horn and Bergamasco [120] summarised their main conclusions regarding ecdysteroid structure-activity relationships in insect systems as follows:

i) and A/B *cis*-ring junction is essential
ii) a 6-oxo-7-ene grouping is essential
iii) a full sterol side-chain is essential
iv) a free 14α-hydroxyl group is essential
v) 2β-, 3β- and 20*R*-hydroxyl groups enhance biological activity
vi) a C-25 hydroxyl diminishes activity

These conclusions were empirically derived from data obtained with a number of different bioassay systems (mainly the *Calliphora, Musca* and *Sarcophaga* assays and the *Chilo* dipping test). On the basis of the above summary, Bergamasco and Horn [171] visualised the ecdysteroid interacting with the ligand-binding pocket of the receptor at 3 major sites: the β-face of the A- and B-rings, the α-face in the region of the 14α-hydroxyl group and the side-chain from C-22 to C-27.

MORE RECENT QSAR/MOLECULAR MODELLING STUDIES

Several major technical developments have taken place in recent years which have permitted a more extensive and sophisticated approach to ecdysteroid QSAR. These are:
i) the ability to identify unambiguously by nmr ecdysteroid samples of less than 1 mg [207].
ii) the availabiltiy of routine HPLC for the verification of purity of ecdysteroid samples.
iii) the development of simple, quick and effective bioassays for the quantitative comparison of the potencies of many ecdysteroid analogues.
iv) the rapid development of computing power and molecular modelling software for non-empirical QSAR methods.

The most informative analysis of ecdysteroid/receptor interaction would be by X-ray crystallography. Although it has been possible to crystallise the LBDs of several nuclear receptors and to determine their structures by X-ray crystallography (ER [208], PR [209], RAR [210], RXR [211], TR [212], USP [213] and VDR [214]), it has not yet been possible to do this for any EcR protein. This probably derives from the fact that EcR only binds ligand with high affinity when it is complexed with USP (RXR), necessitating the co-crystallisation of ligand + EcR LBD + USP LBD. Ultimately it should be possible to achieve this. In the meantime, two approaches are being used to characterise the LBD: i) ligand-based approaches and ii) homology model-based approaches making use of the crystal data for analogous nuclear receptor LBDs which have been crystallised.

Ligand-based molecular modelling

To date, these fall into two categories; those seeking to explain the biological activity of BAHs by developing overlap models (see below) and those using biological activity data for ecdysteroid data sets. The first extensive data set for pure ecdysteroids in a homogeneous assay (B_{II} bioassay) was collected by Dinan et al. [28] and analysed by CoMFA. The potency of ecdysteroids in this assay is believed to reflect their affinity for the ecdysteroid receptor complex. In CoMFA, a template molecule (in this case E) is orientated in a lattice made up of 1Å-sided cubes, according to the conformation derived from its crystal structure. The other ecdysteroids were then aligned according to the 17 carbons of the steroid nucleus. Thus, two models were generated: model A, based on the electrostatic indicator field ($r^2 = 0.903$; $q^2 = 0.631$; 5 components) and model B, based on electrostatic (54% contribution) and steric (37% contribution) fields together with MlogP (9% contribution) ($r^2 = 0.892$; $q^2 = 0.694$; 5 components). Model A revealed that positive charge is favoured on the β-face between the C-2 and C-3 hydroxyls, enveloping C-24 and wrapping round and extending to the end of the side-chain and with smaller regions at C-21 and below C-17. Positive charge is disfavoured most significantly at O-20 and between C-26 and C-27, with less important regions on the β-sides of C-5, C-6 and C-8 and scattered around the end of the side-chain. Model B shows positive charge to be favoured on the near-side of C-24 and extending to H-25, while negative charge is favoured near O-20, on the β-face of C-6 and on the β-face of C-26/C-27 and extending to H-25. Steric bulk is favoured between 11α-H and 11β-H and steric bulk is disfavoured around the perimeter of the side-chain. On the basis of this study, it was possible to put forward a pharmacophore hypothesis for ecdysteroid receptor binding (Fig. {**10**}). In summary, this proposes that ecdysteroid binding is the result of incremental contributions of a number of molecular features, each of which may contribute several -log(EC$_{50}$) units to the binding affinity. These features are heteroatoms at C-2, C-3, C-20, C-22, a large dipole at C-6, and a moderately bulky hydrophobic group attached to C-22 (side-chain), all at approximately the distances and angles shown in Figure (**10**). Other heteroatoms might replace the oxygens. CoMFA does not identify any functionality capable of H-bond donation (but see the section on 4D-QSAR below). It is anticipated that ecdysteroid activity is very sensitive to small changes in the geometry around the C-20 and C-22 dipoles. One or more of the six pharmacophore elements may be absent without full loss of receptor activation.

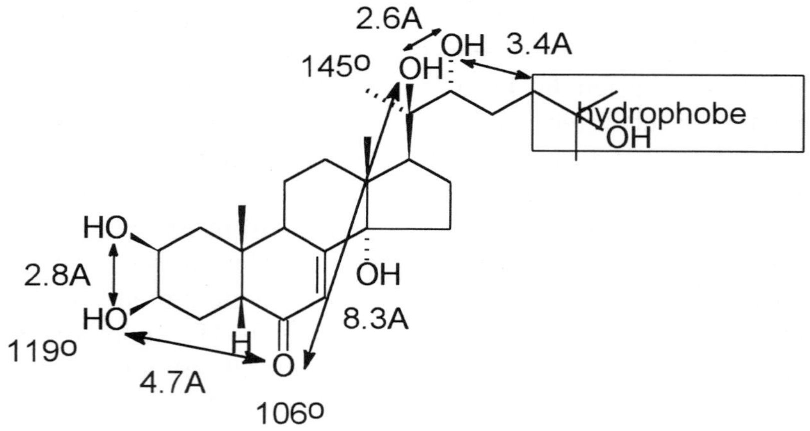

Figure 10: Pharmacophore hypothesis for ecdysteroid potency in the *Drosophila melanogaster* B$_{II}$ bioassay (redrawn from [28]).

The same data set of 71 ecdysteroids (training set) together with a test set of 20 ecdysteroids have been used for a 4D-QSAR study and the outcomes of this compared to the previous CoMFA study, to assess their complementarity [108]. 4D-QSAR has the advantage that it can take into account multiple conformations, whereas CoMFA is based on one conformation of the ligand (usually the crystal structure). Consequently, 4D-QSAR is more structure-activity specific than CoMFA. The initial phase of the 4D-QSAR study was similar to that for the CoMFA study, in that the crystal structure of E was placed into a cubic lattice grid of 1Å resolution. The other ecdysteroids were aligned by fixing the locations of 3 C-atoms (which span across the molecule) in 3D-space and superimposing the structures on the ecdysone template. A set of interaction pharmacophore elements (IPEs) was then assigned to each compound, depending on the type of atom (non-polar, polar/positive, polar/negative, H-bond acceptor, H-bond donor or any type). Each compound was then subjected to conformational ensemble profiling (CEP), which generates information about the conformational flexibility of the molecule. 10000 conformations were sampled for each molecule over a simulation sampling time of 10 ps at 300 K. Each conformation identified from the CEP is placed in the grid cell space according to the alignment. Then, grid cell occupancy descriptors (GCODs) are computed by recording how often a given IPE occupies a given grid cell over the CEP. This generates a vast number of GCODs because of the large number of grid cells and the six IPE types. Partial least squares regression analyses between the dependent variable (biological activity; $-\log[EC_{50}]$) and the set of GCODs are then performed. The 200 most highly weighted GCODs were used as the trial set for deriving the 4D-QSAR models. The final 4D-QSAR models are generated using a genetic algorithm using the 200 PLS-determined descriptors and any other non-GCOD descriptors (e.g. MlogP), which are considered relevant. This process is repeated for each trial alignment and the best models can be combined to create a manifold 4D-QSAR model, which can be used to make consensus predictions of the activities of test compounds. The advantages of this approach are:

- prediction of the active conformation and best alignment of each molecule (which could then be used for CoMFA, if desired)
- the difference between observed and predicted activity can be interpreted in terms of conformational flexiblity (conformational entropy), i.e. an active conformation will occupy the activity-enhancing GCODs, while an inactive conformation will tend to occupy the activity-decreasing GCODs and the CEP (average of the conformations) of the molecule leads to both activity-enhancing and activity-decreasing GCODs being partially occupied.
- 4D-QSAR models can be used as a virtual high-throughput screen to identify further potent ligands.

The best models were based on alignments of C-atoms 6, 12 and 16 (alignment 1) and 2, 3 and 6 (alignment 3), both of which were improved by the inclusion of MlogP (alignment 1 + MlogP: $r^2 = 0.872$, $q^2 = 0.799$; alignment 3 + MlogP: $r^2 = 0.884$, $q^2 = 0.800$). The important GCODs are located near C-2, C-3, C-11, C-12, C-20, C-22 and along the side-chain of the steroid. Many of the GCODs of the "any type" have a negative regression, i.e. cannot be occupied by any ecdysteroid atom without a loss of activity, and this can be interpreted as meaning that this space is occupied by the receptor.

The predictions drawn from the study were:
- C-2 hydroxyl acts as a H-bond acceptor, with restricted space in the receptor binding cavity around this position
- C-3 should be substituted, preferably with a polar negative atom
- C-12 region fits into a hydrophobic region of the receptor
- C-20 hydroxyl is a H-bond donor
- C-22 hydroxyl is a H-bond acceptor
- C-25: not an H-bond acceptor
- the side-chain fits into a non-polar cylindrical binding pocket

In addition, no GCODs were detected for C-6 or C-14, which can be interpreted as meaning that these positions are not involved in receptor binding, or that they are equally involved across the training set. As the training set involved some structural variation at both of these positions, the former possibility is the more likely.

Comparison of the best 4D-QSAR models with the best CoMFA models revealed that they had similar r^2 values, but that the q^2 values are higher for the 4D-QSAR models.

More recently an extended ecdysteroid data set (67-member training set and 33-member test set) and an extensive BAH data set (97-member training set, 19-member test set) have been used for consensus CoMFA/4D-QSAR analysis of 8 possible overlap hypotheses for ecdysteroids and BAHs [Hormann et al., submitted]. A ecdysteroid/BAH superimposition with the BAH with a conformation showing an N-N axis of P-chirality, which is similar to one proposed by Wurtz et al. [107] from homology modelling (see below) was deemed the most probable, but further superimpositions (including one similar to those suggested by Nakagawa et al. [154,156]) were also feasible.

The importance of these models is that they generate testable predictions. For example, we have tested the prediction that there is steric space around C-11 by generating a series of 11α-acyl esters of turkesterone and demonstrating that they retain significant biological activity [143]. We are also preparing suitable derivatives to determine which hydroxyl groups are involved in H-bonding (and how) and to clarify the role of the C-6 ketone. Further, the CoMFA models were used to assess whether ecdysteroid 7,9(11)-dien-6-ones would possess high biological activity before synthesising them as potential photoaffinity labels [148]. Further, such computer models have considerable potential for virtual screening of large data bases of compounds for the identification of active compounds with novel chemistry, either for the identification of novel leads for insecticidal compounds, or for the screening of environmental compounds to identify those which might act as endocrine disruptors in arthropods (see below).

Receptor homology modelling and docking experiments

The structures of the LBDs of several nuclear receptors have been determined by X-ray crystallography [210] and a general structure identified, with conformations which are characteristic for interaction with agonists and antagonists [215]. The common 3D-structure consists of 11 stretches of α-helix (H1 and H3-H12) and 1 region of antiparallel β-sheet, forming a 3-layered α-helical sandwich motif, with H4, H5, H6, H8 and H9 sandwiched between H1 and H3 on one side and H7, H10 and H11 on the other side, creating a hydrophobic binding pocket for the ligand. Helix-12 forms a cap which closes over the binding pocket when an agonistic ligand is bound. However, the precise structure of the LBD is highly dependent on the primary sequence, and these structural changes alter the size and shape of the binding pocket significantly.

Crystallisation of the LBD of EcR and subsequent analysis by X-ray crystallography has not yet been reported. This is not a facile task because EcR only binds ligand with high affinity when it is complexed with USP, and co-crystallisation of the LBDs of EcR and USP and ligand will not be straightforward. Using the 3D-structures of the LBDs of several other nuclear receptor family members as a basis, several groups have developed homology models for EcR; these have focussed on the known crystal structures of RAR and VDR, as these are in the same phylogenetic subfamily as EcR [79], but the sequence conservation is only 25-30%.

Wurtz et al. [107] constructed 2 homology models of the *Chironomus tentans* EcR using the known crystal structures of the human retinoic acid γ (hRARγ) and human vitamin D (hVDR) receptor LBDs as templates, generating the EcRra and EcRvd models, respectively. Both models identify binding pockets which consist of a bulky envelope and shallow tube, but the orientations are different (by 180°) in the two models. They then used these models for docking experiments with 20E and RH5849. In each model, the A-ring of 20E is located in the bulky envelope and the side-chain is located in the shallow tube. The models identify potential H-bond interactions between the receptor and the ecdysteroid. The authors compare the known biological activities of a range of ecdysteroids in the *D. melanogaster* B_{II} bioassay [216] with the expected impact of the associated structural modifications relative to 20E on the molecular space and H-bonding capabilities of the molecules. Neither model accords fully with the biological data; for example, both models predict H-bonding to the C-25 hydroxyl of 20E, but PoA (25-deoxy-20-hydroxyecdysone) has a much higher biological activity (ca. 100-fold) in all arthropod systems, which is not compatible with the C-25 hydroxyl group of 20E H-bonding to the receptor. Similarly, the models do not fully explain the biological activities of some BAHs, although the EcRvd model appears to be better in this respect than the EcRra model. Both models suggest that the *tert*-butyl group of BAHs (which is essential for activity) fits into a non-polar groove, which is not occupied by 20E. Furthermore, a superimposition model is suggested whereby the plane of the BAH A-ring (next to the N-*tert*-butyl) is roughly parallel to the A-ring of the steroid and the B-ring of the BAH superimposes with the beginning of the aliphatic side-chain of the steroid. However, in the EcRra model the A-rings of the steroid and BAH are almost superimposed, while in the EcRvd model they are parallel, but apart from one another.

Santos and Sant'Anna [217] developed homology models for *D. melanogaster* (Diptera) and *Manduca sexta* (Lepidoptera) EcR LBDs based on the common 3D-architecture of nuclear receptors. Their motivation was to seek an explanation for the greater selectivity of certain BAHs (e.g. RH-5992 {**8-2**} and RH-2485 {**8-4**}) for lepidopteran receptors over dipteran receptors. Again, two possible orientations for 20E in the binding pocket were identified; one with the A-ring near the entrance to the pocket (orientation I) and the other with the side-chain near the entrance (orientation II), the more favourable interaction enthalpy being associated with orientation I. This concurs with the EcRvd model of Wurtz et al. [107]. In orientation I, 20E is predicted to form three H-bonds with msEcR (involving C-2, C-6 and C-20 of the steroid) and 4 H-bonds with dmEcR (involving C-2, C-6, C-14 and C-22). In addition the ecdysteroid's methyl groups are in contact with apolar amino-acid side-chains. The BAHs RH-5849 (8-1) and RH-2485 each possessed two possible orientations with each of the receptor LBD models, depending on which phenyl ring is close to the entrance of the binding pocket. Orientation I (B-ring close to the entrance of the binding pocket) is the more energetically favoured, owing to the formation of a greater number of and/or stronger H-bonds. Interestingly, the interaction enthalpy of RH-2485 with msEcR is significantly more negative than for dmEcR, whereas the interaction enthalpies for RH-5849 are similar for msEcR and dmEcR, which is in accord with the greater activity of RH-2485 towards lepidopteran ecdysteroid receptors and the similar activity of RH-5849 with dipteran and lepidopteran receptors. In agreement with the proposal of Wurtz et al. [107], the *tert*-butyl group is located in a hydrophobic cavity which is not occupied when 20E is the ligand. The models also indicate that dmEcR cannot accommodate the 3,5-dimethylphenyl ring (ring-A) of RH-2485 because of steric hindrance with amino acid residues not common to msEcR, forcing the ligand to adopt a different position within the dmEcR binding pocket which is associated with weaker intermolecular interactions than are possible with msEcR.

    Kumar et al. [109] have developed a homology model for the *Choristoneura fumiferana* (Lepidoptera) EcR LBD based on the known crystal structures for the LBDs of the oestrogen receptor, thyroid hormone receptor and vitamin D receptor. Docking experiments with 20E identified two general binding modes, which differed by the ligand being rotated by 180° in the binding pocket. The authors identified amino acid residues which were located within 3.6Å of the docked 20E. Site-directed mutagenesis was then used to modify these amino acids to ala, pro or phe and the mutant receptors were expressed in 3T3 cells and their ability to regulate an ecdysteroid-dependent reporter gene determined in the presence of PoA (2-7) or RG-102240 (GS-E; *N*-[1,1-dimethylethyl]-*N'*-[2-ethyl-3-methoxybenzoyl]-3,5-dimethylbenzohydrazide, 8-5). Most of the mutations significantly reduced both ponA and RG-102240 activity, whereas the mutations A110P and R95A (which are located at the top of the binding cavity) showed a dramatic reduction in ponA activity, but retention of RG-102240 activity. It is believed that these residues make contact with ecdysteroidal ligands, but not with the more compact BAHs (which presumably interact with residues only at the bottom of the binding cavity). Thus, it is possible to separate ecdysteroid activation from BAH activation. This approach is being used to develop modified ecdysteroid receptors as gene switches which can be selectively activated by different ligands, as part of RHeoGene's RHeoplex system (see below).

Thus, EcR LBD homology models based on vertebrate nuclear receptor crystal structures, although instructive for general architecture and residue replacement, are, owing to their low sequence identity, rather speculative when used for ligand docking or as tools for virtual screens [164].

ECDYSTEROID ANTAGONISTS

Antagonistic ecdysteroids

Koreeda et al. [218] reported that the phytoecdysteroid ajugalactone (**4-1**) suppressed the moulting hormone activity of PoA, but not that of 20E, in the *Chilo suppressalis* dipping test. However, no antagonistic activity of ajugalactone was found in the B_{II} bioassay against 20E or PoA. On the contrary, ajugalactone demonstrated agonistic activity ($EC_{50} = 1.6 \times 10^{-7}$ M [28,197]). Approximately 150 pure ecdysteroids have been assessed in the B_{II} bioassay, and until recently none had demonstrated antagonistic activity against 20E (5×10^{-8}M). However, we have recently found that 24-hydroxycathamosterone (**11-1**) possesses antagonistic activity ($EC_{50} = 2.0 \times 10^{-5}$M vs. 5×10^{-8}M 20E) [219]. The stereochemistry of this compound has not yet been fully elucidated, but it will be interesting to determine what makes this compound antagonistic while other structurally closely related compounds are agonistic.

Other classes of ecdysteroid antagonists

Antagonists of vertebrate steroid receptors are proving experimentally and therapeutically very significant [220-222]. It has been known that certain plants contain ecdysteroid receptor agonists (phytoecdysteroids) for over 30 years. It seemed reasonable to assume that plants might also contain compounds which could antagonise the action of ecdysteroids. With the advent of simple, reliable bioassays, it became appropriate to search for such compounds. In the intervening years we have screened over 5000 species of higher plants, a number of fungi, mosses and lichens, and several hundred purified plant natural products to assess their ecdysteroid (ant)agonist activity [113]. The active compounds are summarised in Table (7). The first compounds to be identified as unambiguous ecdysteroid antagonists were cucurbitacins B (11-2) and D (11-3) from *Iberis umbellata* [205]. Antagonistic cucurbitacins and cucurbitanes have also been identified from *Hemsleya carnosiflora* [223], *Cercidiphyllum japonicum* [224] and *Physocarpus opulifolius* [225]. Certain withanolides (e.g. 11-4) are also antagonistic [226], as are certain limonoids (11-6 and 11-7 [227]), stilbenoids (e.g. 11-7, 11-8 and 11-9 [228-230]) and phenylalkanes (e.g. 11-10 [170]. Most of these compounds have been shown to interact with the ligand-binding domain of the ecdysteroid receptor, since they are able to displace [^{3}H]PoA from cell-free receptor preparations at concentrations similar to those found to be active in the B_{II} bioassay. Thus, current and future QSAR studies on these classes of molecules will provide alternative approaches to the characterisation of the LBD of EcR. Remarkably, mono-, di-, tri- and tetrastilbenes all interact with the ecdysteroid receptor ligand binding site with similar affinities (Table {7}), even though they differ considerably in size.

Within the cucurbitacins and withanolides it has been found that certain structural analogues possess agonistic rather than antagonistic activity: hexanorcucurbitacin D (EC_{50} = 2.5 x 10^{-5}M) and withaperuvin D (EC_{50} = 2.5 x 10^{-5}M) [170]. In the case of hexanorcucurbitacin D, this concurs with the hypothesis that antagonistic activity of cucurbitacins is associated with the presence of an α,β-unsaturated ketone in the side-chain [205], which hexanorcucurbitacin D lacks.

The ecdysteroid antagonists identified to date are all of rather low potency. This may be a consequence of them all being plant natural products, which the plant is able to produce in high concentrations at low cost to itself [231].

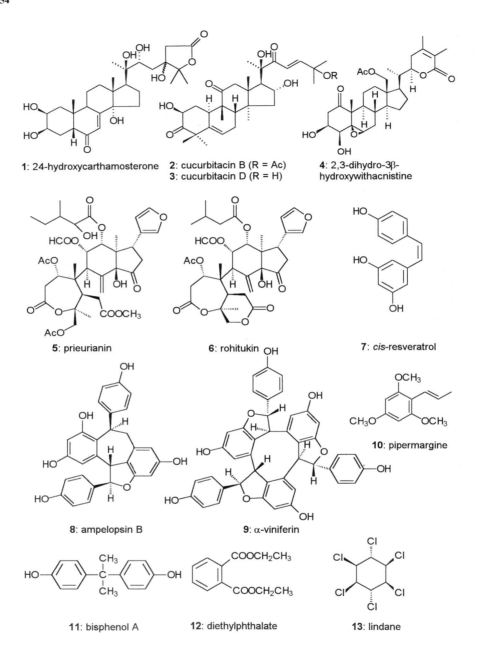

1: 24-hydroxycarthamosterone   **2**: cucurbitacin B (R = Ac)   **4**: 2,3-dihydro-3β-
 3: cucurbitacin D (R = H)   hydroxywithacnistine

5: prieurianin   **6**: rohitukin   **7**: *cis*-resveratrol

8: ampelopsin B   **9**: α-viniferin   **10**: pipermargine

11: bisphenol A   **12**: diethylphthalate   **13**: lindane

Figure 11: Examples of compounds which antagonise ecdysteroid action in the *Drosophila melanogaster* B$_{II}$ cell bioassay.

Table 7: Relative potencies of known ecdysteroid antagonists in the *Drosophila melanogaster* B_{II} bioassay. EC_{50} values were determined in the presence of 5 x 10^{-8}M 20-hydroxyecdysone. Only compounds for which it was possible to determine distinct EC_{50} values are included.

Compound	EC_{50}
Ecdysteroid	
24-hydroxycarthamosterone	$2.0\times10^{-5}M^a$
Cucurbitacins	
cucurbitacin B	$7.5\times10^{-7}M^b$
cucurbitacin C	$1.1\times10^{-6}M^b$
cucurbitacin D	$7.5\times10^{-7}M^b$
cucurbitacin E	$1.1\times10^{-6}M^b$
cucurbitacin F	$8.0\times10^{-7}M^b$
cucurbitacin I	$1.5\times10^{-6}M^b$
25-acetylcucurbitacin F	$1.2\times10^{-5}M^b$
deacetylpicracin	$7.3\times10^{-5}M^b$
3-*epi*-isocucurbitacin D	$7.0\times10^{-6}M^b$
carnosifloside II	$3.4\times10^{-4}M^c$
carnosifloside VI	$1.2\times10^{-4}M^c$
23,24-dihydrocucurbitacin F	$3.0\times10^{-5}M^c$
25-acetyl-23,24-dihydrocucurbitacin F	$3.0\times10^{-5}M^c$
Withanolides	
2,3-dihydro-3β-methoxywithaferin A	$3.5\times10^{-5}M^d$
2,3-dihydro-3β-methoxywithacnistine	$1.0\times10^{-5}M^d$
2,3-dihydro-3β-methoxyiochromolide	$5.0\times10^{-6}M^d$
2,3-dihydro-3β-hydroxywithacnistine	$2.5\times10^{-6}M^d$
Limonoids	
prieurianin	$1.0\times10^{-5}M^e$
rohitukin	$1.3\times10^{-4}M^e$
Stilbenoids	
cis-resveratrol (monostilbene)	$1.2\times10^{-5}M^f$
ampelopsin B (distilbene)	$3.3\times10^{-5}M^g$
α-viniferin (tristilbene)	$1.0\times10^{-5}M^g$
suffruticosol A (tristilbene)	$5.3\times10^{-5}M^f$
suffruticosol B (tristilbene)	$1.4\times10^{-5}M^f$
suffruticosol C (tristilbene)	$2.2\times10^{-5}M^f$
cis-miyabenol C (tristilbene)	$1.9\times10^{-5}M^h$
cis-miyabenol A (tetrastilbene)	$3.1\times10^{-5}M^h$

kobophenol B (tetrastilbene) $3.7 \times 10^{-5} M^h$

Phenylalkanoids
marginatine $5.0 \times 10^{-5} M^b$
croweacin $7.3 \times 10^{-4} M^b$
2,6-dimethoxy-3,4-methylenedioxy-1-(2-propenyl)-benzene $6.0 \times 10^{-4} M^b$
apiole $7.3 \times 10^{-4} M^b$
isoasarone $2.5 \times 10^{-4} M^b$
pipermargine $3.3 \times 10^{-4} M^b$

Alkaloids
isomitraphylline $1.6 \times 10^{-5} M^b$
isopteropodine $1.5 \times 10^{-5} M^b$
pteropodine $1.0 \times 10^{-5} M^b$

Industrial compounds
bisphenol A $1.0 \times 10^{-4} M^i$
diethylphthalate $2.0 \times 10^{-3} M^i$
lindane $3.0 \times 10^{-5} M^i$

a: data taken from [219]
b: data taken from [170]
c: data taken from [205]
d: data taken from [226]
e: data taken from [227]
f: data taken from [228]
g: data taken from [229]
h: data taken from [230]
i: data taken from [240]

APPLICATIONS AND FUTURE PROSPECTS

Agricultural applications

Design of more specific invertebrate pest control agents
Identification of compounds which specifically interfere with the mode of action of ecdysteroids in invertebrates is a very attractive and promising way of controlling pest species while reducing the effects on other animal groups and lowering the environmental burden. The effectiveness of such a strategy has recently been demonstrated by the identification of non-steroidal bisacylhydrazines as agonists of the ecdysteroid receptor complex and the development of several of these compounds as selective

and successful insect control agents [112,152,153]. The ecdysteroids themselves have very limited application in the control of pests, since they do not possess the required properties for successful pesticides; they are too polar, too chemically complicated and too metabolically and environmentally unstable [150]. However, they can contribute significantly as lead compounds. In this sense, the vast diversity of ecdysteroid analogues found in plants provides a natural resource for structure-activity studies (see previous section). This was realised some time ago [189], but the more restricted availablity of ecdysteroid analogues in very pure form, the requirement for larger amounts of each analogue and, most importantly, the complexity of the available bioassay systems then available, restricted the possiblities of this approach at that time. However, the methodological and technical advances of recent years, permitting rapid purification and identification of analogues, extensive QSAR and computerised molecular modelling, will ultimately permit the recognition of further simple, non-steroidal molecules which fit the hypothesis and may lead to novel insecticides. Such approaches can be extended to the taste receptors of insects to identify ecdysteroid analogues (steroidal or non-steroidal) which are potent antifeedants and to provide a further approach to crop protection.

Modification of ecdysteroid levels in crop species
An alternative approach to crop protection based on ecdysteroids is to modify ecdysteroid levels and/or profiles in crop plants (most of which do not contain ecdysteroids) to enhance protection against invertebrate predators. This is based on the assumption that ecdysteroids contribute significantly to the protection of the plants which contain them, at least against non-adapted predators. There is increasing evidence that this is true, but the situation is complicated by the knowledge that there are certain polyphagous insect pests which are unaffected by diets containing very high concentrations of ecdysteroids (reviewed in [26]). However, these tolerant species appear to detoxify the ingested ecdysteroids in specific ways (commonly by acylation with long-chain fatty acids at C-22 [42,43,46,232]) and it seems feasible that plants could be generated which contain ecdysteroid analogues which do not permit the ready biochemical detoxification by predators (e.g. by generating high levels of 22-deoxy ecdysteroids which retain activity on ingestion because they are not a substrate for the detoxification enzymes [233]). The very broad, but

uneven distribution of phytoecdysteroids in the plant world, the fact that low, but detectable levels are present in leaves of 40% of species and that individual ecdysteroid-containing plants can be found in a population of an ecdysteroid-negative species [231] all indicate that the genetic capacity to produce ecdysteroids is retained by most, if not all, plant species. If this is true, modification of the activity of genes which regulate ecdysteroid biosynthesis (either through a traditional selection approach, genetic modification or application of appropriate elicitors) could generate crop plants with elevated ecdysteroid levels. Very few common crop species contain ecdysteroids, with the notable exception of spinach [234]. Thus, ecdysteroids have potential for the GM or non-GM protection of crop species. The apparent non-toxicity of ecdysteroids to mammals [56] makes this potentially very attractive, as it may not be necessary to incorporate tissue- or temporally-specific accumulation of ecdysteroids into the system. The implementation of this strategy requires knowledge of which ecdysteroid analogues are the most effective antifeedants or the most toxic on ingestion (this is likely to vary from insect species to insect species) and a knowledge of the pathway(s) and regulation of phytoecdysteroid biosynthesis, which is currently lacking [3].

QSAR and molecular modelling will contribute signficantly to the characterisation of ligand-binding sites of both intracellular and taste receptors for ecdysteroids.

Improvement of silk yields

Feeding of exogenous ecdysteroids to larvae of the silkworm, *Bombyx mori*, at certain stages of development has been shown to enhance synchronous development of the larvae and, when co-administered with a juvenile hormone analogue, to elevate significantly the yield of silk obtainable from the cocoon [235,236]. Such treatments have considerable potential to improve the efficiency of silkworm rearing by Asian farmers, especially since the hormonal treatments could be based on extracts of appropriate indigenous plant species [237]. No systematic SAR study has yet been performed to identify the most effective ecdysteroid analogues for this application.

Environmental protection

The phenomenon of endocrine disruption by natural and man-made chemicals in the environment is causing considerable concern [238]. Most research has been directed towards aquatic vertebrate species, but it is now realised that invertebrates may also be subject to endocrine disruption [239]. However, research in this area is considerably hampered by the lack of fundamental knowledge which exists about endocrinological processes in invertebrates. However, the ecdysteroid system in insects is relatively well characterised and, thus, it is feasible to determine whether environmental contaminants might interfere with this hormonal system and to assess whether this plays any significant role *in vivo*. In view of the potentially enormous number of substances which need to be assessed, a simple and effective screening system is required. In this context the B_{II} bioassay, which is simple, rapid, robust and has the potential to be automated, offers itself as a preliminary screen to identify compounds which interact with ecdysteroid receptors. A survey of ca. 80 potential environmental contaminants [240] revealed that three (bisphenol A {**11-11**}, diethylphthalate {**11-12**} and lindane {**11-13**}) were quantifiable weak ecdysteroid antagonists. In each case, the biological activity co-chromatographed with the named compound, rather than being associated with any impurity present. The ultimate goal of such screening is, through QSAR and molecular modelling of active classes of compounds, to establish a virtual screen for environmental compounds which would act as a pre-screen to identify the compounds which could potentially interact with the ecdysteroid receptor complex.

Biomedical applications

Gene switching technology
There is considerable interest in gene replacement therapies to correct inborn errors. However, such introduced genes require the correct temporal expression to be fully effective. Thus, systems are being sought which allow these genes to be switched on and off by the availability of an appropriate elicitor. One transgenic system which is receiving attention is the ecdysteroid/ecdysteroid receptor complex [241] (Figure {**9**}), since the ecdysteroid receptor is not a natural component of vertebrate cells, ecdysteroids are believed to be non-toxic to vertebrates and ecdysteroids

do not appear to activate vertebrate steroid hormone receptors, even at very high concentrations. A number of permutations using either all of the EcR protein or the ligand-binding part fused to the DNA-binding domain of another member of the nuclear receptor superfamily are being considered. In insects, EcR naturally forms a heterodimer with USP. The homologue of USP in verbrates is the retinoid X receptor, RXR. When expressed in vertebrate systems, EcR will partner RXR. However, the ecdysteroid binding specificity is considerably altered, such that 20E is largely ineffective at inducing reporter gene activity, while the rare phytoecdysteroid MuA is much more effective, but only at high concentrations (1- 10 μM, whereas 100-fold lower concentrations are effective with EcR/USP heterodimers). PoA is also effective, but concentrations (1 – 10 μM) are required which are 1000-fold greater than those required in insect systems. The non-steroidal bisacylhydrazines (which also have very low mammalian toxicity) are being considered as alternative ligands [109,242]. Gene switches based on ecdysteroid systems are also being developed for plants [243-245]. While these developments are highly interesting, it is clear that they will have to undergo considerable refinement before commercially or medically viable systems are available [56]. Clearly, the identification of much more potent ligands is essential to the medical application of this technology, and QSAR should play an important role in identifying more potent analogues.

Effects of ecdysteroids on vertebrates

There is accumulating evidence that ingested PEs may be beneficial [56]. They appear to be mildly anabolic [60,246], supposedly without the adverse side-effects associated with vertebrate steroids or their analogues [247]. A number of reports suggest that ecdysteroids may be effective in the control of diabetes [248,249]. Recent studies [250,251] have shown that derivatives of PEs exhibit a hepatoprotective action which was related to their effect on liver chromatin and their ability to prevent negative shifts of phospholipid metabolism. Three ecdysteroids from *Ajuga decumbens* (cyasterone, polypodine B and decumbesterone A) have been reported to be potential cancer chemopreventative agents [252]. Thorough clinical trials are required to substantiate and confirm all the above findings.

Although no SAR currently exist for the pharmacological effects of ecdysteroids on any vertebrate species, as these effects become more generally accepted, it will be necessary to conduct such studies to determine which are the most effective analogues to apply.

ABBREVIATIONS

The names of ecdysteroids are abbreviated according to [253]
20E: 20-hydroxyecdysone
20E22Ac: 20-hydroxyecdysone 22-acetate
BAH: bisacylhydrazine
BmEcR: *Bombyx mori* EcR
BmUSP: *Bombyx mori* USP
CoMFA: comparative molecular field analysis
Cya: cyasterone
2dE: 2-deoxyecdysone
25dE: 25-deoxyecdysone
2d20E: 2-deoxy-20-hydroxyecdysone
14dmuA: 14-deoxymuristerone A
3DE: 3-dehydroecdysone
dmEcR: *Drosophila melanogaster* EcR
E: ecdysone
EcR: ecdysteroid receptor
EcRE: ecdysteroid response element
ER: oestrogen receptor
GCOD: grid-cell occupancy descriptor
GM: genetically modified
Ino: inokosterone
IPE: interaction pharmacophore element
LBD: ligand-binding domain
maA: makisterone A
maC: makisterone C
msEcR: *Manduca sexta* EcR
muA: muristerone A
PAGE: polyacrylamide gel electrophoresis
PE: phytoecdysteroid
polB: polypodine B
poA: ponasterone A

PR: progesterone receptor
Pos: poststerone
QSAR: quantitative structure-activity relationships
RAR: retinoic acid receptor
RXR: retinoid X receptor
SAR: structure-activity relationships
USP: ultraspiracle
VDR: vitamin D receptor

ACKNOWLEDGEMENTS

Research by the Insect Biochemistry Group at Exeter University has been/is being supported by the AstraZeneca, BBSRC, INTAS, NERC, RHeoGene LLC, Rohm & Haas and the University of Exeter. I thank the members of the group, past and present, for their contributions to the research and Dr. Anna Barlow for carefully proof-reading the manuscript. I should also like to thank the many researchers who generously provided samples of ecdysteroids and other natural products for biological testing at Exeter.

REFERENCES

[1]   Koolman, J., Ed.; Ecdysone: from chemistry to mode of action, Thieme Verlag, Stuttgart, **1989**.
[2]   Lafont, R.; *Arch. Insect Biochem. Physiol.*, **1997**, 35, 3-20.
[3]   Dinan, L.; *Phytochem.*, **2001**, 57, 325-339.
[4]   Goodwin, T.W.; Horn, D.H.S.; Karlson, P.; Koolman, J.; Nakanishi, K.; Robbins, W.E.; Siddall, J.B.; Takemoto, T.; *Nature*, **1978**, 272, 122.
[5]   Lafont, R.; Horn, D.H.S.; In: *Ecdysone: from Chemistry to Mode of Action*. Thieme Verlag, Stuttgart, **1989**, pp. 39-64.
[6]   Downer, R.G.H.; Laufer, H.; *Endocrinology of Insects*, A.R. Liss Inc., **1983**, 1, pp.
[7]   Koolman, J.; *Zool. Sci.*, **1990**, 7, 563-580.
[8]   Rees, H.H.; *Eur. J. Entomol.*, **1995**, 92, 9-39.
[9]   Thummel, C.S.; *Insect Biochem. Molec. Biol.*, **2001**, 32, 113-120.
[10]   Delbecque, J.-P.; Weidner, K.; Hoffmann, K.H.; *Invert. Reprod. Devel.*, **1990**, 18, 29-42.
[11]   Fugo, H.; Yamauchi, M.; Dedos, S.G.; *Proc. Jpn. Acad. B*, **1996**, 72, 34-37.
[12]   Lagueux, M; Hoffmann, J.A.; Goltzené, F.; Kappler, C.; Tsoupras, G.; Hetru, C.; Luu, B.; In: *Biosynthesis, Metabolism and Mode of Action of Invertebrate Hormones*, Springer Verlag, Berlin, **1984**, pp. 168-180.

[13]   Rees, H.H.; Isaac, R.E.; In: *Biosynthesis, Metabolism and Mode of Action of Invertebrate Hormones*, Springer Verlag, Berlin, **1984**, pp. 181-195.

[14]   Dinan, L.; *Comp. Biochem. Physiol. B*, **1997**, 116, 129-135.

[15]   Hoffmann, K.H.; Espig, W.; Thiry, E.; *GIT Fachz. Lab.*, **1989**, 5, 429-440.

[16]   Ohnishi, E.; *Sericologia*, **1981**, 21, 14-22.

[17]   Thompson, M.J.; Svoboda, J.A.; Weirich, G.F.; *Steroids*, **1984**, 43, 333-341.

[18]   Hagedorn, H.H.; In: *Comprehensive Insect Physiology, Biochemistry and Pharmacology*, **1985**, 8, 205-262.

[19]   Hoffmann, J.A.; Lagueux, M.; In: *Comprehensive Insect Physiology, Biochemistry and Pharmacology*, **1985**, pp.435-460.

[20]   Denlinger, D.L.; In: *Comp. Insect Physiol. Biochem. Pharmacol.*, Pergamon Press, **1985**, 8, 353-412.

[21]   Spindler, K.-D.; Jaros, P.; Weidemann, W.; In: *Reproductive Biology of Invertebrates*, **1998**, X (Part A), pp. 243-269.

[22]   Huberman, A.; *Aquaculture*, **2000**, 191, 191-208.

[23]   Stewart, D.M.; In: *Endocrinology of Selected Invertebrate Types*, A.R. Liss Inc., **1988**, pp. 415-428.

[24]   Joly, R.; Descamps, M.; In: *Endocrinology of Selected Invertebrate Types*, A.R.Liss Inc., **1988**, pp. 429-449.

[25]   Barker, G.C.; Chitwood, D.J.; Rees, H.H.; *Invert. Reprod. Devel.*, **1990**, 18, 1-11.

[26]   Dinan, L.; *Russ. J. Plant Physiol.*, **1998**, 45, 296-305.

[27]   Blackford, M.J.P.; Dinan, L.; *J. Insect Physiol.*, **1997**, 43, 315-327.

[28]   Dinan, L.; Hormann, R.E.; Fujimoto, T.; *J. Comp.-aided Mol. Des.*, **1999**, 13, 185-207.

[29]   Melé, E., Messeguer, J.; Gabarra, R.; Tomás, J.; Coll, J.; Camps, F.; *Entomol. Exp. Appl.*, **1992**, 62, 163-168.

[30]   Mondy, N.; Caïssa, C.; Pitoizet, N.; Delbecque, J.-P.; Corio-Costet, M.-F.; *Arch. Insect Biochem. Physiol.*, **1997**, 35, 227-235.

[31]   Sláma, K.; Abubakirov, N.K.; Gorovits, M.B.; Baltaev, U.A.; Saatov, Z.; *Insect Biochem. Molec. Biol.*, **1993**, 23, 181-185.

[32]   Tanaka, Y.; Takeda, S.; *J. Insect Physiol.*, **1993**, 20, 805-809.

[33]   Robbins, W.E.; Kaplanis, J.N.; Thompson, M.J.; Shortino, T.J.; Joyner, S.C.; *Steroids*, **1970**, 16, 105-125.

[34]   Kubo, I.; Klocke, J.A.; Asano, S.; *J. Insect Physiol.*, **1983**, 29, 307-316.

[35]   Arnault, C.; Sláma, K.; *J. Chem. Ecol.*, **1986**, 12, 1979-1986.

[36]   Camps, F.; Coll, J.; *Phytochem.*, **1993**, 32, 1361-1370.

[37]   Savolainen, V.; Wuest, J.; Lafont, R.; Connat, J.-L.; *Experientia*, **1995**, 51, 596-600.

[38]   Ma, W.-C.; *Ent. exp. appl.*, **1969**, 12, 584-590.

[39]   Tanaka, Y.; Asaoka, K.; Takeda, T.; *J. Chem. Ecol.*, **1994**, 20, 125-133.

[40]   Descoins, C.; Marion-Poll, F.; *J. Insect Physiol.*, **1999**, 871-876.

[41]   Robbins, W.E.; Kaplanis, J.N.; Thompson, M.J.; Shortino, T.J.; Cohen, C.F.; Joyner, S.C.; *Science*, **1968**, 161, 1158-1160.

[42]   Kubo, I.; Komatsu, S.; Asaka, Y.; de Boer, G.; *J. Chem. Ecol.*, **1987**, 13, 785-794.

[43]   Robinson, P.D.; Morgan, E.D.; Wilson, I.D.; Lafont, R.; *Physiol. Entomol.*, **1987**, 12, 321-330.

[44]   Tanaka, Y.; Naya, S.-I.; *Appl. Entomol. Zool.*, **1995**, 30, 285-294.

[45]   Blackford, M.; Clarke, B.; Dinan, L.; *J. Insect Physiol.*, **1996**, 42, 931-936.

64

[46] Blackford, M.; Dinan, L.; *Insect Biochem. Molec. Biol.*, **1997**, 27, 167-177.
[47] Blackford, M.; Dinan, L.; *Entomol. Exp. Appl.*, **1997**, 83, 263-276.
[48] Jones, C.G.; Firn, R.D.; *Phil. Trans.Roy. Soc., Lond. B*, **1991**, 333, 273-280.
[49] Alison, B.; Whiting, P.; Sarker, S.D.; Dinan, L.; Underwood, E.; Šik, V.; Rees, H.H.; *Biochem. System. Ecol.*, **1997**, 25, 255-261.
[50] Schmelz, E.A.; Grebenok, R.J.; Galbraith, D.W.; Bowers, W.S.; *J. Chem. Ecol.*, **1998**, 24, 339-360.
[51] Schmelz, E.A.; Grebenok, R.J.; Galbraith, D.W.; Bowers, W.S.; *J. Chem. Ecol.*, **1999**, 25, 1739-1757.
[52] Schmelz, E.A.; Grebenok, R.J.; Ohnmeiss, T.E..; Bowers, W.S.; *J. Chem. Ecol.*, **2000**, 26, 2883-2896.
[53] Schmelz, E.A.; Grebenok, R.J.; Ohnmeiss, T.E.; Bowers, W.S.; *Arch. Insect Biochem. Physiol.*, **2002**, 51, 204-221.
[54] Simon, P. ; Koolman, J.; In: *Ecdysone: from Chemistry to Mode of Action*, Thieme Verlag, Stuttgart, **1989**, pp.254-259
[55] Sláma, K.; Lafont, R.; *Eur. J. Entomol.*, **1995**, 92, 355-377.
[56] Lafont, R.; Dinan, L.; *J. Insect Sci.*, **2003**, (in press)
[57] Dinan, L.; *Eur. J. Entomol.*, **1995**, 92, 295-300.
[58] Ogawa, S.; Nishimoto, N.; Matsuda, H.; In: *Invertebrate Endocrinology and Hormonal Heterophylly*, Springer Verlag, **1974**, pp. 341-344.
[59] Kosar, K.; Opletal, L.; Vokác, K.; Harmatha, J.; Sovová, M.; Cerovský, J.; Krátký, F.; Dvořák, J.; *Pharmazie*, **1997**, 52, 406-407.
[60] Sláma, K.; Koudela, K., Tenora, J.; Mathova, A.; *Experientia*, **1996**, 52, 702-706.
[61] Chermnykh, N.S.; Shimanovsky, N.L.; Shutko, G.V.; Syrov, V.N.; *Farmakol. Toksikol.*, **1988**, 6, 57-62.
[62] Detmar, M.; Dumas, M.; Bonté, F.; Meybeck, A.; Orfanos, C.E.; *Eur. J. Dermatol.*, **1994**, 4, 558-562.
[63] Okui, S.; Otaka, T.; Uchiyama, M.; Takemoto, T.; Hikino, H.; Ogawa, S.; Nishimoto, N.; *Chem. Pharm. Bull.*, **1968**, 16, 384-387.
[64] Matsuda, H.; Kawaba, T.; Yamamoto, Y.; *Folia Pharmacol. Japon.*, **1970**, 66, 551-563.
[65] Syrov, V.N.; Nabiev, A.N.; Sultanov, M.V.; *Farmakol. Toksikol.*, **1986**, 49, 100-103.
[66] Butenandt, A.; Karlson, P.; *Z. Naturforsch.*, **1954**, 9b, 389-391.
[67] Hocks, P.; Wiechert, R.; *Tetrahedron Lett.*, **1966**, 26, 2989-2993.
[68] Hampshire, F.; Horn, D.H.S.; *J. Chem. Soc. Chem. Commun.*, **1966**, 37-38.
[69] Nakanishi, K.; Koreeda, M.; Sasaki, L.; Chang, M.L.; Hsu, H.Y.; *J. Chem. Soc. Chem. Commun.*, **1966**, 915-917.
[70] Galbraith, M.N.; Horn, D.H.S.; *J. Chem. Soc. Chem. Commun.*, **1966**, 905-906.
[71] Takemoto, T.; Ogawa, S.; Nishimoto, N.; *Yakugaku Zasshi*, **1967**, 1463-1468.
[72] Kaplanis, J.N.; Thompson, M.J.; Robbins, W.E.; Bryce, M.; *Science*, **1967**, 157, 1436-1438.
[73] Heinrich, G.; Hoffmeister, H.; *Experientia*, **1967**, 23, 995.
[74] Imai, S.; Toyosata, T,; Sakai, M.; Sato, Y; Fujioka, S.; Murata, E.; Goto, M.; *Chem. Pharm. Bull.*, **1969**, 17, 335-339.
[75] Báthori, M.; Girault, J.-P.; Kalasz, H.; Mathe, I.; Dinan L.N.; Lafont, R.; *Arch. Insect Biochem. Physiol.*, **1999**, 41, 1-8.

[76]   Dinan, L.; Sarker, S.D.; Bourne, P.; Whiting, P.; Šik, V.; Rees, H.H.; *Arch. Insect Biochem. Physiol.*, **1999**, 41, 18-23.

[77]   Lafont, R.D.; Wilson, I.D.; The ecdysone handbook, 2nd Ed., The Chromatographic Society, Nottingham, U.K., **1996**, pp. 525.

[78]   Lafont, R., Harmatha, J.; Marion-Poll, F.; Dinan, L.; Wilson, I.D.; *Ecdybase*, **2002**, (http://ecdybase.org)

[79]   Laudet, V.; *J. Molec. Endocrinol.*, **1997**, 19, 207-226.

[80]   Kumar, R.; Thompson, E.B.; *Steroids*, **1999**, 64, 310-319.

[81]   Oro. A.E.; McKeown, M.; Evans, R.M.; *Nature*, **1990**, 347, 298-301.

[82]   Yao, T.P.; Forman, B.M.; Jiang, Z.; Cherbas, L.; Chen, J.D.; McKeown, M.; Cherbas, P.; Evans, R.M.; *Nature*, **1993**, 366, 476-479.

[83]   Jones, G.; Jones, D.; *Insect Biochem. Molec. Biol.*, **2000**, 30, 671-679.

[84]   Koelle, M.R.; Talbot, W.S.; Segraves, W.A.; Bender, M.T., Cherbas, P.; Hogness, D.S.; *Cell*, **1991**, 67, 59-77.

[85]   Talbot, W.S.; Swyryd, E.A.; Hogness, D.S.; *Cell*, **1993**, 73, 1323-1337.

[86]   Kapitskaya, M.; Wang, S.; Cress, D.E.; Dhadialla, T.S.; Rhaikel, A.S.; *Molec. Cell. Endocr.*, **1996**, 12, 119-132.

[87]   Cho, W.L.; Kapitskaya, M.Z.; Rhaikel, A.S.; *Insect Biochem. Molec. Biol.*, **1995**, 25, 19-27.

[88]   Palmer, M.J.; Harmon, M.A.; Laudet, V.; *Amer. Zool.*, **1999**, 39, 747-757.

[89]   Swevers, L.; Cherbas, L.; Cherbas, P.; Iatrou, K.; *Insect Biochem. Molecl Biol.*, **1996**, 26, 217-221.

[90]   Imhof, M.O.; Rusconi, S.; Lezzi, M.; *Insect Biochem. Molec. Biol.*, **1993**, 23, 115-124.

[91]   Vögtli, M.; Imhof, M.O.; Brown, N.E.; Rauch, P.; Spindler-Barth, M.; Lezzi, M.; Henrich, V.C.; *Insect Biochem. Molec. Biol.*, **1999**, 29, 931-942.

[92]   Kothapalli, R.; Palli, S.R.; Ladd, T.R.; Sohi, S.S.; Cress, D.; Dhadialla, T.S.; Tzertzinis, G.; Retnakaran, A.; *Devel. Genetics*, **1995**, 17, 319-330.

[93]   Perera, S.C.; Ladd, T.R.; Dhadialla, T.S.; Krell, P.J.; Sohi, S.S.; Retnakaran, A.; Palli, S.R.; *Molec. Cell. Endocr.*, **1999**, 152, 73-84.

[94]   Martinez, A.; Scanlon, D.; Gross, B.; Perera, S.C.; Palli, S.R.; Greenland, A.J.; Windass, J.; Pongs, O.; Broad, P.; Jepson, I.; *Insect Biochem. Molec. Biol.*, **1999**, 29, 915-930.

[95]   Saleh, D.S.; Zhang, J.; Wyatt, G.R.; Walker, V.K.; *Molec. Cell. Endocr.*, **1998**, 143, 91-99.

[96]   Hayward, D.C.; Bastiani, M.J.; Trueman, M.J.; Truman, J.W.; Riddiford L.M.; Ball, E.E.; *Devel. Genes Evolut.*, **1999**, 209, 564-571.

[97]   Hannan, G.N.; Hill, R.J.; *Insect Biochem. Molec. Biol.*, **1997**, 27, 881-897.

[98]   Hannan, G.N.; Hill, R.J.; *Insect Biochem. Molec. Biol.*, **2001**, 31, 771-781.

[99]   Fujiwara, H.; Jindra, M.; Newitt, R.; Palli, S.R.; Hiruma, K.; Riddiford, L.M.; *Insect Biochem. Molec. Biol.*, **1995**, 25, 845-856.

[100]   Jindra, M.; Huang, J.Y.; Malone, F.; Asahini, M.; Riddiford, L.M.; *Insect Molec. Biol.*, **1997**, 6, 41-53.

[101]   Albertsen, M.C.; Brooke, C.D.; Garnaat, C.W.; Roth, B.A.; *Int. Patent Appl. WO 00/15791*, **2000**.

[102]   Rinehart, J.P.; Cikra-Ireland, R.; Flannagan, R.D.; Denlinger, D.L.; *J. Insect Physiol.*, **2001**, 47, 915-921.

66

[103]   Mouillet, J.F.; Delbecque, J.P.; Quennedy, B.; Delachambre, J.; *Eur. J. Biochem.*, **1997**, 248, 856-863.

[104]   Nicolai, M.; Bouhin, H.; Quennedy, B.; Delachambre, J.; *Insect Molec. Biol.*, **2000**, 9, 241-249.

[105]   Durica, D.S.; Wu, X.; Anilkumar, G.; Hopkins, P.M.; Chung, A.C.K.; *Molec. Cell. Endocr.*, **2002**, 189, 59-76.

[106]   Wang, S.F.; Ayer, S.; Segraves, W.A.; Williams, D.R.; Raikhel, A.S; *Molec. Cell. Biol.*, **2000**, 20, 3870-3879.

[107]   Wurtz, J.M.; Guillot, B.; Fagart, J.; Moras, D.; Tietjen, K.; Schindler, M.; *Protein Sci.*, **2000**, 9, 1073-1084.

[108]   Ravi, M.; Hopfinger, A.J.; Hormann, R.E.; Dinan, L.; *J. Chem. Inf. Comput. Sci.*, **2001**, 41, 1587-1604.

[109]   Kumar, M.B.; Fujimoto, T.; Potter, D.W.; Deng, Q.; Palli, S.R.; *Proc. Nat. Acad. Sci., U.S.A.*, **2002**, 99, 14710-14715.

[110]   Wing, K.D.; *Science*, **1988**, 241, 467-469.

[111]   Wing, K.D.; Slawecki, R.A.; Carlson, G.R.; *Science*, **1988**, 241, 470-472.

[112]   Dhadialla, T.S.; Carlson, G.R.; Le, D.P.; *Ann. Rev. Entomol.*, **1998**, 43, 545-569.

[113]   Dinan, L.; Savchenko, T.; Whiting, P.; Sarker, S.D.; *Pestic. Sci.*, **1999**, 55, 331-335.

[114]   Falkenstein, E.; Tillmann, H.-C.; Christ, M.; Feuring, M.; Wehling, M.; *Pharmacol. Revs.*, **2000**, 52, 513-555.

[115]   Tomaschko, K.-H.; *Arch. Insect Biochem. Physiol.*, **1999**, 41, 89-98.

[116]   Schoonhoven, L.M.; Derksen-Koppers, I.; *Ent. exp. appl.*, **1973**, 16, 141-145.

[117]   Ma, W.-C.; *Med. Landbouhogeschool Wageningen*, **1972**, 1-162.

[118]   Marion-Poll, F.; Descoins, C.; *J. Insect Physiol.*, **2002**, 48, 467-476.

[119]   Darazy-Choubaya, D.; Dinan, L.; Nel, P.; Marion-Poll, F.; in preparation

[120]   Horn, D.H.S.; Bergamasco, R.; In: *Comparative Insect Physiology, Biochemistry and Pharmacology, Volume 7*; Kerkut, G.A.; Gilbert, L.I., Eds.; Pergamon Press: Oxford, **1985**; pp. 185-248.

[121]   Kametani, T.; Tsubuki, M.; In: *Ecdysone: from chemistry to mode of action* (Ed. Koolman, J.), Thieme Verlag, Stuttgart, **1989**, pp. 74-96.

[122]   Kovganko, N.V.; *Chem. Nat. Cmpds.*, **1998**, 34, 111-127.

[123]   Siddall, J.B.; Cross, A.D.; Fried, J.H.; *J. Amer. Chem. Soc.*, **1966**, 88, 862-863.

[124]   Furlenmeier, A.; Fürst, A.; Langemann, A.; Waldvogel, G.; Hocks, P.; Kerb, U.; Weichert, P.; *Helv. Chim. Acta*, **1967**, 50, 2387-2396.

[125]   Kerb, U.; Wiechert, R.; Furlenmeier, A.; Fürst, A.; *Tetrahedron Lett.*, **1968**, 9, 4277-4280.

[126]   Mori, H.; Shibata, K.; Tsuneda, K.; Sawai, M.; *Chem. Pharm. Bull.*, **1968**, 16, 563-566.

[127]   Mori, H.; Shibata, K.; *Chem. Pharm. Bull.*, **1969**, 17, 1970-1973.

[128]   Mori, H.; Shibata, K.; Sawai, M.; *Tetrahedron*, **1971**, 27, 1157-1166.

[129]   Kovganko, N.V.; Survilo, V.L.; *Chem. Nat. Cmpds.*, **2000**, 36, 377-380.

[130]   Suksamrarn, A.; Ganpinyo, P.; Sommechai, C.; *Tetrahedron Lett.*, **1994**, 35, 4445-4448.

[131]   Charoensuk, S.; Yingyongnarongkul B.-e.; Suksamrarn, A.; *Tetrahedron*, **2000**, 56, 9313-9317.

[132]   Píš, J.; Girault, J.-P.; Larcheveque, M.; Dauphin-Villement, C.; Lafont, R.; *Steroids*, **1995**, 60, 188-194.

[133]   Mamadalieva, N.Z.; Ramazanov, N.S.; Saatov, Z.; *Khim. Prirod. Soedinen.*, **1999**, 767-770.
[134]   Suksamrarn, A.; Yingyongnarongkul, B.-e.; *Tetrahedron*, **1996**, 52, 12623-12630.
[135]   Suksamrarn, A.; Charoensuk, S.; Yinyongnarongkul, B.-e.; *Tetrahedron*, **1996**, 52, 10673-10684.
[136]   Yingyongnarongkul, B.-e.; Suksamrarn, A.; *Tetrahedron*, **1998**, 54, 2795-2800.
[137]   Suksamrarn, A.; Yingyongnarongkul, B.-e.; *Tetrahedron*, **1997**, 53, 3145-3154.
[138]   Píš, J.; Hykl, J.; Buděšínský M.; Harmatha, J.; *Tetrahedron*, **1994**, 50, 9679-9690.
[139]   Roussel, P.G.; Turner, N.J.; Dinan, L.; *J. Chem. Soc. Chem. Comm.*, **1995**, 933-934.
[140]   Suksamrarn, A.; Pattanaprateep, P.; Tananchatchairatana, T.; Haritakum, W.; Yingyongnarongkul, B.-e., Chimnoi, N.; *Insect Biochem. Molec. Biol.*, **2002**, 32, 193-197.
[141]   Tomás, J.; Camps, F.; Coll, J.; Mele, E.; Pascual, N.; *Tetrahedron*, **1992**, 48, 9809-9818.
[142]   Roussel, P.G.; *Ph.D. Thesis*, University of Exeter, U.K., **1995**, pp. 252
[143]   Dinan, L.; Bourne, P.; Whiting, P.; Tsitsekli, A.; Saatov, Z.; Dhadialla, T.S.; Hormann, R.E.; Lafont, R.; Coll, J.; *J. Insect Sci.* (submitted)
[144]   Dinan, L.; *Arch. Insect Biochem. Physiol.*, **1985**, 2, 295-317.
[145]   Lee, S.-S.; Nakanishi, K.; Cherbas, P.; *J. Chem. Soc. Chem. Comm.*, **1991**, 51-53.
[146]   Harmatha, J.; Buděšínský, M.; Vokáč, K.; *Steroids*, **2002**, 67, 127-135.
[147]   Harmatha, J.; Dinan, L.; Lafont, R.; *Insect Biochem. Mol. Biol.*, **2002**, 181-185.
[148]   Bourne, P.C.; Whiting, P.; Dhadialla, T.S.; Hormann, R.E.; Girault, J.-P.; Harmatha, J.; Lafont, R.; Dinan, L.; *J. Insect Sci.*, **2002**, 2.11, pp. 11.
[149]   Lin, W.; Albanese, C.; Pestell, R.G.; Lawrence, D.S.; *Chem. Biol.*, **2002**, 9, 1347-1353.
[150]   Dinan, L.; In *Ecdysone: from chemistry to mode of action*; Koolman, J., Ed.; Thieme Verlag: Stuttgart, **1989**, pp. 345-354.
[151]   Hsu, A.C.-T.; Fujimoto, T.T.; Dhadialla, T.S.; *ACS Symp. Series*, **1997**, 658, 206-219.
[152]   Carlson, G.R.; *ACS Symp. Series*, **2000**, 767, 8-17.
[153]   Carlson, G.R.; Dhadialla, T.S.; Hunter, R.; Jansson, R.K.; Jany, C.S.; Lidert, Z.; Slawecki, R.A.; *Pest Manag. Sci.*, **2001**, 57, 115-119.
[154]   Nakagawa, Y.; Shimizu, B.-i.; Oikawa, N.; Akamatsu, M.; Nishimura, K.; Kurihara, N.; Ueno, T.; Fujita, T.; *ACS Symposium Series*, **1995**, 606, 288-301.
[155]   Qian, X.; *J. Agric. Food Chem.*, **1996**, 44, 1538-1542.
[156]   Shimizu, B.-i.; Nakagawa, Y.; Hattori, K.; Nishimura, K.; Kurihara, N.; Ueno, T.; *Steroids*, **1997**, 62, 638-642.
[157]   Nakagawa, Y.; Hattori, K.; Shimizu, B.-i.; Akamatsu, M.; Miyagawa, H.; Ueno, T.; *Pestic. Sci.*, **1998**, 53, 267-277.
[158]   Oikawa, N.; Nakagawa, Y.; Nishimura, K.; Ueno, T.; Fujita, T.; *Pestic. Biochem. Physiol.*, **1994**, 48, 135-144.
[159]   Oikawa, N.; Nakagawa, Y.; Nishimura, K.; Ueno, T.; Fujita, T.; *Pestic. Sci.*, **1994**, 41, 139-148.
[160]   Nakagawa, Y.; Nishimura, K.; Oikawa N.; Kurihara, N.; Ueno, T.; *Steroids*, **1995**, 60, 401-405.
[161]   Nakagawa, Y.; Oikawa, N.; Nishimura, K.; Ueno, T.; Fujita, T.; *Pestic. Sci.*, **1995**, 44, 102-105.
[162]   Nakagawa, Y.; Soya, Y.; Nakai, K.; Oikawa, N.; Nishimura, K.; Ueno, T.; Fujita, T.; *Pestic. Sci.*, **1995**, 43, 339-345.

68

[163]   Nakagawa, Y.; Hattori, K.; Minakuchi, C.; Kugimiya, S.; Ueno, T.; *Steroids*, **2000**, 65, 117-123.

[164]   Hormann, R.E.; Dinan, L.; Whiting, P.; *Perspect. Drug Discov. Des.*, in press.

[165]   Hormann, R.E.; Whiting, P.; Dinan L.; in preparation

[166]   Mikitani, K.; *Biochem. Biophys Res. Comm.*, **1996**, 227, 427-432.

[167]   Elbrecht, A.; Chen, Y.; Jurgens, T.; Hensens, O.D.; Zink, D.L.; Beck, H.T.; Balick, M.J.; Borris, R.; *Insect Biochem. Mol. Biol.*, **1996**, 26, 519-523.

[168]   Dinan, L; Whiting, P.; Bourne, P.; Coll, J.; *Insect Biochem. Mol. Biol.*, **2001**, 31, 1077-1082.

[169]   Dixson, J.A.; Elshenawy, Z.M.; Eldridge, J.R.; Dungan, L.B.; Chiu, G.; Wowkun, G.S.; Presentation at the Middle Atlantic Regional ACS Meeting, University of Delaware, **2000**.

[170]   Dinan, L.; Bourne, P.C.; Meng, Y.; Sarker, S.D.; Tolentino, R.B.; Whiting, P.; *Cell. Mol. Life Sci.*, **2001**, 58, 321-342.

[171]   Bergamasco, R.; Horn, D.H.S.; In: *Progress in Ecdysone Research: Developments in Endocrinology, Volume 7*; Hoffmann, J.A., Ed.; Elsevier/Horth-Holland: Amsterdam, **1980**, pp. 299-324.

[172]   Becker, E.; Plagge, E.; *Biol. Zbl.*, **1939**, 59, 326-341.

[173]   Thomson, J.A.; Imray, F.P.; Horn, D.H.S.; *Aust. J. Exp. Biol. Med. Sci.*, **1970**, 48, 321-328.

[174]   Kaplanis, J.N.; Tabor, L.A.; Thompson, M.J.; Robbins, W.E.; Shortino, T.J.; *Steroids*, **1966**, 8, 625-631.

[175]   Adelung, D.; Karlson, P.; *J. Insect Physiol.*, **1969**, 15, 1301-1307.

[176]   Staal, G.B.; *Proc. K. Ned. Akad. Wet. Sect. C*, **1967**, 70, 409-418.

[177]   Watkinson, I.A.; Clarke, B.S.; *PANS*, **1973**, 488-506.

[178]   Suksamrarn, A.; Kumpun, S.; Yingyongnarongkul, B.-e.; *J. Nat. Prod.*, **2002**, 1690-1692.

[179]   Hasegawa, K.; Ata, A.M.; *J. Insect Physiol.*, **1972**, 18, 959-971.

[180]   Sato, Y.; Sakai, M.; Imai, S,; Fujioka, S.; *Appl. Entomol. Zool.*, **1968**, 3, 49-51.

[181]   Píš, J.; Harmatha, J.; Sláma, K.; In: Insect chemical ecology. Academia, Prague, **1991**, pp. 227-234.

[182]   Zolotar, R.M.; Bykhovets, A.I.; Kovganko, N.V.; *Chem. Nat. Cmpds.*, **2001**, 37, 537-539.

[183]   Oberlander, H.; *J. Insect Physiol.*, **1969**, 15, 297-304.

[184]   Chihara, C.J.; Petri, W.H.; Fristrom, J.W.; King, D.S.; *J. Insect Physiol.*, **1972**, 18, 1115-1123.

[185]   Mandaron, P.; *Devel. Biol.*, **1973**, 31, 101-113.

[186]   Ohmori, K.; Ohtaki, T.; *J. Insect Physiol.*, **1973**, 19, 1199-1210.

[187]   Terentiou, P.; Blattmann, M.; Bradbrook, D.; Käuser, G.; Koolman, J;. *Insect Biochem. Molec Biol.*, **1993**, 23, 131-136.

[188]   Eschalier, G.; Ohanessian, A.; *Comptes Rendus D*, **1969**, 268, 1771-1774.

[189]   Cherbas, L.; Yonge, C.D.; Cherbas, P.; Williams, C.M.; *Roux's Arch. Devel. Biol.*, **1980**, 189, 1-15.

[190]   Beckers, C.; Maróy, P.; Dennis, R.; O'Connor, J.D.; Emmerich, H.; *Molec. Cell. Endocr.*, **1980**, 17, 51-59.

[191]   Nakagawa, Y.; Minokuchi, C.; Takahashi, K.; Ueno, T.; *Insect Biochem. Molec. Biol.*, **2002**, 32, 175-180.
[192]   Mikitani, K.; *J. Seric. Sci. Jpn.*, **1995**, 64, 534-539.
[193]   Spindler-Barth, M.; Quack, S.; Rauch, P.; Spindler, K.-D.; *Eur. J. Entomol.*, **1997**, 94, 161-167.
[194]   Cottam, D.M.; Milner, M.J.; *Cell. Mol. Life Sci.*, **1997**, 53, 600-603.
[195]   Nakagawa, Y.; Minokuchi, C.; Ueno, T.; *Steroids*, **2000**, 65, 537-542.
[196]   Clément, C.Y.; Dinan, L.; In: *Insect Chemical Ecology*, Academia Prague, **1991**, pp. 221-226.
[197]   Clément, C.Y.; Bradbrook, D.A.; Lafont, R.; Dinan, L.; *Insect Biochem. Molec. Biol.*, **1993**, 23, 187-193.
[198]   Gateff, E.; Gissman, L.; Shrestha, R.; Plus, N.; Pfister, H.; Schröder, J.; zur Hausen, H.; In: *Invertebrate Systems in vitro*; Kurstak, E.; Maramorosch, K.; Dübendorfer, A., Eds.; Elsevier/North-Holland Biomedical Press: Amsterdam, **1980**; pp. 517-533.
[199]   Saez, E.; Nelson, M.C.; Eshelman, B.; Banayo, E.; Koder, A.; Cho, G.J.; Evans, R.M.; *Proc. Nat. Acad. Sci. USA*, **2000**, 97, 14512-14517.
[200]   Sage, B.A.; Tanis, M.A.; O'Connor, J.D.; *J. Biol. Chem.*, **1982**, 257, 6373-6379.
[201]   Lehmann, M.; Koolman, J.; *Molec. Cell. Endocr.*, **1988**, 57, 239-249.
[202]   Sobek, L.; Böhm, G.-A.; Penzlin, H.; *Insect Biochem. Molec. Biol.*, **1993**, 23, 125-129.
[203]   Deak, P.; Laufer, H.; *Eur. J. Entomol.*, **1995**, 92, 251-257.
[204]   Makka, T.; Seino, A.; Tomita, S.; Fujiwara, H.; Sonobe, H.; *Arch. Insect Biochem. Physiol.*, **2002**, 51, 111-120.
[205]   Dinan, L.; Whiting, P.; Girault, J.-P.; Lafont, R.; Dhadialla, T.S.; Cress, D.E.; Mugat, B.; Antoniewski, C.; Lepesant, J.-A.; *Biochem. J.*, **1997**, 327, 643-650.
[206]   Tomaschko, K.-H.; Guckler, R.; Bückmann, D.; *Nether. J. Zool.*, **1995**, 45, 93-97.
[207]   Girault, J.P.; Lafont, R.; *J. Insect Physiol.*, **1988**, 34, 701-706.
[208]   Brzozowski, A.M.; Pike, A.C.; Dauter, Z.; Hubbard, R.E.; Bonn, T.; Engstrom, O.; Öhmann, L.; Greene, G.L.; Gustafsson, J.A.; Carlquist, M.; *Nature*, **1997**, 389, 753-757.
[209]   Williams, S.P.; Sigler, P.B.; *Nature*, **1998**, 393, 392-396.
[210]   Renaud, J.P.; Rochel, N.; Ruff, M.; Vivat, V.; Chambon, P.; Gronemeyer, H.; Moras, D.; *Nature*, **1995**, 378, 681-689.
[211]   Bourget, W.; Ruff, M.; Chambon, P.; Gronemeyer, H.; Moras, D.; *Nature*, **1995**, 375, 377-382.
[212]   Wagner, R.L.; Apriletti, J.W.; McGrath, M.E.; West, B.L.; Baxter, J.D.; Fletterick, R.J.; *Nature*, **1995**, 378, 690-697.
[213]   Billas, I.M.; Moulinier, L.; Rochel, N.; Moras, D.; *J. Biol. Chem.*, **2001**, 276, 7465-7474.
[214]   Rochel, N.; Wurtz, J.M.; Mitschler, A.; Klaholz, B.; Moras, D.; *Mol. Cell.*, **2000**, 5, 173-179.
[215]   Pike, A.C.W.; Brzozowski, A.M.; Walton, J.; Hubbard, R.E.; Bonn, T.; Gustafsson, J.-Å.; Carlquist, M.; *Biochem. Soc. Trans.*, **2000**, 28, 396-400.
[216]   Harmatha, J.; Dinan, L.; *Arch. Insect Biochem. Physiol.*, **1997**, 35, 219-225.
[217]   Santos, A.C.S.; Sant'Anna, C.M.R.; *J. Molec. Struct. (Theochem)*, **2002**, 585, 61-68.
[218]   Koreeda, M.; Nakanishi, K.; Goto, M.; *J. Am. Chem. Soc.*, **1970**, 92, 7512-7513.
[219]   Harmatha, J.; Dinan, L.; *J. Insect Sci.*, **2002**, 2:16, p10.
[220]   Rosen, J.; Day, A.; Jones, T.K.; Turner Jones, E.T.; Nadzan, A.M.; Stein, R.B.; *J. Med. Chem.*, **1995**, 38, 4855-4874.

70

[221]   Jones, B.B.; Petkovich, M.; *Curr. Pharm. Des.*, **1996**, 2, 155-168.
[222]   Miner, J.N.; Tyree, C.M.; *Vitamins Hormones*, **2001**, 62, 253-280.
[223]   Dinan, L.; Whiting, P.; Sarker, S.D.; Kasai, R.; Yamasaki, K.; *Cell. Mol. Life Sci.*, **1997**, 53, 271-274.
[224]   Sarker, S.D.; Whiting, P.; Lafont, R.; Girault, J.-P.; Lafont, R.; *Biochem. System. Ecol.*, **1997**, 25, 79-80.
[225]   Sarker, S.D.; Whiting, P.; Šik, V.; Dinan, L.; *Phytochem.*, **1999**, 1123-1128.
[226]   Dinan, L.; Whiting, P.; Alfonso, D.; Kapetanidis, I.; *Entomol. Expt. Appl.*, **1996**, 415-420.
[227]   Sarker, S.D.; Savchenko, T.; Whiting, P.; Šik, V.; Dinan, L.; *Arch. Insect Biochem. Physiol.*, **1997**, 35, 211-217.
[228]   Sarker, S.D.; Whiting, P., Dinan, L.; Šik, V.; Rees, H.H.; *Tetrahedron*, **1999**, 55, 513-524.
[229]   Keckeis, K.; Sarker, S.D.; Dinan, L.; *Cell. Mol. Life Sci.*, **2000**, 57, 333-336.
[230]   Meng, Y.; Bourne, P.C.; Whiting, P.; Šik, V.; Dinan, L.; *Phytochem.*, **2001**, 57, 393-400.
[231]   Dinan, L.; Savchenko, T.; Whiting, P.; *Cell. Mol. Life Sci.*, **2001**, 58, 1121-1132.
[232]   Blackford, M.J.P.; Clarke, B.S.; Dinan, L.; *Arch. Insect Biochem. Physiol.*, **1997**, 34, 329-346.
[233]   Connat, J.-L.; Diehl, P.A.; Dumont, N.; Carminati, S.; Thompson, M.J.; *Z. angew. Entomol.*, **1983**, 96, 520-530.
[234]   Grebenok, R.J.; Adler, J.H.; *Phytochem.*, **1991**, 30, 2905-2910.
[235]   Chou, W.S.; Lu, M.S.; In: *Developments in Endocrinology, 7 (Progress in Ecdysone Research)*, Elsevier/North Holland Biomed. Press, Amsterdam, **1980**, pp. 281-297.
[236]   Ninagi, O.; Maruyama, M.; *Jpn. Agric. Res. Quart.*, **1996**, 30, 123-128.
[237]   Chandrakala, M.V.; Maribashetty, V.G.; Jyothi, H.K.; *Current Sci.*, **1998**, 74, 341-346.
[238]   Heffron, J.; *Biochemist*, **1999**, 28-31.
[239]   deFur, P.L.; Crane, M.; Ingersoll, C.; Tattersfield, L., Eds.; Endocrine disruption in invertebrates: endocrinology, testing and assessment; SETAC Press, **1999**.
[240]   Dinan, L.; Bourne, P.; Whiting, P.; Dhadialla, T.S.; Hutchinson, T.H.; *Environ. Toxicol. Chem.*, **2001**, 20, 2038-2046.
[241]   Fussenegger, M.; *Biotechnol. Prog.*, **2001**, 17, 1-51.
[242]   Suhr, S.T.; Gil, E.B.; Senut, M.-C., Gage, F.H.; *Proc. Nat. Acad. Sci. U.S.A.*, **1998**, 95, 7999-8004.
[243]   Jepson, I.; Martinez, A.; Sweetman, J.P.; *Pestic. Sci.*, **1998**, 54, 360-367.
[244]   Martinez, A.; Sparks, C.; Drayton, P.; Thompson, J.; Greenland, A.; Jepson, I.; *Molec. Gen. Genetics*, **1999**, 261, 546-552.
[245]   Martinez, A.; Sparks, C.; Hart, C.A.; Thompson, J.; Jepson, I.; *Plant J.*, **1999**, 19, 97-106.
[246]   Koudela, K.; Tenora, J.; Bajer, J.; Mathova, A.; Sláma, K.; *Eur. J. Entomol.*, **1995**, 92, 349-354.
[247]   Syrov, V.N.; *Biol. Nauki*, **1984**, 11, 16-20.
[248]   Uchiyama, M.; Yoshida, T.; In: *Invertebrate Endocrinology and Hormonal Heterophylly*, Springer Verlag, **1974**, pp. 401-416.
[249]   Najmutdinova, D.K.; Saatov, Z.; *Arch. Insect Biochem. Physiol.*, **1999**, 41, 144-147.
[250]   Badal'yants, K.L.; Nabiev, A.N.; Khushbaktova, Z.A.; Syrov, V.N.; *Dokl. Akad. Nauk Resp. Uzbek.*, **1996**, 46-48.

[251]   Prymak, R.G.; Levitsky, E.L.; Gubsky, Y.I.; Goryushko, H.K.; Kholodova, Y.D.; Vistunova, I.E.; *Ukrain. Biokhim. Zhur.*, **1996**, 68, 84-89.

[252]   Takasaki, M.; Tokuda, H.; Nishino, H.; Konoshima, T.; *J. Nat. Prod.*, **1999**, 62, 972-975.

[253]   Lafont, R.; Koolman, J.; Rees, H.; *Insect Biochem. Molec. Biol.*, **1993**, 207-209.

[254]   Yund, M.A.; In: *Invertebrate Systems in Vitro*, Elsevier/North-Holland Biomedical Press, **1980**, pp. 229-237.

[255]   Cherbas, P.; Trainor, D.A.; Stonard, R.J.; Nakanishi, K.; *J. Chem. Soc. Chem. Commun.*, **1982**, 1307-1308.

[256]   Odnikov, V.N.; Galyautdinov, I.V.; Nedopekin, D.V.; Khalilov, L.M.; Shashkov, A.S.; Kachala, V.V.; Dinan, L.; Lafont, R.; *Insect Biochem. Molec. Biol.*, **2002**, 32, 161-165.

[257]   Sarker, S.D.; Šik, V.; Rees, H.H.; Dinan, L.; *Phytochem.*, **1998**, 49, 2311-2314.

[258]   Meng, Y.; Whiting, P.; Šik, V.; Rees, H.H.; Dinan, L.; *Z. Naturforsch. C*, **2001**, 56, 988-994.

Atta-ur-Rahman (Ed.) *Studies in Natural Products Chemistry, Vol. 29*
73

POTENTIALLY CANCER CHEMOPREVENTIVE AND ANTI-INFLAMMATORY TERPENOIDS FROM NATURAL SOURCES

TOSHIHIRO AKIHISA[a], KEN YASUKAWA[b], and HARUKUNI TOKUDA[c]

[a]*College of Science and Technology, Nihon University, 1-8 Kanda Surugadai, Chiyoda-ku, Tokyo 101-8308, Japan*
[b]*College of Pharmacy, Nihon University, 7-7-1, Narashinodai, Funabashi-shi, Chiba 274-8555, Japan*
[c]*Department of Biochemistry, Kyoto Prefectural University of Medicine, Hirokoji-agaru, Kawaharacho-dori, Kamigyo-ku, Kyoto 602-0841, Japan*

ABSTRACT: Cancer chemoprevention by naturally occurring substances, especially by those occurring in vegetable foods and medicinal plants, seems to be a promising approach since various phytochemicals isolated from these sources have potent chemopreventive activity in animal experiments. Many phytochemicals have the advantage of being non-toxic. A number of terpenoids, which occur as secondary metabolites in all major groups of organisms, from fungi to humans, have been shown to possess chemopreventive activities in animal models. In a recent publication we have discussed the antitumor-promoting and anti-inflammatory activities of naturally occurring triterpenoids and sterols [1]. In this review article we describe the potentially chemopreventive and anti-inflammatory activities of terpenoids of the mono-, sesqui-, di-, sester-, and meroterpenoid classes, and of retinoids. Terpenoids are minor but ubiquitous components of our diet, and are considered relatively non-toxic to humans. These compounds, therefore, have the potential of being used as cancer chemopreventive agents.

INTRODUCTION

Cancer prevention is now one of the most urgent projects in public health. It is evident that an understanding of the mechanisms of the carcinogenesis process is essential for cancer chemoprevention. Most cancer prevention research is based on the concept of multistage carcinogenesis: initiation $\rightarrow$ promotion $\rightarrow$ progression [2–4] [Fig. **(1)**]. Cancer chemoprevention is defined as the use of specific natural and synthetic chemical agents to reverse or suppress carcinogenesis and prevent the development of invasive cancer. There is a growing awareness in that dietary non-nutricial compounds can have important effects as chemopreventive agents, and there have been many studies of the cancer chemopreventive effects of such compounds in animal models. Several classes of natural products are now known to have the potential of being used as chemopreventive agents. These classes include organosulfur compounds, retinoids, carotenoids, curcumin (turmeric yellow), polyphenols, flavonoids, and terpenoids [1,4–12].

Terpenoids occur as secondary metabolites in all major groups of organisms from fungi to humans. Many terpenoids have been shown to possess various biological activities. We have recently reviewed the chemopreventive activities of naturally occurring triterpenoids including sterols [1]. In this paper we review the chemopreventive activities and relevant anti-inflammatory activities of terpenoids belonging to the mono-, sesqui-, di-, sester-, and meroterpenoid classes, and of retinoids.

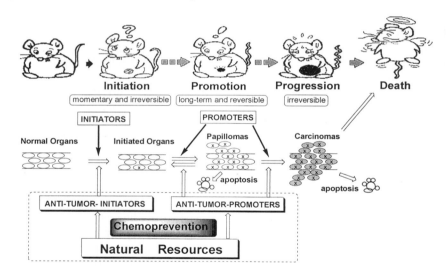

Fig. (1). Multistage chemical carcinogenesis and cancer chemoprevention

RELEVANT BIOACTIVITIES AND BIOASSAY SYSTEMS

The aim of cancer chemoprevention is to reduce the number of initiated cells, inhibit the promotion of initiated cells, or even reverse the promotion itself. There are many strategies that may be useful for cancer prevention [2–4,13].

Inhibition of Initiation

There are strategies that aim to reduce the initiation rate. The anti-initiators are classified as antimutagens, desmutagens, inhibitors of enzymes that activate procarcinogens, or agents that stimulate the metabolization of mutagens to less harmful metabolites. Included are also antioxidants, because a portion of the genetic damage is likely the result of free-radical damage.

Inhibition of Promotion

Carcinogenesis has three stages: initiation, promotion and progression [**Fig. (1)**]. In contrast to both the initiation and progression stage, animal studies indicate that the promotion stage takes a long time to occur and may be reversible, at least in an early stage. Therefore, the inhibition of tumor promotion is expected to be an efficient approach to cancer control [5,14–16]. Biological activities occurring at this stage may act either specifically on the promoted cell to cause it to redifferentiate and hence regain control of its own division or they may act at any of the points in one of the secondary messenger systems or related oncogenes that are frequently implicated as destabilizing agents [13].

Various bioassay systems are available in the literature for the screening of potentially chemopreventive compounds. Table 1 lists the bioassay systems for chemoprevention, as well as anti-inflammatory activities often associated with chemoprevention, reported for terpenoids and described in this review. The assays for anti-carcinogenic compounds might be classified as *in vitro* and *in vivo* primary screening assays, *in vivo* assays of several tumor models for various target organs, and mechanism-based *in vitro* assays.

In vitro Primary Screening Assay

Anti-mutagenic Activity

For screening of anti-mutagenic (AM) and anti-carcinogenic compounds, several tests have been developed that do not take a long time to run [17]. The Ames test [17] is a convenient method for evaluating mutagenic activity and it has been suggested that the mutagenic activity of a number of chemicals can be correlated well with carcinogenic activity [18]. The *umu* test system [19] was developed to evaluate the genotoxic activity of a wide variety of environmental carcinogens and mutagens, using the expression of one of the SOS genes to detect DNA-damaging agents. The

Table 1. Bioassay Systems Related to Cancer Chemoprevention and Anti-inflammatory Activities Described in This Review

Code	Bioassay system
Assay for Anti-carcinogenic Compounds	
In vitro primary screening assay	
AM	Anti-mutagen activity
EBV	Inhibition of Epstein-Barr virus early antigen (EBV-EA) activation induced by 12-*O*-tetradecanoylphorbol-13-acetate (TPA)
In vivo primary screening assay	
CRO	Inhibition of croton oil induced ear edema
TLO	Inhibition of teleosidin induced ear edema
TPA	Inhibition of TPA induced ear edema
In vivo assay	
Colon-1	Inhibition of azoxymethane (AOM) induced colon tumors
Colon-2	Inhibition of *N*-methyl-*N*-nitrosourea (NMU) induced colon tumors
Liver-1	Inhibition of *N*-diethylnitrosamine (DEN)/phenobarbital (PB) induced hepatic tumors
Liver-2	Inhibition of sponateneous hepatoma development in C3H/HeNCrj mice
Lung-1	Inhibition of 4-(methylnitrosoamino)-1-(3-pyridyl)-1-butanone (NNK) induced lung tumors
Lung-2	Inhibition of 4-nitroquinoline-*N*-oxide (4-NQO)/glycerol induced pulmonary tumors
Mam-1	Inhibition of NMU induced mammary tumors
Mam-2	Inhibition of 7,12-dimethylbenz[*a*]anthracene (DMBA) induced mammary tumors
Mam-3	Inhibition of spontaneous mammary tumors (in SHN mice)
Pancr	Inhibition of *N*-nitrobis-(2-hydroxypropyl)amine (BHP) induced pancreatic tumors
Skin-1	Inhibition of DMBA/ TPA induced skin tumors
Skin-2	Inhibition of DMBA/teleocidin B induced skin tumors
Skin-3	Inhibition of DMBA/fumonsin-B1 induced skin tumors
Skin-4	Inhibition of NOR-1/TPA induced skin tumors
Skin-5	Inhibition of DMBA/okadaic acid induced skin tumors
Skin-6	Inhibition of DMBA/mineral oil induced skin tumors
Stoma	Inhibition of *N*-methyl-*N*'-nitro-*N*-nitrosoguanidine (MNNG) induced gastric carinogenesis in rats
DNA	Inhibition of in vivo formation of mammary DMBA-DNA adduct
GSH	Induction of glutathione formation
GST	Induction of glutathione-*S*-transferase (GST), a phase II enzyme
UVB	Inhibition of UVB-induced AP-1 transactivation
Mechanism-based in vitro assay	
AOS	Inhibition of the formation of active oxygen species
AP1	Induction of activated protein 1 (AP-1) activity
APO	Induction of apoptosis
COX	Inhibition of inducible cyclooxygenase (COX-2)
CYP	Inhibition or induction of cytochrome P450 (CYP) enzymes
DIF	Induction of cellular differentiation
GPx	Induction of GPx (glutathione peroxidases) activities
iNOS	Inhibition of inducible NO synthase (iNOS)
5LOX	Inhibition of 5-lipoxygenase
NFAT	Inhibition of release of the transcription factor NF-AT

Table 1. Continued.

Code	Bioassay system
NFkB	Inhibition of nuclear factor kappa B (NF-κB) activity
NO	Inhibition of NO production by NOR-1
ODC	Inhibition of ornithine decarboxylase (ODC)
PKC	Inhibition of protein kinase C (PKC)
PLA	Inhibition of phospholipase A_2
RAS	Inhibition of isoprenylation
TNFa	Inhibition of mouse TNF-α release from BALB/3T3 cells
Assay for Anti-inflammation and Anti-allergic Compounds	
In vivo assay	
AA	Inhibition of arachidonic acid induced edema
BRK	Inhibition of bradykinin induced edeme
CPS	Inhibition of capsaicin induced edema
CRG	Inhibition of carrageenan induced edema
CTN	Innhibition of cotton pellet granuloma
EPP	Inhibition of ethyl phenylpropiolate (EPP) induced edema
HPT	Hepatoprotective activity
HTM	Inhibition of histamine induced edema
PAF	Inhibition of platelet aggregation factor (PAF) acether induced edema
In vitro assay	
ACM	Anticomplementary activity

results of this test are in agreement with those of the Ames test.

Inhibition of EBV-EA Activation

The Epstein-Barr virus (EBV), a herpes virus latently infecting human B-lymphocytes (Raji cells derived from Burkitt's lymphoma), has been thought to be causative of some human cancers [237]. Tumor-promoter-induced EBV activation is used as a means of detecting antitumor-promoting activity. Tumor promoters induce EBV activation possibly through the activation of protein kinase C and mitogen-activated protein kinase [20]. Inhibitory potentials against tumor promoter-induced EBV activation are well correlated with those in some animal models, and the EBV-early antigen (EA) inhibition assay using Raji cells is effective for the primary screening of some inhibitors of tumor promotion [21]. This assay has an advantage since it obtains information on the cytotoxicity from the viability of Raji cells. High viability of these cells is an important factor in developing a compound for the chemoprevention of cancer [22].

In vivo Primary Screening Assay

Assays for the inhibition of croton oil-induced edema, teleosidin-induced edema, and 12-*O*-tetradecanoylphorbol-13-acetate (TPA)-induced edema in animals are used as rapid *in vivo* preliminary tests for screening antitumor-promoting substances [1].

In vivo Assay

Inhibition of Tumor Promotion in Various Target Organs

There are many models for experimental carcinogenesis studies [13,23]. Two-stage carcinogenesis in mouse skin with DMBA (7,12-dimethylbenz[*a*]anthracene) and TPA (Skin-1; Table 1) is used most frequently since skin proves to be a unique target to differentiate between the effect of a test agent as either an anti-initiator or an anti-promoter under appropriate experimental conditions [6]. Further studies using animal models have been performed for carcinogenesis of skin (Skin-2–Skin-6), colon (Colon-1, Colon-2), liver (Liver-1, Liver-2), lung (Lung-1, Lung-2), mammary (Mam-1–Mam-3), pancreas (Pancr), and stomach (Stoma) for terpenoids (Table 1).

Induction of GSH and GST Formation

Glutathione (GSH)-*S*-transferase (GST), one of the phase II enzymes, is responsible for the formation of the conjugates with carcinogens to their deactivation and detoxification. Enhancement of this detoxification may prove to be an important strategy for chemoprevention. GSH is a prototype carcinogen scavenger. It reacts spontaneously or via catalysis by GST with numerous activated carcinogens including N-methyl-N'-nitro-N- nitrosoguanidine (MNNG), aflatoxin B_1 (AFB$_1$), and B[*a*]P-diolepoxide and other activated polycyclic aromatic hydrocarbons. Likewise, GSH protects against mouse skin tumors induced by DMBA/TPA, rat forestomach tumors induced by MNNG, and rat liver tumors induced by AFB$_1$ [23].

Mechanism-based *in vitro* Assay

Inhibtion of the Formation of Active Oxygen Species

There is abundant evidence that activated oxygen species [i.e., singlet oxygen ($^1O_2$), peroxy radicals ($\cdot$OOR), superoxide anion ($\cdot O_2^-$), and hydroxyl radicals ($\cdot$OH)] are involved in carcinogenesis. Potentially, they act both in initiation and in promotion and progression. For example, oxygen radicals can oxidize DNA bases, producing mutagenic lesions. Radicals also cause DNA strand breaks and chromosome deletions and rearrangements. Inflammatory cells produce a wide range of reactive oxygen species. There is evidence associating inflammation with cancers in various tissues including stomach, esophagus, colon/rectum, liver, pancreas, mouth, lung, skin, and bladder. Tumor promoters stimulate the endogeneous production of oxygen radicals in inflammatory cells and keratinocytes, and inhibit endogeneous activities that protect against oxidative damage (e.g., GSH-Px, catalase, and superoxide dismutase). Inhibitors of activated oxygen species have been considered to be potential chemopreventive agents [23,24].

Induction of Apoptosis

Apoptosis (induced programmed cell death) is a well-regulated function of the normal cell cycle. It requires gene transcription and translation. Tumor suppressors have been implicated as inducers of apoptosis. Programmed cell death has been described as the complement to mitosis in the maintenance, growth, and involution of tissues. It is the process by which damaged and excessive cells are eliminated. Apoptosis is inhibited by tumor promoters that stimulate cell proliferation. Induction of apoptosis may inhibit tumor formation. Chemicals that inhibit tumor promotion may act by inducing or preventing inhibition of apoptosis [23].

Inhibition of COX-2 and iNOS

Prostaglandins (PG) and nitric oxide (NO) produced by inducible cyclooxygenase (COX-2) and nitric oxide synthase (iNOS), respectively, have been implicated as important mediators in the processes of inflammation and carcinogenesis. Potential COX-2 and iNOS inhibitors have been considered as anti-inflammatory and cancer chemopreventive agents [25].

Inhibition of ODC

Ornithine decarboxylase (ODC) is the first enzyme in the polyamine biosynthesis pathway. Polyamines play essential roles in cell proliferation and differentiation and participate in macromolecular synthesis. Inhibitors of ODC block aspects of tumor promotion and induce cellular differentiation in several animal carcinogenesis models. Thus induction of ODC has been implicated as being important to carcinogenesis, and ODC activity is an intermediate biomarker of cell proliferation in studies of chemoprevention [26].

Inhibition of NF-κB

Nuclear factor kappa B (NF-κB) serves as a central regulator of the human immune and inflammatory response, and is a family of inducible transcription factors found virtually ubiquitously in all cells and functions in a variety of human diseases including those related to inflammation, cancer, asthma, atherosclerosis, AIDS, septic shock, and arthritis. Due to its role in a wide variety of diseases, NF-κB has become one of the major targets for drug development. Inhibition of NF-κB activity potentially contributes to cancer chemoprevention [27,28].

Inhibition of PKC

Protein kinase C (PKC) actually constitutes a family of several isozymes that are activated by diacylglycerol (DG). The potent tumor promoter, TPA, binds to, and activates PKC, in competition with DG. PKC stimulation results in phosphorylation of regulatory proteins that affect cell proliferation. There is evidence that carcinogenesis may be suppressed by inhibiting this enzyme. Phorbol tumor promoters such as TPA can replace DG in activating PKC. Chemicals that inhibit PKC also inhibit TPA-induced tumor promotion in mouse skin [23].

Inhibition of TNF-α

Tumor necrosis factor (TNF), a cytokine with a molecular weight of 17 kDa, is widely believed to act as a serum factor inducing necrosis of transplanted solid tumors in the mouse. TNF is also known to be important in inflammation, immunoregulation, and mitogenesis. The transformation potential of TNF-α in the initial cells is about 1000 times higher than that of TPA or okadaic acid [5]. These findings raise the possibility that TNF-α, originally discovered and proposed as a tumor necrosis factor, may be an endogeneous tumor promoter in humans. Inhibitors of TNF-α are, therefore, possible antitumor-promoters [5,29].

Assay for Anti-inflammation and Anti-allergic Compounds

Inflammation is one of physiological responses of organisms when they suffer physically or chemically induced stress, and comprises complex processes influenced by chemical mediators [30]. The mediators belong to different chemical classes, such as biologically active amines (histamine, serotonin), proteins and peptides (hydrolytic enzymes, cytokines, growth factors, colony stimulating factors, complement factors, antibodies, kinines), and lipids (PAF, prostanoids, leukotrienes), as well as activated oxygen species mentioned above [31]. Anti-inflammatory compounds can suppress inflammation by inhibiting activity or by interaction with one or several of the above cited chemical mediators. Various assay systems relevant to the physiological responses of organisms are listed in Table 1. As mentioned above, mechanistically, the processes of inflammation and carcinogenesis are closely related [32].

CHEMOPREVENTIVE AND ANTI-INFLAMMATORY EFFECTS OF TERPENOIDS

Tables 2–5 list monoterpenoids (**Mo**), sesquiterpenoids (**Sq**), diterpenoids (**Di**), and retinoids (**Re**), sesterterpenoids (**St**), meroterpenoids and miscellaneous terpenoids (**Me**), respectively, discussed in this review. The bioassay systems in which the compounds exhibited inhibitory effects, together with the major sources of the compounds, are included in the Tables. Most of these terpenoids are isolated from natural sources, and their structures are shown in **Figs. (2)–(5)**. Some terpenoids which exhibit significant and/or a broad range of chemopreventive and anti-inflammatory activities are discussed below.

Monoterpenoids (Mo)

Monoterpenoids have been isolated from the fragrant oils of many plants and are important in the perfumery and flavor industry. A number of these dietary monoterpenoids have chemopreventive activity [33,34].

Limonene (Mo4)

Limonene (**Mo4**), a menthane-type monoterpenoid, occurs naturally in many plant oils, including extracts from orange peel and other citrus fruits, and in mint and in dill, caraway, and celery seed oils. **Mo4** has a broad range of chemopreventive and therapeutic activities including activity against spontaneous and chemically-induced rodent mammary, skin, liver, lung, and forestomach tumors, as well as *ras* oncogene-induced rat mammary tumor [35–56]. In rat mammary carcinogensis models, the chemopreventive effects of limonene are evident at both initiation and promotion [38]. On a molar basis, however, **Mo4** is less potent than some of the other monoterpenoids such as perillyl alcohol (**Mo24**) [38]. In animal studies [37], **Mo4** must be used at doses up to 5% of the diet.

Mo4 is believed to act as an inhibitor of farnesylation of the *ras* oncogene. In addition, it is thought to interfere with the first steps in isoprenylation of hydroxymethylglutarate [35]. In contrast, **Mo4** has been reported to lack chemoprevention effects in a rat multiorgan carcinogenesis model [57].

Perillyl Alcohol (Mo24)

Perillyl alcohol (**Mo24**), a hydroxylated analog of **Mo4** and derived from fruits and vegetables, also possesses a wide range of chemopreventive and chemotherapeutic activities [37,42,49,50,58,59–69]. **Mo24** exhibits chemopreventive activity against liver, mammary gland, pancreas, and colon tumors in rodents [58,59]. This compound appears to function by inhibiting farnesylation of *ras* [42,58,62,66]. Other mechanisms of action have been suggested that are independent of the *ras* pathway, including induction of apoptosis [42,64,67], modulation of AP-1 activity [58,61,63], and induction of cellular differentiation [68]. Another mechanism that has been reported for **Mo24** is the inhibition of the metabolic activation of carcinogens by Phase I and II carcinogen-metabolizing enzymes resulting in the detoxification of carcinogens [58]. Both **Mo4** and **Mo24** appear to serve as prodrugs, their therapeutic effects being attributed to their main metabolites, perillic acid (**Mo31**) and dihydroperillic acid (**Mo32**). **Mo24** was shown to be at least five times more potent than **Mo4** in inhibiting rat mammary carcinogenesis, which might be explained by differences in pharmacokinetics [60,67]. **Mo24** is currently being investigated in clinical trials for the treatment of advanced breast cancer [37]. In addition, use of **Mo24** for chemoprevention of human carcinogenesis has been proposed [60]. There is a cautionary note about the failure of **Mo24** to cause chemoprevention during early rat hepatocarcinogenesis [60]. To study the effects of **Mo24** in early carcinogenesis, putatively initiated cells and preneoplastic foci in hepatocarcinogenesis on male Wistar rats treated with a single dose of *N*-nitrosomorpholine (NNM) were used as a model. **Mo24** exerted no detectable chemopreventive effect in the early stages of rat hepatocarcinogenesis. It rather exerted a phenobarbital-like tumor promoting activity. These data argue against a recommendation of **Mo24** as a chemopreventive agent for healthy humans [60].

Table 2. Monoterpenoids (Mo) from Natural Sources and the Bioassay Systems in which the Compounds Exhibited Inhibitory Activities

Code	Compound	Source & Occurrence	Assay System	References
Acyclic				
Mo1	Geraniol	Geranium oil	ODC, RAS	[59,50]
Mo2	Citral	Lemongrass	Skin-1, TPA	[132,133]
Mo3	Linalool	Orange peel oil	RAS, TPA	[50,133]
Menthane				
Mo4	*d*-Limonene	Essential oils, Orange peel oil	AM, TPA, Liver-1, Lung-1, Mam-1, Mam-2, Stoma, GPx, GST, ODC, NFkB, RAS	[33-35,37-41,43-49, 51,52,54-56, 133-138]
Mo5	*p*-Menth-1-ene	*Rosa rugosa* oil	AM, Lung-1	[49,135]
Mo6	α-Terpinene	*Citrus* spp.	AM	[135]
Mo7	α-Phellandrene	Pimento oil	AM	[135]
Mo8	γ-Terpinene	*Citrus* spp.	AM	[135]
Mo9	Terpinolene	*Citrus* spp.	AM	[135]
Mo10	(+)-*trans*-Isolimonene		AM	[135]
Mo11	(-)-Menthone	Peppermint oil	AM	[135]
Mo12	(-)-Isomenthone	Pennyroyal oil	AM	[135]
Mo13	(+)-Piperitone	*Eucalyptus* spp. oil	AM	[135]
Mo14	(+)-Pulegone	Spearmint oil	AM	[135]
Mo15	Carvone	Spearmint oil	AM, GST, RAS	[42,50,54,135]
Mo16	Piperitenone	*Mentha* spp. oil	AM	[135]
Mo17	Menthol	*Mentha* spp. oil	AM, Mam-2	[134,56,135]
Mo18	(+)-Isomenthol		AM	[135]
Mo19	Terpinen-4-ol	*Melaleuca altemifolia*	AM	[135]
Mo20	(-)-Dihydrocarveol	*Mentha* spp. oil	AM	[135]
Mo21	Carveol	Grapefruit oil	RAS, Mam-2	[34,50,53,137]
Mo22	α-Terpineol	Essential oils	RAS	[50]
Mo23	*p*-Mentha-2,8-dien-1-ol	*Apium graveolens*	GST	[139]
Mo24	Perillyl alcohol	Carraway seed oil	Colon-1, Lung-1, Mam-2, AP1, APO, DIF, NFkB, RAS, UVB	[33,34,36-39,42, 48-50,58,61-69, 134,137]
Mo25	Cuminyl alcohol	*Cuminum cyminum* oil	AM	[135]
Mo26	Sobrerol	Bordeaux terpentine	Mam-2, GST, RAS	[34,53,137,138]
Mo27	Uroterpenol	Constit. of human	Mam-2, RAS	[34,53,137]
Mo28	Limonene-1,2-diol		RAS	[137]
Mo29	Perillaaldehyde	*Siler trilobum*	AM, RAS	[50,135]
Mo30	Cuminaldehyde	Eucalyptus oil	AM	[135]
Mo31	Perillic acid	Metabolite of limonene	NFkB, RAS	[42,134,137]
Mo32	Dihydroperillic acid		RAS	[137]
Mo33	*iso*-Perillic acid	Metabolite of limonene	RAS	[42]
Mo34	Limonen-1,2-oxide	Orange	Lung-1, RAS	[48,49]
Mo35	Anethofuran	*Anethum graveolens*	GST	[54]
Mo36	1,8-Cineole (Eucalyptol)	Eucalyptus oil	TNFa	[140]
Iridoid				
Mo37	8-Acetylharpagide	*Ajuga decumbens*	EBV, Liver-1, Lung-2, Skin-1, Skin-4	[70,71]
Mo38	Harpagide	*A. decumbens*	EBV	[70]
Mo39	Reptoside	*A. decumbens*	EBV	[70]

Table 2. Continued.

Code	Compound	Source & Occurrence	Assay System	References
Mo40	Aucubin	*Plantago major*	TPA	[36]
Mo41	Ipolamiide	*Stachytarpheta cayennensis*	CRG, BRK, HST	[141]
Mo42	Agnuside	*Vitex peduncularis*	COX	[142]
Mo43	Pedunculariside	*V. peduncularis*	COX, *EPP*	[142]
Mo44	Scropolioside A	*Scrophularia*	TPA	[143]
Mo45	Scrovalentinoside	*S. auriculate*	TPA	[143]

Other Menthane-type Monoterpenoids

Carvone (**Mo15**) and perillic acid (**Mo31**), which are possible metabolites of **Mo4**, have more potent inhibitory activity than **Mo4** towards the isoprenylation enzymes in rat brain cytosol [42]. The relative effectiveness of 26 menthane-type monoterpenoids for inhibiting protein isoprenylation and cell proliferation has been determined [50]. Many monoterpenoids were found to be more potent than **Mo4** as inhibitors of small G protein isoprenylation and cell proliferation. The relative potency of limonene-derived monoterpenoids was found to be: monohydroxyl = ester = aldehyde > thiol > acid = diol = epoxide > triol = unsubstituted. The structure-activity relationship observed among the monoterpenoid isoprenylation inhibitors support a role for small G proteins in cell proliferation [50]. The structure-activity relationships among **Mo4** and three hydroxylated derivatives in the prevention of DMBA-induced mammary tumor were studied [53]. Rats were fed control diets or diets with 1% **Mo4**, carveol (**Mo21**), sobrerol (**Mo26**) or uroterpenol (**Mo27**) from 2 weeks before to one week after carcinogen administration. **Mo21**, **Mo26** and **Mo27** significantly prolonged tumor latency and decreased tumor yield. **Mo26** was the most potent of the monoterpenoids tested, decreasing tumor yield to half that of the control, a level previously achieved with 5% **Mo4** diets. **Mo26** is thus 5-fold more potent than **Mo4** both in enhancing carcinogen excretion and in preventing tumor formation. These data demonstrate that hydroxylation of monoterpenoid affects chemopreventive potential, with 2 hydroxyl groups > 1 > 0. Induction of the detoxifying enzyme GST in several mouse target tissues of anethofuran (**Mo35**), **Mo15**, and **Mo4** has been reported [54]. The α, β-unsaturated ketone system in **Mo15** appeared to be critical for the high enzyme-inducing activity.

8-Acetylharpagide (Mo37)

Ajuga decumbens (Labiatae) (flowering whole plant) has been used as a folk medicine for anti-inflammatory, anti-tussive and expectorant effects in China and Japan. 8-Acetylharpagide (**Mo37**), an iridoid glycoside

Monoterpenoid

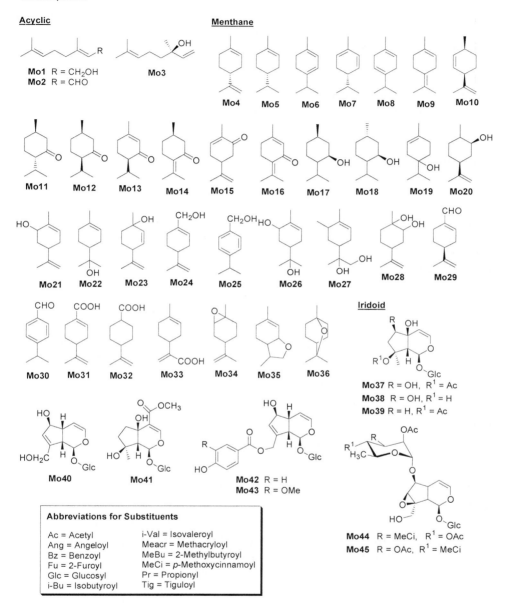

Fig. (2). Structures of monoterpenoids described in this review

derivative obtained from *A. decumbens* (flowering whole plant), exhibited the remarkable inhibitory effect in two-stage carcinogenesis test of mouse skin tumors induced by nitric oxide (NO) donor, NOR-1, as an initiator and TPA as a promoter. Further, **Mo37** exhibited potent antitumor-promoting activity on two-stage carcinogenesis test of mouse hepatic tumor using *N*-diethylnitrosamine (DEN) as an initiator and phenobarbital (PB) as a promoter [70]. **Mo37** showed potent antitumor-promoting activities in mouse skin *in vivo* two-stage carcinogenesis procedure, using DMBA as initiator and TPA as promoter [71]. Furthermore, **Mo37** exhibited potent chemopreventive activity in a mouse pulmonary tumor model [71].

Sesquiterpenoids (Sq)

Sesquiterpenoids (**Sq**) are C_{15} compounds formed by the assembly of three isoprenoid units. They are found in many living systems but particularly in higher plants. There are a large number of sesquiterpenoid carbon skeletons, which arise from the common precursor, farnesyl diphosphate, by various modes cyclization followed, in many cases, by skeletal rearrangement.

Sesquiterpene Lactones

Sesquiterpene lactones (**SLs**) have been isolated from numerous genera of Compositae. They are described as active constituents of some medicinal plants such as *Arnica montana* or *Tanacetum parthenium*. **SLs** are known to possess a wide variety of biological and pharmacological activities. Antitumoral and anti-inflammatory activities as well as allergenic potency have been described. The activities are mediated chemically by α,β-unsaturated carbonyl structures, such as an α-methylene-γ-lactone, an α,β-unsaturated cyclopentenone or a conjugated ester. These structure elements react by Michael-type addition to nucleophiles, especially cysteine sulfhydryl groups. Therefore, exposed thiol groups, such as cysteine residues in proteins, appear to be the primary targets of **SLs**. Many **SLs** inhibit the transcription factor NF-κB by selectively alkylating its p65 subunit probably by reacting with cysteine residues. Twenty-eight **SLs** including **Sq58**, **Sq68**, **Sq69**, **Sq70**, **Sq71**, **Sq72**, **Sq80**, **Sq86**, **Sq87**, **Sq88**, **Sq112**, **Sq120**, **Sq122**, and **Sq124** (Table 3) were assayed for their ability to inhibit NF-κB [72]. The majority of the potent NF-κB inhibitors possess two reactive centers in the form of an α-methylene-γ-lactone group and an α,β- or α,β,γ,δ-unsaturated carbonyl group. Based on computer molecular modeling, a molecular mechanism of action was proposed, which is able to explain the p65 selectivity of the **SLs** and the observed correlation of high activity with bifunctionality of the alkylating agent. A single bifunctional **SL** molecule can alkylate the cysteine residue (Cys38) in the DNA binding loop 1 (L1) and

Table 3. Sesquiterpenoids (Sq) from Natural Sources and the Bioassay Systems in which the Compounds Exhibited Inhibitory Activities

Code	Compound	Source & Occurrence	Assay System	References
Acyclic				
Sq1	*cis*-Nerolidol		EBV	[123]
Sq2	*trans*-Nerolidol		EBV	[123]
Sq3	11-Carboxy-12-*nor*-*trans*-nerolidol		EBV	[123]
Bisabolane				
Sq4	Xanthorrhizol	*Curcuma xanthorrhiza*	COX, iNOS	[25]
Sq5	β-Turmerone	*C. zedoaria*	COX, iNOS	[25]
Sq6	ar-Turmerone	*C. zedoaria*	COX, iNOS	[25]
Cedrane				
Sq7	Cedrol		EBV	[123]
Sq8	3α-Hydroxycedrol		EBV	[123]
Sq9	3β-Hydroxycedrol		EBV	[123]
Sq10	10β-Hydroxycedrol		EBV	[123]
Sq11	12-Hydroxycedrol		EBV	[123]
Drimane				
Sq12	12-Oxo-11-*nor*-drim-8-en-14-oic acid	*Thuja standishii*	EBV	[128]
Sq13	Polygodial	*Polygonum hydropiper*	EBV, Skin-1, CRG, CRO, *AA*, *BRK*, *CPS*, *HTM*, *PAF*	[73,74]
Sq14	Warbruganal	*P. hydropiper*	Skin-1	[74]
Sq15	Cryptoporic acid C	*Cryptoporus volvatus*	PLA	[144]
Sq16	Cryptoporic acid E	*C. volvatus*	AOS, PLA, TLO	[98,144]
Eremophilane				
Sq17	Furanoligularenone	*Ligularia fischeri*	COX, iNOS, NFkB, NO	[145]
Sq18	Petasin	Petasites hybridus	5LOX, PLA, *PAF*	[146,147]
Sq19	Neopetasin	*P. hybridus*	5LOX, PLA	[146]
Sq20	Isopetasin	*P. hybridus*	5LOX, PLA	[146]
Sq21	3-Oxo-8α-hydroxy-10αH-eremo-phila-1,7(11)-dien-12,8β-olide	*Ligularia fischeri*	NO	[145]
Sq22	3-Oxo-8α-methoxy-10αH-eremo-phila-1,7(11)-dien-12,8β-olide	*L. fischeri*	NO	[145]
Eudesmane				
Sq23	β-Eudesmol	Eucalyptus oils	AM	[148]
Sq24	Ilicic acid	*Inula viscosa*	TPA, *AA, EPP*	[149,150]
Sq25	Yomogin	*Artemisia princeps*	iNOS, NO	[151,152]
Sq26	Isohelenin	*A. ludoviciana*	iNOS, NFkB	[79,81,153]
Sq27	Tithofolinolide	*Tithonia diversifolia*	DIF	[155]
Sq28	3β-Acetoxy-8β-isobutyryloxy-reynosin	*T. diversifolia*	Mam-2, DIF	[155]
Sq29	1α-Acetoxy-2α,6β,9β-trifuroyloxy-4β-hydroxy-dihydro-β-agarofuran	*Maytenus cuzcoina*	EBV	[119]
Sq30	1α,2α-Diacetoxy-6β,9β-difuroyloxy-4β-hydroxy-dihydro-β-agarofuran	*M. cuzcoina*	EBV	[119]
Sq31	1α-Acetoxy-6β,9β-difuroyloxy-2α,4β-dihydroxy-dihydro-β-agarofuran	*M. cuzcoina*	EBV	[119]
Sq32	1α-Acetoxy-2α-benzoyloxy-6β,9β-difuroyloxy-4β-dhydro-β-agarofuran	*M. cuzcoina*	EBV	[119]

Table 3. Continued-1.

Code	Compound	Source & Occurrence	Assay System	References
Sq33	1α-Acetoxy-6β,9β-difuroyloxy-2α-propyonyloxy-4β-hydroxy-dihydro-β-agarofuran	*M. cuzcoina*	EBV	[119]
Sq34	1α-Acetoxy-6α,9β-difuroyloxy-2α-(2-methylbutyroyloxy)-4β-hydroxy-dihydro-β-agarofuran	*M. cuzcoina*	EBV	[119]
Sq35	1α,2α,15-Triacetoxy-6β,9β-difuroyloxy-4β-hydroxy-dihydro-β-agarofuran	*M. cuzcoina*	EBV	[119]
Sq36	1α,2α,15-Triacetoxy-6β,9β-dibenzoyloxy-4β-hydroxy-dihydro-β-agarofuran	*M. cuzcoina*	EBV	[119]
Sq37	Eumaytenol	*M. cuzcoina*	EBV	[119]
Sq38	1,2,6,8,15-Pentaacetoxy-9-benzoyloxy-4-hydroxy-dihydro-β-agarofuran		EBV	[121]
Sq39	Triptogelin A-1	*Tripterygium wilfordii*	EBV, Skin-1	[120,121]
Sq40	Triptofordin F-2	*T. wilfordii*	EBV	[121]
Sq41	α-Santonin	*Artemisia* spp.	CRG	[154]
Sq42	Orbiculin H	*Celastrus orbiculatus*	NFkB, NO	[28]
Sq43	Orbiculin I	*C. orbiculatus*	NFkB, NO	[28]
Sq44	Orbiculin D	*C. orbiculatus*	NFkB, NO	[28]
Sq45	Trichothecinol A	*Trichothecium roseum*	EBV	[122]
Sq46	Trichothecinol B	*T. roseum*	EBV	[122]
Sq47	Trichothecinol C	*T. roseum*	EBV	[122]
Sq48	Trichothecin	*T. roseum*	EBV	[122]
Sq49	Euonymine	*Maytenus cuzcoina*	EBV	[119]
Sq50	Atractylon	*Atractylodes japonica*	TPA, Skin-1	[156]
Germacrane				
Sq51	1,4-Dihydroxy-germacra-5*E*-10(14)-diene	*Achillea pannonica*	CRO	[156]
Sq52	Chrysanthediol B		EBV	[123]
Sq53	Chrysanthediol C		EBV	[123]
Sq54	Germacra-4(15),10(14),11-triene-3β,9α-diol		EBV	[123]
Sq55	(7*S*)-4β,5α-Epoxygermacr-1(10)*E*-en-2β,6β-diol	*Santonia chamaecyparissus*	TPA, PLA	[157]
Sq56	(7*S*)-Germacra-4(15)*Z*,9-diene-1α,2β,6β-triol	*S. chamaecyparissus*	TPA, PLA	[157]
Sq57	(7*S*)-Germacra-1(10)*E*,4(15)-dien-2β,5α,6β-triol	*S. chamaecyparissus*	TPA, PLA	[157]
Sq58	Parthenolide	*Tanacetum parthenium*	TPA, iNOS, NFkB, TNFa, CRG	[72,76-78,81-84, 153,154,158-164]
Sq59	Furanodiene	*Curcuma zedoaria*	HPT	[165]
Sq60	Curdione	*C. zedoaria*	HPT	[165]
Sq61	Neocurdione	*C. zedoaria*	HPT	[165]
Sq62	Dehydrocurdione	*C. zedoaria*	CRG, HPT	[165,166]
Sq63	Germacrone	*C. zedoaria*	HPT	[165]
Sq64	13-Hydroxygermacrone	*C. zedoaria*	HPT	[165]

Table 3. Continued-2.

Code	Compound	Source & Occurrence	Assay System	References
Sq65	Costunolide	*Magnolia sieboldii*	AM, Colon-1, iNOS, NFkB, ODC, RAS	[26,85-89,167]
Sq66	3-Acetoxycostunolide	*Podachaenium*	NFkB	[88]
Sq67	7-Hydroxycostunolide	*P. eminens*	NFkB	[88]
Sq68	2α-Acetoxy-14-hydroxy-15-iso-valeroyloxy-9-oxo-costunolide	*Mikania guaco*	NFkB, TNFa	[32,168]
Sq69	14-Hydroxy-15-isovaleroyloxy-9-oxo-melampolide	*M. guaco*	NFkB	[72]
Sq70	1-Methoxymiller-9*Z*-enolide	*Milleria quiqueflora*	NFkB	[72]
Sq71	4β,15-Epoxymiller-9*E*-nolide	*M. quiqueflora*	NFkB, TNFa	[72,168,169]
Sq72	4β,15-Epoxymiller-9*Z*-enolide	*M. quiqueflora*	NFkB	[72,169]
Sq73	9α-Hydroxy-8β-methacryloyloxy-14-oxo-acanthopsermolide-4α,5β-epoxide	*M. quiqueflora*	NFkB	[169]
Sq74	Miller-9*E*-enolide	*M. quiqueflora*	NFkB	[169]
Sq75	Miller-9*Z*-enolide	*M. quiqueflora*	NFkB	[169]
Sq76	15-(2-Hydroxy)-isobutyryloxy-micrantholide	*M. cordifolia*	TNFa	[168]
Sq77	8β-Hydroxy-9α-methacryloyloxy-14-oxo-acanthospermolide	*M. quiqueflora*	NFkB	[169]
Sq78	9α-Hydroxy-8β-methacryloyloxy-14-oxo-acanthospermolide	*M. quiqueflora*	NFkB	[169]
Sq79	15-Acetoxy-9α-methacryloyloxy-8β-hydroxy-14-oxo-acantho-spermolide	*M. quiqueflora*	NFkB	[169]
Sq80	15-Acetoxy-9α-hydroxy-8β-methacryl-oyloxy-14-oxo-acanthospermolide	*M. quiqueflora*	NFkB	[72,169]
Sq81	9α-Acetoxy-miller-1(10)*Z*-enolide	*M. quiqueflora*	NFkB	[169]
Sq82	9α-Acetoxy-4b,15-epoxymiller-1(10)*Z*-enolide	*M. quiqueflora*	NFkB	[169]
Sq83	Vicolide C	*Vicoa indica*	*CTN*	[170]
Sq84	Vicolide D	*V. indica*	*CTN*	[170]
Sq85	2β,5-Epoxy-5,10-dihydroxy-6α-angeloyloxy-9β-isobutyloxy-germacran-8α,12-olide	*Carpesium divaricatum*	iNOS, NFkB	[171]
Sq86	15-Acetoxy-eremantholide B	*Vanillomopis arborea*	NFkB, TNFa	[72,168]
Sq87	Centraherin	*Proteopsis furnensis*	NFkB	[72]
Sq88	Goiazensolide	*Eremanthus mattogrossensis*	NFkB	[72]
Sq89	Diversifolin	*Tithonia diversifolia*	NFkB	[80]
Sq90	Diversifolin methyl ether	*T. diversifolia*	NFkB	[80]
Sq91	Tirotundin	*T. diversifolia*	NFkB	[80]
Guaiane				
Sq92	Dehydrocostus lactone	*Saussurea lappa*	AM	[167]
Sq93	Mokkolactone	*S. lappa*	EBV	[98]
Sq94	Hydroxyachillin	*Tanacetum microphyllum*	TPA, PKC, CRG	[172,173]
Sq95	14-Deoxylactucin	*Achillea setacea*	CRO	[174]

Table 3. Continued-3.

Code	Compound	Source & Occurrence	Assay System	References
Sq96	Rupicolin B	*A. setacea*	CRO	[174]
Sq97	Dehydroleucodin	*Podachaenium eminens*	COX, NFkB	[88,175]
Sq98	14-Hydroxyhypocretenolide	*Leontodon hispidus*	NFkB, CRO	[176]
Sq99	Dehydroleucodin	*Artemisia douglasiana*	COX	[175]
Sq100	3-Chlorodehydroleucodin	*Podachaenium eminens*	NFkB	[88]
Sq101	3,4-Epoxydehydroleucodin	*P. eminens*	NFkB	[88]
Sq102	2-Oxo-8β-methacryloyloxy-10β-hydroxy-guaia-3,11(13)-dien-6α,12-olide	*Viguiera gardneri*	NFkB	[27]
Sq103	2-Oxo-8β-methacryloyloxy-10α-hydroxy-guaia-3,11(13)-dien-6α,12-olide	*V. gardneri*	NFkB	[27]
Sq104	2-Oxo-8β-methacryloyloxy-guaia-1(10),3,11(13)-trien-6α,12-olide	*V. gardneri*	NFkB	[27]
Sq105	4α,10α-Dihydroxy-3-oxo-8β-isobutyryloxyguaia-11(13)-en-12,6α-olide	*Tithonia diversifolia*	DIF	[155]
Sq106	Cynaropicrin	*Saussurea lappa*	NO, TNFa	[177]
Sq107	Procurcumenol	*Curcuma zedoaria*	TNFa	[29]
Sq108	Inuviscolide	*Inula viscosa*	TPA	[149,150]
Sq109	14-Hydroxyhypocretenolide-β-D-glucoside-4',4"-hydroxy-hypocretenoate	*Leontodon hispidus*	iNOS, NFkB, CRO	[176]
Sq110	3-Costoyloxydehydroleucodin	*Podachaenium eminens*	NFkB	[88]
Sq111	Podachaenin	*P. eminens*	NFkB	[88]
Sq112	Cumambrin A	*Eremanthus mattogrossensis*	NFkB	[72]
Sq113	Aerugidiol	*Curcuma zedoaria*	*HPT*	[165]
Sq114	Curcumenol	*C. zedoaria*	*HPT*	[165]
Sq115	Isocurcumenol	*C. zedoaria*	*HPT*	[165]
Pseudoguaiane				
Sq116	11α,13-Dihydrohelenalin acetate	*Arnica montana*	CRO	[178]
Sq117	11α,13-Dihydrohelenalin methacrylate	*A. montana*	NFAT, NFkB, CRO	[178]
Sq118	11α,13-Dihydrohelenalin tigulinate	*A. montana*	NFAT, NFkB	[178]
Sq119	Ergolide	*Inula britannica*	COX, iNOS, NFkB	[179]
Sq120	Helenalin	*Arnica chamissonis*	NFkB, CRG	[72,180-182]
Sq121	Helenalin acetate	*A. montana*	NFAT, NFkB	[178]
Sq122	Helenalin isobutyrate	*A. montana*	NFAT, NFkB	[72,178,180]
Sq123	Helenalin tigulinate	*A. montana*	NFAT, NFkB	[178]
Sq124	Mexicanin I	*A. acaulis*	NFkB	[72,180]
Sq125	Confertdiolide	*Parthenim confertum*	TPA, CRG, EPP	[183]
Sq126	2,3-Dihdydroaromaticin	*Arnica montana*	NFkB	[180]
Sq127	4α-O-Acetyl-pseudoguain-6β-olide	*Parthenium hysterophorus*	TPA	[180]
Sq128	Hysterin	*P. hysterophorus*	TPA	[180]
Sq129	Ambrosanolide	*P. hysterophorus*	TPA, *EPP*	[180]

Table 3. Continued-4.

Code	Compound	Source & Occurrence	Assay System	References
Sq130	Tetraneurin A	*P. hysterophorus*	TPA	[180]
Sq131	Parthenin	*P. hysterophorus*	TPA, *EPP*	[180]
Sq132	Hymenin	*P. hysterophorus*	TPA	[180]
Miscellaneous Sesquiterpenoid				
Sq133	Spathulenol	*Santolia chamaecyparissus*	TPA, *AA* , *PLA*	[157]
Sq134	Lapidin	*Ferula linkii*	CRG	[184]
Sq135	Zerumbone	*Zingiber zerumbest*	EBV, Colon-1, AOS, APO, COX, iNOS, TNFa	[20,24,93,185]
Sq136	8-Hydroxy-α-humulene		EBV	[20]
Sq137	α-Santalol	Sandalwood oil	Skin-1, ODC	[185]
Sq138	β-Santalol	Sandalwood oil	Skin-1, ODC	[185]
Sq139	Celaphanol A	*Celastrus orbiculatus*	NFkB, NO	[28]
Sq140	(1'*E* ,5'*Z*)-2-(2',6'-Dimethylocta-1',5',7'-trienyl)-4-furoic acid	Soft coral *Sinularia* spp.	PLA	[186]
Sq141	(1'E,5'*E*)-2-(2',6'-Dimethylocta-1',5',7'-trienyl)-4-furoic acid	Soft coral *Sinularia* spp.	PLA	[186]
Sq142	Ebenifoline E-1	*Tripterygium wilfordii*	TNFa	[187]
Sq143	Congorinine E-1	*T. wilfordii*	TNFa	[187]
Sq144	Caryophyllene oxide	*Lavandula latifolia*	*CRG*	[188]

another cysteine (Cys120) in the nearby E' region. This cross link alters the position of Tyr36 and additional amino acids in such a way that their specific interactions with the DNA become impossible.

Polygodial (Sq13)

Drymis winteri (Winteraceae) is a plant native to the south of Brazil and some other South American countries. Aqueous alcohol extract obtained from the bark of *D. winteri*, given orally or intraperitoneally (i.p.) to rats, caused dose-related inhibition of paw edema elicited by several mediators of inflammation, such as bradykinin (*BRK*), substance P (SP), or platelet-activating factor (*PAF*-acether), as well as the edema formation caused by ovalbumin (OVA) in animals that had been actively sensitized to this antigen. Polygodial (**Sq13**) is the main constituent present in the bark of *D. winteri*. When assessed *in vitro* it antagonizes, in a concentration-dependent and reversible manner, the contractile responses caused by inflammatory mediators in the guinea-pig ileum and rat portal vein. **Sq13** has been shown to produce a concentration-dependent relaxation in phenylephrine precontracted vessels from guinea pigs and rabbits, by a mechanism largely dependent on the release of NO and stimulation of cyclic guanosine monophosphate (cGMP) pathway [73]. **Sq13** (12.8–128.1 μM/kg, i.p.) 30 min prior, inhibited significantly mouse paw edema induced by PGE_2, *BRK*, SP, dextran, *PAF* or carrageenan (*CRG*) [73]. **Sq13** also inhibited arachidonic acid (*AA*)-, capsaicin (*CPS*)-

and croton oil (CRO)-induced ear edema in mice [73]. **Sq13** (42.7 μM/kg, i.p.), significantly inhibited both exudation and cell influx when assessed in the pleurisy induced by SP and histamine (*HTM*), and to a less extent the inflammatory response caused by *CRG, PAF, BRK* and des-Arg^9-*BRK* [73]. Finally, **Sq13** (4.2–42.7 μM/kg, i.p.) produced dose-related inhibition of paw edema induced by OVA, protecting in a time-dependent manner the anaphylactic shock induced by intravenous administration of OVA in animals which had been actively sensitized by this antigen [73]. These results indicate that the sesquiterpenoid **Sq13**, present as the major component in the bark of *D. winteri*, exerts interesting anti-inflammatory and anti-allergic properties when tested in rats and mice [73]. **Sq13** as well as warburganal (**Sq14**) were evaluated for their inhibitory activity on EBV activation and their *in vivo* antitumor effects on mouse skin [74]. These sesquiterpenoids delayed the formation of papillomas, and markedly reduced both the rate of papilloma-bearing mice and the number of papillomas per mouse on TPA-induced tumor promotion initiated by DMBA [74].

Parthenolide (Sq58)

Tanacetum parthenium (Asteraceae), commonly known as "feverfew," is a popular herbal remedy used as a remedy for fever, arthritis, and migraine. Feverfew products have been shown to exhibit potent anti-inflammatory properties. These anti-inflammatory effects are mainly attributable to sesquiterpene lactones present in the plant. Parthenolide (**Sq58**), the predominant sesquiterpene lactone in feverfew, inhibits the expression of COX-2 and proinflammatory cytokines (TNF-α and IL-1) in the lipopolysaccharide (LPS)-stimulated murine macrophage [75,76]. **Sq58** exerts potent inhibitory effects on the promoter activity of the iNOS gene in THP-1 cells. A tumor-promoting phorbol ester, TPA, significantly increased the iNOS promoter-dependent reporter gene activity, and the TPA-induced increase in iNOS promoter activity was effectively suppressed by **Sq58** which may further explain the anti-inflammatory property of **Sq58** [77]. **Sq58** prevented NF-κB activation [78–81]. This compound also shows anti-inflammatory activity because it inhibits NF-κB activation [82]. Anti-inflammatory activity of **Sq58** against mouse-ear edema induced by TPA [93% edema inhibition at 0.5 mg/ear. ID$_{50}$ (dose of drug inhibiting the edema by 50%) = 0.18 μM/ear] [83], and against acetic acid-induced writhing in mice and carrageenan-induced paw edema (1.2 mg/kg, i.p) in rats, respectively [84], has also been reported.

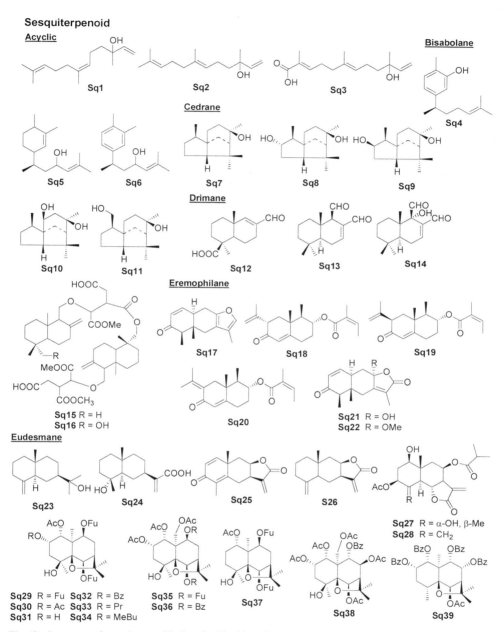

Fig. (3). Structures of sesquiterpenoids described in this review

94

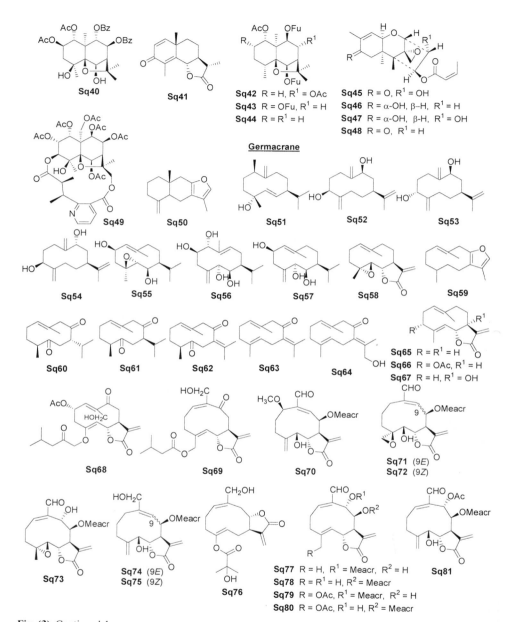

Sq40

Sq41

Sq42 R = H, R¹ = OAc

Sq43 R = OFu, R¹ = H

Sq44 R = R¹ = H

Sq45 R = O, R¹ = OH

Sq46 R = α-OH, β–H, R¹ = H

Sq47 R = α-OH, β-H, R¹ = OH

Sq48 R = O, R¹ = H

Germacrane

Sq49

Sq50

Sq51

Sq52

Sq53

Sq54

Sq55

Sq56

Sq57

Sq58

Sq59

Sq60

Sq61

Sq62

Sq63

Sq64

Sq65 R = R¹ = H

Sq66 R = OAc, R¹ = H

Sq67 R = H, R¹ = OH

Sq68

Sq69

Sq70

Sq71 (9E)

Sq72 (9Z)

Sq73

Sq74 (9E)

Sq75 (9Z)

Sq76

Sq77 R = H, R¹ = Meacr, R² = H

Sq78 R = R¹ = H, R² = Meacr

Sq79 R = OAc, R¹ = Meacr, R² = H

Sq80 R = OAc, R¹ = H, R² = Meacr

Sq81

Fig. (3). Continued-1

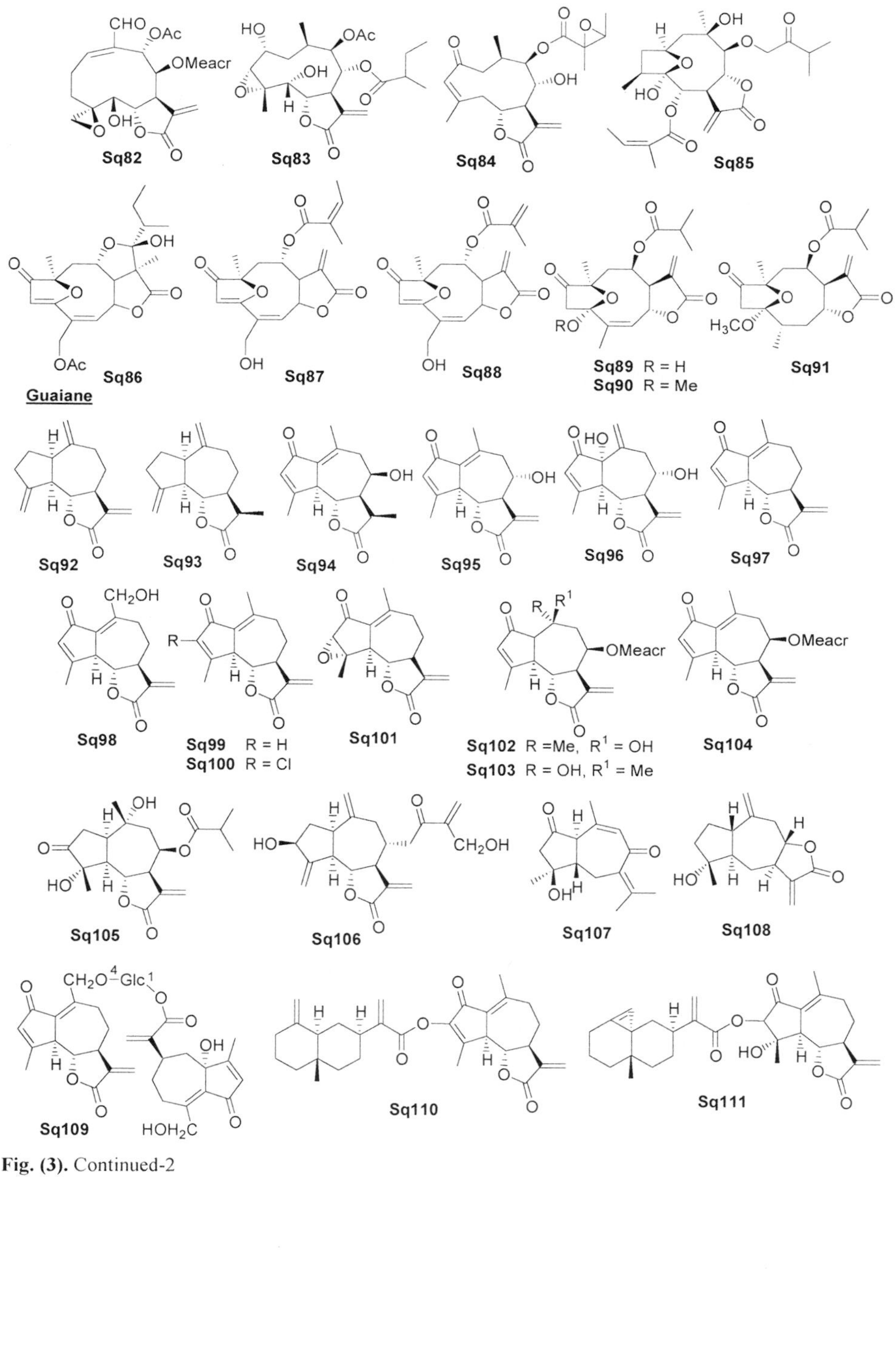

Sq82 Sq83 Sq84 Sq85

Sq86 Sq87 Sq88 Sq89 R = H Sq91
Sq90 R = Me

Guaiane

Sq92 Sq93 Sq94 Sq95 Sq96 Sq97

Sq98 Sq99 R = H Sq101 Sq102 R = Me, R¹ = OH Sq104
Sq100 R = Cl Sq103 R = OH, R¹ = Me

Sq105 Sq106 Sq107 Sq108

Sq109 Sq110 Sq111

Fig. (3). Continued-2

96

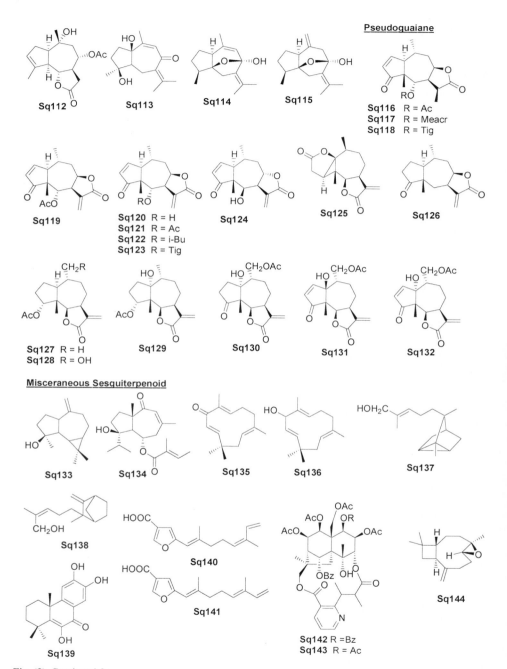

Sq112 **Sq113** **Sq114** **Sq115**

Pseudoguaiane

Sq116 R = Ac
Sq117 R = Meacr
Sq118 R = Tig

Sq119

Sq120 R = H
Sq121 R = Ac
Sq122 R = i-Bu
Sq123 R = Tig

Sq124 **Sq125** **Sq126**

Sq127 R = H
Sq128 R = OH

Sq129 **Sq130** **Sq131** **Sq132**

Misceraneous Sesquiterpenoid

Sq133 **Sq134** **Sq135** **Sq136** **Sq137**

Sq138 **Sq140** **Sq144**

Sq139 **Sq141**

Sq142 R = Bz
Sq143 R = Ac

Fig. (3). Continued-3

Costunolide (Sq65)

Costunolide (**Sq65**) is the predominant sesquiterpene lactone in Saussureae radix [85], and has also been isolated from the stem bark of *Magnolia sieboldii* [86] and *M. grandiflora* [87] as well as from *Podachaenium eminens* [88]. **Sq65** inhibitis significantly the farnesylation process of human lamin-B by farnesyl-proteintransferase (FPTase), in a dose dependent manner *in vitro* (IC_{50} value was calculated as 20 μM) [86]. **Sq65** also exhibits potent chemopreventive effects in carcinogenesis. Cancer preventive azoxymethane (AOM) -induced intestinal carcinogenesis in rats [26,89] and DMBA-induced cheek pouch carcinogenesis in hamsters [90]. These reports demonstrated that **Sq65** suppresses carcinogen-induced cell proliferation in the target tissues. The same compound was shown to exhibit potent inhibitory effects on NOS expression in the endotoxin-activated mouse macrophages [91,92]. Effects of on cellular activation induced by TPA were investigated using a reporter gene assay which was designated to reflect the promoter activity of the iNOS gene in a human monocyte cell line THP-1. iNOS promoter-dependent reporter gene activity was significantly increased by TPA, and the TPA-induced increase of the reporter gene activity was efficiently reduced by **Sq65**, with an IC_{50} of approximately 2 μM. This may further explain the cancer-preventive property of **Sq65** [85]. Inhibition of the transcription factor NF-κB by **Sq65** also has been reported [87,88].

Zerumbone (Sq135)

Zerumbone (**Sq135**) is a food phytochemical that has distinct potentials for use in anti-inflammation, chemoprevention, and chemotherapy. **Sq135** was isolated from the edible rhizomes of *Zingiber zerumbet* (Zingiberaceae). It is a potent inhibitor of tumor promoter TPA-induced EBV activation. The IC_{50} value of **Sq135** (0.14 μM) is noticeably lower than those of α-humulene (2,6,6,9-tetramethyl-1,4,8-cycloundecatriene), lacking the carbonyl group at the 8-position in **Sq135** ($IC_{50} > 100$ μM), and 8-hydroxy-α-humulene ($IC_{50} = 0.95$ μM) [20]. The modifying effects of dietary feeding of **Sq135** on the development of AOM-induced colonic aberrant crypt foci (ACF) were investigated in male F344 rats [93]. Expression of COX-2 in colonic mucosa exposed to AOM and/or **Sq135**, and the effects of **Sq135** on cell proliferation activity of crypts by counting silver-stained nucleolar organizer regions protein (AgNORs) in colonic cryptal cell nuclei also were assayed. To induce ACF rats were given three weekly subcutaneous injections of AOM (15 mg/kg body weight). They were also fed an experimental diet containing 0.01% or 0.05% **Sq135** for 5 weeks, starting one week before the first dosing of AOM. AOM exposure produced 84 ± 13 ACF/rat at the end of the study (week 5). Dietary administration of **Sq135** caused reduction in the frequency of ACF: 72 ± 17 (14% reduction) at a dose of 0.01% and 45 ± 18 (46% reduction, $p<0.001$) at a dose of 0.05%. **Sq135** teeding significantly reduced expression of COX-2 and PG in colonic mucosa. It significantly lowered the number of AgNORs in colonic crypt cell nuclei. These results might suggest possible chemopreventive ability of **Sq135** through suppression of COX-2 expression, cell proliferation activity of colonic mucosa, and induction of phase II detoxification enzymes in the development of carcinogen-induced ACF [93]. Suppressive effects of **Sq135** in a series of *in vitro* bioassays designed to reflect inflammatory and carcinogenic processes, i.e., the anti-inflammatory and chemopreventive potentials in a variety of cell culture experiments, were investigated [24]. **Sq135** effectively suppressed TPA-induced superoxide anion generation from both NADPH oxidase in DMSO-differentiated HL-60 human acute promyelocytic leukemia cells and xanthine oxidase in AS52 Chinese hamster ovary cells. The combined lipopolysaccharide (LPS)- and interferon-γ-stimulated protein expressions of iNOS and COX-2, together with the release of TNF-α, in RAW264.7 mouse macrophages markedly diminished. These suppressive effects were accompanied by a combined decrease in the medium concentration of nitrite and PGE_2, while the expression level of COX-1 remained unchanged. **Sq135** inhibited the proliferation of human colonic adenocarcinoma cell lines in a dose-dependent manner, while the growth of normal human dermal (2F0-C25) and colon (CCD-18 Co) fibroblasts was less affected. It also induced apoptosis in COLO205 cells, as detected by dysfunction of the mitochondria trans-membrane, Annexin V-detected translocation of phosphatidylserine, and chromatin condensation. Intriguingly, α-humulene was virtually inactive in all experiments conducted, indicating that the α,β-unsaturated carbonyl group in **Sq135** may play some pivotal roles in interactions with unidentified target molecule(s) [24].

Diterpenoids (Di)

The diterpenoids constitute a large group of compounds derived from geranylgeranyl diphosphate. They are found in higher plants, fungi, insects and marine organisms.

Table 4. Diterpenoids (Di) from Natural Sources and the Bioassay Systems in which the Compounds Exhibited Inhibitory Activities

Code	Compound	Source & Occurrence	Assay System	References
Abietane				
Di1	Abietic acid	*Pimenta racemosa*	TPA, CRG	[189]
Di2	12-Hydroxydehydroabietic acid	*Picea glehni*	EBV	[127]
Di3	Dehydroabietic acid	*P. glehni*	EBV	[127]
Di4	Methyl dehydroabietate	*P. cocos*	EBV	[131]
Di5	Abieta-8,11,13-trien-7-one	*P. glehni*	EBV	[127]
Di6	Methyl 15-hydroxy-7-oxo-dehydroabietate	*P. glehni*	EBV	[127]
Di7	9α,13α-Epidioxyabiet-8(14)-en-18-oic acid	*P. glehni*	EBV	[127]
Di8	12,15-Dihydroxydehydroabietic acid	*Larix kaempferi*		[129]
Di9	Methyl 15-hydroxy-8,12α-epidioxyabiet-13-en-18-oate	*Abies marocana*	EBV	[126]
Di10	7α,8α,13β,14β-Diepoxyabietan-18-oic acid	*Larix kaempferi*	EBV	[126]
Di11	Methyl 8α,14α,12α,13α-diepoxyabietan-18-oate		EBV	[126]
Di12	Methyl 15-hydroxy-8α,14α,12α,13α-diepoxyabietan-18-oate		EBV	[126]
Di13	18-*nor*-Abieta-8,11,13-triene-4α,7α,15-triol	*L.kaempferi*	EBV	[126]
Di14	Abieta-8.11.13-triene-7α.15.18-triol	*L. kaempferi*	EBV	[126]
Di15	Abieta-8,11,13-triene-7α,15,18-triol 7-acetate		EBV	[126]
Di16	18-*nor*-4α,15-Dihydroxyabieta-8,11,13-trien-7-one	*L. kaempferi*	EBV	[126]
Di17	15,18-Dihydroxyabieta-8,11,13-trien-7-one	*L. kaempferi*	EBV	[126]
Di18	15,18-Dihydroxyabieta-8,11,13-trien-7-one-18-acetate		EBV	[126]
Di19	7α,15-Dihydroxyabieta-8,11,13-trien-18-al	*L. kaempferi*	EBV	[126]
Di20	Karamatsuic acid methyl ester		EBV	[129]
Di21	9,10α-Epoxy-9,10-*seco*-abieta-8,11,13-trien-18-oic acid	*Picea glehni*	EBV	[127]
Di22	12-Hydroxy-6,7-*seco*-abieta-8,11,13-trien-6,7-dial	*Thuja standishii*	EBV	[125]
Di23	7α,8α-Epoxy-6α-hydroxyabieta-9(11),13-dien-12-one	*T. standishii*	EBV	[125]
Di24	12-Methoxyabieta-8,11,13-trien-11-ol	*T. standishii*	EBV	[125]
Di25	Carnosol	Rosemary	DNA, Mam-2, Skin-1, CYP, GST, iNOS, NFkB, NO	[94,95,190-193]
Di26	6α-Hydroxysugiol	*Thuja standishii*	EBV	[125]
Di27	11,12-Dihydroxy-6-oxo-8,11,13-abietatriene		COX	[194]
Di28	Triptoquinone A	*Tripterygium wilfordii*	iNOS	[195]

Table 4. Continued-1.

Code	Compound	Source & Occurrence	Assay System	References
Di29	19-*nor*-Abieta-4(18),8,11,13-tetraen-7-one	*Picea glehni*	EBV	[127]
Di30	Tanshione I		AM	[148]
Di31	13,14-*seco*-13,14-Dioxoabiet-13-en-18-oic acid	*Larix kaempferi*	EBV	[126]
Di32	Larikaempferic acid methyl ester		EBV	[129]
Di33	Triptolide	*Tripterygium wilfordii*	APO, COX, NFkB, TNFa	[96,196-200]
Di34	Standishinal	*Thuja standishii*	EBV	[125]
Di35	Standishinal diacetate	*T. standishii*	EBV	[125]
Di36	Aethiopinone	*Salvia aethiopis*	CRG	[201]
Di37	Tolypodiol	*Tolypothrix nodosa*	TPA	[202]
Atisane				
Di38	*ent*-16α,17-Dihydroxyatisan-3-one	*Euphorbia*	Mam-2	[203]
Di39	*ent*-3β,16α,17-Trihydroxyatisane	*E. quiquecostata*	Mam-2	[203]
Di40	Excoecarin N dimethyl ester	*Excoecaria agallocha*	EBV	[124]
Beyerane				
Di41	*ent*-3β-Hydroxy-15-beyeren-2-one	*Excoecaria agallocha*	EBV	[130]
Di42	Isosteviol		EBV	[123]
Di43	Isosteviol methyl ester		EBV	[123]
Di44	7β-Hydroxyisosteviol		EBV	[123]
Di45	11β-Hydroxyisosteviol		EBV	[123]
Di46	12β-Hydroxyisosteviol		EBV	[123]
Di47	17-Hydroxyisosteviol		EBV	[123]
Di48	17-Hydroxyisosteviol methyl ester		EBV	[123]
Di49	Excoecarin D	*Excoecaria agallocha*	EBV	[124]
Di50	Excoecarin E	*E. agallocha*	EBV	[130]
Cembrane				
Di51	Sarcophytol A	Marine soft coral	EBV, Colon-2, Liver-2, Lung-1, Mam-3, Pancr,	[97-100,204,205]
Di52	Sarcophytol B		Mam-3, Skin-2	[97]
Di53	α-Cembra-2,7,11-triene-4,6-diol	Cigarette smoke	EBV, Skin-1, ODC, PKC	[21,206]
Di54	β-Cembra-2,7,11-triene-4,6-diol	Cigarette smoke	EBV, Skin-1, ODC	[21]
Cleroda				
Di55	Cajucarinolide	*Croton cajucara*	TLO	[207]
Di56	Isocajucarinolide	*C. cajucara*	TLO	[207]
Di57	*trans*-Dehydrocrotonin	*C. cajucara*	CRG	[208]
Dolabrane				
Di58	Akendo 1	*Endospermum*	TPA	[209]
Di59	Akendo 2	*E. diadenum*	TPA, CRG	[209]
Di60	Akendo 3	*E. diadenum*	TPA, 5LOX, CRG	[209]
Kaurane				
Di61	Foliol		NFkB	[210]
Di62	Linearol		NFkB	[210]
Di63	*ent*-Kaur-16-en-19-oic acid	*Helianthus annuus*	EBV, NFkB	[123,210]
Di64	Grandiflorolic acid	*H. annuus*	EBV	[123]
Di65	Grandiflorolic acid angelate	*H. annuus*	EBV	[123]
Di66	Steviol		EBV	[123,211]

Table 4. Continued-2.

Code	Compound	Source & Occurrence	Assay System	References
Di67	Steviol methyl ester		EBV	[123]
Di68	Stevioside	*Stevia rebaudiana*	TPA, Skin-1	[212]
Di69	Rebaudioside A	*S. rebaudiana*	TPA	[212]
Di70	Rebaudioside C	*S. rebaudiana*	TPA	[212]
Di71	Dulucoside A	*S. rebaudiana*	TPA	[212]
Di72	Methyl *ent* -13-hydroxykaur-15-en-19-oate		EBV	[123]
Di73	Stachenone	*Excoecaria agallocha*	EBV	[130]
Di74	Stachenol	*E. agallocha*	EBV	[130]
Di75	Kamebanin	*Isodon japonicus*	NFkB, NO	[213]
Di76	Kamebacetal A	*I. japonicus*	NFkB, NO	[213]
Di77	Excisanin A	*I. japonicus*	NFkB, NO	[213]
Di78	Kamebakaurin	*I. japonicus*	NFkB, NO	[213]
Di79	Cafestol	Green coffee beans	Skin-6, CYP, GSH, GST	[102-105]
Di80	Kahweol	Green coffee beans	Skin-6, CYP, GSH, GST	[102-105]
Di81	Excoecarin M dimethyl ester	*Excoecaria agallocha*	EBV	[124]
Labdane				
Di82	*trans* -Communic acid	*Thuja standishii*	EBV	[125]
Di83	*trans* -Communic acid methyl ester	*T. standishii*	EBV	[128]
Di84	*cis* -Coumaric acid	*Cryptomeria japonica*	CRG	[144]
Di85	15-Oxolabda-8(17),11Z,13E -trien-19-oic acid	*Thuja standishii*	EBV, Skin-1	[125,128]
Di86	12S -Hydroxylabda-8(17),13(16),14-trien-19-oic acid	*T. standishii*	EBV	[125]
Di87	15-Oxolabda-8(17),11Z,13Z -trien-19-oic acid	*T. standishii*	EBV	[128]
Di88	Labda-8(17),13-dien-15,12R -olid-19-oic acid	*T. standishii*	EBV	[125]
Di89	Methyl labda-8(17),13-dien-15,12R -olid-19-oate	*T. standishii*	EBV	[125]
Di90	Methyl 12,13-epoxy-*trans* -	*T. standishii*	EBV	[128]
Di91	Scoparinol	*Scoparia dulcis*	CRG	[235]
Di92	15-Oxolabda-8(17),13Z -dien-19-oic acid	*T. standishii*	EBV	[128]
Di93	Rotundifuran	*Vitex rotundifolia*	APO	[214]
Di94	13-Ethoxylabda-8(17),11,14-trien-19-oic acid N -(methyl)phthalamide	*T. standishii*	EBV	[125]
Di95	Hinokiol	*Cunninghamia*	CRG	[215]
Di96	15-*nor* -14-Oxolabda-8(17),12E -dien-19-oic acid	*T. standishii*	EBV	[128]
Di97	15,16-*bisnor* -13-Hydroxylabda-8(17),11E -dien-19-oic acid	*T. standishii*	EBV	[128]
Di98	15,16-*bisnor* -13-Acetoxylabda-8(17),11E -dien-19-oic acid	*T. standishii*	EBV	[128]
Di99	14,15-*bisnor* -8α-Hydroxylabd-11E -en-13-one	*Picea glehni*	EBV	[127]
Di100	15,16-*bisnor* -13-Oxolabda-8(17),11E -dien-19-oic acid	*T. standishii*	EBV	[128]

Table 4. Continued-3.

Code	Compound	Source & Occurrence	Assay System	References
Di101	Methyl 15,16-*bisnor* -13-oxolabda-8(17),11*E* -dien-19-oate	*T. standishii*	EBV	[128]
Di102	15,16-*bisnor* -13-Oxolabda-8(17)-en-19-oic acid	*T. standishii*	EBV	[128]
Di103	Pinusolide	*Biota orientalis*	CRO	[236]
Di104	19(4→3)*abeo* -8α,13(*S*)-epoxylabda-4(18),14-diene	*Picea glehni*	EBV	[127]
Di105	Borjatriol	*Sideritis mugronensis*	CTN	[216]
Di106	Forskolin		APO	[217]
Di107	13-Epimanoyl oxide	*Picea glehni*	EBV	[127]
Di108	Labdane F2	*Sideritis javalambrensis*	PLA	[218]
Di109	Andalusol	*S. foetens*	TPA, NFkB, NO, CRG, ACM	[107,108,110]
Di110	*ent* -15-Hydroxy-labda-8(17),13*E* -dien-3-one	*Excoecaria agallocha*	EBV	[130]
Di111	Andrographolide	*Andrographis paniculata*	AOS, iNOS, NO	[111,112,219]
Di112	Excoecarin G2	*E. agallocha*	EBV	[124]
Di113	Ribenone	*E. agallocha*	EBV	[130]
Di114	11b-Hydroxyribenone	*E. agallocha*	EBV	[130]
Di115	Ribenol	*E. agallocha*	EBV	[130]
Di116	*ent* -13-*epi* -Manoyl oxide	*E. agallocha*	EBV	[130]
Di117	*ent* -16-Hydroxy-3-oxo-13-*epi* -manoyl oxide	*E. agallocha*	EBV	[130]
Di118	*ent* -13-*epi* -8,13-Epoxy-2-hydoxylabd-1-en-3-one	*E. agallocha*	EBV	[130]
Di119	Manoyl oxide F1	*Sideritis javalambrensis*	PLA	[218]
Di120	Excoecarin A	*E. agallocha*	EBV	[130]
Di121	(13*R* ,14*R*)-*ent* -8α,13;14,15-Diepoxy-13-*epi* -labdan-3β-ol	*E. agallocha*	EBV	[130]
Di122	Excoecarin B	*E. agallocha*	EBV	[130]
Di123	Excoecarin F dimethyl ester	*E. agallocha*	EBV	[124]
Di124	Excoecarin H methyl ester	*E. agallocha*	EBV	[124]
Di125	*ent* -13-*epi* -8,13-Epoxy-2,3-secolabd-14-ene-2,3-dioic acid	*E. agallocha*	EBV	[130]
Di126	*ent* -13-*epi* -8,13-Epoxy-2,3-secolabd-14-ene-2,3-dioic acid 2,3-dimethyl ester	*E. agallocha*	EBV	[130]
Di127	*ent* -13-*epi* -8,13-Epoxy-2,3-secolabd-14-ene-2,3-dioic acid 3-methyl ester	*E. agallocha*	EBV	[130]
Di128	Excoecarin S	*E. agallocha*	EBV, Skin-1	[124]
Di129	*ent* -13-*epi* -8,13-Epoxy-2-oxa-3-oxolabd-14-ene-1*R* -carboxylic acid methyl ester	*E. agallocha*	EBV	[124]
Miscellaneous Diterpenoid				
Di130	17-Hydroxyingenol 20-hexadecanoate	*Euphorbia quiquecostata*	Mam-2	[203]

Table 4. Continued-4.

Code	Compound	Source & Occurrence	Assay System	References
Di131	Ingenol 20-hexadecanoate	*E. quiquecostata*	Mam-2	[203]
Di132	Orthosiphol A	*Orthosiphon stamineus*	TPA	[220]
Di133	Orthosiphol B	*O. stamineus*	TPA	[220]
Di134	Acanthoic acid	*Acanthopanax koreanum*	TNFa	[221,222]
Di135	Totarol	*Thuja standishii*	EBV	[125]
Di136	Lagascatriol		COX, NO	[194]
Di137	Trachyloban-19-oic acid	*Helianthus annuus*	EBV	[123]
Di138	6α-Hydroxyvouacapan-17,7b-lactone	*S. branca*	AOS	[223]
Di139	6α-Acetoxyvouacapan-17,7b-lactone	*S. branca*	AOS	[223]
Di140	6α,7β-Dihydroxyvouacapan-17β-oic acid	*Sucupira branca*	AOS	[223]
Di141	Methyl 6α,7α-dihydroxyvouacapan-17β-oate	*S. branca*	AOS	[223]
Di142	Zahavin A	Marine soft coral	AOS	[224]
Di143	6-Deacetoxy-14,15-deepoxy-xeniculin	Marine soft coral	AOS	[224]
Di144	7,8-Epoxyzahavin A	Marine soft coral	AOS	[224]
Di145	Fuscoside B	Caribbean gorgonian	TPA	[225]

*Carnosol (**Di25**)*

Rosemary (*Rosmarinus officinalis*, Labiatae) is native to southern Europe. Rosemary acts as a mild analgesic and antimicrobial agent in traditional herbal use [95]. The relative amount of carnosol (**Di25**) in dried rosemary leaves is 3.8–4.6%. Among the antioxidant compounds in rosemary leaves, ~90% of the antioxidant activity can be attributed to **Di25** and carnosic acid (11,12-dihydroxy-8,11,13-abietatrien-20-oic acid) [94]. The antioxidant properties of **Di25** and its effects on NO generation, iNOS expression, NF-κB, p38 and p44/42 mitogen-activated protein kinase (MAPK) activation in LPS-stimulated RAW264.7 cells were examined [95]. **Di25** showed potent antioxidant activity in α,α-diphenyl-β-picrylhydrazyl (DPPH) free radicals scavenge and DNA protection from the Fenton reaction [95]. High concentrations of NO are produced by iNOS in inflammation and multiple stage of carcinogenesis [95]. Treatment of mouse macrophage RAW264.7 cell line with **Di25** markedly reduced LPS-stimulated NO production in a concentration-related manner with an IC_{50} of 9.4 μM [95]. Western blot, reverse transcription-polymerase chain reaction, and northern blot analyses demonstrated that **Di25** decreased LPS-induced iNOS mRNA and protein expression [95]. **Di25** treatment showed reduction of NF-κB subunits translocation and NF-κB DNA binding activity in activated macrophages. **Di25** also showed inhibition of iNOS and NF-κB promoter activity in transient transfection assay [95]. These activities were described as down-regulation of inhibitor kappaB (IκB) activity by **Di25** (5 µM), thus inhibition of LPS-induced phosphorylation and degradation of IκB [95]. **Di25** also inhibited LPS-induced p38 and p44/42 MAPK activation at a higher concentration (20 µM) [95]. These results suggest that **Di25** suppresses the NO production and iNOS gene expression by inhibiting NF-κB activation, and provide possible mechanisms for its anti-inflammatory and chemopreventive action [95]. The present data indicate that **Di25** could protect against endotoxin-induced inflammation by blocking NF-κB, p38 and p44/42 MAPK activation, thereby inhibiting the iNOS expression [95]. The effects of dietary intake and intraperitoneal (i.p.) administration of an extract of the spice rosemary and of the rosemary constituent **Di25** on the liver activities of glutathione-S-transferase (GST) and NAD(P)H-quinone reductase (QR) in the female rat were evaluated [96]. These studies indicate that dietary rosemary extract, as well as rosemary extract and **Di25** administered i.p., can substantially enhance live GST and QR activities in female rats. In addition, these studies demonstrate that **Di25** and/or rosemary extract can stimulate the activities of the phase II enzymes GST and QR, can inhibit the activity of mRNA expression of the phase I enzyme cytochrome P4501A1, and can decrease carcinogen-DNA adduct formation in several tissues. This indicates that **Di25** and possibly other constituents of rosemary hold considerable promise as blocking agents that can protect against chemically induced tumors [96]. **Di25** prevents DMBA-induced DNA damage and tumor formation in the rat mammary gland which suggested that this compound has potential for use as a breast cancer chemopreventive agent [94].

*Sarcophytol A (**Di51**)*

Sarcophytol A (**Di51**) is a cembrane-type diterpenoid, isolated from the soft coral, *Sarcophyyton glaucum*. **Di51** inhibits tumor promotion by teleocidin in two-stage carcinogenesis experiments on mouse skin [97]. Treatment with dietary **Di51** inhibited large bowel carcinogenesis with methylnitrosourea, development of spontaneous mammary tumors of SHN mice, and development of spontaneous liver tumors in mice. **Di51** did not show any toxic effects in an acute toxicity test. Sarcophytol B (= **Di51** with an additional hydroxyl group) has a similarly strong inhibitory activity on tumor promotion by teleocidin [98]. **Di51** inhibits the development of spontaneous hepatomas without toxicity, and should be considered as a possible chemopreventive agent for hepatomas in humans [99]. Growth of transplanted human pancreatic cancer cells in nude mice and pancreatic carcinogenesis induced by *N*-nitrobis-(2-hydroxypropyl) amine in Syrian golden hamsters were inhibited by feeding the animals a

Diterpenoid

Abietane

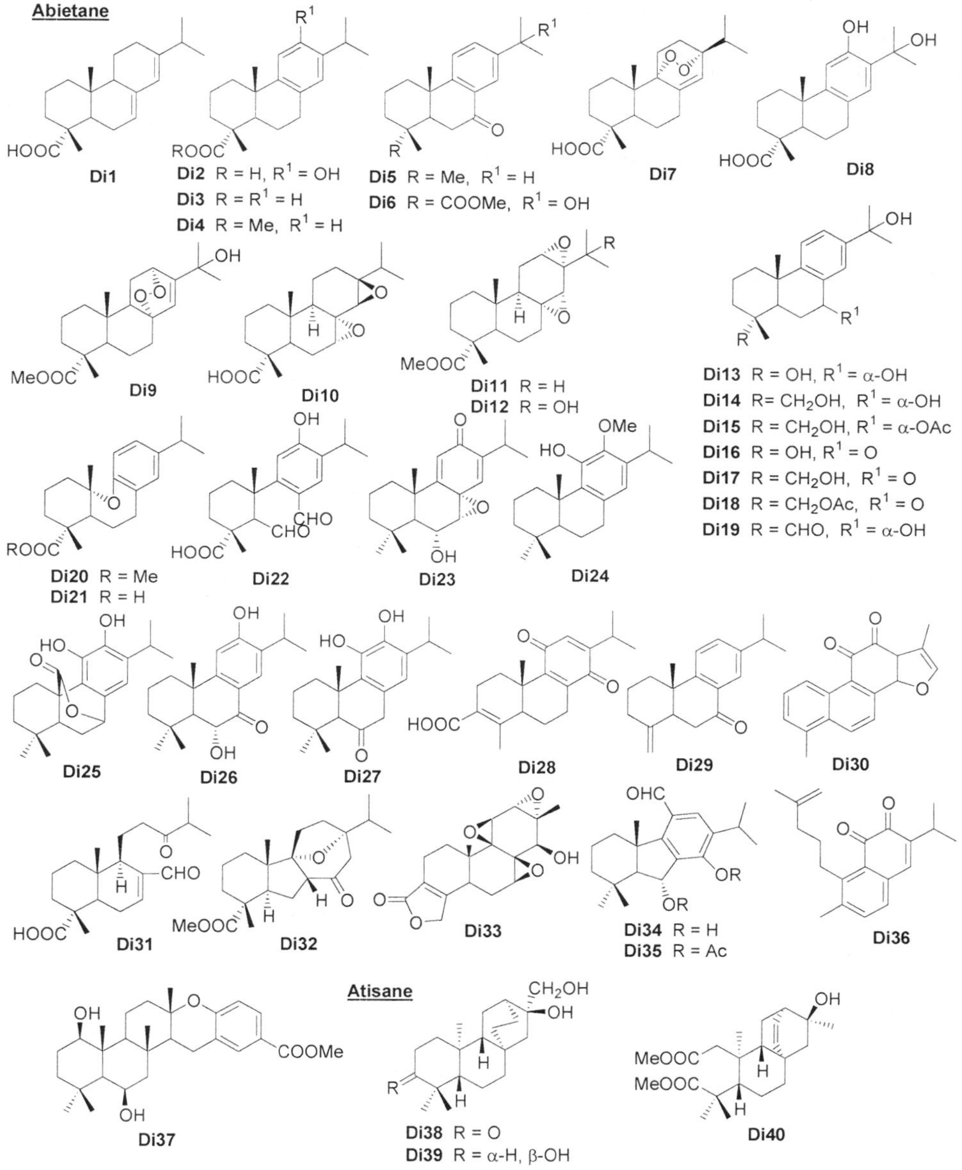

Di1

Di2 R = H, R¹ = OH
Di3 R = R¹ = H
Di4 R = Me, R¹ = H

Di5 R = Me, R¹ = H
Di6 R = COOMe, R¹ = OH

Di7

Di8

Di9

Di10

Di11 R = H
Di12 R = OH

Di13 R = OH, R¹ = α-OH
Di14 R = CH₂OH, R¹ = α-OH
Di15 R = CH₂OH, R¹ = α-OAc
Di16 R = OH, R¹ = O
Di17 R = CH₂OH, R¹ = O
Di18 R = CH₂OAc, R¹ = O
Di19 R = CHO, R¹ = α-OH

Di20 R = Me
Di21 R = H

Di22

Di23

Di24

Di25

Di26

Di27

Di28

Di29

Di30

Di31

Di32

Di33

Di34 R = H
Di35 R = Ac

Di36

Di37

Atisane

Di38 R = O
Di39 R = α-H, β-OH

Di40

Fig. (4). Structures of diterpenoids described in this review

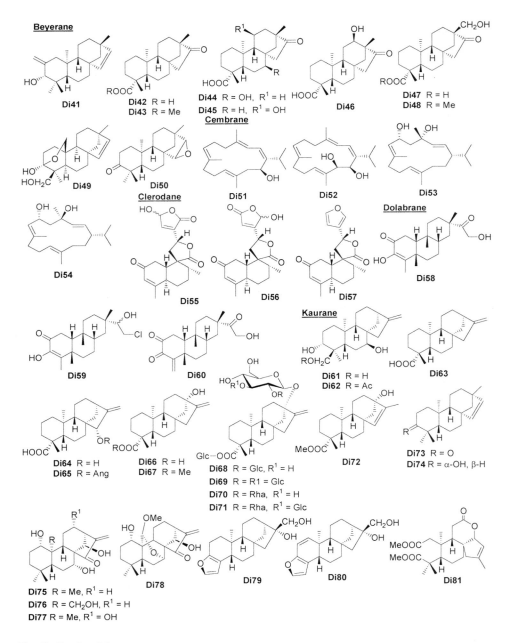

Fig. (4). Continued-1

Labdane

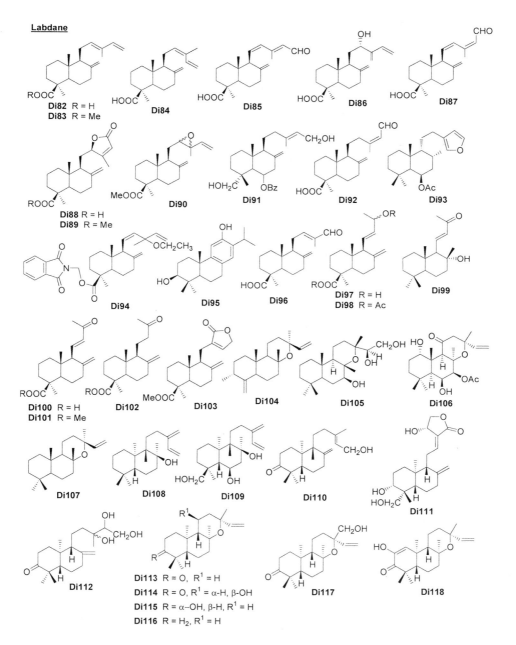

Fig. (4). Continued-2

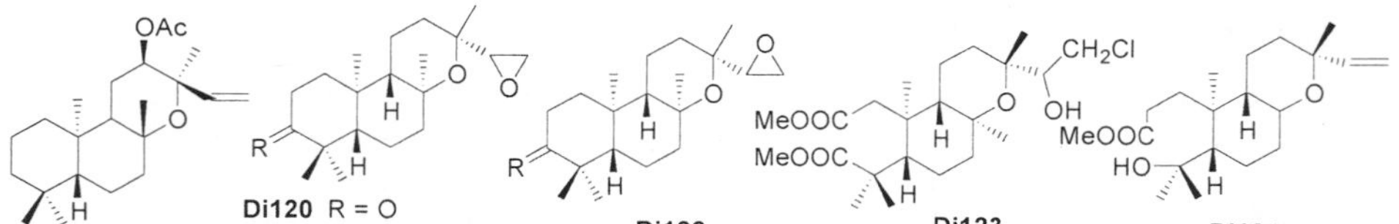

Di119
Di120 R = O
Di121 R = α-OH, β-H
Di122
Di123
Di124

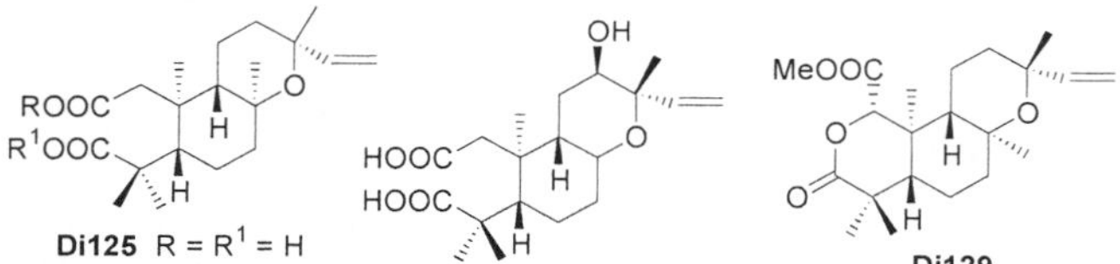

Di125 R = R¹ = H
Di126 R = R¹ = Me
Di127 R = H, R¹ = Me
Di128
Di129

Miscellaneous Diterpenoid

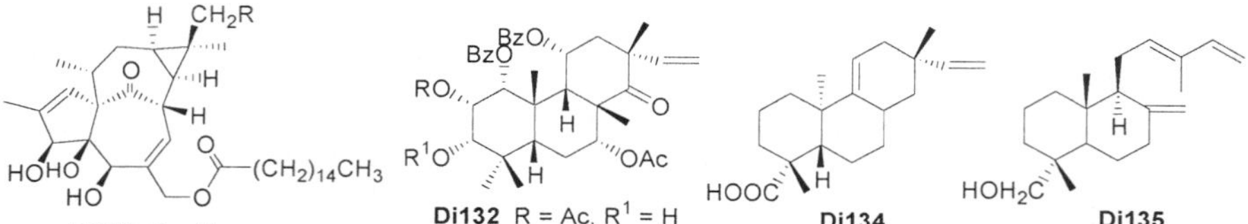

Di130 R = H
Di131 R = OH
Di132 R = Ac, R¹ = H
Di133 R = H, R¹ = Ac
Di134
Di135

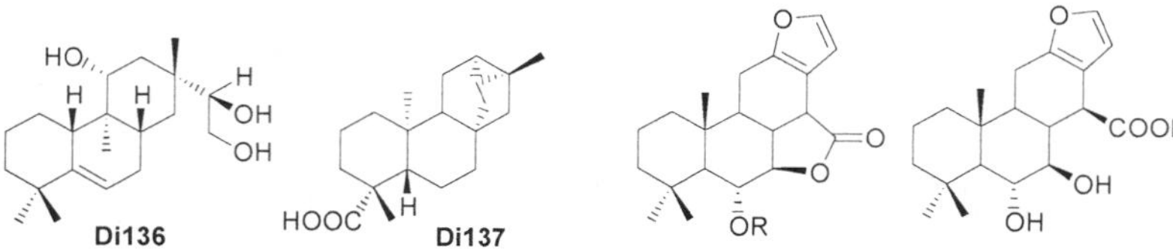

Di136
Di137
Di138 R = H
Di139 R = Ac
Di140 R = H
Di141 R = Me

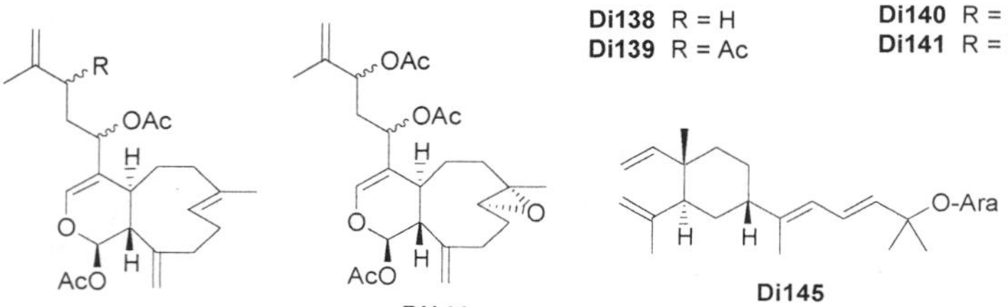

Di142 R = OAc
Di143 R = H
Di144
Di145

Fig. (4). Continued-3

diet containing 0.01% **Di51** [100]. **Di51** also suppresses oxidant formation and DNA oxidation in the epidermis of SENCAR mice exposed to the tumor promoter TPA [101].

*Cafestol (**Di79**) and Kahweol (**Di80**)*

Epidemiological studies have found an inverse association between coffee consumption and the risk of certain types of cancers such as colorectal cancers. In animal models including hamster [102], rat and mouse, the diterpenoids cafestol (**Di79**) and kahweol (**Di80**), that are only found in coffee, produce biological effects compatible with anticarcinogenic properties in liver, kidney, lung and intestinal tissues. The use of several model carcinogens led to the identification of several key mechanisms: (*a*) induction of phase II enzymes involved in carcinogen detoxification [103], (*b*) reduction in the expression of phase I enzyme responsible for carcinogen activation, (*c*) specific inhibition of the enzymatic activity of phase I enzyme responsible for carcinogen activation, and (*d*) stimulation of intracellular antioxidant defense mechanisms. These effects results in inhibition of DNA damage induced by procarcinogens such as AFB_1 [104], B[*a*]P, PhIP, and DMBA. Available information indicates that the **Di79/Di80** effects observed in animal models may be relevant for the human situation. Together, the data on the biological effects of **Di79/Di80** provide a plausible hypothesis to explain some anticarcinogenic effects of coffee observed in human epidemiological studies and in animal experiments [105].

*Andalusol (**Di109**)*

Sidertis species (Lamiaceae) have been extensively used in Spanish folk medicine to treat inflammatory disorders. Andalusol (**Di109**) is a labdane diterpenoid isolated from *S. foetens* with anti-inflammatory activity demonstrated in experimental inflammation model *in vivo* (carrageenan-induced mouse paw edema and TPA-induced mouse ear edema) [**106**]. It also inhibited the release of some inflammatory mediators tested in cellular systems. Thus, **Di109** inhibited NO release in LPS-treated J774 macrophages, granular enzymes secretion and histamine release from mast cells [106–109]. **Di109** inhibited NF-κB activation, a transcription factor necessary for NOS-2 expression in response to LPS and IFN-γ. This compound also inhibited the degradation of IκBα favoring the retention of the inactive NF-κB complexes in the cytosol. These results support a mechanism in which anti-inflammatory properties are mediated by **Di109** [107]. **Di109** has immunosuppressive effects *in vitro* in addition to its anti-inflammatory activity [110].

*Andrographolide (**Di111**)*

Andrographis paniculata (Acanthaceae) has been used medically inmainland China for more than 30 years and has been effective in the treatment of disorders related to bacterial infection (e.g., pneumonia bacteria) and inflammation (e.g., rheumatoid arthritis). Diterpenoids are the primary constituents found in leaves of *A. paniculata*, with andrographlide (**Di111**), a diterpene lactone, as the major constituent. **Di111** has the ability to prevent TPA-induced reactive oxygen species (ROS) production, as well as *N*-formylmethyl-leucyl-phenylalanine (fMLP)-induced adhesion by rat neutrophils. These properties (prevention of ROS production and neutrophils adhesion) suggest that **Di111** might be used as an anti-inflammatory drug [111]. Modulation of the PKC-dependent pathway [23] could give **Di111** the ability to down-regulate macrophage adhesion molecule-1 (Mac-1) up-expression that is essential for neutrophil adhesion and transmigration [112].

Retinoids (Re)

Retinoids (compounds with a 10,15-cyclophytane diterpenoid skeleton) are well-established chemopreventive agents for experimental carcinogenesis of many target organs including mammary glands, urinary bladder, lung, skin, liver, pancreas, colon, and esophagus [113].

Retinoids and Mammary Cancer

The relative efficacy of nontoxic doses of several retinoids in the inhibition of mammary carcinogenesis induced in the rat by both DMBA and MNU (*N*-methyl-*N*-nitrosourea) has been investigated [113]. Retinol acetate (**Re3**) and 4-hydroxyphenyl retinamide (4-HPR) are highly effective in reducing mammary cancer incidence and increasing the latency of induced mammary cancers. In addition, the number of mammary carcinomas is also significantly reduced by the administration of either of these retinoids. Retinol methyl ether was extremely effective against DMBA-induced mammary carcinogenesis [113].

Retinoids and Urinary Bladder Cancer

Synthetic 13-*cis*-retinoic acid not only inhibited the incidence but also reduced the severity of bladder neoplasms induced by the intravesical administration of MNU of female Wister-Lewis rats [113]. In an experimental model with mice treated with *N*-butyl-*N*-(4-hydroxybutyl)-nitrosamine (OH-BBN), retinoic acid (**Re5**), 13-*cis*-retinoic acid, and retinol acetate (**Re2**) show chemopreventive activity. In addition, a quite large number of synthetic *n*-alkyl amide derivatives of **Re2** have a greater activity to toxicity ratio than 13-*cis*-retinoic acid [113].

Table 5. Retinoids (Re), Sesterterpenoids (St), and Miscellaneous Terpenoids from Natural Sources and the Bioassay Systems in which the Compounds Exhibited Inhibitory Activities

Code	Compound	Source & Occurrence	Assay System	References
Retinoid (Re)				
Re1	Retinol		AM, EBV	[36,211,226-228]
Re2	Retinol acetate		AM	[227]
Re3	Retinol palmitate		AM	[227]
Re4	Retinal		AM	[226,227]
Re5	Retinoic acid		AM, EBV, Skin-1, ODC	[21,227]
Sesterterpenoid (St)				
St1	Cyclolinteinone	Marine sponge	COX, iNOS, NFkB	[229]
St2	Manoalide	Marine sponge	PLA	[230]
St3	Mangicol A	*Fusarium heterosporum*	TPA	[231]
St4	Mangicol B	*F. heterosporum*	TPA	[231]
St5	Palauolide	Marine sponge	PLA	[232]
St6	Palauolol	Marine sponge	PLA	[232]
Meroterpenoid and Miscellaneous Terpenoid (Me)				
Me1	Euglobal-G1	*Eucalyptus grandis*	EBV, Lung-2, Skin-1, Skin-3	[117,118]
Me2	Euglobal-G2	*E. grandis*	EBV	[117]
Me3	Euglobal-G3	*E. grandis*	EBV	[117]
Me4	Euglobal-G4	*E. grandis*	EBV	[117]
Me5	Euglobal-G5	*E. grandis*	EBV	[117]
Me6	Euglobal-T1	*E. grandis*	EBV	[117]
Me7	Euglobal-IIc	*E. tereticornis*	EBV	[117]
Me8	Euglobal-Ia1	*E. globulus*	EBV	[117]
Me9	Euglobal-Ia2	*E. globulus*	EBV	[117]
Me10	Euglobal-Bl-1	*E. blakelyi*	EBV	[117]
Me11	Euglobal-Ib	*E. blakelyi*	EBV	[117]
Me12	Euglobal-Ic	*E. blakelyi*	EBV	[117]
Me13	Euglobal-IIa	*E. blakelyi*	EBV	[117]
Me14	Euglobal-IIb	*E. blakelyi*	EBV	[117]
Me15	Euglobal-Am-2	*E. amplifolia*	EBV	[117]
Me16	Euglobal-III	*E. incrassata*	EBV	[117]
Me17	Euglobal-Iva	*E.. globulus*	EBV	[117]
Me18	Euglobal-Ivb	*E. amplifolia*	EBV	[117]
Me19	Euglobal-V	*E. incrassata*	EBV	[117]
Me20	Euglobal-VII	*E. amplifolia*	EBV	[117]
Me21	Euglobal-In-1	*E. incrassata*	EBV	[117]
Me22	Loliolide	*Ajuga decumbens*	EBV	[71]
Me23	Avarol	Mediterranean sponge	TPA, CRG	[233]
Me24	Avarone	Mediterranean sponge	TPA, CRG	[233]
Me25	Canventol	Synthetic compound	Skin-5, RAS, TNFa	[205,234]

112

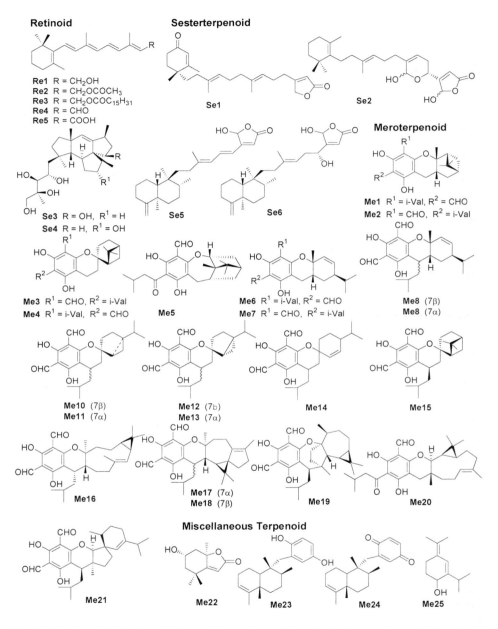

Fig. (5). Structures of retinoids, sesterterpenoids, meroterpenoids, and miscellaneous terpenoids described in this review

Retinol (Re1)

Results from several laboratory and clinical trials suggest that retinoids have chemopreventive effects [114]. In a phase III double-blind, randomized study of **Re1** versus placebo in 2297 patients with moderate to severe AK (actinic keratosis = keratinocytic intraepithelial neoplasia, KIN), daily dietary retinal (**Re5**) (25,000 IU) was associated with a 32% reduction in the risk of developing SCCs (squamous cell carcinomas) of the skin with no significant toxicity [115]. This effect was even more pronounced in patients with eight or more freckles or moles, indicating that **Re1** was especially protective in high-risk subjects. In a subsequent trial of high-risk subjects, 719 patients with at least four prior skin cancers were randomized to receive oral **Re1** (25,000 IU), isotretinoin (5–10 mg) or placebo [116]. In this trial, however, neither **Re1** nor isotretinoin [(13Z)-retinoic acid] was effective in reducing or delaying the occurrence of NMSC (melanoma and non-melanoma skin cancers). These results suggest that **Re1** may be more effective in early promotion and in preneoplastic stages of skin carcinogenesis.

Meroterpenoids (Me)

Euglobal-G1 (Me1) and Other Euglobals

Euglobal-G1 (**Me1**) and -III (**Me16**), isolated from the juvenile leaves of five species of *Eucalyptus* plants (Myrtaceae), exhibited remarkable antitumor-promoting effects on mouse skin tumor promotion in an *in vivo* two-stage carcinogenesis test [117]. Furthermore, upon two-stage carcinogenesis tests, **Me1** exhibited the remarkable inhibitory effect of mouse skin tumors induced by DMBA as an initiator and fumonsin-B1 (a non-TPA-type promoter) as a promoter, and of mouse pulmonary tumors induced by 4-nitroquinoline-*N*-oxide (4-NQO) as an initiator and glycerol as a promoter [118].

INHIBITORY EFFECTS OF TERPENOIDS ON TPA-INDUCED EBV ACTIVATION

The vast majority of cases of African Burkit's lymphoma (BL) are positive for genomic DNA of B-lymphotropic and potentially oncogenic EBV in the tumor cells. EBV does have an aetiological role in BL [237]. Extracts of Euphorbiaceae plants grown in tropical African and southern China contain phorbol esters which enhance EBV-specific events *in vitro* [238]. Most BL cases in Kenya have been diagnosed in the region where *Euphorbia tirucalli* grows densely [239]. The active principle in the extracts of *E. tirucalli* and some other *Euphorbia* spp., which enhances

Table 6. Inhibitory Effects of Some Terpenoids on TPA-induced EBV-EA Activation[a]

Compound Code	IC_{50} (nmol)	References	Compound Code	IC_{50} (nmol)	References	Compound Code	IC_{50} (nmol)	References
Monoterpenoid			Di17	10.0	[126]	Di100	8.7	[128]
Mo37	9.3	[70,71]	Di18	10.3	[126]	Di101	13.4	[128]
Mo38	16.3	[70]	Di19	9.5	[126]	Di102	13.1	[128]
Mo39	9.9	[70]	Di20	18.9	[129]	Di104	10.5	[127]
Sesquiterpenoid			Di21	17.2	[127]	Di107	7.6	[127]
Sq1	18.2	[123]	Di22	17.0	[125]	Di110	3.9	[130]
Sq2	18.6	[123]	Di23	16.0	[125]	Di112	10.9	[124]
Sq3	15.8	[123]	Di24	9.2	[125]	Di113	3.8	[130]
Sq7	10.6	[123]	Di26	8.5	[125]	Di114	10.3	[130]
Sq8	12.9	[123]	Di29	10.4	[127]	Di115	8.1	[130]
Sq9	12.4	[123]	Di31	8.9	[126]	Di116	7.4	[130]
Sq10	15.1	[123]	Di32	11.9	[129]	Di117	3.9	[130]
Sq11	9.3	[123]	Di34	19.8	[125]	Di118	16.1	[130]
Sq12	10.9	[128]	Di35	22.7	[125]	Di120	3.7	[130]
Sq13	6.4	[74]	Di40	9.9	[124]	Di121	4.1	[130]
Sq29	10.5	[119]	Di41	3.0	[130]	Di122	4.2	[130]
Sq30	101	[119]	Di42	12.8	[123]	Di123	8.0	[124]
Sq31	10.1	[119]	Di43	13.0	[123]	Di124	10.1	[124]
Sq32	12.9	[119]	Di44	13.7	[123]	Di125	10.6	[130]
Sq33	10.7	[119]	Di45	9.4	[123]	Di126	16.1	[130]
Sq34	10.9	[119]	Di46	11.2	[123]	Di127	10.1	[130]
Sq35	11.0	[119]	Di47	13.8	[123]	Di128	9.9	[124]
Sq36	14.7	[119]	Di48	14.0	[123]	Di129	11.8	[124]
Sq37	12.5	[119]	Di49	9.9	[124]	Di135	13.5	[130]
Sq45	<0.3	[122]	Di50	9.9	[130]	Di137	9.9	[123]
Sq46	1.6	[122]	Di63	9.9	[123]	*Meroterpenoid*		
Sq47	<0.3	[122]	Di64	9.0	[123]	Me1	9.1	[117]
Sq48	0.3	[122]	Di65	9.0	[123]	Me2	9.5	[117]
Sq49	19.2	[119]	Di66	11.7	[123]	Me3	9.3	[117]
Sq52	9.5	[123]	Di67	13.2	[123]	Me4	9.7	[117]
Sq53	10.6	[123]	Di72	14.0	[123]	Me5	3.8	[117]
Sq54	10.6	[123]	Di73	12.8	[130]	Me6	20.1	[117]
Diterpenoid			Di74	11.6	[130]	Me7	19.7	[117]
Di2	9.0	[127]	Di81	9.3	[124]	Me8	4.0	[117]
Di3	10.9	[127]	Di82	13.5	[125]	Me9	4.0	[117]
Di4	9.6	[131]	Di83	13.8	[128]	Me10	12.5	[117]
Di5	10.3	[127]	Di85	13.4	[125,128]	Me11	12.2	[117]
Di6	9.9	[127]	Di86	15.9	[125]	Me12	13.2	[117]
Di7	8.0	[127]	Di87	13.3	[128]	Me13	13.5	[117]
Di8	9.9	[129]	Di88	19.8	[125]	Me14	4.1	[117]
Di9	9.7	[126]	Di89	20.0	[125]	Me15	9.3	[117]
Di10	13.8	[126]	Di90	13.4	[128]	Me16	10.8	[117]
Di11	12.9	[126]	Di92	13.4	[128]	Me17	3.8	[117]
Di12	10.0	[126]	Di94	16.4	[125]	Me18	4.1	[117]
Di13	13.5	[126]	Di96	13.7	[128]	Me19	16.9	[117]
Di14	10.8	[126]	Di97	13.1	[128]	Me20	8.6	[117]
Di15	10.3	[126]	Di98	13.2	[128]	Me21	22.4	[117]
Di16	11.8	[126]	Di99	9.4	[127]	Me22	9.0	[71]

[a] IC_{50} of β-carotene = 9.4 nmol.

induction of EBV-EA, was identified as a 4-deoxyphorbol ester [240]. The effect of this compound on latent EBV induction was comparable with that of TPA. Tumor-promoter-induced EBV activation has been used for the primary screening of anti-tumor promoters as well as for the screening of TPA-type tumor promoters.

Several terpenoids have been evaluated for their inhibitory effects on EBV-EA activation induced by TPA. Table 6 shows the inhibitory effects of monoterpenoids [70,71], sesquiterpenoids [20,119–123], diterpenoids [21,123–131], and meroterpenoids [117] against TPA (32 pmol, 20 ng)-induced EBV-EA activation in Raji cells. The inhibitory effects were compared with that of β-carotene, a vitamin A precursor that has been studied intensively in cancer chemoprevention using animal models [2,4]. All of the terpenoids tested caused higher viability (60–80%) of Raji cells even at mol ratio of compound to TPA = 1000:1 indicating their very low cytotoxicity at that high concentration (refer to Table 6).

The sesquiterpenoids with a dihydro-β-agarofuran skeleton, isolated from *Maytenus cuzocoina* (Celastraceae), have been tested for their inhibitory effects on EBV-EA activation. Five compounds (**Sq29**, **Sq30**, **Sq31**, **Sq34**, and **Sq35**) were found to be highly active with IC_{50} values = 10–11 nmol [119] which were almost comparable with that of β-carotene (IC_{50} = 9.4 nmol). Four fungal metabolites, trichothecinols A–C (**Sq45**, **Sq46**, and **Sq47**, respectively) and trichothecin (**Sq48**), an antifungal antibiotic, isolated from the fungus *Trichotheium roseum*, exhibited quite high inhibitory effects. For example, **Sq45** and **Sq47** showed IC_{50} values <0.3 nmol [122].

A primary screening of seventeen diterpenoids isolated from the resinous wood of *Excoecaria agallocha* (Euphorbiaceae) has been performed [130]. Seven compounds, viz., **Di41**, **Di110**, **Di113**, **Di117**, **Di120**, **Di121**, and **Di122**, exhibited significant inhibitory effects against EBV-EA activation (IC_{50} values = 3–4 nmol) [130]. A number of other diterpenoids from natural sources are known to be potent anti-tumor-promoters by exhibiting significant inhibitory effects on EBV-EA activation (refer to Table 6).

To search for possible antitumor-promoters, twenty-one compounds with an euglobal structure (= acylphloroglucinol-monoterpenoid or acylphloroglucinol-sesquiterpenoid structure) isolated from *Eucalyptus* plants have been used primary screening [117]. Eublobals-G5 (**Me5**), Ia1 (**Me8**), Ia2 (**Me9**), IIb (**Me14**), Iva (**Me17**), and Ivb (**Me18**) exhibited significant inhibitory effects (IC_{50} values = 3–4 nmol) on EBV activation induced by TPA [117].

CONCLUSION

Both cell cultures and animal studies have shown that many of the naturally occurring mono-, sesqui-, di-, sester-, and meroterpenoids as well as retinoids possess potentially chemopreventive activities. Terpenoids are minor but ubiquitous components of our diet, and have the advantage of being non-toxic or relatively non-toxic to humans. More mechanistic-oriented basic research is needed to elucidate the mechanisms of action. Studies of derivatives of these naturally occurring terpenoids are also necessary to elucidate the structure-activity relationship and to guide the development of novel chemopreventive agents.

ABBREVIATIONS

Refer to Table 1 for abbreviations.

ACKNOWLEDGEMENT

We thank Dr. W. C. M. C. Kokke (Ardmore, Pennsylvania, U.S.A.) for reviewing the manuscript.

REFERENCES

[1]   Akihisa, T.; Yasukawa, K.; In *Studies in Natural Products Chemistry, Bioactive Natural Products*, Atta-ur-Rahman, Ed.; Elsevier Science B.V., Amsterdam, **2002**, Vol. 25, Part F, pp. 43–87.
[2]   Weinstein, I. B.; *Cancer Res.*, **1988**, 48, 4135–4143.
[3]   Piot, H. C.; Dragan, Y. P.; *FASEB J.*, **1991**, 5, 2280–2286.
[4]   Morse, M. A.; Stoner, G. D.; *Carcinogenesis*, **1993**, 14, 1737–1746.
[5]   Murakami, A.; Ohigashi, H.; Koshimizu, K.; *Biosci. Biotech. Biochem.*, **1996**, 60, 1–8.
[6]   Moon, R. C.; Mehta, R. G.; *Prevent. Med.*., **1989**, 18, 576–581.
[7]   Boone, C. W.; Wattenberg, L. W.; *Cancer Res.*, **1994**, 54, 3315–3318.
[8]   Greenwald, P.; In *Chemoprevention of Cancer*, Nixon, D. W., Ed.; CRC Press, Inc., Boca Raton, **1995**, Chapter 1, pp. 1–19.
[9]   Hong, W. K.; Lippman, S. M.; *J. Natl. Cancer Inst. Monographs*, **1995**, 49–53.
[10]   Shidoji, Y.; *J. Jpn. Oil Chem. Soc.*, **1996**, 45, 399–405.
[11]   Ren, S.; Lien, E. J.; *Progr. Drug. Res.*, **1997**, 48, 147–171.
[12]   Kinghorn, A. D.; Cui, B.; Kennelly, E. J.; Luyengi, L.; Chang, L. C.; In *Towards*

Natural Medicine Research in the 21st Century; Ageta, H.; Aimi, N.; Ebizuka, Y.; Fujita, T.; Honda, G., Eds.; Elsevier Science B. V., Amsterdam, **1998**; pp. 125–135.

[13]   Beecher, C.; In *Chemoprevention of Cancer*, Nixon, D. W., Ed.; CRC Press, Inc., Boca Raton, **1995**, Chapter 3, pp. 21–62.

[14]   Sporn, M. B.; *Cancer Res.*, **1976**, 36, 2699–2702.

[15]   Malone, W. F.; Perloff, M.; In *Chemoprevention of Cancer*, Nixon, D. W., Ed.; CRC Press, Inc., Boca Raton, **1995**, Chapter 4, pp. 63–88.

[16]   Iwase, Y.; Takemura, Y.; Ju-ichi, M.; Kawaii, S.; Yano, M.; Okuda, Y.; Mukainaka, T.; Tsuruta, A.; Okuda, M.; Takayasu, J.; Tokuda, H.; Nishino, H.; *Cancer Lett.*, **1999**, 139, 227–236.

[17]   Ames, B. N.; McCann, J.; Yamasaki, E., *Mutat. Res.* **1975**, 31, 347–363.

[18]   McCann, J.; Choi, E.; Yamasaki, E.; Ames, B. N.; *Proc. Natl. Acad. Sci. U.S.A.*, **1975**, 72, 5135–5139.

[19]   Oda, Y.; Nakamura, S.; Oki, I.; *Mutat. Res.*, **1985**, 147, 219–229.

[20]   Murakami, A.; Takahashi, M.; Jiwajinda, S.; Koshimizu, K.; Ohigashi, H.; *Biosci. Biotechnol. Biochem.*, **1999**, 63, 1811–1812.

[21]   Saito, Y.; Takizawa, H.; Konishi, S.; Yoshida, D.; Mizusaki, S.; *Carcinogenesis*, **1985**, 6, 1189–1194.

[22]   Konoshima, T.; *Adv. Exp. Med. Biol.*, **1996**, 404, 87–100

[23]   Kelloff, G. J.; Boone, C. W.; Steele, V. E.; Fay, J. R.; Lubet, R. A.; Crowell, J. A.; Sigman, C. C.; *J. Cell. Biochem., Suppl.*, **1994**, 20, 1–24.

[24]   Murakami, A.; Takahashi, D.; Kinoshita, T.; Koshimizu, K.; Kim, H. W.; Yoshihiro, A.; Nakamura, Y.; Jiwajinda, S.; Terao, J.; Ohigashi, H.; *Carcinogenesis*, **2002**, 23, 795–802.

[25]   Lee, S. K.; Hong, C.-H.; Huh, S.-K.; Kim, S.-S.; Oh, O-J.; Min, H.-Y.; Park, K.-K.; Chung, W.-Y.; Hwang, J.-K.; *J. Environ. Pathol., Toxicol. Oncol.*, **2002**, 21, 141–148.

[26]   Kawamori, T.; Tanaka, T.; Hara, A.; Yamahara, J.; Mori, H.; *Cancer Res.*, **1995**, 55, 1277–1282.

[27]   Schorr, K.; Garcia-Pineres, A. J.; Siedle, B.; Merfort, I.; Da Costa, F. B.; *Phytochem.*, **2002**, 60, 733–740.

[28]   Jin, H. Z.; Hwang, B. Y.; Kim, H. S.; Lee, J. H.; Kim, Y. Ho.; Lee, J. J.; *J. Nat. Prod.*, **2002**, 65, 89–91.

[29]   Jang, M. K.; Sohn, D. H.; Ryu, J. H.; *Planta Med.*, **2001**, 67, 550–552.

[30]   Hirota, M.; Mori, T.; Yoshida, M.; Iriye, R.; *Agric. Biol. Chem.*, **1990**, 54, 1073–1075.

[31]   Safayhi, H.; Sailer, E.-R.; *Planta Med.*, **1997**, 63, 487–493.

[32]   Sporn, M. B.; Roberts, A. B.; *J. Clin. Invest.*, **1986**, 78, 329–332.

[33]   Crowell, P. L.; *J. Nutr.*, **1999**, 129, 775S–778S.

[34]   Crowell, P. L.; *Breast Cancer Res. Treat.*, **1997**, 46, 191–197.

[35]   Bradlow, H. L.; Sepkovic, D. W.; *Anales New York Acad. Sci.*, **2002**, 963, 247–267.

[36]   Stratton, S. P.; Dorr, R. T.; Alberts, D. S.; *European J. Cancer*, **2000**, 36,

1292–1297.

[37]   Gould, M. N.; *Environ. Health Perspect.*, **1997**, 105, 977–979.

[38]   Crowell, P. L.; Siar, A.; Burke, Y. D.; *Adv. Exp. Med. Biol.*, **1996**, 401, 131–136.

[39]   Gould, M. N.; *J. Cell. Biochem., Suppl.*, **1995**, 22, 139–144.

[40]   McNamee, D.; *Lancet*, **1993**, 342, 801.

[41]   Yano, H.; Tatsuta, M.; Iishi, H.; Baba, M.; Sakai, N.; Uedo, N.; *Internat. J. Cancer*, **1999**, 82, 665–668.

[42]   Hardcastle, I. R.; Rowlands, M. G.; Barber, A. M.; Grimshaw, R. M.; Mohan, M. K.; Nutley, B. P.; Jarman, M.; *Biochem. Pharmacol.*, **1999**, 57, 801–809.

[43]   Uedo, N.; Tatsuta, M.; Iishi, H.; Baba, M.; Sakai, N.; Yano, H.; Otani, T.; *Cancer Lett.*, **1999**, 137, 131–136.

[44]   Giri, R. K.; Parija, T.; Das, B. R.; *Oncology Rep.*, **1999**, 6, 1123–1127.

[45]   van Lieshout, E. M.; Ekkel, M. P.; Bedaf, M. M.; Nijhoff, W. A.; Peters, W. H.; *Oncol. Rep.*, **1998**, 5, 959–963.

[46]   van Lieshout, E. M.; Bedaf, M. M.; Pieter, M.; Ekkel, C.; Dijhoff, W. A.; Peters, W. H.; *Carcinogenesis*, **1998**, 19, 2055–2057.

[47]   van Lieshout, E. M.; Posner, G. H.; Woodard, B. T.; Peters, W. H.; *Biochim. Biophys. Acta*, **1998**, 1379, 325–336.

[48]   Karlson, J.; Borg-Karlson, A. K.; Unelius, R.; Shoshan, M. C.; Wilking, N.; Ringborg, U.; Linder, S.; *Anti-Cancer Drugs*, **1996**, 7, 422–429.

[49]   Morse, M. A.; Toburen, A. L.; *Cancer Lett.*, **1996**, 104, 211–217.

[50]   Crowell, P. L.; Ren, Z.; Lin, S.; Vedejs, E.; Gould, M. N.; *Biochem. Pharmacol.*, **1994**, 47, 1405–1415.

[51]   Chander, S. K.; Lannsdown, A. G.; Luqmani, Y. A.; Gomm, J. J.; Coope, R. C.; Gould, N.; Coombes, R. C.; *British J. Cancer*, **1994**, 69, 879–882.

[52]   Gould, M. N.; Moore, C. J.; Zhang, R.; Wang, B.; Kennan, W. S.; Haag, J. D.; *Cancer Res.*, **1994**, 54, 3540–3543.

[53]   Crowell, P. L.; Kennan, W. S.; Haag, J. D.; Ahmad, S.; Vedejs, E.; Gould, M. N.; *Carcinogenesis*, **1992**, 13, 1261–1264.

[54]   Zheng, G. Q.; Kenney, P. M.; Lam, L. K.; *Planta Med.*, **1992**, 58, 338–341.

[55]   Haag, J. D.; Lindstrom, M. J.; Gould, M. N.; *Cancer Res.*, **1992**, 52, 4021–4026.

[56]   Russin, W. A.; Hoesly, J. D.; Elson, C. E.; Tanner, M. A.; Gould, M. N.; *Carcinogenesis*, **1989**, 10, 2161–2164.

[57]   Kimura, J.; Takahashi, S.; Ogiso, T.; Yoshida, Y.; Akagi, K.; Hasegawa, R.; Kurata, M.; Hirose, M.; Shirai, T.; *Jpn. J. Cancer Res.*, **1996**, 87, 589–594.

[58]   Lluria-Prevatt, M.; Morreale, J.; Gregus, J.; Alberts, D. S.; Kaper, F.; Giaccia, A.; Powell, M. B.; *Cancer Epidemiol. Biomark. Prevent.*, **2002**, 11, 573–579.

[59]   Carnesecchi, S. S.; Schneider, Y.; Ceraline J.; Duranton, B.; Gossee, F.; Seiler, N.; Raul, F. J. *Pharmacol. Exp. Therap.*, **2001**, 298, 197–200.

[60]   Loew-Baselli, A.; Huber, W. W.; Kaefer, M.; Bukowska, K.; Schulte-Hermann, R.; Grasl-Kraupp, B.; *Carcinogenesis*, **2000**, 21, 1869–1877.

[61]   Satomi, Y.; Miyamoto, S.; Gould, M. N.; *Carcinogenesis*, **1999**, 20, 1957–1961.

[62]   Ren, Z.; Gould, M. N.; *Carcinogenesis*, **1998**, 19, 827–832.

[63]   Barthelman, M.; Chen, W.; Gensler, H. L.; Huang, C.; Dong, Z.; Bowden, G. T.;

Cancer Res., **1998**, 58, 711–716.

[64]   Reddy, B. S.; Wang, C. X.; Samaha, H.; Lubet, R.; Steele, V. E.; Kelloff, G. J.; Rao, C. V.; *Cancer Res.*, **1997**, 57, 420-425.

[65]   Lantry, L. E.; Zhang, Z.; Gao, F.; Crist, K. A.; Wang, Y.; Kelloff, G. J.; Lubet, R. A.; You, M.; *J. Cell. Biochem. Suppl.*, **1997**, 27, 20–25.

[66]   Ren, Z.; Elson, C. E.; Gould, M. N.; *Biochem. Pharmacol.*, **1997**, 54, 113–120.

[67]   Mills, J. J.; Chari, R. S.; Boyer, I. J.; Coulg, M. N.; Jirtle, R. L.; *Cancer Res.*, **1995**, 55, 979–983.

[68]   Shi, W.; Gould, M. N.; *Cancer Lett.*, **1995**, 95, 1–6.

[69]   Haag, J. D.; Gould, M. N.; *Cancer Chemother. Pharmacol.*, **1994**, 34, 477–483.

[70]   Konoshima, T.; Takasaki, M.; Tokuda, H.; Nishino, H.; *Cancer Lett.,* **2000**, 31, 87–92.

[71]   Takasaki, M.; Tokuda, H.; Nishino, H.; Konoshima, T.; *J. Nat. Prod.,* **1999**, 62, 972–975.

[72]   Ruengeler, P.; Castro, V.; Mora, G.; Goeren, N.; Vichnewski, W.; Pahl, H. L.; Merfort, I.; Schmnidt, T. J.; *Bioorg. Med. Chem.*, **1999**, 7, 2343–2352.

[73]   de Cunha, F. M.; Froede, T. S.; Mendes, G. L.; Malheiros, A.; Echinel, F. V.; Yunes, R. A.; Calixto, J. B.; *Life Sci.*, **2001**, 70, 159–169.

[74]   Matsumoto, T.; Tokuda, H.; In *Antimutagenesis and Anticarcinogenesis Mechanisms II*, Kuroda, Y.; Shankel, D. M.; Waters, M. D., Eds., Prenum Press, New York, **1990**, pp. 423-427.

[75]   Hwang, D.; Fischer, N. H.; Jang, B. C.; Tak, H.; Kim, J. K.; Lee, W.; *Biochem. Biophys. Res. Commun.*, **1996**, 226, 810–818.

[76]   Piela-Smith, T. H.; Liu, X.; *Cell. Immunol.*, **2001**, 209, 89–96.

[77]   Fukuda, K. Hibiya, Y.; Mutoh, M.; Ohno, Y.; Yamashita, K.; Akao, S.; Fujiwara, H.; *Biochem. Pharm.,* **2000**, 60, 595–600.

[78]   Hehner, S. P.; Hofmann, T. G.; Droege, W.; Schmitz, M. L.; *J. Immunol.*, **1999**, 163, 5617–5623.

[79]   Wong, H. R.; Menenzez, I. Y.; *Biochem. Biophys. Res. Commun.*, **1999**, 262, 375–380.

[80]   Ruengeler, P.; Lyss, G.; Castro, V.; Mora, G.; Pahl, H. L.; Merfort, I.; *Planta Med.*, **1998**, 64, 588–593.

[81]   Bork, P. M.; Schnitz, M. L.; Kuhnt, M.; Escher, C.; Heinrich, M.; *FEBS Lett.*, **1997**, 402, 85–90.

[82]   Sheehan, M.; Wong, H. R.; Hake, P. W.; Malhorta, V.; O'Connor, M.; Zingarelli, B.; *Mol. Pharmacol.*, **2002**, 61, 953–963.

[83]   Shinella, G. R.; Giner, R. M.; Recio, M. C.; Mordujovich de Buschiazzo, P.; Rios, J. L.; Manez, S.; *J. Pharm. Pharmacol.*, **1998**, 50, 1069–1074.

[84]   Jain, N. K.; Kulkarni, S. K.; *J. Ethnoparmacol.*, **1999**, 68, 251–259.

[85]   Fukuda, K.; Akao, S.; Ohno, Y.; Yamashita, K.; Fujiwara, H.; *Cancer Lett.*, **2001**, 164, 7–13.

[86]   Park, S. H.; Choi, S. U.; Lee, C. O.; Yoo, S. E.; Yoon, S. K.; Kim, Y. K.; Ryu, S. Y.; *Planta Med.*, **2001**, 67, 358–359.

[87]   Koo, T. H.; Lee, J.-H.; Park, Y. J.; Hong, Y.-S.; Kim, H. S.; Kim, K.-W.; Lee, J. J.;

Planta Med., **2001**, 67, 103–107.

[88] Castro, V.; Ruengeler, P.; Murillo, R.; Hernandez, E.; Mora, G.; Pahl, H. L.; Merfort, I.; *Phytochem.*, **2000**, 53, 257–263.

[89] Mori, H.; Kawamori, T.; Tanaka, T.; Ohnishi, M.; Yamahara, J.; *Cancer Lett.*, **1994**, 83, 171–175.

[90] Ohnishi, M.; Yoshimi, N.; Kawamori, T.; Ino, N.; Hirose, Y.; Tanaka, T.; Yamahara, J.; Miyata, H.; Mori, H.; *Jpn. J. Cancer Res.*, **1997**, 88, 111–119.

[91] Lee, H. J.; Kim, N. Y.; Jang, M. K.; Son, H. J.; Kim, K. M.; Sohn, D. H.; Lee, S. H.; Ryu, J. H.; *Planta Med.*, **1999**, 65, 104–108.

[92] Matsuda, H.; Kagerura, T.; Toguchida, I.; Ueda, H.; Morikawa, T.; Yoshikawa, M.; *Life Sci.*, **2000**, 66, 2151–2157.

[93] Tanaka, T.; Shimizu, M.; Kohno, H.; Yoshitani, S.; Tsukio, Y.; Murakami, A.; Safitri, R.; Takahashi, D.; Yamamoto, K.; Koshimizu, K.; Ohigashi, H.; Mori, H.; *Life Sci.*, **2001**, 69, 1935–1945.

[94] Singletary, K.; MacDonald, C.; Wallig, M.; *Cancer Lett.*, **1996**, 104, 43–48.

[95] Lo, A.-H.; Liang, Y.-C.; Lin-Shiau, S.-Y.; Ho, C.-T.; Lin. J.-K.; *Carcinogenesis*, **2002**, 23, 983–991.

[96] Yang, Y.; Liu, Z.; Tolosa, E.; Yang, J.; Li, L.; *Immunopharmacol.*, **1998**, 40, 139–149.

[97] Fujiki, H.; Suganuma, M.; Suguri, H.; Takagi, K.; Yoshizawa, S.; Ootsuyama, A.; Tanooka, H.; Okuda, T.; Kobayashi, M.; Sugimura, T.; *Basic Life Sci.*, **1990**, 52, 205–212.

[98] Muto, Y.; Ninomiya, M.; Fujiki, H.; *Jpn. J. Clin. Oncology*, **1990**, 20, 219–224.

[99] Yamauchi, O.; Omori, M.; Ninomiya, M.; Okuno, M.; Moriwaki, H.; Suganuma, M.; Fujiki, H.; Muto, Y.; *Jpn. J. Cancer Res.*, **1991**, 82, 1234–1238.

[100] Yokomatsu, H.; Satake, K.; Hiura, A.; Tsutsumi, M.; Suganuma, M.; *Pancreas*, **1994**, 9, 526–530.

[101] Wei, H.; Frenkel, K.; *Cancer Res.*, **1992**, 52, 2298–2303.

[102] Miller, E. G.; McWhorter, K.; Rivera-Hidalgo, F.; Wright, J. M.; Hirsbrunner, P.; Sunahara, G. I.; *Nutr. Cancer*, **1991**, 15, 41–46.

[103] Huber, W. W.; Scharf, G.; Rossmanith, W.; Prustomersky, S.; Sonja, G.-K.; Bettina, P. B.; Turesky, R. J.; Schulte-Hermann, R.; *Archiv. Toxicol.*, **2002**, 75, 685–694.

[104] Cavin, C.; Mace, K.; Offord, E. A.; Schilter, B.; *Food Chem. Toxicol.*, **2001**, 39, 549–556.

[105] Cavin, C.; Holzhaeuser, D.; Scharf, G.; Constable, A.; Huber, W. W.; Schiter, B.; *Food Chem. Toxicol.*, **2002**, 40, 1155–1163.

[106] Navarro, A.; de las Heras, B.; Villar, A.; *Z. Naturforsch.*, **1997**, 52, 844–849.

[107] de las Heras, B.; Navarro, A.; Diaz-Guerra, M. J.; Bermejo, P.; Castrillo, A.; Bosca, L.; Villar, A.; *Br. J. Pharmacol.*, **1999**, 128, 605–612.

[108] Navarro, A.; de las Heras, B.; Villar, A. M.; *Z. Naturforsch. C, J. Biosci.*, **1997**, 52, 844–849.

[109] De las Heras, B.; Navarro, A.; Godoy, A.; Villar, A.; *Pharmacol. Lett.*, **1997**, 7, 111–112.

[110] Navaro, A.; de las Heras, B.; Villar, A.; *Planta Med.*, **2000**, 66, 289–291.

[111] Shen, Y. C.; Chen, C. F.; Chiou, W. F.; *Planta Med.,* **2000**, 66, 314–317.

[112] Shen, Y.-C.; Chen. C.-F.; Chiou, W.-F.; *Br. J. Pharmacol.,* **2002**, 135, 399–406.

[113] Moon, R. C.; Mehta, R. G.; *Preventive Med.,* **1989**, 18, 576–591.

[114] Hong, W. K.; *J. Natl. Cancer Inst. Monogr.,* **1995**, 49–53.

[115] Moon, T. E.; Levine, N.; Cartmel, B.; Bangert, J. L.; Rodney, S.; Dong, Q.; Peng, Y. M.; Alberts, D. S.; *Cancer Epidemiol. Biomarkers Prev.,* **1997**, 6, 949–956.

[116] Levine, N.; Moon, T. E.; Cartmel, B.; Bangert, J. L.; Rodney, S.; Dong, Q.; Peng, Y. M.; Alberts, D. S.; *Cancer Epidemiol. Biomarkers Prev.,* **1997**, 6, 957–961.

[117] Takasaki, M.; Konoshima, T.; Kozuka, M.; Tokuda, H.; *Biol. Pharm. Bull.,* **1995**, 18, 435–438.

[118] Takasaki, M.; Konoshima, T.; Etoh, H.; Pal, S. I.; Tokuda, H.; Nishino, H.; *Cancer Lett.,* **2000**, 155, 61–65.

[119] Gonzalez, A. G.; Tincusi, B. M.; Bazzocchi, I. L.; Tokuda, H.; Nishino, H.; Konoshima, T.; Jimenez, I. A.; Ravelo, A. G.; *Bioorg. Med. Chem.,* **2000**, 8, 1773–1778.

[120] Ujita, K.; Takaishi, Y.; Tokuda, H.; Nishino, H.; Iwashima, A.; Fujita, T.; *Cancer Lett.,* **1993**, 68, 129–133.

[121] Takaishi, Y.; Ujita, K.; Tokuda, H.; Nishino, H.; Iwashima, A.; Fujita, T.; *Cancer Lett.,* **1992**, 65, 19–26.

[122] Iida, A.; Konishi, K.; Kubo, H.; Tomioka, K.; Tokuda, H.; Nishino, H.; *Tetrahedron Lett.,* **1996**, 37, 9219–9220.

[123] Akihisa, T.; Tokuda, H.; Unpublished results.

[124] Konoshima, T.; Konishi, T.; Takasaki, M.; Yamazoe, K.; Tokuda, H.; *Biol. Pharm. Bull.,* **2001**, 24, 1440–1442.

[125] Iwamoto, M.; Ohtsu, H.; Tokuda, H.; Nishino, H.; Matsunaga, S.; Tanaka, R.; *Bioorg. Med. Chem.,* **2001**, 9, 1911–1921.

[126] Ohtsu, H.; Tanaka, R.; In, Y.; Matsunaga, S.; Tokuda, H.; Nishino, H.; *Planta Med.,* **2001**, 67, 655–560.

[127] Kinouchi, Y.; Ohtsu, H.; Tokuda, H.; Nishino, H.; Matsunaga, S.; Tanaka, R.; *J. Nat. Prod.,* **2000**, 63, 817–820.

[128] Tanaka, R.; Ohtsu, H.; Iwamoto, M.; Minami, T.; Tokuda, H.; Nishino, H.; Matsunaga, S.; Yoshitake, A.; *Cancer Lett.,* **2000**, 161, 165–170.

[129] Ohtsu, H.; Tanaka, R.; Matsunaga, S.; Tokuda, H.; Nishino, H.; *Planta Med.,* **1999**, 65, 664–666.

[130] Konishi, T.; Takasaki, M.; Tokuda, H.; Kiyosawa, S.; Konoshima, T.; *Biol. Pharm. Bull.,* **1998**, 21, 993–996.

[131] Ukiya, M.; Akihisa, T.; Tokuda, H.; Hirano, M.; Oshikubo, M.; Nobukuni, Y.; Kimura, Y.; Tai, T.; Kondo, S.; Nishino, H.; *J. Nat. Prod.,* **2002**, 65, 462–465.

[132] Connor, M. J.; *Cancer Lett.,* **1991**, 56, 25–28.

[133] Yasukawa, K.; Takido, M.; Takeuchi, M.; Nakagawa, S.; *Chem. Pharm. Bull.,* **1989**, 37, 1071–1073.

[134] Beaupre, D. M.; Talpaz, M.; Marini, F. C. 3rd.; Cristiano, R. J.; Roth, J. A.; Estrov, Z.; Albitar, M.; Freedman, M. H.; Kurzrock, R.; *Cancer Res.,* **1999**, 59, 2971–2980.

[135] Miyazawa, M.; Okuno, Y.; Nakamura, S.; Kosaka, H.; *J. Agric. Food Chem.*, **2000**, 48, 5440–5443.

[136] Gould, M. N.; *J. Cell. Biochem., Suppl.*, **1993**, 17G, 66–72.

[137] Crowell, P. L.; Gould, M. N.; *Crit. Rev. Oncogen.*, 1994, 5, 1–22.

[138] Elegbede, J. A.; Maltzuman, T. H.; Elson, C. E.; Gould, M. N.; *Carcinogenesis*, **1993**, 14, 1221–1223.

[139] Zheng, G. Q.; Kenney, P. M.; Zhang, J.; Lam, L. K.; *Nutr. Cancer*, **1993**, 19, 77–86.

[140] Juergens, U. R.; Stoeber, M.; Vetter, H.; Eur. *J. Med. Res.*, **1998**, 3, 508–510.

[141] Schapoval, E. E.; Vargas, M. F.; Chaves, C. G.; Bridi, R.; Zuanazzi, J. A.; Henriques, A. T.; *J. Ethnopharmacol.*, **1998**, 60, 53–59.

[142] Suksamrarn, A.; Kumpun, S.; Kirtikara, K.; Yingyongnarongkul, B.; Suksamrarn, S.; *Planta Med.*, **2002**, 68, 72–73.

[143] Giner, R. M.; Villalba, M. L.; Recio, M. C.; Manez, S.; Cerda-Nioklas, M.; Rios, J.; *Eur. J. Pharm.*, **2000**, 389, 243–252.

[144] Wagner, H.; *Planta Med.*, 1989, 55, 235–241.

[145] Hwang, B. Y.; Lee, J.-H.; Koo, T. H.; Kim, H. S.; Hong, Y. S.; Ro, J. S.; Lee, K. S.; Lee, J. J.; *Planta Med.*, **2002**, 68, 101–105.

[146] Thomet, O. A.; Wiesmann, U. N.; Blaser, K.; Simon, H. U.; *Clin. Exp. Allergy*, **2001**, 31, 1310–1320.

[147] Thomet, O. A.; Wiesmann, U. N.; Schapowal, A.; Bizer, C.; Simon, H.; *Biochem. Pharmacol.*, 2001, 61, 1041–1047.

[148] Miyazawa, M.; *J. Jpn. Oil Chem. Soc.*, **1999**, 48, 1057–1066.

[149] Hamandez, V.; del Carmen Recio, M.; Manez, S.; Prieto, J. M.; Giner, R. M.; Rios, J. L.; *Planta Med.*, **2001**, 67, 726–731.

[150] Manez, S.; Recio, M. C.; Gil, I.; Gomez, C.; Giner, R. M.; Waterman, P. G.; Rios, J. L.; *J. Nat. Prod.*, **1999**, 62, 601–604.

[151] Ryu, S. Y.; Oak, M. H.; Kim, K. M.; *Planta Med.*, **2000**, 66, 171–173.

[152] Ryu, J. H.; Lee, H. J.; Jeong, Y. S.; Ryu, S. Y.; Han, Y. N.; *Arch. Pharm. Res.*, **1998**, 21, 481–484.

[153] Wong, H. R.; Menendez, I. Y.; *Biochem. Biophys. Res. Commun.*, **1999**, 262, 375–380.

[154] al-Harbi, M. M.; Qureshi, S.; Ahmed, M. M.; Raza, M.; Miana, G. A.; Shah, A. H.; Jpn. *J. Pharmacol.*, **1994**, 64, 135–139.

[155] Gu, J.-Q.; Gills, J. J.; Eun, J.; Mata-Greenwood, E.; Hawthorne, M. E.; Axelrod, F.; Chavez, P. I.; Fong, H. H. S.; Mehta R. G.; Pezzuto, J. M.; Kinghorn, A. G.; *J. Nat. Prod.*, **2002**, 65, 532–536.

[156] Yu, S.; Yasukawa, K.; Takido, M.; *Phytomedicine*, **1994**, 1, 55–58.

[157] Sala, A.; Recio, M. C.; Giner, R. M.; Manez, S.; Rios, J. L.; *Life Sci.*, **2000**, 66, PL35–PL40.

[158] Reuter, U.; Chiarugi, A.; Bolay, H.; Moskowitz, M. A.; *Ann. Neurol.*, **2002**, 51, 507–516.

[159] Cavallini, L.; Francesoni, M. A.; Zoccarato, F.; Alexandre, A.; *Biochem. Phrmacol.*, **2001**, 62, 141–147.

[160] Kang, B. Y.; Chung, S. W.; Kim, T. S.; *Immunol. Lett.*, **2001**, 77, 159–163.

[161] Kwok, B. H.; Koh, B.; Ndubuisi, M. I.; Elofsson, M.; Crews, C. M.; *Chem. Biol.*, **2001**, 8, 759–766.

[162] Garcia-Pineres, A. J.; Castro, V.; Mora, G.; Schmidt, T. J.; Strunck, E.; Pahl, H. L.; Merfort, I.; *J. Biol. Chem.*, **2001**, 276, 39713–39720.

[163] Sobota, R.; Szwed, M.; Kasza, A.; Bugno, M.; Kordula, T.; *Biochem. Biophys. Res. Commun.*, **2000**, 267, 329–333.

[164] Patel, N. M.; Nozaki, S.; Shortle, N. H.; Bhat-Nakshatri, P.; Newton, T. R.; Rice, S.; Gelfanov, V.; Boswell, S. H.; Goulet, R. J. Jr.; Sledge, G. W. Jr.; Nakshatri, H.; *Oncogene*, **2000**, 19, 4159–4169.

[165] Morikawa, T.; Matsuda, H.; Ninomiya, K.; Yoshikawa, M.; *Biol. Pharm. Bull.*, **2002**, 25, 627–631.

[166] Yoshioka, T.; Fujii, E.; Endo, M.; Wada, K.; Tokunaga, Y.; Shiba, N.; Hohsho, H.; Shibuya, H.; Muraki, T.; *Inflamm. Res.*, **1998**, 47, 476–481.

[167] Kuroda, M.; Yoshida, D.; Kodama, H.; *Agric. Biol. Chem.*, **1987**, 51, 585–587.

[168] Koch, E.; Klaas, C. A.; Ruengeler, P.; Castro, V.; Mora, G.; Vichnewski, W.; Merfort, I.; *Biochem. Pharmacol.*, **2001**, 62, 795–801.

[169] Castro, V.; Murillo, R.; Klaas, C. A.; Meunier, C.; Mora, G.; Pahl, H. L.; Merfort, I.; *Planta Med.*, **2000**, 66, 591–598.

[170] Alam, M.; Susan, T.; Joy, S.; Kundu, A. B.; *Indian J. Exp. Biol.*, **1992**, 30, 38–41.

[171] Kim, E. J.; Jin, H. K.; Kim, Y. K.; Lee, H. Y.; Lee, S. Y.; Lee, K. R.; Zee, O. P.; Han, J. W.; Lee, H. W.; *Biochem. Pharmacol.*, **2001**, 61, 903–910.

[172] Silvan, A. M.; Abad, M. J.; Bermejo, P.; Villar, A.; *Inflamm. Res.*, **1998**, 45, 289–292.

[173] Abad, M. J.; Bermejo, P.; Valverde, S.; Villar, A.; *Planta Med.*, **1994**, 60, 228–231.

[174] Zitterl-Eglseer, K.; Jurenitsch, J.; Korhammer, S.; Haslinger, E.; Sosa, S.; Della Loggia, R.; Kubelka, W.; Franz, C.; *Planta Med.*, **1991**, 57, 444–446.

[175] Wendel, G. H.; Maria, A. O.; Mohamed, F.; Dominguez,, S.; Scardapane, L.; Giordano, O. S.; Guerreiro, E.; Guzman, J. A.; *Pharmaol. Res.*, **1999**, 40, 339–344.

[176] Zidorn, C.; Dirsch, V. M.; Ruengeler, P.; Sosa, S.; Della Loggia R.; Merfort, I.; Pahl, H. L.; Vollmar, A. M.; Stuppner, H.; *Planta Med.*, **1999**, 65, 704–708.

[177] Cho, J. Y.; Baik, K. U.; Jung, J. H.; Park, M. H.; *Eur. J. Pharmacol.*, **2000**, 398, 399–407.

[178] Klass, C. A.; Wagner, G.; Laufer, S.; Sosa, S.; Della Loggia, R.; Bomme, U.; Pahl, H. L.; Merfort, I.; *Planta Med.*, **2002**, 68, 385–391.

[179] Han, J. W.; Lee, B. G.; Kim, Y. K.; Yoon, J. W.; Jin, H. K.; Hong, S.; Lee, H. Y.; Lee, K. R.; Lee, H. W.; Br. *J. Pharmacol.*, **2001**, 133, 503–512.

[180] Lyss, G.; Knorre, A.; Schmidt, T. J.; Pahl, H. L.; Merfort, I.; *J. Biol. Chem.*, **1998**, 273, 33508–33516.

[181] Lyss, G.; Schmidt, T. J.; Merfort, I.; Pahl, H. L.; *Biol. Chem.*, **1997**, 378, 951–961.

[182] Hall, I. H.; Lee, K. H.; Sykes, H. C.; *Planta Med.*, **1987**, 53, 153–156.

[183] Recio, M. C.; Giner, R. M.; Uriburu, L.; Manez, S.; Cerda, M.; De la Ffuente, J. R.; Rios, J. L.; *Life Sci.*, **2000**, 66, 2509–2518.

124

[184] Valencia, E.; Feria, M.; Diaz, J. G.; Gonzalez, A.; Bermejo, J.; *Planta Med.*, **1994**, 60, 395–399.

[185] Murakami, A.; Takahashi, D.; Kinoshita, T.; Koshimizu, K.; Kim, H. W.; Yoshihiro, A.; Nakamura, Y.; Jiwajinda, S.; Terao, J.; Ohigashi, H.; *Carcinogenesis*, **2002**, 23, 795–802.

[186] Williams, D. H.; Faulkner, D. J.; *Tetrahedron*, **1996**, 52, 4245–4256.

[187] Duan, H.; Takaishi, Y.; Momota, H.; Ohmoto, Y.; Taki, T.; Jia, Y.; Li, D.; *J. Nat. Prod.*, **2001**, 64, 582–687.

[188] Shimizu, M.; Shogawa, H.; Matsuzawa, T.; Yonezawa, S.; Hayashi, T.; Arisawa, M.; Suzuki, S.; Yoshizaki, M.; Morita, N.; Ferro, E.; Basualdo, I.; Berganza, L. H.; *Chem. Pharm. Bull.*, **1990**, 38, 2283–2284.

[189] Fernandez, M. A.; Tomos, M. P.; Garcia, M. D.; de las Heras, B.; Villar, A. M.; Saenz, M. T.; *J. Pharm. Pharmacol.*, **2001**, 53, 867–872.

[190] Ho, C. T.; Wang, M.; Wei, G. J.; Huang, T. C.; Huang, M. T.; *BioFactors*, **2000**, 13, 161–166.

[191] Singletary, K. W.; *Cancer Lett.*, **1996**, 100, 139–144.

[192] Offord, E. A.; Mace, K.; Avanti, O.; Pfeifer, A. M. A.; *Cancer Lett.*, **1997**, 114, 275–281.

[193] Chan, M. M.; Ho, C. T.; Huang, H. I.; *Cancer Lett.*, **1995**, 96, 23–29.

[194] de las Heras, B.; Abad, M. J.; Silvan, A. M.; Pascual, R.; Bermejo, P.; Rodriguiez, B.; Villar, A. M.; *Life Sci.*, **2001**, 70, 269–278.

[195] Niwa, M.; Tsutyumishita, Y.; Kawai, Y.; Takahara, H.; Nakamura, N.; Futaki, S.; Takaishi, Y.; Kondoh, W.; Moritoki, H.; *Biochem. Biophys. Res. Commun.*, **1996**, 224, 579–585.

[196] Lin, N.; Sato, T.; Ito, A.; *Arthrit. Rheumat.*, **2001**, 44, 2193–2200.

[197] Jiang, X. H.; Wong, B. C.; Lin, M. C.; Zhu, G. H.; Kung, H. F.; Jiang, S. H.; Yang, D.; Lam, S. K.; *Oncogene*, **2001**, 20, 8009–8018.

[198] Zhao, G.; Vaszar, L. T.; Qiu, D.; Shi, L.; Kao, P. N.; *Am. J. Physiol. Lung Cell. Mol. Physiol.*, **2000**, 279, L959–L966.

[199] Tao, X.; Schulze-Koops, H.; Ma, L.; Cai, M.; Mao, Y.; Lipsky, P. E.; *Arthrit. Rheumat.*, **1998**, 41, 130–138.

[200] Chen, B. J.; *Leukemia & Lymphoma*, **2001**, 42, 253–265.

[201] Hernandez-Perez, M.; Rabanal, R. M.; de la Torre, M. C.; Rodriguez, B.; *Planta Med.*, **1995**, 61, 505–509.

[202] Prinsep, M. R.; Thomson, R. A.; West, M. L.; Wylie, B. L.; *J. Nat. Prod.*, **1996**, 59, 786–788

[203] Kinghorn, A. D.; Fong, H. H. S.; Farnsworth, N. R.; Mehta, R. G.; Moon, R. C.; Moriarty, R. M.; Pezzuto, J. M.; *Current Org. Chem.*, **1998**, 2, 597–612.

[204] Weitberg, A. B.; Corvese, D.; *J. Exp. Clin. Cancer Res.*, **1999**, 18, 433–437.

[205] Fujiki, H.; Suganuma, M.; Komori, A.; Yatsunami, J.; Okabe, S.; Ohta, T.; Sueoka, E.; *Cancer Detect. Prevent.*, **1994**, 18, 1–7.

[206] Saito, Y.; Nishino, H.; Hoshida, D.; Mizusaki, S.; Ohnishi, A.; *Oncology*, **1988**, 45, 122–126.

[207] Ichihara, Y.; Takeya, K.; Hitotsuyanagi, Y.; Morita, H.; Okuyama, S.; Suganuma,

M.; Fujiki, H.; Motidome, M.; Itokawa, H.; *Planta Med.*, **1992**, 58, 549–551.

[208] Carvalho, J. C.; Silva, M. F.; Maciel, M. A.; Pinto, A. C.; Nunes, D. S.; Lima, R. M.; Bastos, J. K.; Sarti, S. J.; *Planta Med.*, **1996**, 62, 402–404.

[209] Paya, M.; Ferrandiz, M. L.; Effadi, F.; Terencio, M. C.; Kijjoa, A.; Pinto, M. M.; Alcaraz, M. J.; *Eur. J. Pharmacol.*, **1996**, 312, 97–105.

[210] Castrillo, A.; de Las Heras, B.; Hortelano, S.; Rodriguez, B.; Villar, A.; Bosca, L.; *J. Biol. Chem.*, **2001**, 276, 15854–15860.

[211] Okamoto, H.; Yoshida, D.; Mizusaki, S.; *Cancer Lett.*, **1983**, 19, 47–53.

[212] Yasukawa, K.; Kitanaka, S.; Seo, S.; *Biol. Pharm. Bull.*, **2002**, 25, 1488–1490.

[213] Hwang, B. Y.; Lee, J. H.; Koo, T. H.; Kim, H. S.; Hong, Y. S.; Ro, J. S.; Lee, K. S.; Lee, J. J.; *Planta Med.*, **2001**, 67, 406–410.

[214] Ko, W. G.; Kang, T. H.; Lee, S. J.; Kim, Y. C.; Lee, B. H.; *Phytotherapy Res.*, **2001**, 15, 535–537.

[215] Du, J.; Wang, M. L.; Chen, R. Y.; Yu, D. Q.; *Planta Med.*, **2001**, 67, 542–547.

[216] Villar, A.; Salom, R.; Alcaraz, M. J.; *Planta Med.*, **1984**, 50, 90–92.

[217] Zhang, L.; Insel, P. A.; *Am. J. Physiol. Cell Physiol.*, **2001**, 281, C1642–C1647.

[218] de las Heras, B.; Villar, A.; Vivas, J. M.; Hoult, J. R.; *Agents Actions*, **1994**, 41, 114–117.

[219] Chiou, W. F.; Chen, C. F.; Lin, J. J.; *Br. J. Pharmacol.*, **2000**, 129, 1553–1560.

[220] Masuda, T.; Masuda, K.; Shiragami, S.; Jitoe, A.; Nakatani, N.; *Tetrahedron*, **1992**, 48, 6787–6792.

[221] Kang, H. S.; Song, H. K.; Lee, J. J.; Pyun, K. H.; Choi, I.; *Mediators Inflamm.*, 1998, 7, 257–259.

[222] Kang, H. S.; Kim, Y. H.; Lee, C. S.; Lee, J. J.; Choi, I.; Pyun, K. H.; *Cell. Immunol.*, **1996**, 170, 212–221.

[223] Di Mascio, P.; Medeiros, M. H. G.; Sies, H.; Bertolotti, S.; Braslavsky, S. E.; Veloso, D. P.; Sales, B. H. L. N.; Magalhaes, E.; Braz-Filho, R.; Bechara, E. J. H.; *J. Photochem. Photobiol. B.: Biology*, **1997**, 38, 169–173.

[224] Hooper, G. J.; Davies-Coleman, M. T.; Schleyer, M.; *J. Nat. Prod.*, **1997**, 60, 889–893.

[225] Jacobson, P. B.; Jacobs, R. S.; *J. Pharmacol. Exp. Therap.*, **1992**, 262, 866–873.

[226] Rauscher, R.; Edenharder, R.; Platt, K. L.; *Mutat. Res.*, **1998**, 413, 129–142.

[227] Brockman, H. E.; Stack, H. F.; Waters, M. D.; *Mutat. Res.*, **1992**, 267, 157–172.

[228] Ong, T.; Whong, W.-Z.; Stewart, J. D.; Brockman, H. E.; *Mutat. Res.*, **1989**, 222, 19–25.

[229] D'acquisto, F.; Lanzotti, V.; Carnuccio, R.; *Biochem. J.*, **2000**, 346, 793–798.

[230] Soriente, A.; De Rosa, M. M.; Scettri, A.; Sodano, G.; Terencio, M. C.; Paya, M.; Alcaraz, M. J.; *Curr. Med. Chem.*, **1999**, 8, 415–431.

[231] Ranner, M. K.; Jensen, P. R.; Fenical W.; *J. Org. Chem.*, **2000**, 65, 4843–4852.

[232] Schmidt, E. W.; Faulkner, D. J.; *Tetrahedron Lett.*, **1996**, 37, 3951–3954.

[233] Ferrandiz, M. L.; Sanz, M. J.; Bustos, G.; Paya, M.; Alcarez, M. J.; De Rosa, S.; *Eur. J. Pharmacol.*, **1994**, 253, 75–82.

[234] Komori, A.; Suganuma, M.; Okabe, S.; Zou, X.; Tius, M. A.; Fujiki, H.; *Cancer Res.*, **1993**, 53, 3462–3464.

[235] Ahmad, M.; Shikha, H. A.; Sadhu, S. K.; Rahman, M. T.; Datta, B. K.; *Pharmazie*, **2001**, 56, 657–660.

[236] Yang, H. O.; Kang, Y. H.; Suh, D. Y.; Kim, Y. C.; Han, B. H.; *Planta Med.*, **1995**, 61, 519–522.

[237] Epstein, M. A.; Achong, B. G.; In *The Epstein-Barr virus*, Epstein, M. A.; Achong, B. G., Eds.; Springer-Verlag, Berlin, **1979**, pp. 321–337.

[238] Mizuno, F.; Koizumi, S.; Osato, T.; Kokwaro, J. O.; Ito, Y.; *Cancer Lett.*, **1983**, 19, 199–205.

[239] Osato, T.; Mizuno, F.; Imai, S.; Aya, T.; Koizumi, S.; Kinoshita, T.; Tokuda, H.; Ito, Y.; Hirai, N.; Hirota, M.; Ohigashi, H.; Koshimizu, K.; Kofi-Tsekpo, W. M.; Were, J. B. O.; Mugambi, M.; *Lancet*, **1987**, 1, 1257–1258.

[240] Fuerstenberger, G.; Hecker, E.; *Tetrahedron Lett.*, **1977**, 925–928.

Atta-ur-Rahman (Ed.) *Studies in Natural Products Chemistry, Vol. 29*

TOTAL SYNTHESIS OF BIOLOGICALLY INTRIGUING DRIMANE-TYPE SESQUITERPENOIDS VIA INTRAMOLECULAR DIELS–ALDER APPROACHES

YOSHIKAZU SUZUKI, KEN-ICHI TAKAO,
and KIN-ICHI TADANO*

Department of Applied Chemistry, Keio University, Hiyoshi, Kohoku-ku, Yokohama 223-8522, Japan

ABSTRACT: The total syntheses of three drimane-type sesquiterpenoids, (–)-mniopetal E, (–)-mniopetal F, and (–)-kuehneromycin A, are described. These natural products inhibit the enzymatic activity of RNA-directed DNA-polymerases (reverse transcriptases) of the human immunodeficiency virus (HIV)-1. Our enantiospecific total syntheses of these target molecules in naturally occurring forms commenced with a known 2,3-anhydro-D-arabinitol derivative, which was prepared using the Sharpless asymmetric epoxidation strategy. A combination of highly stereocontrolled inter- and intramolecular Horner–Emmons carbon elongations led to the two butenolides tethering 1,2,4,9- and 1,4,9-functionalized nona-5,7-diene moieties at the β-carbon. The key step in mniopetal E synthesis is a stereoselective thermal intramolecular Diels–Alder reaction of the former butenolide compound, providing a highly oxygenated tricyclic skeleton with the desired *endo* and π-facial selections. The intramolecular Diels–Alder reaction of the latter butenolide compound for the syntheses of other two drimanes also proceeded with the desired stereoselectivity, which is controlled by a balance of the steric effect and the stereoelectronic effect of a trialkylsilyloxy substituent existing adjacent to the dienophile in accordance with Cieplak's theory. The transformation of the γ-lactone moiety in the cycloadducts to the γ-hydroxy-γ-lactone part was efficiently achieved via a tetracyclic intermediate. Our total syntheses of (–)-mniopetal E, (–)-mniopetal F, and (–)-kuehneromycin A established the unsettled absolute stereochemistries of the antibiotics. In addition, the reported total syntheses by Jauch are briefly reviewed.

INTRODUCTION

In 1994, six novel drimane-type sesquiterpenoids, mniopetals A–F (**1–6**) (Fig. (**1**)), were isolated by Anke and Steglich from the fermentation broth of Canadian *Mniopetalum* sp. 87256 [1, 2]. These natural products are known to inhibit the enzymatic activity of RNA-directed DNA-polymerases (reverse transcriptases) of the human immunodeficiency virus (HIV)-1, avian myeloblastosis virus (AMV), and moloney murine leukemia virus (MMuLV) as shown in Table 1. In addition, mniopetals exhibit moderate antimicrobial and cytotoxic activities. In the following year, a structurally and biologically similar natural product kuehneromniopetal A (**1**)

mniopetal B (**2**)

mniopetal C (**3**)

mniopetal D (**4**)

mniopetal E (**5**)

mniopetal F (**6**)

kuehneromycin A (**7**)

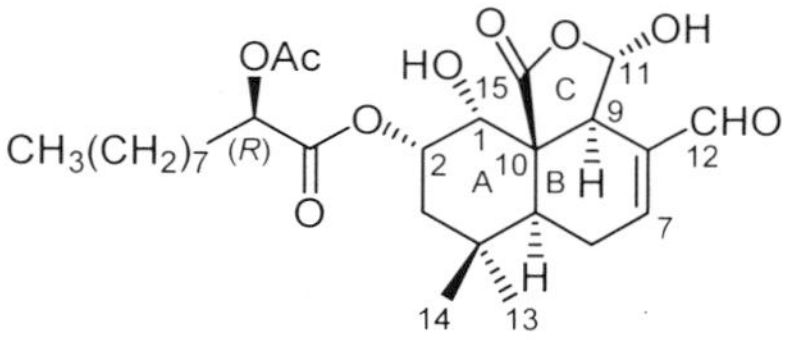

Fig. (1). Structures of mniopetals and kuehneromycin A

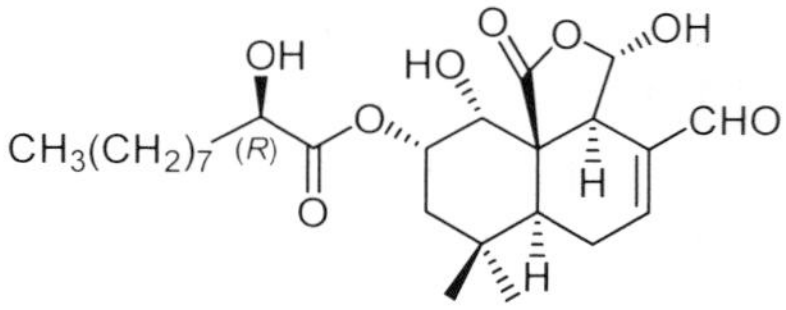

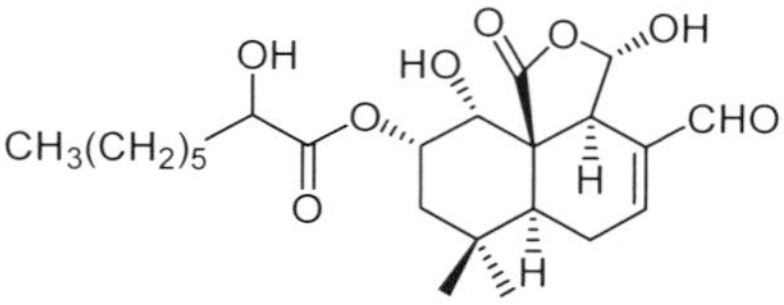

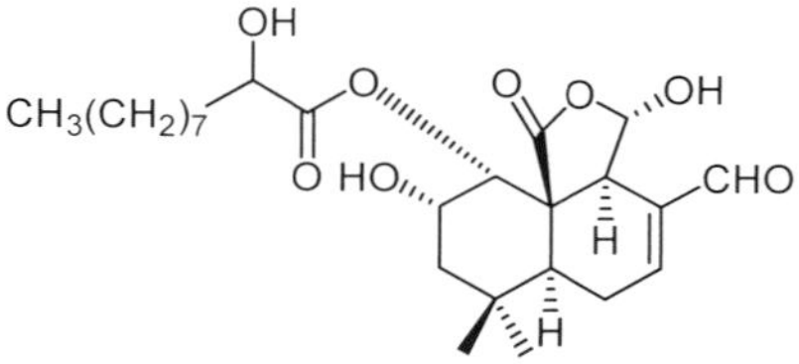

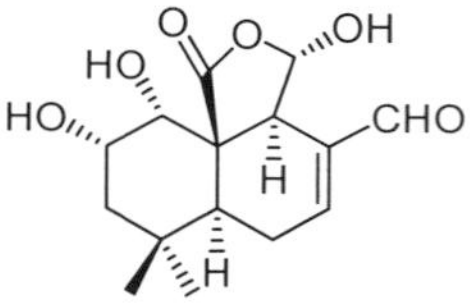

Table 1. Effect of mniopetals and kuehneromycin A of the reverse transcriptases (RT)

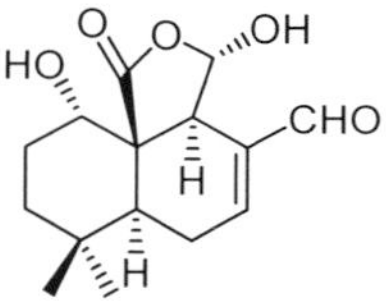

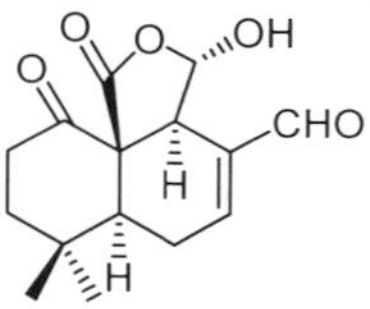

compound		IC_{50} (μM)			
		HIV-1 RT tp 1[a]	HIV-1 RT tp 2[b]	AMV RT tp 1[a]	MMuLV RT tp 1[a]
mniopetal A	(**1**)	> 197	> 197	41	4
mniopetal B	(**2**)	> 223	91	42	1.7
mniopetal C	(**3**)	> 238	190	93	7
mniopetal D	(**4**)	> 223	54	77	6
mniopetal E	(**5**)	> 338	140	> 338	59
mniopetal F	(**6**)	> 320	30	> 320	30
kuehneromycin A	(**7**)	667	54	270	36

[a] poly(A)-$(dT)_{15}$. [b] LTR-template + 18 mer primer.

mycin A (**7**) was found in cultures of Tasmanian *Kuehneromyces* sp. 8758 by the same group [3]. Quite different from previously known inhibitors of HIV reverse transcriptase, these antibiotics do not contain a nucleic acid base nor aromatic ring structures. Instead, they have highly oxygenated drimane-type sesquiterpenoid skeletons that were elucidated by extensive spectroscopic analysis. The structural characteristics of the mniopetal family are (1) a 6–6–5 angularly fused tricyclic framework including a *trans*-fused octahydronaphthalene skeleton (A/B ring), (2) five or six contiguous stereogenic centers including an angular asymmetric quarternary carbon, and (3) a variety of oxygen functionalities such as a γ-hydroxy-γ-lactone ring (C ring). Although the absolute stereochemistry of the mniopetal family had not been determined, it was reasonably assumed to be like those depicted based on the correlation with (+)-1α,15-dihydroxymarasmene (**8**) [4] and (−)-11,12-dihydroxy-7-drimene (**9**) [5, 6] isolated from the same fungus (Fig. (2)) [7]. Kuehneromycin A (**7**) was also presumed to have the same absolute stereochemistry due to a close resemblance between the

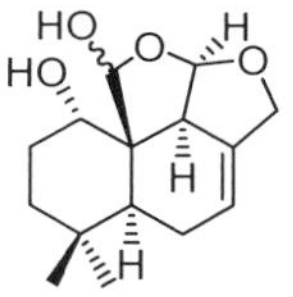

(+)-1α,15-dihydroxymarasmene (**8**) 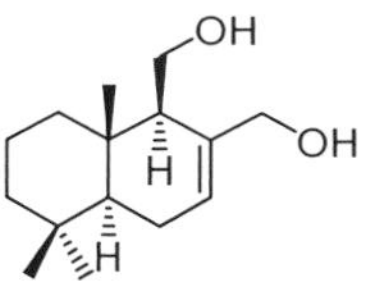   (−)-11,12-dihydroxy-7-drimene (**9**)

Fig. (2). Structures of related natural compounds

CD curves of mniopetals and that of **7**. Because of the biological interest and novelty of their structures, synthetic studies for these sesquiterpenoids have been investigated by several groups including ours.

Regarding the history of natural products chemistry, a large number of drimane-type sesquiterpenoids have been selected as targets of total synthesis. Their functionalized *trans*-decalin moiety and intriguing biological activities have made many organic chemists try to construct them. In particular, antifeedants such as warburganal (**10**) and polygodial (**11**) were synthesized by many groups in the 1980s for agricultural applications as summarized in Fig. (3) [8]. In 1982, Ohno and co-workers reported the first synthesis of (−)-warburganal (**10**), a naturally occurring enantiomeric form starting from natural terpenoid, an abietic acid [9]. Thereafter, several readily available natural products

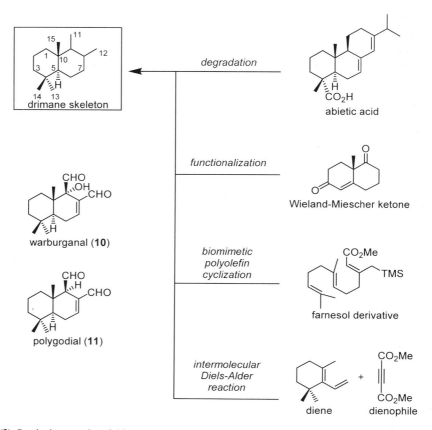

Fig. (3). Synthetic strategies of drimane sesquiterpenoids

were converted into **10** or **11** via the degradation process [10-15]. Wieland–Miescher ketone, prepared by Robinson annulation, was also used as a starting material [13, 16-19]. Biomimetic polyolefin cyclizations of farnesol derivatives realized the formation of the decalin skeleton of drimanes. Many examples demonstrated their effectiveness [20-25]. In addition, Nakanishi and co-workers used the intermolecular Diels–Alder reaction in an efficient total synthesis of (±)-**10** and (±)-**11** [26]. Later, optically active drimanes were synthesized in the Diels–Alder approach by Mori's group [27]. These elegant investigations and efforts seem to complete the development of the synthetic strategy for drimane-type sesquiterpenoid. Considering a practical synthetic scheme of mniopetals in accordance with these methods, however, many oxygenated functional groups of mniopetals might cause problems. In particular, the angular carboxyl group C-10 (the drimane numbering) made us hesitate to use these approaches since the late-stage oxidation of an angular methyl group was expected to be synthetically difficult [28-31]. Regarding natural product synthesis, the construction of an asymmetric quarternary carbon has been recognized as a difficult task, particularly when it is surrounded by other asymmetric carbons. Because the bridgehead carbon at C-10 in mniopetals is just such a situation, synthetic difficulties with the correct control of stereo-selectivities were anticipated in the construction of the decalin ring system of mniopetals. Therefore, we investigated a new synthetic approach to construct the highly oxidized drimane skeleton through synthetic studies of mniopetals. In this article, we discribe our total syntheses of drimane-type sesquiterpenoids, mniopetal E, mniopetal F, and kuehneromycin A, featuring intramolecular Diels–Alder (IMDA) reactions [32-35]. In addition, a similar IMDA approach to their total syntheses by Jauch is also presented [36-41].

Total Synthesis of Mniopetal E

The IMDA reaction of a 1,3,9-decatriene derivative has been demon-strated as a powerful synthetic tool to give a decalin ring system in many studies of total synthesis [42-48]. Previously, we have investi-gated the construction of a decalin ring system equipped with contiguous stereogenic centers including quarternary carbon via an IMDA reaction approach in the synthetic studies of PI-201 (**18**), a novel platelet aggregation inhibitor, as illustrated in Scheme 1 [49, 50]. Two coupling partners, aldehyde **13** and phosphorane **15**, were prepared straightforwardly from ε-caprolactone (6-hexanolide) (**12**) and methyl (*R*)-3-hydroxypentanoate (**14**), respectively. The Wittig reaction of two components led to the IMDA substrate **16**. The key IMDA reaction was carried out under thermal conditions by heating a toluene solution of **16** in a sealed tube at 200 °C for 27 h to give desired *cis*-fused cycloadduct **17** accompanied by other isomers without specified diastereoselectivity. The stereochemistry of the decalin ring fusion and four contiguous stereogenic centers in **17** were confirmed by ^{1}H NMR analysis including an NOE difference experiment. Hydrolysis of ethyl ester and silyl ether gave (−)-PI-201 (**18**) as a natural enantiomeric form. The first total synthesis of **18** verified the unsettled absolute stereochemistry.

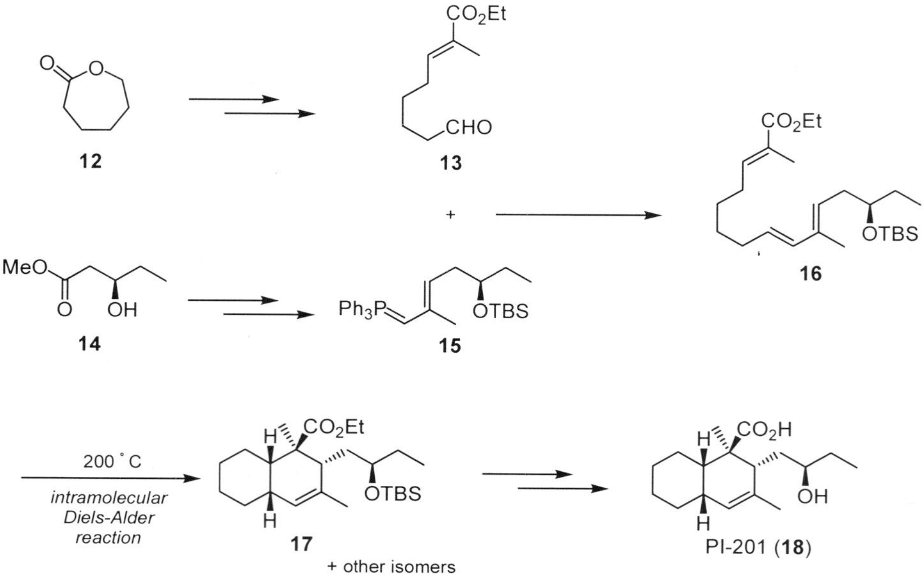

Scheme 1. Total synthesis of PI-201

Mniopetal E (**5**), a prototype of mniopetals A–D, was selected as the initial synthetic target. Our synthetic strategy based on the IMDA approach is depicted in Scheme 2. The substrate of the key IMDA reaction, triene **B**, was designed by making use of all its oxygen functionalities. The oxidation states at C-11 and C-15 were set up as an acid anhydride moiety (X = O) or a butenolide moiety (X = H, H), lowering the LUMO energy of dienophile to activate the reactivity in the cycloaddition step. The C-12 functionality was set up as a protected hydroxymethyl or an aldehyde equivalent. The stereochemistries of α-hydroxyl groups at the C-1 and C-2 positions were anticipated to control the π-facial selectivity of the IMDA reaction. Adjustment of the oxidation states in the cycloadduct **A** would eventually provide the target natural product **5**. Namely, the cycloadduct **A** would be converted into **5** by the modification of the oxidation states at C-11, C-12, and C-15, accompanied by the migration of the carbon–carbon double bond to the conjugate enal form. The synthesis of the substrate **B** would be achieved from 1,4-disubstituted butadiene **C** via the introduction of the butenolide ring by functionalization of the terminal diol moiety. The intermediate **C** would be prepared by the stereoselective introduction of an *E,E*-conjugate diene moiety into a 4,5,6,7-tetrahydroxyheptanal

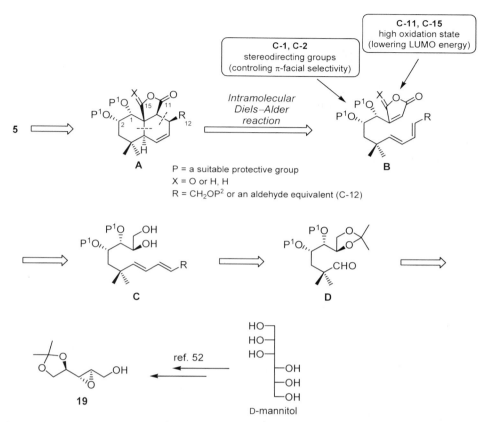

Scheme 2. Synthetic strategy of mniopetal E

derivative **D**. This heptanal **D** could be obtained from the known epoxide **19** through the Payne rearrangement [51], followed by the epoxy ring opening by the attack of an isobutyraldehyde equivalent. The enantiopure **19** had been prepared from D-mannitol by Sharpless and co-workers via the asymmetric epoxidation of (*E*)-1,2-(isopropylidene)-dioxy-3-penten-5-ol [52-55]. According to this retrosynthesis, we embarked on the enantiospecific total synthesis of **5**.

For the synthesis of a heptanal derivative **30** (**D** in the retrosynthesis) from **19**, we investigated two different approaches. The first approach relied on the Wittig olefination of aldehyde **25** (Scheme 3). The second was based on the attack of 2-lithio-2-methylpropionitrile (α-lithiated isobutyronitrile) [56-60] on the Payne rearrangement product **20** (Scheme 4). Our original attempt at the preparation of heptanal **30** from

19 is shown in Scheme 3. Alkaline hydrolysis of **19** according to the reported procedure [52] provided a partially protected D-ribitol **21** via the Payne rearrangement followed by the opening of the resulting terminal epoxy ring by a hydroxide ion. The primary hydroxyl group in **21** was selectively protected as a *tert*-butyldimethylsilyl (TBS) ether providing **22**. The secondary hydroxyl groups in **22** were protected as methoxymethyl (MOM) ethers to afford **23**. Desilylation of **23** with tetrabutylammonium fluoride (TBAF) provided **24**. Oxidation of **24** with Dess–Martin periodinane [61-63] gave an acyclic D-ribose derivative **25**, which was subjected to Wittig olefination with ethylidenetriphenyl-

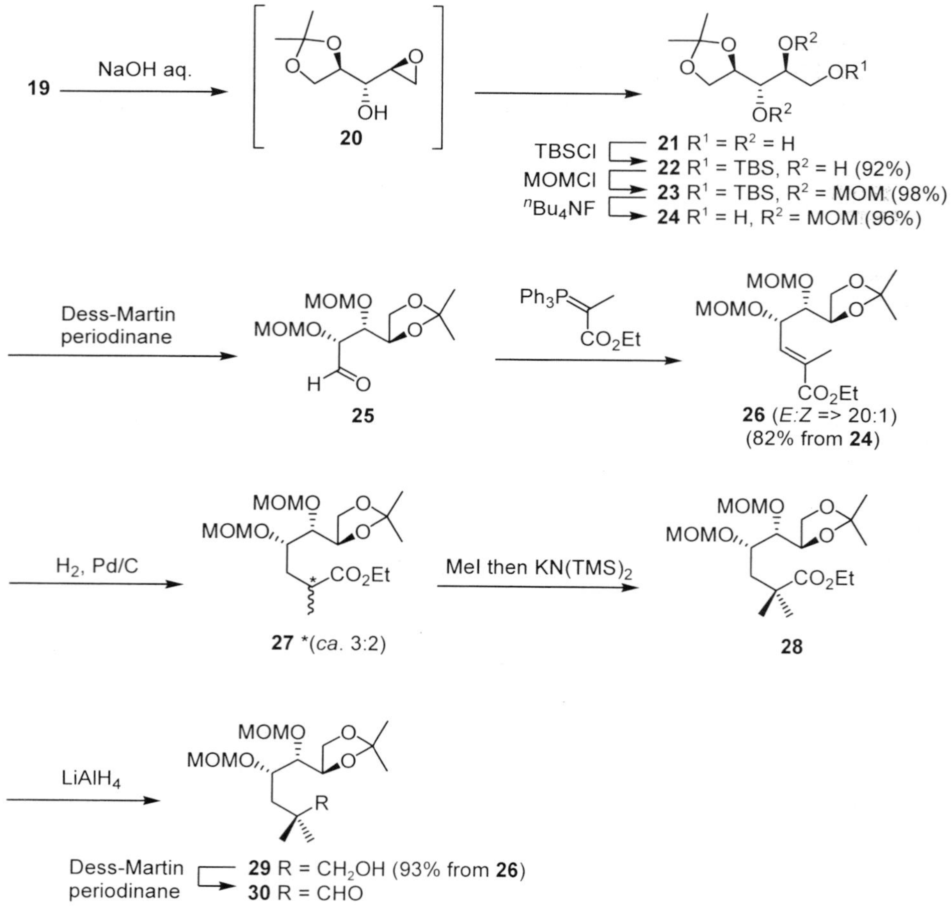

Scheme 3.

phosphorane providing α,β-unsaturated ester **26** with high *E*-selectivity (>20:1, based on the ¹H NMR analysis). Hydrogenation of the double bond in **26** in the presence of Pd on charcoal provided a 3:2 diastereomeric mixture of saturated esters **27**. We did not determine the stereochemistry of the newly created stereogenic centers in the mixture. Next, we focused our attention on constructing the *gem*-dimethyl group by the α-methylation of **27**. The α-methine proton was efficiently deprotonated with KN(TMS)₂, in contrast to recovering the starting material under other basic conditions (LDA, LiN(TMS)₂ or NaH). However the resulting potassium enolate was unstable and immediately decomposed before the addition of MeI. Therefore, we treated **27** with KN(TMS)₂ (3 equiv) in toluene–THF (6:1) at −78 °C in the presence of MeI (4 equiv, internal quenching) to obtain the desired α-dimethyl ester **28** quantitatively. The ester group in **28** was reduced with LiAlH₄, and subsequent Dess–Martin oxidation of the neopentyl alcohol **29** provided **30** in 10 steps from **19**.

In view of the excessive length of the 10-step preparation of the substrate **30** from **19**, we planned to develop a more convenient route to **30**. Our improved route is depicted in Scheme 4. The Payne rearrangement of **19** with KN(TMS)₂ in the presence of 18-crown-6 ether in THF at −18 °C afforded the epoxy-migrated compound **20** in 62% yield along with 30% recovery of **19**. This Payne rearrangement produced the equilibrium mixture of **20** and **19**. We explored this rearrangement by changing the base and reaction temperature. The most

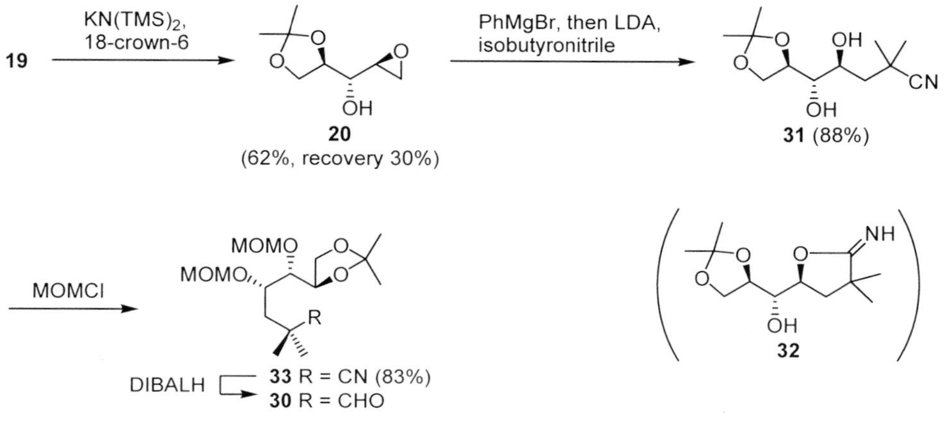

Scheme 4.

effective conditions we found were those using KN(TMS)$_2$ and crown ether. The epoxy ring opening of **20** with 2-lithio-2-methylpropionitrile, derived from isobutyronitrile using LDA as a base, afforded an approximately 1:1 mixture of heptanenitrile **31** and a five-membered cyclic imidate **32**. The latter was presumably formed by the intramolecular attack of the alkoxide, generated after the epoxy ring opening, to the nitrile carbon [57]. As another nucleophile, the dianion of 2-methylpropanoic acid (generated with LDA) [64] was examined in terms of its attack on the 3-*O*-MOM ether of **20**. In this case, the five-membered lactone corresponding to **32** (OMOM in place of OH) was obtained in 86% yield. The lactone formation occurred quite easily during the workup of the reaction mixture. To avoid the formation of **32**, the epoxide **20** was treated with a Grignard reagent prior to the addition of the nucleophile. We expected that the undesired cyclization would be prevented because of the formation of a magnesium chelate between internal dialkoxides. When **20** was treated with PhMgBr (1.5 equiv) at −78 °C for 30 min followed by the addition of 2-lithio-2-methylpropionitrile (3 equiv), to our delight, the desired **31** was obtained in 88% yield. We also explored a one-pot conversion of **19** into **31** [65-67]. However, we could not find any practical conditions for this purpose after examining various bases or additives. The two hydroxyl groups in **31** were protected as MOM ethers providing **33**. Diisobutylaluminium hydride (DIBALH) reduction of the nitrile group in **33** followed by a hydrolytic workup of the resulting imine with dilute aqueous HCl provided **30** in a four-step process from **19**.

The introduction of the conjugated diene unit into **30** (**D** to **C** in the retrosynthesis) was achieved as shown in Scheme 5. Initially, this four-carbon elongation was executed by a sequential Horner–Emmons olefination. Treatment of **30** with the carbanion generated from triethyl phosphonoacetate provided (*E*)-unsaturated ester **34** with complete geometrical stereocontrol. Reduction of **34** with DIBALH gave allylic alcohol **35** in 87% yield from **30**. Manganese dioxide (MnO$_2$) oxidation of **35** afforded unsaturated aldehyde **36**. The second Horner–Emmons olefination of **36** with triethyl phosphonoacetate provided the diene **37** in 91% yield from **35**. The geometrical stereoselection (*E,E* isomer:total of other isomers) was more than 15:1 (^{1}H NMR analysis). Next, we investigated the direct conversion of **30** into **37** by the Horner–Emmons olefination of triethyl 4-phosphonocrotonate. Treatment of **30** with triethyl 4-phosphonocrotonate and LiN(TMS)$_2$ or NaH as a base in THF

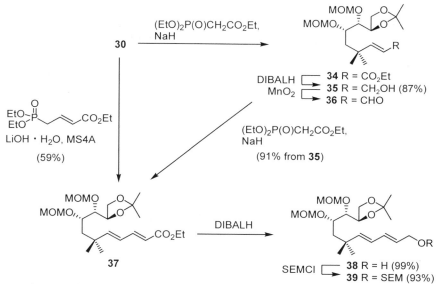

Scheme 5.

failed, resulting in the formation of a mixture of **37** and other isomers in low geometrical selectivity and low yield. On the other hand, the method of Takacs [68] gave the best result. Thus, heating a THF solution of **30** and triethyl 4-phosphonocrotonate in the presence of LiOH·H$_2$O and molecular sieves 4A powder provided **37** with high geometrical selectivity in 59% yield. To make the anticipated IMDA reaction more electronically favorable, the ester group on the diene in **37** was reduced with DIBALH, followed by protection of the resulting allylic alcohol **38** as a 2-trimethylsilylethoxymethyl (SEM) ether providing **39**.

We next explored the construction of the acid anhydride or the butenolide ring into **39** by functionalization of the terminal diol moiety (**C** to **B** in the retrosynthesis). Several initial attempts and the successful route are shown in Scheme 6. Hydrolysis of **39** in aqueous AcOH afforded diol **40**, of which the primary hydroxyl group was protected using any of three protecting groups (R = TBS, Piv, or Ac). The secondary hydroxyl group in thus-obtained silyl ether and two esters was oxidized, providing the respective keto compounds **41–43**. A variety of carbon nucleophiles were examined for their ability to attack the carbonyl groups in **41–43**. For example, (a) the enolate derived from methyl acetate for **41–43** (no reactions), (b) the Peterson-type carbanion

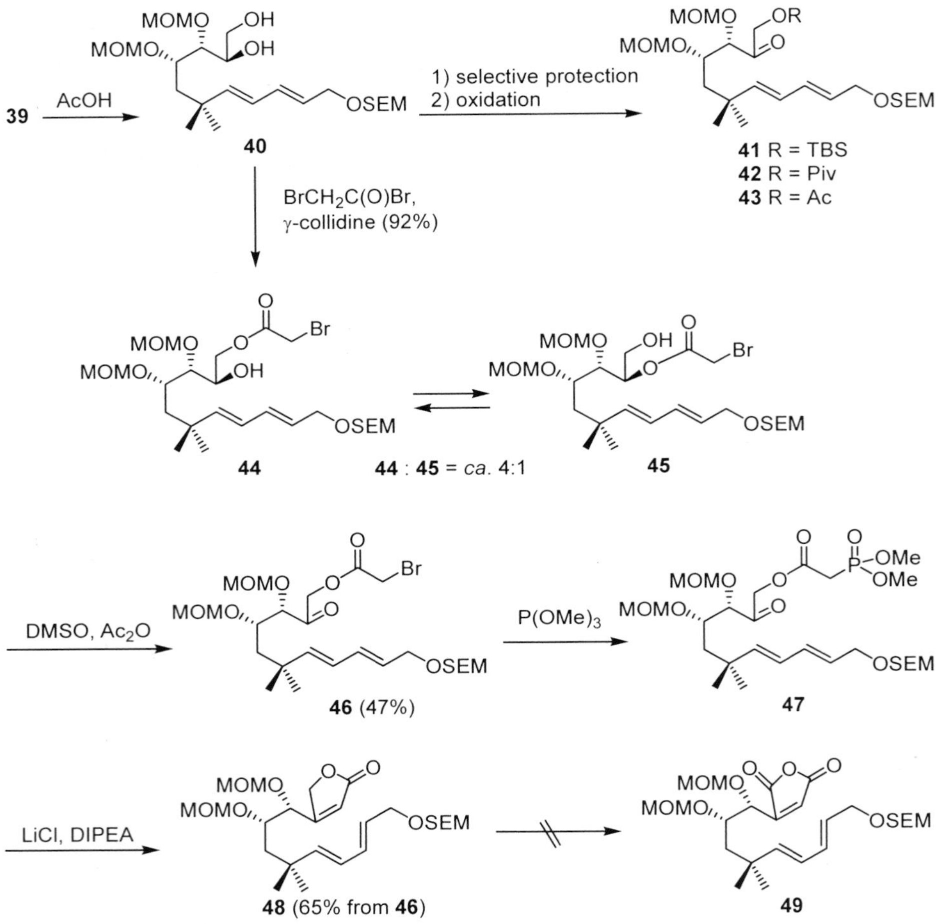

Scheme 6.

of 2-trimethylsilylacetate for **41–43** (no reaction), (c) the Horner–Emmons olefination with triethyl phosphonoacetate for **42** (19% yield of the adduct), and (d) the intramolecular aldol condensation using **43** (LiN(TMS)$_2$ as a base) (no reaction). Unfortunately, we could find no useful results with any of these substrates. For the desired butenolide ring construction, we then focused on the intramolecular Horner–Emmons olefination approach. The primary hydroxyl group in **40** was selectively acylated with bromoacetyl bromide in the presence of γ-collidine [69], affording an approximately 4:1 inseparable mixture of the desired terminal bromoacetate **44** and the regioisomeric ester **45** in a combined yield of 92%. The mixture of **44** and **45** was found to be an equilibrium mixture. Both the ^{1}H NMR spectrum (CDCl$_3$) of regio-isomerically enriched **44** (>90% content) and that of **45** (>90%) revealed the same signal pattern as the 4:1 mixture. The mixture of **44** and **45** was oxidized with DMSO–Ac$_2$O to afford α-bromoacetoxyl ketone **46** in a moderate yield. The ketone **46** was subjected to Arbuzov phosphorylation [70] by being heated in trimethyl phosphite (neat) at 60 °C providing the phosphonoacetate **47**. The intramolecular Horner–Emmons reaction of **47** under Roush–Masamune's conditions [71] proceeded smoothly to give butenolide **48**, the substrate for the IMDA reaction, in 65% yield from **46**. We also examined the conversion of the butenolide **48** into maleic anhydride derivative **49**, which might be a superior dienophile. Using several oxidation procedures, however, we could not find any effective conditions for this conversion.

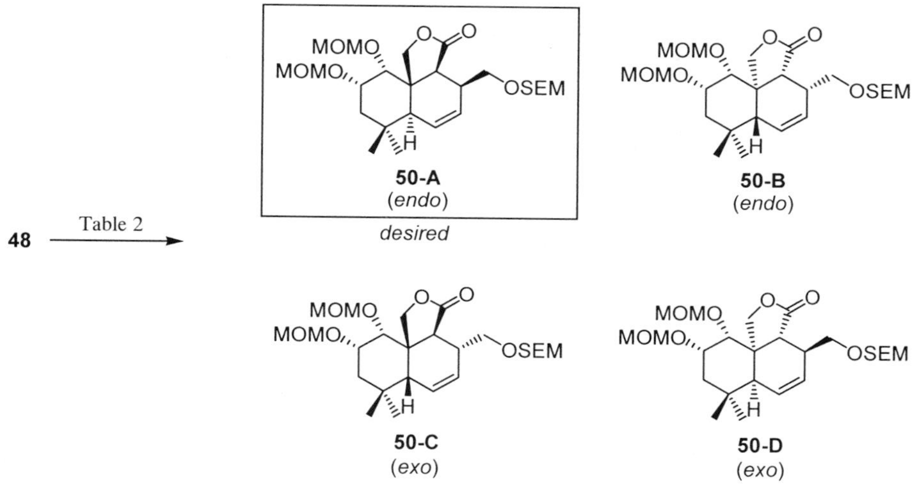

50-A
(*endo*)
desired

50-B
(*endo*)

50-C
(*exo*)

50-D
(*exo*)

48 →(Table 2)→

Scheme 7. Intramolecular Diels–Alder reaction of **48**

Table 2. Intramolecular Diels–Alder reaction of **48**

entry	conditions	results
1	CF$_3$CH$_2$OH (0.03M), 120 °C, 1 h, in a sealed tube	decomposition
2	Et$_2$AlCl, CH$_2$Cl$_2$, -78 °C to 0 °C	recovery of substrate
3	BF$_3$·OEt$_2$, CH$_2$Cl$_2$, -78 °C to 0 °C	decomposition
4	toluene (0.03M), BHT(cat.), 180 °C, 12.5 h, in a sealed tube	**50-A** (54%), **50-B** (22%)

The IMDA reaction of **48** was first conducted in a protonic solvent or Lewis-acid-mediated conditions involving heating it in 2,2,2-trifluoro-ethanol at 120 °C in a sealed tube, or subjecting it to Lewis acid, such as BF$_3$·OEt$_2$ or Et$_2$AlCl in a CH$_2$Cl$_2$ solution (Scheme 7; Table 2, Entries 1–3). None of these conditions provided the desired cyclo-adducts. Eventually, we obtained a successful result by heating a toluene solution of **48** (0.03 M concentration) in a sealed tube at 180 °C for 7.5 h in the presence of a trace amount of 2,6-di-*tert*-butyl-*p*-cresol (BHT) (Entry 4). Under these conditions, two *endo*-cycloadducts **50-A** and **50-B** were isolated in 54% and 22% yields, respectively, after separation on silica gel. Neither *exo*-cycloadduct **50-C** nor **50-D** was found in the reaction mixture. The stereostructures of **50-A** and **50-B** were confirmed by ^{1}H NMR analyses including NOE experiments as shown in Fig. (**4**).

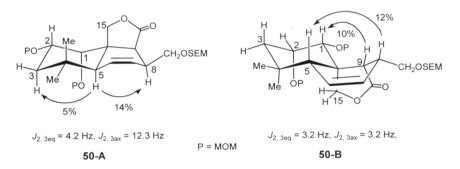

$J_{2, 3eq}$ = 4.2 Hz, $J_{2, 3ax}$ = 12.3 Hz

$J_{2, 3eq}$ = 3.2 Hz, $J_{2, 3ax}$ = 3.2 Hz,

P = MOM

50-A

50-B

Fig. (4). NOE experiments of IMDA adducts

The predominant formation of the *endo*-isomers **50-A** and **50-B** can be explained as follows. Based on the concept of "concerted but asynchronous" cycloaddition, Roush stated that a 1,7,9-decatriene system possessing an electron-withdrawing group at the terminal position of the dienophile part can generally adopt a six-membered chair-like transition state while a six-membered boat-like transition state may be occasionally favored depending on the kind of substituents [45, 46]. It seems reasonable to apply the six-member transition state model to our substrate **48**. Therefore, four chair-like transition states **TS-Chair-A** to **TS-Chair-D** and four boat-like transition states **TS-Boat-A** to **TS-Boat-D** are possible (Fig. (**5**)). All boat-like transition

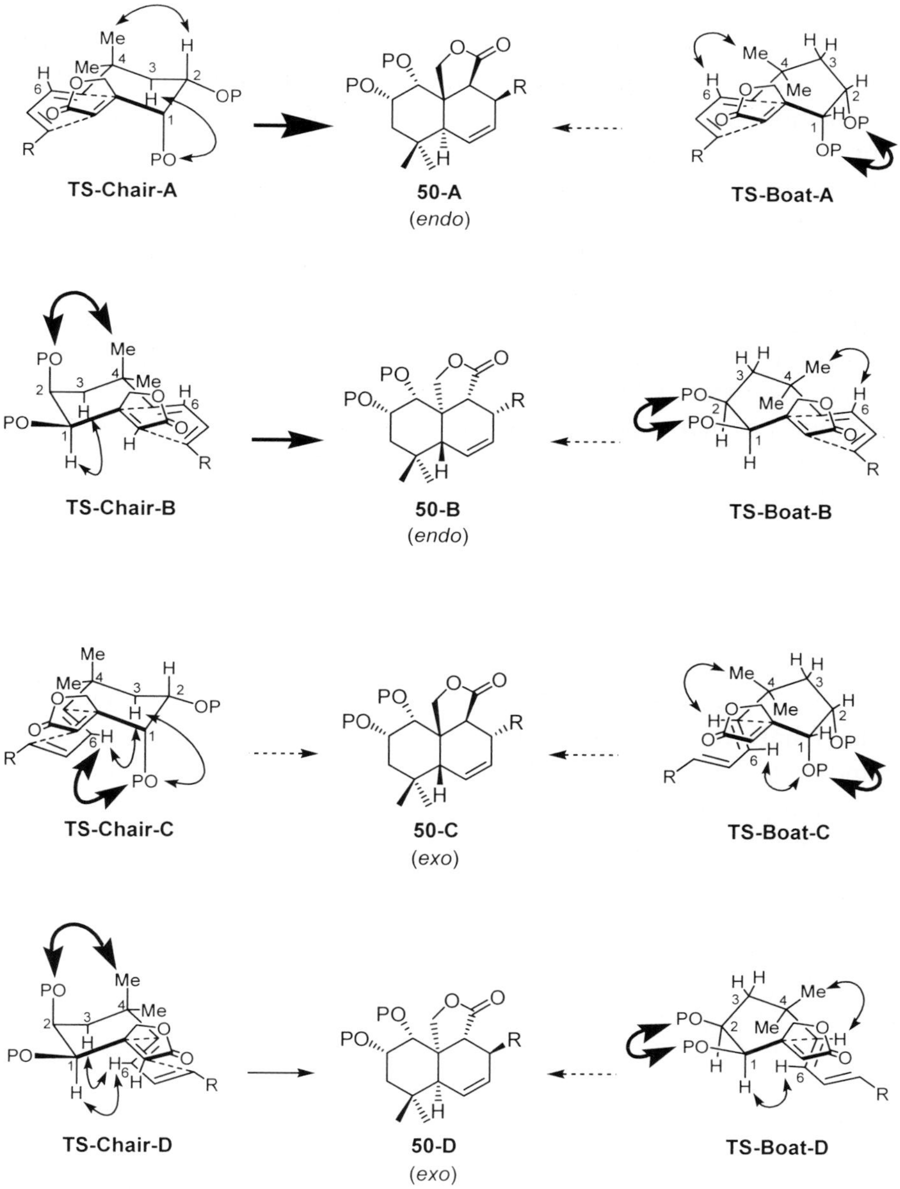

Fig. (5). Plausible transition states

P = MOM ; R = CH₂OSEM

states are disfavored by the strong steric repulsion through a syn-periplaner relationship of the two MOMO groups at C-1 and C-2. Furthermore the *gem*-dimethyl group at C-4 destabilizes the boat-like transition states through 1,3-allylic repulsion with H-6 in the *endo*-mode or 1,2-allylic repulsion with H-5 in the *exo*-mode. These factors allow us to discuss the stereoselectivity of the IMDA reaction without the contribution of boat-like transition states. Two chair-like *endo*-transition states, namely **TS-Chair-A** leading to **50-A** and **TS-Chair-B** leading to **50-B**, seem to be more advantageous compared to the two chair-like *exo*-transition states because of the presence of electronically favorable secondary orbital interactions in the former [72-76]. In addition, a torque generated by twist asynchronicity in the 1,7,9-decatriene system prefers an *endo* approach in the asynchronous transition state [77, 78]. In the two *exo*-transition states, **TS-Chair-C** leading to **50-C** and **TS-Chair-D** leading to **50-D**, severe steric repulsion exists between H-6 and two axial substituents, i.e., OMOM at C-1 and H-3ax. These unfavorable factors certainly impede cyclization. Regarding the π-facial selectivity which affected the ratio of **50-A** and **50-B**, the total magnitude of the 1,3-diaxial repulsions in **TS-Chair-A**, i.e., those occurring between the OMOM at C-1 and H-3 and between H-2 and the axial methyl group at C-4, was likely to be smaller than that expected in **TS-Chair-B** between the OMOM at C-2 and the axial methyl at C-4 and between H-1 and H-3. As a result, the cycloadduct **50-A** was formed preferentially.

Having established an efficient synthetic route to the tricyclic intermediate **50-A** with the correct stereochemistry for mniopetal

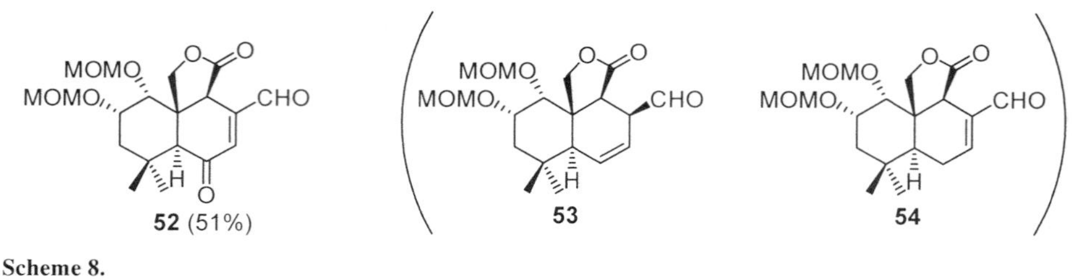

Scheme 8.

synthesis, we investigated the introduction of the α,β-unsaturated aldehyde moiety in the B-ring by oxidation of the C-12 hydroxyl group in **51**, which was prepared from **50-A** by desilylation (Scheme 8). Contrary to our expectations, neither the desired aldehyde **53** nor the conjugated enal **54** was obtained by the pyridinium chlorochromate (PCC) oxidation or the *o*-iodoxybenzoic acid (IBX) oxidation [61, 62, 79] of **51**. Rather, γ-keto α,β-unsaturated aldehyde **52** was obtained in 51% yield with the former oxidant and in 19% yield with the latter (40% recovery of **51**). The structure of **52** was established by ^{1}H and ^{13}C NMR analyses. Compound **52** was formed, probably by successive oxidation of the intermediary enal **54**. This facile oxidation at the allylic position of **54** can be explained by the high acidity of the pseudoaxial H-6β in **54**. Steglich reported that the pseudoaxial H-6β in mniopetal F (**6**) was smoothly exchanged by deuterium during NMR measurement in CD$_3$OD [2]. The conformational change of the tricyclic framework accompanying migration of C-6–C-7 double bond to C-7–C-8 may enable the C-12 carbon to be oxidized readily. At this stage, we concluded that it was advisable to prepare a substrate, which is equipped with an aldehyde group or its synthetic equivalent as the diene terminal, for the IMDA reaction.

We next designed dithiolane derivative **62** as the substrate possessing a synthetic equivalent to the aldehyde group in **5**. The preparation of **62** from **38** is depicted in Scheme 9. MnO$_2$ oxidation of **38** provided α,β:γ,δ-unsaturated aldehyde **55**. The aldehyde group was protected as the 1,3-dithiolane under the standard conditions, providing **56** in 60% yield along with the 33% yield of de-*O*-isopropylidene derivative **57**. Acid hydrolysis of **56** in aqueous acetic acid provided additional **57**. In the preliminary experiment, the 1,3-dioxolane derivative corresponding to **56** was also prepared. However, the 1,3-dioxolane moiety was strongly sensitive to acidic conditions and was difficult to treat. We expected that the introduction of the butenolide part into **57** could be achieved using the reaction sequence similar to that used for **40**. However, the four-step conversion of **57** to **62**, i.e., (1) bromo-acetylation of the primary hydroxyl group, (2) DMSO oxidation of the secondary hydroxyl group, (3) the Arbuzov reaction with P(OMe)$_3$, and (4) the intramolecular Horner–Emmons reaction, provided **62** in less effective yields of 8% to 22% without reproducibility. In some cases, the 1,3-dithiolane group was not compatible with these conditions. We next explored the butenolide construction using the direct introduction of the phosphonoacetyl group into the primary hydroxyl group. The

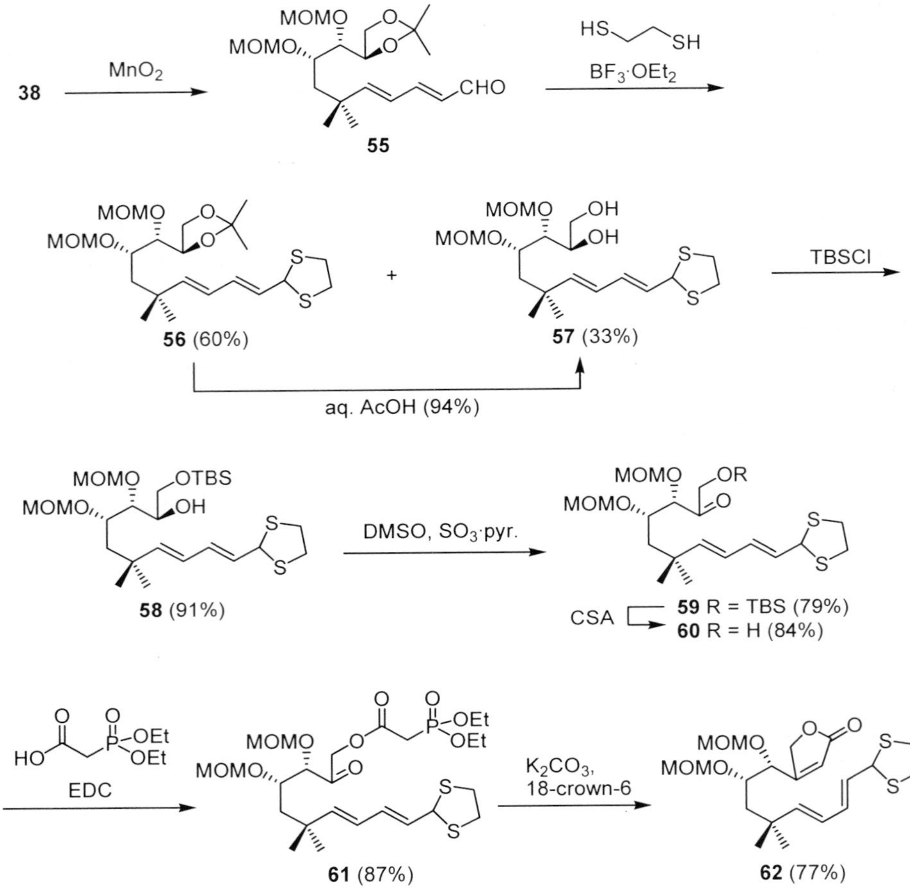

Scheme 9.

primary hydroxyl group in **57** was protected temporarily as TBS ether, and the secondary hydroxyl group in the resulting **58** was oxidized. Desilylation of the resulting ketone **59** was achieved in 84% yield by treatment with camphorsulfonic acid (CSA) in MeOH affording α-hydroxyketone **60**. Esterification of **60** with diethylphosphonoacetic acid was achieved efficiently in the presence of a 1-[3-(dimethylamino)-propyl]-3-ethylcarbodiimide (EDC). Intramolecular Horner–Emmons olefination of the resulting phosphonoacetate **61** provided butenolide **62** in 77% yield by the action of K_2CO_3 and 18-crown-6 ether in toluene [80]. Under Roush–Masamune's conditions [71], the formation of **60** accompanied the desired olefination.

We examined the IMDA reaction of **62** under Lewis-acid-mediated conditions (Scheme 10). As in the case of the Lewis-acid-mediated IMDA reaction of the substrate **48**, no good results were obtained (Table 3, Entries 1–3). Treatment of **62** with 5.0 M LiClO$_4$ in Et$_2$O [81] resulted in the formation of a complex mixture (Entries 4, 5). We found that the IMDA reaction of **62** proceeded under thermal conditions as a 0.02 M toluene solution in a sealed tube in the presence of BHT (Entry 6). In contrast to the case of **48**, the IMDA reaction of **62** required a longer heating time for completion. Two *endo*-cycloadducts **63-A** and **63-B** were isolated in 62% and 21% yields, respectively, after separation on silica gel. Furthermore, an *exo*-cycloadduct **63-D** was isolated in a trace amount of 2%. The stereochemistries of the

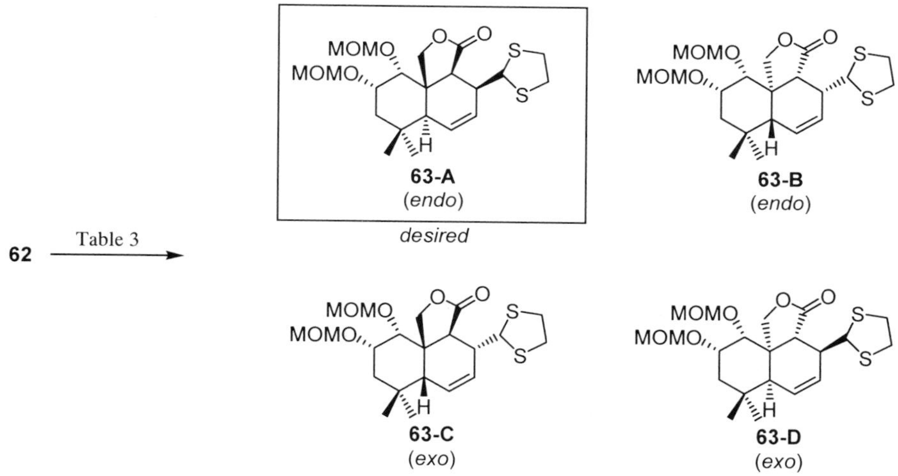

Scheme 10. Intramolecular Diels–Alder reaction of **62**

Table 3. Intramolecular Diels–Alder reaction of **62**

entry	conditions	results
1	Et$_2$AlCl, CH$_2$Cl$_2$, -78 °C to rt	no reaction
2	EtAlCl$_2$, CH$_2$Cl$_2$, -78 °C to rt	decomposition
3	BF$_3$·OEt$_2$, CH$_2$Cl$_2$, -78 °C	decomposition
4	5.0M LiClO$_4$ / Et$_2$O	no reaction
5	5.0M LiClO$_4$ / Et$_2$O, 10 mol% CSA	decomposition
6	toluene (0.02M), BHT, 180 °C, 34h	**63-A** (62%), **63-B** (21%), **63-D** (2%)

cycloadducts **63-A**, **B**, and **D** were determined unambiguously by ^{1}H NMR analysis including NOE difference spectroscopy for **63-A** and **63-D** and NOESY spectroscopy for **63-B** as shown in Fig. (**6**).

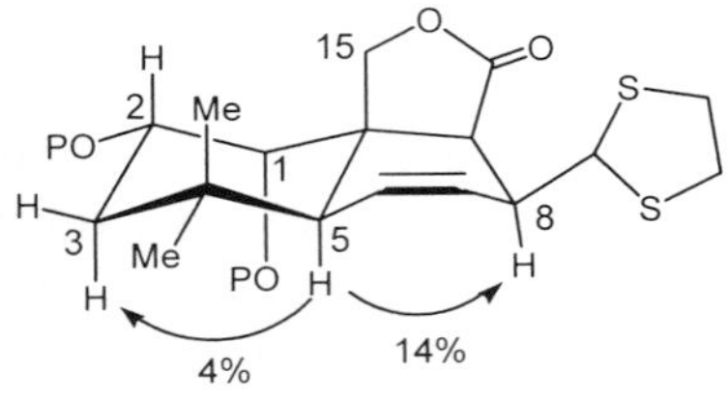

$J_{2,3eq}$ = 4.0 Hz, $J_{2,3ax}$ = 12.5 Hz

63-A

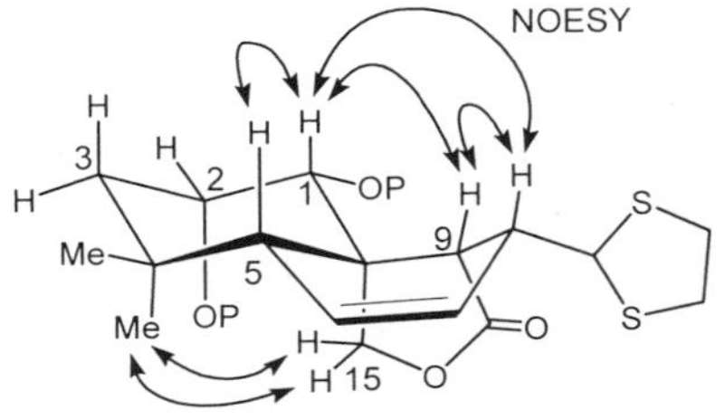

$J_{2,3eq}$ = 3.2 Hz, $J_{2,3ax}$ = 3.2 Hz,

63-B

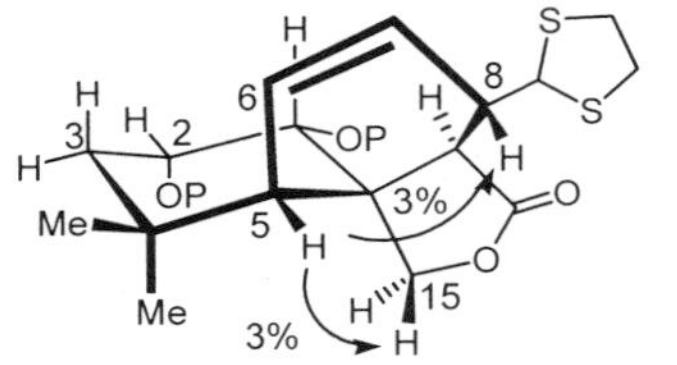

$J_{2,3eq}$ = 3.4 Hz, $J_{2,3ax}$ = 6.1 Hz

P = MOM

63-D

Fig. (6). NOE experiments of IMDA adducts

Having practical access to the desired cycloadduct **63-A** in hand, the remaining tasks for the total synthesis of **5** were the adjustment of the oxidation states in the γ-lactone moiety, migration of the carbon–carbon double bond, and deprotection. These tasks were completed as shown in Schemes 11 and 12. We decided to transform 1,3-dithiolane **63-A** first into dimethyl acetal **64** because of the probable incompatibility of the dithioacetal with advanced oxidation reactions. This conversion was carried out using Hg(ClO$_4$)$_2$·3H$_2$O [82] (Scheme 11). Treatment of **64** with Na$_2$RuO$_4$ as a 1.0 M aqueous NaOH solution [83, 84] in a mixture of 1.0 M aqueous KOH and *t*-BuOH at 50 °C provided γ-hydroxy-γ-lactone **65** as an inseparable diastereomeric mixture (ca. 9:1) on the hemiacetal carbons. The oxidation of **64** to **65** presumably proceeded through saponification of the γ-lactone and subsequent oxidation of the primary hydroxyl group followed by ring closure of the resulting acyclic β-formyl carboxylate. To obtain the regioisomeric γ-hydroxy-γ-lactone

67, we attempted the regioselective reduction of the right-hand carbonyl (C-11) of the succinic anhydride moiety in anhydride **66**, which was obtained by Jones oxidation of **65**. In the case of DIBALH reduction of

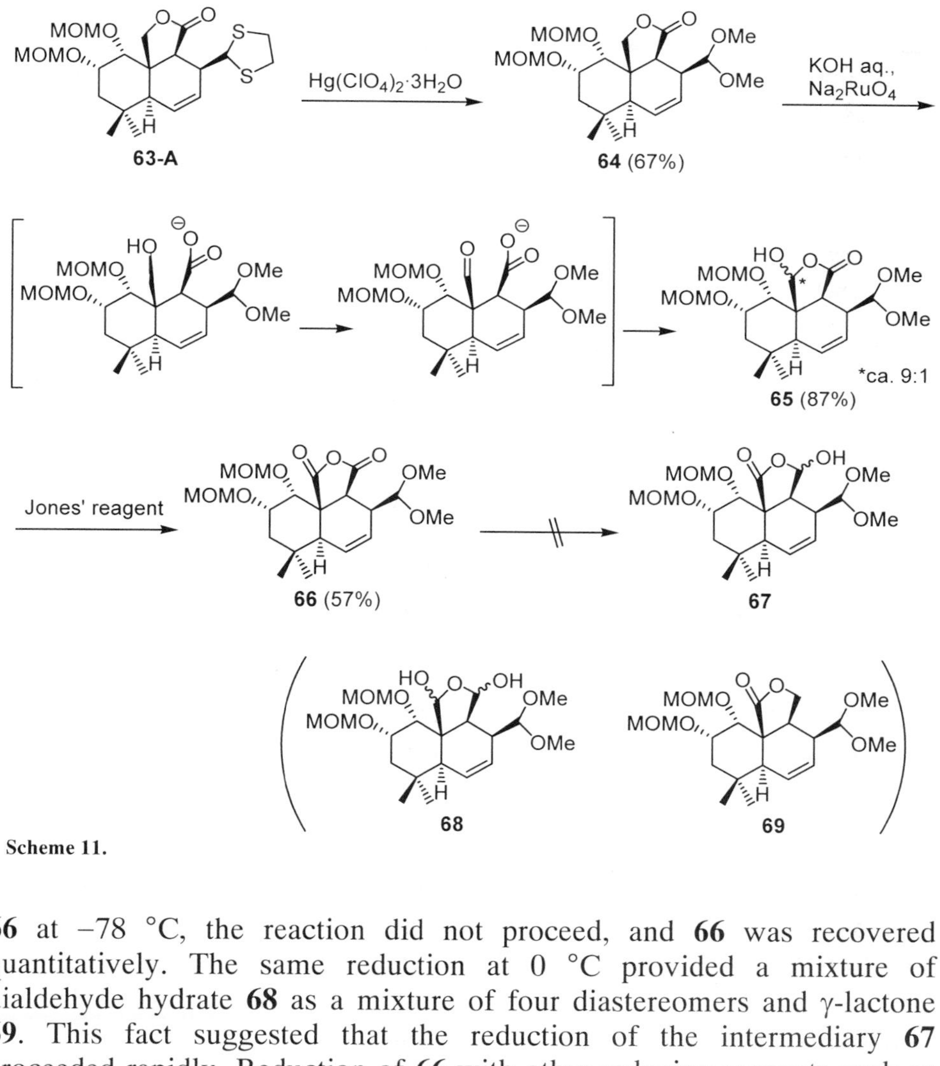

Scheme 11.

66 at −78 °C, the reaction did not proceed, and **66** was recovered quantitatively. The same reduction at 0 °C provided a mixture of dialdehyde hydrate **68** as a mixture of four diastereomers and γ-lactone **69**. This fact suggested that the reduction of the intermediary **67** proceeded rapidly. Reduction of **66** with other reducing reagents such as Li(t-BuO)$_3$AlH [85], Na$_2$Fe(CO)$_4$ [86], or L-Selectride also gave unfruitful results.

The successful conversion of **65** into mniopetal E (**5**) was eventually achieved as shown in Scheme 12. The DIBALH reduction of **65** at −78

°C provided an inseparable diastereomeric mixture of **68** in 67% yield along with 30% recovery of **65**. Attempts at selective protection of the less-hindered hydroxyl group at C-11 in **68** by silylating or acylating

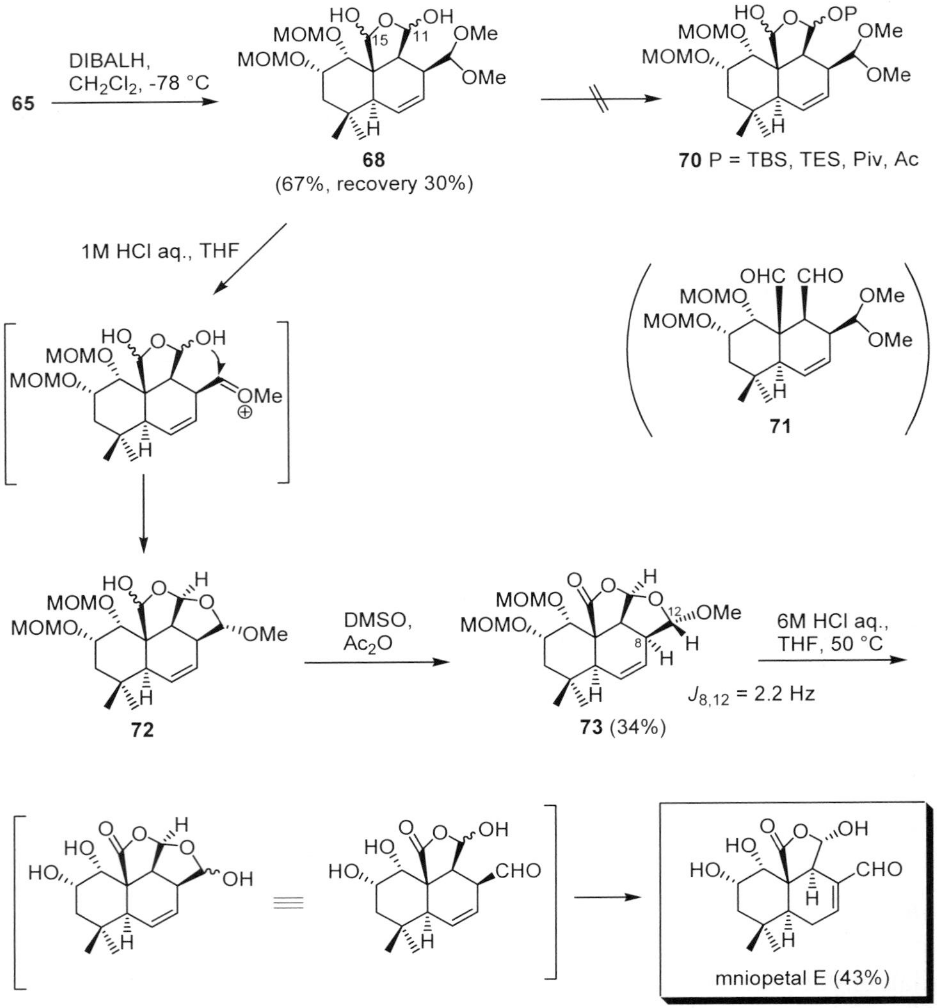

Scheme 12. Completion of the total synthesis of mniopetal E

reagents failed to provide mono-protected **70**. Only dialdehyde **71** was obtained in some cases. Fortunately, brief treatment of **68** with a trace of 1 M aqueous HCl in THF provided a tetracyclic methyl acetal **72** as a diastereomeric mixture regarding the hemiacetal carbons. This reaction presumably proceeded as shown in brackets via intramolecular attack of the right-hand hemiacetal-hydroxyl group to the oxocarbenium cation generated by elimination of 1 equiv of methanol from **68**. Consequently, the right-hand hemiacetal-hydroxyl group in **68** was selectively protected. Oxidation of **72** with DMSO–Ac$_2$O provided tetracyclic γ-lactone **73** as a single diastereomer. Based on ^{1}H NMR analysis, the stereochemistry of C-12 in **73** was confirmed as depicted. The coupling constant of H-8 and H-12 was 2.2 Hz, indicating their *trans*-relationship. The two-step yield of **73** from **68** was 34%. Treatment of **73** with a 1:1 mixed solution of 6 M aqueous HCl and THF at 50 °C caused deprotection of the MOM groups, hydrolysis of methyl acetal, and simultaneous double-bond migration to provide (–)-mniopetal E (**5**) in 43% yield. Once the double bond migrates, the resulting α,β-unsaturated aldehyde is not subject to intramolecular hemiacetal formation. The spectroscopic data of synthetic **5** were well matched with those of natural **5** kindly provided by Professor Steglich. Comparison of the optical rotation of synthetic **5** ($[\alpha]_D^{29.5}$ –58) with that of the natural product ($[\alpha]_D^{20}$ –57) established the absolute stereo-chemistry as depicted.

In summary, we completed the first total synthesis of (–)-mniopetal E (**5**), a prototype of mniopetals A–D, in its natural form. Our synthesis featured the following aspects: (1) the substrate **62** for the IMDA reaction was synthesized in enantiopure form from the known building block **19**, and (2) the stereoselective IMDA reaction of **62** under thermal conditions realized practical access to the desired *endo*-cycloadduct **63-A** possessing the entire carbon skeleton with the correct stereo-chemistry. Our total synthesis of **5** as the natural form established the unsettled absolute configuration.

Our Total Syntheses of Mniopetal F and Kuehneromycin A

As the total synthesis of mniopetal E (**5**) was compleated, we next focused our attention on mniopetal F (**6**) and kuehneromycin A (**7**) as synthetic targets. The retrosynthesis of **6** and **7** is shown in Scheme 13. We anticipated that the IMDA reaction of the substrate **G** would proceed favorably via a chair-like *endo*-transition state **F**, which would lead to the desired adduct **E**. In general, a substituent neighboring the dienophile part would dispose of the sterically less-demanding equatorial orientation in the transition state. However, the trialkylsilyloxy group prefers an axial orientation as a result of the maximum hyperconjugative interaction between σ^*_{C-O} and $\pi^*_{C=C}$, which lowers the dienophile LUMO [87-89]. Accordingly, we designed the substrate **G**, carrying a trialkylsilyl ether at C-1. The synthesis of **G** would be achieved from aldehyde **I** via **H** by the introduction of the diene and dienophile parts. The intermediate **I** would be prepared from the known 2-deoxy-D-threitol derivative **74**, which had been prepared from **19** by Sharpless et al. [90].

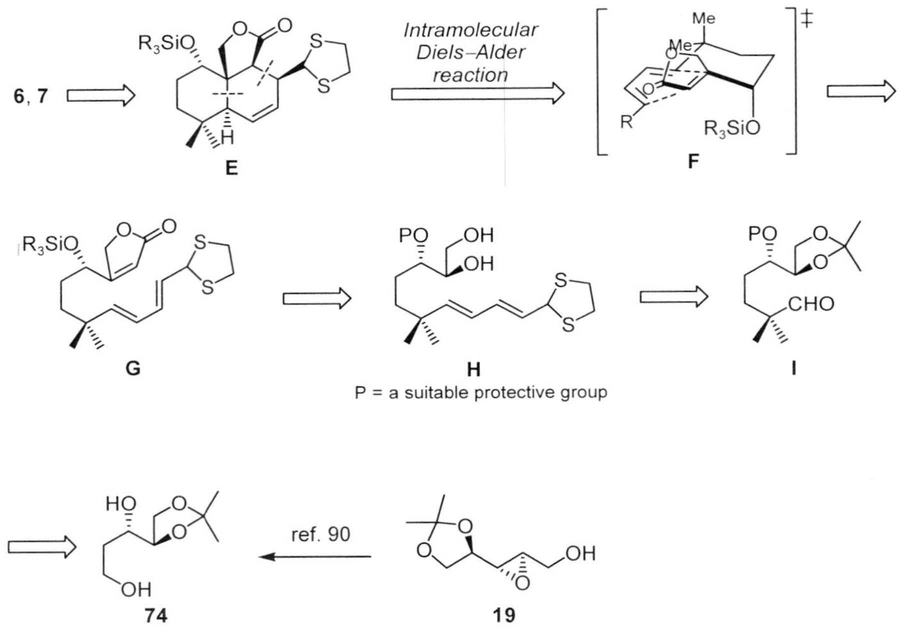

Scheme 13. Synthetic strategy of mniopetal F and kuehneromycin A

The synthesis of substrates **88–96** for the IMDA reaction is depicted in Schemes 14 and 15. Treatment of **74** with I_2, PPh_3, and imidazole [91] provided iodide **75**. The substitution of the iodo group in **75** by an anion generated from 2-methylpropionitrile [56-60] provided a heptanitrile derivative **76** in 95% yield. Then the secondary hydroxyl group in **76** was protected as an MOM ether to provide **77**. Reduction of **77** with DIBALH followed by acidic hydrolysis gave aldehyde **78**. The Horner–Emmons reaction of **78** with triethyl 4-phosphonocrotonate in the presence of LiOH·H$_2$O and molecular sieves 4A powder [68] provided $\alpha,\beta{:}\gamma,\delta$-unsaturated ester **79**. The geometric ratio (E,E-isomer: other isomers) was determined to be >20:1 by ^{1}H NMR analysis. DIBALH reduction of **79** followed by MnO$_2$ oxidation of the resulting

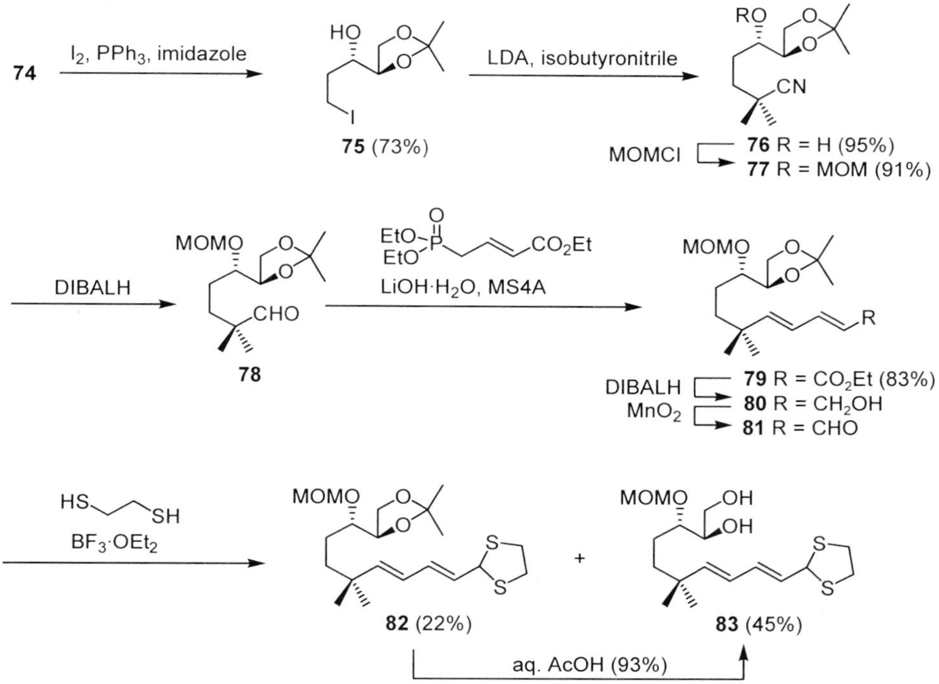

Scheme 14.

80 gave aldehyde **81**. The aldehyde group in **81** was protected as the 1,3-dithiolane, providing **82** along with the de-O-isopropylidene derivative **83**. Additional **83** was obtained by acid hydrolysis of **82**.

Selective protection of the primary hydroxyl group in **83** afforded the TBS ether **84** (Scheme 15). Oxidation of **84** and subsequent desilylation of the resulting keto derivative **85** with CSA provided α-hydroxy ketone **86**. Esterification of **86** by the action of EDC provided α-phosphono-acetate **87**. The intramolecular Horner–Emmons reaction of **87** in the presence of K$_2$CO$_3$ and 18-crown-6 ether [80] produced a butenolide **88**. The MOM group in **88** was converted into a variety of trialkylsilyl groups to investigate the substituent effect on the π-facial selectivity in an IMDA reaction. Thus, treatment of **88** with aqueous HCl, followed by reprotection of the hydroxyl group in the resulting **89** with a variety of silylating reagents, provided five trialkylsilyl ethers **90–94**. The preparation of substrate **92** from **89** by silylation with TBSOTf in pyridine was accompanied by the transformation of the butenolide moiety into 2-*tert*-butyldimethylsilyloxyfuran. Obtaining the substrate **92** required acid treatment of this intermediate. Acetate **95** and pivaloate **96** were also prepared by respective acylation.

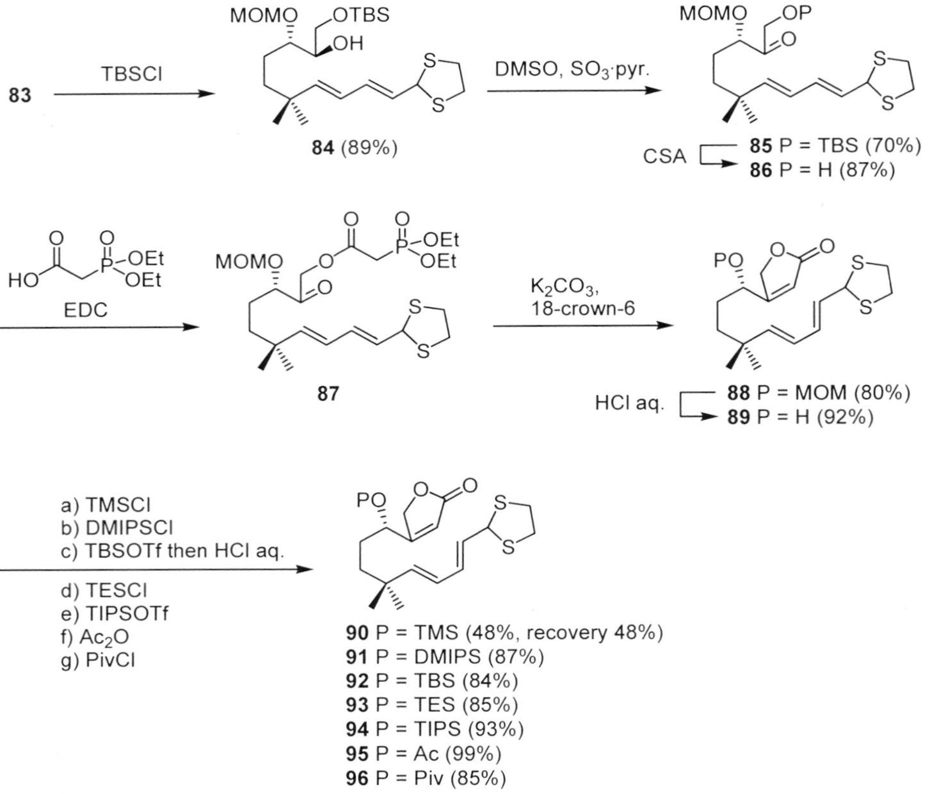
83
TBSCl
MOMO
OTBS
OH
S
S
84 (89%)
DMSO, SO_3·pyr.
MOMO
OP
O
CSA
85 P = TBS (70%)
86 P = H (87%)
O
O
P
OEt
OEt
HO
EDC
MOMO
87
K_2CO_3,
18-crown-6
PO
HCl aq.
88 P = MOM (80%)
89 P = H (92%)
a) TMSCl
b) DMIPSCl
c) TBSOTf then HCl aq.
d) TESCl
e) TIPSOTf
f) Ac_2O
g) PivCl
90 P = TMS (48%, recovery 48%)
91 P = DMIPS (87%)
92 P = TBS (84%)
93 P = TES (85%)
94 P = TIPS (93%)
95 P = Ac (99%)
96 P = Piv (85%)

Scheme 15.

The results of the thermal IMDA reactions using the substrates **88–96** are summarized in Scheme 16 and Table 4. We did not conduct the IMDA reactions of **88–96** using a Lewis acid catalyst because of our previous unfruitful results with Lewis acids obtained in the mniopetal E synthesis. Each solution of the substrate in toluene was heated at 210 °C in a sealed tube in the presence of a catalytic amount of BHT. After purification of the reaction mixture by silica gel column chromatography, the diastereomeric ratios (*endo*/*exo* and **A/B**) of the

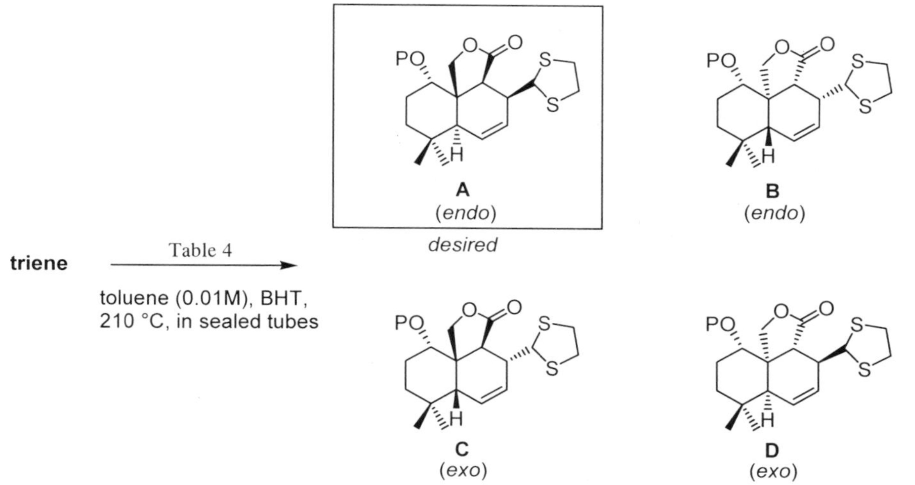

Scheme 16. Intramolecular Diels-Alder reactions of **88–96**

Table 4. Intramolecular Diels-Alder reactions of **88–96**

entry	substrate	products	P	time (h)	yield (%)	endo : exo [a]	A : B [a]
1	88	97	MOM	40	86	> 20 : 1	8 : 12
2	89	98	H	40	66	> 20 : 1	8 : 12
3	90	99	TMS	40	70	> 20 : 1	13 : 7
4	91	100	DMIPS	70	81	> 20 : 1	13 : 7
5	92	101	TBS	50	84	> 20 : 1	13 : 7
6	93	102	TES	40	78	> 20 : 1	12 : 8
7	94	103	TIPS	40	50 [b]	> 20 : 1	10 : 10
8	95	104	Ac	40	74	> 20 : 1	5 : 15
9	96	105	Piv	40	72	> 20 : 1	7 : 13

[a] determined by 300 MHz ^{1}H NMR [b] recovery of substrate ca. 30%

cycloadducts were determined by ^{1}H NMR analysis.

As shown in Fig. (**7**), the relative configurations for the diastereomeric adducts were determined by NOESY correlations and coupling constants for ring protons. All the IMDA reactions proceeded with high *endo/exo* selectivity (**A+B/C+D**), while the π-facial selectivity of the *endo*-adducts (**A/B**) depended on the protecting group at C-1. The C-1 alkyl- and acyl-protected substrates (**88**, **95**, or **96**) and non-protected substrate **89** provided the undesired cycloadduct **B** preferentially

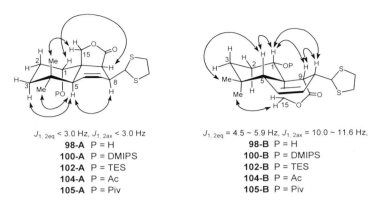

$J_{1,\,2eq} < 3.0$ Hz, $J_{1,\,2ax} < 3.0$ Hz

98-A P = H
100-A P = DMIPS
102-A P = TES
104-A P = Ac
105-A P = Piv

$J_{1,\,2eq} = 4.5 \sim 5.9$ Hz, $J_{1,\,2ax} = 10.0 \sim 11.6$ Hz,

98-B P = H
100-B P = DMIPS
102-B P = TES
104-B P = Ac
105-B P = Piv

Fig. (7). NOESY correations of IMDA adducts

(Entries 1, 2, 8, or 9). On the other hand, the silyl ethers **90–94** provided the desired *endo*-adduct **A** predominantly (Entries 3–7). In the case of the bulkier silyl ethers **93** and **94**, the π-facial selectivity significantly decreased (Entries 6 and 7 compared to 3–5). As anticipated, by employing the silyl ethers **90–92** for the IMDA reaction, the desired cycloadduct **A** for the syntheses of mniopetal F and kuehneromycin A was obtained stereoselectively.

To explain the diastereoselectivities, the analogous transition states used for the cycloaddtion of triene **48** can be depicted as shown in Fig. **(8)**. Although the steric repulsion by the synperiplaner relationship between C-1 and C-2 is reduced, the four boat-like transition states **TS-Boat-A, B, C**, and **D** are still disfavored by the allylic repulsion of H-5 and H-6. Thus, we again discuss the stereoselectivity in consideration of four chair-like transition states. The high *endo* selectivity in all cases is explained by the same three factors (secondary orbital interaction [72-76], twist asynchronicity [77, 78] and steric repulsion), which are not affected by the protecting groups. We assume that the π-facial selectivity (the ratio of **A/B**) is determined by the balance of a steric effect and a stereoelectronic effect directed by the stereochemistry at C-1. The steric effect originates from 1,3-diaxial repulsion between PO and H-3 in **TS-Chair-A** and 1,3-allylic repulsion between PO and H-9 in **TS-Chair-B**. Considering the decrease of π-facial selectivity in the case of the bulkier silyl group, 1,3-diaxial repulsion is more influenced by the steric bulkiness of the PO group compared with 1,3-allylic repulsion. Therefore, the steric effect of the PO group prefers

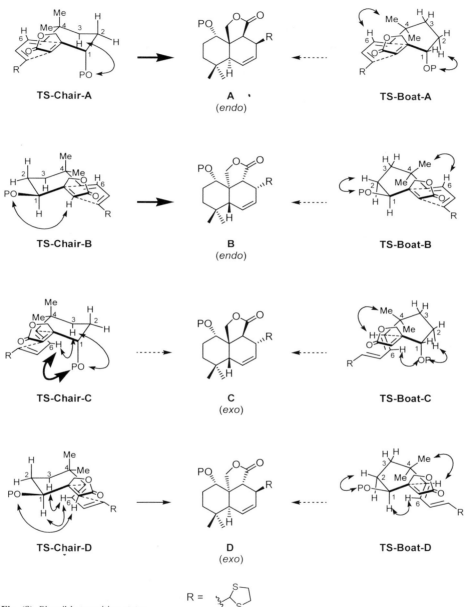

Fig. (8). Plausible transition states

to adopt **TS-Chair-B**. On the other hand, the stereoelectronic effect leads the PO group to an axial orientation favoring **TS-Chair-A**. This direction is enlarged in the case of the trialkylsilyloxy group. In previous reports, it was supposed that this stereoelectoronic effect is a result of the maximum hyperconjugative interaction between axial σ^*_{C-O} and $\pi^*_{C=C}$, which lowers the dienophile LUMO [87-89] as shown in Fig. (**9-A**). This effect is explained more reasonably by Cieplak's theory [92-98], i.e., it is a result of the maximum hyperconjugative interaction between the partial antibonding $\sigma_{\ddagger}^*$ and equatorial σ_{C-H} as shown in Fig. (**9-B**).

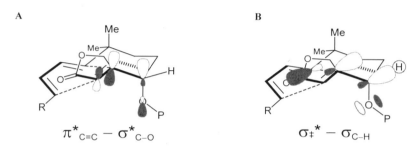

$$\pi^*_{C=C} - \sigma^*_{C-O} \qquad\qquad \sigma_{\ddagger}^* - \sigma_{C-H}$$

Fig. (9). Stereoelectoronic effect. (**A**):Funk model; (**B**):Cieplak model

As depicted in Scheme 17, total synthesis of mniopetal F (**6**) was achieved from the dimethylisopropylsilyl (DMIPS) ether-protected cycloadduct **100** by the analogous reaction sequence used for mniopetal E synthesis. Desilylation of **100** with TBAF afforded a diastereomeric mixture **98**. Treatment of **98** with Hg(ClO$_4$)$_2$·3H$_2$O [82] provided dimethylacetals **106-A** (44%) and **106-B** (25%), which were separated by chromatography on silica gel. At this stage, the enantiopurity of **106-A** (>98% ee) was confirmed by ^{1}H NMR analyses of (*R*)- and (*S*)-*O*-acetylmandelate derivatives of **106-A**. No racemization of the C-1 in **91** occurred under the IMDA conditions. The diastereomerically pure **106-A** was protected as an MOM ether. The γ-lactone in the resulting **107** was hydrolyzed, and the resulting γ-hydroxy carboxylate was oxidized with Na$_2$RuO$_4$ [83, 84], affording γ-hydroxy-γ-lactone **108**. Treatment of **108** with two equivalents of DIBALH gave dialdehyde hydrate **109** (44%) along with **107** (21%) and **108** (21%). Brief treatment of **109** with *p*-toluenesulfonic acid (TsOH) provided a tetracyclic intermediate **110**, which was oxidized to lactone **111**. Acid

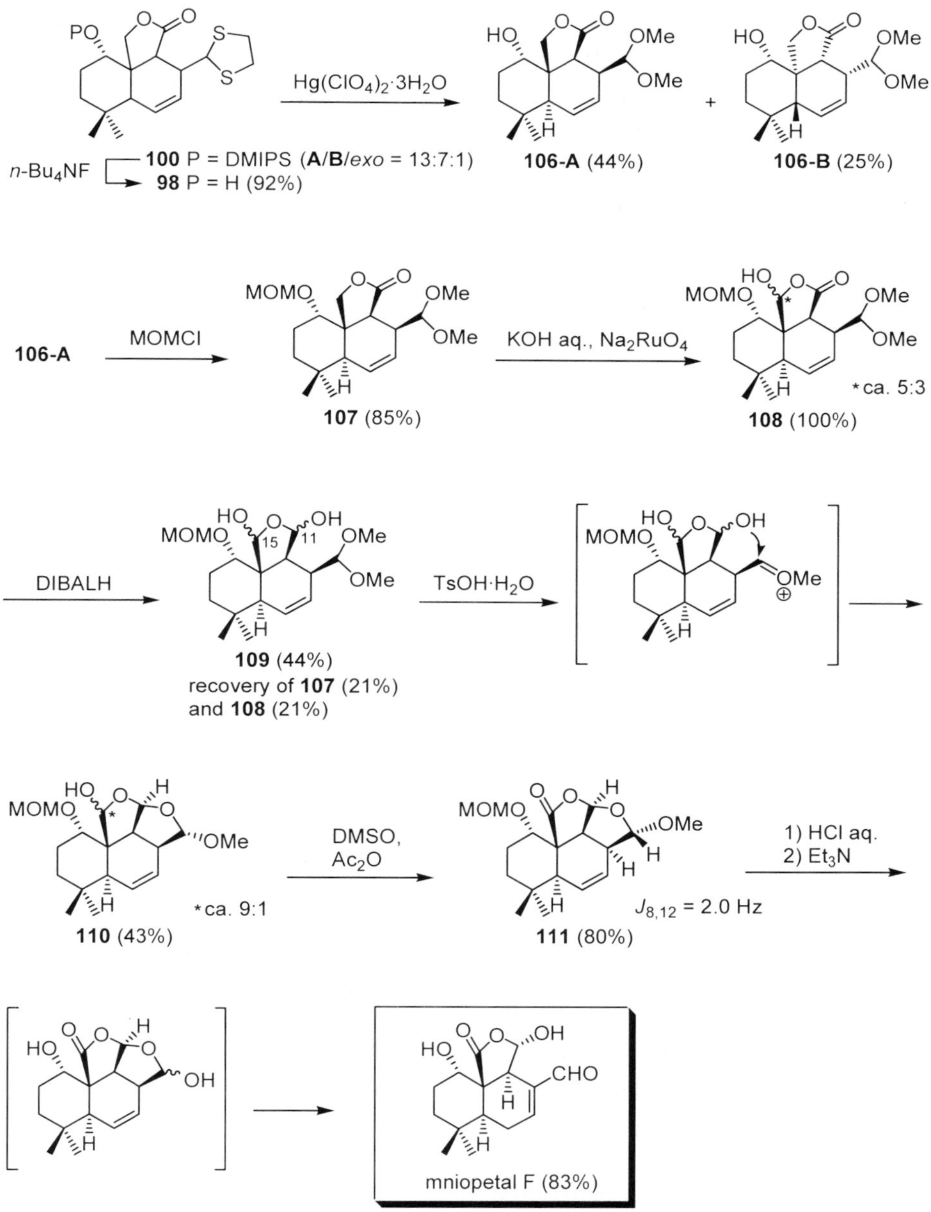

Scheme 17. Completion of the total synthesis of mniopetal F

hydrolysis of the acetal moiety and the subsequent base-mediated double-bond migration provided (−)-mniopetal F (**6**). The spectroscopic data (^{1}H and ^{13}C NMR, IR, MS, and CD) of the synthetic **6** were well matched with those for natural **6**.

We next investigated the total synthesis of kuehneromycin A (**7**). Partial hydrolysis of **111** in trifluoroacetic acid (TFA) gave alcohol, which was oxidized to keto-lactone **112** (Scheme 18). Treatment of **112** with TFA followed by triethylamine caused unexpected decarboxylation, providing a mixture of kuehneromycin B (**113**) and panudial (**114**), which are known as natural products related to **7** [3] . The synthesis of

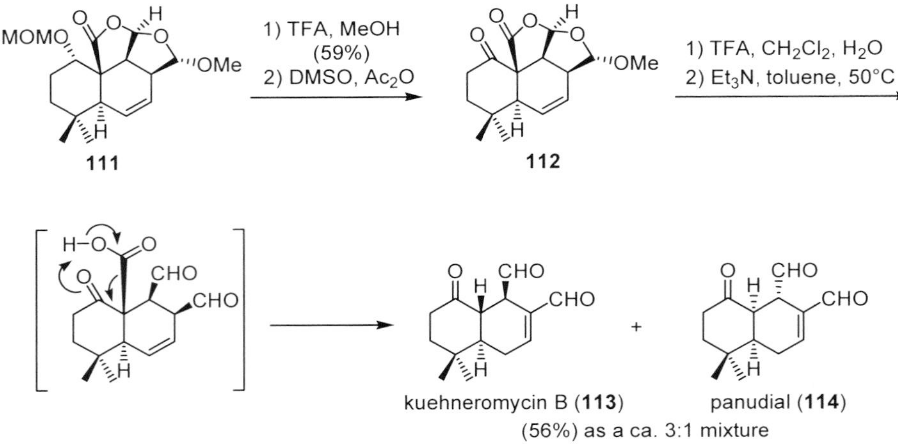

Scheme 18.

7 was also achieved from mniopetal F (**6**) (Scheme 19). Selective protection of the C-11 hydroxyl group in **6** with the triethylsilyl (TES) group and following Dess–Martin oxidation provided **116** as an anomeric mixture at C-11 (the ratio was ca. 1:1). Acidic hydrolysis of the TES group in **116** provided (−)-kuehneromycin A (**7**) as a sole product. Our synthetic (−)-**7** was identical in all respects to a natural sample ([α]$_D$, ^{1}H and ^{13}C NMR, IR, and MS). The stereochemistry at C-11 in synthetic **7** was regulated as only the α-hydroxy form, even though **116** was an anomeric mixture. It is likely that the α-hydroxy form of kuehneromycin A (**7**) is stabilized by the intramolecular hydrogen bond between C-11 hydroxyl and C-12 carbonyl groups indicated by a absorption band at 3400 cm^{-1} in the IR spectrum of **7**. The same intramolecular hydrogen bond seems to regulate the stereochemistry at C-11 in structurally similar natural products, mniopetals (**1–6**).

In summary, we completed the total syntheses of (–)-mniopetal F (**6**) and (–)-kuehneromycin A (**7**) in their natural forms. Our total synthesis feature a the highly *endo*-selective IMDA reaction using some substrates carrying a trialkylsilyloxy group at C-1. The π-facial selectivity in the IMDA reaction was controlled by the stereoelectronic effect of the silyloxy group adjacent to the dienophile part.

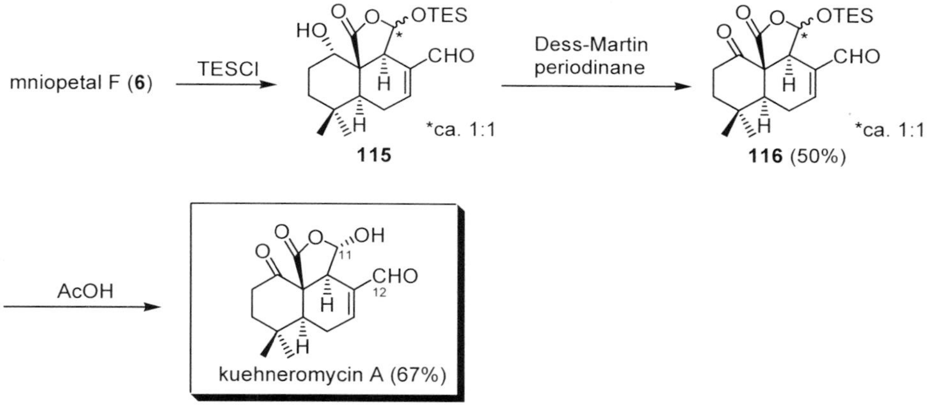

Scheme 19. Completion of the total synthesis of kuehneromycin A

Total Syntheses of Mniopetals E and F and Kuehneromycin A by Jauch

Jauch achieved the short and elegant total synthesis of mniopetals during almost the same period in which our total syntheses were reported [36-41]. We briefly introduce Jauch's achievements on this subject. The key step of his synthesis was the IMDA reaction of 2-substituted butenolide **124** (Scheme 20), which was prepared by the diastereoselective PhSeLi-induced Baylis–Hillman reaction of **122** and Feringa's butenolide **123** [99]. The cycloadduct of this IMDA reaction **125** was transformed into kuehneromycin A and mniopetals E and F, respectively.

The synthesis started with the Horner–Emmons reaction of 2,2-di-methyl-4-pentenal **117** [100, 101] and trimethyl 4-phosphonocrotonate providing **118** in 85% yield. The DIBALH reduction of ester **118** gave alcohol **119**, which was protected as *tert*-butyldiphenylsilyl (TBDPS) ether **120**. Chemoselective hydroboration of the terminal double bond in **120** with 9-BBN and subsequent oxidation with H_2O_2/NaOH gave alcohol **121**. Oxidation of primary hydroxyl group in **121** by TEMPO/PhI(OAc)$_2$ [102] provided aldehyde **122**, the substrate for the planned Baylis–Hillman reaction. Since Feringa's butenolide is highly

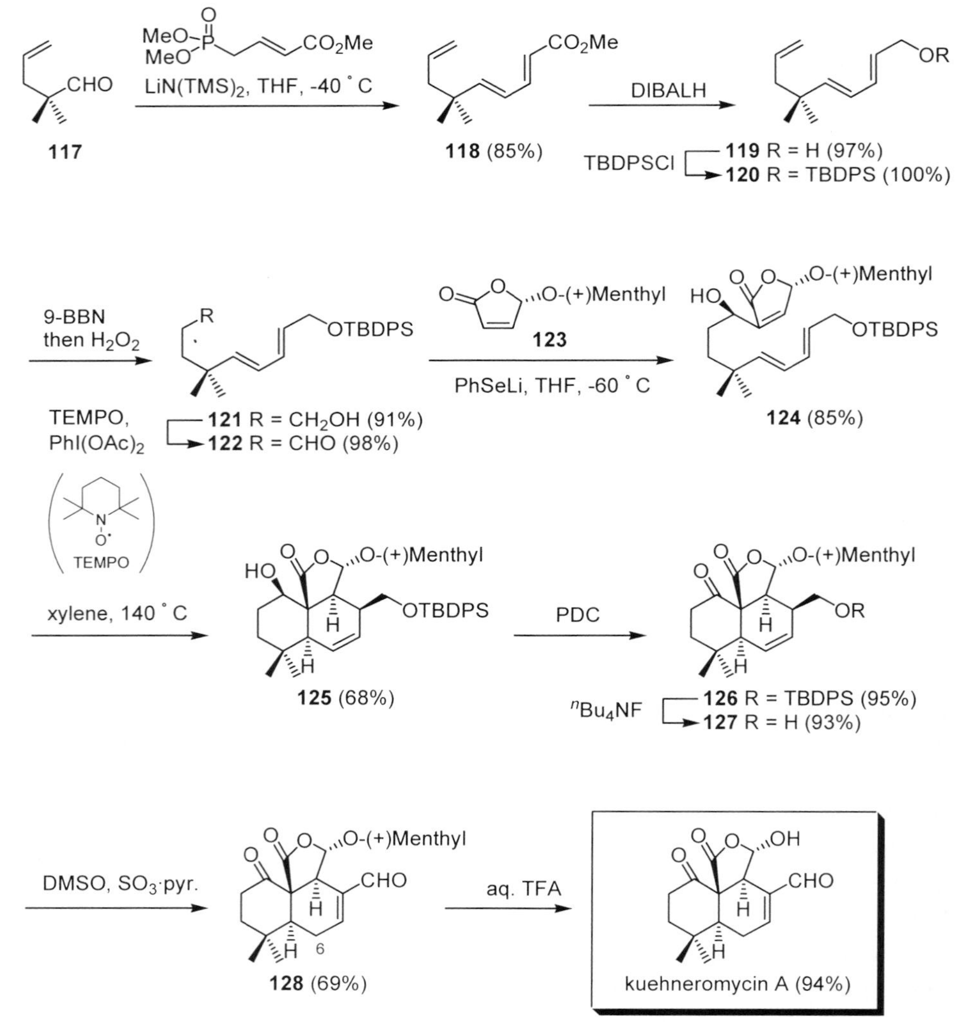

Scheme 20.

base-sensitive, Jauch developed a new variant of the Baylis–Hillman reaction using weakly basic and strongly nucleophilic PhSeLi [103]. The treatment of aldehyde **122** and Feringa's butenolide **123** with PhSeLi at −60 °C gave trienolide **124** in 88% yield with a high level of diastereo-selection, which was explained by a Zimmermann–Traxler-like transition state in the aldol step. The key IMDA reaction was carried out under thermal conditions by heating a xylene solution of **124** to 140 °C for 60 h to give the desired *cis*-fused cycloadduct **125** in 68% yield along with some unidentified side products and about 20% recovered starting material. The boat-like transition state depicted in Fig. (**10**) was proposed for the stereoselectivity obserbed in the IMDA reaction [36].

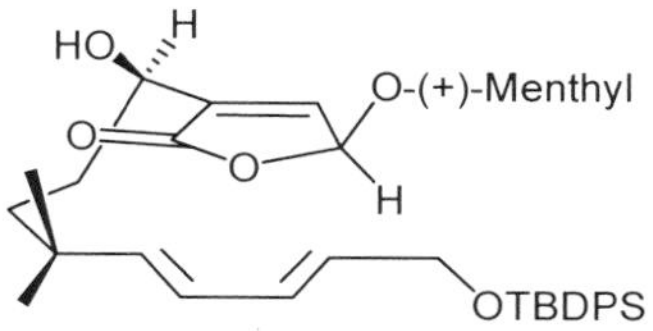

Fig. (10). A possible transition state

The transformation of **125** to kuehneromycin A (**7**) was as follows. The equatorial hydroxyl group in **125** was oxidized to the ketone **126** with PDC in the presence of MS3A. Cleavage of the TBDPS ether gave primary alcohol **127**. Oxidation of **127** to **128** was difficult because of the concomitant peroxidation of the C-6 position in **128** (as in our case with **51**; see Scheme 8). However, Parikh–Doering oxidation [104] was found to work well for the desired transformation. In our case, Parikh–Doering oxidation of **51** gave no reaction. In the final step, the menthyl residue was removed by the action of 80% aqueous TFA solution to provide kuehneromycin A (**7**). Synthetic **7** was identical in all respects, including optical rotation, to natural **7**.

The total synthesis of mniopetal E (**5**) was achieved in a six-step sequence from the cycloadduct **125** (Scheme 21). The treatment of **125** with trifluoromethanesulfonic anhydride (Tf₂O) afforded triflate **129**. Elimination of trifluoromethanesulfonic acid to introduce a double bond was accomplished in 2,6-lutidine at 100 °C, giving **130** in 84% yield. Deprotection of TBDPS ether was followed by Parikh–Doering oxidation of the resulting **131** by the same procedure used in the case of **127**. Taking advantage of the electron-rich double bond between C-1 and C-2 in **132**, dihydroxylation with OsO$_4$/NMO proceeded stereo-selectively from the *si*-face leading to *cis*-diol **133** in 62% yield. Finally, hydrolysis of menthyl acetal by exposure to TFA/H$_2$O/acetone (1:1:1) provided mniopetal E (**5**) in quantitative yield as a natural enantiomeric form.

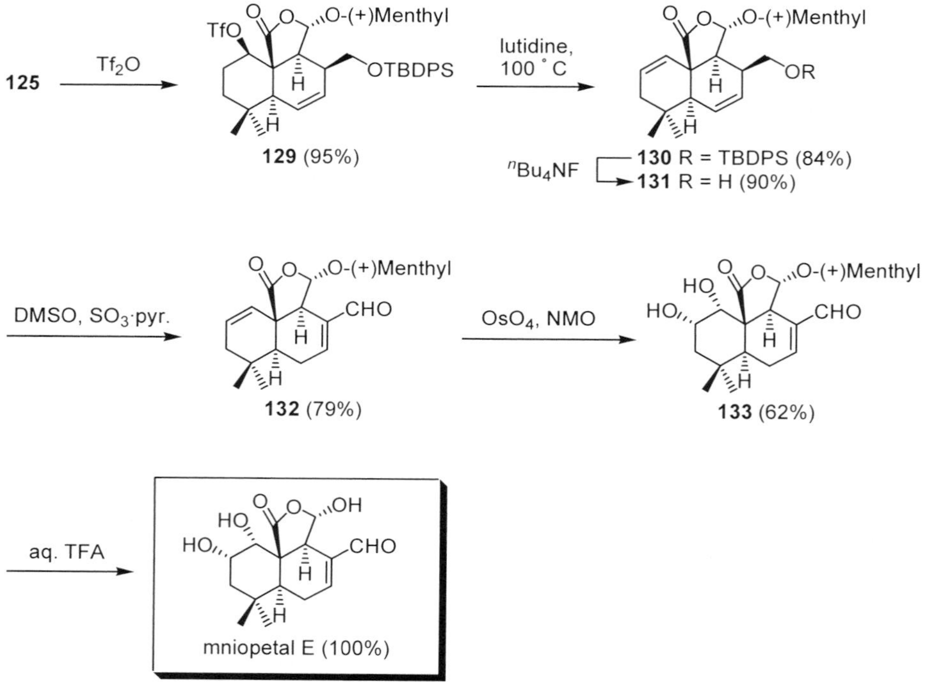

Scheme 21.

The transformation of **125** into mniopetal F (**6**) requires the inversion of a secondary hydroxyl group at the C-1 position, which has the opposite configuration of **6**. Although several attempts failed, Jauch found ultimately that a sterically small nucleophile attacks the C-1 position from the axial direction by an S$_N$2 mechanism (Scheme 22). Namely, **125** was transformed into triflate **129**, and subsequent treatment with KNO$_2$, 18-crown-6 ether, and a trace amount of water at 40 °C [105] provided the desired inverted alcohol **134** in 50% yield along with 25% recovered starting material (66% corrected yield). The hydroxyl group in **134** was protected to give *p*-nitrobenzoate **135**.

Desilylation of **135** afforded alcohol **136**, which was oxidized to enal **137** by Parikh–Doering oxidation in 62% yield. Removal of the *p*-nitrobenzoyl group and the menthyl group resulted in completion of the total synthesis of mniopetal F (**6**).

Thus, Jauch effectively synthesized kuehneromycin A, mniopetal E, and mniopetal F by a strategy employing a new variant of the Baylis-Hillman reaction.

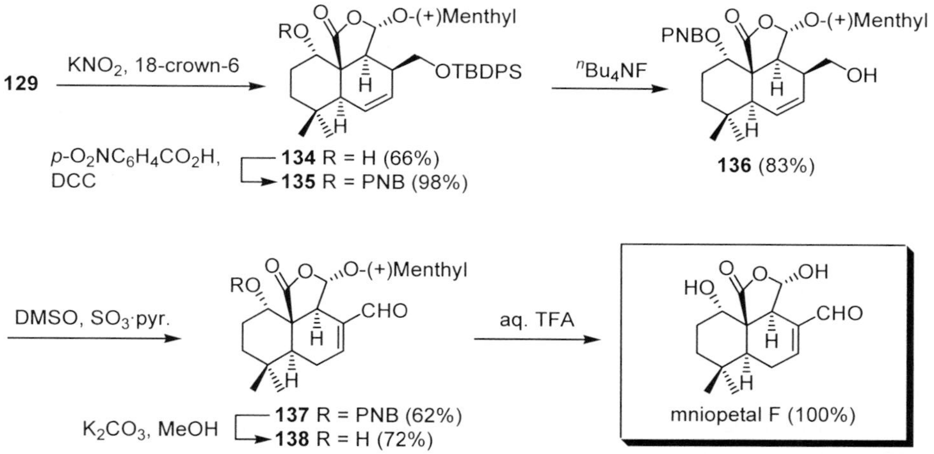

Scheme 22.

ACKNOWLEDGEMENTS

We thank our co-workers whose names are cited in the references. They are largely responsible for the synthetic achievements described in this article, and their distinguished ability and patience made the completion of the total syntheses possible. We are grateful to Professor W. Steglich (University of München) for sending us copies of the spectra of natural mniopetals and kuehneromycin A. We gratefully acknowledge the Japan Interaction in Science and Technology Forum (JIST) and Taisho Pharmaceutical Co., Ltd. for their financial support.

REFERENCES

[1]   Kuschel, A.; Anke, T.; Velten, R.; Klostermeyer, D.; Steglich, W.; König, B. *J. Antibiot.* **1994**, *47*, 733–739.

[2]   Velten, R.; Klostermeyer, D.; Steffan, B.; Steglich, W.; Kuschel, A.; Anke, T. *J. Antibiot.* **1994**, *47*, 1017–1024.

[3]   Erkel, G.; Lorenzen, K.; Anke, T.; Velten, R.; Gimenez, A; Steglich, W. *Z. Naturforsch. C* **1995**, *50*, 1–10.

[4]   Ayer, W. A.; Craw, P. A. *Can. J. Chem.* **1989**, *67*, 1371–1380.

[5]   Mori, K.; Watanabe, H. *Tetrahedron* **1986**, *42*, 273–281.

[6]   He, J.-F.; Wu, Y.-L. *Tetrahedron* **1988**, *44*, 1933–1949.

[7]   Velten, R.; Steglich, W.; Anke, T. *Tetrahedron: Asymmetry* **1994**, *5*, 1229–1232.

[8]   Jansen, B. J.; De Groot, A. *Natural Product Reports* **1991**, 319–337.

[9]   Okawara, H.; Nakai, H.; Ohno, M. *Tetrahedron Lett.* **1982**, *23*, 1087–1090.

[10]   Manna, S.; Yadagiri, P.; Falck, J. R. *J. Chem. Soc., Chem. Commun.* **1987**, 1324–1325.

[11]   Urones, J. G.; Marcos, I. S.; Martin, D. D. *Tetrahedron* **1988**, *44*, 4547–4554.

[12]   Ayer, W. A.; Talamas, F. X. *Can. J. Chem.* **1988**, *66*, 1675–1685.

[13]   Goldsmith, D. J.; Kezar III, H. S.; *Tetrahedron Lett.* **1980**, *21*, 3543–3546.

[14]   Razmillic, I.; Sierra, J.; Lopez, J.; Cortes, M. *Chem. Lett.* **1985**, 1113–1114.

[15]   Oyarzum, M. L.; Cortes, M.; Sierra, *J. Synth. Commun.* **1982**, 12, 951–958.

[16]   Banerjee, A. K.; Laya-Mimo, M. In *Studies in Natural Products Chemistry*; Atta-ur-Rahman, Ed.; Vol. 24, Elsevier Science B. V: Amsterdam, **2000**, vol. *24*, pp. 175–213.

[17]   Kende, A. S.; Blacklock, T. J. *Tetrahedron Lett.* **1980**, *21*, 3119–3122.

[18]   Jansen, B. J. M.; Sengers, H. H. W. J. M.; Bos, H. J. T.; De Groot, A. *J. Org. Chem.* **1988**, *53*, 855–859.

[19]   Banerjee, A. K.; Vera, W. *Synth. Commun.* **2000**, *30*, 4375–4385.

[20]   Nakata, T.; Akita, H.; Naito, T.; Oishi, T. *J. Am. Chem. Soc.* **1979**, *101*, 4400–4401.

[21]   Akita, H.; Naito, T.; Oishi, T. *Chem. Lett.* **1979**, 1365–1368.

[22]   Akita, H.; Naito, T.; Oishi, T. *Chem. Pharm. Bull.* **1980**, *28*, 2166–2171.

[23]   Nakata, T.; Akita, H.; Naito, T.; Oishi, T. *Chem. Pharm. Bull.* **1980**, *28*, 2172–2177.

[24]   Kato, T.; Suzuki, T.; Tanemura, M.; Kumanireng, A. S.; Ototani, N.; Kitahara, Y. *Tetrahedron Lett.* **1971**, 1961–1964.

[25]   Ohsuka, A.; Matsukawa, A. *Chem. Lett.* **1979**, 635–636.

[26]   Tanis, P. A.; Nakanishi, K. *J. Am. Chem. Soc.* **1979**, *101*, 4398–4400.

[27]   Mori, K.; Takaishi, H. *Liebigs Ann. Chem.* **1989**, 695–697.

[28]   Tamai, Y.; Mizutani, Y.; Hagiwara, H.; Uda, H.; Harada, N. *J. Chem. Res.* **1985**, (S) 148–149; (M) 1746–1787.

[29]   Hagiwara, H.; Uda, H. *J. Chem. Soc., Chem. Commun.* **1987**, 1351–1353.

[30]   Hanselmann, R.; Benn, M. *Synth. Commun.* **1996**, *26*, 945–961.

[31]   Toyooka, N.; Nishino, A.; Momose, T. *Tetrahedron* **1997**, *53*, 6313–6326.

[32]   Murata, T.; Ishikawa, M.; Nishimaki, R.; Tadano, K. *Synlett* **1997**, 1291–1293.

[33]   Suzuki, Y.; Nishimaki, R.; Ishikawa, M.; Murata, T.; Takao, K.; Tadano, K. *Tetrahedron Lett.* **1999**, *40*, 7835–7838.

[34]   Suzuki, Y.; Nishimaki, R.; Ishikawa, M.; Murata, T.; Takao, K.; Tadano, K. *J. Org. Chem.* **2000**, *65*, 8595–8607.

[35]   Suzuki, Y.; Ohara, A.; Sugaya, K.; Takao, K.; Tadano, K. *Tetrahedron* **2001**, *57*, 7291–7301.

[36]   Jauch, J. *Synlett* **1999**, 1325–1327.

[37]   Reiser, U.; Jauch, J.; Herdtweck, E. *Tetrahedron: Asymmetry* **2000**, *11*, 3345–3349.

[38]   Jauch, J. *Angew. Chem., Int. Ed. Engl.* **2000**, *39*, 2764–2765.

[39]   Jauch, J. *Eur. J. Org. Chem.* **2001**, 473–476.

[40]   Jauch, J. *Synlett* **2001**, 87–89.

[41]   Reiser, U.; Jauch, *J. Synlett* **2001**, 90–92.

[42]   Fukumoto, K. *J. Syn. Org. Chem. Jpn.* **1994**, *52*, 2–18.

[43]   Fallis, A. G. *Acc. Chem. Res.* **1999**, *32*, 464–474.

[44]   Brocksom, T. J.; Nakamura, J.; Ferreira, M. L.; Brocksom, U. *J. Braz. Chem. Soc.* **2001**, *12*, 597–622.

[45]   Roush, W. R. In *Advances in Cycloaddition*; Curran, D. P., Ed.; JAI Press: Greenwich, **1990**; vol. *2*, pp. 91–146.

[46]   Roush, W. R. In *Comprehensive Organic Synthesis*; Trost, B. M.; Fleming, I.; Paquette, L. A., Eds.; Pergamon Press: Oxford, **1991**; vol. *5*, pp. 513–550.

[47]   Kappe, C. O.; Murphree, S. S.; Padwa, A. *Tetrahedron* **1997**, *42*, 14179–14233.

[48]   Suzuki, Y.; Murata, T.; Takao, K.; Tadano, K. *J. Syn. Org. Chem. Jpn.* **2002**, *60*, 679–690.

[49]   Tadano, K.; Murata, T.; Kumagai, T.; Ogawa, S. *Tetrahedron Lett.* **1993**, *34*, 7279–7282.

[50]   Murata, T.; Kumagai, T.; Ishikawa, M.; Tadano, K.; Ogawa, S. *Bull. Chem. Soc. Jpn.* **1996**, *69*, 3551–3561.

[51]   Payne, G. B. *J. Org. Chem.* **1962**, *27*, 3819–3822.

[52]   Katsuki, T.; Lee, A. W. M.; Ma, P.; Martin, V. S.; Masamune, S.; Sharpless, K. B.; Tuddenham, D.; Walker, F. J. *J. Org. Chem.* **1982**, *47*, 1373–1380.

166

[53] Schmid, C. R.; Bryant, J. D. *Org. Synth.* **1993**, *72*, 6–13.

[54] Takano, S.; Kurotaki, A.; Takahashi, M.; Ogasawara, K. *Synthesis* **1986**, 403–406.

[55] Takano, S.; Ogasawara, K. *J. Syn. Org. Chem. Jpn.* **1987**, *45*, 1157–1170.

[56] Borgne, J. F. L.; Cuvigny, T.; Larcheveque, M.; Normant, H. *Synthesis* **1976**, 238–240.

[57] Murata, S.; Matsuda, I. *Synthesis* **1978**, 221–222.

[58] Matsuda, I.; Murata, S.; Ishii, Y. *J. Chem. Soc., Perkin Trans. 1* **1979**, 26–30.

[59] Smith, J. G. *Synthesis* **1984**, 629–656.

[60] Semmelhack, M. F.; Harrison, J. J.; Thebtaranonth, Y. *J. Org. Chem.* **1979**, *44*, 3275–3277.

[61] Dess, D. B.; Martin, J. C. *J. Org. Chem.* **1983**, *48*, 4155–4156.

[62] Dess, D. B.; Martin, J. C. *J. Am. Chem. Soc.* **1991**, *113*, 7277–7287.

[63] Ireland, R. E.; Liu, L. *J. Org. Chem.* **1993**, *58*, 2899–2899.

[64] Creger, I. I. *J. Org. Chem.* **1972**, *37*, 1907–1918.

[65] Behrens, C. H.; Ko, S. Y.; Sharpless, K. B.; Walker, F. J. *J. Org. Chem.* **1985**, *50*, 5687–5696.

[66] Behrens, C. H.; Sharpless, K. B. *J. Org. Chem.* **1985**, *50*, 5696–5704.

[67] Page, P. C. B.; Rayner, C. M.; Sutherland, I. O. *J. Chem. Soc., Perkin Trans. 1* **1990**, 1375–1382.

[68] Takacs, J. M.; Jaber, M. R.; Clement, F.; Walters, C. *J. Org. Chem.* **1998**, *63*, 6757–6760.

[69] Ishihara, K.; Kurihara, H.; Yamamoto, H. *J. Org. Chem.* **1993**, *58*, 3791–3793.

[70] Bhattacharya, A. K.; Thyagarajan, G. *Chem. Rev.* **1981**, *81*, 415–430.

[71] Blanchette, M. A.; Choy, W.; Davis, J. T.; Essenfeld, A. P.; Masamune, S.; Roush, W. R.; Sakai, T. *Tetrahedron Lett.* **1984**, *25*, 2183–2186.

[72] Hoffmann, R.; Woodward, R. B. *J. Am. Chem. Soc.* **1965**, *87*, 4388–4389.

[73] Hoffmann, R.; Woodward, R. B. *J. Am. Chem. Soc.* **1965**, *87*, 4389–4390.

[74] Ginsburg, D. *Tetrahedron* **1983**, *39*, 2095–2135.

[75] García, J.; Mayoral, J.; Salvatella, L. *Acc. Chem. Res.* **2000**, *33*, 658–664.

[76] Arrieta, A.; Cossío, F. P.; *J. Org. Chem.* **2001**, *66*, 6178–6180.

[77] Wu, T.-C.; Houk, K. N. *Tetrahedron Lett.* **1985**, *26*, 2293–2296.

[78] Brown, F. K.; Houk, K. N. *Tetrahedron Lett.* **1985**, *26*, 2297–2300.

[79] Frigero, M.; Santagostino, M.; Sputore, S.; Palmisano, G. *J. Org. Chem.* **1995**, *60*, 7272–7276.

[80] Aristoff, P. A. *J. Org. Chem.* **1981**, *46*, 1954–1957.

[81] Saito, N.; Grieco, P. A. *J. Syn. Org. Chem. Jpn.* **2000**, *58*, 39–49.

[82] Fujita, E.; Nagao, Y.; Kaneko, K. *Chem. Pharm. Bull.* **1978**, *26*, 3743–3751.

[83] Lee, D. G.; Hall, D. T.; Cleland, J. H. *Can. J. Chem.* **1972**, *50*, 3741–3743.

[84]   Corey, E. J.; Myers, A. G. *J. Am. Chem. Soc.* **1985**, *107*, 5574–5576.

[85]   Canne, P.; Pamondon, J.; Akssira, M. *Tetrahedron* **1988**, *44*, 2903–2912.

[86]   Thuring, J. W. J. F.; Nefkens, G. H. L.; Schaafstra, R.; Zwanenburg, B. *Tetrahedron* **1995**, *51*, 5047–5056.

[87]   Funk, R. L.; Zeller, W. E. *J. Org. Chem.* **1982**, *47*, 180–182.

[88]   Müller, G.; Jas, G. *Tetrahedron Lett.* **1992**, *33*, 4417–4420.

[89]   Kanoh, N.; Ishihara, J.; Murai, A. *Synlett* **1995**, 895–897.

[90]   Ma, P.; Martin, V. S.; Masamune, S.; Sharpless, K. B.; Viti, S. M. *J. Org. Chem.* **1982**, *47*, 1380–1381.

[91]   Garegg, P. J.; Samuelsson, B. *J. Chem. Soc., Perkin Trans.1* **1980**, 2866–2869.

[92]   Mehta, G.; Uma, R. *Acc. Chem. Res.* **2000**, *33*, 278–286.

[93]   Coxon, J. M.; Froese, R. D. J.; Ganguly, B.; Marchand, A. P.; Morokuma, K. *Synlett*, **1999**, 1681–1703.

[94]   Coxon, J. M.; McDonald, D. Q. *Tetrahedron Lett.* **1992**, *33*, 651–654.

[95]   Ishida, M.; Aoyama, T.; Kato, S. *Chem. Lett.* **1989**, 663–666.

[96]   Ishida, M.; Benia, Y.; Inagaki, S.; Kato, S. *J. Am. Chem. Soc.* **1990**, *112*, 8980–8982.

[97]   Ishida, M.; Aoyama, T.; Benia, Y.; Yamabe, S.; Kato, S.; Inagaki, S. *Bull. Chem. Soc. Jpn.* **1993**, *66*, 3430–3439

[98]   Ishida, M.; Inagaki, S. *J. Syn. Org. Chem. Jpn.* **1994**, *52*, 649–657.

[99]   Feringa, B. L.; De Jong, J. C. *Bull. Chem. Soc. Chim. Belg.* **1992**, *101*, 627–640.

[100]   Brannock, K. C. *J. Am. Chem. Soc.* **1959**, *81*, 3379–3383.

[101]   Salomon, G.; Ghosh, S. *Org. Synth.* **1984**, *62*, 125–133.

[102]   De Mico, A.; Marfarita, R.; Parlanti, L.; Vescovi, A.; Piacantelli, G. *J. Org. Chem.* **1997**, *62*, 6974–6977.

[103]   Jauch, J. *J. Org. Chem.* **2001**, *66*, 609–611.

[104]   Tidwell, T. T. In *Organic Reactions*; Paquette, L. A., Ed.; Wiley: New York, **1991**, vol. *39*, p. 297 and references cited therein.

[105]   Moriarty, R. M.; Zhuang, H.; Penmasta, R.; Liu, K.; Awasthi, A. K.; Tuladhar, S. M.; Rao, M. S. C.; Singh, V. K. *Tetrahedron Lett.* **1993**, *34*, 8029–8032.

Atta-ur-Rahman (Ed.) *Studies in Natural Products Chemistry, Vol. 29*

SYNTHESIS OF BIOACTIVE DITERPENES

AJOY K. BANERJEE*, PO. S. POON NG AND MANUEL S. LAYA

Centro de Química, Instituto Venezolano de Investigaciones Científicas (IVIC),Apartado 21827, Caracas 1020-A, Venezuela

ABSTRACT: This article reviews the literature published dealing with the synthesis of some bioactive diterpenes. It describes the biological activity and synthesis of only four diterpenes: pisiferic acid, carnosic acid, triptolide and miltirone. This review excludes the discussions of Taxodione, a bioactive diterpene, because it has already been reviewed [85]. The utility of several reagents in the total synthesis of terpenoid compounds has been documented. It can be observed that several routes have been developed for the synthesis of a single diterpene.

INTRODUCTION

Among the various classes of natural products, the terpenoids have sometime occupied a special position and received chemist's attention. A large number of these compounds exhibit significant biological activities such as antifungal, antibacterial, antiviral, antitumor and insecticidal properties. The interest in this type of natural products is manifested by a large number of articles dealing with their isolation and synthesis that appear in scientific literature [1]. In recent decades an enormous effort has been focused on the development of new methods for the stereoselective synthesis of the above-mentioned diterpenes.

This review aims to present the most relevant aspects of the synthesis of the above mentioned bioactive diterpenes. It also describes the preparation of several potential intermediates, which have been utilized for the total synthesis of these bioactive diterpenes.

---

* Address correspondence to this author at Centro de Química , Instituto Venezolano de Investigaciones Científicas(IVIC), Apartado 21827, Caracas 1020-A; Venezuela; Tel 0582-12-5041324, Fax0582-12-5041350; E-mail abanerje@quimica.ivic.ve

SYNTHESIS OF BIOACTIVE DITERPENES

Pisiferic Acid

Isolation And Biological Activity

Pisiferic acid (1), an abietane type diterpene acid, was first isolated [2] from the methanol extract of leaves and twigs of *chamaecyparis pisifera Endle* (Japanese name: Sawara) whose timber has been found to be one of the best materials for rice chests. The isolation and structural studies of related diterpenoids possessing angular oxygenated methyl groups, pisiferol (4), o-methylpisiferic acid (2), methylpisiferate (3), and o-methyl-pisiferate (5) were also reported by Yatagai [3,4] Fig (1). Both pisiferic acid (1) and o-methylpisiferic acid (2) exhibit antibacterial activity against Gram-positive bacteria Staphylococcus aureus and Bacillus subtilis [2]. Kibayashi and Nishino [5] reported that pisiferic acid (1) shows antifungal activity against Pyricularis cryzae and cytotoxicity against HeLA cells. They also observed that pisiferic acid (1) exhibits antibacterial activity against Gram-negative bacteria, which was not reported by Egawa et al. [1]. Methyl pisiferate (3) is active against S. aureus, B. subtilis, P. cryzae and HeLA cellas but inactive against Gram-negative bacteria, and shows significant cytotoxicity against HeLA cells. In the mode of action of the pisiferic acid species on the bacterial macromolecule peptidoglycan synthesis is predominantly inhibited in B. subtilis while non-specific inhibition is observed in P. vulgaris [6]. Pisiferic acid inhibits predominantly DNA synthesis in HeLA cells as compared with RNA and protein synthesis. Several pisiferic acid species show inhibitory action HeLA DNA polymerase α and inhibitory activity is about 1/20 of aphidicolin [7].

The antimicrobial activity of three analogues of pisiferic acid, namely ferruginol (6), dehydroabietic acid (7) and podocarpic acid (8), which have aromatic C-ring, has also been studied. Ferruginol (6) shows significant activities against all Gram-positive bacteria tested, while dehydroabietic acid (7) and podocarpic acid (8) have less and no activity against the bacteria, respectively. These observations may lead one to the following conclusions: (a) the isopropyl group on aromatic ring is responsible for the biological activity of pisiferic acid and ferruginol, (b) an enhancement in biological activity has been observed when the isopropyl and hydroxyl group are ortho to each other, (c) the presence of a carboxyl group in ring A decreases the activity of diterpenoids, (d) the

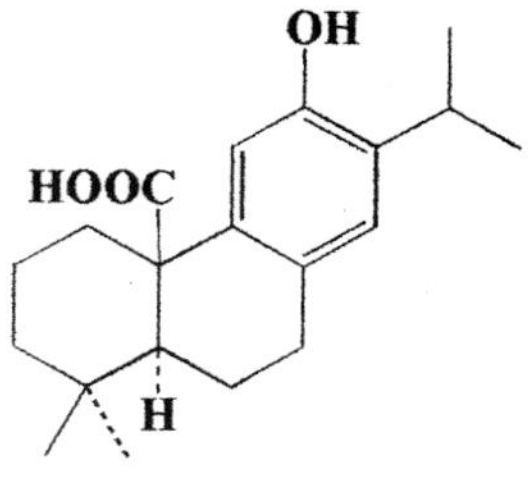

Fig (1) The diterpenoid compounds (1) to (5) were isolated by Yatagai et al [2,3], from the leaves of Chamaceyparis Pisifera endle. Ferruginol (6) is diterpene phenol, while dehydroabietic acid (7) and Podocarpic acid (8) are the resin acids. It is recommendable to consult the reference[1], to have important informations of compounds (6), (7) and (8).

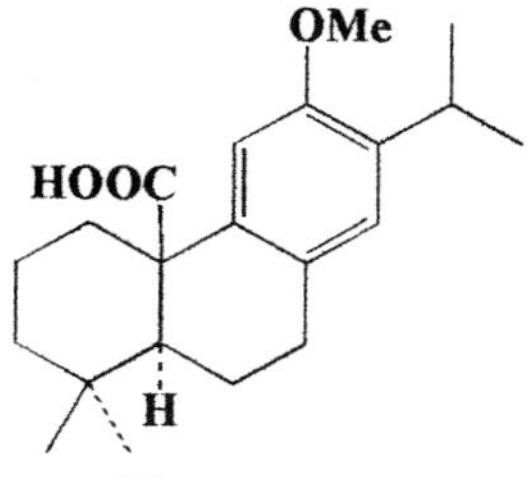

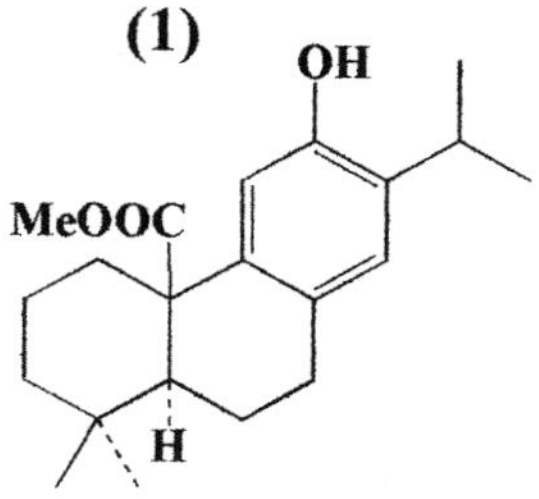

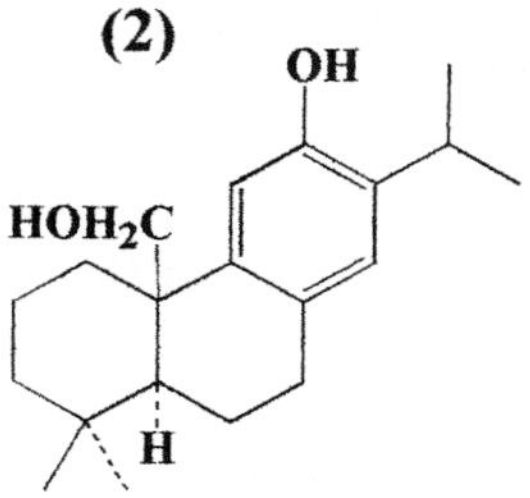

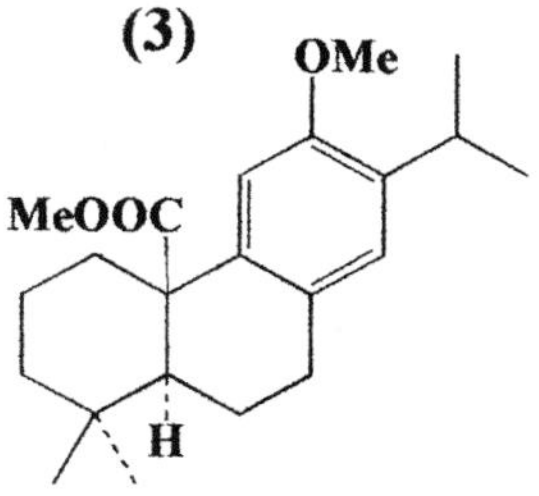

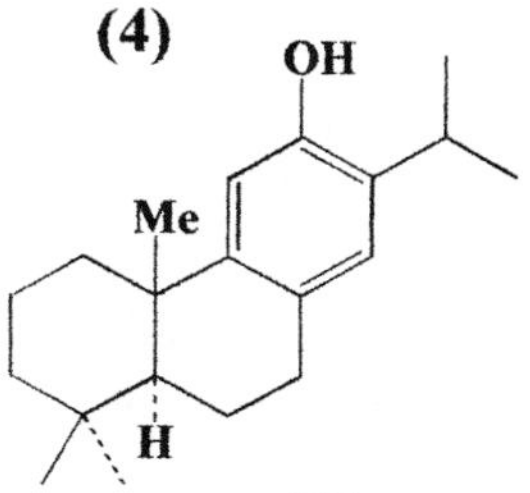

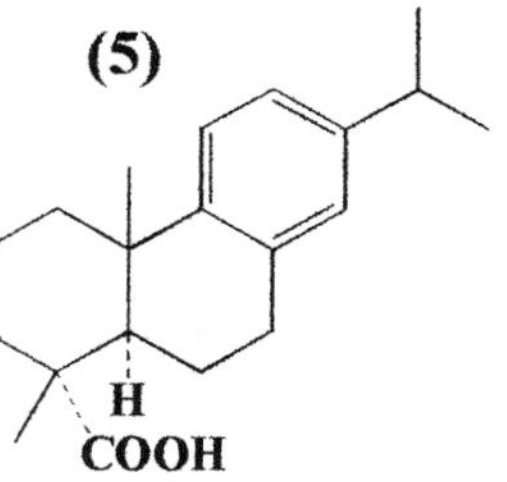

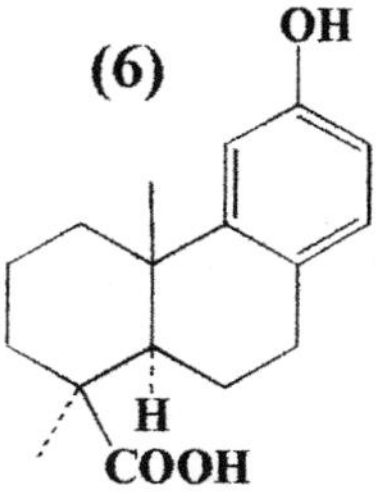

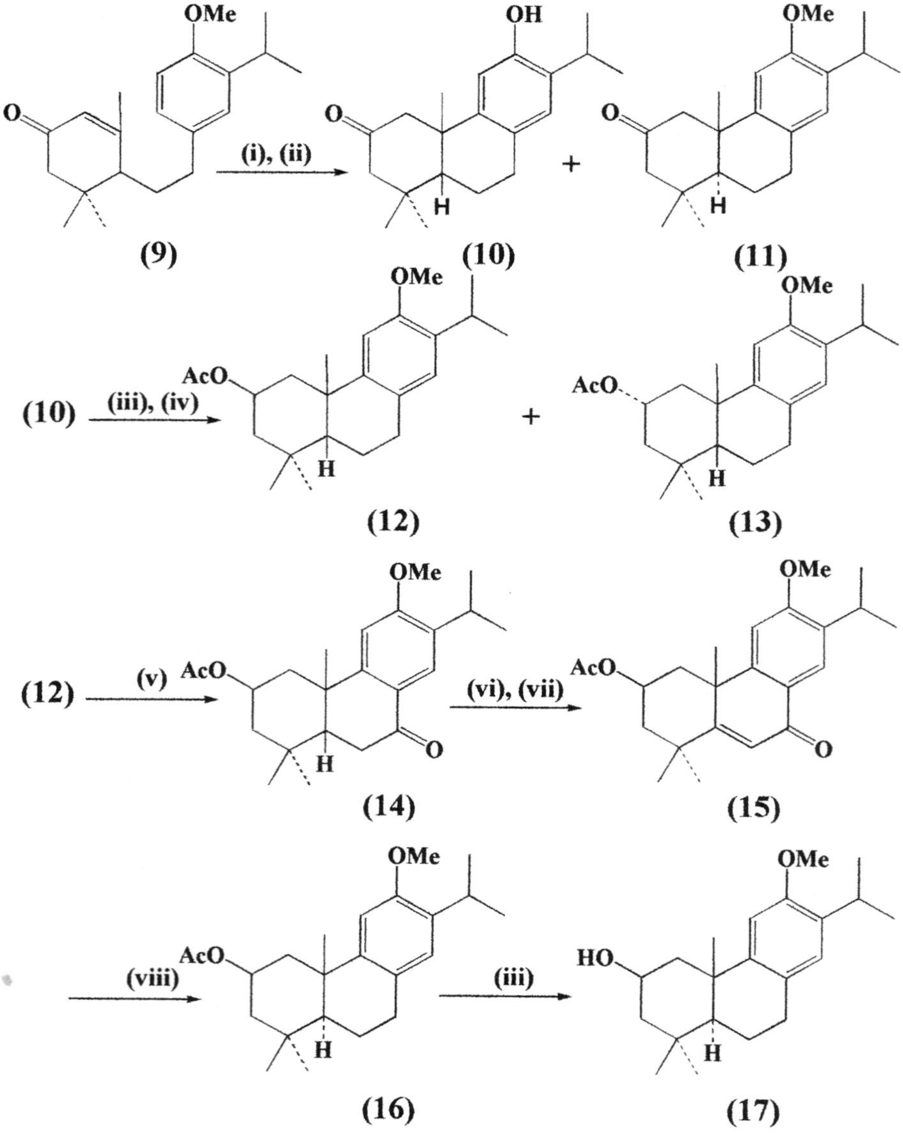

(9) → (i), (ii) → (10) + (11)

(10) → (iii), (iv) → (12) + (13)

(12) → (v) → (14) → (vi), (vii) → (15)

→ (viii) → (16) → (iii) → (17)

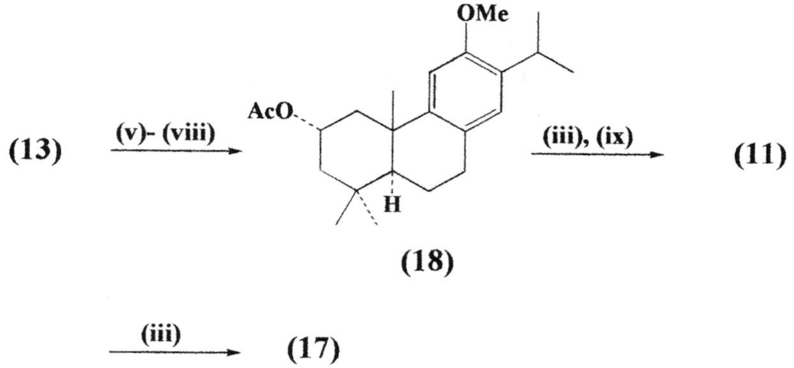

(13) $\xrightarrow{\text{(v)- (viii)}}$ **(18)** $\xrightarrow{\text{(iii), (ix)}}$ **(11)**

$\xrightarrow{\text{(iii)}}$ **(17)**

Fig. (2). The cyclization of enone (9), gives origin of two Cyclized products (10) and (11). Ketone (10), Ketone (10) is converted to the saturated ketone (14)under standard organic reactions.Bromination and dehydrobromination of ketone (14) yields the α,β-unsaturated ketone (15), which on subjection to catalytic hydrogenation affords (16) and this on reduction, produces alcohol (17). The compound (13) yields (18) by standard reactions that are used for the transformation of (12) to (16).Reduction with metal hydride followed by oxidation affords ketone (11), which is converted to alcohol (17)

Reagents: (i) $BF_3.Et_2O$, C_2H_5SH; (ii) $HgCl_2$, MeCN; (iii) $LiAlH_4$, THF; (iv) Ac_2O, Pyr; (v) CrO_3-H_2SO_4; (vi) $C_5H_6Br_3N$, CH_2Cl_2; (vii) Li_2CO_3, LiBr, DMF; (viii) Pd-C, $HClO_4$; (ix) PCC

presence of the C-10 carboxylic acid is a necessary factor for the activity against Gram-negative bacteria, and (e) the C-12 hydroxyl group may play a key role in the antifungal activity because o-methylpisiferic acid (2) and o-methylpisiferate (5) are inactive.

Synthesis of Pisiferic Acid

Matsumoto and Usui [8] reported the first total synthesis of (±)-pisiferic acid (1) and (±)-methylpisiferate (3) and this is depicted in Fig. (2). In this synthesis the fundamental skeleton of pisiferic acid has been constructed by the cyclization of enone (9). The introduction of C-10 carboxylic acid in pisiferic acid was accomplished by transannular oxidation. The depicted synthesis has been divided in two parts, Fig. (2) describes the preparation of the alcohol (17) and Fig. (3) deals with the conversion of the alcohol (17) to pisiferic acid (1).

Cyclization of enone (9) in hexane with boron trifluorideetherate in presence of 1,2-ethanedithiol, followed by hydrolysis with mercury (II) chloride in acetonitrile, yielded the cis-isomer (10) (16%) and trans-isomer (11) (28%). Reduction of (10) with lithium aluminium hydride in tetrahydrofuran followed by acetylation with acetic anhydride and pyridine gave two epimeric acetates (12) (32%) and (13) (52%) whose configuration was determined by NMR spectroscopy. Oxidation of (12) with Jones reagent afforded ketone (14) which was converted to the α,β-unsaturated ketone (15) by bromination with pyridinium tribromide in dichloromethane followed by dehydrobromination with lithium carbonate and lithium bromide in dimethylformamide. Ketone (15), on catalytic hydrogenation with Pd-C in the presence of perchloric acid, produced compound (16) (72%) and (14) (17%). The compound (16) was converted to alcohol (17) by reduction with lithium aluminium hydride.

Following the procedure for the transformation of acetate (12) to (16), the acetate (13) was converted to trans-acetate (18). Reduction of (18) with metal hydride followed by oxidation of the resulting alcohol with pyridinum chlorochromate produced ketone (11), which afforded alcohol (17) by reduction with lithium aluminium hydride.

Next step of this synthesis consisted in the conversion of alcohol (17) to pisiferic acid (1) and this has been described in Fig. (3). The alcohol (17) in hexane was treated with Pb(OAc)$_4$ in presence of iodine at room temperature to obtain the epoxy triene (19) (51%) whose structure was confirmed by spectroscopy. Treatment of (19) with acetyl p-toluene-sulfonic in dichloromethane yielded an olefinic acetate (20) and this was hydrogenated to obtain (21). The compound (22) could be isolated from (21) on subjection to reduction, oxidation and esterification respectively. The conversion of (22) to (23) was accomplished in three steps (reduction with sodium borohydride, immediate dehydration in dichloromethane and catalytic hydrogenation). Demethylation of (23) with anhydrous aluminium bromide and ethanethiol at room temperature produced pisiferic acid (1). Similar treatment of (23) with aluminium chloride and ethanethiol in dichloromethane yielded methylpisiferate (3).

The most interesting aspects of the present synthesis are (i) preparation of ketones (10) and (11), (ii) high yield in most of the steps, (iii) use of transannular oxidation for the introduction of C-10 carboxylic acid.

An alternative synthesis of (±)-pisiferic acid (1) was developed by Banerjee and collaborators [9] and this is depicted in Fig. (4).

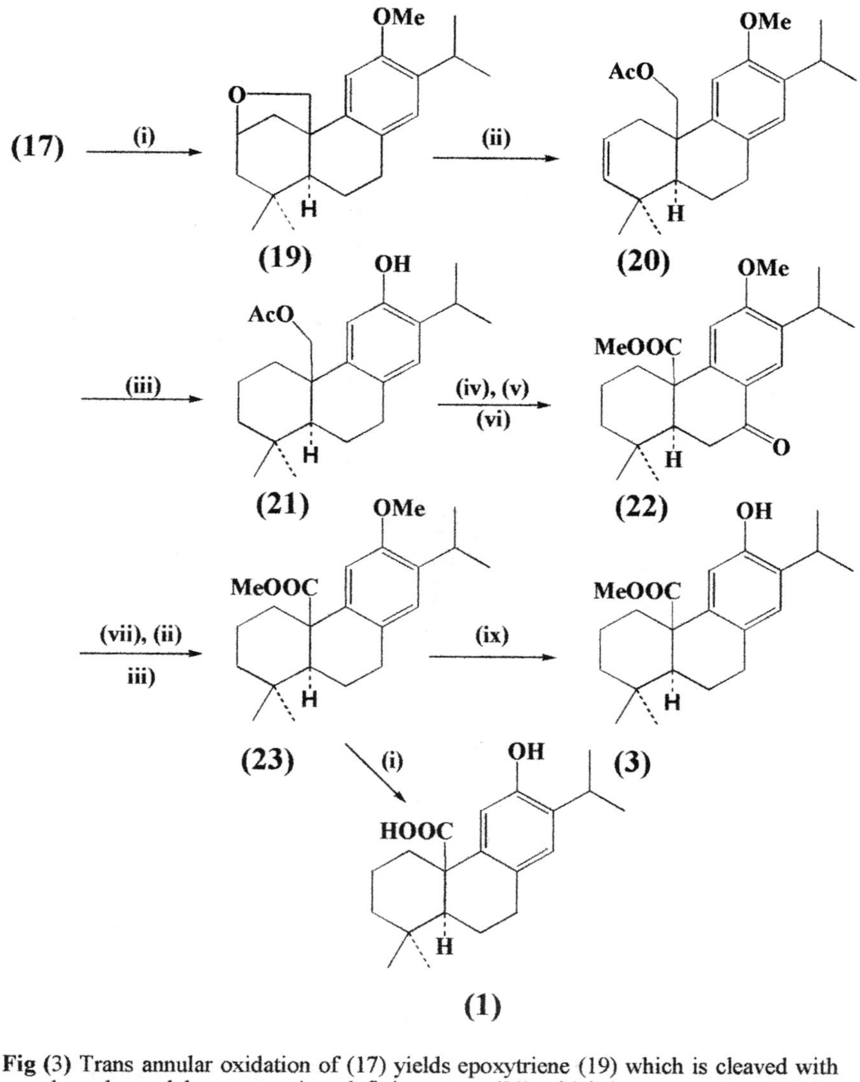

Fig (3) Trans annular oxidation of (17) yields epoxytriene (19) which is cleaved with acetyl p-toluensulphonate to give olefinic acetate (20) which is converted to (22) by hydrogenation, reduction, oxidation respectively. Its conversion to (23) is carried out by standard organic reactions described in Fig.(2). Demethylation of (23) with aluminium chloride, leads the formation of methyl pisiferate (3).

Reagents: (i) Pb(OAc)$_4$, I$_2$, hexane; (ii) acetyl p-toluenesulfonate; (iii) H$_2$, Pd-C; (iv) LiAlH$_4$, THF; (v) CrO$_3$-H$_2$SO$_4$; (vi) CH$_2$N$_2$; (vii) NaBH$_4$·EtOH; (viii) AlBr$_3$, C$_2$H$_5$SH; (ix) AlCl$_3$, C$_2$H$_5$SH.

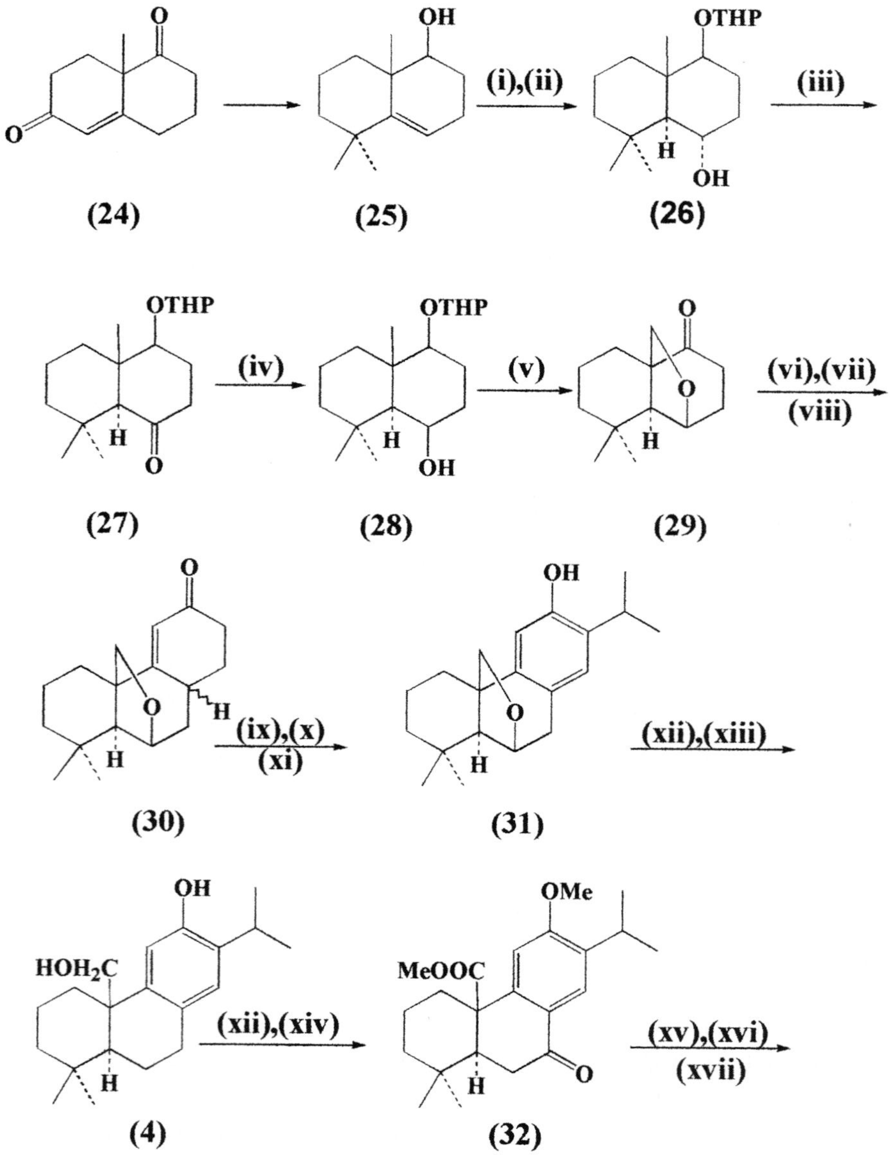

(24) → (25) (i),(ii) → (26) (iii) →

(27) (iv) → (28) (v) → (29) (vi),(vii) (viii) →

(30) (ix),(x) (xi) → (31) (xii),(xiii) →

(4) (xii),(xiv) → (32) (xv),(xvi) (xvii) →

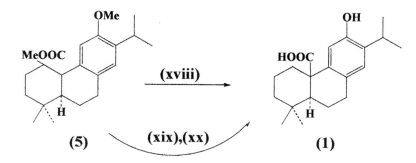

(5)   (xviii)   (xix),(xx)   (1)

Fig (4) The transformation of the ketone (24) to the cyclic ether (9) applying the standard organic reactions is described It wa subjected to three sequencial reactions with reagents mentioned for the conversion to cyclic ether (30). Isopropylation and by aromatization, it produces the phenol (31), which is converted to pisiferol (4). This on subjection to oxidation, esterification and deoxygenation respectively, furnish O-methyl pisiferate (5) and this is easily converted to pisiferic acid (1).

Reagents: (i) DHP, p-TsOH, (ii) $BF3.Et_2O$, $NaBH_4$,H_2O_2 (30%), KOH; (iii) CrO_3, Py; (iv) $LiALH_4$, THF; (v) hn, 250W, $Pb(OAc)_4$, $I2,C_6H_{12}$, $CrO_3\cdot H_2SO_4$; (vi) NaH, HCO_2Et; (vii) CH_2=CH.COMe; (viii) NaOMe, MeOH; (ix) $CO(COOEt)_2$, NaH, DME; (x) MeLi, Et_2O; (xi)HCl, MeOH; (xii) Me_2SO_4, NaOH; (xiii) Zn, ZnI_2, MeCOOH; (xiv) CH_2N_2, Et_2O; (xv) $NaBH_4$-MeOH; (xvi) TsCl,Py; (xvii) NaI/Zn dust, (xviii) $AlBr_3$, C_2H_5SH; (xix) Quinoline, MeCOOH; (xx) BBr_3, CH_2Cl_2

The starting material for the present synthesis was Wieland-Miescher ketone (24), which was converted to the known alcohol (25) by the published procedure [10]. Tetrahydropyranylation of alcohol (25) followed by hydroboration-oxidation afforded the alcohol (26), which on oxidation produced ketone (27). Reduction of (27) with metal hydride gave the alcohol (28) (56%). This in cyclohexane solution on irradiation with lead tetraacetate and iodine produced the cyclic ether that was oxidized to obtain the keto-ether (29). Subjection of the keto-ether (29) to three sequential reactions (formylation, Michael addition with methyl vinyl ketone and intramolecular aldol condensation) provided tricyclic ether (30) whose NMR spectrum showed it to be a mixture of C-10 epimers. The completion of the synthesis of pisiferic acid (1) did not require the separation of epimers and thus the tricyclic ether (30) was used for the next step. The conversion of (30) to tricyclic phenol (31) was accomplished in three steps (ethoxycarbonylation, Grignard reaction with methyl-lithium, and acid-catalyzed dehydration).

The phenol (31) was methylated and treated with zinc, zinc iodide and acetic acid to yield pisiferol (4). Its methyl derivative on oxidation with Jones reagent at room temperature followed by esterification furnished the keto ester (32). Its conversion to o-methylpisiferate (5) was accomplished by subjection to three sequential reactions (reduction with metal hydride, tosylation and elimination). Its identity was confirmed by comparing its spectral data and melting point with an authentic specimen [8]. The transformation of o-methylpisiferate (5) to pisiferic acid (1) was achieved by treatment with aluminium bromide and ethanethiol. This transformation was also achieved by heating methylpisiferate (5) with quinoline and acetic acid followed by treatment of the resulting product with boron tribromide in dichloromethane. The yield of pisiferic acid (1) was not very high probably due to the decarboxylation that occurred during heating with quinoline and acetic acid.

Another approach [9] was also made to accomplish the synthesis of pisiferic acid and is depicted in Fig. (5).

The starting material for the present synthesis is the keto-alcohol (34), which was prepared from the methyl analog of Wieland-Miescher ketone (33)[11,12]. Tetrapyranylation of alcohol (34) followed by reduction with sodium borohydride yielded a mixture of alcohols whose p-tosylsulphonyl derivative on heating with lithium bromide, lithium carbonate and dimethylformamide afforded the oily olefin (35). These conditions not only caused the dehydrosulphonation but also hydrolysis of the tetrahydropyranyl group thus shortening the reaction sequence by one step. The oily olefin (35) was oxidized to ketone (36), which was formylated and subjected to Robinson annelation with methyl vinyl ketone prepared in situ following the procedure of Howell and Taylor [13]. The resulting adduct without purification was heated by boiling with sodium methoxide in methanol to obtain the tricyclic ketone (37). The yield was improved when the adduct was heated with sodium hydride in dimethoxyethane. This was converted to methoxyabietatriene (38) in four steps (ethoxycarbonylation, Grignard reaction withy methyl-lithium, acid-catalized dehydration and methoxylation). Hydroboration-oxidation of (38) followed by oxidation of the resulting alcohol furnished the ketone (39). It was reduced with lithium aluminium hydride in tetrahydrofuran to yield the alcohol (40) whose structural assignment was derived from ^{1}H NMR spectroscopic data. Irradiation of a mixture of alcohol (40) and lead

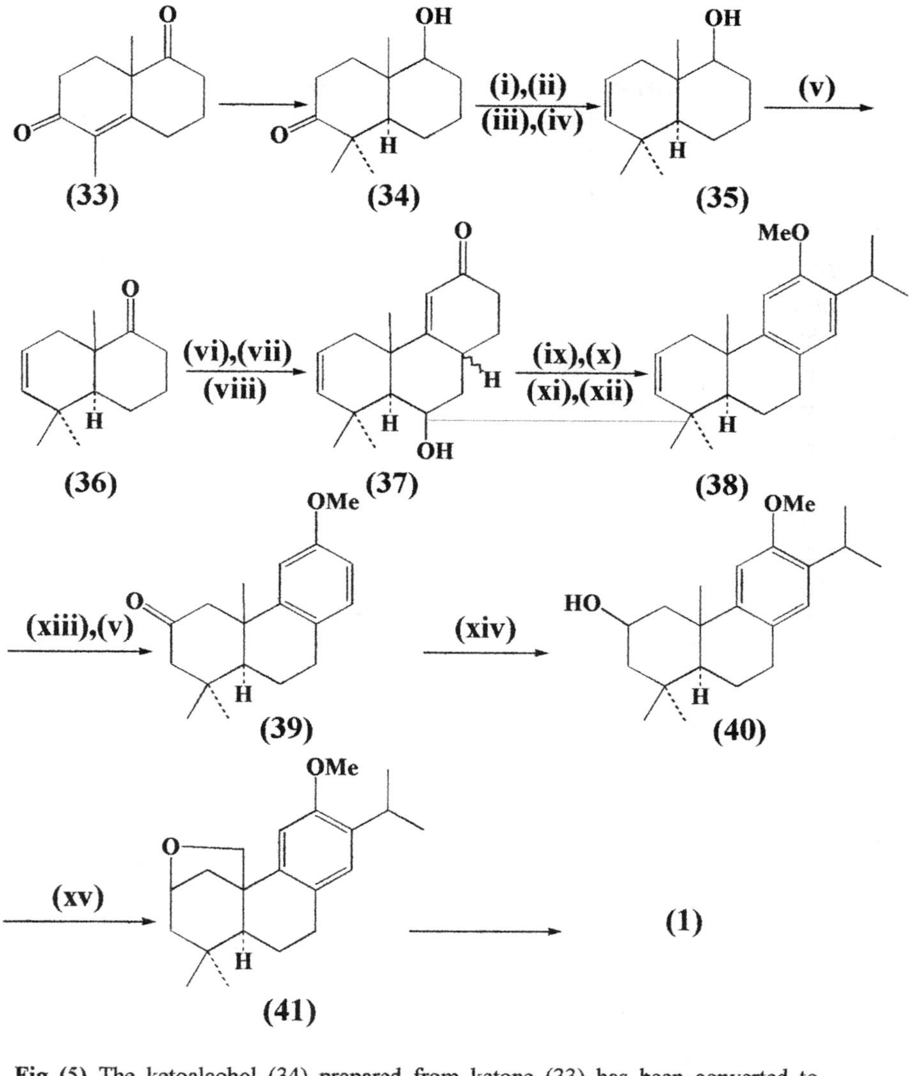

Fig (5) The ketoalcohol (34) prepared from ketone (33) has been converted to methoxyabietatriene(38), by standard organic reactions. Hydroboration, oxidation of (38), followed by oxidation, yielded the ketone (39), which is converted to the alcohol (40), by metal hydride reduction. On subjection to transanular oxidation with lead tetracetate and benzene alcohol (40) furnishes (41) whose conversion to pisiferic acid has already been described.

Reagents: (i), DHP, TsOH; (ii) NaBH$_4$, MeOH; (iii) TsCl, Py; (iv) LiBr, Li$_2$CO$_3$, DMF; (v) CrO$_3$·H$_2$SO$_4$; (vi) NaH, HCO$_2$Et; (vii) CH$_2$=CHCOMe; (viii) NaH, DME; (ix) CO(COOEt)$_2$; (x) MeLi, Et$_2$O; (xi)HCl, MeOH; (xii) Me$_2$SO$_4$, K$_2$CO$_3$, Me$_2$CO; (xiii) BH$_3$·THF,H$_2$O$_2$, KOH; (xiv) LiAlH$_4$; (xv) Pb(OAc)$_4$, C$_6$H$_{12}$

tetraacetate in benzene with a 250 W tungsten lamp yielded methoxyabietatriene (41) in 50% yield whose identity was established by direct comparison with an authentic specimen (i.r., t.l.c., mixed m.p.). When the transannular oxidation was carried out in cyclohexane solution with lead tetraacetate and iodine the yield of the abietatriene (41) was very poor and a large amount of iodocyclohexane and other unidentified products were obtained. As the abietatriene (41) has been previously converted [8] into pisiferic acid, the detailed description of the transformation of abietatriene (41) to pisiferic acid (1) has been omitted.

In the present synthesis, the preparation of two cyclic ethers (31) and (41) and their transformations to pisiferic acid (1) are described. In this synthesis, the introduction of carboxylic acid at angular position has been accomplished by transannular oxidation.

Pal and Mukerjee [14] developed another total formal synthesis of racemic pisiferic acid (1) and this is described in Fig. (6).

Tetralone (42) on subjection to Reformatsky reaction with methyl-4-bromocrotonate followed by catalytic hydrogenation of the resulting product produced the ester (43) which was converted to keto-ester (44) by oxidation with chromic acid. This on bromination and dehydrobromination followed by methylation respectively furnished the ester (45). It was converted to tricyclic ketone (46) by treatment with methylmagnesium iodide followed by intramolecular cyclization of the resulting carbonol and then oxidation with pyridinium chlorochromate. The hydroxymethylene derivative of (46) was treated with alkaline hydrogen peroxide and then esterified to provide diester (47). Its transformation to diazomethyl ketone (48) was effected in three steps (partial hydrolysis, reaction with ethylchloroformate and then treatment of the resulting anhydride with diazomethane). Intramolecular cyclization of (48) effected with trifluoroacetate acid in dichloromethane produced the enedione (49). Its conversion to methyl ($\pm$)-o-methylpisiferate (5) was effected by catalytic hydrogenation, thioacetalisation and desulfurisation with Raney nickel respectively. As methyl ($\pm$)-o-methylpisiferate has already been converted to ($\pm$)-pisiferic acid (1), an alternative method for the synthesis of (5) constitutes a total formal synthesis of ($\pm$)-pisiferic acid.

The synthesis of methyl ($\pm$)-o-methylpisiferate (5) developed by Pal and Mukerjee possesses some interesting features: (i) the smooth conversion of tetralone (42) to diazomethyl ketone (48), (ii) aryl

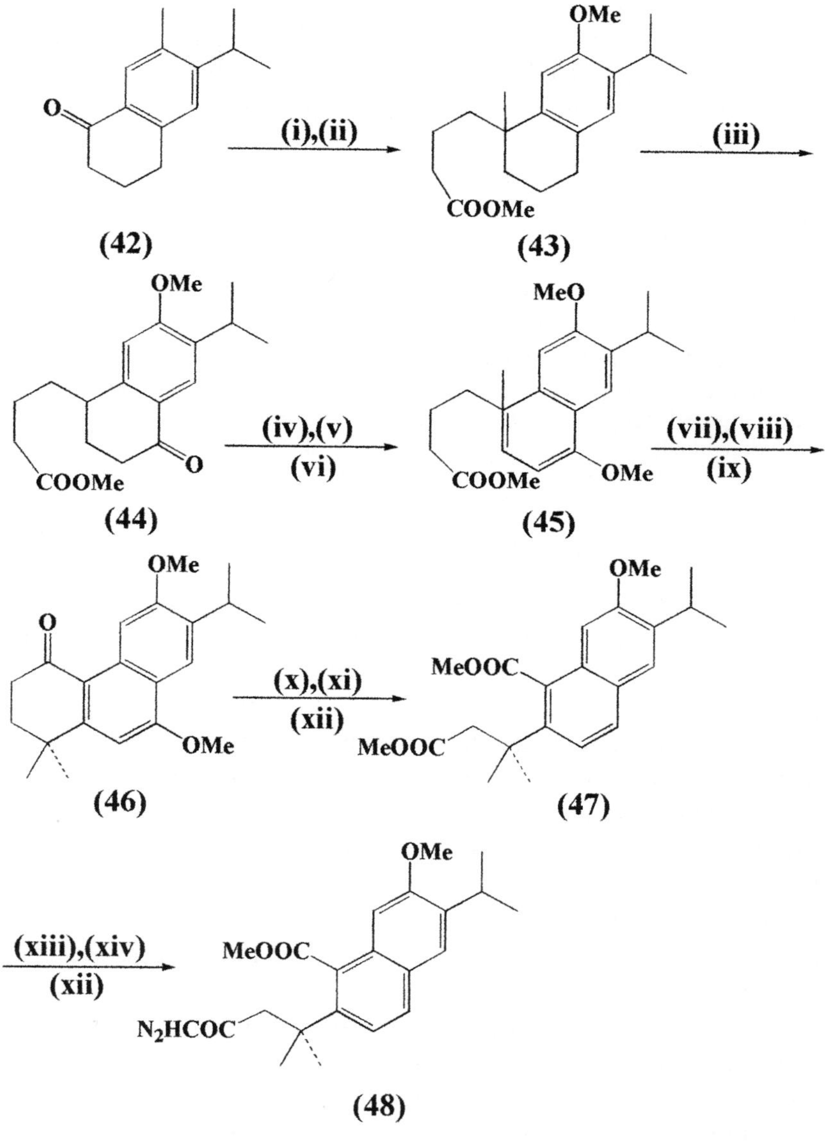

(42)   (i),(ii) → (43)   (iii) →

(44)   (iv),(v) (vi) → (45)   (vii),(viii) (ix) →

(46)   (x),(xi) (xii) → (47)

(xiii),(xiv) (xii) → (48)

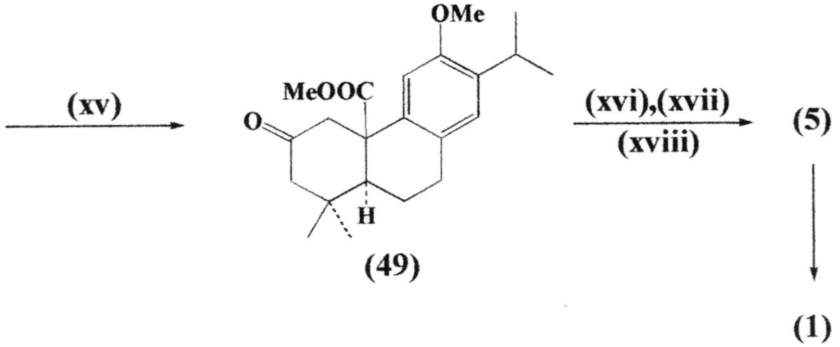

Fig (6)The transformation of the tetralone (42) to the ester (45) is described. Its tranformation to the tricyclic ketone (46) involves aromatization, subjection to Grignard reaction and intramolecular cyclisation and oxidation. Its conversion to diazomethylketone (48) is carried out in three steps and this on acid catalysed cyclisåtion yields the enedione (49), which is converted to methy (±)-methylpisiferate (5) by standard organic reactions.

Reagents: (i) BrCH$_2$CH=CO$_2$Me, Zn; (ii) H$_2$, $_{10}$% Pd-C; (iii) CrO$_3$, AcOH; (iv) Br$_2$,Et$_2$O; (v) LIBr, Li$_2$CO$_3$, DMF; (vi) MeI, K$_2$CO$_3$, Me$_2$CO; (vii) MeMgI, Et$_2$O; (viii) P$_2$O$_5$, H$_3$PO$_4$; (ix) PCC, CH$_2$Cl$_2$; (x) NaH,HCO$_2$Et; (xi) H$_2$O$_2$; (xii) CH$_2$N$_2$,Et$_2$O; (xiii) KOH,THF; (xiv) ClCOOEt, Et$_3$N; (xv) TFA, CH$_2$Cl$_2$; (xvi) H$_2$ $_{10}$% Pd-C; (xvii) (CH$_2$SH)$_2$, BF$_3$.Et$_2$O; (xviii) Raney-Nickel

participation in intramolecular cyclisation of (48) to enedione (49), (iii) selective transformation of (49) into methyl-o-methylpisiferate (5).

The synthesis is lengthy. We believe that the transannular reaction for the introduction of carboxylic acid at annular position is more convenient than the method adopted by Pal and Mukerjee [14].

Mori and Mori [15] have developed a total synthesis of both enantiomers of o-methylpisiferic acid, and this is reported in Fig. (7) and Fig. (8).

Reduction of 2,2-dimethyl-cyclohexane-1,3-dione (50) with Baker's yeast gave alcohol (ee 98.3%) whose tetrahydropyranyl derivative on methoxycarbonylation produced (51) quantitatively. Michael addition of (51) with methyl vinyl ketone followed by heating the adduct under reflux with pyrrolidine in benzene yielded (52) in 85% yield as stereoisomeric mixture whose separation presented problems. In order to eliminate the complexity due to a chiral center in tetrahydropyranyl protective group, deprotection of (52) was achieved by treatment with p-toluenesulphonic acid in methanol. The product obtained was a mixture of the lactone (53) and hydroxy ester (54). Probably the stereoisomer of

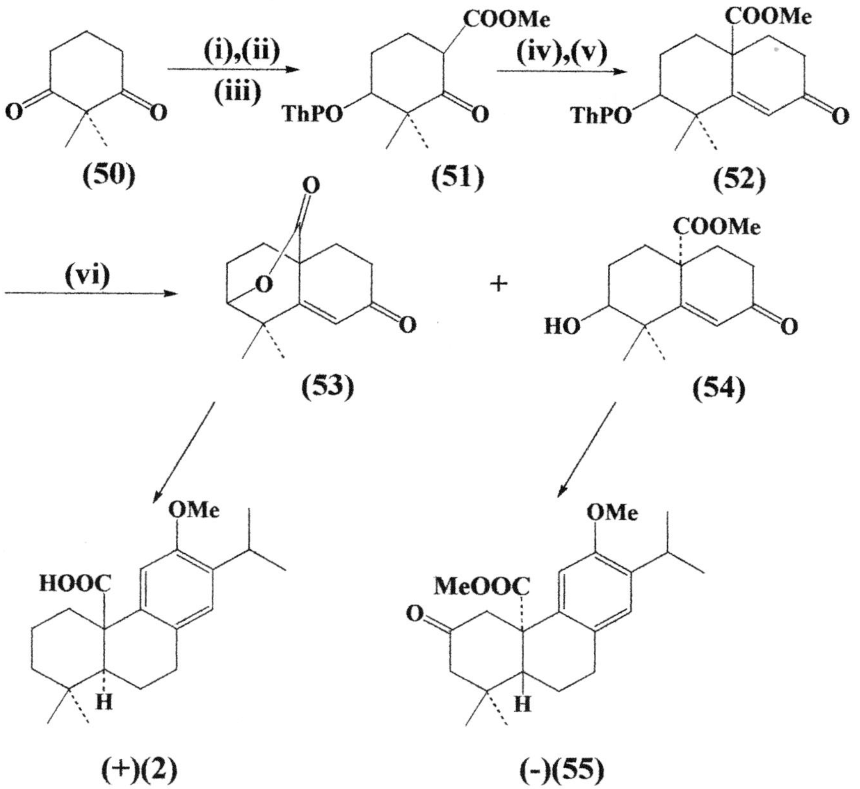

Fig (7) The transformation of cyclohexene-1,3-dione (50) to ketone (52) is described. This involves reduction, terahydropyranilation, methoxy-carbonylation and Michael addition, with methyl vinyl ketone. Treatment of (52) with p-TsOH in methanol, leads the formation of lactone (53) and hydroxy ester (54). The lactone (53) is regarded as an appropiate intermediate for natural o-methyl pisiferic acid (+)(2)and hydroxi ester (54) is intermediate for (-)(55).

Reagents: (i) Baker's yeast; (ii) DHP, p-TsOH; (iii) CO(COOMe)$_2$, NaH; (iv) CH$_2$CHCOMe; (v) p-TsOH, C$_6$H$_6$; (vi) p-TsOH, MeOH

(52) with a β-oriented CO$_2$Me group was easily converted into the lactone (53) by intramolecular acid-catalyzed reaction with β-OH obtained by deprotection of (52). The stereochemistries of (53) and (54) were evident from the S-absolute configuration of the OH group of the alcohol obtained by the reduction of (50). The lactone (53) can therefore be regarded as appropriate intermediate to the natural methylpisiferic acid (+) (2) and while the hydroxy ester (54) is a potential intermediate for the pisiferic acid (-) (55) Fig. (7). In Fig. (8) the transformation of lactone (53) to natural pisiferic acid (+) (2) is described.

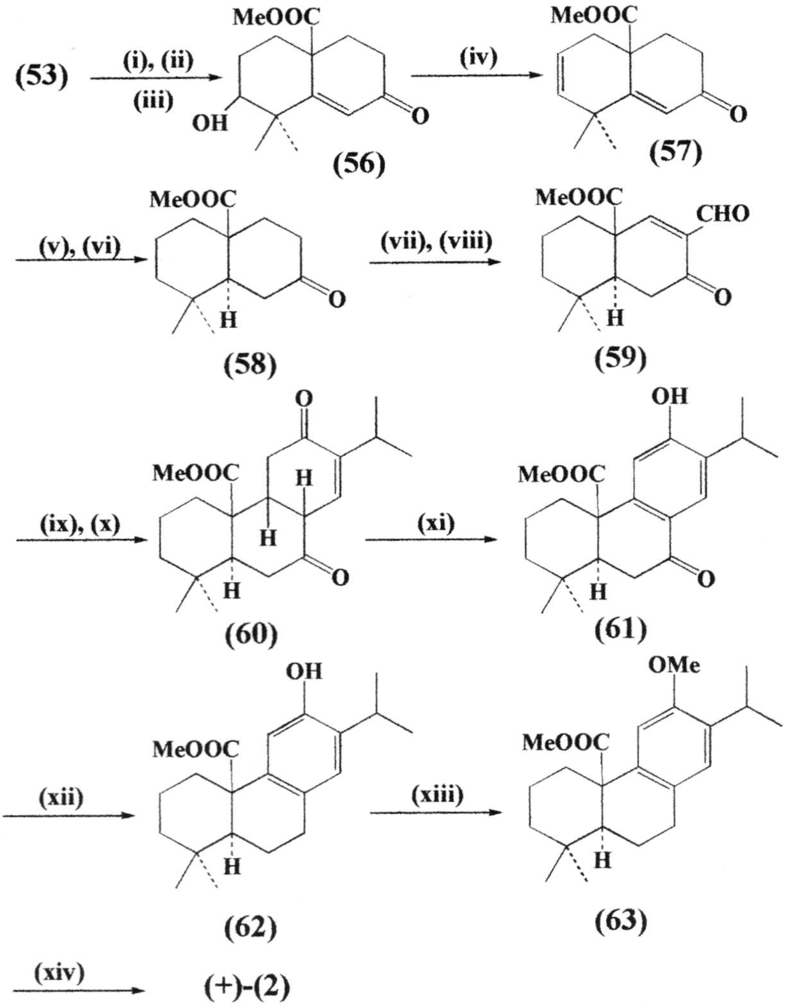

Fig (8) The transformation of lactone (53) to keto ester (58) is described. The unsaturated aldehyde (59) is converted to tricyclic ketone (60) by two steps (Michael addition, and intramolecular aldol condensation). And this on subjection to aromatization and hydrogenation respectively leads the formation of (62) whose transformation to (+)O-methyl pisiferic acid (2) is accomplished by methylation and hydrolisis.

Reagents: (i) KOH, MeOH; (ii) AcOH; (iii) CH_2N_2; (iv) TfCl, DMAP; (v) PtO_2, EtOAc; (vi) CrO_3, H_2SO_4; (vii) HCO_2Me, NaH, C_6H_6; (viii) DDQ; (ix) Na enolate, $Me_2CHCH_2COCH_2COOC_4H_9$; (x) TsOH, AcOH; (xi) $C_6H_5NHBr_3$, AcOH; (xii) H_2, Pd-C, EtOAc-H_2SO_4; (xiii) Me_2SO_4, K_2CO_3; (xiv) t-BuOK, DMSO.

Alkaline hydrolysis of (53), followed by acidification and esterification produced the ester (56), whose optical purity was estimated to be 100 %. This on treatment with 2.5 eq TfCl and 5 eq of 4,-N,N-dimethylaminopyridine (DMAP) in dichloromethane yielded (57) (89%). Hydrogenation of (57) with PtO_2 not only reduced the double bond but also the carbonyl group producing the hydroxyester which on oxidation with Jones CrO_5 reagent afforded the ketone (58). When the hydrogenation of (57) was carried out with Pd-C, the major product was not the expected (58) but a hydrogenolysis product without carbonyl group. Formylation of (58) followed by dehydrogenation with 2,3-dichloro-5,6-dicyano benzoquinone (DDQ) led to the formation of the unsaturated aldehyde (59). Addition of sodium enolate of t-butyl isovalerylacetate to (59) afforded an adduct which on treatment with p-toluenesulphonic acid in refluxing acetic acid underwent cyclization yielding the tricyclic ketone (60) (84%). The trans-syn-cis-stereochemistry of (60) was assigned on the basis of Meyer's work on analogous work of tricyclic compounds [16]. Ketone (60) on treatment with pyridinium tribromide in acetic acid underwent aromatization affording (61) in quantitative yield which suffered hydrogenolysis on hydrogenation with palladium and carbon in ethylacetate and sulphuric acid. The resulting product (62) on methylation with dimethyl sulfate and potassium carbonate yielded (63). Its transformation to (+)-o-methyl pisiferic acid (2) was accomplished by heating with potassium t-butoxide in dimethylsulfoxide. Following the similar procedure the hydroxy ester (54) was converted to (-)-o-methyl pisiferic acid (55).

Geiwiz and Hasslinger [17] have developed an attractive synthesis of (+)-pisiferic acid (1) from dehydroabietic acid (64a). This is depicted in Fig. (9).

Metal hydride reduction of methyl dehydroabietate (64b) afforded alcohol (65) whose tosyl derivative on heating with sodium iodide and zinc in dimethylformamide yielded (66). The use of hexamethylphosphoramide gave an inferior yield of (66). Regio and stereoselective acetoxylation with $Pb(OAc)_4$ in acetic acid at 100°C gave only 30% yield of (67) but the same experiment realized with $Pb(OAc)_4$ using a Hg medium-pressure lamp at room temperature yielded (67) in 74%. The ^1H NMR spectrum of (67) showed that the OAc group was introduced in 7α-position. This on subjection to acid-catalyzed β-elimination (EtOH/10% HCl) produced (68) in quantitative yield.

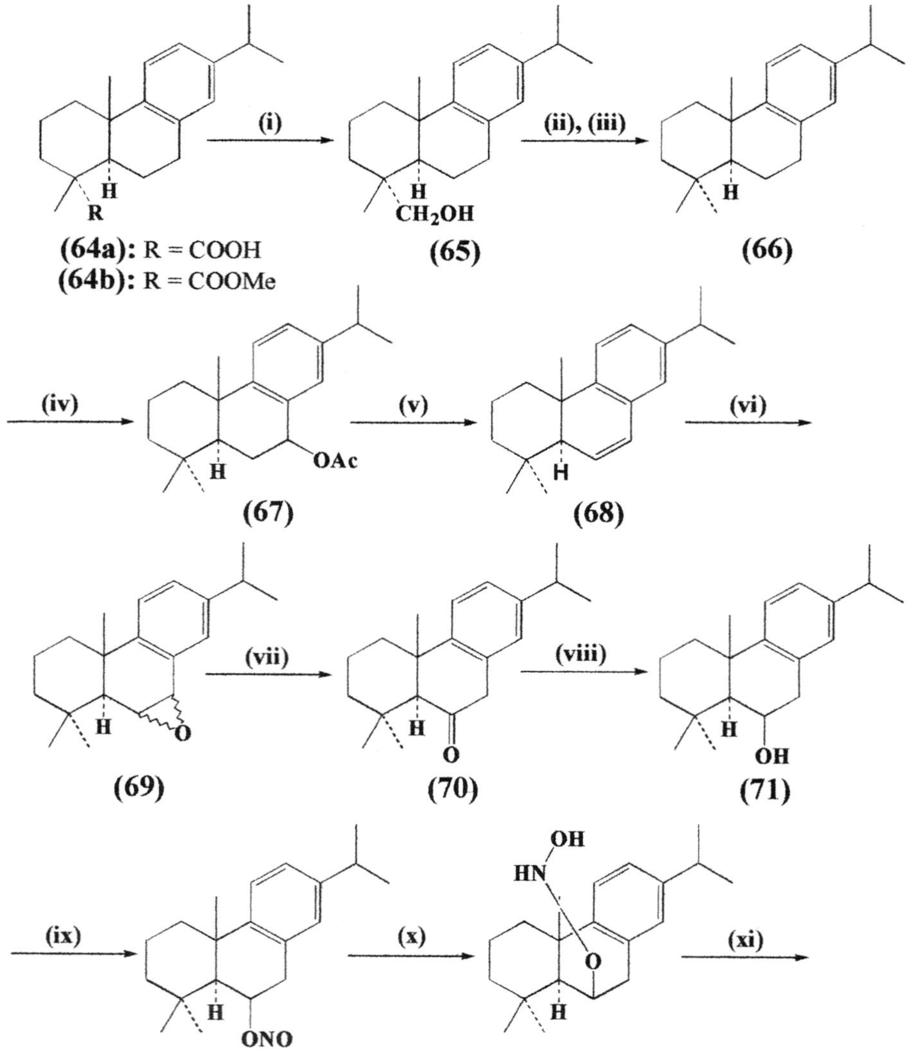
(i)
(ii), (iii)
(iv)
(v)
(vi)
(vii)
(viii)
(ix)
(x)
(xi)
H
R
CH2OH
OAc
O
OH
ONO
HN
(64a): R = COOH
(64b): R = COOMe
(65)
(66)
(67)
(68)
(69)
(70)
(71)
(72)
(73)

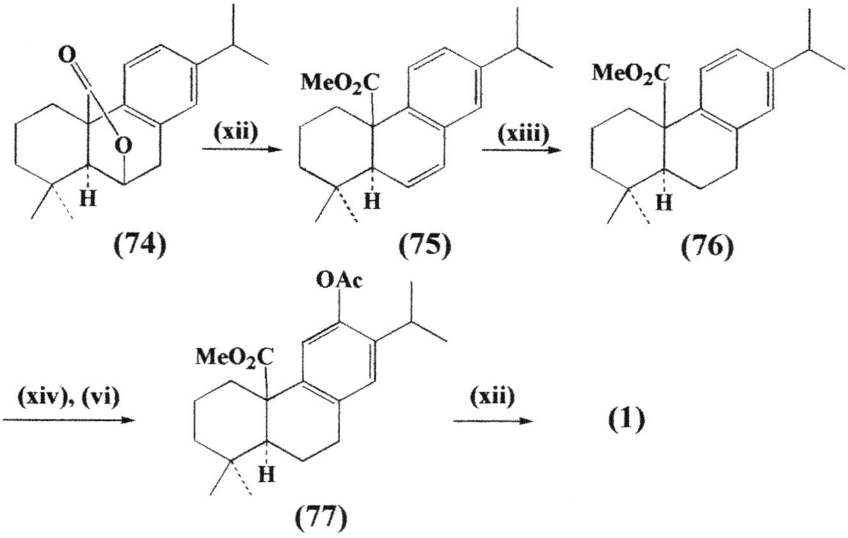

Fig (9) Methyl dihydroabietate (64b) has been converted to compound (68). The standard organic reactions have been utilized for its conversion to nitrite (72) which on irradiation followed by oxidation yields lactone (74). Cleavage of lactone ring followed by hydrogenation and esterification produced (76) which was converted to (77) by Friedel-Crafts acylation and Bayer-Villiger oxidation. This yields psiferic acid (1) by hydrolysis.

Reagents: (i) LiAlH$_4$, THF; (ii) TsCl, Py; (iii) NaI, Zn, DMF; (iv) Pb(OAc)$_4$; (v) EtOH, 10%HCl; (vi) MCPBA, CH$_2$Cl$_2$; (vii) TsOH, C$_6$H$_6$; (viii) LiAH$_4$, THF; (ix) NOCl, Py; (x) hv, Hg lamp; (xi) PDC; (xii) lithium methanethiolate; (xiii) H$_2$, Pd-C; (xiv) AcCl, AlCl$_3$.

Epoxidation of (68) with m-chloroperbenzoic acid in dichloromethane gave a mixture of epoxides (69) which on treatment with toluene-4-sulfonic acid in benzene produced the ketone (70). The alcohol (71), obtained by metal hydride reduction, was treated with nitrosyl chloride in pyridine to yield the nitrite (72). Irradiation of the nitrite (72) in benzene under argon and at room temperature with a high-pressure Hg lamp using a solidex glass filter. The resulting product, probably a dimer [18,19], after treatment with boiling i-PrOH under argon provided the compound (73) (65%) and this on oxidation with pyridinium dichromate (PDC) produced lactone (74) which on cleavage with lithium methanethiolate followed by esterification produced (75). Compound (76), obtained by catalytic hydrogenation of (75), on subjection to Friedel-Craft acylation with AcCl and AlCl$_3$ and then Bayer-Villiger oxidation with m-chloroperbenzoic acid afforded (+)-12-o-acetyl pisiferate (77). This was hydrolysed with lithium methanethiolate to obtain (+)-pisiferic acid (1).

The synthesis of (+)-pisiferic acid (1) was also achieved from abietic acid (78) [17] Fig. (**10**). Its methyl derivative (79) forms iron carbonyl complex (80) with Fe(CO)$_5$ which on metal hydride reduction, tosylation and reduction afforded (81). Its transformation to allylic alcohol (82) was carried out by treatment of iodine and potassium bicarbonate, which was converted to α,β-unsaturated ketone (83) with acetic anhydride in dimethylsulfoxide, and this underwent aromatization on treatment with bromine in acetic acid followed by quenching with potassium bicarbonate yielding the compound (84). It was methylated and the resulting product was converted to the nitrite (85) following the similar procedure utilized for the conversion of (68-73). Photolysis of nitrite (85) in benzene following the procedure mentioned above, did not form a dimer as before. The resulting product after treatment with I-PrOH afforded the hydroxyl amine (86) which was oxidized to lactone (87) (77%). The cleavage of the lactone ring with lithium methanethiolate as described before was not possible. The lactone ring of (87) survived even at high temperatures and the methoxy group was cleaved. This difficulty was overcome by the use of aluminium tribromide and tetrahydrothiophene. Lactone (87) on stirring with this reagent for 24 hr at room temperature yielded (+)-pisiferic acid (1) (82%) and the olefin (88) (10%).

The synthesis of (+)-pisiferic acid (1) presents many important aspects: (i) functionalization of the angular methyl group to lactone, (ii) cleavage of lactone ring to carboxylic acid, and (iii) conversion of (78) to allylic alcohol (81) via iron carbonyl complex

Carnosic Acid

Isolation And Biological Activity

The abietane diterpene carnosic acid (89), a derivative of ferruginol, is found in the popular Labiatae herb, sage and rosemary and is considered a precursor of other diterpenoid constituents in the herb [20,21]. Wenkert et al [22] have established the structure of carnosic acid (89). Carnosic acid (89) and related diterpenes such as carnasol (90) and rosemanol (91) Fig. (**11**) possess powerful antioxidant activities [23] but carnosic acid (89) is the most powerful potency among these diterpenes. Carnosic acid (89)

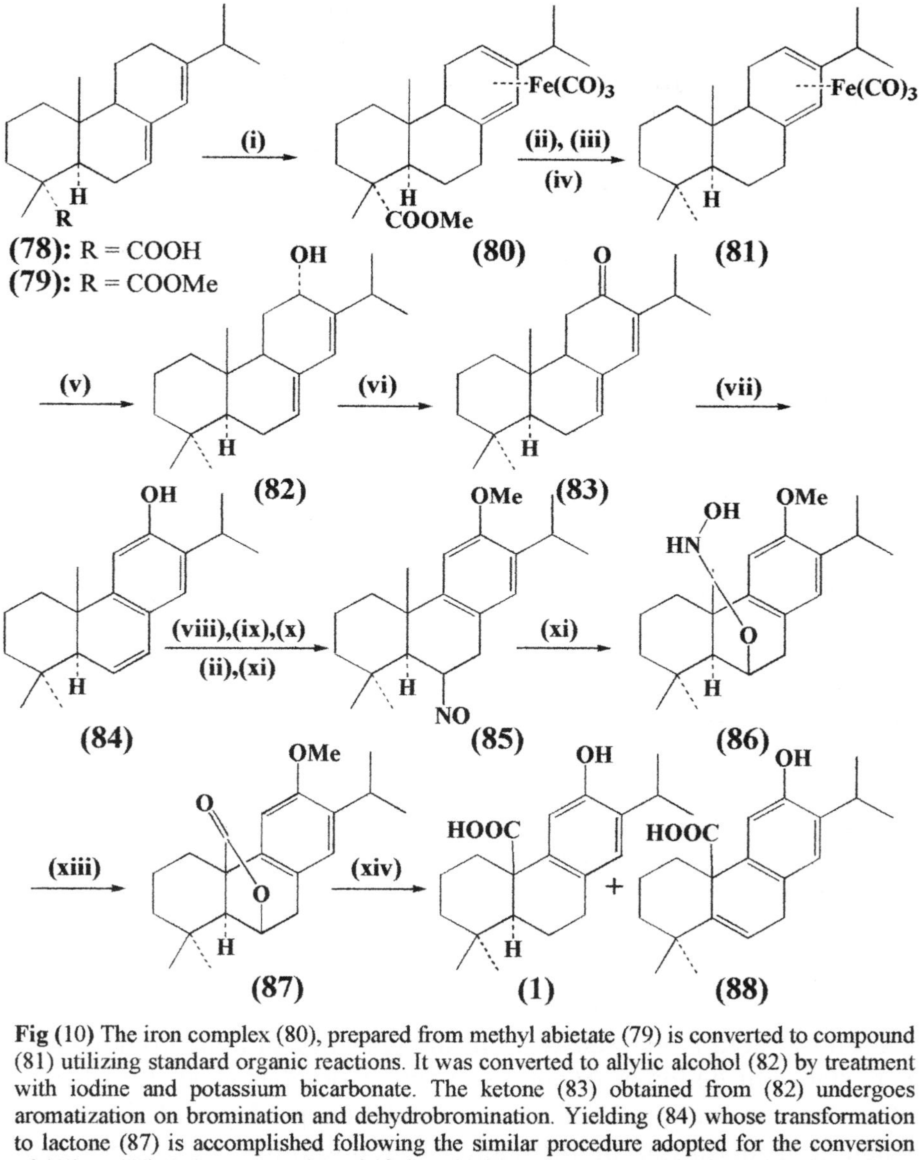

Fig (10) The iron complex (80), prepared from methyl abietate (79) is converted to compound (81) utilizing standard organic reactions. It was converted to allylic alcohol (82) by treatment with iodine and potassium bicarbonate. The ketone (83) obtained from (82) undergoes aromatization on bromination and dehydrobromination. Yielding (84) whose transformation to lactone (87) is accomplished following the similar procedure adopted for the conversion of (68) to (74). It is converted to pisiferic acid (1) by treatment with aluminium bromide in C_4H_8S.

Reagents: (i) CH_2N_2, $Fe(CO)_5$; (ii) $LiAlH_4$, THF; (iii) TsCl, Py; (iv) NaI, Zn, DMF; (v) I_2, $KHCO_3$;(vi) DMSO, Ac_2O, $KHCO_3$, H_2O; (vii) AcOH, Br_2, Et_2O; (viii) Me_2SO_4, K_2CO_3; (ix) MCPBA, CH_2Cl_2;(x) TsOH, C_6H_6; (xi) NOCl, Py; (xii) hv, C_6H_6; (xiii) PDC; (xiv) $AlBr_3$, C_4H_8S.

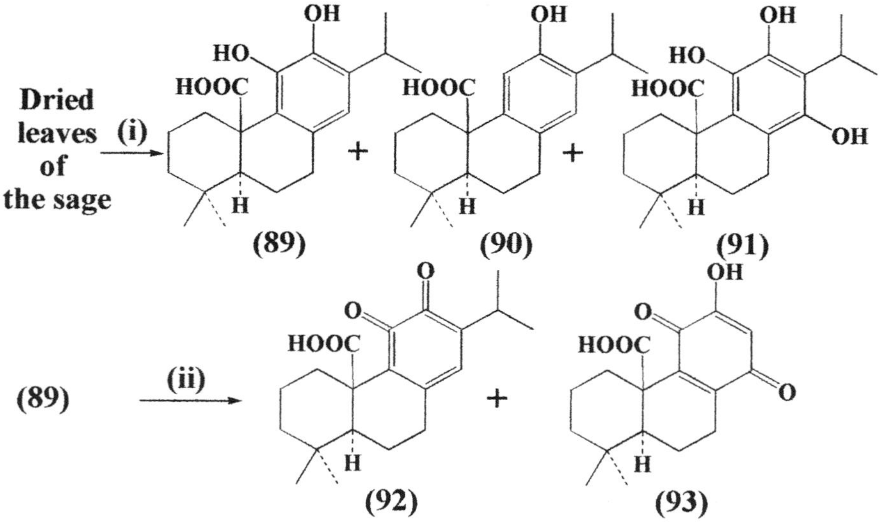

Fig (11) Carnosic acid (89) is an abietane diterpene and has been isolated from the leaves of sage as pale-yellow powder, m.p. 194.5 - 195°C (crystallized from acetic acid). Its oxidation products have been identified as (92), an orange powder, m.p. 129°C and (93), a yellow powder, m.p. 172 - 173°C.

Reagents: (i) MeOH, MeCOOH; (ii) MeCN, FeCl₃.

posseses o-diphenol structure and undergoes oxidation easily. It is necessary to mention that most diphenol compounds exhibit potent antioxidant activity in food systems [24]. An oxidation and isomerization pathway was suggested by Wenkert [22] to explain the formation of carnasol (90) from carnosic acid (89) via an o-quinone intermediate. Recently Masuda et al. [25] have proposed an antioxidant mechanism for carnosic acid. This is based on the chemical structures of oxidation products, which are radical termination compounds. These are produced from carnosic acid (89) in the antioxidation process.

Carnosic acid is very unstable in the presence of aqueous solvents, but it is stable in nonpolar organic solvents. Curvelier et al. [26] have also observed the formation of several oxidation products of carnosic acid when it is dissolved in methanol. The instability of carnosic acid is probably due to air oxidation catalyzed by a transition metal ion such as iron, which often exists as impurity in polar solvents. It has been observed by Masuda et al. [25] that addition of a catalyst amount of ferric chloride to a solution of carbonic acid in acetonitrile leads the formation of products (92) and (93) in low yield. It has been found that the oxidation products do not show any activity. This observation confirms that they may be regarded as termination compounds on the basis of their stability against radicals.

Carnosic acid has two reactive phenolic groups in the aromatic ring. Inatani et al. [23] suggested by that the phenolic group at C-11 is more important for its antioxidant activity on the basis of structure-activity relationship studies using the structurally related rosemanol derivative. It has been shown recently by Turk et al. [27] that carnosic acid shows the strongest inhibitory effect against HIV-1 protease which is regarded as potential target for developing new drugs useful in the treatment of AIDS.

Synthesis of Carnosic Acid

The first total synthesis of carnosic acid (89). was accomplished by Meyer et al. [28]. Meyer developed two different routes for the synthesis of carnosic acid. This is exhibited in Figs. (**12**) and (**13**).

In Fig. (**12**) keto ester (**94**) was selected as starting material. It was converted to the formyl derivative (**95**) which yielded α,β-unsaturated aldehyde (**96**) by treatment with DDQ. Michael addition of the sodium enolate of tert-butyl- isovalerylacetate to aldehyde (**96**) afforded the adduct (**97**) as a mixture of C-11 diastereomers. By fractional crystallization one of the adducts could be separated but for the synthetic purpose the mixture was not separated. Treatment of the adduct (**97**) with p-toluenesulfonic acid in glacial acetic acid caused t-butyl ester cleavage, decarboxylation and cyclodehydration leading the formation of tricyclic enedione (**98**) in 80% yield. This approach was previously utilized by Meyer in the synthesis of nimbiol [29]. Treatment of (**98**) with pyridinium bromide perbromide, followed by hydrogenolysis with palladium and carbon caused aromatization of (**98**) leading the formation of the phenolic ester (**99**).

In order to introduce a second hydroxyl group al C-11 of the ester (**99**) its sodium salt was treated with p-nitrobenzene diazonium chloride in methanol. The resulting product on methylation and then reduction with potassium dithionate afforded the amino ester (**100**). Diazotization and methanolysis produced ethyl (±)-carnosate dimethyl ether (**101**), which on saponification produced carnosic acid dimethyl ether (**102**). Wenkert [22] accomplished demethylation of dextrorotatory carnosic dimethyl ether with boron tribromide. Thus the work of Meyer lacks only the

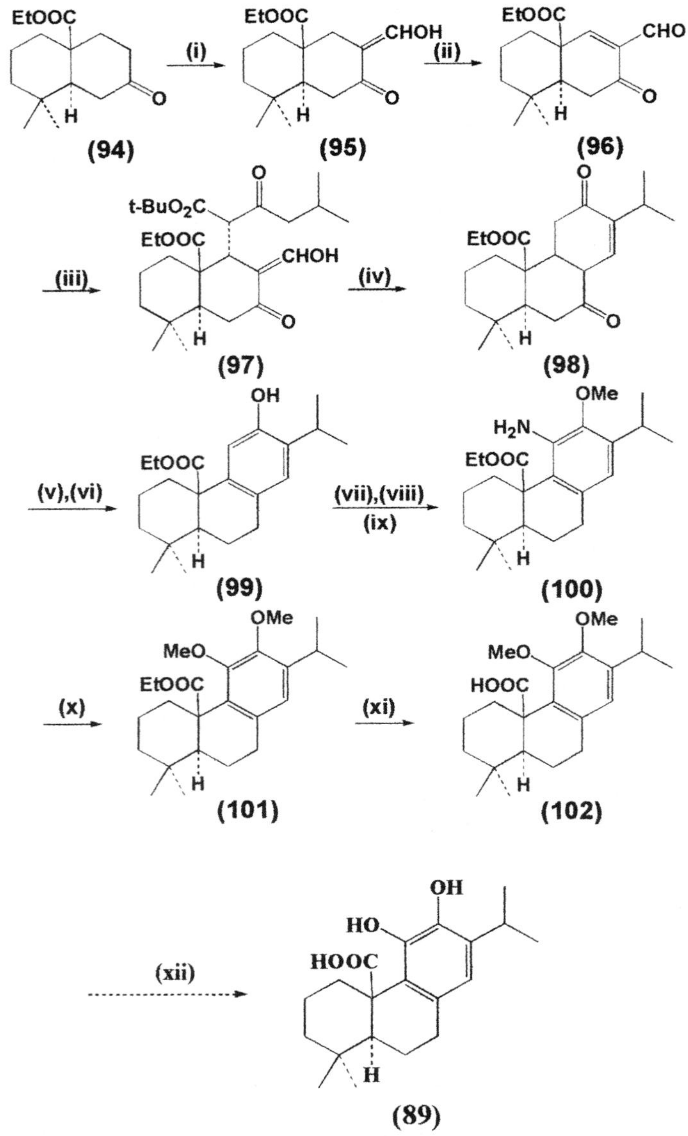

Fig (12) Transformation of keto ester (94) to (96) is described. Michael addition leads the formation of the adduct (97) which is subjected to cyclization, aromatization and hydrogenolysis to obtain the phenol (99). This on diazotization, methylation and reduction afforded the amino ether (100). Further diazotization, methanolysis and saponification produce ethyl (±)-carnosic acid dimethylether (102).

Reagents: (i) HCO_2Et; (ii) DDQ; (iii) $C_{11}H_{20}O_3$; (iv) PTS/MeCOOH; (v) Py, HBr; (vi) Pd/C, (vii) ArN_2Cl; (viii) Me_2SO_4; (ix) $Na_2S_2O_4$; (x) $NaNO_2$, MeOH; (xi) C_4H_9OK/Me_2SO; (xii) BBr_3.

resolution of racemic dimethoxy acid to constitute a formal total synthesis of carnosic acid.

Meyer also developed another route for the synthesis of carnosic acid (89) and this is described in Fig. (13).

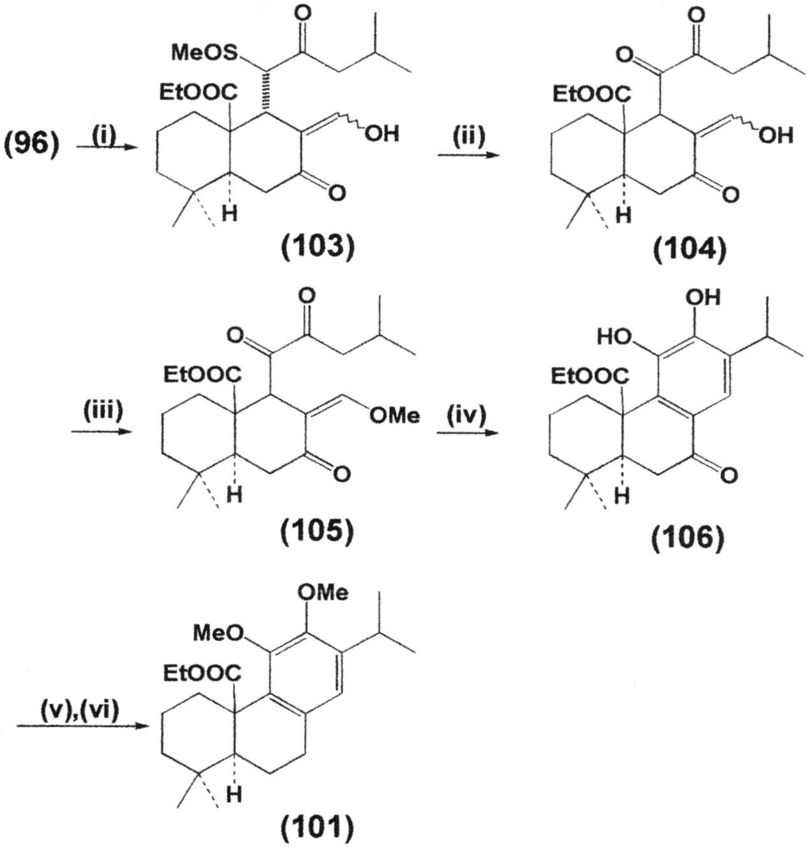

Fig (13) The adduct (103) prepared from (96) is converted to diketone (104) by Pummerer rearrangement. Treatment of (104) with p-TsOH and methanol affords enol ether (105) which on treatment with sodium methoxide in methanol yields the catechol (106) which was converted to ethyl (±)-carnosate dimethylether (101) by methylation and hydrogenolysis respectively.

Reagents: (i) MeSOCH$_2$COCH$_2$CHMe$_2$; (ii) HOAc, H$_2$O; (iii) MeOH, TsOH; (iv) MeONa, MeOH; (v) Me$_2$SO$_4$; (vi) Pd/H$_2$

In this approach the keto aldehyde (96) was subjected to Michael addition with β-keto sulphide to obtain the adduct (103) in quantitative yield. These adducts differ in configuration only at C-11 or at sulfur because both diastereomers lead to the same diketone (104). Pummerer rearrangement of the adduct (103) under mild acid conditions converted the sulfide group to carbonyl group affording the diketone (104) which could not be cyclized with p-toluenesulfonic acid in acetic acid. Treatment of the diketone (104) with p-toluenesulfonic and methanol afforded hydroxymethylene enol ether (105). This underwent cyclization, elimination and aromatization when refluxed with sodium methoxide. The resulting keto catechol (106) was converted to the previously described ethyl (±)-carnosate dimethyl ether (101) by methylation and hydrogenolysis.

The synthesis of carnosic acid dimethyl ether by Meyer et al records some interesting aspects. Michael addition of ketoaldehyde with two new reagents proved very useful in the synthesis of tricyclic ketone. In the second approach a very easy method has been described for the introduction of hydroxyl group at C-11 position.

Banerjee et al. [30] have accomplished a formal total síntesis of (±)-carnosic acid dimethyl ether and this is described in Fig. (**14**).

The olefin (107), prepared by the published procedure [31], was methylated to obtain (108) which was treated with N-bromosuccinimide in 1,4-dioxane containing perchloric acid. A mixture of bromohydrine was obtained, which without purification was oxidized with Jones reagent. [32] The resulting product on purification yielded the bromoketone (110). This indicated that one of the bromohydrin was (109). The bromoketone (110) was converted to alcohol (111) by reductive dehalogenation with zinc dust and acetic acid followed by reduction with sodium borohydride. The β-configuration of the hydroxyl group was confirmed by NMR spectroscopic analysis. [33] Oxidation of the alcohol (111) with lead tetraacetate and iodine in cyclohexane produced the cyclic ether (112) which was converted into ketoether (113) by (i) treatment with acetic anhydride and boron trifluoride, (ii) hydrolysis of the resulting product with 1,8-diazabicyclic 5.4.0 undecane (DBU) and (iii) oxidation with Jones reagent. It was allowed to react with dimethyl carbonate to give (114) whose transformation to the enone (115) was achieved by (i) annelation with methyl vinyl ketone, (ii) heating of the resulting product with lithium chloride in hexamethylphosphoric triamide, and (iii) heating of the resulting diketone with sodium hydride in dimethoxyethane. The β-ketoester (116), obtained from (115), gave on treatment with methyllithium in diethyl ether tertiary alcohol which without isolation was heated under reflux with methanolic hydrochloric

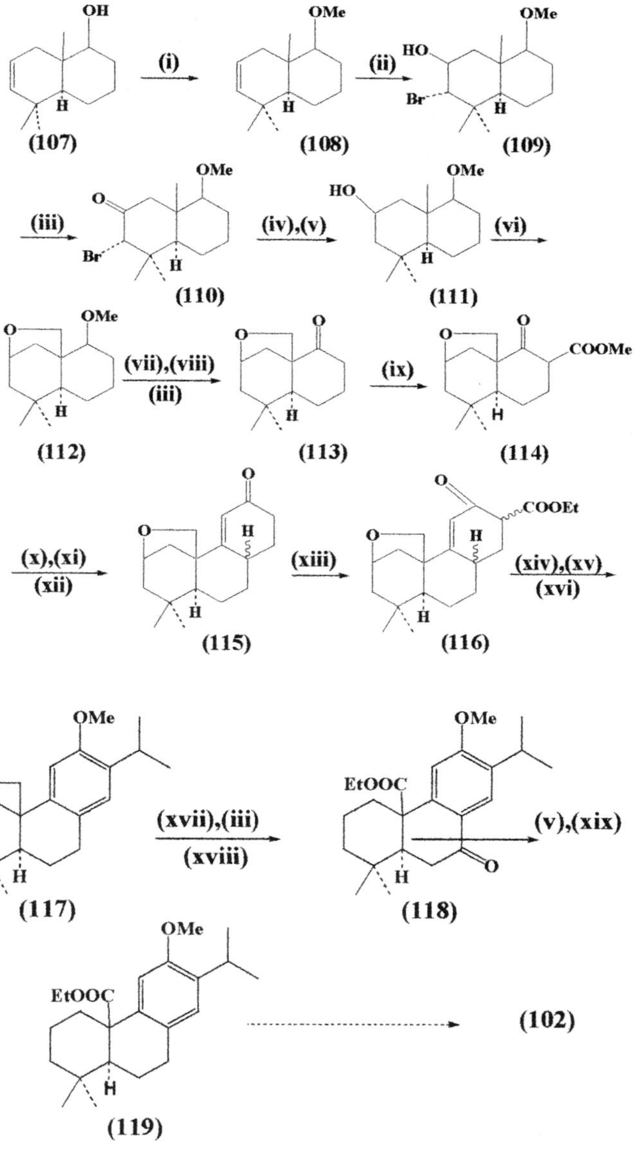

Fig (14) Olefin (107) has been converted to cyclic ether (114) by standard reactions. Its transformation to enone (115) is accomplished by annelation with methyl vinyl ketone and heating the resulting diketone with sodium hydride in dimethoxyethane. The ketoester (116) is subjected to Grignard reaction with methyllithium, aromatization and methylation to obtain the cyclic ether (117). Its transformation to phenolic ester (119) has been achieved by reduction, oxidation and esterification and deoxygenation.

Reagents: (i) MeI,NaH; (ii) NBS, 1,4-dioxane, HClO$_4$; (iii) CrO$_3$-H$_2$SO$_4$; (iv) Zn dust, MeCOOH; (v) NaBH$_4$, EtOH; (vi) Pb(OAc)$_4$, C$_6$H$_{12}$, I$_2$, hυ ; (vii) Ac$_2$O, BF$_3$ Et$_2$O; (viii) DBU; (ix) CO(OMe)$_2$; (x) CH$_2$CHCOMe; HMPA; (xii) NaH, DME; (xiii) CO(OEt)$_2$; (xiv) MeLi, Et$_2$O;(xv) HCl; (xvi) Me$_2$SO$_4$, NaOH; (xvii) Zn, ZnI$_2$; (xvii EtI, K$_2$CO$_3$, Me$_2$CO; (xix) P$_2$I$_4$, C$_6$H$_6$.

acid (1%) followed by methylation with dimethyl sulphate and sodium hydroxide to afford the cyclic ether (117). Its conversion to ketoester (118) was accomplished in three steps (reduction with zinc iodide, oxidation with Jones reagent and esterification with ethyl iodide and potassium carbonate in acetone). Reduction of (118) with sodium borohydride followed by heating the resulting alcohol with diphosphorous tetraiodide in benzene yielded the phenolic ester (119) whose transformation to carnosic acid dimethyl ether (102) has already been described by Meyer et al. [28]. Thus an alternative synthesis of the phenolic ester (119) constitutes a formal total synthesis of carnosic acid dimethyl ether.

Oommen [34] has reported the synthesis of a potential intermediate of carnosic acid and this is described in Fig. (15).

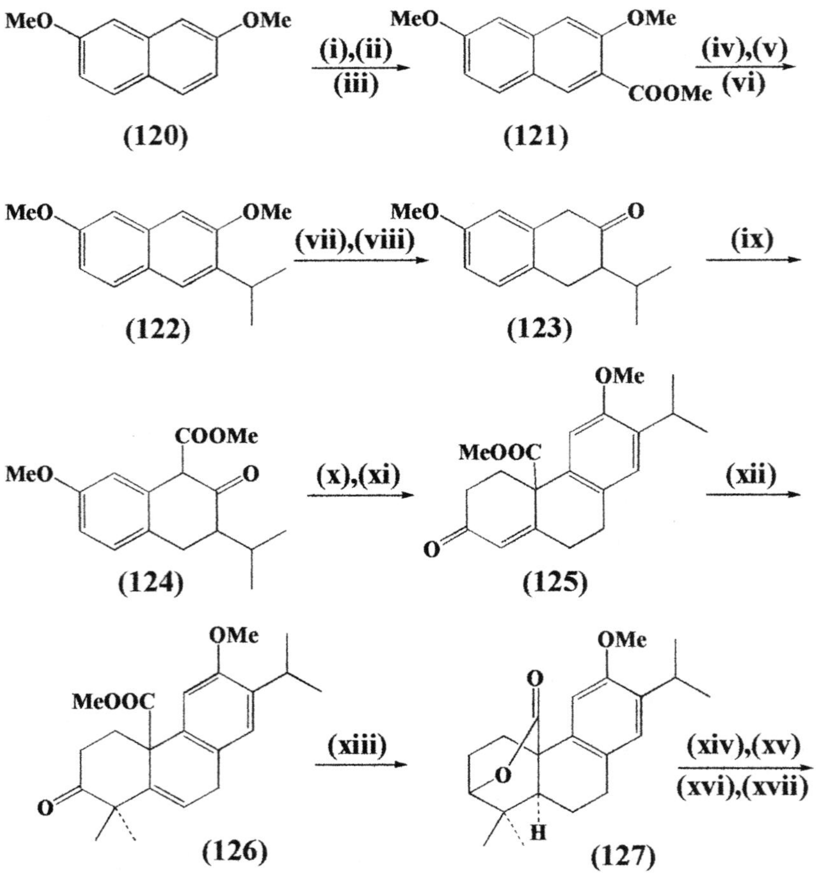

Fig (15) Tetralone (123) is prepared from 2,7-dimethoxynapthalene (120) using the standard organic reactions. It is carboxymethylated with dimethylcarbonate. The resulting compound (124) on subjection to Michael condensation yeilds tricyclic ketone (125) which on methylation followed by thioketalisation produces (128). This on desulfurisation with Raney nickel is converted to (129).

Reagents: (i) t-BuLi; (ii) CO $_2$; (iii) MeOH, H $_2$SO $_4$; (iv) MeMgI, Et $_2$O; (v) MeCOOH; (iv) H $_2$, Pd/C; (vii) l liq, NH $_3$; (viii) 3M HCl, Me $_2$CO; (ix) CO(COOMe) $_2$, NaH; (x) CH $_2$CHCOMe, Triton B; (xi) PTS/toluer (xii) MeI, C $_4$H $_9$OK, t-BuOH; (xiii) H $_2$ at high pressure (xiv) (CH $_2$-SH) $_2$; (xv) BF $_3$ Et $_2$O; (xvi) CH $_2$N $_2$, (xvii) Raney nickel, EtOH; (xviii) Raney nickel, EtOH, H$_2$.

2,7-dimethoxynaphthalene (120) on lithiation followed by carbonation with dry ice yielded an acidic material, which on esterification gave (121). It was converted to (122) in three steps (Grignard reaction with MeMgI, dehydration with glacial acetic acid and hydrogenation with 5% Pd-C at 60 psi pressure). Birch reduction of (122) followed by acid hydrolysis produced the tetralone (123) which was carboxymethylated with dimethyl carbonate and sodium hydride. The resulting product (124) was made to react with methyl vinyl ketone in presence of triton B methoxide. The product obtained was heated with p-toluenesulfonic acid in toluene to synthesize the tricyclic ketone (125). It was methylated with methyl iodide and potassium t-butoxide according the procedure of Woodward-Barton [35]. The crude methylation product not only contained (126) but had minor quantities of polymethylation product as was evidenced by the mass spectrum.

Attempted hydrogenation of (126) at atmospheric pressure in presence of 5% palladium-on-charcoal yielded a complex mixture which contained products of reduction of the 5,6 double bond and probably of 3-keto group. Hydrogenation of (126) at higher pressure produced the saturated lactone (127). In order to avoid the participation of the 3-keto group during hydrogenation, the compound (126) was converted to the thioketal (128) by standard method. The partial demethylation of the C-20 ester group occurred during thioketalization and therefore the crude product was esterified with diazomethane to obtain the thioketal derivative (128) in a pure form. An alcoholic solution of the thioketal (128) was heated with an active variety of Raney-nickel prepared by the method of Burgstahler [36]. Desulphurisation occurred along with the partial reduction of the 5,6-double bond. The complete reduction of the 5,6-double bond was achieved by refluxing the mixture obtained from the previous experiment with a large excess of the same variety of Raney-nickel in a current of hydrogen. This operation yielded methyl-o-methyl-11-desoxycarnosate (129). Meyer et al. [28] have reported the process of oxygenation of the C-11 position of 11- desoxycarnosate. They have also reported [28] the method of demethylation of the C-20 ester and C-12 methyl ether. Thus the total synthesis of (129) constitutes the synthesis of a potential intermediate for the synthesis of dl-carnosic acid.

Tiam et al. [37] have developed the synthesis of a model compound for the synthesis of carnosic type diterpenes by an expedient synthetic route. Instead of following the standard A→B→C approach as was done by previous authors [28,30], the present authors have developed an efficient convergent synthetic approach (AC→ABC). This is described in Fig. (16).

The starting material for the present synthesis was the 5,5-dimethyl-1,3-cyclohexandione (130) which was made to react with isopropyl alcohol to the enone (131). In order to build the C-ring of carnosic acid, β-phenethyl alcohol (132) was iodated in presence of triphenyl phosphine and imidazole in acetonitrile to produce (133). The enone (131) was alkylated with iodide (133) in THF in presence of LDA. The resulting compound (134) was treated with Me_3SiCN in presence of ZnI_2. The product (135) on heating under reflux with hydrochloric acid and alcohol underwent cyclization along with hydrolysis of the nitrile group to carboxylic acid affording the tricyclic compound (136) in 47% yield.

The present work reports an excellent method for the introduction of carboxylic acid on the angular position. In addition Tian et al. [37] have developed a very interesting procedure for the construction of the fundamental skeleton of carnosic acid.

Triptolide

Isolation and Biological Activity

Kupchan et al. [38] first isolated triptolide, a highly oxygenated diterpene from an ethanolic extract of the Chinese medicinal plant Triptergium

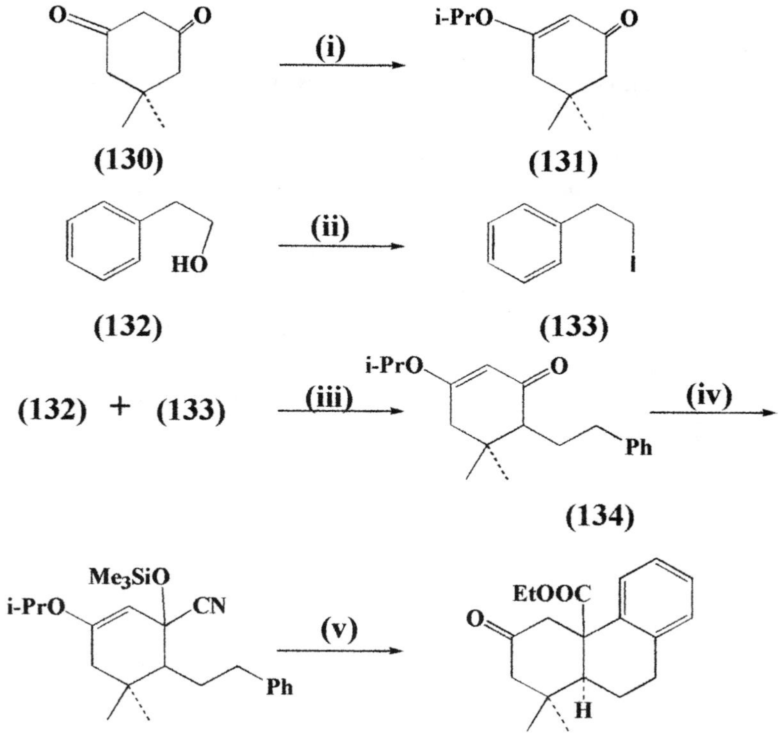

Fig (16) The enone (131), prepared from (130) is alkylated with iodide (133) to obtain the product (134). It reacts with trimethylsilyl cyanide and zinc iodide. The resulting product (135) is converted to tricyclic compound (136) by heating with hydrochloric acid and ethanol.

Reagents: (i) isopropyl alcohol, PTS, benzene, reflux; (ii) I_2, Ph_3P, imidazole, MeCN; (iii) LDA, THF, -78°C; (iv) Me_3SiCN, ZnI_2, benzene; (v) HCl, EtOH, reflux.

Wilfordil Hook F (celastraceae) on the basis of bioassay-directed fractionation. The ethanol extract was concentrated in an ethylacetate layer of an ethylacetate-water participation. The ethylacetate extract was eluted on silica gel with chloroform and 5% methanol in chloroform. The latter fraction was further chromatographed on silica AR-CC-7 with chloroform to yield a triptolide-enriched fraction. This was further chromatographed on silica AR CC-7 to yield triptolide (137) which on crystallization from CH_2Cl_2-Et_2O was obtained as white needles, m.p.

226-227°C, $[\alpha]^{25}$D-154°C (c 0.369, CH$_2$Cl$_2$); uv(max)(EtOH) 218 nm (ε 14,000), mass spectrum m/s 360.1600 (M$^+$), (Calcd 360.1573).

Triptolide (149) possesses a 9.11-epoxy-14 β–hydroxy system which is important for biological activity [39]. Triptolide exhibits impressive activity in vivo against murine L-1210 leukemia. The survival time of mice bearing 1615 leukemia is prolonged by the use of triptolide. [40] It shows intense cytotoxicity with cultivated KB cells. These data indicate that the triptolide (148) is useful as an antitumor agent. It has been reported that many biological effect of Triptergium Wilfordii has been observed owing to the presence of triptolide. [41]. Its immunosuppressive and inflammatory properties in human peripheral blood lymphocytes, jurkat T-cells and human bronchial epithelial cells has been reported. [42]. Triptolide exhibits in vitro cytotoxic activity against several types of carcinoma. [43] Recently it has been reported [44] that triptolide exhibits a remarkable in vitro cytotoxicity and in vitro growth inhibition against chalangiocarcinoma in hamsters model. Triptolide demonstrate potent cytotoxicity with a panel of cultured mammalian cell lines. This cytotoxic response profile can be compared with an activity profile that has been observed with various cancer chemotherapeutic drugs such as etoposide, taxol, etc. In addition modest antitumor activity has been observed in athymic mice carrying human breast tumors. Triptolide is not mutagenic toward Salmonella typhimurium strain TM 667 either in the absence or presence of a metabolic activating system. It is very strange because α,β-unsaturated carbonyl and the epoxide functionalities are each known to induce mutagenic effects. [45,46] Absence of mutagenic potential suggests that triptolide does not have covalently interest with DNA. These observations perfectly support the previous findings [47] that triptolide is found to inhibit DNA synthesis in L-1210 leukemia cells without directly damaging DNA. Triptolide has been reported to inhibit colony formation of several human cancer lines. [48]

Synthesis of Triptolide

Berchtold et al. [49] have reported the first total synthesis of racemic triptolide (149) and this is described in Fig. (**17**).

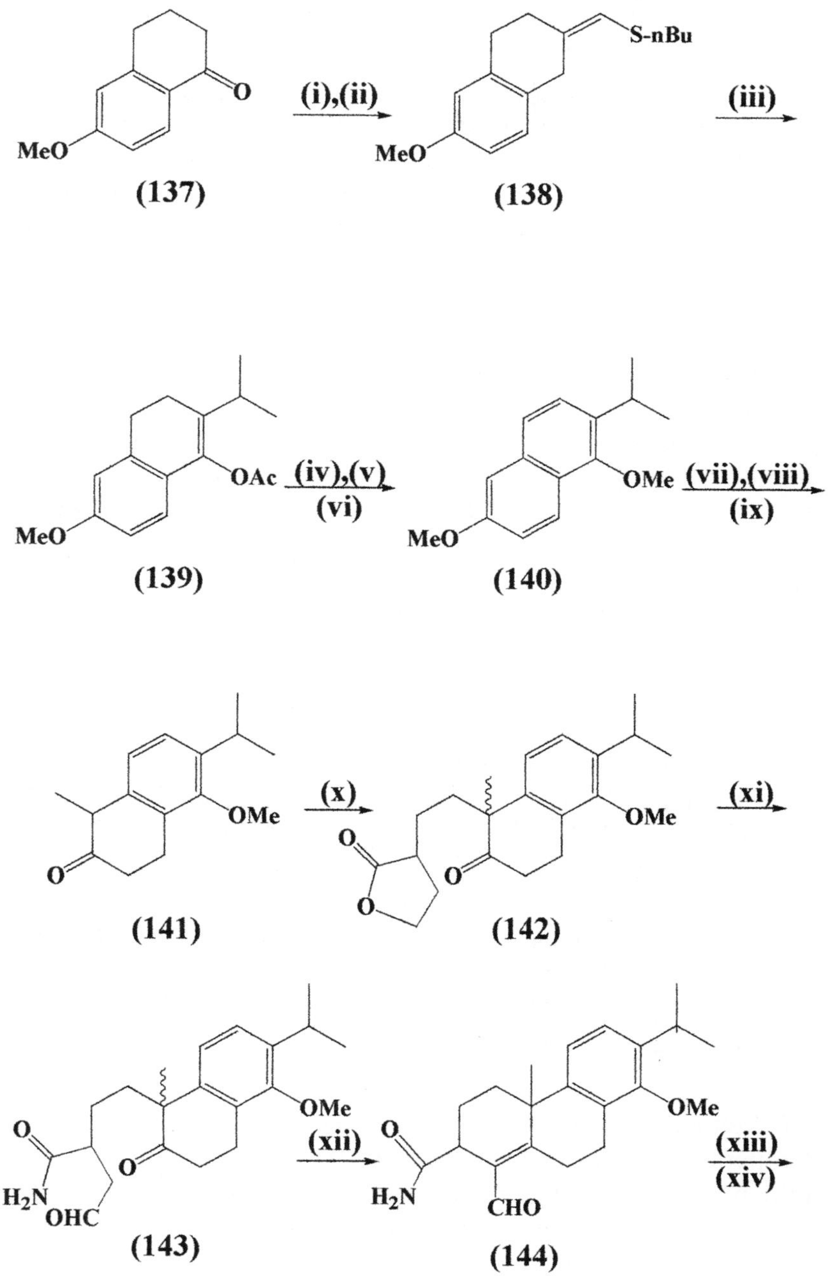

(137) → (i),(ii) → (138) → (iii) →

(139) → (iv),(v) (vi) → (140) → (vii),(viii) (ix) →

(141) → (x) → (142) → (xi) →

(143) → (xii) → (144) → (xiii) (xiv) →

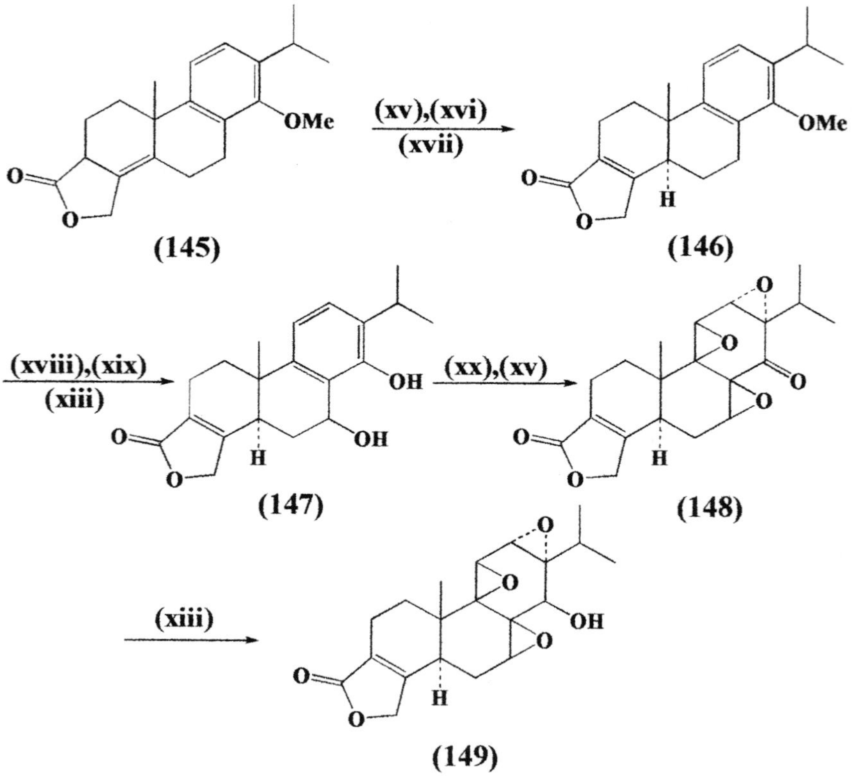

Fig (17) Transformation of 6-methoxy-α-tetralone (137) is described. Alkylation of (137) followed by cleavage of lactone and aldol condensation provided (144), which is converted to (146) by reduction, epoxidation, elimination and hydrogenation respectively. It is converted to phenol (147). Oxidation of (147) yields triptonide (148) which on reduction gives triptolide (149).

Reagents: (i) HCO₂Et, NaH; (ii) n-BuSH/p-TsOH; (iii) LiMe₂Cu/Ac₂O; (iv) S, Δ; (v) H₂SO₄, MeOH; (vi) Me₂SO₄, Ba(OH)₂, DMF; (vii) Na, EtOH; (viii) (COOH)₂; (ix) C₄H₉N, MeI; (x) C₆H₉O₂I, NaH; (xi) Me₂NH, CrO₃; (xii) Al₂O₃, EtOAc, p-TsOH; (xiii) NaBH₄; (xiv) HCl; (xv) MCPBA; (xvi) Et₃N, Me₂SO₄; (xvii) Pd-C, H₂; (xviii) CrO₃, HOAc; (xix) BBr₃, CH₂Cl₂; (xx) Na₅IO₆, MeOH.

6-Methoxy-α-tetralone (137), which on formylation at room temperature followed by treatment with n-butanethiol in presence of p-toluenesulfonic acid gave the compound (138), which was converted with lithium dimethylcopper and acetic anhydride to (139). Subjection of (139) to dehydrogenation, hydrolysis and methylation, respectively, yielded (140) which was converted to tetralone (141) by reduction, hydrolysis and methylation of pyralidine enamine. Alkylation of (141) with 2-

(iodoethyl)butyrolactone provided (142). Cleavage of the lactone ring of (142) with diethylamine and subsequent oxidation of the resulting primary alcohol afforded aldehyde (143) as a mixture of epimers which underwent alumina catalyzed aldol condensation to afford (144). It was reduced and treated with acid to obtain β,γ-lactone (145) as single isomer. This was converted to the C-5α compound (146) by three steps (epoxidation, elimination and hydrogenation). Its conversion to phenol (147) was accomplished by benzylic oxidation, demethylation and metal hydride reduction respectively. Triptonide (148) was obtained by oxidation of (147) with sodium periodate in aqueous methanol, and then epoxidation. This on reduction followed by purification on silica gel produced triptolide (149).

The synthesis of triptolide (149) by Berchtold[49], provides many interesting information for organic chemists. An excellent method has been developed for the construction of butenolide ring. The periodate oxidation of o-hydroxymethylphenoles appears to be a convenient method for the stereospecific construction of the C-ring functionality in triptolide and related substances.

Van Tamelen and coworkers have developed three syntheses of triptolide (149) and these are described in Fig. (**18**).

In the first synthesis [50] phenol (150) prepared from dehydroabietic acid as starting material [51] and this was converted to trifluoroacetate (151). The azide (152) prepared from (151), underwent Curtis rearrangement yielding isocyanate (153). Reduction of (153) followed by heating the resulting material with formic acid and formaldehyde provided the tertiary amine (154). Its conversion to ketone (155) was accomplished in three steps: (a) oxidation with m-cloroperbenzoic acid, (b) Cope elimination and (c) oxidative cleavage.

Hydroxymethylation of ketone (155) was followed by protection of the aliphatic hydroxy group (2-methoxypropyl ether) and addition of an α-benzyloxymethylene group at C-4. Acidic workup at the last stage of the reaction sequence produced (156). Its transformation to aldehyde (157) was carried out by successive treatment with methoxypropyl ether, acetic anhydride and pyridine, hydrochloric acid and methanol, and finally chromic acid, pyridine and hydrochloric acid. Dehydration of (157) led to the formation of (158) in 20% yield. Reagents other than the mentioned produced appreciable quantities of the cis A/B isomer. The butenolide (159) was finally synthesized by oxidation and hydrogenolysis. In order to complete the synthesis of triptolide it was necessary to introduce the

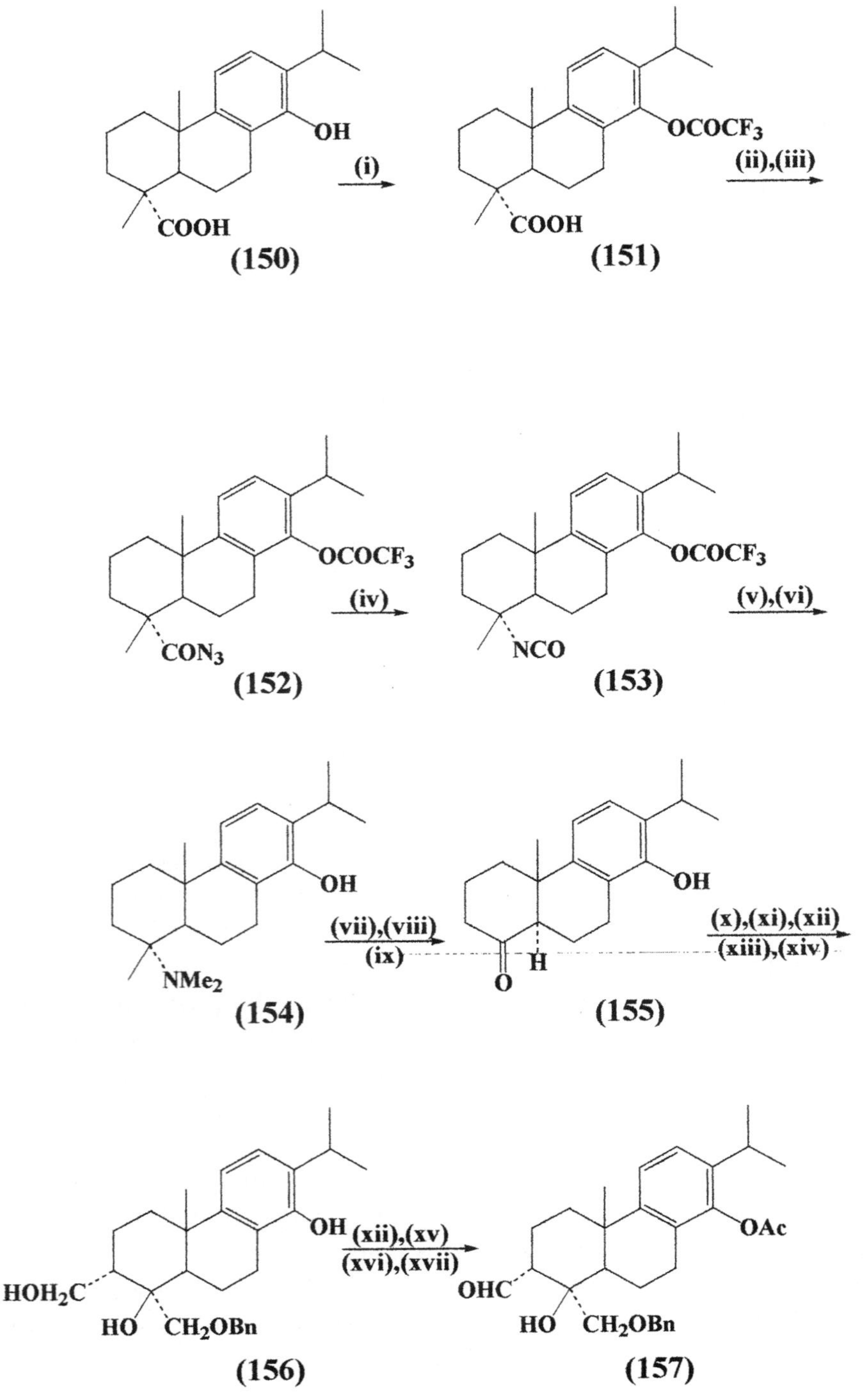

$(150) \xrightarrow{(i)} (151) \xrightarrow{(ii),(iii)}$

$(152) \xrightarrow{(iv)} (153) \xrightarrow{(v),(vi)}$

$(154) \xrightarrow[(ix)]{(vii),(viii)} (155) \xrightarrow[(xiii),(xiv)]{(x),(xi),(xii)}$

$(156) \xrightarrow[(xvi),(xvii)]{(xii),(xv)} (157)$

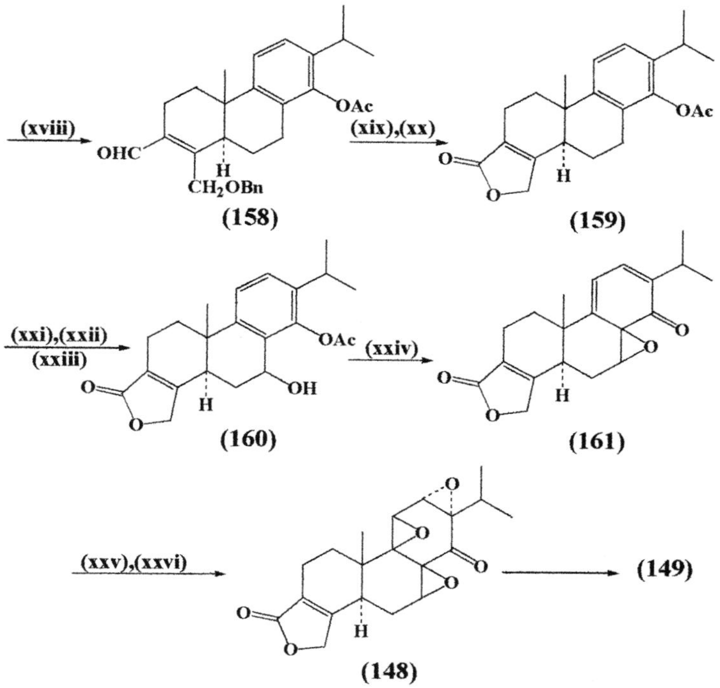

(158)   **(159)**

(160)   **(161)**

(148)

Fig (18) A series of reactions are carried out for the conversion of dehydroabietic acid to ketone (155) to hydroxymethylation, protection of aliphatic hydroxyl group, addition of an -benzyloxymethyl group and acid treatment provides (156) which is converted to (158). The butenolide function is constructed by successive oxidation and debenzylation to yield (159) which is converted to triptolide (149) by conventional methods.

Reagents: (i) (CF$_3$CO)$_2$O; (ii) SOCl$_2$; (iii) NaN$_3$; (iv) toluene, 100 $^\circ$ C ; (v) LiAlH$_4$; (vi) H$_2$ CH$_2$O; (vii) MCPBA; (viii) Reflux, 30 min , CHCl$_3$; (ix) OsO$_4$, NaIO$_4$; (x)LDA; (xi) CH$_2$O; (MeOCMe=CH$_2$, AcOH; (xiii) PhCH$_2$OCH$_2$Li; (xiv) HCl; (xv) Ac$_2$O, Py; (xvi) HCl, MeOH; (xvii CrO$_3$, Py, HCl; (xviii) O-C$_6$H$_4$(NH$_2$)$_2$, HCl; (xix) NaClO$_2$, HOSO$_2$NH$_2$; (xx) H$_2$, Pd; (xxi) (AcOH; (xxii) KOH; (xxiii) NaBH$_4$; (xxiv) NaIO$_4$; (xxv) H$_2$O$_2$, KOH; (xxvi) 3,5-(NO$_2$)$_2$C$_6$H$_3$CO$_3$H.

highly oxygenated functional groups at B/C ring of triptolide and therefore the compound (159) was subjected to oxidation, hydrolysis and metal hydride reduction respectively to yield alcohol (160). Oxidation produced epoxy dienone (161). Basic epoxidation and then peracid epoxidation afforded triptonide (148) which was already converted by Berchtold [49] to triptolide (149).

The second synthesis of triptolide (149) by van Tamelen [52] requires fewer steps and proceeds in an overall yield 40-50 times than that realized earlier [50], and this is described in Fig. (**19**).

Bicyclic diketone monoethylene ketal (162) prepared following the procedure of Kitahara et al. [53] was converted into octalin ketal (163) in one pot sequence which involves (a) reduction with liq. NH$_3$, (b)

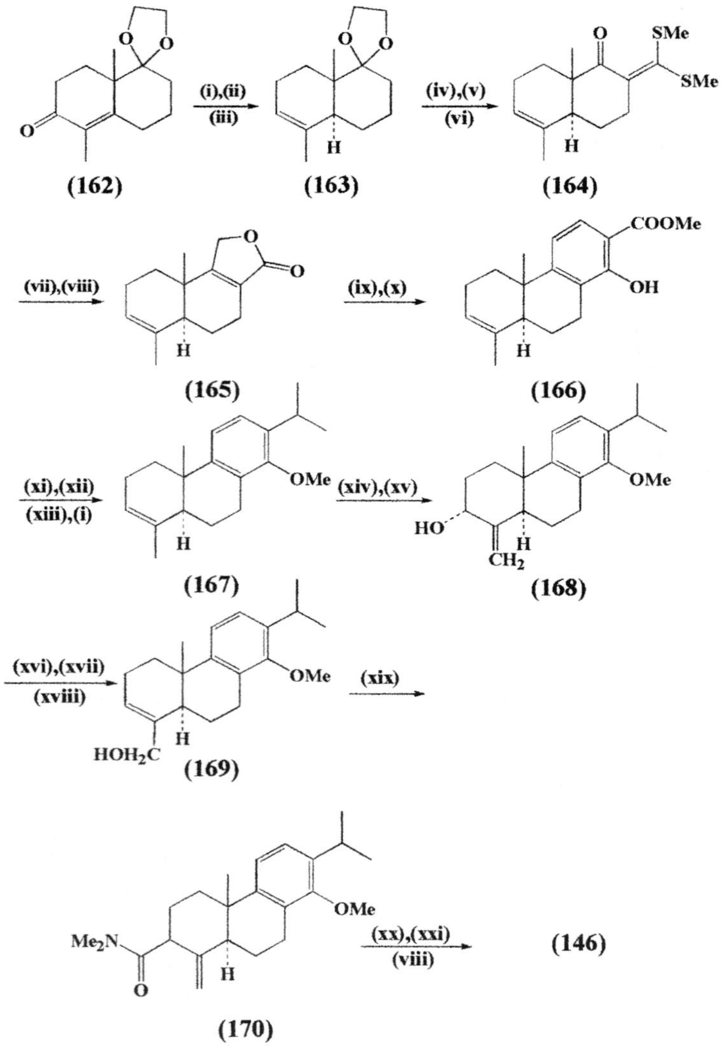

206

(162)   (163)   (164)

(165)   (166)

(167)   (168)

(169)

(170)   (146)

Fig (19) Octalin ketal (163) is converted to kete dithioacetal (164) by the cleavage of ketal function and condensation with carbon disulfide and methyl iodide. Subjection of (164) to the action of dimethylsulfonium methylide and acid hydrolysis leads to the formation of unsaturated lactone (165).Its furan silyl ether derivative is caused to undergo Diels-Alder reaction with methyl acrylate to obtain salicyclic ester (166) which is converted by standard organic reactions to abietane ether (167). It is converted to allylic alcohol (168) by epoxidation and elimination. Alcohol (169) obtained from (168) yields orthoamide which undergoes transformation to amide (170). Its conversion to the previously reported intermediate has been achieved by epoxidation, elimination and hydrolysis.

Reagents: (i) Li/liq. NH$_3$; (ii) Et$_2$POCl; (iii) LiEtNH$_2$, THF, t-BuOH; (iv)3M H$_2$SO$_4$; (v) CS$_2$, Me(C$_4$H$_9$)$_2$PhOLi; (vi) MeI; (vii) CH$_2$=SMe$_2$; (viii) HCl; (ix) LDA, TBDMSCl; (x) MeOC(Me)=CH$_2$; (xi) MeI, NaH; (xii) MeLi; (xiii) MeSO$_2$Cl, Et$_3$N; (xiv) MCPBA, CH$_2$Cl$_2$; (xv) LDA, THF; (xvi) SOCl$_2$, Py; (xvii) NaOAc; (xviii) NaOMe, MeOH; (xix) CH(OMe)$_2$NMe$_2$; (xx) MCPBA; (xxi) LiN(SiMe$_3$)$_2$.

subsequent reaction with diethyl chlorophosphate and (c) reduction of enol phosphate. The product obtained after the acid treatment of (163) was subjected to reaction with carbon disulphide followed by the addition of methyl iodide yielded ketone dithioacetal (164). It was made to react with dimethylsulfonium methylide [54] and the resulting product on acid hydrolysis afforded the unsaturated lactone (165). The furanal silyl ether derivative of (165) on subjection to Diels-Alder reaction with methyl acrylate produced salicylic ester (166). Its conversion to abietane ether (167) was accomplished by standard organic reactions. It was converted to allylic alcohol (168) by epoxidation and base promoted elimination. Allylic chloride from (168) produced acetate, which on methanolysis produced alcohol (169). When the alcohol was heated with dimethylformamide dimethylacetal the orthoamide was first formed which underwent α-elimination followed by cyclopropanation and rearrangement. The resulting amide (170) on epoxidation and lithium hexamethyldisililazide induced β-elimination and acid hydrolysis efforded the previously reported[49] intermediate (146), which proved potential intermediate for the synthesis of triptolide (149).

The third synthesis of triptolide by van Tamelen [55] features a biomimetic approach to the ring system. This is documented in Fig. (**20**).

2-Isopropyl anisole (171) was converted to bromide (172) by metalation, formylation and bromination. Alkylation with cyclopropyl ketoester produced (173) whose transformation to alcohol (174) was achieved by saponification, decarboxylation and reduction.. Its conversion to homoallylic bromide (175) was accomplished by the method of Julia et al. [56]. Alkylation of ethyl acetoacetate with bromide (175) furnished β-ketoester (176). It was subjected to cyclization with stannic chloride in dichloromethane. The resulting tricyclic alcohol provided the olefinic ester (177) by treatment with mesylchloride and triethylamine. Epoxidation followed by elimination led to the previously reported intermediate (146) whose conversion to triptolide (149) has already been described.

The present synthesis of (146) requires only twelve steps from the starting material. In this synthesis purification of only four intermediates is required. The overall yield is about 15%. This is superior to the previous methods reported by van Tamalen [50,52], which involves 20-30 steps and the yield ranged from 0.3-15%.

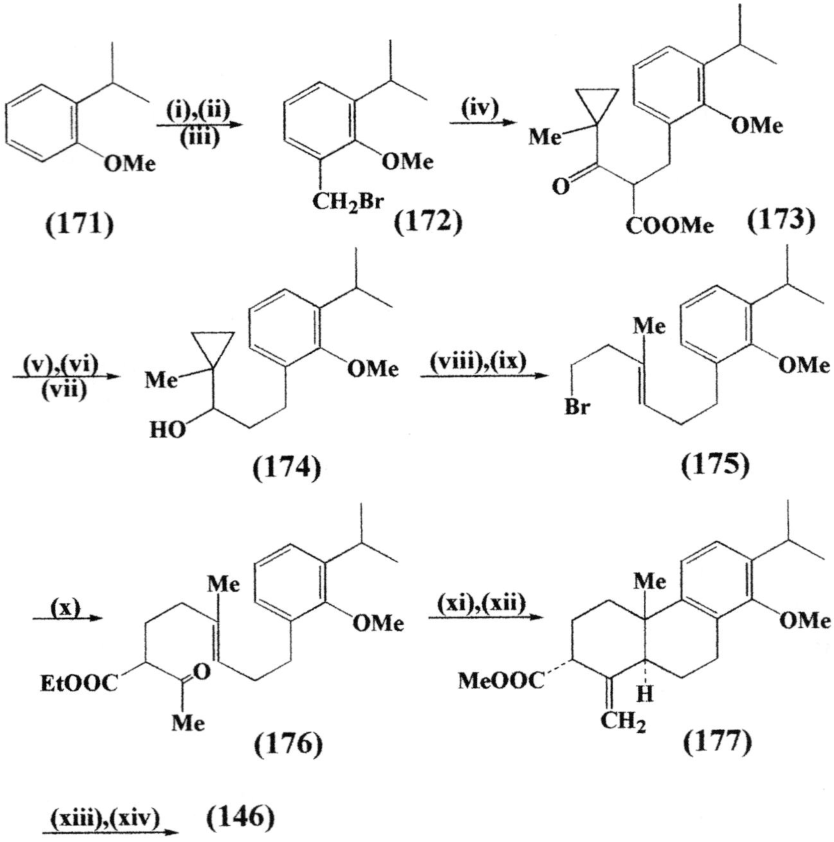

(171) → (i),(ii),(iii) → (172) → (iv) → (173)

(174) → (v),(vi),(vii) ← ... (viii),(ix) → (175)

(176) → (x) ← ... (xi),(xii) → (177)

(xiii),(xiv) → (146)

Fig (20) Bromide (172), prepared from alcohol (171) on alkylation yields (173) which is converted to (174) without any difficulty. Conversion of (174) to homoallylic bromide (175) is accomplished by the method of Julia. On alkylation followed by cyclization generated the tricyclic alcohol which undergoes dehydration to yield unsaturated ester (177) which is converted to unsaturated lactone (146) by epoxidation and elimination.

Reagents: (i) n-BuLi; (ii) CH_2O; (iii) PBr_3; (iv) $MeCH(CH_2)_3COCH_2COOMe$, NaH; (v) $Ba(OH)_2$; (vi) H_2O, Et_2O, $90^{\circ}C$; (vii) LAH , Et_2O; (viii) PBr_3, LiBr, collidine; (ix) $ZnBr_2$ in Et_2O; (x) $MeCOCH_2COOEt$; (xi) $SnCl_4$; (xii) $MeSO_2Cl$, Et_3N; (xiii) MCPBA ; (xiv) n-BuLi, i-C_3H_7NH.

Banerjee and Azócar [57] have developed an alternative synthesis of compound (167) whose transformation to triptolide (149) has already been described. [52] The synthesis of (167) is described in Fig. (**21**).

Tetralone (178) [58] was chosen as starting material. Its condensation with 1-chloro-3-pentanone yielded the already reported [59] tricyclic ketone (179) in 58% yield. This method constitutes an alternative method

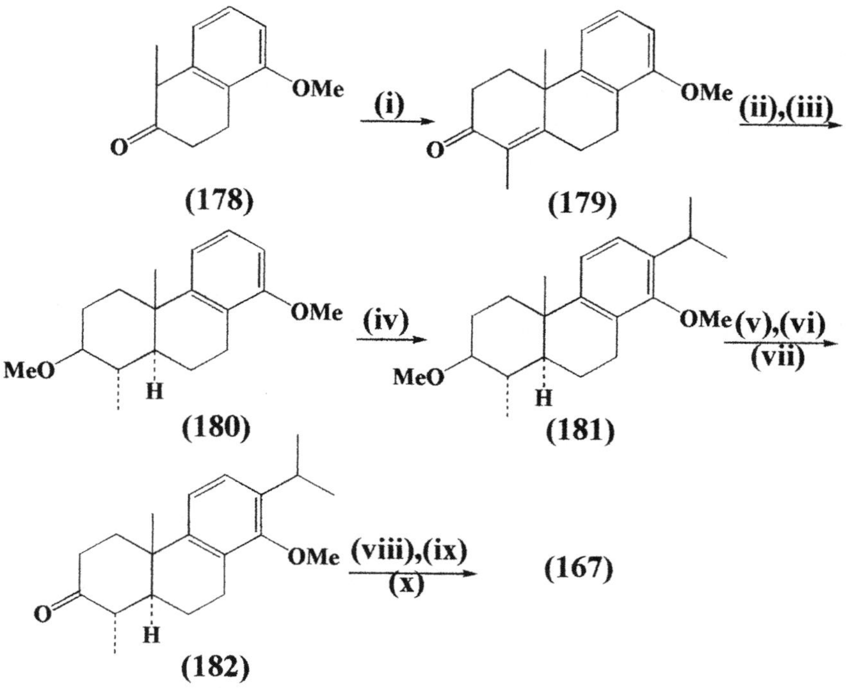

Fig (21) Ketone (179) on reduction and methylation produces (180). It is alkylated with isopropanol to obtain (181). Subjection of (181) to demethylation, methylation and oxidation yields ketone (182) which is converted to olefin (167) by reduction, tosylation and detosylation.

Reagents: (i) C$_4$H$_7$OCl, PTS, C$_7$H$_8$; (ii) Na, n-propanol; (iii) NaH, MeI, THF; (iv) isopropanol, PPA; (v) Me$_3$SiCl$_3$, NaI, MeCN; (vi) CrO$_3$, H$_2$SO$_4$; (vii) Me$_2$SO$_4$, MeCOMe, K$_2$CO$_3$; (viii) NaBH$_4$, MeOH; (ix) TsCl, Py; (x) LiBr, DMF.

for ketone (179). Reduction of (179) with sodium and n-propanol afforded an alcohol which on treatment with methyl iodide and sodium hydride in tetrahydrofuran afforded the dimethoxy derivative (180). The trans ring juncture, the α-configuration of 1-Me and the β-configuration of 2-OMe of (180) were assumed on the basis of previous works. [60,61]. Treatment of (180) with isopropanol and polyphosphoric acid [62] afforded the alkylated product (181). The alkylation experiment tried with isopropanol and sulphuric acid [63] and isopropanol and boron trifluoride [64] did not afford satisfactory yield. Treatment of (181) with MeSiCl$_3$ and sodium iodide in acetonitrile [65] afforded an oily compound, which on oxidation and methylation furnished ketone (182) in

45% yield. Subjection of ketone (182) to reduction yielded an alcohol whose tosyl derivative was converted to hexahydrophenanthrene (167) by standard organic reagents in 55% yield. As it was already been converted to triptolide [52] the present method constitutes an additional approach towards the synthesis of triptolide (149).

The present approach for the synthesis of hexahydrophenanthrene (167) is very convenient on the basis of less complicated experimental procedures and acceptable yield in most of the steps.

Recently Yang et al. [66] have developed a concise total synthesis of triptolide (149) and this is described in Fig. (22).

Compound (184), prepared from 2-isopropylphenol (183), on subjection to benzylic deprotonation and coupling of the mixed cuprate with 2-methyl-2-vinyloxirane following the method of Lipshutz[67] afforded allylic alcohol (185). Its bromide derivative was treated with methyl acetoacetate. The resulting compound on subjection to radical crystallization [68] furnished β-ketoester (186) (40%). The radical cyclization was highly stereoselecticve because four stereocenters were set up in one step. The conversion of (186) to lactone (187) was achieved [69] in three steps: (a) formation of vinyl triflate, (b) reduction and (c) Pd-catalyzed carbonylation. Deprotection of (187) led to the formation of diol (188) along with its epimer. It was converted to monoepoxide (189) following the procedure of Alder [70,71]. It was again subjected to epoxidation by two different processes to obtain triptonide (148). Epoxidation was carried out with (a) methyl(trifluoromethyl) dioxirane and (b) basic H_2O_2. Complete stereocontrol and high efficiency, were offered by these two oxidation processes. Reduction of triptonide (148) with sodium borohydride in methanol in presence of Ru(fod)$_3$ yielded triptolide (149) (47%) along with its α-hydroxy epimer. In the absence of Eu(fod)$_3$ the formation of the α-hydroxy epimer was obtained in major amount.

The method described for the synthesis of triptolide is concise and affords good yield in most of the steps. We believe that the experimental procedures are complicated and requires skilled experimentalists. Yang et al [72] have also accomplished an enantioselective total synthesis of (-)-triptolide.

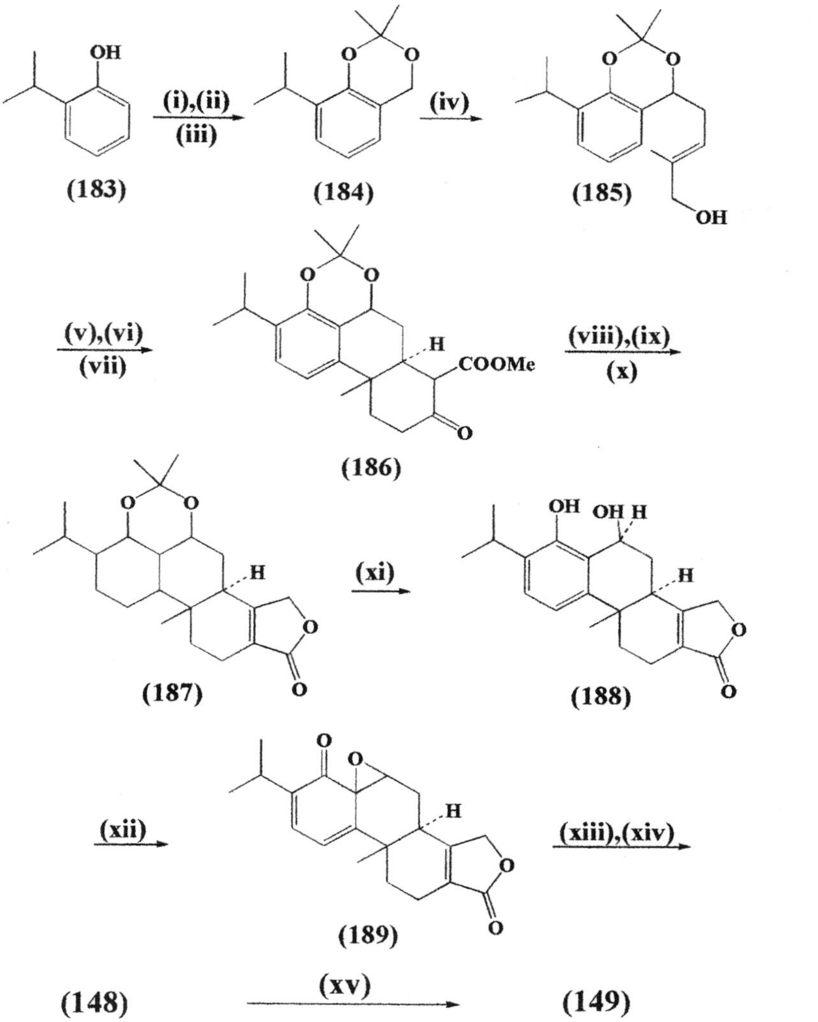

Fig (22) 2-isopropylphenol (183) is converted to allylic alcohol (185). Its bromide derivative reacts with methyl acetoacetate. The resulting compound undergoes radical cyclization yields ketoester (186) whose vinyl triflate on reduction and carbonylation furnishes lactone (187). Diol (188) obtained from (187), is converted to monoepoxide (189) which on further epoxidation produces triptonide (148). Its conversion to triptolide (149) is accomplished by reduction.

Reagents: (i) (CH₂O)n, SnCl₄, 2,6-lutidone; (ii) NaBH₄, MeOH; (iii) NMe₂COMe₂, TsOH; (iv) n-BuLi, THF, CuCN, 2-thienyllithium, 2-methyl-2-vinyloxirane; (v) PPh₃, CBr₄, CH₂Cl₂; (vi) MeCOCH₂COOMe, NaH, n-BuLi, THF; (vii) Mn(OAc)₃·2H₂O; (viii) KHMDS, PhNTf₂, H₂O; (ix) DIBAL-H, THF; (x) LiCl, n-BuN, Pd(PPh₃)₄, CO; (xi) PPTS, MeCN, H₂O; (xii) NaIO₄, TsOH, MeCN, H₂O; (xiii) CF₃COMe, oxone, NaHCO₃, MeCN, H₂O; (xiv) H₂O₂, NaOH, MeOH; (xv) Eu(fod)₃, NaBH₄, MeOH.

Miltirone

Isolation and Biological Activity

Tanshen (Salvia miltiorrhiza Bung), a medicinal plant, has been used in traditional Chinese medicine for its tranquilizing, sedative, circulation-promoting and bacteriocidal effects. [73], It has proven to be a rich source of abietane o-quinone diterpenoids. Miltirone (197) is a tricyclic diterpenoid quinone which has been isolated from the roots of salvia miltiorrhiza Bung. The isolation of miltirone constitutes a new addition to naturally occurring quinines related to tanshinones [74,75] isolated from the same source.

Dry clean tanshen rhizomes were powdered and extracted with hexane for three days at room temperature. The hexane solution was kept overnight and then filtered. After removal of the solvent a residue was obtained which was separated into seven colored fractions by column chromatography with silica gel. Miltirone was isolated by preparative tlc from fraction 1 (light red) using hexane:ethyl acetate (4:1) followed by benzene-acetone (20:1). The product obtained was recrystallized from ethylacetate, m.p. 100-101°C. Its structure was confirmed by mass spectrum, NMR, IR and UV spectra which agree quite closely with those of Ho et al [76]. Miltirone showed antioxidant behavior comparable to that of the commonly used phenolics BHT and BEA [77]. The antioxidant activity of miltirone in lard at 100°C was determined with a Rancimat. Miltirone and other related compounds may have the potential of being used as natural antioxidants in food and cosmetics.

Synthesis of Miltirone

Nasipuri and Mitra [63] reported the first total synthesis of miltirone (197) and this is described in Fig. (23).

4-Bromo-2-isopropylanisole (191) was prepared from p-bromoanisole (190). It was caused to react with magnesium and ethylene oxide to obtain the alcohol (192). Its bromoderivative on heating with diethylmalonate and sodium ethoxide followed by hydrolysis with alcoholic potassium hydroxide yielded the substituted malonic acid. This on heating furnished acid (193). Its acid chloride in benzene underwenr cyclization with aluminium chloride to yield tetralone (194). It was

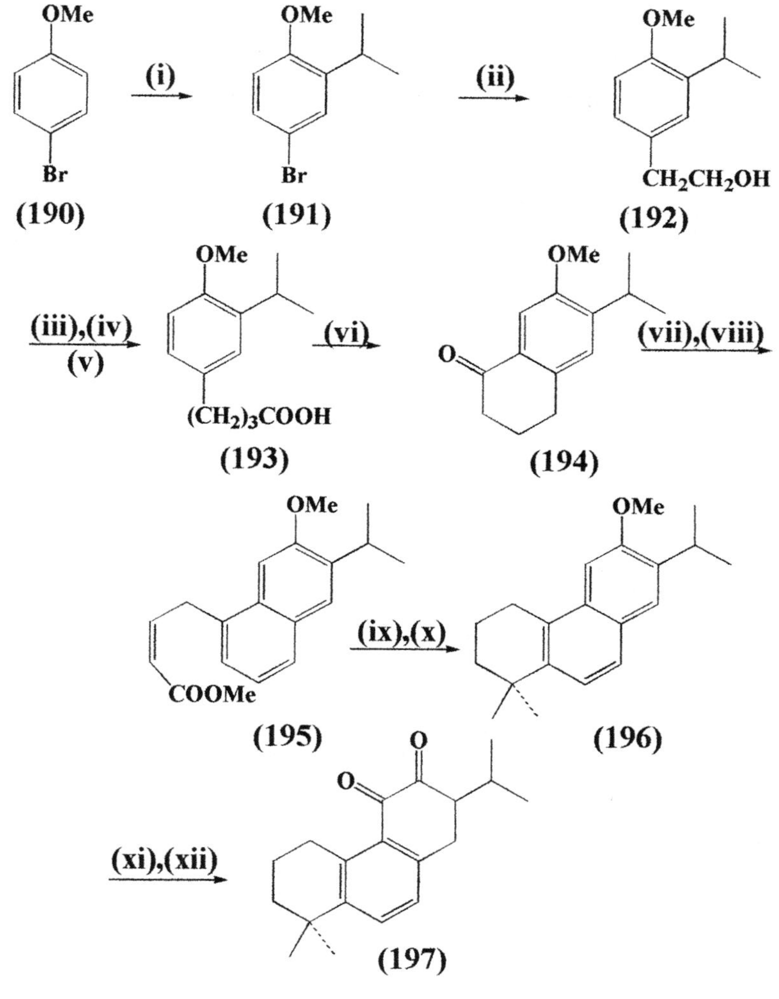

Fig (23) Acid (193) prepared from bromoanisole (190) by standard organic reactions, undergoes cyclization with polyphosphoric acid leading the formation of tetralone (194). A Reformatsky reaction on compound (194) with methyl g-bromocrotonate followed by aromatization produces compound (195). It is converted to compound (196) by treatment with methylmagnesium and cyclization. The corresponding phenol was oxidized to miltirone (197).

Reagents: (i) i-PrOH-H$_2$SO$_4$; (ii) Mg, (CH$_2$)$_2$O; (iii) PBr$_3$, CCl$_4$; (iv) CH$_2$(COOEt)$_2$, NaOEt; (v) KOH, EtOH (20%); (vi) PCl$_5$, AlCl$_3$, C$_6$H$_6$; (vii) BrCH$_2$CHCHCOOMe; (viii) Pd black, CO$_2$; (ix) MeMgI; (x) PPA; (xi) Py, HCl, 210-220°C; (xii) air or Fremy's salt.

subjected to a Reformatsky reaction with methyl-γ-bromocrotonate. The resulting compound on heating with palladium black furnished ester

(195). The transformation of ester (195) to phenanthrene (196) was carried out in two steps: (i) treatment with excess of methylmagnesium bromide and (ii) heating the resulting alcohol with polyphosphoric acid at 170°C. Demethylation of (196) followed by oxidation provided miltirone (197).

The synthesis of miltirone (197) by Nasipuri follows the route similar to Thomson's [74] and Kasikawa's [78] approach to tanshinones bearing the benzofuran unit.

Knapp and Sharma [79] reported a very short synthesis of miltirone (197) and this is described in Fig. (24).

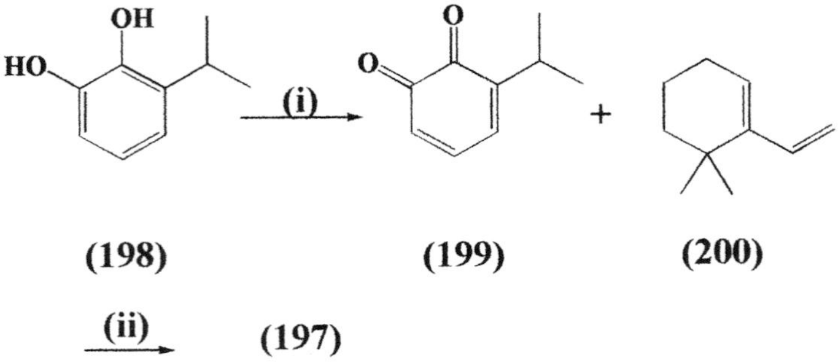

(198)   **(199)**   **(200)**

(ii) → **(197)**

Fig (24) Oxidation of (198) gives o-quinone (199) which on heating with vinylcyclohexene (200) gives miltirone (197).

Reagents: (i) Pb(OAc)$_4$; (ii) heat.

O-quinone (199) prepared from catechol (198) was heated with vinylcyclohexene (200) in refluxing ethanol. Miltirone (197) was isolated by column chromatography in 28-30% yield. About 5% of (200) could be recovered Lower temperatures or longer reaction times damaged the reaction.

Snyder et al [80] developed a novel synthesis of miltirone (197) and this is described in Fig. (25).

Snyder observed the o-quinone dienophile reported by Knapp et. al. Sharma [79] was very unstable and thus the yield of miltirone was poor. In order to improve the yield of miltirone, it was planned to prepare o-quinone dienophile in situ. This plan was carried out by heating a mixture of 3-isopropyl-1,2-dihydroxybenzene (201), 6´,6-methyl-1-vinylcyclohexene (200) and silver oxide in anhydrous ethanol under

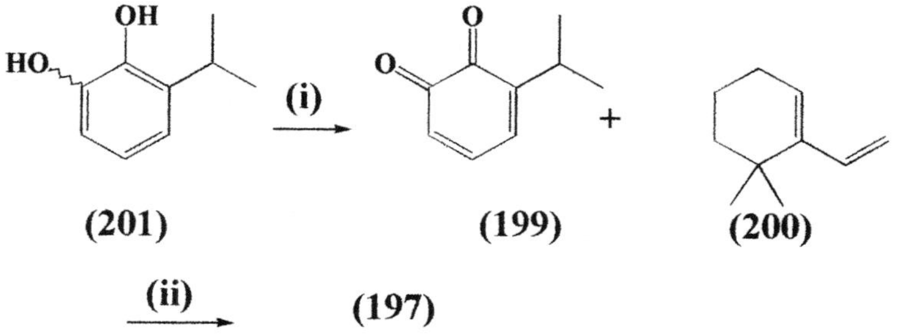

Fig (25) Catechol (201) on oxidation with silver oxide generates 3-isopropyl-o-benzoquinone (199) which undergoes ultrasound-promoted cycloaddition with 6,6-dimethyl-1-vinylcyclohexene (200) yielding the synthesis of miltirone (197).

Reagents: (i) Ag$_2$O; (ii) ultrasound.

ultrasonication. Miltirone (197) was obtained in 93% yield. In this process isopropylbenzene (201) is converted to isopropyl-o-benzoquinone (199) by silver oxide oxidation and this is subjected to Diels-Alder reaction using ultrasound to accelerate the addition.

Thus it can be observed that ultrasound-promoted cycloaddition has proven to be an expedient route for the synthesis of miltirone. We believe this approach would prove very valuable for the synthesis of related abietane natural products.

Majetich et al [81] have reported an excellent synthesis of miltirone (197) using cyclialkylation-based strategy. This is described in Fig. (26).

Cyclohexanone (202) was converted to compound (203) whose transformation to cyclohexanone (204) was accomplished in three steps. It underwent cyclialkylation with boron trifluoride etherate affording the cyclized product (205) (R=R,=OMe) in 64% yield along with naphthalene (206) (R=R$_1$= H,H). Compound (205) on heating under reflux with DDQ in benzene produced ketone (207) whose tosylhydrazone on treatment with sodium cyanoborohydride afforded reduced product (208). Deprotection of the aryl methyl ethers and oxidation with ceric ammonium nitrate led to the formation of miltirone (197).

The most interesting part of this synthesis is the cyclialkylation-based strategy for the synthesis of functionalized hydrophenanthrene. The utility of this methodology is featured in the synthesis of many diterpenoids.

Banerjee et al [82] have synthesized an important intermediate for the synthesis of miltirone (197) and this is described in Fig. (27).

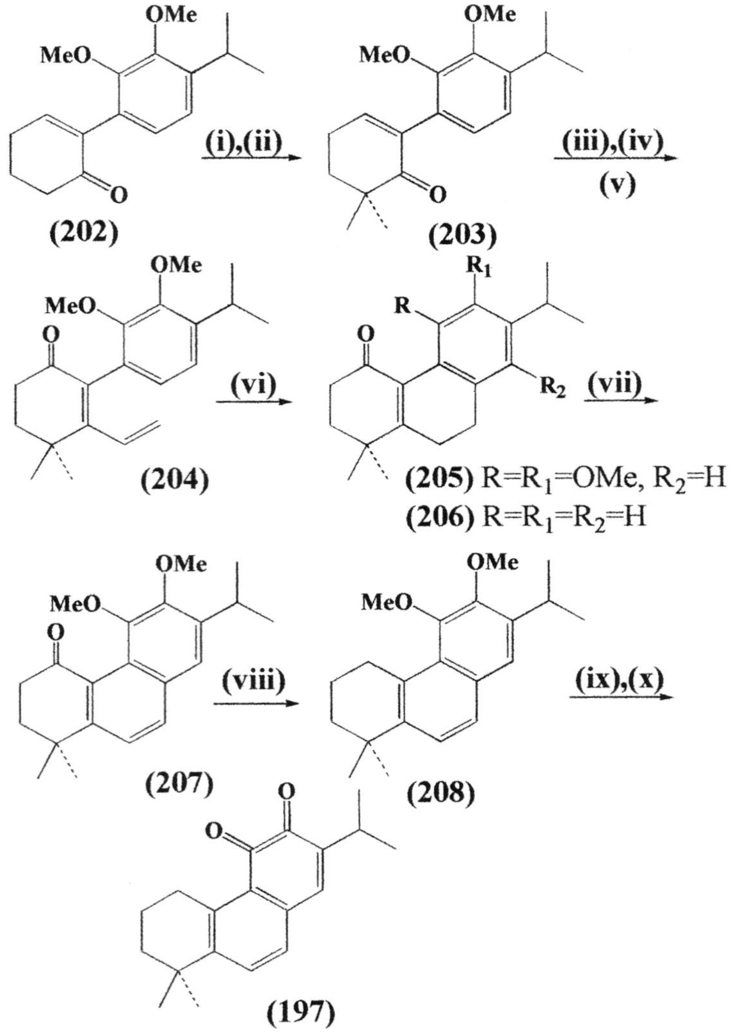

Fig (26) Methylation of (202) produces (203) which is converted to dienone (204) which on subjection to cyclicalkylation produces the cyclized product (205). Dehydrogenation of (205) produces ketone (207) whose transformation to miltirone (197) was achieved by deoxygenation, demethylation and oxidation respectively.

Reagents: (i) LDA/MeI; (ii) LDA, CH$_3$I; (iii) vinyllithium; (iv) H$_3$O$^+$; (v) PDC; (vi) BF$_3$·Et$_2$O, CCl$_4$; (vii) DDQ; (viii) NH$_2$NHTS, NaBH$_3$CN; (ix) BBr$_3$; (x) (NH$_4$)$_2$Ce(NO$_3$)$_6$.

In 1992 the synthesis of ketoester (209) was reported [30] and this was utilized as starting material for the present work. Reduction of ketoester

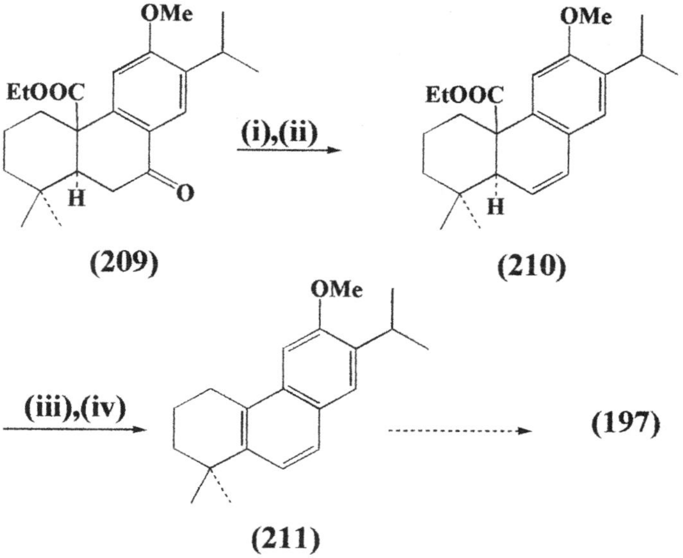

Fig (27) Reduction of ketoester (209), followed by dehydration produced olefin (210, which was heated with quinoline and acetic acid. The resulting acidic material on treatment with dimethylformamide, pyridine and lead tetraacetate afforded tetrahydrophenanthrene (211).

Reagents: (i) NaBH$_4$, MeOH; (ii) TsOH, SiO$_2$, C$_6$H$_6$; (iii) Quinoline, acetic acid; (iv) DMF, C$_5$H$_5$N, Pb(OAc)$_4$

(209) followed by dehydration afforded olefin (210). It was heated with quinoline and acetic acid [83]. The resulting acidic product without purification was treated with lead tetraacetate and pyridine in DMF at room temperature [84]. This process caused decarboxylation generating a mixture of olefins, which on chromatographic purification furnished the tetrahydrophenanthrene (211) in 55% whose m.p. and spectroscopic data were identical with that reported [78]. The conversion of (211) to miltirone (197) has already been reported [78] and thus the present approach for the synthesis of (211) constitutes the synthesis of the potential intermediate for miltirone.

CONCLUSIONS

We have described the synthesis of several bioactive diterpenes. Several methods varying in efficiency and operational simplicity have been developed for the synthesis of the above mentioned bioactive diterpenes.

We believe that in the near future many appropriate synthetic approaches will be developed for large scale production of these diterpenes.

ACKNOWLEDGEMENTS

We thank all the people who have provided published and unpublished information regarding the synthesis of the described diterpenes. We are grateful to Fondo Nacional de Ciencia, Tecnología e Innovación (FONACIT) and the Instituto Venezolano de Investigaciones Científicas (IVIC) for financial support to carry out some of the work presented in this review.

REFERENCES

[1] Goldsmith, D. in: J. ApSimon (Eds.), *Total Synthesis of Natural Products*, Vol. 8, John Wiley & Sons, New York, **1992**, p. 1-243.

[2] Fukui, H.; Koshimizu, K.; Egawa, H. *Agric. Biol. Chem.* **1978**, *42*, 1419-1423.

[3] Yatagai, M.; Takahashi, T. *Phytochemistry* **1979**, *18*, 176-179.

[4] Yatagai, M.; Takahashi, T. *Phytochemistry* **1980**, *19*, 1149-1151.

[5] Kobayashi, K.; Nishino, C. *Agric. Biol. Chem.* **1986**, *50*, 2405.

[6] Kobayashi, K.; Nishino, C.; Fukushima, M.; Shiobara, Y.; Kodama, M. *Agric. Biol. Chem.* **1988**, *52*, 77-83.

[7] Kobayashi, K.; Kuroda, K.; Shinomiya, T.; Nishino, C.; Ohya, J.; Sato, S. *Int. J. Biochem.* **1989**, *21*, 463-468.

[8] Matsumoto, T.; Usui, S. *Bull. Chem. Soc. Jpn.* **1982**, *55*, 1599-1604.

[9] Banerjee, A. K.; Hurtado, H.; Laya, M.; Acevedo, J. C.; Alvárez, J. *J. Chem. Soc. Perkin Trans I*, **1988**, 931-938.

[10] Sondheimer, F.; Elad, D. *J. Amer. Chem. Soc.*, **1957**, *79*, 5542-5546.

[11] Dutcher, J.S.; Mac Millan, J.G.; Heath, C.N. *J. Org. Chem.* **1976**, *41*, 2663-2669.

[12] Banerjee, A.K.; Pita-Boente, M.I. *Heterocycles* **1985**, *23*, 5-10.

[13] Howell, F.H.; Taylor, D.A.H. *J. Chem. Soc.* **1958**, 1248-1254.

[14] Pal, S.K.; Mukerjee, D. *Tetrahedron Lett.* **1998**, *39*, 5831-5832.

[15] Mori, K.; Mori, H. *Tetrahedron* **1986**, *42*, 5531-5538.

[16] Meyer, W.L.; Manning, R.A.; Schroeder, P.G.; Shew, D.C. *J. Org. Chem.*, **1977**, *42*, 2754-2769.

[17] Geiwiz, J.; Haslinger, E. *Helv. Chim. Acta,* **1995**, *78*, 818-832.

[18] Nusslaum, A.K.; Robinson, C.H. *Tetrahedron,* **1962**, *17*, 35-59.

[19] Baumgarten, H.E.; Staklis, A.; Miller, E.M. *J. Org. Chem.* **1965**, *30*, 1203-1206.

[20] Curvelier, M.-E.; Richard, H.; Berset, C. *J. Amer. Oil Chem. Soc.* **1996**, *73*, 645-652.

[21] Tada, M. *Food Ingred. J. Jpn.* **2000**, *184*, 33-39.

[22]   Wenkert, E.; Fuchs, A.; McChesney, J.D. *J. Org. Chem.* **1965**, *30*, 2931-2934.

[23]   Inatani, E.; Nakatani, N.; Fuwa, H. *Agric. Biol. Ch.* **1983**, *47*, 521-528.

[24]   Shahidi, F.; Janitha, P.K.; Vanasundara, P.D. *Crit. Rev. Food Sci. Nutr.* **1992**, *32*, 67-103.

[25]   Masuda, T.; Inaba, Y.; Takeda, Y. *J. Agric. Food Chem.* **2001**, *49*, 5560-5565.

[26]   Curvelier, M.-E.; Berset, C.; Richard, H. *J. Agric. Food Chem.* **1994**, *42*, 665-669.

[27]   Paris, A.; Struker, B.; Renko, M.; Turk. V.; Puki, M.; Umek, A.; Korant, B.D. *J. Nat. Prods.* **1993**, *56*, 1426-1430.

[28]   Meyer, W.L.; Manning, R.A.; Schindler, E.; Schroeder, R.S.; Shew, D.C. *J. Org. Chem.* **1976**, *41*, 1005-1016.

[29]   Meyer, W.L.; Clemans, G.B.; Manning, R.A. *J. Org. Chem.* **1975**, *40*, 3686-3688.

[30]   Banerjee, A.K.; González, N.C.; Peña, C.A. *J. Chem. Res (S)* **1992**, 50.

[31]   Banerjee, A.K.; Hurtado, H.S.; Laya, M.L.; Acevedo, A.J.; Alvárez, J. *J. Chem. Soc., Perkin Trans I.* **1988**, 931-938.

[32]   Bowers, A.; Halsall, T.G.; Jones, E.R.H.; Lemin, A.J. *J. Chem. Soc.* **1953**, 2548-2560.

[33]   Bhaca, N.S.; William, D.H. *Application of NMR Spectroscopy in Organic Chemistry,* Holden-Day, San Francisco, **1964**, p. 79-102.

[34]   Oomen, P.K. *Bull. Chem. Soc. Jpn.* **1976**, *49*, 1985-1988.

[35]   Woodward, R.B.; Patchett, A.A.; Barton, D.H.R.; Ives, D.A.J.; Kelly, R.B. *J. Amer. Chem. Soc.* **1954**, *76*, 2852-2853.

[36]   Burgstahler, A.W.; Abdel-Rahman, N.O. *J. Amer. Chem. Soc.* **1963**, *85*, 173-180.

[37]   Tian, Y.; Chen, N.; Wang, H.; Pan, X.F.; Hao, X.J.; Chen, C.X. *Syn. Communs.,* **1997**, *27*, 1577-1582.

[38]   Kupchan, S.M.; Court, W.A.; Dailey, R.G. Jr.; Gilmore, C.J.; Bryan, R.F. *J. Amer. Chem. Soc.* **1972**, *94*, 7194-7195.

[39]   Kupchan, S.M.; Schubert, R.M. *Science* **1974**, *185*, 791-793.

[40]   Shang, T.-M.; Chen, Z.-Y.; Linc, C. *Acta Pharmacol. Sin.* **1981**, *2*, 128-130.

[41]   Yang, S.-X.; Gao, N.-L.; Xie, S.-S.; Zhang, W.R.; Long, Z.-Z.. Int. *J. Inmunopharmacol.* **1992**, *14*, 963-969.

[42]   Qiu, D.; Zhao, G.; Aoki, Y.; Shi, L.; Uyei, A.; Nazarian, S.; Ng, J.C.H.; Kao, P.N. *J. Biol. Chem.* **1999**, *274*, 13443-13450.

[43]   Shamon, L.A.; Pezzuto, J.M.; Graves, J.M.; Mehta, R.R.; Wangcharoentrakul, S.; Sangsuwan, R.; Chaichana, S.; Tuchinda, P.; Cleason, F.; Reutrakul, V. *Cancer Lett.* **1996**, *112*, 113-117.

[44]   Tengchaisri, T.; Chawengkirttikul, R.; Rachaphaew, N.; Reutrakul, V.; Sangsuwan, R.; Sirisinha, S. *Cancer lett.* **1998**, *133*, 169-175.

[45]   McCann, J.; Choi, E.; Yamasaki, E.; Ames, B.N. *Proc. Nat. Acad. Sci. USA* **1975**, *72*, 5135-5139.

[46]   Kupchan, S.N. *Fed. Proc.* **1974**, *33*, 2288-2295.

[47]   Xu, J.-N.; Li, C.-C.; Huang, Z.-Q. *Euro J. Pharmacol.* **1999**, *183*, 1700-1701.

[48]   Wei, Y.-S.; Adachi, I. *Acta Pharmacol. Sin.* **1991**, *12*, 406-410.

[49]   Buckanin, R.S.; Chen, S.J.; Frieze, D.M.; Sher, F.T.; Berchtold, S.A. *J. Amer. Chem. Soc.* **1980**, *102*, 1200-1201.

220

[50]   van Tamelen, E.E.; Demers, J.F.; Taylor, E.G.; Koller, K. *J. Amer. Chem. Soc.* **1980**, *102*, 5424-5425.
[51]   Tahara, A.; Akita, H. *Chem. Pharm. Bull.* **1975**, *23*, 1976-1983.
[52]   Garver, L.C.; van Tamelen, E.E. *J. Amer. Chem. Soc.* **1982**, *104*, 867-869.
[53]   Kitahara, Y.; Yoshikoshi, A.; Oida, S. *Tetrahedron Lett.* **1964**, 1763-1770.
[54]   Corey, E.J.; Chaykovsky, M. *J. Amer. Chem. Soc.* **1965**, *87*, 1353-1364.
[55]   van Tamelen, E.E.; Leiden, T.M. *J. Amer. Chem. Soc.* **1982**, *104*, 1785-1786.
[56]   Julia, M.; Julia, S.; Tchen, S.-Y. *Bull. Soc. Chim. Fr.* **1961**, 1849-1853.
[57]   Banerjee, A.K.; Azócar, J.A. *Synth. Communs.* **1999**, *29*, 249-256.
[58]   Cornforth, J.W.; Robinson, R. *J. Chem. Soc.* **1949**, 1855-1865.
[59]   Shiozaki, M.; Mori, K.; Natsui, M. *Agric. Biol. Chem.* **1972**, *36*, 2539-2546.
[60]   Chitty, S.L.; Rao, S.S.K.; Dev, S.; Banerjee, D.K. *Tetrahedron* **1966**, *22*, 2311-2318.
[61]   Banerjee, A.K.; Canudas-González, N.; Nieto, G.C.; Peña-Matheud, C.A. *J. Chem. Res.* **1990**, 266-267.
[62]   Wolinsky, J.; Lau, R.; Kamsher, J.; Cimarusti, C.M. *Synth. Communs.* **1972**, *2*, 327-330.
[63]   Nasipuri, D.; Mitra, A.K. *J. Chem. Soc. Perkin Trans I.* **1972**, 285-287.
[64]   McKenna, J.F.; Shaw, F.J.J. *J. Amer. Chem. Soc.* **1937**, *59*, 470-471.
[65]   Olah, G.A.; Hysain, A.; Balaram Gupta, B.G.; Narang, S.C. *Angew. Chem. Int. Ed. Engl.* **1981**, *20*, 690-691.
[66]   Yang, D.; Ye, X.-Y.; Xu, M.; Pang, K.'W.; Zou, N.; Letcher, R.M. *J. Org. Chem.* **1998**, *63*, 6446-6447.
[67]   Lipshutz, B.N.; Ellisworth, E.L.; Behling, J.R.; Campbell, A.L. *Tetrahedron Lett.* **1988**, *29*, 893-896 and other references cited therein.
[68]   Snider, R.H.; Nohan, R.; Kates, S.A. *Tetrahedron Lett.* **1987**, *28*, 841-844 and other references cited therein.
[69]   Crisp, G.T.; Meyer, A.D. *J. Org. Chem.* **1992**, *57*, 6972.
[70]   Adler, E.; Brasen, S.; Miyake, H. *Acta Chem. Scand.* **1971**, *25*, 2055-2069.
[71]   Becker, N.D.; Bremholt, T.; Adler, E. *Tetrahedron Lett.* **1972**, *41*, 42054208.
[72]   Yang, D.; Ye, X.-Y.; Xu, N. *J. Org. Chem.* **2000**, *65*, 2208-2217.
[73]   Fang, C.-N.; Chang, P.-L.; Hsu, T.-P. *Acta Chinn. Sin.* **1976**, *34*, 197-209.
[74]   Baillie, A.C.; Thomson, R.N. *J. Chem. Soc.* **1968**, 48-52.
[75]   Kakisawa, H.; Hayashi, T.; Yamazaki, T. *Tetrahedron Lett.* **1969**, 301-304.
[76]   Zhang, K.Q.; Bao, Y.; Wu, P.; Rosen, R.T.; Ho, C.T. *J. Agric. Food Chem.* **1990**, *38*, 1194-1197.
[77]   Ruschig, H.; Korger, G.; Aumuller, W.; Wagner, R.; Weyer, R.; Bander, A.; Scholza, *J. Arzenim.-Forsch.* **1958**, *8*, 448-453.
[78]   Tateishi, M.; Kusumi, T.; Kakisawa, H. *Tetrahedron* **1971**, *27*, 237-244.
[79]   Knapp, S.; Sharma, S. *J. Org. Chem.* **1985**, *50*, 4996-4998.
[80]   Lee, J.; Mei, H.S.; Snyder, J.K. *J. Org. Chem.* **1990**, *55*, 5013-5016.
[81]   Majetich, G.; Liu, S.; Fang, J.; Siesel, D.; Zhang, Y. *J. Org. Chem.* **1997**, *62*, 6928-6951.
[82]   Banerjee, A.K.; Azócar, J.A.; Carrasco, M.C.; Laya, M. *Synth. Communs.* **2002**, *31*, 2471-2478.
[83]   Aranda, G.; Fetizon, M. *Synthesis* **1975**, 330-332.

[84]   Murai, A.; Teketsuru, H.; Masamune, T. *Bull. Chem. Soc. Jpn.* **1980**, *53*, 1049-1056.

[85]   Banerjee, A. K.; Carrasco, M. C. In *Studies in Natural Products Chemistry*, (Atta-Ur-Rahman), Elsevier, Amsterdam, **1994**, Vol *14* pp. 667-702.

Atta-ur-Rahman (Ed.) *Studies in Natural Products Chemistry, Vol. 29*

SEARCH FOR BIOACTIVE NATURAL PRODUCTS FROM UNEXPLOITED MICROBIAL RESOURCES

MASAMI ISHIBASHI

Graduate School of Pharmaceutical Sciences, Chiba University, 1-33 Yayoi-cho, Inage-ku, Chiba 263-8522, Japan

ABSTRACT: The Myxomycetes (true slime molds) are an unusual group of primitive organisms that may be assigned to one of the lowest classes of eukaryotes. As their fruit bodies are very small and it is very difficult to collect much quantity of slime molds, few studies have been made on the chemistry of myxomycetes. Cultivation of the plasmodium of myxomycetes in a practical scale for natural products chemistry studies is known only for very limited species such as *Physarum polycephalum*. We recently studied the laboratory-cultivation of myxomycetes and several species have been successfully cultured in agar plates. Chemical constituents of cultured plasmodia of several species of myxomycetes of the genera *Didymium* and *Physarum* were examined to obtain several sterols, new lipid, or pyrroloiminoquinone derivatives. Previous studies on the chemistry of the secondary metabolites of myxomycetes by other groups are also described here.

1. INTRODUCTION

Natural Products continue to provide a great structural diversity and offer major opportunities for finding novel low-molecular-weight lead compounds. No more than 10% of the earth's biodiversity has been examined for biological activity tests [1]. How to access this unexplored natural diversity is quite an important problem. Among the many unexplored organisms, here we focus on "Myxomycetes".

The Myxomycetes (true slime molds) are an unusual group of primitive organisms that may be assigned to one of the lowest classes of eukaryote [2]. In the assimilative phase of their life cycle (Figure 1)

224

they form a free-living, multinucleate, acellular, mobile mass of protoplasm, called a plasmodium, which feeds on living bacteria. Under certain conditions, the plasmodium undergoes sporulation to develop into small, fungus-like fruit bodies which often have unique structures and colors. Spores, released from fruit bodies, germinate into protozoan-like myxamoeba, which mate to form a zygote which develops into the plasmodial stage. Although the myxomycetes have been recognized for a long time, chemical studies on the secondary metabolites of the myxomycetes are limited so far, which may be mainly attributable to extreme paucity of materials and a lack of adequate knowledge of their laboratory cultivation.

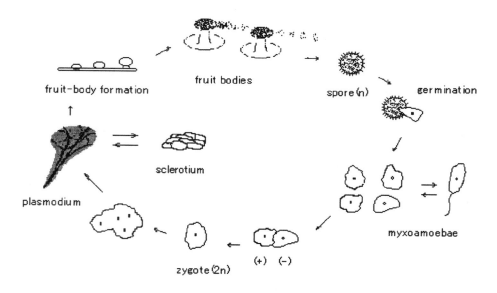

Figure 1. Lifecycle of myxomycetes

Steglich [3] and Asakawa's group [4] have isolated several types of metabolites from field-collected samples of several myxomycete species. Cultivation of the plasmodium of myxomycetes in a considerable scale to carry out chemical studies is known only for very limited species such as *Physarum polycephalum* [5,6]. Cultivation of other myxomycetes for natural products chemistry studies had never been reported. In this chapter we describe the previous studies on the chemistry of secondary metabolites of myxomycetes as well as our recent results of the basic studies on laboratory-culture of myxomycetes.

2. Chemical Constituents of Field-Collected Myxomycetes

2. 1. Bisindoles

Fruit bodies of wild myxomycetes often have bright colors. In 1980 Steglich *et al.* isolated red and yellow pigments from methanol extracts of 2 g of the red fruit bodies of *Arcyria denudata* [7]. Structure of these pigments were elucidated by spectral data as a series of bisindole maleimides, and they were named arcyriarubin B (**1**), C (**2**), arcyriaflavin B (**3**), C (**4**), and arcyrioxepin A (**5**). **1**, **2**, and **5** were red and **3** and **4** were pale yellow pigments. Arcyriaflavins were structurally related to staurosporin, an anitibiotic and antihypertensive isolated from *Streptomyces staurospreus* (vide infra), while compound **1**, **2**, and **5** exhibited medium inhibiting action against *Bacillus brevis* and *B. subtilis*in the platelet diffusion assay. From the same myxomycetes, four other bisindole maleimides, arcyriarubin A (**6**), arcyriaflavin A (**7**),

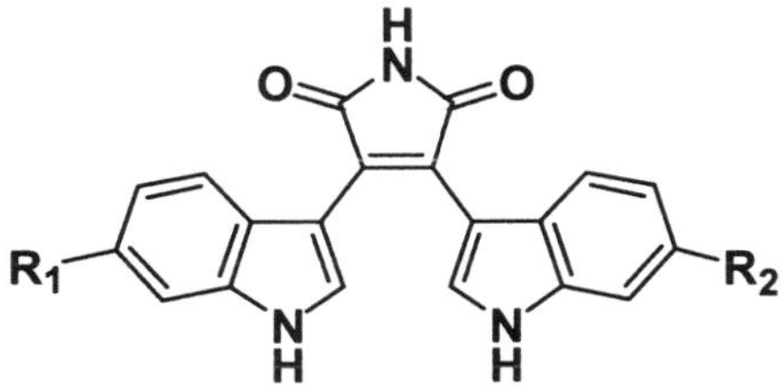

1 R₁=OH, R₂=H
2 R₁=R₂=OH
6 R₁=R₂=H

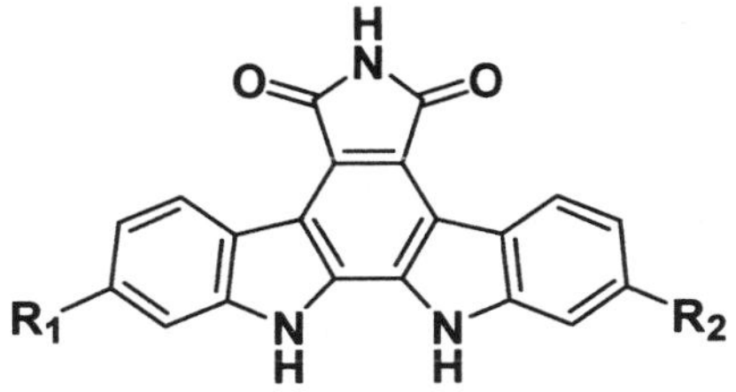

3 R₁=OH, R₂=H
4 R₁=R₂=OH
7 R₁=R₂=H

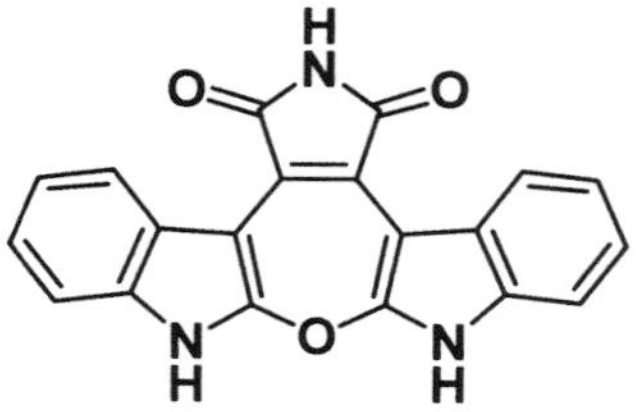

5 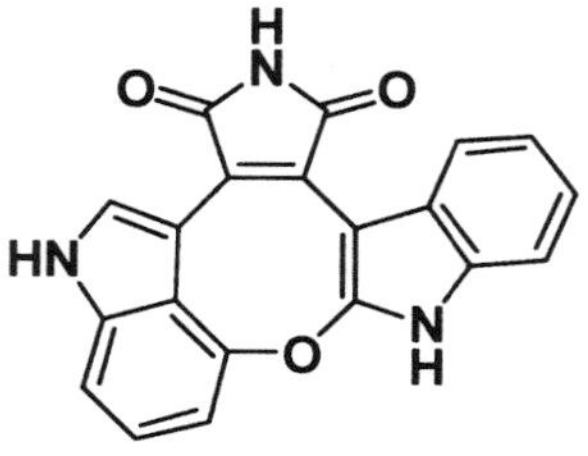   **8** 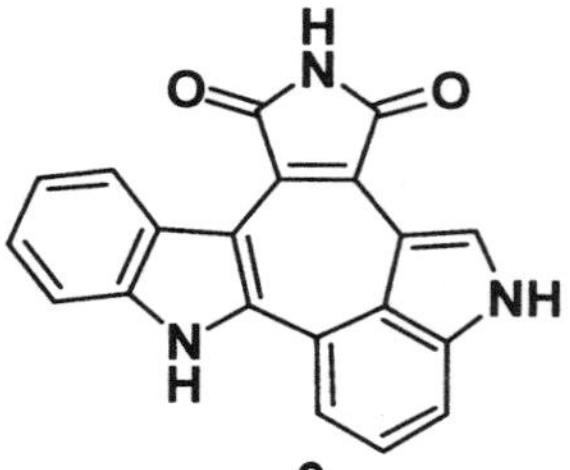   **9**

arcyriaoxocin A (**8**), and arcyriacyanin A (**9**) were also isolated [3].

The structures of these bisindole maleimides were confirmed by synthesis. The reaction of indolyl magnesium bromide (**10**) with 2,3-dibromo-*N*-methylmaleimide (**11**) in toluene led to bisindolylmaleimide (**12**), which was converted into arcyriarubin A (**6**) through alkaline hydrolysis followed by heating with ammonium acetate (Scheme 1). The reaction of **10** and **11** in THF yielded monosubstitution product (**13**) which after protection of the indole NH group with the Boc residue was used to prepare unsymmetrically substituted bisindolylmaleimide, arcyriarubin B (**1**) [8].

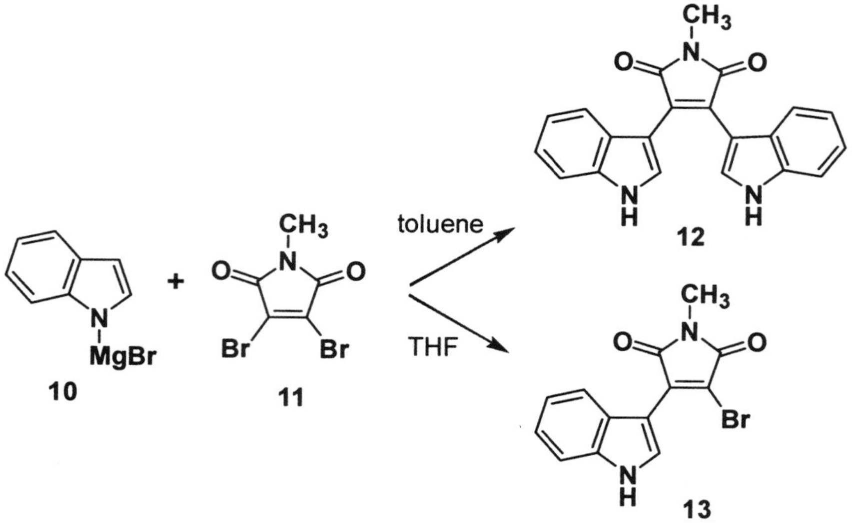

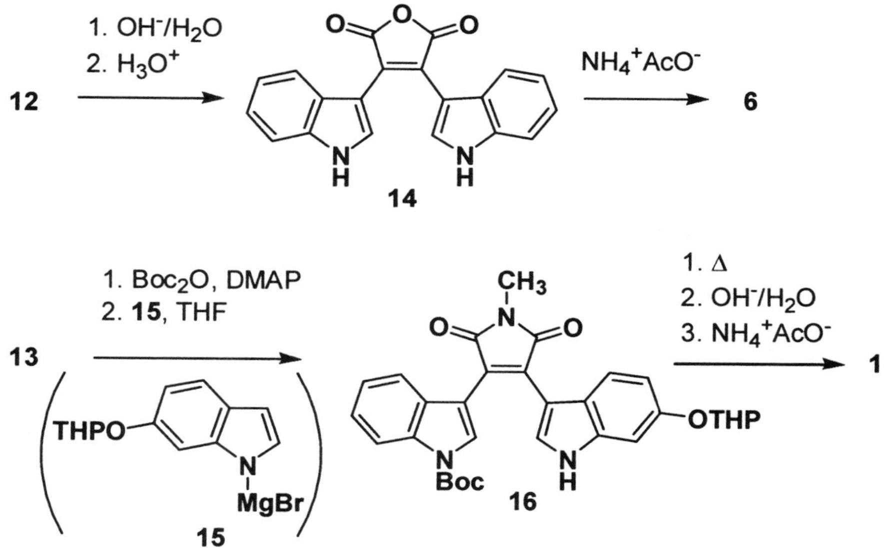

Scheme 1

Arcyriaoxocin A (**8**) was also prepared from the bisindolylmaleimide derivative (**17**) through acid catalyzed oxidative cyclization (Scheme 2). **17** was obtained by the similar procedures to those in Scheme 1 from reactions with a bromomaleimide and an indolyl magnesium salt [9].

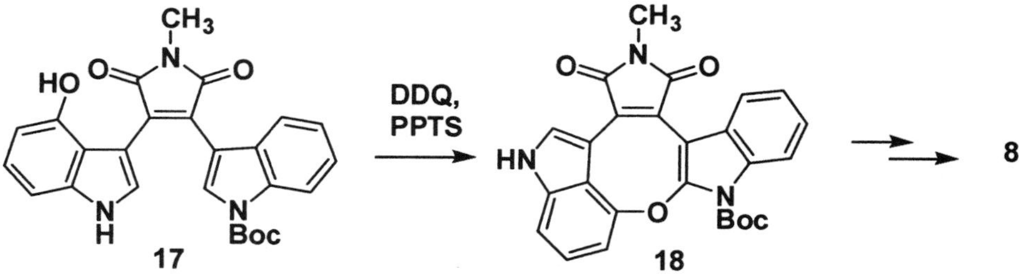

Scheme 2

Evidence for the structure of arcyriacyanin A (**9**) was provided by synthesis through three different routes (Scheme 3) [10]. In the first

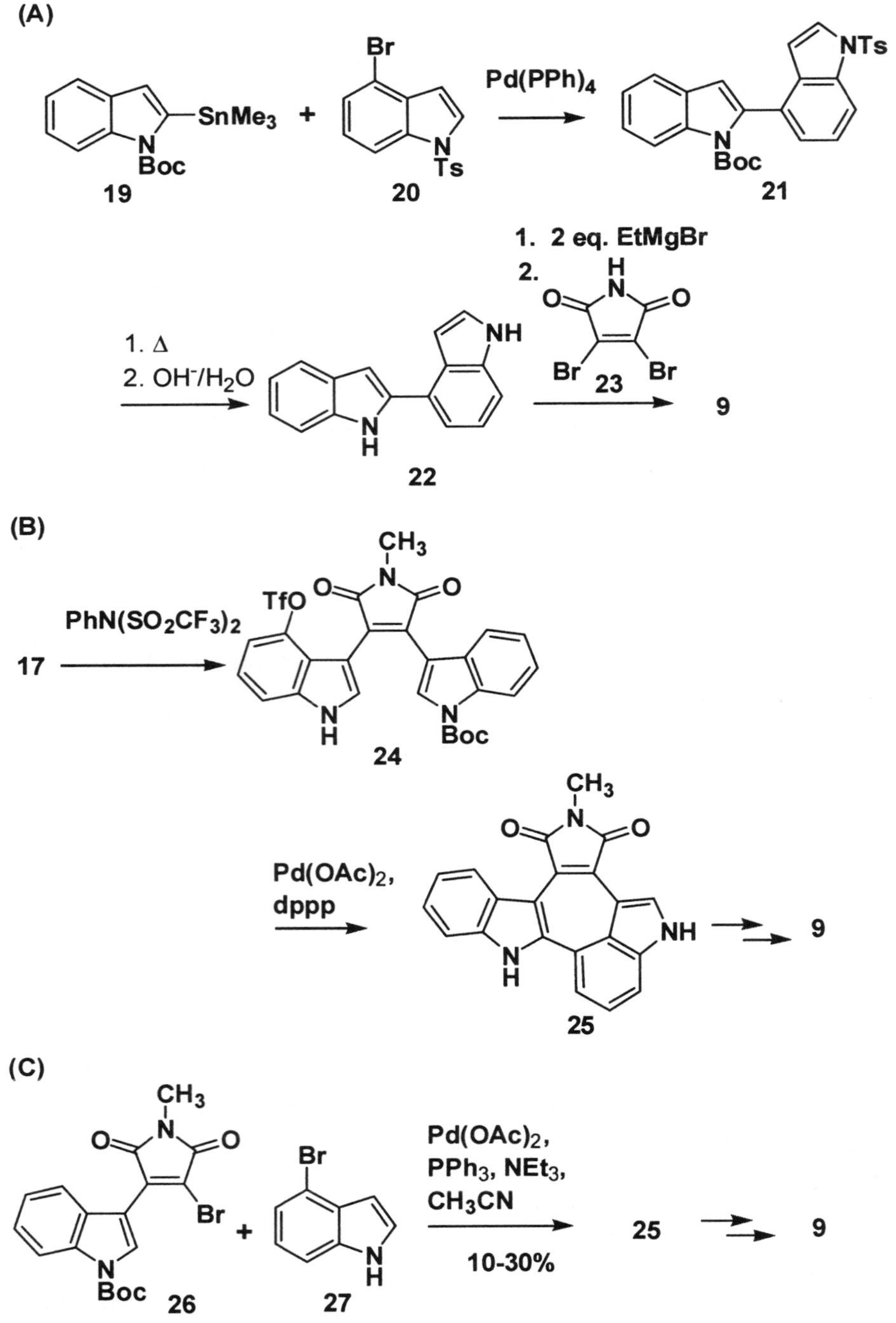

(A)

(B)

(C)

Scheme 3

synthesis (Scheme 3, (A)), Stille coupling of stannylindole (**19**) with 1-tosyl-4-bromoindole (**20**) followed by removal of the *N*-protecting groups from the coupling product (**21**) by alkaline hydrolysis afforded a 2,4'-biindole (**22**), the bisbromomagnesium salt of which was treated with 3,4-dibromomaleimide (**23**) to give arcyriacyanin A (**9**). In the second synthesis (Scheme 3, (B)) the hydroxybisindole compound (**17**, Scheme 2) was converted into its triflyl derivative (**24**) with *N*-phenyltriflimide and DMAP, and the Heck cyclization of **24** took place in DMF with catalytic amounts of palladium (II) acetate and 1,3-bis(diphenylphosphino)propane (dppp), and a large excess of triethylamine to yield *N*-methylarcyriacyanin A (**25**), which was readily converted to arcyriacyanin A (**9**) by alkaline hydrolysis, acidic workup, and reaction of the resulting anhydride with hexamethyldisilazane. The key step of in the third approach (Scheme 3, (C)) was a domino Heck reaction between bromo(indolyl)maleimide (**26**) and 4-bromoindole (**27**), which afforded *N*-methylarcyriacyanin A (**25**) in up to 33% yield.

Murase *et al.* studied the biological activity of synthetic arcyriacyanin A (**9**) [11]. The compound **9** was tested for inhibitory activity against a panel of 39 human cancer cell lines. The effective inhibitory dose of **9** was rather high (> 10^{-5} M), however, it was noted that the compound proved to have unique selective inhibition when compared with the data base of anticancer reagents developed to date, suggesting that arcyriacyanin A (**9**) might be a useful leading compound for anticancer drugs with a novel mode of action. Compound **9** also exhibited inhibitory activity against protein kinases, which were critical in regulating cell growth and differentiation and encoded by many oncogenes. The compound inhibited protein kinase C (PKC) and protein tyrosine kinase (PTK) over 1 µg/ml, but not protein kinase A (PKA) or calmodulin-dependent protein kinaseC (CAMP) in the range of 1-100 µg/ml.

Recently considerable attention has been focused on the metabolites belonging to bisindolylmaleimides such as staurosporine (**28**) [12], UCN-01 (**29**) [13], rebeccamycin (**30**) [14], which were produced by the family of *Streptomyces*, *Actinomycetes*, and *Saccharothrixes*. These metabolites cause topoisomerase I mediated DNA cleavage, potent inhibition of protein kinase C and cell-cycle-regulating cyclin-dependent kinase (CDK), and cell-cycle checkpoint inhibition [15]. It seems interesting that myxomycetes also contain the bisindole metabolites having related structures to **28 – 30**.

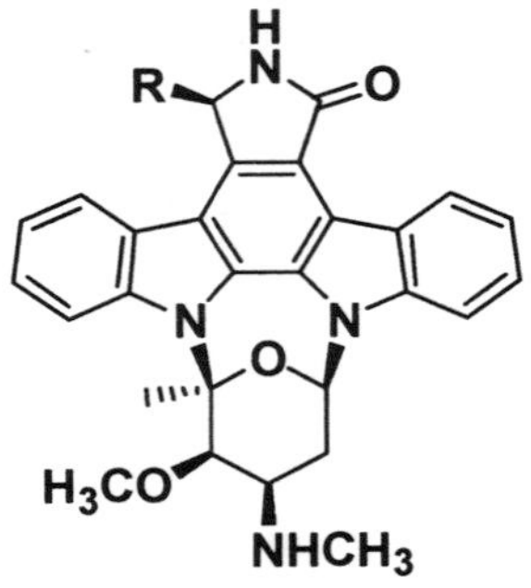

28 R=H
29 R=OH

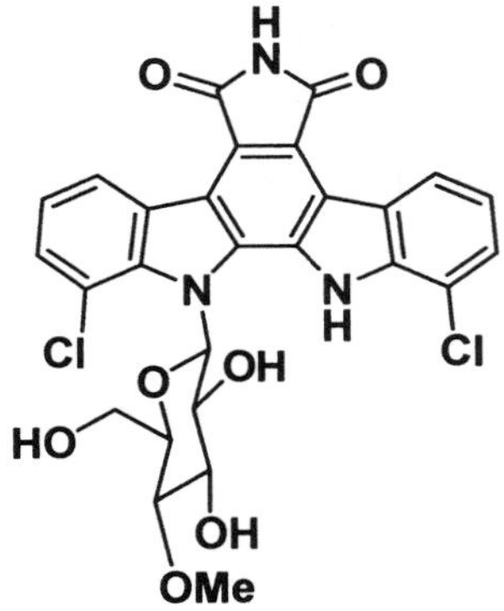

30

Asakawa *et al.* collected 1.65 kg of a different species of myxomycete, *Lycogala epidendrum*, and from the ethyl acetate extract of this myxomycete, three dimethyl pyrroledicarboxylates attached to two indoles, named lycogarubins A-C (**31-33**), were isolated [16]. Their structures have been established by a combination of 2D NMR spectroscopy, chemical degradation, and X-ray crystallographic analysis of the pentamethylated compound (**34**). Lycogarbins A-C (**31-33**) did not show antimicrobial activity at 400 μg/mL, while lycogarbin C (**33**) showed anti HSV-1 virus activity (IC$_{50}$ 17.2 μg/ml) *in vitro*.

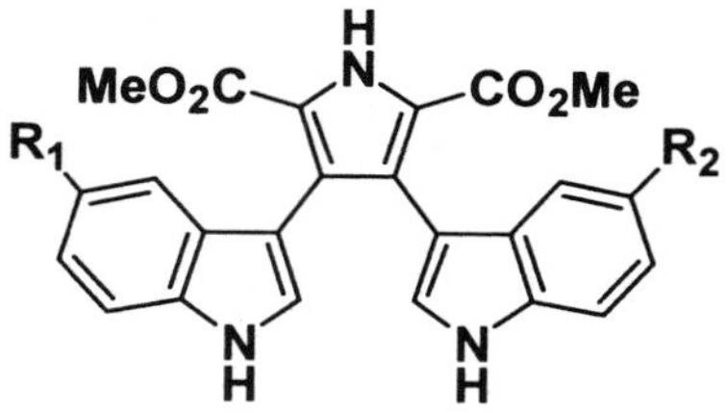

31 R$_1$=R$_2$=OH
32 R$_1$=OH, R$_2$=H
33 R$_1$=R$_2$=H

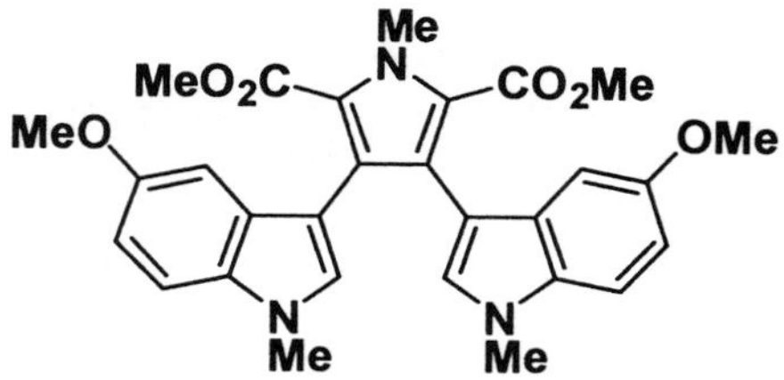

34

Steglich *et al.* also studied the constituents of the fruit bodies of *Lycogala epidendrum* [17] and isolated not only lycogarubins A-C (**31-33**) but also lycogalic acid A (**35**) [18], staurosporinone (**36**) [19], and arcyriarubin A (**6**) and arcyriaflavin A (**7**) previously isolated from *Arcyria denudata*.

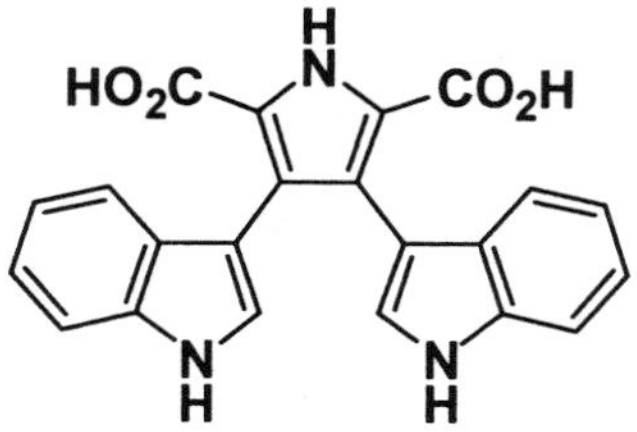

35 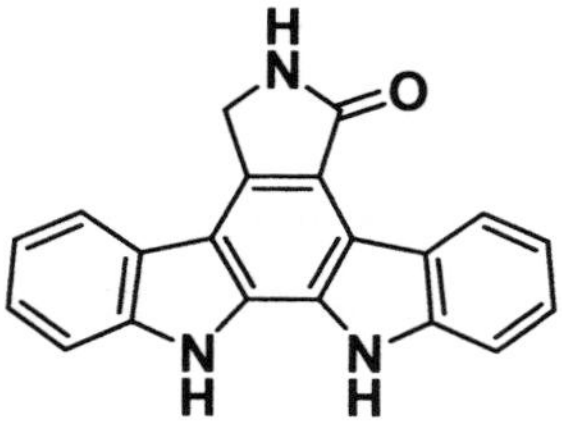   36

Hoshino *et al.* described that in cultures of *Chromobacterium violaceum*, lycogalic acid (**35**, = chromopyrrolic acid) was derived from tryptophan [18].   It was therefore assumed that lycogalic acid (**35**) was formed by oxidative dimerization of 3-(indol-3-yl)pyruvic acid followed by reaction of the resulting 1,4-dicarbonyl intermediate with an equivalent of ammonia.   On the basis of this idea, lycogarubin C (**33**) was synthesized from methyl 3-(indol-3-yl)pyruvate (**37**) in a simple one-pot process (Scheme 4).

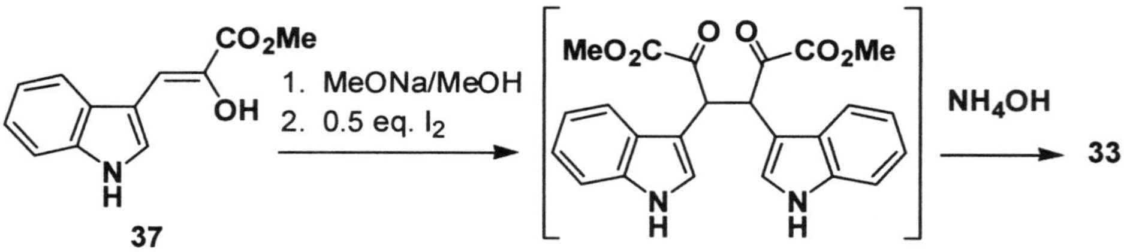

Scheme 4

2.2. Naphtoquinones

In addition to the bisindole alkaloids described above, some naphtoquinones are known as myxomycetes pigments. Steglich *et al.* isolated two naphtoquinone compounds, trichione (**38**) and homotrichione (**39**), as the red main pigments of fruit bodies of the myxomycete *Trichia floriformis* (= *Metatrichia floriformis*) [20]. Trichione (**38**) contains a 2,5-dihydroxynaphtoquinone chromophore, substituted in position 3 by an unsaturated side chain carrying a terminal hemimalonate moiety, while homotrichione (**39**) differs by the presence of two additional CH_2 groups in the side chain. In addition, this myxomycete contained the bisindole pigments, arcyriaflavin B (**3**) and C (**4**), previously isolated from *Arcyria denudata*.

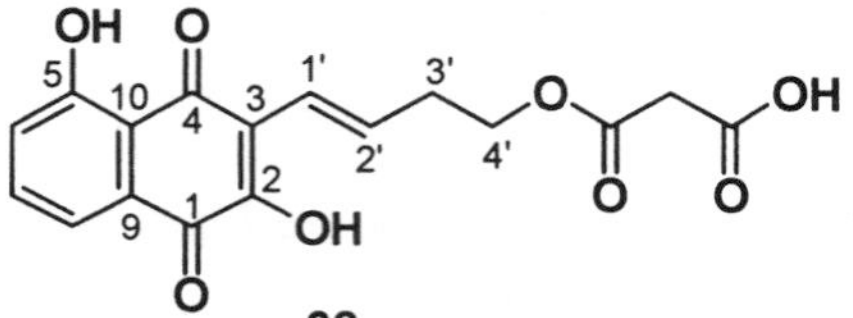

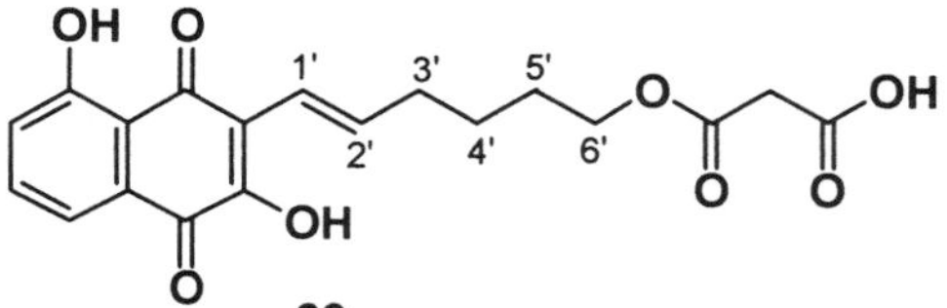

From the myxomycete of the same genus, *Metatrichia vesparium*, vesparione (**40**), a naphtha[2,3-b]pyrandione derivative was isolated as a red pigment of its fruit bodies, and this compound exhibited antibiotic properties [21].

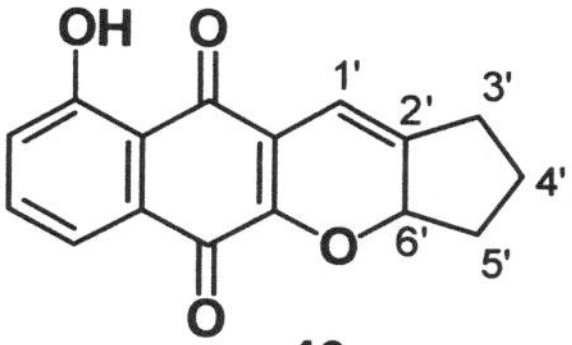

40

Vesparione (**40**) may originate from homotrichione (**39**) by cyclization of the side chain, and its structure was established by synthesis through reaction of naphtoquinone (**41**) and cyclopentanecarboxyaldehyde (**42**) followed by oxidative cyclization in the presence of DDQ (Scheme 5).

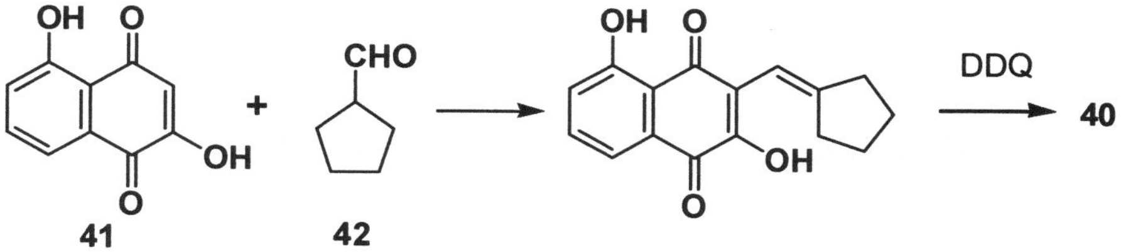

Scheme 5

2.3. Lipids

From the fruit bodies of *Lycogala epidendrum*, Asakawa *et al.* isolated, in addition to lycogarubins A-C (**31-33**), seven unusual polyacetylene triglycerides, named lycogarides A-G (**43-49**) [22,23]. These compounds possess acetylene- and epoxide-containing fatty acid residues, and their relative stereochemistries were established by a combination of two-dimensional NMR spectroscopyand chemical degradation. The absolute structures were elucidated from the modified Mosher's methods using MTPA or 2NMA [24] esters as well as CD exiton chirality methods of their dibenzoate derivatives. lycogarides A-C (**43-45**) completely inhibited germination of the root

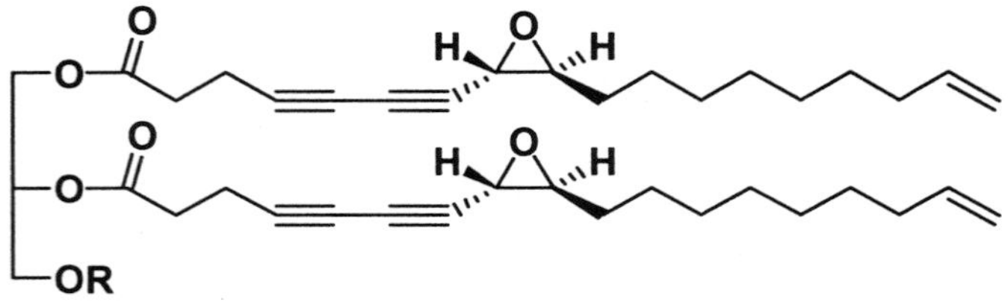

43 R=

44 R= $(CH_2)_nCH_3$ n = 14 ~ 20

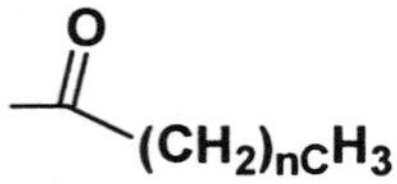

45 R=

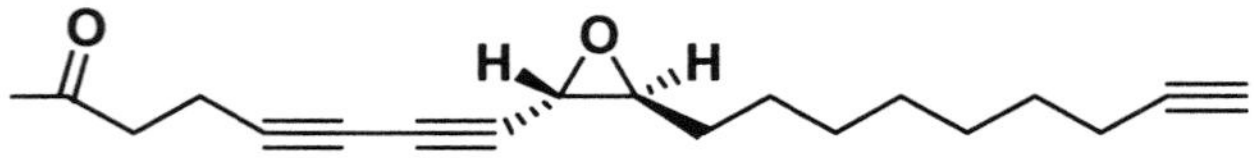

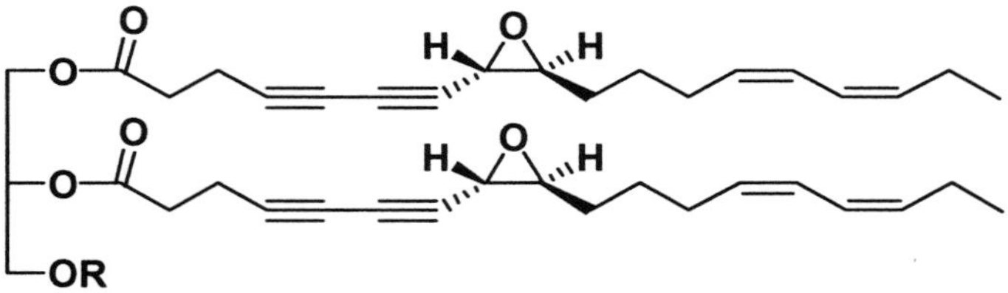

46 R=

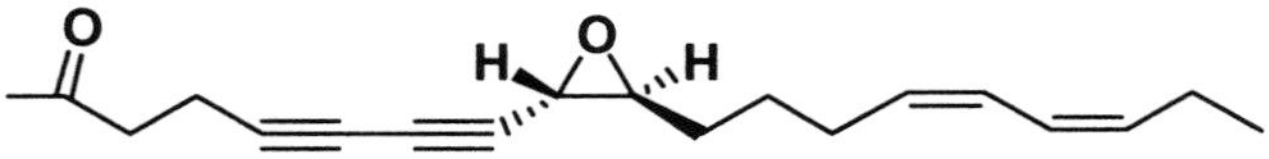

47 R= C_{16}-C_{20} saturated and unsaturated fatty acid unit

48 R= H

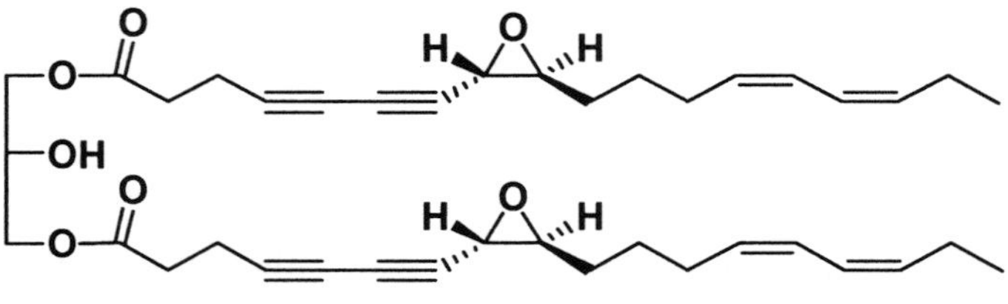

49

and second coleoptile of rice in husk at 100 ppm.

Řezanka analysed the fatty acid composition of several species of myxomycetes. In addition to the common fatty acids, polyunsaturated and methylene non-interrupted polyunsaturated fatty acids, for example, with 5, 9- and/or 5, 11-double bonds, were identified by GS-mass spectrometry of their corresponding oxazolines [25]. Multibranched polyunsaturated fatty acid and its four glycosides (**50 - 54**) were isolated from seven different myxomycetes [26]. The absolute configurations of hydroxyl groups were determined by modified Mosher's method, and the glycosides were revealed to contain glucose, mannose, and rhamnose. It may be interesting that these fatty acid constituents contained by the seven species of myxomycetes were different (Table 1).

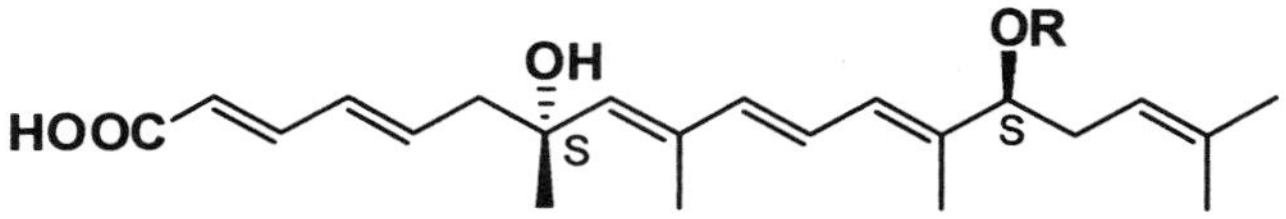

50 R=H
51 R=α-D-Glu

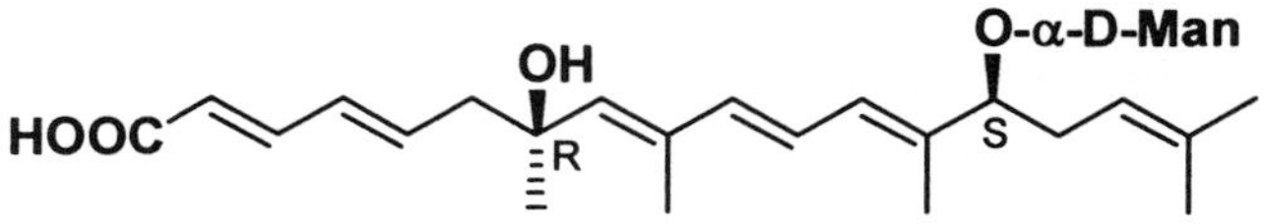

52

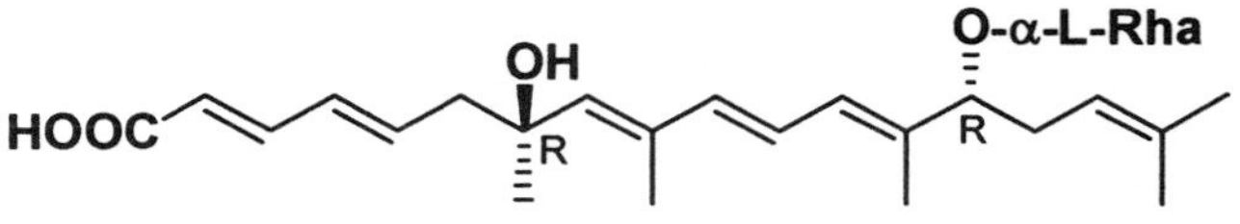

53

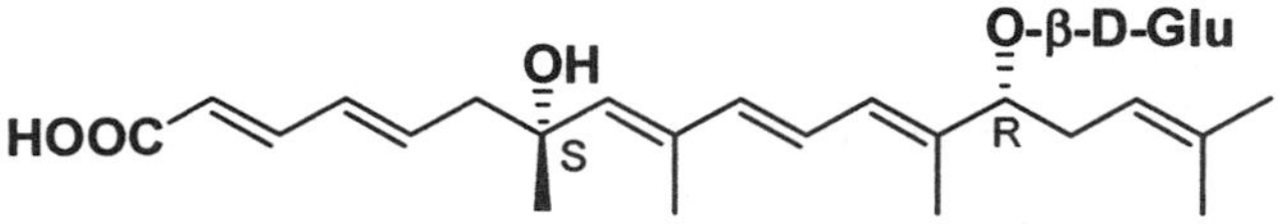

54

Table 1. Distribution of the fatty acid derivatives identified in the seven species of myxomycetes

Species	Order and Family	fatty acid constituents
Arcyria cinerea	Trichiales, Arcyriaceae	**50, 51**
Arcyria denudata		**50, 51**
Arcyria nutans		**50, 51**
Trichia varia	Trichiales, Trichiaceae	**51**
Fuligo septica	Physarales, Physaraceae	**52**
Physarum polycephalum		**53**
Lycogala epidendrum	Liceales, Enteridiaceae	**54**

2.4. Constituents of Wild Plasmodium

Compounds described above are all those isolated from wild myxomycetes in the stage of fruit bodies in their lifecycle. In the lifecycle of myxomycetes (Figure 1), there is a stage of plasmodium before that of fruit body. A few studies of chemical constituents of plasmodial stages of myxomycetes have been described.

One of the most common myxomycetes is *Fuligo septica*, which form yellowish slime masses (plasmodia) and can be often found in woods after damp weather. Methanol extracts of the plasmodia or aethalia of *F. septica* was studied by Steglich *et al.* to yield a yellow solution which turned orange on acidification with hydrochloric acid. A greater part of the pigments was present in the form of salts, from which the free pigments were obtained on acidification. Separation by chromatographies on Sephadex LH-20 with methanol and/or acetone/methanol (4:1), in the absence of light using a cooled column, afforded a main pigment fuligorubin A (**55**) [27]. Spectral studies revealed that fuligorubin A (**55**) possessed a tetramic acid moiety with undecapentaene chain coupled via a carbonyl group. The UV spectrum of **55** showed absorption maxima at λ 243 and 425 nm, while on addition of alkali the long wavelength absorption shifts hypsochromically to λ 377 nm, and a color change from orange to lemon-yellow occurred. The absolute configuration of **55** was determined on the basis of comparison of the CD spectral data with related known tetramic acid derivatives. Fuligorubin A (**55**) was thought to be involved in photoreceptor and energy conversion processes during the lifecycle of *F. septica*.

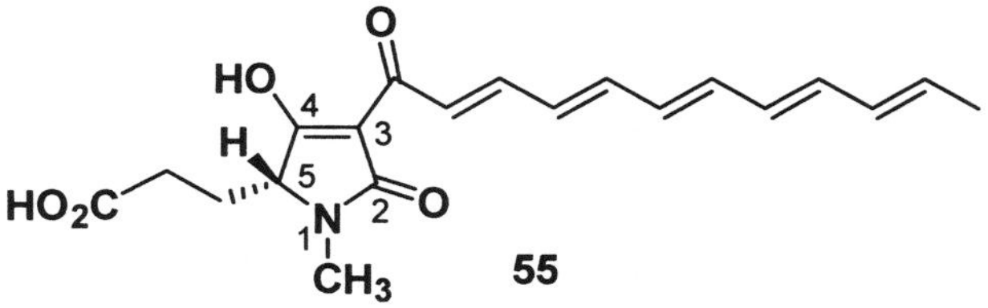

Further structural confirmation of fuligorubin A (**55**) was provided by its short, efficient total synthesis, achieved by Ley *et al.* using coupling of t-butyl 4-diethylphosphono-3-oxobutanethioate (**56**) with deca-2,4,6,8-tetraenal (**57**) and subsequent substitution with a glutamic acid derivative (**58**) followed by Dieckmann cyclisation (Scheme 6) [28]. Therefore, it was suggested that in the biosynthesis of fuligorubin A (**55**) D-glutamic acid was condensed with a heptaketide chain.

From plasmodia of another common myxomycetes *Ceratiomyxa fruticulosa*, five 6-alkyl-4-methoxypyran-2-ones, ceratiopyron A (**60**) ~ D (**63**), and ceratioflavin A (**64**) were isolated [29]. Their structures were determined by spectroscopic and chemical invesitigations, and five compounds possessed side chains with varying

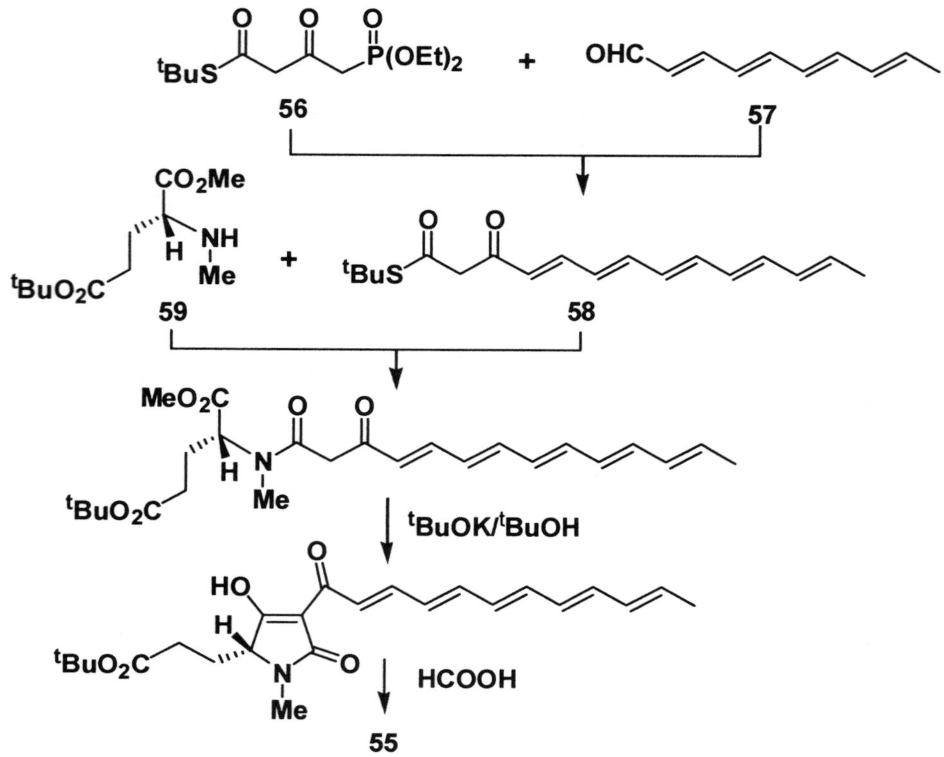

Scheme 6

degrees of unsaturation. The positions of the double bonds were defined by mass spectral analyses of the silylenol ethers of the OsO$_4$ oxidation products.

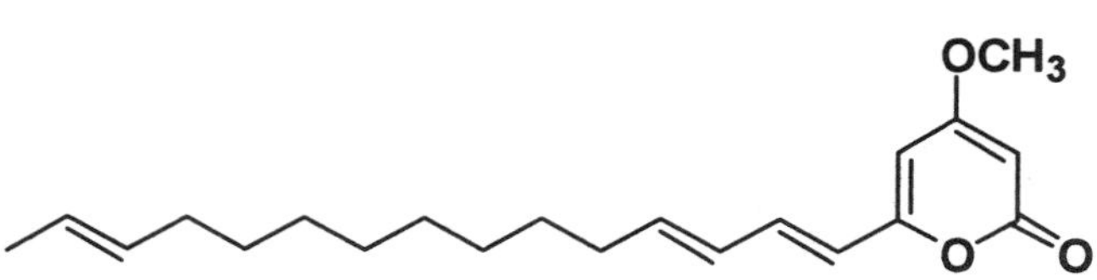

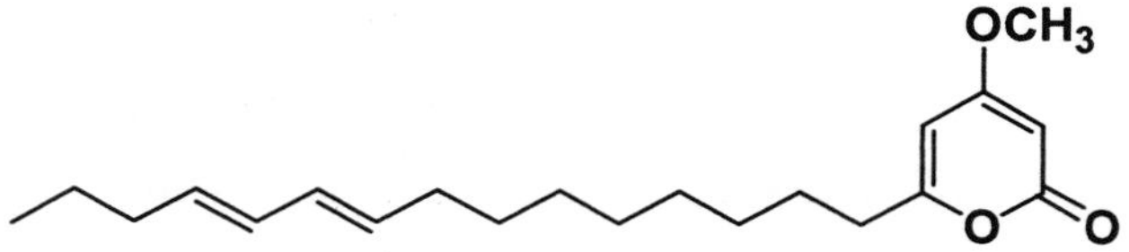

61

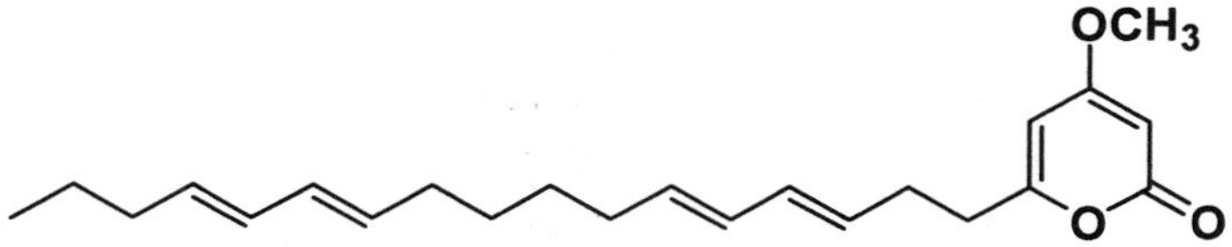

62

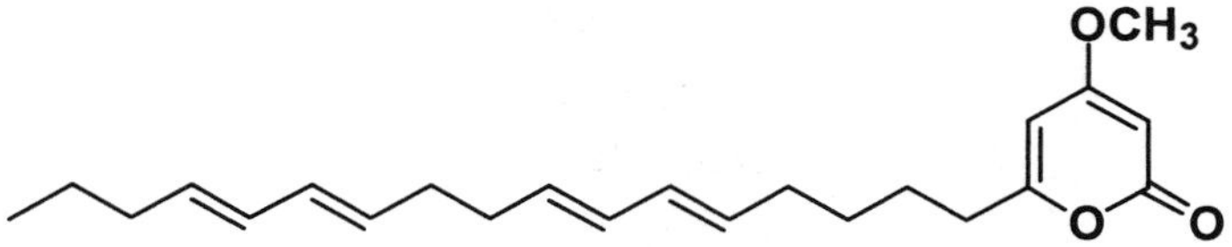

63

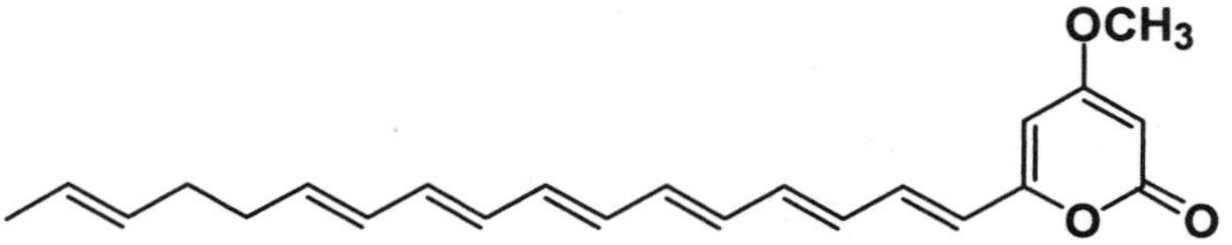

64

3.   Chemical Constituents of *Physarum polycephalum*

3.1.   Plasmodial constituents of *P. polycephalum*

Cultivation of the plasmodium of myxomycetes in a practical scale has been known only for very limited species.   *Physarum polycephalum* may be the most familiar species of myxomycetes because of the ease with which its plasmodium may be grown in the laboratory.   It has been therefore used extensively in physiological, biochemical, and genetic studies as well as in schools for model organisms.   Mass culture of the plasmodia of *Physarum polycephalum* was possible to collect enough quantity of materials, and sterol contents of the plasmodia of this species had been studied since 1950's [30-32]. The plasmodia of this species are noticeably yellow-colored and can reach the size of up to 2 m^2 in its natural habitat. Steglich and Steffan's group were interested in the plasmodial yellow pigments of *P. polycephalum*, which were considered to act as photoreceptors, and they cultured the plasmodia on oat flakes and harvested after 3 days. From the methanol extracts of the plasmodia, physachrome A (**65**) was isolated by chromatography on Sephadex LH 20 with methanol in the dark. Its structure was established as all-*trans* *N*-[11-(2-acetylamino-3-hydroxyphenyl)-2,4,6,8,10-undecapentaenoyl]-*S*-glutamine (**65**) on the basis of spectroscopic evidence and hydrogenation to a decahydro derivative. The *S*-configuration of the glutamine residue of **65** was determined by hydrolysis of its decahydro derivative with 6N HCl which yielded *S*-glutamic acid, which was proved by GC analysis of its *N*-pentafluoropropionyl diisopropylester on a Chirasil-D-Val column [33]. Physachrome A (**65**) showed some structural similarities to fuligorubin A (**55**), a plasmodial pigment from the myxomycete *Fuligo septica*.

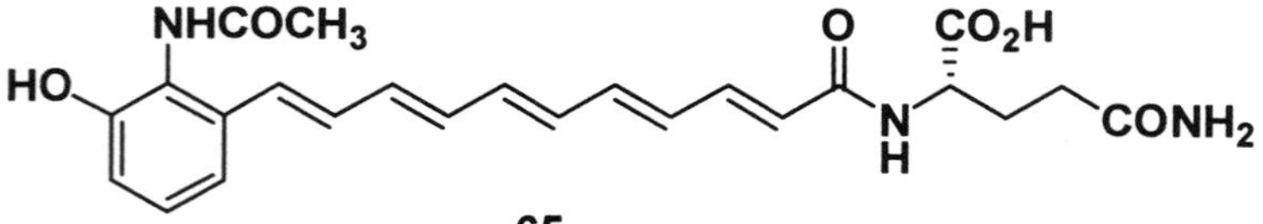

65

Plasmodia of *P. polycephalum* can also be collected by liquid suspension culture. The extract of the suspension culture of plasomodia of *P. polycephalum* was subjected to repeated chromatographies in the dark at low temperature (4 °C) on Sephadex LH-20 with methanol and methanol/water (19:1) as eluents to yield an amorphous orange-red powder. This pigment showed, however, strong line broadening in the ^{1}H NMR spectrum because of binding with calcium and other metals. The pigment was therefore partitioned between citrate/HCl buffer and ethyl acetate, and the organic layer was extracted with aqueous EDTA solution and then evaporated to yield a purified pigment, named phsarorubinic acid A (**66**), which showed narrow signals with definite splitting in the ^{1}H NMR in CD$_3$OD/CDCl$_3$ solution [34]. The absolute configuration of C-5 position of **66** was deduced as *S* from comparison of its CD spectrum with that of a synthetic model compound (**67**), which was prepared from *N*-methyl-L-serine. Although this pigment (**66**) had a related structure to fuligorubin A (**55**), previously obtained from *Fuligo septica*, the configurations of C-5 position were different from each other; this result was confirmed by the fact that the CD spectrum of these two compounds showed opposite Cotton effect. Further investigations on the pigments of *P. polycephalum* carried out by Steffan *et al.* led to isolation of an analog of **66**, physarorubinic acid B (**68**), containing one less double bond [35].

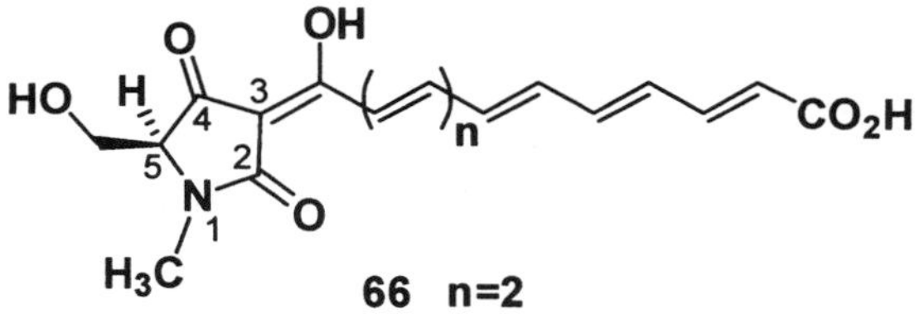

66 n=2
68 n=1

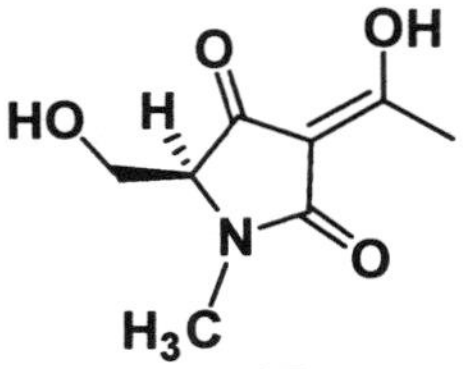

67

Steffan *et al.* studied the difference in the pigments of the suspension cultured plasmodia of *P. polycephalum* under dark and light conditions. On the basis of HPLC analyses, physarorubinic acids A (**66**) and B (**68**) were contained in both cultures under dark and light conditions. Apparently it was revealed that light stimulated the formation of two metabolites. Because these metabolites were very sensitive to light, all steps from incubation and extraction to chromatography were carried out under exclusion of light and at low temperature (4 °C). After careful purification procedures, two pigments, polycephalin B (**69**) and C (**70**) were obtained in pure form. These two pigments possessed a cyclohexene ring with two side-chains of polyeneacyltetramic acid residues [35]. The absolute configurations at the C-5 and C-5' positions of the two tetramic acid units were determined through comparison of CD spectra of the hydrogenated natural compound and a synthetic tetramic acid (**67**). The relative configuration of the C-3" and C-4" in the cyclohexene moiety was suggested as trans by the coupling constants (*J*=8.8 Hz). The absolute configuration of the chiral centers in the cyclohexene ring was supposed to be 3"R,4"R as found by comparison of CD spectra of the natural product (**70**) and synthetic model compounds [36].

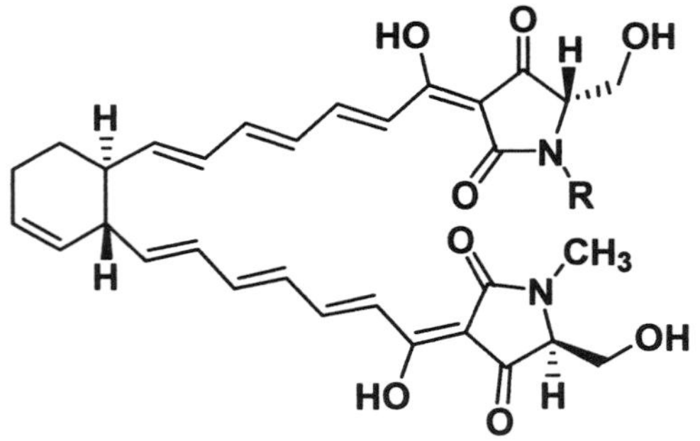

69 R = H
70 R = CH₃

Steffan *et al.* further investigated the yellow pigments in the extract of suspension-cultured plasmodia of *P. polycephalum* and isolated another yellow pigment, chrysophysarin A (**71**) [37]. This compound contained an imidazole ring, a tetraene system, and a leucine residue whose carboxy group was modified to an acetyl moiety. The absolute configuration of the modified leucine residue was assigned by the synthesis of *N*-(3,3-dimethylacryloyl) derivative of (*S*)-leucine (**72**) and by comparison of the corresponding CD spectra. The biosynthesis of chrysophysarin A (**71**) was elucidated to feeding labeled acetate to plasmodia, and L-leucine, acetate, and histidine were proposed as biosynthetic building blocks.

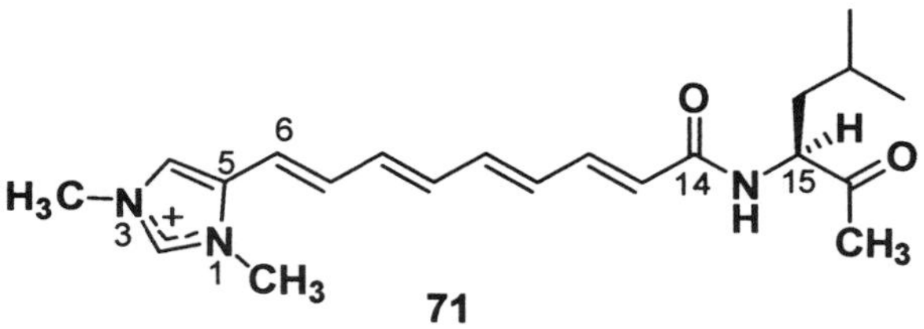

3.2.   Constituents of myxoamoebae of *P. polycephalum*

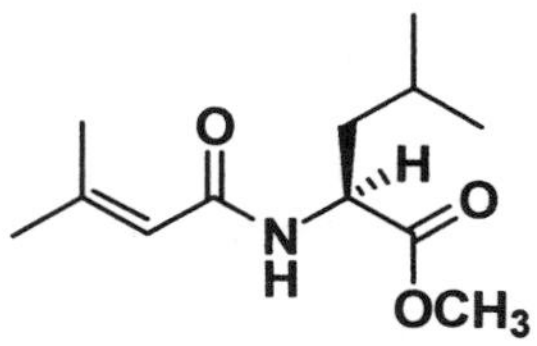

The plasmodial stage of *Physarum polycephalum* can be cultured as described above. The myxoamoeboid stage of this species can be also cultured on a lawn of bacteria *Aerobacter aerogenes* on agar medium in the dark at 24 °C. In this stage, *Physarum* grows as unicellular and motile myxoamoebae, which recognize different mating types and fuse to form zygotes then differentiate into multinuclear, acellular plasmodia (Fig. 1). Murakami-Murofushi *et al.* examined the sterol contents of the myxoamoebae of *P. polycephalum* and found that the myxoamoebae contained poriferasterol, Δ^5-ergostenol, and 22-dihydroporiferasterol in the ratio of 80:15:5 [38]. Murakami-Murofushi *et al.* further studied the lipid fraction of this myxomycete and succeeded in isolating novel lysophosphatidic acid (PHYLPA) (73) [39]. This compound, composed of cyclic phosphate and cyclopropane-containing hexadecanoic acid, inhibited more than

80% of the affinity-purified calf thymus DNA polymerase α activity at a concentration of 10 μg/mL. Inhibition was observed for DNA polymerase α but not for DNA polymerase β or γ from various eukaryotic species, nor did it inhibit DNA polymerase I from *E. coli*.

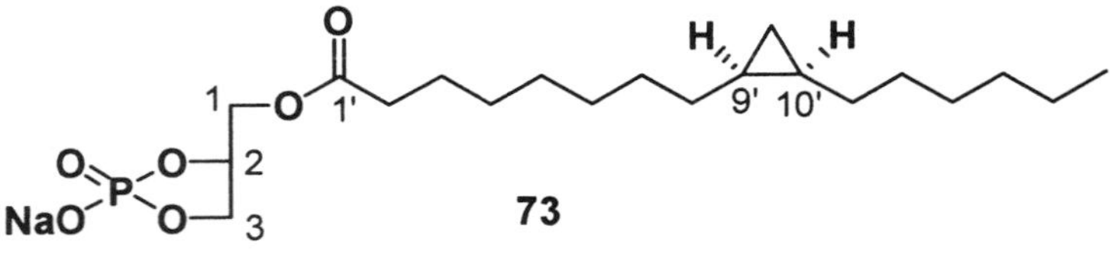

73

Four possible stereoisomers of PHYLPA (**72**) were synthesized in enantioselective manners by Kobayashi *et al.* [40] Preparation of one diastereomer of cyclopropane-containing hexadecanoic acid (**81**) was summarized in Scheme 7, starting with enzymatic hydrolysis of *meso* diester (**74**).

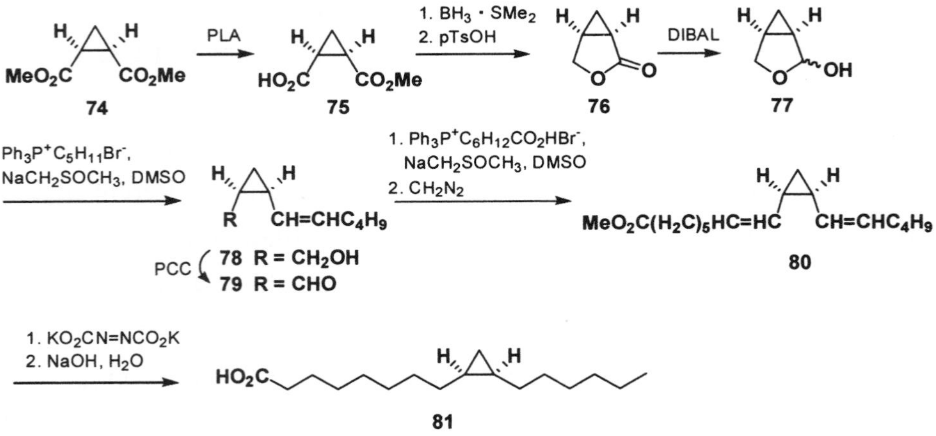

Scheme 7

Coupling of the acid (**81**) with (*R*)-glycerol acetonide (**82**), followed by formation of cyclic phosphate by reacting with tris(1,2,4-triazole) phosphate to afford one isomer of PHYLPA (**73**) (Scheme 8).

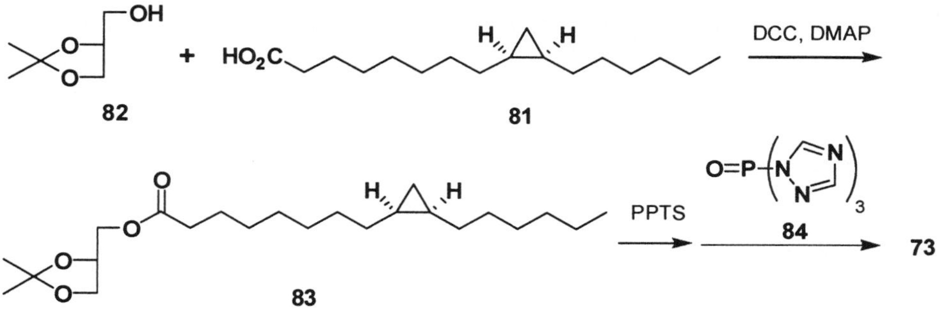

Scheme 8

Three other isomers (**85**, **86**, and **87**) were also prepared in a similar manner. Discrimination of the isomers by spectroscopic analysis was found difficult because ^{1}H and ^{13}C NMR spectra of four

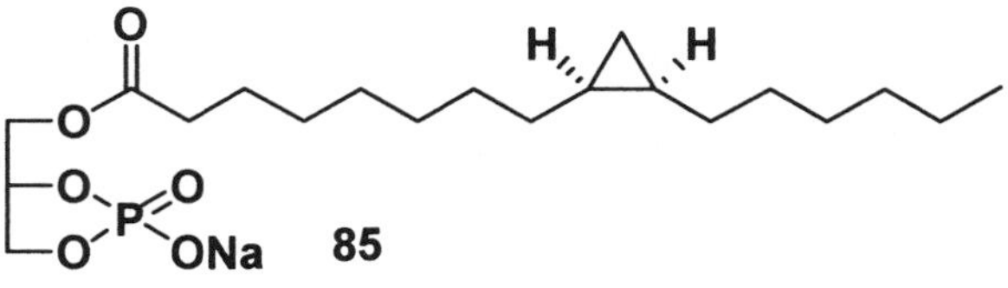

85

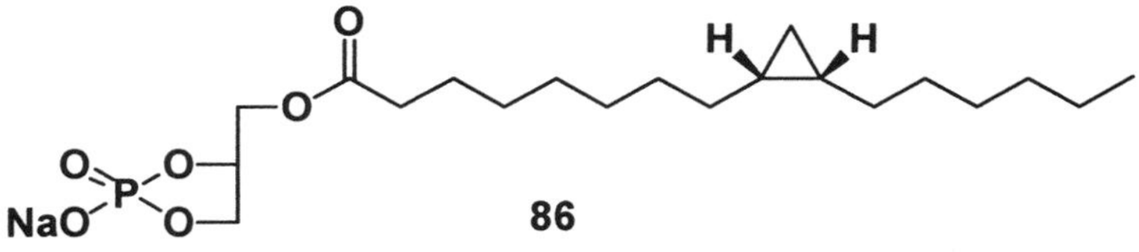

86

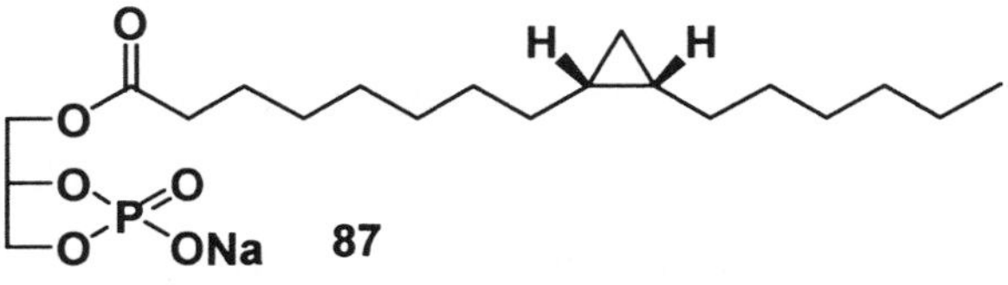

87

isomers (**73**, **85**, **86**, and **87**) and natural PHYLPA were indistinguishable. However, inhibition activity of each isomer for immunoaffinity-purified calf thymus DNA polymerase α was examined, and only **73** with 2*R*, 9'*S*,10'*R*-configurations exhibited the comparable activity as natural PHYLPA. Relative activities of other isomers compared to natural PHYLPA were estimated to be ca 1/3, <1/10, and <1/10 for **85**, **86**, and **87**, respectively. The structure of natural PHYLPA was therefore concluded as **73** (sodium 1-*O*-[(9'*S*,10'*R*)-9',10'-methanohexadecanoyl]-*sn*-glycerol 2,3-cyclic phosphate). A series of synthetic derivatives of PHYLPA were prepared and tested for their ability to inhibit tumor cell invasion and metastasis. Among these, 1-palmitoyl cyclic lysophosphatidic acid (**88**) was most potent in inhibiting invasion, with 93.8% inhibition at the concentration of 25 μM [41].

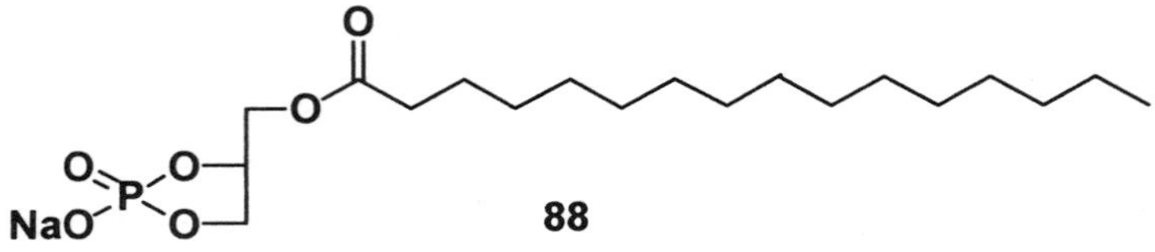

88

4. Recent Our Studies on the Chemical Constituents of Myxomycetes

4.1. Construction of culture collection of myxomycetes

The explorative studies of myxomycetes chemistry described above demonstrated that myxomycetes developed a rather unique secondary metabolism which offers a wide field of further studies. Only a few of more than 500 known species have been investigated so far. Therefore, Steglich described that for further progress the development of methods for the cultivation of myxomycetes is crucial to overcome the paucity of material [3]. Being consistent with these words, we initiated studies on laboratory culture of myxomycetes in Japan a few years ago. Cultivation of the plasmodium of myxomycetes in a considerable scale to carry out chemical studies has been known only for very limited species such as *Physarum polycephalum*. Cultivation of other myxomycetes for natural products chemistry studies had never been reported. We recently collected the fruit bodies of the myxomycetes (133 species, 361 strains) in Japan, and laboratory-cultivation of these myxomycetes were investigated.

Attempts on the laboratory-cultures of myxomycetes were initiated by spreading spores, which were contained in the fruit bodies, on an agar medium together with a suspension of *Escherichia coli*. After germination of the spores into myxamoebae, which were visualized by plaque formation in the *E. coli* culture (Figure 2). The myxamoebic plaques were then repetitively transferred to new agar plates several times to remove contaminating organisms. Some

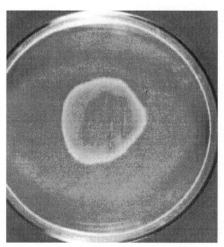

Figure 2. Myxoamoebae of *Didymium squamulosum*

248

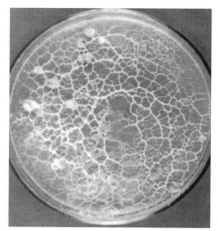

Figure 3.   Cultured plasomodium of *Didymium squamulosum*

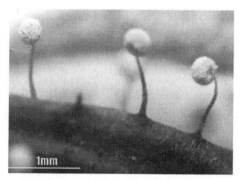

Figure 4.   Fruit bodies of *Didymium iridis* (wild)

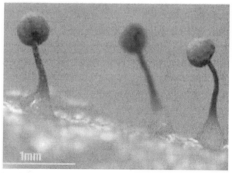

Figure 5.   Fruit bodies of *Didymium iridis* (cultured)

cultures proceeded to develop into the plasmodial stage, and these plasmodial cultures (Figure 3) were then able to grow by adding oatmeal to the agar medium in the absence of *E. coli*. Some strains showed formation of fruit-bodies on the oatmeal agar plates from the plasmodial cultures in the presence or absence of light (Figures 4 and 5).

In 1998, the spore-germination experiment was carried out on non-nutrient medium under dark condition at 25 degree, and only 3.5% sample showed spore germination. These experiments were continued in 1999-2002, and ratio of the samples which showed spore germination increased every year (Table 2). Table 3 shows the relationship between spore germination ratio and the genus. There are genuses which showed high germination ratio such as *Didymium*, *Physarum*, and

Table 2 Results of spore germination experiments

year	Conditions				Results	
	Culture medium	tempe-rature	Light or dark	Test strains (species)	Spore-germina-tion	Spore-germina-tion(%)
1998	Non nutrient/*E. coli*	25°C	Dark	57 (33)	2 (2)	3.5%
1999	Non nutrient/*E. coli* CM/2	25°C	Light (16h/day) or dark	68 (36)	4 (4)	6%
2000	Non nutrient/*E. coli* CM/2 LP/*E. coli*	22°C	Light (16h/day) or dark	65 (42)	13 (13)	21%
2001	Non nutrient/*E. coli* LP/*E. coli* Agar/*E. coli* A/*E. coli*	22°C	Light (16h/day) or dark	102 (65)	62 (37)	61%
2002	LP/*E. coli* Agar/*E. coli*	22°C	Dark	69 (55)	24 (22)	35%

[a]Ingradients of culture medium: "Non nutirient": KH_2PO_4 1.45%, $Na_2HPO_4 \cdot 12H_2O$ 2.4%, Agar 2.0%. "CM/2": Cornmeal agar 0.9%, Agar 1.0%. "LP": Lactose 0.1%, Peptone 0.1%, KH_2PO_4 0.205%, $Na_2HPO_4 \cdot 12H_2O$ 0.083%, Agar 1.5%. "A": Glucose 0.5%, Peptone 0.5%, Yeast extract 0.05%, KH_2PO_4 0.225%, $Na_2HPO_4 \cdot 12H_2O$ 0.137%, $MgSO_4 \cdot 7H_2O$ 0.05%, Agar 1.5%. "Agar": Agar 1.5%.

Table 3. Relationship between genus and spore-germination ratio

genus	1998 strains	%	1999 strains	%	2000 strains	%	2001 strains	%	2002 strains	%	1998-2002 strains	%
Ceratiomyxa	0 (4)		0 (4)		1 (3)	33%	2 (3)	67%	1 (3)	33%	4 (17)	24%
Cribraria	0 (2)		0 (6)		0 (11)		2 (7)	29%	0 (8)		2 (34)	6%
Tubifera	0 (1)		0 (1)		0 (1)		1 (2)	50%	0 (3)		1 (8)	13%
Arcyria	0 (6)		0 (12)		2 (8)	25%	5 (9)	56%	2 (8)	25%	9 (43)	21%
Perichaena					1 (1)	100%	1 (2)	50%			2 (3)	67%
Hemitrichia	0 (2)		0 (2)		0 (2)		1 (2)	50%	1 (2)		2 (10)	20%
Badhamia	0 (1)						1 (3)	33%			1 (4)	25%
Craterium	0 (4)				2 (2)	100%	11 (11)	100%	2 (3)	67%	15 (20)	75%
Physarum	0 (8)		0 (6)		1 (7)	14%	12 (16)	75%	7 (13)	50%	20 (50)	40%
Diderma	0 (1)		0 (2)				3 (4)	75%	3 (3)	100%	6 (10)	60%
Didymium	2 (4)	50%	4 (5)	80%	5 (7)	71%	17 (17)	100%	5 (5)	80%	33 (38)	87%
Lamproderma			0 (1)				1 (1)	100%			1 (2)	50%
Stemonaria							1 (1)	100%			1 (1)	100%
Stemonitopsis	0 (1)		0 (4)		1 (3)	33%	2 (5)	40%	0 (1)		3 (14)	21%

Craterium, and there are also genuses which showed low germination ratio such as *Cribraria* and *Arcyria*.

Out of 361 strains (133 species) of wild fruit bodies of myxomycetes collected in 1998-2002, spore germination was observed for 105 strains on an agar medium. Among them 40 strains (20 *Didymium* sp., 7 *Craterium* sp., and 7 *Physarum* sp.) were able to develop into the plasmodial stage, and 19 strains (14 *Didymium* sp., 4 *Physarum* sp., and 1 *Craterium* sp.) successfully showed formation of fruit bodies from the cultured plasmodia in a plate agar medium in the presence or absence of light, implying that spore-to-spore cultivation, i.e., rotation of one life cycle, was realized on agar-plates.

4.2. Constituents of *Didymium minus*

For a preliminary examination by TLC and antimicrobial activity tests, a small quantity of the cultured plasmodia of five myxomycetes of the genus *Didymium* (*Didymium squamulosum* [42], *D. minus*, *D. marineri*, *D. iridis*, and *D. flexuosum*) were collected [43], and these five species were revealed to contain almost similar constituents on the basis of TLC examination.   The major constituents of *D. minus* proved to be clionasterol (**89** = a 24-epimer of β-sitosterol), poliferasterol (**90**), 22,23-dihydrobrassicasterol (**91**), isofucosterol (**92**), clerosterol (**93**), and 24-methylenecholesterol (**94**), together with a glycoside glucoclionasterol (**95**) [44].   Isolation of glucoclionasterol (**95**) has been only reported from *Arthrocnenum glaucum* [45].

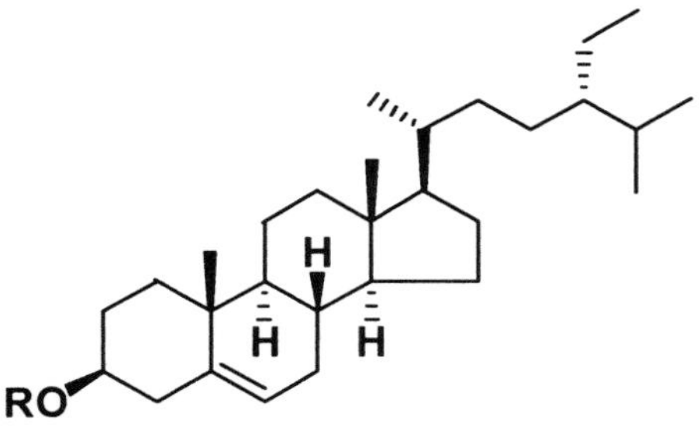

RO

89 R = H
95 R = Glc

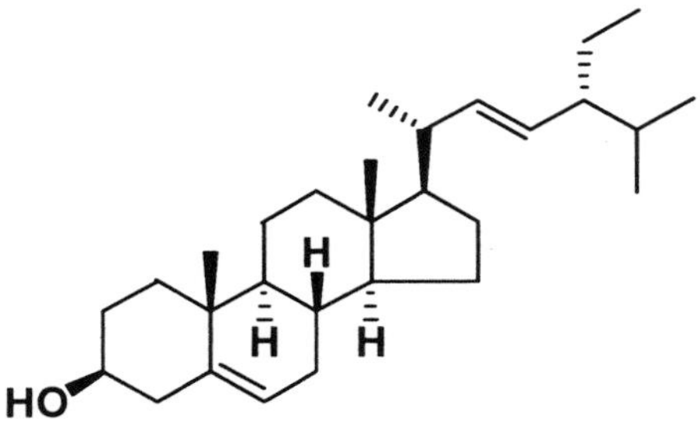

HO

90

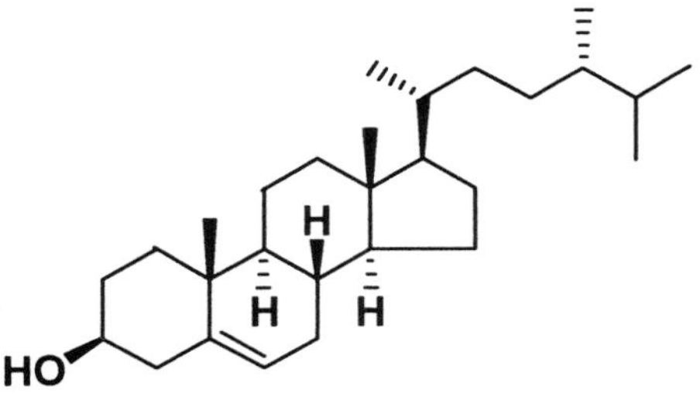

HO

91

HO

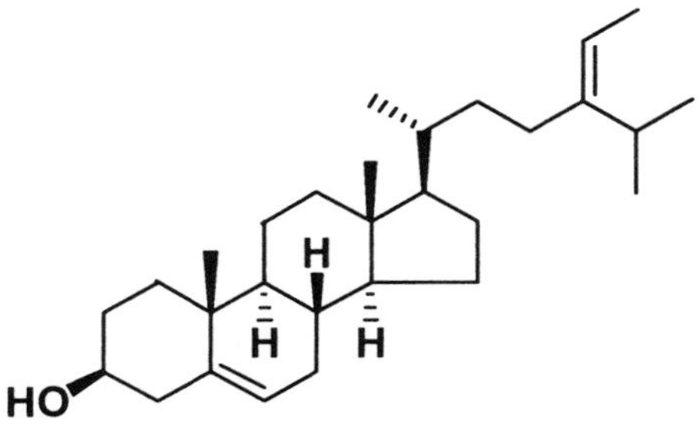

92

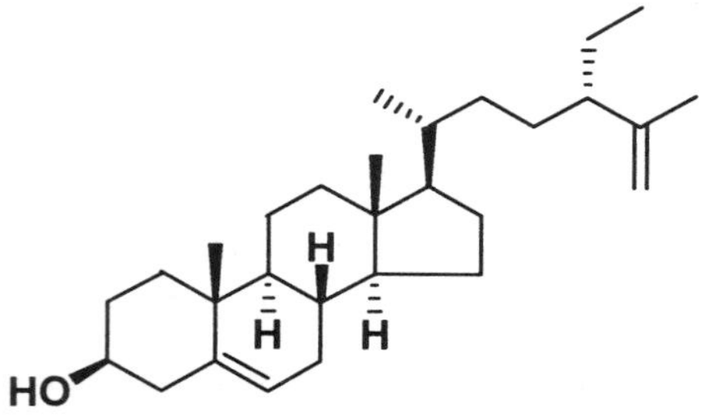

93 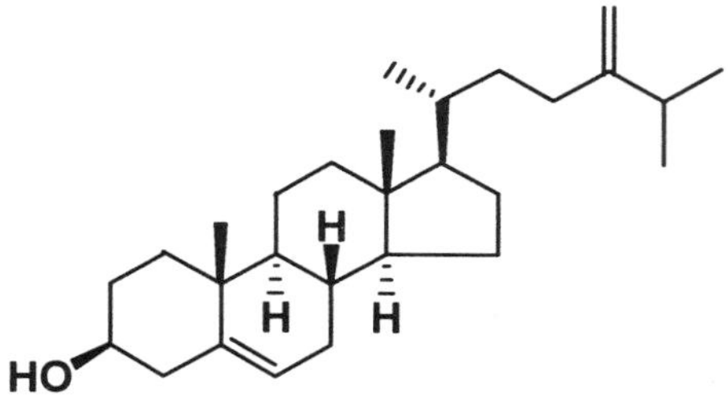   94

4.3. Constituents of *Didymium bahiense* var. *bahiense*

The fruit bodies of the myxomycetes *Didymium bahiense* var. *bahiense* (Order Physarales; Family Didymiaceae) were collected at Ina, Nagano Prefecture, Japan, in August, 1999. The spores contained in the fruit bodies were applied on an agar plate (lactose 0.1%, peptone 0.1%, KH$_2$PO$_4$ 0.205%, Na$_2$HPO$_4$·12H$_2$O 0.083%, agar 1.5%) with a suspension of *Escherichia coli* (0.1 mL in Nutrient media or Heart Infusion media, DIFCO). After static incubation at 25 °C in the dark condition for 4-5 days, myxamoebic plaque appeared, and the plaque was transferred several times to new agar plates containing the same media as above until the plasmodial formation was observed. The plasmodia were then mass cultured in agar plates (the same media as above) with oatmeal (*ca.* 0.2 g / plate, autoclaved prior to use) for 1-2 weeks at 25 °C in the dark condition. The harvested plasmodial cells were extracted with 90% MeOH, and the BuOH-soluble material of the extract was subjected to silica gel column chromatography, followed by separation by gel filtration on Sephadex LH-20. The fractions containing a Ninhydrin-positive spot on TLC were further purified by silica gel and Sephadex LH-20 chromatographies to give makaluvamine A (**96**), which was identified by ESIMS (*m/z* 202 (M+H)$^+$), and ^{1}H and

^{13}C NMR spectral data [46]. Makaluvamine B (**97**) was also isolated and identified from the same extract.

Makaluvamine A (**96**), possessing a unique pyrroloiminoquinone structure, was first isolated from a Fijian marine sponge *Zyzzya* cf. *marsailis* as a highly cytotoxic topoisomerase II inhibitor [47], and presumably biogenetically derived from a tryptophan. Makaluvamine A (**96**) was first isolated from myxomycetes in this study, and it seems to be noteworthy that a marine sponge isolate was obtained from a terrestrial organism, a myxomycete, which ostensibly appear phylogenetically unrelated.

Bahiensol (**98**), a new glycerolipid has been isolated from a cultured plasmodium of myxomycete *Didymium bahiense* var. *bahiense*. This new lipid was obtained from an antimicrobial fraction against *Bacillus subtilis*. The molecular formula of **98** was suggested as $C_{19}H_{40}O_5$ by the HRFABMS data (m/z 349.2950, [M+H]$^+$, Δ –0.4 mmu) and its planar structure was elucidated by spectral data. Analyses of the ^{1}H-^{1}H COSY and HMBC spectra suggested the presence of a glycerol unit, and that C_{16} aliphatic chain is connected to C-3' of glycerol unit through an ether linkage. Two secondary hydroxyl groups [δ_H 3.72 (1H, m) and 3.51 (1H, m)] were included in the C_{16} aliphatic chain, and one of the two secondary hydroxyl groups was shown to be located on C-3 position from the COSY (H$_2$-1/H$_2$-2 and H$_2$-2/H-3) and HMBC correlations (H$_2$-1/C-2, H$_2$-1/C-3, H$_2$-2/C-1, and H$_2$-2/C-3). The position of the other secondary hydroxyl group (position 'ω') was implied to be between C-6 to C-13, and MS/MS and HMQC-TOCSY experiments were carried out to determine the exact position 'ω', but were unsuccessful at this time. From these results, structure of bahiensol (**1**) was concluded to be 3-*O*-(3,ω-dihydroxyhexadecanyl)-glycerol.

Constituents in fruit bodies of *D. bahiense* var. *bahiense* were analyzed by TLC examination and were found to be different from those of the plasmodial stage.

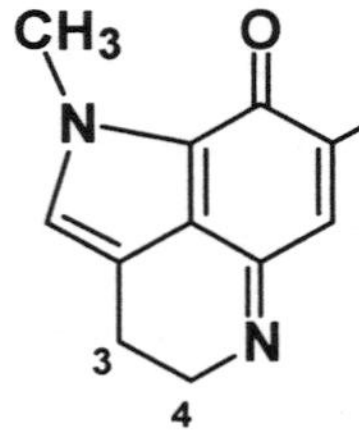

96

97 Δ^{3,4}

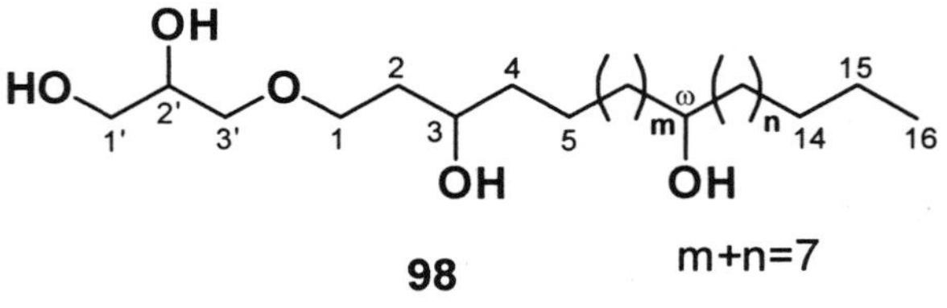

98 m+n=7

4.4. Constituents of plasmodia of other species

Mass culture of plasmodia of a few other species were also carried out. Among them, constituents of the plasmodium of *Physarum rigidum* were investigated to find that this plasmodium contain a complex mixture of yellow pigments. One of the yellow pigments from the complex mixture was isolated as almost pure state and spectral studies revealed that this pigment [**99**, ESIMS (positive) *m/z* 459 (M+Na)$^+$; ESIMS (negative) *m/z* 435 (M-H)$^-$] possesses a polyene structure having a methoxyphenyl acetamide and a glycine unit.

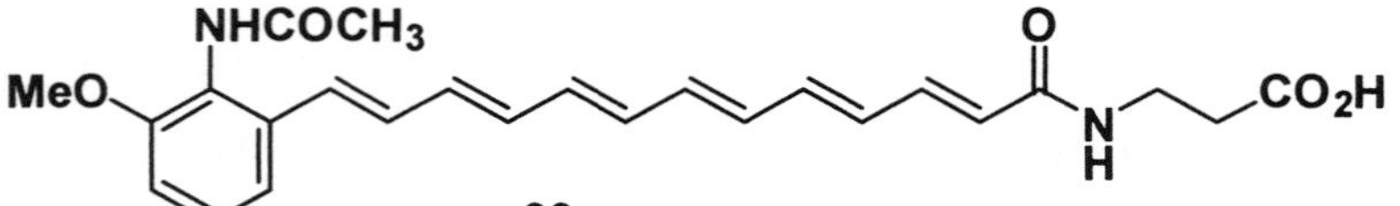

99

4.5. Constituents of field-collected samples

We also investigated field-collected samples of myxomycetes fruit bodies. TLC spot tests as well as antimicrobial activity tests were carried out on 13 species of wild myxomycetes. Chemical constituents of the field-collected samples of fruit bodies of *Lindbladia tubulina*, which (#21615) was collected at Takamagahara, Ohtsu, Kochi-shi, in Kochi Prefecture, Japan, in August 2001, were examined to isolate naphtoquinone pigments, lindbladione (**100**) and its *O*-methyl derivatives (**101** and **102**). Lindbladione (**100**) was obtained as dark brown solids, and shown to have the molecular formula $C_{16}H_{14}O_7$ by the HRFABMS data (*m/z* 319.0816, $[M+H]^+$, Δ –0.1 mmu). The UV spectrum of **100** showed absorption maxima at 265 and 367 nm, which were shifted to 271 and 379 nm, respectively, on addition of alkali (NaOH), indicating the presence of phenol group(s). The ^{13}C NMR spectrum of **100** showed signals for three carbonyls (δ_C 206.4, 189.6, and 184.1), ten sp^2 olefinic or aromatic carbons, and three sp^3 ones (δ_C 42.4, 20.2, and 14.3). The ^{1}H NMR spectrum of **100** in CD_3OD showed only six signals due to one aromatic proton (δ_H 6.97, 1H, s), one *E*-olefin (δ_H 8.07 and 7.45, each 1H, d, *J*=16.0 Hz), and one *n*-propyl group [δ_H 2.53 (2H, t, *J*=7.6 Hz), 1.60 (2H, m), 0.90 (3H, t, *J*=7.6 Hz). The presence of *n*-propyl group was confirmed by its ^{1}H-^{1}H COSY spectrum. Since eight out of ten unsaturation degrees were accounted for from ^{13}C NMR data, **100** was inferred to have a two rings. The presence of *n*-propyl group was further corroborated by the HMBC correlations (from H_3-16 to C-15 and C-14, from H_2-15 to C-16 and C-14, and from H_2-14 to C-16 and C-15), and this n-propyl group was shown to be attached on the carbonyl group resonating at δ_C 206.4 (C-13) by the HMBC correlations (from H_2-14 to C-13 and from H_2-15 to C-13). In the HMBC spectrum of **100**, one olefinic proton at δ_H 8.07 (H-11) showed long-range connectivities with two carbonyl carbons at δ_C 206.4 (C-13) and 189.6 (C-4) and also with sp^2 carbons at δ_C 176.4 (C-2) and 122.5 (C-12), while another olefinic proton at δ_H 7.45 (H-12) showed correlations with the carbonyl carbon at δ_C 206.4 (C-13) and sp^2 carbon at δ_C 113.3 (C-3). From these observations, a 3-oxo-hex-1-enyl group was inferred to be attached on the sp^2 carbon resonating at δ_C

113.3 (C-3). The aromatic proton at δ_H 6.97 (H-8) showed HMBC correlations to carbonyl carbon at δ_C 184.1 (C-1) and sp^2 carbons at δ_C 141.5 (C-6), 149.5 (C-7), 125.0 (C-9), and 112.2 (C-10). By interpreting these spectral data, a naphtoquinone nucleus with four hydroxyl groups at C-2, C-5, C-6, and C-7 was constructed for compound **100**. There was no possibility of inverse location of substituents on C-2 and C-3 on the basis of observation of the HMBC correlations (^3J) from H-8 to the C-1 carbonyl group and from H-11 to the C-4 carbonyl group. Thus, the structure of lindbladione was concluded as 2,5,6,7-tetrahydroxy-3-(3-oxo-hex-1-enyl)-[1,4]naphtho-quinone (**100**). Steglich said that lindbladione (**100**) was responsible for the color change from dark to red on treatment of the plasmodia of *Lindbladia tubulina* with mineral acid [3]. However, no spectral data and detailed structure elucidation of **100** were not found in the literature. Lindbladione (**100**) was reisolated here together with its methoxy derivatives (**101** and **102**) and fully characterized. The cytotoxic activity of compound **100** against murine leukemia P388 cells was examined, but it proved to be inactive at 25 μg/mL.

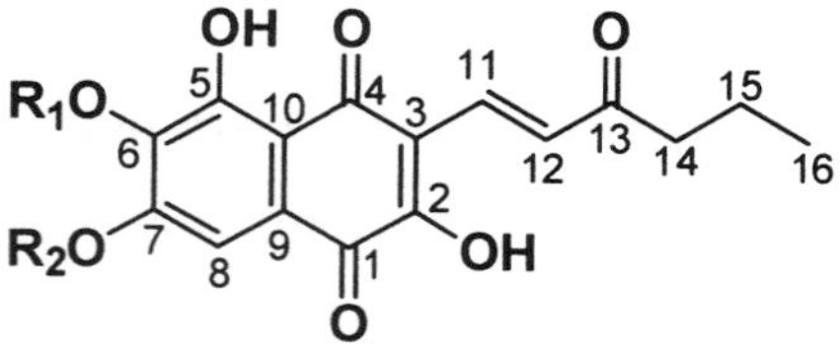

100	$R_1=R_2=$ H	
101	$R_1=$ H, $R_2=$ Me	
102	$R_1=R_2=$ Me	

Pigments contained in the fruit bodies of *Cribraria intricata* (#21517), collected at Takamagahara, Ohtsu, Kochi-shi, in Kochi Prefecture, Japan, in August 2001, were also examined recently, and the extract of this myxomycete was subjected to chromatography using ODS eluted with acetonitrile and water to afford lindbladione (**100**) as a major pigment constituent (0.3% yield). Myxomycetes, *Lindbladia tubulina* and *Cribraria intricata*, belong to different genus but to the same family Cribrariaceae. It may be possible that lindbladione (**100**) is one of the common red pigments contained in myxomycetes of the family Cribrariaceae.

From another species of the genus *Cribraria*, *Cribraria purpurea*, which (#22101) was collected at Mt. Shiraga, Monobe-mura in Kochi Prefecture, Japan, in November 2001, a new naphthoquinone pigment, cribrarione A (**103**) was obtained as dark purple powder. The molecular formula of **103** was revealed as $C_{13}H_{10}O_7$ by ^{13}C NMR aided with the DEPT and HMQC data along with the HRFABMS data (*m/z* 279.0494, $[M+H]^+$, Δ −1.1 mmu). The UV spectrum of **103** showed absorption maxima at 275, 316, and 510 nm, which were shifted to 310, 537, and 568 nm, respectively, on addition of alkali (NaOH), indicating the presence of phenol group(s). The ^1H NMR spectrum of **103** in CDCl$_3$ showed signals due to one aromatic proton, three hydroxyl protons, one methoxy group, one oxymethine, and one oxymethylene groups. The ^{13}C NMR spectrum of **103** showed signals for two carbonyls, eight other sp^2 carbons, and each one sp^3 oxymethine, oxymethylene, and methoxy carbons. Since six out of nine unsaturation degrees were accounted for from ^{13}C NMR data, **103** was inferred to have three rings. By interpreting 2D NMR spectral data, a naphthoquinone nucleus with two hydroxyl groups at C-5 and C-8 and a methoxy group on C-2 was constructed for compound **103**. It should be noteworthy that in the HMBC spectrum of **103** (J_{C-H}=8 Hz) cross-peaks were clearly observed from OH-5 (δ_H 13.2) to C-4 (δ_C 180.9) and C-3 (δ_C 110.1). These HMBC correlations may be attributable to 2- and 3-bond J_{C-H} couplings, respectively, through a rigid intramolecular hydrogen bond between the hydroxyl proton (OH-5) and the oxygen of the C-4 carbonyl group, which further corroborated the proposed naphtoquinone structure. The ^{13}C chemical shift of C-7 (δ_C 158.0) implied that this carbon bore an oxygen atom, and the unsaturation degree of **103** had suggested the presence of one more ring other than the naphtoquinone. The C-7, therefore, had to be connected with the sp^3 methylene carbon (C-12) through an ether-oxygen atom to give rise to a dihydrofuran ring moiety. From these results, the structure of cribrarione A was concluded as **103**. Crude extract of *Cribraria*

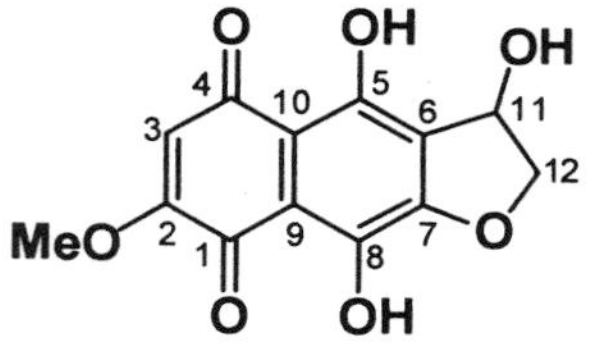

103

purpurea exhibited antimicrobial activity against *Bacillus subtilis*, and this activity was revealed to be ascribable to cribrarione A (**103**), since **103** was substantially active against *B. subtilis* with a diameter of inhibition zone 11 mm at 5 μg per paper disc (8 mm in diameter).

Dihydroarcyriarubin C (**104**), a new bisindole alkaloid, has been isolated from fruit bodies of a myxomycete *Arcyria ferruginea*, which was collected at Hao, Yasu-cho, Kochi Prefecture, together with two known bisindoles, arcyriarubin C (**2**) and arcyriaflavin C (**4**), and the structure of **104** was elucidated by spectral data. Arcyriaflavin C (**4**) was also obtained, together with arcyriaflavin B (**3**), from fruit bodies of *Tubifera casparyi*, which (#16839) was collected at Mt. Miune, Monobe-mura, Kochi Prefecture, in November 1997. Arcyriaflavin C (**4**) was revealed to exhibit cell cycle inhibition activity on HeLa cells on the basis of flow cytometry studies. Fruit bodies of *Fuligo candida*, which (#23446) was collected at Motoyama-cho, Kochi Prefecture, in August 2002, were revealed to contain a relatively high quantity of cycloanthranilylproline (**105**).

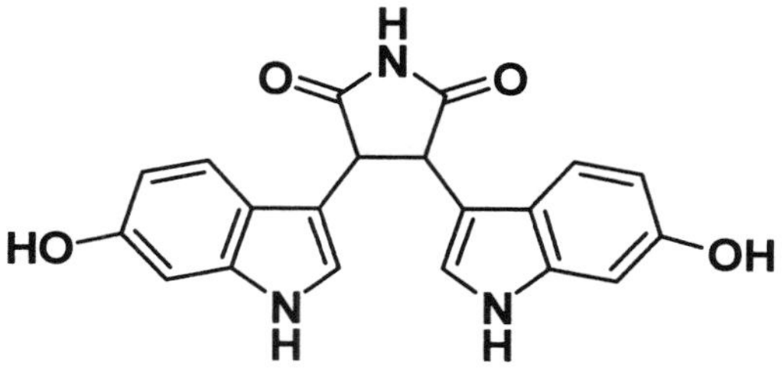

104

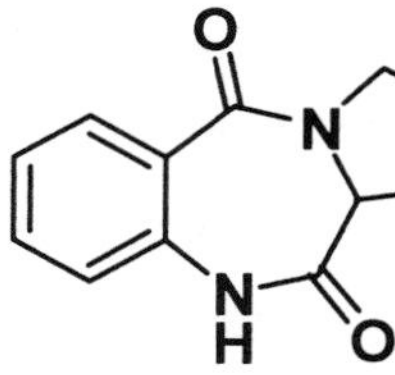

105

ACKNOWLEDGEMENT. The author is grateful to coworkers (Satomi Nakatani, Ayano Naoe, Yuka Misono, Yae Ishikawa, Tomoko Iwasaki, Dai Iwata, Yasuko Kono, Satomi Imai, and Mana Mitamura) for their hard working with thousand of culture plates, and Dr. Akira Ito, Kyorin Pharmaceutical Co., Ltd., for his valuable discussions and encouragements throughout this work, and the members of the Japanese Society of Myxomycetology since the organisms were collected on the occasion of their collection tour, and particularly to Yukinori Yamamoto (Kochi Kita High School) for collection of various wild myxomycetes and Dr. Jun Matsumoto (Keio University) for identification and collections of the myxomycetes. We also thank Dr. Kentaro Yamaguchi and Dr. Hiroko Seki of our university for instrumental analyses and Professor Naoto Yamaguchi for helpful discussions and guidance of cell cycle inhibition studies. This work was partly supported by a Grant-in-Aid from the Ministry of Education, Culture, Sports, Science and Technology of Japan, and by a Grant-in-Aid from the Uehara Memorial Foundation, the Japan Securities Scholarship Foundation, the Naito Foundation, and the Nagase Science and Technology Foundation.

260

REFERENCES

[1]   Harvey, A. *Drug Discovery Today*, **2000**, *5*, 294-300.

[2]   Martin, G. W.; Alexopoulos, C. J. The Myxomycetes, Iowa City: University of Iowa Press (1969).

[3]   Steglich, W. *Pure Appl. Chem.* **1989**, *61*, 281-288.

[4]   Hashimoto T., Yasuda A., Akazawa K., Takaoka S., Tori M., Asakawa, Y. *Tetrahedron Lett.* **1994**, *35*, 2559-2560.

[5]   Murakami-Murofushi, K.; Shioda, M.; Kaji, K.; Yoshida, S.; Murofushi, H. *J. Biol. Chem.* **1992**, *267*, 21512-21517.

[6]   Eisenbarth, S.; Steffan, B. *Tetrahedron* **2000**, *56*, 363-365 and refereces cited therein.

[7]   Steglich, W.; Steffan, B.; Kopanski, L.; Eckhardt, G. *Angew. Chem. Int. Ed. Engl.* **1980**, 459-460.

[8]   Brenner, M.; Rexhausen, H.; Steffan, B.; Steglich, W. *Tetrahedron* **1988**, *44*, *44*, 2887-2892.

[9]   Mayer, G.; Wille, G.; Steglich, W. *Tetrahedron Lett.* **1996**, *37*, 4483-4486.

[10]   Brenner, M.; Mayer, G.; Terpin, A.; Steglich, W. *Chem. Eur. J.* **1988**, *3*, 70-74.

[11]   Murase, M.; Watanabe, K.; Yoshida, T.; Tobinaga, S. *Chem. Pharm. Bull.* **2000**, *48*, 81-84.

[12]   Ōmura, S.; Iwai, Y.; Hirano, A.; Nakagawa, A.; Awaya, J.; Tsuchiya, H.; Takahashi, T.; Masuma, R. *J. Antibiot.* **1977**, *30*, 275-282.

[13]   I. Takahashi, E. Kobayashi, K. Asano, M. Yoshida, H. Nakano, *J. Antibiot.*, **40**, 1782-1784 (1987).

[14]   Nettleton, D. E.; Doyle, T. W.; Krishnan, B.; Matsumoto, G. K.; Clardy, J. *Tetrahedron Lett.* **1985**, *26*, 4011-4014.

[15]   *e.g.*) Toogood, P. L. *Curr. Opin. Chem. Biol.* **2002**, *6*, 472-478.

[16]   Hashimoto, T.; Yasuda, A.; Akazawa, K.; Takaoka, S.; Tori, M.; Asakawa, Y. *Tetrahedron Lett.* **1994**, *35*, 2559-2560.

[17]   Fröde, R.; Hinze, C.; Josten, I.; Schmidt, B.; Steffan, B.; Steglich, W. *Tetrahedron Lett.* **1994**, *35*, 1689-1690.

[18]   Hoshino, T.; Komiya, Y.; Hayashi, T.; Uchiyama, T.; Kaneko, K. *Biosci.*

Biotech. Biochem. **1993**, *57*, 775-781.

[19]   Yasuzawa, T.; Iida, T.; Yoshida, M.; Hirayama, N.; Takahashi, M.; Shirahata, K.; Sano, H. *J. Antibiot.* **1986**, *39*, 1072-1078.

[20]   Kopanski, L.; Li, G.-R.; Besl, H.; Steglich, W. *Liebigs Ann. Chem.* **1982**, 1722-1729.

[21]   Kopanski, L.; Karbach, D.; Selbitschka, G.; Steglich, W. *Liebigs Ann. Chem.* **1987**, 793-796.

[22]   Hashimoto, T.; Akazawa, K.; Tori, M.; Kan, Y.; Kusumi, T.; Takahashi, H.; Asakawa, Y. *Chem. Pharm. Bull.* **1994**, *42*, 1531-1533.

[23]   Buchanan, M. S.; Hashimoto, T.; Asakawa, Y. *Phytochemistry* **1996**, *41*, 791-794.

[24]   Kusumi, T.; Takahashi, H.; Xu, P.; Fukushima, T.; Asakawa, Y.; Hashimoto, T.; Kan, Y.; Inouye, Y. *Tetrahedron Lett.* **1994**, *35*, 4397-4400.

[25]   Řezanka, T. *Phytochemistry* **1993**, *33*, 1441-1444.

[26]   Řezanka, T. *Phytochemistry* **2002**, *60*, 639-646.

[27]   Casser, I.; Steffan, B.; Steglich, W. *Angew. Chem. Int. Ed. Engl.* **1987**, *26*, 586-587.

[28]   Ley, S. V.; Smith, S. C.; Woodward, P. R. *Tetrahedron Lett.* **1988**, *29*, 5829-5832.

[29]   Velten, R.; Josten, I.; Steglich, W. *Liebigs Ann.* **1995**, 81-85.

[30]   Emanuel, C. F. *Nature* **1958**, *182*, 1234.

[31]   Lenfant, M.; Lecompte, M. F.; Farrugia, G. *Phytochemistry* **1970**, *9*, 2529-2535.

[32]   Bullock, E.; Dawson, C. J. *J. Lip. Res.* **1976**, *17*, 565-571.

[33]   Steffan, B.; Praemassing, M.; Steglich, W. *Tetrahedron Lett.* **1987**, *28*, 3667-3670.

[34]   Nowak, A.; Steffan, B. *Liebigs Ann.* **1997**, 1817-1821.

[35]   Nowak, A.; Steffan, B. *Angew. Chem. Int. Ed.* **1998**, *37*, 3139-3141.

[36]   Blumenthal, F.; Polborn, K.; Steffan, B. *Tetrahedron* **2002**, *58*, 8433-8437.

[37]   Eisenbarth, S.; Steffan, B. *Tetrahedron* **2000**, *56*, 363-365.

[38]   Murakami-Murofushi, K.; Nakamura, K.; Ohta, J.; Yokota, T. *Cell Struct. Funct.* **1987**, *12*, 519-534.

[39]   Murakami-Murofushi, K.; Shinoda, M.; Kaji, K.; Yoshida, S.; Murofushi, H. *J. Biol. Chem.* **1992**, *267*, 21512-21517.

[40]   Kobayashi, S.; Tokunoh, R.; Shibasaki, M.; Shinagawa, R.; Murakami-Murofushi, K. *Tetrahedron Lett.* **1993**, *34*, 4047-4050.

[41]   Mukai, M.; Imamura, F.; Ayaki, M.; Shinkai, K.; Iwasaki, T.; Murakami-Murofushi, K.; Murofushi, H.; Kobayashi, S.; Yamamoto, T.; Nakamura, H.; Akedo, H. *Int. J. Cancer* **1999**, *81*, 918-922.

[42]   Ishibashi, M.; Mitamura, M.; Ito, A. *Nat. Med.* **1999**, *53*, 316-318.

[43]   Ishikawa, Y.; Misono, Y.; Nakatani, S.; Ito, A.; Matsumoto, J.; Ishibashi, M. *J. Nat. Hist. Mus. Inst., Chiba* **2002**, *7*, 1-4.

[44]   Ishikawa, Y.; Kono, Y.; Iwasaki, T.; Misono, Y.; Nakatani, S.; Ishibashi, M.; Ito, A.; Matsumoto, J. *Nat. Med.* **2001**, *55*, 312.

[45]   Falsone, G.; Catane, F.; Kadlecech, V.; Pichler, B.; Wintersteiger, R.; Haslinger, E.; Presser, A.; Birkofer, L. *Pharm. Pharmacol. Lett.* **1998**, *4*, 184-187.

[46]   Ishibashi, M.; Iwasaki, T.; Imai, S.; Sakamoto, S.; Yamaguchi, K.; A. Ito, *J. Nat. Prod.* **2001**, *64*, 108-110.

[47]   Radisky, D. C.; Radisky, E. S.; Barrows, L. R.; Copp, B. R.; Kramer, R. A.; Ireland, C. M. *J. Am. Chem. Soc.* **1993**, *115*, 1632-1638.

[48]   Ishikawa, Y.; Ishibashi, M.; Yamamoto, Y.; Hayashi, M.; Komiyama, K. *Chem. Pharm. Bull.* **2002**, *50*, 1126-1127.

Atta-ur-Rahman (Ed.) *Studies in Natural Products Chemistry, Vol. 29*

POLYHYDROXY-*p*-TERPHENYLS AND RELATED *p*-TERPHENYLQUINONES FROM FUNGI: OVERVIEW AND BIOLOGICAL PROPERTIES

VALERIA CALÌ, CARMELA SPATAFORA
AND CORRADO TRINGALI[*]

*Dipartimento di Scienze Chimiche, Viale A. Doria 6,
I-95125 Catania, Italy.*

ABSTRACT: Natural occurrence of products with a *p*-terphenyl core is essentially restricted to fungi and lichens. Recently, an increasing number of metabolites based on the *p*-terphenyl nucleus and showing interesting biological properties or unusual structures has been reported. The literature on polyhydroxy-*p*-terphenyls and *p*-terphenylquinones isolated from 35 different fungal species (both from fungal cultures or from fruiting bodies of basidiomycetes) has been reviewed, and a total of 115 fungal metabolites are reported here, with emphasis on their biological properties. Some semisynthetic analogues or lichen metabolites are also included. An array of biological activity has been reported for these metabolites, and in particular for fully aromatic polyhydroxy-*p*-terphenyls, including cytotoxic activity against tumoral cells and other antiproliferative properties, antibacterial activity, antioxidant or radical scavenging activity, anti-inflammatory activity and other properties of biomedical or agronomical interest. Simple *p*-terphenyls with a tricyclic C-18 basic skeleton, including their acyclic *O*-alkyl or *O*-acyl derivatives are treated in Section 2. Those showing a polycyclic C-18 skeleton are discussed in Section 3, while Section 4 includes those possessing an alkylated skeleton (C-19 or more). The main biosynthetic pathways and synthetic methods for some of the cited metabolites are summarised in Sections 5 and 6, respectively. Globally considered, the data here reported confirm the important properties of this peculiar group of metabolites and open promising perspectives in the search for further bioactive polyhydroxy-*p*-terphenyls.

1   INTRODUCTION

Natural occurrence of products with a *p*-terphenyl core is essentially restricted to fungi and lichens. In particular, *p*-terphenylquinones were known as fungal pigments for a long time. Indeed, the colouring principles in mushrooms and toadstools attracted the interest of scientists since the late XIX century and the pioneering work in this field has been reviewed in the older literature [1]. A detailed historical report on fungal *p*-terphenylquinones – including the basic steps of determination of the structures of polyporic acid (**3**[#]) and atromentin (**18**) (see Section 2.4) by

---

[*] To whom correspondence should be addressed. E-mail: ctringali@dipchi.unict.it
[#] Compounds are numbered according to their citation in the following Sections

Kögl – was compiled by Thomson in 1957 [2]. Fully aromatic polyhydroxy-*p*-terphenyls have generally been reported later than the related terphenylquinones. Recently, an increasing number of *p*-terphenyl metabolites showing interesting biological properties and/or unusual structures has been reported. Bioactive metabolites with a *p*-terphenyl core have been obtained both from cultured fungi or fruiting bodies of basidiomycetes, and showed cytotoxic, antibacterial, anti-inflammatory, antioxidant, anti-insect, and other promising biological properties. Thus, we wish to review here the chemical and biological literature on polyhydroxy-*p*-terphenyls and terphenylquinones of fungal origin, with emphasis on the biological properties of the reported compounds.

Compounds will be discussed in Sections based on structural analogy, generally related to their biosynthetic origin. Thus, simple *p*-terphenyls with a tricyclic C-18 basic skeleton, including their acyclic *O*-alkyl or *O*-acyl derivatives are treated in Section 2, normally in order of increasing number of oxygenated functions. Those showing a policyclic C-18 skeleton are discussed in Section 3, while Section 4 includes those possessing an alkylated skeleton (C-19 or more). The main biosynthetic pathways and selected synthetic methods are summarised in Sections 5 and 6, respectively. For the sake of completeness, terphenyls from lichens are shortly reported in Section 2.6.

Various numbering systems of the terphenyl nucleus have been used in the reviewed literature, so that a different number may refer to the same substituent position, as assigned by Authors.

Taxonomical names of the species are cited as reported in the original article; in some cases we report here in parentheses more recent systematic names or older synonyms, to facilitate retrieving a specific fungal species.

Excellent books and reviews devoted to fungal metabolites, or more specifically to pigments of fungal origin, and including *p*-terphenylquinones or related polyhydroxy-*p*-terphenyls, have been previously published. Among them, we wish cite here the well-known Turner's two volumes on 'Fungal Metabolites' [3,4], the reviews of Steglich [5,6] Gill and Steglich [1], and Gill [7–9]. Two reviews devoted to bioactive metabolites from Japanese mushrooms have been recently compiled by Hashimoto and Asakawa [10] and by Jikai [11]. The biosynthesis of bioactive fungal metabolites has been the subject of a specific review, including some terphenyl derivatives [12].

2   TRICYCLIC *p*-TERPHENYLS (C-18 BASIC SKELETON)

2.1   Terphenyls bearing three oxygenated functions

Only two fungal *p*-terphenyls bearing three oxygenated functions are known, namely terferol (**1**), and its methylated analogue **2**.

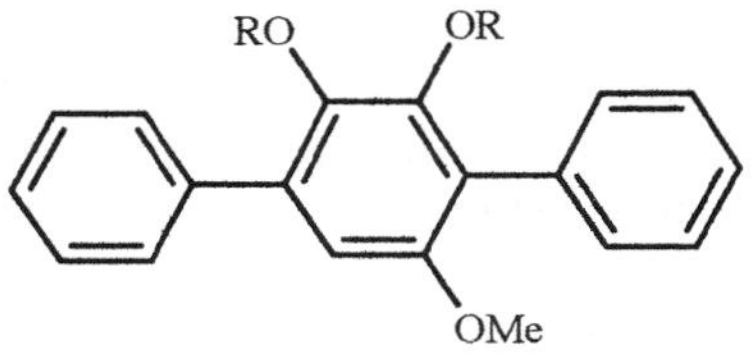

	R
1	H
2	Me

The fully aromatic terferol was isolated from cultures of *Streptomyces showdoensis* through a fractionation guided by enzyme inhibition assays on cyclic adenosine 3',5'-monophosphate phosphodiesterase (cAMP-PDE) [13-15]. Levels of cAMP, controlled by adenylate cyclase and cAMP-PDE, are abnormal in a number of disease states. Terferol showed inhibitory activity not only against cAMP-PDE but also against cyclic guanosine 3',5'-monophosphate phosphodiesterase (cGMP-PDE) for various rat tissues. Very recently, terferol was also identified as a selective acetylcholinesterase (AchE) inhibitor (N98-1,021A) through a screening of 2141 strains of microorganisms. [16] Also 2',3',5'-trimethoxy-*p*-terphenyl (**2**) had been obtained from fungal cultures, in this case of the basidiomycete *Peniophora gigantea (Phlebiopsis gigantea)* [17].

2.2   Terphenyls bearing four or five oxygenated functions

The early isolated *p*-terphenylquinones were obtained mostly from fruiting bodies of basidiomycetes or lichens. The prototype of this class of fungal metabolites is polyporic acid (**3**), originally isolated by Stahlschmidt from *Polyporus nidulans* [18]. This simple terphenylquinone occurs in a few Aphyllophorales, among them *Lopharia papyracea, Polyporus rutilans (P. nidulans, Hapalopilus nidulans, H. rutilans)* and *Peniophora filamentosa (Phanerochaete filamentosa)* [1]. It was identified in a mycelial culture of *Polyporus nidulans* by direct injection of a fragment of the mycelium into a mass spectrometer [19], and isolated also from lichens of *Sticta* genus [20]. Polyporic acid was more than once recognized to possess antibacterial activity [21,22], and was also reported as an antileukemic principle, significantly prolonging the life-span of mice inoculated with lymphocitic leukemia [23]. This compound also induces contraction of isolated rabbit ileum suspended in oxygenated Tyrode solution [24]. The effects of 13 quinones, including **3** and **18**, on the mitochondrial dihydrorotate dehydrogenase (DHO-DH) from potato tuber parenchyma have been reported [25]. Inhibitors of DHO-DH have been studied for their antitumor and immunomodulating effects. Polyporic acid from *Hapalopilus rutilans* has $IC_{50} = 10^{-4}\text{-}10^{-3}$ M. [26] The authors investigated the possible relationship between poisoning due to ingestion of *H. rutilans* and DHO-DH inhibiton by **3**, and concluded that the intoxication symptoms are due to the high content of polyporic acid.

The dimethoxy derivative of polyporic acid, betulinan A (**4**), is another of the few known *p*-terphenyls with unsubstituted side rings. It has recently been isolated from fruiting bodies of *Lenzites betulina*, together with the related betulinan B (**113**, see Section 4.1) [27]. Betulinan A and B are potent inhibitors of lipid peroxidation in rat liver microsomes (thiobarbituric acid method), and **4** showed the highest activity ($IC_{50} = 0.46$ µg/mL).

	R
3	H
4	Me

The first *p*-terphenylquinone isolated from cultures of micromycetes is volucrisporin (**5**), obtained from *Volucrispora aurantiaca*, a member of Hyphomycetes [28]. Volucrisporin has a simple symmetrical structure with two *meta*-hydroxylated side rings, and was the subject of early biosynthetic studies on *p*-terphenylquinones (See Section 5) [29]. Another example of terphenylquinone isolated from fungal cultures is the 4-hydroxy derivative of polyporic acid, ascocorynin (**6**), obtained from *Ascocoryne sarcoides* (whose conidial state is reported as *Coryne dubia = Pirobasidium sarcodides*) [30]. In the serial dilution test on fourteen Gram-positive and Gram-negative bacteria, **6** inhibited the growth of several Gram-positive bacteria, and in particular showed a MIC of 20-30 µg/mL against *Bacillus stearothermophilus*. Gram-negative bacteria were not affected at a concentration up to 100 µg/mL. In the plate diffusion test on five fungal strains, **6** (100 µg) was not active.

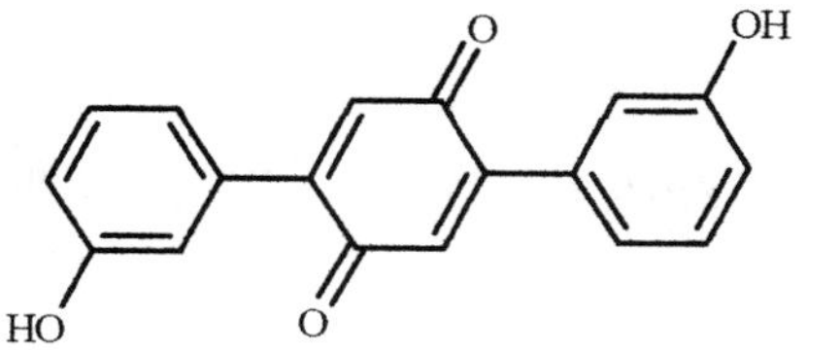

5

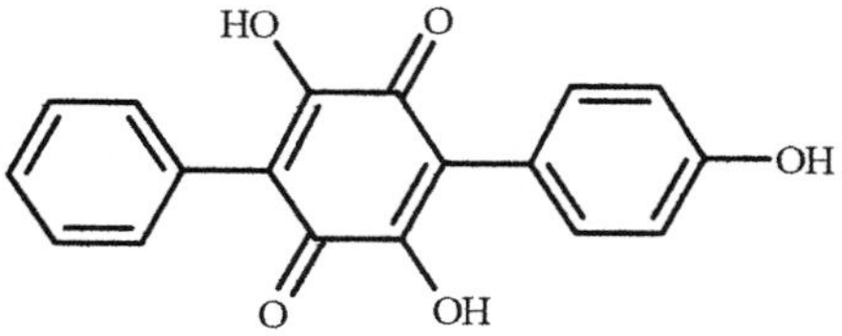

6

2.3 Terphenyllin, terprenin and analogues

Several *p*-terphenyl metabolites have been obtained from cultures of the filamentous fungus *Aspergillus candidus* and all those based on the simple C-18 tricyclic skeleton are discussed here together, apart from the number of oxygenated functions. In early biosynthetic studies on cultures of *A. candidus*, the fully aromatic terphenyllin (reported as 1,4-dimethoxy-2,4',4"-trihydroxy-*p*-terphenyl, **7**) was isolated [31-33]. An independent study by Takahashi *et al.* on *A. candidus* metabolites toxic for tumoral HeLa cells reported the isolation of **7** as a cytotoxic principle (toxin A, ID_{50} = 10 µg/mL) [34] together with the related 'compound E' (deoxyterphenyllin, **8**) [35]. Cytotoxicity to HeLa cells of the metabolites of *A. candidus* as well as fifteen related compounds of synthetic or semisynthetic origin and atromentin (**18**) was also reported. Some tested compounds induced different morphological changes and the authors suggests that methylation or removal of the 4"-hydroxy group may lead to the loss of a typical 'R type' change induced by **7**.

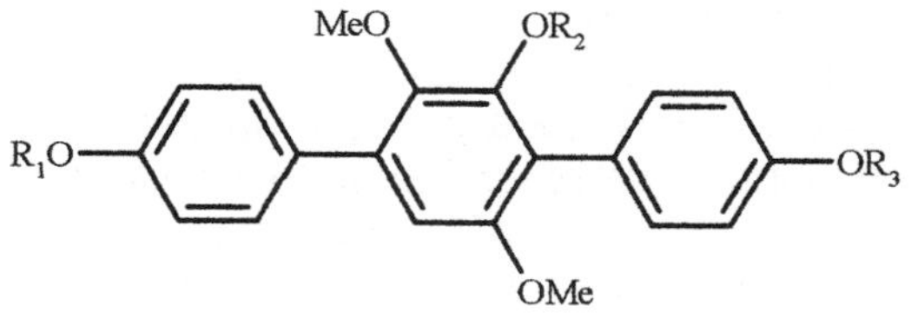

	R_1	R_2	R_3
7	H	H	H
9a	Me	Me	H
9b	H	Me	Me

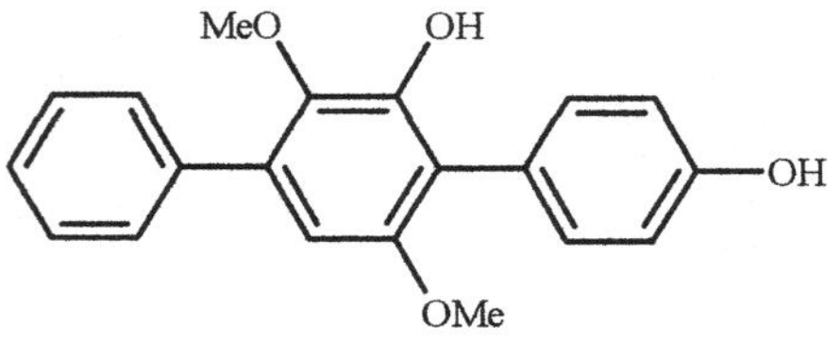

MeO OH

OMe

8

RO O

MeO OMe

O

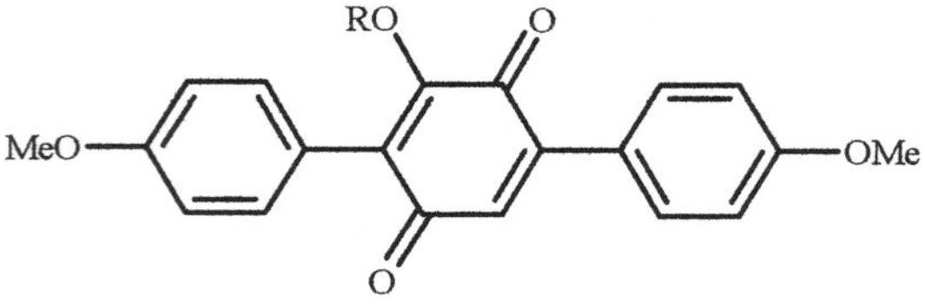

	R
10	H
11	Me

The majority of compounds showed cytotoxicity in the range 3-10 μg/mL. The terphenyllin dimethylether **9** (two alternative structures **9a** and **9b** are reported) and the terphenylquinones **10** and **11** showed mild activity even at at 1.0 μg/mL.

From a culture of *A. candidus* isolated from a mouldy unbleached flour, two metabolites with plant growth inhibiting properties were isolated, namely **7** and the new hydroxyterphenyllin (**12**) [36]. Both compounds significantly ($P < 0.01$) inhibited wheat coleoptile (*Triticum aestivum*) growth at 10^{-3} M, (respectively 35% and 100% relative to control), but **12** proved to be considerably more active than **7**, inhibiting wheat coleoptile growth also at 10^{-4} M (42%) and 10^{-5} M (8%). The semisynthetic hydroxyterphenyllin tetraacetate was tested but revealed inactive. According to the authors, these data suggest that biological activity is a function of number and position of the free hydroxy groups. Both **7** and **12** were independently obtained by other workers from cultured *A. candidus* and the second, considered a new metabolite, was named 3-hydroxyterphenillin [37]. A further study on *A. candidus* cultures, following the isolation of candidusins A and B (**72**, **73** see Section 3.1) [38], afforded **12** and the 3,3"-dihydroxyterphenyllin **13** [39]. The isolated compounds were tested for inhibitory activity against the development of sea urchin (*Hemicentrotus pulcherrimus*) embryos, as a step of a search for new metabolic inhibitors of cell proliferation. Compounds **12** and **13** showed a similar inhibitory pattern and complete block of first cleavage was obtained at 50 μg/mL for **12** and at 20 μg/mL for **13**. At lower dosages (5-10 μg/mL) the compounds induced irregular shaped and(or) odd-numbered cells in the embryos.

In a recent screening of moulds, *A. candidus* showed the highest antioxidant activity [40]. A subsequent study allowed identification and characterization of **12**, **13** and **73** as major components of broth filtrate [41]. The antioxidative properties of the purified metabolites were evaluated with various methods and compared to those of butylhydroxyanisol (BHA) and α-tocopherol. Safety studies were also made to evaluate their cytotoxicity, genotoxicity and mutagenicity. In the linoleic acid peroxidation system, the inhibition of peroxidation (IP%) for the three compounds in the range 12,5 –200 μg/mL was greater than 95% and significantly higher than that of α-tocopherol. The authors suggest that the strong antioxidant activity of the cited compounds could be related to the presence of a methoxy substituent *ortho* to a hydroxyl. Using the Rancimat method in lard, **13** showed the highest protection factor (7.82), substantially higher than those of BHA (5.58) and α-tocopherol (4.29). The greater activity of **13** appears related to the presence of a second *ortho*-hydroxy group. Also the scavenging effects on α,α-diphenyl-β-picrylhydrazyl (DPPH) radicals was evaluated and both **12** and **13** exhibited marked scavenging effects (94.7 and 96.0%, respectively), similar to those of BHA and α-tocopherol. Safety studies showed that the three compounds were neither cyto- nor genotoxic toward human intestine cells, nor mutagenic toward *Salmonella typhimurium*. Very recently, a further study on the antioxidant activity of mycelia extracts obtained from submerged cultures of *A. candidus* (CCRC 31543) confirmed the presence of **12**, **13** and **73** [42].

A number of *Aspergillus* spp. have been submitted to a chemical screening, pointing to the identification of possible chemotaxonomic indicators [43]. The results indicated that *A. campestris* and *A. taichungensis*, producing both **7** and **13**, should be placed in section *Candidi*.

A demethylated analogue of **13**, namely **14** (3,3"-dihydroxy-6'-desmethylterphenyllin), was obtained from the sclerotia of *Penicillium raistrickii*, whose extracts displayed potent anti-insect activity [44]. As reported in Section 3.1, **14** was obtained together with the candidusin-related metabolites **75** and **76**, a xanthone derivative, and two well-known fungal metabolites, griseofulvine and its 6-desmethyl analogue. All compounds were evaluated for activity in dietary assay against the corn earworm *Helicoverpa zea* and the fungivorous beetle *Carpophilus hemipterus*. Griseofulvin was responsible for most of the activity of the extract, nevertheless **14**, at 500 ppm, induced a 28% reduction in growth rate in *H. zea* larvae. This compound also showed a moderate antibacterial activity against *Staphylococcus aureus* (13-mm inhibition zone in standard disk assay at 200 μg/disk), whereas it was inactive when tested at 200 μg/disk against *Candida albicans*.

Probably the more promising lead compounds in the search of new drugs of natural origin among *p*-terphenyl fungal metabolites are the prenylated terphenyllin analogues recently obtained from the strain RF-5762 of *Aspergillus candidus*. Indeed, a screening program to find novel immunosuppressant, anticancer or anti-inflammatory agents from microbial fermentation afforded the new terprenin (**15**), 3-methoxyterprenin (**16**) and 4"-deoxyterprenin (**17**) together with the previously isolated terphenyllins **8**, **9**, **12** and **13** [45,46].

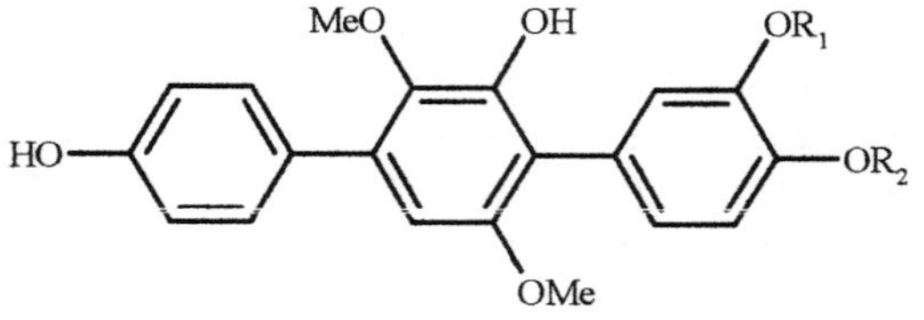

	R$_1$	R$_2$
12	H	H
15	H	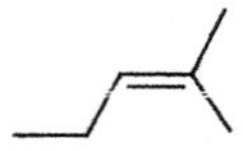 (prenyl)
16	Me	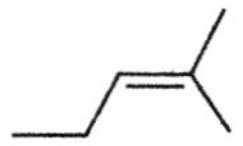 (prenyl)

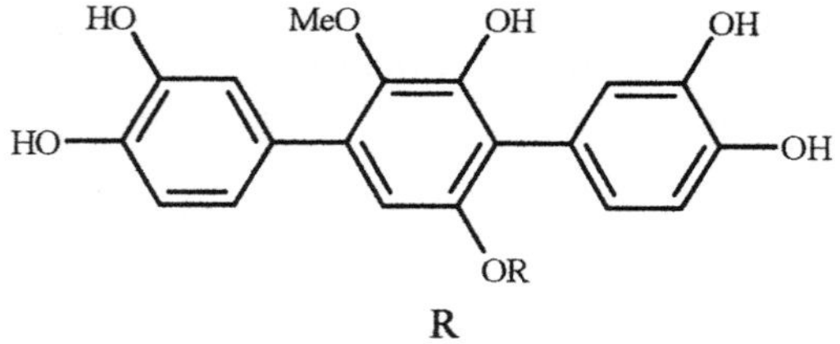

	R
13	Me
14	H

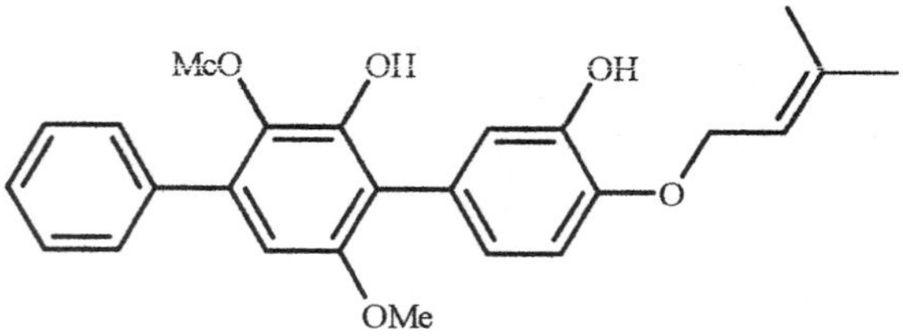

17

The Authors report a potent antiproliferative activity of **15** against human lung cancer Lu-99, human leukaemia CCFR-CEM, and mouse leukaemia P388 cells (IC$_{50}$ values: 1.0, 0.2 and 12.0 ng/mL, respectively). All terprenins showed strong inhibitory activity against mouse spleen lymphocytes stimulated with concanavalin A (Con A) or lipopolysaccharide (LPS). With Con A (or LPS) the IC$_{50}$ values were 1.2 (4.5), 2.0 (8.0) and 5.6 (15.6) ng/mL for **15**, **16** and **17**, respectively. Terprenins had no antimicrobial activity against bacteria and fungi, and no immunosuppressive activity was found for the strictly related terphenyllins **8**, **9**, **12** and **13**. In further works, a highly efficient and practical total synthesis of terprenin has been reported (see Section 6) [47-49], as well as data on the highly potent immunosuppressive effect of **15** on the in vitro Immunoglobulin E (IgE) production of human lymphocytes stimulated with anti-CD40 and IL-4 (IC$_{50}$ = 0.18 nM). This is particularly noticeable in comparison with the little inhibition showed by the strong immunosuppressant FK506, even at high concentration. Terprenin was also found to suppress antigen-specific IgE production in mice by oral administration in a dose-dependent manner, reaching a significant suppressive effect at 20 and 40 mg kg^{-1}, without any toxicological signs. These results are promising for future development of new antiallergic drugs. In a subsequent paper, 4"-deoxyterprenin (**17**) has been reported as a potent and selective cytotoxic metabolite isolated from *A. candidus* F23967 [50]. This compound has potent cytotoxic activity against nineteen tumour or hyper-proliferative cell lines, in particular mouse keratinocytes (MK) cells (IC$_{50}$ = 14 nM). A study of the action mechanism suggests that its potent cytotoxic property is related to selective inhibition of pyrimidine biosynthesis.

2.4 Terphenyls bearing six oxygenated functions: atromentin and analogues

Atromentin (**18**), the 4,4"-dihydroxy analogue of polyporic acid, accounts for the reddish-brown colour of the external parts of *Paxillus atrotomentosus*, from which it was originally isolated by Thörner [2]. It was later obtained from an array of basidiomycetes belonging to the genera *Paxillus, Hydnellum (Hydnum), Lampteromyces, Leccinum, Leucogyrophana, Omphalotus, Rhizopogon, Suillus, Xerocosmus* [1], and from cultures of *Clitocybe subilludens* [51]. Among them, it is worth of mentioning here the isolation from *Hydnellum diabolous* as an anticoagulant principle [52,53] as well as the *in vivo* studies using rabbit ileum, showing that **18** possesses a significant smooth muscle stimulant activity [54]. In an antimicrobial screening of mushroom metabolites, **18** showed MIC in the range 25-100 µg/mL [21]. More recently, a study on the antibacterial activity of some naturally occurring quinones, including

3 and **18**, has been reported [22]. The results suggest that the antibiotic activity depends on the *para* substituents of the benzoquinone ring, the activity being inversely proportional to the polarity of the metabolite. In particular, the activity against *Bacillus subtilis* is noticeably higher for 4,4"-dimethoxyatromentin (**19**, MIC = 5 μg/mL) with respect to atromentin (MIC = 500 μg/mL). The study on effect of **18** on DHO-DH from potato tuber has been cited above [25].

Two acetylated derivatives of atromentin have been reported, namely diacetylatromentin (**20**), isolated from fruiting bodies of *Anthracophyllum archeri* [55], and acetylatromentin (**21**), obtained from sporophores of *Albatrellus cristatus, Paxillus atrotomentosus, P. panuoides* and *Omphalotus olearius* [56]. In this work, the semisynthetic methoxyatromentin **22** was prepared, a compound later isolated as a natural constituent from fruiting bodies of *Thelephora ganbajun* collected in Yunnan (China) [57].

Flavomentins A – D (**23** – **26**) are orange-yellow terphenyl quinones possessing the basic skeleton of atromentin and obtained from fresh or dried fruiting bodies of *P. atrotomentosus* and *P. panuoides* [56]. These pigments are clearly related to the colourless 'leucomentins', reported below, co-occurring in *P. atrotomentosus*. Both groups of metabolites are peculiar in bearing unsaturated acyl residues. In particular, flavomentin A (**23**), and flavomentin B (**24**) bear respectively two or one acyl residue of (2*Z*,4*S*,5*S*)-4,5-epoxy-2-hexenoic acid (EPH) on the central ring, while flavomentin C (**25**) and flavomentin D (**26**) are monoesters respectively of (2*Z*,4*E*)-2,4-hexadienoic and 4,5-dihydroxy-2-hexenoic acids. It is worth noting here that **26**, during its hydrolytic conversion into **18**, affords a lactone form (osmundalactone) from the acid residue. Flavomentin B was also isolated from *P. panuoides* var. *ionipus*. The Authors mention the surprising NMR equivalence of the *p*-hydroxyphenyl residues attached to the central ring in **24**, explained as the effect of a rapid and synchronous migration of the acyl residue and the OH proton. In the ^{13}C NMR spectra of **24** and **21** at room temperature, the peaks of quinone ring are not discernible, due to the pronounced broadening caused by the cited shifts of the acyl group.

The 3,6-dibenzoylatromentin, named aurantiacin (**27**) was first isolated as a dark red pigment from *Hydnum (Hydnellum) aurantiacum* [58] and subsequently obtained from various *Hydnellum* spp. [1]. Aurantiacin was included in the above cited antimicrobial screening of fungal metabolites, but was inactive (MIC > 100 μg/mL) [21].

Very recently, ganbajunin A (**28**), a further orange pigment related to atromentin, the methoxy atromentin **22** and other fully aromatic *p*-terphenyls (reported here in the following) were isolated from fruiting bodies of the Chinese mushroom *Thelephora ganbajun* [57].

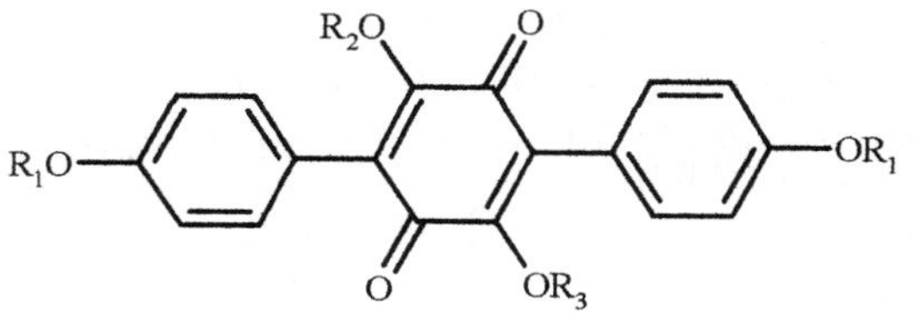

	R_1	R_2	R_3
18	H	H	H
19	Me	H	H
20	H	Ac	Ac
21	H	Ac	H
22	H	Me	H
23	H	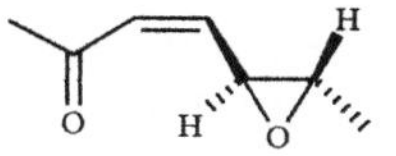 (EPH residue)	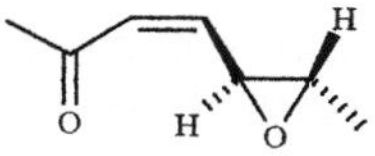 (EPH residue)
24	H	H	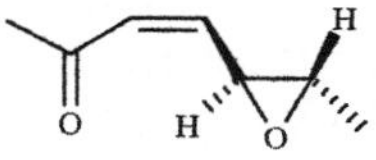 (EPH residue)
25	H	H	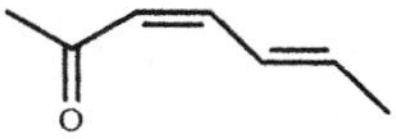 (dienone residue)
26	H	H	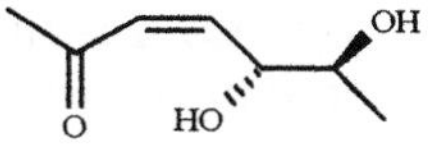 (dihydroxy residue)
27	H	Bz	Bz
28	Bn	Bn	Me

The first leuco-derivatives of atromentin were isolated by Steglich and co-workers from the acetone extract of sporophores of *Paxillus atrotomentosus* and named 'leucomentins' [59]. These are esters of 'leuco-atromentin' (**29**), the hypothetical precursor of atromentin (**18**), according to Thörner [1]. Steglich isolated leucomentin 2, (**30**) 3 (**31**) and 4 (**32**), all bearing acyl EPH residues, and showed that the leuco-precursor of atromentin is the unstable **31**. This metabolite, present in relatively large amount in fruiting bodies of *P. atrotomentosus* and co-occurring with minor amounts of **32**, is easily degraded to **30** during chromatography, and affords atromentin by alkaline hydrolysis. The absolute configuration of the oxirane ring in the EPH moiety has been established by its conversion into (+)-*O*-acetylosmundalactone on acid catalysed acetylation of **31**, affording also tetra-*O*-acetylatromentin. The mechanism of the acid-catalysed cleavage of the acyl groups has been defined by means of ^{18}O labeling.

More recently, in a search for free radical scavengers from basidiomycetes, leucomentins **30** and **32** were isolated through bioguided fractionation of extracts from fruiting bodies of *P. panuoides* [60]. The continuing study of the same fungus afforded also the new leucomentin 5 (**33**) and leucomentin 6 (**34**) [61], differing from **32** for the replacement of one EPH group with an acetyl group or a 2,4-hexadienoyl group, respectively. Very recently, leucomentin 3 (**31**) was obtained from the Japanese inedible basidiomycete *P. atrotomentosus* var. *bambusinus*, together with a novel dimeric lactone, bis-osmundalactone [62]. Compounds **30, 32, 33** and **34** exhibited strong inhibitory activity against lipid peroxidation in rat liver microsomes, with IC_{50} values of 0.06, 0.11, 0.11 and 0.06 μg/mL, respectively [61]. It is worth noting that **30** and **34** are twenty times as active as the control, vitamin E (IC_{50} = 1.5 μg/mL). From the fruiting bodies of *P. curtisii* the new leucoatromentin-related esters curtisian A - D (**35 - 38**) were isolated [63]. Also these polyhydroxy-*p*-terphenyls, bearing acetyl, benzoyl, phenylbutyril, 3-hydroxybutyryl and 3-acetoxybutyryl as substituents on the central aromatic ring, exhibited inhibitory activity against lipid peroxydation in rat liver microsomes, with IC_{50} values of 0.15 (**35**), 0.17 (**36**), 0.24 (**37**) and 0.14 (**38**) μg/mL. These dose-dependent activities were 10-20 times higher than that of vitamin E (IC_{50} = 2.5 μg/mL), used as a control. In addition, **36** and **38** were superoxide radical scavengers comparable to BHA, with IC_{50} = 36.2 and 28.3, respectively, in xanthine/xanthine oxidase system. Curtisians were not active as DPPH radical scavengers, and consequently **36** and **38** are not proton-donating antioxidants. Very recently, a further leucoatromentin analogue, thelephorin A (**39**), has been isolated from fruiting bodies of *Thelephora vialis* collected in Japan [64]. The symmetrical polyhydroxy-*p*-terphenyl core of **39** is asymmetrically substituted with benzoyl and phenylacetyl groups. Thelephorin A was tested for free radical scavenging activity against DPPH and showed EC_{50} = 0.028 (amount of compound necessary to decrease the initial DPPH concentration by 50%). Ascorbic acid, used as positive control, was ten times less potent (EC_{50} = 0.27). Compound **39** was not cytotoxic, when tested against HeLa tumor cells at 50 μg/mL.

As cited above, also the fruiting bodies of *Thelephora ganbajun* afforded leuco-derivatives of atromentin, namely ganbajunins C – G (**40 – 44**), together with the polycyclic ganbajunin B (**66**, see Section 3.1) [57,65]. It is worth noting here that *T. ganbajun* is one of the most favourite edible mushrooms in Yunnan (China), and afforded terphenyls metabolites with a yield of approximately 800 mg/Kg dried material. Further work on Yunnan basidiomycetes gave another acylated leuco-atrometin, 2,3-diacetoxy-4',4",5,6-tetrahydroxy-*p*-terphenyl (**45**), isolated

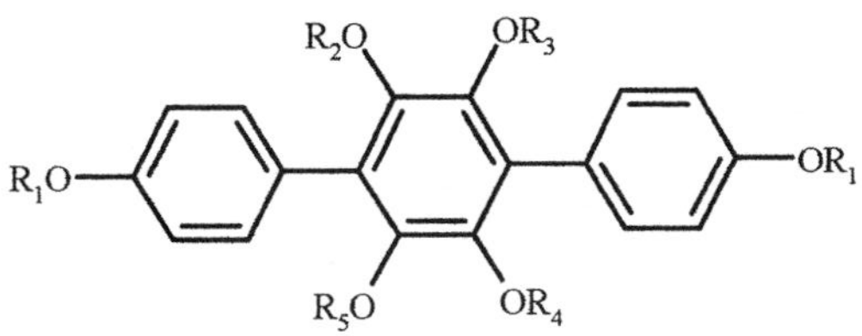

	R_1	R_2	R_3	R_4	R_5
29	H	H	H	H	H
30	H	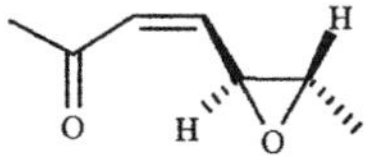	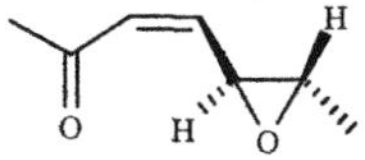	H	H
31	H	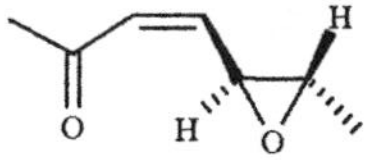	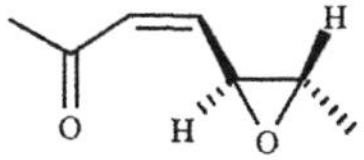	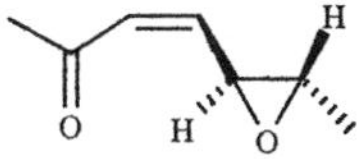	H
32	H	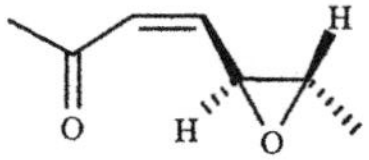	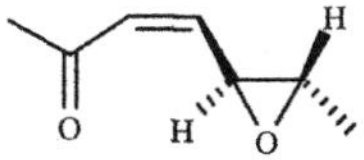	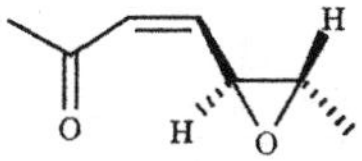	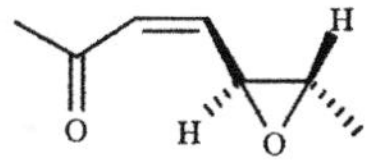
33	H	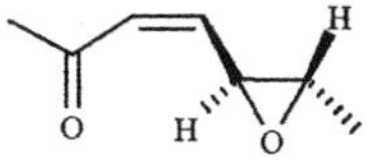	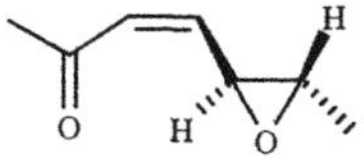	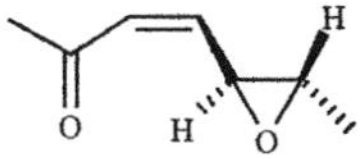	Ac
34	H	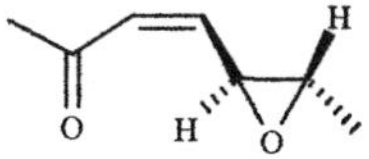	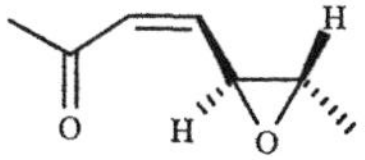	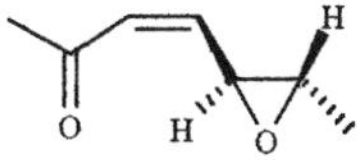	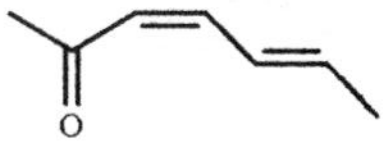
35	H	Ac	Ac	Bz	Ac
36	H	Ac	Ac	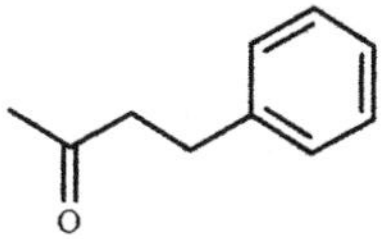	Ac

276

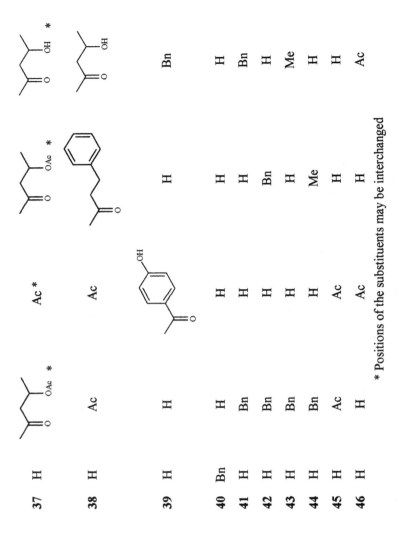

37	H	Ac*	H	H	*
38	H	Ac	Ac	H	Bn
39	H	H	H	H	H
40	Bn	H	H	H	H
41	H	Bn	H	H	Bn
42	H	Bn	H	Bn	H
43	H	Bn	H	H	Me
44	H	Bn	H	Me	H
45	H	Ac	Ac	H	H
46	H	H	Ac	H	Ac

* Positions of the substituents may be interchanged

from fruiting bodies of *Boletopsis grisea (Polyporus griseus = P. repandus)* [66]. Another simple leuco-atromentin analogue, kynapcin-12 (**46**) has been recently isolated from fruiting bodies of the Korean mushroom *Polyozellus multiplex* [67], as a continuation of previous studies on the same source which had afforded polyozellin (**84**, see Section 3.2). Both *Polyozellus* metabolites are new non-peptidyl and small molecular weight inhibitors of prolyl endopeptidase (PEP), a serine protease playing an important role in degradation of proline-containing neuropeptides and possibly involved in Alzheimer's disease. PEP inhibitors could be useful to develop anti-dementia drugs. Kynapcin-12, in particular, had IC_{50} = 1.25 µM and inhibited more than 90% of PEP activity at 40 ppm (97.6 µM).

2.5 Terphenyls bearing seven oxygenated functions

The structure of the 4,3",4"-trihydroxy derivative of polyporic acid was assigned in 1942 by Akagi to the brown pigment leucomelone (**47**) isolated from the black edible mushroom *Polyporus leucomelas* (*Boletopsis leucomelaena*) collected in Japan [68]. From the same fungus, Akagi obtained also a colourless compound named protoleucomelone (**48**), established as the leuco-peracetate of **47**.

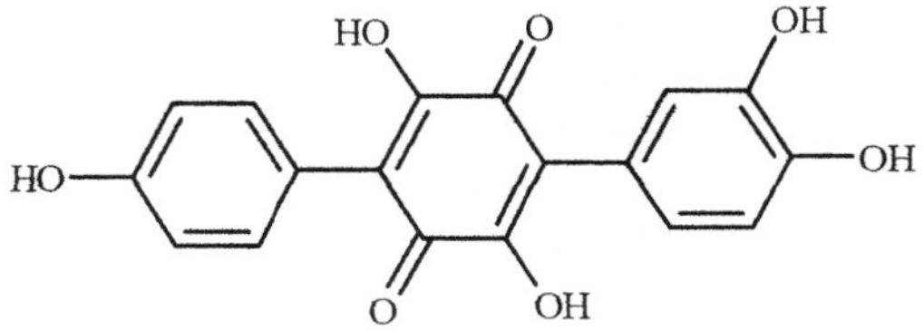

47

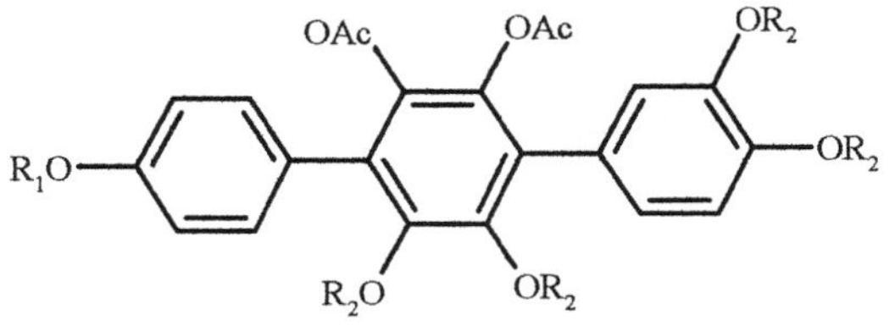

	R_1	R_2
48	Ac	Ac
49	Ac	H
50	H	H

In 1987 Steglich and co-workers [69] observed that the properties of Akagi's leucomelone and protoleucomelone correspond to those of cycloleucomelone **57** and its leuco-peracetate **58** (see Section 3.1), and suggested that **47** could be a biogenetical precursor of **57**. Actually, a compound possessing the structure of Akagi's protoleucomelone (**48**) has been obtained as a semisynthetic peracetate of the polyhydroxy-*p*-terphenyls **49** and **50**, isolated as antimicrobial principles through a bioguided fractionation of an aqueous acetone extract from the fruiting bodies of *Sarcodon leucopus* (*S. laevigatus* = *Hydnum leucopus*). Both metabolites had mild activity against both Gram-positive and Gram-negative bacteria [70]. In a more recent re-examination of the same fungus (see Section 3.3) [71], cytotoxicity data for **49** and **50** against KB tumoral cells have been reported, showing that activity for **50** (ED_{50} = 2.2 µg/mL) is significantly higher than for its 4"-acetoxy analog **49** (ED_{50} = 22.0 µg/mL).

2.6 Terphenyls from lichens

To the best of our knowledge, the few chemical studies reporting isolation and characterization of terphenyls from natural sources not belonging to the Kingdom of Fungi are confined to lichens, and are shortly reported here.

Polyporic acid (**3**), the above cited terphenylquinone occurring in several species of basidiomycetes, has also been found in lichens of the *Sticta* genus [19]. Also thelephoric acid (**82**, see Section 3.2), widely distributed within Thelephoraceae, has been found in lichens of *Lobaria* genus [3]. From the lichen *Relicina connivens* (*Parmelia butleri*) the butlerins A-F (**51 - 56**) were isolated [72, 73]. These polyporic acid-related metabolites have not yet been found in fungi.

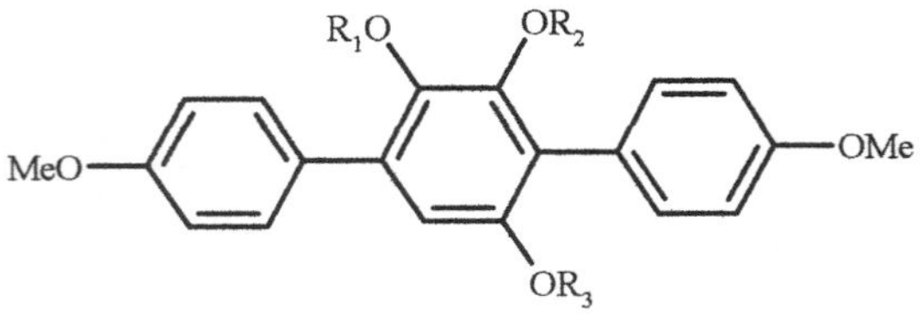

	R_1	R_2	R_3
51	Me	Ac	Me
52	Ac	Me	Me
53	Me	Ac	Ac

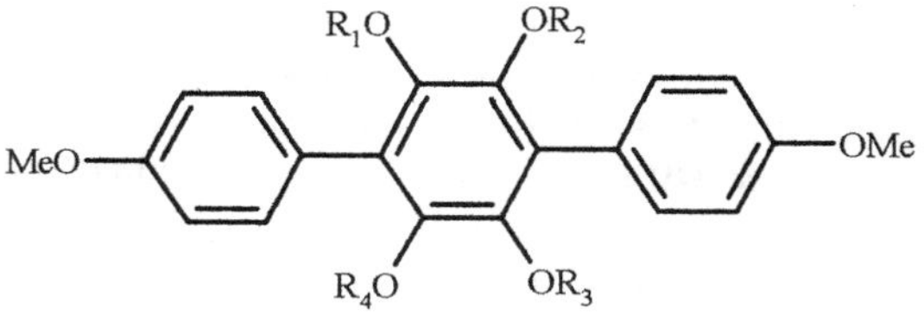

R₁O, OR₂ ... MeO — ... — OMe ... R₄O, OR₃

	R₁	R₂	R₃	R₄
54	Me	Me	Me	Ac
55	Me	Ac	Me	Ac
56	Me	Ac	Ac	Me

3   POLYCYCLIC *p*-TERPHENYLS (C-18 BASIC SKELETON)

This section deals with fungal *p*-terphenyls maintaining the basic C-12 skeleton but possessing more than three cycles, formed through ether bridges or other functionalization involving oxygenated carbons. Polycyclic *p*-terphenyls belong mainly to two groups, including one or two benzofuranoid moieties, respectively analogues of cycloleucomelone (**57**) and thelephoric acid (**82**) (Sections 3.1 and 3.2, respectively). Further less common *p*-terphenyls are included here on the basis of their polycyclic structure.

3.1   Benzofuranoid terphenyls

A number of benzofuranoid terphenyls have been obtained from basidiomycetes. The above cited cycloleucomelone (**57**) was obtained by Steglich and co-workers from the fruiting bodies of *Boletopsis leucomelaena* (*B. leucomelas* = *Polyporus leucomelas*) [69]. From the same fungus the related, fully aromatic cycloleucomelone leuco-peracetate (**58**), cycloleucomelone leuco-tetraacetates **59** and **60**, cycloleucomelone leuco-triacetate **61** and the blue pigment **62** were also isolated, co-occurring with the leuco peracetate of thelephoric acid (**82**, see Section 3.3). In this work, Steglich showed that leucomelone (**47**) and protoleucomelone (**48**) originally obtained by Akagi (see Section 2.5) are actually **57** and its leuco-peracetate **58**. Cycloleucomelone has been also obtained from *Paxillus atrotomentosus* [56], *Anthracophyllum discolor* (from Chile), *A. archeri* (from Australia), and an unnamed *Anthracophyllum* sp. (from Papua New Guinea), where it co-occurs with its monoacetates **63** and **64** [55]. From these species the yellow pigment anthracophyllin (**65**), namely the diacetate of **57**, has also been obtained. More recently, **57** was isolated also from the basidiomycete *Thelephora ganbajun* collected in Yunnan, together with the   fully aromatic benzofuranoid ganbajunin B (**66**), and the above cited ganbajunin A (**28**) (see Section 2.4) and ganbajunins C-G (**40** – **44**, see Section 2.4) [65]. Further compounds related to the quinol form of **57** and named Bl-I, – Bl-V, (**67** – **71**), were obtained from the MeOH extract of the fruiting bodies of *Boletopsis leucomelas* [74]. These metabolites, differing each other for the number of acetoxy group, had an inhibitory effect on 5-lipoxygenase, a key enzyme in the synthesis of leukotrienes, involved in various inflammatory and allergic diseases. In particular, the IC_{50} of Bl-III (**69**) was 0.35 μM, comparable to that of a therapeutic drug for asthma (AA-861).

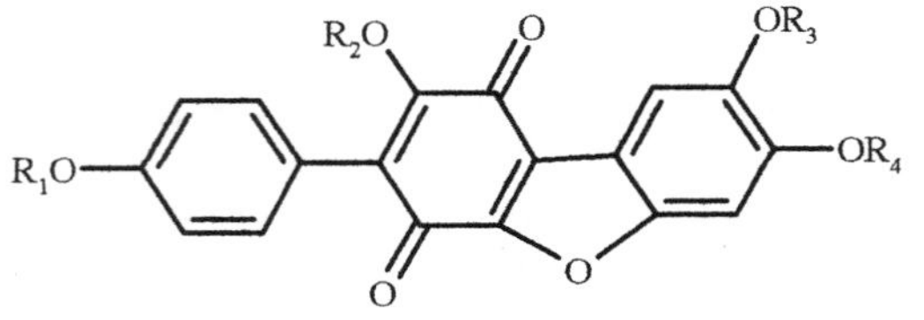

	R_1	R_2	R_3	R_4
57	H	H	H	H
63	H	Ac	H	H
64	H	H	Ac	H
65	H	Ac	Ac	H

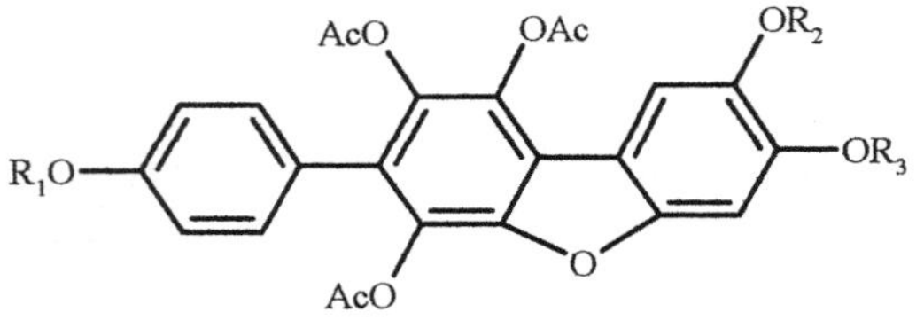

	R_1	R_2	R_3
58	Ac	Ac	Ac
59	H	H	Ac
60	H	Ac	H
61	H	H	H

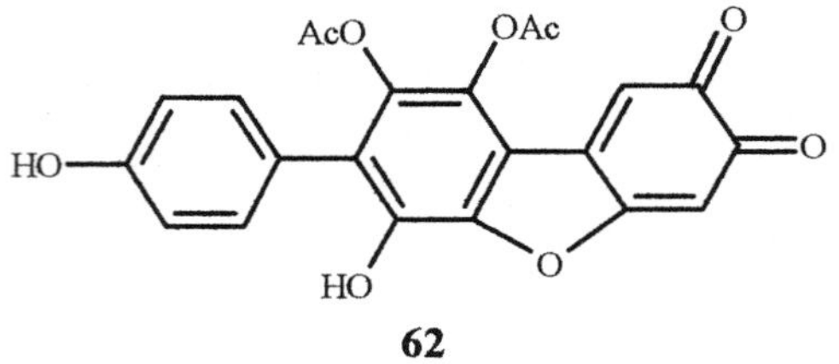

62

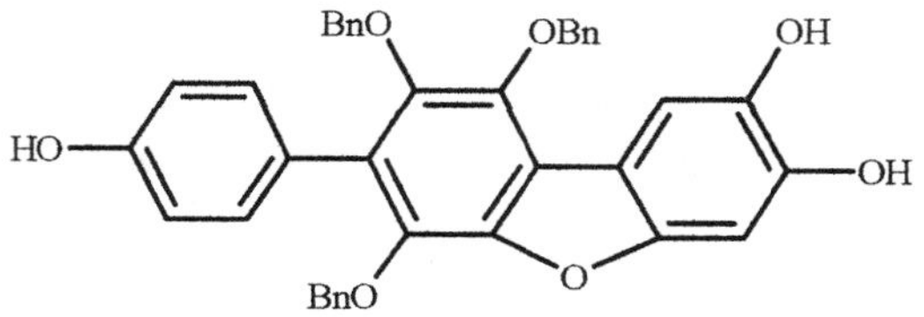

66

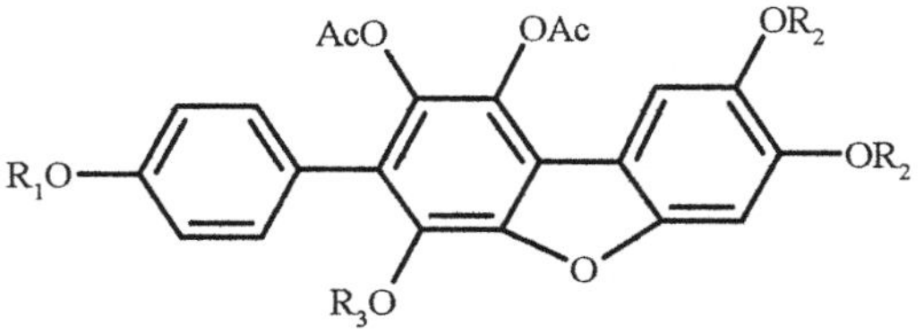

	R₁	R₂	R₃
67	Ac	Ac	Ac
68	H	Ac	Ac
69	Ac	H	Ac
70	H	H	Ac
71	H	H	H

Candidusin A (**72**) and candidusin B (**73**) were the first isolated fully aromatic benzofuranoid terphenyls and were obtained from cultures of *Aspergillus candidus* [38]. These metabolites were cytotoxic on sea urchin embryos (total block of the initial cleavage) at 10^{-4} M and 5 x 10^{-5} M, respectively. Candidusin B was also studied for inhibitory effect on DNA, RNA and protein synthesis, and proved active at 10 μg/mL (2.7 x 10^{-5} M) in inhibiting DNA and RNA synthesis, but not protein synthesis. Both candidusins were clearly active at 50 μg/mL in antibacterial assays against *Bacillus subtilis*. More recently, **73** was isolated as antioxidative component of broth filtrate from *Aspergillus candidus*, together with **12** and **13** [41]. As reported in Section 2.3, the antioxidant properties of these three metabolites were comparable or higher as those of BHA and α-tocopherol. In particular, **73** inhibited linoleic acid peroxidation more than 95%, and was moderately active in Rancimat method in lard and DPPH radical scavenging assay. No cytotoxicity or mutagenicity was found for **73**, in contrast with the above cited data on DNA and RNA, thus inviting further studies on polyhydroxyterphenyls as potential antioxidants in food applications. In a very recent chemotaxonomical screening of *Aspergillus* spp., the isolation of the new candidusin C (**74**, 2,2'-epoxyterphenyllin) from the chemically unexplored *A. campestris* has been reported, in addition to **8**, **12** and chlorflavonin [43].

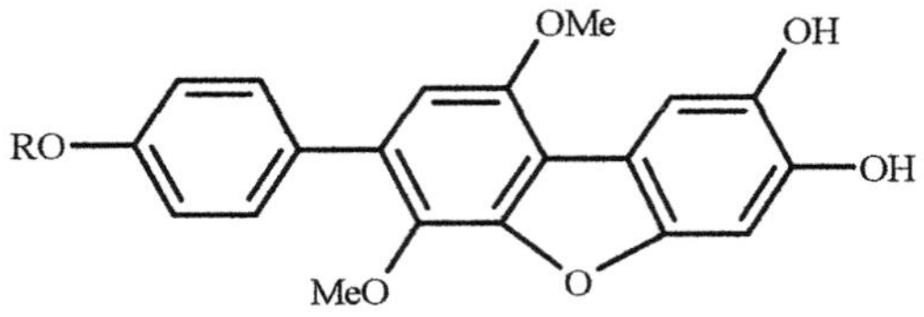

	R
72	H
74	Me

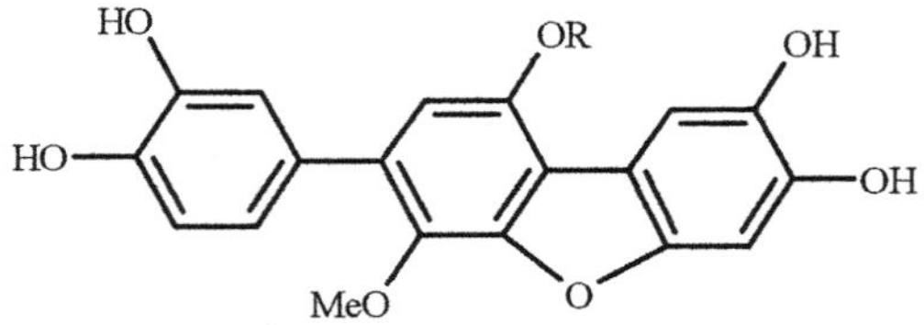

	R
73	Me
76	H

Two further *p*-terphenyls related to candidusins, namely 3'-demethoxy-6'-desmethyl-5'-methoxycandidusin B (**75**) and 6'-desmethylcandidusin B (**76**) were more recently obtained from sclerotia of *Penicillium raistrickii* together with **14** and other metabolites, among them griseofulvin (see Section 2.3) [44].

A further benzofuranoid, arenarin A (**77**), was isolated from the sclerotia of *Aspergillus arenarius*, where co-occurred with arenarin B and arenarin C (**116** and **117**, See section 4.1) [75]. These three prenylated terphenyls possess mild antifeedant activity against the fungivorous beetle *Carpophilus hemipterus*. In particular, **77** induced a 13% reduction in feeding rate at 100 ppm dietary level. Arenarins A and B were also tested against NCI's 60 human tumour cells, **77** displaying average GI_{50} = 4.8 µg/mL.

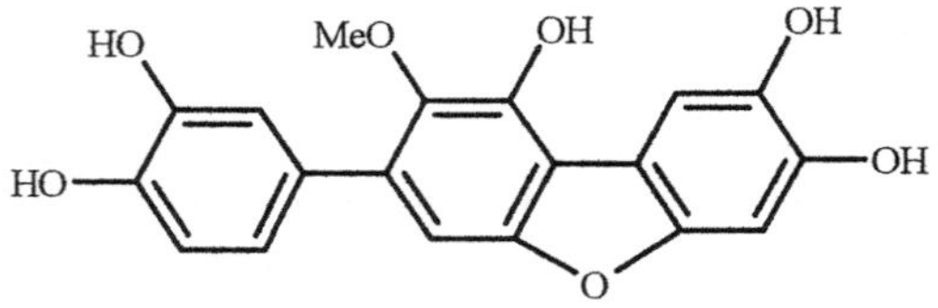

75

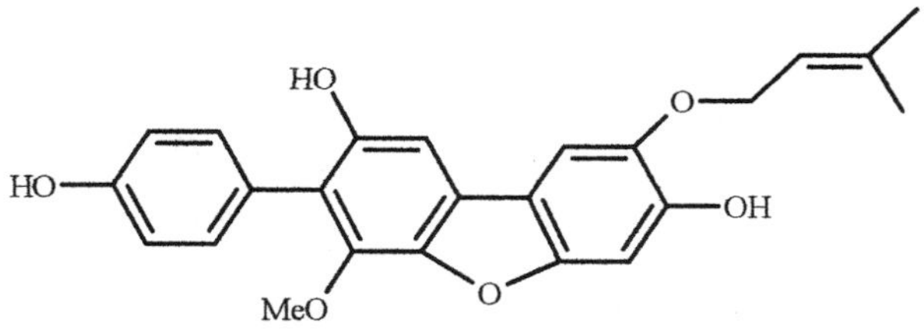

77

The hydroxylated analogue of cycloleucomelone, **78**, previously isolated as the first benzofuranoid terphenyl, was obtained from *Boletus elegans* (*Suillus grevillei*) and originally reported as 'pigment C$_2$' [76]. On standing in solution, particularly in acetone, **78** was easily converted into thelephoric acid (**82**). Thus, pigment C$_2$ was reported as a probable precursor of **82**, itself isolated from *S. grevillei* [77]. More recently, Gill and Steglich proposed for **78** the name cyclovariegatin, by analogy with cycloleucomelone [1], since its hypothetical biogenetic precursor is known as variegatin (**79**) [76]. Two acetylated pigments related to **78** were obtained from an *Anthracophyllum* sp. collected in Papua New Guinea, namely 2-*O*-acetylcyclovariegatin (**80**) and 2,3',8-tri-*O*-acetylcyclovariegatin (**81**) [55]. When dissolved in methanol, **80** is rapidly converted into thelephoric acid.

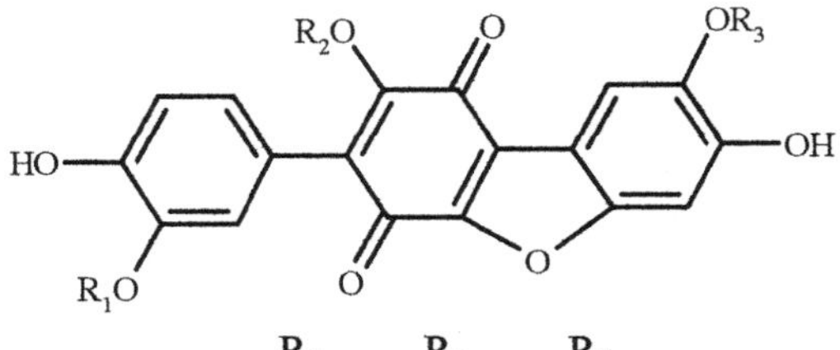

	R$_1$	R$_2$	R$_3$
78	H	H	H
80	H	Ac	H
81	Ac	Ac	Ac

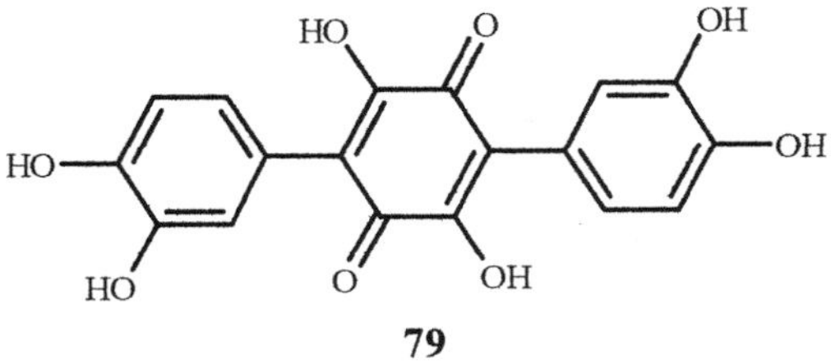

79

3.2 Bisbenzofuranoid Terphenyls

A further group of polycyclic p-terphenyls is characterized by a bisbenzofuranoid core constituting a pentacyclic, rigid structure well represented by one of the early isolated terphenylquinones, thelephoric acid (82). Later isolated by Gripenberg from sporophores of *Hydnum suaveolens*, its structure was correctly established [78]. Thelephoric acid is now known to be a common pigment in basidiomycetes belonging to the Thelephoraceae family [1]. Recently, 82 has been isolated from fruiting bodies of the Korean toadstool *Polyozellus multiplex* in the course of a screening for prolyl endopeptidase (PEP) inhibitors, and found to have IC_{50} = 0.157 ppm (446 nM). The same screening afforded also the related unusual *meta*-terphenyl kynapcin-9 (83), [IC_{50} = 0.087 ppm (212 nM)] [79]. In a previous study on PEP inhibitors from *P. multiplex*, the leuco diacetate of thelephoric acid, polyozellin (84), was obtained [80]. Polyozellin inhibited PEP in a dose-dependent fashion with an IC_{50} value of 2.72 μM, but its activity was lower than that of poststatin. In antimicrobial assays at 1 mg/mL, no activity was observed.

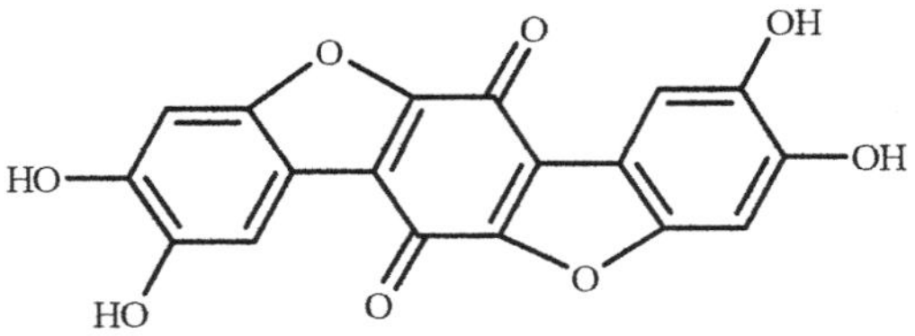

82

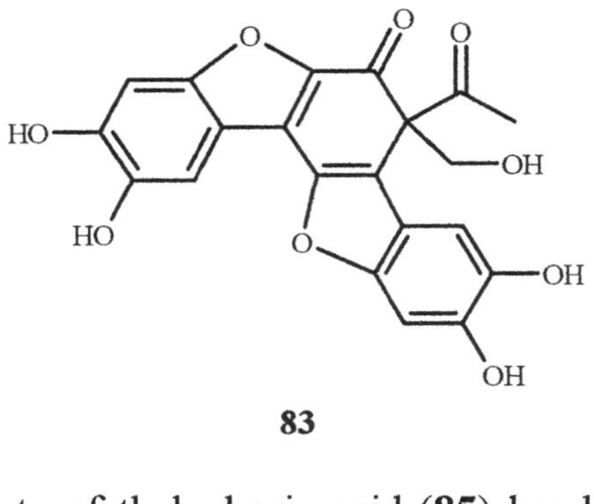

83

The leuco peracetate of thelephoric acid (85) has been isolated from *Boletopsis leucomelaena* [69]. The leuco permethyl ether of thelephoric acid (86) was obtained in minute amounts from mycelial mat of *Pulcherricium caeruleum (Corticium caeruleum)*, a basidiomycete characterized by an indigo blue colour of its hymenial surface, and its structure established by X-ray crystal analysis [81,82]. An independent study on the blue pigments of this fungus confirmed a relationship with leuco-thelephoric acid derivatives but suggested on the basis of MS data they could be polymeric [83,84]. In a subsequent research on cultured *C. caeruleum,* the crude pigment was obtained by washing the surface of the culture with chloroform. From this, through direct chromatographic separation or by preliminary conversion to their acetates or leucoacetates, the *p*-terphenyl benzobisbenzofurans corticin A (**87**) and B (**88**), clearly related to the quinol form of thelephoric acid, and the *m*-terphenyl corticin C (**89**), were obtained [85]. The structures of corticins were proposed on the basis of the available spectral data (essentially ^{1}H NMR) and that of corticin B was confirmed by synthesis. The synthesis of benzobisbenzofuranoid derivatives and their MS data were also reported (See Section 6). Gill and Steglich observed [1] that structure **89** appears biogenetically improbable, nevertheless to date this is not the only reported fungal *meta*-terphenyl, a further example being **83** [79]. In this connection it is worth noting here that hydroxylation of a *m*-terphenyl has been obtained in a fermentation process employing *Aspergillus parasiticus* [86].

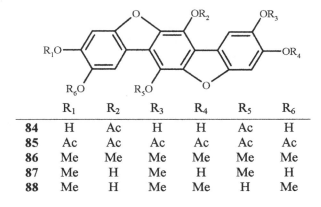

	R_1	R_2	R_3	R_4	R_5	R_6
84	H	Ac	H	H	Ac	H
85	Ac	Ac	Ac	Ac	Ac	Ac
86	Me	Me	Me	Me	Me	Me
87	Me	H	Me	H	Me	H
88	Me	H	Me	Me	H	Me

89

3.3 Other polycyclic terphenyls

The unusual epoxyenedione **90** was obtained from cultures of the AAA566 *Streptomyces* strain. This compound showed weak antibiotic activity against *Bacillus subtilis* [87].

Phlebiarubrone (**91**), the first isolated polycyclic *p*-terphenyl, is an *ortho*-quinone strictly related to polyporic acid (**3**), originally obtained as a red pigment from a culture of *Phlebia strigosozonata (Punctularia strigosozonata)* [88, 89]. Phlebiarubrone lacks of free hydroxy groups, the 5',6'-oxygens being linked through a methylene bridge. The violet pigments 4'-hydroxyphlebiarubrone (**92**), 4',4"-dihydroxyphlebiarubrone (**93**) and 3',4',4"-trihydroxyphlebiarubrone (**94**) were later isolated, together with **91,** from cultures of *Punctularia atropurpurascens* [90]. In biological assays **91, 92** and **94** showed weak antibacterial and cytotoxic activity, whereas **93** was inactive. The production of terphenylquinones **91** - **94** in cultures of different strains and species of the genus *Punctularia* has been cited in a work on the isolation of bioactive kauranes from *P. atropurpurascens* [91].

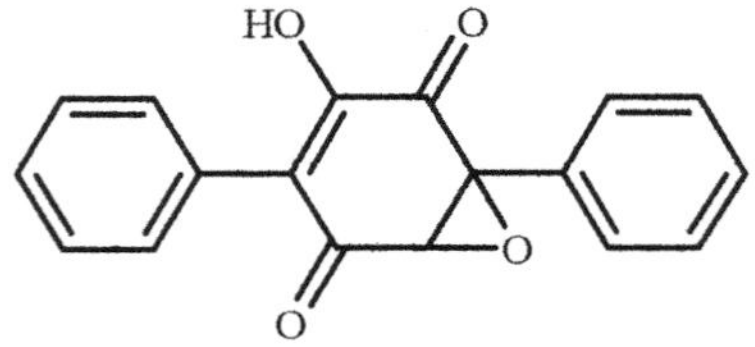

90

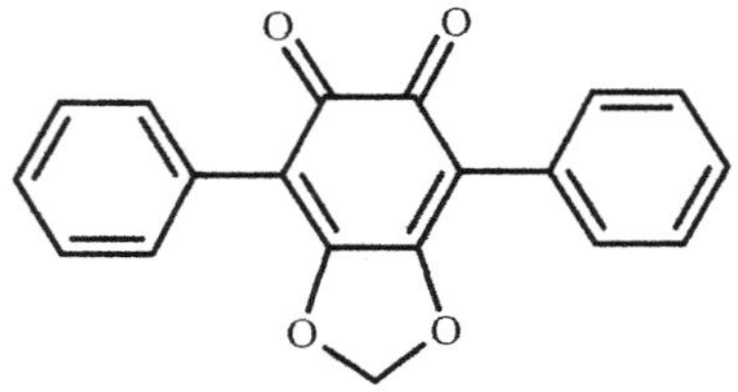

91

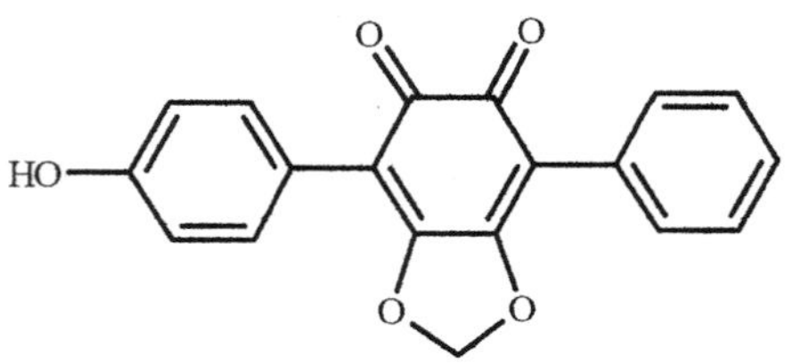

92

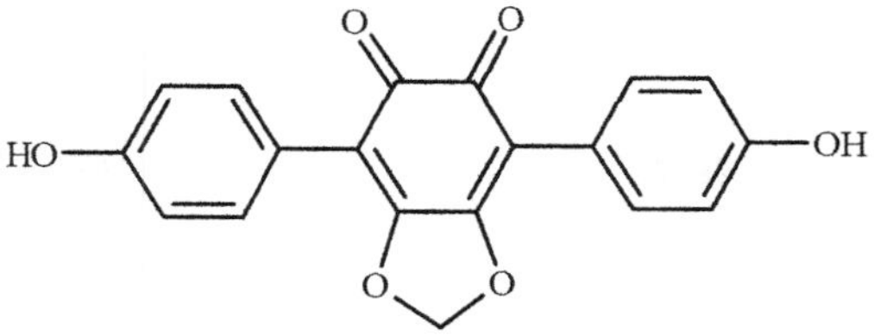

93

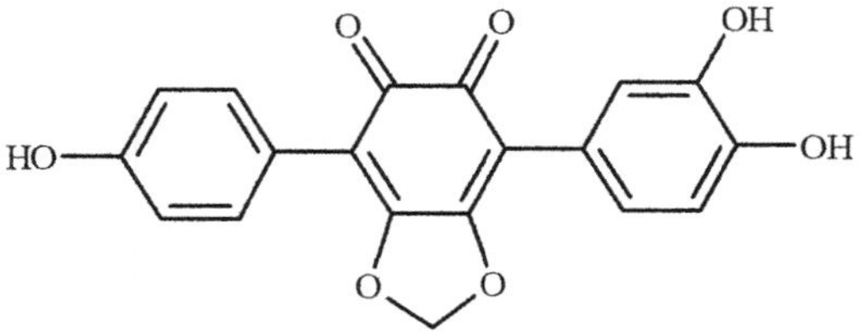

94

In an extensive study of *Paxillus atrotomentosus* and *P. panuoides* carried out by Steglich and co-workers, a new group of *p*-terphenyls possessing unique γ- and δ-lactone-acetal spiro structures linked to a 4,5-dihydroxy-1,2-benzoquinone core, the spiromentins A - D, (**95 - 98**) were isolated [56]. The structures of these violet pigments, established on the basis of spectral data, were confirmed by the synthesis of model compounds and biomimetic conversion of flavomentin B (**24**) into **96** and **97**. (See section 6). This easy conversion suggests that spiromentins could be artefacts formed from flavomentins during work-up of the fungus.

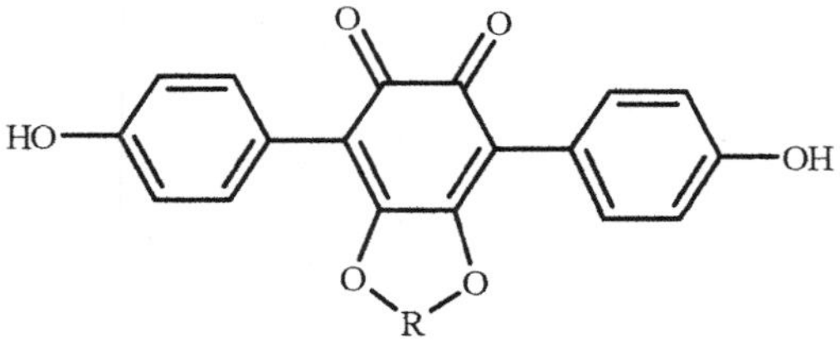

R
_____

95

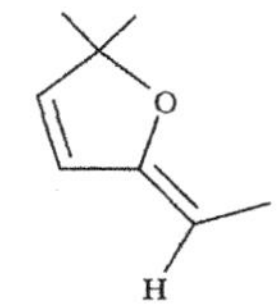

288

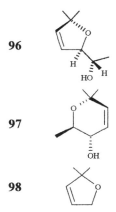

96

97

98

Spiromentin B and C (but no leucomentin or flavomentin) were subsequently obtained from the fresh fruiting bodies of *P. atrotomentosus* collected in Japan, together with six new terphenylquinol metabolites, the colourless spiromentins E - J (**99 – 104**), with a symmetrical structure including seven cycles [92]. It is worth noting here that assignment of the relative stereochemistry at the spiro asymmetric carbons was based on the observed equivalence of chemical shifts of H-2'/H-2" and H-3'/H-3". The same study reports the isolation from *P. atrotomentosus* of (+)-osmundalactone and other three γ-lactones related to the spiromentins structures.

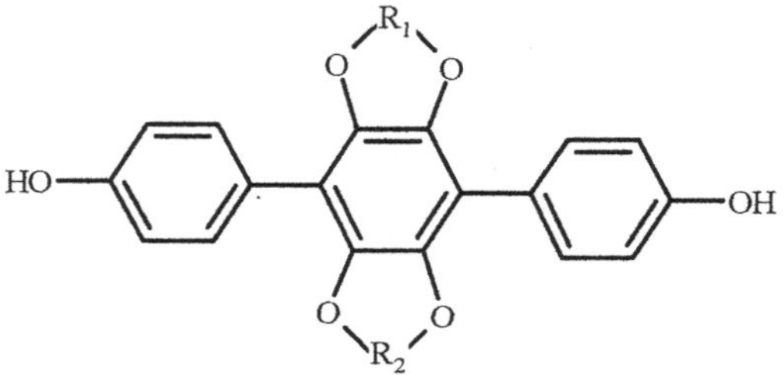

	R₁	R₂
99	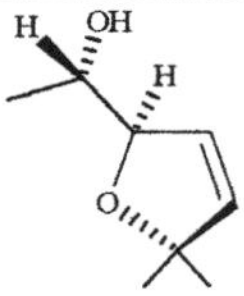	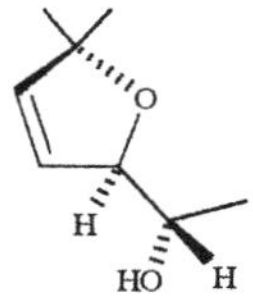

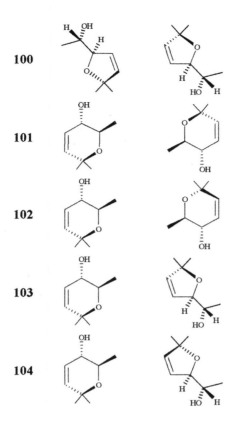

A recent re-examination of *Sarcodon leucopus* afforded an unique nitrogenous metabolite with a *p*-terphenyl core, sarcodonin (**105**) [71]. This relatively unstable metabolite, clearly related to the co-occurring and previously reported **49** and **50** [70], could be isolated through a modified extraction/isolation protocol, based on ethyl acetate extraction of freeze-dried fruiting bodies followed by acetyl polyammide or diol-silica gel column chromatography. Sarcodonin is unique among terphenyl metabolites, in possessing a *N*-oxide function and an hydroxamic acid moiety in a 1,4-diazine ring bonded to positions 3 and 4 of the external dihydroxylated ring of the *p*-terphenyl core. Determination of the structure of this metabolite was achieved through a careful study of the more stable peracetate, by use of an array of spectroscopic techniques including ^{15}N NMR, reverse-detected 2D NMR, ROESY data and Molecular Mechanics calculation as well as chemical degradation. Sarcodonin exhibited moderate cytotoxicity toward KB (ED$_{50}$ = 10.0 µg/mL) and P-388 (ED$_{50}$ = 27.0 µg/mL) cultured tumor cells.

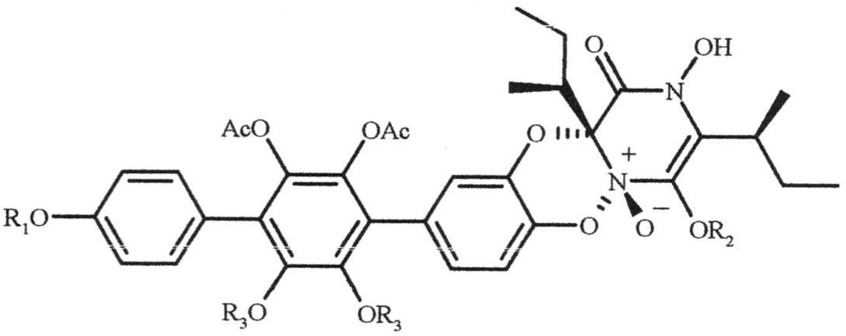

	R_1	R_2	R_3
105	Ac	H	H
106	H	H	H
107	H	H	Ac
108	Ac	Ac	H

Very recently, a continuation of the work on fruiting bodies of *Sarcodon leucopus* afforded six new sarcodonins, three of them, respectively sarcodonin α (106), sarcodonin β (107) and sarcodonin γ (108), differing from 105 for the acetylation pattern of the hydroxy groups, and three 1β-epimers of metabolites 106 – 108, namely episarcodonin (109), episarcodonin α (110), and episarcodonin β (111). The related violet pigment with *ortho*-quinone structure, sarcoviolin α (112) was also obtained [93]. In the primary anticancer assay carried out at NCI (Bethesda), compounds 105, 108, 109 and 110 were tested against NCI-H460 (Lung), MCF7 (Breast) and SF-268 (CNS) tumour cell cultures. All compounds proved active at concentration 5×10^{-5} M. In particular, the highest cytotoxicity was observed towards SF-268 cells, with 105, 108, and 109 showing respectively 96%, 93% and 95% killed cells.

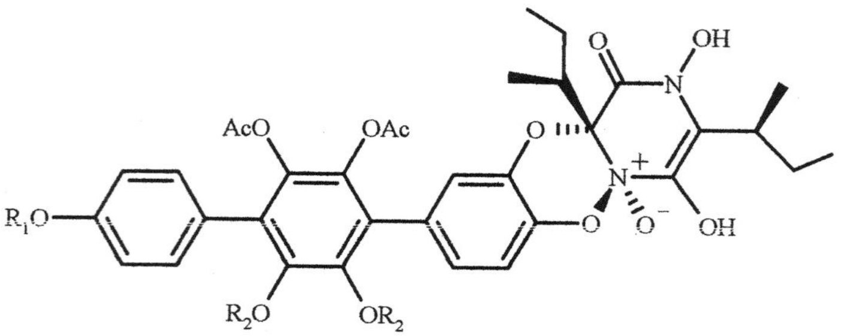

	R_1	R_2
109	Ac	H
110	H	H
111	H	Ac

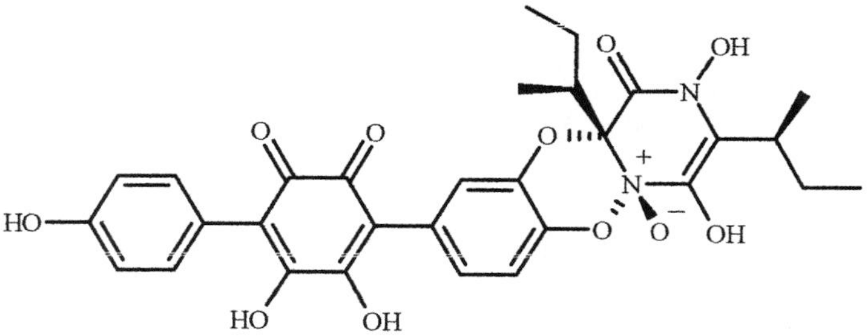

112

4 ALKYLATED *p*-TERPHENYLS

In this section metabolites bearing one or more alkyl groups directly attached to the C-18 *p*-terphenyl core will be reviewed, including the xylerythrin group, which is originated through insertion of a third phenylpropanoid unit (See Section 5).

4.1 Terphenyls with an alkylated C-18 basic skeleton

The alkylated derivative of polyporic acid, betulinan B (**113**), co-occuring with betulinan A (**4**), was obtained from *Lenzites betulina*. Both compounds were tested for inhibitory activity against lipid peroxidation in rat liver microsomes (See Section 2), **113** showing an $IC_{50} = 2.88$ µg/mL [27]. Gripenberg *et al.* isolated a series of alkylated *p*-terphenyls from decaying wood attacked by the fungus *Peniophora sanguinea* (see Section 4.2). The simplest among them is the yellow pigment penioflavin (**114**), a benzofuranoid aldehyde [94].

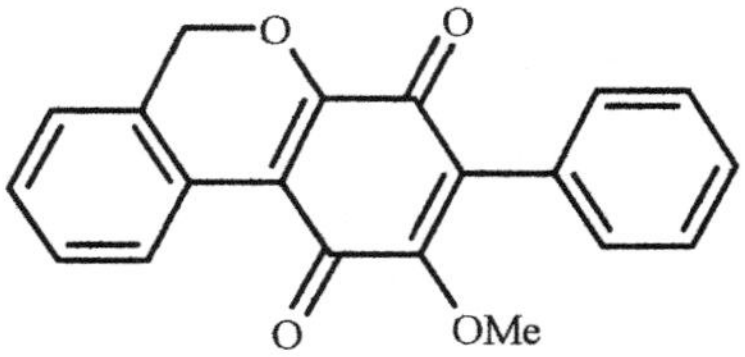

113

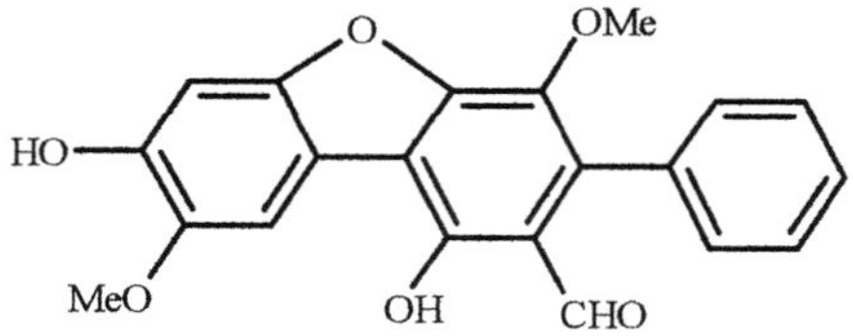

114

Among the early isolated terphenylquinones, we wish to mention here muscarufin (**115**), reported by Kögl and Erxleben in 1930 as a pigment from the cap skin of fly agaric *Amanita muscaria*, a common poisonous basidiomycete with a peculiar red cap [95]. Although muscarufin became popularly known as the colouring principle of *A. muscaria*, more recent studies failed to validate Kögl's results and the main pigments of fly agaric were positively identified as members of the class of betalains [96].

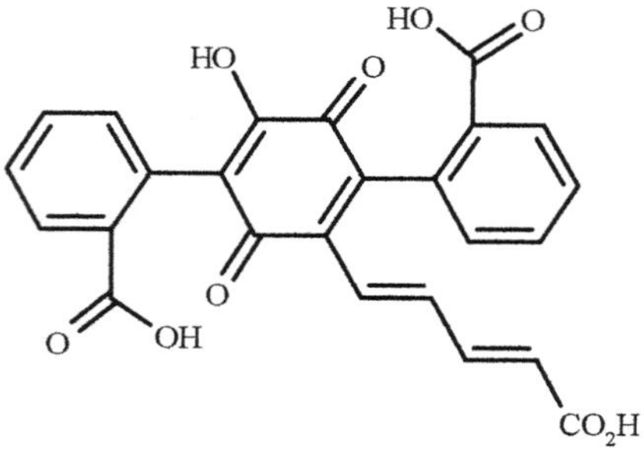

115

Two prenylated terphenylquinones, arenarin B (**116**) and C (**117**), with an unprecedented additional carbocyclic ring attached to the terphenyl core, have recently been isolated from the sclerotia of *Aspergillus arenarius*, produced by solid fermentation on corn kernels [75]. The extract of the sclerotia was highly active against the agriculturally important corn pest *Helicoverpa zea* and the fungivorous dried fruit beetle *Carpophilus hemipterus*. Fractionation of this extract afforded **116** and **117** together with arenarin A (**77**, see Section 3.1). These metabolites have mild antifeedant activity against *Carpophilus hemipterus*, **116** and **117** showing respectively a 20% and 13% reduction in feeding rate at 100 ppm dietary level. Arenarins were also submitted to the NCI's 60 human tumour cells panel, **116** having average $GI_{50} = 3.8$ $\mu g/mL$.

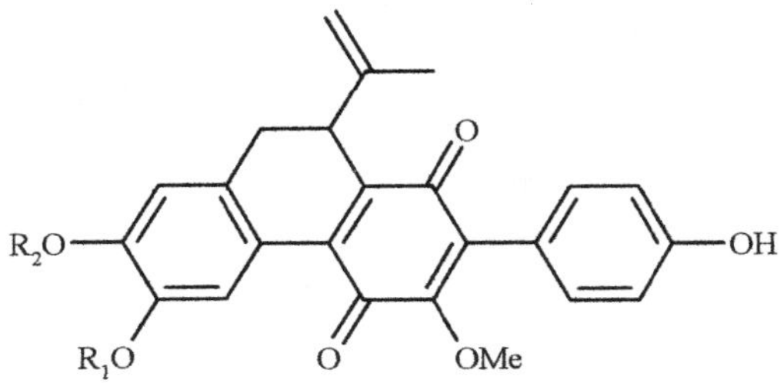

	R₁	R₂
116	H	(2-methyl-pent-2-enyl, OH)
117	Me	(2-methyl-pent-2-enyl, OH)

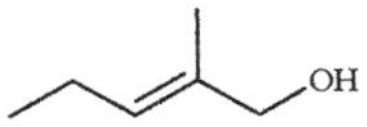

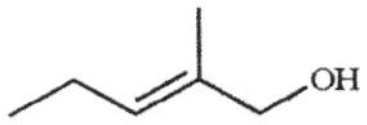

4.2 Xylerythrin and analogues

The early work by Gripenberg et al. on *P. sanguinea* infected wood afforded the dark red pigment xylerithrin (**118**) and its simple derivative 5-*O*-methylxylerythrin (**119**) [97, 98]. The synthesis of **118** was carried out also by other authors [99]. Subsequently, the related peniophorin (**120**) [100] and peniophorinin (**121**) were isolated [101]. Both compounds were subjected to further confirmative X-ray studies, allowing the unambiguous location of all hydroxy group in the former [101], and a revision of the structure for the latter [102, 103].

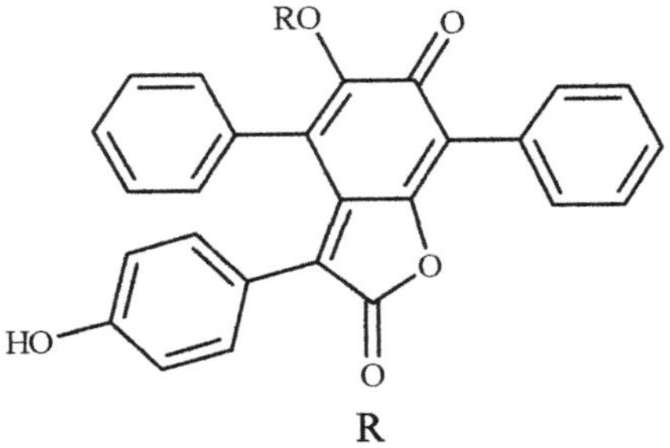

	R
118	H
119	Me

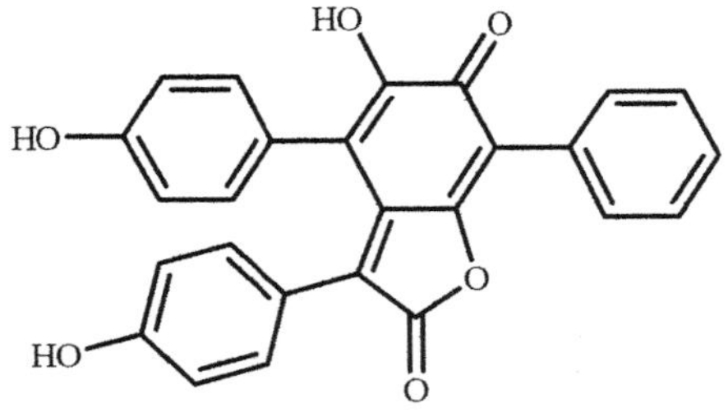

120

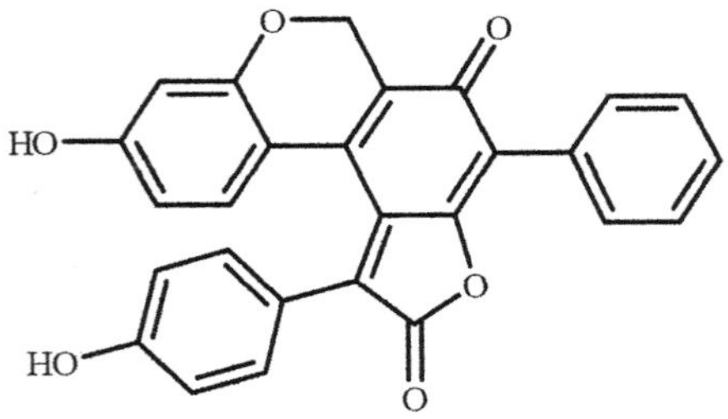

121

Further work on the same source afforded peniosanguin (**122a** or **122b**) and its methylether (**123a** or **123b**) [104], in which structures location of an hydroxyl group could not be established without ambiguity. A further related compound isolated from wood infected by *P. sanguinea* was xylerythrinin (**124**) whose structure was established by X-ray analysis of its methylether [105]. A TLC study on pigments from *P. sanguinea* has been also reported [106].

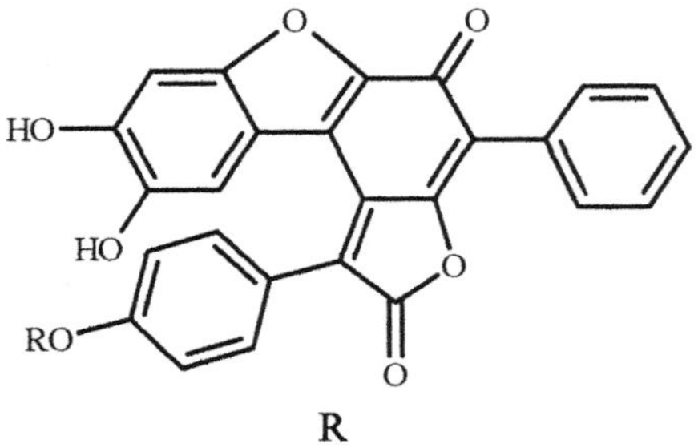

	R
122a	H
123a	Me

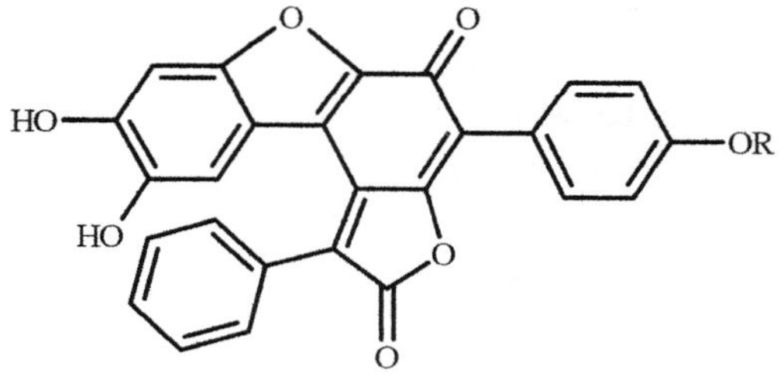

	R
122b	H
123b	Me

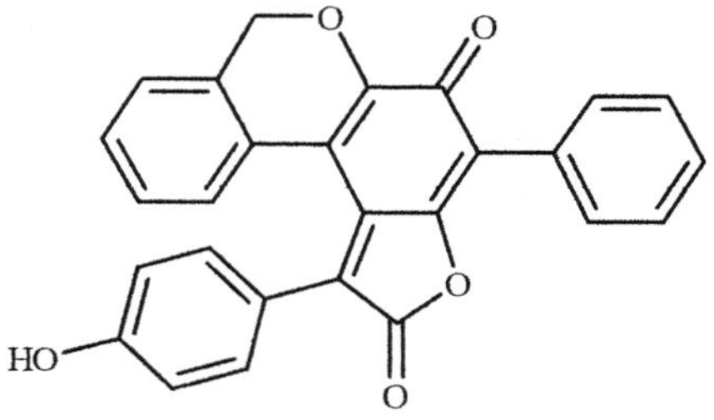

124

5   BIOSYNTHESIS

Polyhydroxyterphenyls and terphenylquinones are generally included among compounds derived from arylpiruvic acids (shikimate-chorismate pathway). Scheme 1 summarizes the biosynthetic ways to some natural polyhydroxy-*p*-terphenyls, as suggested by some authors and corroborated by experimental evidences [5, 29, 51]. Read and Vining were the first to suggest that natural terphenylquinones could originate by condensation of two molecules of an unbranched phenylpropanoid precursor [107]. This was confirmed by their biogenetic studies with ^{14}C-labelled compounds, carried out on cultures of the hyphomycete *Volucrispora aurantiaca*, showing that shikimic acid, phenylalanine, phenyllactic acid and *m*-tyrosine are precursors of the terphenylquinone volucrisporin (**5**), whereas cinnamic acid, *m*-hydroxy cinnamic acid and tyrosine are not incorporated into **5**.

These results indicated also that *meta* hydroxylation in volucrisporin occurs before the formation of the *p*-terphenyl core [29]. Further studies supported the hypothesis that **5** can be biosynthesized via phenylpiruvic acid and *m*-hydroxyphenylpiruvic acid [108].

Condensation of two phenylpiruvate units should produce the simple polyporic acid (3), not bearing hydroxyl groups on side rings. Some studies were devoted to clarifying the hydroxylation mechanism. The isolation of the 4-hydroxy derivative of polyporic acid, ascocorynin (6) from mycelial cultures of *Ascocoryne sarcoides*, where neither polyporic acid (3) nor atromentin (18) were detected, was considered a further evidence in favour of a biosynthesis of hydroxylated terphenyls through direct condensation of phenylpiruvic acid and hydroxyphenylpiruvic acid, rather than by enzymatic hydroxylation of 3 [29]. Atromentin is a key intermediate for further conversion to the more highly hydroxylated terphenyls [1]. The occurrence of atromentin and thelephoric acid (82) in cultures of *Clitocybe subilludens (Omphalotus subilludens)* corroborated the hypothesis that 3 may act as a biogenetic precursor of 82, since a simultaneous disappearance of the former and appearance of the latter was observed [51].

The first biosynthetic study on a fully aromatic polyhydroxy-*p*-terphenyl was carried out on cultures of A*spergillus candidus* and led to the isolation of terphenyllin (7), which incorporated ^{14}C-labelled phenylalanine and methionine [31]. Phenylalanine is employed to form the *p*-terphenyl core, whereas methionine supplies methylation of the hydroxyls of the central ring. Acetate and cinnamate were not incorporated, thus confirming that the terphenyl system arises by condensation of phenylpiruvate precursors [33]. According to Steglich *et al.*, the easy conversion of leucomentin 3 (31) to atromentin by alkaline hydrolysis suggests that leuco-derivatives of atromentin may accumulate in fruiting bodies of basidiomycetes as precursor of 18 or related pigments isolated in small amounts [59].

The possibility of enzymatic hydroxylation subsequent to the formation of the terphenyl nucleus cannot be completely ruled out, on the basis of some studies on the biosynthesis of the simple, not hydroxylated *ortho*-terphenylquinone phlebiarubrone (91), formed in cultures of *Phlebia strigosozonata (Punctularia strigosozonata)*. Early studies employing ^{13}C-labelled precursors showed that the terphenyl core is derived from phenylalanine, whereas the methylene bridge comes from formate or from methionine [109,110], and these results further substantiated the first steps of the biogenetic pathway outlined in Scheme 1. Nevertheless, more recent studies, based on administration of ^{13}C-phenylalanine and ^{13}C-tyrosine as well as of ^{14}C-phenylalanine and ^{14}C-phenylpiruvate to cultures of *Punctularia atropurpurascens*, confirmed that 91 is biosynthesized from initial condensation of two molecules of either phenylpiruvic acid or phenylalanine but showed that its hydroxylated derivatives 92, 93 and 94, not incorporating tyrosine, are obtained through subsequent consecutive hydroxylations [111, 112].

Scheme 1 – Biosynthetic ways for some polyhydroxy-*p*-terphenyls.

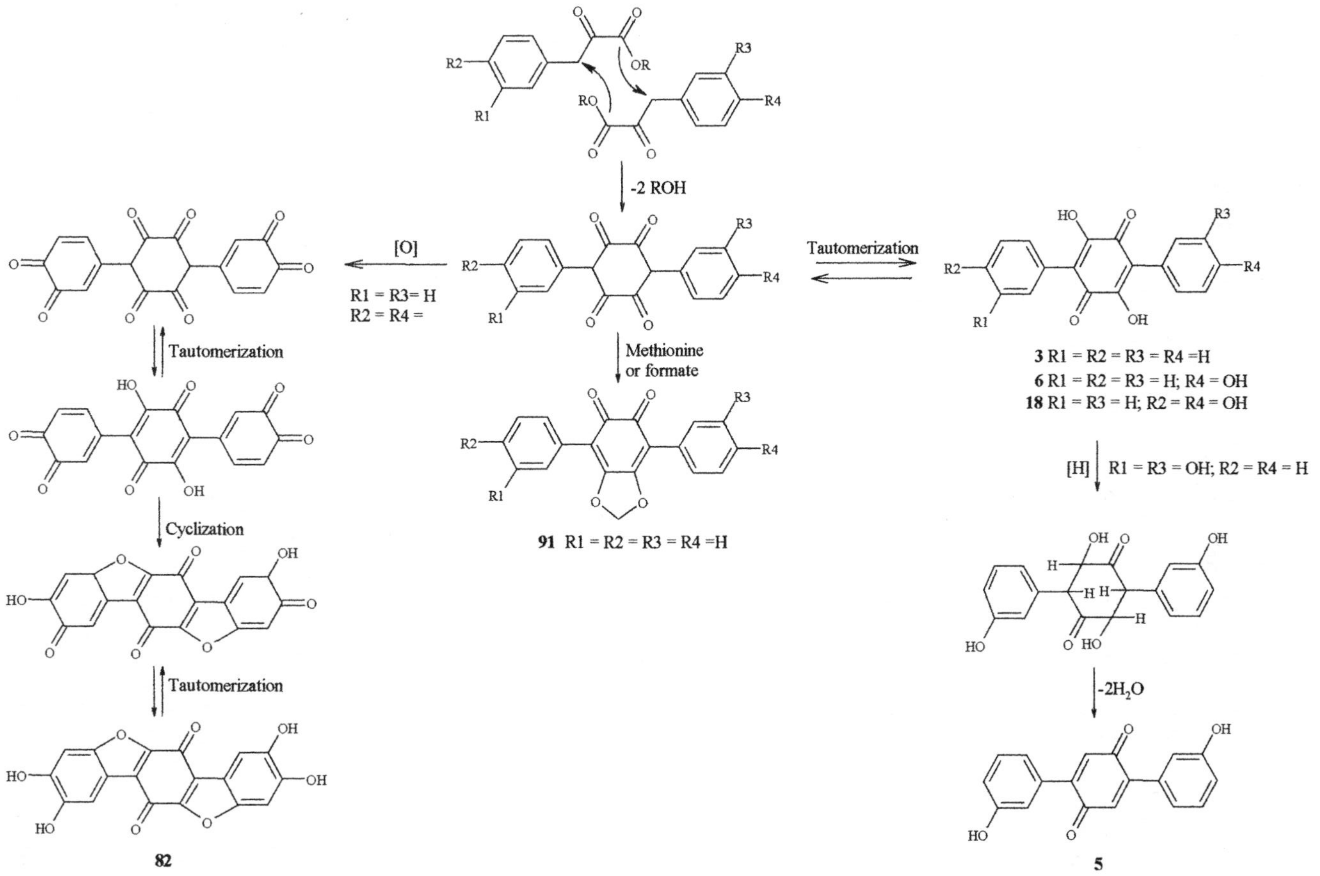

R1
R2
R3
R4
OR
RO
-2 ROH
[O]
R1 = R3= H
R2 = R4 =
Tautomerization
Methionine
or formate
3 R1 = R2 = R3 = R4 =H
6 R1 = R2 = R3 = H; R4 = OH
18 R1 = R3 = H; R2 = R4 = OH
[H]
R1 = R3 = OH; R2 = R4 = H
Cyclization
-2H₂O
91 R1 = R2 = R3 = R4 =H
82
5

Simple alkylated polyhydroxyterphenyls like penioflavin (**114**), isolated from *Peniophora sanguinea*, can be derived formally from a C-18 terphenylquinone as precursor and subsequent formation of a new carbon-carbon bond [94]. The biosynthesis of xylerithrin (**118**) and related pigments from *P. sanguinea*, according to the studies of Von Massow [113, 114], requires the insertion of a third unbranched phenylpropanoid unity on a pulvinic acid intermediate derived from a terphenylquinone, as indicated in Scheme 2.

Scheme 2 – Biosynthesis of xylerythrins

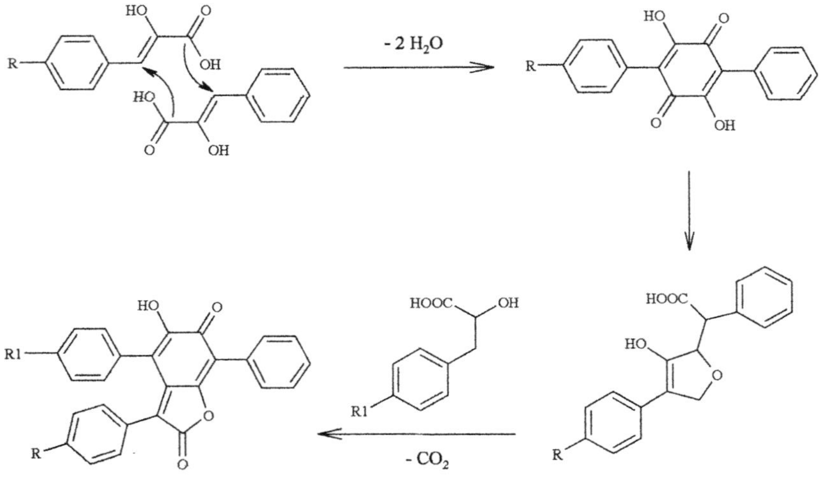

118 R = OH; R1 = H
120 R = R1 = OH

6   SYNTHESIS

Many natural polyhydroxy-*p*-terphenyls have also been obtained by total or partial synthesis. The cited reviews of Gill and Steglich [1, 8] include a number of early synthetic methods and related references. For the sake of brevity, we report here only some general schemes.

Among the early synthetic works, we wish to mention here the synthesis of polyporic acid (**3**) [115] and thelephoric acid (**82**) [78]. As summarised in Scheme 3a, several terphenylquinones have been synthesized with moderate yields starting from 2,5-dichlorobenzoquinone by arylation with *N*-nitrosoacetanilides or diazonium salts and subsequent alkaline hydrolysis. This method allowed preparation of symmetrical and unsymmetrical terphenylquinones, this latters with low yields.

Scheme 3a – Synthesis of *p*-terphenylquinones. For details see ref. [116]

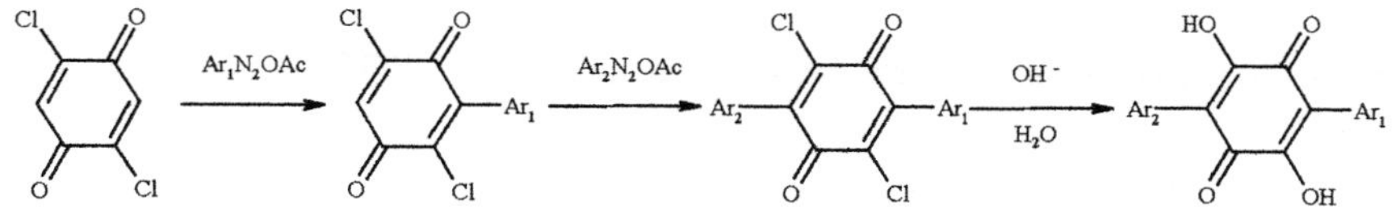

The synthesis of benzobisbenzofuranoid derivatives starting from chloranil (2,3,5,6-tetrachlorobenzoquinone) was also reported (Scheme 3b) [85]. Chloranil is made to react with two equivalents of a suitable phenol in the presence of sodium ethoxide. Phenoxyde substitutes chlorine in a first stage and then ring closure afford the benzobisfuranoid product, but other products are also formed.

Scheme 3b – Synthesis of *p*-terphenylquinones. For details see ref. [85]

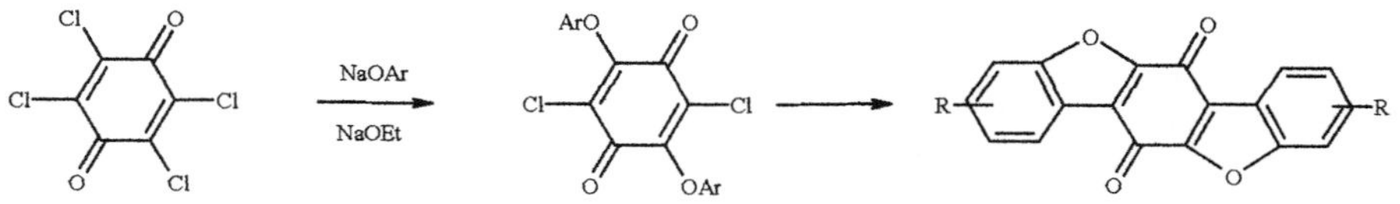

A general synthesis based on a biogenetic model has been developed according to Scheme 4a [117, 118]. In this versatile synthetic route a bis-benzylaciloin (**I**) is a key intermediate to grevillins (**II**), which are isomerised to terphenylquinones (**III**) by ethoxide catalysed rearrangement. Based on this general way with various modifications, a number of terphenylquinones have been prepared, among them polyporic acid (**3**) ascocorynin (**7**) and leucomelone (**47**) [116].

Scheme 4a – Synthesis of terphenylquinones. For details see ref. [117, 118]

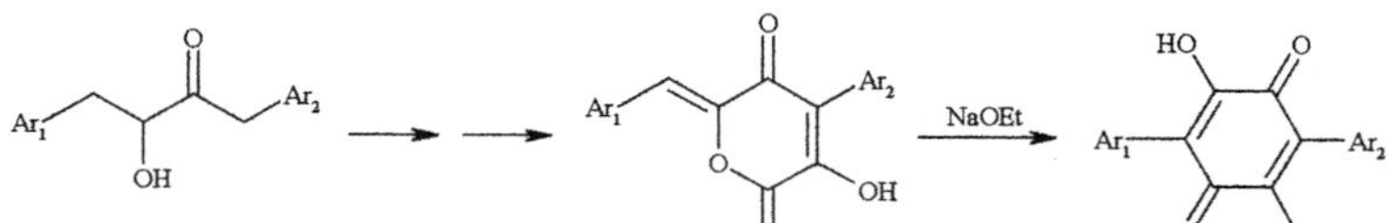

| I | II | II |

This method allowed also the synthesis of xylerythrins (**IV**, Scheme 4b), through a Perkin-type condensation of the terphenylquinone with arylacetic acids, thus supporting the proposed common biosynthetic way to different groups of fungal pigments [117].

Scheme 4b – Synthesis of xylerythrins. For details see ref. [117]

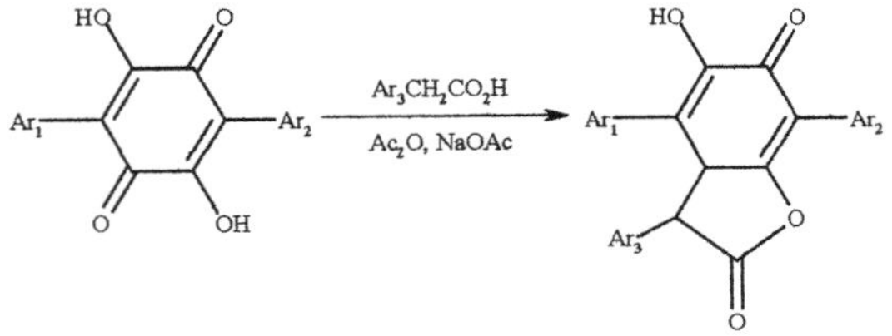

A total synthesis of xylerithrin was also reported, starting from 4,5-dimethoxy-3,6-diphenyl-o-benzoquinone and a p-tosyl derivative of phenylacetic methylester [99], via 2-oxo-3-(p-hydroxyphenyl)-4,7diphenyl-5,6-dihydroxybenzofuran.

Another example revealing the chemical relationships between different groups of terphenylquinones and possibly miming their biogenesis is the reported conversion of the flavomentin C (**25**) dimethylether to a flavomentin B (**24**) derivative and this latter in turn to derivatives of spiromentins B (**96**) and C (**97**) [1].

Fully aromatic polyhydroxyterphenyls have been generally obtained through catalytic reduction of a quinone form, this latter prepared according to previously reported methods, as for example in the case of butlerins A-C [119] or terphenyllin derivatives [35]. More recently, a different approach to the synthesis of this kind of polyhydroxyterphenyls has been developed to obtain terprenin (**15**), and reported as a promising methodology for the development of new antiallergic drugs [47,48]. In this highly efficient and practical synthesis the p-terphenyl skeleton is constructed through two Suzuki reactions, and the side chain is added as last step. Appropriate protecting groups are used to discriminate oxygenated functions. Actually, two different synthetic ways affording **15** have been reported. In one case a stepwise process is proposed (Scheme 6, route a), starting from 3-bromo-2,5-dimethoxybenzaldehyde (**V**), which undergoes a Suzuki reaction with an appropriate phenyl boronic acid to give a biphenyl derivative (**VI**). A further Suzuki reaction on the bromobiphenyl **VII** led to the terphenyl structure **VIII**, subsequently affording **15** with 40% overall yield. The second synthesis (Scheme 6, route b) is a one-pot process starting from the same bromobenzaldehyde, and employing a bromine-idodine-substituted intermediate **IX** to give **15** via Suzuki reactions.

Scheme 6 – Synthesis of terprenin (Routes a and b). For details see ref. [47, 48]

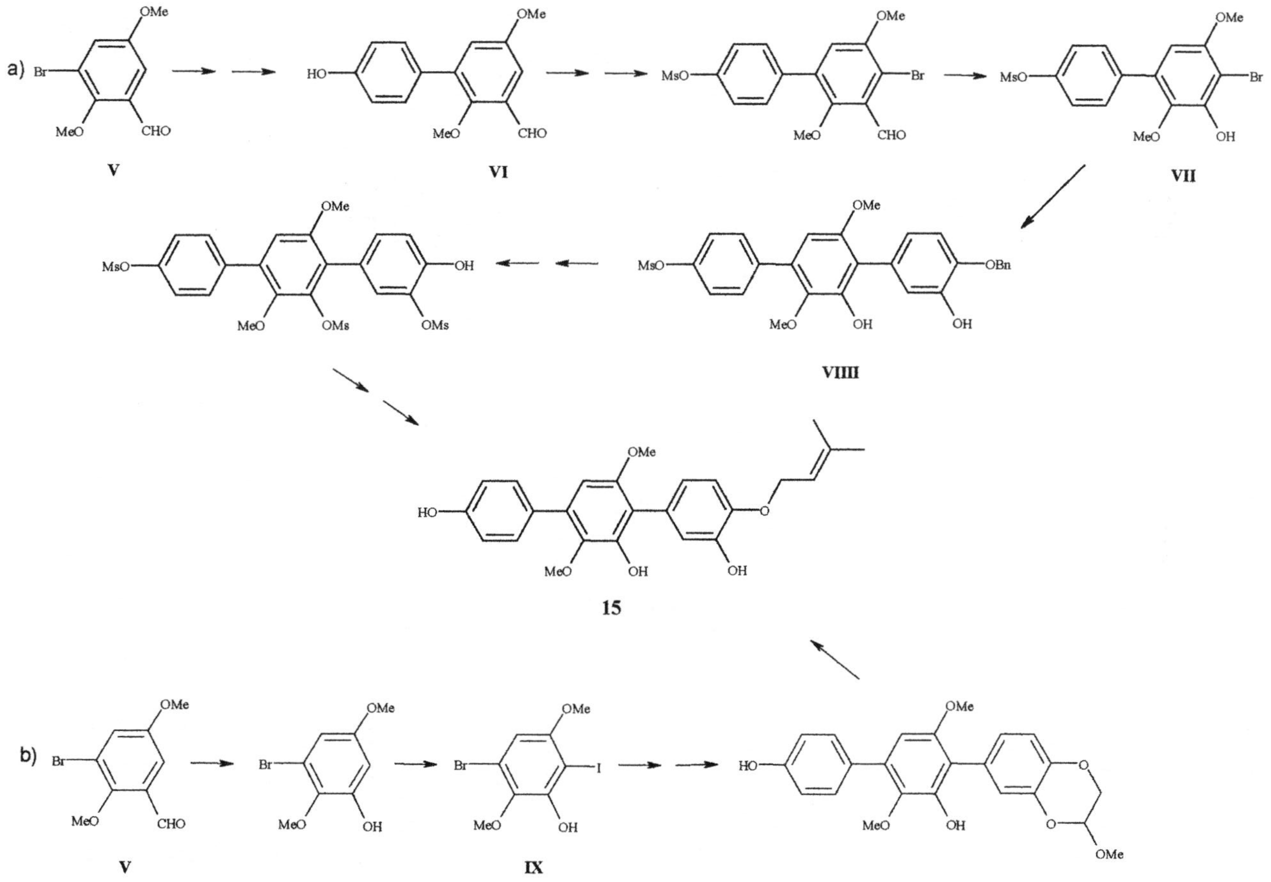

7   CONCLUDING REMARKS

The literature on polyhydroxy-*p*-terphenyls and *p*-terphenylquinones isolated from 35 different fungal species (both from fungal cultures or from fruiting bodies of basidiomycetes) has been reviewed, and a total of 115 fungal metabolites are reported here, with emphasis on their biological properties. A few semisynthetic analogues or lichen metabolites are also included. Although terphenyl metabolites are found essentially within the Kingdom of Fungi or in lichens, some of them are well-known as quinone pigments of basidiomycetes. Less known is the array of biological activity that has been reported for these metabolites, and in particular for fully aromatic polyhydroxy-*p*-terphenyls, including cytotoxic activity against tumoral cells and other antiproliferative properties, antibacterial activity, antioxidant or radical scavenging activity, anti-inflammatory activity and other properties of biomedical or agronomical interest. Among these, we wish to mention here the potent antiproliferative and immunosuppressive activity of the fully aromatic terprenin (**15**), a metabolite produced in cultures of *Aspergillus candidus*. Some of the metabolites produced by this filamentous fungus have also been reported for strong antioxidant properties, like hydroxyterphenyllin (**12**), dihydroxyterphenyllin (**13**) and candidusin B (**73**). Strong inhibition of lipid peroxidation was also reported for leucomentins, in particular leucomentin 2 (**30**) and leucomentin 6 (**34**), isolated from fruiting bodies of *Paxillus atrotomentosus* and curtisians A – D (**35 – 38**), obtained from sporophores of *P. curtisii*. Telephorin A (**39**), recently obtained from *Thelephora vialis*, is a potent free radical scavenger against DPPH, not cytotoxic against HeLa tumour cells. It is worth noting here that safety studies carried out on **12**, **13** and **73** showed that these compounds were neither cytotoxic nor mutagenic. These data contrast with other results reporting cytotoxicity or inhibitory activity on DNA/RNA synthesis of terphenyl metabolites and suggest that further studies could be useful to evaluate the safety of these compounds for possible use as food antioxidants. In this connection, it should be noted that some *p*-terphenyls are accumulated even in relatively large amounts in fruiting bodies of basidiomycetes considered edible or at least not particularly toxic.

As a further example of the promising properties of *p*-terphenyl metabolites, we wish to mention here the benzofuranoid compounds isolated from the fruiting bodies of *Poliporus leucomelas*, and in particular Bl-III (**69**), displaying an inhibitory effect on 5-lipoxygenase comparable to that of a therapeutic drug for asthma.

Structure-activity relationships of *p*-terphenyls have scarcely been investigated, and the reported data cannot be easily compared, so that further studies are needed to achieve conclusive indications. In a study on the cytotoxic activity of terphenyllin (**7**) and fourteen natural or semisynthetic analogues, the Authors suggest that methylation or removal of the 4"-hydroxy group may lead to the loss of a typical 'R type' change induced by **7**. All compounds showed cytotoxicity in the range 3-10 µg/mL, but the level of toxicity is not clearly correlated with the structure. The cytotoxicity of compound **50** (2',3'-diacetoxy-3,4,5',6',4"-pentahydroxy-*p*-terphenyl) is approximately ten times higher than that of its 4"-acetoxy analog **49**. Terprenin (**15**) and its analogues **16** and **17** showed comparable immunosuppressive activity, whereas no activity was found for the strictly related terphenyllins **8**, **9**, **12** and **13**. In a work reporting the plant growth inhibiting activity of **8**, **13** and their peracetate, biological activity was higher for the compound bearing the greater number of free hydroxyl groups. The antioxidant activity of **12**, **13** and **73** has been correlated with the presence of a methoxy substituent *ortho* to hydroxyl. In a study on the antibacterial activity of natural occurring quinones, including polyporic acid (**3**) and atromentin (**18**), the activity was inversely proportional to the polarity of the metabolite, and in particular dimethoxyatromentin (**19**) was noticeably more active than **18**.

Globally considered, the data here reported confirm the important properties of this peculiar group of metabolites and open promising perspectives in the search for further natural polyhydroxy-*p*-terphenyls and new synthetic methods, as well as for structure-activity relationship and safety studies.

ABBREVIATIONS

Ac	= Acetyl
Ar	= Aryl
Bn	= Benzyl
Bz	= Benzoyl
DNA	= Desoxyribonucleic Acid
EC_{50}	= Effective Concentration 50%
ED_{50}	= Effective Dose 50%
GI_{50}	= Growth Inhibition 50%
IC_{50}	= Inhibitory Concentration 50%
ID_{50}	= Inhibitory Dose 50%
Me	= Methyl
MIC	= Minimum Inhibitory Concentration
MS	= Mass Spectrometry
Ms	= Methansulfonyl, mesyl
NCI	= National Cancer Institute
NMR	= Nuclear Magnetic Resonance
RNA	= Ribonucleic Acid
ROESY	= Rotation Overhauser Enhancement SpectroscopY

TLC   = Thin Layer Chromatography

ACKNOWLEDGEMENTS

The authors gratefully acknowledge Prof. M. Piattelli (University of Catania, Italy) for helpful suggestions. Financial support was obtained from the Ministero della Istruzione, Università e Ricerca (MIUR, Roma, Italy – grant PRIN 2001) and from the Università degli Studi di Catania (Catania, Italy – Progetti di Ricerca di Ateneo).

REFERENCES

[1]   Gill, M.; Steglich, W.; *Fort. Chem. Org. Nat.*, **1987**, *51*, 1-317.
[2]   Thomson, R.H. In *Naturally Occurring Quinones*; Butterworths Scientific Publications, London, **1957**.
[3]   Turner, W.B. In *Fungal metabolites*; Academic Press, London, **1971**.
[4]   Turner, W.B; Aldridge, D.C.; In *Fungal metabolites*; Academic Press, London, **1983**
[5]   Steglich, W.; *Pilzfarbstoffe. Chemie in unserer Zeit*, **1975**, *9*, 117-123.
[6]   Steglich, W.; Pigments of Higher Fungi (Macromycetes) In *Pigments in Plants*; Czygan, F.C., Ed; 2nd ed., Stuttgart: G. Fischer, **1980**, pp. 393-412.
[7]   Gill, M.; *Nat. Prod. Rep.*, **1994**, 67-90.
[8]   Gill, M.; *Nat. Prod. Rep.*, **1996**, 513-28.
[9]   Gill, M.; *Nat. Prod. Rep.*, **1999**, *16*, 301-317.
[10]   Hashimoto, T.; Asakawa, Y.; *Heterocycles*, **1998**, *47*, 1067-1110.
[11]   Jikai, L.; *Heterocycles*, **2002**, *57*, 157-167.
[12]   Steglich, W.; Elzenhoefer, T.; Casser, I.; Steffan, B.; Rabe, U.; Boeker, R.; Knerr, H. J.; Anke H.; Anke, T.; *DECHEMA Monogr.*, **1993**, *129*, 3-13.
[13]   Atsushi N. *et al.* (Sankyo Co., Ltd., Japan).   Jpn. Kokai Tokkyo Koho JP 57024395, **1982**.
[14]   Nakagawa, F.; Enokita, R.; Naito, A.; Iijima, Y.; Yamazaki, M.; *J. Antibiot.*, **1984**, *37*, 6-9.
[15]   Nakagawa, F.; Takahashi, S.; Naito, A.; Sato, S.; Iwabuchi, S.; Tamura, C.; *J. Antibiot.*, **1984**, *37*, 10-12.
[16]   Dong, Y.; Zheng, Z.; Zhang, Q.; Zhang, H.; Lu, X.; Shu, W.; Shan, Y.; Ma, Y.; Mo, Y.; He, B.; *Zhongguo Kangshengsu Zazhi* **2002**, *27*, 260-263.
[17]   Briggs, L.H.; Cambie, R.C.; Dean, I.C.; Dromgoole, S.H.; Fergus, B.J.; Ingram, W.B.; Lewis, K.G.; Small, C.W.; Thomas, R.; Walker, D.A.; *N. Z. J. Sci.*, **1975**, *18*, 565-576.
[18]   C. Stahlschmidt, *Liebigs Ann. Chem.* **1877,** *187*, 177 cfr. Ref. [1] and [20]
[19]   Mosbach, K.; Guilford, H.; Lindberg, M.; *Tetrahedron Lett.*, **1974**, *17*, 1645-1648.
[20]   Murray, J.; *J. Chem. Soc.* **1952**, 1345-50.
[21]   Benedict, R.G.; Brady, L.R.; *J. Pharm. Sci.*, **1972**, *61*, 1820-1822.
[22]   Brewer, D.; Jen, W.C.; Jones, G.A.; Taylor, A.; *Can. J. Microbiol.*, **1984**, *30*, 1068-1072.
[23]   Burton, J. F.; Cain, B.F.; *Nature*, **1958**, 1326-1327.
[24]   Jirawongse, V.; Ramstad, E.; Wolinsky, J.; *J. Pharm. Sci.*, **1962**, *51*, 1108-1109
[25]   Miersch, J.; *Biochem. Physiol. Pflanz.*, **1986**, *181*, 405-410.

[26] Kraft, J.; Bauer, S.; Keilhoff, G.; Miersch, J.; Wend, D.; Riemann, D.; Hirschelmann, R.; Holzhausen, H. J.; Langner, J.; *Archives of Toxicology*, **1998**, *72*, 711-721.

[27] In-Kyoung, L.; Bong-Sik, Y.; Soo-Muk, C.; Won-Gon, K.; Jong-Pyong, K.; In-Ja, R.; Hiroyuki, K.; Ick-Dong, Y.; *J. Nat. Prod.*, **1996**, *59*, 1090-1092.

[28] Divekar, P.V.; Read, G.; Vining, L.C.; Haskins, R.H.; *Can. J. Chem.*, **1959**, *37*, 1970-1976.

[29] Read, G.; Vining, L.C.; Haskins, R.H.; *Can. J. Chem.*, **1962**, *40*, 2357-2361.

[30] Quack, W.; Scholl, H.; Budzikiewicz, H.; *Phytochemistry*, **1982**, *21*, 2921-2923.

[31] Marchelli, R.; Vining, L.C.; *J. Chem. Soc., Chem. Commun.*, **1973**, *15*, 555-556.

[32] Andreetti, G. D.; Bocelli, G.; Sgarabotto, P.; *Cryst. Struct. Commun.*, **1974**, *3*, 145-149.

[33] Marchelli, R.; Vining, L. C.; *J. Antibiot.*, **1975**, *28*, 328-331.

[34] Takahashi, C.; Yoshihira, K.; Natori, S.; Umeda, M.; Ohtsubo, K.; Saito, M., *Experientia*, **1974**, *30*, 529-530.

[35] Takahashi, C.; Yoshihira, K.; Natori, S.; Umeda, M.; *Chem. Pharm. Bull.*, **1976**, *24*, 613-620.

[36] Cutler, H.G.; LeFiles, J.H.; Crumley, F.G.; Cox, R.H.; *J. Agric. Food Chem.*, **1978**, *26*, 632-635.

[37] Kurobane, I.; Vining L.C.; McInnes A.G.; Smith D.G.; *J. Antibiot.*, **1979**, *32*, 559-564.

[38] Kobayashi, A.; Takemura, A.; Koshimizu, K.; Nagano, H.; Kawazu, K.; *Agric. Biol. Chem.*, **1982**, *46*, 585-589.

[39] Kobayashi, A.; Takemoto, A.; Koshimizu, K.; Kawazu, K.; *Agric. Biol. Chem.*, **1985**, *49*, 867-868.

[40] Yen, G.C.; Lee, C.A.; *J. Food Prot.*, **1996**, *59*, 1327-1330.

[41] Yen, G.C.; Chang, Y.C.; Sheu, F.; Chiang, H.C.; *J. Agric. Food Chem.*, **2001**, *49*, 1426-1431.

[42] Yen, G.C.; Chang, Y.C.; Chen, J.P.; *J. Food Sci.* **2002**, *67*, 567-572.

[43] Rahbaek, L; Frisvad, J.C.; Christophersen, C.; *Phytochemistry.*, **2000**, *53*, 581-586.

[44] Belofsky, G.N.; Gloer, K.B.; Gloer, J.B.; Wicklow, D.T.; Dowd, P.F.; *J. Nat. Prod.*, **1998**, *61*, 1115-1119.

[45] Kamigauchi, T.; Suzuki, R.; (Shionogi & Co., Ltd.) PCT Int. Appl. WO97/39999, **1997**.

[46] Kamigauchi, T.; Sakazaki, R.; Nagashima, K.; Kawamura, Y.; Yasuda, Y.; Matsushima, K.; Tani, H.; Takahashi, Y.; Ishii, K.; Suzuki, R.; Koizumi, K.; Nakai, H.; Ikenishi, Y.; Terui, Y.; *J. Antibiot.*, **1998**, *51*, 445-450.

[47] Kawada, K.; Arimura, A.; Tsuri, T.; Fuji, M.; Komurasaki, T.; Yonezawa, S.; Kugimiya, A.; Haga, N.; Mitsumori, S.; Inagaki, M.; Nakatani, T.; Tamura, Y.; Takechi, S.; Taishi, T.; Kishino, J.; Ohtani, M.; *Angew. Chem., Int. Ed.*, **1998**, *37*, 973-975.

[48] Yonezawa, S.; Komurasaki, T.; Kawada, K.; Tsuri, T.; Fuji, M.; Kugimiya, A.; Haga, N.; Mitsumori, S.; Inagaki, M.; Nakatani, T.; Tamura, Y.; Takechi, S.; Taishi, T.; Ohtani, M.; *J. Org. Chem.*, **1998**, *63*, 5831-5837.

[49] Kawada, K.; Ohtani, M.; Suzuki, R.; Arimura, A.; (Shionogi & Co., Ltd., Japan); PCT Int. Appl. WO 98/04508, **1998**.

[50] Stead, P.; Affleck, K.; Sidebottom, P.J.; Taylor, N.L.; Drake, C.S.; Todd, M.; Jowett, A.; Webb, G.; *J. Antibiot.*, **1999**, *52*, 89-95.

[51] Sullivan, G.; Garrett, R.D.; Lenehan, R.F.; *J. Pharm. Sci.*, **1971**, *60*, 1727-1729.

[52] Euler, K.L.; Tyler, V.E.; Brady, L.E.; Malone, M.H; *Lloydia*, **1965**, *28*, 203-206.

306

[53]   Khanna, J.M.; Malone, M. H.; Euler K.L, Brady, L.R; *J Pharm. Sci.*, **1965**, *54*, 1016-1020.
[54]   Sullivan, G.; Guess, W.L.; *Lloydia*, **1969**, *32*, 72-75.
[55]   Jagers, E.; Hillen-Maske, E.; Schmidt, H.; Steglich, W.; Horak, E.; *Z. Naturforsch. B: Chem. Sci.*, **1987**, *42*, 1354-1360.
[56]   Besl, H.; Bresinsky, A.; Geigenmueller, G.; Herrmann, R.; Kilpert, C.; Steglich, W.; *Liebigs Ann. Chem.*, **1989**, 803-810.
[57]   Hu, L.; Gao, J.M, Liu, J.K.; *Helv. Chim. Acta*, **2001**, *84*, 3342-3349.
[58]   Gripenberg, J.; *Acta Chem. Scand.*, **1956**, *10*, 1111-1115.
[59]   Holzapfel, M.; Kilpert, C.; Steglich, W.; *Liebigs Ann. Chem.*, **1989**, 797-801.
[60]   Yun, B.S.; Lee, I.K.; Kim, J.P.; Yoo, I.D.; *J. Microbiol. Biotechn.*, **2000**, *10*, 233-237.
[61]   Yun, B.S.; Lee, I.K.; Kim, J.P; Yoo, I.D.; *J. Antibiot.*, **2000**, *53*, 711-713.
[62]   Hashimoto, T.; Arakawa, T.; Tanaka, M.; Asakawa, Y.; *Heterocycles*, **2002**, *56*, 581-588.
[63]   Yun, B.S.; Lee, I.K.; Kim, J.P.; Yoo, I.D.; *J. Antibiot,.* **2000**, *53*, 114-122.
[64]   Tsukamoto, S.; Macabalang, A.D.; Abe, T.; Hirota, H., Ohta, T.; *Tetrahedron*, **2002**, *58*, 1103-1105.
[65]   Hu, L.; Liu, J.K.; *Z. Naturforsch. C*, **2001**, *56*, 983-987.
[66]   Hu, L.; Don, Z.J.; Liu, J.K.; *Chin. Chem. Lett.*, **2001**, *12*, 335-336.
[67]   Lee, H.-J.; Rhee, I.-K.; Lee, K.–B; Yoo, I.-D.; Song, K.–S.; *J. Antibiot.*, **2000**, *53*, 714-719.
[68]   Akagi, M.; *J. Pharm. Soc. Japan*, **1942**, *62*, 129-134.
[69]   Jaegers, E.; Hillen-Maske, E.; Steglich, W.; *Z. Naturforsch., B: Chem. Sci.*, **1987**, *42*, 1349-1353.
[70]   Tringali, C.; Piattelli, M.; Geraci, C.; Nicolosi, G.; Rocco, C.; *Can. J. Chem.*, **1987**, *65*, 2369-2372.
[71]   Geraci, C.; Neri, P.; Paternò, C.; Rocco, C.; Tringali, C.; *J. Nat. Prod.*, **2000**, *63*, 347-351.
[72]   Elix, J.A.; Gaul, K.L.; Hockless, D.C. R.; Wardlaw, J.H.; *Aust. J. Chem.*, **1995**, *48*, 1049-1053.
[73]   Elix, J.A.; Ernst-Russell, M.A.; *Aust. J. Chem.*, **1996**, *49*, 1247-1250.
[74]   Takahashi, A.; Kudo, R.; Kusano. G.; Nozoe, S.; *Chem. Pharm. Bull.*, **1992**, *40*, 3194-3196.
[75]   Oh, H.; Gloer, J.B.; Wicklow, D.T.; Dowd, P.F.; *J. Nat. Prod.*, **1998**, *61*, 702-705.
[76]   Edwards, R.L.; Gill, M.; *J. Chem. Soc. Perkin Trans. 1*, **1975**, *4*, 351-354.
[77]   Edwards, R.L.; Gill, M.; *J. Chem. Soc. Perkin Trans. 1*, **1973**, 1921-1929.
[78]   Gripenberg, J.; *Tetrahedron*, **1960**, *10*, 135-143.
[79]   Kwak, J.–Y.; Rhee, I.–K.; Lee, K.–B.; Hwang, J.–S.; Yoo, I–D.; Song, K.–S.; *J. Microbiol. Biotechnol.*, **1999**, *9*, 798-803.
[80]   Hwang, J.–S.; Song, K.–S.; Kim, W.-G; Lee, T.-H; Koshino, H.; Yoo, I.–D.; *J. Antibiot.*, **1997**, *50*, 773-777. 1877
[81]   Weisgraber, K.; Weiss, U.; Milne, G.W.A.; Silverton, J.V.; *Phytochemistry*, **1972**, *11*, 2585-2587.
[82]   Silverton, J.V.; *Acta Crystalogr.*, **1973**, B29, 293- 298.
[83]   Neveu, A.; Baute, R.; Bourgeois, G.; Deffieux, G.; *Bull. Soc. Pharm. Bordeaux*, **1974**, *113*, 121- 129.
[84]   Neveu, A.; Baute, R.; Deffieux, G.; *Bull. Soc. Pharm. Bordeaux*, **1974**, *113*, 77-85.
[85]   Briggs, L.H.; Cambie, R.C.; Dean, I.C.; Hodges, R.; Ingram, W.B.; Rutledge, P.S.; *Aust. J. Chem.*, **1976**, *29*, 179-190.

[86]   Salvo, J.J.; Mobley, D.P.; Brown, D.W.; Caruso, L.A.; Yake, A.P.; Spivack, J.L.; Dietrich, D.K.; *Biotechnol. Prog.*, **1990**, *6*, 193-197.
[87]   Colson, K.L.; Jackman, L.M.; Jain, T.; Simolike, G.; Keeler, J.; *Tetrahedron Lett.*, **1985**, *26*, 4579-4582.
[88]   McMorris, T.C.; Anchel, M.; *Tetrahedron Lett.*, **1963**, *5*, 335-337.
[89]   McMorris, T.C.; Anchel, M.; *Tetrahedron*, **1967**, *23*, 3985-3991.
[90]   Anke, H.; Casser, I.; Herrmann, R.; Steglich, W.; *Z. Naturforsch., C. Biosci.*, **1984**, *39C*, 695-698.
[91]   Anke, H.; Casser, I.; Steglich, W.; Pommer, E. H.; *J. Antibiot.*, **1987**, *40*, 443-449.
[92]   Buchanan, M.S.; Hashimoto, T.; Takaoka, S.; Asakawa, Y.; *Phytochemistry*, **1995**, *40*, 1251-1257.
[93]   Calì, V.; Spatafora, C.; Tringali, C.; unpublished results
[94]   Gripenberg, J.; *Acta Chem. Scand., Ser. B*, **1978**, *B32*, 75-76.
[95]   Kögl, F.; Erxleben, H.; *Liebigs Ann. Chem*, **1930**, *479*, 11-25.
[96]   Musso, H., *Tetrahedron*, **1979**, *35*, 2843-2853.
[97]   Gripenberg, J.; *Acta Chem. Scand.*, **1965**, *19*, 2242-2243.
[98]   Gripenberg, J.; Martikkala, J.; *Acta Chem. Scand.*, **1969**, *23*, 2583-2588.
[99]   Wanzlick, H. W.; Jahnke, U.; *Chem. Ber.*, **1968**, *101*, 3753-3760.
[100]   Gripenberg, J.; Martikkala, J.; *Acta Chem. Scand.*, **1970**, *24*, 3444-3448.
[101]   Gripenberg, J.; *Acta Chem. Scand.*, **1970**, *24*, 3449-3354.
[102]   Gripenberg, J.; Hiltunen, L.; Pakkanen, T.; Pakkanen, T.; *Acta Chem. Scand., Ser. B*, **1980**, *B34*, 575-578.
[103]   Gripenberg, J.; Hiltunen, L.; Niinisto, L.; Pakkanen, T.; Pakkanen, T.; *Acta Chem. Scand., Ser. B*, **1979**, *B33*, 1-5.
[104]   Gripenberg, J.; *Acta Chem. Scand.*, **1971**, *25*, 2999-3005.
[105]   Gripenberg, J.; Hiltunen, L.; Pakkanen, T.; Pakkanen, T.; *Acta Chem. Scand., Ser. B*, **1979**, *B33*, 6-10.
[106]   Von Massow, F.; *J. Chromatogr.*, **1975**, *105*, 391-392.
[107]   Read, G.; Vining, L.C.; *Chem. & Ind.* **1959**, 1547-8.
[108]   Chandra, P.; Read, G.; Vining, L.C.; *Can. J. Biochem.*, **1966**, *44*, 403-413.
[109   Bose, A.K.; Khanchandani, K.S.; Funke, P.T.; Anchel, M.; *J. Chem. Soc. D*, **1969**, *22*, 1347-1348.
[110]   Anchel, M.; Bose, A.K.; Khanchandani, K.S.; Funke, P.T.; *Phytochemistry*, **1970**, *9*, 2335-2338.
[111]   Gill, M.; Steglich, W.; *Prog. Chem. Org. Nat. Prod.*, **1987**, *51*, 286 pp.
[112]   Boeker, R.; Anke, T.; Casser, I.; Steglich, W.; *Dechema Biotechnology Conferences*, **1990**, 4, 225-228.
[113]   Von Massow, F.; *Phytochemistry*, **1977**, *16*, 1695-1698.
[114]   Von Massow, F.; Noppel, H.E., *Phytochemistry*, **1977**, *16*, 1699-1701.
[115]   Cain, B.F.; *J. Chem. Soc. C*, **1966**, 1041-1045.
[116]   Pattenden, G.; Pegg, N.A.; Kenyon, R.W.; *J. Chem. Soc., Perkin Trans. 1*, **1991**, 2363-2372.
[117]   Lohrisch, H.J.; Schmidt, H.; Steglich, W.; *Liebigs Ann. Chem.*, **1986**, 195-204.
[118]   Pattenden, G.; Pegg, N.A.; Kenyon, R.W.; *Tetrahedron Lett.*, **1987**, *28*, 4749-4752.
[119]   Ernst-Russell, M.A.; Chai, C.L. L.; Hockless, D.C.R.; Elix, J.A.; *Aust. J. Chem.*, **1998**, *51*, 1037-1043.

Atta-ur-Rahman (Ed.) *Studies in Natural Products Chemistry, Vol. 29*

HALOGEN-CONTAINING ANTIBIOTICS FROM STREPTOMYCETES

ŘEZANKA, T. AND SPÍŽEK, J.

Institute of Microbiology, Academy of Sciences of the Czech Republic, Vídeňská 1083, 142 20, Prague 4, Czech Republic

ABSTRACT. Thousands of new structures of organic compounds have so far been described in the genus *Streptomyces,* the best-known producer of antibiotics. Streptomycetes are the most thoroughly studied microorganisms producing biologically active compounds; out of them several tens contain a halogen atom in their molecules. Mainly is chlorine, however, brominated and fluorinated compounds have also been described. The main attention was devoted to strains producing industrially manufactured antibiotics such as chloramphenicol, chlortetracycline, or the group of antibiotics of vancomycin type. However, other halogenated metabolites also exhibit interesting biological effects. It will be the aim of this chapter to describe the occurrence, isolation, structure and occasionally also biosynthesis of these metabolites.

INTRODUCTION

For a long time it was assumed that halogen containing compounds are only exceptionally detected in living organisms. This assumption began to change at the end of the sixties when bioorganic chemistry experienced a stormy progress due to the development of modern physico-chemical methods. As a result numbers of compounds discovered during the last 30 years increased by several orders of magnitude. Halogenated metabolites found in both terrestrial and aquatic organisms were studied mainly with respect to their potential pharmacological activity. Unfortunately, the number of newly described compounds steadily increases, whereas that of compounds applied in the practice grows only relatively slowly and the difference between the two groups of compounds permanently increases.

Chlorinated derivatives are the most frequent halogenated metabolites found in the nature. This is mainly due to the fact that chlorine is the most common halogen found on the Earth. Chloramphenicol, isolated from the soil bacterium *Streptomyces venezuelae,* was the very first halogenated metabolite detected. This important commercial antibiotic will be dealt with in chapter 1. Other chlorinated metabolites, *e.g.* chlortetracycline, will be discussed in chapter 2.

The three remaining halogens, *i.e.* fluorine, bromine and iodine, are only rarely present in soil microorganisms, *e.g.* streptomycetes. Compounds with those halogens were mainly prepared by directed biosynthesis by adding a respective anion to the cultivation medium (*e.g.* addition of bromine yielding bromotetracycline). Fluorine derivatives such as fluoroacetate and 4-fluorothreonine were also detected. Iodine-containing metabolites have not yet been described in streptomycetes. The chemical structure of halogenated metabolites found in streptomycetes is highly varied. The simplest derivative (*e.g.* fluoroacetate described above) contains only two carbon atoms, whereas certain chemically complex antibiotics contain a substantially higher number of carbon atoms.

In the following subchapters we tried to describe individual compounds according to their chemical structure with a special attention paid to important industrial antibiotics.

Chloramphenicol

In 1947 a strain of *S. venezuelae* [1-4] was isolated from a soil sample collected in Venezuela. It produced a wide-spectrum antibiotic called chloramphenicol (**1**) according to its chemical structure. Due to its simple chemical structure it is still used. It is the only antibiotic produced by the chemical synthesis rather than biosynthetically. Nevertheless, *S. venezuelae* is used as a model organism in biosynthetic and genetic experiments.

Thus for instance its biosynthesis [5] was investigated with the aid of stable isotopes (^{2}H, ^{13}C), in experiments in which acetic acid or dichloroacetic acid served as precursors. In streptomyces it was found that the addition of amino acids such as phenylalanine, or P-nitrophenylserine increases the chloramphenicol production. In further experiments NaCl was replaced with NaBr and bromo and chloroderivatives could be isolated from the cultivation broth [6].

Chloramphenicol is produced by *S. venezuelae via* the following pathway: chorismic acid → *p*-aminophenylalanine → *p*-aminophenylserine → dichloroacetyl → *p*-aminophenylserine → chloramphenicol.

Chloramphenicol is a relatively simple compound containing two asymmetric carbon atoms (4 isomers exist). Biological activity of all of them was investigated. The results obtained can be summarized as follows:

1. Nitro group is not important for the biological activity.
2. Substitution of the phenyl group by another aromatic or heterocyclic ring decreases the activity.

3.   There is at least one hydrogen on the nitrogen atom. Substitution of this hydrogen atom by any other group results in the loss of the biological activity.

4.   Esterification of the primary alcohol group also leads to the loss of the biological activity.

5.   Out of four possible isomers (1-4) only the D-threo isomer (1) is active. The activity of the remaining isomers is about 1 %.

Chloramphenicol has an outstanding bacteriostatic effect against a number of Gram-positive and Gram-negative bacteria. It inhibits growth of the genera *Aerobacter, Staphylococcus, Streptococcus, Diplococcus, Proteus, Bacillus, Vibrio, etc.* It also inhibits rickettsiae, including those causing louse-borne typhus, and some large viruses (causing trachoma, veneral lymphogranuloma, atypical pneumonia, and some other diseases). The antibiotic is a specific drug for the treatment of typhoid and paratyphoid fevers, dysentery, brucellosis, toxic dyspepsia, trachoma, and other diseases. Chloramphenicol is very effective against bacillary dysentery in children. When given perorally administered, it is well absorbed and is carried by blood to tissue and tissue fluids.

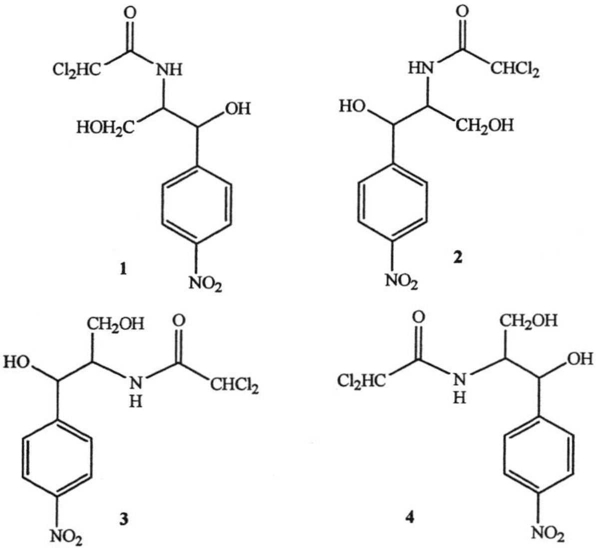

Fig. (1). Four possible isomers of chloramphenicol

Tetracyclines

Tetracyclines [7-12] are probably the most important group of halogen containing antibiotics. They are produced primarily by *S. aureofaciens*, however, further streptomycetes producing these antibiotics were described. Similar structures have recently been found also in other microorganisms (see below). Chlortetracycline (**5**) is the most common derivative. About one hundred of other halogenated tetracyclines are known at present. In addition, organic chemists prepared thousands of synthetic and semisynthetic tetracyclines. Chlortetracycline is used as a wide-spectrum antibiotic. Microorganisms sensitive to chlortetracycline develop resistance, but this resistance develops only slowly, much more slowly than resistance to streptomycin. The excessive use of chlortetracycline in medicine increases the occurrence of resistant microorganisms. Chlortetracycline should preferably be used to treat diseases whose causative agents are resistant against penicillin or streptomycin. It is successfully used for the treatment of bacterial pneumonia, brucellosis, tularemia, pertussis, scarlet fever, anthrax, and other bacterial diseases. Moreover, chlortetracycline can be used to treat various forms of louse-borne typhus, various rickettsioses, and also some viral diseases (trachoma, lymphogranuloma). Doses not exceeding 500 mg are applied perorally three- to four-times a day.

Biosynthesis of tetracycline was studied extensively and a number of various intermediates were isolated from the cultivation broth (**6-37**) [13-16], *e.g.* isochlortetracycline (**6**). High amounts of chlortetracycline are used in agriculture, *e.g.* in food additives on fish farms or also in pig, calf, sheep and poultry breeding.

The addition of NaBr to the medium leads to the production of a brominated derivative (bromotetracycline) (**7**) instead of the chlorinated one [17]. Other chlortetracyclines such as demethylchlortetracycline (**8**) [18] or anhydrotetracycline (**9**) [19] were isolated from mutants of *S. aureofaciens* and have similar antibiotic effects.

The structure-function relationships can be summarized as follows.

1) Substitution in positions 5, 6 and 9 is not important. For instance, the exchange of chlorine with bromine in position 7 yields compounds with very similar antibiotic effects.

2) Substitutions in positions 2, 11a and 12a result in biologically active compounds only if the original compound can be easily released in the target organism.

3) Replacement of the dimethylamino group leads to a decreased biological activity or its complete loss.

4) Changes in the spatial arrangement (5-epimer) give rise to biologically inactive compounds.

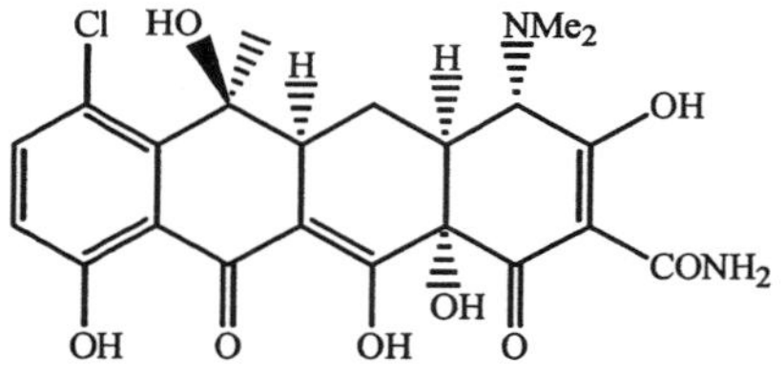

5

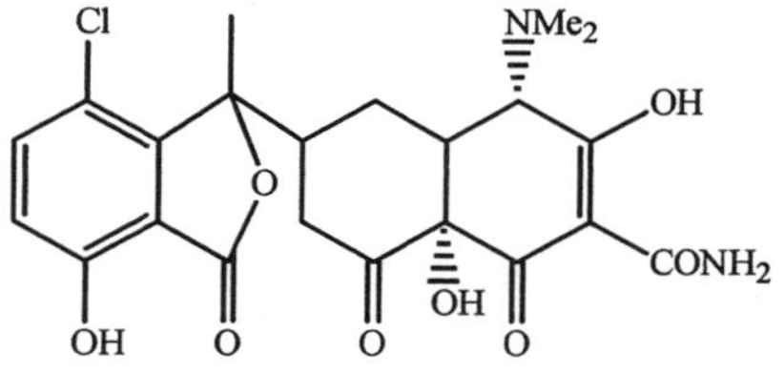

6

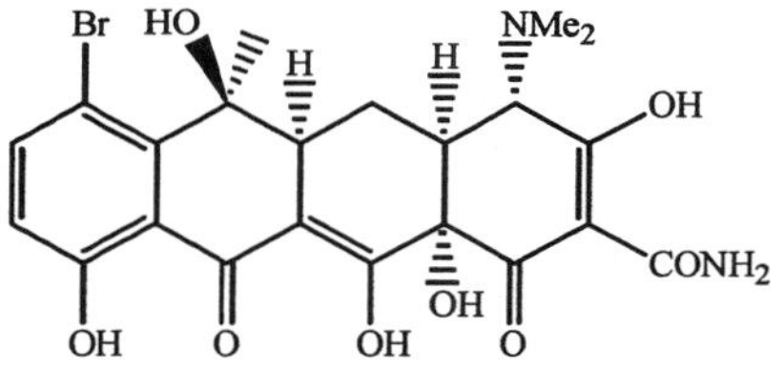

7

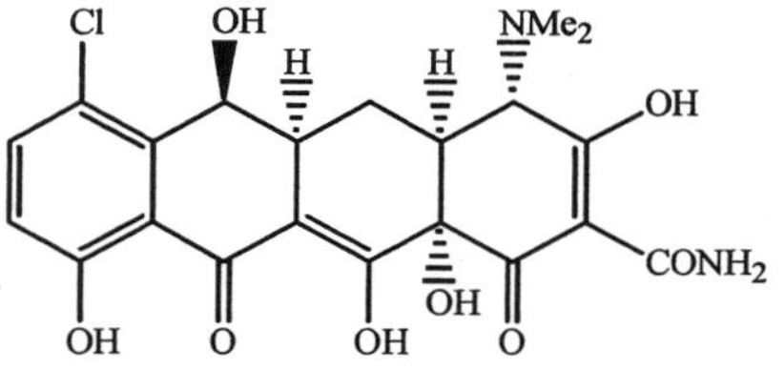

8

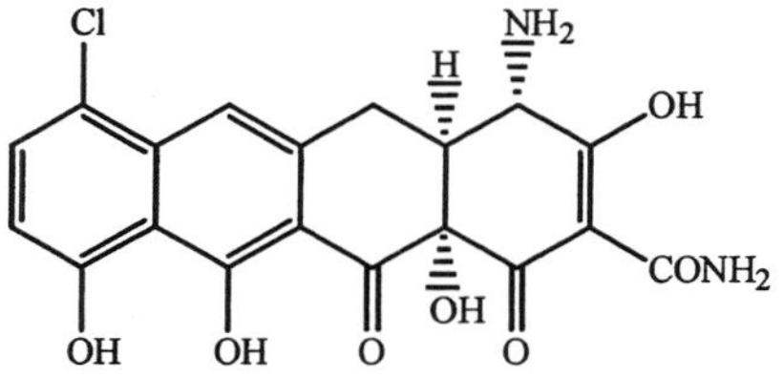

9

Fig. (2). Tetracyclines

314

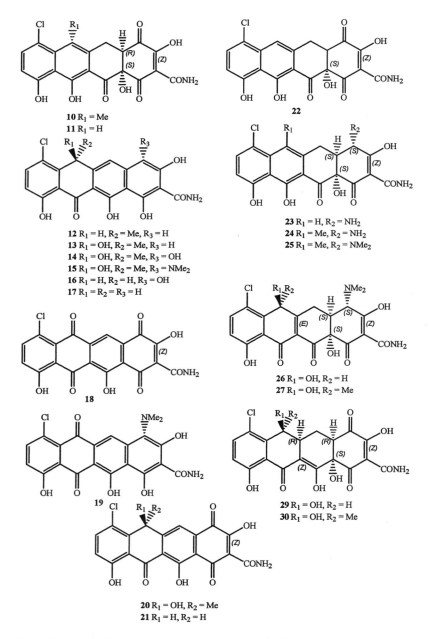

Fig. (3). Some important tetracycline metabolites

No	Name
10	4-oxoanhydrochlortetracycline
11	Methylchlorpretetramide
12	Methylhydroxychlorpretetramide
13	Chloraureovocidin
14	methylchlortetramide-blue
15	6-demethyl-6-deoxy-7-chloraureovocidin
16	Chlorpretetramide
17	chloro-A-C-diquinone
18	chlortetramide-blue
19	chlortetramide-green
20	4-oxoanhydrodemethylchlortetracycline
21	6-demethyl-6-deoxy-7-chlortetramid-green
22	4-oxoanhydrodemethylchlortetracycline
23	4-aminoanhydrodemethylchlortetracycline
24	4-aminoanhydrochlortetracycline
25	anhydrochlortetracycline
26	dehydrodemethylchlortetracycline
27	dehydrochlortetracycline
28	4-oxodemethylchlortetracycline
29	4-oxochlortetracycline
30	4-oxodehydrochlortetracycline

More recently, dactylocyclines A (31), B (32), D (33) and E (34) have been isolated from a *Dactylosporangium* strain ATCC 53693 [20-24]; these have the opposite stereochemistry at C6. Dactylocyclinones *i.e.* 8-methoxychlortetracycline (Sch 36969) (35), 2′-N-Methyl-8-methoxychlortetracycline (Sch 33256) (36) and 4a-hydroxy-8-methoxychlortetracycline (Sch 34164) (37) are the parent members of the family that are the 6-epimers of 8-methoxychlortetracycline from *Actinomadura brunnea* [25,26]. The dactylocyclinones contain unusual sugar moieties, with dactylocycline A being the biologically most active member. A group of tetracyclines including SF 2575 (38) has been isolated from *Streptomyces* species containing C-glycosidic bond at C9 and a salicyloyl ester at C4.

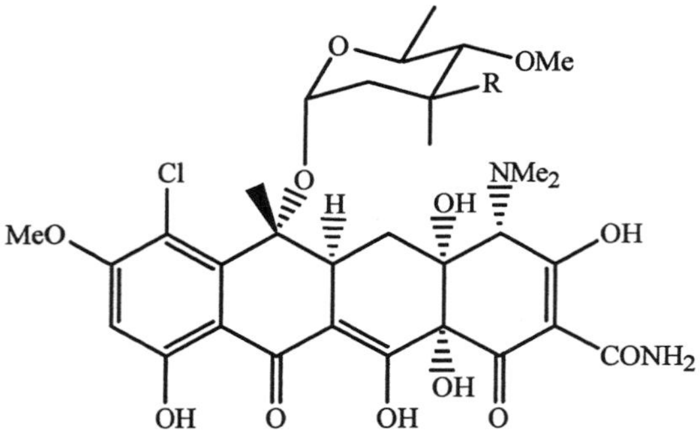

31 R = NHOH
32 R = NO$_2$
33 R = NHOAc
34 R = OH

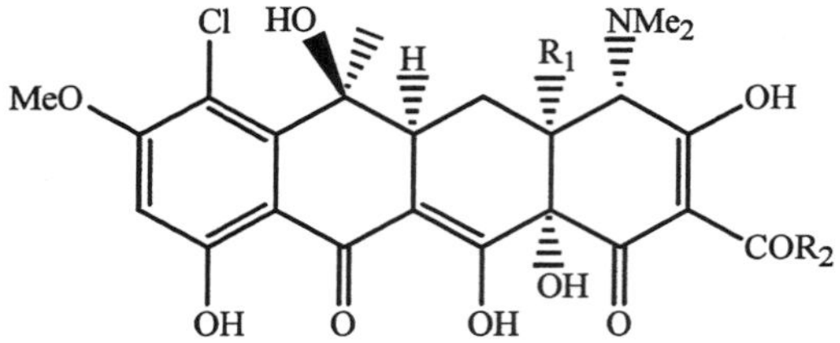

35 R$_1$ = H, R$_2$ = NH$_2$
36 R$_1$ = H, R$_2$ = NHMe
37 R$_1$ = OH, R$_2$ = NH$_2$

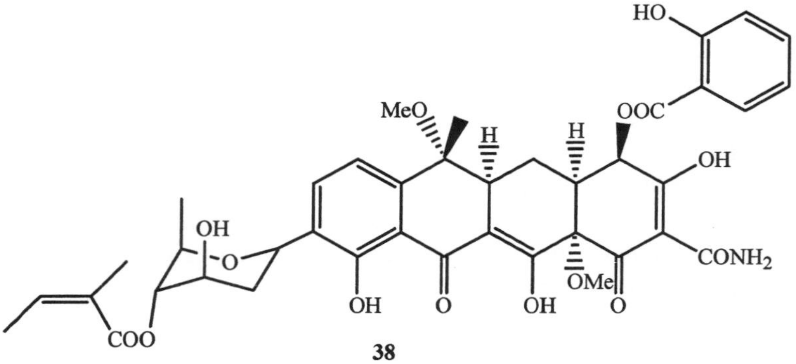

38

Fig. (4). Dactylocyclines and dactylocyclinones.

Oxygen compounds

S. malachitofuscus produces X14766A, the only chlorine containing polyether antibiotic (39) described so far. The producer was isolated from the Mexican soil in 1976. The antibiotic is active against Gram-positive bacteria forming strong complexes with both monovalent and divalent metals (K, Na, Rb, Cs, Ba, Sr, Ca, Mg, Li) [27,28].

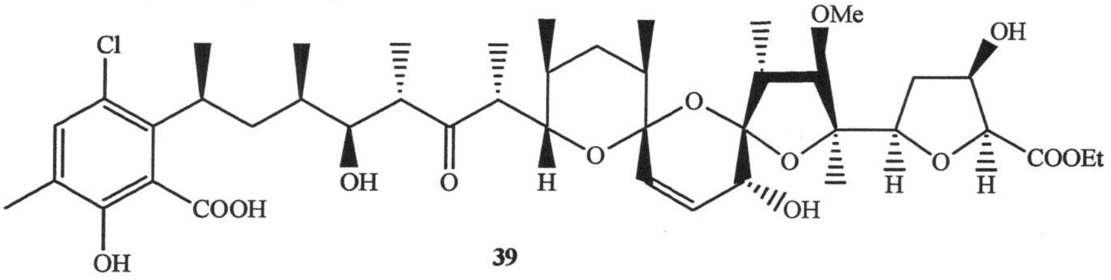

Fig. (5). Structure of polyether antibiotic-X14766A

Lysolipins I (40) and X (41) [29] were isolated from *S. violaceoniger*. The compounds inhibit both Gram-positive and Gram-negative bacteria inducing probably the lytic activity of bacterial cells. The antibiotic was one of the first antibiotics whose biosynthesis was studied with the aid of labeled stable precursors. A highly interesting experiment was performed including the cultivation of *S. violaceoniger* in the atmosphere of isotopically labeled oxygen (^{18}C) as shown in Fig. 6. Compounds BE19412A (47) BE19412B (48) active against implanted Ehrlich cancer cells were isolated [30] from *Streptomyces* sp.

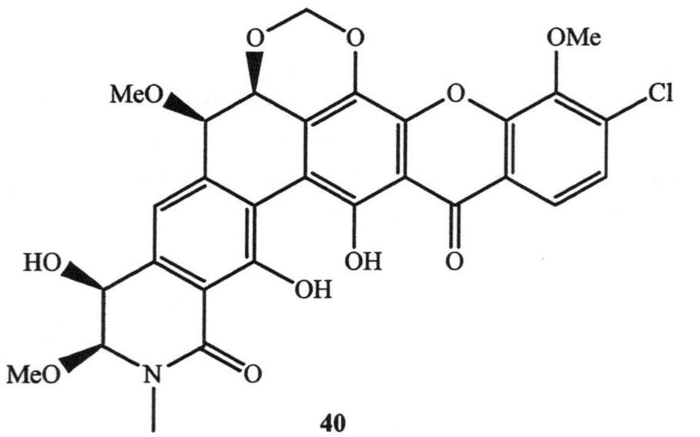

40

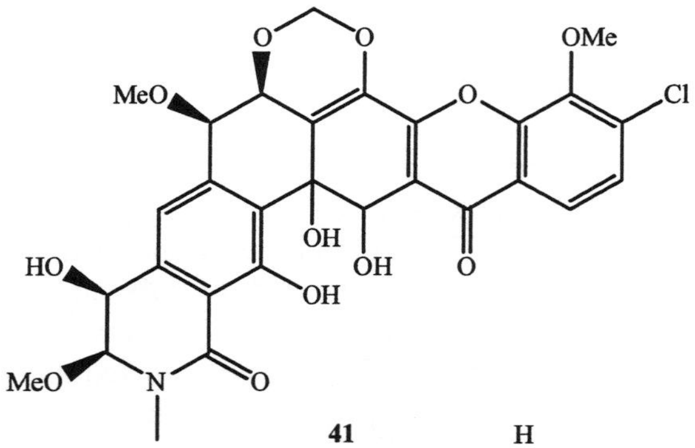

41

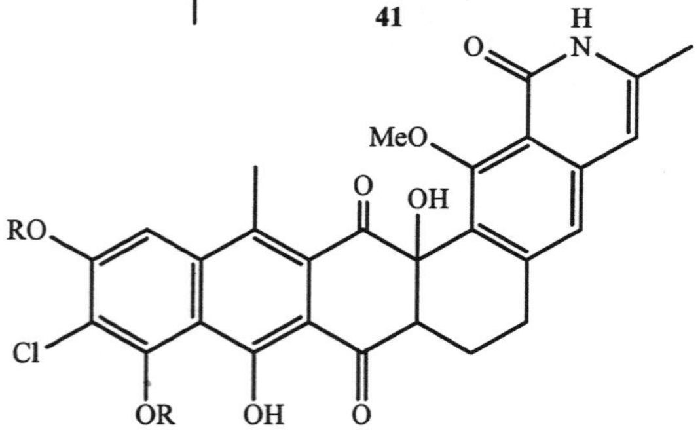

42 R = H
43 R = Me

Fig. (6). Lysolipin like compounds

A known strain of *S. antibioticus* was found to produce a compound designated simocyclinon D8 (**44**) [31]. The compound exhibits antibiotic activity against Gram-positive bacteria and cytostatic effects against various cancer cells. Its structure and biosynthesis were clarified on the basis of physical-chemical methods, mainly by 1D and 2D NMR and ESI-MS. It was found that during its biosynthesis the incorporation of ^{18}O into the aminocoumarine residue is probably catalyzed by a monooxygenase.

Naphtocyclinone (**45**) was isolated from the mycelium of *S. arenae* [32]. Another similar metabolite (naphtomevalin, *i.e.* **46**) was isolated from an Australian soil sample at 2.5 μg/ml.

S. aculeolatus produces a group of antibiotics [33] with biological activity against Gram-positive bacteria (**47-50**). The structure of these antibiotics [34] was clarified by physico-chemical methods, mainly by 1D and 2D NMR. It was found that the production in a 50 L fermentor reaches up to several tens of milligrams of pure compounds.

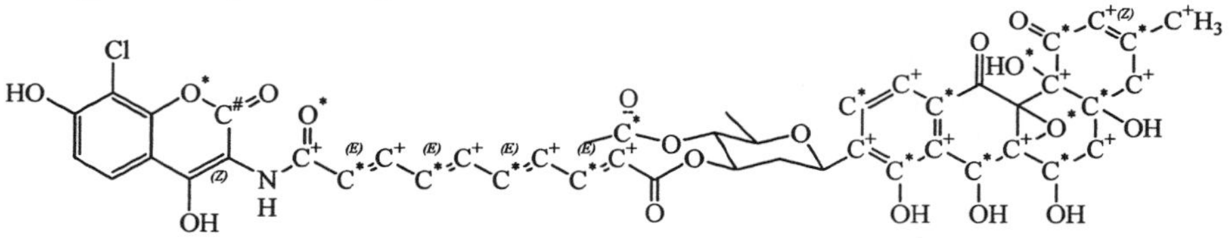

C^* [1,3-$^{13}C_2$]malonic acid, C^+ [2-^{13}C]malonic acid, $C^\#$ L-[1-^{13}C]tyrosine, * $^{18}O_2$

50

Fig. (7). Precursors in biosynthesis of simocyclinon D8

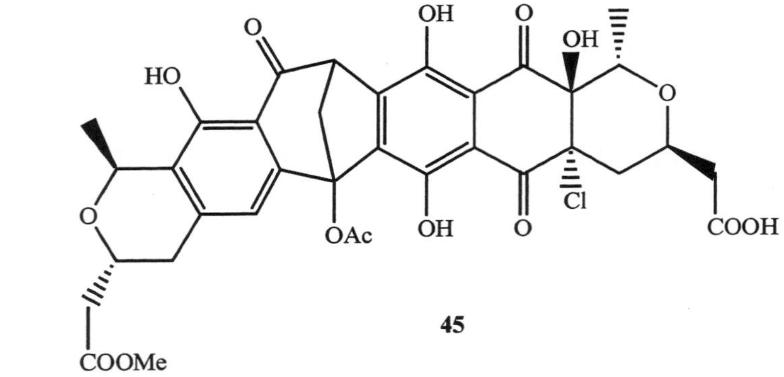

45

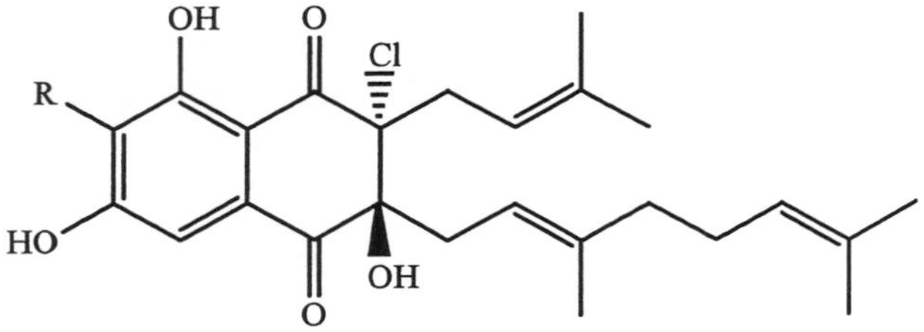

46 R = H
49 R = Me

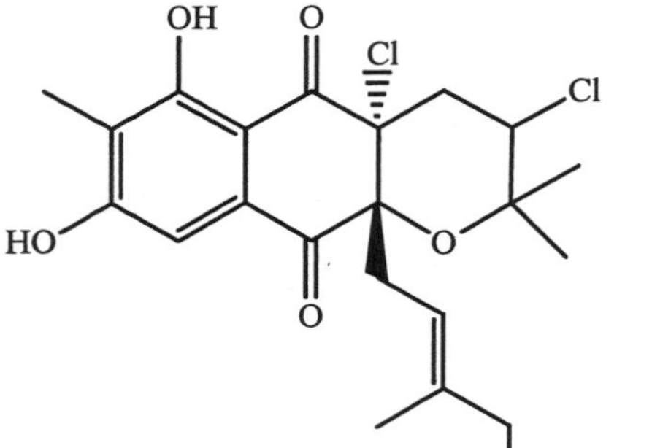

50

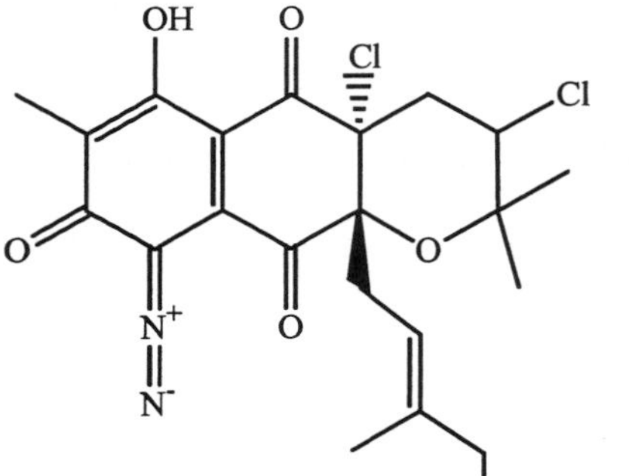

48

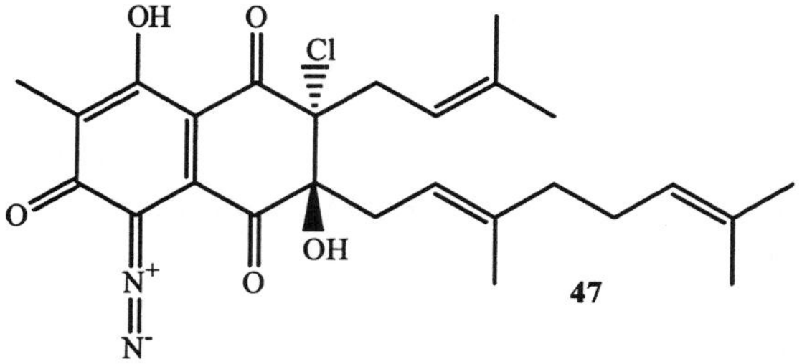

47

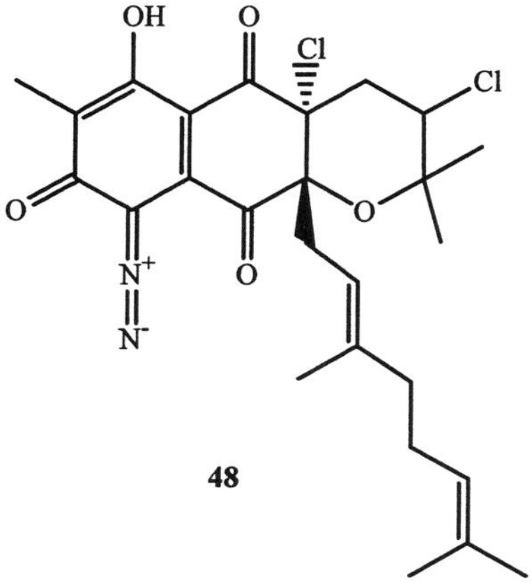

48

Fig. (8). Naphtocyclinones

Macrolide antibiotics are frequent in the genus *Streptomyces*. During the last 50 years several thousands of new antibiotics were described [35]. However, macrolides with heteroatoms in their structure, *e.g.* a halogen, are much less frequent. A 34-membered macrolide antibiotic colubricidin A (**51**) isolated from a new *Streptomyces* species can serve as example. The compound exhibits a very low activity on Gram-positive and Gram-negative bacteria. On the other hand, it has an excellent nematocidal activity against *Caenorhabditis elegans*.

Another compound, ansamitocin P3 (**52**), was isolated [36] from lichens of the family *Thuidiaceae*. However, it may be assumed that the lichens do not synthesize this compound but that it is rather a product of streptomycetes living on surface of these lichens. Ansamitocin P3 exhibits an excellent cytotoxicity against human tumors, leukemia and carcinoma in cell cultures.

Roseophilin (**53**) was isolated [37] from *S. griseoviridis* and found to exhibit cytotoxicity against leukemic and epidemoid cells at a concentration of IC=0.5 μM.

Ansamycin antibiotics are probably the most complex organic compounds produced by the genus *Streptomyces*. In addition to the antibacterial, they also exhibit antiviral effects. Ansamitocin P-3 shows a potent cytotoxicity against the human solid tumor cell lines A-549 and HT-29. The compound exhibits a significant activity against P-388 lymphocytic leukemia in mice and both 9PS (murine lymphocytic leukemia) and 9KB (human nasopharyngeal carcinoma) in cell culture systems. [38].

Naphtomycin A (**54**) was isolated from *S. collinus* and its structure was clarified by means of spectral methods. Naphtomycin A is a strong vitamin K antagonist. In addition to naphtomycins A and H (**55**), a geometric isomer of naphtomycin B (**56**), was produced by a streptomycete isolated from the Indian soil [39].

Naphtomycins B and C, *i.e.* thio- and chloroansamycins, were isolated [40] and found to exhibit the antibiotic activity [41]. As already mentioned above, their structure is very complex and has been revised several times [42,43]. Their biosynthesis was investigated using labeled isotopes [44].

S. antibioticus produces another unusual macrolide antibiotic [45] called chlorothricin (**57**) containing, in addition to the aglycone (modified methylsalicylic acid), saccharides, dideoxyhexoses in the first place. Their biosynthesis was investigated by the incorporation of stable isotopes [45-47]. The compounds were only active in a synthetic medium and inactive in a complex one [48]. Compounds designated MC031-034 (**58-61**) are similar to chlorothricin mentioned above and were isolated from the cultivation broth of *Streptomyces* sp. collected in Japan. Another compound, 2-hydroxychlorothricin (**62**), with antitumor activity, was isolated [66] from *Streptomyces* K818.

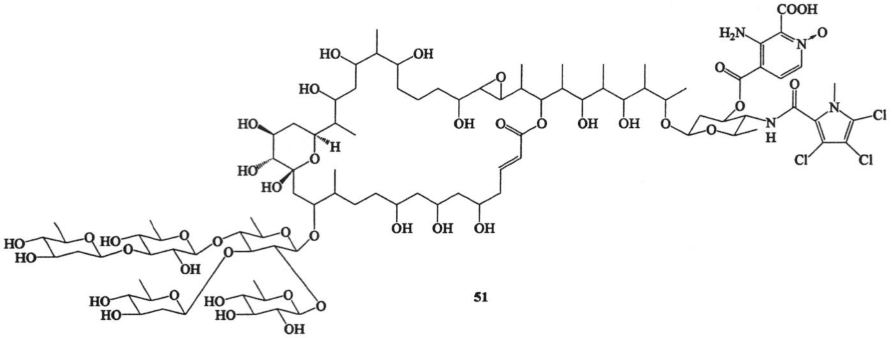

51

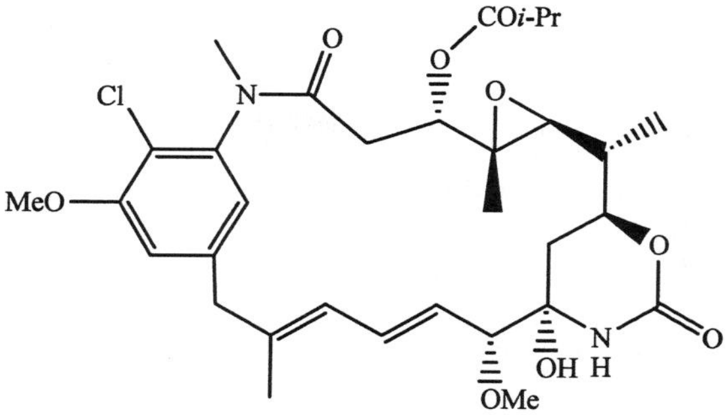

52

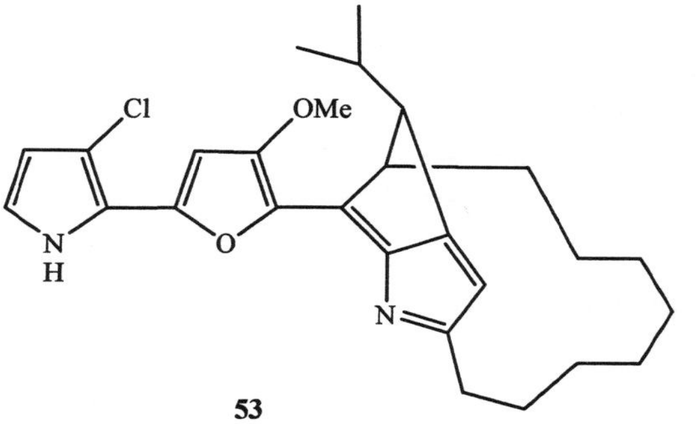

53

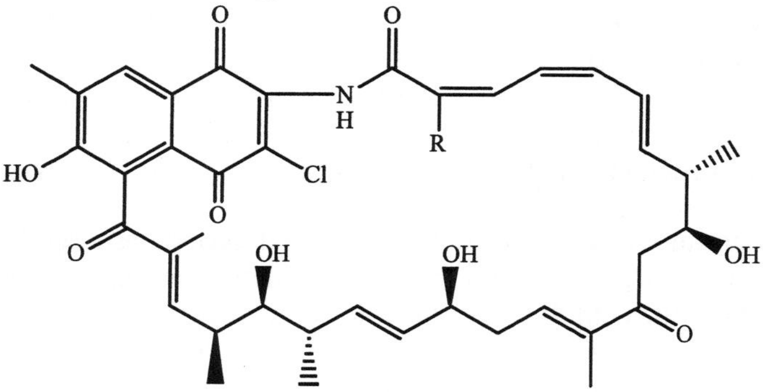

54 R = Me
55 R = H

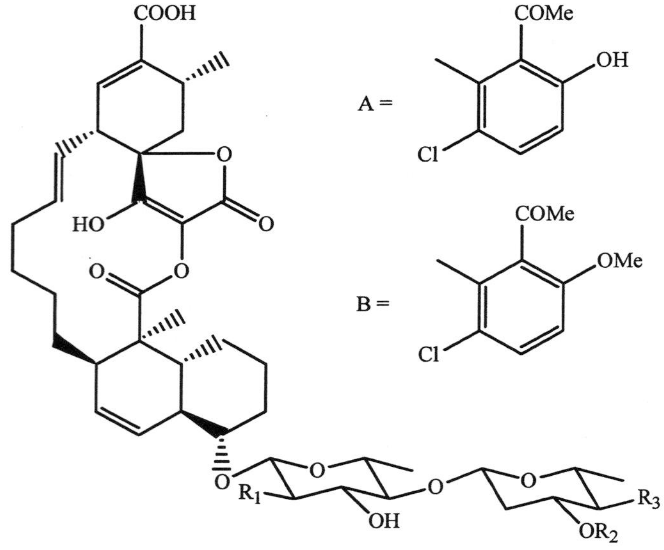

Fig. (9). Chemical structures of macrocyclic antibiotics

57 R₁ = H, R₂ = B, R₃ = OH (may be also Br)
58 R₁ = H, R₂ = A, R₃ = OH
59 R₁ = OH, R₂ = A, R₃ = OH
60 R₁ = H, R₂ = H, R₃ = OA
61 R₁ = OH, R₂ = H, R₃ = OA
62 R₁ = OH, R₂ = B, R₃ = OH

A large group of antibiotics called orthomycins was described as a product of different streptomycetes [50]. The group includes flambamycin, curamycin and avilamycin. These antibiotics exhibit a very low if not negligible activity against Gram-negative bacteria, yeasts and fungi. On the other hand, they have a significant effect on staphylococci and streptococci.

Flambamycin exhibits a very low toxicity and shows an interesting activity against Gram-positive bacteria, *Neisseria,* Gram-negative bacteria and some Gram-positive bacilli. It is practically inactive against Gram-negative bacilli, yeasts and filamentous fungi. This selectivity coupled with excellent therapeutic activity in mice infected experimentally with *Staphylococcus aureus*, *Streptococcus pyogenes haemolyticus*, or *Neisseria meningitidis* encouraged structural investigation of flambamycin.

In the course of studies [51-53] on the production of antimicrobial agents by microorganisms, a new antibiotic flambamycin (**57**) (also known as 21190 RP), was discovered in the culture broth of *S. hygroscopicus* DS 23230. The strain was isolated from a soil sample collected in Great Britain and its antibiotic properties were demonstrated by classical methods. It exhibits all the main morphological and biochemical characteristics of the species *S. hygroscopicus*, especially the tight spiral sporophores, the dark grey color of the normally sporulated mycelium, and also in ageing cultures the production of dark patches and of an exudate on surface of colonies on an agar medium.

South-American soil was a good source [54] of *S. curacoi* producing the antibiotic curamycin (**58**). This antibiotic also contains unusual saccharides, *i.e.* L-lyxose and 4-O-methyl-D-fucose.

Avilamycin A (**59**) isolated [55] from *S. viridochromogenes* collected from a Venezuelan soil sample, exhibits similar effects. Its structure was elucidated by NMR methods and two unusual saccharides – evalose and evermicose – were identified in the sugar moiety.

Antitumor antibiotics of the enediyne group, containing a highly unusual structure of the 9-membered unsaturated ring, were isolated from an unidentified streptomycete [56,57] designated L585-6 (**60**). These compounds were found to be highly active primarily against known cell lines [58] but almost inactive against Gram-negative bacteria. They form strong complexes with DNA preventing its replication and, as a result, exhibit a high activity against different cancer types.

57 R_1 = OH, R_2 = CO*i*-Pr, R_3 = Ac
58 R_1 = H, R_2 = R_3 = Ac
59 R_1 = H, R_2 = CO*i*-Pr, R_3 = Ac

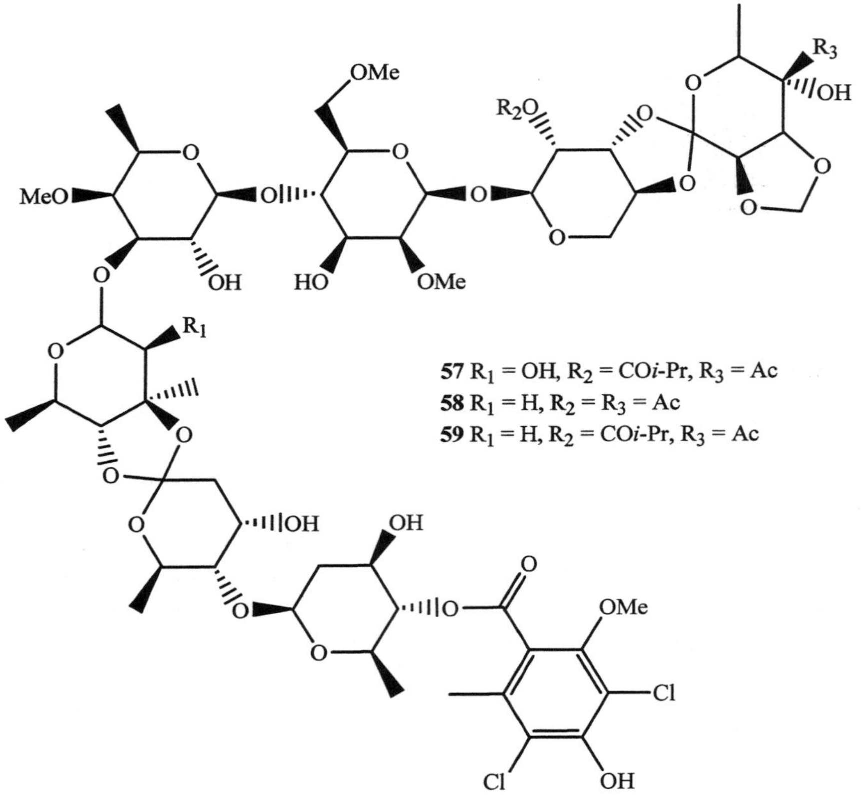

Fig. (10). Chemical structure of orthosomycin antibiotics.

Antibiotic C-1027 (**61**) was isolated from the cultivation broth of *S. globisporus* [59] and its molecular weight was determined to be 15 000 [60]. The chromophore part of this antibiotic [60] contains a 9-membered ring incorporated into the 16-membered macrocycle [61]. Complexity of its structure was described several times [61,62] and its absolute stereochemistry was determined on the basis of not only 2D NMR but also of CD spectra. Its high cytotoxicity exceeds even that of doxorubicin [63]. *S. graminofaciens* was several times erratically described [64,65], as a producer of chlorinated flavone derivatives (**62,63**).

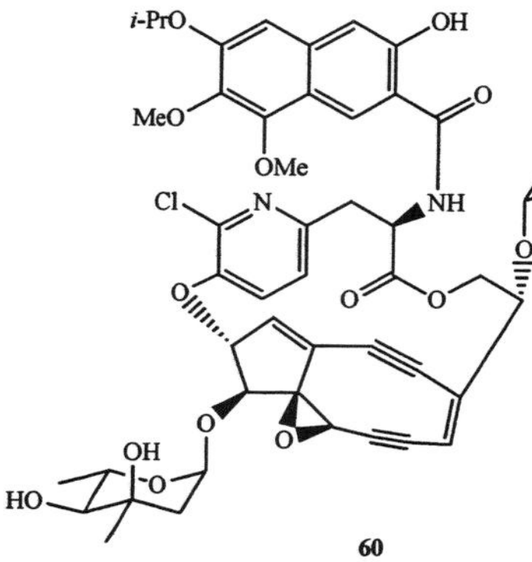

Fig. (11). Unusual structures of antibiotics with enediyne group

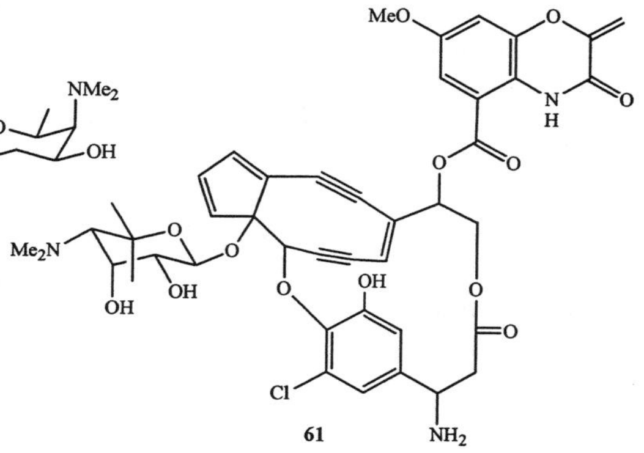

It was shown later that the streptomycete can only modify rather than biosynthesize these compounds [66] and that all such derivatives, *i.e.* even genisteins **(64,65)** come from the soy meal or cotton seed and are not produced by streptomycetes. These results demonstrate clearly that flavonoids isolated from streptomycetes cultivated on media containing plant components, such as soy meal or cotton seed, always come from the cultivation medium and are not of microbial origin. The same effect was observed in *S. griseus* [65] producing two additional chlorinated genisteins **(66,67)**. This also holds true [67] for isoflavone **(68)** isolated from Japanese soil.

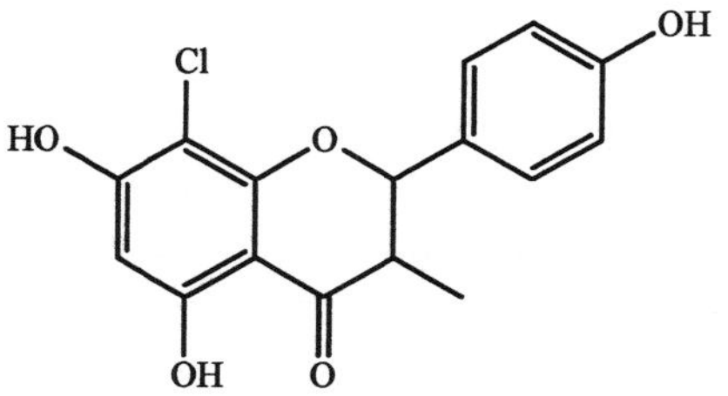

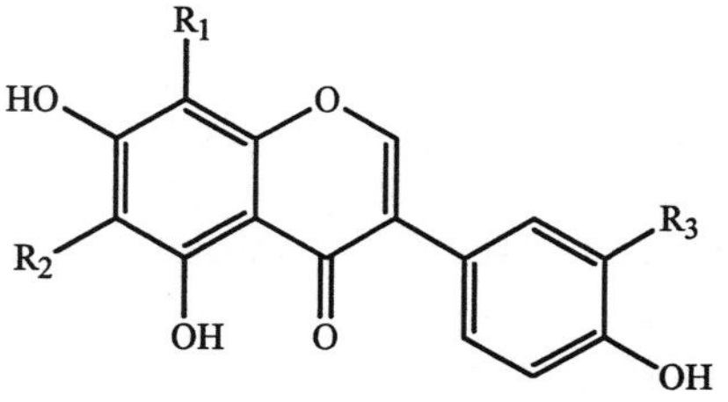

62 2(*S*):3(*S*)
63 racemic mixture of 2(*S*):3(*R*) and 2(*R*):3(*S*)

64 R_1 = Cl, R_2 = H, R_3 = H
65 R_1 = Cl, R_2 = Cl, R_3 = H
66 R_1 = H, R_2 = Cl, R_3 = H
67 R_1 = H, R_2 = Cl, R_3 = Cl
68 R_1 = Cl, R_2 = H, R_3 = OH

Fig. (12). Derivatives of flavones metabolized by *Streptomyces*.

Acidic lipophilic antibiotics pentalenolactones (**69,70**), active against Gram-positive and Gram-negative bacteria including acid-fast bacteria, were isolated [68] from *Streptomyces* sp.

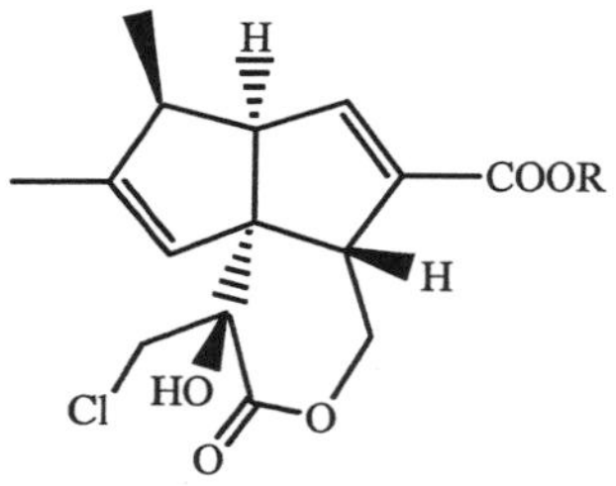

69 R = H
70 R = Me

Fig. (13). Pentalenolactones from *Streptomyces*

Nitrogen compounds

An L-leucine antimetabolite called AL-719 was isolated [69] from the cultivation broth of different *Streptomyces* species. In this metabolite one of two methyl groups is substituted by chlorine (**71**).

A glutamic acid analogue (**72**) was isolated [70] from a production *Streptomyces* strain. It was found to be partially effective against *Micrococcus luteus*.

Another species of the genus *Streptomyces* [70], identified as *S. xanthocidicus*, was found to produce a compound called FR900148 (**73**). This compound inhibits the growth of both Gram-positive and Gram-negative bacteria. It is assumed to inhibit the cell wall biosynthesis. Using physico-chemical methods its structure was determined as 1-*N*-valyl-3-chloro-2,5-dihydro-5-oxo-1*H*-pyrrole-2-carboxylic acid.

4-Chlorothreonine (**74**) was isolated [71] from the streptomycete designated OH5093. It exhibits the antibiotic activity against the yeast *Candida*.

Armentomycin (**75**) [72] inhibits the growth of bacteria on a synthetic medium. Biosynthesis of this non-protein amino acid is catalyzed by a peroxidase incorporating chlorinated substituents into the molecule without a simultaneous removal of other functional groups.

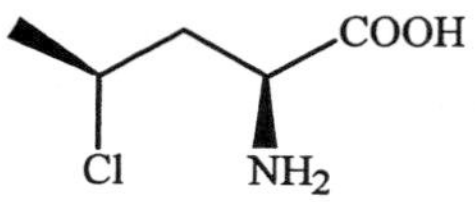

Fig. (14). Amino acid metabolites

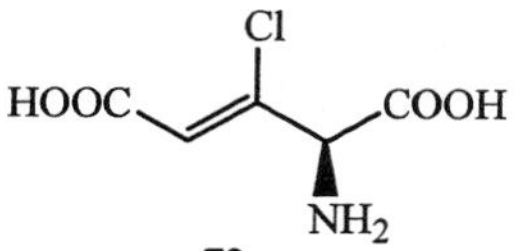

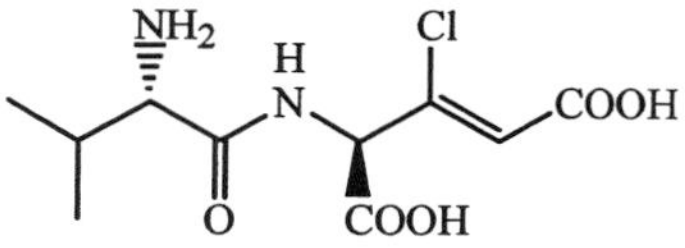

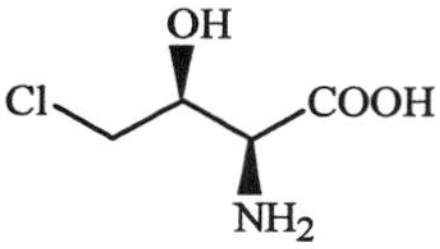

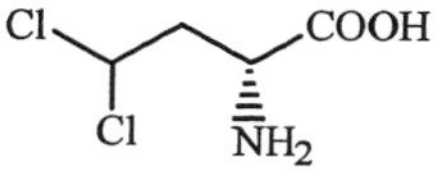

The antimetabolic activity of di- tri- and oligopeptides has been generally recognized. Compound (**76**) was isolated [73] from the cultivation broth of *Streptomyces* sp. 372A. The compound inhibits growth of Gram-positive and Gram-negative bacteria, however, only on a chemically defined medium. The addition of L-glutamine abolishes its effect.

Chlorocardin (**77**) was isolated as a β-lactam antibiotic [74] and exhibits biological activity against bacteria of the genus *Pseudomonas in vitro.*

A new herbicide resormycin (**78**) [75] was isolated from the streptomycete *S. platensis* exhibiting a significant biological effect against the unicellular green alga *Selenastrum capricornutum.* Resormycin inhibits primarily rather the growth in the dark than in the light. The effect of the compound on weeds in crops was also investigated in field experiments.

A complex of actinomycins Z (Z1-Z5) with only two chlorine containing β-depsipeptides (**79**=Z3, **80**=Z5) was isolated from the well-known *S. fradiae* [76]. Both actinomycins (Z3 and Z5) exhibit significant cytotoxicity against cancer cells.

Pepticinnamines (A-F) (**81**=E) inhibiting pharnesyl transfer and belonging to the protein family were isolated [77] from the streptomycete OH 4652. Their structure [78] was determined by means of NMR and found to contain five unusual amino acids and O-pentenylcinnamic acid.

The complete structure of RP 18,631 **(82)**, a new chlorine containing antibiotic related to novobiocin, has been determined [79] using a combination of degradative and NMR techniques. Of particular interest is the long-range couplings observed in the pyrrole ring present in the molecule.

76

77

79 3-OH-5MePro
80 5-MePro

78

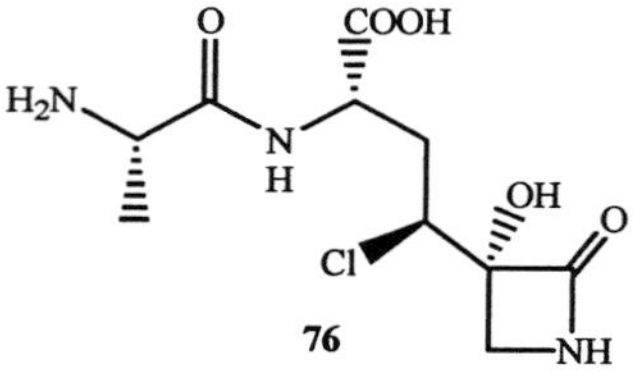

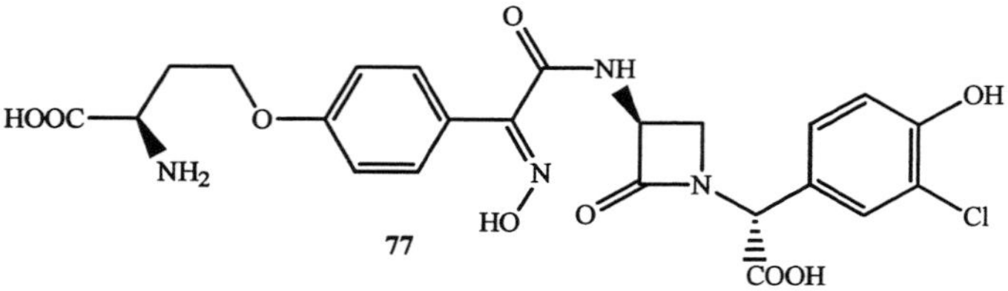

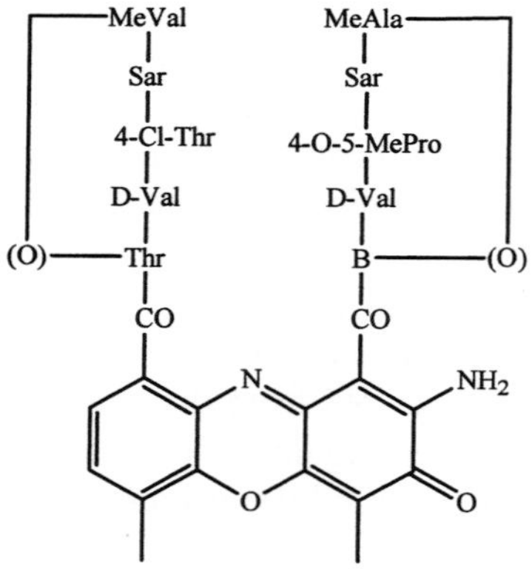

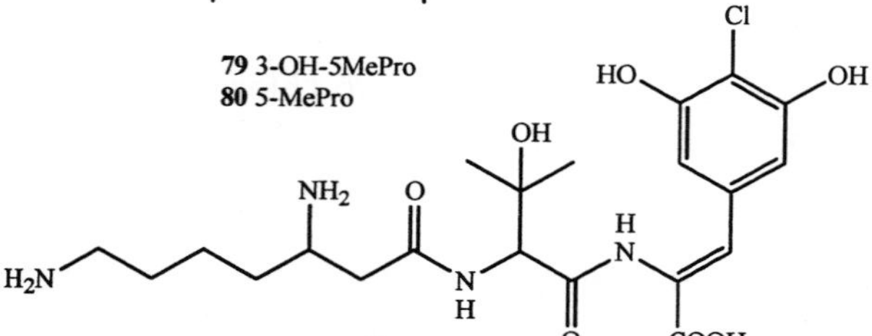

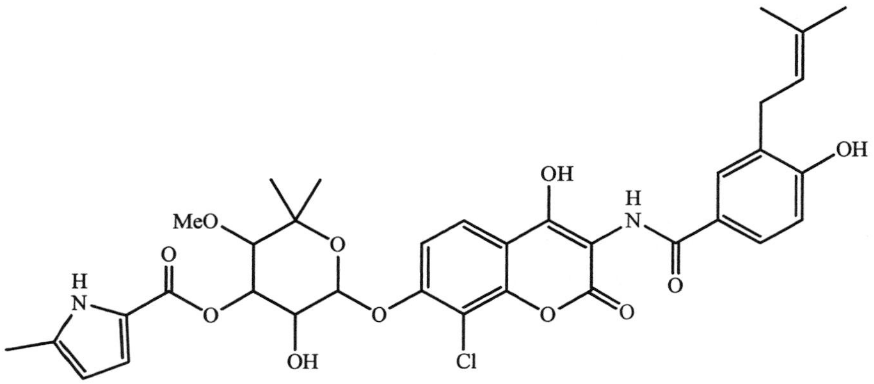

Fig. (15). Other oligopeptides.

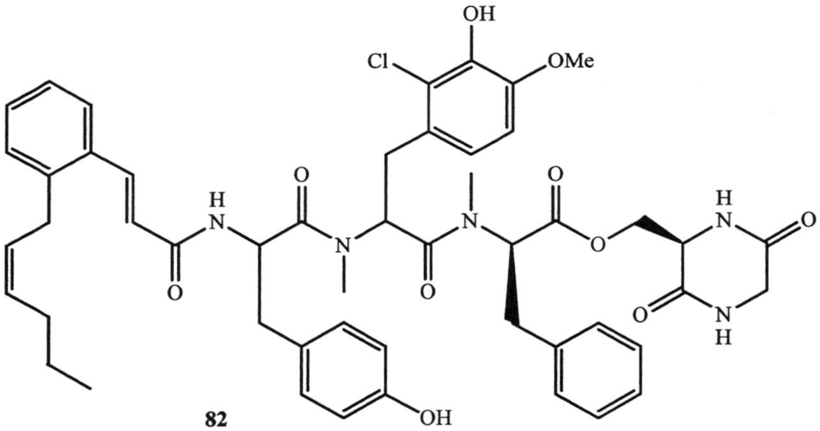

Vancomycins were first marketed more than 35 years ago. It was found that they exhibit an excellent activity against Gram-positive bacteria. They still remain the antibiotic of choice for the treatment of infections caused by *Staphylococcus aureus*, of methicillin-resistant strains in particular. Vancomycin is not absorbed from the digestive tract and is hence preferentially used for the treatment of intestinal infections. The whole group of glycopeptide antibiotics includes more than 200 different chemical structures including antibiotics such as A42867, A82846, A8350, chloroorienticin, decaplanin, eremomycin, MM 45289, MM 477611, OA-7653, orienticin and UK72051.

As a matter of fact, vancomycin (**82**) is produced by *Amycolatopsis orientalis* [80], it is only mentioned here since the strain was previously called *S. orientalis*.

Avoparcin, a mixture of highly similar glycopeptides avoparcin α (**83**) and β (**84**) belongs to commercially available antibiotics used mainly in veterinary medicine. The aglycone called avoparcin ε (**85**) was later isolated from the cultivation broth of *Streptomycete*.

S. fungicidus [81] produces the antibiotic enduracidin including again two compounds, *viz.* enduracidins A (**86**) and B (**87**). Both enduracidins exhibit a high *in vivo* and *in vitro* antibacterial activity against Gram-positive bacteria, including bacteria resistant to other known antibiotics. Additional compounds structurally related with vancomycin (OA7653A=**88** and OA7653B=**89**) were isolated [82] from *S. hygroscopicus*. Their chemical structure was determined mainly by MS and NMR. The compounds in question contain 7 amino acid residues.

Other complex glycopeptide antibiotics (**90-96**) produced by *S. virginiae* [83] contain galactose in their chemical structure.

Two neuroprotectins (A=**97**, B=**98**) including the previously described complestatin were isolated from the fermentation broth of *Streptomyces* Q27107. Neuroprotectins protect primary cultured chick telencephalic neurons from glutamate- and kainate-induced excitotoxicities in a dose-dependent fashion

Complestatin (**99**) [84,85] was isolated from the mycelium of *S. lavendulae* and its structure was determined by NMR. It contains two unusual amino acids as shown in Fig. 16. The compound inhibits hemolysis of sensitized sheep erythrocytes mediated by guinea pig and human complement at concentrations of 0.4 and 0.7 µg/ml, respectively.

Chlorpeptin I (**100**) [86,87] was isolated from *Streptomyces* sp. It was found that it mainly contains aromatic amino acids and exhibits strong anti-HIV effects. Another glycopeptide antibiotic (**101**) [88] designated LY264826, structurally related with vancomycin but containing a C-methylated sugar in its chemical structure, is effective against enterococci. Other glycopeptides (**102-110**) derived from teicoplanin were prepared [89] and their activity against Gram-positive bacteria was investigated.

The aglycone was obtained by enzymatic hydrolysis of the side chain containing fatty acids and its additional reductive alkylation yielded products exhibiting a much higher activity against both staphylococci and enterococci (the most efficient compounds were active at 0.25 − 2 µg/ml).

Chloptosin (**111**) [90] also containing unusual amino acids was found to exhibit apoptotic activity against human adenocarcinoma.

Glycopeptide antibiotic A35512B (**112**) [91] was isolated from a soil streptomycete.

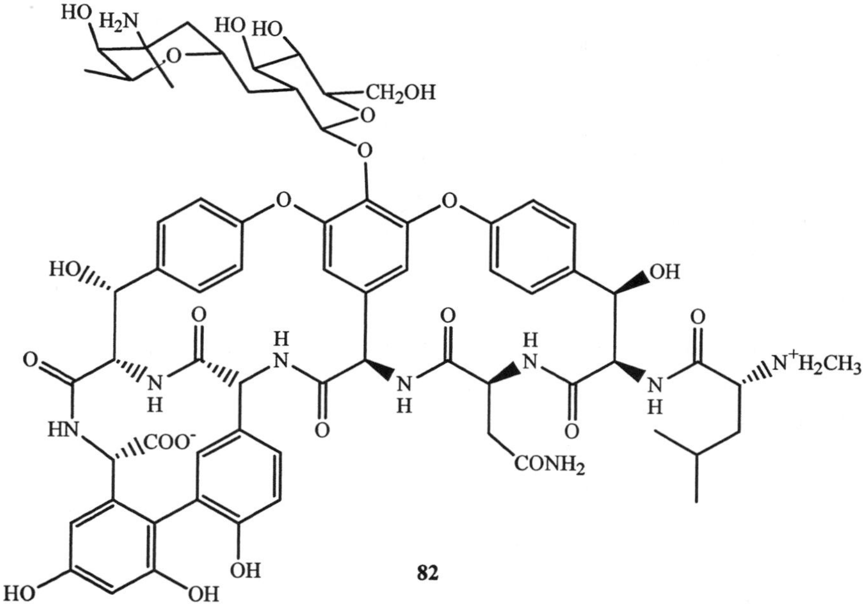

82

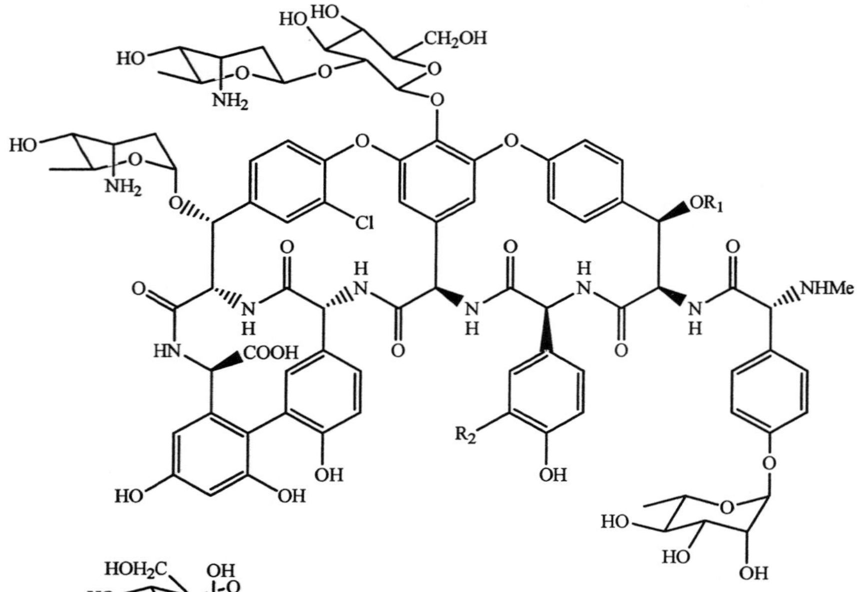

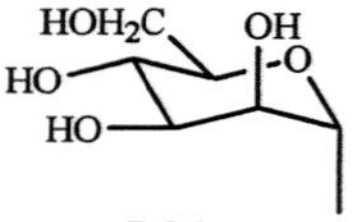

83 R_1 = α-D-Man, R_2 = H
84 R_1 = α-D-Man, R_2 = Cl
85 R_1 = H, R_2 = Cl

α–D-Man

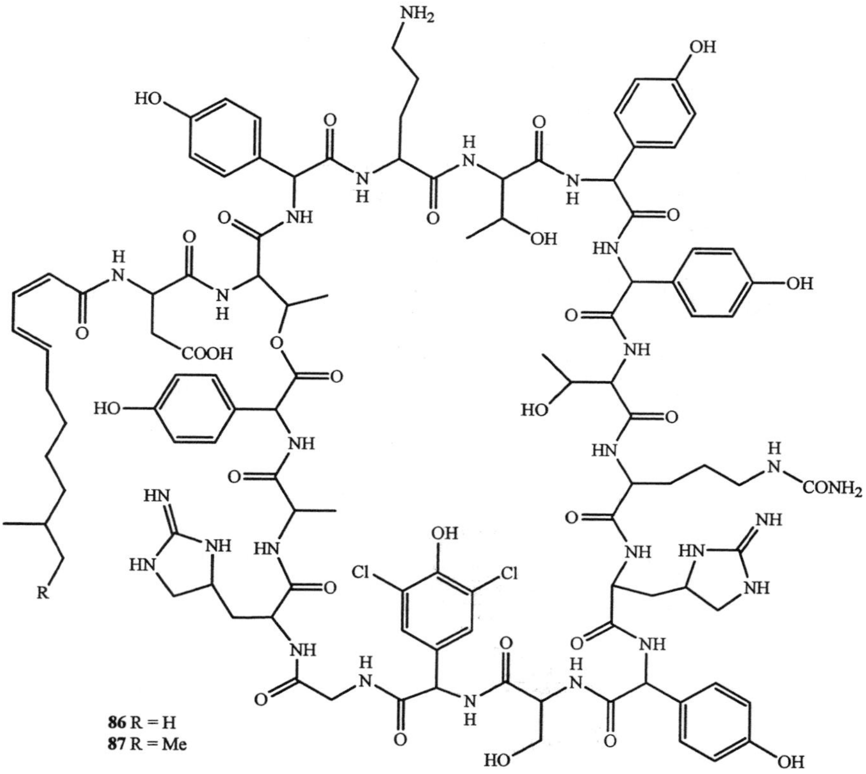

86 R = H
87 R = Me

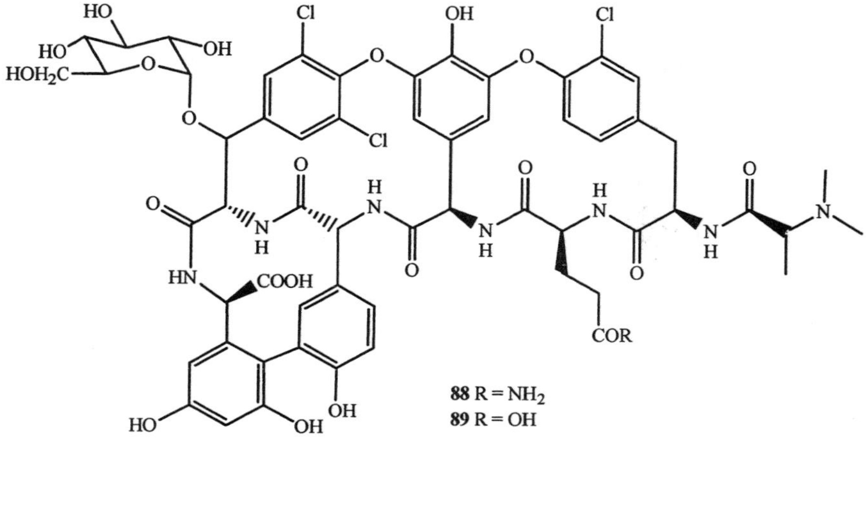

88 R = NH₂
89 R = OH

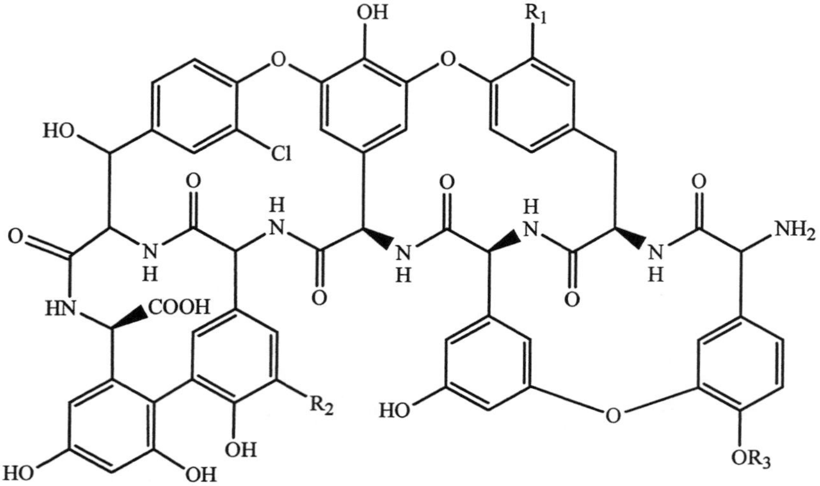

	R1	R2	R3		R1	R2	R3
90	Cl	Cl	H	**94**	H	H	H
91	Cl	H	H	**95**	Cl	Cl	Gal-Gal
92	Cl	Cl	Gal	**96**	Cl	Cl	Gal-Gal
93	H	Cl	H				

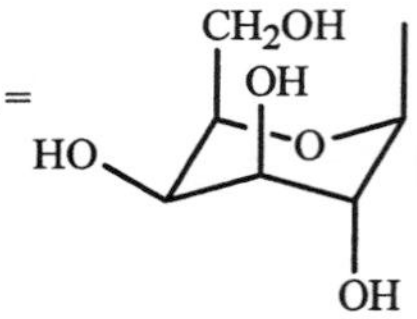

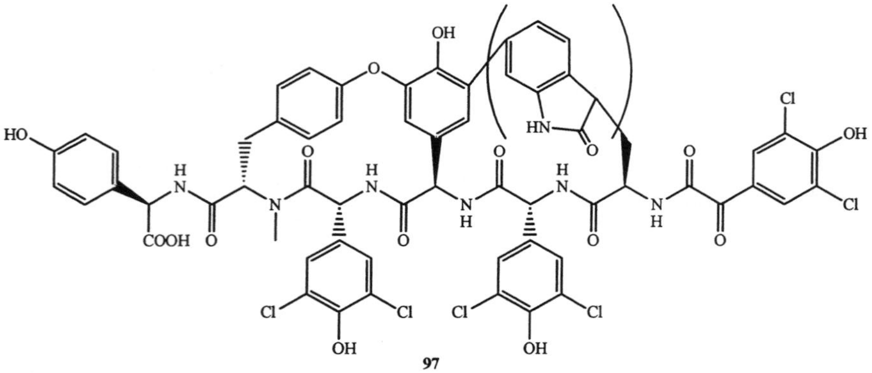

97

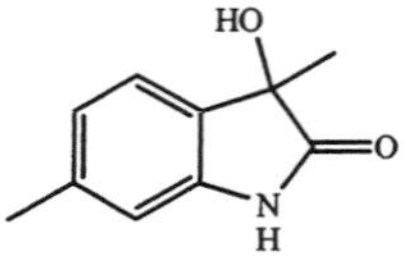

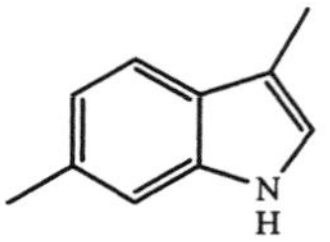

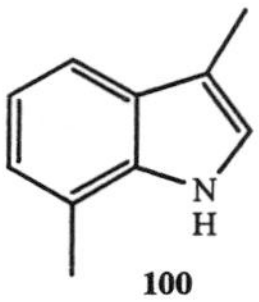

98   **99**   **100**

101

111

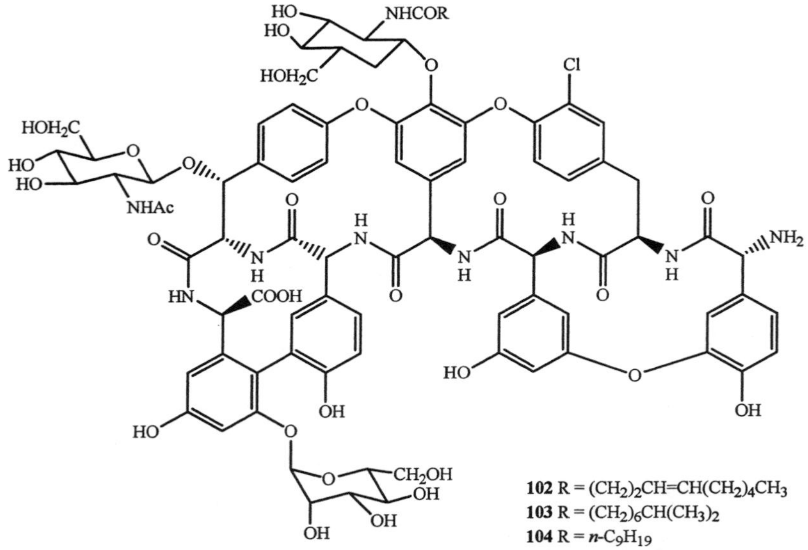

102 R = (CH$_2$)$_2$CH=CH(CH$_2$)$_4$CH$_3$
103 R = (CH$_2$)$_6$CH(CH$_3$)$_2$
104 R = n-C$_9$H$_{19}$
105 R = (CH$_2$)$_6$CH(CH$_3$)CH$_2$CH$_3$
106 R = (CH$_2$)$_7$CH(CH$_3$)$_2$
107 R = (CH$_2$)$_8$CH(CH$_3$)$_2$
108 R = n-C$_{11}$H$_{23}$
109 R = (CH$_2$)$_4$CH(CH$_3$)CH$_2$CH$_3$
110 R = n-C$_8$H$_{17}$

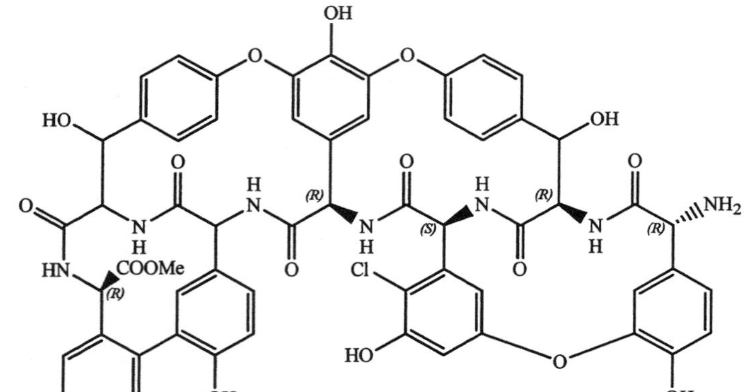

112

Fig. (16). Glycopeptide antibiotics

The chemical structure of new anticancer antibiotics duocarmycin C1 (113) and C2 (114), isolated from the cultivation broth of *Streptomyces* sp. [92,93] was elucidated. The antibiotic was found to exhibit a high antibacterial activity against Gram-positive bacteria (0.01μg/ml). Both duocarmycins are active against lymphocyte leukemia and sarcoma in mice [94-96]. Duocarmycins B1 (115) and B2 (116) [97] were isolated after addition of bromine to the cultivation medium. Additional seven duocarmycins were isolated, however, only duocarmycins C and B contained a halogen atom [98]. Pyrindamycins A and B, found later to be identical with duocarmycins C1 and C2, were isolated from a streptomycete designated SF2582 [99]. Pyrroindomycin A, which does not contain chlorine and pyrroindomycin B with chlorine in its structure (117) are the main components of the antibiotic complex isolated from the cultivation broth of *S. rugosporus* [100]. When clarifying their structure by means of physico-chemical methods it was found that they contain a trisaccharide and exhibit an excellent activity against Gram-positive bacteria [101]. A semisynthetic derivative of pyrroindomycin B (pyrroindomycin B-AC-2) (118) has an outstanding activity against the exponential phase cells, however, it does not affect the stationary phase cells. The antibiotic is very efficient against *Staphylococcus aureus*. As the compound really exhibits significant antibiotic effects, optimization of the cultivation process was investigated [102]. Thus for instance at a glucose concentration higher than 7.5 g/l its yield decreases, however, the concentration of glucose must not decrease below 5 g/l as the antibiotic is not produced any more under these conditions. The effect can be reversed by increasing the content of nitrogenous compounds such as ammonium chloride, arginine or glutamine. The effect of other carbon compounds was also investigated, however, it was found that some of them, *e.g.* sucrose or starch, are poorly metabolized. Other compounds, such as biotin or L-tryptophan, increase the production of pyrroindomycin.

Virantmycin (119) isolated from *S. nitrosporeus* [103] exhibits an antiviral activity [104] and a weak antifungal activity. It prevents peroxidation of lipids in rat liver microsomes. In the cell assay, benzastatins C and D inhibited glutamate toxicity in N18-RE-105 cells with EC_{50} values of 2.0 and 5.4 μM, respectively [105]. Pyrroxamycin (98) isolated from an unidentified streptomycete [106] was found to inhibit Gram-positive bacteria and dermatophytes.

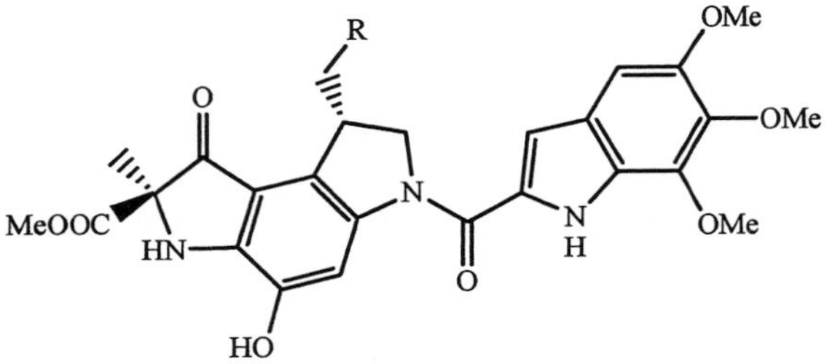

114 R = Cl
116 R = Br

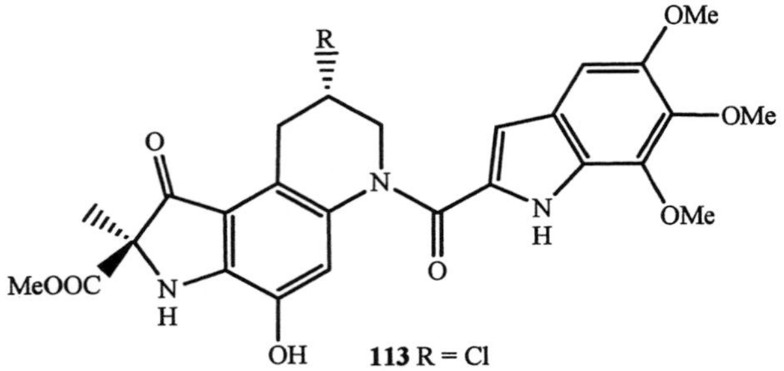

113 R = Cl
115 R = Br

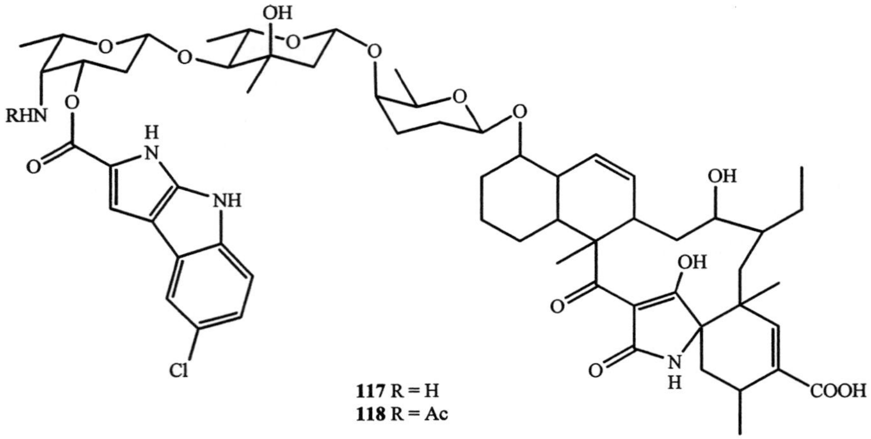

117 R = H
118 R = Ac

Fig. (17). Indol antibiotics.

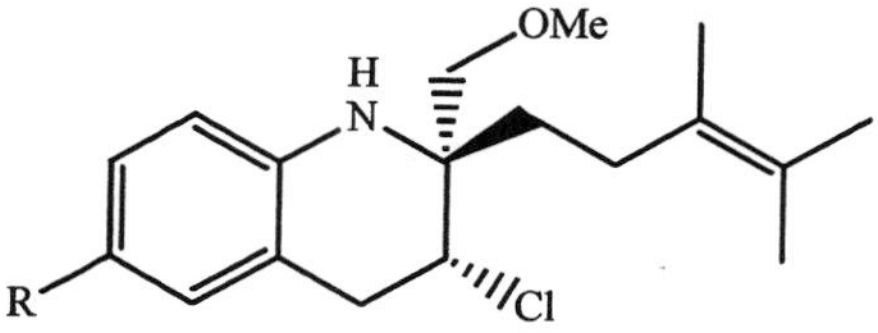

119 R = COOH
120 R = CONH$_2$

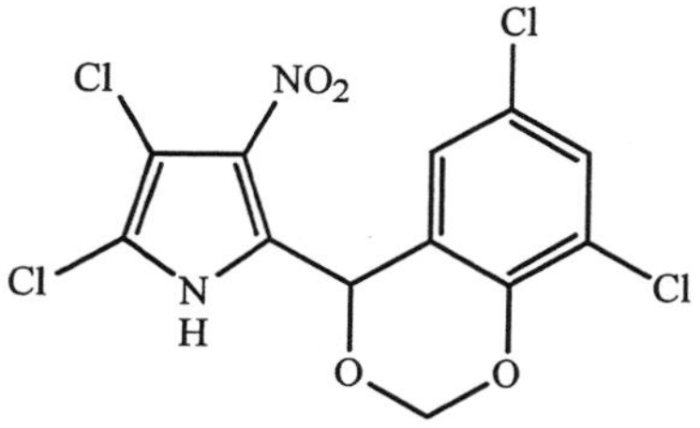

121

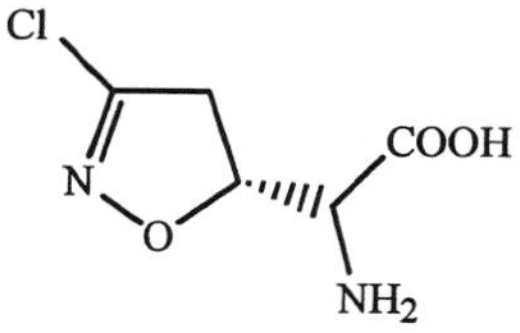

122

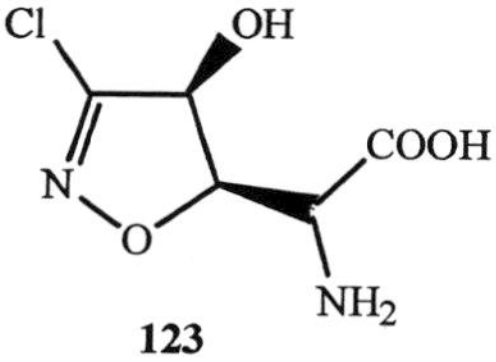

123

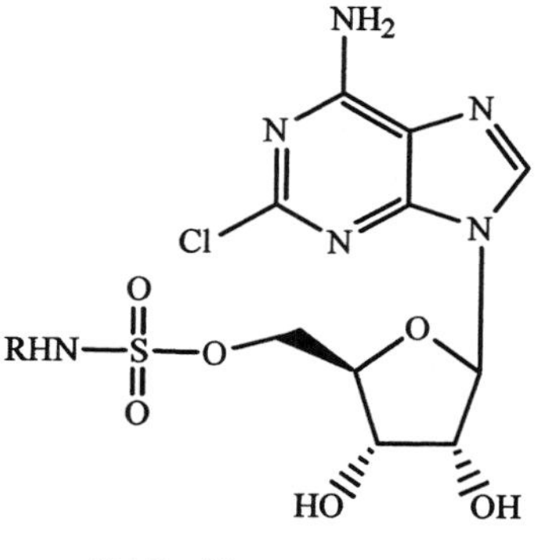

124 R = H
125 R = COCH(CH$_3$)NH$_2$

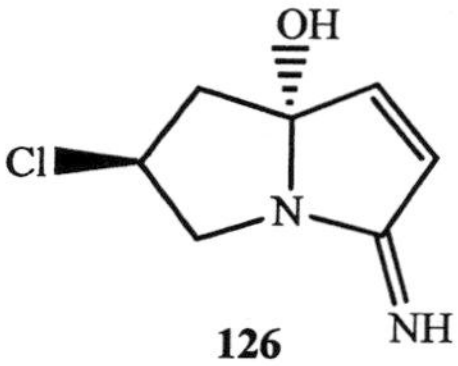

126

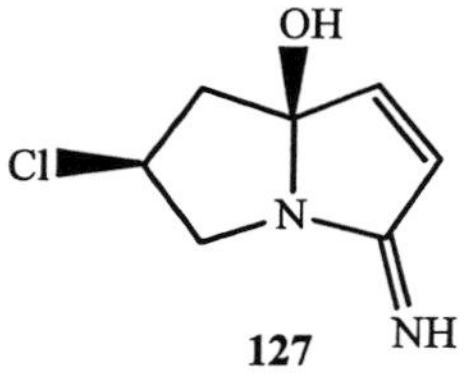

127

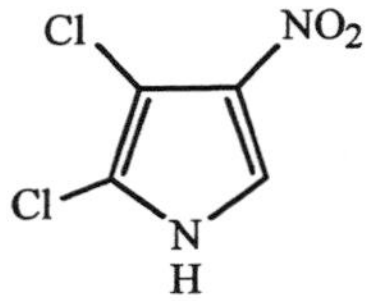

	R₁	R₂	R₃	R₄
	R_1	R_2	R_3	R_4
131	Cl	H	Pr	OH
132	Cl	H	Et	OH
133	Cl	Cl	Pr	OH
134	Cl	H	Pr	OMe
135	Br	H	Pr	OH
136	Cl	Cl	Et	OH
137	Cl	H	Bu	OH
138	Cl	H	Et	OMe
139	Cl	Cl	Pr	OMe

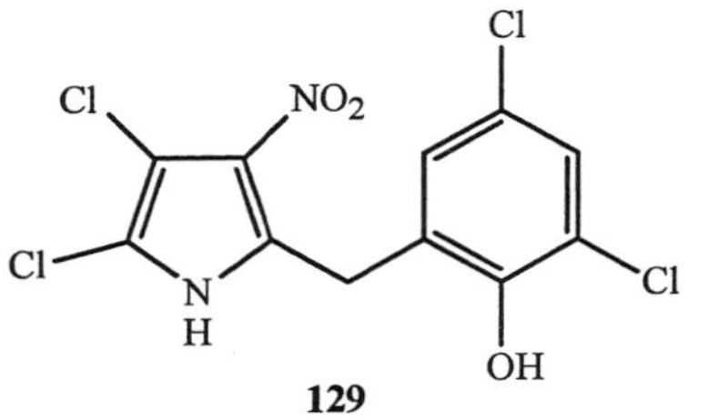

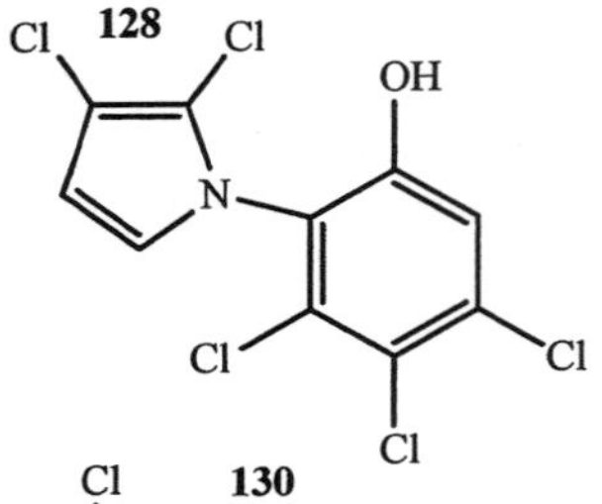

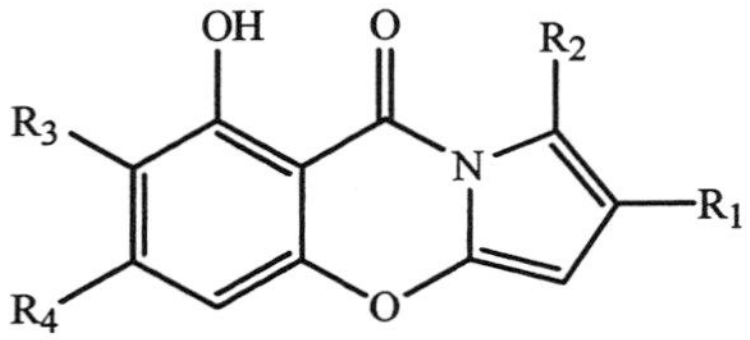

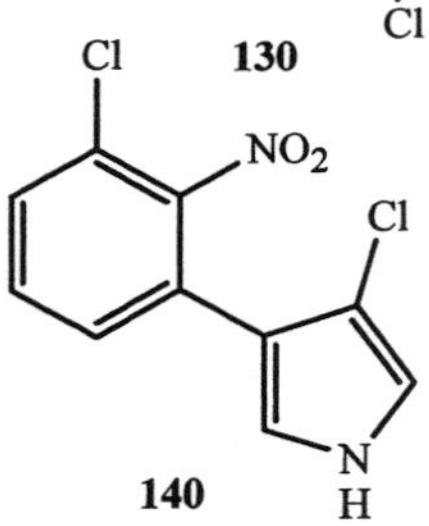

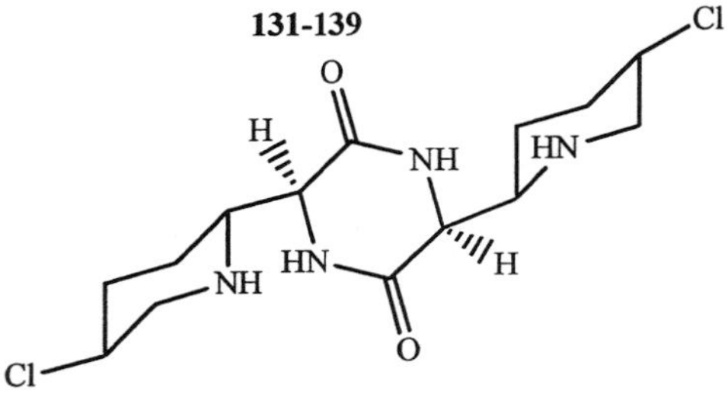

Fig. (18). Other N-heterocycles

From a cultivation of *S. sviceus* in a 250 l fermentor almost 0.25 kg of the antibiotic U42126 (**122**) was isolated [107]. The antimetabolite antibiotic U-42126 significantly increased the life span of tumor-bearing L1210 leukemia mice at low drug levels with no demonstrable signs of toxicity to the hosts.

Even higher quantities of an antibiotic (**123**) exhibiting anticancer effects were isolated [108] with a total of 150 g of pure compound from 16 000 l.

S. rishirensis produces [109] a nucleoside antibiotic designated AT265 (**124**). Its structure was determined to be a 5′-O-sulphamoyl derivative stimulating phosphate groups [110]. Because of its unionized nature, the molecule can cross cell membranes. A derivative of this compound (**125**) designated ascamycin exhibits similar biological effects.

Clazamycins A and B (**126,127**) were isolated [111] from *S. ponicerus* (similar to *S. cinereoruber*) [112].

Two wide-spectrum antifungal compounds, *viz.* pyrrolomycin A and B (**128,129**) were isolated from a streptomycete [113,114]. Their structure was clarified on the basis of physico-chemical [114] and spectroscopic (X-ray) [115] properties.

Neopyrrolomycin (**130**) inhibiting growth of Gram-positive and Gram-negative bacteria and of various fungi was isolated from a streptomycete [116].

A series of benzoxazines (**131-139**) were isolated [117] from fermentation broth of *S. rimosus*. These compounds were found to inhibit bacterial histidine kinase and were produced by a controlled cultivation with added NaBr and NaI. As mentioned above, streptopyrol (**131**) inhibits the nitrogen regulator II histidine kinase from *Escherichia coli* with IC_{50} of 20 μM and exhibits antimicrobial activity against a wide range of bacteria and fungi.

S. pyrocinia produces [118] an antifungal antibiotic pyrrolnitrin (**140**) active against mycobacteria. The antimycobacterial activity can be well correlated with the presence of the halogen and nitro group in the aromatic ring.

A compound designated 593A (**141**) was isolated [119] from *S. griseoluteus*. It was found that this compound exhibits high anticancer, antileukemic effects.

Fluorinated compounds

Practically only three fluorinated derivatives were described. As early as in 1957 nucleocidin (**142**) was isolated from the cultivation broth of *S. calvus* producing [120]. It is frustrating that the recent attempts at the isolation of this antibiotic from *S. calvus* were unsuccessful [120-123]. *S. cattleya* produces mM concentrations of 4-fluorothreonine (**143**) and fluoroacetate (**144**) when cultivated in a medium with fluorine ions.

All other fluorinated antibiotics were either prepared by a partial chemical synthesis or by cultivations to which fluorine ions were added. All metabolites prepared in this way and initially considered very promising with respect to their increased biological activity were an absolute failure, in spite of the fact that just the contrary was often described in the patent literature.

Several synthetic compounds can be presented as examples: 5-fluoro-N_b-acetyltryptamine (145) and 6-fluoro-N_b-acetyltryptamine (146) and the novel metabolites 5-fluoro- and 6-fluoro-substituted alkaloids (147) and (148) [124], 5-fluoro-β-hydroxy-N_b-acetyltryptamine (149) and 6-fluoro-β-hydroxy-N_b-acetyltryptamine (150) from the ethyl acetate extract of S. staurosporeus cultures after addition of tryptamine, 5-fluorotryptamine and 6-fluorotryptamine [125]. 5''-Fluoropactamycin (151) and 5''-fluoropactamycate (152) were also prepared by S. pactum, after supplementation of the medium with 3-amino-5-fluorobenzoic acid [126]. The macrolides FK-506 and FK-520 were obtained [127] by fermentation of two streptomycetes. The parent compounds were obtained from S. tsukubaensis No 9993 and from S. hygroscopicus subsp. ascomyceticus ATCC 14891. The following derivatives (153) [127] were prepared by a partial chemical synthesis, where the sugar is fluoro substituted β-D-galactopyranose or α-D-arabinofuranose.

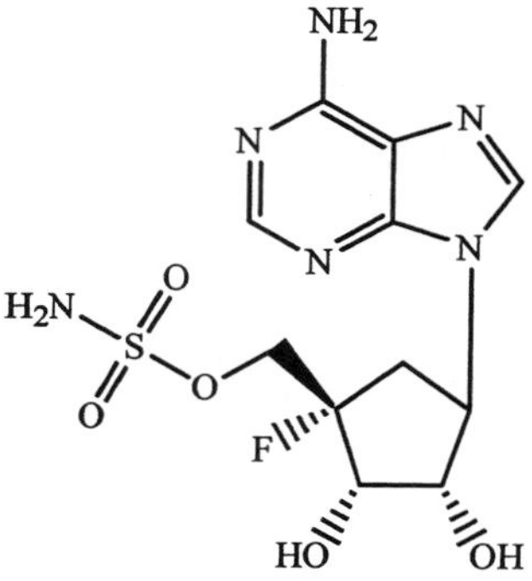

142

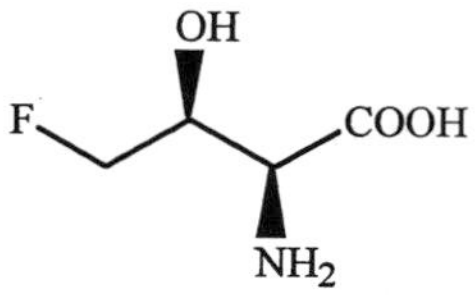

143

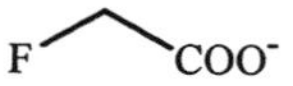

144

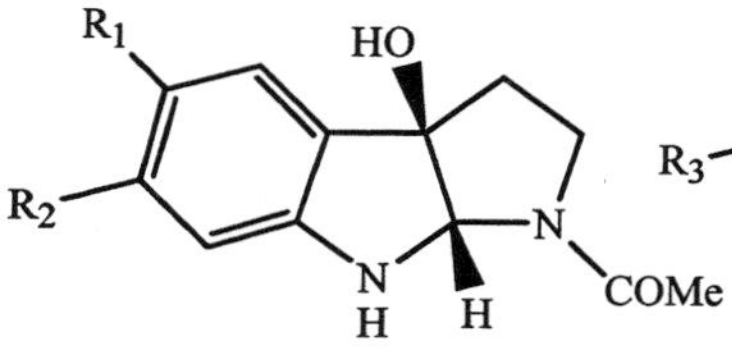

147 $R_1 = F, R_2 = H$
148 $R_1 = H, R_2 = F$

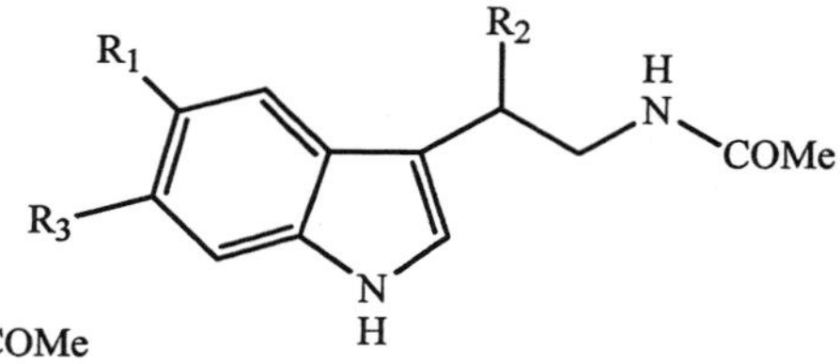

145 $R_1 = F, R_2 = R_3 = H$
146 $R_1 = R_2 = H, R_3 = F$
149 $R_1 = F, R_2 = OH, R_3 = H$
150 $R_1 = H, R_2 = OH, R_3 = F$

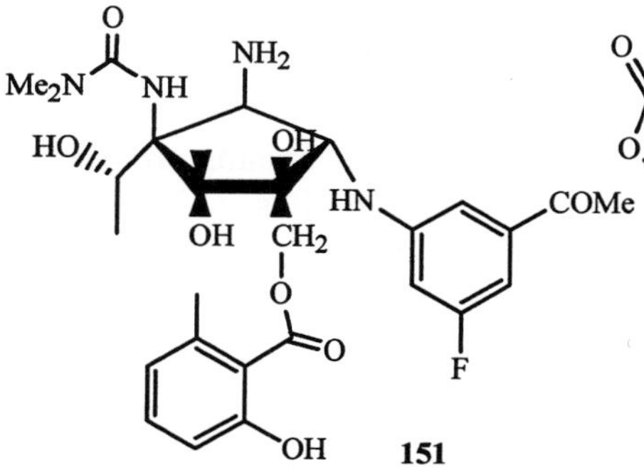

151

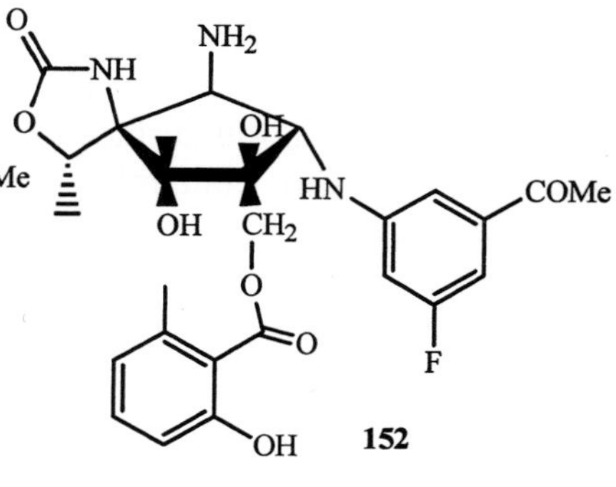

152

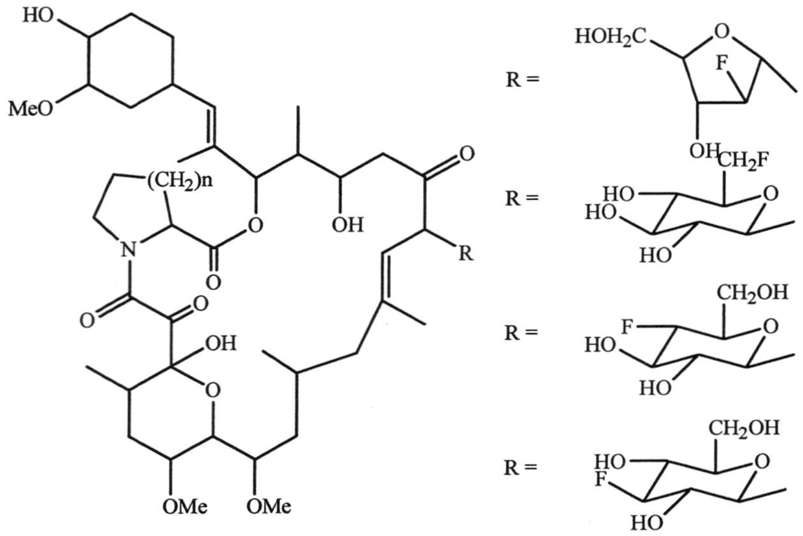

Fig. (19). F containing compounds

CONCLUSION

Halogen containing secondary metabolites exhibit interesting biological effects, such as antibiotic, anticancer and others. New producers of secondary metabolites such as streptomycetes isolated from unusual locations, *e.g.* from salty soils, lakes with increased salt concentrations, or marine streptomycetes or streptomycetes growing on surface or inside marine organisms are expected to produce new halogen containing secondary metabolites [128]. Molecular biological and genetic methods together with combinatorial biochemistry of streptomycetes also represent a promising approach towards the isolation of new halogen containing biologically active compounds [129].

REFERENCES

[1] Gottlieb, D.; Bhattacharyya, P.K.; Anderson, H.W.; Carter, H.E.; *J. Bacteriol.*, **1948**, *55*, 409-417.

[2] Bartz, Q.R.; *J. Biol. Chem.*, **1948**, *172*, 445-450.

[3] Malik V.S.; *Advances in Applied Microbiology*, **1972**, *15*, 297-336.

[4] Vining L.C.; Westlake D.W.S.; *Biotechnology of industrial antibiotics*, **1984**, 387-412.

[5] Simonsen, J.N.; Paramasigamani, K.; Vining, L.C.; McInnes, A.G.; Walter, J.A.; Wright, L.C.; *Can. J. Microbiol.*, **1978**, *24*, 136-142.

[6] Smith, C.G.; *J. Bacteriol.*, **1958**, *75*, 577-583.

[7] Labeda, D.P.; *Int. J. Syst. Bacteriol.*, **1995**, *45*, 124-127.

[8] Durckheimer, W.; *Angew. Chem.*, **1975**, *87*, 751-764.

[9] Podojil, M.; Blumauerova, M.; Cudlin, J.; Vanek, Z.; *Biotechnology of industrial antibiotics*, **1984**, 259-280.

[10] Hunter, I.S.; Hill, R.A.; *Biotechnology of antibiotics*, **1997**, 659-682

[11] Behal, V.; Hunter, I.S.; *Genetics Biochemistry of antibiotic production*, **1995**, 359-385.

[12] Behal, V.; *Critical Review Biotechnol.*, **1987**, *5*, 275-317.

[13] Vanek, Z.; Cudlin, M.; Blumauerova, M.; Hostalek, Z.; *Folia Microbiol.*, **1971**, *16*, 225-240.

[14] Abulnaja, K.O.; *Alexandria J. Pharm. Sci.*, **1997**, *11*, 17-21.

[15] Bertan, M.; Revallova, V.; Vrublova, K.; Martincer, L.; Timcova, J., *Czech Patent* No. 282681, **1997**.

[16] Ryan, M.J.; Lotvin, J.A.; Strathy, N.; Fantini, S.E.; *US Patent* No. 5589385, **1996**.

[17] Lepetit, S.P.A. ; *GB Patent* No. 772149, **1957**.

[18] McCormick, J.R.D.; Sjoler, N.O.; Hirsch, U.; Jensen, E.R.; Doerschuk, A.P.; *J. Am. Chem. Soc.*, **1957**, *79*, 4561-4563.

[19] McCormick, J.R.D.; Jensen, E.R.; *J. Am. Chem. Soc.*, **1969**, *91*, 206-206.

[20] Ryan, M.J.; US Patent No. 5965429, **1999**.

[21] Wells, J.S.; Osullivan, J.; Aklonis, C.; Ax, H.A.; Tymiak, A.A.; Kirsch, D.R.; Trejo, W.H.; Principe, P.; *J. Antibiot.*, **1992**, *45*, 1892-1898.

[22] Tymiak, A.A.; Ax, H.A.; Bolgar, M.S.; Kahle, A.D.; Porubcan, M.A.; Andersen, N.H.; *J. Antibiot.*, **1992**, *45*, 1899-1906.

[23] Devasthale, P.V.; Mitscher, L.A.; Telikepalli, H.; Vervelde, D.; Zou, J.Y.; Ax, H.A.; Tymiak, A.A.; *J. Antibiot.*, **1992**, *45*, 1907-1913.

[24] Tymiak, A.A.; Aklonis, C.; Bolgar, M.S.; Kahle, A.D.; Kirsch, D.R.; Osullivan, J.; Porubcan, M.A.; Principe, P.; Trejo, W.H.; Ax, H.A.; Wells,

348

J.S.; Andersen, N.H.; Devasthale, P.V.; Telikepalli, H.; Vervelde, D.; Zou, J.Y.; Mitscher, L.A.; *J. Org. Chem.*, **1993**, *58*, 535537.

[25] Patel, M.; Gullo, V.P.; Hegde, V.R.; Horan, A.C.; Gentile, F.; Marquez, J.A.; Miller, G.H.; Puar, M.S.; Waitz, J.A.; *J. Antibiot.*, **1987**, *40*, 1408-1413.

[26] Smith, E.B.; Munayyer, H.K.; Ryan, M.J.; Mayles, B.A.; Hegde, V.R.; Miller, G.H.; *J. Antibiot.*, **1987**, *40*, 1419-1425.

[27] Liu, C.M.; Hermann, T.E.; Prosser, B.L.T.; Palleroni, N.J.; Westley, J.W.; Miller, P.A.; *J. Antibiot.*, **1981**, *34*, 133-138.

[28] Westley, J.W.; Evans, R.H.; Sello, L.H.; Troupe, N.; Liu, C.M.; Blount, J.F.; Pitcher, R.G.; Williams, T.H.; Miller, P.A.; *J. Antibiot.*, **1981**, *34*, 139-147.

[29] Drautz, H.; Keller-Schierlein, W.; Zahner, H.; *Arch. Microbiol.*, **1975**, *106*, 175-190.

[30] Tsukamoto, M.; Nakajima, S.; Arakawa, H.; Sugiura, Y.; Suzuki, H.; Hirayama, M.; Kamiya, S.; Teshima, Y.; Kondo, H.; Kojiri, K.; Suda, H.; *J. Antibiot.*, **1998**, *51*, 908-914.

[31] Schimana, J.; Fiedler, H.P.; Groth, I.; Sussmuth, R.; Beil, W.; Walker, M.; Zeeck, A.; *J. Antibiot.*, **2000**, *53*, 779-787.

[32] Krone, B.; Zeeck, A.; *Liebigs Ann. Chem.*, **1983**, 471-502.

[33] Shomura, T.; Gomi, S.; Ito, M.; Yoshida, J.; Tanaka, E.; Amano, S.; Watabe, H.; Ohuchi, S.; Itoh, J.; Sezaki, M.; Takebe, H.; Uotani, K.; *J. Antibiot.*, **1987**, *40*, 732-39.

[34] Gomi, S.; Ohuchi, S.; Sasaki, T.; Itoh, J.; Sezaki, M.; *J. Antibiot.*, **1987**, *40*, 740-749.

[35] Omura, S.; Macrolide antibiotics: chemistry, biology, and practice (Ed. by Satoshi Omura), 2nd ed. Amsterdam, Elsevier Science, 2002.

[36] Suwanborirux, K.; Chang, C.J.; Spjut, R.W.; Cassady, J.M.; *Experientia* **1990**, *46*, 117-120.

[37] Hayakawa, Y.; Kawakami, K.; Seto, H.; Furihata, K.; *Tetrahedron Lett.*, **1992**, *33*, 2701-2704.

[38] Williams, T.H.; *J. Antibiot.*, **1975**, *28*, 85-86.

[39] Mukhopadhyay, T.; Franco, C.M.M.; Reddy, G.C.S.; Ganguli, B.N.; Fehlhaber, H.W.; *J. Antibiot.* **1985**, *38*, 948-951

[40] Kellerschierlein, W.; Meyer, M.; Zeeck, A.; Damberg, M.; Machinek, R.; Zahner, H.; Lazar, G.; *J. Antibiot.*, **1983**, *36*, 484-492.

[41] Hooper, A.M.; Rickards, R.W.; *J. Antibiot.*, **1998**, *51*, 845851.

[42] Brufani, M.; Cellai, L.; *J. Antibiot.*, **1979**, *32*, 167-168.

[43] Kellerschierlein, W.; Meyer, M.; Cellai, L.; Cerrini, S.; Lamba, D.; Segre, A.; Fedeli, W.; Brufani, M.; *J. Antibiot.*, **1984**, *37*, 1357-1361.

[44] Hooper, A.M.; Rickards, R.W.; *J. Antibiot.*, **1998**, *51*, 958-962.

[45] Lee, J.J; Lee, J.P.; Keller, P.J.; Cottrell, C.E.; Chang, C.J.; Zahner, H.; Floss, H.G.; *J. Antibiot.*, **1986**, *39*, 1123-1134.

[46] Holzbach, R.; Pape, H.; Hook, D.; Kreutzer, E.F.; Chang, C.J.; Floss, H.G.; *Biochemistry*, **1978**, *17*, 556-560.

[47] Mascaretti, O.A.; Chang, C.J.; Hook, D.; Otsuka, H.; Kreutzer, E.F.; Floos, H.G.; *Biochemistry*, **1981**, *20*, 919-924.

[48] Keller-Schierlein, W.; Muntwyler, R.; Pache, W.; Zahner, H.; *Helv. Chim. Acta*, **1969**, *52*, 127-142.

[49] Yamamoto, I.; Nakagawa, M.; Hayakawa, Y.; Adachi, K.; Kobayashi, E.; *J. Antibiot.*, **1987**, *40*, 1452-1454.

[50] Wright, D.E.; *Tetrahedron* **1979**, *35*, 1207-1237.

[51] Ninet, L.; Benazet, F.; Charpentie, Y.; Dubost, M.; Florent, J.; Lunet, J.; Mancy, D.; Preud´Homme, J.; *Experientia* **1974**, *30*, 1270-1272.

[52] Ollis, W.D.; Smith, C.; Sutherl, I.O.; Wright, D.E.; *J. Chem. Soc. Chem. Comm.*, **1976**, 350.

[53] Ollis, W.D.; Smith, C.; Sutherl, I.O.; Wright, D.E.; *Tetrahedron* **1979**, *35*, 105-127.

[54] Galmarini, O.L.; Deulofeu, V.; *Tetrahedron* **1961**, *15*, 76-86.

[55] Keller-Schierlein, W.; Heilman, W.; Ollis, W.D.; Smith, C.; *Helv. Chim. Acta*, **1979**, *62*, 7-20.

[56] Lam, K.S.; Hesler, G.A.; Gustavson, D.R; Crosswell, A.R.; Veitch, J.M.; Forenza, S.; Tomita, K.; *J. Antibiot.*, **1991**, *44*, 472-478.

[57] Hofstead, S.J.; Matson, J.A.; Malacko, A.R.; Marquardt, H.; *J. Antibiot.*, **1992**, *45*, 1250-1254.

[58] Zein, N.; Casazza, A.M.; Doyle, T.W.; Leet, J.E.; Schroeder, D.R.; Solomon, W.; Nadler, S.G.; *Proc. Natl. Acad. Sci. USA*, **1993**, *90*, 8009-8012.

[59] Hu, J.; Xue, Y.C.; Xie, M.Y.; Zhang, R.; Otani, T.; Minami, Y.; Yamada, Y.; Marunaka, T.; *J. Antibiot.*, **1988**, *41*, 1575-1579.

[60] Otani, T.; Minami, Y.; Marunaka, T.; Zhang, R.; Xie, M.Y.; *J. Antibiot.*, **1988**, *41*, 1580-1585.

[61] Iida, K.; Ishii, T.; Hirama, M.; Otani, T.; Minami, Y.; Yoshida, K.; *Tetrahedron Lett.*, **1993**, *34*, 4079-4082.

[62] Minami, Y.; Yoshida, K.; Azuma, R.; Saeki, M.; Otani, T.; *Tetrahedron Lett.*, **1993**, *34*, 2633-2636.

[63] Zhen, Y.; Ming, X.; Yu, B.; Otani, T.; Saito, H.; Yamada, Y.; *J. Antibiot.*, **1989**, *42*, 1294-1298.

[64] Kondo, H.; Nakajima, S.; Yamamoto, N.; Okura, A.; Satoh, F.; Suda, H.; Okanishi, M.; Tanaka, N.; *J. Antibiot.*, **1990**, *43*, 1533-1542.

[65] Konig, W.A.; Krauss, C.; Zahner, H.; *Helv. Chim. Acta*, **1977**, *60*, 2071-2078.

[66] Anyanwutaku, I.O.; Zirbes, E.; Rosazza, J.P.N.; *J. Nat. Prod.*, **1992**, *55*, 1498-1504.

[67] Komiyama, K.; Funayama, S.; Anraku, Y.; Mita, A.; Takahashi, Y.; Omura, S.; Shimasaki, H.; *J. Antibiot.*, **1989**, *42*, 1344-1349.

[68] Aizawa, S.; Akutsu, H.; Satomi, T.; Kawabata, S.; Sasaki, K.; *J. Antibiot.*, **1978**, *31*, 729-731.

[69] Narayanan, S.; Iyengar, M.R.S.; Ganju, P.L.; Rengaraju, S.; *J. Antibiot.*, **1980**, *33*, 1249-1255.

[70] Kuroda, Y.; Okuhara, M.; Goto, T.; Yamashita, M.; Iguchi, E.; Kohsaka, M.; Aoki, H.; Imanaka, H.; *J. Antibiot.*, **1980**, *33*, 259-266.

[71] Yoshida, H.; Arai, N.; Sugoh, M.; Iwabuchi, J.; Shiomi, K.; Shinose, M.; Tanaka, Y.; Omura, S.; *J. Antibiot.*, **1994**, *47*, 1165-1166.

[72] Argoudelis, A.D.; Herr, R.R.; Mason, D.J.; Pyke, T.R.; Zieser, J.F.; *Biochemistry*, **1967**, *6*, 165-170.

[73] Scannell, J.P.; Pruess, D.L.; Blount, J.F.; Ax, H.A.; Kellett, M.; Weiss, F.; Demmy, T.C.; Williams, T.H.; Stempel, A.; *J. Antibiot.*, **1975**, *28*, 1-6.

[74] Nisbet, L.J.; Mehta, R.J.; Oh, Y.; Pan, C.H.; Phelen, C.G.; Polansky, M.J.; Shearer, M.C.; Giovenella, A.J.; Grappel, S.F.; *J. Antibiot.*, **1985**, *38*, 133-138.

[75] Igarashi, M.; Nakamura, H.; Naganawa, H.; Takeuchi, T.; *J. Antibiot.*, **1997**, *50*, 1026-1031.

[76] Lackner, H.; Bahner, I.; Shigematsu, N.; Pannell, L.K.; Mauger, A.B.; *J. Nat. Prod.*, **2000**, *63*, 352-356.

[77] Omura, S.; Verpyl, D.; Inokoshi, J.; Takahashi, Y.; Takeshima, H.; *J. Antibiot.*, **1993**, *46*, 222-228.

[78] Shiomi, K.; Yang, H.; Inokoshi, J.; Verpyl, D.; Nakagawa, A.; Takeshima, H.; Omura, S.; *J. Antibiot.*, **1993**, *46*, 229-234.

[79] Dolak, L.A.; *J. Antibiot.*, **1973**, *26*, 121-125.

[80] Nagarajan, R.; *J. Antibiot.*, **1993**, *46*, 1181-1195.

[81] Iwasaki, H.; Horii, S.; Asai, M.; Mizuno, K.; Ueyanagi, J.; Miyake, A.; *Chem. Pharm. Bull.*, **1973**, *21*, 1184-1191.

[82] Ang, S.G.; Williamson, M.P.; Williams, D.H.; *J. Chem. Soc. Perkin Trans. I*, **1988**, 1949-1956.

[83] Boeck, L.D.; Mertz, F.P.; Clem, G.M.; *J. Antibiot.*, **1985**, *38*, 1-8.

[84] Kaneko, I.; Kamoshida, K.; Takahashi, S.; *J. Antibiot.*, **1989**, *42*, 236-241.

[85] Seto, H.; Fujioka, T.; Furihata, K.; Kaneko, I.; Takahashi, S.; *Tetrahedron Lett.*, **1989**, *30*, 4987-4990.

[86] Matsuzaki, K.; Ikeda, H.; Ogino, T.; Matsumoto, A.; Woodruff, H.B.; Tanaka, H.; Omura, S.; *J. Antibiot.*, **1994**, *47*, 1173-1174.

[87] Tanaka, H.; Matsuzaki, K.; Nakashima, H.; Ogino, T.; Matsumoto, A.; Ikeda, H.; Woodruff, H.B.; Omura, S.; *J. Antibiot.*, **1997**, *50*, 58-65.

[88] Rodriguez, M.J.; Snyder, N.J.; Zweifel, M.J.; Wilkie, S.C.; Stack, D.R.; Cooper, R.D.G.; Nicas, T.I.; Mullen, D.L.; Butler, T.F., Thompson, R.C.; *J. Antibiot.*, **1998**, *51*, 560-569.

[89] Snyder, N.J.; Cooper, R.D.G.; Briggs, B.S.; Zmijewski, M.; Mullen, D.L.; Kaiser, R.E.; Nicas, T.I.; *J. Antibiot.*, **1998**, *51*, 945-951.

[90] Umezawa, K.; Ikeda, Y.; Uchihata, Y.; Naganawa, H.; Kondo, S.; *J. Org. Chem.*, **2000**, *65*, 459-463.

[91] Hunt, A.H.; *J. Am. Chem. Soc.*, **1983**, *105*, 4463-4468.

[92] Yasuzawa, T.; Iida, T.; Muroi, K.; Ichimura, M.; Takahashi, K.; Sano, H.; *Chem. Pharm. Bull.*, **1988**, *36*, 3728-3731.

[93] Takahashi, I.; Takahashi, K.; Ichimura, M.; Morimoto, M.; Asano, K.; Kawamoto, I.; Tomita, F.; Nakano, H.; *J. Antibiot.*, **1988**, *41*, 1915-1917.

[94] Nagamura, S.; Saito, H.; *Khimiya Geterotsiklicheskikh Soedinenii*, **1998**, 1636-1657.

[95] Cacciari, B.; Romagnoli, R.; Baraldi, P.G.; Da Ros, T., Spalluto, G.; *Expert Opinion Therapeutic Patents*, **2000**, *10*, 1853-1871.

[96] Ishii, S.; Nagasawa, M.; Kariya, Y.; Yamamoto, H.; Inouye, S.; Kondo, S.; *J. Antibiot.*, **1989**, *42*, 1713-1717.

[97] Ogawa, T.; Ichimura, M.; Katsumata, S.; Morimoto, M.; Takahashi, K.; *J. Antibiot.*, **1989**, *42*, 1299-1301.

[98] Yasuzawa, T.; Muroi, K.; Ichimura, M.; Takahashi, I.; Ogawa, T.; Takahashi, K.; Sano, H.; Saitoh, Y.; *Chem. Pharm. Bull.*, **1995**, *43*, 378-391.

[99] Ohba, K.; Watabe, H.; Sasaki, T.; Takeuchi, Y.; Kodama, Y.; Nakazawa, T.; Yamamoto, H.; Shomura, T.; Sezaki, M.; Kondo, S.; *J. Antibiot.*, **1988**, *41*, 1515-1519.

352

[100] Ding, W.D.; Williams, D.R.; Northcote, P.; Siegel, M.M.; Tsao, R.; Ashcroft, J.; Morton, G.O.; Alluri, M.; Abbanat, D.; Maiese, W.M.; Ellestad, G.A.; *J. Antibiot.*, **1994**, *47,* 1250-1257.

[101] Singh, M.P.; Petersen, P.J.; Jacobus, N.V.; Mroczenskiwildey, M.J.; Maiese, W.M.; Greenstein, M.; Steinberg, D.A.; *J. Antibiot.*, **1994**, *47,* 1258-1265.

[102] Abbanat, D.; Maiese, W.; Greenstein, M.; *J. Antibiot.*, **1999**, *52*, 117-126.

[103] Pearce, C.M.; Sers, J.K.M.; *J. Chem. Soc. Perkin Trans. I*, **1990**, 409-411.

[104] Omura, S.; Nakagawa, A.; *Tetrahedron Lett.*, **1981**, *22*, 2199-2202.

[105] Kim, W.G.; Kim, J.P.; Yoo, I.D.; *J. Antibiot.*, **1996**, *49*, 26-30.

[106] Yano, K.; Oono, J.; Mogi, K.; Asaoka, T.; Nakashima, T.; *J. Antibiot.*, **1987**, *40*, 961-969.

[107] Martin, D.G.; Duchamp, D.J.; Chidester, C.G.; *Tetrahedron Lett.*, **1973**, *27*, 2549-2552.

[108] Martin, D.G.; Chidester, C.G.; Mizsak, S.A.; Duchamp, D.J.; Baczynskyj, L.; Krueger, W.C.; Wnuk, R.J.; Meulman, P.A.; *J. Antibiot.*, **1975**, *28*, 91-93.

[109] Takahashi, E.; Beppu, T.; *J. Antibiot.*, **1982**, *35*, 939-947.

[110] Isono, K.; Uramoto, M.; Kusakabe, H.; Miyata, N.; Koyama, T.; Ubukata, M.; Sethi, S.K.; McCloskey, J.A.; *J. Antibiot.*, **1984**, *37*, 670-672.

[111] Miura, K.; Hamada, M.; Takeuchi, T.; Umezawa, H.; *J. Antibiot.*, **1979**, *32*, 762-765.

[112] Dolak, L.A.; DeBoer, C.; *J. Antibiot.*, **1980**, *33*, 83-84.

[113] Ezaki, N.; Shomura, T.; Koyama, M.; Niwa, T.; Kojima, M.; Inouye, S.; Ito, T.; Niida, T.; *J. Antibiot.*, **1981**, *34*, 1363-1365.

[114] Koyama, M.; Kodama, Y.; Tsuruoka, T.; Ezaki, N.; Niwa, T.; Inouye, S.; *J. Antibiot.*, **1981**, *34,* 1569-1576.

[115] Kaneda, M.; Nakamura, S.; Ezaki, N.; Iitaka, Y.; *J. Antibiot.*, **1981**, *34*, 1366-1368.

[116] Nogami, T.; Shigihara, Y.; Matsuda, N.; Takahashi, Y.; Naganawa, H.; Nakamura, H.; Hamada, M.; Muraoka, Y.; Takita, T.; Iitaka, Y.; Takeuchi, T.; *J. Antibiot.*, **1990**, *43*, 1192-1194.

[117] Trew, S.J.; Wrigley, S.K.; Pairet, L.; Sohal, J.; Shanu-Wilson, P.; Hayes, M.A.; Martin, S.M.; Manohar, R.N.; Chicarelli-Robinson, M.I.; Kau, D.A.; Byrne, C.V.; Wellington, E.M.H.; Moloney, J.M.; Howard, J.; Hupe, D.; Olson, E.R.; *J. Antibiot.*, **2000**, *53*, 1-11.

[118] Di Santo, R.; Costi, R.; Artico, M.; Massa, S.; Lampis, G.; Deidda, D.; Pompei, R.; *Biorg. Med. Chem. Lett.*, **1998**, *8*, 2931-2936.

[119] Gitterman, C.O.; Rickes, E.L.; Wolf, D.E.; Madas, J.; Zimmerman, S.B.; Stoudt, T.H.; Demny, T.C.; *J. Antibiot.*, **1970**, *23,* 305-310.

[120] O'Hagan, D.; Harper, D.B.; *J. Fluorine Chem.*, **1999**, *100,* 127-133.

[121] Morton, G.O.; Lancaster, J.E.; Van Lear, G.E.; Fulmor, W.; Meyer, W.E.; *J. Am. Chem. Soc.*, **1969**, *91*, 1535-1537.

[122] Jenkins, I.D.; Verheyden, J.P.H.; Moffatt, J.G.; *J. Am. Chem. Soc.*, **1976**, *98*, 3346-3357.

[123] Maguire, A.R.; Meng, W.D.; Roberts, S.M.; Willetts, A.J.; *J. Chem. Soc. Perkin Trans. I*, **1993**, 1795-1808.

[124] Yang, S.W.; Cordell, G.A.; *J. Nat. Prod.*, **1997**, *60*, 44-48.

[125] Yang, S.W.; Cordell, G.A.; *J. Nat. Prod.*, **1997**, *60*, 230-235.

[126] Adams, E.S.; Rinehart, K.L.; *J. Antibiot.*, **1994**, *47*, 1456-1465.

[127] Koch, K.; *US Patent* No. 5612316, **1997**.

[128] Fenical, W.; *Chem. Rev.* **1993**, *93*, 1673-1683.

[129] Hutchinson, C.R.; McDaniel, R.; *Curr Opin Investig Drugs.* **2001**, *2,* 1681-1690.

Atta-ur-Rahman (Ed.) *Studies in Natural Products Chemistry, Vol. 29*

NATURAL BRIDGED BIARYLS WITH AXIAL CHIRALITY AND ANTIMITOTIC PROPERTIES

OLIVIER BAUDOIN AND FRANÇOISE GUÉRITTE

*Institut de Chimie des Substances Naturelles, CNRS,
avenue de la Terrasse, 91198 Gif-sur-Yvette cedex, France
baudoin@icsn.cnrs-gif.fr / gueritte@icsn.cnrs-gif.fr*

ABSTRACT: Numerous molecules containing a biaryl unit have been found in Nature. Many of them exhibit a variety of biological actions. In this review, we will focus only on natural bridged biaryls that possess axial chirality and exhibit antimitotic activities. Thus, we will first present the occurrence and structure of three families of natural compounds that display these particular structural and biological features: the allocolchicinoids, steganes and rhazinilam-type compounds. We will then describe the semi-synthetic and synthetic approaches to biaryls belonging to these three series. Their interaction with tubulin, a heterodimeric protein that is critical for the formation of microtubules and consequently of the mitotic spindle, will be discussed.

1. INTRODUCTION

Natural biaryls, bridged or unbridged, show remarkable variations in their structure and biological activity [1]. Some of these exhibit antimitotic properties, inhibiting cell division, and are thus potential anticancer compounds. Antimitotic agents interact with tubulin or its assembled form, microtubules, which are involved in the major process of cell division [2]. The drugs designed to interact with these targets have proved to be extremely useful in cancer chemotherapy. The vinca alkaloids (vinblastine, vincristine and vinorelbine) together with the taxoids (paclitaxel and docetaxel) represent two of the main series of clinically used antimitotic agents. This review will focus on the three families of axially chiral natural biaryls that are known to interact with tubulin (allocolchicinoids, steganes and rhazinilam-type compounds) represented by the lead compounds allocolchicine **1**, steganacin **2** and rhazinilam **3**, respectively. These bridged biaryls possess, respectively, a seven-, eight- and nine-membered median ring.

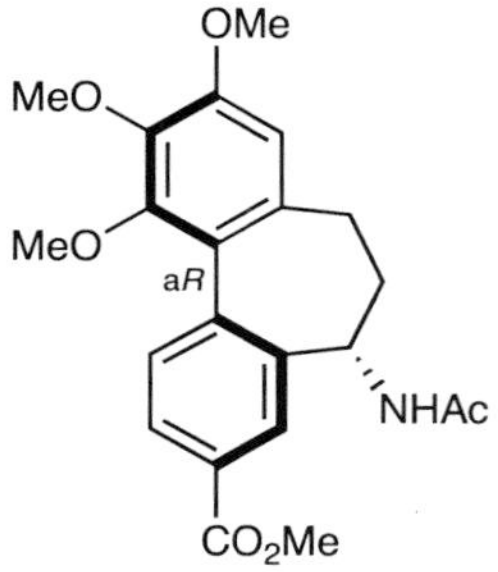

1: (–)-allocolchicine 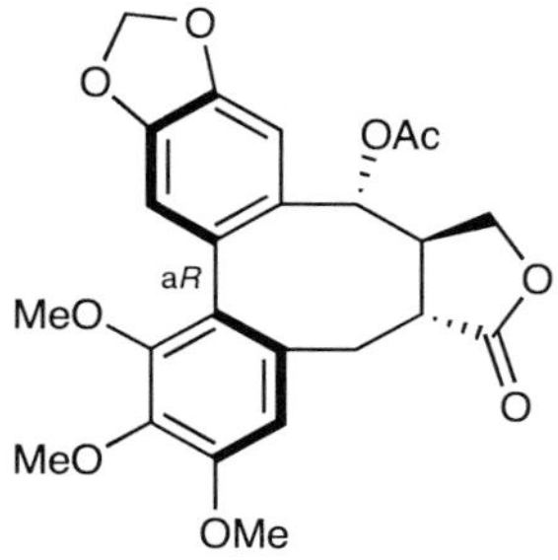   2: (–)-steganacin 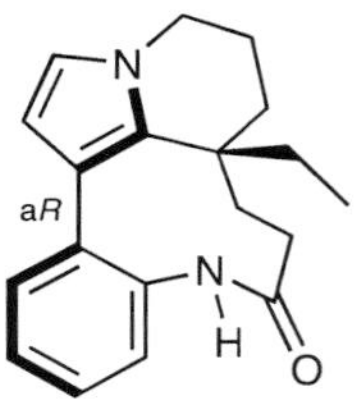   3: (–)-rhazinilam

The common structural feature of these natural substances is the presence of a chiral biaryl axis arising from the presence of the bridging ring and, in the case of allocolchicinoids and steganes, from the steric hindrance generated by the methoxy substituents located *ortho* to the biaryl axis. This particular substitution pattern hinders or prevents free rotation around the biaryl axis, leading to atropisomerism. The axial configuration of (–)-allocolchicine **1** is the same as that of (–)-steganacin **2** and (–)-rhazinilam **3** and is designated as a*R* or *M* (see Ref [1] for the stereochemical descriptors of chiral biaryl axis).

This review, after covering general aspects concerning the structure, occurrence and biological activity in the allocolchicinoid, stegane and rhazinilam series, will focus on recent work related to the semi-synthesis and total synthesis of these molecules together with structure-activity relationships.

2. STRUCTURE, OCCURRENCE AND BIOLOGICAL ACTIVITY

2.1. Allocolchicinoids

Allocolchicine **1** is the oldest known natural allocolchicinoid, isolated originally from *Colchicum autumnale* L. as substance O [3,4], then from *Colchicum kesselringii* Rgl. as substance K3 [5]. The former plant also produces colchicine **4** as the major alkaloid, which has been thought to be a biogenetic precursor of **1** (see Section **3.1.1**). Although colchicine **4** is not, *stricto sensu*, a biaryl compound (the tropolone ring does not present the character of aromaticity), it is important to underline its role not only in the determination of the configurational assignment of chiral bridged biaryl systems, but also in the discovery of tubulin, the protein subunit of microtubules.

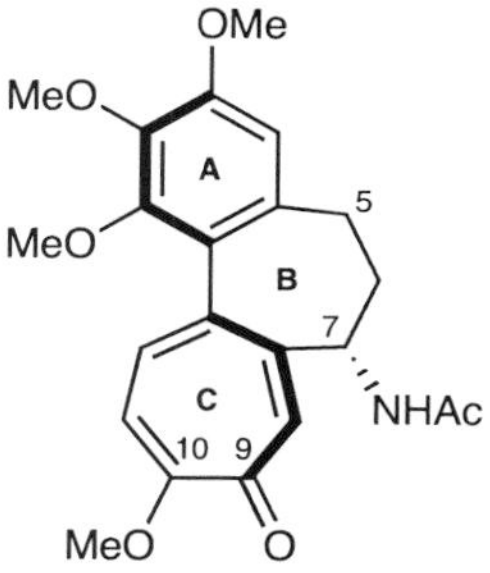

4: (–)-(aR,7S)-colchicine   5: (+)-(aS,7R)-colchicine

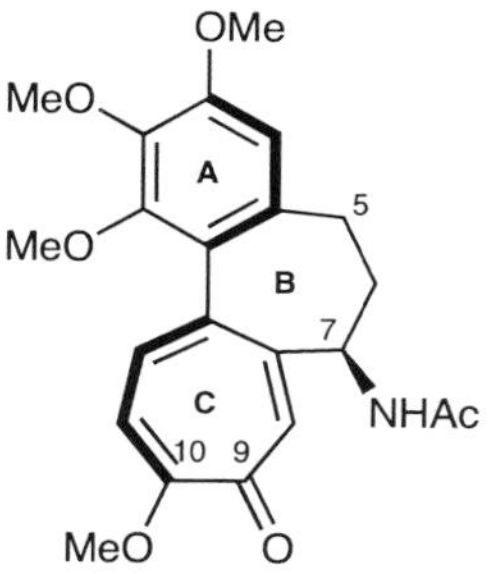

Isolated first in 1820 [6], colchicine **4** was recognized as the active principle of *C. autumnale* whose extracts have been used over the past 2000 years in the treatment of gout. Colchicine, still used today as an antigout agent, has also a unique place among the antimitotic agents since it was the first tubulin-binding agent to be discovered. Thanks to the work of Taylor, Borisy and Shelanski, the antimitotic property of colchicine mentioned by Pernice in 1889 [7] was linked to its interaction with tubulin in the late sixties [8,9]. Colchicine was shown to be an inhibitor of tubulin assembly into microtubules, but too toxic to be used as an anticancer drug. Since the discovery of colchicine as the first agent interacting with tubulin, it was used extensively as a tool in the preparation of purified tubulin as well as in the study of mitotic processes and antimitotic drug action [10,11,12]. The process of colchicine binding to tubulin as well as the determination of structure-activity relationships (SAR) at the colchicine binding site has been extensively reviewed [13,14,15]. To summarize briefly its antimitotic effect, natural (–)-colchicine binds in an irreversible manner to a high affinity site on the tubulin heterodimer inducing a conformational change within the protein [16]. Formation of the tubulin-colchicine complex leads to the inhibition of tubulin assembly and, consequently, to potent antimitotic activity.

From a structural point of view, natural (–)-colchicine **4** is a tricyclic compound possessing a trimethoxyphenyl ring, a seven-membered bridging ring substituted with an acetamide group and a tropolone ring. The relative configuration of natural (–)-colchicine was determined by X-ray analysis [17,18] and showed two degrees of chirality: the stereogenic center at C-7 and the axial chirality due to the restricted rotation around the tropolone-phenyl bond. The absolute configuration at C-7 was deduced by chemical degradation to a known chiral amino acid [19] and confirmed by the X-ray analysis of 2-acetyl-2-demethylthiocolchicine [20]. A number of X-ray structures has been obtained for various colchicinoids [14] but the absolute configuration of the tropolone-phenyl axis of natural colchicine **4** was only firmly confirmed by the X-ray analysis of a urea derivative of N-acetylcolchinol, a synthetic member of the allocolchicinoid series, and the three-dimensional structure was correlated to NMR and circular dichroism data [21]. Long described as (a*S*, 7S), the configuration of natural (−)-colchicine was recently corrected as being (a*R*, 7S), following the *CIP* rules adopted by *IUPAC* [22,23,24]. Therefore, the axial configuration of all (−)-colchicinoids, as well as (−)-allocolchicinoids described before 1999 as (a*S*) has to be changed to (a*R*). Extensive studies in the colchicinoid series have highlighted the crucial influence of the absolute configuration of the biaryl axis and of the conformation of the seven-membered bridging ring on the binding to tubulin [13,14,15]. For example, the dextrogyre unnatural (+)-(a*S*, 7*R*)-colchicine **5** as well as its deacetamido derivative were found to be less active in inhibiting tubulin assembly than natural (−)-colchicine **4** [25,26,27].

To the best of our knowledge few natural (−)-allocolchicinoids have been reported. As mentioned above, natural (−)-allocolchicine **1** was isolated from *C. autumnale* [3,4] and *C. kesselringii* [5]. 3-Demethylallocolchicine **6**, identified previously as the methyl ester of 2-demethylcolchicinic acid, was also isolated from the latter plant [5]. Other natural allocolchicinoids, also named dibenzocycloheptylamines, have been reported. Colchibiphenyline **7** was isolated from *Colchicum ritchii* R. Br. [28] and (−)-androbiphenyline **8** from *Androcymbium palaestinum* (Boiss.) Bak. [29], *C. ritchii* [28] and *Colchicum decaisnei* Boiss. [30]. The latter plant also produces (−)-jerusalemine **9**, (−)-salimine and (−)-suhailamine [30], though the exact structure of salimine was later questioned [31] and suhailamine had a reported structure identical to that of allocolchicine.

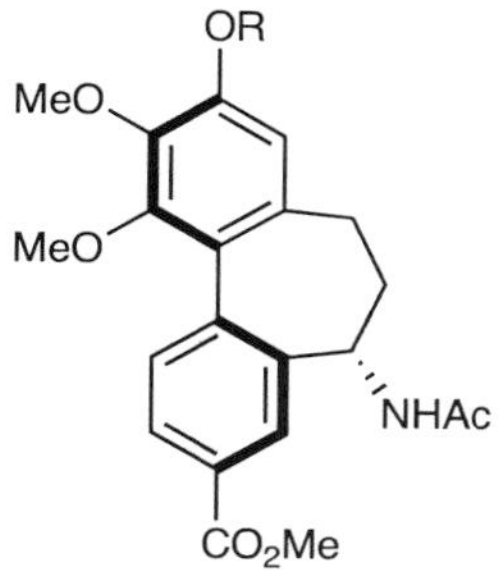

1 (R= Me)
6 (R= H)

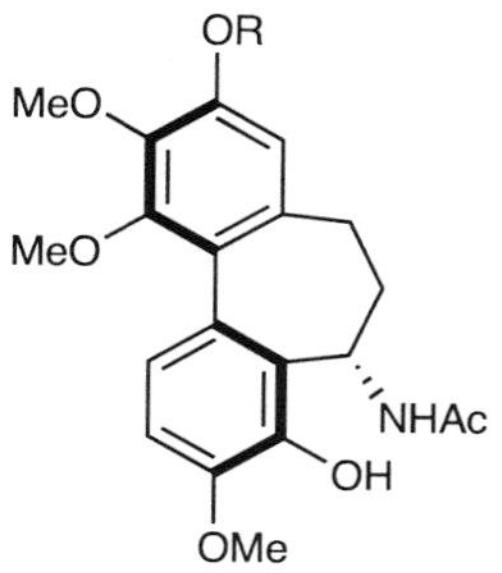

7: (R= H) colchibiphenyline
8: (R= Me) androbiphenyline

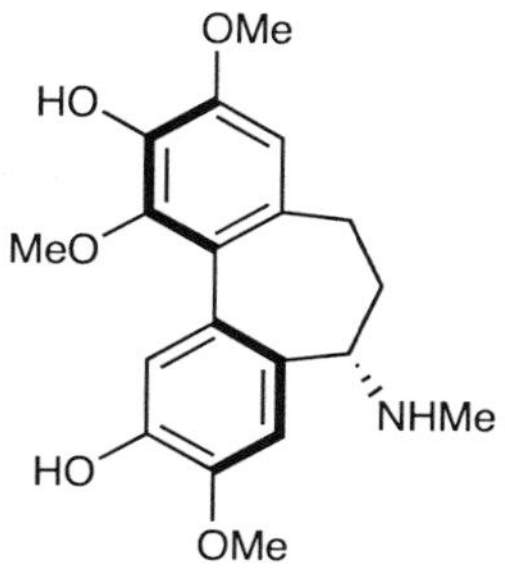

9: jerusalemine

The X-ray analysis of (–)-allocolchicine **1**, Fig. (**1**), showed a similar conformation to that of (–)-colchicine **4** [32]. The seven-membered ring adopts a slightly more flattened boat conformation than that of the solid-state conformation of colchicine, and the dihedral angle between the two phenyl rings is 48.7°. Natural (–)-allocolchicine **1** was shown to bind to the colchicine binding site on tubulin with a higher affinity than colchicine [33,34]. The SAR studies in the allocolchicinoid series, using semi-synthetic or synthetic analogues of allocolchicine, will be reviewed in section **3.1.1**.

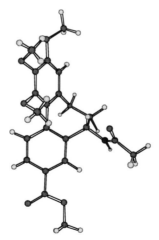

Fig. (1). X-ray crystal structure of (–)-allocolchicine **1** [32]

2.2. Steganes

The steganes are natural antimitotic compounds that structurally belong to the dibenzocyclooctadiene lignans. Several reviews have underlined the occurrence, biogenesis and structural variety of this family encompassing more than 100 substances [1,35]. Among these, the members of the stegane series share in common the presence of a lactone ring fused to an eight-membered bridge linking two polyoxygenated phenyl units. To our knowledge, the Apiaceae *Steganotaenia araliaceae* Hochst. is the only source of the dibenzocyclooctadiene lactones. Four steganes were first isolated by purification of a cytotoxic extract of *S. araliaceae*: (–)-steganacin **2**, (–)-steganangin **11**, (–)-steganone **12** and (–)-steganol **13** [36]. The absolute configuration of these natural steganes was originally described as the antipodal structures of **2**, **11-13** [36]. Thus, structure **14** was first attributed to (–)-steganacin. This was later corrected by two independent asymmetric syntheses of natural (–)-steganone [37] and unnatural (+)- and natural (–)-steganacin [38,39] (see Section **3.2.2.**).

(–)-Steganacin, the major dibenzocyclooctadiene lignan isolated from *S. araliaceae*, was shown to inhibit tubulin assembly and to bind to the colchicine binding site of tubulin [40,41]. This compound also possesses significant cytotoxicity and *in vivo* antitumor activity [36]. The antitubulin activity as well as the cytotoxicity of the steganes were linked to the axial chirality which is similar to that of allocolchicinoids [42,43]. A few other dibenzocyclooctadiene lactone lignans have been isolated from *S. araliaceae*: araliangin **15** [44], neoisostegane **16** [45,46,47] steganolides A **17** [48], B **18** and C **19** [49], episteganangin **20** [50] and 10-demethoxystegane **21** [51]. In addition, steganoates **22** and **23**, possessing an opened lactone ring, have also been found in the same species [50].

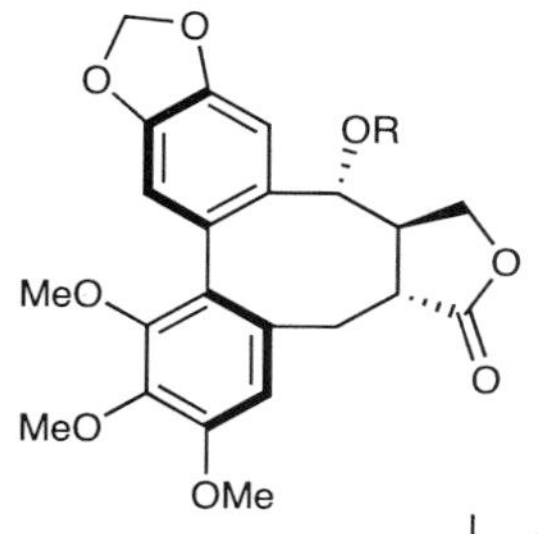

2: steganacin (R=Ac)
11: steganangin (R= -OC...)
13: steganol (R=H)

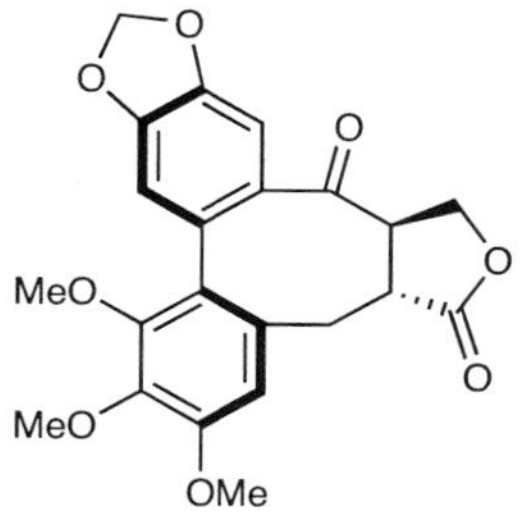

12: steganone

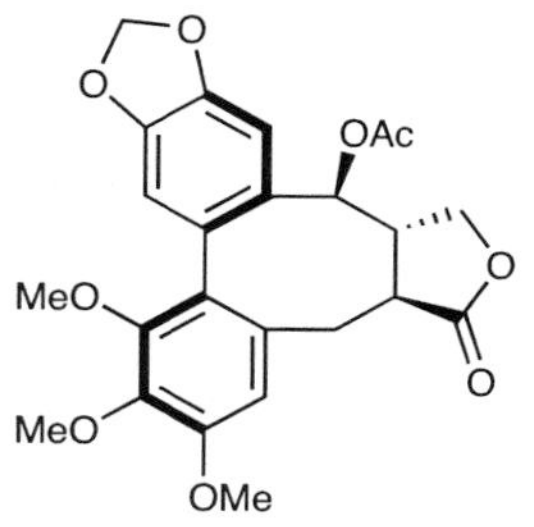

14

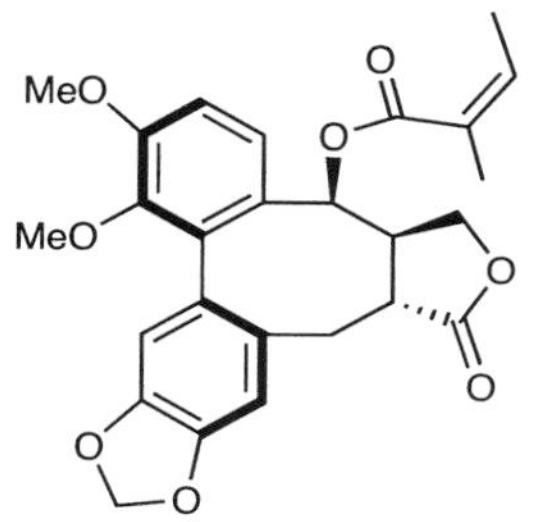

15: araliangin

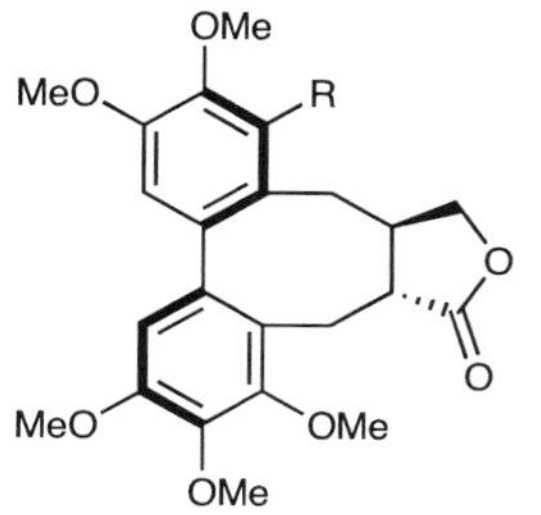

16: neoisostegane (R=H)
17: steganolide A (R=Me)

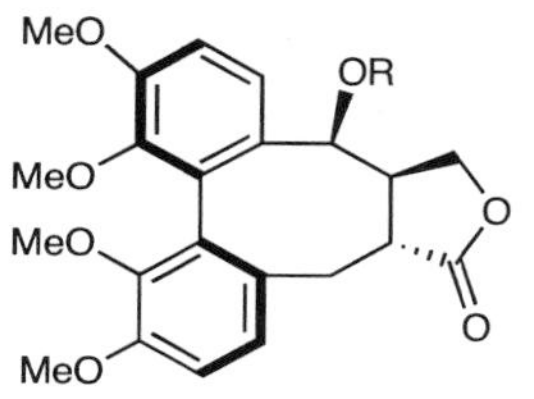

18: steganolide B (R= -OC...)

19: steganolide C (R=-OC...)

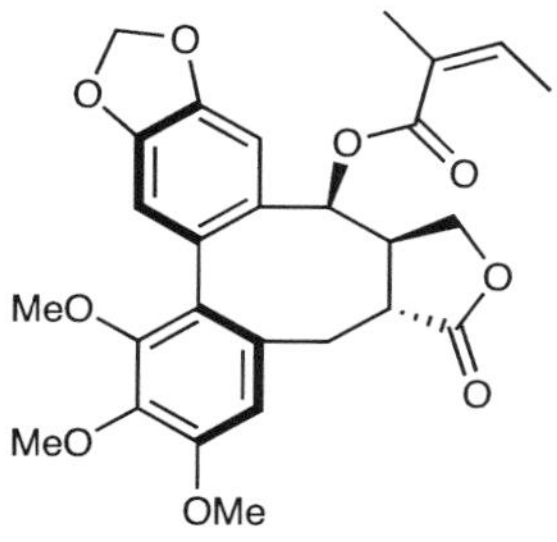

20: episteganangin

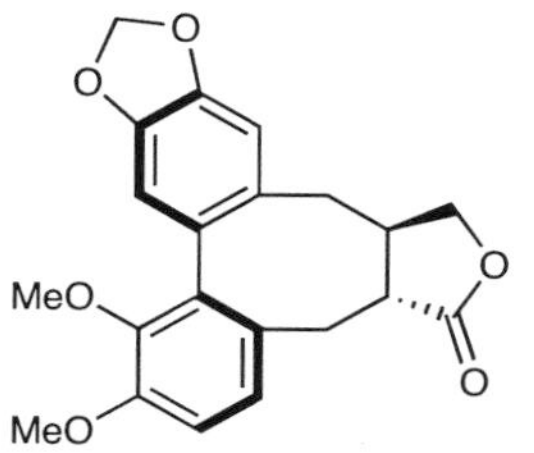

21: 10-demethoxystegane

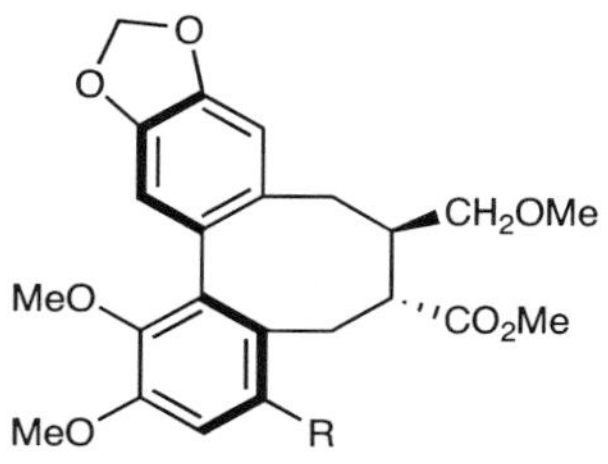

22: steganoate A (R=H)
23: steganoate B (R=OMe)

The first published X-ray structure of a stegane was that of (−)-9-episteganol obtained from natural (−)-steganone by sodium borohydride reduction [36], but no crystallographic data were provided. Fig. (**2**) shows the three-dimensional structure of one unit of the dimeric distegyl ether **24** formed by coupling of two molecules of (−)-steganol [52]. The eight-membered ring adopts a twist-boat conformation and the dihedral angle between the two phenyl rings is about 71°.

362

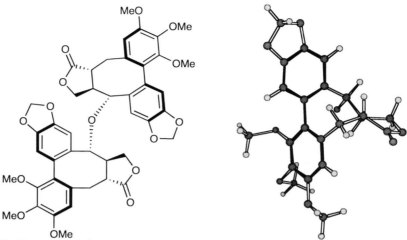

Fig. (2). X-ray crystal structure of distegyl ether **24** (one steganol subunit has been omitted for clarity) [52]

The X-ray analysis of other steganes allowed to show that the conformation of the median ring depends on the axial chirality with regard to the other stereogenic centers. Thus, for compounds with opposite axial configuration to that of levogyre steganes, such as **16** and **17**, the median ring adopts a twist-boat-chair conformation [1].

2.3. Rhazinilam

(–)-Rhazinilam **3** is an axially chiral phenyl-pyrrole compound which was first isolated from *Melodinus australis* (F. Muell.) Pierre [53]. It has also been found in other Apocynaceae such as *Rhazya stricta* Decaisne [54,55], *Aspidosperma quebracho-blanco* Schlecht. [56,57], *Leuconitis eugenifolia* A. DC [58,59] and *Kopsia singapurensis* [60]. More recently, (–)-rhazinilam was isolated from intergeneric somatic hybrid cell cultures of two Apocynaceae, *Rauvolfia serpentina* Benth. Ex Kurz and *Rhazia stricta* Decaisne [61].

(–)-Rhazinilam **3** is characterized by the presence of four rings: the phenyl A-ring, the nine-membered lactam B-ring, the pyrrole C-ring and the piperidine D-ring. According to X-ray data, the A-C dihedral angle of (–)-rhazinilam **3** is 90°, the amide bond possesses a *cis* conformation, and the median ring exists in a boat-chair conformation, Fig. **(3)** [56]. The absolute configurations at C-20 and at the phenyl-pyrrole axis were deduced by semi-synthesis from (+)-1,2-didehydroaspidospermidine [62] (See Section **3.1.3.**).

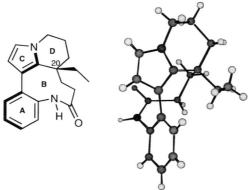

Fig. (3). X-ray crystal structure of (–)-rhazinilam **3** [56]

Rhazinilam was found to be an oxidative artefact derived from the unstable natural 5,21-dihydrorhazinilam **25** [58,59,60]. Other compounds of the (–)-rhazinilam series have been isolated from various Apocynaceae. (–)-3-Oxorhazinilam **26**, also called rhazinicine, was founded in *Kopsia dasyrachis* Ridl. [63] and in hybrid plant cell cultures [64].

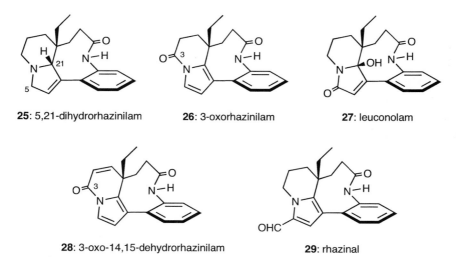

25: 5,21-dihydrorhazinilam   **26**: 3-oxorhazinilam   **27**: leuconolam

28: 3-oxo-14,15-dehydrorhazinilam   **29**: rhazinal

Leuconolam **27** was isolated from different species of *Leuconotis* [58,59], 3-oxo-14,15-dehydrorhazinilam **28** was found in cell suspension cultures of *A. quebracho-blanco* [65], and (–)-rhazinal **29** was obtained from a *Kopsia* sp. [66].

The tubulin-binding properties of (–)-rhazinilam were discovered through screening of a number of Malaysian plant extracts [60]. Natural (–)-rhazinilam induces tubulin spiralization, inhibiting tubulin assembly in the same way as vinblastine-like alkaloids, and protects microtubules from cold disassembly such as with paclitaxel [67]. This effect has never been observed with other microtubule poisons. For this reason, and despite the *in vivo* inactivity of (–)-rhazinilam [67], a number of analogues have been prepared by semi-synthesis and total synthesis (see Sections **3.1.3.** and **3.2.3.**) in order to improve the pharmacological properties of this molecule.

3. SYNTHESIS OF NATURAL COMPOUNDS AND ANALOGUES

3.1. Biogenesis and semi-synthesis

3.1.1. Allocolchicinoids

Nucleophile-induced contraction of the tropolone ring of colchicine

Degradation studies of the tropolone C-ring of (–)-colchicine **4** were extensively conducted in the mid 20th century [for reviews: 14,68,69]. In particular, treatment of colchicine or isocolchicine with sodium methoxide in refluxing methanol produces allocolchicine **1** [70,71]. This finding, along with the fact that both alkaloids occur in the same plant, support the hypothesis that allocolchicine **1** is biosynthetically derived from colchicine **4**.

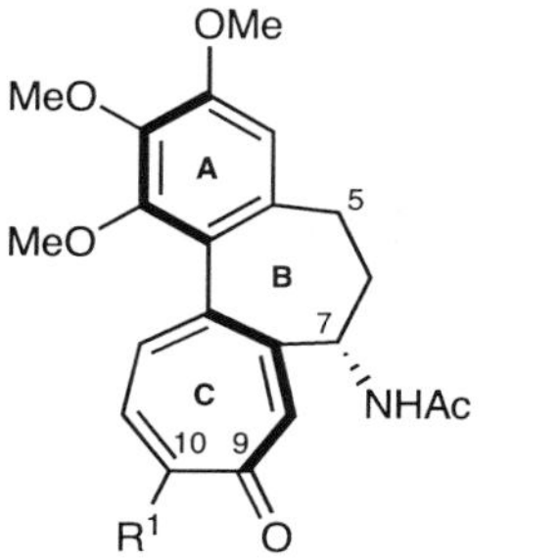

4: (R^1=OMe) colchicine
33: (R^1=OH) colchiceine
34: (R^1=SMe) thiocolchicine

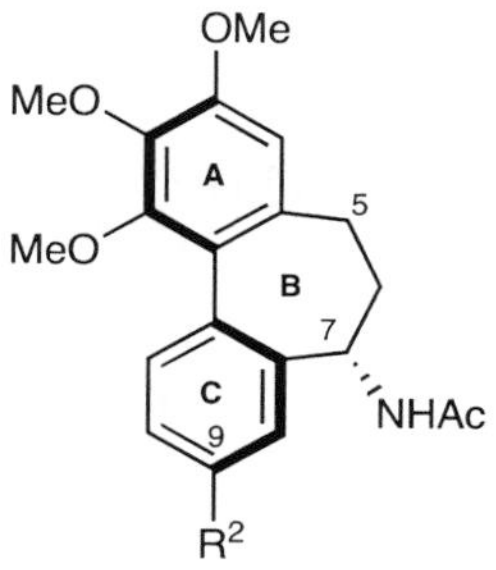

1: (R^2=CO$_2$Me) allocolchicine
30: (R^2=CO$_2$H) allocolchiceine
31: (R^2=OH) N-acetylcolchinol
32: (R^2=OMe) N-acetylcolchinol
methyl ether

According to the authors, the tropolone ring contraction giving **1** may arise from rearrangement of the methoxide adduct **A** [70]. Alternatively, one could propose that **A** would be prone to form the bicyclo[4.1.0] intermediate **B** (by analogy with intermediate **C**, *vide infra*), which would then rearrange to give **1**.

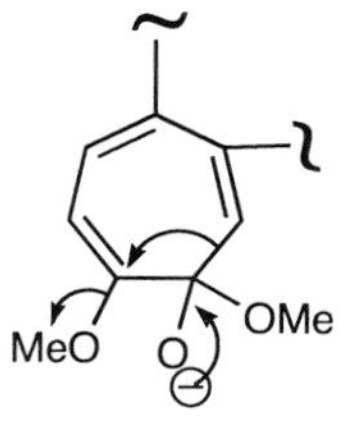

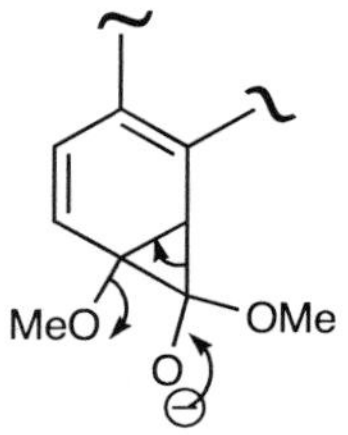

Allocolchicine and its congeners such as allocolchiceine **30**, N-acetylcolchinol **31** and its methyl ether **32** have played an important role in the investigation of the SAR in the colchicine series and have been used as standards for the tubulin assay [for reviews: 11,13]. These congeners were also obtained semi-synthetically from colchicinoid precursors. For instance, treatment of colchiceine **33**, obtained from acid hydrolysis of **4**, with hydrogen peroxide in alkaline medium yielded N-acetylcolchinol **31**, which was methylated with diazomethane to give the methyl ether **32** [72]. Alternatively, **31** was produced from allocolchiceine **30** by Schmidt rearrangement (NaN$_3$, concd. H$_2$SO$_4$), diazotization of the intermediate aniline with nitrous acid followed by hydrolysis [70]. More recently, colchicine has been directly converted to *N*-acetylcolchinol methyl ether **32** by treatment with 30% aqueous hydrogen peroxide [73]. This oxidation was proposed to proceed *via* bicyclo [4.1.0] intermediate **C**, and subsequently **D**, which underwent decarboxylation to afford **32**.

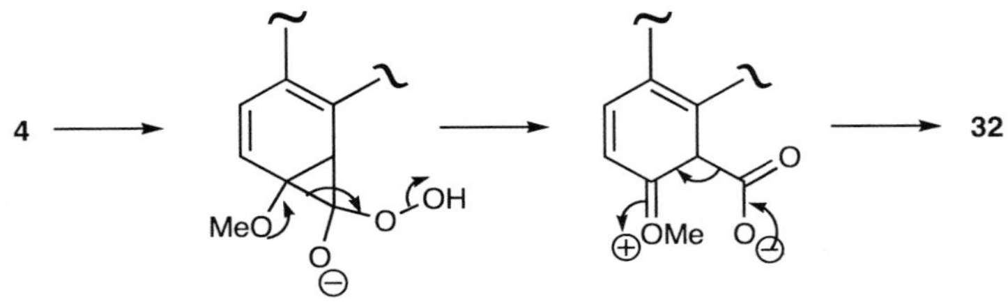

These simple semi-synthetic procedures have been extensively used in the preparation of new allocolchicinoids for biological screening purposes [11]. Recently, these molecules found a new interest when it was shown by scientists at Angiogene Pharmaceuticals that *N*-acetylcolchinol phosphate **35**, termed ZD6126, is a prodrug of *N*-acetylcolchinol **31** which causes the selective destruction of tumour vasculature [74]. This compound, obtained semi-synthetically from colchicine, is currently under phase I clinical trials. Another patent from Brandeis University described methyl ketone **36** as a reversible inhibitor of tubulin polymerization (in contrast to colchicine), with 100-fold cytotoxicity compared to colchicine toward a number of cell lines [75].

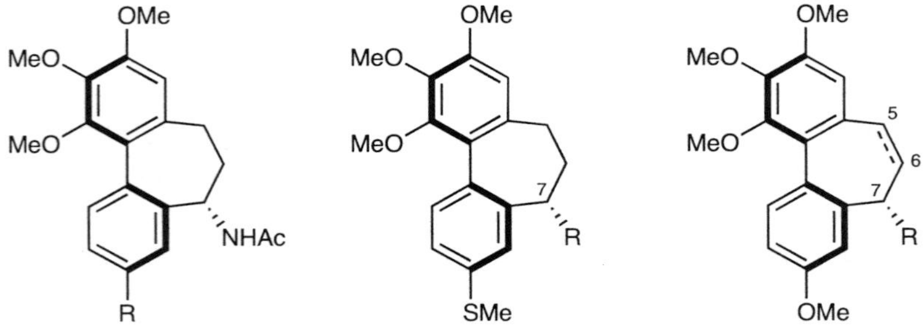

35: R=OPO$_3$H$_2$
36: R=COCH$_3$

37: R=(=O)
38: R=OH

39: R=(=O)
40: R=(=O), $\Delta^{5,6}$
41: R=NHCOCH$_2$Cl

The Brossi and Lee groups have extensively studied SAR of allocolchicinoids prepared by semi-synthesis [11,14]. In particular, a series of allo*thio*colchicinoids were semi-synthetically obtained starting from thiocolchicine **34** by aniline-induced ring contraction of thiocolchicone[†] [76,77]. Among them, the C-7 analogues **37-38** showed strong inhibition of tubulin assembly and cytotoxicity toward cancer cells. For instance, ketone **37** had an IC_{50} of 1.7 µM for microtubules assembly inhibition (compared to 2.1 µM for reference compound **32**), and nanomolar IC_{50} values for several cancer cell lines (2.5 to 4-fold more potent than colchicine). More recently, it was found that the allocolchicine analogues **39** and **40**, obtained semi-synthetically from **32**, display antimitotic properties equipotent to **32** and potent cytotoxicity toward various cancer cell lines [78]. Moreover, chloroacetamide analogue **41**, designed to form a covalent bond with tubulin, showed potent cytotoxicity against KB-VIN, a multi-drug resistant cell line [79].

Other oxidations of the tropolone ring

Schröder and co-workers studied the addition of formaldehyde-*O*-oxide, generated *in situ* at low temperature by ozonolysis of ketene diethylacetal, to the tropolonic ring of various colchicine analogues, Fig. (**4**) [80]. The intermediate ozonide **E** formed from **4** decomposed upon warming to give *N*-acetylcolchinol methyl ether **32** *via* the presumed intermediate **F**, in 81% yield.

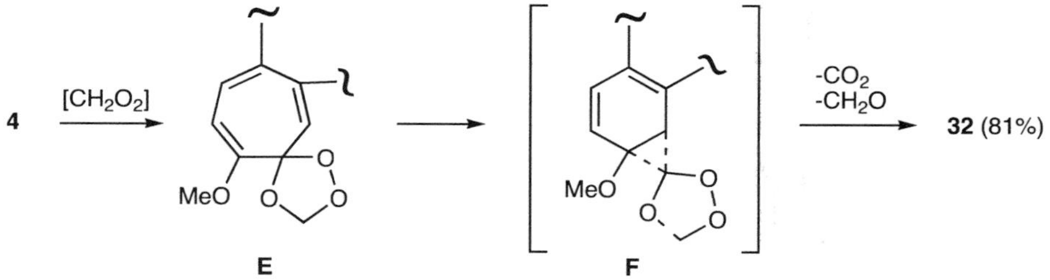

Fig. (4). Oxidation of colchicine with formaldehyde-*O*-oxide [80]

The photooxidation of colchicine with singlet oxygen has been reported by Seitz and co-workers, Fig. (**5**) [22]. This [4+2] cycloaddition produced the endoperoxide **42** stereoselectively, which could be isolated in 85% yield. Endoperoxide ring opening with triphenylphosphine

---

[†] thiocolchicine having a keto group in the C-7 position

triggered a sequence of domino reactions leading to *N*-acetylcolchinol methyl ether **32** in 40% yield from **4**. Alternatively, methanolysis of **42** in the presence of silica afforded the natural allocolchicinoid androbiphenyline **8** in 60% overall yield. This impressive chemical process was exploited to generate novel allocolchicinoids having an eight membered B-ring [81].

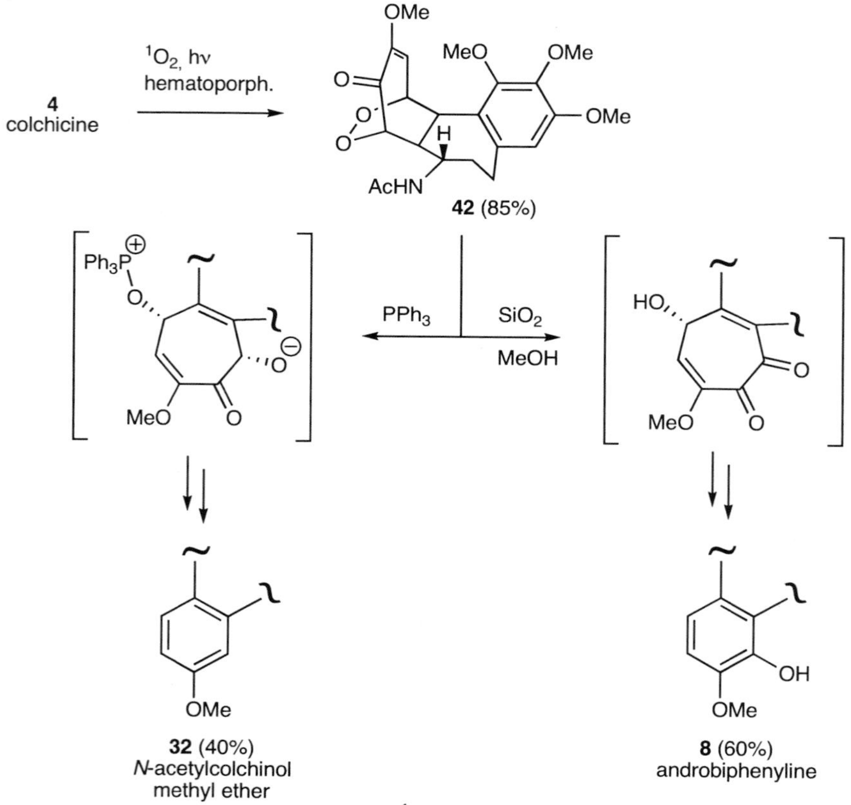

Fig. (5). Photooxidation of colchicine with $^1O_2$ [22]

From the above literature studies, it seems clear that semi-synthesis has turned out to be the method of choice to generate a whole range of novel allocolchicinoid compounds from readily available colchicine precursors. Some of these synthetic allocolchicinoids have very potent antimitotic properties, and they have been used as tools to study tubulin-ligand interactions. The recent discovery of phosphate **35** as a prodrug with anti-angiogenesis properties illustrates the ongoing interest in this long-studied class of biaryl molecules.

3.1.2. Steganes

The stegane lignans are thought to originate biogenetically from the same pathway as the podophyllotoxin lignans, a common precursor being the dibenzylbutyrolactone yatein, Fig. (**6**) [35].

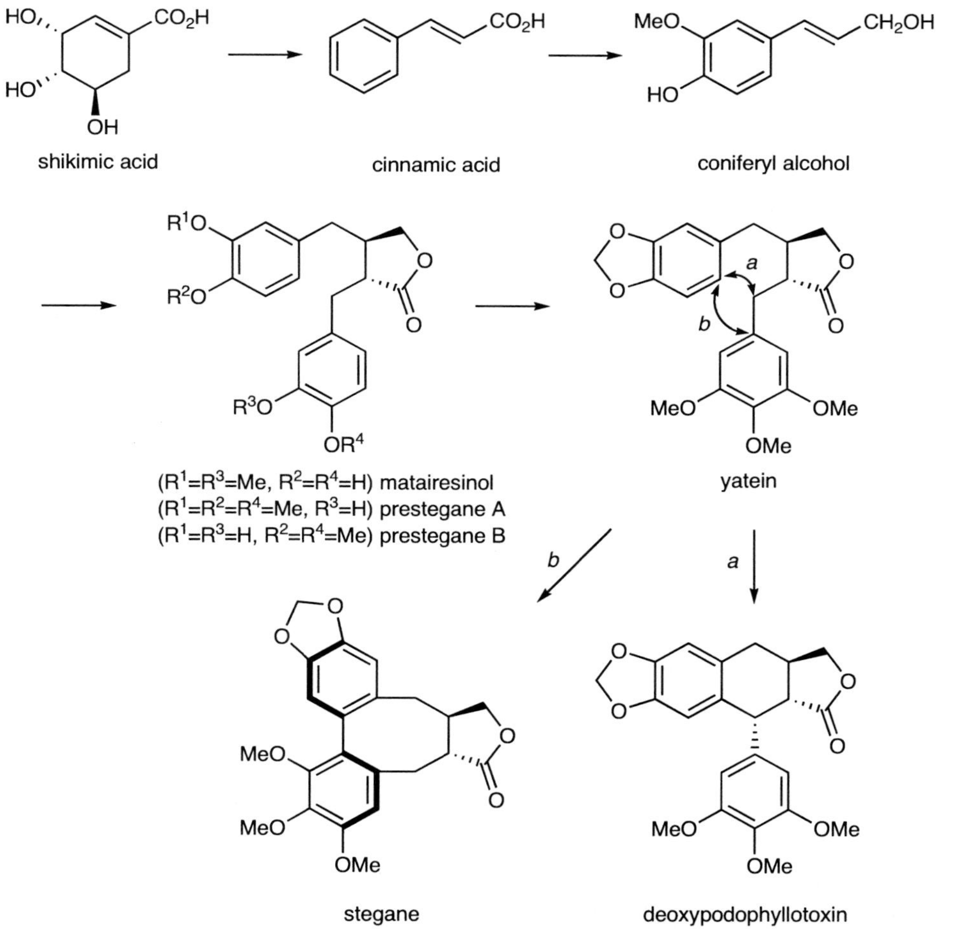

shikimic acid   cinnamic acid   coniferyl alcohol

$(R^1=R^3=Me, R^2=R^4=H)$ matairesinol
$(R^1=R^2=R^4=Me, R^3=H)$ prestegane A
$(R^1=R^3=H, R^2=R^4=Me)$ prestegane B

yatein

stegane   deoxypodophyllotoxin

Fig. (6). The shikimic acid-cinnamic acid pathway to steganes and podophyllotoxins

This lactone derives from the shikimic acid-cinnamic acid pathway, *via* oxidative dimerization of coniferyl alcohol. Whereas the presence of yatein itself was not detected in the plants producing steganes, presteganes A and B were isolated in *Steganotaenia araliaceae* and constitute therefore viable biogenetic precursors [82,83].

Another clue for the biogenesis of steganes was given by the early work of Schlessinger and co-workers, who reported the biomimetic total synthesis of (±)-isostegane, Fig. (7) [84].

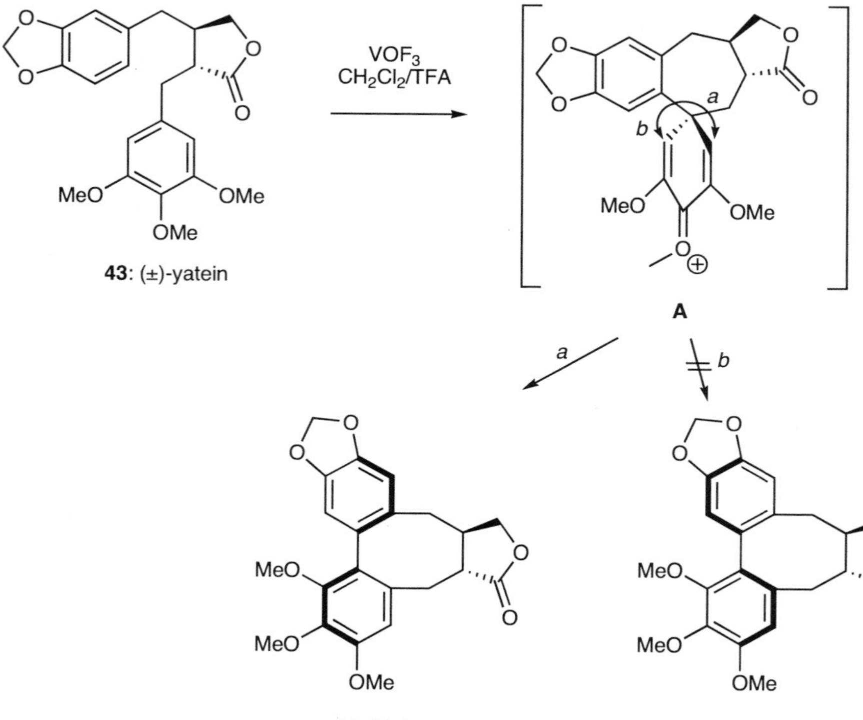

Fig. (7). Biomimetic synthesis of racemic isostegane [84] (TFA = trifluoroacetic acid)

Racemic yatein **43** was obtained by Michael addition of the anion of piperonaldehyde dithiomethyl acetal to 5*H*-furan-2-one (butenolide), followed by trapping of the resulting enolate with 3,4,5-trimethoxybenzyl bromide (see section 3.2.2). This process gave **43** with the desired *trans* stereochemistry at the butyrolactone. Oxidative coupling of the two benzenoid units was accomplished using VOF$_3$ in fluoro-acid medium. The presumed spirodienone intermediate **A** undergoes a stereoselective phenyl group migration following path *a*, to furnish racemic isostegane **44**, having the shown relative stereochemistry between the biaryl axis and the lactone ring. The formation of stegane **45**, which would result from migration on the opposite face following path *b*, was not observed.

The biogenetic relevance of spirodienone intermediate **A** is supported by the isolation of other spirodienones from *Eupomatia laurina*, namely the eupodienones [85], which were found to rearrange under acid conditions to give dibenzocyclooctadiene-type lignans [86].

A number of biomimetic semi- or total syntheses using an oxidative coupling of a yatein or matairesinol-type intermediate to form stegane or isostegane lignans such as in Fig. (7) have been reported [35,87]. The major contributions will be reviewed in section 3.2.2.

3.1.3. Rhazinilam

Shortly after the report of the structure elucidation of (–)-rhazinilam **3**, Smith published in the same paper the semi- and total synthesis of this alkaloid, shedding light on its plausible biogenetic origin [62]. Indeed, the naturally-occurring indole alkaloid (+)-1,2-didehydroaspidospermidine **46**, Fig. (**8**), when treated first with *m*CPBA, then with ferrous sulphate, gave **3** in ca. 30% yield. The mechanism of this stepwise conversion was first proposed by Smith, and later refined by our own group [88,89]. Aspidosperma alkaloid **46** was obtained quantitatively from the more readily available (+)-vincadifformine **47** by concd. hydrochloric acid-induced decarboxylation. Treatment of **46** with *m*CPBA at –20°C in methylene chloride afforded 5,21-dihydrorhazinilam *N*-oxide **48** in 65% isolated yield, *via* the postulated bis-oxidized intermediate **A**. In the presence of Fe(II), path *a*, **48** was reduced in 30 min to a 9:1 mixture of rhazinilam **3** and 5,21-dihydrorhazinilam **25**, the latter giving rhazinilam upon exposure to air for several days. This slow conversion **25→3** suggested that rhazinilam was formed directly from **48** *via* a Polonovski-type reaction. Indeed, submission of **48** to the usual Polonovski conditions (Ac$_2$O, Et$_3$N), path *b*, afforded rhazinilam in 81% yield. The reaction sequence **46→3** could also be performed in 50% yield in a one-pot fashion by careful monitoring by TLC.

In the plant, the unstable dihydrorhazinilam **25** is most certainly the direct biogenetic precursor of rhazinilam, which is an isolation artefact. Indeed, **25** was isolated together with **3** from the Apocynaceae *Leuconotis eugenifolia* and *Leuconotis griffithii* [58,59] and *Kopsia singapurensis* [60]. The question whether (+)-1,2-didehydroaspidospermidine **46** is the actual biogenetic precursor of **25** is still open.

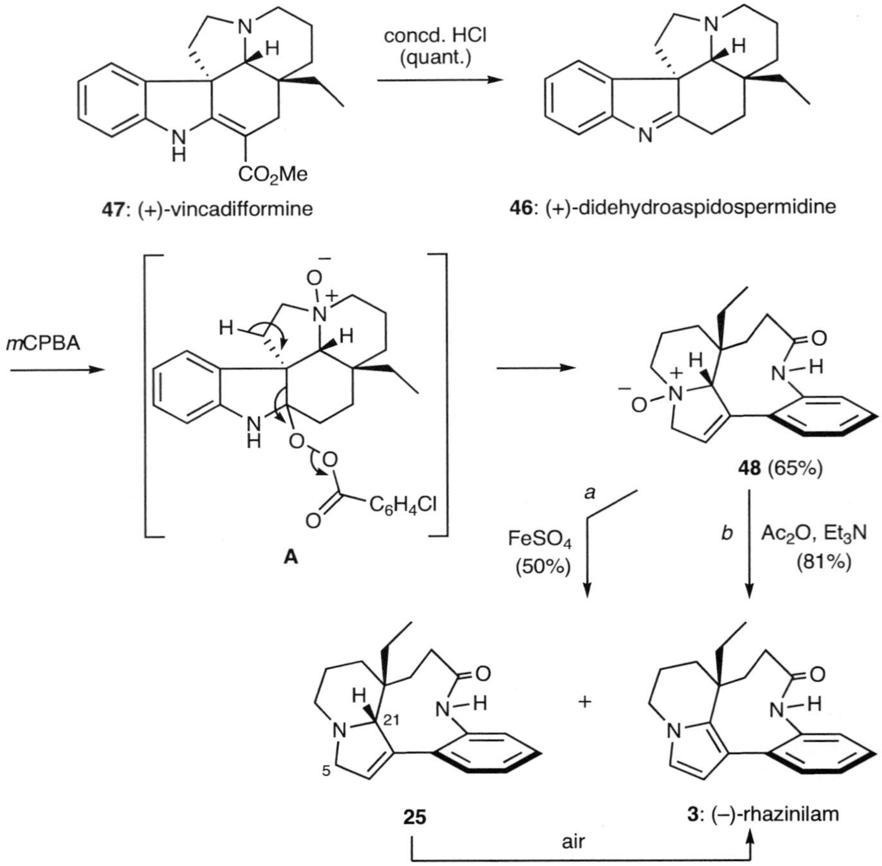

Fig. (8). Biomimetic semi-synthesis of rhazinilam [88,89] (*m*CPBA = *m*-chloroperbenzoic acid)

The preceding semi-synthetic scheme has been utilized by our group in order to conduct SAR studies [88,89]. In Fig. **(9)** is represented the rhazinilam scaffold bearing new substituents or modified chemical functions. Most analogues were markedly less active than rhazinilam in inhibiting microtubules disassembly and less cytotoxic on KB cells. The most active compound in both assays was **49**, bearing a C-14/C-15 double bond. The *dextro* enantiomer of rhazinilam, having opposite configuration at both C-20 and the biaryl axis, was synthesized in the same manner as above, starting from the *l e v o* enantiomer of 1,2-didehydroaspidospermidine, which itself can be obtained from (–)-tabersonine [60]. In contrast to the natural enantiomer, (+)-rhazinilam is biologically inactive [88].

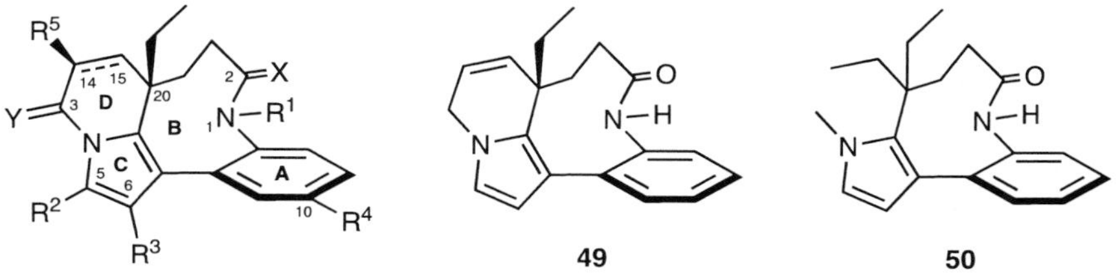

X = O, S, (H,H); Y = (H,H), (H,OH), O
R^1 = H, Me, carbonyl; R^2 = H, carbonyl
R^3 = H, CHO; R^4 = H, Br, NHAc
R^5 = H, Et, Bn

Fig. (9). Analogues of (–)-rhazinilam prepared by semi-synthesis [88,89,91]
(Bn = benzyl)

Finally, one has to mention the work of Lévy and co-workers who described the semi-synthesis of D-ring *seco*-rhazinilam analogues from (–)-tabersonine using an Emde degradation and *m*CPBA oxidation [90]. The *seco*-analogue **50** is only twice less active than rhazinilam on the inhibition of microtubules disassembly, illustrating that the absence of the D-ring is not strictly deleterious to the antitubulin activity [91,92].

3.2. Total Synthesis

3.2.1. Allocolchicinoids

After the pioneering work of Rapoport and Cook in the early 1950's, aimed at elucidating the structure of major colchicine degradation products, there were only a few reported total syntheses of allocolchicines, in part due to their easy availability by semi-synthesis. These syntheses are reviewed below, along with new synthetic approaches giving rise to analogues.

Ring enlargement of phenanthrenes

In 1950, Rapoport reported the first total synthesis of an important allocolchicine congener, (±)-colchinol methyl ether **51**, Fig. (**10**) [93,94].

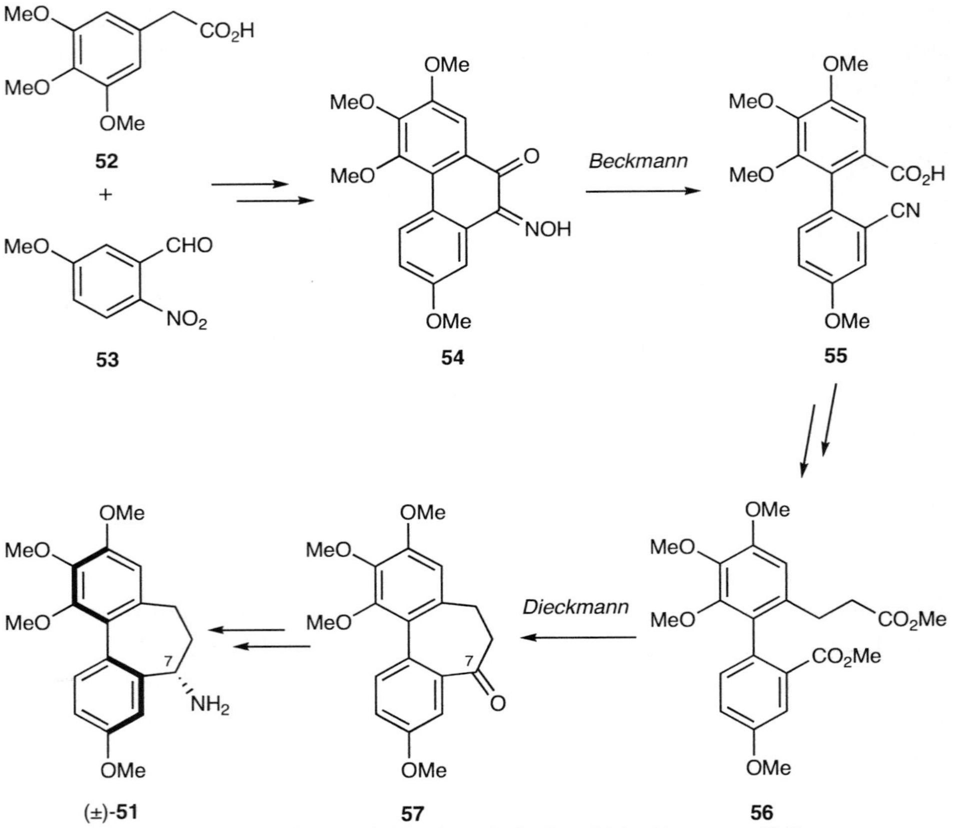

Fig. (10). Total synthesis of (±)-colchinol methyl ether **51** by Rapoport [94]

The phenanthrenequinone oxime **54** was built in four steps from the two benzenoid precursors **52** and **53**. Beckmann rearrangement of **54** furnished the cyano-acid **55**. The latter, after reduction to the corresponding cyano-aldehyde, was homologated by Knoevenagel condensation with malonic acid to give, after reduction, hydrolysis and esterification, the diester **56**. This compound underwent Dieckmann condensation, installing the seven-membered C-7 ketone **57** in 69% yield after hydrolysis and decarboxylation of the intermediate β-ketoester.

Oxime formation and catalytic hydrogenation provided racemic colchinol methyl ether **51**.

Shortly after this report, Cook and co-workers disclosed a short total synthesis of (–)-colchinol methyl ether **51** and its *N*-acetyl derivative (–)-**32** through elaboration of phenanthrenes, Fig. **(11)** [95]. Phenanthrene **58** was submitted to osmium tetroxide dihydroxylation and lead tetraacetate C-C bond cleavage to give **59**, which was cyclized in acid medium to form the seven-membered cyclic enone **60**. This compound was hydrogenated to give the corresponding ketone **57**, which was converted to racemic colchinol methyl ether **51** as previously through oxime formation. While in the previous report Rapoport and co-workers had failed to perform the resolution of **51**, Cook et al. succeeded in this operation using (+)-6,6'-dinitrodiphenic acid, giving the bio-active enantiomer (–)-**51** and by extension its *N*-acetyl derivative (–)-**32**.

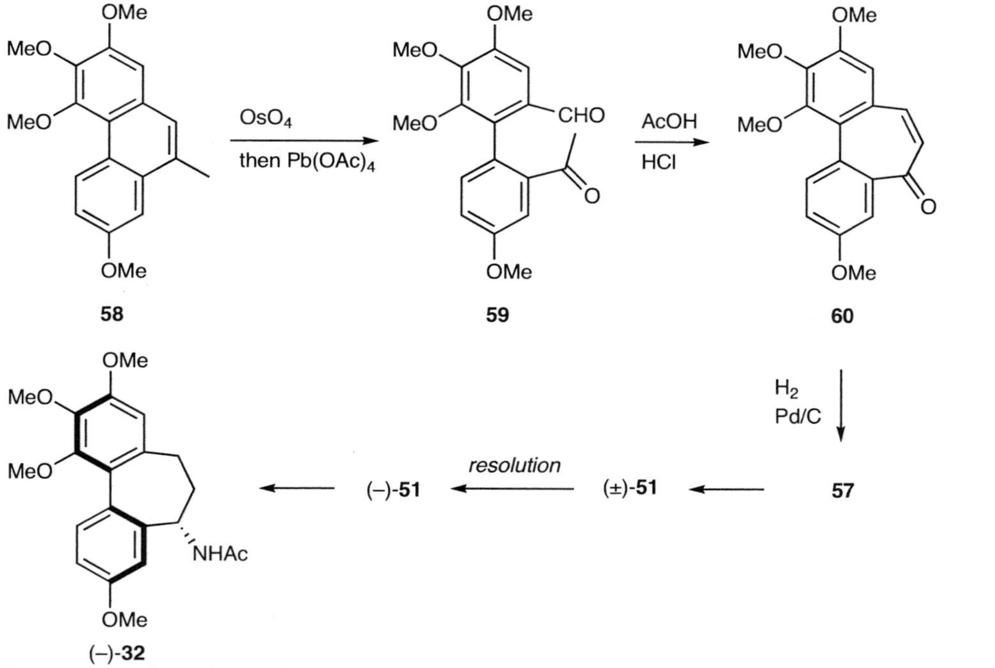

Fig. (11). Total synthesis of (–)-*N*-acetylcolchinol methyl ether **32** by Cook [95]

Intramolecular oxidative biaryl coupling

Phenolic or non-phenolic oxidative coupling methods have been extensively used since the 1970's to synthesize polyoxygenated bridged biaryl compounds, as will be illustrated later in the case of steganes. Remarkably enough, the use of transition metal oxidants is particularly suited to perform *intramolecular* biaryl coupling whereas most other methods are better suited to *intermolecular* coupling.

In the case of allocolchicinoids, a non-phenolic intramolecular oxidative coupling was used by Macdonald for the total synthesis of (±)-*N*-acetylcolchinol **31**, Fig. (**12**) [96].

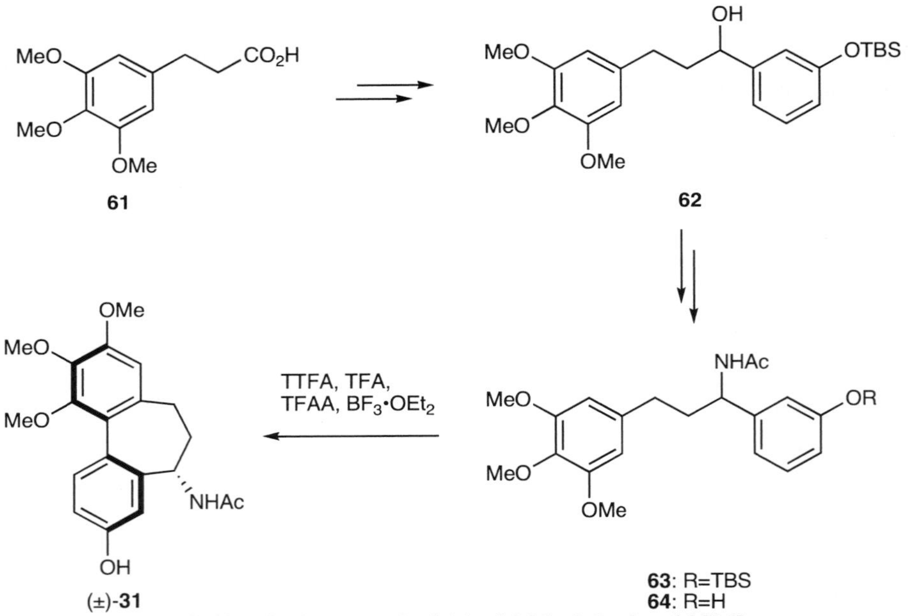

Fig. (12). Total synthesis of (±)-*N*-acetylcolchinol **31** by Macdonald [96] (TBS = *tert*-butyldimethylsilyl; TTFA = thallium(III) trifluoroacetate; TFAA = trifluoroacetic anhydride)

The carboxylic acid **61** was converted to its aldehyde counterpart which, upon condensation with *m*-*tert*-butyldimethylsilyloxyphenyl-magnesium bromide, gave the alcohol **62**. Conversion of the alcohol to the acetamide **63** was effected by a one-pot tosylation/sodium azide displacement, followed by hydrogenation (H$_2$/Pd(C)/Florisil) and acetylation. Treatment of **63** by thallium(III) trifluoroacetate (TTFA) afforded the target colchinol **31** in 71% isolated yield. It was shown that the cyclization occurred through *non-phenolic* coupling, followed by desilylation of the cyclized product, rather than desilylation followed by *phenolic* oxidative coupling. Indeed, treating unprotected alcohol **64** under the same conditions or with other oxidants did not give colchinol **31**.

Another example of intramolecular oxidative biaryl coupling in the synthesis of allocolchicine-type structures was provided by the total synthesis of the allocolchicinoid jerusalemine **9**, Fig. (**13**) [31].

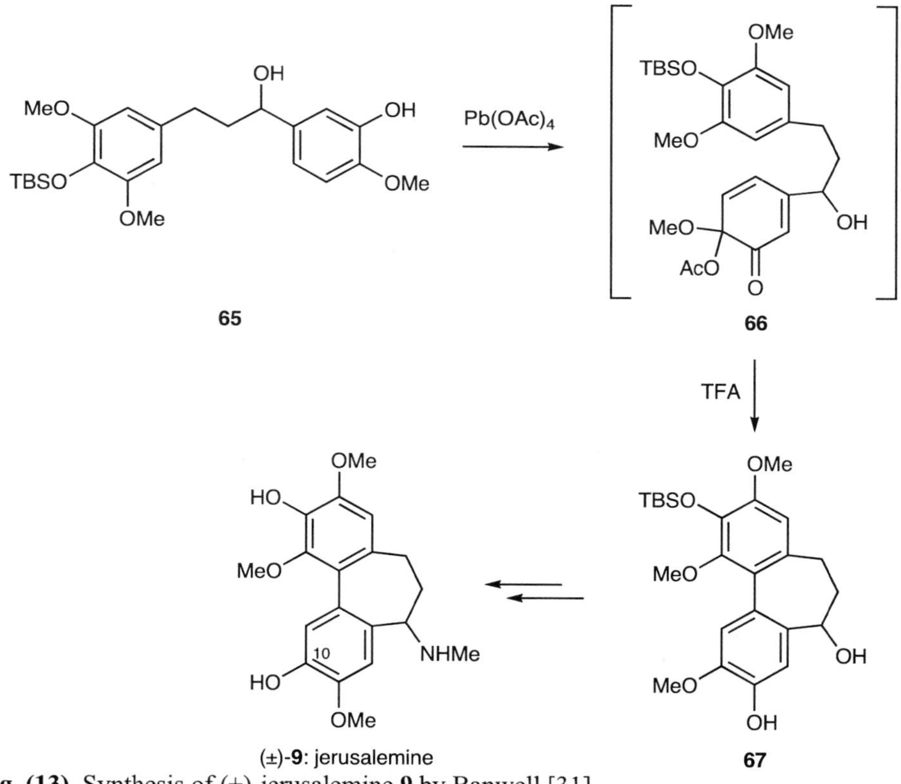

Fig. (13). Synthesis of (±)-jerusalemine **9** by Banwell [31]

Compared to allocolchicine, jerusalemine bears an extra C-10 hydroxyl group which raises regioselectivity issues. These were solved using a formal phenolic oxidative coupling of the two benzenoid moieties in phenol **65**. The coupling was accomplished using the Umezawa cyclization [97] by treating **65** with lead tetraacetate to give the intermediate *p*-quinone acetate **66** which underwent cyclization in the presence of TFA to give biaryl **67**. The latter, upon manipulation of oxygenated functions at C-9 and C-10, was converted to jerusalemine **9**. A similar synthetic pathway was utilized previously by Banwell and co-workers in their synthesis of colchicine, which involved an allocolchicine-type intermediate similar to **67** (having a methyl instead of a TBS group) [98,99].

Other methods

The allocolchicine positional isomers **70-73**, Fig. (**14**), were synthesized by Boyé and Brossi by cyclization of the open-chain biphenyl carboxylic acid **68** in acid medium [14,100,101]. Whereas the allocolchicinol methyl ether analogues **70** and **71** were inactive as tubulin polymerization inhibitors, the deamino analogues **72** and **73** were found to be equipotent with allocolchicine.

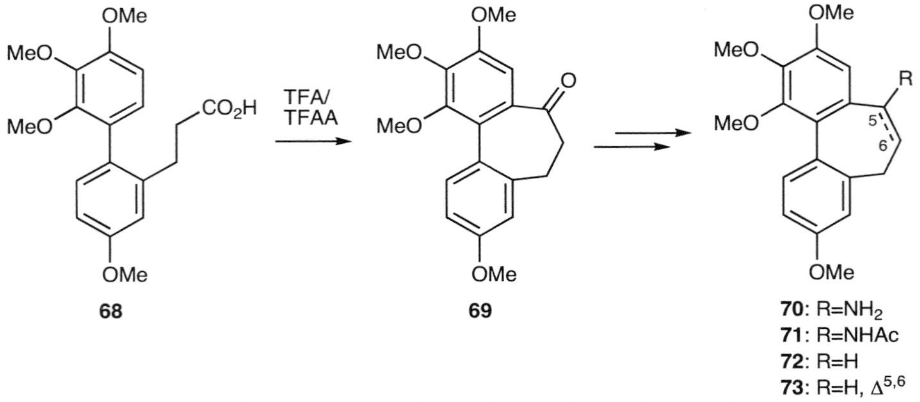

Fig. (14). Synthesis of allocolchicine analogues by Boyé and Brossi [14]

In the course of their synthesis of colchicine [102], Banwell and co-workers found that the tricyclic intermediate **74**, Fig. (**15**), aromatized in the presence of *p*-toluenesulfonic acid to give the colchinol analogue **75** [103]. The methyl ester **76** and the previously described methyl ether **72** obtained from **75** were found to be potent inhibitors of tubulin polymerization.

74

75: R=OH
72: R=OMe
76: R=CO2Me

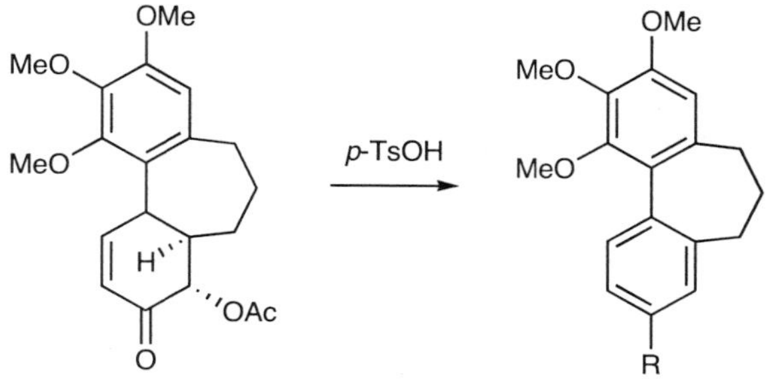

Fig. (15). Synthesis of allocolchicine analogues by Banwell [103] (*p*-TsOH = *p*-toluenesulfonic acid)

More recently, a stereocontrolled approach to the synthesis of allocolchicinoids was disclosed by Wulff and co-workers using a diastereoselective benzannulation of chromium carbene complexes, Fig. (**16**) [104,105].

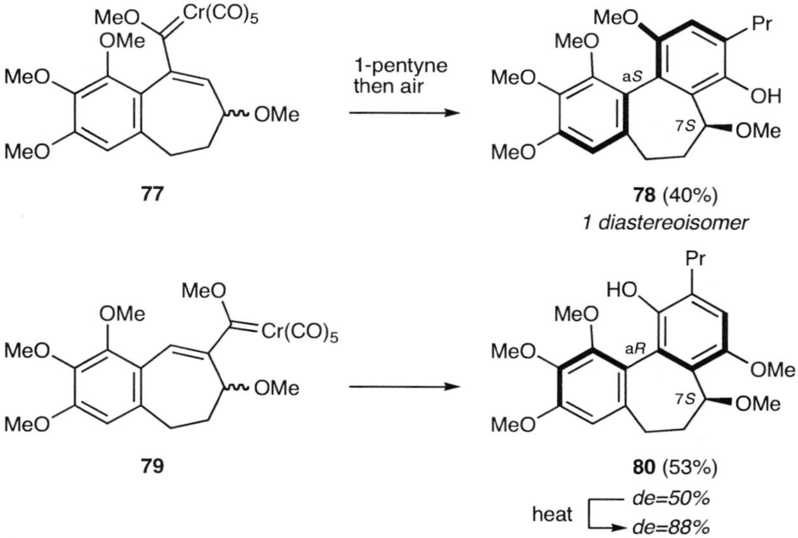

Fig. (16). Diastereoselective synthesis of allocolchicinoids by Wulff [104,105]

Thus, reaction of chromium carbene **77** with 1-pentyne followed by exposure to air provided biaryl **78** having the (a*S*,7*S*) relative configuration in 40% yield and as a single diastereoisomer. Starting from regioisomeric complex **79** resulted in the formation of a mixture of both diastereoisomers of **80**, in favor (3:1) of that having the (a*R*,7*S*) relative configuration. Heating the mixture in toluene at 120°C caused equilibration in strong favor of the (a*R*,7*S*)-configured compound.

Contrary to most allocolchicinoids which lack a C-11 substituent, compounds **78** and **80** show stable axial chirality. The above study constitutes the first example in the allocolchicinoid series where both configurations at the biaryl axis can be obtained from a given stereochemistry at C-7. In the natural allo series, free rotation around the biaryl axis is often possible at room temperature, and the configuration of the biaryl axis is controlled by the stereochemistry at C-7 due to conformational constraints in the C-ring [14]. While several total enantioselective syntheses of colchicine address the control of the stereochemistry at C-7 [106], no direct enantioselective synthesis of natural allocolchicines has been reported to date.[†]

3.2.2. Steganes

Due to the interesting biological activity of several of its members and to the synthetic challenge raised by the presence of its stereogenic elements, the stegane family has attracted considerable attention in the organic synthesis community for the past thirty years [for reviews: 1,35,107,108]. The control of the relative and absolute stereochemistry of the butyrolactone ring and the biaryl axis is a key issue that was partially or fully addressed during the course of total syntheses. Most of these benefit from the possible isomerization at both the biaryl axis and the α-position of the lactone ring.

Isomerization studies

It was shown early on by Kende and co-workers that isosteganone **81** can be thermally isomerized to steganone **12**, Fig. (**17**) [109]. This operation does not involve simple biaryl bond rotation since isosteganacin is unable to isomerize thermally to steganacin. It occurs rather through ring opening of the butyrolactone to form intermediate **A**, which

---

[†] in Banwell's total synthesis of (−)-colchicine an enantiomerically pure allocolchicinoid intermediate was synthesized, see [99]

undergoes rotation around the biaryl bond and recyclization. The driving force of this process is provided by the movement of the keto group of **81** (IR band at 1707 cm^{-1}) into coplanarity with the piperonyl ring to give steganone (IR band at 1667 cm^{-1}).

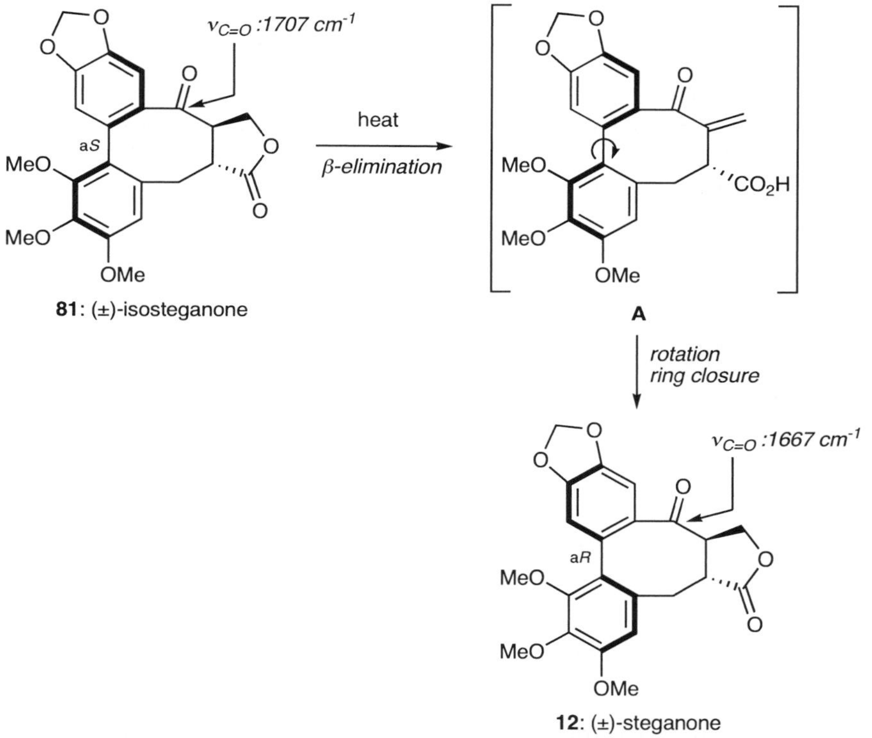

Fig. (17). Thermal isomerization of isosteganone to steganone [109]

Further studies described the interconversion of stegane diastereoisomers *via* thermal (for the biaryl axis) or acid/base-induced (for the α-lactone position) processes, Fig. **(18)** [110,111,112]. These isomerization possibilities were largely exploited in the different total syntheses discussed in the following pages.

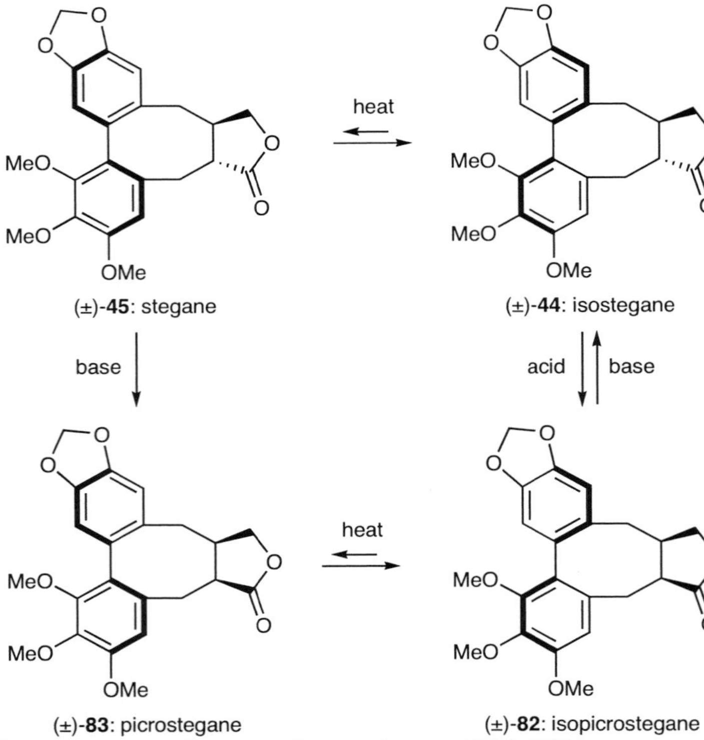

(±)-**45**: stegane   (±)-**44**: isostegane

(±)-**83**: picrostegane   (±)-**82**: isopicrostegane

Fig. (18). Interconversion of stegane diastereoisomers [110-112]

Ring enlargement of phenanthrenes

Starting from 1976, the group of Raphael published a series of papers describing the synthesis of (±)-steganone and (±)-steganacin *via* the intermediacy of phenanthrene **86**, Fig. (**19**) [113,114,115,116]. This compound was originally synthesized from the enamine **84** (available in a few steps) *via* a conrotatory photocyclization followed by the elimination of HBr (65% yield) [113,114]. The availability of **86** was later improved using the POCl₃-mediated dehydrative cyclization of amide **85** [115,116]. Expansion from a 6 to an 8-membered ring was performed by [2+2] addition with dimethyl acetylenedicarboxylate followed by electrocyclic rearrangement of the bicyclo[4.2.0] intermediate, furnishing **87** in high yield.

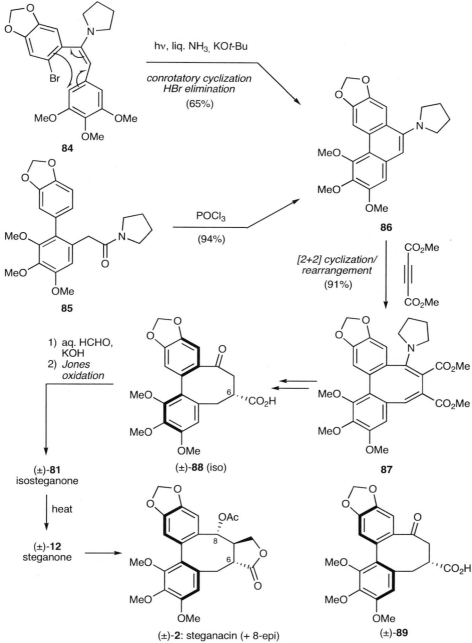

Fig. (19). Total synthesis of (±)-steganone and (±)-steganacin by Raphael [113-116]

Chemical manipulations at the cyclooctatetraene ring, including hydrolysis of the enamine group and decarboxylation, led to the keto-acid **88** which possesses the *iso*stegane relative configuration at the biaryl axis and C-6. Introduction of the butyrolactone ring was performed by treatment with aqueous formaldehyde in the presence of potassium hydroxide followed by Jones oxidation (of the C-8 alcohol resulting from cross-Cannizzaro reaction) to give racemic isosteganone **81**. The latter was isomerized in refluxing xylenes to produce racemic steganone **12** as depicted previously in Fig. (**17**). Alternatively, heating keto-acid **88** produced a 1:1 mixture of the latter and its atropisomer **89**, the latter giving steganone after the same reaction sequence as that producing isosteganone. Reduction of steganone with sodium borohydride gave a 1:1 mixture of steganol **13** and 8-episteganol, whereas reduction with lithium tri-*t*-butoxyaluminum hydride gave a 5:2 mixture in favor of steganol. The latter was easily acetylated to give racemic steganacin **2**. In a later report, the resolution of keto-acid **88** could be performed through amide formation with (–)-(*S*)-2-amino-3-phenylpropan-1-ol, furnishing (+)-**88** which was converted as described previously with (±)-**88** to optically pure (–)-steganone [116]. Finally it can be noted that modifications of the Raphael synthesis have been conducted by other groups, namely Krow [117] and Narasimhan [118,119].

Intramolecular oxidative biaryl coupling

This biomimetic method has proved particularly suited to the construction of electron-rich polyoxygenated biaryls and has therefore constituted the principal synthetic strategy to produce steganes. As described in section 3.1.2., the Schlessinger synthesis of (±)-isostegane **44** has pioneered this field in the sense that the construction of the *trans*-fused lactone yatein **43** and its cyclization to form isostegane were both diastereoselective (see Fig. (**7**)). This approach, which gave also a formal access to the stegane series *via* isomerization (see Fig. (**18**)), has inspired a number of different total syntheses.

Another milestone synthesis was reported by the group of Kende the same year as the Schlessinger and the Raphael syntheses, Fig. (**20**) [120]. The synthetic sequence arrived at the same keto-acid intermediate **89** as in the Raphael synthesis. The oxidative coupling precursor **90** was obtained from condensation of a malonic ester onto 3,4,5-trimethoxybenzyl bromide.

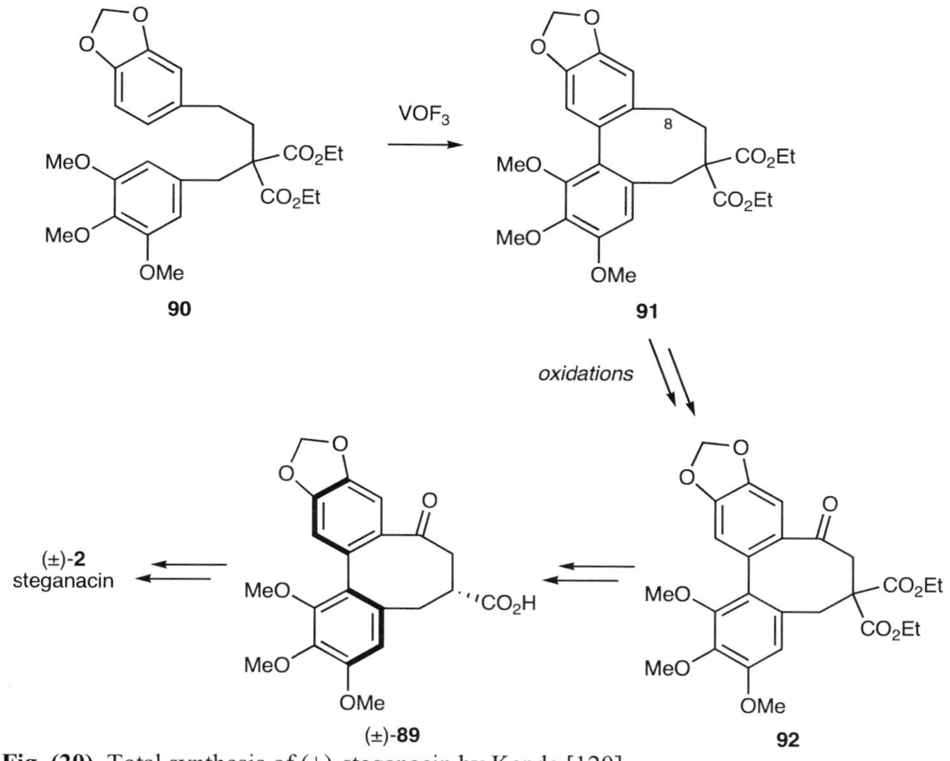

Fig. (20). Total synthesis of (±)-steganacin by Kende [120]

Vanadium(V)-mediated intramolecular coupling of **90** furnished the dibenzocyclooctadiene diester **91** (45% yield), which underwent sequential oxidations at the benzylic C-8 position (*via* the bromide and the alcohol) to install the requisite keto group. Saponification of both esters in **92** followed by decarboxylation at 200°C furnished keto-acid **89**, which was treated with formaldehyde as described above to give racemic steganone **12** and eventually steganacin **2**.[†] From a later report [109] and the work of Raphael, it seems obvious that under the decarboxylation conditions **89** was produced in an equimolar mixture with its atropisomer

---

[†] as well as its C-8 epimer, episteganacin, arising from the non-selective sodium borohydride reduction of steganone

88 (see Fig. (**19**)). The resulting contamination of steganone with isosteganone was indeed observed [109].

The first enantioselective synthetic efforts, led by the Koga and the Robin groups in the late 1970's, were directed at determining the absolute configuration of steganacin. An enantiospecific synthesis of (+)- and (−)-steganacin was proposed by Koga and co-workers, using the Schlessinger synthetic pathway [38,39,121,122,123]. In order to control the two possible absolute configurations of the *trans*-fused lactone of the intermediate (+)- and (−)-yatein **43**, three different approaches were envisaged, Fig. (**21**).

The optically pure γ-lactone synthons **93** (R=Bn or Tr) were obtained from *L*-glutamic acid and were used as chiral templates in three different asymmetric sequences. In paths *a* and *b*, a 1,3-asymmetric induction from the γ to the α-lactone position was performed in two different manners, whereas in path *c* a 1,2-asymmetric induction from the γ to the β-lactone position was described.

In path *a* [121,123], lactone **93** (R=Bn) was aldolized with piperonal to give after dehydration the (*E*)-olefin **94** which was hydrogenated to give **95** with 64% de resulting from 1,3-asymmetric induction. After translocation of the lactone carbonyl group at the γ-position, the second aryl group was introduced *via* completely diastereoselective alkylation, to give (+)-yatein **43** (also named deoxypodorhizon), the optical purity of which could be improved by recrystallization. In path *b* [122,123], the lithium enolate of **93** (R=Tr) was alkylated with piperonyl bromide to give **96** with 58% de from 1,3-asymmetric induction. Carbonyl translocation and alkylation as in path *a* gave optically enriched (−)-yatein **43**. Path *c* [38,39] started with butenolide **97** (R=Bn or Tr) obtained from **93** by α-selenoxide elimination. 1,4-Addition of the dithiane anion of 3,4,5-trimethoxybenzaldehyde to **97** followed by reductive desulfurization gave lactone **98** with 96% de for R=Bn and 100% de for R=Tr, resulting from 1,2-asymmetric induction. Compound **98** was alkylated with piperonyl bromide to give lactone **99**. Conversion of **99** to optically pure (+)-yatein **43** was effected as before by manipulation of the lactone ring.

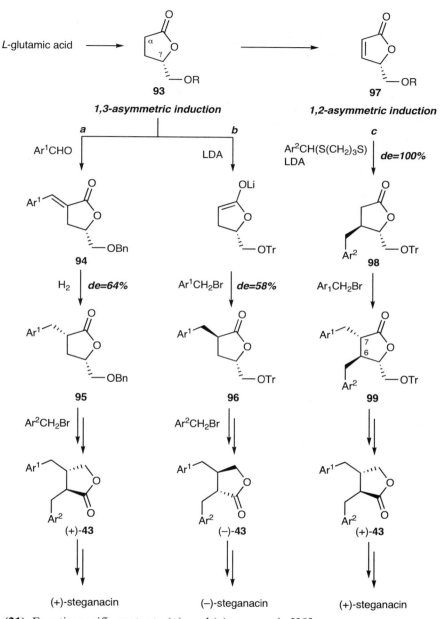

Fig. (21). Enantiospecific routes to (+)- and (−)-steganacin [39]
(Ar^1 = piperonyl; Ar^2 = 3,4,5-trimethoxyphenyl; R = Tr or Bn; Tr = triphenylmethyl; Bn
= benzyl; LDA = lithium diisopropylamide)

The elaboration of (+)- and (–)-yatein to (+)- and (–)-steganacin, respectively, was operated as follows. (+)-Yatein **43** was the first enantiomer obtained in a completely diastereoselective fashion from path *c* [38]. Nonphenolic oxidative coupling according to Schlessinger (see Fig. (**7**)) gave (–)-isostegane **44**, which was thermally isomerized to give 39% of (+)-stegane **45** and 61% of recovered (–)-**44** (see Fig. (**18**)). Benzylic oxidation of (+)-**45** with DDQ in acetic acid gave directly (+)-steganacin **2** in 11% yield. This report allowed the first determination of the absolute configuration of (+)-steganacin and by extension of the natural *levo* enantiomer.

On the other hand, pure (–)-yatein **43**, which was obtained from recrystallization of the optically enriched material furnished by path *b*, was sequentially converted as above to (+)-isostegane **44** and (–)-stegane **45**, the latter being transformed into (–)-steganacin **2** in 61% yield by stepwise bromination, hydrolysis and acetylation [39].

Since the publication of these pioneering and particularly elegant studies, several groups have reported modifications or innovations in order to increase the level of asymmetric induction in the synthesis of (+)- and (–)-yatein and, by extension, of steganes [108,124]. Most recently, Enders and co-workers disclosed an original enantioselective synthesis of (–)-isostegane, Fig. (**22**) [125,126]. The asymmetric Strecker reaction of piperonal with the enantiomerically pure secondary amine **100** and KCN gave compound **101**, which upon deprotonation with LDA and Michael addition to butenolide gave the lactone **102** in 79% yield and 98% de after purification. Alkylation of **102** with 3,4,5-trimethoxybenzyl bromide followed by cleavage of the chiral auxiliary with silver nitrate afforded lactone **103** in 78% yield, complete diastereoselection and 97% ee. Conversion of the latter to (+)-yatein **43** was effected by reduction with NaBH$_4$ and catalytic hydrogenolysis. Finally, (–)-isostegane **44** was produced in good yield by oxidative coupling mediated by thallium oxide/BF$_3$•OEt$_2$ in TFA.

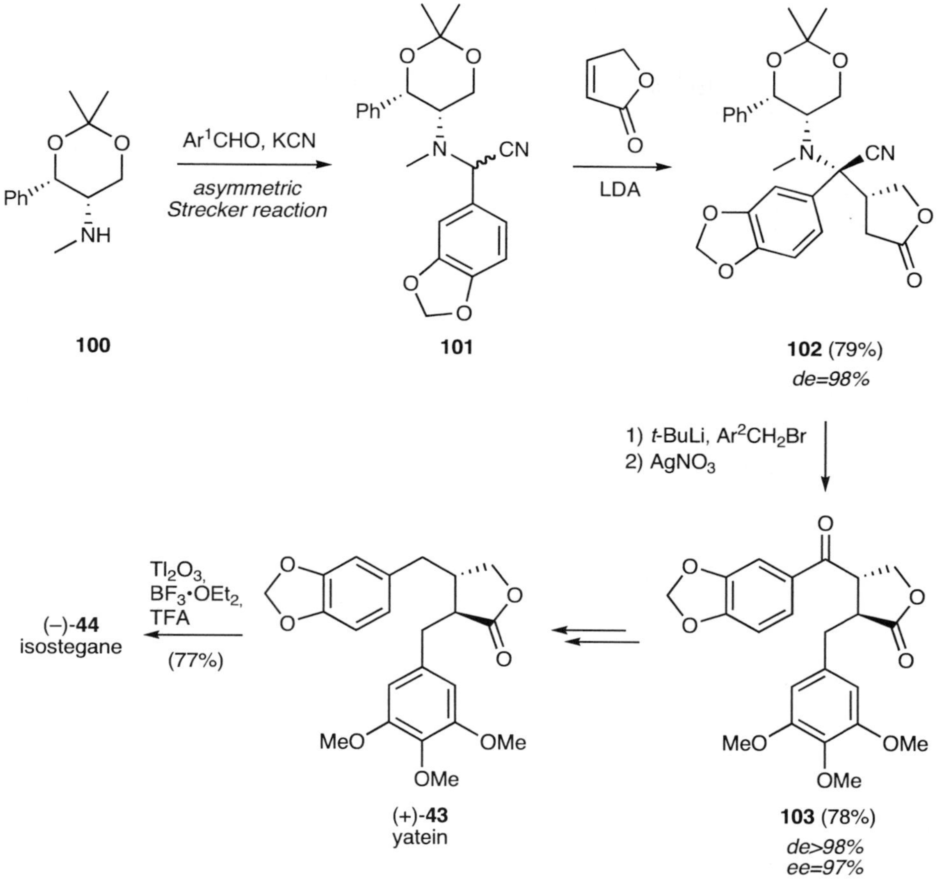

Fig. (22). Enantioselective synthesis of (−)-isostegane by Enders [125,126]
(Ar^1 = piperonyl, Ar^2 = 3,4,5-trimethoxyphenyl)

All total syntheses discussed up to this point arrive at intermediates having the *iso*stegane relative configuration of the biaryl axis and the eight-membered ring substituents, and necessitate further isomerization at the biaryl axis to produce the stegane skeleton. In the particular case of approaches based on intramolecular oxidative biaryl couplings, this relative configuration arises from the stereoselectivity of the aryl group migration depicted previously in Fig. (7).

In their total synthesis of (±)-steganone, Magnus and co-workers solved this problem *via* the stereoconvergent synthesis of the bicyclo[5.1.0] intermediate **106**, Fig. (23) [127,128].

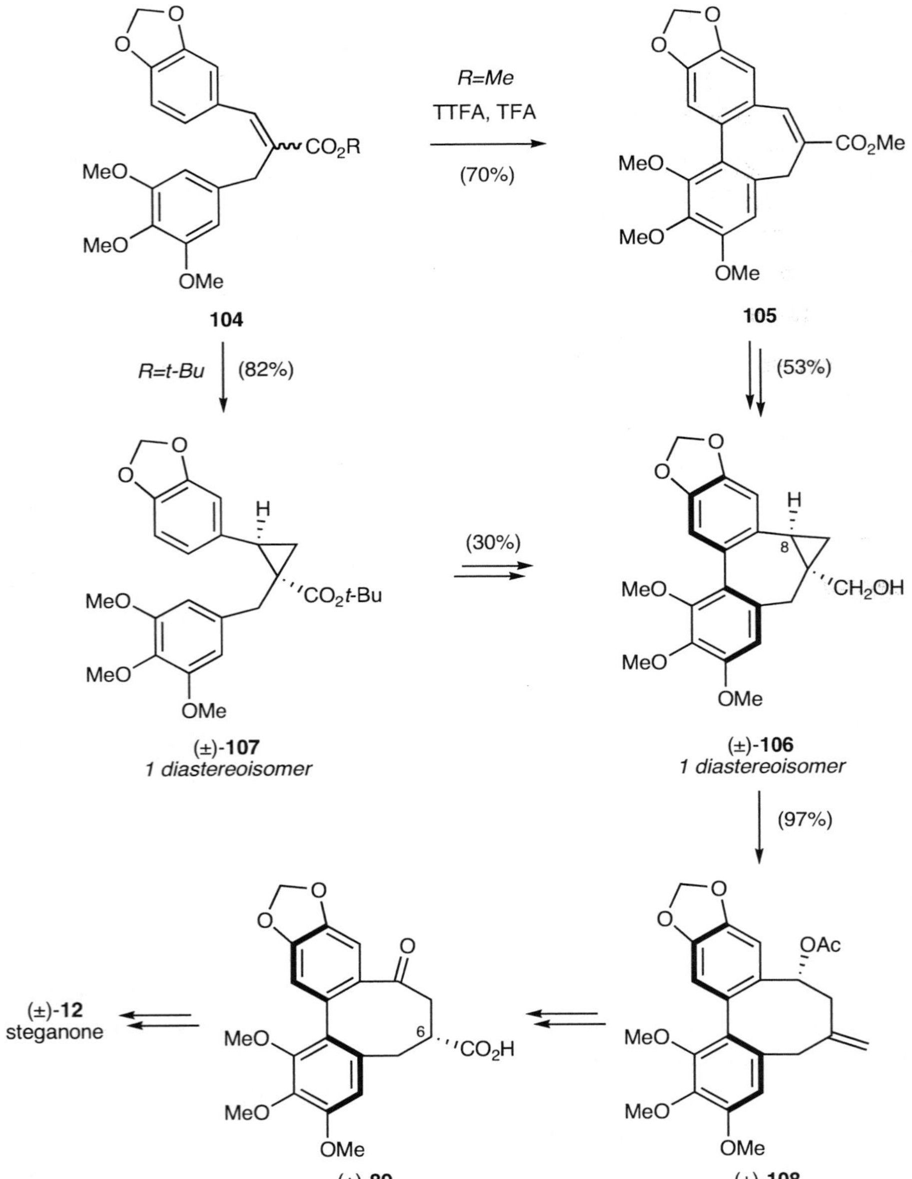

Fig. (23). Total synthesis of (±)-steganone by Magnus [127,128]

Cyclopropane **106** was obtained by two routes: in the first route, the conjugated ester **104** (R=Me, >95% *E*-isomer) underwent oxidative biaryl coupling in the presence of TTFA to give biaryl cycloheptene **105** in good yield and in a stereoconvergent manner (both E and Z-isomers gave independently **105**). After reduction of the ester to the primary alcohol, a Simmons-Smith cyclopropanation occurred in a completely diastereoselective fashion from the most accessible face of the double bond, furnishing **106** in 53% overall yield from **105**. Alternatively, cyclopropanation of **104** (R=t-Bu, 2:1 mixture of E:Z isomers) using the sulfoxonium ylide method gave cyclopropane **107** as a single diastereoisomer in a stereoconvergent manner (both E and Z isomers gave **107**). After conversion to the methyl ester, oxidative biaryl coupling and reduction, compound **106** was obtained in 30% overall yield. Rearrangement of the bicyclo[5.1.0] to the cyclooctadiene skeleton occurred upon solvolysis with AcOH/NaOAc/HClO$_4$, with configuration inversion at C-8 to give compound **108** in quantitative yield. The latter was converted to the known racemic keto-acid **89** *via* hydroboration and Jones oxidation. Compound **89**, which has the correct relative configuration of the biaryl axis and C-6, was converted without isomerization to racemic steganone using the conditions developed by Raphael (aq. HCHO/KOH then Jones oxidation, see Fig. (**19**)).

It can be noted that since the initial use of vanadium(V) in acid medium for the biaryl oxidative coupling by Kende and Schlessinger, improvements have been made using other metal oxides, including manganese(III), cerium(IV), ruthenium(IV) and thallium(III) as illustrated above in the Magnus and Enders syntheses [87,129].

Ullmann biaryl coupling

The application of the Ullmann coupling to the synthesis of dibenzocyclooctadiene lignans was described in the late 1970's by the groups of Ziegler and of Brown and Robin.

The former group [130,131,132] used a modified low-temperature Ullmann procedure to produce biaryl intermediate **111**, Fig. (**24**): the thiadioxolane **109** (X=Br), prepared from piperonal, was submitted to halogen-lithium exchange with *n*-BuLi, and the resulting organolithium intermediate (X=Li) was treated with CuI•P(OEt)$_3$ to give the corresponding organocopper reagent. Coupling of the latter with iodide **110** occurred upon warming from −78 to 25°C providing, after hydrolysis of the imine, aldehyde **111** in very good overall yield.

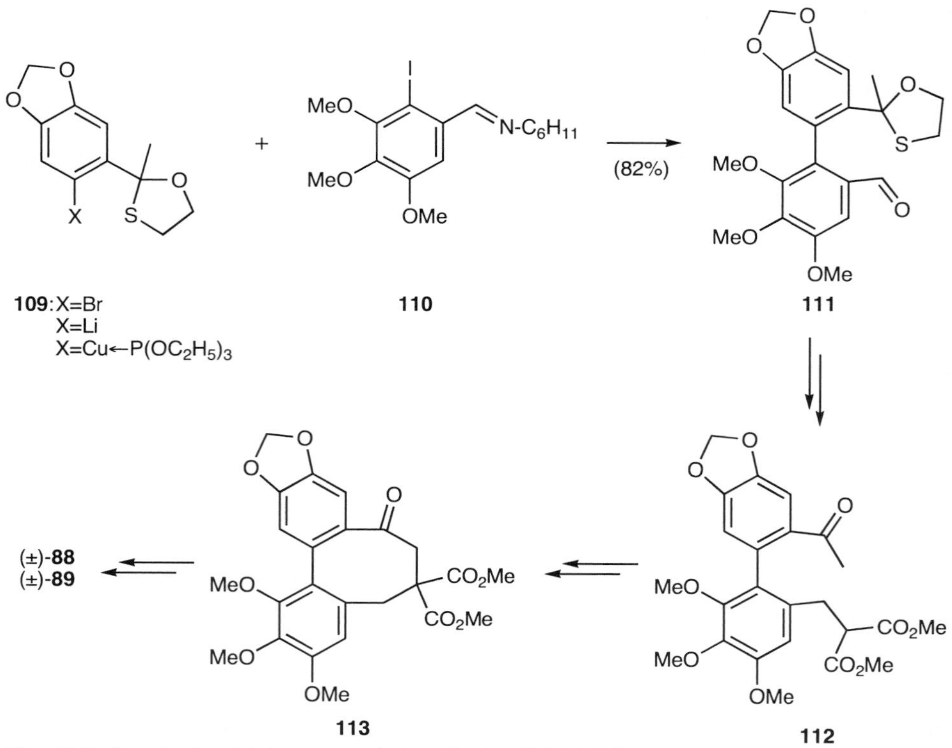

Fig. (24). Synthesis of (±)-steganacin by Ziegler [130,132]

Transformation of **111** into keto-diester **112** was performed uneventfully in three steps, as was the cyclization to form the cyclooctadiene ring *via* formation of the α-bromoketone (with pyridinium tribromide) and deprotonation of the malonate with *t*-BuOK. The so-formed keto-diester **113** is analogous to the one previously described in the Kende synthesis (compound **92**, Fig. (**20**)). Saponification and decarboxylation accordingly provided a mixture of the atropisomeric keto-acids **88** and **89**. Similarly to the preceding studies, this mixture led to (±)-steganacin *via* isomerization of the isosteganone/steganone mixture to (±)-steganone. It can be noted that attempts at intramolecular Ullmann coupling to form intermediate **113** failed [131].

The total synthesis of racemic and later of natural (−)-steganone was described by Brown and Robin using a classical Ullmann coupling at a later stage, Fig. (**25**) [37,133,134,135,136]. Lactone **114** was first synthesized using the Koga 1,3-asymmetric induction method (see path *b*, Fig. (**21**)).

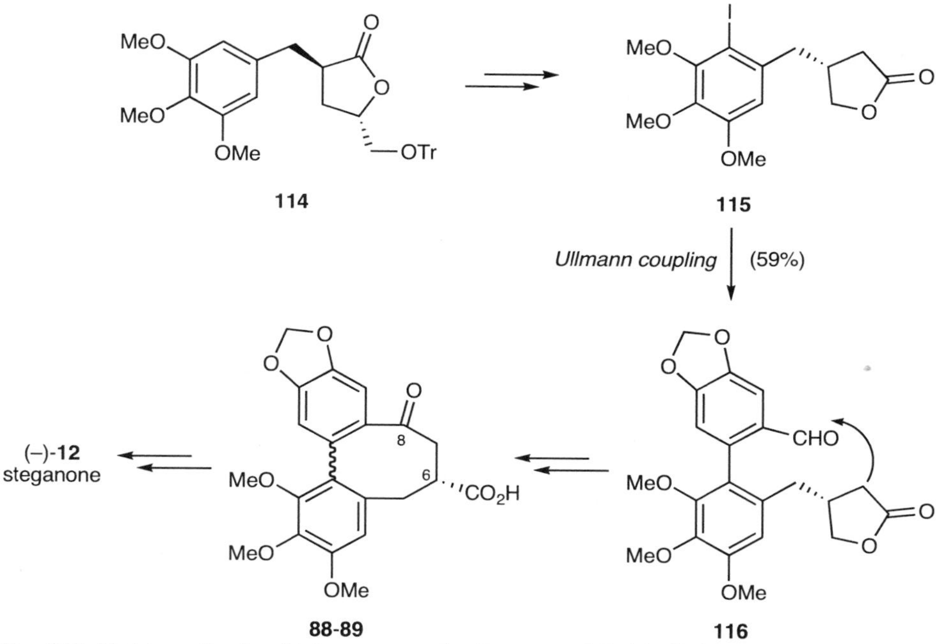

Fig. (25). Total synthesis of (–)-steganone by Brown and Robin [37]

After elaboration of the lactone ring, the iodination of the benzene ring gave optically pure iodide **115**. Classical Ullmann coupling of the latter with 2-bromopiperonal in the presence of copper at 225°C afforded the biaryl aldehyde **116** in 59% yield. Condensation of the lactone lithium enolate on the aldehyde group, followed by Jones oxidation (to install the C-8 ketone), decarboxylation with barium hydroxide, and Jones oxidation (to form the C-6 carboxylic acid), gave an atropisomeric mixture of **88** and **89** having the correct absolute configuration at C-6. This mixture was converted to (–)-steganone **12** as previously described *via* isomerization of the (+)-isosteganone/(–)-steganone mixture. This report constituted the first total synthesis of (–)-steganone [37], which, together with the synthesis of (+)-steganacin by Koga [38], confirmed its absolute configuration.

Grignard/oxazoline (Meyers) biaryl coupling

In the early 1980's, Meyer and co-workers developed a novel method for synthesizing axially chiral biaryl molecules based on the diastereoselective substitution of an aryl methoxy group *ortho* to a chiral oxazoline by an aryl Grignard reagent [137,138]. Its application to the asymmetric synthesis of (–)-steganone was subsequently disclosed, Fig. (**26**) [139]. The reaction of Grignard reagent **117** with chiral oxazoline **118** proceeded with good diastereoselection (75% de), affording the optically pure biaryl intermediate **119** after removal of the undesired atropisomer by flash chromatography. The diastereoselectivity of this reaction originates from the formation of an intermediate bearing a highly coordinated magnesium atom [138]. The functionalization of **119**, which is configurationally stable due to the bulkiness of the acetal and oxazoline substituents, was performed carefully in order to avoid epimerization at the biaryl bond. Deprotection of the acetal group and immediate reaction of the so-formed configurationally unstable aldehyde with methylmagnesium bromide yielded the configurationally stable alcohol **120** as a 1:1 mixture of alcohol epimers. Protection of the alcohol with allyl iodide, careful cleavage of the oxazoline group and rapid homologation of the so-formed benzylic alcohol yielded compound **121**, which was elaborated *via* the C-8 ketone to give the dibenzocyclooctadiene **113**, already known from the Ziegler synthesis (Fig. (**24**)), but this time in enantiomerically pure form. Decarboxylation at the C-6 position followed by esterification with diazomethane gave the methyl ester **122** having the desired *R* configuration at C-6, but as a mixture of a*R* and a*S* diastereoisomers due to thermal epimerization at the decarboxylation step. The desired a*R* atropisomer was separated and converted as described previously to (–)-steganone. Analysis of the specific rotation of the latter indicated that 10-12% racemization had occurred, presumably during the elaboration of **121** which produced a configurationally labile intermediate.

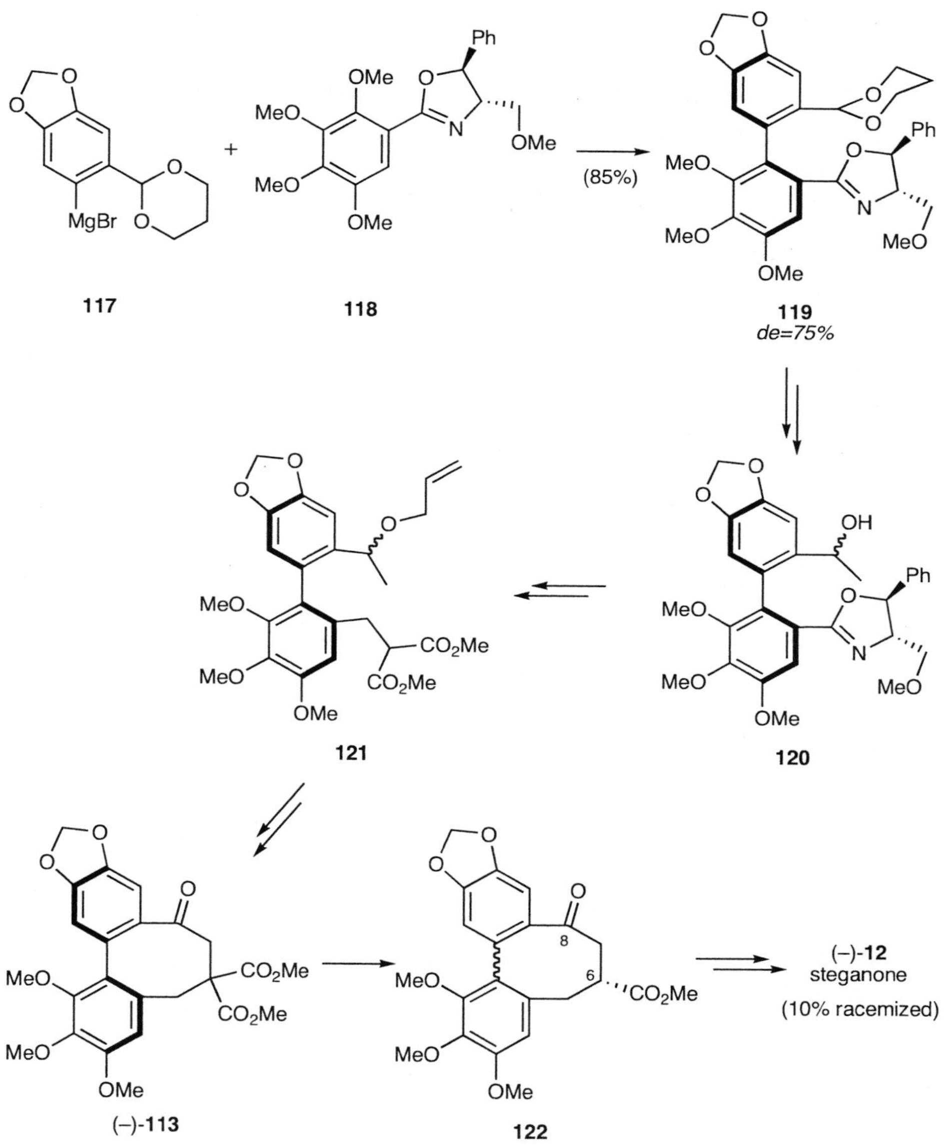

Fig. (26). Total synthesis of (–)-steganone by Meyers [139]

This synthetic plan, which initially involved an efficient asymmetric biaryl synthesis, suffered from two weaknesses, namely in the manipulation of configurationally fragile intermediates and in the epimerization at the biaryl axis during the decarboxylation step. These two issues were recently addressed in the Uemura and the Molander syntheses.

Suzuki biaryl coupling with (arene)chromium complexes

Another diastereoselective strategy for synthesizing axially chiral biaryl compounds was developed by the Uemura group using the cross-coupling reaction of planar chiral (arene)chromium complexes with arylboronic acids [for review: 140]. The illustration of its usefulness was provided *inter alia* by the formal synthesis of (−)-steganone, Fig. (27) [141,142]. The optically pure (arene)chromium tricarbonyl complex **123** was prepared by thermal complexation of the corresponding chiral aryl acetal with $Cr(CO)_6$. As its benzene ring contains a plane of symmetry, complex **123** does not have planar chirality. The directed *ortho*-lithiation of **123** with *n*-BuLi in toluene followed by quenching with $BrCF_2CF_2Br$ gave complex **124**, this time having planar chirality, in 54% yield and with high diastereoselectivity (de=90%) in favor of the pR-configured diastereoisomer. Recrystallization and acetal hydrolysis furnished enantiomerically pure arene(chromium) complex **125**. Suzuki-Miyaura cross-coupling of chromium-complexed aryl halides with arylboronic acids occurs in a particularly smooth fashion since the oxidative insertion of palladium is favored by the electron-deficient character of the carbon-halogen bond induced by the complexation with chromium. Indeed, cross-coupling of **125** with boronic acid **126** occurred under mild conditions, furnishing biaryl **127** having the aR axial configuration in completely diastereoselective fashion. In this case, the chromium tricarbonyl residue literally blocks the β face of aryl **125** and the aldehyde group of **126** takes position on the α face, opposite from the chromium. Functionalization of the aldehyde and the alcohol groups and oxidative removal of the chromium tricarbonyl group led uneventfully to biaryl compound **121**, previously described in the Meyers synthesis (see Fig.(26)).

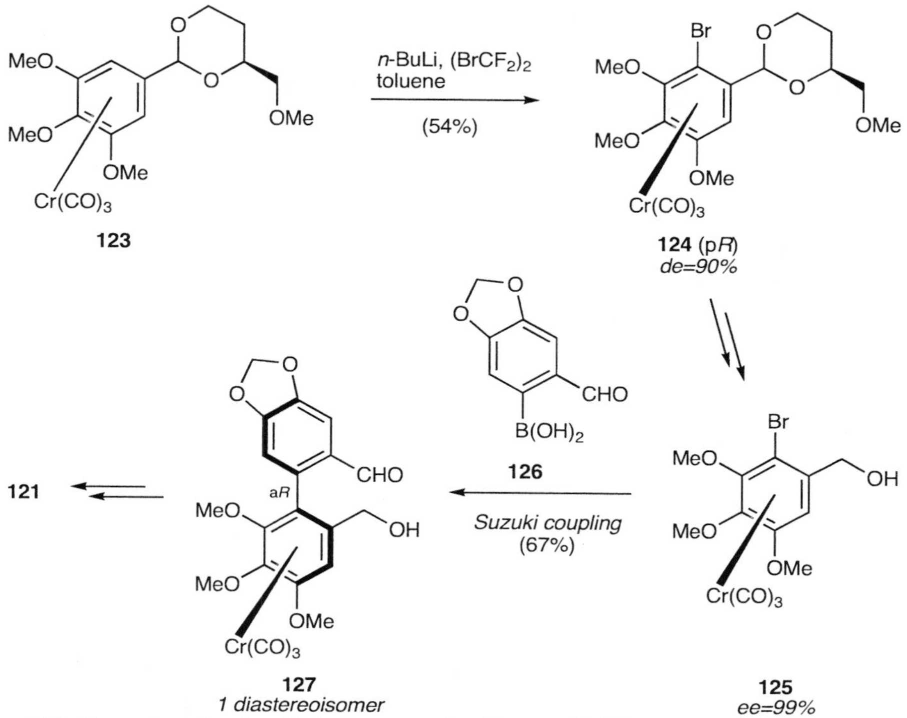

Fig. (27). Formal synthesis of (–)-steganone by Uemura [141]

This synthesis resolved one of the issues encountered in the Meyers synthesis, *i. e.* the epimerization of configurationally labile intermediates, since in this case the rotation around the biaryl axis is precluded by the presence of the bulky chromium tricarbonyl group.

The second stereochemical issue, *i. e.* the epimerization of the biaryl bond during functionalization of the eight-membered ring, was solved recently by Molander and co-workers in their synthesis of (–)-steganone, Fig. (**28**) [143]. The chromium complex **127** obtained above in the Uemura synthesis was transformed into the benzylic bromide **128** which underwent an unusually efficient Stille coupling with stannane **129** to give compound **130**. An original samarium(II) iodide-promoted 8-endo ketyl-olefin radical cyclization yielded the chromium-complexed isopicrosteganol **131** in good yield and with complete diastereoselection. In order to perform stereochemical inversion at C-7, the C-8 alcohol was first oxidized to the ketone with sodium acetate-buffered PCC. In the event, decomplexation occurred as well and picrosteganone **132** was produced as a 2:1 mixture of interconverting atropisomers. Upon treatment with DBU in refluxing THF, the mixture was converted to enantiomerically pure (–)-steganone *via* stereochemical inversion at C-7 and biaryl-bond isomerization to form the most thermodynamically stable (a*R*,6*R*,7*R*) relative configuration.

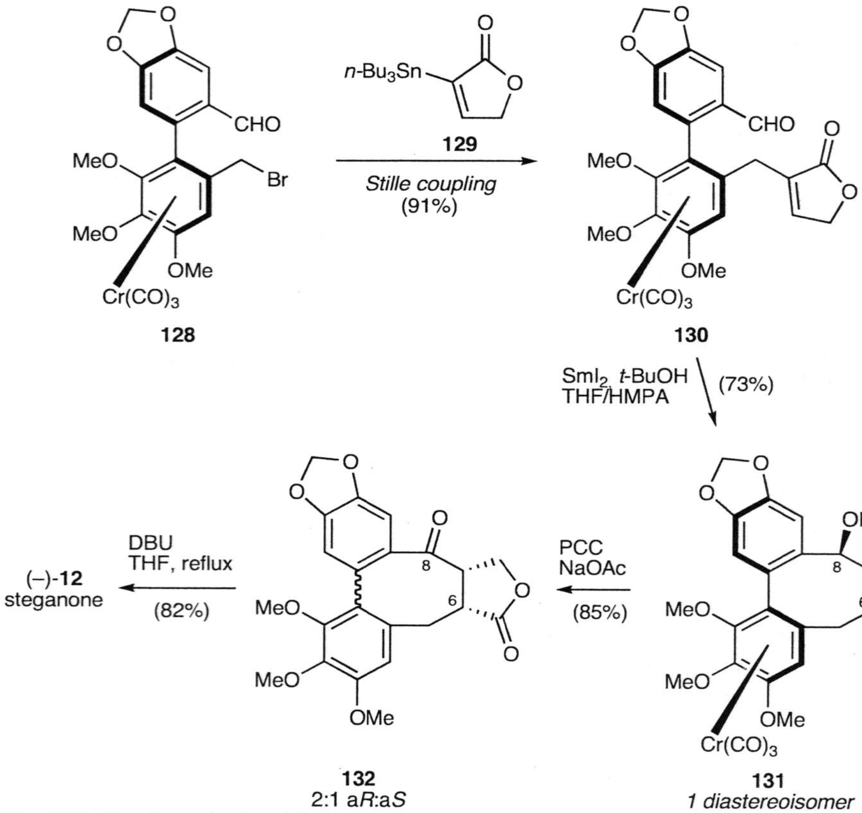

Fig. (28). Total synthesis of (–)-steganone by Molander [143]
(PCC = pyridinium chlorochromate, DBU = 1,8-diazabicyclo[5.4.0]undec-7-ene)

Other methods

An original diastereoselective approach to the synthesis of steganone analogues was reported by Motherwell and co-workers, Fig. (**29**) [144]. The diyne **135** was obtained from bromide **133** and lactone **134** *via* diastereoselective alkylation and deprotection of the alkyne groups. A cobalt-mediated [2+2+2] cycloaddition of **135** with bis(trimethylsilyl)acetylene using photochemical irradiation provided the dibenzocyclooctadiene compound **136** as the minor product (19% yield) in a diastereoselective fashion. Removal of the acetal and trimethylsilyl groups gave the racemic steganone analogue **137**, lacking the dioxolane ring but having the same (aR,6R,7R) relative configuration.

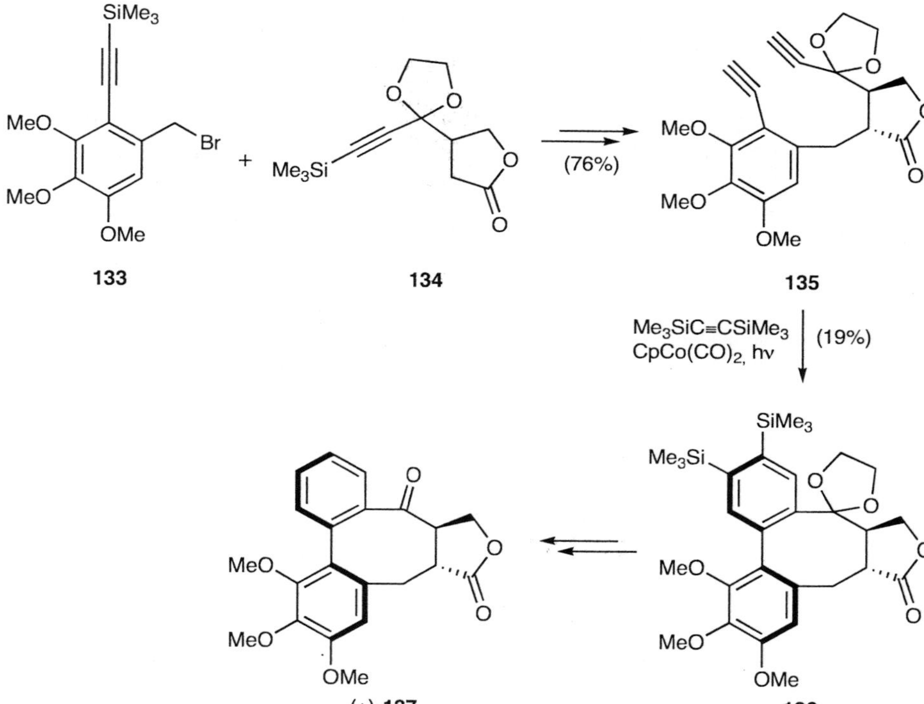

Fig. (29). Synthesis of steganone analogues by Motherwell [144]

Biological activity

Given the large number of publications regarding the total synthesis of steganes, relatively few SAR studies have been undertaken in these chemical series [for review: 145]. SAR have been investigated mainly by the Koga and Tomioka and by the Brown and Robin groups using total synthesis and isomerization reactions. The inhibition of microtubules assembly [40,41,42] as well as the cytotoxicity towards the human nasopharynx carcinoma KB cell line [43] have been reported for natural products and analogues, Table 1, showing the influence of each stereogenic element.

Table 1. Antitubulin activity and cytotoxicity of selected steganes [42,43,149]

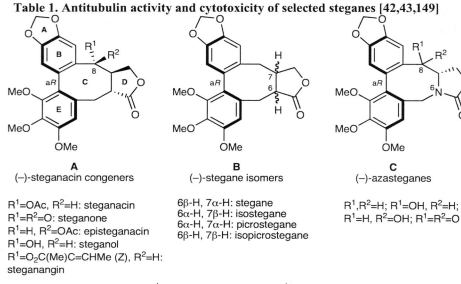

A	**B**	**C**
(–)-steganacin congeners	(–)-stegane isomers	(–)-azasteganes

R^1=OAc, R^2=H: steganacin
R^1=R^2=O: steganone
R^1=H, R^2=OAc: episteganacin
R^1=OH, R^2=H: steganol
R^1=O_2C(Me)C=CHMe (Z), R^2=H: steganangin

6β-H, 7α-H: stegane
6α-H, 7β-H: isostegane
6α-H, 7α-H: picrostegane
6β-H, 7β-H: isopicrostegane

R^1,R^2=H; R^1=OH, R^2=H;
R^1=H, R^2=OH; R^1=R^2=O

Compound	Cytotoxicity (KB cells) IC_{50}, µg/mL	Inhibition of microtubules assembly (racemates) IC_{50}, µM
(–)-steganacin	<0.3	3.5
(+)-steganacin	1.75	
(–)-steganone	6.2	125
(–)-episteganacin	16.6	inact
(–)-steganol	0.53	110
(–)-steganangin	0.4	
(–)-stegane	0.49	60
(+)-stegane	86	
(–)-isostegane	0.47	inact
(–)-picrostegane	8.8	inact
(–)-isopicrostegane	<0.3	5
(–)-azasteganes	<0.3	

The aR absolute configuration of the biaryl axis is crucial since all aS-configured compounds are only weakly cytotoxic (see for instance (+)-steganacin and (+)-stegane). The configuration at C-8 is also important, since episteganacin is markedly less active than steganacin. (–)-Steganacin is the most active congener of C-8 substituted compounds of structure **A**, with an IC_{50}<0.3 µg/mL on KB cells [43]. The exact IC_{50} value given by Kinghorn and co-workers is 0.1 µg/mL (0.2 µM) for KB cells and 0.1-5.0 µg/mL (0.2-10 µM) on a number of other cell lines [50].

As shown with stegane isomers of structure **B**, different relative configurations at C-6 and C-7 give good activities, (–)-isopicrostegane (with a *cis*-lactone ring on the α face) being the most cytotoxic compound. The inhibition of tubulin polymerization by steganes, measured with racemic mixtures, accounted at least in part for the observed cytotoxicities, Table 1 [42]. Thus, the lowest IC_{50} values were again obtained with (±)-steganacin and (±)-isopicrostegane and were comparable to the IC_{50} value of (–)-colchicine (2.5 μM).

Introduction of a bulky sugar moiety at the C-8 position [52,146] or demethylation of the aryl methoxy groups [147] led to reduced biological activity. By contrast, the synthesis of aza-isopicrosteganes of structure **C** proved particularly fruitful [148,149]. Indeed, IC_{50} values lower than that of steganacin were found for these compounds on KB cells. It was shown that the eight-membered C-ring of these azaisopicrosteganes adopts a boat-chair conformation which proved essential for the activity.

3.2.3. Rhazinilam

Several structural features of (–)-rhazinilam **3** raise interesting synthetic challenges: the axially chiral phenyl-pyrrole A-C biaryl bond, the fused pyrrole-piperidine C-D rings, the stereogenic quaternary carbon (C-20) *ortho* to the phenyl-pyrrole axis, the nine-membered lactam B-ring. Three racemic (Smith, Sames, Magnus) and one asymmetric (Sames) total syntheses have been published to date, which all proceed *via* construction of the pyrrole ring and diastereoselective control of the axial chirality by the central chirality at C-20.

Total syntheses with construction of the pyrrole ring

In the three different approaches envisaged for the total syntheses of rhazinilam, the construction of the pyrrole ring was operated in two different manners: a Knorr-type reaction in the Smith synthesis and a 1,5-electrocyclization of an allyl-iminium compound in the Sames and Magnus syntheses.

Soon after the structure elucidation of rhazinilam, Smith and co-workers published its semi-synthesis and racemic total synthesis in the same report [62]. The total synthesis, Fig. (**30**), started with a Knorr-type construction of the phenyl-pyrrole scaffold from keto-acid **138** and aminoacetal **139** in the presence of sodium acetate in acetic acid [150]. A methyl ester protecting group was installed in the 2-position of the pyrrole ring by Vilsmeier formylation, oxidation and esterification, furnishing compound **140** in 10.5% yield from **138**. *N*-Alkylation of phenyl-pyrrole **140** with racemic lactone tosylate **142**, obtained in six steps in 22% yield from keto-diester **141**, followed by aluminum trichloride-mediated cyclization and hydrogenation of the nitro group gave amino-acid **143**, having the A-C-D ring system of rhazinilam (39% overall yield). The installation of the lactam B-ring was performed by cyclization of **143** in the presence of DCC to give **144** in >95% yield. The final steps of the synthesis proceeded uneventfully through hydrolysis of the methyl ester and decarboxylation.

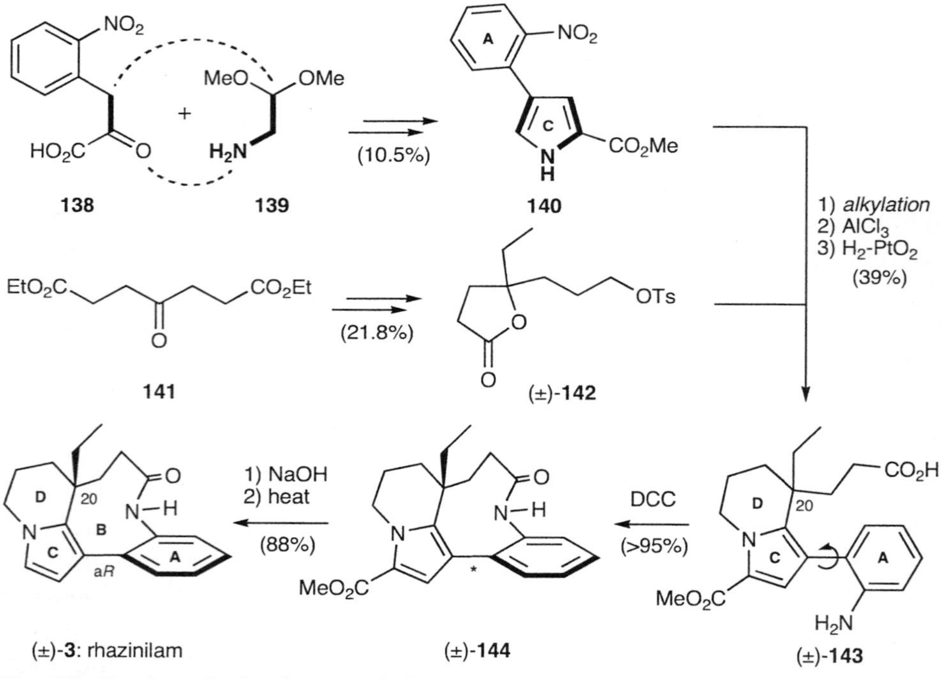

Fig. (30). Total synthesis of (±)-rhazinilam by Smith [62] (DCC = 1,3-dicyclohexylcarbodiimide)

As compound **144** may only exist in the (a*R*,20*R*) relative configuration, the quantitative conversion of amino-acid **143** to **144** indicates that free rotation around the phenyl-pyrrole bond occurs in **143** at room temperature. Indeed, if **143** existed as a mixture of non-interconverting atropisomers, (a*R*,20*R*) and (a*S*,20*R*), **144** would be obtained in maximum 50% yield together with the unreacted (a*S*,20*R*)-configured amino-acid. Therefore, the absolute configuration of the biaryl axis can be completely controlled by the absolute configuration at C-20, a property that was further exploited in the Sames enantioselective synthesis.

In 2000 and 2002, Sames and co-workers described a total synthesis of (±)- and (−)-rhazinilam using an original strategy to build the quaternary C-20 stereogenic center, Fig. (**31**) [151,152]. Phenyl-pyrrole **146**, containing the A-C-D ring system of rhazinilam as well as a quaternary C-20 carbon atom bearing a prochiral diethyl group, was synthesized from allyl-iminium salt **145** using the Grigg methodology [153]. Indeed, refluxing **145** in toluene in the presence of silver carbonate induced the formation of the zwitterionic intermediate **A**, which underwent a 1,5-electrocyclization to form the corresponding dihydropyrrole ring. The latter was oxidized *in situ* to form the pyrrole ring, giving **146** in 70% yield. Protection of the 2-pyrrole position followed by reduction of the nitro group afforded compound **147** in high yield. Desymmetrization of the two enantiotopic ethyl groups at C-20 to form the ethyl-vinyl-bearing chiral compound **151** was effected by metal-induced C-H activation (dehydrogenation) according to the following protocol: after Schiff base formation with 2-benzoylpyridine, the platinum complex **148** was formed by complexation with the dimethylplatinum reagent [Me$_2$Pt(μ-SMe$_2$)]$_2$. Addition of one equiv of triflic acid caused liberation of methane and formation of cationic complex **149**, the thermolysis of which generated the platinum alkene-hydride complex **150**. Decomplexation of the platinum with potassium cyanide followed by hydrolysis of the Schiff base provided the chiral phenyl-pyrrole **151** in good overall yield.

145 A 146 147 148 149 150 (±)-151

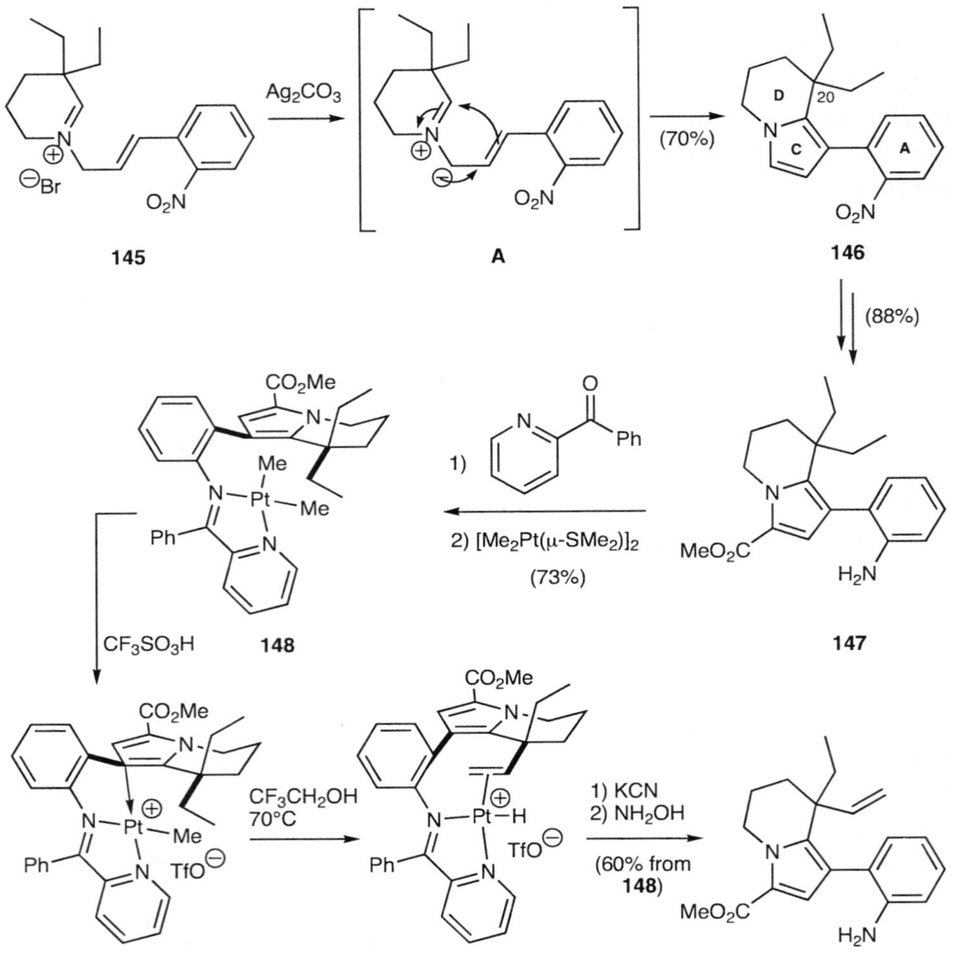

Fig. (31). Total synthesis of (±)-rhazinilam by Sames [151]

The asymmetric version of this methodology was published recently, Fig. (**32**) [152]. Schiff base formation between aniline **147** and chiral oxazolines **152** followed by formation of the platinum complex as above afforded chiral complexes **153** (R=Ph, *i*-Pr, Cy or *t*-Bu). Submitting these complexes to the reaction sequence described above generated the Schiff bases **154** with moderate yields but with good diastereoselectivities, R=Cy being optimal. The C-H activation step was highly sensitive to the temperature, higher temperatures providing better yields but lower de.

After separation of the desired major diastereoisomer **154**, the removal of the chiral auxiliary furnished vinyl compound **151** in enantiomerically pure form. The latter was directly converted to the 9-membered lactam **144** in 58% yield *via* a palladium-catalyzed carbonylation (10 atm CO, HCOOH, DME, 150°C). Removal of the methyl ester as previously described furnished (–)-rhazinilam. This elegant work constitutes the first asymmetric total synthesis of the natural product.

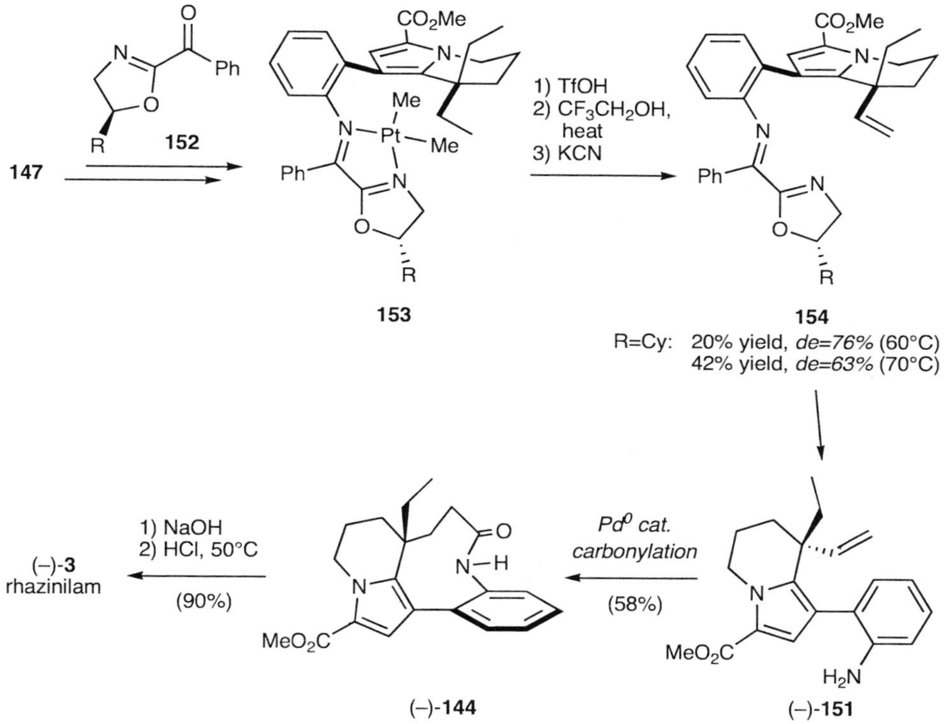

Fig. (32). Total synthesis of (–)-rhazinilam by Sames [152]
(TfOH = triflic acid = trifluoromethanesulfonic acid)

In 2001, Magnus and co-workers reported a straightforward synthesis of racemic rhazinilam by initial construction of the pyrrole ring from a piperidone, Fig. (**33**) [154]. The sequential alkylation of piperidone **155** with iodoethane and allyl bromide furnished piperidone **156**, having the requisite C-20 substitution of rhazinilam. After formation of the thiophenyl iminoether **157**, *N*-alkylation with allyl bromide **158** furnished the corresponding iminium intermediate which underwent 1,5-

electrocyclization/thiophenol elimination in basic medium to form phenyl-pyrrole **159** in good yield. Homologation of the vinyl group by sequential hydroboration and oxidations furnished carboxylic acid **160**, which was reduced catalytically (Raney Ni) to the corresponding aminoacid. The cyclization of the latter was performed in the presence of 2-chloro-1-methylpyridinium iodide to give racemic rhazinilam, with an overall yield of 8% for nine steps. Compared to the two previous total syntheses, this approach was significantly shorter for the following reasons: 1. the conversion of the piperidone **156** to the phenyl-pyrrole **159** was effected without changing the oxidation state of the piperidine ring; 2. the pyrrole ring protection/deprotection steps were avoided.

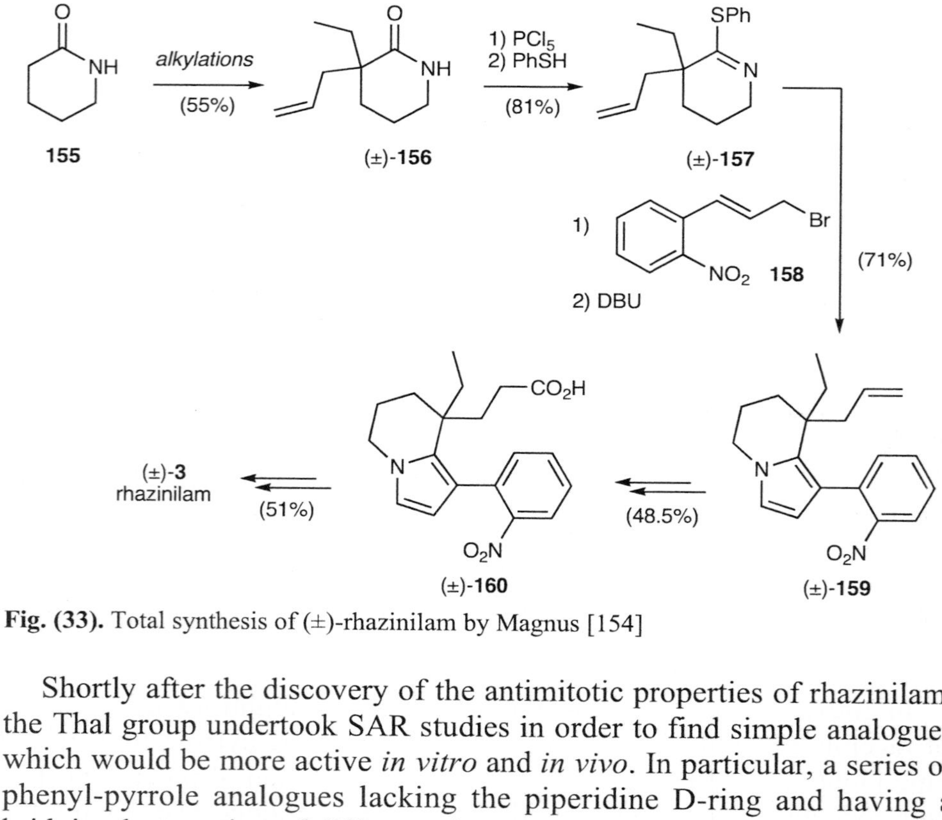

Fig. (33). Total synthesis of (±)-rhazinilam by Magnus [154]

Shortly after the discovery of the antimitotic properties of rhazinilam, the Thal group undertook SAR studies in order to find simple analogues which would be more active *in vitro* and *in vivo*. In particular, a series of phenyl-pyrrole analogues lacking the piperidine D-ring and having a bridging lactam ring of different sizes were synthesized and biologically evaluated, Fig. **(34)** [92,155,156]. The substituted phenyl-pyrroles **161** were prepared using the Barton-Zard or the Gupton methodologies [150]. After functionalization of the pyrrole ring to form intermediate **162**, the phenyl-pyrrole rhazinilam analogues **163** having a six- to nine-membered lactam ring were obtained. All compounds, including those with a nine-membered lactam ring, were found to be less active than rhazinilam by more than two orders of magnitude. This result provided the first clue that the substitution at the C-20 atom is essential for the lactam ring to adopt the rigid boat-chair conformation which is probably the active conformation.

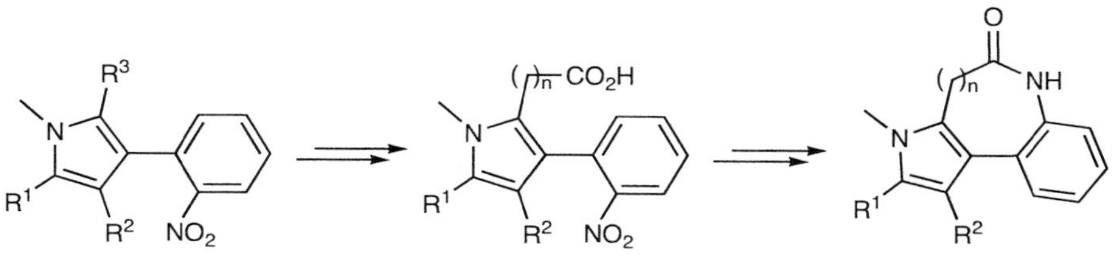

161
R^1=H, n-Bu; R^2=H, Me;
R^3=H, CO$_2$Et

162
n=0-3

163
n=0-3; R^1=H, n-Bu;
R^2=H, Me

Fig. (34). Synthesis of phenyl-pyrrole analogues of rhazinilam by Thal [92]

Synthesis of analogues by palladium-catalyzed biaryl coupling

In the continuation of the SAR studies initiated by the Thal group, Guéritte and co-workers have synthesized a series of biphenyl analogues of rhazinilam, Fig. (35) [157,158,159]. These compounds of general structure **164** have the A-B-C ring system of rhazinilam, with a nine-membered B lactam, lactone, carbamate or urea ring, and a mono- or disubstituted C-9 carbon atom (homologous to C-20 in rhazinilam). They were synthesized from a 2-substituted phenyl halide **166** and a 2-substituted phenylboronic acid or phenylstannane **165** *via* palladium-catalyzed Suzuki or Stille coupling, functionalization at C-9 and cyclization. The IC$_{50}$ values of selected analogues for cytotoxicity towards KB cells and microtubules disassembly inhibition are listed in Table 2 [159,160]. The influence of the substitution at C-9 is shown with compounds **164a-c**: the disubstitution by alkyl groups of a minimum size such as ethyl gives better activity. As shown with compound **164d**, the lactam function can be advantageously replaced by a carbamate function but compounds with a lactone or a urea group (**164e-f**) are less active. Carbamate **164g** having the six-membered D-ring of rhazinilam is less active than carbamate **164d** lacking this D-ring. The most active biphenyl analogue **164d** has, like rhazinilam, stable axial chirality and two atropisomeric enantiomers could be separated by chiral HPLC. It was found that, as with rhazinilam, the *levo*, aR-configured enantiomer is responsible for the antimitotic activity of the racemic mixture. In addition, (–)-**164d** was found to be twice as active as (–)-rhazinilam in inhibiting both microtubules assembly and disassembly, and showed comparable cytotoxicities on human cancer cell lines [159,161].

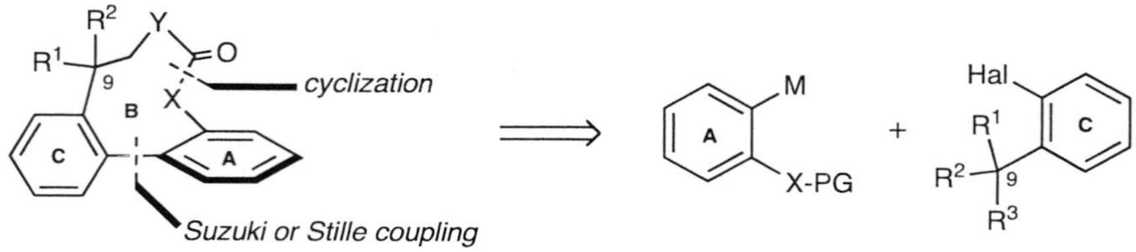

164
R^1,R^2=alkyl groups;
X=NH,O; Y=CH$_2$, NH,O

165
M=B(OH)$_2$, Sn-nBu$_3$
X=NH / PG=t-Boc
X=O / PG=Me

166
Hal=Br,I

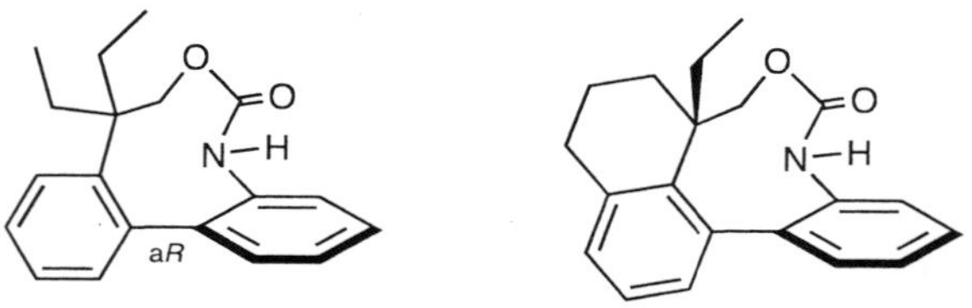

(–)-**164d**   **164g**

Fig. (35). Retrosynthetic analysis of rhazinilam biphenyl analogues **164** [158,159] (t-Boc = *tert*-butyloxycarbonyl, PG = protecting group)

Table 2. Antitubulin activity and cytotoxicity of selected biphenyl analogues of rhazinilam [159,160]

Compound	Cytotoxicity (KB cells) IC$_{50}$, µM	Inhibition of microtubules disassembly IC$_{50}$, µM
(–)-rhazinilam	2	3
164a: R^1=Et, R^2=H, X=NH, Y=CH$_2$	inact	63
164b: R^1=R^2=Et, X=NH, Y=CH$_2$	22	24
164c: R^1=R^2=Me, X=NH, Y=CH$_2$	53	51
164d: R^1=R^2=Et, X=NH, Y=O	5	3
(–)-**164d**	2	1.5
(+)-**164d**	10	inact
164e: R^1=R^2=Et, X=Y=NH	80	45
164f: R^1=R^2=Et, X=O, Y=CH$_2$	22	inact
164g	21	9

Biphenyl-carbamate **164d**, the most active analogue of rhazinilam to date, was taken as a lead and its synthesis was re-examined for the purpose of adaptation to the construction of a library of analogues, Fig. (**36**) [161].

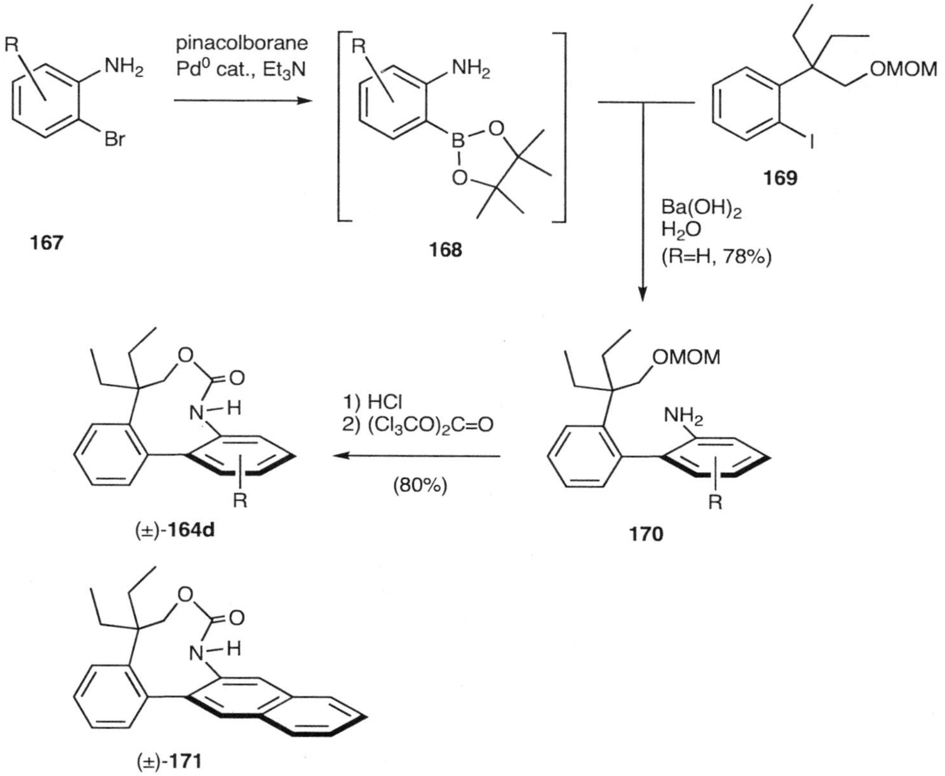

Fig. (36). Synthesis of racemic biaryl-carbamate analogues of rhazinilam by Baudoin-Guéritte [161]
(MOM = methoxymethyl)

The palladium(0)-catalyzed borylation/Suzuki coupling protocol was developed by Baudoin and co-workers in order to synthesize sterically hindered 2,2'-disubstituted biaryl compounds such as **170** in high yield and in a convenient manner [162,163]. Indeed, the catalytic borylation of 2-bromoaniline **167** (R=H) gave the corresponding pinacolboronic ester **168** which was reacted in a one-pot fashion with phenyl iodide **169** in the presence of barium hydroxide to give the functionalized biphenyl **170** in 78% yield. Cleavage of the MOM group and cyclization in the presence of triphosgene furnished biphenyl-carbamate analogue **164d** in 62% overall yield. Various compounds bearing substituents (R) on the A-ring were synthesized according to this scheme, starting from the corresponding substituted anilines **167**. The biological evaluation of these new analogues in the cytotoxicity and tubulin assays revealed that they were all less active than their unsubstituted counterpart, except the naphthyl-phenyl analogue **171** which showed a better cytotoxicity toward the breast adenocarcinoma MCF7 cell line.

Other phenyl-pyrrole and phenyl-pyridine analogues of rhazinilam, having seven to nine-membered B-rings, have been synthesized using a Suzuki-Miyaura biaryl coupling to install the biaryl bond, Fig. (37).

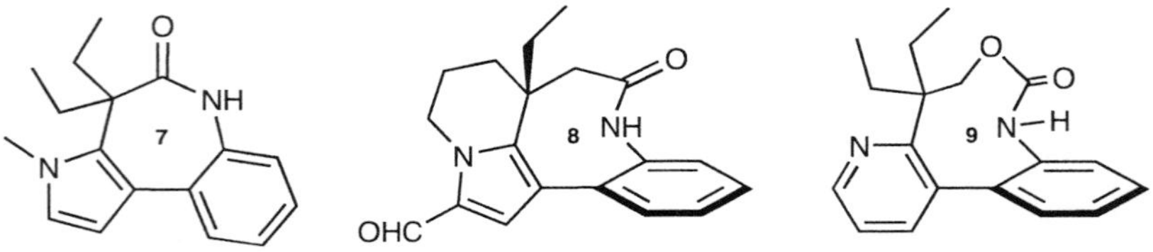

Fig. (37). Rhazinilam analogues obtained by Suzuki-Miyaura biaryl coupling [164-166]

The phenyl-pyrrole **172**, having a seven-membered lactam B-ring, showed cytotoxicity towards KB cells with an IC_{50} of 7 µM [164]. By contrast with rhazinilam, which is an inhitor of both microtubules assembly and disassembly, **172** showed only a moderate inhibition of microtubules assembly with an IC_{50} of 27 µM, suggesting another mode of action. B-Norrhazinal **173**, having an eight-membered B-ring, showed a slightly lower activity than rhazinilam in the cytotoxicity and tubulin assays [165]. A series of phenyl-pyridine analogues of biphenyl carbamate **164d** were synthesized by Queguiner and co-workers using picolinic metalation as a key step [166]. Carbamate **174** was the most active analogue, but it was less active than rhazinilam and biphenyl **164d**, with an IC_{50} of 18 µM for microtubules disassembly inhibition.

These extensive SAR studies have shown that the rigid biaryl/nine-membered lactam structure of rhazinilam adopting a boat-chair conformation is an essential feature for its antimitotic properties. Moreover, the absolute a*R* configuration of the biaryl axis is, as in the allocolchicine and stegane series, absolutely required for biological activity. The possibility of synthesizing more active analogues will thus reside in the fine-tuning of the drug-tubulin interactions while respecting these structural requirements.

4. CONCLUSION

In the case of chiral natural substances, it is always surprising to note that Nature often chooses to produce the correct enantiomer or diastereoisomer that will fit with a given cellular receptor and will lead to biological activity. This feature also concerns the natural bridged biaryls with axial chirality possessing antimitotic activity, which we have described in this review. *Colchicum autumnale*, *Steganotaenia araliaceae* and *Rhazia stricta* led respectively to active (–)-colchicine, (–)-steganacine and (–)-rhazinilam. Their (+)-enantiomers were not found in Nature but were chemically prepared to demonstrate the importance of the 3D structure for the interaction with tubulin. The discovery of natural (–)-colchicine, (–)-steganacin and (–)-rhazinilam was the starting point for chemists to prepare a number of analogues in these three series. Thus, important advances were made in the knowledge of SAR and in the search for therapeutically useful agents. In addition, new synthetic methodologies, especially in asymmetric synthesis, were developed to obtain these structurally challenging molecules.

ABBREVIATIONS

t-Boc	=	*tert*-butoxycarbonyl
*m*CPBA	=	*meta*chloroperbenzoic acid
Cy	=	cyclohexyl
DBU	=	1,8-diazabicyclo[5.4.0]undec-7-ene
DCC	=	1,3-dicyclohexylcarbodiimide
DDQ	=	2,3-dichloro-5,6-dicyano-1,4-benzoquinone
de	=	diastereomeric excess
ee	=	enantiomeric excess
equiv	=	equivalent
HPLC	=	high performance liquid chromatography
IC$_{50}$	=	50% Inhibition Concentration: for the tubulin assay, the concentration of compound required to inhibit 50% of

the rate of microtubules assembly or disassembly; for cytotoxicity towards cancer cells, the concentration of compound causing 50% growth inhibition after incubation

LDA	=	lithium diisopropylamide
MOM	=	methoxymethyl
PCC	=	pyridinium chlorochromate
SAR	=	Structure-Activity Relationships
TBS	=	*tert*-butyldimethylsilyl
TFA	=	trifluoroacetic acid
TFAA	=	trifluoroacetic anhydride
TfOH	=	trifluoromethanesulfonic (triflic) acid
TTFA	=	thallium(III) trifluoroacetate
p-TsOH	=	*p*-toluenesulfonic acid

ACKNOWLEDGEMENTS

The authors gratefully thank Dr. Robert H. Dodd for assistance in the preparation of the manuscript and the Centre National de la Recherche Scientifique and Institut National de la Santé et de la Recherche Médicale for financial support.

REFERENCES

[1] Bringmann, G.; Günther, C.; Ochse, M.; Schupp, O.; Tasler, S. In *Progress in the Chemistry of Organic Natural Products*; Herz, W., Falk, H., Kirby, G.W., Moore, R.E., Eds.; Springer-Verlag: Wien, **2001**.

[2] Jordan, A.; Hadfield, J.A.; Lawrence, N.J.; McGown, A.T.; *Med. Res. Rev.,* **1998**, *18*, 259.

[3] Santavy, F.; Macak, V.; *Chem. Listy.,* **1953**, *47*, 1214.

[4] Santavy, F.; Macak, V.; *Collect. Czech. Chem. Commun.,* **1954**, *19*, 805.

[5] Yusupov, M.K.; Sadykov, A.S.; *Zh. Obsch. Khim.,* **1964**, *34*, 1677.

[6] Pelletier, P.J.; Caventou, J.; *Ann. Chim. Phys.,* **1820**, *14*, 69.

[7] Pernice, B.; *Silicia Med.,* **1889**, 265.

[8] Borisy, G.G.; Taylor, E.W.; *J. Cell. Biol.,* **1967**, *34*, 525.

[9] Shelanski, M.L.; Taylor, E.W.; *J. Cell. Biol.,* **1967**, *34*, 549.

[10] Dustin, P. In *Microtubules*, 2nd Ed., Springer-Verlag: New York, **1984**.

[11] Shi, Q.; Chen, K.; Morris-Natschke, S.L.; Lee, K.-H.; *Curr. Pharm. Des.,* **1998**, *4*, 219.

[12] Correia, J.J.; Lobert, S.; *Curr. Pharm. Des.*, **2001**, *7*, 1213.

[13] Capraro, H-G.; Brossi, A. In *The Alkaloids*; Brossi, A. Ed.; Academic Press: New York, **1984**, *23*, 1.

[14] Boyé, O.; Brossi, A. In *The Alkalois*; Brossi, A., Cordell, G.A., Eds.; Academic Press: New York, **1992**, *41*, 125.

[15] Hastié, S.B.; *Pharmac. Ther.*, **1991**, *51*, 377.

[16] Downing, K.H.; Nogales, E.K.; *Eur. Biophys. J.*, **1998**, *27*, 431.

[17] Morrisson, J. D.; *Acta Cryst.*, **1951**, *4*, 69.

[18] King, M. V.; de Vries, J. L.; Pepinsky, R.; *Acta Cryst.*, **1952**, *5*, 437.

[19] Corrodi, H.; Hardegerr, E.; *Helv. Chim. Acta*, **1955**, *38*, 2030.

[20] Silverton, J.V.; Dumont, R.; Brossi, A.; *Acta. Cryst.*, **1987**, *C43*, 1802.

[21] Brossi, A.; Boyé, O.; Muzaffar, A.; Yeh, H.J.C.; Toome, V.; Wegrzynski, B.; George, C.; *FEBS Lett.*, **1990**, *262*, 45.

[22] Brecht, R.; Haenel, F.; Seitz, G.; *Liebigs Ann./Recueil*, **1997**, *11*, 2275.

[23] Berg, U.; Bladh, H.; *Helv. Chim. Acta.*, **1999**, *82*, 323.

[24] Brossi, A.; Lee, K.-H.; Yeh, H.J.C.; *Helv. Chim. Acta.*, **1999**, *82*, 1223.

[25] Roesner, M.; Capraro, H.-G.; Jacobson, A.E.; Atwell, L.; Brossi, A.; Iorio, M.A.; Williams, T.H.; Sik, R.H.; Chignell, C.F.; *J. Med. Chem.*, **1981**, *24*, 257.

[26] Brossi, A.; Yeh, H.J.C.; Chrzanowska, M.; Wolff, J.; Hamel, E.; Lin, C.; Quin, F.; Suffness, M.; Silverton, J.; *Med. Res. Rev*, **1988**, *8*, 77.

[27] Berg, U.; Deinum, J.; Lincoln, P.; Kvassman, J.; *Bioorg. Chem*, **1991**, *19*, 53.

[28] Al-Tel, T.H.; Abu-Zarga, M.H.; Sabri, S.S.; Frayer, A.J.; Sharma, M.; *J. Nat. Prod.* **1990**, *53*, 623.

[29] Tojo, E.; Abu Zarga , M.H.; Freyer, A.; Shamma, M.; *J. Nat. Prod.*, **1989**, *52*, 1163.

[30] Abu Zarga, M.H.; Sabri, S.S.; Al-Tel, T.H.; Atta-Ur-Rahman, Z.; Shah, Z., Feroz, M.; *J. Nat. Prod.*, **1991**, *54*, 936.

[31] Banwell, M.G.; Fam, M.A.; Gable, R.W.; Hamel, E.; *J. Chem. Soc., Chem. Commun.,* **1994**, 2647.

[32] Mackay, M.F.; Lacey, E.; Burden, P.; *Acta. Cryst.*, **1989**, *C45*, 799.

[33] Fitzgerald, T.J.; *Biochem. Pharmacol.*, **1976**, *25*, 1383.

[34] Itoh, Y.; Brossi, A.; Hamel, E.; Lin, C.M.; *Helv. Chim. Acta*, **1988**, *71*, 1199.

[35] Ayres, D.C.; Loike, J.D.; In *Lignans. Chemical, Biological and Clinical Properties; Chemistry and Pharmacology of Natural Products*; Phillipson, J.D., Ayres, D.C., Baxter, H., Eds.; Cambridge University Press: Cambridge (UK), **1990**.

[36] Kupchan, S.M.; Britton, R.W.; Ziegler, M.F.; Gilmore, C.J.; Restivo, R.J.; Bryan, R.F.; *J. Am. Chem. Soc.*, **1973**, *95*, 1336.

[37] Robin, J.-P.; Gringore, O.; Brown, E.; *Tetrahedron Lett.*, **1980**, *21*, 2709.

[38] Tomioka, K.; Ishiguro, T.; Koga, K.; *Tetrahedron Lett.*, **1980**, *21*, 2973.

[39] Tomioka, K.; Ishiguro, T.; Iitaka, Y.; Koga, K.; *Tetrahedron*, **1984**, *40*, 1303.

[40] Wang, R. W.-J.; Rebhun, L.I.; Kupchan, S.M.; *Cancer Res.*, **1977**, *37*, 3071.

[41] Schiff, P.B.; Kende, A.; Horwitz, S.B.; *Biochem. Biophys. Res. Commun.,* **1978**, 737.

[42] Zavala, F.; Guénard, D.; Robin, J.-P.; Brown, E.; *J. Med. Chem.*, **1980**, *23*, 546.

[43] Tomioka, K.; Ishiguro, T.; Mizuguchi, H.; Komeshima, N.; Koga, K.; Tsukagoshi, S.; Tsuruo, T.; Tashiro, T.; Tanida, S.; Kishi, T.; *J. Med. Chem*, **1991**, *34*, 54.

414

[44] Taafrout, M.; Rouessac, F.; Robin, J.-P.; *Tetrahedron Lett.*, **1983**, *24*, 197.

[45] Taafrout, M.; Rouessac, F.; Robin, J.-P.; *Tetrahedron Lett.*, **1983**, *24*, 2983.

[46] Hicks, R.P.; Sneden, A.T.; *Tetrahedron Lett.*, **1983**, *24*, 2987.

[47] Taafrout, M.; Rouessac, F.; Robin, J.-P.; Hicks, R.P.; Shillady, D.D.; Sneden, A.T.; *J. Nat. Prod*, **1984**, *47*, 600.

[48] Taafrout, M.; Landais, Y.; Robin, J.-P.; *Tetrahedron Lett.*, **1986**, *24*, 1781.

[49] Robin, J.-P.; Davoust, D.; Taafrout, M.; *Tetrahedron Lett.*, **1986**, *27*, 2871.

[50] Wickramaratne, D.B.M.; Pengsuparp, T.; Mar, W.; Chai, H.-B.; Chagwedera, T.E.; Beecher, C.W.W.; Farnsworth, N.R.; Kinghorn, A.D.; Pezzuto, J.M.; Cordell, G.A.; *J. Nat. Prod.*, **1993**, *56*, 2083.

[51] Meragelman, K.M.; McKee, T.C.; Boyd, M.R.; *J. Nat. Prod.*, **2001**, *64*, 1480.

[52] Houlbert, N.; Brown, E.; Robin, J.-P.; Davoust, D.; Chiaroni, A.; Prangé, T.; Riche, C.; *J. Nat. Prod.*, **1985**, *48*, 345.

[53] Linde, H.H.A.; *Helv. Chim. Acta.*, **1965**, *48*, 1822.

[54] Banerji, A.; Majumder, P.L.; Chatterjee, A.; *Phytochem.*, **1970**, *9*, 1491.

[55] De Silva, K.T.; Ratcliffe, A.H.; Smith, G.F.; Smith, G.N. *Tetrahedron Lett.*, **1972**, *10*, 913.

[56] Abraham D.J.; Rosenstein, R.D.; Lyon, R.L.; Fong, H.H.S. *Tetrahedron Lett.*, **1972**, *10*, 909.

[57] Lyon, R.L; Fong, H.H; Farnsworth, N.R; Svoboda, G.H.; *J. Pharm. Sciences*, **1973**, *62*, 218.

[58] Goh, S.H.; Razak Mohd Ali, A.; *Tetrahedron Lett.*, **1986**, *27*, 2501.

[59] Goh, S.H.; Razak Mohd Ali, A.; Wong, W.H.; *Tetrahedron*, **1989**, *45*, 7899.

[60] Thoison, O.; Guénard, D.; Sévenet, T.; Kan-Fan, C.; Quirion, J.-C.; Husson, H.-P.; Deverre, J.-R.; Chan, K.C.; Potier, P.; *C. R.. Acad. Sc. Paris II*, **1987**, *304*, 157.

[61] Sheludko, S.; Gerasimenko, I.; Platonova, O.; *Planta Med.*, **2000**, *66*, 656.

[62] Ratcliffe, A.H.; Smith, G.F.; Smith, G.N.; *Tetrahedron Lett.*, **1973**, *52*, 5179.

[63] Kam, T.-S.; Subramaniam, G.; Chen, W.; *Phytochem.*, **1999**, *51*, 159.

[64] Gerasimenko, I.; Sheludko, Y.; Stöckigt, J.; *J. Nat. Prod*, **2001**, *64*, 114.

[65] Aimi, N.; Uchida, N.; Ohya, N.; Hosokawa, H.; Takayama, H.; Sakai, S.-I.; Mendoza, L.A.; Polz, L.; Stöckigt, J.; *Tetrahedron Lett.*, **1991**, *32*, 4949.

[66] Kam, T.S.; Tee, Y.M.; Subramaniam, G.; *Nat. Prod. Lett.*, **1998**, *12*, 307.

[67] David, B.; Sévenet, T.; Morgat, M.; Guénard, D.; Moisand, A.; Tollon, Y.; Thoison, O.; Wright, M.; *Cell. Motil. Cytoskeleton*, **1994**, *28*, 317.

[68] Wildman, W.C. In *The Alkaloids*; Manske, R.H.F., Ed.; Academic Press: New York, **1952**; *6*, 261.

[69] Santavy, F.; *Planta Med., Suppl.*, **1968**, 46.

[70] Fernholz, H.; *Liebigs Ann. Chem.*, **1950**, *568*, 63.

[71] Santavy, F.; *Helv. Chim. Acta*, **1948**, *31*, 821.

[72] Cech, J.; Santavy, F.; *Collect. Czech. Chem. Commun.*, **1949**, *14*, 532.

[73] Iorio, M.; *Heterocycles*, **1984**, *22*, 2207.

[74] Dougherty, G. (Angiogene Pharmaceuticals Ltd., UK); PCT Int. Appl. WO 99/02166, **1999**.

[75] Timasheff, S.M.; Gorbunoff, M.J.; Perez-Ramirez, B. (Brandeis University, USA); United States Patent US5760092, **1998**.

[76] Shi, Q.; Chen, K.; Brossi, A.; Verdier-Pinard, P.; Hamel, E.; McPhail, A.T.; Tropsha, A.; Lee, K.-H.; *J. Org. Chem.*, **1998**, *63*, 4018.

[77] Shi, Q.; Chen, K; Brossi, A.; Verdier-Pinard, P.; Hamel, E.; McPhail, A.T.; Lee, K.-H.; *Helv. Chim. Acta*, **1998**, *81*, 1023.

[78] Guan, J.; Zhu, X.-K.; Brossi, A.; Tachibana, Y.; Bastow, K.F.; Verdier-Pinard, P.; Hamel, E.; McPhail, A.T.; Lee, K.-H.; *Collect. Czech. Chem. Commun.*, **1999**, *64*, 217.

[79] Han, S.; Hamel, E.; Bastow, K.F.; McPhail, A.T.; Brossi, A.; Lee, K.-H.; *Bioorg. Med. Chem. Lett.*, **2002**, *12*, 2851.

[80] Dilger, U.; Franz, B.; Röttele, H.; Schröder, G.; *J. Prakt. Chem.*, **1998**, *340*, 468.

[81] Brecht, R.; Seitz, G.; Guénard, D.; Thoret, S.; *Bioorg. Med. Chem.*, **2000**, *8*, 557.

[82] Taafrout, M.; Rouessac, F.; Robin, J.-P.; *Tetrahedron Lett.*, **1983**, *24*, 3237.

[83] Taafrout, M.; Rouessac, F.; Robin, J.-P.; *Tetrahedron Lett.*, **1984**, *25*, 4127.

[84] Damon, R.E.; Schlessinger, R.H.; Blount, J.F.; *J. Org. Chem.*, **1976**, *41*, 3772.

[85] Bowden, B.F.; Read, R.W.; Taylor, W.C.; *Aust. J. Chem.*, **1980**, *33*, 1823.

[86] Bowden, B.F.; Read, R.W.; Taylor, W.C.; *Aust. J. Chem.*, **1981**, *34*, 799.

[87] Ward, R.S.; Hughes, D.D.; *Tetrahedron*, **2001**, *57*, 4015 and references therein.

[88] David, B.; Sévenet, T.; Thoison, O.; Awang, K.; Païs, M.; Wright, M.; Guénard; *Bioorg. Med. Chem. Lett.*, **1997**, *7*, 2155.

[89] Dupont, C.; Guénard, D.; Tchertanov, L.; Thoret, S.; Guéritte, F.; *Bioorg. Med. Chem.*, **1999**, *7*, 2961.

[90] Lévy, J.; Soufyane, M.; Mirand, C.; Döé de Maindreville, M.; Royer, D.; *Tetrahedron: Asymmetry*, **1997**, *8*, 4127.

[91] Soufyane, M.; *Ph. D. Thesis*; Reims University (France); **1993**.

[92] Alazard, J.-P.; Millet-Paillusson, C.; Guénard, D.; Thal, C.; *Bull. Soc. Chim. Fr.*, **1996**, *133*, 251.

[93] Rapoport, H.; Williams, A.R.; Cisney, M.E.; *J. Am. Chem. Soc.*, **1950**, *72*, 3324.

[94] Rapoport, H.; Williams, A.R.; Cisney, M.E.; *J. Am. Chem. Soc.*, **1951**, *73*, 1414.

[95] Cook, J.W.; Jack, J.; Loudon, J.D.; Buchanan, G.L.; MacMillan, J.; *J. Chem. Soc.*, **1951**, 1397.

[96] Sawyer, J.S.; Macdonald, T.L.; *Tetrahedron Lett.*, **1988**, *29*, 4839.

[97] Hara, H.; Shinoki, H.; Hoshino, O.; Umezawa, B.; *Heterocycles*, **1983**, *20*, 2155.

[98] Banwell, M.G.; Lambert, J.N.; Mackay, M.F.; Greenwood, R.J.; *J. Chem. Soc., Chem. Commun.*, **1992**, 974.

[99] Banwell, M.G.; *Pure Appl. Chem.*, **1996**, *68*, 539.

[100] Boyé, O.; Itoh, Y.; Brossi, A.; *Helv. Chim. Acta*, **1989**, *72*, 1690.

[101] Boyé, O.; Brossi, A.; Yeh, H.J.C.; Hamel, E.; *Can. J. Chem.*, **1992**, *70*, 1237.

[102] Banwell, M.G.; Lambert, J.N.; Corbett, M.; Greenwood, R.J.; Gulbis, J.M.; Mackay, M.F.; *J. Chem. Soc., Perkin Trans. 1*, **1992**, 1415.

[103] Banwell, M.G.; Cameron, J.M.; Corbett, M.; Dupuche, J.R.; Hamel, E.; Lambert, J.N.; Lin, C.M.; Mackay, M.F.; *Aust. J. Chem.*, **1992**, *45*, 1967.

[104] Vorogushin, A.V.; Wulff, W.D.; Hansen, H.-J.; *Org. Lett.*, **2001**, *3*, 2641.

[105] Vorogushin, A.V.; Wulff, W.D.; Hansen, H.-J.; *J. Am. Chem. Soc.*, **2002**, *124*, 6512.

[106] Lee, J.C.; Cha, J.K.; *Tetrahedron*, **2000**, *56*, 10175 and references therein.

416

[107] Ward, R.S.; *Chem. Soc. Rev.*, **1982**, *11*, 75.

[108] Ward, R.S.; *Tetrahedron*, **1990**, *46*, 5029.

[109] Kende, A.; Liebeskind, L.S.; Kubiak, C.; Eisenberg, R.; *J. Am. Chem. Soc.*, **1976**, *98*, 6389.

[110] Brown, E.; Robin, J.-P.; *Tetrahedron Lett.*, **1978**, 3613.

[111] Tomioka, K.; Mizuguchi, H.; Koga, K.; *Tetrahedron Lett.*, **1979**, 1409.

[112] Tomioka, K.; Mizuguchi, H.; Ishiguro, T.; Koga, K.; *Chem. Pharm. Bull.*, **1985**, *33*, 121.

[113] Hughes, L.R.; Raphael, R.A.; *Tetrahedron Lett.*, **1976**, 1543.

[114] Becker, D.; Hughes, L.R.; Raphael, R.A.; *J. Chem. Soc., Perkin Trans. 1*, **1977**, 1674.

[115] Larson, E.R.; Raphael, R.A.; *Tetrahedron Lett.*, **1979**, 5041.

[116] Larson, E.R.; Raphael, R.A.; *J. Chem. Soc., Perkin Trans. 1*, **1982**, 521.

[117] Krow, G.R.; Damodaran, K.M.; Michener, E.; Wolf, R.; Guare, J.; *J. Org. Chem.*, **1978**, *43*, 3950.

[118] Narasimhan, N.S.; Aidhen, I.S.; *Tetrahedron Lett.*, **1988**, *29*, 2987.

[119] Aidhen, I.S.; Narasimhan, N.S.; *Indian J. Chem., Section B*, **1993**, *32*, 211.

[120] Kende, A.S.; Liebeskind, L.S.; *J. Am. Chem. Soc.*, **1976**, *98*, 267.

[121] Tomioka, K.; Mizuguchi, H.; Koga, K.; *Tetrahedron Lett.*, **1978**, 4687.

[122] Tomioka, K.; Koga, K.; *Tetrahedron Lett.*, **1979**, 3315.

[123] Tomioka, K.; Mizuguchi, H.; Koga, K.; *Chem. Pharm. Bull.*, **1982**, *30*, 4304.

[124] van Oeveren, A.; Jansen, J.F.G.A.; Feringa, B.L.; *J. Org. Chem.*, **1994**, *59*, 5999 and references therein.

[125] Enders, D.; Kirchhoff, J.; Lausberg, V.; *Liebigs Ann. Chem.*, **1996**, 1361.

[126] Enders, D.; Lausberg, V.; Del Signore, G.; Berner, O.M.; *Synthesis*, **2002**, 515.

[127] Magnus, P.; Schultz, J.; Gallagher, T.; *J. Chem. Soc., Chem. Commun.*, **1984**, 1179.

[128] Magnus, P.; Schultz, J.; Gallagher, T.; *J. Am. Chem. Soc.*, **1985**, *107*, 4984.

[129] Planchenault, D.; Dhal, R.; Robin, J.-P.; *Tetrahedron*, **1993**, *49*, 5823 and references therein.

[130] Ziegler, F.E.; Fowler, K.W.; Sinha, N.D.; *Tetrahedron Lett.*, **1978**, 2767.

[131] Ziegler, F.E.; Schwartz, J.A.; *J. Org. Chem.*, **1978**, *43*, 985.

[132] Ziegler, F.E.; Chliwner, I.; Fowler, K.W.; Kanfer, S.J.; Kuo, S.J.; Sinha, N.D.; *J. Am. Chem. Soc.*, **1980**, *102*, 790.

[133] Brown, E.; Robin, J.-P.; *Tetrahedron Lett.*, **1977**, 2015.

[134] Brown, E.; Dhal, R.; Robin, J.-P.; *Tetrahedron Lett.*, **1979**, 733.

[135] Dhal, R.; Brown, E.; Robin, J.-P.; *Tetrahedron*, **1983**, *39*, 2787.

[136] Robin, J.-P.; Dhal, R.; Brown, E.; *Tetrahedron*, **1984**, *40*, 3509.

[137] Meyers, A.I.; Lutomski, K.A.; *J. Am. Chem. Soc.*, **1982**, *104*, 879.

[138] Meyers, A.I.; Himmelsbach, R.J.; *J. Am. Chem. Soc.*, **1985**, *107*, 682.

[139] Meyers, A.I.; Flisak, J.R.; Aitken, R.A.; *J. Am. Chem. Soc.*, **1987**, *109*, 5446.

[140] Kamikawa, K.; Uemura, M.; *Synlett*, **2000**, 938.

[141] Uemura, M.; Daimon, A.; Hayashi, Y.; *J. Chem. Soc., Chem. Commun.*, **1995**, 1943.

[142] Kamikawa, K.; Watanabe, T.; Daimon, A.; Uemura, M.; *Tetrahedron*, **2000**, *56*, 2325.

[143] Monovich, L.G.; Le Huérou, Y.; Rönn, M.; Molander, G.A.; *J. Am. Chem. Soc.*, **2000**, *122*, 52.

[144] Bradley, A.; Motherwell, W.B.; Ujjainwalla, F.; *Chem. Commun.*, **1999**, 917.

[145] Sackett, D.; *Pharmac. Ther.*, **1993**, *59*, 163.

[146] Hicks, R.P.; Sneden, A.T.; *J. Nat. Prod.*, **1985**, *48*, 357.

[147] Tomioka, K.; Kawasaki, H.; Koga, K.; *Chem. Pharm. Bull.* **1990**, *38*, 1899.

[148] Tomioka, K.; Kubota, Y.; Kawasaki, H.; Koga, K.; *Tetrahedron Lett.*, **1989**, *30*, 2949.

[149] Kubota, Y.; Kawasaki, H.; Tomioka, K.; Koga, K.; *Tetrahedron*, **1993**, *49*, 3081.

[150] for a recent review of the synthesis of pyrroles: Ferreira, V.F.; de Souza, M.C.B.V.; Cunha, A.C.; Pereira, L.O.R.; Ferreira, M.L.G.; *Org. Prep. Proced. Int.*, **2001**, *33*, 411.

[151] Johnson, J.A.; Sames, D.; *J. Am. Chem. Soc.*, **2000**, *122*, 6321.

[152] Johnson, J.A.; Li, N.; Sames, D.; *J. Am. Chem. Soc.*, **2002**, *124*, 6900.

[153] Grigg, R.; Myers, P.; Somasunderam, A.; Sridharan, V.; *Tetrahedron*, **1992**, *48*, 9735.

[154] Magnus, P.; Rainey, T.; *Tetrahedron*, **2001**, *57*, 8647.

[155] Alazard, J.-P.; Millet-Paillusson, C.; Boyé, O.; Guénard, D.; Chiaroni, A.; Riche, C.; Thal, C.; *Bioorg. Med. Chem. Lett.*, **1991**, *1*, 725.

[156] Alazard, J.-P.; Boyé, O.; Gillet, B.; Guénard, D.; Beloeil, J.-C.; Thal, C.; *Bull. Soc. Chim. Fr.*, **1993**, *130*, 779.

[157] Pascal, C.; Guéritte-Voegelein, F.; Thal, C.; Guénard, D.; *Synth. Commun.*, **1997**, *27*, 1501.

[158] Pascal, C.; Dubois, J.; Guénard, D.; Guéritte, F.; *J. Org. Chem.*, **1998**, *63*, 6414.

[159] Pascal, C.; Dubois, J.; Guénard, D.; Tchertanov, L.; Thoret, S.; Guéritte, F.; *Tetrahedron*, **1998**, *54*, 14737.

[160] Pascal, C.; *Ph. D. Thesis*; Paris XI Orsay University (France); **1997**.

[161] Baudoin, O.; Claveau, F.; Thoret, S.; Herrbach, A.; Guénard, D.; Guéritte, F.; *Bioorg. Med. Chem.*, **2002**, *10*, 3395.

[162] Baudoin, O.; Guénard, D.; Guéritte, F.; *J. Org. Chem.*, **2000**, *65*, 9268.

[163] Baudoin, O.; Cesario, M.; Guénard, D.; Guéritte, F.; *J. Org. Chem.*, **2002**, *67*, 1199.

[164] Dupont, C.; Guénard, D.; Thal, C.; Thoret, S.; Guéritte, F.; *Tetrahedron Lett.*, **2000**, *41*, 5853.

[165] Banwell, M.; Edwards, A.; Smith, J.; Hamel, E.; Verdier-Pinard, P.; *J. Chem. Soc., Perkin Trans. 1*, **2000**, 1497.

[166] Pasquinet, E.; Rocca, P.; Richalot, S.; Guéritte, F.; Guénard, D.; Godard, A.; Marsais, F.; Quéguiner, G.; *J. Org. Chem.*, **2001**, *66*, 2654.

Atta-ur-Rahman (Ed.) *Studies in Natural Products Chemistry, Vol. 29*
© 2003 Elsevier B.V. All rights reserved.

SYNTHETIC STUDIES ON BIOLOGICALLY ACTIVE ALKALOIDS STARTING FROM LACTAM-TYPE CHIRAL BUILDING BLOCKS

NAOKI TOYOOKA* AND HIDEO NEMOTO

Faculty of Pharmaceutical Sciences, Toyama Medical and Pharmaceutical University, Sugitani 2630, Toyama 930-0194, Japan

Abstract: This review describes synthetic studies on biologically active alkaloids starting from three lactams. A stereodivergent process for the synthesis of 3-piperidinol alkaloids and the total synthesis of cytotoxic marine alkaloids, clavepictines A, B, pictamine, and lepadin B, were achieved starting from our original lactam-type chiral building block **1**. The synthesis of the proposed structure for dart-poison frog alkaloid **223A** and a stereodivergent process for the synthesis of decahydroquinoline-type dart-poison frog alkaloids were achieved from lactam **31**. Finally, synthesis of the revised structure for the above alkaloid **223A** and dart-poison frog alkaloid **207I** was also achieved starting from lactam **68**.

INTRODUCTION

Construction of versatile chiral building blocks would provide us with powerful tools for the synthesis of certain natural products. For instance, 6-membered lactam is an attractive building block for the synthesis of alkaloids containing a piperidine ring system. We designed lactam **1** as a promising chiral building block for the synthesis of biologically active alkaloids containing a 3-piperidinol ring core. Both enantiomers of **1** were synthsized using biocatalysis (lipase and bakers' yeast), and elaboration of **1** resulted in the first total synthesis of the marine alkaloids clavepictines A, B, pictamine, and lepadin B. A stereodivergent process for the synthesis of α,α'-disubstituted 3-piperidinol alkaloid has also been established. The synthesis of the 3-piperidinol alkaloids, prosafrinine, iso-6-cassine, prosophylline, and prosopinine, was achieved based upon the above process. On the other hand, we established a stereodivergent process for the synthesis of decahydroquinoline-type dart-poison frog alkaloids, and the synthetic route to the proposed structure for the alkaloid **223A** was explored starting from known lactam **31**. Finally, synthesis of the revised structure for the alkaloid **223A** and dart-poison frog alkaloid **207I** was also achieved starting from *cis*-substituted lactam **68**.

1. SYNTHETIC STUDIES ON BIOLOGICALLY ACTIVE ALKALOIDS STARTING FROM LACTAM 1

1.1. Synthesis of both enantiomers of lactam 1

A number of α, α'-disubstituted 3-piperidinol alkaloids have been found in *Cassia* or *Prosopis* species,[1] and, quite recently, alkaloids including this structural unit have also been isolated from ascidian.[2] Many of these alkaloids showed interesting pharmacological activities such as anesthetic, analgestic, and antibiotic activities. Clavepictines A and B, and pictamine, isolated from tunicate, by Cardellina, II[3] or Faulkner[4] and co-workers, possess 3-piperidinol structure. Lepadins A, B, and C, isolated from tunicate, by Steffan[5] and Andersen[6] and co-workers, also have this structural unit. We designed lactam **1** as a useful

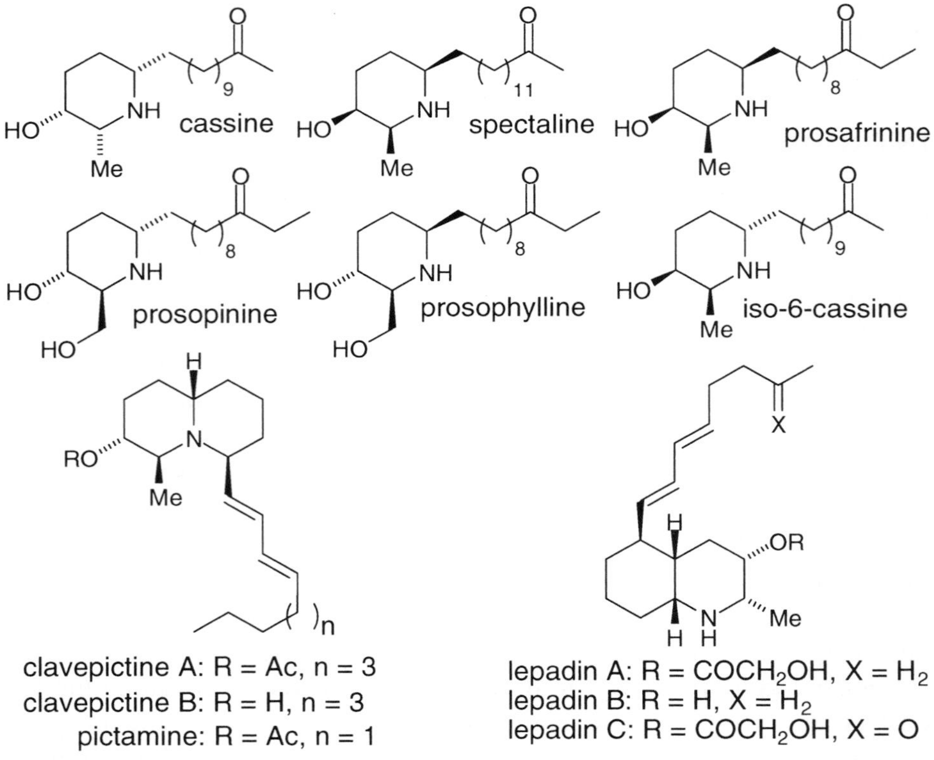

clavepictine A: R = Ac, n = 3
clavepictine B: R = H, n = 3
pictamine: R = Ac, n = 1

lepadin A: R = COCH₂OH, X = H₂
lepadin B: R = H, X = H₂
lepadin C: R = COCH₂OH, X = O

Fig. (1). Representative alkaloids possessing the α,α'-disubstituted 3-piperidinol nuclei

chiral building block for the synthesis of the above alkaloids containing the 3-piperidinol nuclei. First, we examined the synthesis of both enantiomers of **1** starting from known β-keto ester **2** [7]. The lipase-mediated transesterification of (±)-**1**, derived from **2**, afforded the acetate **3** in 47% yield (>99% ee) along with (-)-**1** (52%, 91% ee). Hydrolysis of **3** with K$_2$CO$_3$ gave rise to (+)-**1**. The enantiomer (-)-**1** was found to be derived more effectively from the bakers' yeast reduction of **2** in high optical yield (98% ee). Direct recrystallization of the crude reduction product resulted in the enantiomerically pure (-)-**1** in 88% isolated yield.

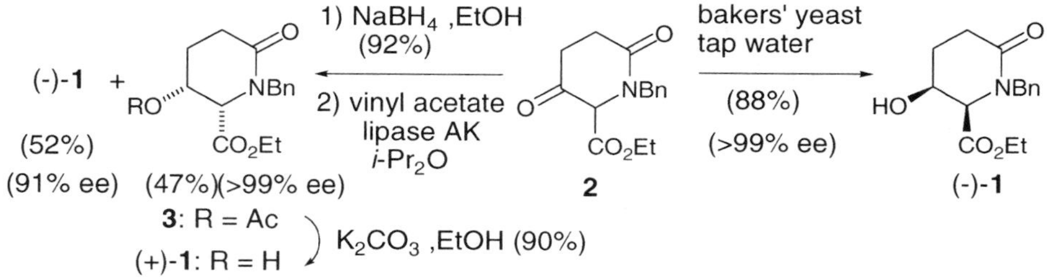

Scheme 1. Synthesis of both enantiomers of lactam **1**

The absolute stereochemistry of (-)-**1** was determined to be 2*R*, 3*S* by conversion of (-)-**1** to known piperidine **5**[8] as shown in Scheme 2.

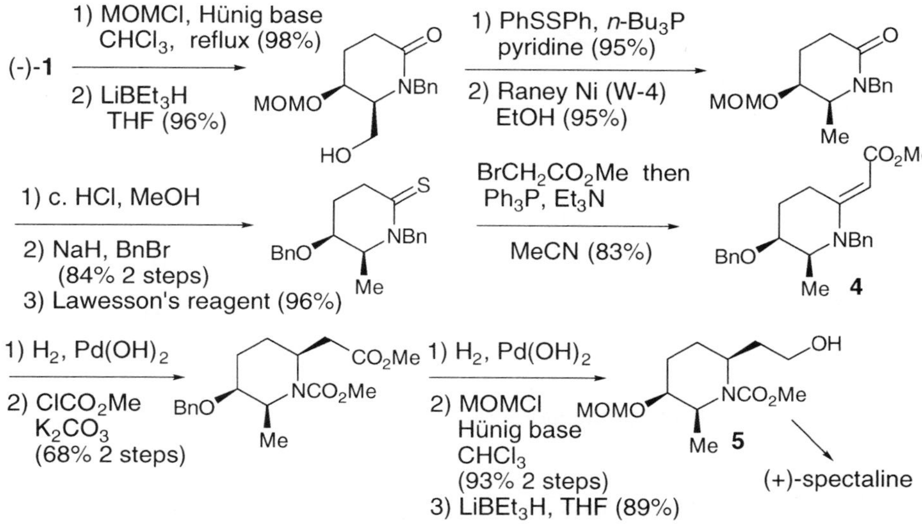

Scheme 2. Determination of the absolute stereochemistry of (-)-**1**

The stereoselectivity of the reduction of vinylogous urethane **4**, as shown in Scheme 2, may be attributed to steric hindrance, by which the catalytic hydrogenation occurs from the less hindered site (α-face) of **4** to give an all *cis*-substituted reduction product (Fig 2).

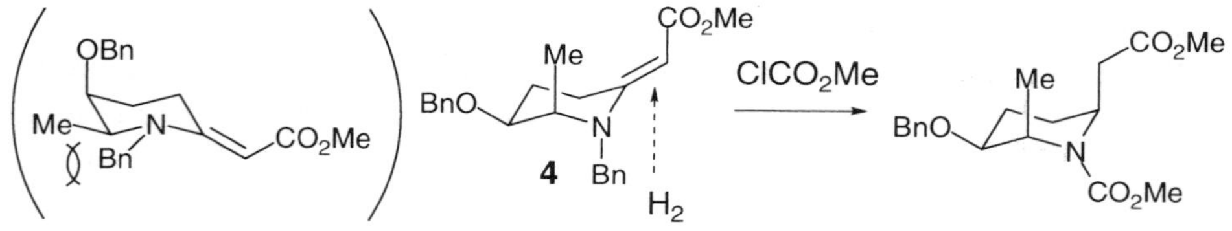

Fig. (2). Stereochemical course of the reduction of **4**

1.2. Stereodivergent synthesis of 3-piperidinol alkaloids

With the enantiomeric pair of **1** in hand, we examined the stereodivergent synthesis of α,α'-disubstitued 3-piperidinol alkaloids. The basic strategy we used to prepare the building blocks (**I-IV**) from (-)-**1** is presented in Figure 3.

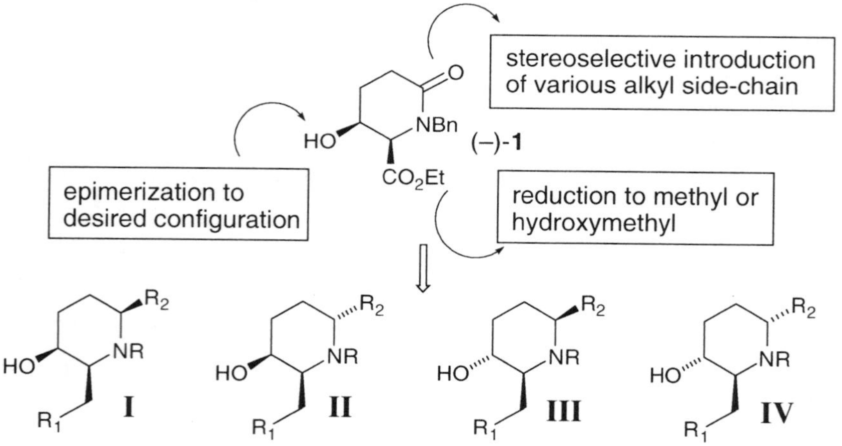

(R = H or Me, R$_1$ = H or OH, R$_2$ = various alkyls)

Fig. (3). Basic strategy for the stereodivergent synthesis of 3-piperidinol alkaloids

The C-6 side chain was installed by using the lactam moiety with stereochemical control. In addition, the hydroxyl group at the 3-position was readily epimerized to the desired configuration for the synthesis of the corresponding alkaloids.

The chiral building block of type **I** was synthesized as shown in Scheme 2.

The chiral building block of type **II** was synthesized as follows. Reduction of **4**, prepared in Scheme 2, with NaBH₃CN in the presence of trifluoroacetic acid provided a 14 : 1 mixture of *trans*-(2,6)- and *cis*-(2,6)-piperidines. Because it was difficult to isolate the major , desired *trans*-isomer in a pure state, the epimeric mixture was used for subsequent transformation. Hydrogenation of the mixture over Pd(OH)₂ followed by treatment of the resulting amine with ClCO₂Me gave diastereopure *trans*-piperidine in 68% isolated yield, which was transformed into the alcohol **6** in 92% yield.

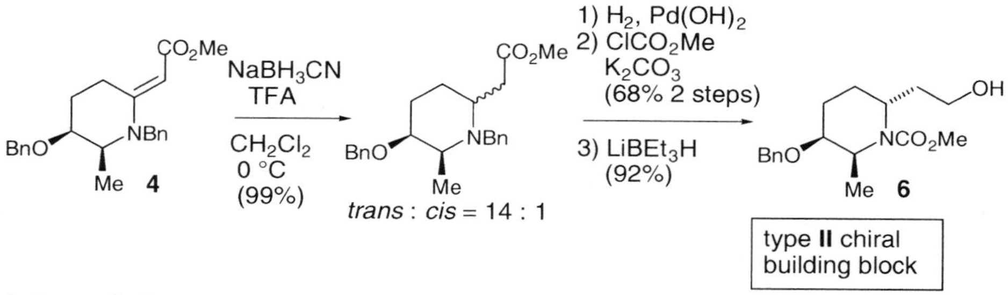

Scheme 3. Synthesis of type **II** chiral building block **6**

The *trans*-selectivity observed in the above reduction is explained by the following factors. Conformer **A** for the iminium salt, generated from **4** under acidic conditions is favored relative to **B** because of $A^{(1,2)}$ strain between the *N*-benzyl and the methyl groups at the α-position, so the hydride attacks from the preferred β-axial direction, leading to a chairlike transition state to give rise to *trans*-(2,6)-piperidine selectively.

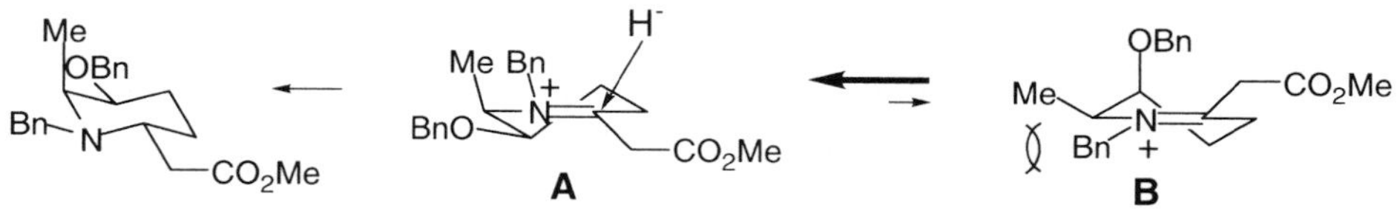

Fig. (4). Selectivity of the iminium reduction derived from **4**

Another *trans*-type of chiral building block **III** was synthesized in a similar iminium reduction of the corresponding vinylogous urethane of *trans*-(2,3)-congener **11**. Protection of the hydroxyl group with BnBr gave benzyl ether, whose MOM protecting group was deprotected with conc. HCl in MeOH. Oxidation of the resulting alcohol with PCC in the presence of NaOAc afforded the ketone **7** in 95%ee, which was recrystallized from *i*-Pr$_2$O to give rise to an enantiomerically pure compound in 83% isolated yield. When the oxidation was performed under Swern conditions, complete racemization occurred. Reduction of the keto alcohol obtained from hydrogenolysis of **7** with NaB(OAc)$_3$H provided diol **8** with complete stereochemical control. Both hydroxyl groups in **8** were protected as an acetonide **9**, which was subjected Eschenmoser's contraction-sulfide extrusion reaction[9], after conversion to thiolactam **10**, to provide vinylogous urethane **11**. Reduction of **11** with NaBH$_3$CN gave *cis*-(2,6)-piperidine **12**, which was reduced with LiAlH$_4$ to give rise to piperidine **13**.

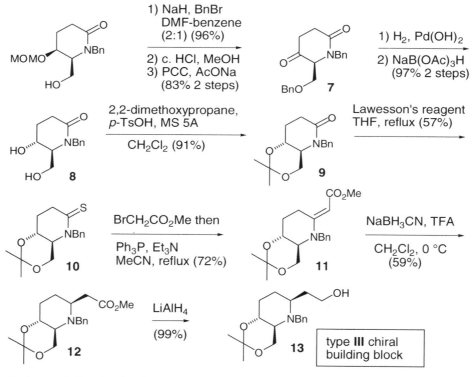

Scheme 4. Synthesis of type **III** chiral building block **13**

Finally, the type **IV** chiral building block **15** was synthesized via the iminium reduction of the corresponding iminiun salt derived from the vinylogous urethane **14** as shown in Scheme 5. With all four stereoisomers of the chiral building blocks (**I~IV**) for synthesis of 3-piperidinol alkaloid in hand, we examined the synthesis of prosafrinine, iso-6-cassine, prosophylline, and prosopinine. The alkaloid prosafrinine was synthesized from type **I**, iso-6-cassine was synthsized from type **II**, prosophylline was synthesized from type **III**, and prosopinine was synthesized from type **IV** chiral building block, respectively, as shown in Scheme 6 and 7[10].

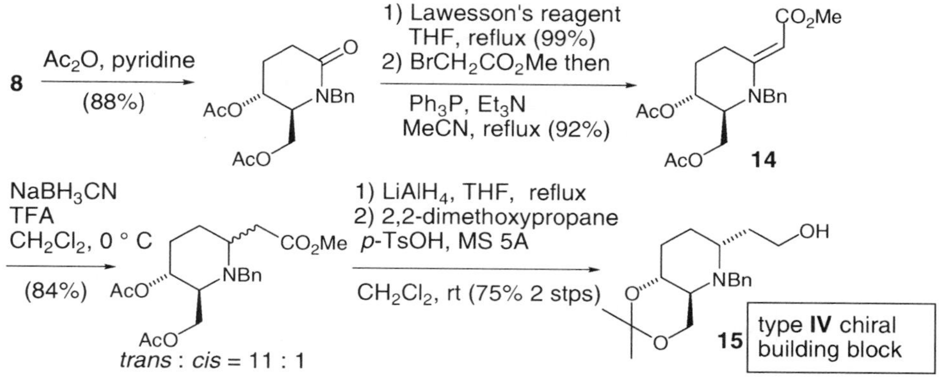

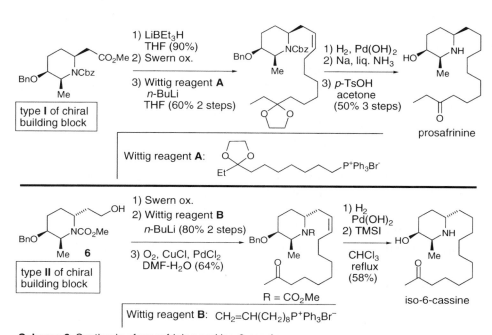

Scheme 5. Synthesis of type **IV** chiral building block **15**

Scheme 6. Synthesis of prosafrinine and iso-6-cassine

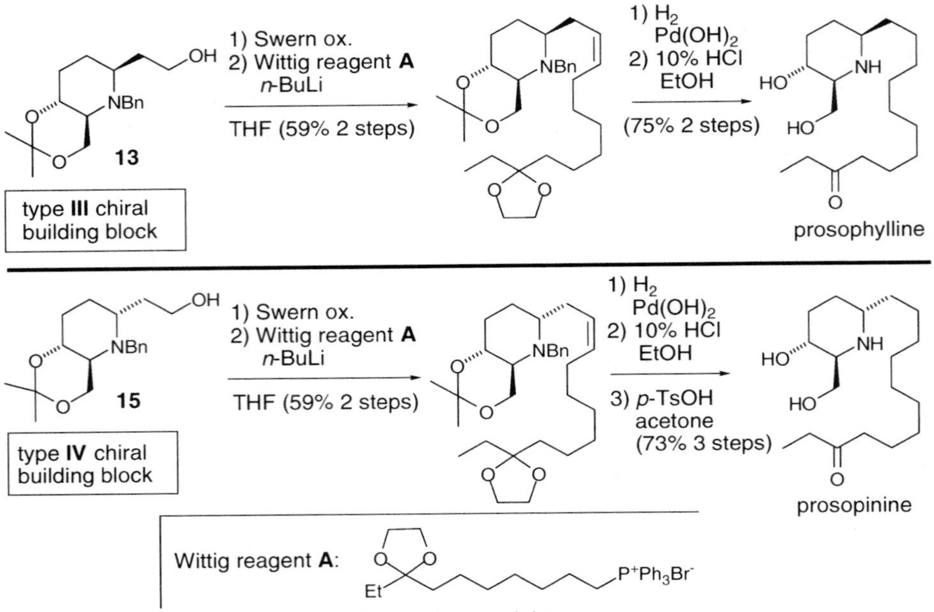

Scheme 7. Synthesis of prosophylline and prosopinine

1.3. Enantioselective total synthesis of marine alkaloids

In 1991, Cardellina, II and co-workers isolated clavepictines A and B from the tunicate *Clavelina* picta, and these alkaloids showed substantial cyctotoxic activity against human solid cell lines. In the same year, Faulkner and co-worker isolated pictamine, the bis-nor congener of clavepictine A, from the same marine species. Although the relative stereochemistry of these alkaloids was determined on the basis of extensive NMR studies for clavepictine A in conjunction with X-ray diffraction analysis for clavepictine B, the absolute stereochemistry of these alkaloids was unknown.

On the other hand, Steffan isolated lepadin B from the tunicate *Clavelina* lepadiformis, and, four years later, Andersen and co-workers isolated lepadin B, and C along with lepadin A from the same tunicate. Professor Andersen also reported lepadin A and B showed significant cytotoxic activity toward a variety of murine and human cancer cell lines. The absolute stereochemistry of these alkaloids was also unknown.

Here, we achieved the first enantioselective total synthesis of clavepictines A and B, pictamine, and lepadin B starting from the common lactam-type chiral building block **1**, and determined the absolute stereochemistry of these interesting alkaloids.

Swern oxidation of alcohol **15**, prepared in Scheme 5, followed by Wittig-Horner reaction of the resulting aldehyde afforded the α,β-unsaturated ester, which was converted to alcohol **16** in a 3-step sequence. Swern oxidation of **16** and Wittig-Horner reaction of the resulting aldehyde gave conformationally constrained piperidine **17**.

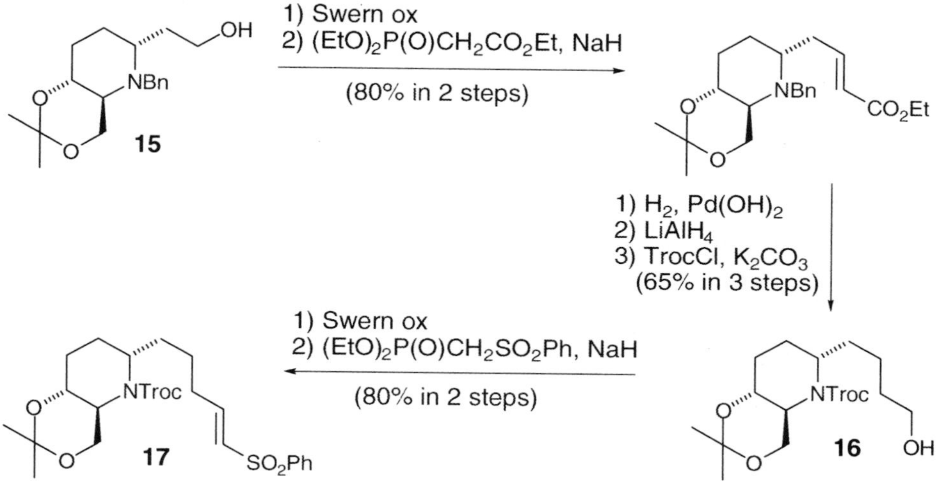

1) Swern ox
2) (EtO)2P(O)CH2CO2Et, NaH
(80% in 2 steps)
OH
NBn
15
CO2Et
1) H2, Pd(OH)2
2) LiAlH4
3) TrocCl, K2CO3
(65% in 3 steps)
1) Swern ox
2) (EtO)2P(O)CH2SO2Ph, NaH
(80% in 2 steps)
NTroc
17
SO2Ph
16
OH

Scheme 8. Synthesis of conformationally constrained piperidine **17**

Treatment of **17** with a Cd-Pb complex[11] in aqueous THF furnished the quinolizidine **18** in 92% yield as the only cyclized product. Stereochemistry of this quinolizidine was assigned by the NOE between H_a and H_b, in addition to the coupling constant of H_c. Finally, this assignment was confirmed by X-ray diffraction analysis.

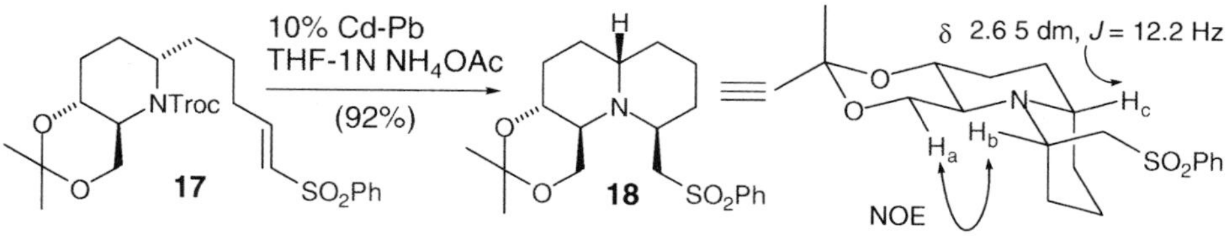

10% Cd-Pb
THF-1N NH4OAc
(92%)
NTroc
17
SO2Ph
H
N
18
SO2Ph
δ 2.6 5 dm, J = 12.2 Hz
Hc
Ha
Hb
NOE
SO2Ph

Scheme 9. Key Michael-type quinolizidine ring closure reaction of piperidine **17**

This high-kinetic stereoselectivity can be rationalized as shown below. In the two kinds of folded chairlike transition states **A** and **B**, steric repulsion between H_a and H_b protons in **B** is present, so the cyclization occurs via transition state **A** to afford the desired *cis*-product.

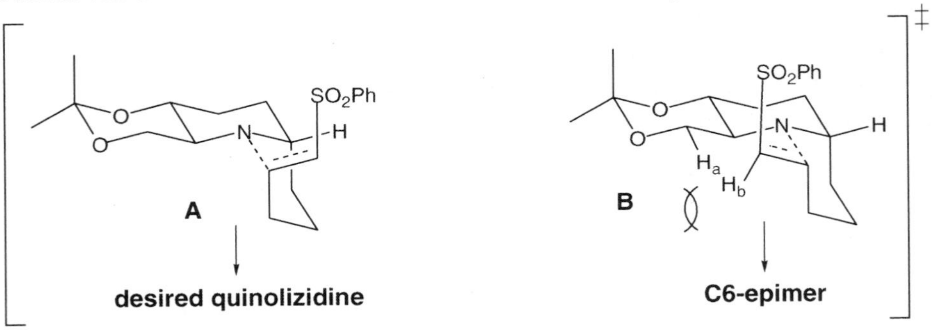

A

desired quinolizidine

B

C6-epimer

Fig. (5). Stereochemical course of key cyclization reaction of **17**

Completion of the total synthesis of clavepictines A and B is shown in Scheme 10. Conversion of **18** to the alcohol **19** via 4 steps and reduction of the hydroxyl group in **19** via iodide gave the quinolizidine **20**. Finally, the dienyl moiety was constructed by Julia coupling, and deprotection furnished clavepictine B. Acetylation of the hydroxyl group of clavepictine B gave clavepictine A.

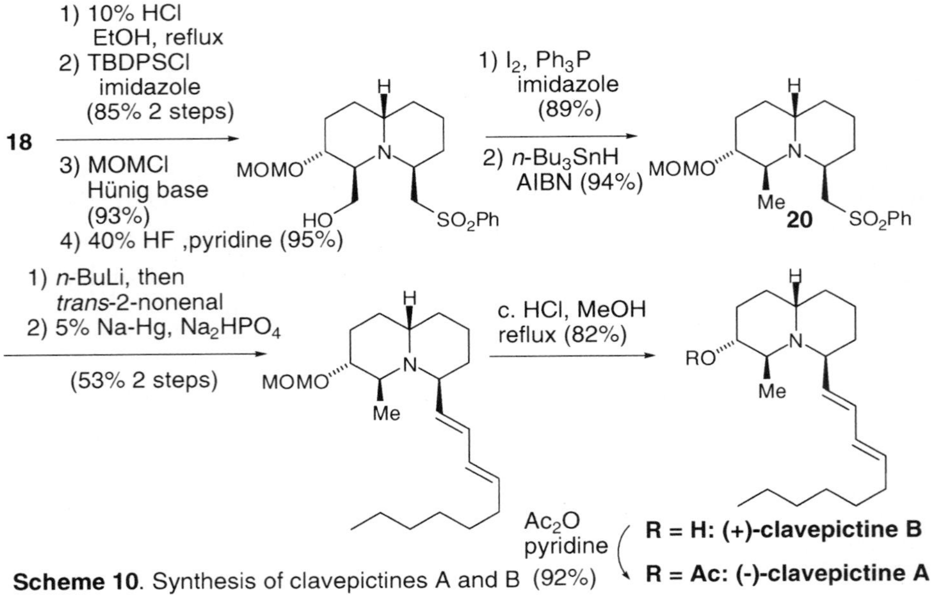

Scheme 10. Synthesis of clavepictines A and B (92%)

R = H: (+)-clavepictine B
R = Ac: (-)-clavepictine A

Thus, we achieved the first enantioselective total synthesis of clavepictines A and B using the highly stereoselective Michael-type quinolizidine ring closure reaction as the key step[12]. Cha's group also completed the total synthesis of clavepictines A and B independently[13].

By the same procedure, we also completed the total synthesis of (-)-pictamine, and determined the absolute stereochemistry of this alkaloid as depicted in Scheme 11.

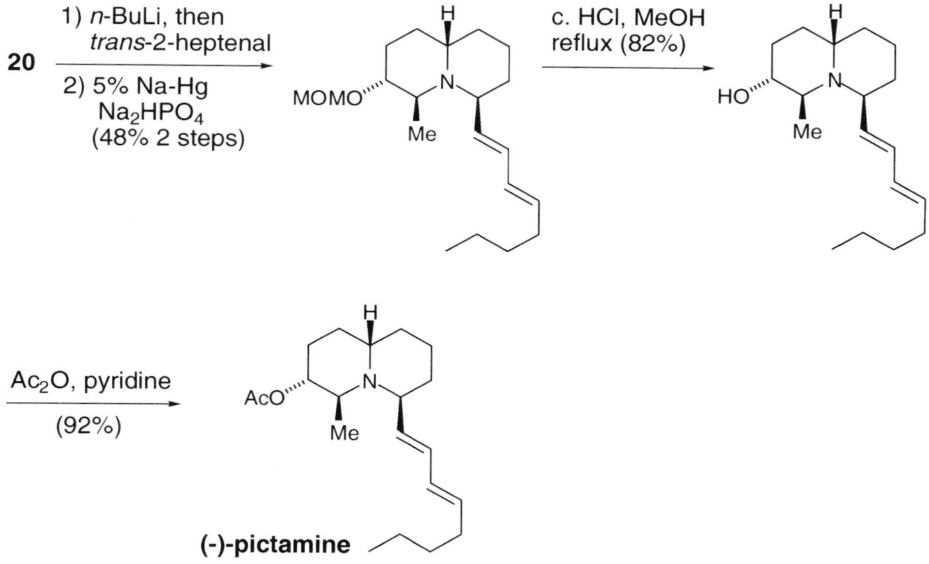

20

1) *n*-BuLi, then
   trans-2-heptenal

2) 5% Na-Hg
   Na$_2$HPO$_4$
   (48% 2 steps)

MOMO'''' ... Me

c. HCl, MeOH
reflux (82%)

HO'''' ... Me

Ac$_2$O, pyridine

(92%)

AcO'''' ... Me

(-)-pictamine

Scheme 11. Synthesis of pictamine

Next, we examined the total synthesis of lepadin B starting from common lactam **1**. The piperidone, derived from **1**, was converted to enoltriflate using the Comins triflating reagent[14], followed by Pd-catalyzed CO insertion reaction[15] of the resulting triflate to afford the enaminoester **21**. The Michael-type conjugate addition reaction of **21** proceeded smoothly to yield the adduct **22** as the single stereoisomer. Carbon chain elongation at the C-2 position of **22** was performed by Arndt-Eistert sequence to provide the homologated ester **23**. This ester was transformed into the methyl ketone **25** via Weinreb's amide[16] **24**, and subsequent oxidative cleavage of the terminal alkene yielded the keto aldehyde **26**.

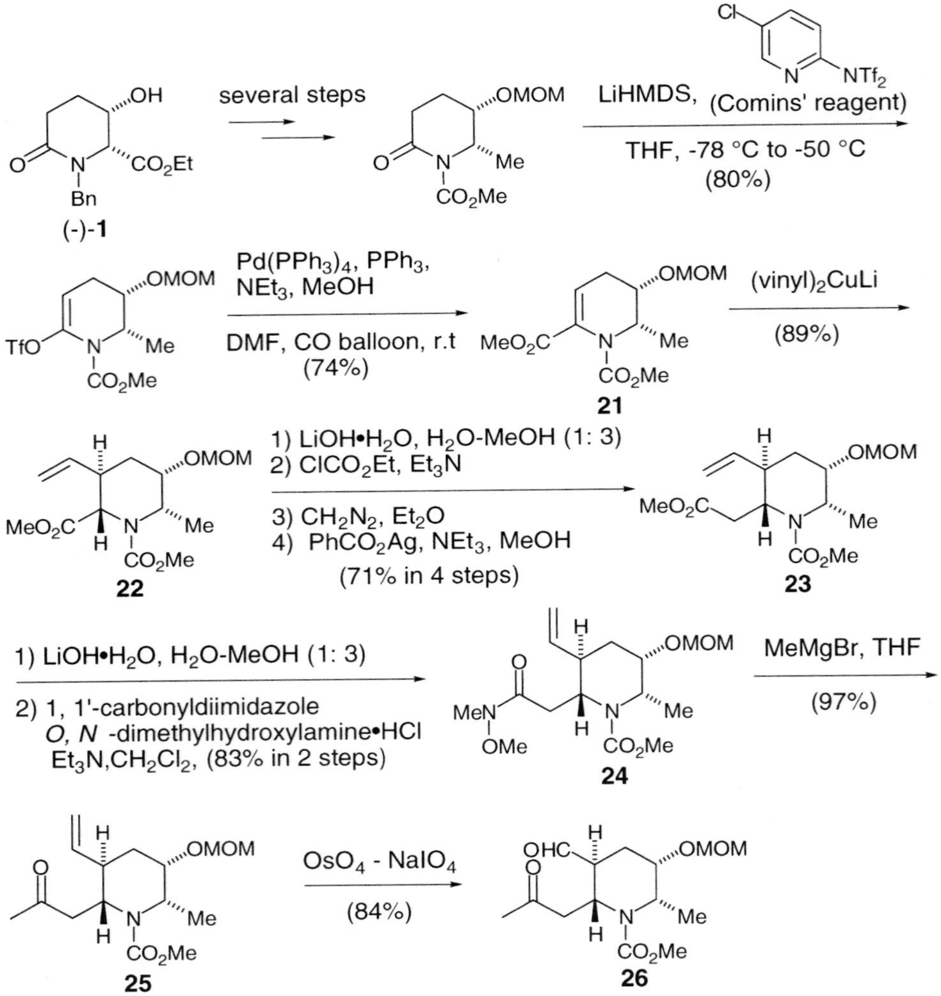

Scheme 12. Synthesis of key intermediate **26** for the aldol cyclization

With the requisite aldehyde **26** in hand, the stage was now set for the key intramolecular aldol cyclization. Thus, treatment of **26** with 4 equivalents of DBU in refluxing benzene gave the cyclized product in a ratio of 14 : 1, and the major product **27** was isolated in 60% yield by flash column chromatography. The major product in this key reaction was *cis*-octahydroquinolinone as expected. The stereochemistry of the major product was determined by the NOE experiment shown in Scheme 13.

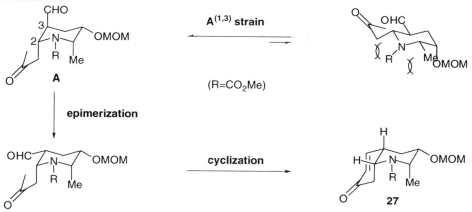

Scheme 13. Key intramolecular aldol type of cyclization of **26**

Conformation of the keto aldehyde **26** should be **A** by severe $A^{(1,3)}$ strains[17], but in this conformation both appendages in 2 and 3 positions were not cyclizable *trans* diaxial.   So, the epimerization at the 3 position occurs first, and then cyclization will proceed to afford the desired *cis*-enone.

Fig. (6). Reaction pathway of the aldol type of cyclization reaction of **26**

The phenylsulfonylmethyl group was introduced at the 5-position of **27** stereoselectively in a 2-step sequence for the final Julia coupling. The carbonyl group in the ketone **28** was removed by Barton's procedure[18] to give decahydroquinoline derivative **29**. After conversion of **29** to Boc derivative **30**, Julia coupling was performed using *trans* 2-heptanal to construct the desired diene system.   Finally, deprotection of both protective groups furnished the lepadin B. The spectral data for trifluoroacetate salt of synthetic lepadin B were identical with those for trifluoroacetate salt of the natural product. From the optical rotation, the absolute stereochemistry of natural lepadin B was determined, unambiguously[19].

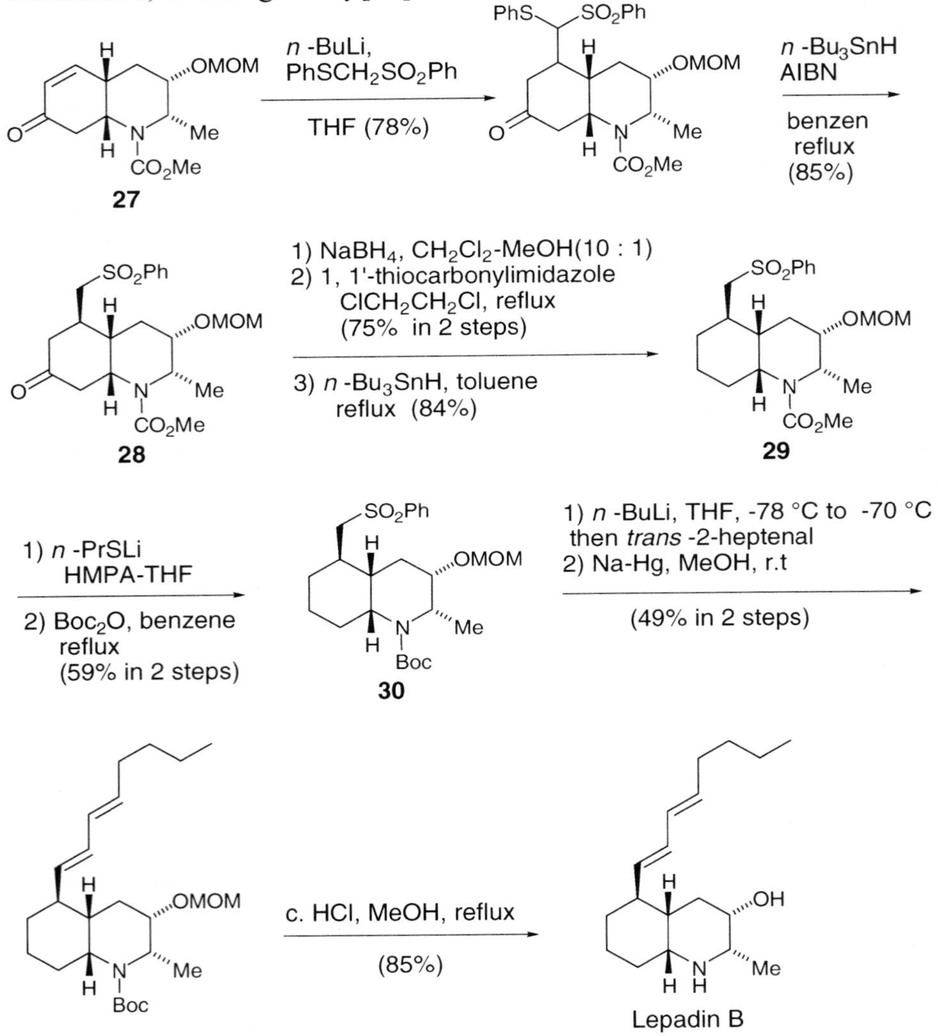

Scheme 14. Completion of the synthesis of lepadin B

Kibayashi's group also completed the total synthesis of lepadins A, B, and C[20].

2. SYNTHETIC STUDIES ON BIOLOGICALLY ACTIVE ALKALOIDS STARTING FROM LACTAM 31

2.1. Stereodivergent process for synthesis of the decahydroquinoline type of *Dendrobates* alkaloids

The neotropical dart poison frogs contain a remarkable diversity of alkaloids, and the 2,5-disubstituted decahydroquinolines represent, a major class of these amphibian alkaloids[21]. Isolation of these alkaloids from some ants strengthens a dietary hypothesis for the origin of the above alkaloids that have been detected in extracts of frog skin[22]. In addition, these alkaloids containing both *cis* and *trans* ring fusion have been identified as well as diastereomers at the C-2 and C-5 position.

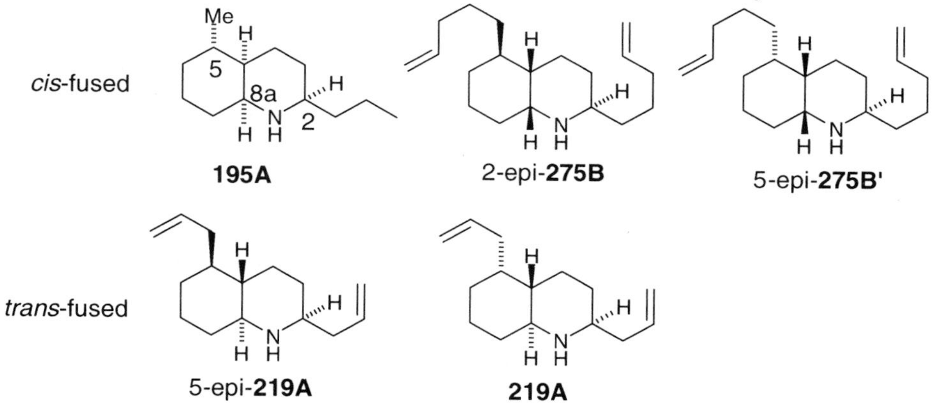

Fig. (7). Representative decahydroquinoline-type dart-poison frog alkaloids

The structural diversity and pharmacological activity associated with this class of alkaloids have stimulated synthetic activity by numerous groups[23]. However, the route which can be applicable to the synthesis of both *cis*- and *trans*-fused ring systems has been reported only to a limited extent[24]. Moreover, no methodology for the divergent synthesis of the 2,8a-*cis* and -*trans* substituted ring system has been reported to date.

The common starting material **32**, derived from known lactam **31**[25], was treated with divinyl lithium cuprate to afford the adduct **33** as a single isomer. The stereoselectivity of this addition reaction can be explained by $A^{(1,3)}$ strain and stereoelectronic effect[26] the same as our recent investigations. On the other hand, deprotection of the TBDPS group with TBAF followed by oxazolizinone ring formation using sodium hydride as a base afforded the oxazolizinone **34**. Conjugate addition reaction of **34** proceeded smoothly to provide the adduct **35** as a single isomer again. The stereochemistry of **35** was determined by the NOE between H_a and H_b, and half-height width ($W_{1/2}$, 11.5 Hz) of H_c in **35** as shown in Scheme 15.

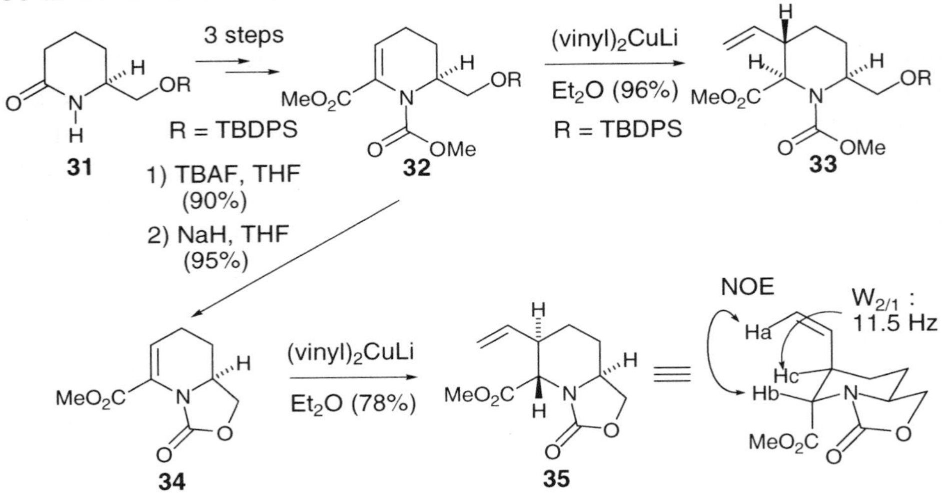

Scheme 15. Synthesis of diastereomers of trisubstituted piperidines **33** and **35**

The stereoselectivity of conjugate addition reaction of **34** can be rationalized as follows. The conformation of **34** is restricted to **A** by the oxazolizinone ring, and the vinyl anion attacks from stereoelectronically preferred β-axial direction to give rise to the adduct **35** exclusively.

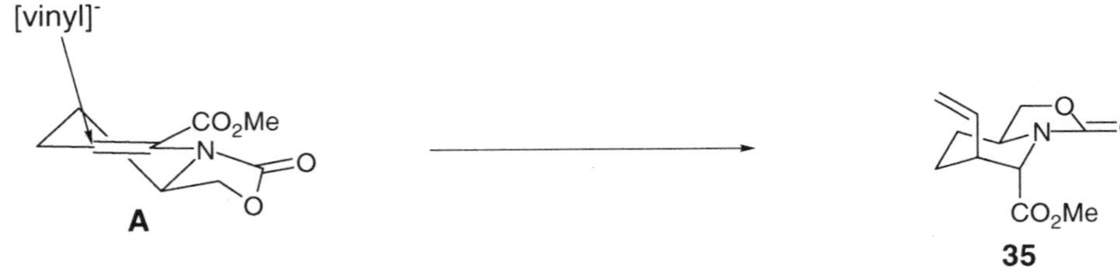

Fig. (8). Stereochemical course of conjugate addition reaction of **34**

The carbon-chain at the α-position on both adducts **33** and **35** was elongated by Arndt-Eistert sequence to provide the homologated esters **36** and **37**. The esters **36** and **37** were transformed into the methyl ketones **40** and **41** via Weinreb's amides **38** and **39**.

Oxidative cleavage of the terminal olefin in **40** and **41** gave the keto aldehydes **42** and **43**.

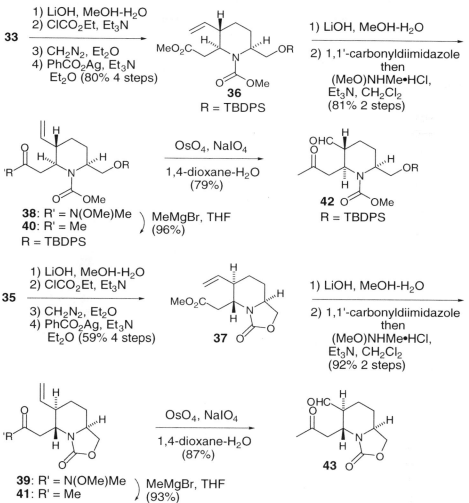

Scheme 16. Synthesis of the aldehydes **42** and **43**

The aldehydes **42** and **43** were subjected to the key intramolecular aldol-type cyclization to afford the *cis*-fused quinolinone **44** selectively (*cis:trans*=6.5:1, 52% isolated yield) or **45** exclusively (51% yield). The stereochemistry of **44** and **45** was determined by NOE experiments as shown in Scheme 17.

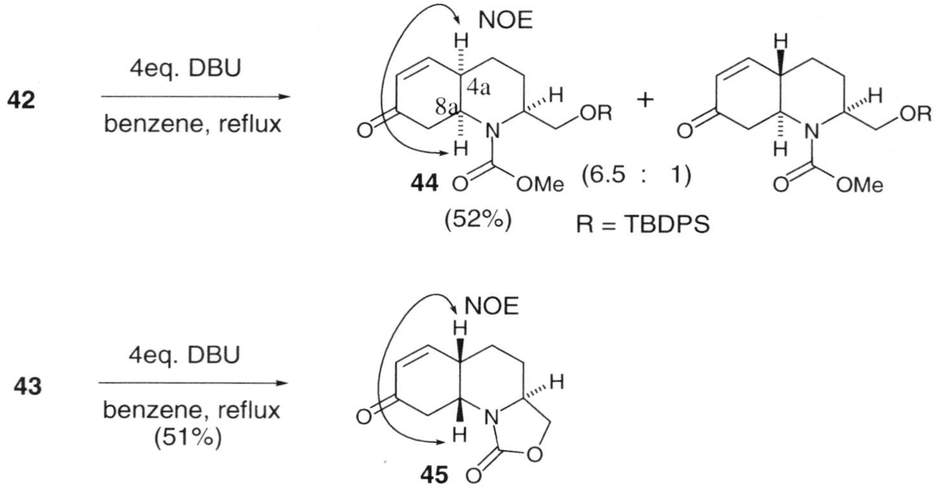

Scheme 17. Key intramolecular aldol-type cyclization reaction of **42** and **43**

Next, we examined the synthesis of a 4a,8a-*trans*-fused ring system such as **49** using the same aldol-type cyclization reaction of **48**. The methyl ketone **9** was converted to alcohol **46**, which was treated with sodium hydride to afford the oxazolizinone **47**. The terminal olefin in **47** was cleaved oxidatively to give rise to the keto aldehyde **48**. However, the aldol-type cyclization reaction of **48** under the same reaction conditions as for **42** or **43** was very messy and no cyclized product was isolated.

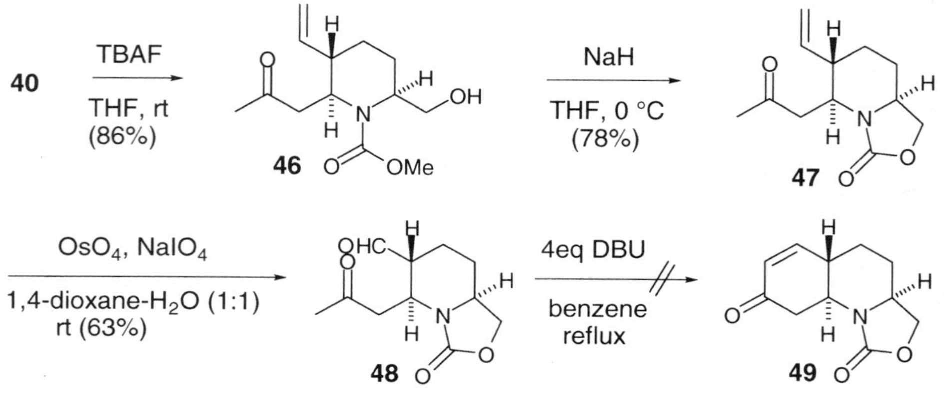

Scheme 18. Attempt to synthesize *trans*-fused octahydroquinolinone **49**

According to the above observation, we anticipate that the aldol-type cyclization reaction of **42** should proceed via not *trans*-enone **T** but epimeric aldehyde **E** as shown in Fig. 9.

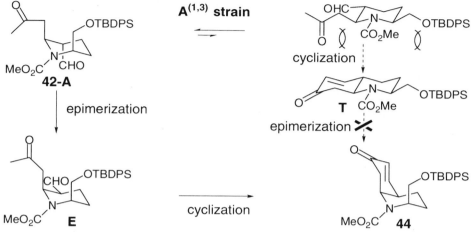

Fig. (9). Reaction mechanism for the aldol-type cyclization reaction of **42**

For the synthesis of *trans*-fused quinolinone, we were forced to develop an alternative route to **49**. We designed the ring-closing methathesis reaction of diene **52**. Deprotection of the TBDPS group in **36** with TBAF followed by treatment of the resulting alcohol with NaH gave the oxazolizinone **50**. Reduction of **50** with Super-Hydride provided the alcohol **51**. Swern oxidation of **51** followed by Grignard reaction and PCC oxidation of the resulting secondary alcohol afforded the vinyl ketone **52**. Ring-closing methathesis reaction of **52** using Grubbs' catalyst[27] gave rise to the *trans*-fused quinolinone **49** in good yield.

Scheme 19. Synthesis of *trans*-fused octahydroquinolinone **49**

438

We next examined installation of the alkyl side chain at the C-5 position. Thus, treatment of the above enones with dimethyl lithium cuprate afforded the adducts **53** and **55** exclusively or **56** selectively (α:β=3:1). The stereochemistry of the adducts **53** and **55** was determined to be 4a,5-*cis* relationship by the NOE between H_a and the methyl group on the C-5 position of **54** or between H_b and the methyl group on the C-5 position of **55**, respectively. On the other hand, the stereochemistry of the major product on the conjugate addition reaction of **49** was determined to be **56** by the large (13 Hz) coupling constant of H-5 in minor adduct **57** suggesting the axial (α) position of this proton. The C-5 epimer of **53** was synthesized by means of conjugate addition followed by Ito-Saegusa oxidation[28] of the resulting silyl enol ether and catalytic hydrogenation of the enone **58** to give rise to the epimeric octahydroquinolinone **59**[29].

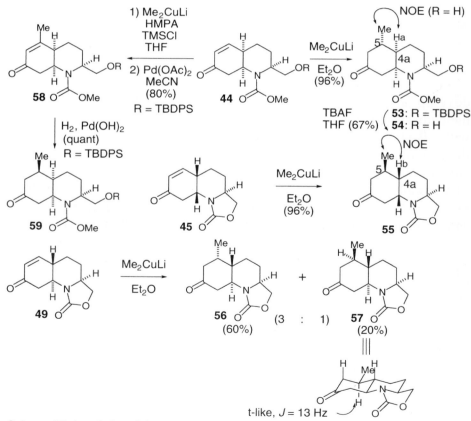

Scheme 20. Instalation of the alkyl group at the 5 position of enones **44**, **45**, and **49**

2.2. Synthesis of a proposed structure of alkaloid 223A

The alkaloid **223A (P)**, the first member of a new trialkyl-substituted indolizidine class of amphibian alkaloids, was isolated from a skin extract of a Panamanian population of the frog *Dendrobates pumilio* Schmidt (Dendrobatidae) in 1997 along with three higher homologs of **P**[29]. The structure of this alkaloid has been established to be **P** based upon GC-MS, GC-FTIR, and ¹H-NMR spectral studies. In this paper, we would like to report the first synthesis of alkaloid **223A** by sequential use of our original Michael-type conjugate addition reaction to enaminoesters (**i, ii**) as the key step.

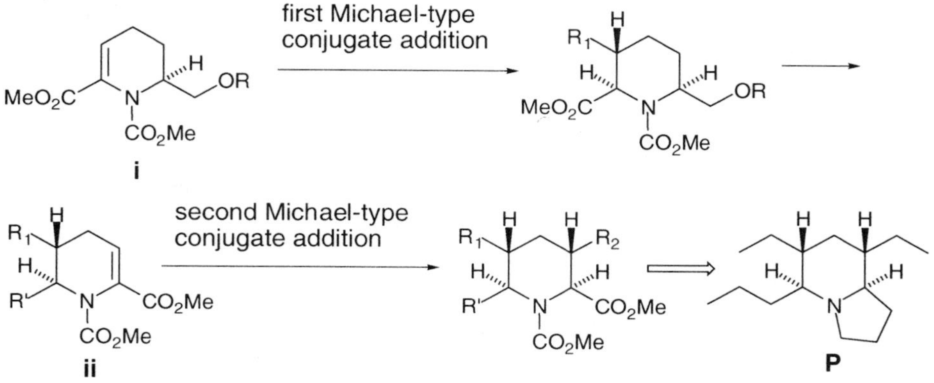

Fig. (10). Synthtic design for the proposed structure of alkaloid **223A (P)**

The synthesis began with (2S)-amide **31**, which was converted to the cyclic enaminoester **60** using triflation of the intermediate imidourethane with Comins' reagent followed by a palladium-catalyzed CO insertion reaction of the resulting enol triflate. The first Michael-type conjugate addition reaction of **60** proceeded smoothly giving the vinyl adduct as a single isomer which was transformed into the methyl urethane **62** *via* the alcohol **61**. The alcohol **62** was converted to the methyl ester, which was transformed into the cyclic enaminoester **63** using Rubio's protocol[30]. The second Michael-type conjugate addition of **63** proceeded to afford adduct **64**, again as a single isomer. The stereochemistry of **64** was determined by the coupling constants and NOE of the oxazolizinone derivative **65**.

440

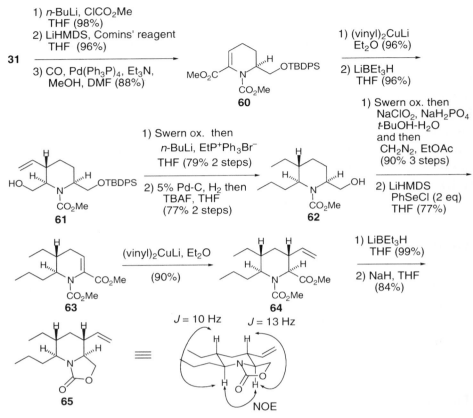

Scheme 21. Synthesis of 2,3,5,6-tetrasubstituted piperidine ring system **64**

The stereoselectivity of the second and key Michael-type conjugate addition reaction can be rationalized as follows. The conformation of **63** will be restricted to **63-A** due to $A^{(1,3)}$ strain between the N-methoxycarbonyl and n-propyl groups in **63-B**. Attack of the vinyl anion from the stereoelectronically favored α-axial direction provides the adduct **64** exclusively. It is noteworthy that the stereochemical course of the above reaction is controlled by the stereoelectronic effect in spite of severe 1,3-diaxial steric repulsion between the axial ethyl group at the 5-position and the incoming vinyl anion. This remarkable stereoselectivity can be also explained by Cieplak's hypothesis[31]. On the preferred conformation **63-A**, the developing σ^* of the transition state is stabilized by the antiperiplanar donor σ_{C-H} at the C-4 position.

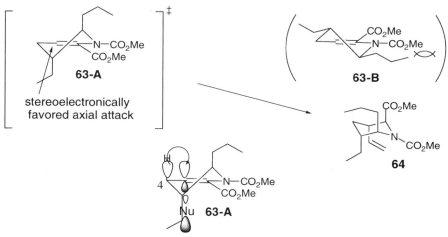

Fig. (11). Explanation for the stereoselectivity of the conjugate addition reaction of **63**

Elaboration of the adduct **64** into the indolizidine **P**, previously proposed as the structure for natural **223A**, is shown in Scheme 22. Reduction of **64** with Super-Hydride followed by Swern oxidation and Wittig-Horner reaction of the resulting aldehyde afforded the α,β-unsaturated ester **66**. Hydrogenation of **66** and reduction of the resulting saturated ester gave the corresponding alcohol, whose hydroxyl group was protected as the MOM ether **67**. Finally, deprotection of the methoxycarbonyl group and cleavage of the MOM ether followed by indolizidine formation furnished **P**. However the ¹H-, ¹³C-NMR and IR spectra of **P** were not identical with those for the natural product, nor was the GC retention time[32].

Scheme 22. Synthesis of proposed structure for alkaloid **223A (P)**

3. SYNTHETIC STUDIES ON BIOLOGICALLY ACTIVE ALKALOID STARTING FROM LACTAM 68

3.1. Synthesis of a revised structure of alkaloid 223A

The close similarity of the Bohlmann bands in the vapor phase FTIR spectra of **P** and natural **223A** indicated the same 5,9-Z configuration for both compounds. Based upon a detailed comparison of the ^{1}H- and ^{13}C-NMR spectra, we concluded that natural **223A** differed from **P** only in the 6-position configuration. Therefore, we commenced the synthesis of the diastereomeric indolizidine **R** from piperidone **68**[33]. The piperidone **68** was converted to cyclic enaminoester **69** in the same manner as described in Scheme 21. Addition of divinylcuprate to **69** provided the adduct **70** as a single isomer in excellent yield. The indicated coupling constant and NOE experiments of oxazolizinone **72** derived from **70** *via* the alcohol **71** confirmed that the stereogenic centers of the key intermediate **70** were correct for the synthesis of the target indolizidine **R**.

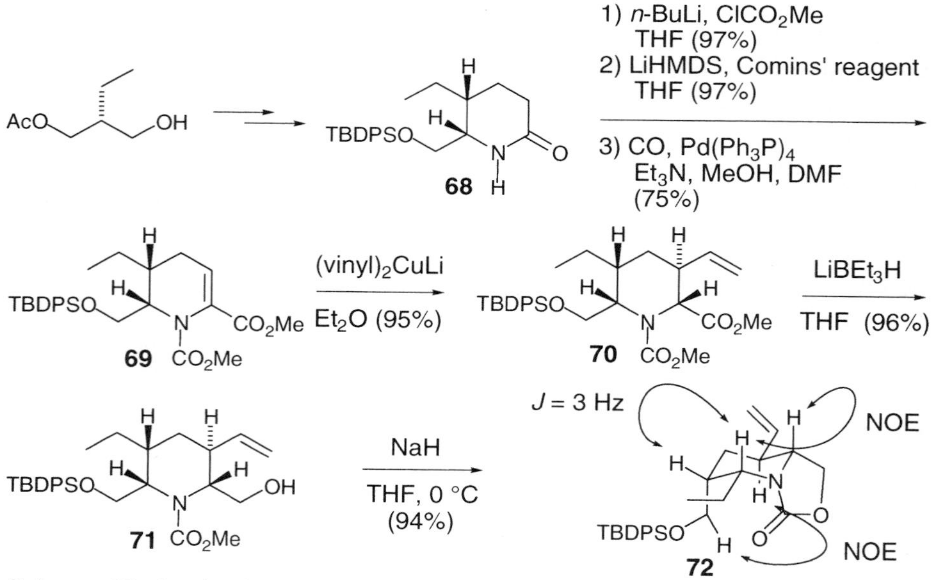

Scheme 23. Synthesis of the C-6 epimer (**71**) of tetrasubstituted piperidine **64**

Stereoselectivity of this Michael-reaction can again be explained by A$^{(1,3)}$ strain and a stereoelectronic effect as shown in Figure 2.

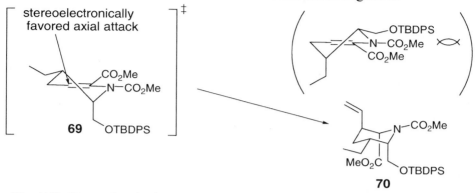

Fig. (12). Stereochemical course of conjugate addition reaction of **69**

The synthesis of **R** was accomplished *via* the alcohols **73** and **74** in the same manner as in the synthesis of **P** in Scheme 22. The spectral data for **R** (¹H-, ¹³C-NMR, GC-FTIR, GC-EIMS) were completely identical with those for the natural product. Thus, the structure of natural **223A** is revised to **R**, and the relative stereostructure of this alkaloid was determined to be 5$R*$, 6$R*$, 8$R*$, 9$S*$ by the present synthesis[32].

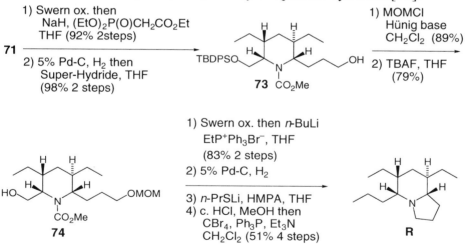

Scheme 24. Completion of the synthesis of the revised structure for alkaloid **223A** (**R**)

3.2. Synthesis of dart-poison frog alkaloid 207I

The 5,8-disubstituted indolizidines and 1,4-disubstituted quinolizidines are the more common structural patterns found in amphibian skin[21]. None of these alkaloids has so far been reported from any other source. In addition, the biological activity of only a few 5,8-disubstituted indolizidines has been investigated due to the isolation in minute quantities from the skin. Among them, the relative stereochemistry of quinolizidine **207I** was anticipated to be **75** by our chiral synthesis of **76**[35] followed by stereocontrolled synthesis of **75**[36]. A sample of synthetic racemate of **75** had produced the best separations on GC analysis with β-dextrin chiral column[36].

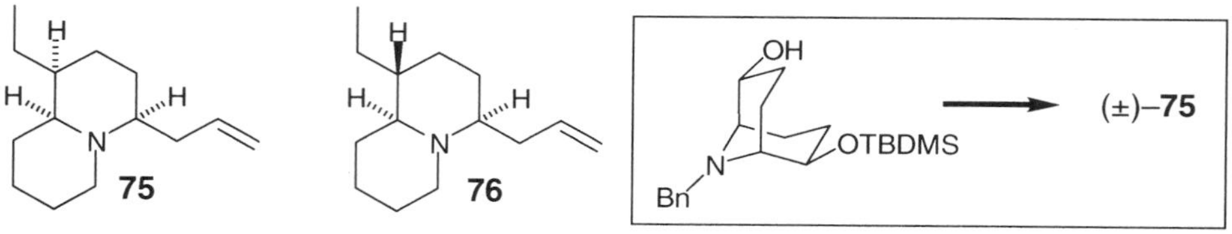

Fig (13). Structure of natural alkaloid **207I**

We planned the enantioselective synthesis of **75** to determine the absolute stereochemistry of natural quinolizidine **207I**.

The synthesis began with enantiomerically pure triflate **69**, which was converted to **77** using palladium-catalyzed Sonogashira-type coupling reaction[37] with propagyl ether in good yield. Catalytic hydrogenation of **77** over 5% Rh-C in EtOAc under medium pressure (4 atm) gave the desired reduction product **78** in 90% yield. The stereochemistry of **78** was determined to be that of the desired all *cis*-trisubstituted piperidine on the basis of the observation of NOE on the oxazolizinone **79** as shown in Scheme 25. After deprotection of the silyl group with TBAF, the carbon-chain at the 2 position was homologated by Swern oxidation followed by Wittig reaction of the resulting aldehyde to afford the homologated ether **80**. Hydrogenation and deprotection of the tetrahydropyranyl group with PPTS gave rise to alcohol **81**. This was transformed into the terminal olefin **82** using Sharpless'[38] and Grieco's[39] procedure. Finally, quinolizidine ring-closure was performed by a three-step sequence to provide the quinolizidine (+)-**75**.

The GC analysis of our synthetic (+)-**75** revealed that the absolute stereochemistry of natural **207I** is 1*S*, 4*S*, 10*S* as shown in Scheme 25. On the GC analysis using β–Dex-120 cyclodextrin-based column (Supelco Inc., Bellefonte, PA; 30 m, 0.25 mm i. D., 25um film thickness), synthetic (+)-**75** was coeluted with the shorter retention time peak of racemate and the natural product was coeluted with the longer retention time peak of racemate[40].

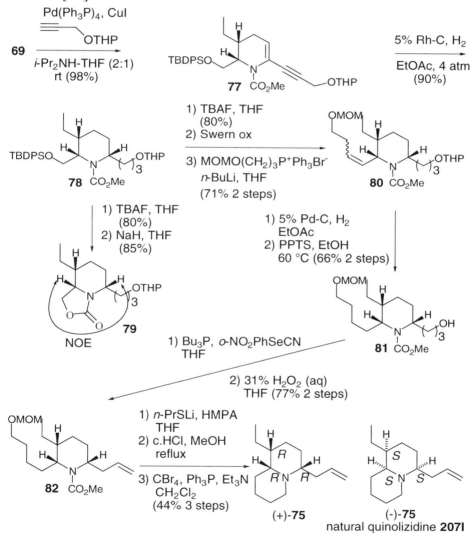

Scheme 25. Synthesis of the enantiomer of natural **207I**

ACKNOWLEDGEMENTS

We are grateful to Dr. John H. Cardellina, II, National Cancer Institute, Prof. D. John Faulkner, University of California, San Diego, and Prof. Raymond J. Andersen, University of British Columbia, for kindly providing us with ^{1}H, ^{13}C NMR spectra of natural clavepictines A, B, Pictamine, and lepadin B. We are also indebted to Dr. Thomas F. Spande, National Institute of Health, for measurement of the GC analysis of synthetic, racemic, and natural quinolizidine **207I** with β-dextrin chiral column, and to Prof. André Rassat, Ecole Normale Supérieure, for kindly providing us with racemic quinolizidine **207I**.

REFERENCES AND NOTES

[1]   Strunz, G.M.; Findlay, J.A. In The Alkaloids; Brossi, A., Ed.; Academic Press: New York, 1985; Vol. 26, pp 89-183.

[2]   McCoy, M.C.; Faulkner, D.J.; *J. Nat. Prod.*, **2001**, *64*, 1087-1089.

[3]   Raub, M.F.; Cardellina, II, J.H.; Choudhary, M.I.; Ni, C.-Z.; Clardy, J.; Alley, M.C.; *J. Am. Chem. Soc.*, **1991**, *113*, 3178-3180.

[4]   Kong, F.; Faulkner, J. D.; *Tetrahedron Lett.*, **1991**, *32*, 3667-3668.

[5]   Steffan, B.; *Tetrahedron*, **1991**, *47*, 8729-8732.

[6]   Kubanek, J.; Williams, D.E.; de Silva, E.D.; Allen, T.; Andersen, R.J.; *Tetrahedron Lett.*, **1995**, *36*, 6189-6192.

[7]   Bonjoch, J.; Serret, I.; Bosch, J.; *Tetrahedron*, **1984**, *40*, 2505-2511.

[8]   Momose, T.; Toyooka, N.; *Tetrahedron Lett.*, **1993**, *34*, 5785-5786.

[9]   Roth, M.; Dubs, P.; Götschi, E.; Eschenmoser, A.; *Helv. Chim. Acta*, **1971**, *54*, 710-734.

[10]   Toyooka, N.; Yoshida, Y.; Momose, T.; *Tetrahedron Lett.*, **1995**, *36*, 3715-3718; Toyooka, N.; Yoshida, Y.; Yotsui, Y.; Momose, T.; *J. Org. Chem.*, **1999**, *64*, 4914-4919.

[11]   Dong, Q.; Anderson, C.E.; Ciufolini, M.A.; *Tetrahedron Lett.*, **1995**, *36*, 5681-5682.

[12]   Toyooka, N.; Yotsui, Y.; Yoshida, Y.; Momose, T.; *J. Org. Chem.*, **1996**, *61*, 4882-4883; Toyooka, N.; Yotsui, Y.; Yoshida, Y.; Momose, T.; Nemoto, H.; *Tetrahedron*, **1999**, *55*, 15209-15224.

[13]   Ha, J.D.; Lee, D.; Cha, J.K.; *J. Org. Chem.*, **1997**, *62*, 4550-4551; Ha, J.D.; Cha, J.K.; *J. Am. Chem. Soc.*, **1999**, *121*, 10012-10020.

[14]   Comins, D.L.; Dehghani, A.; *Tetrahedron Lett.*, **1992**, *33*, 6299-6302.

[15]   Cacchi, S.; Morera, E.; Orter, G.; *Tetrahedron Lett.*, **1985**, *26*, 1109-1112.

[16]   Nahm, S.; Weinreb, S.M.; *Tetrahedron Lett.*, **1981**, *22*, 3815-3818.

[17]   Hoffmann, R.W.; *Chem. Rev.*, **1989**, *89*, 1841-1873.

[18]   Barton, D.H.R.; McCombie, S.W.; *J. Chem. Soc., Perkin Trans. 1*, **1975**, 1574-1585.

[19]   Toyooka, N.; Okumura, M.; Takahata, H.; *J. Org. Chem.*, **1999**, *64*, 2182-2183; Toyooka, N.; Okumura, M.; Takahata, H.; Nemoto, H.; *Tetrahedron*, **1999**, *55*, 10673-10684.

[20]   Ozawa, T.; Aoyagi, S.; Kibayashi, C.; *Org. Lett.*, **2000**, *2*, 2955-2958; Ozawa, T.; Aoyagi, S.; Kibayashi, C.; *J. Org. Chem.* **2001**, *66*, 3338-3347.

[21]   Daly, J.W.; Garraffo, H.M.; Spande, T.F.; In *Alkaloids: Chemical and Biological Perspectives*; Pelletier, S.W., Ed.; Pergamon Press: New York, **1999**, Vol. *13*, pp 1-161; Daly, J.W.; In *The Alkaloids*; Cordell, G.A., Ed.; Academic Press: New York, **1998**, Vol. *50*, pp 141-169.

[22]   Spande, T.F.; Jain, P.; Garraffo, H.M.; Pannell, L.K.; Yeh, H.J.C.; Daly, J.W.; Fukumoto, S.; Imamura, K.; Tokuyama, T.; Torres, J.A.; Snelling, R.R.; Jones, T.H.; *J. Nat. Prod.*, **1999**, *62*, 5-21.

[23]   Aoyagi, S.; Hirashima, S.; Saito, K.; Kibayashi, C.; *J. Org. Chem.*, **2002**, *67*, 5517-5526; Smith, C.J.; Holmes, A.B.; Press, N.J.; *Chem. Commun.*, **2002**, 1214-1215; Davis, F.A.; Chao, B.; Rao, A.; *Org. Lett.*, **2001**, *3*, 3169-3171; Kim, G.; Jung, S.; Kim, W.-J.; *Org. Lett.*, **2001**, *3*, 2985-2987; Tan, C.-H.; Holmes, A.B.; *Chem. Eur. J.*, **2001**, *7*, 1845-1854; Comins, D.L.; Huang, S.; McArdle, C.L.; Ingalls, C.L.; *Org. Lett.*, **2001**, *3*, 469-471; Wei, L.-L.; Hsung, R.P.; Sklenica, H.M.; Gerasyuto, A.I.; *Angew. Chem., Int. Ed.*, **2001**, *40*, 1516-1518; Shu, C.; Alcudia, A.; Yin, J.; Liebeskind, L.S.; *J. Am. Chem. Soc.*, **2001**, *123*, 12477-12487; Oppolzer, W.; Flaskamp, E.; Bieber, L.W.; Hel. Chim. Acta, **2001**, 84, 141-145; Akashi, M.; Sato, Y.; Mori, M.; J. Org. Chem., **2001**, 66, 7873-7874; Back, T.G.; Nakajima, K.; *J. Org. Chem.*, **2000**, *65*, 4543-4552; Michael, J.P.; Gravestock, D.; *J. Chem. Soc., Perkin Trans. 1*, **2000**, 1919-1928; Enders, D.; and Thiebes, C.; *Synlett*, **2000**, 1745-1748; Williams,G.M.; Roughley, S.D.; Davies, S.D.; Holmes, A.B.; Adams, J.P.; *J. Am. Chem. Soc.*, **1999**, *63*, 9910-9918; Riechers, T.; Krebs, H.C.; Wartchow, R.; Habermehl, G.; *Eur. J. Org. Chem.*, **1998**, 2641-2646; Bardou, A.; Celerier, J.-P.; Lhommet, G.; *Tetrahedron Lett.*, **1998**, *39*, 5189-5192; Back, T.G.; Nakajima, K.; *J. Org. Chem.*, **1998**, *63*, 6566-6571; Pearson, W.H.; Suga, H.; *J. Org. Chem.*, **1998**, *63*, 9910-9918; Comins, D.L.; Lamunyon, D.H.; Chen, X.H.; *J. Org. Chem.*, **1997**, *62*, 8182-8187; Toyota, M.; Asoh, T.; Fukumoto, K.; *Tetrahedron Lett.*, **1996**, *37*, 4401-4404; Davies, S.G.; Bhalay, G.; *Tetrahedron: Asymmetry*, **1996**, *7*, 1595-1596; Naruse, M.; Aoyagi, S.; Kibayashi, C.; *Tetrahedron Lett.*, **1994**, *35*, 9213-9216; Naruse, M.; Aoyagi, S.; Kibayashi, C.; *J. Chem. Soc., Perkin Trans. 1*, **1996**, 1113-1124; Murahashi, S.; Sasao, S.; Saito, E.; Naota, E.; *J. Org. Chem.*, **1992**, *57*, 2521-2523; Murahashi, S.; Sasao, S.; Saito, E.; Naota, E.; *Tetrahedron*, **1993**, *49*, 8805-8826 and references cited therein.

[24]   Comins', Schultz's, and Kunz's approaches could be applicable to the divergent synthesis of *cis*- and *trans*-fused decahydroquinoline ring system. See: Comins, D.L.; Dehghani, A.; *J. Chem. Soc., Chem. Commun.*, **1993**, 1838-1839; Comins, D.L.; Dehghani, A.; *J. Org. Chem.*, **1995**, *60*, 794-795; Schultz, A.G.; McCloskey, P.J.; Court, J.J.; *J. Am. Chem. Soc.*, **1987**, *109*, 6493-6502; McCloskey, P.J.; Schultz, A.G.; *J. Org. Chem.*, **1988**, *53*, 1380-1383;

448

Weymann, M.; Schultz-Kukula, M.; Kunz, H.; *Tetrahedron Lett.*, **1998**, *39*, 7835-7838.

[25] Hodgkinson, T.J.; Shipman,M.; Synthesis, 1998, 1141-1144.

[26] Deslongchamps, P. *Stereoelectronic Effects in Organic Chemistry*; Pergamon: New York, **1983**; pp 209-290.

[27] Schwab, P.; France, M.B.; Ziller, J.W.; Grubbs, R.H.; *Angew. Chem. Int. Ed. Engl.*, **1995**, *34*, 2039-2041.

[28] Ito, Y.; Hirao, T.; Saegusa, T.; *J. Org. Chem.*, **1978**, *43*, 1011-1013.

[29] Toyooka, N.; Okumura, M.; Nemoto, H.; *J. Org. Chem.*, **2002**, *67*, 6078-6081.

[30] Garraffo, H.M.; Jain, P.; Spande, T.F.; Daly, J.W.; *J. Nat. Prod.*, **1997**, *60*, 2-5.

[31] Ezquerra, J.; Escribano, A.; Rubio, A.; Remuinan, M.J.; Vaquero, J.J.; *Tetrahedron: Asymmetry*, **1996**, *7*, 2613-2626.

[32] Cieplak, A.S.; *J. Am. Chem. Soc.*, **1981**, *103*, 4540-4542.

[33] Toyooka, N.; Fukutome, A.; Nemoto, H.; Daly, J.W.; Spande, T.F.; Garraffo, H.M.; Kaneko, T.; *Org. Lett.*, **2002**, *4*, 1715-1717.

[34] This piperidone was synthesized from known (2R)-2-(hydroxymethyl)butyl acetate. See: Izquierdo, I.; Plaza, M.P.; Rodriguez, M.; Tamayo, J.; *Tetrahedron: Asymmetry*, **1999**, *10*, 449-455.

[35] Momose, T.; Toyooka, N.; *J. Org. Chem.*, **1994**, *59*, 943-945; Toyooka, N.; Tanaka, K.; Momose, T.; Daly, J.W.; Garraffo, H.M.; *Tetrahedron*, **1997**, *53*, 9553-9574.

[36] Michel, P.; Rassat, A.; *Chem. Commun.*, **1999**, 2281-2283; P. Michel, P.; Rassat, A.; Daly, J.W.; Spande, T.F.; *J. Org. Chem.*, **2000**, *65*, 8908-8918.

[37] Okita, T.; Isobe, M.; *Synlett*, **1994**, 589; T. Okita T.; Isobe, M.; *Tetrahedron*, **1995**, 51, 3737-3744; Foti, C.J.; Comins, D.L.; *J. Org. Chem.*, **1995**, *60*, 2656-2657; Nicolaou, K.C.; Shi, G.-Q.; Namoto, K.; Bernal, F.; *Chem. Commun.*, **1998**, 1757-1758.

[38] Sharpless, K.B.; Young, M.W.; *J. Org. Chem.*, **1975**, *40*, 947-949.

[39] Grieco, P.A.; Gilman, S.; Nishizawa, M.; *J. Org. Chem.*, **1976**, *41*, 1485-1486.

[40] Toyooka, N.; Nemoto, H.; *Tetrahedron Lett.*, **2003**, *44*, 569-570.

Atta-ur-Rahman (Ed.) *Studies in Natural Products Chemistry, Vol. 29*
© 2003 Elsevier B.V. All rights reserved.

ADVANCES IN CHEMICAL SYNTHESIS OF CARBASUGARS AND ANALOGUES

GLORIA RASSU*[a], LUCIANA AUZZAS[a], LUIGI PINNA[b],
LUCIA BATTISTINI[c], and CLAUDIO CURTI[c]

[a]*CNR, Istituto di Chimica Biomolecolare - Sezione di Sassari, I-07100 Sassari, Italy;* [b]*Università degli Studi di Sassari, Dipartimento di Chimica, I-07100 Sassari, Italy;* [c]*Università degli Studi di Parma, Dipartimento Farmaceutico, I-43100 Parma, Italy*

ABSTRACT: Carbasugars, carbocyclic structures where a carbon atom – usually a methylene – replaces oxygen in the heterocyclic motif of the carbohydrates, represent an important class of natural and synthetic compounds that exhibit far-reaching biological effects. In this article the carbasugars have been divided into three main categories, those embodying a cyclopentane motif (furanoid carbasugars), those incorporating a cyclohexane motif (pyranoid carbasugars), and those bearing rather rare cyclopropane, cyclobutane, cycloheptane, and cyclooctane cores (contracted and expanded carbasugars).

In the three major sections of this article, the most representative syntheses of the carbasugars and their closest analogues are analyzed, and the key carbocycle forming manoeuvres are focused on in particular detail. Reference to the observed biological activity is also included, where sufficient data is available. Only recent literature from the 90s onwards was considered, although the most pioneer works on this matter have also been remembered.

Lack of space did not permit us to cover this field completely; however, in order to offer the reader a wider coverage of this subject matter, two tables listing important review articles and papers dealing with carbasugars, not discussed in the text, have been included in the final section of this account.

1. INTRODUCTION

The term carbasugars (alias pseudosugars) is used to identify an ample and varied family of polyhydroxylated cyclic organic compounds that share a homomorphic relationship with structures that fall into the class of the carbohydrates [1].

Strictly speaking, the basic motif of a carbasugar possesses a carbohydrate counterpart, be it a real molecule or only hypothetically obtainable. To be precise, the "carba" counterpart of a monosaccharide arises from the replacement of the hemiacetal ring oxygen in carbohydrates by a carbon atom (usually a CH_2), leaving the rest of the structure and chirality intact. In literature, however, this distinction does not appear to be so clear and often polyhydroxylated carbocyclic compounds that do not bare a true correlation with existing sugar motifs, but only a formal similarity, are grouped together with the carbasugars.

In this article, which deals with carbasugars, we chose to concentrate our efforts on the carbasugar targets whose structural core motif finds a precise correspondence in one of the existing subclasses of the monosaccharide cycles. In order to help the reader in this journey through the realm of the carbasugars, a chart (Chart 1) has been laid-out to permit an immediate view of the compounds dealt with in this article whilst also indicating the section where their synthesis and eventually their biological functions will be discussed.

Chart 1. Relevant Structures Covered in this Article.

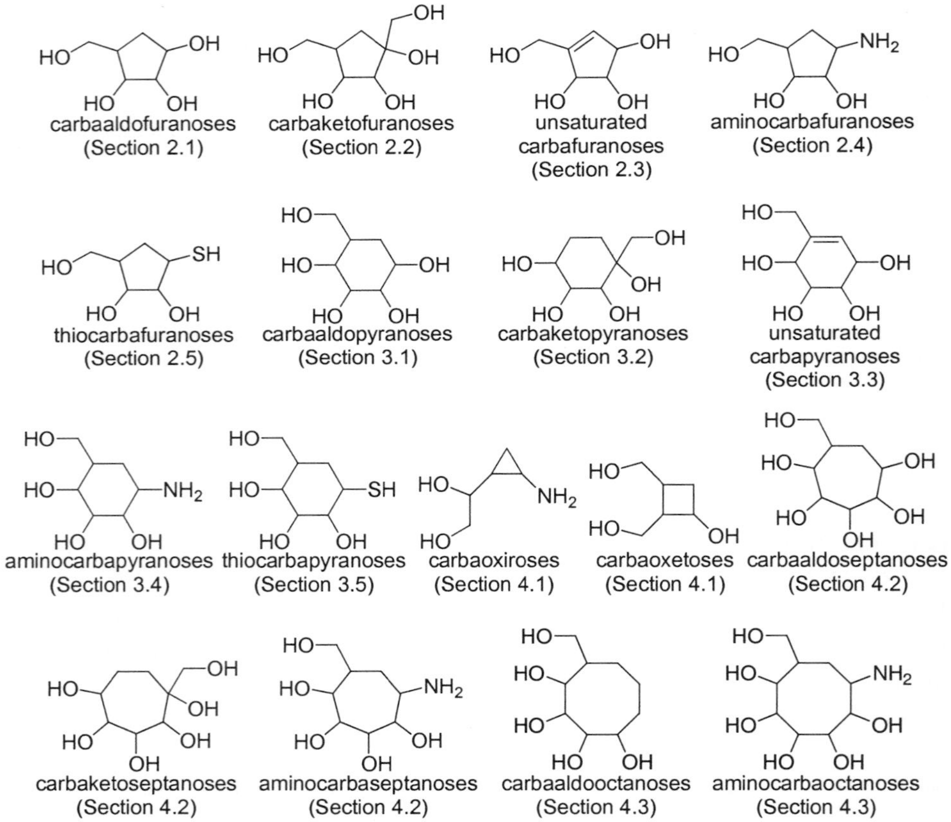

carbaaldofuranoses
(Section 2.1)

carbaketofuranoses
(Section 2.2)

unsaturated
carbafuranoses
(Section 2.3)

aminocarbafuranoses
(Section 2.4)

thiocarbafuranoses
(Section 2.5)

carbaaldopyranoses
(Section 3.1)

carbaketopyranoses
(Section 3.2)

unsaturated
carbapyranoses
(Section 3.3)

aminocarbapyranoses
(Section 3.4)

thiocarbapyranoses
(Section 3.5)

carbaoxiroses
(Section 4.1)

carbaoxetoses
(Section 4.1)

carbaaldoseptanoses
(Section 4.2)

carbaketoseptanoses
(Section 4.2)

aminocarbaseptanoses
(Section 4.2)

carbaaldooctanoses
(Section 4.3)

aminocarbaoctanoses
(Section 4.3)

To clearly see where the line was drawn in choosing the literature reviewed, a chart (Chart 2) has been prepared summarizing analogous high profile structures that will not be dealt with in this article.

Although this choice criterion gave rise to a somewhat restricted representation of the molecular diversity present in the carbasugar domain, the repertoire of the syntheses proved, nonetheless, formidable. In this article, for reasons of concision, only a selected number of manuscripts will be analyzed giving precedence to the most appealing synthetic achievements encountered in literature from the early 90s onwards.

Chart 2. Relevant Polyhydroxylated Carbocycles Uncovered.

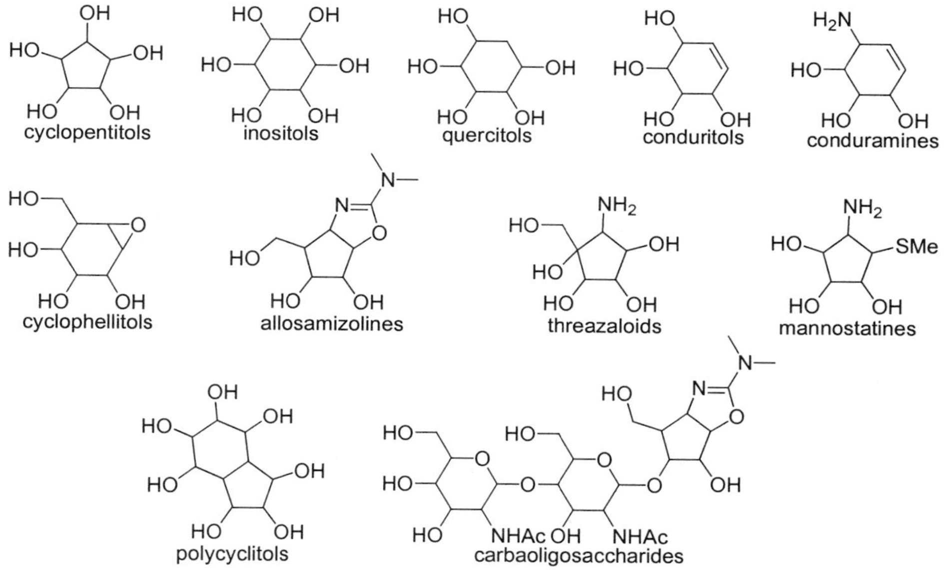

cyclopentitols   inositols   quercitols   conduritols   conduramines

cyclophellitols   allosamizolines   threazaloids   mannostatines

polycyclitols   carbaoligosaccharides

Only several pioneer works, from earlier literature, will be taken into consideration to recall the historical and scientific impact they created. Nonetheless, an appendix (Chapter 5) has been compiled at the end of this article to help the reader, which lists both reviews and commentaries (Table 1), together with research articles which have not necessarily been analyzed on this occasion (Table 2), but the contents of which are pertinent to the theme of this manuscript.

2. FURANOID CARBASUGARS

Carbocyclic analogues of furanoid sugars have represented a relevant topic in the realm of carbasugars ever since the discovery of their occurrence as substructures in natural bioactive compounds, such as carbocyclic nucleosides and certain carbocyclic sugar conjugates [2], at the end of '60s. It was soon recognised that their lack of the acetal function coupled with their resistance toward chemical or enzymatic degradation could offer a tool of biological relevance, allowing these carbafuranosidic compounds to mimic the furanose counterparts in biological systems. After the first synthesis of a carbocyclic analogue of adenosine [3], the isolation of natural carbanucleosides such as aristeromycin and neplanocin A [4], with relevant antiviral and anticancer activities, launched a series of studies directed towards the synthesis of carbafuranose entities in quite diverse stereochemical and constitutional variants.

According to the classical sugar classification, this Chapter is subdivided into five sections; carbaaldofuranoses (2.1), carbaketofuranoses (2.2), unsaturated carbafuranoses (2.3), aminocarbafuranoses (2.4) and thiocarbafuranoses (2.5).

2.1. Carbaaldofuranoses

The synthetic strategies designed to access 4a-carbaaldofuranoses can be grouped in two main categories: those utilizing enantiopure poyhydroxylated synthons, such as carbohydrates [5] and those moving from non-sugar racemic or achiral precursors. Apart from some appealing examples of target-oriented (linear) syntheses, the majority of the approaches are focused on stereocontrolled, variable methodologies which guarantee access to numerous stereoconstitutional variants of a given carbafuranose.

A nice example of the diversity-oriented synthesis of carbafuranoses was recently reported by Griengl and co-workers [6], starting from enantiomerically pure norborn-5-en-2-one 1 (Scheme 1). The Authors successfully synthesized all the stereochemical variants of the 4a-carbapentofuranose family in both the α- and β-pseudoanomeric configurations. The Bayer-Villiger norbornenone products, 2 and 3, were envisioned as the right intermediates with which ribo-, lyxo-, arabino- and xylo-configured carbafuranoses could be obtained. As an example, Scheme 1 illustrates the chiral synthesis of 4a-carba-α-D-xylofuranose 9, starting from *cis*-fused bicyclic furanone 3. *Cis*-hydroxylation of 3 and subsequent protection of the diol formed as an acetonide yielded dioxolane 4 whose lactone frame was then reductively cleaved to produce cyclopentane tetraol 5 in high yields. After regioselective bromination to 6, one carbon shortening of the side chain was performed, via debromination to olefin 7 and subsequent dihydroxylation, oxidative fission, and NaBH$_4$ reduction, resulting in the formation of protected carbasugar **8**. Target α-D-xylo carbafuranose **9** was obtained after acidic deprotection of **8** in a 9% yield over 11 steps from **3**.

Scheme 1

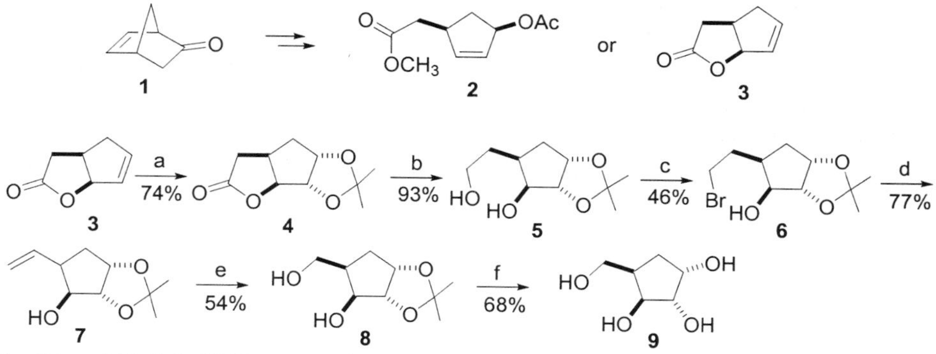

Conditions: (a) i: OsO$_4$, NMO, acetone; ii: DMP, *p*-TsOH; (b) LiAlH$_4$, Et$_2$O; (c) Ph$_3$P, Br$_2$, Et$_3$N, CH$_2$Cl$_2$; (d) i: 2-nitrophenylselenocyanate, NaBH$_4$, EtOH, 0°C to rt; ii: H$_2$O$_2$, EtOH; (e) i: OsO$_4$, NaIO$_4$, Et$_2$O/H$_2$O; ii: NaBH$_4$, MeOH; (f) i: 80% aq AcOH, reflux; ii: Ac$_2$O, pyridine, DMAP, CH$_2$Cl$_2$; iii: MeONa, MeOH.

The development of flexible and efficient methodologies to assemble *de novo* complex molecules endowed with varied substitutions and chiralities, is one of the major concerns of synthetic organic chemistry; and this concept is especially pressing when compounds with potential biological activities are involved.

Scheme 2

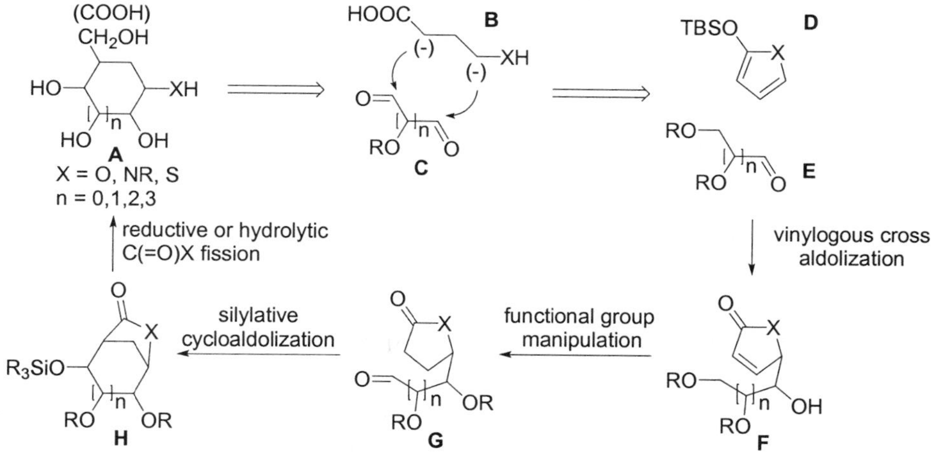

In line with this reasoning, Rassu and colleagues recently implemented a truly efficient synthetic plan for carbasugars endowed with a quite remarkable potential of flexibility and adaptability [7]. As summarized in the prospect of Scheme 2, a generic carbasugar **A** is forged via a sort of cycloadditive manoeuvre involving two complementary substructures, the dianion of a γ-heterosubstituted butanoic acid **B**, and the hydroxylated α,ω-dialdehyde **C**. In practice, dienoxy heterocycles **D** served as synthetic equivalents of dianion **B**, whilst aldehydo sugars **E** were the surrogates of α,ω-dicarbonyl moieties **C**. In the synthetic direction, the overall construction strategy features three key transformations, that is (1) a vinylogous cross aldolization between donor **D** and acceptor **E** to deliver **F**, (2) an unprecedented silylative intramolecular aldol reaction (**G** to **H**), and (3) a reductive or hydrolytic fission of the C(=O)X bond to arrive at carbasugar targets **A**.

Scheme 3

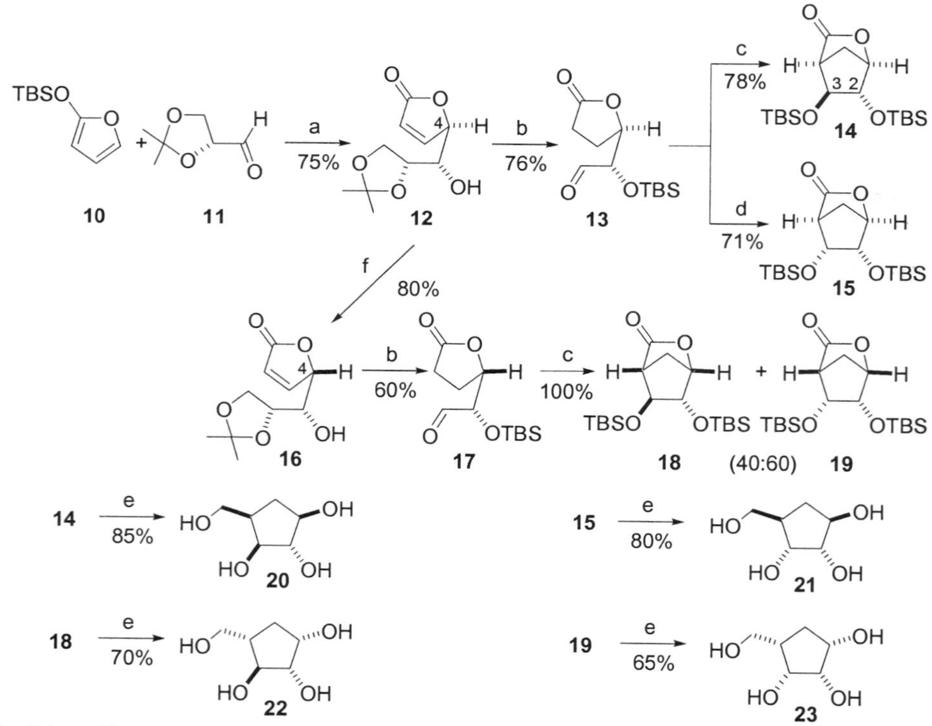

Conditions: (a) BF₃·OEt₂, CH₂Cl₂, -80°C; (b) i: NaBH₄, NiCl₂, MeOH, 0°C to rt; ii: TBSOTf, 2,6-lutidine, CH₂Cl₂; iii: 80% aq AcOH, 50°C; iv: aq NaIO₄, SiO₂, CH₂Cl₂; (c) DIPEA (3 equiv), TBSOTf (3 equiv), CH₂Cl₂, 25°C; (d) DIPEA (3 equiv), TBSOTf (3 equiv), CH₂Cl₂, -90°C; (e) i: LiBH₄, THF, 0°C to rt; ii: 6N HCl/THF/MeOH (1:2:2); (f) Et₃N, 3 cycles.

An example highlighting this synthetic tactic is outlined in Scheme 3, where 4a-carbafuranoses **20**, **21** and **22**, **23** were the chosen target compounds [7b,c]. The opening move was the BF₃-assisted vinylogous aldol reaction between furan-based silyloxy diene **10** and 2,3-*O*-isopropylidene-D-glyceraldehyde **11**, giving (4*R*)-configured unsaturated lactone **12** with a high yield and good stereoselectivity. Lactone **12** was then transformed into the six-carbon long aldehyde **13** by a clean, four-step sequence including (1) olefin saturation, (2) silylation of the free secondary hydroxyl, (3) acetonide deblocking, and (4) oxidative breakage of the terminal diol moiety. Aldehyde **13** was thus ready for the crucial annulative reaction. After an extensive optimization study, it was found that exposure of aldehyde **13** to an excess of the TBSOTf/DIPEA reagent couple resulted in formation of a mixture of separable 2,3-*trans*-configured bicycle **14** (thermodynamically favoured), and 2,3-*cis*-disposed isomer **15**, whose ratio strongly depended upon the reaction conditions. In a divergent way, isomeric bicycle adducts **18** and **19** were similarly prepared, starting from (4*S*)-configured lactone **16**, in turn, obtained from base-promoted epimerization of the previously mentioned lactone **12**. In parallel, bicycloheptanoids **14**, **15** and **18**, **19** were subjected to reductive opening of the γ-lactone framework furnishing, after global deprotection, 4a-carba-β-D-xylofuranose (**20**), 4a-carba-β-D-ribofuranose (**21**), 4a-carba-β-L-arabinofuranose (**22**), and 4a-carba-β-L-lyxofuranose (**23**), respectively, in remarkable high yields.

In a rather comprehensive work aimed at synthesizing diverse carbapentofuranoses, Désiré and Prandi [8] exploited a cobalt-catalyzed radical cyclization/oxygenation of a wide number of 6-iodohex-1-enitols to assemble the carbocyclic frame of the targets. As an example, Scheme 4 displays how 4a-carba-α-D-arabinofuranose **27** was prepared by starting with readily available glucopyranoside **24**.

Scheme 4

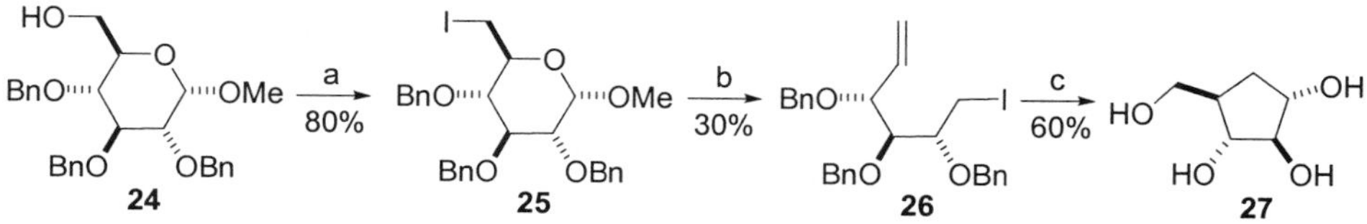

Conditions: (a) PPh$_3$, I$_2$, imidazole, toluene, reflux; (b) i: Zn, 94% aq EtOH, reflux; ii: NaBH$_4$, EtOH; iii: TsCl, pyridine, iv: NaI, HMPA, 60°C; (c) i: 10N NaOH, NaBH$_4$, Co(salen) catalyst, EtOH, 40°C; ii: H$_2$, Pd/C, EtOH.

In the event, Mitsunobu-type displacement of the unprotected hydroxyl of **24** delivered the iodinated sugar **25**, which was then elaborated into acyclic iodohexenitol **26** via ring opening using activated zinc in refluxing ethanol, followed by reduction, tosylation and iodination. The pivotal radical cyclization/oxygenation of **26** was finally carried out in the presence of a catalytic amount of cobalt(salen) complex in air. This produced a protected carbasugar intermediate which was fully liberated to **27** by hydrogenolysis. In a parallel way, a remarkable set of cyclopentane analogues including α-D-ribo, β-L-ribo, and β-D-arabino-configured isomers was engineered.

An intramolecular radical cyclization on carbohydrate derivatives was the method developed by Lundt and Horneman in a highly stereocontrolled synthesis of functionalized cyclopentanes [9]. Due to the neutral conditions required for the cyclization step, which prevented competing side reactions such as β-elimination and base-promoted epimerization, the potential of the employed radical intramolecular reaction versus the classical carbanionic carbon-carbon bond forming reactions was demonstrated.

Starting from C_4-epimeric C_2-substituted α,β-unsaturated bromolactones **28** and **33**, readily available from 7-bromo-7-deoxy-D-glycero-D-galacto-heptono-1,4-lactone, the radical cyclization was performed by using tributyltin hydride/AIBN mixture (Scheme 5).

Scheme 5

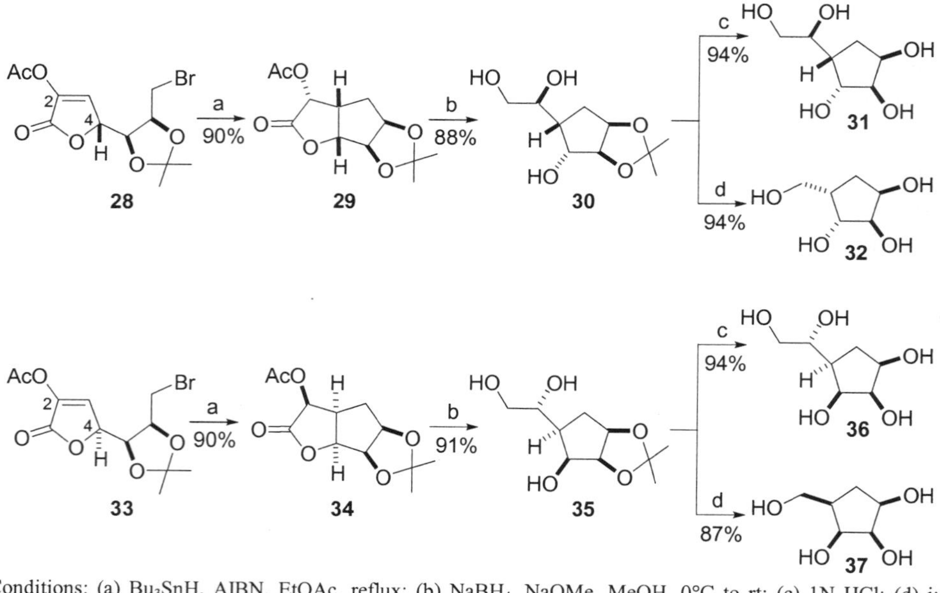

Conditions: (a) Bu$_3$SnH, AIBN, EtOAc, reflux; (b) NaBH$_4$, NaOMe, MeOH, 0°C to rt; (c) 1N HCl; (d) i: NaIO$_4$, H$_2$O, 0°C; ii: NaBH$_4$, H$_2$O; iii: 1N HCl.

The 5-*exo*-trig radical cyclization resulted in exclusive formation of *cis*-annulated bicyclic compounds **29** and **34**, which were in parallel transformed into cyclitols **30** and **35** by lactone reductive opening and deacetylation. In a divergent manner, by preserving the integrity of the carbon backbone, or by sacrificing one-carbon atom, either 4a-carbahexofuranose **31** or 4a-carbapentofuranose **32** were obtained via simple chemistry. By exactly paralleling the above mentioned conservative or degradative protocols, carbafuranose analogues **36** and **37** were also assembled starting from **35**.

Aiming at the discovery of novel inhibitors of cell wall arabinan portion biosynthesis, Lowary and Callam [10] designed an elegant route to methyl carbafuranosides **43** and **44**, employing, as the key step, a ring closing metathesis reaction.

Scheme 6

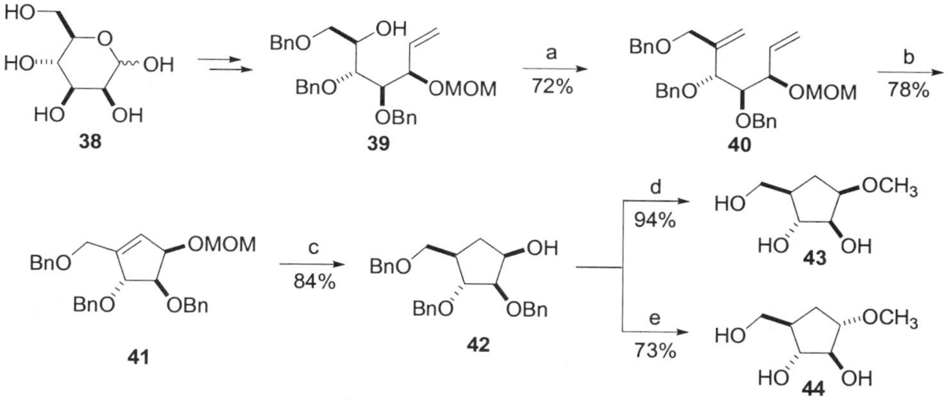

Conditions: (a) i: PCC, NaOAc, 4Å MS, CH₂Cl₂; ii: Ph₃PCH₃Br, BuLi, THF, -78°C to rt; (b) (PCy₃)(C₃H₄N₂Mes₂)Cl₂Ru=CHPh, (20 mol%), toluene, 60°C; (c) i: (Ph₃P)₃RhCl (30 mol%), H₂, toluene; ii: trace HCl, CH₃OH; (d) i: CH₃I, NaH, THF; ii: Pd/C, H₂, CH₃OH, AcOH; (e) i: DEAD, PPh₃, p-NO₂-C₆H₄CO₂H, toluene; ii: NaOCH₃, CH₃OH; iii: CH₃I, NaH, THF; iv: Pd/C, H₂, CH₃OH, AcOH.

As shown in Scheme 6, the synthesis started from the D-mannose derived alcohol **39** which was transformed to diene **40** by hydroxyl oxidation to a ketone, followed by Wittig olefination. Ring closing metathesis, promoted by a ruthenium-based catalyst, allowed the desired cyclopentene **41** to be obtained, which was reduced with Wilkinson's catalyst to alcohol **42**. Conversion of **42** to methyl 4a-carba-β-D-arabinofuranoside (**43**) was finally executed by a two-stage sequence including methylation and hydrogenolytic removal of the benzyl protecting groups. Alternatively, Mitsunobu inversion at the pseudoanomeric position of compound **42** provided methyl 4a-carba-α-D-arabinofuranoside (**44**) in good yield.

Branched-chain sugars constitute a widely represented subclass of carbohydrates which have long attracted the attention of synthetic organic chemists due to their possible incorporation into sugar-modified nucleosides [11]. In the carbasugar realm, instead, C-branched furanoid motifs have been scarcely considered, and only a few reports have been published on this subject.

Monti's research group very recently described an enantioselective and divergent approach towards certain 3-methyl 4a-carbaaldofuranoses and their 2-deoxy analogues, capitalyzing on a well established chemoenzymatic resolution of racemic cyclopentene precursors [12].

Scheme 7

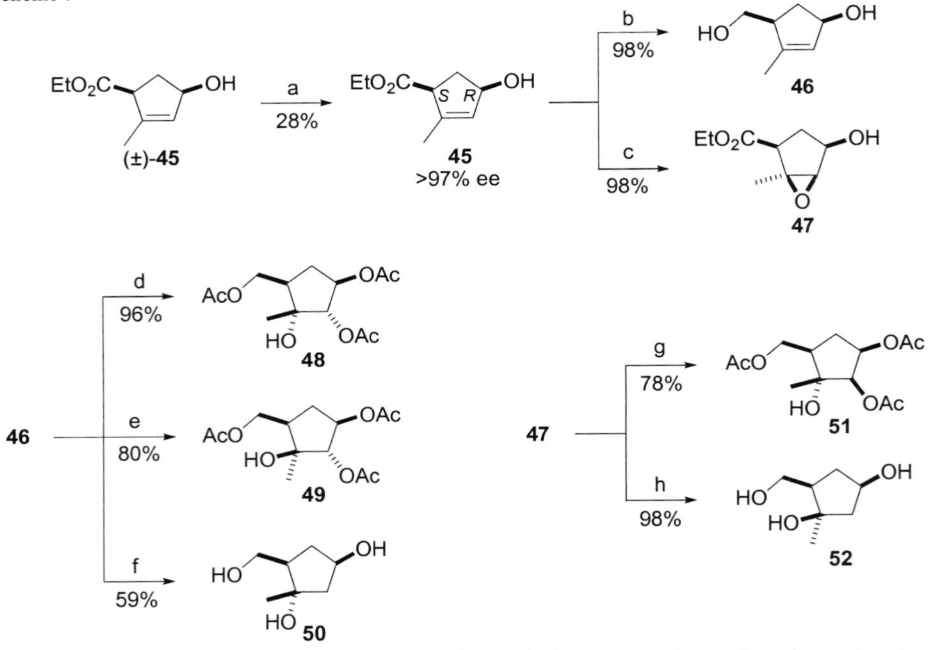

Conditions: (a) i: Porcine Pancreas lipase (PPL); (b) LiAlH₄, Et₂O, -20°C; (c) VO(acac)₂, Bu^{t}O₂H, C₆H₆; (d) i: OsO₄, NMO, acetone/H₂O; ii: Ac₂O, pyridine; (e) i: DMDO, acetone, -20°C; ii: HClO₄, H₂O; iii: Ac₂O, pyridine; (f) i: NaH, BnBr, TBAI, Et₂O, 0°C; ii: DMDO, acetone, -20°C; iii: LiAlH₄, Et₂O/THF (2:1), reflux; iv: H₂, Pd/C, MeOH; (g) i: HClO₄, H₂O; ii: LiAlH₄, THF; iii: Ac₂O, pyridine; (h) LiAlH₄, Et₂O, reflux.

As shown in Scheme 7, lipase-assisted resolution of the racemic hydroxy ester (±)-**45** allowed for isolation of chiral non-racemic enantiomer **45**, which served to arrive at both diol **46** (via ester reduction) and all-*cis* epoxide **47** (via *tert*-butylhydroperoxide treatment). In a remarkably flexible fashion, cyclopentene **46** was then used as the direct precursor of 3-methyl-branched β-D-ribocarbafuranose **48**, β-D-xylocarbafuranose **49** and 2-deoxy analogue **50**, whilst epoxide **47** was the precursor of C₃-branched congeners **51** and **52**.

In a continuing effort directed towards the assembly of a varied repertoire of carbasugar entities, Rassu and co-workers exploited enantiopure 4,5-*threo*-configured butenolide **12** (Scheme 8) as a matrix to forge C_2 methyl-branched carba-analogues **58** and **59** [13].

Scheme 8

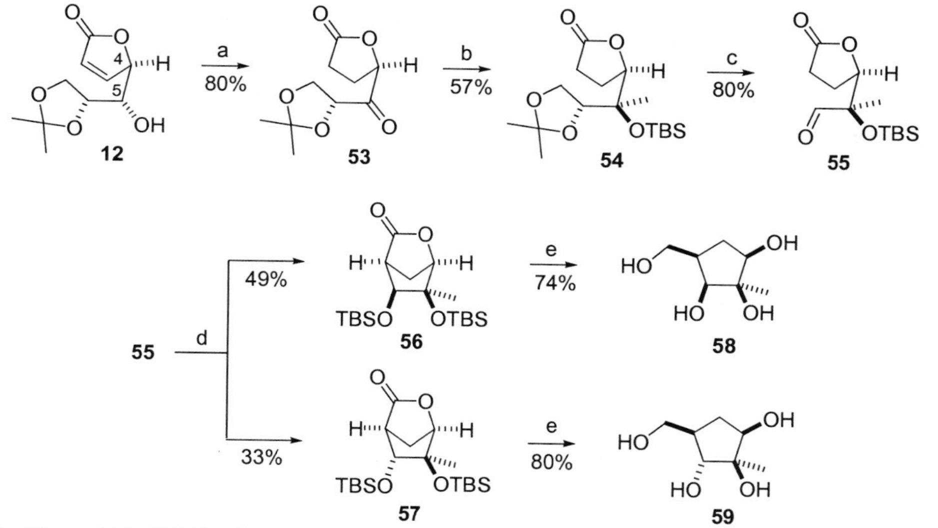

Conditions: (a) i: NaBH$_4$, NiCl$_2$, MeOH, 0°C to rt; ii: (COCl)$_2$, DMSO, CH$_2$Cl$_2$, -78°C, then Et$_3$N; (b) i: MeMgCl, THF, -30°C; ii: TBSOTf, 2,6-lutidine, CH$_2$Cl$_2$; (c) i: 80% aq AcOH, 50°C; ii: aq NaIO$_4$, SiO$_2$, CH$_2$Cl$_2$; (d) TBSOTf, DIPEA, CH$_2$Cl$_2$; (e) i: LiBH$_4$, THF, 0°C to rt; ii: 6N HCl/THF/MeOH (1:2:2).

Reduction of the olefinic bond in **12** and Swern oxidation of the free carbinol function provided ketone **53**, onto which installation of the methyl group was performed by reaction with methylmagnesium chloride in THF. After protection of the resulting tertiary alcohol as a TBS-ether, fully protected triol **54** was obtained with a useful 80% diastereomeric excess. Acetonide deblocking and oxidative fission of the diol formed led to aldehyde **55**, ready for the planned cyclization step. Exposure of aldehyde **55** to TBSOTf/DIPEA reagent system smoothly resulted in formation of the desired bicyclic adducts **56** and **57** which were isolated in a 82% combined yield (60:40 ratio).

In parallel, individual, pure bicycloheptanoids **56** and **57** were then subjected to reductive opening of the γ-lactone framework (LiBH$_4$, THF) followed by acidic removal of the silyl protecting groups. This resulted in completion of the syntheses, with 2*C*-methyl-4a-carba-β-D-lyxoaldofuranose (**58**) formed in a 74% isolated yield and 2*C*-methyl-4a-carba-β-D-arabinoaldofuranose (**59**) formed in a good 80% yield.

2.2. Carbaketofuranoses

To date, only a few reports dealing with carba-analogues of ketohexofuranoses exist [14], among which only two present the synthesis of enantiomerically pure targets [14a,b,e].

In the pioneering work by Wilcox and Gaudino, a straightforward route to the carbocyclic analogue of D-fructofuranose, **64**, and its 6-phosphate derivative was delineated [14a,b]. As shown in Scheme 9, the first move consisted of Wittig olefination of benzyl-protected arabinose **60** with carboxy-*tert*-butylmethylene triphenyl phosphorane to deliver unsaturated ester **61**, which was then cleverly elaborated into dibromide **62** via a reaction cascade encompassing Swern oxidation of the secondary OH, ester hydrolysis, diastereoselective addition of dibromomethyl lithium, and carboxylic acid methylation.

Scheme 9

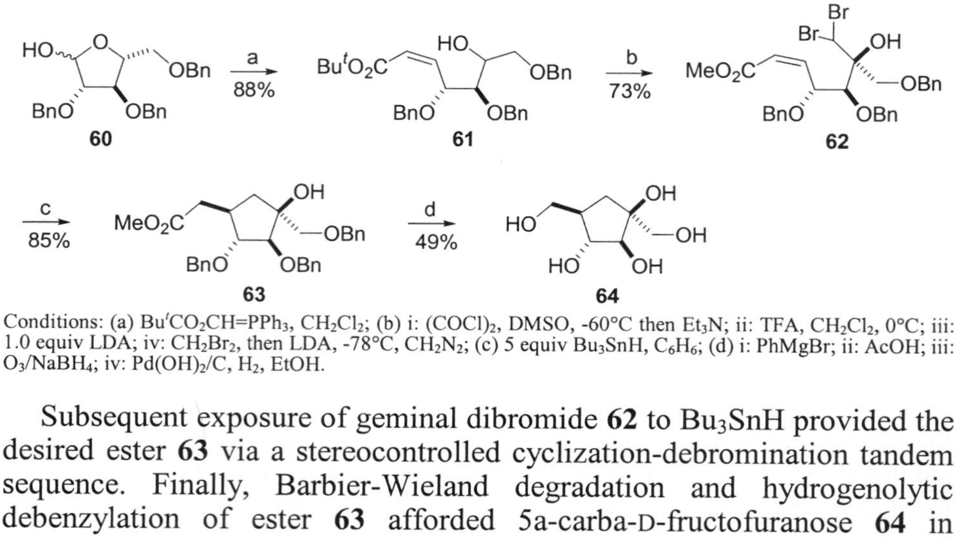

Conditions: (a) $Bu^tCO_2CH=PPh_3$, CH_2Cl_2; (b) i: $(COCl)_2$, DMSO, -60°C then Et_3N; ii: TFA, CH_2Cl_2, 0°C; iii: 1.0 equiv LDA; iv: CH_2Br_2, then LDA, -78°C, CH_2N_2; (c) 5 equiv Bu_3SnH, C_6H_6; (d) i: PhMgBr; ii: AcOH; iii: $O_3/NaBH_4$; iv: $Pd(OH)_2/C$, H_2, EtOH.

Subsequent exposure of geminal dibromide **62** to Bu_3SnH provided the desired ester **63** via a stereocontrolled cyclization-debromination tandem sequence. Finally, Barbier-Wieland degradation and hydrogenolytic debenzylation of ester **63** afforded 5a-carba-D-fructofuranose **64** in excellent yields. In a preliminary enzymological study, its 6-phosphate derivative showed to be an interesting substrate for the enzymes of the glycolysis pathway.

Motivated by the potential of fructose analogues as useful pharmacophores *en route* to allosteric modulators of phosphofructokinase-2 enzymes, Al-Abed and Seepersaud [14e] recently devised a practical, concise approach to carbafructose **64**, relying on a ring closing metathesis reaction governed by Schrock's catalyst (Scheme 10). Moving from unsaturated ketose **65**, easily obtained from 2,3,5-tri-*O*-

benzyl-D-arabinose **60**, diene **66** was first assembled via vinylmagnesium addition, followed by benzylation of the tertiary hydroxyl group. Transformation of **66** into the requisite diallyl alcohol **67** required two further steps, namely SeO$_2$/*tert*-butylhydroperoxide oxidation and benzylation of the primary hydroxyl formed.

Scheme 10

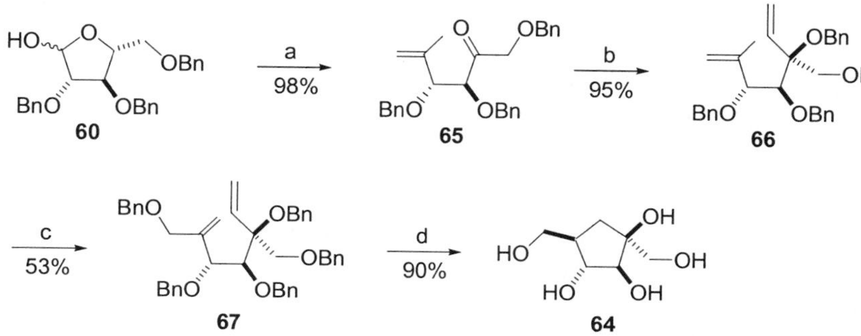

Conditions: (a) i: [O], TEMPO, NaOCl; ii: MeMgBr, THF; iii: Ac$_2$O, DMAP, EtOAc; iv: SOCl$_2$, pyridine, NaOMe; v: DMSO, (COCl)$_2$, -78°C, then Et$_3$N; (b) i: vinylmagnesium bromide, THF; ii: BnBr, DMF, NaH; (c) i: SeO$_2$, Bu^{t}O$_2$H, CH$_2$Cl$_2$; ii: BnBr, DMF, NaH; (d) i: Schrock's catalyst, hexane, reflux; ii: Pd, H$_2$, EtOH.

Facing the decisive ring-closure step, Schrock's molybdenum catalyst was used to promote the stereocontrolled annulative metathesis of diene **67**, which ultimately yielded carbafructofuranose **64** after exhaustive debenzylation.

As a whole, this procedure encompasses 11 steps from D-arabinose with a 45% global yield, as compared to an 18% yield, 12-step historical synthesis by Wilcox and Gaudino.

2.3. Unsaturated Carbafuranoses

Carbafuranoses possessing an unsaturation between C$_4$ and C$_{4a}$ positions are often exploited as scaffolds for the assembly of nucleoside carba-analogues related to the neplanocin family, many of which exhibit attractive antiviral properties [2].

As part of a drug discovery programme, Agrofoglio and collaborators [15] reported an easy route to enantiomerically pure cyclopentene **71**, by emphasizing a ring closing metathesis reaction as the pivotal manoeuvre (Scheme 11). Starting from tetra-*O*-benzyl-D-galactopyranose **68**, installation of the first olefin terminus was achieved through Wittig reaction affording, after oxidation of the resulting alcohol, L-tagatose (L-lyxohexulose) derivative **69**. A second olefination reaction delivered diene **70**, ready for the ring closing metathesis step.

Thus, exposure of **70** to a catalytic amount of Schrock's molybdenum complex smoothly provided the desired unsaturated carbasugar **71**.

Scheme 11

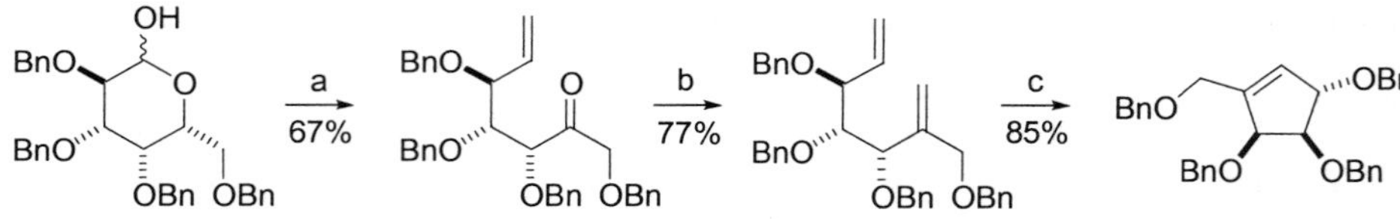

Conditions: (a) i: Ph₃PCH₃Br, BuLi, THF, -78°C to rt; ii: PCC, NaOAc, 4Å MS, CH₂Cl₂; (b) Ph₃PCH₃Br, BuLi, THF, -78°C to rt; (c) Schrock's catalyst (12 mol%), C₆H₆, 80°C, dry box.

A simple and conceptually different strategy was devised by Metha and Mohal, exploiting a norbornyl framework as a useful cyclopentanoid equivalent [16]. The key operation in this approach was the setting up of a Grob-like bottom-to-top fragmentation of suitably crafted norbornane derivatives unveiling the cyclopentanoid motif with its integral functionality complement.

Scheme 12

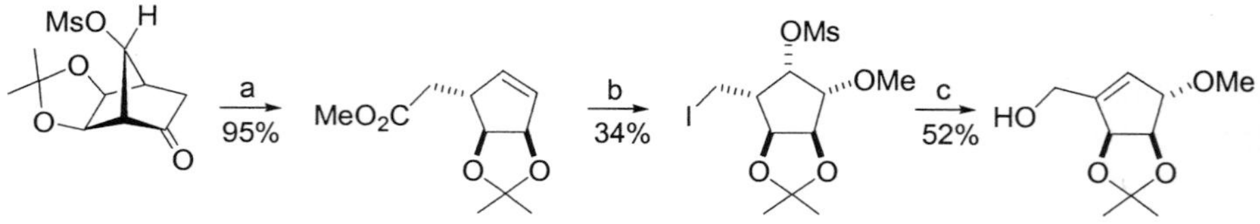

Conditions: (a) NaOMe, MeOH; (b) i: OsO₄, NMO, acetone/H₂O (4:1), rt; ii: MeI, Ag₂O, 4Å MS; iii: DIBAL-H, CH₂Cl₂, -78°C; iv: PhI(OAc)₂, I₂, cyclohexane, hv, 500W; v: Na₂CO₃, MeOH; vi: MsCl, Et₃N, CH₂Cl₂, -10°C to 0°C; (c) i: NaOAc, DMF, 100°C; ii: KOH, MeOH, rt.

As an example (Scheme 12), exposure of racemic **72** to sodium methoxide resulted in a smooth fragmentation of the norbornyl backbone to achieve olefinic methyl ester **73**, whose elaboration then afforded densely functionalized cyclopentane framework **74**. Sodium acetate-mediated displacement of the iodo functionality within **74**, followed by elimination of the mesylate group and base-promoted deacetylation finally provided the unsaturated carbasugar **75**, which nicely served as a building block for the preparation of a set of neplanocin analogues.

2.4. Aminocarbafuranoses

Aminocarbafuranoses have been widely utilized as building blocks for the construction of carbocyclic nucleosides whose heterocyclic base is forged upon the pre-existing amine substituent of the carbasugar frame. In this context, several methodologies to access carbafuranosylamines have been developed, and a punctual compilation exists on this subject matter [2c,d]. Nonetheless, to update this issue, very recent reports on the synthesis of carbafuranosylamine derivatives are here analyzed.

In planning a diversity-oriented synthetic panel of constitutionally and stereochemically diverse anomeric 4a-carbafuranosylamines, our research group remained faithful to the general plan delineated in Scheme 2 (*vide supra*), by utilizing the pyrrole-based dienoxy silane **76** as the pivotal amine source [7b,e].

Scheme 13

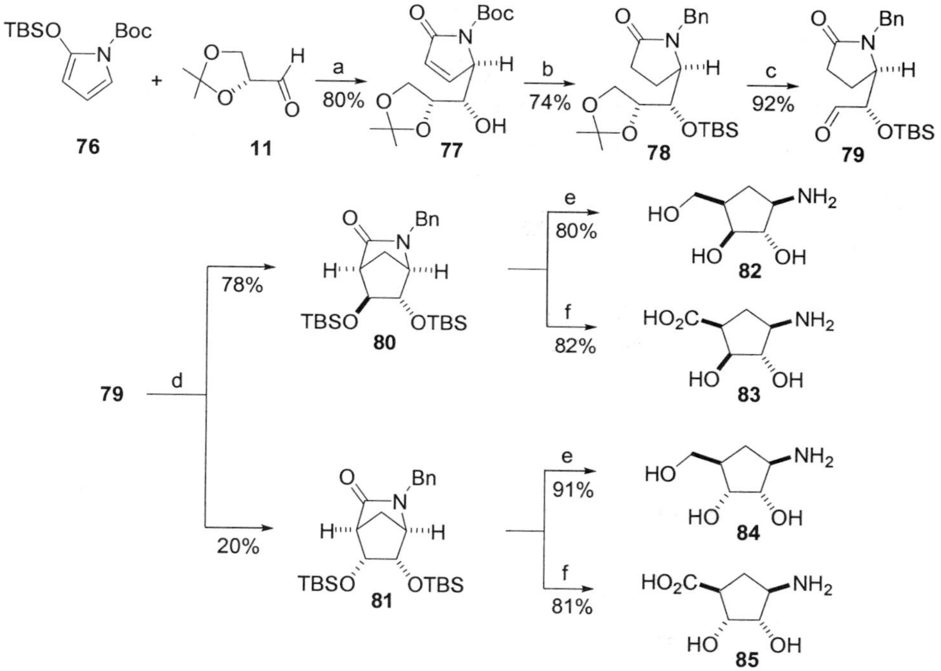

Conditions: (a) SnCl$_4$, Et$_2$O, -80°C; (b) i: NaBH$_4$, NiCl$_2$, MeOH, 0°C to rt; ii: TBSOTf, 2,6-lutidine, CH$_2$Cl$_2$; iii: TBSOTf, DIPEA, CH$_2$Cl$_2$; iv: BnCl, KH, THF, 70°C; (c) i: 80% aq AcOH, 50°C; ii: aq NaIO$_4$, SiO$_2$, CH$_2$Cl$_2$; (d) TBSOTf, DIPEA, CH$_2$Cl$_2$; (e) i: Na, liq NH$_3$, THF, -78°C; ii: Boc$_2$O, CH$_3$CN, DMAP; iii: NaBH$_4$, wet THF; iv: 6N HCl/THF/MeOH (1:2:2), then Dowex 50Wx8; (f) i: Na, liq NH$_3$, THF, -78°C; ii: Boc$_2$O, CH$_3$CN, DMAP; iii: aq LiOH, THF; iv: 6N HCl/THF/MeOH (1:2:2), then Dowex 50Wx8.

As depicted in Scheme 13, the SnCl$_4$-promoted vinylogous aldol reaction between silyloxy pyrrole **76** and D-glyceraldehyde **11** afforded the unsaturated lactam **77** in a good yield and excellent diastereoselectivity. Saturation of the carbon-carbon double bond within **77** using the NiCl$_2$/NaBH$_4$ mixture, followed by protection of the free secondary carbinol group as a TBS-ether and N-Bn for N-Boc protecting group exchange gave lactam **78** in 74% yield from **77**. Conversion of the seven-carbon lactam **78** to six-carbon aldehyde **79** was then effected via acetonide deblocking and subsequent $NaIO_4$-promoted oxidative fragmentation of the terminal diol. Exposure of aldehydo lactam **79** to the TBSOTf/DIPEA reagent system resulted in chemoselective formation of silyl-protected isomeric cycloaldols **80** and **81** in a 98% combined yield (80:20 ratio). Working in parallel, bicycles **80** and **81** were transformed into 1-deoxy-1-amino-4a-carbaaldopentofuranoses **82** and **84** by a sequence of four operations namely, liquid ammonia debenzylation, N-Boc protection, reductive lactam opening, and overall deprotection. Noteworthily, bicycloheptanoids **80** and **81** could also be conveniently elaborated to rare carbasugar amino acids **83** and **85** by hydrolytic breakage of the C(=O)N linkage.

In a structure-based drug-discovery programme in the field of influenza virus neuraminidase inhibitors, researchers at BioCryst Pharmaceuticals discovered aminocarbafuranose sialyl mimetic **89** as a very potent candidate against influenza A and B viruses [17]. To arrive at this highly substituted compound in a non-racemic format, the viable route of Scheme 14 was executed.

Scheme 14

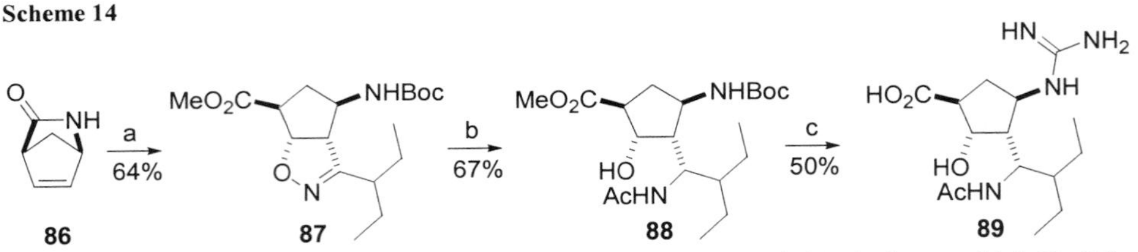

Conditions: (a) i: HCl, MeOH; ii: (Boc)₂O, Et₃N; iii: PhNCO, Et₃N, 2-ethyl-1-nitrobutane; (b) i: H₂, PtO₂, MeOH, HCl (100 psi); ii: Ac₂O, Et₃N; (c) i: HCl, Et₂O; ii: pyrazole carboxamide·HCl, DIPEA; iii: NaOH.

Commercially available lactam **86** was chosen as the chiral non-racemic starting material. The delicate issue of introducing vicinal aminoalkyl and hydroxyl substituents in **86** with the right stereoconfiguration was solved by converting **86** into isoxazoline **87** through a high-yielding, three-step sequence involving (1) methanolysis of the lactam bond, (2) protection of the amino functionality as a carbamate, and (3) [3+2] dipolar cycloaddition using the nitrile oxide obtained from 2-ethyl-1-nitrobutane. Reductive cleavage of the isoxazoline ring of **87** followed by acetylation of the amino group delivered N-acetylderivative **88** which was finally transformed into amino acid **89** via deblocking of the pseudo-anomeric amino function, guanylation, and final ester hydrolysis.

2.5. Thiocarbafuranoses

Apart from a short communication [18] where carbafuranose mercapto derivatives were prepared to be used as intermediates in the formal synthesis of certain carbanucleosides, only one report has been published dealing with the synthesis of thiocarbafuranoses [7c].

As shown in Scheme 15, the thiophene-based silyloxy diene **90** was adopted as an homologative reagent in the opening vinylogous aldol coupling to glyceraldehyde **11**.

Schema 15

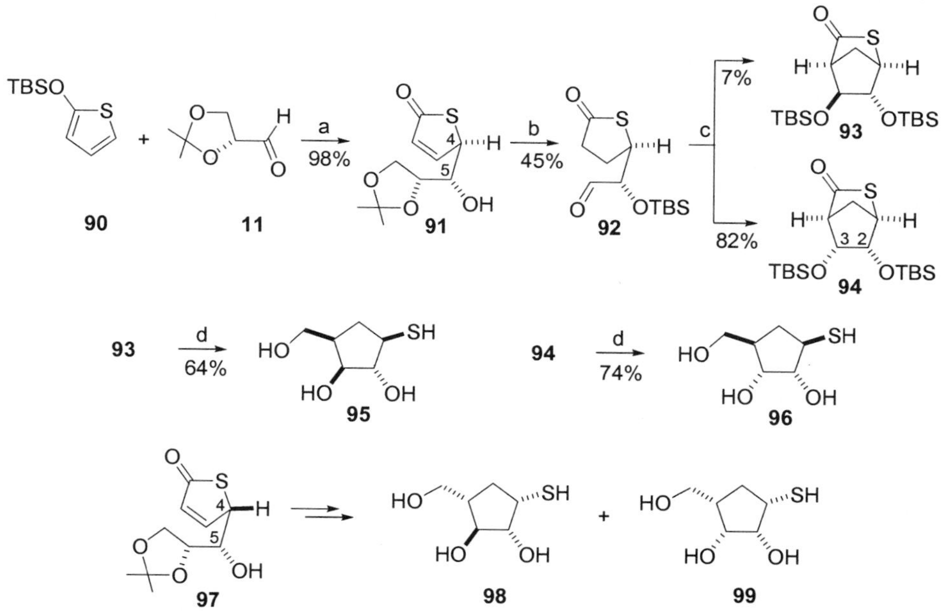

Conditions: (a) BF$_3$·OEt$_2$, CH$_2$Cl$_2$; (b) i: NaBH$_4$, NiCl$_2$, MeOH, 0°C to rt; ii: TBSOTf, 2,6-lutidine, CH$_2$Cl$_2$; iii: 80% aq AcOH, 50°C, iv: aq NaIO$_4$, SiO$_2$, CH$_2$Cl$_2$; (c) TBSOTf, DIPEA, CH$_2$Cl$_2$; (d) i: LiBH$_4$, THF, 0°C to rt; ii: 6N HCl/THF/MeOH (1:2:2).

The 4,5-*threo*-thiobutenolide **91** thus obtained was then reduced and shortened by one carbon atom to arrive at aldehyde **92** onto which a highly stereoselective cyclopentane-forming annulation was performed via a TBSOTf/DIPEA-promoted silylative protocol. In the event, a major 2,3-*cis*-configured bicycle educt **94** was isolated in a 82% yield, accompanied by only 7% of *trans*-isomer **93**. Completion of the synthesis entailed the reductive cleavage of the thiolactone bond within **93** and **94**, followed by acidic deprotection, giving 1-deoxy-1-thio-4a-carba-β-D-xylofuranose (95) and its β-D-ribo-configured analogue 96 in moderately good yields.

Paralleling this procedure, two further β-L-configured mercapto derivatives, namely arabinocarbafuranose 98 and lyxocarbafuranose 99, were synthesized starting from 4,5-erythro-disposed butenolide 97.

3. PYRANOID CARBASUGARS

It was the pyranoid form of naturally occurring aldohexoses that first inspired McCasland, Furuta and Duhram in devising a new class of sugar analogues in which the acetalic ring-oxygen was replaced by carbon [19]. They coined the term "pseudosugar" presently out of date to designate such carba-analogues of carbohydrates. Targeting racemic 5a-carba-talopyranose (103) (Scheme 16), the Authors utilized the racemic methyl ester 101 prepared in a three step sequence from keto acid 100.

Scheme 16

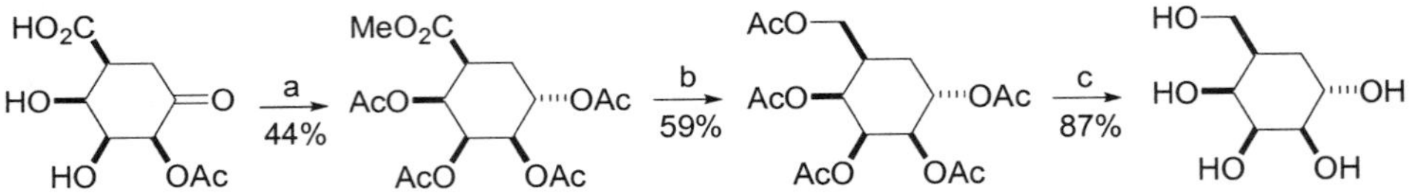

Conditions: (a) i: NaBH₄, H₂O; ii: TFA, MeOH, reflux to rt; iii: Ac₂O, cat H₂SO₄, 90-100°C; (b) i: LiAlH₄, THF, reflux; ii: Ac₂O, cat H₂SO₄; (c) 2M HCl, EtOH, reflux.

Reduction of the ester function of 101 and subsequent protection of the resulting hydroxymethyl group gave peracetate 102 which, on acidic hydrolysis, produced talopyranose 103, the first exponent of the carbasugar family.

It was then postulated and soon recognised that carbasugars, by virtue of their close resemblance to parent monosaccharides, could possess interesting biological activities. Therefore, after McCasland's pioneering work, there has been extreme interest in the synthesis of carbasugars and, in particular, of carbapyranoid motifs.

McCasland's expectation of creating a novel class of compounds featuring high biological potential by means of a slight structural modification represents a lesson of how to rationally approach drug design in medicinal chemistry. The first proof of exactness of this intuition came from synthetic 5a-carba-α-D-galactopyranose [20], isolated some years later from a fermentation broth of *Streptomyces* sp. MA-4145 and found to possess weak antibiotic properties [21].

After this, an ever-increasing number of carbapyranoses from natural sources was discovered (e.g. validamine, acarbose, gabosines, etc.), whose interesting biological activities were soon ascertained.

Nowadays, an enormous amount of bibliographic data attests a great variety of biological properties for natural and synthetic carbapyranoses, spanning the field of antibiotics, sweeteners, insuline release mediators, cellular messengers, protein anchors, and so on [22]. Most of these properties are mainly related to a common mechanism of action, namely the inhibition of several glycosidase and glycosyltransferase enzymes which translates into diverse therapeutic applications [23,24].

In this chapter we review recent achievements in the chemical synthesis of pyranoid carbasugars, pointing out relevant biological implications where claimed.

3.1. Carbaaldopyranoses

Several efficient methodologies towards enantiopure carbaaldopyranoses have been reported which exploited diverse natural and synthetic chiral sources [5,25].

A recent desymmetrization strategy of achiral dienylsilanes was developed by Angelaud and Landais [26], affording direct access to various classes of potent glycosidase inhibitors. The methodology is founded on functionalization of 1,4-dienylsilanes – such as **104** – by diastereocontrolled and enantioselective Sharpless dihydroxylation (Scheme 17). This desymmetrization step could be followed by two alternative routes towards carbapyranoses, diverging on the stereoselective introduction of hydroxymethyl function on the remaining allylsilane double bond.

Using the silicon group as a proton equivalent, access to 1,5-*cis*-configured carbaaldopyranoses was guaranteed. Thus, exposure of **104** to asymmetric dihydroxylation mixture followed by benzylation afforded enantiopure diol **105** which was subjected to diastereoselective cyclopropanation giving rise to bicycle **106**. Electrophilic ring opening of cyclopropane **106** occurred *anti* to silicon group, giving homoallylic halide **107**, having the required 1,5-*cis* configuration. Conversion of iodide **107** into polyol **108** was achieved through a two step sequence involving silylation and *anti*-selective dihydroxylation. Subsequent oxidation of the C_6-Si bond and acetylation gave the known carba-β-D-altropyranose (**109**) in a 23% overall yield.

Noteworthy, if otherwise elaborated, **107** could diverge into the 6-deoxy-analogue **111**, an isomer of the biologically valuable carbafucopyranose.

Scheme 17

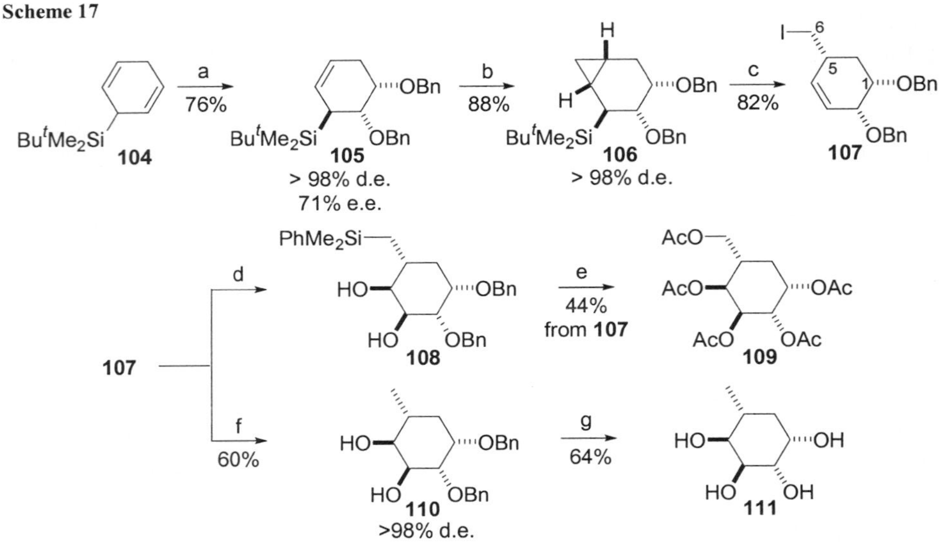

Conditions: (a) i: $K_2OsO_2(OH)_4$, $(DHQ)_2PYR$, Bu^tOH/H_2O, K_2CO_3, $K_3Fe(CN)_6$, 0°C; ii: NaH, BnBr; (b): $ZnEt_2$, CH_2I_2, DCE; (c) NIS, MeCN; (d) i: Bu^tLi, $PhMe_2SiCl$, THF, -100°C; ii: OsO_4, NMO, THF; (e) i: $Hg(OAc)_2$, AcO_2H; ii: Ac_2O, pyridine; iii: Pd/C, H_2; iv: Ac_2O, pyridine; (f) i: BuLi, THF, -80°C; ii: OsO_4, THF, NMO; (g) Pd/C, H_2, EtOH.

In a complementary approach (Scheme 18), using the silicon group as a latent hydroxy function, 1,5-*trans*-configured carbapyranose **115** was obtained. In the event, oxidation of the C-Si bond of isopropylidene protected diol **112** occurred with retention of configuration at C_3 to form **113**, which was finally transformed into 5a-carba-α-D-galactopyranose derivative **115** via [2,3]-Wittig sigmatropic rearrangement followed by dihydroxylation and peracetylation.

Scheme 18

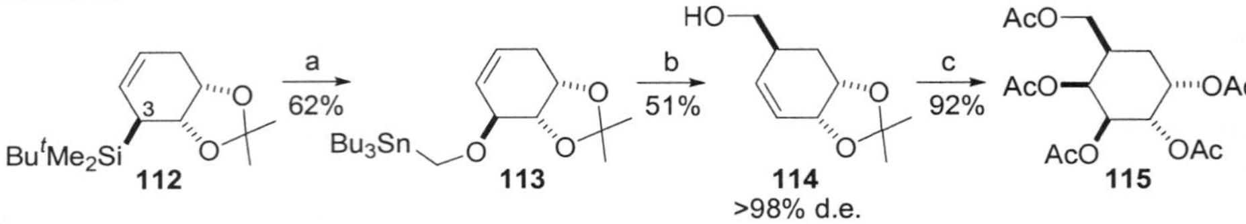

Conditions: (a) i: H_2O_2, KF, $KHCO_3$, DMF, 60°C; ii: KH, THF, Bu_3SnCH_2I; (b) BuLi, THF, -60°C; (c) i: Ac_2O, pyridine; ii: OsO_4, NMO, THF; iii: aq AcOH, THF, 80°C; iv: Ac_2O, pyridine.

Clever entries to enantiopure carbapyranoses exploited optically pure or meso *cis*-diene diols easily obtained from aromatic compounds by microbial transformations [27]. In the impressive work described by Pingli and Vandewalle [28] optically active building-block **116** (Scheme 19) was prepared by enantiotoposelective lipase-catalyzed hydrolysis of the corresponding *meso* dibutyrate.

Starting from **116**, the synthesis of all possible D- and L-configured 5a-carbaaldohexoses possessing a 2,3-*cis* substitution pattern (gulo, talo, manno, and allo series) was indeed realized in a brilliant, divergent way.

Scheme 19

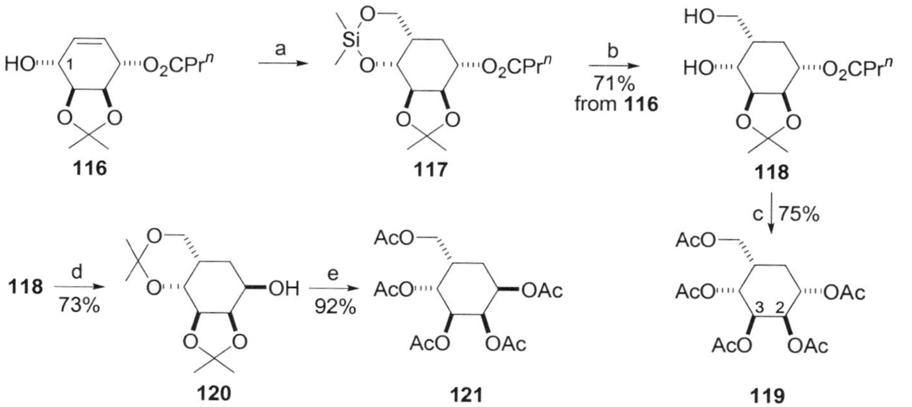

Conditions: (a) i: BrCH$_2$(Me)$_2$SiCl, Et$_3$N, DMAP, CH$_2$Cl$_2$, 0°C; ii: Bu$_3$SnH, AIBN, C$_6$H$_6$, reflux to rt; (b) i: KF, KHCO$_3$, H$_2$O$_2$ (35%), THF/MeOH (1:1); ii: Na$_2$SO$_3$, 0°C; (c) i: KHCO$_3$, MeOH; ii: *p*-TsOH, MeOH; iii: Ac$_2$O, pyridine; (d) i: DMP, DMF, PPTS; ii: KHCO$_3$, MeOH; iii: (COCl)$_2$, DMSO, CH$_2$Cl$_2$, -78°C, then Et$_3$N; iv: NaBH$_4$, THF/MeOH (1:1), -78°C; (e) i: *p*-TsOH, MeOH; ii: Ac$_2$O, pyridine.

To access 5a-carba-L-gulopyranose derivatives **119** and **121**, Stork cyclization clearly transformed cyclohexene **116** into cyclic silylether **117** which, after oxidative cleavage of the C-Si bond, led to **118**, the direct precursor of β-L-configured carbahexose **119**. The α-L-anomeric counterpart **121** was obtained from the common intermediate **118** by a short sequence centered upon inversion of configuration of the pseudoanomeric carbon. Inversion of the configuration at C$_1$ in **116** and application of the same protocol ensured preparation of both anomers of the D-talo-configured carbasugars.

Targeting α- and β-D-mannocarbapyranoses **125** and **127** (Scheme 20), the same researchers envisioned the C$_5$ hydroxymethyl group to be installed via [2,3]-Wittig rearrangement on advanced intermediate **123**.

Indeed, starting from the densely hydroxylated compound **122**, formation of stannane **123** and [2,3]-sigmatropic rearrangement under Still's condition easily gave cyclohexene **124**, whose hydroboration-oxidation led to protected carbapyranose **125**. On the other hand, silylation of the hydroxymethyl moiety and hydration of double bond in

124 furnished **126**. Selective manipulation of the pseudoanomeric hydroxyl group in **126** provided 5a-carba-β-D-mannopyranose **127** in a good overall yield (52%) from **122**.

Scheme 20

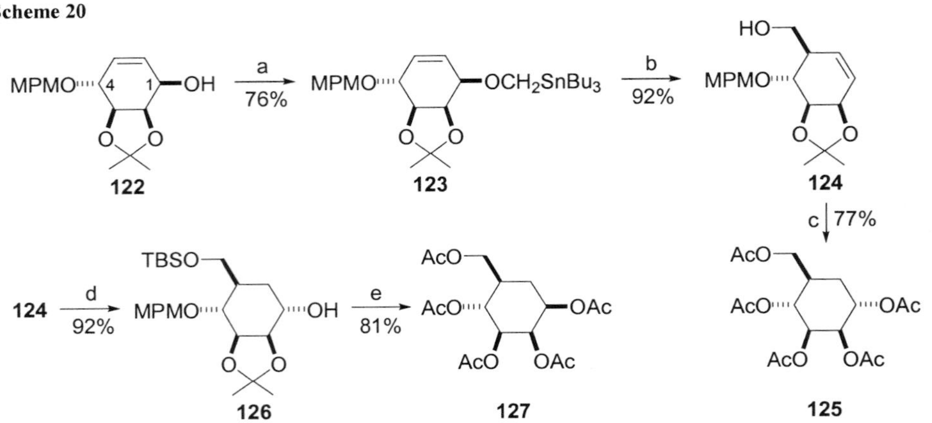

Conditions: (a) KH, ICH$_2$SnBu$_3$, THF, 0°C; (b) BuLi, THF, -78°C; (c) i: BH$_3$, THF, -78°C; ii: H$_2$O$_2$, NaOH; iii: Pd/C, H$_2$, MeOH; iv: *p*-TsOH, MeOH; v: Ac$_2$O, pyridine; (d) i: TBSCl, imidazole, DMF; ii: BH$_3$, THF -78°C; iii: H$_2$O$_2$, NaOH, 0°C; (e) i: (COCl)$_2$, DMSO, -78°C, then Et$_3$N; ii: NaBH$_4$, MeOH, 0°C; iii: Pd/C, H$_2$, MeOH; iv: *p*-TsOH, MeOH; v: Ac$_2$O, pyridine.

Quite remarkably, by switching the allyl alcohol protection from C$_4$ to C$_1$ in the intermediary compound **122**, construction of α- and β-D-allocarbapyranoses was also completed.

One of the most utilized routes to carbaaldopyranoses is represented by a furan-based Diels-Alder approach [25]. Several works centered upon this approach have appeared in literature, among which one of the most versatile has been recently proposed by Metha and Mohal [29]. In their norbornyl route to polyoxygenated cyclohexanes and cyclopentanes, they recognised bicyclo[2.2.1]heptane (norbornyl) motifs of type **128** (Scheme 21) as locked carbasugars, in which the carbapyranose backbone could easily be retrieved through top-to-bottom excision of the C$_1$-C$_7$ bond. Alternatively, a carbafuranose skeleton could be unmasked by bottom-to-top fragmentation of the C$_1$-C$_2$ linkage (*vide supra*) [16]. *Endo*-hydroxy-7-norbornenone ketal **128**, readily available through Diels-Alder reaction between 5,5-dimethoxy-1,2,3,4-tetrachlorocyclopentadiene and vinyl acetate, was thus chosen as an ideal scaffold to access pyranose-type carbasugars. As an example of this strategy, Scheme 21 highlights the divergent synthesis of carbapyranoses **115** and **134** in a racemic format [29]. After transformation of racemic ketal **128** into ketone **129**, top-to-bottom fragmentation was performed with sodium in methanol, giving ester **130**. Stereoselective dihydroxylation of **130** followed by functional group manipulation delivered the desired 5a-carba-α-D,L-galactopyranose (**115**). Alternatively, ester **131** could be transformed into fuco-configured 6-deoxy-carba-analogue **134** via exhaustive reduction of the C$_5$ ester moiety.

Scheme 21

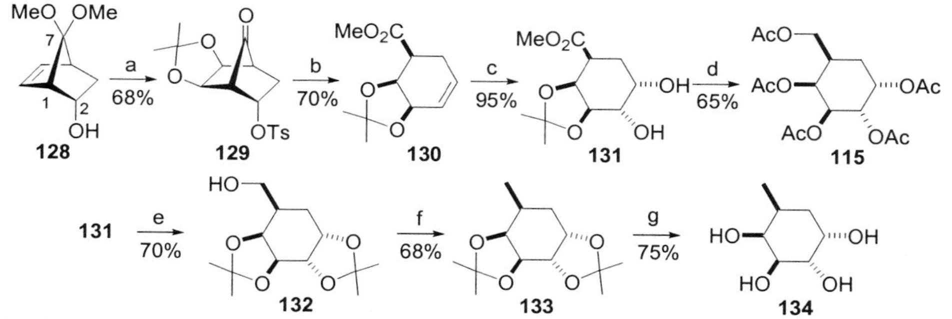

Conditions: (a) i: TsCl, pyridine, DMAP, CH$_2$Cl$_2$; ii: OsO$_4$, NMO, acetone/H$_2$O (4:1); iii: Amberlist-15, acetone/H$_2$O; (b) NaOMe, MeOH; (c) OsO$_4$, NMO; (d) i: LiAlH$_4$, THF; ii: Amberlist-15, MeOH/H$_2$O; iii: Ac$_2$O, pyridine; (e) i: acetone, Amberlist-15, 4Å MS; ii: LiAlH$_4$, THF, 0°C; (f) i: TsCl, pyridine, CH$_2$Cl$_2$; ii: NaBH$_4$, DMSO; (g) Amberlist-15, MeOH/H$_2$O.

The majority of methodologies resulting in reliable diastereoselective routes to carbaaldopyranoses suitably exploits natural carbohydrates or their derivatives as chiral starting sources [5].

Scheme 22

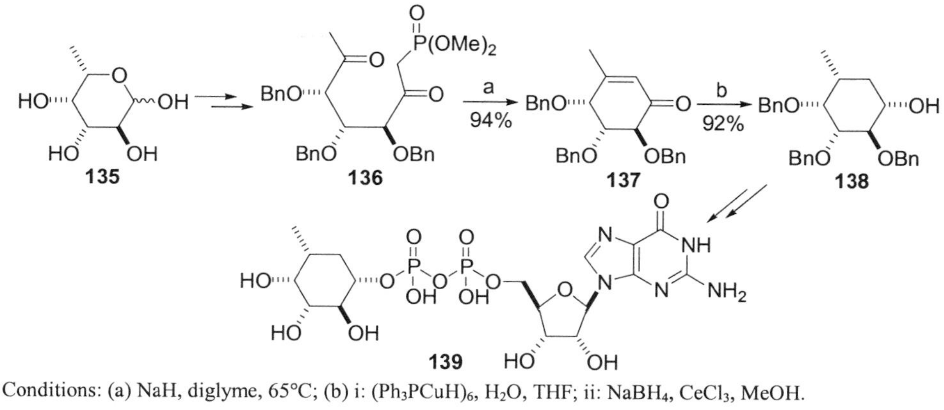

Conditions: (a) NaH, diglyme, 65°C; (b) i: (Ph$_3$PCuH)$_6$, H$_2$O, THF; ii: NaBH$_4$, CeCl$_3$, MeOH.

In 1992 Toyokuni synthesized 5a-carba-α-D-fucopyranose **138** (Scheme 22) by a straightforward route involving an intramolecular Emmons-Horner-Wadsworth olefination of dioxophosphonate **136**, readily available from L-fucose **135** [30]. The intramolecular olefination of **136** proceeded easily to give inosose derivative **137** which, after stereoselective reduction of the carbon-carbon double bond and subsequent carbonyl reduction, led to α-alcohol **138**. In this protected form, **138** was employed as the scaffold for the assembly of a carbocyclic analogue of the nucleoside diphosphosugar guanosine 5'(β-L-fucopyranosyl diphosphate) (GDP-fucose) **139**. Interestingly, carba-GDP-fucose **139** showed high inhibitory activity against α(1→ 3/4)Fuc-T of human colonic adenocarcinoma cells origin, even stronger than natural GDP-fucose. Analogously, in a related work the carbasugar analogue of UDP-galactose was reported to be a competitive inhibitor of β-(1,4)-galactosyltransferase with activity similar to that of UDP-galactose [31].

Another clever application of carbohydrate chemistry for the synthesis of carbaaldopyranoses is described in a recent publication of Sudha and Nagarajan (Scheme 23) [32]. Inspired by a previously reported protocol by Büchi and Powell [33], the pivotal step of their divergent strategy to carbasugars is represented by smooth conversion of glycal intermediate **140** into cyclohexene **141**, via a thermal [3,3]-sigmatropic Claisen rearrangement, followed by reduction of the resulting aldehyde group.

Scheme 23

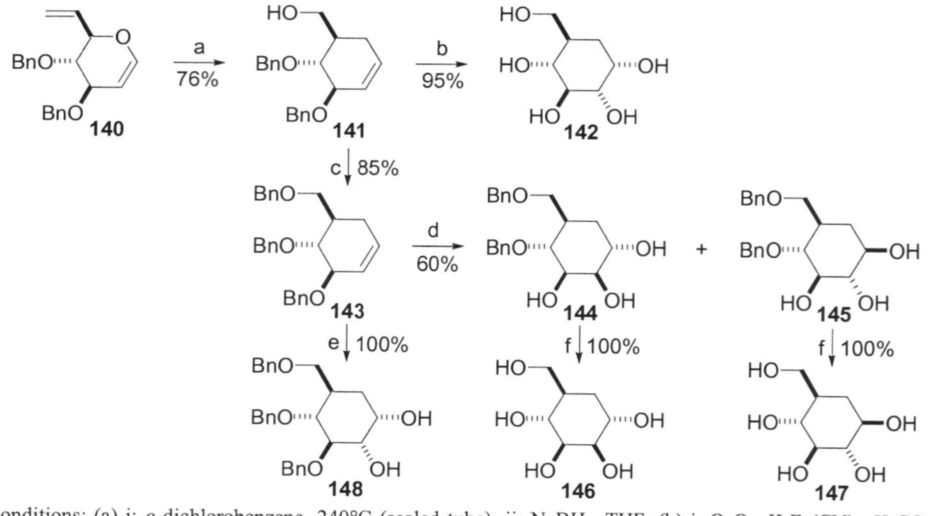

Conditions: (a) i: *o*-dichlorobenzene, 240°C (sealed tube); ii: NaBH₄, THF; (b) i: OsO₄, K₃Fe(CN)₆, K₂CO₃, Bu^{t}OH/H₂O; ii: 20% Pd(OH)₂/C, H₂, 55 psi; (c) NaH, DMF, BnBr; (d) MCPBA, H₂O, 10% H₂SO₄; (e) AcOH/H₂O, AgOAc, I₂, cat Na, MeOH; (f) 20% Pd(OH)₂/C, H₂, 55 psi.

Controlled *cis*-dihydroxylation of the olefinic double bond of alcohol **141** allowed preparation of 5a-carba-α-D-glucopyranose (**142**), whilst 5a-carba-α-D-mannopyranose (**146**) and 5a-carba-β-D-glucopyranose (**147**) were prepared through epoxidation, ring opening, and final deprotection. Unfortunately, an attempt to obtain the β anomer of **146** by *cis*-

dihydroxylation of **143** under Woodsward's conditions (*syn* dihydroxylation on the more hindered face) failed, yielding only the α-D-gluco-configured carbasugar **148**.

In planning the synthesis of a series of carbapyranoses in their chiral non racemic format, Zanardi and co-workers [7d] utilized the seven-carbon lactone **149**, in turn prepared by elaboration of butenolide **12** (*vide supra*, Scheme 3).

Scheme 24

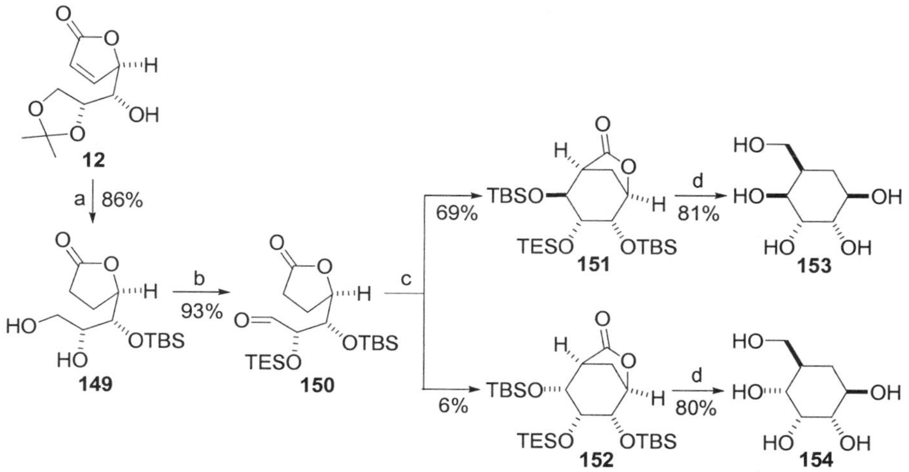

Conditions: (a) i: NaBH₄, NiCl₂, MeOH, 0°C to rt; ii: TBSOTf, 2,6-lutidine, CH₂Cl₂; iii: 80% aq AcOH, 50°C; (b) i: TESOTf, pyridine, DMAP; ii: (COCl)₂, DMSO, CH₂Cl₂, -78°C to -30°C, then Et₃N, -78°C; (c) TBSOTf, DIPEA, CH₂Cl₂, (d) i: LiBH₄, THF, 0°C to rt; ii: 6N HCl/THF/MeOH (1:2:2).

Thus, according to Scheme 24, diol **149** was converted into protected aldehyde **150** via persilylation and Swern oxidation. Subsequent intramolecular silylative aldolization smoothly provided epimeric bicyclooctanoids **151** and **152** as a separable 92:8 mixture. Bicyclic adducts **151** and **152** were then in parallel elaborated to unmask the hydroxymethyl and the pseudoanomeric functions, leading to 5a-carba-β-D-gulopyranose (**153**) and 5a-carba-β-D-allopyranose (**154**), respectively.

Scheme 25

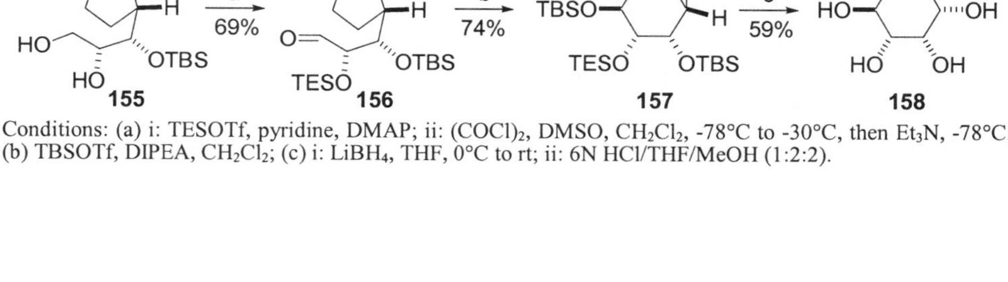

Conditions: (a) i: TESOTf, pyridine, DMAP; ii: (COCl)₂, DMSO, CH₂Cl₂, -78°C to -30°C, then Et₃N, -78°C; (b) TBSOTf, DIPEA, CH₂Cl₂; (c) i: LiBH₄, THF, 0°C to rt; ii: 6N HCl/THF/MeOH (1:2:2).

Application of the same protocol to lactone **155** (Scheme 25), smoothly obtained from epimeric butenolide **16** (*vide supra*, Scheme 3) resulted in formation of β-L-manno-configured carbapyranose **158** through the intermediacy of aldehyde **156** and bicycle **157**.

3.2. Carbaketopyranoses

In the past, ketose-type pyranose carbasugars were mainly investigated in the search for sucrose mimics to employ as non-nutritive sweeteners [25a,b]. Probably due to this seemingly unappealing potential and the inherent difficulties related to their stereoconstitutional features, up until 1990 only one report was found to have been published dealing with the asymmetric synthesis of D- and L-carba-analogues of fructopyranose [34].

A more efficient approach was proposed by Shing who demonstrated the synthetic versatility of natural quinic acid as the starting material for the preparation of carbapyranoses of both aldo- and keto-typologies [35].

Scheme 26

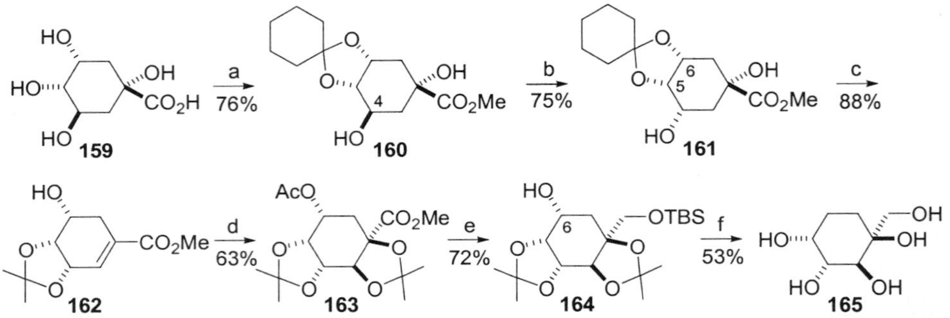

Conditions: (a) i: cyclohexanone, C_6H_6, DMF, Dowex 50Wx8, reflux; ii: NaOMe/MeOH, 0°C; (b) i: PCC, 3Å MS, pyridine, CH_2Cl_2; ii: NaBH$_4$, MeOH, 0°C; (c) acetone, *p*-TsOH, 88%; (d) i: Ac$_2$O, pyridine; ii: OsO$_4$, TMNO, pyridine, Bu^{t}OH/H$_2$O; iii: 2-methoxypropene, CSA, CH_2Cl_2; (e) i: DIBAL-H, THF, 0°C; ii: TBSCl, imidazole, DMAP, CH_2Cl_2; (f) i: PhOC(S)Cl, pyridine, DMAP, CH_2Cl_2; ii: Bu$_3$SnH, AIBN, toluene; iii: TFA/H$_2$O (1:1).

The synthesis of 6a-carba-β-D-fructopyranose (**165**) (Scheme 26) comprised four key steps, namely the inversion of C_4 configuration in quinic ester **160** to obtain **161**, the shift of the ketal protection from C_5-C_6 to C_4-C_5 to provide **162**, the stereocontrolled dihydroxylation of *epi*-shikimate **162**, and the C_6-deoxygenation reaction of advanced intermediate **164**. Carbasugar **165** was obtained in an overall 12% yield from quinic acid **159** for a 13-step sequence.

An improved preparation of the same target **165** was executed by Maryanoff's group [36] exploiting a strictly analogous chemistry (Scheme 27). Here, protected carbocyclic ester **162** was first deoxygenated to **166**, which was next subjected to dihydroxylation to provide, after full deprotection, fructopyranose **165**. Noteworthily, protected carbaketose **167** also served as a scaffold for the synthesis of sulfamate **168**, the carba-analogue of clinically useful antiepileptic drug topiramate.

Scheme 27

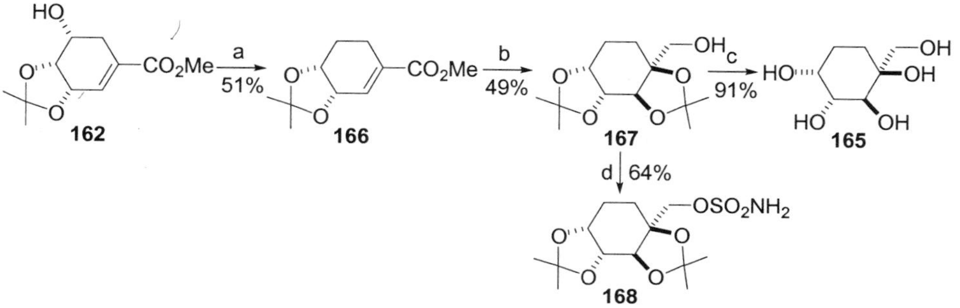

Conditions: (a) i: PhOC(S)Cl, pyridine, DMAP, CH_2Cl_2; ii: Bu_3SnH, $(Bu^tO)_2$, toluene, reflux; (b) i: OsO_4, TMNO, pyridine, Bu^tOH/H_2O, reflux; ii: 2-methoxypropene, CSA, CH_2Cl_2; iii: DIBAL-H, THF, -20°C to rt; (c) TFA/H_2O (1:1); (d) H_2NSO_2Cl, Et_3N, 0°C.

Topiramate analogue **168** was enlisted into a comprehensive structural-activity relationship study [37], along with about a hundred varied synthetic analogues, which allowed identification of a new promising antiepileptic lead. From this study, carba-analogue **168** resulted three times more active than its corresponding carbohydrate-based drug.

3.3. Unsaturated Carbapyranoses

A variety of unsaturated pyranoid carbasugars populates the arena of carbasugar-related enzyme inhibitors. In the most promising representatives the double bond is located between C_5 and C_{5a}, and, strictly speaking, these motifs cannot be related to any corresponding existing or potentially existing sugars. Nonetheless, the relevant biological properties of unsaturated carbapyranoses enticed us to consider this important structural typology within this compilation.

Among simple unsaturated carbaaldopyranoses, one of the most well-known examples is represented by MK7607, an effective herbicidal substance isolated from the fermentation broth of *Corvularia eragrostidis* [38]. Preparation of its unnatural antipode **174** was recently accomplished by Singh who exploited (-)-shikimic acid **169** as the chiral source (Scheme 28) [39]. As with quinic acid, the shikimate structure can be regarded as a quasi-carbasugar, embodying most of the structural requisites of a variety of carbapyranose targets.

Scheme 28

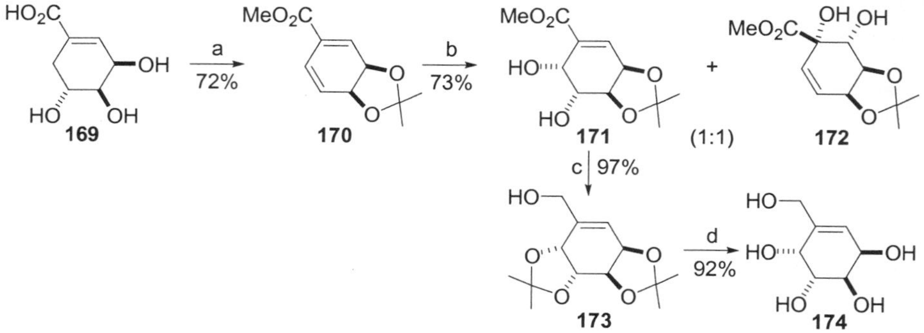

Conditions: (a) i: CSA, MeOH, reflux; ii: DMP, CSA; iii: Tf₂O, DMAP, pyridine, CH₂Cl₂, -20°C; iv: CsOAc, DMF; (b) OsO₄, NMO, BuᵗOH/H₂O (10:1); (c) i: DMP, CSA, ii: DIBAL-H, THF, -10°C; (d) TFA/H₂O (6:1).

Thus, according to Scheme 28, (-)-shikimic acid **169** was converted to cyclohexadiene derivative **170** via esterification of carboxyl group, protection of the *cis*-disposed hydroxyls, and elimination of the remaining carbinol moiety. Catalytic dihydroxylation of **170** gave unsaturated esters **171** and **172** as a separable 1:1 mixture. The 5,5a-unsaturated isomer **171** was finally elaborated into the desired (-)-MK7607 (**174**) via simple protection-reduction-deprotection sequence.

Scheme 29

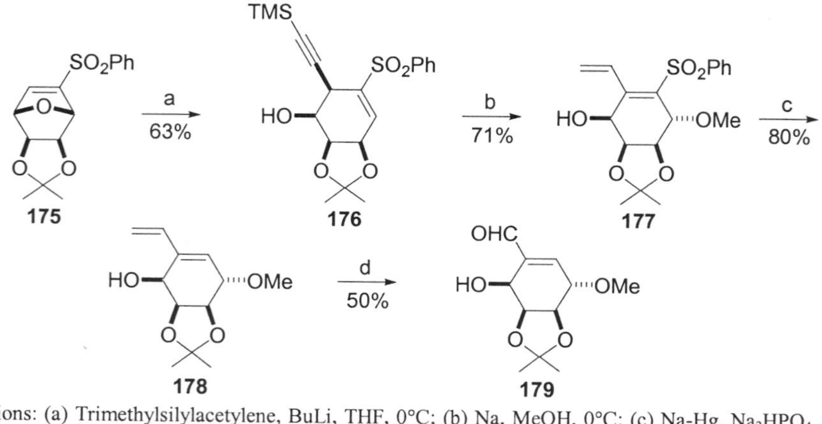

Conditions: (a) Trimethylsilylacetylene, BuLi, THF, 0°C; (b) Na, MeOH, 0°C; (c) Na-Hg, Na₂HPO₄, MeOH, -20°C; (d) i: NaIO₄, MeCN/CCl₄/H₂O (1:1:1.5); ii: RuCl₃·H₂O.

A second typology of unsaturated carbapyranoses is represented by the rancinamycins, a group of secondary metabolites produced by *Streptomyces lincolnensis*, whose important antibiotic activity was attested [40]. An interesting way to arrive at these constructs was outlined by Arjona starting from the chiral non racemic Diels-Alder adduct **175** (Scheme 29) [41]. The pivotal manoeuvre of the synthesis was the unprecedented transformation of alkynyl derivative **176** into dienylsulfone **177** through a one-pot transformation involving a methoxide Michael addition, followed by vinylsulfone isomerisation, and triple bond reduction. From **177**, the synthesis was completed by desulfonylation to **178** and subsequent oxidation of the exocyclic double bond to provide aldehyde **179**, the protected form of rancinamycin III. Conveniently, optically active diene **177** also served as the precursor of several aldopyranose carbasugars.

Gabosines represent a large group of secondary metabolites of great biological interest isolated from different *Streptomycetes* strains and traditionally they are referred to as ketocarbasugars due to the presence of a keto functionality on pyranose skeleton [42].

Scheme 30

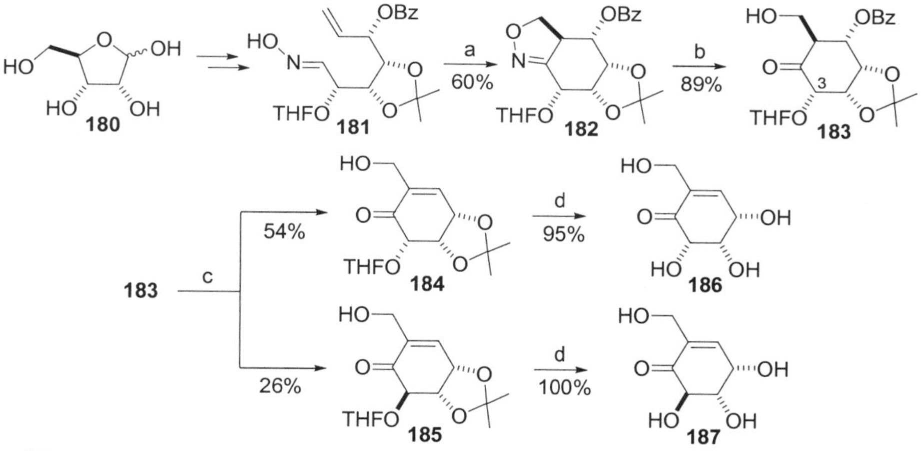

Conditions: (a) NaOCl, Et$_3$N, CH$_2$Cl$_2$; (b) H$_2$, Ni-Raney, EtOH, AcOH; (c) DABCO, THF; (d) TFA, CH$_2$Cl$_2$.

Working on the assumption that the enone moiety is a basic requisite for the biological activity of gabosine C-related compounds, Lygo and co-workers envisaged a synthetic strategy to arrive at unsaturated carbapyranoses **186** (the enantiomeric form of natural gabosine C) and **187** (gabosine E) (Scheme 30) [43]. Central to this strategy was an efficient and stereoselective intramolecular nitrile oxide cycloaddition reaction involving protected ω-unsaturated oxime **181** (ex D-ribose) to form cycloadduct **182**.

Reductive cleavage of the isoxazole ring within **182** installed the requested carbonyl and hydroxymethyl functionalities giving densely substituted cyclohexanone **183** in high yields. Exposure of **183** to DABCO resulted in the formation of a separable 2:1 mixture of enones **184** and **185** presumably arising from an irreversible C$_3$-epimerization, elimination and C$_3$-epimeric re-equilibration. The major isomer **184** was the direct precursor of unnatural gabosine C (**186**), whilst epimer **185** served as the precursor of natural gabosine E (**187**).

In a more recent synthesis, Lubineau and Billault [44] developed a novel method to access gabosine-related cyclohexene structures by exploiting a Nozaki-Kishi cyclization as the key step (Scheme 31).

Scheme 31

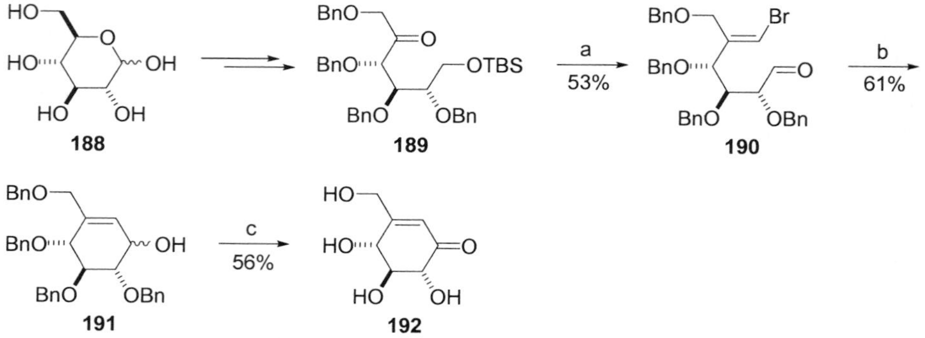

Conditions: (a) i: Ph$_3$PCHBr, THF; ii: TBAF, THF; iii: (COCl)$_2$, DMSO, CH$_2$Cl$_2$, -80°C, then Et$_3$N; (b) CrCl$_2$, 0.1% NiCl$_2$, DMF; (c) i: PCC, AcONa, 4Å MS, CH$_2$Cl$_2$; ii: BCl$_3$, CH$_2$Cl$_2$.

Thus, the glucose-derived L-sorbose intermediate **189** was transformed into aldehyde-vinylalogenide **190** through a sequence involving a Wittig reaction, desilylation, and Swern oxidation. Exposure of **190** to catalytic CrCl$_2$/NiCl$_2$ mixture promoted the crucial intramolecular cyclization giving a mixture of allylic alcohols **191**, whose oxidation and deprotection led to gluco-configured gabosine I (**192**).

Two main attributes are ascribed to natural shikimic acid; the first, of practical nature, is related to its use as a chiral source for asymmetric synthesis, the second, of biochemical prominence, is connected to the key role it exerts in the production of benzenoid rings of natural aromatic amino acids and other important metabolites [45]. The biological relevance of shikimic acid and the challenging nature of its multichiral structure have motivated an active search for the development of viable asymmetric syntheses of this compound and novel structural variants [46].

A practical approach to unnatural (+)-shikimic acid derivative **197** (Scheme 32) was proposed by Vandewalle, based on extensive transformation of the desymmetrized building-block **193** [47]. The synthesis began with a clean Mitsunobu-type elimination to form, after hydrolysis of the ester moiety, the cyclohexene triol **194**.

Scheme 32

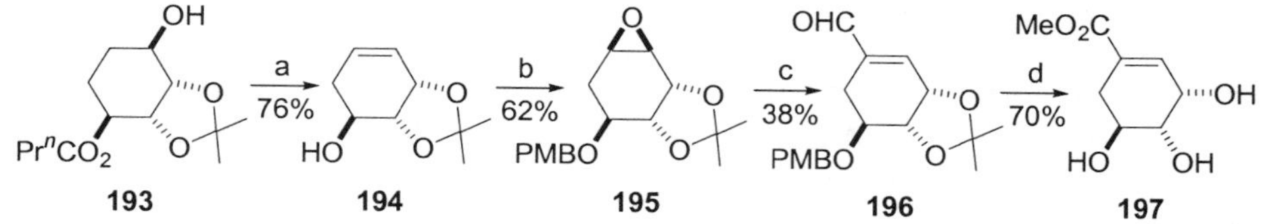

Conditions: (a) i: DEAD, PPh₃, THF, reflux; ii: K₂CO₃, MeOH; (b) i: PMBBr, KOBuᵗ, THF; ii: MCPBA, CH₂Cl₂, THF; (c) i: 1,3-dithiane, BuLi, HMPA, THF, -20°C; ii: HgO, BF₃·OEt₂, THF/H₂O (5:1), 40°C; iii: MsCl, Et₃N, CH₂Cl₂; (d) i: NaClO₂, 2-methyl-2-butene, aq NaH₂PO₄ (pH 3), BuᵗOH; ii: CH₂N₂, Et₂O, 0°C; iii: DDQ, CH₂Cl₂, H₂O; iv: p-TsOH, MeOH, 50°C.

Protection of **194** as a *p*-methoxybenzylether and subsequent epoxydation led to the *trans*-epoxide **195**, which was transformed into the unsaturated aldehyde **196** by a three-reaction sequence, including regioselective oxirane opening with a 1,3-dithiane anion, hydrolysis of the dithioacetal formed, and dehydration. Chlorite promoted aldehyde oxidation, methyl ester formation, and removal of the hydroxyl protections delivered methyl (+)-shikimate **197** in a remarkable 12% yield from **193**.

Pericosine B, a secondary metabolite produced by a strain of *Periconia bissoides* [48], proved to be an interesting target compound due to its significant antitumoral activity. From a structural point of view, pericosine B (**201**, Scheme 33) can be considered as an unsaturated carbauronic acid or an hydroxylated shikimic acid derivative.

Scheme 33

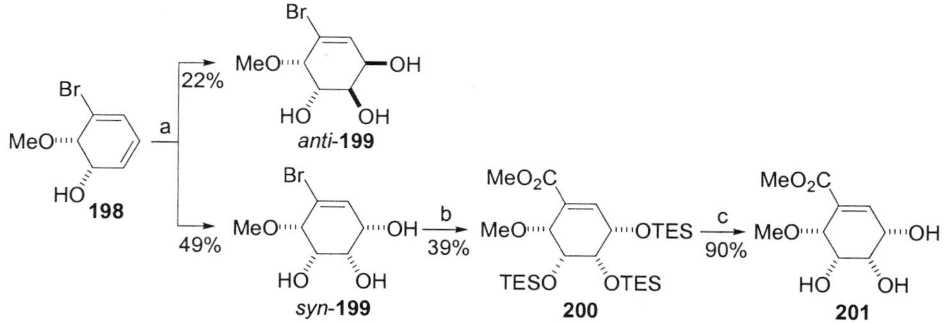

Conditions: (a) OsO₄, quinuclidine, CH₂Cl₂, -78°C; (b) i: TESOTf, Et₃N, ii: BuᵗLi, ClCO₂Me, Et₂O, -78°C; (c) TFA, CH₂Cl₂.

In a work aimed at confirming the stereochemical assignment of pericosine B, Donohoe and co-workers [49] delineated a short route, based on direct dihydroxylation of the enantiomerically pure brominated cyclohexadiendiol derivative **198** (Scheme 33). Exposure of **198** to a stoichiometric OsO$_4$/quinuclidine system, gave the desired "contrasteric" product *syn*-**199** as the major product, accompanied by a minor amount of anti-configured isomer *anti*-**199** (2.2:1 ratio). After protection of the free hydroxyl groups of *syn*-**199** as triethylsilyl ethers, the synthesis was completed by introduction of a carboxymethyl group to furnish **200**, and final deprotection. The methyl ester **201** formed proved to be identical in all respects to natural pericosine B, thus corroborating the original structural assignment.

3.4. Aminocarbapyranoses

Aminocarbapyranoses (carbaglycosylamines) represent one of the most interesting groups of natural and synthetic carbasugars by virtue of the relevant biological profile they display, and the challenging chemical and stereochemical diversity they embody. This carbasugar subgroup includes potent glycosidase inhibitors, some of which have been exploited as lead compounds to design useful drug candidates.

The first evidence for aminocarbapyranoses occurring in Nature dates back to early '70s with the discovery of validamine **202** (Chart 3), a carbasugar constituent of Validamycins, an antibiotic complex which shows growth inhibition activity against bacterial diseases of rice plants [50]. Shortly after, validamine **202**, along with valiolamine **203** and valienamine **204**, (Chart 3), were isolated from the fermentation broth of *Streptomyces hygroscopicus* var. *limoneus* [50].

Chart 3

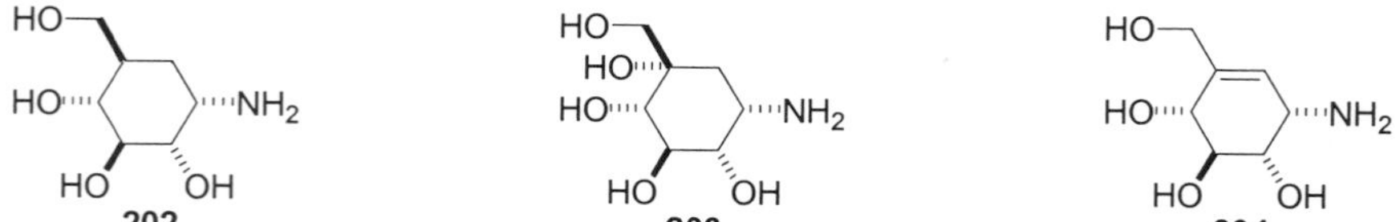

The majority of naturally occurring aminocarbasugars bears the amino functionality located at the pseudoanomeric position; however, replacement of one or more hydroxyl groups by amino functions on the carbasugar backbone may offer great opportunities for structural and stereochemical variation. This section is subdivided into three subsections, dealing with the synthesis of validamine, valiolamine, and valienamine-type carbaglycosyl amine structures.

3.4.1. Validamine and analogues

The most popular chemical entries to carbapyranosyl amines of the validamine subgroup exploit two major ouvertures, namely Diels-Alder annulation and ex-chiral pool-derived precursors homologation.

A nice stereoselective synthesis of validamine **202** was accomplished by Yoshikawa's group [51], capitalizing on the longstanding familiarity with cyclization of nitrofuranose derivatives [52]. The focal intermediate of the synthesis was nitrofuranose adduct **206** (Scheme 34), which was obtained via a high-yielding multistep transformation of D-glucuronolactone **205**.

Scheme 34

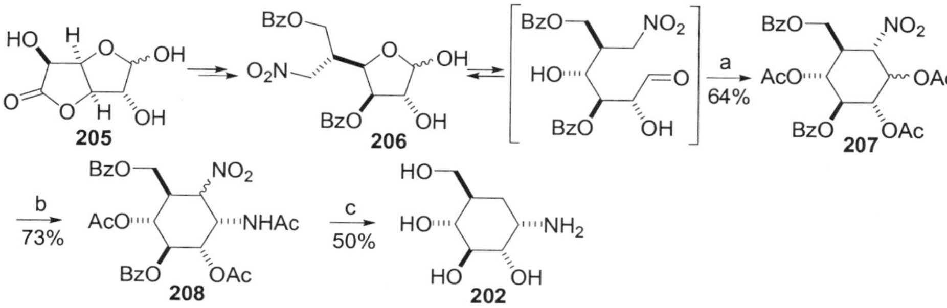

Conditions: (a) i: CsF, DMF; ii: *p*-TsOH, Ac₂O; (b) i: liq NH₃, THF, -78°C; ii: *p*-TsOH, Ac₂O; (c) i: Bu₃SnH, AIBN, toluene, 110°C; ii: 1% NaOMe/MeOH; iii: Ac₂O, pyridine; iv: 1% NaOMe/MeOH; v: 80% aq NH₂NH₂.

Treatment of **206** with cesium fluoride in DMF clearly promoted an intramolecular nitro-aldol reaction furnishing, after acetylation, nitrocyclitol **207** as an α,β-anomeric mixture. Exposure of **207** to liquid ammonia in THF resulted in formation of kinetically favoured α-acetamido derivative **208**, probably from a Michael-type addition of ammonia to a nitroolefin intermediate. Reductive elimination of nitro-group in **208** and subsequent deprotection gave validamine **202**, in a global 6% yield from glucuronolactone **205**.

In a semisynthetic study directed towards the preparation of potentially bioactive validamine stereoanalogues, Kameda and co-workers exploited, as a useful chiral source, validamine itself, conveniently obtained by microbial degradation of Validamycins [53]. Scheme 35 illustrates how natural validamine derivative **209** was simply transformed into manno- and galacto-configured aminocarbasugars **211** (2-*epi*-validamine) and **214** (4-*epi*-validamine). Synthetic analogues **211** and **214** were investigated in inhibitory assays against several mannosidases and galactosidases, as well as N-linked oligosaccharide-processing mannosidases. Remarkably, while galacto-validamine **214** proved to be a weak inhibitor of α- and β-galactosidases, manno-validamine **211** showed potent competitive inhibition of various α-mannosidase enzymes.

Scheme 35

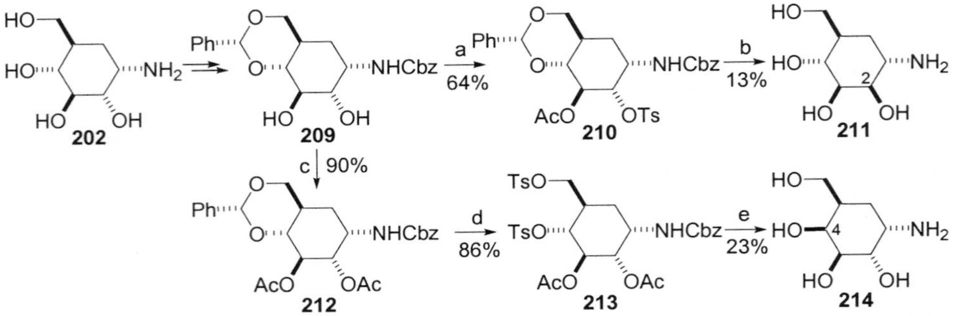

Conditions: (a) i: *p*-TsCl, CH₂Cl₂, pyridine, 0°C to rt; then Ac₂O, 0°C to rt; (b) i: NaOBz, DMF, 140°C; ii: 1N HCl, MeOH, reflux; iii: 10% Ba(OH)₂, 80°C; (c) Ac₂O, pyridine; (d) i: 0.5N HCl, MeOH, reflux; ii: *p*-TsCl, CH₂Cl₂, pyridine, 0°C to rt; (e) i: NaOBz, DMF, 140°C; ii: 10% Ba(OH)₂, 80°C.

In a target-oriented investigation aimed at the synthesis of novel fucosidase inhibitory compounds, Ogawa utilized dibromide **218** (Scheme 36) as a pivotal intermediary compound [54].

Scheme 36

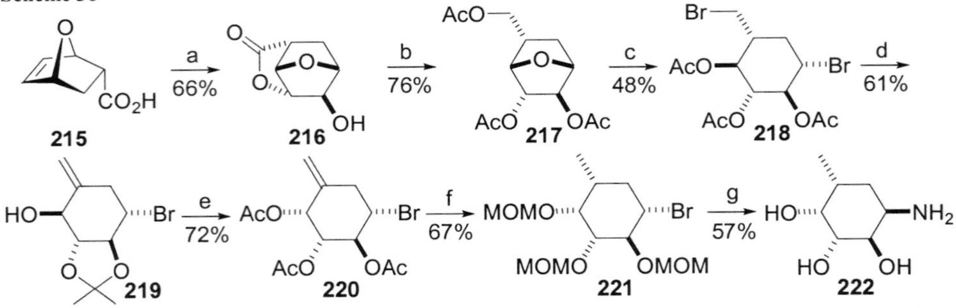

Conditions: (a) 90% aq HCO₂H, 35% H₂O₂, 70°C; (b) i: LiAlH₄, THF; ii: Ac₂O, pyridine; (c) 20% HBr/AcOH, 85°C (sealed tube); (d) i: AgF, pyridine; ii: 4N HCl/THF, 60°C; iii: DMP, *p*-TsOH, DMF, 60°C; (e) i: MsCl, pyridine, 0°C; ii: 60% aq AcOH; iii: Ac₂O, pyridine; iv: 60% aq AcOH, 60°C; v: Ac₂O, pyridine; (f) i: 4N HCl/THF; ii: MOMCl, DIPEA, 40°C; iii: H₂, Wilkinson catalyst, C₆H₆, MeOH; (g) i: NaN₃, DMF, 90°C; ii: 4N HCl/THF; iii: Ph₃P, 5% aq THF, 60°C, then Dowex 50Wx2.

Starting from optically resolved carboxylic acid **215**, the synthesis of bromide **218** encompassed three steps including performic acid treatment to **216**, reductive fission of the lactone moiety to obtain **217**, and double bromination. A three-step sequence involving mono-debromination, deacetylation and selective 2,3-*O*-isopropylidene protection converted **218** into *exo*-methylene precursor **219**, on which C₄ inversion to **220** was performed via mesylate. Changing the acetyl groups for methoxymethyl ether protection and subsequent stereoselective double bond saturation resulted in the formation of bromide **221**, which was finally elaborated into 5a-carba-α-L-fucopyranosylamine (**222**) via azidolysis and azide reduction. Compound **222** showed relevant antifucosidase activity against α-L-fucosidase from bovine kidney, with a more than 3000-fold increase in activity in comparison to the parent 5a-carba-α-L-fucopyranose.

As part of a wide investigation directed towards the preparation of a small repertoire of aminated carbasugars and amino acids thereof, Rassu and co-workers used dienoxy silane **76** as a homologative nitrogen-containing synthon [7a,b,e].

Scheme 37

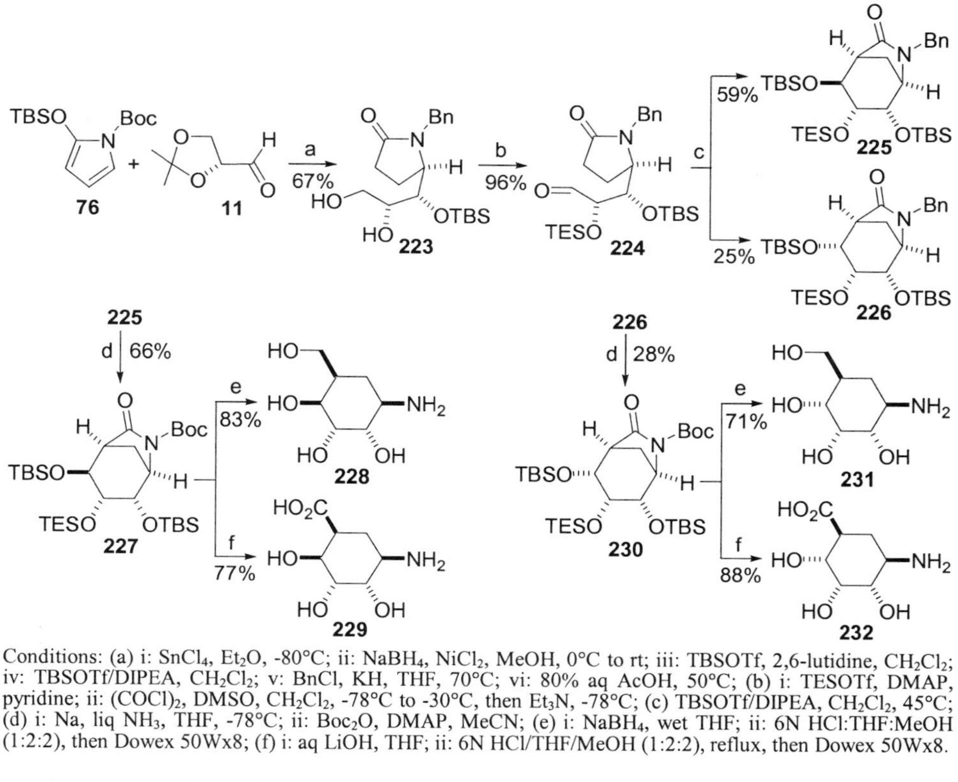

Conditions: (a) i: SnCl₄, Et₂O, -80°C; ii: NaBH₄, NiCl₂, MeOH, 0°C to rt; iii: TBSOTf, 2,6-lutidine, CH₂Cl₂; iv: TBSOTf/DIPEA, CH₂Cl₂; v: BnCl, KH, THF, 70°C; vi: 80% aq AcOH, 50°C; (b) i: TESOTf, DMAP, pyridine; ii: (COCl)₂, DMSO, CH₂Cl₂, -78°C to -30°C, then Et₃N, -78°C; (c) TBSOTf/DIPEA, CH₂Cl₂, 45°C; (d) i: Na, liq NH₃, THF, -78°C; ii: Boc₂O, DMAP, MeCN; (e) i: NaBH₄, wet THF; ii: 6N HCl:THF:MeOH (1:2:2), then Dowex 50Wx8; (f) i: aq LiOH, THF; ii: 6N HCl/THF/MeOH (1:2:2), reflux, then Dowex 50Wx8.

As summarized in Scheme 37, gulo- and allo-configured 1-amino-5a-carba-β-D-hexopyranoses **228** and **231**, as well as the related amino acid counterparts, **229** and **232**, sprang from the common lactam intermediate **224**. The preparation of aldehyde **224** from silyloxypyrrole **76** and D-glyceraldehyde **11** entailed a Lewis acid promoted vinylogous cross aldolization followed by functional group manipulation, including olefin hydrogenation, silylation of hydroxyl, acetonide breakage, and Swern oxidation. The key manoeuvre which installs the carbapyranose backbone was represented by a high-yielding annulative aldosilylation, which led to the desired bicyclic lactams **225** and **226**. After switching the nitrogen protection to N-Boc, individual carbamates **227** and **230** were in parallel converted to carbapyranosyl amines **228** and **231** by reductive lactam fission followed by global deprotection. In a divergent way, **227** and **230** were transformed into amino acids **229** and **232** by hydrolytic cleavage and deprotection.

3.4.2. Valiolamine and analogues

Among all the aminocarbasugars obtained by chemical or microbial degradation of validamycins [50], valiolamine is the strongest α-glucosidase inhibitor. Its N-[2-hydroxy-1-(hydroxymethyl)ethyl]-derivative, coded as AO-128, displays higher activity than the parent valiolamine and is undergoing clinical trial for treatment of diabetes [55].

Scheme 38

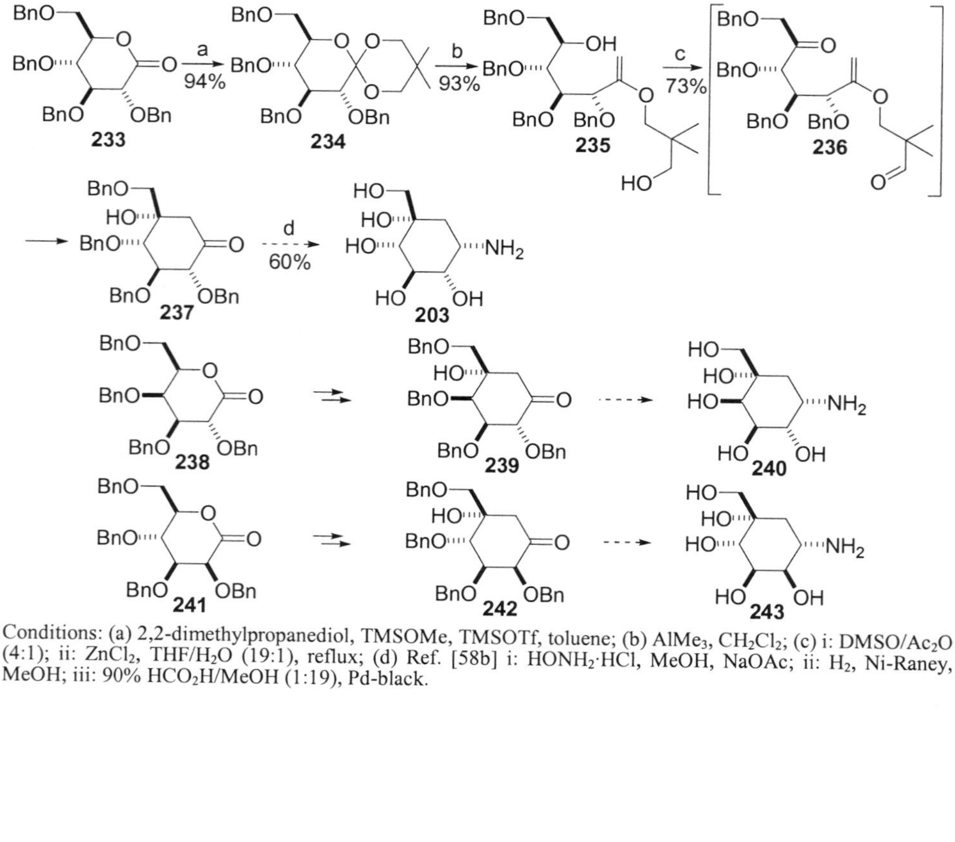

Conditions: (a) 2,2-dimethylpropanediol, TMSOMe, TMSOTf, toluene; (b) AlMe₃, CH₂Cl₂; (c) i: DMSO/Ac₂O (4:1); ii: ZnCl₂, THF/H₂O (19:1), reflux; (d) Ref. [58b] i: HONH₂·HCl, MeOH, NaOAc; ii: H₂, Ni-Raney, MeOH; iii: 90% HCO₂H/MeOH (1:19), Pd-black.

In spite of the great biological potential of valienamine and its derivatives, only a few methods have been reported for their synthesis [56].

A chemically efficient, virtual synthesis of valiolamine **203**, and its D-galacto- and D-manno-configured congeners **240** and **243** (Scheme 38) was recently developed by Ikegami and co-workers, based on two strategic operations; a novel enol ether formation from sugar ortho esters and a stereoselective aldocyclization of alkyl enol ethers [57]. Starting from tetra-*O*-benzylgluconolactone (**233**), preparation of the ortho ester **234** was secured by exposure to 2,2-dimethylpropanediol under acid-base conditions. Treatment of **234** with a 5-fold molar excess of AlMe₃ gave the enol ether **235** in a 93% yield, via double fragmentation of both the pyran and dioxane rings. Oxidation of the secondary hydroxyl within **235** delivered ketone **236** which was directly converted to D-gluco inosose **237** by ZnCl₂-assisted aldol cyclization. Inosose **237** was the same advanced intermediate upon which Fukase and Horii based their total synthesis of **203** [58].

The application of the same protocol to D-galacto- and D-manno-configured lactones **238** and **241** resulted in the construction of two further inosose variants, **239** and **242**, the proximate precursors of galacto- and manno-valiolamines **240** and **243**.

Searching for a simple, divergent methodology to access valiolamine **203** and certain structurally varied relatives, Shing and Wang [59] utilized the unsaturated scaffold **244** whose preparation from quinic acid **159** was already established [60].

Scheme 39

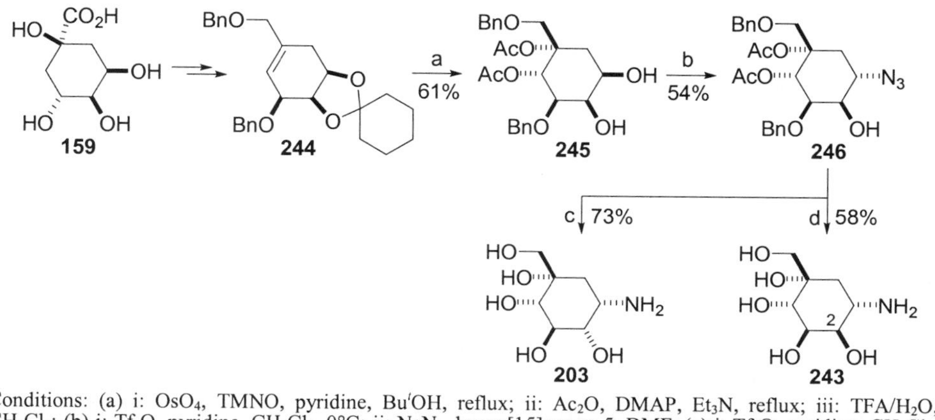

Conditions: (a) i: OsO₄, TMNO, pyridine, BuᵗOH, reflux; ii: Ac₂O, DMAP, Et₃N, reflux; iii: TFA/H₂O, CH₂Cl₂; (b) i: Tf₂O, pyridine, CH₂Cl₂, 0°C; ii: NaN₃, benzo[15]crown-5, DMF; (c) i: Tf₂O, pyridine, CH₂Cl₂, 0°C; ii: Bu₄NOAc, THF; iii: K₂CO₃, MeOH; iv: H₂, Pd(OH)₂, EtOH; (d) i: K₂CO₃, MeOH; ii: H₂, Pd(OH)₂, EtOH.

As can be seen in Scheme 39, alkene **244** was first dihydroxylated and then manipulated into partially protected cyclitol **245**, a flexible intermediate of this synthetic plan. Incorporation of the nitrogen at the pseudo-anomeric position was then carried out, as usual, by azide S_N2 displacement of the C_1 hydroxyl. There was obtained azido-carbapyranose **246**, which nicely served to prepare both valiolamine **203** and its C_2 epimer **243**.

By adding flexibility to the project, carbasugar **245** also acted as the scaffold with which β-D-manno-, and β-D-glucovalienamine, as well as 2-amino-2-deoxy-β-D-glucovalienamine were constructed.

3.4.3. Valienamine and analogues

Valienamine **204** represents the cardinal subunit of relevant naturally occurring carbaoligosaccharides including validamycin-type antibiotics and the potent α-amilase inhibitor acarbose [61].

Scheme 40

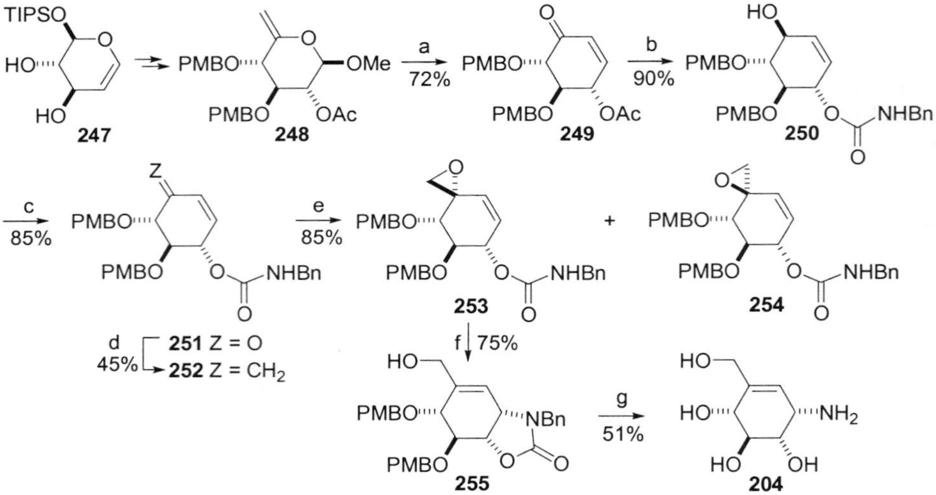

Conditions: (a) i: Hg₂Cl₂, acetone/H₂O, reflux; ii: MsCl, DMAP, pyridine; (b) i: NaBH₄, CeCl₃, EtOH, -78°C; ii: PhCH₂NCO, C₆H₆, reflux; iii: K₂CO₃, MeOH; (c) PDC, AcOH, EtOAc; (d) CH₂I₂, Zn, TiCl₄, THF; (e) MCPBA, NaHCO₃, CH₂Cl₂, (f) KHMDS, 18-crown-6, THF, -78°C to rt; (g) i: Na, liq NH₃, THF, -78°C; ii: LiOH, 30% aq EtOH, reflux.

One of the most interesting routes to **204** was developed in 1994 by Park and Danishefsky (Scheme 40), where a brilliant tethered S_N2' reaction was executed to install, in one go, both the C_1 carbasugar nitrogen and the C_5 hydroxymethyl terminus [62].

The journey began with D-glucal **247**, which was elaborated into the densely hydroxylated cyclohexene **250** by a multistep sequence. Oxidation of the C_5 hydroxyl of **250** to ketone **251** and subsequent methylenation then gave rise to the *exo*-methylene compound **252**, which was subjected to *m*-chloroperbenzoic acid epoxydation. A separable mixture of C_5 epimeric epoxides **253** and **254** was obtained, where the 5*R*-configured isomer **253** predominated (2:1 ratio). Facing the crucial intramolecular amination, compound **253** was exposed to potassium hexamethyldisilazide to achieve a clean conversion to cyclohexylamine **255**. Finally, the synthesis was completed by global deprotection of **255** giving valienamine **204** in a good isolated yield.

In continuing with this section, we will highlight a further class of compounds, namely the carba-analogues of naturally occurring sialic acid, which represent a class of potential inhibitors of viral neuraminidase enzymes [63].

A short, scalable synthesis of carbapyranuronic acid **259**, a sialyl mimetic compound related to the antiinfluenza drug Tamiflu®, was developed by Kim's group [64] from (-)-shikimic acid **169** (Scheme 41).

Scheme 41

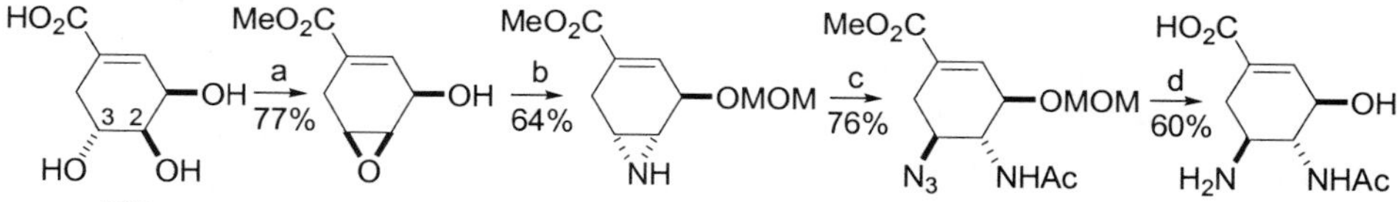

Conditions: (a) i: 0,5% aq HCl, MeOH, reflux; ii: Ph₃P, DEAD; THF, 0°C to rt; (b) i: MOMCl, DIPEA, CH₂Cl₂, reflux; ii: NaN₃, NH₄Cl, MeOH/H₂O, reflux; iii: MsCl, Et₃N, CH₂Cl₂, 0°C to rt; iv: Ph₃P, THF, 0°C to rt; v: Et₃N, H₂O; (c) i: NaN₃, NH₄Cl, DMF, 65-75°C; ii: AcCl, pyridine; (d) i: H₂, Lindlar catalyst, EtOH; ii: 0.48 N KOH, THF; iii: TFA, CH₂Cl₂.

Transformation of the C_2 and C_3 hydroxyls in **169** to *trans*-disposed C_2 and C_3 amino groups in **259** relied on conversion of the starting acid into an aziridine followed by azide-ion fragmentation. Thus, shikimic acid **169** was first elaborated into **257** by a well experienced sequence consisting of formation of epoxide **256** and epoxide-to-azide conversion via NaN₃ oxirane opening followed by azide reduction, and cyclization. Then, the aziridine ring was opened regioselectively by exposure to sodium azide in DMF to deliver *trans*-disposed amino-azide **258**, the close precursor of unsaturated sialyl mimetic **259**.

An equally efficient preparation of a new carbocyclic influenza sialidase inhibitor, namely the guanidine substituted cyclohexene derivative **265**, has been recently reported by Bianco and co-workers [65], choosing (-)-quinic acid **159** as a starting precursor (Scheme 42). Moving from methyl shikimate **260**, fully protected sulfite **261** was cleanly prepared by thionyl chloride treatment followed by acetylation of the remaining free hydroxyl group.

Scheme 42

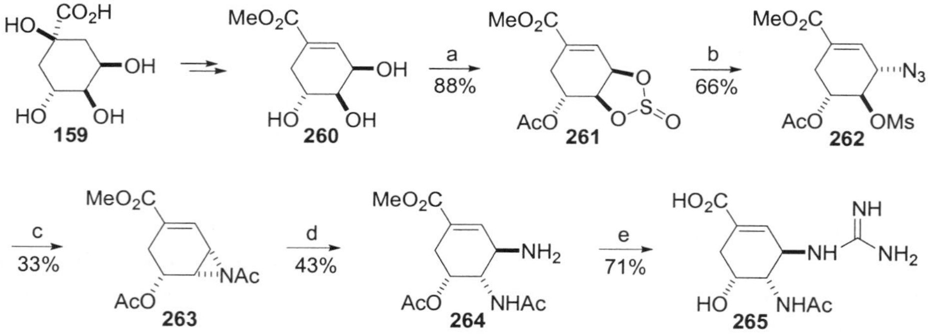

Conditions: (a) i: SOCl₂, Et₃N, THF, 0°C to rt; ii: Ac₂O, pyridine; (b) i: NaN₃, DMF; ii: MsCl, Et₃N, CH₂Cl₂, 0°C; (c) i: Ph₃P, THF; ii: AcCl, CH₂Cl₂, 0°C; iii: NaH, THF; (d) i: NaN₃, NH₄Cl, DMF; ii: H₂, Lindlar catalyst, EtOH; (e) i: N,N'-bis-(t-Boc)-thiourea, HgCl₂, Et₃N, DMF, 0°C to rt; ii: 1N KOH, THF.

Exposure of sulfolane **261** to sodium azide and mesylation regioselectively afforded mesylate **262**, which was transformed into aziridine **263** by a three-step sequence involving azide to amine Staudinger reduction, acetylation, and ring closure. The *trans* 1,2-diamino derivative **264** was quickly obtained from **263** by azide opening of the aziridine ring, followed by reduction.

Finally, methyl ester **264** was elaborated into the zanamivir-related guanidine derivative **265** by guanidylation and selective removal of the hydroxyl and carboxylic acid protections.

3.5. Thiocarbapyranoses

Thiocarbapyranoses are a quite rare compound class in the scenario of pyranoid carbasugars, although replacement of a hydroxyl group with a sulfurized moiety is a classical isosteric modification in medicinal chemistry.

Lew and Kim [66] recently synthesized a sulfide isostere of carbocyclic influenza neuraminidase inhibitor **268**, moving from trityl protected aziridine **266**, in turn prepared from (-)-quinic acid **159** (Scheme 43). Ring opening of **266** with 1-propanethiol promoted by BF₃ etherate, followed by acetylation, gave sulfide **267** which was finally hydrogenated to diamine **268**, in a good chemical yield. As the authors expected, mercapto-derivative **268** resulted to be a strong inhibitor of influenza virus neuraminidase, with an activity comparable to that of the corresponding hydroxy analogue.

Scheme 43

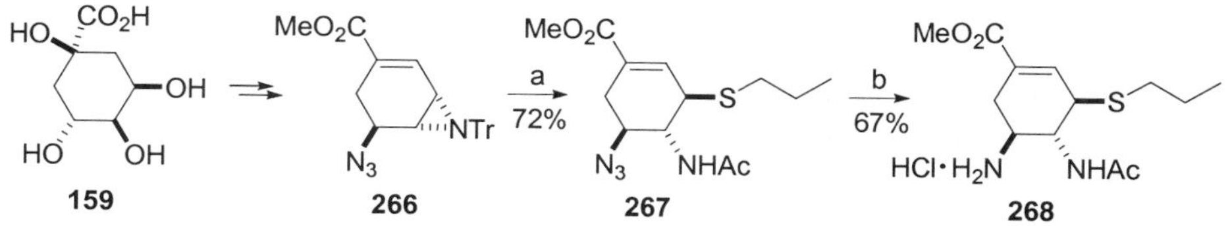

Conditions: (a) i: 1-propanethiol, BF$_3$·OEt$_2$; ii: AcCl, Et$_3$N, CH$_2$Cl$_2$; (b) i: H$_2$, Lindlar catalyst, EtOAc; ii: 1N HCl.

The exquisite synthetic potential of thiophene-based dienoxy silane **90** (Scheme 44) as a nucleophilic homologative reactant *vis-à-vis* protected hydroxy aldehydes, such as **11**, was recently exploited by Rassu and co-workers to enter the scantly investigated domain of sulfur-containing carbasugars [7d].

Scheme 44

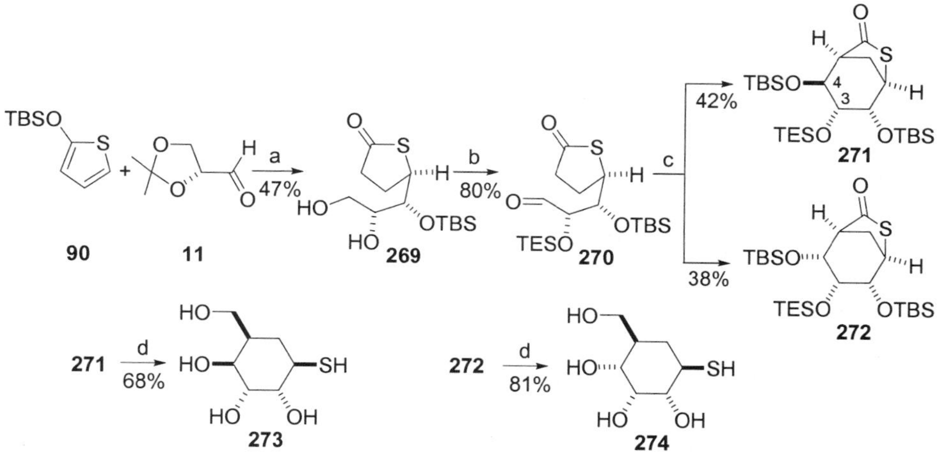

Conditions: (a) i: BF$_3$·OEt$_2$, CH$_2$Cl$_2$, -80°C; ii: NaBH$_4$, NiCl$_2$, MeOH, 0°C to rt; iii: TBSOTf, 2,6-lutidine, CH$_2$Cl$_2$, iv: 80% aq AcOH, 50°C; (b) i: TESOTf, pyridine, DMAP; ii: (COCl)$_2$, DMSO, CH$_2$Cl$_2$, -78°C to -30°C, then Et$_3$N, -78°C; (c) TBSOTf/DIPEA, CH$_2$Cl$_2$; (d) i: LiBH$_4$, THF, 0°C to rt; ii: 6N HCl/THF/MeOH (1:2:2).

As already mentioned (Section 2.1), the preparation of β-D-gulo- and β-D-allo-configured carbapyranosyl thiols **273** and **274** utilized thiolactone **269** as an advanced chiral scaffold. Approaching the decisive annulation step, which creates the carbapyranose ring, the required aldehyde **270** was implemented from **269** via silylation and Swern oxidation.

Exposure of **270** to the TBSOTf/DIPEA reagent couple once more triggered a cascade of reactions, namely chemoselective thiolactone enolization, intramolecular aldolization, and hydroxyl silylation, which ultimately gave rise to 3,4-*trans* and 3,4-*cis* bicyclic thiolactones **271** and **272**. Final unmasking of the pseudoanomeric thiol function and terminal hydroxymethyl moiety was performed on the individual precursors **271** and **272** to deliver, as expected, the corresponding thiocarbasugars **273** and **274** in high isolated yields.

4. "CONTRACTED" AND "EXPANDED" CARBASUGARS

Just as Nature has done with the carbohydrates, where the favoured conformations assume furanoid and pyranoid motifs, organic chemistry has also favoured five- and six-membered carbocyclic targets within the parallel realm of carbasugars, dedicating only a marginal interest to small- and medium-sized structures. In truth, small-sized carbasugars (carbaoxiroses, carbaoxetoses) and medium-sized carbasugars (carbaseptanoses and carbaoctanoses) hardly ever have cyclic monosaccharide counterparts as they are mainly found in the furanoid and pyranoid formats for entropic/enthalpic reasons. Only with particular strategies or protection intervention can contracted or expanded rings be stabilized and isolated.

This section groups together a limited number of diastereoselective syntheses of carbaoxirose and carbaoxetose (4.1), as well as several approaches to ring-expanded carbaseptanose (4.2) and carbaoctanose (4.3) derivatives.

4.1. Carbaoxiroses and Carbaoxetoses

During a wide study directed towards the asymmetric synthesis of cyclopropyl carbocyclic nucleosides, Chu and collaborators [67] targeted 1-deoxy-1-amino-2a-carbatetrooxirose derivatives **279** and **280** (Scheme 45) which were envisioned as the key intermediates for the preparation of purine and pyrimidine nucleosides. As an example, the total asymmetric synthesis of cyclopropane nucleosides **281** and **282** began from the Wittig olefination of 2,3-*O*-isopropylidene-D-glyceraldehyde (**11**) with carboxylmethylmethylene triphenyl phosphorane, furnishing unsaturated ester **275**. Reduction of the ester moiety and silylation provided intermediate **276** which was subjected to Simmons-Smith cyclopropanation to afford, after TBAF-promoted desilylation, cyclopropyl alcohol **277** in excellent yields and diastereoselectivity.

Scheme 45

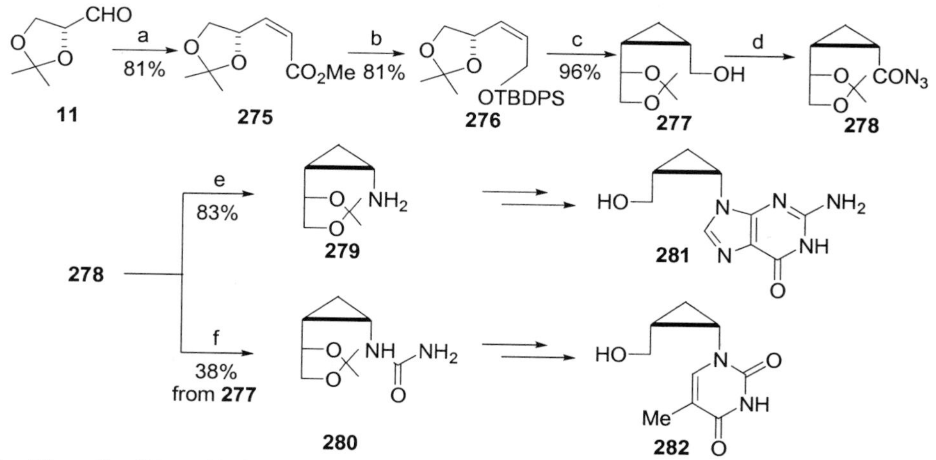

Conditions: Conditions: (a) Ph$_3$P=CHCO$_2$Me; (b) i: DIBAL-H, -78°C; ii: TBDPSCl, imidazole; (c) i: Zn(Et)$_2$, ICH$_2$Cl; ii: TBAF; (d) i: RuO$_2$, NaIO$_4$; ii: ClCO$_2$Et, Et$_3$N; iii: NaN$_3$; (e) i: toluene, BnOH, 100°C; ii: H$_2$, Pd/C; (f) i: toluene, 100°C; ii: NH$_3$.

Conventional chemistry converted carbinol **277** into acyl azide **278**, which was treated under Curtius rearrangement conditions to obtain either amine **279** or urea derivative **280**, the respective precursors of guanine derivative **281** and thymidine analogue **282**.

A similar approach was also applied by the same authors [68], who addressed a series of carbocyclic cyclopropyl nucleosides of the L-series by utilizing L-gulonic γ-lactone as the chiral source.

In the search for novel therapeutic agents against HIV and herpes viruses, attention was focused on synthesis of carba-analogues of natural nucleosides. In this perspective, researchers at Bristol-Myers Squibb [69] planned the preparation of a series of carbaoxetose nucleoside analogues which represent new structures closely related to anti-HIV natural compound oxetanocin (Scheme 46). Racemic cyclobutane diester **283** served as the precursor to arrive at the targeted compounds. Thus, **283** was first saponified to diacid **284** and then resolved chemically via crystallization of the corresponding bis-glycinamides. Both amide moieties of **285** were converted to hydroxymethyl groups to produce enantiomerically pure diol **287** in good yields. Transformation of **287** to key intermediary compound **288** required three further steps namely, benzoylation, hydrolysis of the ketal function, and carbonyl reduction. By employing similar chemistry, enantiomer *ent*-**288** was also prepared by starting with the amide **286**.

Scheme 46

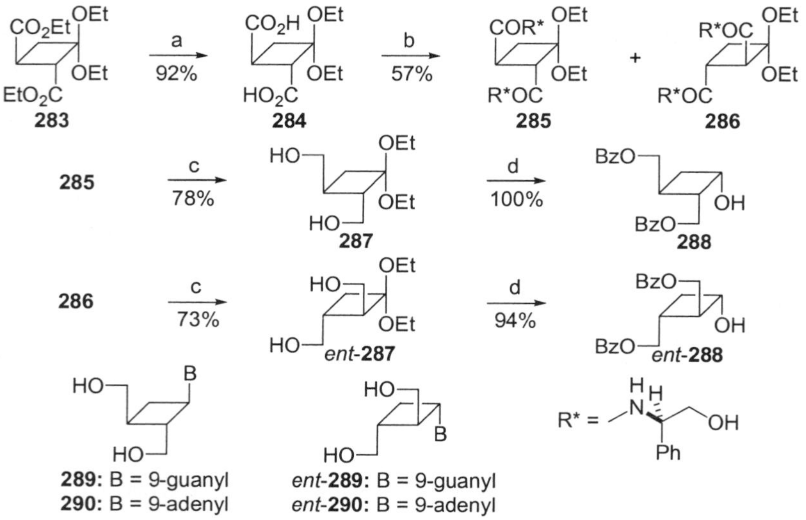

Conditions: (a) KOH, MeOH, H$_2$O; (b) i: *R*-(-)-2-phenylglycinol, DCC; ii: crystallization; (c) i: TBSCl, imidazole; ii: N$_2$O$_4$, NaOAc, CCl$_4$; iii: LiBH$_4$, THF; (d) i: BzCl, pyridine; ii: H$_2$SO$_4$, MeCN, H$_2$O; iii: LS-selectride.

By conventional chemistry, enantiopure branched-chain 3a-carbatetrooxetose derivatives **288** and *ent*-**288** were finally incorporated into carbaoxetanocin analogues **289**, *ent*-**289**, **290**, and *ent*-**290**, whose efficacy against a range of herpes viruses and cytomegalovirus infections was assayed.

Scheme 47

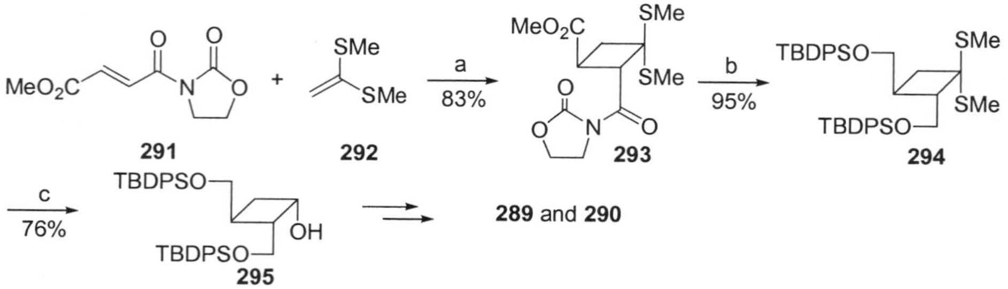

Conditions: (a) Ti(Pr^{i}O)$_2$Cl$_2$, (2*S*,3*S*)-2,3-*O*-(1-phenylethylidene)-1,1,4,4-tetraphenylbutane-1,2,3,4-tetraol (10 mol%); (b) i: (MeO)$_2$Mg, MeOH, 0°C; ii: LiAlH$_4$, Et$_2$O, 0°C; iii: TBDPSCl, Et$_3$N, DMAP; (c) i: NCS-AgNO$_3$; ii: DIBAL-H, toluene.

A total enantioselective synthesis of antiviral carbocyclic oxetanocins **289** and **290** was also completed by the Ichikawa group by utilizing a noticeable asymmetric [2+2] cycloaddition reaction to forge the core cyclobutane ring of the targets [70]. As shown in Scheme 47, exposure of a 1:1 mixture of oxazolidinone **291** and 1,1-bis-(methylthio)ethylene (**292**) to a catalytic amount of a chiral Ti(IV)-catalyst (10 mol%) delivered cyclobutane derivative **293** in 83% yield and in >98% enantiomeric excess. The compound thus obtained was converted to silylether **294** by sequential treatment with dimethoxymagnesium in methanol followed by LiAlH$_4$ reduction and silylation. Treatment of **294** with *N*-chlorosuccinimide-silver nitrate followed by DIBAL-H finally gave rise to protected 3a-carbaoxetose derivative **295** in a 76% yield for the two steps. Compound **295** served as the common precursor with which optically pure carbocyclic oxetanocins **289** and **290** were constructed. Noticeably oxetanocin **290** displayed a high efficacy against hepatitis B virus and HIV (IC$_{50}$, 0.024 and 0.03 μg·mL^{-1}, respectively), whilst compound **289** revealed consistent activity against herpes simplex virus and human cytomegalovirus (IC$_{50}$, 0.047 and 0.08 μg·mL^{-1}, respectively).

4.2. Carbaseptanoses

Even though medium-sized carbocycles are structural motifs often encountered in naturally occurring organic substances, richly hydroxylated structures reminiscent of the carbohydrate counterparts are rare, and efforts made to obtain these derivatives via synthesis have not been particularly intense.

The total synthesis of carbaseptanose **300**, and those of the aminated derivatives **304**, *ent*-**304**, **305**, and *ent*-**305**, recently developed and carried out by our group, represent a remarkable success story in this field [7f]. As already mentioned in the previous chapters dealing with the synthesis of furanoid and pyranoid carbasugars, the strategy adopted here exploits the dual nucleophilic reactivity of furan- and pyrrole-based dienoxysilanes towards suitable sugar-derived aldehyde acceptors. To enter optically pure septanose **300** (β-D-glycero-D-gulo configuration), the authors started with the boron trifluoride-assisted vinylogous aldol reaction between L-threose aldehyde **296** and 2-[(*tert*-butyldimethylsilyl)oxy]furan (**10**) (Scheme 48), to prepare the butenolide adduct **297**, which contains the entire carbon skeleton of the target. Simple manipulation of **297** providing aldehyde **298** proceeded smoothly, as indicated, setting the stage for the crucial septanose forming reaction. In the event, exposure of **298** to TBSOTf/DIPEA reagent mixture at room temperature furnished silylated tricyclic compound **299**, with little, if any, diastereomeric contamination. Thus, having secured the construction of the seven-membered carbon ring, synthesis of **300** was completed by reductive opening of the lactone moiety, followed by deprotection.

Scheme 48

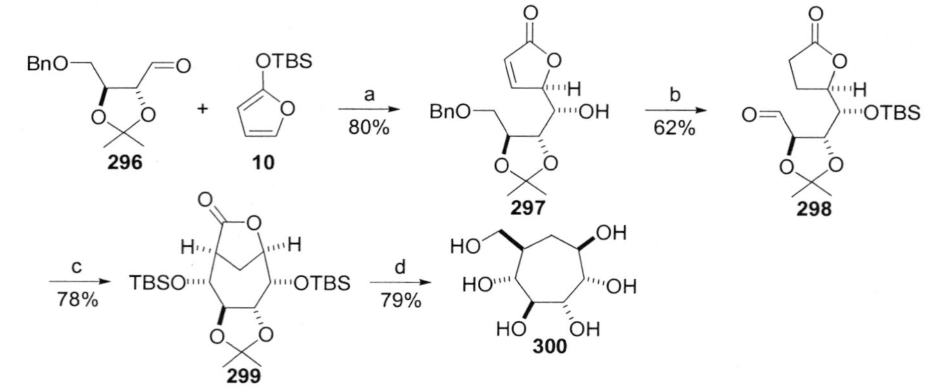

Conditions: (a) BF₃·OEt₂, CH₂Cl₂, -80°C; (b) i: NaBH₄, NiCl₂, MeOH, 0°C to rt; ii: TBSOTf, 2,6-lutidine, CH₂Cl₂; iii: H₂, Pd(OH)₂; iv: (COCl)₂, DMSO, CH₂Cl₂, -78°C, then Et₃N; (c) TBSOTf, DIPEA, CH₂Cl₂; (d) i: LiBH₄, THF, 0°C to rt; ii: 6N HCl/THF/MeOH (1:2:2), then Dowex 50Wx8.

Chiral non-racemic 6a-carba-β-D-glycero-D-guloheptoseptanose **300** was synthesized via a highly productive and diastereoselective eight-steps sequence with a good 31% global yield from threose derivative **296**.

Scheme 49

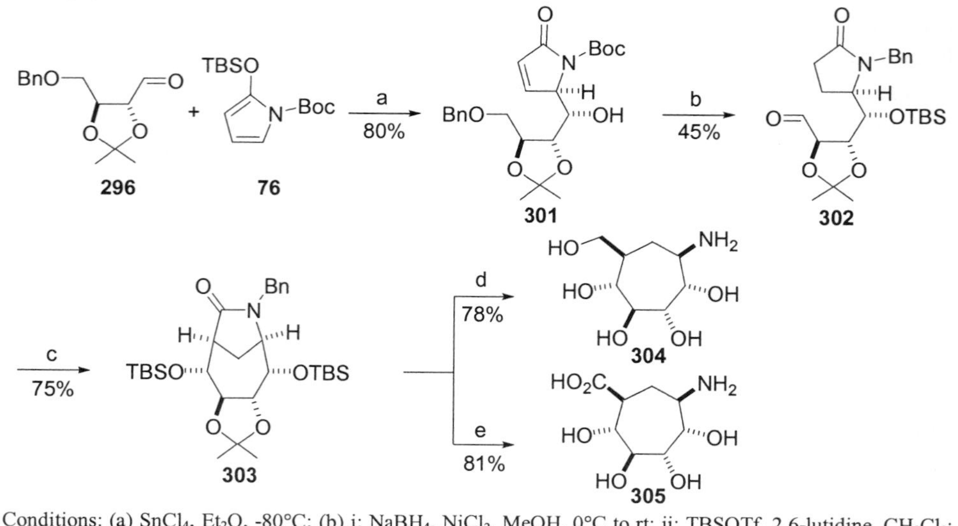

Conditions: (a) SnCl₄, Et₂O, -80°C; (b) i: NaBH₄, NiCl₂, MeOH, 0°C to rt; ii: TBSOTf, 2,6-lutidine, CH₂Cl₂; iii: CAN, MeCN; iv: BnCl, KH, THF, 60°C; v: H₂, Pd(OH)₂; vi: (COCl)₂, DMSO, CH₂Cl₂, -78°C, then Et₃N; (c) TBSOTf, DIPEA, CH₂Cl₂; (d) i: Na, liq NH₃, THF, -78°C; ii: (Boc)₂O, DMAP, MeCN; iii: NaBH₄, wet THF; iv: 6N HCl/THF/MeOH (1:2:2), then Dowex 50Wx8; (e) i: Na, liq NH₃, THF, -70°C; ii: 3N HCl, THF, reflux, then Dowex 50Wx8.

The aminated nature of **304** and **305** called for the use of pyrrole-based silane **76**. At first, aldol coupling between *N*-(*tert*-butoxycarbonyl)-2-[(*tert*-butyldimethylsilyl)oxy]pyrrole (**76**) and aldehyde **296** under SnCl₄ guidance furnished lactam **301** in 80% yield and >98% diastereoselectivity (Scheme 49). Almost paralleling the previously disclosed protocol of **300**, compound **301** was converted into *N*-benzyl protected aldehyde **302**, ready for the requisite cyclization reaction. Exposure of **302** to the TBSOTf/DIPEA mixture resulted in exclusive formation of lactam **303**, which admirably served to obtain both aminocarbasugar **304** and amino acid **305** in a divergent way. Thus, reductive opening of **303** provided 1-deoxy-1-amino-6a-carba-β-D-glycero-D-guloheptoseptanose (**304**), while hydrolytic breakage of the lactam moiety of **303** produced aminouronic derivative **305**.

Scheme 50

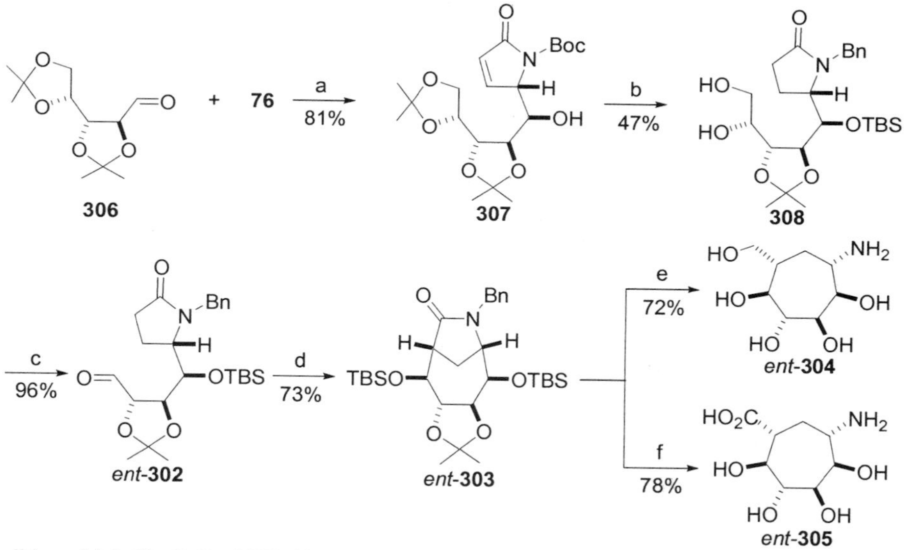

Conditions: (a) SnCl₄, Et₂O, -80°C; (b) i: NaBH₄, NiCl₂, MeOH, 0°C to rt; ii: TBSOTf, 2,6-lutidine, CH₂Cl₂; iii: CAN, MeCN; iv: BnCl, KH, THF, 60°C; v: 80% aq AcOH, 50°C; (c) aq NaIO₄, SiO₂, CH₂Cl₂; (d) TBSOTf, DIPEA, CH₂Cl₂; (e) i: Na, liq NH₃, THF, -78°C; ii: (Boc)₂O, DMAP, MeCN; iii: NaBH₄, wet THF; iv: 6N HCl/THF/MeOH (1:2:2), then Dowex 50Wx8; (f) i: Na, liq NH₃, THF, -78°C; ii: 3N HCl, THF, reflux, then Dowex 50Wx8.

During the same study, the Italian researchers aiming at the synthesis of carbaoctanose compounds (*vide infra*), synthesized a nine-carbon long intermediate **308** which served both to access the octanose and the septanose targets. As shown in Scheme 50, the route to **308** entailed vinylogous aldol coupling between pyrrole **76** and D-arabinose derived aldehyde **306** producing lactam **307**, the immediate precursor of **308**. To target amino septanose compounds *ent*-**304** and *ent*-**305**, a one carbon atom sacrifice was required, which led to the eight-carbon long aldehyde *ent*-**302**, enantiomer of the ex L-threose compound **302** (*vide supra*). Paralleling the above discussed procedure, *ent*-**302** was clearly converted to *ent*-**304** and *ent*-**305** via the intermediacy of tricyclic compound *ent*-**303**.

It should be noted at this point, that the delicate issue of constructing medium-sized carbocycles found an elegant solution in this synthesis, based on an unprecedented silylative cycloaldolization protocol.

A common procedure used to forge carbocyclic systems is represented by the ring-closing metathesis reactions exploiting various Grubbs-type carbene catalysts. Based on this cyclization technology, Skaanderup and Madsen [71], on the way to polyhydroxy-nortropane calystegines, reported the asymmetric synthesis of aminated 6a-carbahexoseptanose derivatives **311**.

Scheme 51

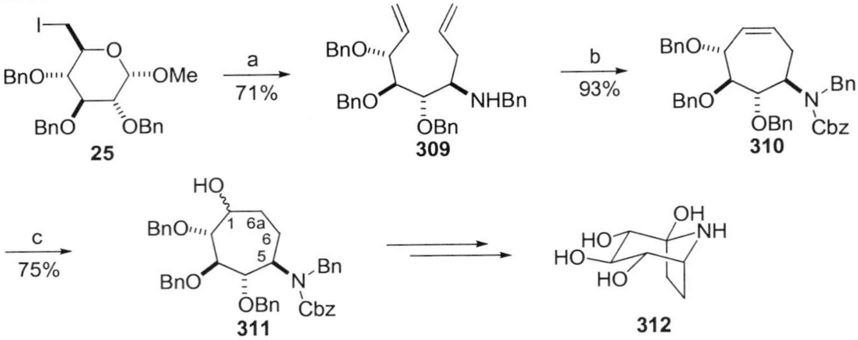

Conditions: (a) Zn, BnNH$_2$, CH$_2$=CHCH$_2$Br, THF, sonication, 40°C; (b) i: CbzCl, KHCO$_3$, EtOAc, H$_2$O; ii: 2% (PCy$_3$)(C$_3$H$_4$N$_2$Mes$_2$)Cl$_2$Ru=CHPh; (c) i: BH$_3$·THF, THF; ii: NaOH, H$_2$O$_2$, 0°C.

As shown in Scheme 51, for the synthesis of calystegine B$_2$ **312**, the starting material was the benzyl protected glucopyranoside **25**. Sonicating a mixture of **25** and zinc dust in THF caused a reductive fragmentation of the pyranose ring to generate an unsaturated aldehyde intermediate which was trapped as the corresponding benzyl imine. Slow addition of allyl bromide led to allylation of the imine to give amino diene **309** as the major diastereoisomer. After Cbz-protection of the amino group, compound **309** was metathesized into cycloheptene **310** using the Grubbs carbene catalyst. Hydroboration-oxydation of the double bond within **310** furnished alcohols **311** as the major regioisomers, having the requisite 5-deoxy-5-amino-6a-carbahexoseptanose configuration. Oxidation of the free C$_1$ hydroxyl to a keto group and subsequent spontaneous intramolecular amination allowed preparation of calystegine B$_2$ **312**, a potent inhibitor of glucosidase enzymes. The same triple domino rearrangement/ring-closing metathesis approach was independently and almost simultaneously exploited by Boyer and Hanna [72] to synthesize chiral calystegine **312**. Here, the chemistry parallels exactly the reaction sequence of Scheme 51 with yields and stereochemical efficiency comparing well with those reported in the Madsen's work [71].

Quite recently, Marco-Contelles [73] has made use of free radical cyclization and ring-closing metathesis in order to develop useful synthetic protocols to access a number of chiral non-racemic, densely oxygenated medium-sized carbocycles from carbohydrate precursors.

Scheme 52

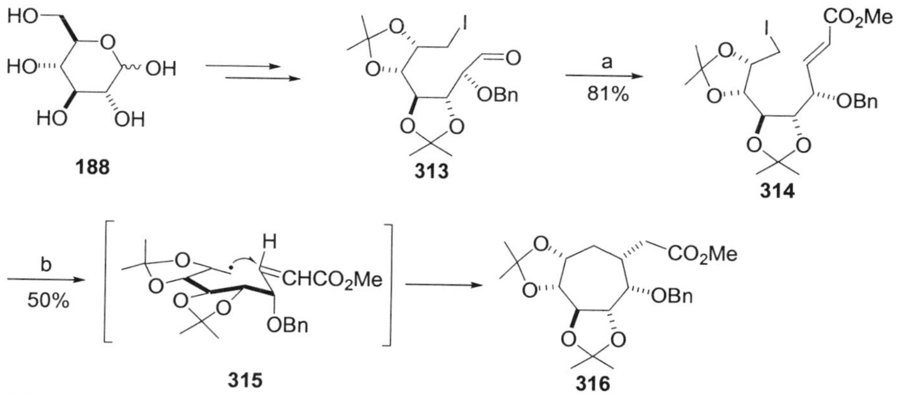

Conditions: (a) Ph$_3$P=CHCO$_2$Me, CH$_2$Cl$_2$; (b) AIBN, Bu$_3$SnH, toluene, 80°C, slow addition.

As an highlighting example, conversion of D-glucose **188** into methyl 6a-carbaoctoseptanuronate **316** is displayed in Scheme 52. Wittig elongation of aldehyde **313**, readily obtainable from D-glucose, with carboxymethylmethylene triphenyl phosphorane gave rise to E-configured unsaturated ester **314**, ready for the free radical cyclization. Exposure of **314** to AIBN-Bu$_3$SnH under slow addition conditions, remarkably resulted in ring annulation to produce septanuronic ester **316** in reasonable yield.

Mechanistically, in the transition state leading to **316**, conformer **315** bearing most of the substituents in a pseudoequatorial orientation should be operative, resulting in formation of the carbocycle with the substituent at the newly formed stereocenter located in the α-orientation.

During the same study, the Spanish researchers also succeeded in the construction of a variety of cycloheptitols, including the calystegine precursor **310** which proved identical to the compound obtained by Madsen and Hanna during their synthesis of calystegine B$_2$ (*vide supra*).

The synthesis of homochiral (+)- and (-)-calystegines B$_2$ was also addressed by two French research groups. Depezay and co-workers [74,75] utilized 6-deoxy-6-vinyl-D-glucopyranose **317**, readily accessible from α-D-methylglucoside, as the key reaction intermediate (Scheme 53).

Scheme 53

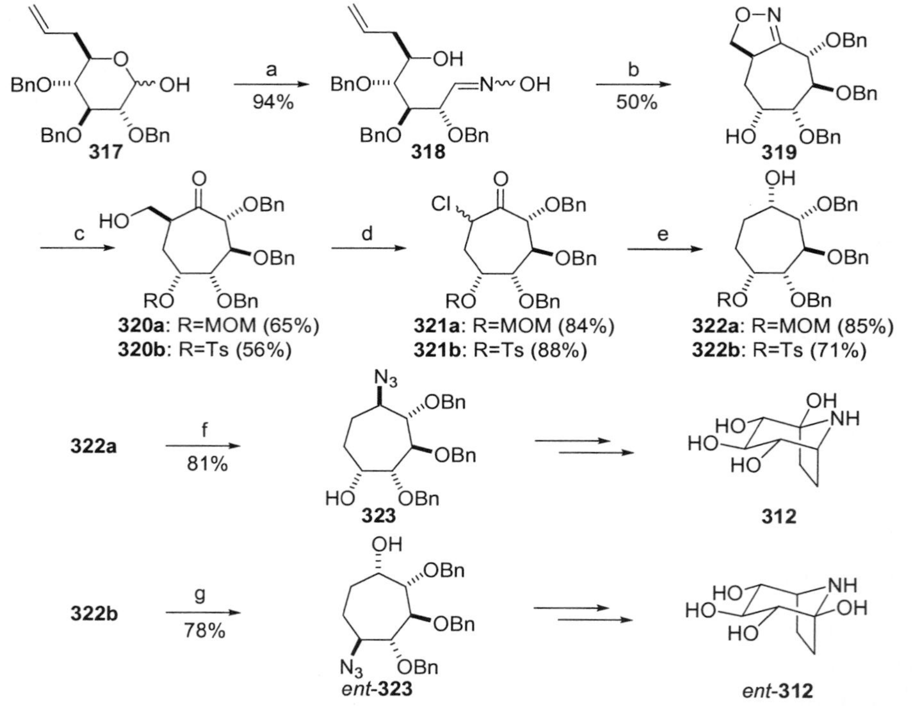

Conditions: (a) NH$_2$OH·HCl, MeONa, MeOH; (b) aq NaOCl, CH$_2$Cl$_2$, 20°C; (c) i: for **320a**: MOMCl, DIPEA; for **320b**: TsCl, pyridine; ii: H$_2$, Ni-Raney, B(OH)$_3$, MeOH/H$_2$O; (d) excess (COCl)$_2$, DMSO, -60°C, then Et$_3$N; (e) i: Zn, TMEDA, AcOH, EtOH; ii: DIBAL-H, Et$_2$O, -50°C; (f) i: ZnN$_6$·2Py, PPh$_3$, DIAD; ii: MeOH, H$^+$; (g) NaN$_3$, DMF, 80°C.

Pyranose was then converted into the oximes **318** by treatment with hydroxylamine, which underwent intramolecular cycloaddition using aqueous NaOCl to afford isoxazoline **319** with excellent diastereocontrol. With intelligent planning, the hidden symmetry of **319** was used to divergently convert this intermediate into the two quasi-meso compounds **322a** and **322b**, in turn precursors to (+)- and (-)-calystegine B$_2$ (**312** and *ent*-**312**). Thus **319** was either protected as its methoxymethyl ether, or tosylated and the products formed independently reduced to form keto-alcohols **320a** and **320b**. The Swern oxidation with an excess of reagents converted **320a** and **320b** into α-chloroketones **321a** and **321b** in excellent yields, probably via a reaction cascade consisting of oxidation, electrophilic chlorination, and chlorine-assisted retro-Claisen deformylation. Reductive dechlorination of **321a** and **321b**, followed by carbonyl reduction, delivered the compounds pair **322a** and **322b** (6a-carba-α-D-glucohexoseptanose configuration). In order to synthesize each of the two calystegine B_2 enantiomers, **322a** and **322b** were converted into the azido alcohol enantiomers **323** and *ent-***323** which served to arrive at the planned targets.

The idea of exploiting the latent symmetry of an intermediate to arrive at the target enantiomers in a divergent manner lay at the base of the strategy developed by Boyer and Lallemand [76] for the total synthesis of **312** and *ent-***312**. As shown in Scheme 54, the starting point was the exomethylene sugar **324** (ex D-glucose) which was subjected to Ferrier rearrangement to afford cyclohexanone **325**.

Scheme 54

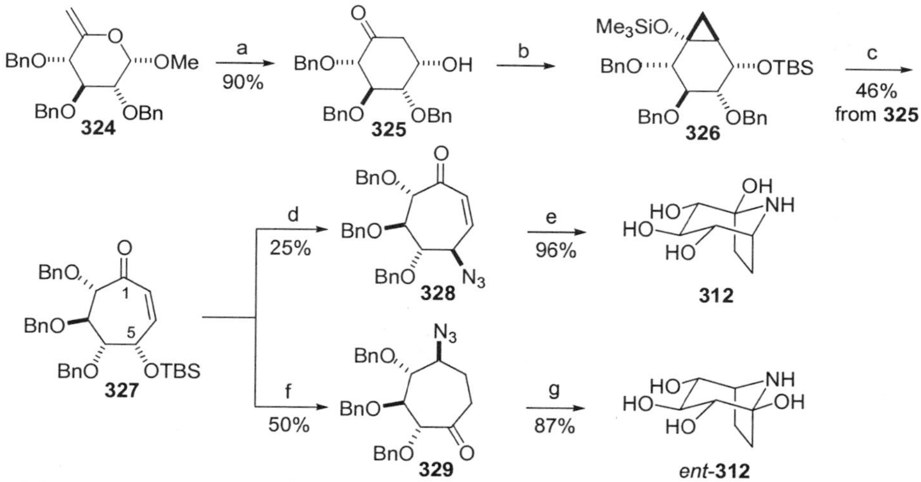

Conditions: (a) Hg(OAc)$_2$, acetone, 90% aq AcOH (1%); (b) i: TBSOTf, 2,6-lutidine, CH$_2$Cl$_2$; ii: LDA, TMSCl, THF, -70°C; iii: Et$_2$Zn, CH$_2$I$_2$, toluene, 0°C; (c) i: FeCl$_3$, DMF, 70°C; ii: NaOAc, MeOH, reflux; (d) i: TBAF, THF; ii: MsCl, pyridine; iii: DIBAL-H, Et$_2$O, -60°C; iv: NaN$_3$, DMF; v: Dess-Martin reagent, pyridine, CH$_2$Cl$_2$; (e) i: H$_2$, Pd/C, AcOH, H$_2$O; ii: Permutite 50, aq NH$_3$; (f) i: H$_2$, Pd/C, EtOH; ii: DIBAL-H, Et$_2$O, -60°C; iii: MsCl, DMAP, pyridine; iv: NaN$_3$, DMF, 80°C; v: TBAF, THF; vi: PCC, CH$_2$Cl$_2$; (g) i: H$_2$, Pd/C, AcOH, H$_2$O; ii: Permutite 50, aq NH$_3$.

A sequence of conventional transformations allowed cyclopropane bicycle **326** to be prepared which was quickly enlarged to cycloheptenone **327** by treatment with iron trichloride in DMF and subsequent dehydrochlorination. Ketone **327** represented the branching point of the synthesis due to its masked symmetric nature. In fact, when the azido group was installed on C_1 and a carbonyl group was put in on C_5

compound **329** was obtained, precursor of (-)-calystegine B$_2$ (*ent-***312**); when however the azido group was implemented at C$_5$ leaving the C$_1$ carbonyl unscathed, compound **328** was obtained, precursor of (+)-calystegine B$_2$ (**312**). These manoeuvres were carried out as indicated and enabled the synthesis to be completed in style.

A clever total synthesis of both enantiomers of aminated 6a-carbahexoseptanoses **336** and *ent-***336**, the immediate precursors of the tropane alkaloids calystegines A$_3$ (**337** and *ent-***337**), was accomplished by Johnson and Bis [77], which centered upon the enzymatic desymmetrization of meso aminotropanediol **333**.

Scheme 55

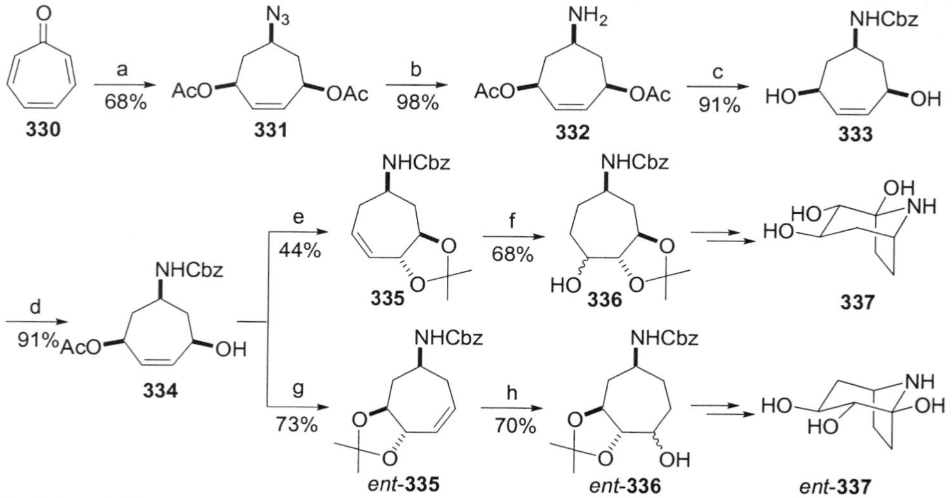

Conditions: (a) i: NaBH$_4$, MeOH, -15°C; ii: LiOAc·2H$_2$O, Pd(OAc)$_2$, MnO$_2$, AcOH, benzoquinone; iii: MsCl, Et$_3$N, 0°C iv: NaN$_3$, DMF, 75°C; (b) H$_2$, Lindlar catalyst, EtOH; (c) i: ClCO$_2$Bn, Na$_2$CO$_3$, EtOAc/H$_2$O; ii: K$_2$CO$_3$, MeOH; (d) Amano P-30 lipase, isopropenyl acetate, 50°C; (e) i: TBSCl, imidazole; ii: NaCN, MeOH; iii: MsCl, Et$_3$N; iv: NaBH$_4$, Ph$_2$Se$_2$; v: H$_2$O$_2$, THF, CH$_2$Cl$_2$; vi: HF, CH$_3$CN; vii: acetone, Amberlyst 15; (f) BH$_3$·DMS, Et$_2$O, -20°C to 0°C, then 30% H$_2$O$_2$, 2N NaOH; (g) i: MsCl, Et$_3$N, 0°C; ii: NaBH$_4$, Ph$_2$Se$_2$, 0°C; iii: H$_2$O$_2$, THF, CH$_2$Cl$_2$, -78°C to rt; iv: K$_2$CO$_3$, MeOH; v: 2,2-dimethoxypropane, *p*-TsOH, acetone; (h) Thexyl·BH$_2$, Et$_2$O, -30°C to -15°C, then 30% H$_2$O$_2$, 2N NaOH.

Starting with tropone **330**, the azido compound **331** was first synthesized, as shown in Scheme 55. Then, compound **331** was chemoselectively reduced to unsaturated amine **332** by Lindlar catalyst, and this material was elaborated to the meso carbamate **333**, ready for enzymatic asymmetrization. Treatment of **333** with Amano P-30 lipase in the presence of isopropenyl acetate resulted in formation of the enantiomerically pure (>98% ee) monoacetate **334**, the common intermediate to both calystegines **337** and *ent-***337**. Using conventional chemistry, elaboration of the functional groups within tropane **334** resulted in formation of protected diol **335**, which was subjected to hydroboration-oxidation to deliver triol **336** as a mixture of isomers. Alternatively, **334** was employed to produce *ent*-**336** via the intermediacy of carbamate *ent*-**335**. Carbasugars **336** and *ent*-**336** were not far from the targeted alkaloids and these transformations were effected without trouble.

A unique member of the seven carbon carbaketose family, namely hept-2-uloseptanosylamine **342**, was assembled by Mandal [78,79] during a remarkable total synthesis of chiral carbocyclic nucleosides from D-glucose. Here, a diastereoselective intramolecular nitrone cycloaddition reaction was the decisive move to install the carbaseptanose ring system (Scheme 56).

Scheme 56

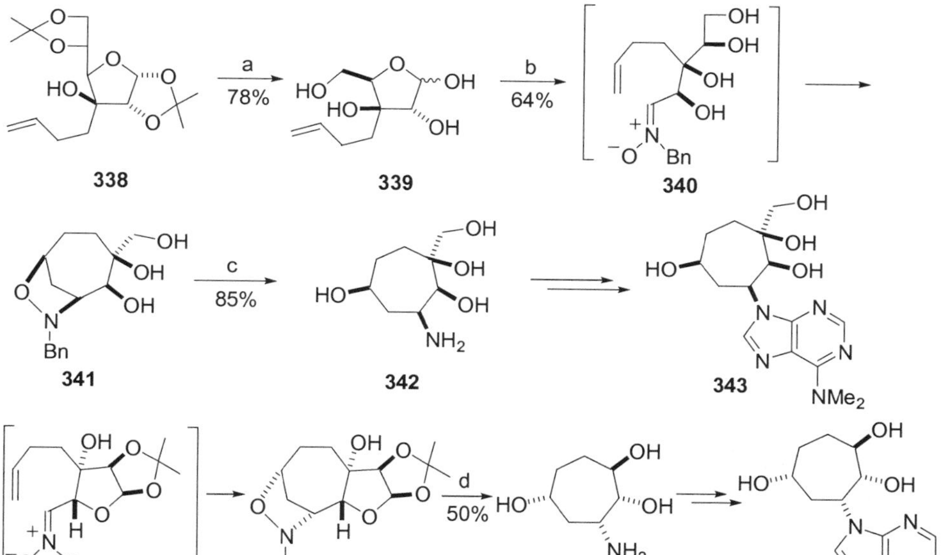

Conditions: (a) i: AcOH/H$_2$O (1:1), 60°C; ii: NaIO$_4$, EtOH, H$_2$O; iii: NaBH$_4$, MeOH; iv: 4% aq H$_2$SO$_4$, dioxane, H$_2$O; (b) i: BnNHOH, 2-fluoroethanol; ii: NaIO$_4$, EtOH, H$_2$O; iii: NaBH$_4$, MeOH; (c) Pd/C, cyclohexene, reflux; (d) i: 4% H$_2$SO$_4$, MeCN/H$_2$O; ii: NaIO$_4$, EtOH, H$_2$O, 10°C; iii: NaBH$_4$, MeOH, 10°C; iv: Pd/C, cyclohexene, EtOH, reflux.

The synthesis began with conversion of D-glucose-derived compound **338** to aldehydofuranose **339**. Reaction of **339** with benzylhydroxylamine in 2-fluoroethanol formed the corresponding nitrone **340** which spontaneously underwent 1,3-dipolar cycloaddition furnishing, after sequential NaIO$_4$ and NaBH$_4$ treatment, the isoxazolidinocarbocycle **341**.

Hydrogenolytic cleavage of the N,O-bond within **341** and debenzylation afforded carbaketose **342**, which nicely served to build up the purine carbocyclic nucleoside **343**.

In the same manner, isoxazolidine **345**, obtained from enose-nitrone substrate **344**, was used to create the 6a-carbahexoseptanosylamine **346**, the core component of the expanded adenosine congener **347**.

4.3. Carbaoctanoses

Polyhydroxylated cyclooctane compounds with a substitution pattern reminiscent of that of the cyclic octanose sugars are fairly rare in current scientific literature. This is probably related to the high density of functional groups and contiguous stereocentres together with the known difficulty of constructing large eight-membered rings.

Scheme 57

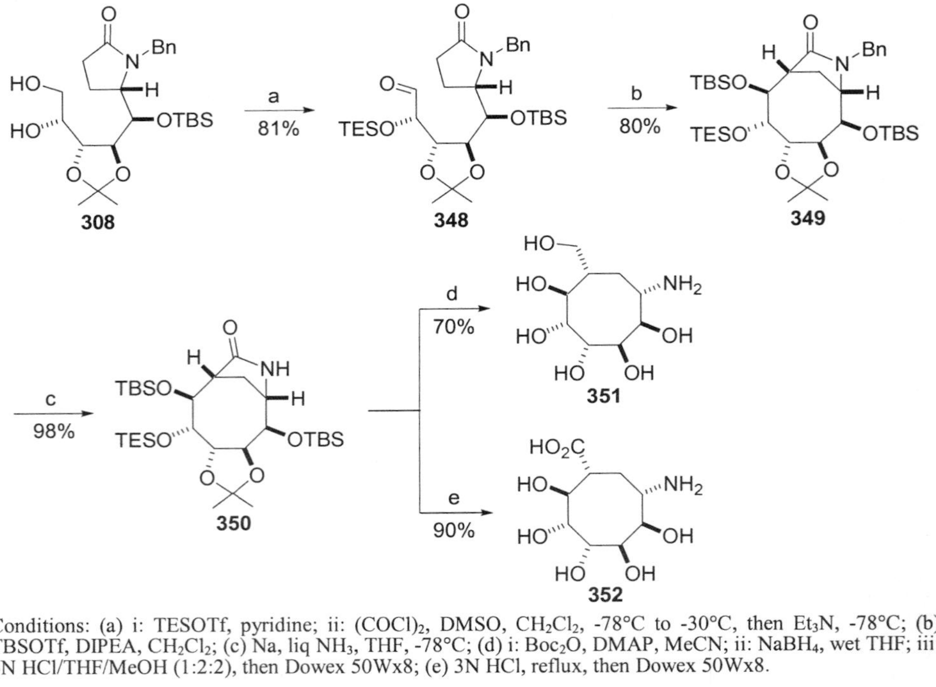

Conditions: (a) i: TESOTf, pyridine; ii: (COCl)$_2$, DMSO, CH$_2$Cl$_2$, -78°C to -30°C, then Et$_3$N, -78°C; (b) TBSOTf, DIPEA, CH$_2$Cl$_2$; (c) Na, liq NH$_3$, THF, -78°C; (d) i: Boc$_2$O, DMAP, MeCN; ii: NaBH$_4$, wet THF; iii: 6N HCl/THF/MeOH (1:2:2), then Dowex 50Wx8; (e) 3N HCl, reflux, then Dowex 50Wx8.

Aiming at the construction of enantiopure carbaoctanose amine compounds, such as β-L-erythro-D-manno-configured 7a-carbaoctanosyl amines **351** and **352**, Rassu and co-workers [7f] exploited the above described advanced intermediate **308**, which embodied the complete nine-carbon skeleton of the targeted compounds. As shown in Scheme 57, after persilylation, an adaptation of the Swern oxidation was chosen to arrive at the aldehyde **348**. At this point, the formation of the eight-membered ring awaited the decisive step of ring closure. Exposure of **348** to the TBSOTf/DIPEA reagent mixture pleasingly afforded tricycle **349** as the only detectable product. Excision of the *N*-benzyl linkage proceeded smoothly to deliver lactam **350**, which could be readily transformed into either **351** or **352**. Thus, completion of 7a-carba-1-deoxy-1-aminooctooctanose **351** was achieved via a reaction sequence involving *N*-Boc activation, reductive cleavage of the amide bond, and deprotection. Alternatively, the preparation of unprecedented γ-amino acid **352** was accomplished straightforwardly by exposure of **350** to refluxing aqueous hydrochloric acid.

The glycoside-to-carbocycle transformation represents an attractive route for the synthesis of a variety of carbocycles, starting from available sugar precursors. Among the various techniques to implement such operations, the reductive Claisen rearrangement of exomethylene C-vinyl glycopyranosides governed by triisobutyl aluminium (TIBAL) proved to be a powerful tool to obtain cyclooctanose carbasugars in a simple way. In recent remarkable works, Sinaÿ and collaborators [80,81] utilized this strategy to synthesize four cyclooctanose mimetics, namely the methyl 5,6-dideoxy-5-hydroxymethyl-7a-carbaheptooctanosides **353**, *ent*-**353**, **354**, and *ent*-**354** in Chart 4.

Chart 4

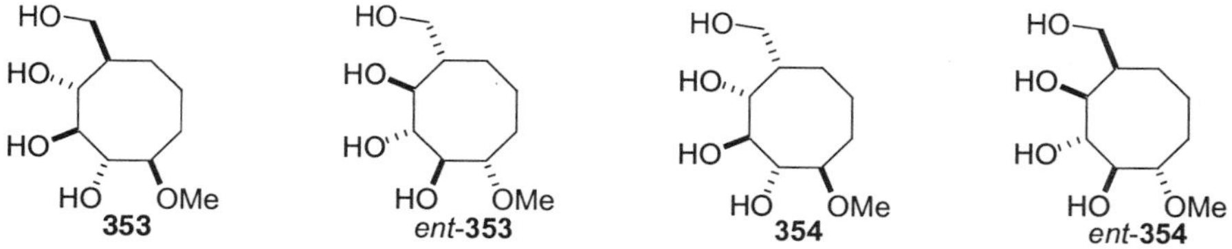

To highlight the procedure, the preparation of D-gluco- and L-ido-configured mimetics **353** and **354** is displayed in Scheme 58.

The opening move was the transformation of the known glucopyranoside **355** into exomethylene vinylpyranose **356**. TIBAL-promoted Claisen rearrangement of **356** provided the cyclooctene derivative **357** almost quantitatively, which was then transformed to protected cyclooctanose **358** by methylation followed by hydroboration-oxidation. Installation of the hydroxymethyl function at C_5 required three further operations; oxidation of the C_5-hydroxyl, Tebbe methylenation, and hydroboration-oxidation. In the event, a mixture of epimeric cyclooctanoids **359** and **360** was formed, which were efficiently elaborated into the targeted carbasugars **353** and **354**.

Scheme 58

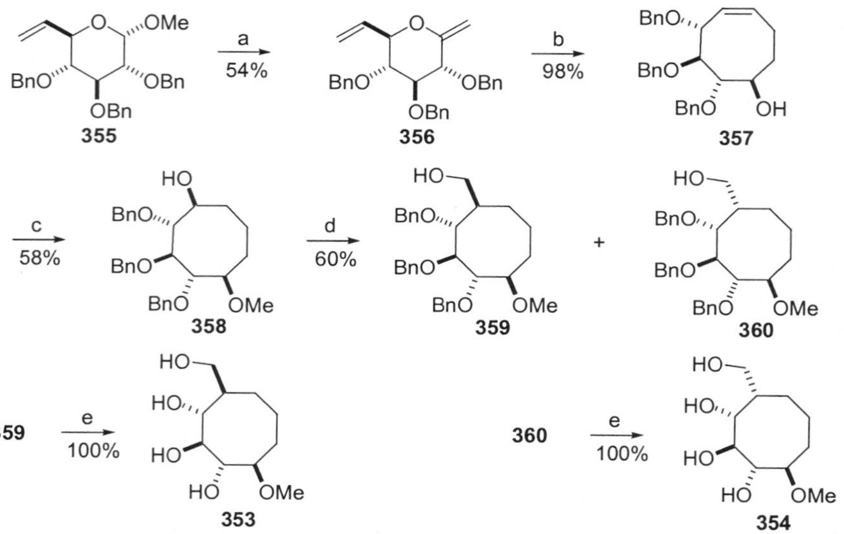

Conditions: (a) i: TfOH, AcOH, H₂O, 80°C; ii: PCC, 4Å MS, CH₂Cl₂, 0°C to rt; iii: Tebbe reagent, pyridine/THF (1:1), -78°C to rt; (b) TIBAL, toluene, 50°C; (c) i: NaH, MeI, DMF; ii: BH₃·THF, then 11% NaOH, 35% H₂O₂, 0°C to rt; (d) i: PCC, MS, CH₂Cl₂, 0°C; ii: Tebbe reagent, pyridine, THF, -78°C to rt; iii: BH₃·THF, then 11% NaOH, 35% H₂O₂, 0°C to rt; (e) H₂, Pd/C, EtOAc, MeOH.

By adopting the same noteworthy TIBAL-mediated carbocyclization technique, preparation of unsaturated octanoid **363** (Scheme 59) was successfully accomplished by the van Boom [82] group, during a study directed towards the synthesis of conformationally locked L-idose analogues.

Scheme 59

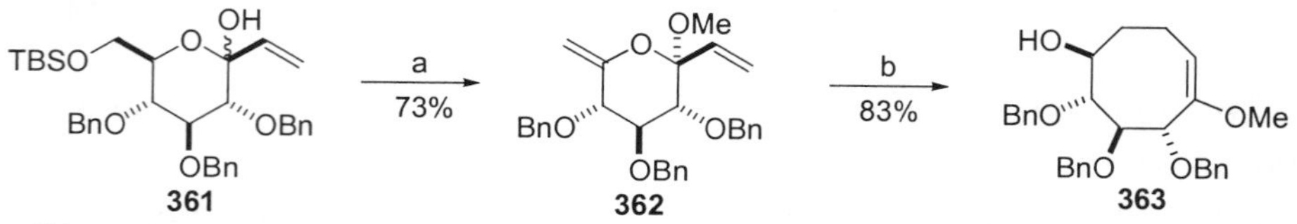

Conditions: (a) i: MeOH, K-10 clay, MS; ii: TBAF, THF; iii: I₂, imidazole, PPh₃, toluene; iv: NaH, DMF; (b) TIBAL, toluene.

Thus, ketose **361** was first methylated and then subjected to a three step transformation including desilylation followed by iodination and hydride-promoted HI elimination. This furnished vinyl ketoside **362** whose treatment with excess of TIBAL ensured smooth Claisen rearrangement to afford cyclooctenic carbocycle **363** in excellent yield.

As already mentioned, ring closing metathesis represents one of the most popular methods for the construction of carbocyclic systems from open chain dienes. Quite often, however, unfavourable thermodynamic factors impede the synthesis of medium-sized rings and, in particular, eight-membered rings. By using proper chain constraints, that is, dioxolane moieties or other ring blockages, carbocyclization may be easier. Indeed, Hanna and Ricard [83] reported a short synthesis of medium-sized carbocyclic rings from galactose, by utilizing olefin metathesis assisted by Grubbs' ruthenium catalyst.

Scheme 60

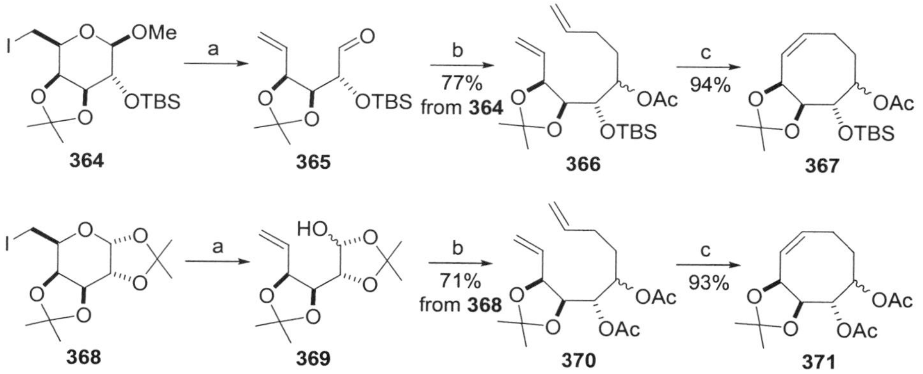

Conditions: (a) Zn, THF, H_2O, sonication; (b) i: butenylmagnesium bromide, Et_2O; ii: Ac_2O, pyridine; (c) Grubbs' catalyst, CH_2Cl_2, reflux.

As shown in Scheme 60, dienes **366** and **370**, respectively prepared from galactopyranosides **364** and **368**, underwent efficient ring closing metathesis furnishing the corresponding 5,6-dideoxy-5,6-didehydro-7a-carbaheptooctanoses **367** and **371**, which were isolated as mixtures of *cis/trans* isomers.

5. COROLLARY

This last section groups together a number of contributions in the field of carbasugar synthesis which, for reasons of space, could not be dealt with in detail in various sections of the text. For the readers comfort, the publications have been sorted into two tables; Table 1 lists reviews and accounts about chemistry and biology of carbasugars and Table 2 enumerates recent reaserch papers focusing on the synthesis of variously-sized carbocyclic carbasugars.

Table 1. Recent Reviews and Accounts about Carbasugars and Their Analogues

Author	Title	Year[ref]
Agrofoglio, L. et al.	*Synthesis of Carbocyclic Nucleosides.*	1994 [2c]
Balci, M. et al.	*Conduritols and Related Compounds.*	1990 [84]
Barco, A. et al.	*D-(-)-Quinic Acid: a Chiron Store for Natural Product Synthesis.*	1997 [85]
Baumgartner, J. et al.	*Chemo-enzymatic Approaches to Enantiopure Carbasugars and Carbanucleosides.*	1998 [86]
Borthwick, A.D. et al.	*Synthesis of Chiral Carbocyclic Nucleosides.*	1992 [2b]
Campbell, M.M. et al.	*The Biosynthesis and Synthesis of Shikimic Acid, Chorismic Acid, and Related Compounds.*	1993 [45]
Carless, H.A.J.	*The Use of Cyclohexane-3,5-diene-1,2-diols in Enantiospecific Synthesis.*	1992 [27a]
Chrétien, F. et al.	*A Concise Route to Carba-Hexopyranoses and Carba-Pentofuranoses from Sugar Lactones.*	1998 [87]
Compain, P. et al.	*Carbohydrate Mimetics-Based Glycosyltransferase Inhibitors.*	2001 [24]
Crimmins, M.T.	*New Developments in Enantioselective Synthesis of Cyclopentyl Carbocyclic Nucleosides.*	1998 [2d]
Dalko, P. et al.	*Recent Advances in the Conversion of Carbohydrate Furanosides and Pyranosides into Carbocycles.*	1999 [88]
Douglas, K.T. et al.	*Chemical Basis of the Activity of Glyoxalase I, an Anticancer Target Enzyme.*	1985 [89]
Dumortier, L. et al.	*Chemo-Enzymatic Total Synthesis of Some Conduritols, Carba-sugars and (+)-Fortamine.*	1998 [90]
Ferrier, R.J. et al.	*The Conversion of Carbohydrate Derivatives into Functionalized Cyclohexanes and Cyclopentanes.*	1993 [5]
Hudlicky, T. et al.	*Toluene-Dioxygenase-Mediated cis-Dihydroxylation of Aromatics in Enantioselective Synthesis. Iterative Glycoconjugates Coupling Strategy and Combinatorial Design for the Synthesis of Oligomers of nor-Saccharides, Inositols and Pseudosugars with Interesting Molecular Properties.*	1996 [91]
Hudlicky, T. et al.	*Modern Methods of Monosaccharide Synthesis from Non-Carbohydrate Sources.*	1996 [25c]
Jiang, S. et al.	*Chemical Synthesis of Shikimic Acid and Its Analogues.*	1998 [46]
Kiefel, M.J. et al.	*Recent Advances in the Synthesis of Sialic Acid Derivatives and Sialylmimetics as Biological Probes.*	2002 [63b]
Landais, Y.	*Desymmetrisation of Dienylsilanes. Stereoselective Access to Cyclitols and Carba-Sugars.*	1998 [92]
Lew, W. et al.	*Discovery and Development of GS 4104 (Oseltamivir): an Orally Active Influenza Neuraminidase Inhibitor.*	2000 [63a]
Lillelund, V.H. et al.	*Recent Developments of Transition-State Analogue Glycosidase Inhibitors of Non-Natural Product Origin.*	2002 [23]
Marquez, V.E. et al.	*Carbocyclic Nucleosides.*	1986 [2a]
Martìnez-Grau, A. et al.	*Carbocycles from Carbohydrates via Free Radical Cyclization: New Synthetic Approaches to Glycomimetics.*	1998 [93]
Ogawa, S.	*Synthetic Studies on Glycosidase Inhibitors Composed of 5a-Carba-Sugars.*	1998 [22]
Sears, P. et al.	*Carbohydrate Mimetics: a New Strategy for Tackling the Problem of Carbohydrate-Mediated Biological Recognition.*	1999 [94]
Sinnott, M.L.	*Catalitic Mechanisms of Enzymic Glycosyl Transfer.*	1990 [95]
Suami, T.	*Chemistry of Pseudo-Sugars.*	1990 [25a]
Suami, T. et al.	*Chemistry of Carba-Sugars (Pseudo-Sugars) and Their Derivatives.*	1990 [25b]
Truscheit, E. et al.	*Chemistry and Biochemistry of Microbial α-Glucosidase Inhibitors.*	1981 [61]
Vasella, A. et al.	*Glycosidase Mechanisms.*	2002 [96]
Vogel, P. et al.	*Stereoselective Synthesis of C-Disaccharides, Aza-C-Disaccharides and C-Glycosides of Carbapyranoses Using the "Naked Sugars".*	1998 [97]

Table 1 *(continued)*

Vogel, P.	*Synthesis of Rare Carbohydrates and Analogues Starting from Enantiomerically Pure 7-Oxabycyclo[2.2.1]heptyl Derivatives ("Naked Sugars").*	2000 [98]
Vogel, P.	*Monosaccharide and Disaccharide Mimics: New Molecular Tools for Biology and Medicine.*	2001 [99]

Table 2. Recent Research Articles Dealing with Synthesis and Biology of Carbasugars and Their Analogues

Author	Title	Sect.	Year[ref]
Aceña, J.L. et al.	*Strain-Directed Bridge Cleavage of (Phenylsulfonyl)-7-oxabicyclo[2.2.1]heptane Derivatives: Application to the Total Synthesis of Carba-α-DL-glucopyranose.*	3.1	1992 [100]
Aceña, J.L. et al.	*A Stereodivergent Access to Naturally Occurring Aminocarbasugars from (Phenylsulfonyl)-7-oxabicyclo[2.2.1]heptane Derivatives. Total Synthesis of Penta-N,O-acetyl-(±)-validamine and Its C$_1$ and C$_2$ Stereoisomers.*	3.4.1	1994 [101]
Afarinkia, K. et al.	*A Novel and Concise Synthesis of (±)-2-epi-Validamine.*	3.4.1	1999 [102]
Agrofoglio, L. et al.	*Synthesis of a New Exocyclic Amino Carbocyclic Nucleoside with Potential Antiviral Activity.*	2.4	1993 [103]
Angelaud, R. et al.	*Stereocontrolled Access to Carba-C-disaccharides Via Functionalized Dienylsilanes.*	3.1	1997 [104]
Bach, G. et al.	*Gabosines, New Carba-Sugars from Streptomyces.*	3.3	1993 [42a]
Baumgartner, H. et al.	*Partial Synthesis of a Carbocyclic Nokkomycin Analogue.*	2.4	1992 [105]
Bennett, S.M. et al.	*Studies on the Chemoselectivity and Diastereoselectivity of Samarium(II) Iodide Mediated Transformations of Carbohydrate Derived ω-Halo-α,β-Unsaturated Esters.*	2.1	1998 [106]
Béres, J. et al.	*Stereospecific Synthesis and Antiviral Properties of Different Enantiomerically Pure Carbocyclic 2'-Deoxyribonucleoside Analogues Derived from Common Chiral Pools: (+)-(1R,5S)- and (-)-(1S,5R)-2-Oxabicyclo[3.3.0]oct-6-en-3-one.*	2.4	1990 [107]
Bodenteich, M. et al.	*Synthesis of Protected (±)-α- and β-Carba-psicofuranose.*	2.2	1992 [14c]
Borthwick, A.D. et al.	*An Efficient Synthesis of a Chiral Carbocyclic 2'-Deoxyribonucleoside Synthon by Directed Reduction.*	2.1	1994 [108]
Boyer, S.J. et al.	*Carbocyclic Nucleoside Analogs. 1. Concise Enantioselective Synthesis of Functionalized Cyclopentanes and Formal Total Synthesis of Aristeromycin.*	2.4	1997 [109]
Bray, B.L. et al.	*Improved Procedures for the Preparation of (+)-(1R,2S,4R)-4-Amino-2-hydroxy-1-hydroxymethyl Cyclopentane.*	2.4	1995 [110]
Campbell, J.A. et al.	*Chirospecific Syntheses of Precursors of Cyclopentane and Cyclopentene Carbocyclic Nucleosides by [3+3]-Coupling and Transannular Alkylation.*	2.4	1995 [111]
Chandler, M. et al.	*Approaches to Carbocyclic Analogues of the Potent Neuraminidase Inhibitor 4-Guanidino-Neu5Ac2en. X-Ray Molecular Structure of N-[(1S,2S,6R)-2-azido-6-benzyloxy-methyl-4-formylcyclohex-3-enyl]acetamide.*	3.4.1	1995 [112]
Chun, B.K. et al.	*Asymmetric Synthesis of Carbocyclic C-Nucleoside, (-)-9-Deazaaristeromycin.*	2.1	1999 [113]
Comin, M.J. et al.	*First Synthesis of (-)-Neplanocin C.*	2.3	2000 [114]
Csuk, R. et al.	*Synthesis of Racemic Carbocyclic Cyclopropanoid Nucleoside Analogues.*	4.1	1995 [115]
Dong, H. et al.	*Biosynthesis of the Validamycins: Identification of Intermediates in the Biosynthesis of Validamycin A by Streptomices hygroscopicus var. limoneus.*	3.4	2001 [50]

Table 2 *(continued)*

Draths, K.M. et al.	*Shikimic Acid and Quinic Acid: Replacing Isolation from Plant Sources with Recombinant Microbial Biocatalysis.*	3.3	1999 [116]
Elliott, R.D. et al.	*Phosphonate Analogs of Carbocyclic Nucleotides.*	2.4	1994 [117]
Entwistle, D.A. et al.	*Synthesis of Pseudosugars from Microbial Metabolites.*	3.1	1995 [118]
Ezzitouni, A. et al.	*(1S,2R)-[(Benzyloxy)methyl]cyclopent-3-enol. A Versatile Synthon for the Preparation of 4',1'a-Methano- and-1',1'a-Methanocarbocyclic Nucleosides.*	2.4	1997 [119]
Fukase, H. et al.	*Synthesis of a Branched-Chain Inosose Derivative, a Versatile Synthon of N-Substituted Valiolamine Derivatives from D-Glucose.*	3.4.2	1992 [58a]
Gallos, J.K. et al.	*Facile Synthesis of Enantiomerically Pure Carbafuranoses: Precursors of Carbocyclic Nucleosides.*	2.1	2001 [120]
Gómez, A.M. et al.	*Regio- and Stereocontrolled 6-Endo-Trig Radical Cyclization of Vinyl Radicals: a Novel Entry to Carbasugars from Carbohydrates.*	3.1	1998 [121]
Gravier-Pelletier, C. et al.	*Efficient Access to Azadisaccharide Analogues.*	4.2	2001 [122]
Hanessian, S. et al.	*Design and Synthesis of Mimics of S-Adenosyl-L-Homocysteine as Potential Inhibitors of Erythromycin Methyltransferases.*	3.1	2000 [123]
Horneman, A.M. et al.	*Stereoselective Radical Induced Cyclisation of Unsaturated Aldonolactones: Synthesis of Highly Functionalized, Enantiomerically Pure Cyclopentane Derivatives.*	2.1	1995 [124]
Horneman, A.M. et al.	*Highly Functionalised Cyclopentanes by Radical Cyclisation of Unsaturated Bromolactones. I. Preparation of 5-Deoxycarbahexofuranoses.*	2.1	1997 [125]
Horneman, A.M. et al.	*Highly Functionalised Cyclopentanes by Radical Cyclisation of Unsaturated Bromolactones III. Preparation of Carbaaldohexofuranoses. Determination of the Relative Configuration at C-4/C-5 of 2,3-Unsaturated Heptono-1,4-lactones by Means of ^{1}H NMR Spectroscopy.*	2.1	1999 [126]
Hsiao, C.-N. et al.	*Efficient Syntheses of Protected (2S,3S)-2,3-Bis(hydroxy-methyl)cyclobutanone, Key Intermediates for the Synthesis of Chiral Carbocyclic Analogues of Oxetanocin.*	4.1	1990 [127]
Ito, H. et al.	*Zirconium-Mediated, Highly Diastereoselective Ring Contraction of Carbohydrate Derivatives: Synthesis of Highly Functionalized, Enantiomerically Pure Carbocycles.*	4.1	1993 [128]
Jeanneret, V. et al.	*Carbaxylosides of 4-Ethyl-2-oxo-2H-benzopyran-7-yl as Non-Hydrolizable, Orally Active Venous Antithrombotic Agents.*	3.1	1998 [129]
Jiang, S. et al.	*Enantiospecific Synthesis of (-)-5-epi-Shikimic Acid and (-)-Shikimic Acid.*	3.3	1997 [130]
Jiang, S. et al.	*Synthesis of 3-Deoxy-3,3-difluoroshikimic Acid and Its 4-Epimer from Quinic Acid.*	3.3	1999 [131]
Johansen, S.K. et al.	*Synthesis of Carbasugars from Aldonolactones: Ritter-Type Epoxide Opening in the Synthesis of Polyhydroxylated Aminocyclopentanes.*	2.1 2.4	1999 [132]
Jotterand, N. et al.	*Total Asymmetric Synthesis of Doubly Branched Carba-hexopyranoses and Amino Derivatives Starting from the Diels-Alder Adducts of Maleic Anhydride to Furfuryl Ester.*	3.4.3	1999 [133]
Kapeller, H. et al.	*Synthesis of Methyl 5-Azido-5-deoxy-2,3-O-isopropylidene-carba-α-D-allo-hexafuranuronate, the Sugar Part of Carbapolyoxins and Carbanikkomycins.*	2.1	1997 [134]
Karpf, M. et al.	*New, Azide-Free Transformation of Epoxides into 1,2-Diamino Compounds: Synthesis of the Anti-Influenza Neuraminidase Inhibitor Oseltamivir Phosphate (Tamiflu).*	3.4.4	2001 [135]
Katagiri, N. et al.	*A Highly Efficient Synthesis of the Antiviral Agent (+)-Cyclaradine Involving the Regioselective Cleavage of Epoxide by Neighboring Participation.*	2.4	1997 [136]

Table 2 *(continued)*

Kim, C.U. et al	*Structure-Activity Relationship Studies of Novel Carbocyclic Influenza Neuraminidase Inhibitors.*	3.4.4	1998 [137]
Knapp, S. et al.	*Intramolecular Amino Delivery Reactions for the Synthesis of Valienamine and Analogues.*	3.4.3	1992 [138]
Lee, K. et al.	*Synthesis Using Ring Closure Metathesis and Effect on Nucleoside Transport of a (N)-Methanocarba S-(4-Nitrobenzyl)Thioinosine Derivative.*	2.3	2001 [139]
Le Merrer, Y. et al.	*A Concise Route to Carbasugars.*	4.2	1999 [140]
Ley, S.V. et al.	*Microbial Oxidation in Synthesis: Preparation of Pseudo-α-D-Glucopyranose from Benzene.*	3.1	1992 [141]
Lew, W. et al.	*A New Series of C_3-Aza Carbocyclic Influenza Neuraminidase Inhibitors: Synthesis and Inhibitory Activity.*	3.4.4	1998 [142]
Lew, W. et al.	*Carbocyclic Influenza Neuraminidase Inhibitors Possessing a C_3-Cyclic Amine Side Chain: Synthesis and Inhibitory Activity.*	3.4.4	2000 [143]
Marco-Contelles, J. et al.	*6-Exo Free Radical Cyclization of Acyclic Carbohydrate Intermediates: A New Synthetic Route to Enantiomerically Pure Polyhydroxylated Cyclohexane Derivatives.*	3.1	1992 [144]
Marco-Contelles, J. et al.	*Cycloheptenols from Carbohydrates.*	4.2	2000 [145]
Marco-Contelles, J. et al.	*Synthesis of Cycloheptenols from Carbohydrates by Ring-Closing Metathesis.*	4.2	2000 [146]
Marschner, C. et al.	*Synthesis of α- and β-D-Carbaribofuranose from (+)-norborn-5-en-2-one.*	2.1	1990 [147]
Marschner, C. et al.	*Synthesis of α- and β-D,L-ribo-Carbahex-2-ulofuranose.*	2.2	1993 [14d]
Martínez, L.E. et al.	*Highly Efficient and Enantioselective Synthesis of Carbocyclic Nucleoside Analogs Using Selective Early Transition Metal Catalysis.*	2.4	1996 [148]
Maryanoff, B.E.	*Structure-Activity Studies on Anticonvulsivant Sugar Sulfamates Related to Topiramate. Enhanced Potency with Cyclic Sulfate Derivatives.*	3.2	1998 [37]
Matsugi, M. et al.	*Highly Stereoselective Synthesis of Carbocycles via a Radical Addition Reaction Using 2,2'-Azobis(2,4-dimethyl-4-methoxyvaleronitrile) [V-70L].*	2.1	1999 [149]
Maudru, E. et al.	*Radical Cyclisation of Carbohydrate Alkynes: Synthesis of Highly Functionalised Cyclohexanes and Carbasugars.*	3.1	1998 [150]
Maycock, C.D. et al.	*An Application of Quinic Acid to the Synthesis of Cyclic Homochiral Molecules: A Common Route to Some Interesting Carbocyclic Nucleoside Precursor.*	2.4	1993 [151]
McNulty, J. et al.	*Selective Addition of Grignard Reagents to 2,3-O-Isopropylidene Bis-Weinreb Tartaric Acid Amide.*	4.3	2001 [152]
Metha, G. et al.	*Norbornyl Route to Polyoxygenated Cyclohexanes. A Facile Entry into Carbasugars and Shikimic Acid.*	3.1 3.3	1998 [153]
Metha, G. et al.	*A Norbornyl Route to Cyclohexitols: Stereoselective Synthesis of Conduritol-E, allo-Inositol, MK 7607 and Gabosines.*	3.3	2000 [154]
Metha, G. et al.	*A General Norbornyl Based Synthetic Approach to Carbasugars and "Confused" Carbasugars.*	3.1	2001 [155]
Metha, G. et al.	*A Norbornyl Route to Aminocyclohexitols: Syntheses of Diverse Aminocarbasugars and "Confused" Aminocarbasugars.*	3.4.1	2002 [156]
Mirza, S. et al.	*Synthesis of Shikimic Acid and its Phosphonate Analogue Via Knoevenagel Condensation.*	3.3	1991 [157]
Neufellner, E. et al.	*A Novel Access to Derivatives of 3-Azido-3-deoxy-4a-carba-α-DL-ribofuranose, Potential Intermediates for the Synthesis of Carbachryscandin and Carbapuromycin.*	2.4	1998 [158]

Table 2 *(continued)*

Norimine, Y. et al.	*Enantio- and Diastereoselective Synthesis of N-[(1R,2R,3R,4R)-2,3-Diacetoxy-4-(acetoxymethyl)cyclopentyl]acetamide, a Synthetic Key Intermediate of (+)-Cyclaradine.*	2.4	1998 [159]
Obara, T. et al.	*New Neplanocin Analogues. 7. Synthesis and Antiviral Activity of 2-Halo Derivatives of Neplanocin A.*	2.3	1996 [160]
Ogawa, S. et al.	*Synthesis of DL-2-Amino-2-deoxyvalidamine and Its Three Diastereoisomers.*	3.4.1	1990 [161]
Ogawa, S. et al.	*Synthesis and Glycosidase Inhibitory Activity of 5a-Carba-α-DL-fucopyranosylamine and -galactopyranosylamine.*	3.4.1	2000 [162]
Ohira, S. et al.	*Synthesis of (-)-Neplanocin A via C-H Insertion of Alkylidenecarbene.*	2.3	1995 [163]
Ovaa, H. et al.	*A Versatile Approach to the Synthesis of Highly Functionalised Carbocycles.*	3.3 3.4	1999 [164]
Park, K.H. et al.	*Enantioselective Synthesis of (1R,4S)-1-Amino-4-(hydroxymethyl)-2-cyclopentene, a Precursor for Carbocyclic Nucleoside Synthesis.*	2.4	1994 [165]
Parry, R.J. et al.	*Biosynthesis of Aristeromycin: Evidence for the Intermediacy of a 4β-Hydroxymethyl-1α,2α,3α-trihydroxycyclopentanetriol.*	2.1	1990 [166]
Parry, R.J. et al.	*Investigations of Aristeromycin Biosynthesis: Evidence for the Intermediacy of a 2α,3α-Dihydroxy-4β-(hydroxymethyl)cyclopentane-1β-amine.*	2.4	1991 [167]
Parry, R.J. et al.	*Synthesis of 1α-Pyrophosphoryl-2α,3α-dihydroxy-4β-cyclopentanemethanol-5-phosphate, a Carbocyclic Analog of 5-Phosphoribosyl-1-pyrophosphate (PRPP).*	2.1	1993 [168]
Patil, S.D. et al.	*(±)-Carbocyclic 5'-Nor-2'-deoxyguanosine and Related Purine Derivatives: Synthesis and Antiviral Properties.*	2.4	1992 [169]
Posner, G.H.	*New Synthetic Methodology Using Organosulfur Compounds.*	3.3	1990 [170]
Redlich, H. et al.	*Radical Cyclisation of Hept-1-enitols.*	3.1	1992 [171]
Renaut, P. et al.	*5a-Carba-β-D-, 5a-Carba-β-L-, and 5-Thio-β-L-xylopyranosides as New Orally Active Venous Antithrombotic Agents.*	3.1	1998 [172]
Roberts, S.M. et al.	*Radical Cyclisation Reactions Leading to Polyhydroxylated Cyclopentane Derivatives: Synthesis of (1R,2R,3S,4R)- and (1S, 2S,3R,4S)-4-Hydroxyethylcyclopentane-1,2,3-triol.*	2.1	1992 [173]
Rohloff, J.C. et al.	*Practical Total Synthesis of the Anti-influenza Drug GS-4104.*	3.4.4	1998 [174]
Seepersaud, M. et al.	*Studies Directed toward the Synthesis of Carba-D-arabinofuranose.*	2.1	2000 [175]
Shing, T.K.M. et al.	*(-)-Quinic Acid in Organic Synthesis. 1. A Facile Synthesis of 2-Crotonyloxymethyl-(4R,5R,6R)-4,5,6-Trihydroxycyclohex-2-enone.*	3.3	1990 [176]
Shing, T.K.M. et al.	*(-)-Quinic Acid in Organic Synthesis. 3. Stereocontrolled Syntheses of Pseudo-α-D-glucopyranose and Pseudo-α-D-mannopyranose.*	3.1	1992 [60]
Shing, T.K.M. et al.	*Enantiospecific Syntheses of Penta-N,O,O,O,O-acetylvalidamine and Penta-N,O,O,O,O-acetyl-2-epi-validamine.*	3.4.1	1995 [177]
Shing, T.K.M. et al.	*Enantiospecific Syntheses of Valienamine and 2-epi-Valienamine.*	3.4.3	1999 [178]
Shoberu, K.A. et al.	*Synthesis of Pseudo-Ribofuranoses by Stereocontrolled Reactions on 4-Hydroxycyclopent-2-enylmethanol Derivatives.*	2.1	1992 [179]
Siddiqui, S.M. et al.	*Search for New Purine- and Ribose-Modified Adenosine Analogues as Selective Agonists and Antagonists at Adenosine Receptors.*	2.4	1995 [180]

Table 2 *(continued)*

Soulié, J. et al.	*General Access to Polyhydroxylated Nortropane Derivatives through Hetero Diels-Alder Cycloaddition. Part II: Synthesis of (±)-Calystegine B₂.*	4.2	1996 [181]
Takahashi, T. et al.	*A New Synthetic Approach to Pseudo-Sugars by Asymmetric Diels-Alder Reaction. Synthesis of Optically Pure Pseudo-β-D-mannopyranose, 1-Amino-1-deoxypseudo-α-D-manno-pyranose and Pseudo-α-L-mannopyranose Derivatives.*	3.1	1990 [182]
Tang, Y.-Q. et al.	*Gabosines L, N and O: New Carba-Sugars from* Streptomyces *with DNA-Binding Properties.*	3.3	2000 [42b]
Trost, B.M. et al.	*Total Synthesis of (±)-Valienamine and (+)-Valienamine via a Strategy Derived from New Palladium-Catalyzed Reactions.*	3.4.3	1998 [183]
Tsunoda, H. et al.	*Synthesis of Some 5a-Carbaglycosylamides, Glycolipid Analogs of Biological Interests.*	3.4.1	1994 [184]
van Hooft, P.A.V. et al.	*Stereoselective Transformations on* D-Glucose-Derived *Eight-Membered Ring Carbocycles.*	4.3	2001 [185]
Verduyn, R. et al.	*Synthesis of Carba-β-DL-Fucose and Carba-β-DL-Galactose from* myo-*Inositol.*	3.1	1996 [186]
Vorwerk, S. et al.	*Carbocyclic Analogues of* N-*Acetyl-2,3-didehydro-2-deoxy-D-Neuraminic Acid (Neu5Ac2en, DANA): Synthesis and Inhibition of Viral and Bacterial Neuraminidase.*	3.4.4	1998 [187]
Yamakoshi, Y.N. et al.	*High Pressure Mediated Asymmetric Diels-Alder Reaction of Chiral Sulfinylacrylate Derivatives with Furan and 2-Methoxyfuran.*	3.3	1996 [188]
Yoshikawa, M. et al.	*Synthesis of (-)-Aristeromycin from* D-*Glucose.*	2.4	1990 [189]
Yoshikawa, M. et al.	*Facile Syntheses of Pseudo-α-D-arabinofuranose, and Two Pseudo-D-arabinofuranosylnucleosides, (+)-Cyclaradine and (+)-1-Pseudo-β-D-arabinofuranosyluracil, from* D-*Arabinose.*	2.1	1994 [190]
Wang, Z.X. et al.	*Asymmetric Epoxidation by Chiral Ketones Derived from Carbocyclic Analogues of Fructose.*	3.2 3.4.2	2001 [191]
Werschkun, B. et al.	*From* D-*Glucose to a New Chiral Cyclooctenone.*	4.3	1997 [192]
Williams, M.A. et al.	*Structure-Activity Relationships of Carbocyclic Influenza Neuraminidase Inhibitors.*	3.4.4	1997 [193]
Wood, H.B. et al.	*Synthesis of (-)-3-Homoshikimic Acid and (-)-3-Homoshikimate-3-phosphate.*	3.3	1993 [194]
Zhou, Z. et al.	*Samarium(II) Iodide Mediated Transformations of Carbohydrate Derived Alkenyl Iodides.*	2.1	1997 [195]

6. CONCLUSIONS

In this article we have analized a number of papers dedicated to the synthesis of selected carbocyclic polyols which bare an authentic resemblance to one of the many subclasses of the carbohydrate family. This criterion lead us to exclude a whole world of biologically important carbocyclic compounds from this compilation which, although strictly speaking do not exhibit a direct corrispondence to existing carbohydrates, continue to be included in the family of the carbasugars by many authors. Thus, important cyclopentitols and cyclohexitols, including inositols, quercitols, conduritols, cyclophellitols, threhazoloids, and allosamizolines, have been regrettably neglected. In listing the syntheses,

512

a weighted contribution has been given by the furanosidic and pyranosidic carbasugars, with a brief section being dedicated to the less represented 3, 4, 7, and 8 membered ring congeners.

A minor space has been dedicated to biological aspects. Studies of biological activity have been commented on only when explicitly carried out by the authors in their original manuscripts. Although *avant-gard* techniques of bibliographic research were adopted to thoroughly scour the domain of carbasugar literature important contributions may have been unvoluntarily overlooked and we wish to apologize to any authors whose work has escaped our attention.

ACKNOWLEDGEMENTS

The authors are grateful to all of their co-workers for the invaluable contribution to the projects carried out in the laboratories mentioned in this review. Special thanks are due to Prof. Giovanni Casiraghi, Prof. Franca Zanardi, and Dr. Lucia Marzocchi for their help given in compiling this account. The authors are also grateful to the sources that funded their work in the carbasugar area; Ministero dell'Istruzione, dell'Università e della Ricerca (MIUR): Programmi di Ricerca Scientifica di Rilevante Interesse Nazionale, Anni 2000 and 2002, and Regione Autonoma della Sardegna.

REFERENCES

[1]   McNaught, A.D.; *Carbohydr. Res.*, **1997**, *297*, 1-92.

[2]   a) Marquez, V.E.; Lim, M.-I.; *Med. Res. Rev.*, **1986**, *6*, 1-40. b) Borthwick, A.D.; Biggadike, K.; *Tetrahedron*, **1992**, *48*, 571-623. c) Agrofoglio, L.; Suhas, E.; Farese, A.; Condom, R.; Challand, S.R.; Earl, R.A.; Guedj, R.; *Tetrahedron*, **1994**, *50*, 10611-10670. d) Crimmins, M.T.; *Tetrahedron*, **1998**, *54*, 9229-9272.

[3]   Shealy, Y.F.; Clayton, J.D.; *J. Am. Chem. Soc.*, **1966**, *88*, 3885-3887.

[4]   a) Kusaka, T.; Yamamoto, H.; Shibata, M.; Muroi, M.; Kishi, T.; Mizuno, K.; *J. Antibiot.*, **1968**, *21*, 255-271. b) Yagimuna, S.; Muto, N.; Tsujino, M.; Sudate, Y.; Hayashi, M.; Otani, M.; *J. Antibiot.*, **1981**, *34*, 359-366.

[5]   Ferrier, R.J.; Middelton, S.; *Chem. Rev.*, **1993**, *93*, 2779-2831.

[6]   Marschner, C.; Baumgartner, J.; Griengl, H.; *J. Org. Chem.*, **1995**, *60*, 5224-5235.

[7]   a) Rassu, G.; Auzzas, L.; Pinna, L.; Zanardi, F.; Battistini, L.; Casiraghi, G.; *Org. Lett.*, **1999**, *1*, 1213-1215. b) Rassu, G.; Auzzas, L.; Pinna, L.; Battistini, L.; Zanardi, F.; Marzocchi, L.; Acquotti, D.; Casiraghi, G.; *J. Org. Chem.*, **2000**, *65*, 6307-6318. c) Rassu, G.; Auzzas, L.; Pinna, L.; Zambrano, V.; Battistini, L.;

Zanardi, F.; Marzocchi, L.; Acquotti, D.; Casiraghi, G.; *J. Org. Chem.*, **2001**, *66*, 8070-8075. d) Zanardi, F.; Battistini, L.; Marzocchi, L.; Acquotti, D.; Rassu, G.; Pinna, L.; Auzzas, L.; Zambrano, V.; Casiraghi, G.; *Eur. J. Org. Chem.*, **2002**, 1956-1965. e) Rassu, G.; Auzzas, L.; Pinna, L.; Zambrano, V.; Zanardi, F.; Battistini, L.; Marzocchi, L.; Acquotti, D.; Casiraghi, G.; *J. Org. Chem.*, **2002**, *67*, 5338-5342. f) Rassu, G.; Auzzas, L.; Pinna, L.; Zambrano, V.; Zanardi, F.; Battistini, L.; Marzocchi, L.; Curti, C.; Casiraghi, G.; *J. Org. Chem.*, **2003**, submitted.

[8]   Désiré, J.; Prandi, J.; *Eur. J. Org. Chem.*, **2000**, 3075-3084.

[9]   Horneman, A.M.; Lundt, I.; *J. Org. Chem.*, **1998**, *63*, 1919-1928.

[10]   Callam, C.S.; Lowary, T.L.; *J. Org. Chem.*, **2001**, *66*, 8961-8972.

[11]   a) Walton, E.; Jenkins, S.R.; Nutt, R.F.; Zimmerman, M.; Holly, F.W.; *J. Am. Chem. Soc.*, **1966**, *88*, 4524-4525. b) Rosenthal, A.; Sprinzl, M.; *Can. J. Chem.*, **1969**, *47*, 3941-3946.

[12]   Audran, G.; Acherar, S.; Monti, H.; *Eur. J. Org. Chem.*, **2003**, *92*, 92-98.

[13]   Rassu, G.; Auzzas, L.; Pinna, L.; Zambrano, V.; Battistini, L.; Curti, C.; unpublished results.

[14]   a) Wilcox, C.S.; Gaudino, J.J.; *J. Am. Chem. Soc.*, **1986**, *108*, 3102-3104. b) Gaudino, J.J.; Wilcox, C.S.; *Carbohydr. Res.*, **1990**, *206*, 233-250. c) Bodenteich, M.; Marquez, V.E.; *Tetrahedron*, **1992**, *48*, 5961-5968. d) Marschner, C.; Penn, G.; Griengl, H.; *Tetrahedron*, **1993**, *49*, 5067-5078. e) Seepersaud, M.; Al-Abed, Y.; *Org. Lett.*, **1999**, *1*, 1463-1465.

[15]   Gillaizeau, I.; Charamon, S.; Agrofoglio, L.A.; *Tetrahedron Lett.*, **2001**, *42*, 8817-8819.

[16]   Metha, G.; Mohal, N.; *Tetrahedron Lett.*, **2001**, *42*, 4227-4230.

[17]   Babu, Y.S.; Chand, P.; Bantia, S.; Kotian, P.; Dehghani, A.; El-Kattan, Y.; Lin, T.-H.; Hutchinson, T.L.; Elliot, A.J.; Parker, C.D.; Ananth, S.L.; Horn, L.L.; Laver, G.W.; Montgomery, J.A.; *J. Med. Chem.*, **2000**, *43*, 3482-3486.

[18]   Gallos, J.K.; Massen, Z.S.; Koftis, T.V.; Dellios, C.C.; *Tetrahedron Lett.*, **2001**, *42*, 7489-7491.

[19]   McCasland, G.E.; Furuta, S.; Durham, L.J.; *J. Org. Chem.*, **1966**, *31*, 1516-1521.

[20]   McCasland, G.E.; Furuta, S.; Durham, L.J.; *J. Org. Chem.*, **1968**, *33*, 2835-2844.

[21]   Miller, T.W.; Arison, B.H.; Albers-Schonberg, G.; *Biotechnol. Bioeng.*, **1973**, *15*, 1075-1080.

[22]   Ogawa, S. In *Carbohydrate Mimics: Concepts and Methods*; Chapleur, Y., Ed.; Wiley-VCH: Weinheim, **1998**; Chapter 5, pp. 87-106.

[23]   Lillelund, V.H.; Jensen, H.H.; Liang, X.; Bols, M.; *Chem. Rev.*, **2002**, *102*, 515-553.

[24]   Compain, P.; Martin, O.R.; *Bioorg. Med. Chem. Lett.*, **2001**, *9*, 3077-3092.

514

[25] a) Suami, T.; *Top. Curr. Chem.*, **1990**, *154*, 257-283. b) Suami, T.; Ogawa, S.; *Adv. Carbohydr. Chem. Biochem.*, **1990**, *48*, 21-90. c) Hudlicky, T.; Entwistle, D.A.; Pitzer, K.; Thorpe, A.J.; *Chem. Rev.*, **1996**, *96*, 1195-1220.

[26] Angelaud, R.; Landais, Y.; *Tetrahedron Lett.*, **1997**, *38*, 8841-8844.

[27] a) Carless, H.A.J.; *Tetrahedron: Asymmetry*, **1992**, *3*, 795-826. b) Hudlicky, T.; Thorpe, A.J.; *Chem. Commun.*, **1996**, 1993-2000.

[28] Pingli, L.; Vandewalle, M.; *Tetrahedron*, **1994**, *50*, 7061-7074.

[29] Metha, G.; Mohal, N.; Lakshminath, S.; *Tetrahedron Lett.*, **2000**, *41*, 3505-3508.

[30] Cai, S.; Stroud, M.R.; Hakomori, S.; Toyokuni, T.; *J. Org. Chem.*, **1992**, *57*, 6693-6696.

[31] Yuasa, H.; Palcic, M.M.; Hindsgaul, O.; *Can. J. Chem.*, **1995**, *73*, 2190-2195.

[32] Sudha, A.V.R.L.; Nagarajan, M.; *Chem. Commun.*, **1998**, 925-926.

[33] Büchi, G.; Powell, J.E.; *J. Am. Chem. Soc.*, **1967**, *89*, 4559-4560.

[34] Ogawa, S.; Uematsu, Y.; Yoshida, S.; Sasaki, N.; Suami, T.; *J. Carbohydr. Chem.*, **1987**, *6*, 471-478.

[35] Shing, T.K.M.; Tang, Y.; *Tetrahedron*, **1991**, *47*, 4571-4578.

[36] McComsey D.F.; Maryanoff, B.E.; *J. Org. Chem.*, **1994**, *59*, 2652-2654.

[37] Maryanoff, B.E.; Costanzo, M.J.; Nortey, S.O.; Greco, M.N.; Shank, R.P.; Schupsky, J.J.; Ortegon, M.P.; Vaught, J.L.; *J. Med. Chem.*, **1998**, *41*, 1315-1343.

[38] Yoshikawa, M.; Chiba, N.; Mikawa, T.; Ueno, S.; Harimaya, K.; Iwata, M.; *Chem. Abstr.*, **1985**, *122*, 185533e.

[39] Song, C.; Jiang, S.; Singh, G.; *Synlett*, **2001**, 1983-1985.

[40] a) Argoudelis, A.D.; Pyke, T.R.; Sprague, R.W.; *J. Antibiot.*, **1976**, *29*, 777-786. b) Argoudelis, A.D.; Sprague, R.W.; Mizsak, S.A.; *J. Antibiot.*, **1976**, *29*, 787-796.

[41] Arjona, O.; Iradier, F.; Medel, R.; Plumet, J.; *Tetrahedron: Asymmetry*, **1999**, *10*, 3431-3442.

[42] a) Bach, G.; Breiding-Mack, S.; Grabley, S.; Hammann, P.; Hütter, K.; Thiericke, R.; Uhr, H.; Wink, J.; Zeeck, A.; *Liebigs Ann. Chem.*, **1993**, 241-250. b) Tang, Y.-Q.; Maul, C.; Höfs, R.; Sattler, I.; Grabley, S.; Feng, X.-Z.; Zeeck, A.; Thiericke, R.; *Eur. J. Org. Chem.,* **2000**, 149-153.

[43] Lygo, B.; Swiatyj, M.; Trabsa, H.; Voyle, M.; *Tetrahedron Lett.*, **1994**, *35*, 4197-4200.

[44] Lubineau, A.; Billault, I.; *J. Org. Chem.*, **1998**, *63*, 5668-5671.

[45] Campbell, M.M.; Sainsbury, M.; Searle, P.A.; *Synthesis*, **1993**, 179-193.

[46] Jiang, S.; Singh, G.; *Tetrahedron*, **1998**, *54*, 4697-4753.

[47] Dumortier, L.; Van der Eycken, J.; Vandewalle, M.; *Synlett*, **1992**, 245-246.

[48] Numata, A.; Iritani, M.; Yamada, T.; Minoura, K.; Matsumura, E.; Yamori, T.; Tsuruo, T.; *Tetrahedron Lett.*, **1997**, *38*, 8215-8218.

[49] Donohoe, T.J.; Blades, K.; Helliwell, M.; Waring, M.J.; *Tetrahedron Lett.*, **1998**, *39*, 8755-8758.

[50] Dong, H.; Mahmud, T.; Tornus, I.; Lee, S.; Floss, H.G.; *J. Am. Chem. Soc.*, **2001**, *123*, 2733-2742.

[51] Yoshikawa, M.; Murakami, N.; Yokokawa, Y.; Inoue, Y.; Kuroda, Y.; Kitagawa, I.; *Tetrahedron*, **1994**, *50*, 9619-9628.

[52] a) Yoshikawa, M.; Cha, B.C.; Nakae, T.; Kitagawa, I.; *Chem. Pharm. Bull.*, **1988**, *36*, 3714-3717. b) Kitagawa, I.; Cha, B.C.; Nakae, T.; Okaichi, Y.; Takinami, Y.; Yoshikawa, M.; *Chem. Pharm. Bull.*, **1989**, *37*, 542-544.

[53] Kameda, Y.; Kawashima, K.; Takeuchi, M.; Ikeda, K.; Asano, N.; Matsui, K.; *Carbohydr. Res.*, **1997**, *300*, 259-264.

[54] Ogawa, S.; Maruyama, A.; Odagiri, T.; Yuasa, H.; Hashimoto, H.; *Eur. J. Org. Chem.*, **2001**, 967-974.

[55] Horii, S.; Fukase, H.; Matsuo, T.; Kameda, Y.; Asano, N.; Matsui, K.; *J. Med. Chem.*, **1986**, *29*, 1038-1046.

[56] a) Ogawa, S.; Shibata, Y.; *Chem. Lett.*, **1985**, 1581-1582. b) Hayashida, M.; Sakairi, N.; Kuzuhara, H.; *J. Carbohydr. Chem.*, **1988**, *7*, 83-94.

[57] Ohtake, H.; Li, X.-L.; Shiro, M.; Ikegami, S.; *Tetrahedron*, **2000**, *56*, 7109-7122.

[58] a) Fukase, H.; Horii, S.; *J. Org. Chem.*, **1992**, *57*, 3642-3650; b) Fukase, H.; Horii, S.; *J. Org. Chem.*, **1992**, *57*, 3651-3658.

[59] Shing, T.K.M.; Wang, L.H.; *J. Org. Chem.*, **1996**, *61*, 8468-8479.

[60] Shing, T.K.M.; Cui, Y.-X.; Tang, Y.; *Tetrahedron*, **1992**, *48*, 2349-2358.

[61] Truscheit, E.; Frommer, W.; Junge, B.; Müller, L.; Schmidt, D.D.; Wingender, W.; *Angew. Chem. Int. Ed.*, **1981**, *20*, 744-761.

[62] Park, T.K.; Danishefsky, S.J.; *Tetrahedron Lett.*, **1994**, *35*, 2667-2670.

[63] a) Lew, W.; Chen, X.; Kim, C.U.; *Curr. Med. Chem.*, **2000**, *7*, 663-672. b) Kiefel, M.J.; Von Itzstein, M.; *Chem. Rev.*, **2002**, *102*, 471-490.

[64] Kim, C.U.; Lew, W.; Williams, M.A.; Liu, H.; Zhang, L.; Swaminathan, S.; Bischofberger, N.; Chen, M.S.; Mendel, D.B.; Tai, C.Y.; Laver, W.G.; Stevens, R.C.; *J. Am. Chem. Soc.*, **1997**, *119*, 681-690.

[65] Bianco, A.; Brufani, M.; Manna, F.; Melchioni, C.; *Carbohydr. Res.*, **2001**, *332*, 23-31.

[66] Lew, W.; Williams, M.A.; Mendel, D.B.; Escarpe, P.A.; Kim, C.U.; *Bioorg. Med. Chem. Lett.*, **1997**, *7*, 1843-1846.

[67] Zhao, Y.; Yang, T.; Lee, M.; Lee, D.; Newton, M.G.; Chu, C.K.; *J. Org. Chem.*, **1995**, *60*, 5236-5242.

[68] Lee, M.G.; Du, J.F.; Chun, M.W.; Chu, C.K.; *J. Org. Chem.*, **1997**, *62*, 1991-1995.

[69] Bisacchi, G.S.; Braitman, A.; Cianci, C.W.; Clark, J.M.; Field, A.K.; Hagen, M.E.; Hockstein, D.R.; Malley, M.F.; Mitt, T.; Slusarchyk, W.A.; Sundeen, J.E.; Terry, B.J.; Tuomari, A.V.; Weaver, E.R.; Young, M.G.; Zahler, R.; *J. Med. Chem.*, **1991**, *34*, 1415-1421.

516

[70] Ichikawa, Y.; Narita, A.; Shiozawa, A.; Hayashi, Y.; Narasaka, K.; *Chem. Commun.*, **1989**, 1919-1921.

[71] Skaanderup, P.R.; Madsen, R.; *Chem. Commun.*, **2001**, 1106-1107.

[72] Boyer, F.-D.; Hanna, I.; *Tetrahedron Lett.*, **2001**, *42*, 1275-1277.

[73] Marco-Contelles, J.; de Opazo, E.; *J. Org. Chem.*, **2002**, *67*, 3705-3717.

[74] Duclos, O.; Mondange, M.; Duréault, A.; Depezay, J.-C.; *Tetrahedron Lett.*, **1992**, *33*, 8061-8064.

[75] Duclos, O.; Duréault, A.; Depezay, J.-C.; *Tetrahedron Lett.*, **1992**, *33*, 1059-1062.

[76] Boyer, F.-D.; Lallemand, J.-Y.; *Tetrahedron*, **1994**, *50*, 10443-10458.

[77] Johnson, C.R.; Bis, S.J.; *J. Org. Chem.*, **1995**, *60*, 615-623.

[78] Roy, A.; Chakrabarty, K.; Dutta, P.K.; Bar, N.C.; Basu, N.; Achari, B.; Mandal, S.B.; *J. Org. Chem.*, **1999**, *64*, 2304-2309.

[79] Bar, N.C.; Roy, A.; Achari, B.; Mandal, S.B.; *J. Org. Chem.*, **1997**, *62*, 8948-8951.

[80] Wang, W.; Zhang, Y.; Sollogoub, M.; Sinaÿ, P.; *Angew. Chem. Int. Ed.*, **2000**, *39*, 2466-2467.

[81] Wang, W.; Zhang, Y.; Zhou, H.; Blériot, Y.; Sinaÿ, P.; *Eur. J. Org. Chem.*, **2001**, 1053-1059.

[82] van Hooft, P.A.V.; van der Marel, G.A.; van Boeckel, C.A.A.; van Boom, J.H.; *Tetrahedron Lett.*, **2001**, *42*, 1769-1772.

[83] Hanna, I.; Ricard, L.; *Org. Lett.*, **2000**, *2*, 2651-2654.

[84] Balci, M.; Sütbeyaz, Y.; Seçen, H.; *Tetrahedron*, **1990**, *46*, 3715-3742.

[85] Barco, A.; Benetti, S.; De Risi, C.; Marchetti, P.; Pollini, G.P.; Zanirato, V.; *Tethrahedron: Asymmetry*, **1997**, *8*, 3515-3545.

[86] Baumgartner, J.; Griengl, H. In *Carbohydrate Mimics: Concepts and Methods*; Chapleur, Y., Ed.; Wiley-VCH: Weinheim, **1998**; Chapter 12, pp. 223-237.

[87] Chrétien, F.; Khaldi, M.; Chapleur, Y. In *Carbohydrate Mimics: Concepts and Methods*; Chapleur, Y., Ed.; Wiley-VCH: Weinheim, **1998**; Chapter 8, pp. 143-156.

[88] Dalko, P.I.; Sinaÿ, P.; *Angew. Chem. Int. Ed.*, **1999**, *38*, 773-777.

[89] Douglas, K.T.; Shinkai, S.; *Angew. Chem. Int. Ed.*, **1985**, *24*, 31-44.

[90] Dumortier, L.; Liu, P.; Van der Eycken, J.; Vandewalle, M. In *Carbohydrate Mimics: Concepts and Methods*; Chapleur, Y., Ed.; Wiley-VCH: Weinheim, **1998**; Chapter 11, pp. 209-221.

[91] Hudlicky, T.; Abboud, K.A.; Entwistle, D.A.; Fan, R.; Maurya, R.; Thorpe, A.J.; Bolonick, J.; Myers, B.; *Synthesis*, **1996**, 897-911.

[92] Landais, Y.; *Chimia*, **1998**, *52*, 104-111.

[93] Martínez-Grau, A.; Marco-Contelles, J.; *Chem. Soc. Rev.*, **1998**, *27*, 155-162.

[94] Sears, P.; Wong, C.-H.; *Angew. Chem. Int. Ed.*, **1999**, *38*, 2300-2324.

[95] Sinnott, M.L.; *Chem. Rev.*, **1990**, *90*, 1171-1202.

[96] Vasella, A.; Davies, G.J.; Böhm, M.; *Curr. Opin. Chem. Biol.*, **2002**, *6*, 619-629.

[97]   Vogel, P.; Ferritto, R.; Kraehenbuehl, K.; Baudat, A. In *Carbohydrate Mimics: Concepts and Methods*; Chapleur, Y., Ed.; Wiley-VCH: Weinheim, **1998**; Chapter 2, pp. 19-48.

[98]   Vogel, P.; *Curr. Org. Chem.*, **2000**, *4*, 455-480.

[99]   Vogel, P.; *Chimia*, **2001**, *55*, 359-365.

[100]   Aceña, J.L.; Arjona, O.; de la Pradilla, R.F.; Plumet, J.; Viso, A.; *J. Org. Chem.*, **1992**, *57*, 1945-1946.

[101]   Aceña, J.L.; Arjona, O.; de la Pradilla, R.F.; Plumet, J.; Viso, A.; *J. Org. Chem.*, **1994**, *59*, 6419-6424.

[102]   Afarinkia, K.; Mahmood, F.; *Tetrahedron*, **1999**, *55*, 3129-3140.

[103]   Agrofoglio, L.; Condom, R.; Guedj, R.; Challand, R.; Selway, J.; *Tetrahedron Lett.*, **1993**, *34*, 6271-6272.

[104]   Angelaud, R.; Landais, Y.; Parra-Rapado, L.; *Tetrahedron Lett.*, **1997**, *38*, 8845-8848.

[105]   Baumgartner, H.; Marschner, C.; Pucher, R.; Singer, M.; Griengl, H.; *Tetrahedron Lett.*, **1992**, *33*, 6443-6444.

[106]   Bennett, S.; Biboutou, R.K.; Zhou, Z.; Pion, R.; *Tetrahedron*, **1998**, *54*, 4761-4786.

[107]   Béres, J.; Sági, Gy.; Tömösközi, I.; Gruber, L.; Baitz-Gács, E.; Ötvös, L.; De Clerq, E.; *J. Med. Chem.*, **1990**, *33*, 1353-1360.

[108]   Borthwick, A.D.; Crame, A.J.; Exall, A.M.; Weingarten, G.G.; *Tetrahedron Lett.*, **1994**, *35*, 7677-7680.

[109]   Boyer, S.J.; Leahy, J.W.; *J. Org. Chem.*, **1997**, *62*, 3976-3980.

[110]   Bray, B.L.; Dolan, S.C.; Halter, B.; Lackey, J.W.; Schilling, M.B.; Tapolczay, D.J.; *Tetrahedron Lett.*, **1995**, *36*, 4483-4486.

[111]   Campbell, J.A.; Lee, W.K.; Rapoport, H.; *J. Org. Chem.*, **1995**, *60*, 4602-4616.

[112]   Chandler, M.; Conroy, R.; Cooper, A.W.J.; Lamont, R.B.; Scicinski, J.J.; Smart, J.E.; Storer, R.; Weir, N.G.; Wilson, R.D.; Wyatt, P.G.; *J. Chem. Soc., Perk. Trans. 1*, **1995**, 1189-1197.

[113]   Chun, B.K.; Chu, C.K.; *Tetrahedron Lett.*, **1999**, *40*, 3309-3312.

[114]   Comin, M.J.; Rodriguez, J.B.; *Tetrahedron*, **2000**, *56*, 4639-4649.

[115]   Csuk, R.; von Scholz, Y.; *Tetrahedron*, **1995**, *51*, 7193-7206.

[116]   Draths, K.M.; Knop, D.R.; Frost, J.W.; *J. Am. Chem. Soc.*, **1999**, *121*, 1603-1604.

[117]   Elliott, R.D.; Rener, G.A.; Riordan, J.M.; Secrist III, J.A.; Bennett, L.L., Jr.; Parker, W.B.; Montgomery, J.A.; *J. Med. Chem.*, **1994**, *37*, 739-744.

[118]   Entwistle, D.A.; Hudlicky, T.; *Tetrahedron Lett.*, **1995**, *36*, 2591-2594.

[119]   Ezzitouni, A.; Russ, P.; Marquez, V.E.; *J. Org. Chem.*, **1997**, *62*, 4870-4873.

[120]   Gallos, J.K.; Dellios, C.C.; Spata, E.E.; *Eur. J. Org. Chem.*, **2001**, 79-82.

[121]   Gómez, A.M.; Danelón, G.O.; Valverde, S.; López, J.C.; *J. Org. Chem.*, **1998**, *63*, 9626-9627.

[122] Gravier-Pelletier, C.; Maton, W.; Lecourt, T.; Le Merrer, Y.; *Tetrahedron Lett.*, **2001**, *42*, 4475–4478.

[123] Hanessian, S.; Sgarbi, P.W.M.; *Bioorg. Med. Chem. Lett.*, **2000**, *10*, 433-437.

[124] Horneman, A.M.; Lundt, I.; Søtofte, I.; *Synlett*, **1995**, 918-920.

[125] Horneman, A.M.; Lundt, I.; *Tetrahedron*, **1997**, *53*, 6879-6892.

[126] Horneman, A.M.; Lundt, I.; *Synthesis*, **1999**, 317-325.

[127] Hsiao, C.-N.; Hannick, S.M.; *Tetrahedron Lett.*,**1990**, *31*, 6609-6612.

[128] Ito, H.; Motoki, Y.; Taguchi, T.; Hanzawa, Y.; *J. Am. Chem. Soc.*, **1993**, *115*, 8835-8836.

[129] Jeanneret, V.; Vogel, P.; Renaut, P.; Millet, J.; Theveniaux, J.; Barberousse, V.; *Bioorg. Med. Chem. Lett.*, **1998**, *8*, 1687-1688.

[130] Jiang, S.; McCullough, K.J.; Mekki, B.; Singh, G.; Wightman, R.H.; *J. Chem. Soc., Perkin Trans. I*, **1997**, 1805-1814.

[131] Jiang, S.; Singh, G.; Boam, D.J.; Coggins, J.R.; *Tetrahedron: Asymmetry*, **1999**, *10*, 4087-4090.

[132] Johansen, S.K.; Kørno, H.T.; Lundt, I.; *Synthesis*, **1999**, 171-177.

[133] Jotterand, N.; Vogel, P.; Schenk, K.; *Helv. Chim. Acta*, **1999**, *82*, 821-847.

[134] Kapeller, H.; Griengl, H.; *Tetrahedron*, **1997**, *53*, 14635-14644.

[135] Karpf, M.; Trussardi, R.; *J. Org. Chem.*, **2001**, *66*, 2044-2051.

[136] Katagiri, N.; Matsuhashi, Y.; Kokufuda, H.; Takebayashi, M.; Kaneko, C.; *Tetrahedron Lett.*, **1997**, *38*, 1961-1964.

[137] Kim, C.U.; Lew, W.; Williams, M.A.; Wu, H.; Zhang, L.; Chen, X.; Escarpe, P.A.; Mendel, D.B.; Laver, W.G.; Stevens, R.C.; *J. Med. Chem.*, **1998**, *41*, 2451-2460.

[138] Knapp, S.; Naughton, A.B.J.; Dhar, T.G.M.; *Tetrahedron Lett.*, **1992**, *33*, 1025-1028.

[139] Lee, K.; Cass, C.; Jacobson, K.A.; *Org. Lett.*, **2001**, *3*, 597-599.

[140] Le Merrer, Y.; Gravier-Pelletier, C.; Maton, W.; Numa, M.; Depezay, J.-C.; *Synlett*, **1999**, 1322-1324.

[141] Ley, S.V.; Yeung, L.L.; *Synlett*, **1992**, 291-292.

[142] Lew, W.; Wu, H.; Mendel, D.B.; Escarpe, P.A.; Chen, X.; Laver, W.G.; Graves, B.J.; Kim, C.U.; *Bioorg. Med. Chem. Lett.*, **1998**, *8*, 3321-3324.

[143] Lew, W.; Wu, H.; Chen, X.; Graves, B.J.; Escarpe, P.A.; MacArthur, H.L.; Mendel, D.B.; Kim, C.U.; *Bioorg. Med. Chem. Lett.*, **2000**, *10*, 1257-1260.

[144] Marco-Contelles, J.; Pozuelo, C.; Jimeno, M.L.; Martinez, L.; Martinez-Grau, A.; *J. Org. Chem.*, **1992**, *57*, 2625-2631.

[145] Marco-Contelles, J.; de Opazo, E.; *Tetrahedron Lett.*, **2000**, *41*, 2439-2441.

[146] Marco-Contelles, J.; de Opazo, E.; *J. Org. Chem.*, **2000**, *65*, 5416-5419.

[147] Marschner, C.; Penn, G.; Griengl, H.; *Tetrahedron Lett.*, **1990**, *31*, 2873-2874.

[148] Martínez, L.E.; Nugent, W.A.; Jacobsen, E.N.; *J. Org. Chem.*, **1996**, *61*, 7963-7966.

[149] Matsugi, M.; Gotanda, K.; Ohira, C.; Suemura, M.; Sano, A.; Kita, Y.; *J. Org. Chem.*, **1999**, *64*, 6928-6930.

[150] Maudru, E.; Singh, G.; Wightman, R.H.; *Chem. Commun.*, **1998**, 1505-1506.

[151] Maycock, C.D.; Barros, M.T.; Santos, A.G.; Godinho, L.S.; *Tetrahedron Lett.* **1993**, *34*, 7985-7988.

[152] McNulty, J.; Grunner, V.; Mao, J.; *Tetrahedron Lett.*, **2001**, *42*, 5609-5612.

[153] Metha, G.; Mohal, N.; *Tetrahedron Lett.*, **1998**, *39*, 3285-3286.

[154] Metha, G.; Lakshminath, S.; *Tetrahedron Lett.*, **2000**, *41*, 3509-3512.

[155] Metha, G.; Talukdar, P.; Mohal, N.; *Tetrahedron Lett.*, **2001**, *42*, 7663-7666.

[156] Metha, G.; Lakshminath, S.; Talukdar, P.; *Tetrahedron Lett.*, **2002**, *43*, 335-338.

[157] Mirza, S., Harvey, J.; *Tetrahedron Lett.*, **1991**, *32*, 4111-4114.

[158] Neufellner, E.; Kapeller, H.; Griengl, H.; *Tetrahedron*, **1998**, *54*, 11043-11062.

[159] Norimine, Y.; Hayashi, M.; Tanaka, M.; Suemune, H.; *Chem. Pharm. Bull.*, **1998**, *46*, 842-845.

[160] Obara, T.; Shuto, S.; Saito, Y.; Snoeck, R.; Andrei, G.; Balzarini, J.; De Clercq, E.; Matsuda, A.; *J. Med. Chem.*, **1996**, *39*, 3847-3852.

[161] Ogawa, S.; Tonegawa, T.; *Carbohydr. Res.*, **1990**, *204*, 51-64.

[162] Ogawa, S.; Sekura, R.; Maruyama, A.; Yuasa, H.; Hashimoto, H.; *Eur. J. Org. Chem.*, **2000**, 2089-2093.

[163] Ohira, S.; Sawamoto, T.; Yamato, M.; *Tetrahedron Lett.*, **1995**, *36*, 1537-1538.

[164] Ovaa, H.; Codée, J.D.C.; Lastdrager, B.; Overkleeft, H.S.; van der Marel, G.A.; van Boom, J.H.; *Tetrahedron Lett.*, **1999**, *40*, 5063-5066.

[165] Park, K.H.; Rapoport, H.; *J. Org. Chem.*, **1994**, *59*, 394-399.

[166] Parry, R.J.; Haridas, K.; De Jong, R.; Johnson, C.R.; *Tetrahedron Lett.*, **1990**, *31*, 7549-7552.

[167] Parry, R.J.; Haridas, K.; De Jong, R.; Johnson, C.R.; *J. Chem. Soc., Chem. Commun.*, **1991**, 740-741.

[168] Parry, R.J.; Haridas, K.; *Tetrahedron Lett.*, **1993**, *34*, 7013-7016.

[169] Patil, S.D.; Koga, M.; Schneller, S.W.; *J. Med. Chem.*, **1992**, *35*, 2191-2195.

[170] Posner, G.H.; *Pure Appl. Chem.*, **1990**, *62*, 1949-1956.

[171] Redlich, H.; Sudau, W.; Szardenings, A.K.; Vollerthun, R.; *Carbohydr. Res.*, **1992**, *226*, 57-78.

[172] Renaut, P.; Millet, J.; Sepulchre, C.; Theveniaux, J.; Barberousse, V.; Jeanneret, V; Vogel, P.; *Helv. Chim. Acta*, **1998**, *81*, 2043-2052.

[173] Roberts, S.M.; Shoberu, K.A.; *J. Chem. Soc., Perkin Trans. I*, **1992**, 2625-2632.

[174] Rohloff, J.C.; Kent, K.M.; Postich, M.J.; Becker, M.W.; Chapman, H.H.; Kelly, D.E.; Lew, W.; Louie, M.S.; McGee, L.R.; Prisbe, E.J.; Schultze, L.M.; Yu, R.H.; Zhang, L.; *J. Org. Chem.*, **1998**, *63*, 4545-4550.

[175] Seepersaud, M.; Al-Abed, Y.; *Tetrahedron Lett.*, **2000**, *41*, 7801-7803.

[176] Shing, T.K.M.; Tang, Y.; *Tetrahedron*, **1990**, *46*, 6575-6584.

[177] Shing, T.K.M.; Tai, V.W.-F.; *J. Org. Chem.*, **1995**, *60*, 5332-5334.

520

[178] Shing, T.K.M.; Li, T.Y.; Kok, S.H.-L.; *J. Org. Chem.*, **1999**, *64*, 1941-1946.

[179] Shoberu, K.A.; Roberts, S.M.; *J. Chem. Soc., Perkin Trans. I*, **1992**, 2419-2425.

[180] Siddiqui, S.M.; Jacobson, K.A.; Esker, J.L.; Olah, M.E.; Ji, X.; Melman, N.; Tiwari, K.N.; Secrist III, J.A.; Schneller, S.W.; Cristalli, G.; Stiles, G.L.; Johnson, C.R.; Ijzerman, A.P.; *J. Med. Chem.*, **1995**, *38*, 1174-1188.

[181] Soulié, J.; Faitg, T.; Betzer, J.-F.; Lallemand, J.-Y.; *Tetrahedron*, **1996**, *52*, 15137-15146.

[182] Takahashi, T.; Kotsubo, H.; Iyobe, A.; Namiki, T.; Koizumi, T.; *J. Chem. Soc., Perkin Trans. I*, **1990**, 3065-3072.

[183] Trost, B.M.; Chupak, L.S.; Lübbers, T.; *J. Am. Chem. Soc.*, **1998**, *120*, 1732-1740.

[184] Tsunoda, H.; Ogawa, S.; *Liebigs Ann. Chem.*, **1994**, 103-107.

[185] van Hooft, P.A.V.; Litjens, R.E.J.; van der Marel, G.A.; van Boeckel, C.A.A.; van Boom, J.H.; *Org. Lett.*, **2001**, *3*, 731-733.

[186] Verduyn, R.; van Leeuwen, S.H.; van der Marel, G.A.; van Boom, J.H.; *Recl. Trav. Chim. Pays-Bas*, **1996**, *115*, 67-71.

[187] Vorwerk, S.; Vasella, A.; *Angew. Chem. Int. Ed.*, **1998**, *37*, 1732-1734.

[188] Yamakoshi, Y.N.; Ge, W.-Y.; Sugita, J.; Okayama, K.; Takahashi, T.; Koizumi, T.; *Heterocycles*, **1996**, *42*, 129-133.

[189] Yoshikawa, M.; Okaichi, Y.; Cha, B.C.; Kitagawa, I.; *Tetrahedron*, **1990**, *46*, 7459-7470.

[190] Yoshikawa, M.; Yokokawa, Y.; Inoue, Y.; Yamaguchi, S.; Murakami, N.; *Tetrahedron*, **1994**, *50*, 9961-9974.

[191] Wang, Z.X.; Miller, S.M.; Anderson, O.P.; Shi, Y.; *J. Org. Chem.*, **2001**, *66*, 521-530.

[192] Werschkun, B.; Thiem, J.; *Angew. Chem. Int. Ed.*, **1997**, *36*, 2793-2794.

[193] Williams, M.A.; Lew, W.; Mendel, D.B.; Tai, C.Y.; Escarpe, P.A.; Laver, W.G.; Stevens, R.C.; Kim, C.U.; *Bioorg. Med. Chem. Lett.*, **1997**, *7*, 1837-1842.

[194] Wood, H.B.; Ganem, B.; *Tetrahedron Lett.*, **1993**, *34*, 1403-1406.

[195] Zhou, Z.; Bennett, S.M.; *Tetrahedron Lett.*, **1997**, *38*, 1153-1156.

Atta-ur-Rahman (Ed.) *Studies in Natural Products Chemistry, Vol. 29*

521

THE BIOSYNTHESIS AND PROPERTIES
OF ANTI-CARBOHYDRATE ANTIBODIES

JOHN H. PAZUR

Dept. of Biochemistry and Molecular Biology, The Pennsylvania State University, University Park, PA 16802 USA

ABSTRACT: Antibodies are an important group of bioactive natural products and are of the globulin class of proteins. Anti-carbohydrate antibodies are produced by immunization of animals with carbohydrate containing antigens. The antigens may be a polysaccharide, glycoprotein, glycolipid or a synthetic conjugate of carbohydrate and protein. The antibody that is produced is polyclonal and has specificity for the carbohydrate residue of the antigen. Many types of anti-carbohydrate antibodies have been isolated in pure form by affinity chromatography. These are specific for mono- or disaccharide units of antigens *(e.g.* glucose, galactose, mannose, rhamnose, fucose, glucuronic acid, N-acetyl glucosamine, lactose, isomaltose, lactosamine, neuraminic-galactose, glucuronic-galactose, mannose-glucuronic, rhamnose-glucuronic, arabinose-glucuronic, me glucuronic-galactose). The chemical and biological properties of these antibodies have been determined. To illustrate, molecular weight, light and heavy chain type, immunoglobulin class, inhibition, ultracentrifugation, electrophoresis, isoelectrofocusing and specificity are recorded. Many of these anti-carbohydrate antibodies have been useful in applications. Some have been used in medicine for the identification of pathogenic microorganisms, some in studies to determine the structure of carbohydrate polymers, some for the analysis of additives to processed food and beverages and some in pharmaceutical uses to prepare vaccines for treatment of diseases.

INTRODUCTION

Antibodies are an important group of bioactive natural products which are produced in the serum of vertebrates in response to activation of the immune system by foreign substances. Since there are numerous activating substances, there are a large number of antibodies with each displaying specificity for the immunizing substance. The antibodies are the principal agents for defense against diseases caused by pathogenic microorganisms, viruses and animal cells which have undergone transformation into diseased cells. The function of the antibody is to direct the invading and injurious substance to the macrophage cells where the substance is decomposed by enzymes from the cytoplasmic vesicles. The list of antigens includes compounds from proteins, glycoproteins, polysaccharides, nucleoproteins, lipoproteins and numerous synthetic conjugates linking carbohydrates to proteins.

In 1923 antibodies specific for a polysaccharide in the pneumococcal cell wall were detected in the serum from rabbits immunized with non-viable pneumococcal cells by Heidelberger and Avery [1]. Cell wall polysaccharides from other bacteria as reported by Lancefield [2], McCarty [3], Krause [4], Pazur [5], and Karakawa [6] activated the plasma cells in the serum of animals to initiate the biosynthesis of antibodies which were specific for the carbohydrates. Such antibodies are appropriately classified as anti-carbohydrate antibodies. The antibodies which are synthesized on immunization are polyclonal antibodies and consist of various numbers of isoforms. The antibodies which are synthesized by fusion of immune spleen cells and myeloma cells are monoclonal antibodies.

With the advent of affinity chromatography to isolate enzymes [7] it was possible to adapt the method for the isolation of antibodies [8]. The purification of many types of anti-carbohydrate antibodies with specificity for carbohydrate residues of macromolecules has been achieved. The antibodies were isolated by affinity chromatography on adsorbents with proper ligands. The immunodeterminants of the antigens have been found to be mono- or disaccharide residues of the antigen. The anti-carbohydrate antibodies with unique properties are useful in a variety of medicinal and technological applications. In this article the emphasis is on polyclonal antibodies specifically the preparation, properties and uses of anti-carbohydrate antibodies.

GENERAL CONSIDERATIONS

Structure of Antibodies

The antibodies are similar in structure to other globulin proteins which are present in serum of vertebrates [9]. The antibody molecule consists of two light polypeptide chains and two heavy polypeptide chains [10]. The amino acids are present in all the chains but the number of residues and the sequence will vary in different antibody multiforms [9]. Light chains contain approximately 220 amino acid residues and heavy chains about 450 residues. The complete sequence of the chains of human IgG myeloma protein has been determined by Edelman [11]. The chains are held together in the unique conformational structure of the antibody molecule by a few covalent disulfide bonds between the chains and many electrostatic bonds between the amino groups of one chain and the hydroxyl groups of another chain. The covalent bonds are represented in Formula 1, 2 and 3 [peptide, disulfide and glycosidic] and the electrostatic bonds in Formula 4 may be numerous and depend on the type of antibody generated in response to the antigen. The molecule contains a few carbohydrate chains attached by glycosidic bonds to heavy peptide chains. From the NH_2 terminal end each chain has a variable segment and then a constant segment of amino acids constituting about half of the chain [12]. A combination of amino acids in the variable contains the active site of the molecule. More detailed information on the antibody structure is recorded in the Nobel Lectures by Dr. Porter and Dr. Edelman [13, 14].

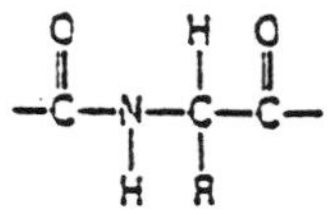

C—H···O
C—H···C⁻
C—H···N
N—H···O
N⁺—H···O
N—H···N
4

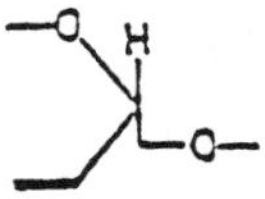

-C-N-C-C-
  | |
  H   R
1

-S-S-
2

3

The homogeneity and molecular size of the purified antibody preparations were determined by ultracentrifugation in a Spinco Model E centrifuge. The results of this analysis on anti-lactose antibodies are illustrated in the photographs of the ultracentrifuge patterns shown in Fig. (1). The

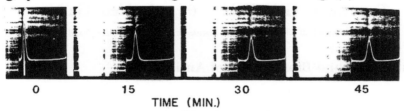

0   15   30   45
TIME (MIN.)

Fig. (1). Spinco Model E ultracentrifugation pattern for anti-lactose antibodies.

sedimentation velocity and the sedimentation equilibrium data were obtained. Similar results were obtained with other antibodies. The sedimentation coefficient was calculated from the data by the method described by Schachman [15] and was found to be 7.1S. The molecular weight was calculated as 150,000 daltons. The molecular weight was also determined to be the same by the sucrose density ultracentrifugation method described in a later section.

Most antibodies have carbohydrate residues in the structure. In acid hydrolysates the following have been identified as fucose, mannose, galactose, glucosamine and neuraminic acid identified by paper chromatography, Fig. (2), and by specific colorimetric tests [16]. The amino sugars were N-acetylated prior to hydrolysis. All residues occur as oligosaccharide in chains attached to hydroxyl groups of amino acids of the peptide chains of the constant region [17]. The carbohydrate content of different antibodies ranges from 3-12% [9].

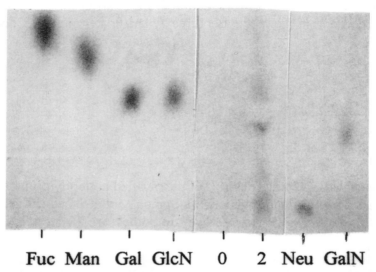

Fuc Man Gal GlcN   0   2 Neu GalN

Fig. (2).   Paper chromatograph of reference monosaccharides and acid hydrolosate (0 and 2 h) of antibodies.

Immunization and Biosynthesis of Antibodies

Antibody preparations are polyclonal or monoclonal depending on the use of immunization methods or fusion techniques. In this article emphasis is on the polyclonal antibodies, their preparation, properties and uses. Substances that are foreign to the circulatory system of an animal or diseased cells which are produced by transformation of healthy cells can activate the immune system to produce antibodies which ultimately destroy the substances and the diseased cells. Selected antigens containing carbohydrate can be used for the production of a desired type of antibody for immunizing the animals. The synthesis of polyclonal antibodies is illustrated in Fig. (3A). A solution of carbohydrate containing antigen is mixed with an equal volume of Freunds complete adjuvant and is injected interdermally at multi-sites on the back of a rabbit. The process is repeated weekly for 15 weeks. Blood samples are drawn weekly after the 4th injection and serum is prepared by a standard method. Agar double diffusion tests performed as described below showed that the serum contained antibodies, Fig. (**3B**). The antibodies and antigen yielded a precipitin band. With longer periods of immunization it has been found that some antigens possess more than one determinant and two or more types of antibodies are produced, Fig. (**3C**).

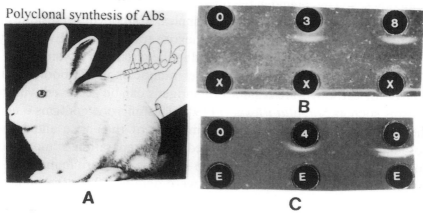

Fig. (3). Immunization of a rabbit for synthesis of polyclonal antibodies and agar diffusion of sera from rabbits immunized with different antigens (xanthan and erythropoietin).

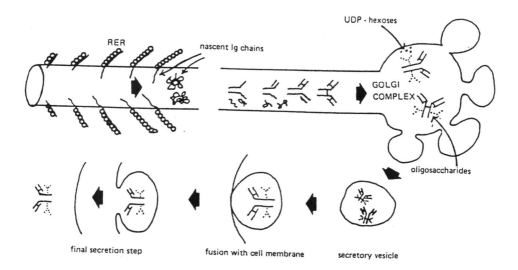

Fig. (4). Diagram of reactions for *in vivo* synthesis and secretion of antibodies; scheme for cleavage of leader, chain assembly, intracellular transport, and secretion [18].

The series of reactions and events for the *in vivo* synthesis of antibodies occurring in the plasma cells [18] are shown in Fig. (**4**). The polysome with one messenger RNA is attached to endoplasmic reticulum. Polypeptides are synthesized as dictated by the genetic information for synthesis of light or heavy chains. These chains are moved along the reticulum and aligned into the antibody shape by folding in the proper conformation. Bonds are formed to hold the chains together in the golgi complex. Carbohydrate residues from sugar nucleotide are added to chains by hexose transferases. Eventually the completed antibody molecule is secreted in the circulatory system.

The synthesis of anti-carbohydrate monoclonal antibodies having specificity for carbohydrate residues of antigens can be achieved with many organisms such as *Shigella Flexneri* [19]. This microbe contains 0 antigen in the cell wall and the antigen is composed of rhamnose and N-acetyl glucosamine [20]. The series of steps used in the preparation of monoclonals is illustrated in Fig. (**5**). In the first step BALD/c mice are immunized with nonviable bacterial cells. Spleen cells were obtained from the immunized mice and fused with Sp2/0 plasmycytoma cells (Institute for Medical Research, Camden, NJ) with the propylene glycol technique. Hybrids were selected by an ELISA assay for 0-antigen positive clones. These were selected and then grown in amounts needed for immunizing mice to obtain ascite fluid. The fluid contains the monoclonal antibodies. A photograph of the agar diffusion plate for the monoclonals from *Shigella Flexneri* is shown in the inset of Fig. (**5**).

A diagram of the general structure of an antibody molecule is reproduced in Fig. (**6**) [21].

Of the procedures for detecting antibodies, the double diffusion method is reliable [22]. The procedure is performed with agarose on microscope slides. The slides are first covered with hot agarose 0.5% solution in 0.2M phosphate buffer of pH 7.2. When the agarose solidifies wells are cut in the desired arrangement. The antigen and the serum or purified antibodies are placed in adjacent wells and placed in a moist chamber. Agar diffusion is allowed to proceed until development of immunobands between the antigen and antibody, the time varying from 6 to 36 h. Some animals yield one type of antibody with an antigen, others yield two or more types with the same antigen. The bands on the plates were photographed, Fig. (**3**) for a permanent record and for comparisons with results obtained in other tests.

Monoclonal Synthesis of Abs

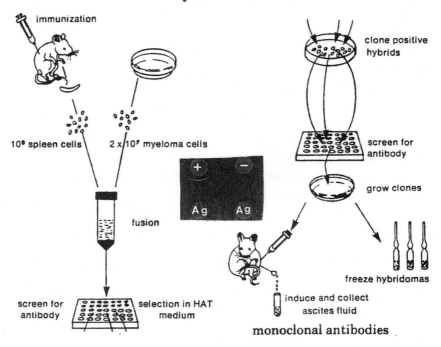

Fig. (5). Immunization of mouse and a method of preparation of monoclonal antibodies and agar diffusion of positive and negative clones with *S. flexneri* antigen.

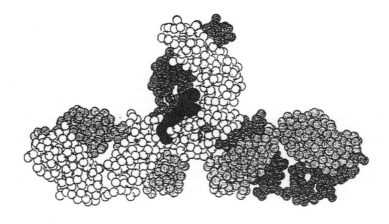

Fig. (6). A possible structural arrangement of antibody. Space-filling view of the molecule. One complete heavy chain is in white and the other is dark gray; the two light chains are lightly shaded. The large black spheres represent the individual hexose units of the complex carbohydrate [21].

Preparation of Antigens

Polysaccharides

The cell walls of microbes contain antigenic polysaccharides. The preparation of cell walls was from freshly grown organisms and is described [23]. The polysaccharides were extracted from the cells with a solution of 0.2 N HCl and 0.2 N KCl [24]. The extract was subjected to dialysis to remove inorganic salts and with chloroform-butyl alcohol (4:1) to remove the protein by centrifugation. The filtrate was subjected to chromatography on P-60 acrylamide and the polysaccharides eluted with phosphate buffer pH 7.0. The polysaccharide was precipitated with ethyl alcohol (5 vol) from the eluate and collected [24].

Glycoproteins

The preparation methods for the glycoprotein antigens were varied. Myeloma protein was obtained from ascite fluid [25]. The glucoamylase was from A niger [26], the erythropoietin was from recombinant hampster cells [27] and tumor glycoprotein was from colon tumor tissue [28].

Glycoconjugates and Immunoadsorbents

The reaction used for coupling p-aminophenyl α or β glycosides and bovine serum albumin was the carbodiimide reaction [29]. A sample of 40 mg of 2-p-aminophenyl glycoside was dissolved in 2 ml of water and added to a 10% BSA solution. After adjusting the pH to 4.5, 0.2 g of CMC was added and the mixture was stirred at room temperature for 16 h in which time coupling of the reactants occurred. The mixture was then dialysed for 24 h and taken to dryness by lyophilization. Appropriate amounts of the products were used for immunizations. The immunoadsorbents with carbohydrate ligands were synthesized from CNBr activated Sepharose or from AH-Sepharose [30] and carboxy compound using the carbodiimide reaction for the latter. In a typical experiment samples of 4 g of CNBr activated Sepharose-4B and 0.05 g of the p-aminophenyl glycoside were used. The product was placed in a column (1 cm X 20 cm) and equilibrated with 0.02 M phosphate buffer of pH 7 and used for affinity chromatography. AH-Sepharose which has a free amino group at one end of 6-carbon spacer for coupling carbohydrate ligands with a free carboxy group is used to synthesize this adsorbent by the carbidiimide method.

PURIFICATION AND CRITERIA OF PURITY

Affinity Chromatography for Isolating Antibodies

Affinity chromatography was used for purifying anti-lactose and other anti-carbohydrate antibodies [31]. An affinity is performed on columns of insoluble adsorbents consisting of a support to which are attached ligands which bind the antibody being purified. The method requires the preparation of a suitable adsorbent bearing ligands that form a complex with the substance being isolated. An eluting system which releases the desired substance from the ligand needs to be available or readily prepared. The formation of the complex should be specific and reversible, with the latter being necessary for the release of the antibody at a later stage. The serum containing the antibody is prepared by immunization of selected animals with the antigen. The serum obtained is passed through the column of adsorbent containing ligands specific for the antibody with a buffer solution of pH 7. The compounds in the serum which do not bind to the ligand will pass unretarded through the adsorbent, and those which bind to the ligand are immobilized on the support. The antibody is released later from the complex by a solution containing an agent with a structure complementary to the active site of the antibody or by a nonspecific chaotropic agent. The eluate from the column was monitored continuously for UV absorbing compounds at 280 nm. A typical elution pattern for the anti-lactose antibodies is shown in Fig. (7). The eluate which was obtained with the ligand was collected and

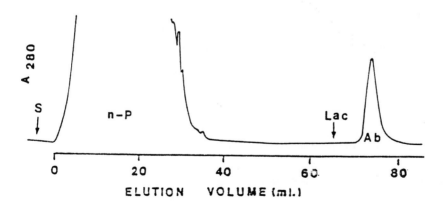

Fig. (7). Affinity chromatography pattern for isolation of anti-lactose antibodies. n-p, non-proteins; Ab , antibody.

mixed with an equal volume of saturated ammonium sulfate. The mixture was maintained at 4° overnight. The precipitate which formed was collected by centrifugation and dissolved in 0.2 ml of phosphate buffer and saline of pH 7.2. Antibodies from a number of such preparations of single antibodies were combined and the resulting solution was used for the antibody tests.

Electrophoresis

Polyacrylamide gel electrophoresis is conducted utilizing a published procedure [32]. Samples of approximately 50 μg of the antibody solution were subjected to electrophoresis. The buffer solution was of pH 8.3 and consisted of 0.005 M Tris and 0.04 M glycine. Electrophoresis was conducted at a constant current of 2.5 ma/gel for periods of 4 to 6 hrs. The finished gels were stained with 0.02% Coomassie Blue G-250 to reveal the protein components. The results for immune serum and the purified anti-glucose antibodies were photographed, Fig. (**8A**). The non-antibody protein components in the serum have been removed by the affinity method.

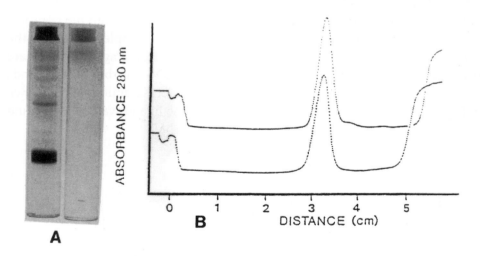

Fig. (8). Gel electrophoresis patterns of immune serum and affinity purified antibodies [A]. Density gradient ultracentrifugation (B), top pattern glucose oxidase, lower pattern mesquite antibody.

Density Gradient Ultracentrifugation

Another method for determining homogeneity and molecular size of antibodies is density gradient ultracentrifugation [33]. A sample of anti-mesquite antibodies was centrifuged in a SW-65 rotor at 65,000 RPM for 16 h in a Beckman L-75 ultracentrifuge. At the end of this time the tubes were removed and the gradient solutions were fractionated by means of an ISCO Density Gradient Fractionator (ISCO Inc., Lincoln, NE). The UV absorbance of the solution from the density gradient columns was measured continuously at 280 nm during fractionation. The sedimentation patterns are reproduced in Fig. (8B). The sedimentation data for the anti-gum mesquite antibodies and glucose oxidase were used to calculate the molecular weight of the antibodies using the relationship (34):$(D_1/D_2)=(M_1/M_2)^{\frac{2}{3}}$ [33] D_1 and M_1 are distance and molecular weight for the unknown and D_2 and M_2 for the standard. Glucose oxidase was the standard and is of molecular weight of 150,000 as determined by ultracentrifugation in a Spinco centrifuge [15]. The antibodies are of the same molecular weight.

Inhibition

The tests for antibody activity in the presence of inhibitors were performed in capillary tubes following the method of McCarty and Lancefield [35]. Quantitative data for levels of antigen-antibody complex at different concentrations of potential inhibitors were obtained [36] and a plot of the results is shown in Fig. (9A).

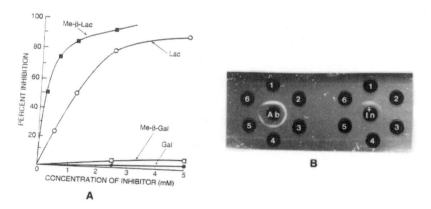

Fig. (9). Inhibition of anti-lactose antibodies by increasing amounts of methyl β-lactoside and lactose (A) and by the agar diffusion method (B).

The specificity of the antibodies can be verified by inhibition tests using agar diffusion data [35]. The results of such experiments with anti-lactose antibodies are shown in Fig. (**9B**). A sample of the antibody is first incubated with the potential inhibitor for 2 h at room temperature. Appropriate amounts of the digest and the pure antibody sample are placed in the separate center wells of an agar plate. Decreasing amounts of antigen ($20 \rightarrow 1$ μg) are placed in outer wells (1 to 6). Agar diffusion is used to detect the formation and the number and intensity of precipitin bands. A comparison of the amount of antigen required to give the same amount of precipitin as the native antibody and the antibody treated with inhibitor permits calculation of the percentage of inhibition.

Specificity

To determine whether the carbohydrate residues or other moieties of the antigen activate plasma cells to synthesize an antibody, a series of agar diffusion tests with lactosyl conjugates was conducted with anti-lactose antibodies. These antigens were tested also with the anti-BSA antibodies, Fig. (**10B**). Fig. (**10A**) shows that all conjugates yielded a precipitin with anti-lactose antibodies but Fig. (**10B**) shows that anti-BSA antibody reacts only with BSA and those conjugates containing a BSA moiety. These results show that the carbohydrate moiety is the immunodeterminant of antigen and specifies the type of antibodies that will be synthesized.

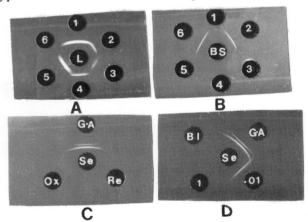

Fig. (10). Agar diffusion of different proteins and a lipid conjugated with lactose against anti-lactose antibodies (well L) and against anti-BSA antibodies (Well BS); wells 1-6 contained Lac-poly, Lac-BSA, Lac-sphingosine, Lac-ORA, Lac-HGG and BSA; Chemical modification of the antigen by periodate oxidation or borohydride reduction can effect an agar diffusion against anti-gum arabic antibodies; (Se), GA=gum arabic BI=blank.

Chemical Modification of the Antigens

Chemical modifications of the antigens were achieved by oxidation of the monosaccharide residues with periodate and reduction of the carboxyl groups of the uronic acid residues by the carbodiimide and borohydride methods, Fig. (10C). The periodate oxidation was performed by a procedure described in the literature [37]. The reduction of the uronic acid residues of the antigen was performed by the carbodiimide (CMC) and sodium borohydride method [38]. The oxidized and reduced types of antigens no longer reacted with the antibodies, Fig. (10C).

Acid hydrolysis of gum arabic was effected at several concentrations of mineral acid and for different time periods. One hydrolysis was conducted in dilute acid (0.01 N HCl) for short periods (10 min). Aliquots of the hydrolysate were removed at various intervals, neutralized with barium carbonate, and analyzed for reducing sugars by paper chromatography in the solvent system of n-butyl alcohol:pyridine:water (6:4:3 by volume). A portion of each hydrolysate was also used in agar diffusion tests to determine the effects of acid hydrolysis on the antigenicity of the sample antigen. Other samples of the antigen were then subjected to hydrolysis in more concentrated acid for longer periods (1.0 N HCl for 2 h). The products in both hydrolysates were subjected to agar diffusion against homologous antiserum, Fig. (10D). It is noted that hydrolysis in dilute acid destroyed antigenicity for one type of anti-gum arabic antibody but concentrated acid destroyed reactivity for both types of antibodies

Isoelectric Focusing

Gel isoelectric focusing of the antiserum and antibody preparation was performed in 10% polyacrylamide gels in a 2% ampholine-sucrose solution of pH gradient 5-8. The procedure employed was essentially that described in the literature [39]. The analysis was performed on anti-lactose preparation from three different rabbits immunized in an identical fashion. The finished gels were stained for proteins by the Coomassie stain. The results are shown in Fig. (11). Animals of the same breed and immunized in an identical fashion with the same antigen synthesize 13, 16 and 21 different types of isoforms (lactose antibodies) which are specific for the same immunodeterminant of the antigen. It is remarkable that different molecular entities are produced reacting with the same antigenic determinant in the immune response. This no doubt is due to different genome constitutions.

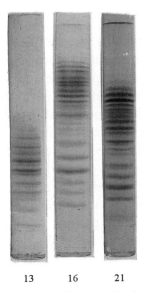

13   16   21

Fig. (11). Isoelectric focusing pattern for anti-lactose antibodies composed of 13, 16 and 21 Isoforms.

Liquid isoelectric focusing [40] was performed using an LKB model 8100 (110 ml) electrofocusing column. All procedures were performed according to the manufacturer, LKB, 12221 Parklawn Dr., Rockville, MD. A sample of 10 mgs of anti-lactose antibodies to be focused was first dialyzed at 4°C for 48 hrs against a 1% glycerol solution. A stabilizing gradient of sucrose 5% to 50% containing 2% w/v carrier ampholytes with a pH gradient of 6.0 to 8.0 was employed.

At the start of electrofocusing the maximum power did not exceed 1.0 watt and was gradually increased to 2.0 watts during the initial 24 hrs. After focusing for 65 hrs, the power was stopped and the column emptied. One ml fractions were collected, maintaining the fractions at 4°C. The pH of each fraction was measured using a combination electrode and a Corning pH meter. The isoelectric points were calculated at the peak of each fraction and are recorded in Table 1. The 280 nm absorbance of the eluate was measured using a Beckman Du-2 spectrophotometer. The pattern is shown in a later section.

Table 1.   **Isoelectric Points for Anti-Lactose Isoforms**

Ab	1	2	3	4	5	6	7	8	9	10	11	12	13
pI	8.1	7.9	7.8	7.4	7.2	6.8	6.6	6.5	6.4	6.2	6.1	5.9	5.8

TYPES OF ANTI-CARBOHYDRATE ANTIBODIES

General Comments

The carbohydrate residues of polysaccharide antigens of the cell walls of bacteria which activate the immune system are the immunodeterminants of antigens. It has been found that the carbohydrate residues of glycoconjugates and glycoproteins are also immunodeterminants. For example the serum of animals immunized with the conjugate glucose-BSA contained antibodies which reacted with the glucose of the glycoconjugate. The serum of rabbits immunized with a glycoprotein glucoamylase contained antibodies which reacted with the mannose residues of glucoamylase. With the development of adsorbents with carbohydrate ligands it became possible to purify anti-carbohydrate antibodies by affinity chromatography. The method has been used to purify eighteen such antibodies with specificity for monosaccharide and oligosaccharide residues. This method of affinity chromatography was developed for purifying enzymes [7] and then it was used for purifying anti-polysaccharide antibodies [41, 42].

A number of species of the Streptococci bacteria cause major human diseases. Important studies on the identification and classification of the strains and groups of Streptococci are due to Lancefield [43], Heidelberger [44] and collaborators. The classification system is based on the antigenic carbohydrates. A scheme for pathogenicity of the groups of Streptococci is recorded in Table 2. Such information is useful for identification of the organism causing the disease.

Table 2. **Pathogenicity of Serological Groups of** *Streptococci*

Animal	Pathogenic	Occasionally Pathogenic	Non-Pathogenic
Man	A	B, C, D, F, G	K, L
Cattle	B, C	A, G	D, E, H, L
Horse	C		
Monkey		A, G	C
Dog	G, L, M		C
Rabbit	C	A, B	

Specificity for Residues of Polysaccharides

N-acetyl Glucosamine, Rhamnose, Lactose, Gum Arabic, Gum Xanthan, Gum Guar, Gum Mesquite.

The polysaccharide of *Streptococcus pyogenes* consists of rhamnose and N-acetyl glucosamine residues. Antibodies have been isolated from *S. pyogenes* by affinity chromatography by the technique described in an earlier section. The immune serum of *S. pyogenes* was provided by Dr. McCarty, Rockefeller University. The affinity pattern for the isolation of the antibodies is reproduced in Fig. (12) [45].

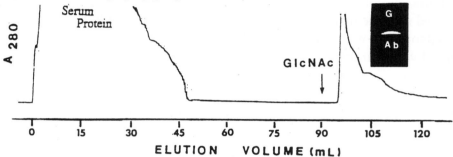

Fig. (12). Purification of anti-group A streptococcal antibodies by affinity chromatography; inset is agar-diffusion pattern.

The adsorbent used for the isolation of the antibodies was synthesized from CNBr-activated Sepharose and *p*-aminophenyl 2-acetamido-2-deoxy-α-D-glucopyranoside by the method described previously. The UV-absorbing material which eluted with *N*-acetylglucosamine was collected and concentrated. Agar diffusion showed that the sample yielded a precipitin band with the group A polysaccharide from the cell wall of *S. pyogenes* (inset of Fig. (12). The constituent units of this polysaccharide are *N*-acetylglucosamine (GlcNAc) and rhamnose (Rha) [46]. Inhibitions with these monosaccharides and the antibodies are shown in Fig. (13A).

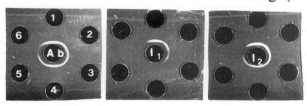

Fig. (13A). Inhibition patterns for anti-S. pyogenes antibodies. Ab=antibody, I₁ =Ab + GlcNAc, I₂=Ab + Rha. Wells 106 and other outside wells contain decreasing amounts (20-1 µg) of antigen (group A polysaccharide).

On isoelectrofocusing, the antibody preparation was shown to consist of 10 isoantibodies, and each member possessed anti-GlcNAc activity. The latter was shown by the results of coupling electrofocusing with agar diffusion. In this procedure, duplicate samples of antibody preparation are subjected to identical isoelectrofocusing. One finished gel is stained for protein with Coomassie Blue and the other gel is embedded in fluid agar. When the agar solidifies, a trough is cut in the agar about 2 cm from the gel and a 2% solution of the antigen is placed into the trough. Diffusion is allowed to proceed for 12 to 48 h and the precipitin band which formed was observed opposite each multiform of the antibody.

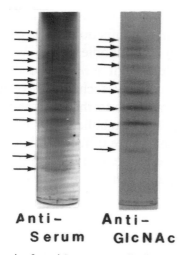

Anti– A nti–
S erum GlcNAc

Fig. (13B). Electrofocusing for anti-S. pyogenes antibodies and serum.

Anti-carbohydrate antibodies have also been isolated by affinity chromatography of immune sera obtained from rabbits immunized with vaccines of *Streptococcus pneumoniae* type 32, or *Streptococcus mutans*, strain KI-R, Fig. (**14A**) [47]. The antibodies of *S. penumoniae* were directed at a polysaccharide of L-rhamnose and D-glucose in the cell wall and antibodies against *S. mutans* against a capsular polysaccharide of L-rhamnose D-galactose and D-glucose. Both types of antibodies were specific for L-rhamnose units of the antigens but were immunologically distinct, Fig. (**14B and C**). In spite of modern antimicrobial agents, *Streptococcus pneumoniae*, remains a leading cause of mortality in persons of all ages. It is the cause of pneumonia but also causes meningitis and septicema.

538

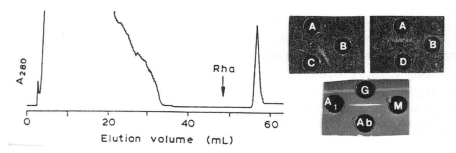

Fig. (14).   Affinity pattern , agar diffusions (C and D), A=*S. pneumoniae*, B=*S. mutans,* and immunoglobulin class (Ab) of anti-*S. mutans* or anti-*S. pneumoniae* antibodies (Ab), A$_1$, G, M, globulins..

Streptococcus mutans is the organism that is the cause of dental caries which constitutes the majority of dental infections. The organism causes a plaque that forms on the teeth.   Antibodies have been prepared by immunizing rabbits with non-viable cells of *S. mutans* and are anti-dextran antibodies. The organism synthesizes a polysaccharide dextran from glucose in food [48]. The polysaccharide adheres to the teeth and forms a plaque where other organisms can adhere. The metabolic cycles of the bacteria lead to formation of acids which etch the surface of teeth and dental caries form. The antibodies can be used to identify the *S. mutans* in the oral cavity and preventative methods can be instigated. The antibodies are of the IgG class of immunoglobulins, Fig. **(14D)**.

The cell wall of *Streptococcus faecalis* strain N contains a polysaccharide of glucose and galactose, which has been isolated and characterized [49]. The polymer has many lactose units as side chains. The lactose chains are the immunodeterminant of this antigen which activates the immune system to synthesize anti-lactose antibodies. Fig. **(15A)** shows the rate of synthesis of the antibodies on successive immunization. The anti-lactose antibodies have been used most extensively in studies relating to structural aspects and biological properties.   The methods used to characterize the anti-lactose antibodies are applicable to the other anti-carbohydrate antibodies which are discussed in this article. The antibodies were prepared from the cell wall polysaccharide by the method described in an earlier section, Fig. (7). Agar diffusion tests showed that the serum contained antibodies which formed a precipitin complex with the polysaccharide. The antibodies were isolated from serum by affinity chromatography on the Sepharose with lactose ligands. The affinity pattern is reproduced in Fig. (7). Several lines of evidence establish the antibodies are specific for the lactose units of the antigen. First the adsorption on lactosyl adsorbent and elution with lactose, second oxidized polysaccharide with periodate no longer gave a precipitin with the anti-lactose antibody [37]. It is well known that periodate oxidizes carbohydrate residues. Third, oxidation of terminal galactose residues by galactose oxidase [50, 51] and the product no longer yielded a precipitin with the antibody. Fourth, lactose inhibits the formation of the precipitin product between the antibody and polysaccharide. Fifth, a precipitin formed with four different lactose conjugates showing the lactosyl unit of the conjugate reacted with the antibody and not BSA the other unit of the conjugate, Fig. **(10)**.

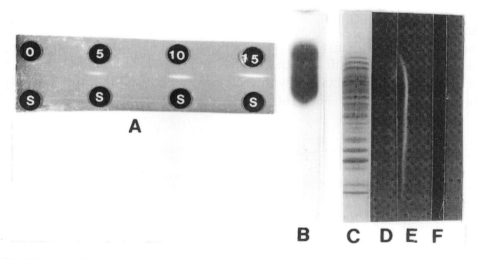

Fig. (15).   Agar diffusion (A), electrophoresis (B) and isoelectric focusing (C-F) of anti-lactose antibodies.

Gel electrophoresis yielded a broad band and indicated several protein components, Fig. **(15B)**. It was necessary to establish whether the preparation contained several types of antibodies or only anti-lactose antibodies. The preparation was subjected to electrofocusing [39]. Duplicate gels were made by electrofocusing. One gel was stained for protein and showed the presence of multiproteins, Fig. **(15C)**. The unstained gel was examined by the coupled method and yielded a precipitin which was opposite each band in the stained gel, Fig. **(15D-F)**. These results showed that the purified antibody preparation consisted of 13 protein species each with antibody activity directed at lactose-containing antigen.

An LKB liquid isoelectrofocusing apparatus was used to separate isoforms of antibodies [52]. The anti-lactose antibodies have been separated into individual components by this procedure [53]. A sucrose-stabilized pH 6-8 gradient of ampholine was used as the liquid column. The gradient solution (110 ml) was introduced into the apparatus, and the unit was prefocused for 24 hr. A sample of 10 mg of the anti-lactose antibodies was then introduced into the gradient column. Further isoelectrofocusing was performed at 800 V for 65 hr. At the end of this time, the gradient was fractionated in 1 ml fractions, and the UV absorbance of eluate and of each fraction was measured at 280 nm. A plot of the data is presented in Fig. (16A) and this shows that 13 individual components are present in the anti-lactose antibodies. Gel electrophoresis was performed on a sample of each peak fraction. The stained gel shows that each sample yielded a positive test with the protein reagent, Fig. (16B). The peak samples of fractions 1, 6, and 10 were used in agar diffusion tests with the polysaccharide antigen and a precipitin band was obtained with each sample verifying that the fractions have antibody activity, Fig. (16C).

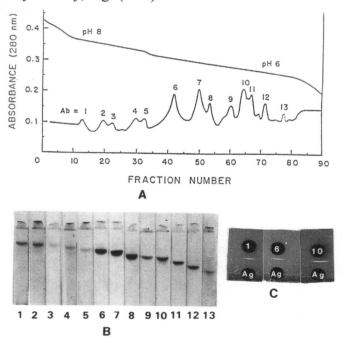

Fig. (16). Liquid electrofocused pattern (A) electrophorised in gels stained by protein reagent (B) and agar diffusion (C) of anti-lactose antibodies.

The method for preparing the light and heavy chains from immunoglobulins has been published [54]. The method was used for preparing the chains of anti-lactose antibodies. A sample of 20 mg anti-lactose antibodies was dissolved in 1 ml 0.02 M phosphate buffer and dialyzed for 48 hr and used for the analysis. The antibodies in the solution were treated with an equal volume of 0.01 M solution of dithiothreitol in buffer of pH 8.2 for 2 hr at room temperature to separate chains. Next, the reduced antibodies were treated with 5 ml of 0.12 M iodoacetamide in the same buffer for 15 min. During the reduction by dithiothreitol and alkylation by iodoacetamide, the pH of the solution was maintained at 8.2 by addition of M Tris buffer. Finally, the reduced and alkylated antibodies were dialyzed against M propionic acid at 4°C overnight and then fractionated by chromatography on Sephadex G-100 using 0.5 M propionic acid as the eluting solvent. Forty fractions of 5 ml were collected from the column and the UV-absorbing fractions were located by absorbance measurements. Three peaks of UV-absorbing material were obtained corresponding to fractions 12-15, fractions 17-21, and fractions 24-31. Each group of fractions was combined separately and dialyzed against distilled water for 24 h. The fractions were taken to dryness by lyophilization. The amount of heavy chains (fractions 17-21) obtained was 8 mg and of light chains (fractions 24-31) the amount was 5 mg. The third fraction was undissociated

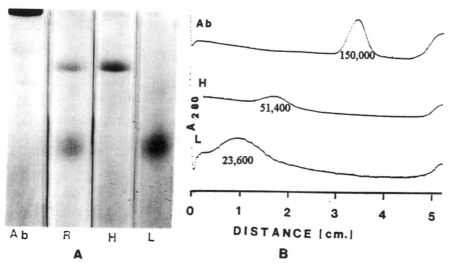

Fig. (17). Gel electrophoresis (A) and sedimentation on density gradient ultra centrifugation (B) of anti-lactose antibodies and dissociated antibodies. Antibody (Ab), reaction mixture (R), heavy chain (H), light chain (L).

antibody. The native antibody, the reaction mixture and the light and heavy chains were subjected to gel electrophoresis, Fig. (**17A**). It is noted that the light and the heavy chains each yielded only one band which stained with the protein reagent. Fig. (**17B**) shows that the native antibody and chains sedimented to yield single UV absorption bands at 280 nm. The electrophoresis migration rates and the molecular weights are recorded in the figure. However, electrofocusing yielded the results in Fig. (**18**). The preparation of native antibodies and the light and heavy chains each yielded 13 protein bands [53].

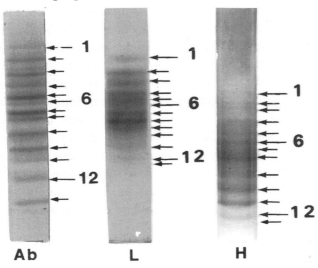

Fig. (18). Isoelectric focusing of anti-lactose antibodies (Ab) and the Heavy (H) and the Light (L) chains.

In the present studies, it has been shown that the synthesis of a set of isoantibodies occurs by a mechanism in which one type of light chain always combines with a complementary heavy chain. It is suggested that the synthesis of a set occurs in a group of plasma cells, all of which possess the same type of receptor substance on the cell surface. These cells are therefore stimulated by the same structural group of the antigen and apparently produce only one type of light chain and one type of heavy chain. These chains are assembled by a single-chain pairing mechanism into antibodies of identical light- and heavy-chain composition. Since synthesis of the anti-lactose antibodies occurs in different cells activated by the same determinant group of the antigen, a set of antibodies with structural differences but with identical specificity is produced.

A number of anti-carbohydrate antibodies were prepared with antigens which are polysaccharide gums. The industrial use of polysaccharide gums is in the formulation of processed foods, beverages, adhesives, pharmaceuticals and personal care products. Gum arabic, gum xanthan, gum guar and gum mesquite are used in these products. Antibodies against the gums have been produced by immunizing rabbits with a suspension of the individual gum mixed with Freunds complete adjuvant. The antibodies specific for the individual gums were isolated by affinity chromatography on an adsorbent with gum ligands and elution with ammonium thiocyanate. The results for anti-gum arabic in Fig. (19) show that two types of antibodies are produced on immunization with this gum [55]. The gum is composed of the following monosaccharide residues galactose, rhamnose, glucuronic acid and arabinose but in a complicated array of residues. The gum contains two types of side chains which are the immunodeterminants of the gum and two types of antibody are produced. Mild acid hydrolysis released arabinofuranose. The ability of the modified antigen to form precipitin with the antibody is destroyed for one type of antibody, Fig. (10D). Periodate oxidation and borohydride reduction of the antigen destroys the activity of both determinants, Fig. (10C).

Fig. (19). Agar diffusion of pre-immune (P) and immune serum (Se) against gum arabic (Ar).

A two-column affinity chromatography method, Fig. (20), was devised and used to separate the anti-arabic antibodies into two sets by using two types of adsorbents [56]. For the preparation of one adsorbent a sample of gum arabic was hydrolyzed in 0.01 M HCl for 10 min and the acid hydrolyzed gum was coupled to AH-Sepharose 4B. The second adsorbent was synthesized by coupling native gum arabic to AH-Sepharose 4B. The columns were attached in a series with the hydrolyzed gum arabic ligand-column preceding the native gum arabic ligand-column. In order to decrease

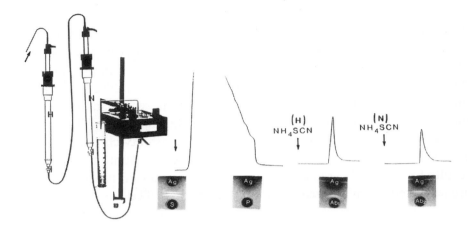

Fig. (20). Two column affinity chromatography of anti-arabic antibodies, two types of antibody. H=hydrolized gum column, N=native gum column. S=serum, P=non-adsorbed protein, Ab₁ and Ab₂=antibodies.

the amount of extraneous protein, the immune serum was treated with an equal amount of saturated $(NH_4)_2SO_4$ for precipitating the globulin fraction. The resulting precipitate was collected by centrifugation and redissolved in phosphate-buffered saline, pH 7.2. The globulin solution from the serum was then applied to the column bearing the hydrolyzed gum arabic ligands and the eluate from this column was passed through the column bearing native gum arabic ligands. After all non-adsorbed proteins had been eluted from the two columns, the columns were separated and each was eluted using 1 M ammonium thiocyanate. During elution the eluates from the columns were monitored for UV absorbance and the UV-absorbing fractions were collected. These fractions were treated with an equal volume of saturated $(NH_4)_2SO_4$ to precipitate the antibodies. The antibodies were recovered by centrifugation and were dissolved in phosphate buffer and saline of pH 7.2. The individual antibody types and the serum were subjected to electrofocusing [56] and the results are shown in Fig. (21).

It is noted in the figure that gel 1 contains four isomeric antibodies. Gel 2 contains nine isomeric antibodies. Gel 3 is a mixture of the two types of antibodies. The results of the coupling method of analysis show that all of the isomers possess antibody activity and two separate precipitin bands were obtained for the two types of antibodies.

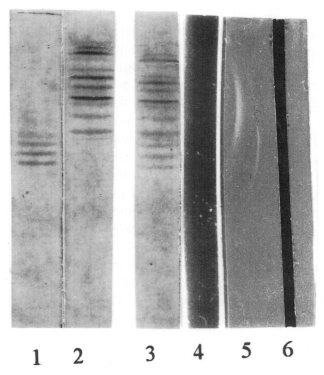

1 2 3 4 5 6

Fig. (21). Electrofocusing of anti-gum arabic antibodies.

Xanthan is an exocellular polysaccharide gum produced by the bacterium *Xanthomonas campestris*. It has been used to produce antibodies by immunization of rabbits. In structure, the polysaccharide is composed of mannose, glucose, 6-acetyl-mannose, 4,6-pyruvate mannose and glucuronic acid [57]. Xanthan has some remarkable rheological properties and can function as thickening, stabilizing and emulsifying agent in aqueous solutions. Anti-xanthan antibodies have been isolated from serum of immunized rabbits with xanthan by chromatography on Sepharose with xanthan ligands, Fig. **(22A)**. The antibodies are unique in that pyruvate is an essential unit of the immunodeterminant, Fig. **(22B)** [58]. Xanthan reduced by borohydride or oxidized by periodate is non-reactive with the antibodies, Fig. **(22C)**.

Guar is a polysaccharide of galacto-mannan type [59]. Guar was used to immunize rabbits to produce anti-carbohydrate antibodies. Inhibition experiments showed that galactose was an inhibitor but the mannose was not. Guar was subjected to periodate oxidation [37] and to oxidation by galactose oxidase [51], and in both cases the oxidized guar did not react with the antibody, Fig. (**23A and B**). Thus, terminal galactose residues of the antigen are the immunodeterminants of guar. Fig. (**24**) shows the anti-guar antibodies react with β-galactose-BSA conjugate. Therefore the galactose residues are linked by β-galactoside linkage to the main mannose chain of guar. Further guar does not react with anti-α-Gal-BSA but reacts with β-gal-BSA showing the glycosidic linkage is β. A report has been published that the linkage is α on the basis of methylation results but this is not substantiated by the antibody and α-galactose binding lectin results [60].

Gum mesquite is obtained principally from the mesquite tree (Prosopis Juliflora) and is a complex acidic polysaccharide. The gum consists of 4-methyl glucuronic acid, D-galactose and L-arabinose [61]. The immune system is activated by 4-methyl-glucuronic acid -(1→6)-galactose. The properties of the antibodies are comparable to that of other gum antigen antibodies. For example, the molecular weight is 150,000 calculated from data of ultracentrifugation, Fig. (**8B**) using the density gradient method and glucose oxidase as the standard [62, 63]. The immunodeterminant is established by the agar diffusion procedure. The gum on periodate oxidation no longer reacts with the purified anti-mesquite antibodies.

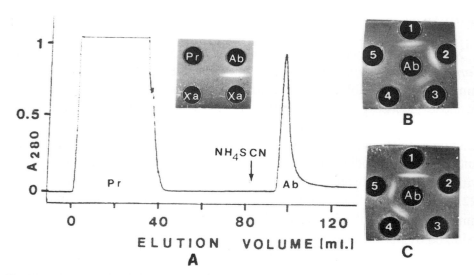

Fig. (22). Some properties of anti-xanthan antibodies, affinity chromatography (A) agar diffusion (inset) diffusion with xanthan of different pyruvate content (6,4,2.8,0.5,0.2%) in wells 1-5 (B) xanthan reference, reduced, oxidized, deacetylated and mannosidase treatment. (C).

Fig. (23). Oxidation of Guar (Gu) by Periodate (POx) and by galactose oxidase (EOx).

Fig. (24). Reaction of guar (Gu), and α-galactose-BSA(α) and β-galactose-BSA (β) with anti-guar antibodies (A), anti β-Galactose Abs (B) and anti α-galactose binding lectin (C).

Specificity for Carbohydrate Residues of Glycoproteins

Erythropoietin, Glucoamylase A, Myeloma protein, Tumor, Monoclonal.

There are many glycoproteins which can activate the synthesis of anti-carbohydrate antibodies. Erythropoietin (EPO) is a hormone with a glycoprotein structure and is immunogenic [64]. It is produced primarily in the kidney and is an essential hormone for initiating and regulating the production of red blood cells. Recombinant erythropoietin (r-EPO) has been produced by the introduction of the human erythropoietin gene into Chinese hamster ovary cells [65, 66]. The molecule is composed of approximately 40% carbohydrate (N-acetyl neuraminic acid, N-acetyl glucosamine, fucose, mannose and galactose) and 60% protein (common amino acids) [67]. There are several types of carbohydrate chains in this glycoprotein.

Antibodies directed against recombinant erythropoietin have been obtained from rabbits immunized with the hormone and Freunds complete adjuvant [64]. Two sets of antibodies are present in the serum of rabbits that have been immunized, Fig. (**3C**). The results of oxidation of the erythropoietin with periodate, Fig. (**25A**), reaction of the antibodies with lectins of known carbohydrate specificity, Fig. (**25B**) and inhibition of the antibodies with the structural monosaccharide residues of the hormone, Fig. (**26**), have established the types of antibodies to be anti-carbohydrate antibodies and the immunodeterminants for these antibodies are oligosaccharides with the following structure: N-acetyl neuraminyl α(2→3)

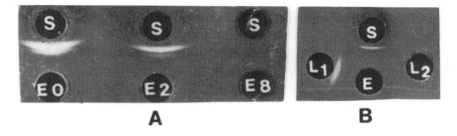

Fig. (25). Periodate oxidation (A) and reaction with lectins (B) of anti-erythropoietin antibodies (S), E0, E2, E8= hrs of oxidation, L1 = Maackia, L2 = Sambucus.

galactose and galactosyl $\beta(1\rightarrow4)$N-acetyl glucosamine (lactosamine) [68]. These antibodies should be of value medically for monitoring the levels of EPO used in the treatment of anemia. The antibodies also should be useful for detecting abuses of the hormone by athletes for enhancing performance in athletic competitions. The EPO is being used to stimulate red blood cell synthesis in several disease situations.

Fig. (26). Inhibition of erythropoietin (E), reaction with serum (S) by N-acetyl-neuraminic acid, (SN), and by lactosamine, (SNL).

Glucoamylase is an enzyme capable of hydrolyzing starch quantitatively to glucose [69] which in turn is enzymatically converted to yield a high fructose syrup [70]. This sweetener is used in processed foods, beverages and confectionaries. Glucoamylase is needed for the first step of this industrial process. The enzyme is produced by many types of fungi but the glucoamylase from *Aspergillus niger* is commonly used. Glucoamylase was found to be a glycoprotein containing the carbohydrates mannose, glucose and galactose `and is antigenic. Since the fungus produced several anti-carbohydrate hydrolytic enzymes it was necessary to purify the glucoamylase.

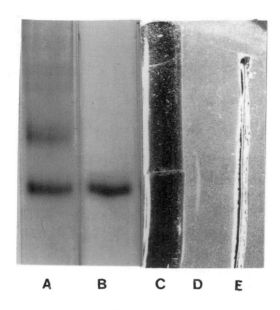

Fig. (27). Gel electrophoresis of glucoamylase in filtrate and purified glucoamylase, enzyme gel unstained and coupled with anti-glucoamylase antibodies.

The purification was by chromatography on DEAE cellulose. Electrophoresis of the purified enzyme and the original filtrate are shown in Fig. (27) [71]. Rabbits were immunized with the pure glucoamylase [72]. Antibodies in the serum were purified by affinity chromatography. The affinity pattern for glucoamylase purification is shown in Fig. (28). Electrofocusing and assay by the coupled method for antibodies is shown in Fig. (29). The glucoamylase antibodies consist of 8 isomers [73].

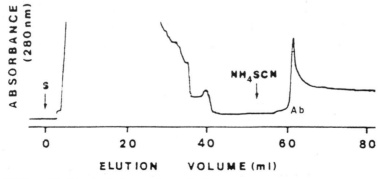

Fig. (28). Affinity purification of anti-glucoamylase antibodies.

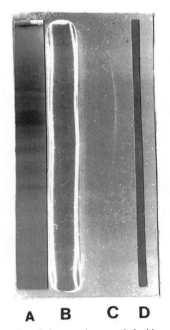

Fig. (29). Isoelectric focusing of glucoamylase coupled with agar diffusion.

Many of the myeloma proteins exhibit antibody-like activity and some react selectively with carbohydrate antigens [74]. Myeloma proteins which are reported to be homogeneous in molecular structure have been used extensively for elucidating the structure of different types of immunoglobulins [11, 12]. A myeloma protein preparation from ascitic fluid

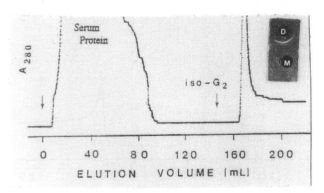

Fig. (30). Affinity chromatography of myeloma protein on isomaltosyl-Sepharose, elution with isomaltose. Inset of agar-diffusion plate.

from BALB/c mice bearing W3129 plasma cell tumors was isolated by affinity chromatography [75]. The protein has been obtained in highly purified form by adsorption on isomaltosyl-Sepharose and elution with isomaltose solution, Fig. (**30**). Results of isoelectrophoresis and agar diffusion on the affinity purified sample revealed that the purified myeloma protein consisted of 6 isomeric proteins, Fig. (**31**).

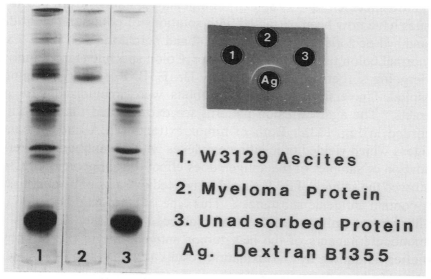

1. W3129 Ascites

2. Myeloma Protein

3. Unadsorbed Protein

Ag. Dextran B1355

Fig. (31). Electrophoresis of multiple myeloma fluid W3129 (1) myeloma protein adsorbed and eluted, (2) and non-adsorbed protein (3) and agar diffusion against dextran B1355 (Ag).

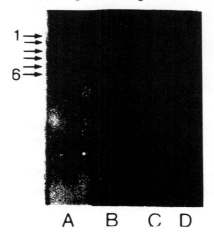

A   B   C   D

Fig. (32). Electric focusing of affinity purified myeloma protein.

By electrofocusing and agar diffusion each isomer possesses anti-dextran activity with the dextran B-1355, Fig. (32). The use of myeloma proteins as homogeneous immunoglobulins does not always yield unambiguous results

A tumor antigen, carcinoembryonic antigen (CEA)was isolated early by extraction of colon tumor tissue with perchloric acid. It was reported to be a marker for colorectal cancer [76]. Subsequent studies did not confirm this absolute specificity [77, 78, 79]. Antigenic substances from colon tumor tissues have now been extracted with solutions of potassium chloride 0.45 M and hydrochloric acid 0.01 M of pH 2 and perchloric acid [80]. Extracts of normal colon tissue were also made. Colon specimens from individuals undergoing surgery were provided by Dr. E. B. Rosenberg (Mount Sinai Hospital, University of Miami). Rabbits were immunized with all the extracts. The agar diffusion patterns for extracts and the antibodies are recorded in Fig. (33). Both of tumor extracts (CEA and E) contained antigens which yielded precipitin complexes with the antibodies. Periodate oxidation of samples of the antigen does effect the antigenicity of CEA but destroyed the activity of E. It should be noticed extracts of normal tissue did not contain antigenic substances. This type of periodate oxidation occurs with lactose-polysaccharide and glucose oxidase antigens. The immunodeterminants of the new tumor antigens are in the carbohydrate moieties as shown by the oxidation results in contrast to CEA which is reported to have determinants in the protein moiety [81].

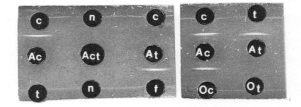

Fig. (33). Agar diffusion plate of colon and normal tissue, n = normal tissue extract, c = CEA perchloric acid extract, t=KCl-HCl extract, At = immune serum against tumor, Ac = immune serum against CEA, Act=mixture of sera.

As revealed by gel electrophoresis and agar diffusion, two new antigenic substances were obtained as the major components of extracts of human colon tumors with the KCl-HCl solution. The coupled electrophoresis-agar diffusion analysis showed that the components reacted with immune serum from rabbits immunized with new antigens. These antigens are present in colon tumor tissue but were not present in normal colon tissue. The

electrophoresis coupled with agar diffusion, was devised to check for the formation of precipitin complexes by the antibody preparations and the antigens. In this procedure, duplicate samples of antibody preparation are subjected to identical electrophoresis. One finished gel is stained for protein with Coomassie Blue and the other gel is embedded in fluid agar. When the agar solidifies, a trough is cut in the agar about 2 cm from the gel and a 2% solution of the antigen is placed into the trough. Diffusion is allowed to proceed for 12 to 48 h and plates which develop precipitin bands are photographed. Results are recorded in Fig. (**34**).

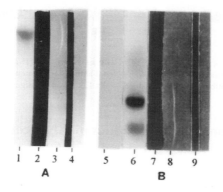

Fig. (34). Coupled electrophoresis - agar diffusion analysis of tumor extracts and immune serum A; Perchloric acid extract (CEA, 1-4); normal extract (5); B,KCl-HCl extract (6-9).

Many types of monoclonal antibodies have been prepared including anti-carbohydrate monoclonals. The coupled inhibition-agar diffusion method was used to determine the nature of carbohydrate residues which inhibited a monoclonal antibody. The monoclonal antibodies were obtained from Dr. Bundle, University Alberta [19]. The scheme in preparing antibody involves fluid from primed BALB/c mice that were injected with appropriate hybridoma cells, Fig. (**4**) . The O-antigen contains 2 types of carbohydrate constituents, namely *N*-acetyl-glucosamine and rhamnose [20]. Both carbohydrates were tested as potential inhibitors of the precipitin reaction [36]. The results of these experiments are shown in Fig. (**35**). It can be seen that the O-antigen yielded strong precipitin bands with the monoclonal antibodies at concentrations of antigen in wells 1, 2 and 3 but did not yield a precipitin band at a concentration of antigen (A) (well 4). Similar results were obtained with antibodies pre-incubated with rhamnose (C). However antibodies incubated with *N*-acetyl-glucosamine (B) did not react as readily with comparable amounts of antigen and no precipitin bands were obtained with antigen in wells 3 and 4. Further bands of reduced intensities were obtained with the antigen (wells 1 and 2). The immunodeterminant of this monoclonal is *N*-acetyl-glucosamine.

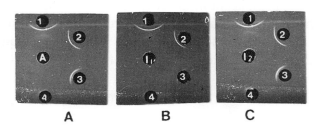

Fig. (35). Agar diffusion patterns of monoclonal antibodies with an O-antigen of N-acetyl-glucosamine and rhamnose from *Shigella* variant Y and potential inhibitors: A, monoclonal antibodies, B, nonoclonal antibodies with N-acetyl-glucosamine, C, monoclonal antibodies with rhamnose, wells 1,2,3, and 4 antigen from *Shigella* at concentrations of 0.25, 0.05, 0.01 and 0.005%.

Specificity for Carbohydrate Residues of Glycoconjugates
α-D-Glucose, β-D-Glucose, α-L-Fucose, β-D-Mannose, β-D-Galactose, *Staphylococcus Aureus* Polysaccharide

Glycoconjugates have been used to immunize rabbits and to obtain anti-carbohydrate antibodies. The conjugates were synthesized from α or β-p-aminophenol glucosides and cyanogen bromide activated Sepharose or AH Sepharose using the carbodiimide method [30, 82]. Conjugates of α and β glucose with BSA were used to immunize different rabbits [83]. Affinity chromatography was used on columns of α or β glucosyl Sepharose adsorbents for isolating the antibodies from the serum of the rabbits immunized with the glycoconjugates. These are shown in Fig. (**36**) [83]. The columns were eluted with α methyl and β glucoside in the order indicated on the patterns. It is noted in the figure that methyl-α-glucoside elutes a UV absorbing compound from α-column but methyl-β-glucoside did not. From the β column the methyl-β-glucoside eluted a UV compound, α-glucoside did not. Agar diffusion of the antibodies against α and β glucose-BSA and α- or β-glucose horse globulin are recorded in Fig. (**37**). Gel electrophoretic patterns show that the antibodies were free of other serum proteins, Fig. (**38A**). The figure also contains electrofocusing results, showing that preparations consist of multi-molecular proteins with the α having 6 members and the β having 9 members, Fig. (**38B**). Inhibition studies show that α antibodies are inhibited with me α-glucoside but not by me β-glucoside. The β antibodies are inhibited by me-β-glucoside and not the α glucoside, Fig. (**39**).

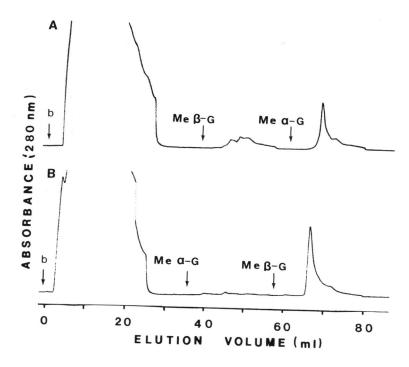

Fig. (**36**). Affinity patterns for purification of α-glucose Abs and β glucose Abs.

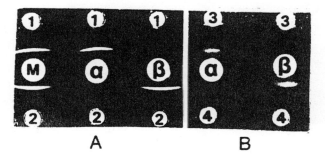

Fig. (**37**). Agar diffusion of anti-α-glucose Abs and β-glucose Abs. M is a mixture of α and B Abs; 1 = α-glucose-BSA; 2 = β-glucose-BSA; 3 = α-glucose horse globulin, 4 = β-glucose horse globulin.

556

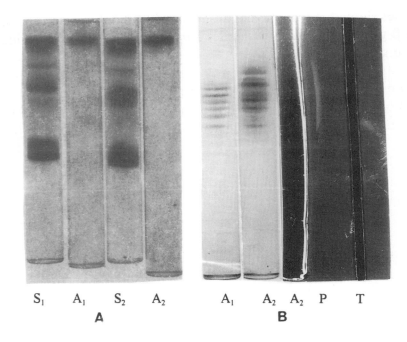

S_1 A_1 S_2 A_2 A_1 A_2 A_2 P   T

A   B

Fig. (38). Electrophoresis (A) of a serum (S_1) and α-Abs (A_1) and β-serum (S_2) and β Abs (B_2). Electrofocusing with agar diffusion (B); P = precipitin; T = trough of antigen..

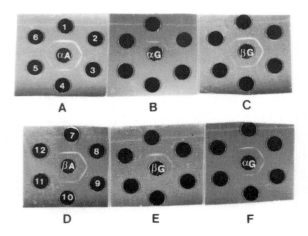

Fig. (39). Inhibition of α and β-glucose Abs by α and β methyl glucosides.

The glycoconjugate α-L-fucose-BSA was used to immunize rabbits. The immunization was performed interdermally at multisites on the back of the neck and repeated weekly for 10 weeks. Blood samples were collected weekly and immune sera samples were prepared. It was found in later experiments that on immunization by this glycoconjugate two types of antibodies were produced. One type was specific only for α-L-fucose and the other for BSA and the conjugate with a BSA moiety, Fig (**40**).

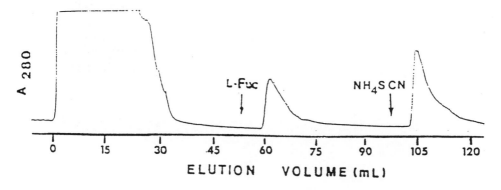

Fig. (40). Two column affinity chromatography of immune serum from rabbits immunized with α-L-fucose-BSA (see the text).

A two column affinity chromatography method was used to separate the antibodies [84]. One adsorbent was α-L-fucose-Sepharose-4B and the other was BSA-Sepharose-4B prepared as outlined in the method in a previous section. The columns were attached in series with the fucose Sepharose column preceeding the BSA-Sepharose column. The immune serum was applied to the first column. The columns were eluted with a solution of α-L-fucose and then with a solution of NH$_4$SCN. The eluates from the columns were monitored for UV absorbance at 280 nM and the fractions with UV absorbing components were collected separately. The affinity pattern is reproduced in Fig. (**40**). Each fraction was mixed with an equal volume of saturated ammonium sulfate. The precipitates were collected by centrifugation and dissolved in phosphate buffer of pH 7. Agar diffusion of the immune serum and the isolated antibodies is shown in Fig. (**41**). It is noted in the figure that the fucose had eluted antibodies that reacted with the glycoconjugate but not with BSA. The NH$_4$SCN eluted the second antibodies that reacted with BSA but also with the conjugate because the latter contained a BSA moiety.

Fig. (41). Agar diffusion pattern; B = BSA, A = immune serum, P = purified anti-fucose Abs, C = purified anti-BSA Abs, F = fucose-BSA.

A micro method has been used previously for measuring inhibition of antigen-antibody complex formation by potential inhibitors. Fig. **(42)** shows

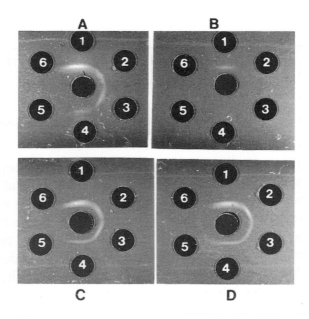

Fig. (42). Inhibition of anti-fucose Abs (A and B) and anti-BSA Abs (C and D) by α-L fucose.

that fucose inhibits the anti-Fuc antibodies but not the anti-BSA antibodies. Gel isoelectrofocusing results show that anti-fucose antibodies are composed of eleven isomers and anti-BSA antibodies of seven different isomers, Fig. **(43)**.

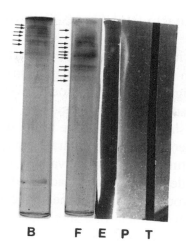

B   F   E   P   T

Fig. (43). Isoelectrofocusing and agar diffusion of anti-Fuc and anti-BSA antibodies.

Anti-α-D-mannose antibodies have been isolated from the serum of rabbits immunized with the glycoconjugate of α-D-mannose and bovine serum albumin [85]. The antibodies were purified by affinity chromatography with adsorption on a mannosyl-Sepharose column and elution with α-D-mannose or methyl α-D-mannoside. Such antibodies should be especially useful for studying the detection of diseases due to the appearance of abnormal glycoproteins that contain mannose polymers.

Antibodies having specificity for the galactose moiety of glycoconjugates have been isolated from the sera of rabbits immunized with vaccine of β-D-galactosyl-bovine serum albumin [82]. The antibodies were isolated by affinity chromatography on adsorbents bearing galactose ligands. That the antibodies are specific for the galactose moiety and not the BSA moiety is shown by a periodate oxidation and agar diffusion experiment, Fig. (44).

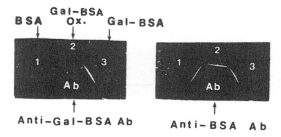

Fig. (44). Anti-galactose-BSA and anti-BSA antibodies.

USES OF ANTI-CARBOHYDRATE ANTIBODIES

Antibodies with specificity for specific carbohydrate residues of antigens have been isolated from serum of rabbits immunized with carbohydrate containing antigens. Eighteen such antibodies were purified by affinity chromatography on adsorbents with ligands of carbohydrate residues. Electrofocusing results showed that all the antibodies occurred in multi-protein forms. A number of anti-carbohydrate antibodies have had useful applications. In the future additional advances with anti-carbohydrate antibodies should be made and these can lead to developments of new and improved products and processes. Some of the applications are summarized:
1.   An early use was development of a serological method based on anti-carbohydrate antibodies which is used to identify bacteria causing infectious diseased in humans. A listing of pathogenic groups of streptococci is presented in Table 2.
2.   It has been possible to produce vaccines to use for immunization against certain bacterial infections and other human diseases. *Staphylococcus aureus* capsular polysaccharide in combination with a carrier protein has been used to prepare monovalent vaccines specific for *S. aureus* [86].   The vaccine was administered to groups of healthy adults and to patients with end-stage renal disease. The antibodies are directed at the polysaccharide moiety of the glycoconjugate.   The data of this study show that conjugate-induced antibodies to *S. aureus* can provide partial protection against *S. aureus* bacteremia, Fig. (45).

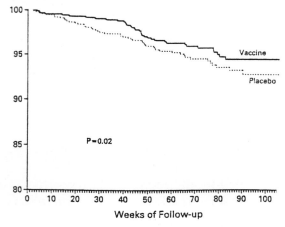

Fig. (45). Survival curves for *Staphlococcus aureus* bacteremia. The P value is for the difference between the two groups at 40 weeks.

3.   An analytical method employing specificity of anti-carbohydrate antibodies has been developed for determination of D and L forms and of α and β glycosidic bonds [87]. All of the center wells of the plates contain the same amount of antibodies. The outer wells contain decreasing amounts of antigen. The antibodies are first incubated with isomers of glucose, fucose or galactose, Fig. (46). The first set of wells on each plate is the standard. One isomer combines with the antibody and reduces the amount of precipitin. The number of precipitin bands in comparison to the number of bands in the standard sample is observed. To test the procedure, galactose constituents of two polysaccharides were determined. Poly 1 is a lactosyl polysaccharide and poly 2 is from flaxseed. On hydrolysis and testing with D-galactose-BSA antibodies it can be seen in the figure that poly 1 constituent is D-galactose and poly 2 constituent is L-galactose, Fig. (46). Anti-glucose antibodies can be used to identify terminal glucosidic linkage shown in the last plate.

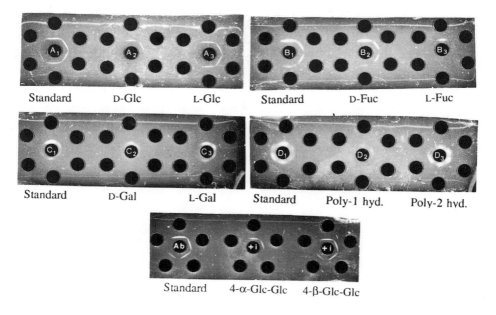

Fig. (46). Analysis of D and L monosaccharides and α and β terminal glycoside linkages in oligo and polysaccharides.

4.   Antibodies specific for a hormone erythropoietin, are being tested as a monitor for hormone therapy of anemia. The antibody can be used to identify the misuse of the hormone by athletes to enhance performance in competitive sports.

562

5. Antibodies that are specific for polysaccharide gums are used for detection and identification of gum additives in processed foods, beverages and confectionaries, Fig. (47). The tests are performed with a control of gum and its complementary antibody with food products and the antibody. The products tested were ice cream, soup, candy and salad dressing. All but the last yielded positive tests with the specific antibodies that were used.

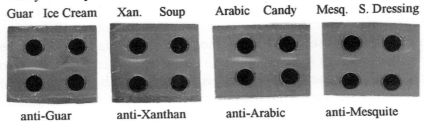

Guar Ice Cream   Xan. Soup   Arabic Candy   Mesq. S. Dressing

anti-Guar   anti-Xanthan   anti-Arabic   anti-Mesquite

Fig. (47). The identification of gum polysaccharides used as additives in processed foods and beverages by use of precipitin reaction with specific antibodies.

6. New antigenic compounds from colon tumor tissue have been isolated and antibodies have been produced. These antigens are not detectable in extracts of normal colon tissue. The antigens may be suitable as markers of colon cancer.

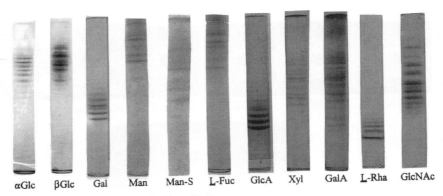

αGlc   βGlc   Gal   Man   Man-S   L-Fuc   GlcA   Xyl   GalA   L-Rha   GlcNAc

Fig. (48). The isoforms of anti-monosaccharide antibodies.

All the anti-carbohydrate antibodies purified by affinity chromatography are isomers of a different number of isoforms. The forms can be separated by electrofocusing and all have been found to have combining activity for the immunodeterminant of the same antigen by the method of coupled analysis of electrofocusing and agar diffusion [88]. The isoforms of anti-monosaccharide antibodies are shown in Fig. (48).

REFERENCES

[1]   Heidelberger, M.; Avery, O.T.; *J. Exptl. Med.*, **1923**, 38, 73-79.

[2]   Lancefield, R.C.; Harvey Lectures, **1940-1941**, Ser. 36, 251-290.

[3]   McCarty, M.; *J. Exptl. Med.*, **1958**, 108, 311-323.

[4]   Krause, R.M.; *Bacteriol. Rev.*, **1963**, 27, 369-380.

[5]   Pazur, J.H.; *Adv. Carbohydr. Chem. Biochem.*, **1998**, 53, 201-261.

[6]   Karakawa, W.W.; Wagner, J.E.; Pazur, J.H.; *J. Immunol.*, **1971**, 107, 554-562.

[7]   Cuatrecasas, P.; Wilchek, M.; Anfinsen, C.B.; *Proc. Nat. Acad. Sci. U.S.A.*, **1968**, 2, 636-642.

[8]   Pazur, J.H.; Miller, K.B.; Dreher, K.L.; Forsberg, L.S.; *Biochem. Biophys. Res. Commun.*, **1976**, 70, 545-550.

[9]   Eisen, H.N.   In *Immunology, An Introduction to Molecular and Cellular Principles of the Immune Responses*, Harper & Row, New York, **1980**, 2nd ed., pp. 338-378.

[10]   Fleischman, J.B.; Porter, R.R.; Press, E.M.; *Biochem. J.*; **1963**, 88, 220-227.

[11]   Edelman, G.M.; *Biochemistry*, **1970**, 9, 3197-3204.

[12]   Kabat, E.A.; Wu, T.T.; *Ann. NY Acad. Sci.*, **1974**, 190, 382-390.

[13]   Porter, R.R.; *Science*, **1973**, 180, 713-716.

[14]   Edelman, G.M.; *Science*, **1973**, 180, 830-839.

[15]   Stellwagen, E.; Schachman, H.K.; *Biochemistry*, **1962**, 1, 1056-1069.

[16]   French, D.; Knapp, D.W.; Pazur, J.H.; *J. Am. Chem. Soc.*, **1950**, 72, 5148-5152.

[17]   Kornfeld, R.; Kellar, J.; Baenziqer, J.; Kornfeld, S.; *J. Biol. Chem.*, **1971**, 246, 3259-3268.

[18]   Uhr, J.W.; *Cellular Immunology*, **1970**, 1, 228-244.

[19]   Carlin, N.I.A.; Gidney, M.A.J.; Lindberg, A.A.; Bundle, D.R.; *J. Immunol.*, **1986**, 137, 2361-2366.

[20]   Kenne, L.; Lindberg, B.; Petersson, K.; Katzenellenbogen, E.; Romanowska, E.; *Eur. J. Biochem.*, **1978**, 91, 279-284.

[21]   Silverton, E.W.; Navia, M.A.; Davies, D.R.; *Proc. Nat. Acad. Sci.*, **1977**, 74, 5142-5148.

[22]   Ouchterlony, O.; *Acta Path. Microbiol.*, **1949**, 26, 507-515.

[23]   Bleiweis, A.S.; Karakawa, W.W.; Krause, R.M.J.; *J. Bacteriol.*, **1964**, 88, 1198-1202.

[24]   Pazur, J.H.; Kane, J.A.; Dropkin, D.J.; Jackman, L.M.; *Archives Biochem. Biophys.*, **1972**, 150, 382-381.

[25]   Potter, M.; *Federation Proceedings*, **1970**, 29, 85-91.

[26]   Pazur, J.H. In *Method of Enzymology*, **1972**, 28, 931-934.

[27]   Takeuchi, M.; Kobata, A.; *Glycobiology*, **1991**, 1, 337-346.

[28]   Pazur, J.H.; *J. Chromatography*, **1995**, 663, 51-57.

[29]   Khorana, H.G.; *Chem. Industry*, **1955**, 1087-1088.

[30]   Pharmacia Affinity Chromatography, Principles & Methods, Laboratory Separation Division, S-751 82, Uppsala, Sweden, 1986-1988.

[31]   Pazur, J.H.; Dreher, K.L.; Forsberg, L.S.; *J. Biol. Chem.*, **1978**, 253, 1832-1837.

564

[32]   Davis, B.J.; *Annals N.Y. Academy Sci.*, **1964**, 121, 404-427.
[33]   Pazur, J.H.; Reed, A.S.; Scrano, R.S.; *Carb. Res.*, **2001**, 336, 195-201.
[34]   Martin, R.G.; Ames, B.N.; *J. Biol. Chem.*, **1961**, 236, 1372-1379.
[35]   McCarty, M.; Lancefield, R.C.; *J. Expt. Med.*, **1955**, 102, 11-28.
[36]   Pazur, J.H.; Kelly, S.A.; *J. Immunol. Methods*, **1984**, 75, 107-116.
[37]   Pazur, J.H.; Cepura, A.; Kane, J.; Hellerquist, C.G.; *J. Biol. Chem.*, **1973**, 248, 279-284.
[38]   Taylor, R.L.; Conrad, H.E.; *Biochemistry*, **1972**, 11, 1383-1388.
[39]   Doerr, P.; Chrambach, A.; *Anal. Biochem.*, **1971**, 42, 96-107.
[40]   Valmet, E.; *Science Tools*, **1968**, 15, 8-11.
[41]   Pazur, J.H.; Miller, K.B., Dreher, K.L. Forsberg, L.S.; *Biochem. Biophys. Res. Commun.*, **1976**, 70, 545-550.
[42]   Pazur, J.H.; Dreher, K,L.; *Biochem. Biophys. Res. Commun.*, **1977**, 74, 818-824.
[43]   Lancefield, R.C.; *Harvey Lectures*, **1941**, 36, 251-290.
[44]   Heidelberger, M. In *Research in Immunochemistry and Immunobiology;* Kwapinski, J.B.G.; Day, E., Eds.; University Park Press: Baltimore London, Tokyo, **1973**; Vol. *3*, pp. 1-40.
[45]   Pazur, J.H.; Erikson, M.S.; Tay, M.E.; Allen, P.Z.; Carbohydr. Res. **1983**, 124, 253-263.
[46]   Coligan, J.E.; Schnute, W.C.; Kindt, T.J.; *J. Immunol.*, **1975**, 114, 1654-1658.
[47]   Gahnberg, L.; Krasse, B.; *Infect. Immun.*, **1981**, 33, 697-703.
[48]   Taubman, M.A.; Smith, D.J.; *Infection and Immunity*, **1974**, 9, 1079-1091.
[49]   Pazur, J.H.; Anderson, J.S.; Karakawa, W.W.; *J. Biol. Chem.*, **1971**, 246, 1793-1798.
[50]   Avigad, G.; Amaral, D.; Asensio, C.; Horecker, B.L.; *J. Biol. Chem.*, **1962**, 237, 2736-2743.
[51]   Pazur, J.H.; Knull, H.R.; Chevalier, G.E.; *J. Carbohydrates, Nucleosides, Nucleotides*, **1977**, 4, 129-146.
[52]   Pazur, J.H.; Dreher, K.L.; Tay, M.E.; Pazur, B.A.; *Biochem. Biophys.*, **1953**, 113, 555-561.
[53]   Pazur, J.H.; Tay, M.E.; Pazur, B.A.; Miskiel, F.J.; *J. Protein Chem.*, **1987**, 6, 387-399.
[54]   Edelman, G.M.; Olins, D.E.; Gally, J.A.; Zinder, N.D.; *Proc. Natl. Acad. Sci., USA*, **1963**, 50, 753-761.
[55]   Miskiel, F.J.; Pazur, J.H.; *Carbohydrate Polymers*, **1991**, 16, 17-35.
[56]   Pazur, J.H.; Miskiel, F.J.; Witham, T.F.; Marchetti, N.; *Carbohydrate Research*, **1991**, 214, 1-10.
[57]   Melton, L.D.; Mindt, L.; Rees, D.A.; Sanderson, G.R.; *Carbohydrate Research*, **1976**, 46, 245-257.
[58]   Pazur, J.H.; Miskiel, F.J.; Marchetti, N.T.; *Carbohydrate Polymers*, **1995**, 27, 85-91.
[59]   Heyne, E.; Whistler, R.L.; *J. Amer. Chem. Soc.*, **1948**, 70, 2249-2252.
[60]   Hayes, C.H.; Goldstein, I.J.; *J. Biol. Chem.*, **1974**, 249, 1904-1914.
[61]   Smith, F.; *J. Amer. Chem. Soc.*, **1951**, 2646-2652.

[61] Smith, F.; *J. Amer. Chem. Soc.*, **1951**, 2646-2652.
[62] Pazur, J.H.; Kleppe, K.; Anderson, J.S.; *Biochim. Biophys. Acta*, **1962**, 65, 369-372.
[63] Pazur, J.H.; Kleppe, K.; *Biochemistry*, **1964**, 3, 578-583.
[64] Pazur, J.H.; Jensen, P.J.; Murray, A.K.; *Pro. Chem.*, **2000**, 19, 629-631.
[65] Jacobs, K. *et al.*; *Nature*, **1985**, 313, 806-810.
[66] Lin, F.-K. *et al.*, *Proc. Natl. Acad. Sci. USA*, **1985**, 82, 7580-7584.
[67] Dordal, M.S.; Wang, F.F.; Goldwasser, E.; *Endocrinology*, **1985**, 116, 2293-2299.
[68] Pazur, J.H.; Murray, A.K.; *Glycoconjugate Journal*, **2001**, 18, 85-86.
[69] Pazur, J.H.; Ando, T.; *J. Biol. Chem.*, **1959**, 234, 1966-1970.
[70] Antrim, R.L.; Colilla, W.; Schnyder, B.J.; *Appl. Biochem. Bioeng.*, **1979**, 2, 97-155.
[71] Pazur, J.H.; Ando, T.; *J. Biol. Chem.*, **1960**, 235, 297-302.
[72] Pazur, J.H.; Forry, K.R.; Tominaga, Y.; Ball, E.M.; *Biochem. Biophys. Res. Commun.*, **1981**, 100, 420-426.
[73] Pazur, J.H.; Liu, B.; Miskiel, F.J.; *Biotech. Appl. Biochem.*, **1990**, 12, 63-78.
[74] Potter, M.; Boyce, C.R.; *Nature (London)*, **1962**, 93, 1086-1087.
[75] Pazur, J.H.; Tay, M.E.; Rovnak, S.E.; Pazur, B.A.; *Immunology Letters*, **1982**, 5, 285-291.
[76] Krupey, J.; Gold, P.; Freedman, S.O.; *J. Expt. Med.*, **1968**, 128, 387-398.
[77] Martin, F.; Martin, M.C.; *Int. J. Cancer*, **1970**, 6, 352-356.
[78] Lo Gerfo, P.; Krupey, J.; Hansen, H.J.; *New Eng. J. Med.*, **1971**, 285, 138-145.
[79] Neville, A.M.; *Tumor Biol.*, **1988**, 9, 61-62.
[80] Pazur, J.H.; *J. Chromatography B.*, **1995**, 663, 51-57.
[81] Hammarstrom, S. *et al.*, *Proc. Nat. Acad. Sci. USA*, **1975**, 72, 1528-1532.
[82] Pazur, J.H.; *Carbohydr. Res.*, **1982**, 107, 243-254.
[83] Pazur, J.H.; Miskiel, F.J.; Marchetti, N.T.; Shiels, H.R.; *Pharm. Pharmacol. Lett.*, **1993**, 2, 232-235.
[84] Pazur, J.H.; Liu, B.; Witham, T.F.; *J. Prot. Chem.*, **1994**, 13, 59-66.
[85] Pazur, J.H.; Liu, B.; Li, N-Q.; *Natural Product Lett*, **1992**, 1, 51-57.
[86] Shinefield, H. *et al.*; *N. Engl. J. Med.*, **2002**, 346, 491-496.
[87] Pazur, J.H.; Reed, A.M.; Li, N-Q.; *Carbohydr. Polymers*, **1994**, 24, 171-175.
[88] Pazur, J.H.; *Anal. Chima Acta*, **1999**, 383, 127-136.

Atta-ur-Rahman (Ed.) *Studies in Natural Products Chemistry, Vol. 29*

PROTEIN AND NON-PROTEIN PROTEASE INHIBITORS FROM PLANTS

GIDEON M. POLYA

Department of Biochemistry, La Trobe University, Bundoora, Melbourne, Victoria 3086, Australia

ABSTRACT: Plants are consumed by bacteria, fungi and animals and utilization of plant proteins requires their hydrolysis, a process which is catalyzed by proteases. Plants defend themselves against other organisms by elaborating physical and chemical defenses, the latter including protein and non-protein inhibitors of proteases. Plant protease inhibitor proteins can be either constitutive or inducible as a result of wounding or pathogen invasion. Proteases are classified into the aspartic proteases, cysteine proteases, metalloproteases and serine proteases on the basis of the involvement of aspartate, cysteine, metal ions and serine, respectively, in the catalytic mechanism. Extracellular proteases are involved in digestion, blood clotting, inflammatory responses to invasion, and extracellular matrix digestion required for angiogenesis and tissue re-modelling. Intracellular proteolysis must be exquisitely regulated to avoid autolysis and intracellular proteases are involved in proprotein processing, lysosome- and proteasome-mediated protein destruction, cell division and apoptosis. A large variety of plant protease inhibitor proteins and peptides have been resolved including aspartic protease inhibitor proteins, cysteine protease inhibitory phytocystatins, metallocarboxypeptidase inhibitor proteins and serine protease inhibitor proteins such as the Bowman-Birk, cereal bifunctional, Kunitz, potato type I, potato type II, mustard family, squash family, serpin and other protease inhibitors. Plant protease inhibitor proteins have potential transgenic applications for crop plant defense. A variety of non-protein protease inhibitors have also been resolved from plants. Plant protease inhibitors have potential for pharmaceutical development especially in relation to Alzheimer's disease, angiogenesis, cancer, inflammatory disease and viral and protozoal infection.

INTRODUCTION

Plants defend themselves against herbivores and pathogenic micro-organisms (notably fungi) by elaborating bioactive secondary metabolites and defensive proteins. Production of such defensive agents may be constitutive or induced by wounding or pathogen infection. Many such non-protein and protein defensive bioactives inhibit plant protein digestion by the attacking organisms. The hydrolysis of peptide bonds in peptides and polypeptides is catalyzed by proteases which are classified into the aspartic, cysteine, metallo- and serine proteases on the basis of the involvement of aspartate, cysteine, divalent metal ions and serine, respectively, in the catalytic mechanism [1-6]. A variety of animal proteins, protease autoinhibitory domains and plant defensive proteins act as inhibitory substrate analogues by binding to the active site. Key residues about the scissile peptide bond contribute to inhibitor specificity and are denoted thus: (N-terminal side)-P2-P1-(peptide bond to be hydrolysed)-P1′-P2′ [1].

The destructive potential of proteases means that their activity has to be tightly regulated to prevent autolysis of protease-producing cells. Thus digestive proteases such as chymotrypsin, pepsin and trypsin are produced as inactive zymogens (proenzymes) and are subsequently activated after secretion. The serine protease-catalysed process of blood clotting involves a cascade of successive proteolytic activations of the blood clotting factor proteases involved [2-6].

Proteases that function intracellularly also require strict regulation. Thus proteolysis of endocytotically ingested proteins is compartmented within lysosomes. Cytosolic protein destruction is achieved via proteasome complexes, the protein substrates being marked for proteolytic destruction by covalent linkage to the protein ubiquitin. The net level of cellular proteins depends upon the balance between such proteolysis and *de novo* synthesis through gene expression. Thus the critical process of cell division is regulated in part by the *de novo* synthesis and proteolytic degradation of cell cycle stage-specific cyclin proteins. The process of apoptosis or programmed cell death involves tight regulation of a cascade of cysteine proteases called caspases. Finally, proteins are typically made as proproteins and specific proteases are required to process newly-synthesized proproteins [2-6].

Proteases are regulated *in vivo* by autoinhibition (as zymogens), proteolytic activation, turnover and by endogenous protease inhibitor proteins. The following brief sketch of protease complexity is accompanied by succinct reference to the physiological context and hence the potential pharmacological relevance of plant-derived protease inhibitors to be described later in this review.

Aspartic proteases include the gastric proteases pepsin A, pepsin B, gastricsin and chymosin that are secreted as inactive zymogens and are activated through removal of autoinhibitory domains in the low pH conditions of the stomach [7-9]. Oesophagitis is caused by undue exposure of the oesophagus to gastric aspartic proteases [10]. Overexpression of the aspartic protease cathepsin D in breast cancer cells is linked to increased metastasis [11]. β-Secretase is a transmembrane aspartic protease involved in the generation of the β-amyloid involved in the amyloid deposition in Alzheimer's disease [12]. The aspartic protease renin cleaves angiotensinogen to yield angiotensin I which is subsequently converted to the vasoconstrictive hormone angiotensin II by the metalloprotease angiotensin I converting enzyme (ACE) [13]. Human immunodeficiency virus 1 (HIV-1) protease is an aspartic protease required for replication of the acquired immunodeficiency syndrome (AIDS)-causing HIV-1 virus and has accordingly become a major pharmacological target for anti-HIV-1 drugs [8, 14]. Major haemoglobinases of *Plasmodium falciparum* include aspartic proteases which represent major potential targets for anti-malarial therapy [15, 16].

Cysteine proteases from plants include papain and the papain-like proteases actinidin, aleurain, bromelain, caricain, chymopapain and ficin. Animal papain-like cysteine proteases include the lysosomal cathepsins B, C, H, K, L and S. Cathepsin C is a multimeric dipeptidyl aminopeptidase but the other papain-related proteases are monomeric endopeptidases. Cathepsin B also has dipeptidyl carboxypeptidase activity and cathepsin H has aminopeptidase activity [1, 17, 18]. Type I pyroglutamyl peptidases are cytosolic cysteine peptidases [19]. The lysosomal cathepsins are compartmented in lysosomes and are also inhibited by cystatin proteins such as type 1 cystatins (stefins A, B and D), type 2 cystatins (cystatins C, D and S) and the cystatin-like domains on kininogens that are released by kallikrein during wounding and microbial invasion [17, 20]. The calpains are heterodimeric, Ca^{2+}-regulated cysteine proteases involved in Ca^{2+}-mediated signalling, apoptosis and cyclin turnover and are inhibited by calpastatin. Calpains have been implicated in Alzheimer's disease, type 2 diabetes and gastric cancer [21-26]. Caspases cleave peptide bonds on the C-terminal side of Asp (i.e. c-asp-ases) and are activated in protease cascades leading to apoptosis [24, 25, 27]. Cysteine proteases of *Plasmodium falciparum* are potential chemotherapeutic targets [28].

Metalloproteases include carboxypeptidase [1, 29, 30], ACE (a dipeptidyl carboxypeptidase) [13, 31] and a variety of matrix metalloproteases (matrixins or MMPs) [1, 32, 33]. ACE catalyzes the formation of the vasoconstrictive hormone angiotensin II from angiotensin I, some ACE inhibitors being important anti-hypertensive drugs [13, 31].

The MMPs catalyze the hydrolysis of collagen and of extracellular matrix proteoglycans and glycoproteins and accordingly are of importance in angiogenesis, embryogenesis, wound healing, inflammation and tumour growth. The MMPs include MMP-1, MMP-8, MMP-13 and MMP-18 (collagenases), MMP-2 and MMP-9 (gelatinases A and B, respectively), MMP-3, MMP-10 and MMP-11 (stromelysins 1, 2 and 3, respectively), MMP-7 (matrilysin), MMP-12 (metalloelastase) and MMP-14, MMP-15, MMP-16 and MMP-17 (membrane-type MMPs MT1-MMP, MT2-MMP, MT3-MMP and MT4-MMP, respectively). The MMPs are regulated by turnover and by inhibitory proteins called TIMPs (tissue inhibitors of MMPs) [32, 33].

Serine proteases include the blood clotting factors V, VII, VIII, IX, X, XI, XII and XIII, cathepsin G, chymotrypsin, chymase, granzymes A, B, D and F, kallikrein, plasmin, prolyl endopeptidases, proteases A-D, subtilisin, thrombin, trypsin, tryptase and urokinase type plasminogen activator (uPA). Chymotrypsin cleaves on the C-terminal side of a hydrophobic amino acid whereas trypsin cleaves on the C-terminal side of a basic amino acid residue. Elastase cleaves on the C-terminal side of a small amino acid residue. The serine proteases are variously involved in angiogenesis, blood clotting, cytosolic proteolysis, digestion, inflammation, proprotein processing and tissue remodelling [1-6]. Prolylendopeptidases are serine oligopeptidases involved in amnesia, depression, diabetes and blood pressure regulation [34].

PLANT NON-PROTEIN PROTEASE INHIBITORS

Non-protein aspartic protease inhibitors

A number of plant natural products have been isolated that inhibit an aspartic protease secreted by *Candida albicans* [35, 36], the most potent of which (IC_{50} values 8-14 µM) are the phenolics ellagic acid [35] and mattucinol-7-O-[4″,6″-O-(S)-hexahydroxydiphenyl]-β-glucopyranoside [35] and the triterpene betulinic acid [36] (Table 1). A macromolecule from *Anchusa* is a potent inhibitor of pepsin (K_i 20 nM) [37] (Table 1).

The devastating global impact of HIV-1/AIDS has led to the discovery of compounds interfering with key enzymes involved in HIV-1 replication, notably HIV-1 protease [8, 14, 38-41]. HIV-1 protease inhibitors are major components of anti-HIV-1 therapy [41]. Many plant natural products have been isolated that inhibit HIV-1 protease [42-57] (Table 1) of which the most potent is the abietane diterpene carnosolic acid (IC$_{50}$ 0.2 μM) [54]. Other relatively active plant HIV-1 protease inhibitors (IC$_{50}$ values 8-14 μM) include the phenolics corilagin [41], condensed tannins [45], kaempferol [42, 43, 47], α-mangostin [48, 49] and γ-mangostin [48]; the triterpenes betulinic acid [53], oleanolic acid [49, 53, 55, 57], ursolic hydrogen maleate [53] and uvaol [53]; and the abietane diterpenes 7-O-ethylrosmanol [54] and rosmanol [54] (Table 1).

Table 1. Non-protein aspartic protease inhibitors
Some representative examples are given of plant sources of the cited compounds (further plant sources can be readily accessed via the Web). IC$_{50}$ (concentration for 50% inhibition) values are given in round brackets. K$_d$ (dissociation constant) or K$_i$ (enzyme-inhibitor dissociation constant) values are given in square brackets. For convenience compounds are grouped into alkaloids (also encompassing N-containing aromatic pseudoalkaloids), phenolics, terpenes and other compounds and are listed alphabetically within these four groupings.

Compound (class)	Selected plant sources (Family)	Protease specificity (IC$_{50}$) [K$_d$, K$_i$] (other proteases inhibited)	Ref.
Candida albicans secreted aspartic protease; Pepsin			
Phenolics			
Ellagic acid (polyphenol)	Widespread, ellagitannin product; *Fragaria* spp. (Rosaceae)	*Candida albicans* secreted aspartic protease (11 μM); Pepsin (37 μM)	[35]
Mattucinol-7-O-[4″,6″-O-(S)-hexahydroxy-diphenoyl]-β-D-glucopyranoside (polyphenol glycoside)	*Miconia myriantha* (Melastomataceae)	*Candida albicans* secreted aspartic protease (8 μM); Pepsin (67 μM)	[35]
3-Geranyl-2,4,6-trihydroxy-benzophenone (phenolic ketone)	*Tovomita krukovii* (Clusiaceae)	*Candida albicans* secreted aspartic protease (109 μM)	[36]
1,3,5,7-Tetrahydroxy-8-isoprenylxanthone (xanthone)	*Tovomita krukovii* (Clusiaceae)	*Candida albicans* secreted aspartic protease (46 μM)	[36]
1,3,5-Trihydroxy-8-isoprenylxanthone (xanthone)	*Tovomita krukovii* (Clusiaceae)	*Candida albicans* secreted aspartic protease (80 μM)	[36]
Terpenes			
Betulinic acid (lupane triterpene)	Widespread; *Rhododendron arboreum* (Ericaceae)	*Candida albicans* secreted aspartic protease (14 μM) (HIV-1 protease)	[36]
Unknown structure			
Anchusa Pepsin inhibitor (63 kDa macromolecule)	*Anchusa strigosa* (Boraginaceae) [root]	Pepsin [20 nM]	[37]
HIV-1 protease			
Phenolics			
Acacetin (= 5,7-Dihydroxy-4′-methoxyflavone)	*Ammi visnaga* (Apiaceae), *Ginkgo biloba* (Ginkgoaceae),	HIV-1 protease (> 176 μM)	[42]

(flavone)	*Agastache foeniculum* (Lamiaceae)		
Apigenin (= 5,7,4'-Trihydroxyflavone) (flavone)	*Ocimum sanctum* (Lamiaceae), ferns	HIV-1 protease (60 µM)	[43-45]
Amariin (hydrolysable tannin)	*Phyllanthus amarus* (Euphorbiaceae)	HIV-1 protease (< 53 µM)	[42]
Baicalein (= 5,6,7-Trihydroxyflavone) (flavone)	*Oroxylum indicum* (Bignoniaceae)	HIV-1 protease (480 µM)	[43]
Butein (chalcone)	*Robinia pseudoacacia, Vicia faba* (Fabaceae)	HIV-1 protease (< 184 µM)	[42]
(+)-Catechin (flavan-3-ol)	Widespread (esp. tannin constituent); *Gossypium* sp. (Malvaceae)	HIV-1 protease (~ 172 µM)	[42]
Chrysin (= 5,7-Dihydroxyflavone) (flavone)	Widespread; *Passiflora coerulea* (Passifloraceae), *Pinus* spp. (Pinaceae)	HIV-1 protease (125 µM)	[43]
Condensed tannins (containing epicatechin & epiafzelechin) (polyphenol)	*Xanthoceras sorbifolia* (Sapindaceae)	HIV-1 protease (4 µM)	[46]
Corilagin (hydrolysable tannin)	*Phyllanthus amarus* (Euphorbiaceae)	HIV-1 protease (21 µM)	[42]
Demethylated Gardenin A (= 5,6,7,8,3',4',5'-Nonahydroxyflavone) (flavonol)	*Ocimum* sp. (Lamiaceae)	HIV-1 protease (11 µM)	[43]
3,2'-Dihydroxyflavone (flavone)	Semi-synthetic	HIV-1 protease (12 µM)	[43]
7,8-Dihydroxyflavone (flavone)	Semi-synthetic	HIV-1 protease (100 µM)	[43]
Epigallocatechin- (4β→8, 2β→*O*-7)-epicatechin (tannin)	*Xanthoceras sorbifolia* (Sapindaceae)	HIV-1 protease (121 µM)	[46]
Fisetin (= 5-Deoxy-quercetin; 3,7,3',4'-Tetrahydroxyflavone) (flavonol)	*Rhus cotinus, R. rhodantherma* (Anacardiaceae), *Acacia* spp. (Fabaceae)	HIV-1 protease (50 µM) (Neutral endopeptidase)	[43]
Fortunellin (= 5-Hydroxy-7-glucosyl-rhamnosyl-4'-methoxyflavone) (glycosylated flavone)	See Acacetin	HIV-1 protease (< 84 µM)	[42]
Galangin (= 3,5,7-Trihydroxyflavone) (flavonol)	*Escallonia* spp. (Saxifrageaceae), ferns, *Alpinia officinarum* (Zingiberaceae)	HIV-1 protease (520 µM)	[43]
Gardenin A (flavone)	*Ocimum* sp. (Lamiaceae)	HIV-1 protease (190 µM)	[43]
Geraniin (hydrolysable tannin)	*Phyllanthus* (Euphorbiaceae), *Geranium* (Geraniaceae), *Fuchsia* (Onagraceae)	HIV-1 protease (< 79 µM)	[42]
Gossypin (= Gossypetin 8-*O*-glucoside; 3,5,7,8,3',4'-Hexahydroxyflavone 8-*O*-glucoside)	*Gossypium indicum, Hibiscus vitifolis* (Malvaceae)	HIV-1 protease (~104 µM)	[42]

(flavonol O-glycoside)			
Isoquercitrin (= Quercetin 3-O-glucoside)	Widespread; *Gossypium herbaceum* (Malvaceae), *Morus alba* (Moraceae)	HIV-1 protease (< 108 μM)	[42]
Kaempferol (flavonol)	Widespread; *Cuscuta reflexa* (Convolvulaceae), *Pisum sativum* (Fabaceae)	HIV-1 protease (7 μM)	[42, 43, 47]
Luteolin (= 5,7,3',4'-Tetrahydroxyflavone) (flavone)	Widespread; *Apium graveolens* (Apiaceae); widespread as glycosides	HIV-1 protease (< 175 μM) (ACE, Neutral endopeptidase)	[42]
α-Mangostin (prenylated xanthone)	*Garcinia mangostana* (Clusiaceae)	HIV-1 protease (5 μM)	[48, 49]
γ-Mangostin (prenylated xanthone)	*Garcinia mangostana* (Clusiaceae)	HIV-1 protease (5 μM)	[48]
Morin (= 3,5,7,2',4'-Pentahydroxyflavone) (flavonol)	*Morus alba, M.* spp., *Chlorophora tinctoria* (Moraceae)	HIV-1 protease (24 μM)	[43]
Myricetin (= 3,5,7,3',4',5'-Hexahydroxyflavone) (flavonol)	*Azadirachta indica, Soymida febrifuga* (Meliaceae)	HIV-1 protease (22 μM) (Neutral endopeptidase)	[43]
Myricitrin (flavonol)	*Bauhinia microstachya* (Fabaceae), *Myrica rubra* (Myricaceae)	HIV-1 protease (~ 157 μM)	[42]
Naringin (= 2,3-Dihydroapigenin 7-O-rhamnosyl-glucoside) (flavanone O-glycoside)	*Adiantum* spp., *Ceterach officinarum* (Adiantaceae), *Origanum vulgare* (Lamiaceae), *Citrus aurantium, C. limon, C. paradisi, C. sinensis* (Rutaceae)	HIV-1 protease (220 μM)	[43]
$N^1,N^4,N^7,N^{10},N^{13}$-Penta-$p$-coumaroyl-spermidine (coumaroyl amide)	Semi-synthetic - cf. N^1,N^5,N^{10}-Tri-p-coumaroylspermidine	HIV-1 protease (33 μM)	[50]
Quercetagetin (= 6-Hydroxyquercetin; 3,5,6,7,3',4'-Hexahydroxyflavone) (flavonol)	*Eupatorium gracile* (Asteraceae), other Asteraceae; glycosides in *Tagetes erecta* (marigold) (Asteraceae)	HIV-1 protease (1000 μM)	[43]
Quercetin (= 3,5,7,3',4'-Pentahydroxyflavone) (flavonol)	Widespread; *Oenothera biennis* (Onagraceae)	HIV-1 protease (20 μM; 36 μM; 59 μM; 66 μM)	[43, 47]
Repandusic acid (hydrolysable tannin)	*Phyllanthus amarus* (Euphorbiaceae)	HIV-1 protease (13 μM)	[42]
Rhoifolin (= 5, 4'-Dihydroxy -7- rhamnosyl-glucosyl-flavone) (glycosylated flavone)	*Evodiopanax innovans* (Araliaceae)	HIV-1 protease (< 87 μM)	[42]
Robinin (= Kaempferol 3-O-galactosyl-rhamnosyl-7-O-rhamnoside) (flavonol O-glycoside)	*Vinca minor* (Apocynaceae), *Pueraria* spp., *Robinia pseudoacacia, Vigna* spp. (Fabaceae)	HIV-1 protease (< 68 μM)	[42]
Rutin (= Quercetin 3-O-rutinoside; Rutoside) (flavonol O-glycoside)	Widespread; *Sophora japonica* (Fabaceae)	HIV-1 protease (< 82 μM; 500 μM)	[42, 43]
Tangeretin (= 5,6,7,8,4'-Pentahydroxyflavone) (flavone)	*Citrus reticulata, C.* spp. (Rutaceae)	HIV-1 protease (~134 μM)	[42]

N^1,N^5,N^{10},N^{14}-Tetra-*p*-coumaroylspermidine (phenolic amide)	Semi-synthetic - cf. N^1,N^5,N^{10}-Tri-*p*-coumaroylspermidine	HIV-1 protease (34 µM)	[50]
N^1,N^5,N^{10}-Tri-*p*-coumaroylspermidine (phenolic amide)	*Artemisia caruifolia* (Asteraceae)	HIV-1 protease (91 µM)	[50]
Terpenes			
Absinthin (guaine sesquiterpene dimer)	*Artemisia caruifolia* (Asteraceae)	HIV-1 protease (< 50% inhibition at 202 µM)	[51]
Absintholide (guaine sesquiterpene dimer)	*Artemisia caruifolia* (Asteraceae)	HIV-1 protease (< 50% inhibition at 202 µM)	[51]
Agastanol (diterpene)	*Agastache rugosa* (Lamiaceae)	HIV-1 protease (360 µM)	[52]
Agastaquinone (diterpene)	*Agastache rugosa* (Lamiaceae)	HIV-1 protease (87 µM)	[52]
α-Amyrin (= α-Amyrenol; Viminalol) (ursene triterpene)	*Alstonia boonei* (Apocycaceae), *Balanophora elongata* (Balanophoraceae)	HIV-1 protease (80 µM) (Collagenase)	[53]
β-Amyrin (ursene triterpene)	*Cycnomorium songaricum* (Cynomoriaceae)	HIV-1 protease (> 100 µM) (Collagenase)	[53]
Anabsin (guaine sesquiterpene dimer)	*Artemisia caruifolia* (Asteraceae)	HIV-1 protease (< 50% inhibition at 202 µM)	[51]
Anabsinthin (guaine sesquiterpene dimer)	*Artemisia caruifolia* (Asteraceae)	HIV-1 protease (< 50% inhibition at 200 µM)	[51]
Betulinic acid (lupene triterpene)	Widespread; *Rhododendron arboreum* (Ericaceae)	HIV-1 protease (9 µM)	[53]
Carnosol (abietane diterpene)	*Rosmarinus officinalis, Salvia officinalis* (Lamiaceae)	HIV-1 protease (> 30 µM)	[54]
Carnosolic acid (abietane diterpene)	*Rosmarinus officinalis* (Lamiaceae)	HIV-1 protease (0.2 µM)	[54]
Caruifolin A (germacranolide sesquiterpene)	*Artemisia caruifolia* (Asteraceae)	HIV-1 protease (< 50% inhibition at 376 µM)	[51]
Caruifolin B (guaine sesquiterpene dimer)	*Artemisia caruifolia* (Asteraceae)	HIV-1 protease (293 µM)	[51]
Caruifolin C (guaine sesquiterpene dimer)	*Artemisia caruifolia* (Asteraceae)	HIV-1 protease (< 50% inhibition at 202 µM)	[51]
Caruifolin D (guaine sesquiterpene dimer)	*Artemisia caruifolia* (Asteraceae)	HIV-1 protease (< 50% inhibition at 202 µM)	[51]
2α,19α-Dihydroxy-3-oxo-12-ursen-28-oic acid (ursane triterpene)	*Geum japonica* (Rosaceae) [plant]	HIV-1 protease (72 % inhibition at 37 µM)	[49, 55]
10′,11′-Epiabsinthin (guaine sesquiterpene dimer)	*Artemisia caruifolia* (Asteraceae)	HIV-1 protease (< 50% inhibition at 202 µM)	[51]
Epipomolic acid (triterpene)	*Geum japonica* (Rosaceae)	HIV-1 protease (42 % inhibition at 38 µM)	[49, 55]
Escin Ia	*Aesculus chinensis*	HIV-1 protease (35 µM)	[56]

(triterpene saponin)	(Hippocastanaceae)		
Escin Ib (triterpene saponin)	*Aesculus chinensis* (Hippocastanaceae)	HIV-1 protease (50 µM)	[56]
Escin IVc (triterpene saponin)	*Aesculus chinensis* (Hippocastanaceae)	HIV-1 protease (35 % inhibition at 100 µM)	[56]
Escin IVd (triterpene saponin)	*Aesculus chinensis* (Hippocastanaceae)	HIV-1 protease (34 % inhibition at 100 µM)	[56]
Escin IVe (triterpene saponin)	*Aesculus chinensis* (Hippocastanaceae)	HIV-1 protease (16 % inhibition at 100 µM)	[56]
Escin IVf (triterpene saponin)	*Aesculus chinensis* (Hippocastanaceae)	HIV-1 protease (13 % inhibition at 100 µM)	[56]
7-*O*-Ethylrosmanol (abietane diterpene)	Semi-synthetic from Carnosolic acid	HIV-1 protease (5 µM)	[54]
Euscaphic acid (= 3-α-hydroxy isomer of Tormentic acid) (triterpene)	*Geum japonica* (Rosaceae)	HIV-1 protease (0 % inhibition at 37 µM) (cf. Tormentic acid)	[49, 55]
Isoescin Ia (triterpene saponin)	*Aesculus chinensis* (Hippocastanaceae)	HIV-1 protease (39 % inhibition at 100 µM)	[56]
Isoescin Ib (triterpene saponin)	*Aesculus chinensis* (Hippocastanaceae)	HIV-1 protease (15 % inhibition at 100 µM)	[56]
Maslinic acid (triterpene)	*Geum japonica* (Rosaceae)	HIV-1 protease (100% inhibition at 38 µM)	[49, 55]
Oleanolic acid (oleanene triterpene)	*Luffa cylindrica* (Cucurbitaceae), *Rosmarinus officinalis* (Lamiaceae)	HIV-1 protease (8 µM; 22 µM) (C3-convertase, Elastase)	[46]
3-Oxotirucalla-7,24-diene-21-oic acid (triterpene)	*Xanthoceras sorbifolia* (Sapindaceae)	HIV-1 protease (~40 µM)	[46]
7-*O*-Methylrosmanol (abietane diterpene)	Semi-synthetic from Carnosolic acid	HIV-1 protease (4 µM)	[54]
Rosmanol (abietane diterpene)	Semi-synthetic from Carnosolic acid	HIV-1 protease (2 µM)	[54]
Tormentic acid (triterpene)	*Geum japonica* (Rosaceae)	HIV-1 protease (49 % inhibition at 37 µM)	[49, 55]
Ursolic acid (= Malol; Malolic acid; Micromerol; Prunol; Urson) (ursene triterpene)	Widespread; *Cynomorium songaricum* (Cynomoriaceae), *Arctostaphylos uva-ursi*, (Ericaceae)	HIV-1 protease (8 µM; 85% inhibition at 39 µM) (Elastase)	[49, 53, 55, 57]
Ursolic acid hydrogen malonate (= 3-*O*-Malonyl ursolic acid hemiester) (triterpene)	*Cynomorium songaricum* (Cynomoriaceae)	HIV-1 protease (6 µM)	[53]
Ursolic acid methyl ester (ursene triterpene)	Semi-synthetic from Ursolic acid	HIV-1 protease (14 µM)	[53]
Uvaol (= Urs-12-ene-3,28-diol) (ursene triterpene)	*Crataegus pinatifida* (Rosaceae)	HIV-1 protease (6 µM)	[57]

Non-protein inhibitors of cysteine proteases

Triterpene sulphates from the fungus *Fusarium compactum* inhibit rhinovirus 3C protease [58], synthetic cysteine protease inhibitors are undergoing clinical trials against the common cold (rhinovirus) and rheumatoid arthritis [59] and a variety of plant proteins are cysteine protease inhibitors [20]. However the only non-protein cysteine protease inhibitor of plant origin found in this survey was an oligosaccharide gum (> 2 million Da) from *Hakea gibbosa* (Proteaceae) that inhibits the cysteine peptidase pyroglutamate aminopeptidase [60]. In contrast, a large number of plant-derived compounds are pro-apoptotic through activation of caspases [61], an example being ginsenoside Rh-2, a triterpene glycoside from *Panax ginseng* (Araliaceae) [62].

Non-protein inhibitors of metallopeptidases and metalloproteases

A variety of plant bioactives inhibit the metallopeptidase aminopeptidase N [63-68], the most potent inhibitors (variously active at 1-7 μM) being the phenolics curcumin, phloretin and quercetin [64-67] and the lupane triterpene betulinic acid (IC_{50} value 7 μM) [68] (Table 2).

Inhibitors of angiotensin I converting enzyme (ACE) are of major interest as potential anti-hypertensive and anti-diabetic agents [13, 31, 69]. A variety of plant alkaloids inhibit ACE, namely a range of bisbenzylisoquinolines such as (+)-tetrandine [70] and the azetidine derivative nicotianamine (IC_{50} 0.3 μM) [71, 72] (Table 2). Of a wide range of plant phenolic ACE inhibitors [73-81] the most potent inhibitors found are the condensed tannins procyanidin B-2 3,3′-di-*O*-gallate, procyanidin B-5 3,3′-di-*O*-gallate and procyanidin C-1 3,3′,3″-tri-*O*-gallate (IC_{50} values 2, 1 and 2 μM, respectively) [80-82] (Table 2). The iridoid monoterpenes oleacein and sambuceins I-III inhibit ACE (IC_{50} values about 30 μM) [83] (Table 2).

Endothelin converting enzyme and neutral endopeptidase are also involved in cardiovascular homeostasis through regulation of endothelin levels [84]. The pterocarpinoid phytoalexin daleformis inhibits the metallopeptidase endothelin converting enzyme (IC_{50} 9 μM) [85]. Various phenolics inhibit the metallopeptidase neutral endopeptidase [76, 77], the most effective being the flavonol myricetin (IC_{50} 42 μM) [76] (Table 2).

Major metalloproteases include the zinc-containing proteases carboxypeptidase A and thermolysin [1, 29, 30, 86]. A variety of matrix metalloproteases (MMPs), including collagenases and gelatinases, are of major importance in relation to extracellular matrix degradation and hence angiogenesis, development, tumour growth, wound healing, inflammation and tissue remodelling [32, 33]. A variety of plant-derived phenolics inhibit collagenase and related MMPs [87-96] of which the most potent (with IC_{50} values of about 0.2 μM) are the phenylpropanoids cimicifugic acids A-C [89], fukinolic acid [89] and rosmarinic acid [89]. The gallotannin (-)-epigallocatechin-3-gallate inhibits collagenase and gelatinases A and B [87, 90-94] (Table 2). The ursane triterpene α-amyrin and its fatty acid esters inhibit collagenase [97]. Tropolone monoterpenes variously inhibit carboxypeptidase A, thermolysin and collagenase [98, 99], hinokitiol and γ-thujaplicin inhibiting all three enzymes [98] (Table 2).

Table 2. Plant non-protein metallopeptidase inhibitors
For details see the legend to Table 1.

Compound (class)	Selected plant sources (Family)	Protease specificity (IC_{50}) [K_d, K_i] (other proteases)	Ref.
Aminopeptidase			
Phenolics			
Apigenin (= 5,7,4'-Trihydroxyflavone) (flavone)	*Apium, Daucus* (Apiaceae), *Achillea, Artemisia* (Asteraceae), *Mentha, Thymus* (Lamiaceae) spp.; ferns	Aminopeptidase N (42% inhibition at 300 μM)	[63]
Baicalein (= 5,6,7-Trihydroxyflavone) (flavone)	*Scutellaria* spp. (Lamiaceae), *Oroxylum indicum* (Bignoniaceae)	Aminopeptidase N (57% inhibition at 300 μM)	[63]
Chrysin (= 5,7-Dihydroxyflavone) (flavone)	*Populus* (Salicaceae), *Escallonia* (Saxifragaceae) spp.	Aminopeptidase N (49% inhibition at 300 μM)	[63]
Curcumin Curcumin (= Diferuloylmethane; Turmeric yellow) (phenylpropanoid)	*Curcuma longa, C. aromatica, C. xanthorrhiza, C. zedoaria, Zingiber officinale* (Zingiberaceae)	Aminopeptidase N (75-89% inhibition at 3 μM)	[64, 67]
Diosmetin (= 5,7, 3'-Trihydroxy-4'-methoxyflavone) (flavone)	*Arnica* spp. (Asteraceae), *Salvia tomentosa* (Lamiaceae), *Stemodia viscosa* (Scrophulariaceae)	Aminopeptidase N (45% inhibition at 300 μM)	[63]
Genistein Genistein (= 4',5,7-Trihydroxy-isoflavone) (isoflavone)	*Prunus* spp. (Rosaceae), *Genista* spp., *Phaseolus lunatus, Trifolium subterraneum* (Fabaceae)	Aminopeptidase N (39-76% inhibition at 100 μM)	[64, 67]
Isorhamnetin (= 3,5,7,3',4'-Pentahydroxyflavone 3'-methyl ether; (flavonol)	Widespread; aglycone & glycoside in *Arnica, Artemisia dracunculus, Haplopappus* (Asteraceae) spp.	Aminopeptidase N (49% inhibition at 300 μM)	[63]
Myricetin (= 3,5,7,3',4',5'-Hexahydroxyflavone) (flavonol)	*Haplopappus canescens* (Asteraceae), *Soymida febrifuga* (Meliaceae)	Aminopeptidase N (48% inhibition at 300 μM)	[63]
Phloretin (= 2',4,4',6'-Tetrahydroxy-dihydrochalcone) (dihydrochalcone)	Aglycone & glycoside in *Malus domestica* (Rosaceae)	Aminopeptidase N (72-91% inhibition at 5 μM)	[64, 67]

Quercetin (= 3,5,7,3′,4′-Pentahydroxyflavone) (flavonol)	Widespread; *Podophyllum peltatum* (Berberidaceae), *Allium cepa* (Liliaceae)	Aminopeptidase N (56-79% inhibition at 1 μM)	[64, 67]
Rhamnetin (= 3,5,3′,4′-Tetrahydroxy-7-methoxyflavone) (flavonol)	*Cistus* spp. (Cistaceae); many Asteraceae & Lamiaceae	Aminopeptidase N (54% inhibition at 300 μM)	[63]
Terpenes			
Betulinic acid (lupane triterpene)	Widespread; *Tovomita krukovii* (Clusiaceae), *Diospyros perigrina* (Ebenaceae)	Aminopeptidase N (7 μM)	[68]
Angiotensin I converting enzyme (ACE)			
Alkaloids			
Cycleahomine (bisbenzylisoquinoline)	*Stephania tetrandra* (Menispermaceae)	ACE	[70]
2,2′-N,N-Dimethyl-tetrandinium dichloride (bisbenzylisoquinoline)	*Stephania tetrandra* (Menispermaceae)	ACE	[70]
Fangchinoline (bisbenzylisoquinoline)	*Isopyrum thalictroides, Pachygone dasycarpa, Stephania erecta* (Menispermaceae)	ACE	[70]
Fenfangjine A (bisbenzylisoquinoline)	*Stephania tetrandra* (Menispermaceae)	ACE	[70]
Fenfangjine B (bisbenzylisoquinoline)	*Stephania tetrandra* (Menispermaceae)	ACE	[70]
Fenfangjine C (bisbenzylisoquinoline)	*Stephania tetrandra* (Menispermaceae)	ACE	[70]
Fenfangjine D (bisbenzylisoquinoline)	*Stephania tetrandra* (Menispermaceae)	ACE	[70]
2′-N-Methyl-tetrandinium chloride (bisbenzylisoquinoline)	*Stephania tetrandra* (Menispermaceae)	ACE	[70]
Nicotianamine (= N-[N-(3-Amino-3-carboxypropyl)-3-amino-3-carboxypropyl]-azetidine-2-carboxylic acid) (azetidine carboxylic acid)	*Glycine max* (soybean) (Fabaceae) (fermented soybean), *Angelica keiskei* (Apiaceae)	ACE (0.3 μM)	[71, 72]
(+)-Tetrandine (bisbenzylisoquinoline)	*Cissampelos pareira, Stephania tetrandra* (Menispermaceae)	ACE	[70]
(+)-Tetrandine-2′-N-α-oxide (bisbenzylisoquinoline)	*Stephania tetrandra* (Menispermaceae)	ACE	[70]
(+)-Tetrandine-2′-N-β-oxide (bisbenzylisoquinoline)	*Stephania tetrandra* (Menispermaceae)	ACE	[70]
Phenolics			
Amentoflavone (= 3′,8″-Biapigenin) (biflavone)	*Cycas revoluta* (Cycadaceae), *Podocarpus montanus* (Podocarpaceae)	ACE (~75% inhibition at 500 μM)	[73, 82]
Areca II-5-C (tannin)	*Areca catechu* (Palmae)	ACE	[74]
Astragalin (= Kaempferol-3-O-β-D-glucoside) (flavonol glycoside)	*Dipladenia martiana* (Apocynaceae)	ACE (401 μM)	[75]
Caffeic acid	Widespread; *Coffea arabica*	ACE (65% inhibition at 500	[76]

(phenylpropanoid)	(Rubiaceae)	μM)	
Catechin (flavan-3-ol)	Widespread tannin component; *Hypericum perforatum* (Hypericaceae), *Vitis vinifera* (Vitaceae)	ACE (40% inhibition at 300 μM)	[76]
Diosmin (= 3′,5,7-Trihydroxy-4′-methoxy-7-O-rutoside) (flavone-O-glycoside)	*Diosma crenulata* (Rutaceae)	ACE (44% inhibition at 300 μM)	[76]
Ellagic acid (= Benzoaric acid; Lagistase) (phenolic acid lactone)	Widespread, ellagitannin product; *Psidium guajava* (Myrtaceae), *Fragaria* spp. (Rosaceae)	ACE (34% inhibition at 300 μM)	[76]
(-)-Epicatechin (= (2R,3R)-5,7,3′,4′-Tetrahydroxyflavan-3-ol) (flavan-3-ol)	Widespread; *Podocarpus nagi* (Podocarpaceae), *Crataegus monogyna* (Rosaceae), *Camellia sinensis* (Theaceae)	ACE (34% inhibition at 1137 μM)	[78]
(-)-Epicatechin 3-O-gallate (hydrolysable tannin)	Widespread; *Rheum palmatum* (Polygonaceae)	ACE (80 μM)	[81]
(-)-Epigallocatechin 3-O-gallate (hydrolysable tannin)	Widespread; *Rheum palmatum* (Polygonaceae)	ACE (70 μM)	[81]
Eriocitrin (= 5,7,3′,4′-Tetrahydroxyflavanone-7-O-rutinoside) (flavanone O-glycoside)	*Mentha piperita* (Lamiaceae), *Myoporum tenuifolium* (Myoporaceae), *Citrus* spp. (Rutaceae)	ACE (28% inhibition at 300 μM)	[76]
Eriosema compound B (prenylated xanthone)	*Eriosema tuberosum* (Fabaceae)	ACE (195) (NEP)	[77]
Hesperidin (= 3′,5,7-Trihydroxy-4′-methoxyflavanone-7-rutinoside) (flavanone-O-glycoside)	*Hyssopus, Mentha* (Lamiaceae), *Citrus* spp., *Poncirus trifoliata* (Rutaceae)	ACE (23% inhibition at 300 μM)	[76]
Hypericum compound H8 (prenylated xanthone)	*Hypericum roeperanum* (Hypericaceae)	ACE (104) (NEP)	[77]
Fisetin (= 3,7,3′,4′-Tetrahydroxyflavone) (flavonol)	*Rhus cotinus, R. rhodantherma* (Anacardiaceae), *Acacia* spp., *Glycine max, Robinia pseudoacacia* (Fabaceae)	ACE (23% inhibition at 300 μM)	[76]
Isoorientin (= Luteolin 6-C-glucoside) (flavone C-glycoside)	*Polygonum orientale* (Polygonaceae)	ACE (48% inhibition at 736 μM)	[78]
Isoquercitrin (= Quercetin 3-O-glucoside) (flavonol-O-glycoside)	*Gossypium herbaceum, Morus alba* (Moraceae)	ACE (32% [53%] inhibition at 711 [646] μM)	[75, 78]
Isovitexin (= Apigenin 6-C-glucoside) (flavone C-glycoside)	Widespread; *Vitex lucens* (Verbenaceae)	ACE (46% inhibition at 763 μM)	[78]
Kaempferol-3-O-(2″-O-galloyl)-glucoside (flavonol-O-glycoside)	*Diospyros kaki* (Ebenaceae), *Euphorbia pekinensis* (Euphorbiaceae)	ACE (466 μM)	[75]
Luteolin (= 5,7, 3′,4′-Tetrahydroxyflavone) (flavone)	Widespread in leaves; *Ammi, Cuminum, Daucus* (Apiaceae), *Lavandula, Mentha, Ocimum, Origanum, Rosmarinus, Thymus*	ACE (55% inhibition at 300 μM)	[76]

	(Lamiaceae) spp.		
(+)-Mesquitol (= 3,7,3',4'-Tetrahydroxyflavan) (flavan)	*Prosopis glandulosa* (Fabaceae)	ACE 9 (~75% inhibition at 1100 µM)	[73, 82]
Morin (= 3,5,7,2',4'-Pentahydroxyflavone) (flavonol)	*Artocarpus heterophyllus, Chlorophora tinctoria, Morus alba* (Moraceae)	ACE 9 (64% inhibition at 1092 µM)	[73, 76, 79]
Myricetin (= 3,5,7,3',5',7'-Hexahydroxyflavone) (flavonol)	*Haplopappus canescens* (Asteraceae), *Soymida febrifuga* (Meliaceae)	ACE (26% inhibition at 300 µM)	[76]
Naringenin (= 5,7,4'-Trihydroxyflavanone) (flavanone)	*Artemisia, Baccharis, Centaurea, Dahlia* spp. (Asteraceae)	ACE (22% inhibition at 300 µM)	[76]
Naringin (= Naringenin-7-O-(2'-O-rhamnosyl)glucoside) (flavanone glycoside)	*Adiantum* spp., *Ceterach officinarum* (Adiantaceae), *Origanum vulgare* (Lamiaceae), *Citrus aurantium, C.* spp. (Rutaceae)	ACE (37% inhibition at 300 µM)	[76]
Orientin (= Luteolin-8-C-glucoside) (flavone C-glycoside)	Widespread; *Polygonum orientale* (Polygonaceae)	ACE (20% inhibition at 736 µM)	[78]
Procyanidin B1 (dimeric flavan-3-ol)	*Lespedeza capitata* (Fabaceae), *Hypericum perforatum* (Hypericaceae), *Vitis vinifera* (Vitaceae)	ACE (58% inhibition at 571 µM)	[79]
Procyanidin B2 (dimeric flavan-3-ol)	*Hypericum perforatum* (Hypericaceae), *Vitis vinifera* (Vitaceae)	ACE (25% inhibition at 570 µM)	[78, 82]
Procyanidin B-2 3,3'-di-O-gallate (condensed tannin)	*Rheum palmatum* (Polygonaceae)	ACE (2 µM)	[81]
Procyanidin B3 (dimeric flavan-3-ol)	*Lespedeza capitata* (Fabaceae), *Hypericum perforatum* (Hypericaceae), *Vitis vinifera* (Vitaceae)	ACE (62% inhibition at 571 µM)	[79, 82]
Procyanidin B6 (dimeric flavan-3-ol)	*Lespedeza capitata* (Fabaceae), *Vitis vinifera* (Vitaceae)	ACE (37% inhibition at 571 µM)	[79, 82]
Procyanidin C1 (trimeric flavan-3-ol)	*Hypericum perforatum* (Hypericaceae), *Vitis vinifera* (Vitaceae)	ACE (45% inhibition at 570 µM)	[78]
Procyanidin C2 (trimeric flavan-3-ol)	*Lespedeza capitata* (Fabaceae), *Vitis vinifera* (Vitaceae)	ACE (44% inhibition at 196 µM)	[79, 82]
Procyanidin B-5 3,3'-di-O-gallate (condensed tannin)	*Rheum palmatum* (Polygonaceae)	ACE (1 µM)	[81]
Procyanidin C-1 3,3',3'''-tri-O-gallate (condensed tannin)	*Rheum palmatum* (Polygonaceae)	ACE (2 µM)	[81]
Procyanidin polymer (flavan-3-ol polymer)	*Pistacia lentiscus* (Anacardiaceae)	ACE (at ~10 µM)	[80]
Quercetin (= 3,5,7,3',4'-Pentahydroxyflavone) (flavonol)	Widespread; *Podophyllum peltatum* (Berberidaceae), *Allium cepa* (Liliaceae)	ACE (23% inhibition at 300 µM)	[76]
Quercetin-3-O-(2''-O-galloyl)-glucoside	*Diospyros kaki* (Ebenaceae)	ACE (48% inhibition at 487 µM)	[75]

(flavonol-*O*-glycoside)			
Quercitrin (= Quercetin-3-rhamnoside) (flavonol *O*-glycoside)	Widespread; *Polygonum* spp. (Polygonaceae), *Quercus tinctoria* (Fagaceae)	ACE (51% inhibition at 300 µM)	[76]
Rhamnetin (= 7-*O*-Methylquercetin) (flavonol)	*Cistus* spp. (Cistaceae)	ACE (23% inhibition at 300 µM)	[76]
Rosmarinic acid (phenylpropanoid)	*Anethum, Levisticum, Sanicula, Astrantia* (Apiaceae), *Symphytum* (Boraginacaeae), *Melissa, Mentha, Rosmarinus* (Lamiaceae) spp.	ACE (27% inhibition at 300 µM)	[76]
Terpenes			
Oleacein (iridoid monoterpene)	*Jasminum grandiflorum* (Oleaceae)	ACE (36 µM)	[83]
Oleuropein (seco-iridoid glycoside)	*Olea europaea* (Oleaceae)	ACE (38% inhibition at 300 µM)	[76]
Sambucein I (iridoid monoterpene)	*Jasminum azoricum* (Oleaceae)	ACE (28 µM)	[83]
Sambucein II (iridoid monoterpene)	*Jasminum azoricum* (Oleaceae)	ACE (26 µM)	[83]
Sambucein III (iridoid monoterpene)	*Jasminum azoricum* (Oleaceae)	ACE (~30 µM)	[83]
Endothelin-converting enzyme (ECE)			
Daleformis (pterocarpinoid phytoalexin)	*Dalea filiciformis* (Fabaceae)	ECE (9 µM)	[85]
Neutral endopeptidase (NEP)			
Eriosema compound B (prenylated xanthone)	*Eriosema tuberosum* (Fabaceae)	NEP (50 µM)	[77]
Fisetin (= 5-Deoxy-quercetin; 3,7,3',4'-Tetrahydroxyflavone) (flavonol)	*Rhus cotinus* (Anacardiaceae), *Acacia* spp., *Glycine max* (Fabaceae)	NEP (220 µM)	[76]
Hypericum compound H8 (prenylated xanthone)	*Hypericum roeperanum* (Hypericaceae)	NEP (81 µM)	[77]
Luteolin (= 5,7,3',4'-Tetrahydroxyflavone) (flavone)	Widespread; *Apium graveolens* (Apiaceae)	NEP (127 µM)	[76]
Myricetin (= 3,5,7,3',4',5'-Hexahydroxyflavone) (flavonol)	*Azadirachta indica, Soymida febrifuga* (Meliaceae), *Haplopappus canescens* (Asteraceae)	NEP (42 µM)	[76]
Quercetin (= 3,5,7,3',4'-Pentahydroxyflavone) (flavonol)	Widespread; *Oenothera biennis* (Onagraceae), *Koelreuteria henryi* (Sapindaceae)	NEP (192 µM)	[76]
Other Metalloproteases			
Baicalein (= 5,6,7-Trihydroxyflavone) (flavone)	*Scutellaria* spp. (Lamiaceae), *Oroxylum indicum* (Bignoniaceae)	MMP-2 (2 µM), MMP-9 (6 µM)	[87]
Catechin (= Catechinic acid; Catechuic acid) (flavan-3-ol)	Widespread; *Gossypium* sp. (Malvaceae), *Agrimonia eupatoria* (Rosaceae), *Salix caprea* (Salicaceae)	Collagenase (1800 µM)	[88]

Cimicifugic acid A (phenylpropanoid ester)	*Cimicifuga* spp. (Ranunculaceae)	Collagenase (47% inhibition at 0.2 μM)	[89]
Cimicifugic acid B (phenylpropanoid ester)	*Cimicifuga* spp. (Ranunculaceae)	Collagenase (64% inhibition at 0.2 μM)	[89]
Cimicifugic acid C (phenylpropanoid ester)	*Cimicifuga* spp. (Ranunculaceae)	Collagenase (47% inhibition at 0.2 μM)	[89]
Cimicifugic acid D (phenylpropanoid ester)	*Cimicifuga* spp. (Ranunculaceae)	Collagenase (20% inhibition at 0.2 μM)	[89]
Cimicifugic acid E (phenylpropanoid ester)	*Cimicifuga* spp. (Ranunculaceae)	Collagenase (26% inhibition at 0.2 μM)	[89]
Cimicifugic acid F (phenylpropanoid ester)	*Cimicifuga* spp. (Ranunculaceae)	Collagenase (37% inhibition at 0.2 μM)	[89]
Delphinidin (anthocyanidin)	*Punica granatum* (Punicaceae), *Solanum tuberosum* (Solanaceae)	MMP-2 (3 μM), MMP-9 (13 μM)	[87]
(-)-Epicatechin-3-gallate (flavan-3-ol, gallotannin)	*Camellia sinensis* (Theaceae)	Collagenase (< 110 μM), MMP-2 (Gelatinase A) (95 μM), MMP-9 (Gelatinase B) (28 μM), MMP-12 (< 1 μM)	[87, 90, 91]
(-)-Epigallocatechin (flavan-3-ol, gallotannin)	*Davidsonia pruriens* (Davidsoniaceae), *Hamamelis virginiana* (Hamamelidaceae), *Camellia sinensis* (Theaceae)	MMP-2 (Gelatinase A) (160 μM; 450 μM), MMP-9 (Gelatinase B) (390 μM)	[87, 92]
(-)-Epigallocatechin-3-gallate (flavan-3-ol, gallotannin)	*Davidsonia pruriens* (Davidsoniaceae), *Hamamelis virginiana* (Hamamelidaceae), *Camellia sinensis* (Theaceae)	Collagenase (< 110 μM), MMP-2 (Gelatinase A) (6 μM; 8 μM; 15 μM), MMP-9 (Gelatinase B) (0.3 μM; 13 μM; 30 μM), MMP-12 (< 1 μM), Type IV collagenase (10 μM)	[87, 90-94]
Fisetin (flavonol)	*Acacia* spp. (Fabaceae)	MMP-2 (8 μM), MMP-9 (160 μM)	[87]
Fukiic acid (phenolic acid)	Semi-synthetic from Fukinolic acid	Collagenase (20% inhibition at 0.2 μM)	[89]
Fukinolic acid (phenylpropanoid ester)	*Cimicifuga* spp. (Ranunculaceae)	Collagenase (51% inhibition at 0.2 μM)	[89]
Honokiol (lignan)	*Magnolia obovate, M. officinalis* (Magnoliaceae)	MMP-9 (~ 0.1 μM)	[95]
Magnolol (lignan)	*Magnolia obovate, M. officinalis* (Magnoliaceae)	MMP-9 (~ 0.1 μM)	[95]
Morin (flavonol)	*Morus alba, Chlorophora tinctoria* (Moraceae)	MMP-2 (25 μM), MMP-9 (300 μM)	[87]
Myricetin (flavonol)	*Soymida febrifuga* (Meliaceae)	MMP-2 (10 μM), MMP-9 (12 μM)	[87]
Pelargonidin (anthocyanidin)	*Pelargonium* spp. (Geraniaceae), *Fragaria* spp. (Rosaceae),	MMP-2 (200 μM), MMP-9 (30 μM)	[87]
Phloretin (dihydrochalcone)	*Malus domestica* (Rosaceae)	MMP-2 (20 μM)	[87]
Procyanidins (polyphenolic oligomers)	*Vitis vinifera* (Vitaceae)	Collagenase (38 μM)	[87, 96]
Rosmarinic acid (phenylpropanoid ester)	*Anethum, Levisticum, Sanicula, Astrantia* (Apiaceae), *Symphytum* (Boraginacaeae),	Collagenase (51% inhibition at 0.2 μM)	[89]

	Rosmarinus (Lamiaceae) spp.		
Taxifolin (dihydroflavonol)	*Engelhardtia chrysolepis* (Juglandaceae), *Pinus maritima* (Pinaceae)	MMP-2 (50 μM), MMP-9 (55 μM)	[87]
Theaflavin (polycyclic benzopyran)	*Camellia sinensis* (Theaceae)	Type IV collagenase (20 μM)	[94]
Theaflavin digallate (polycyclic benzopyran)	*Camellia sinensis* (Theaceae)	Type IV collagenase (30 μM)	[94]
Terpenes			
α-Amyrin (= α-Amyrenol; Viminalol) (ursane triterpene)	*Alstonia boonei* (Apocynaceae), *Balanophora elongata* (Balanophoraceae)	Collagenase (< 100 μM)	[97]
α-Amyrin linoleate (= α-Amyrin *cis*-9,*cis*-12-octadecadienoic acid ester) (ursane triterpene fatty acid ester)	Semi-synthetic from α-Amyrin	Collagenase (< 100 μM)	[97]
α-Amyrin palmitate (= α-Amyrin hexadecanoic acid ester) (ursane triterpene fatty acid ester)	*Lobelia inflata* (Campanulaceae); Semi-synthetic from α-Amyrin	Collagenase (< 100 μM)	[97]
β-Dolabrin (tropolone monoterpene)	*Thujopsis dolobrata, T. plicata* (Cupressaceae)	Carboxypeptidase A (20 μM), Collagenase (89 μM)	[98]
Hinokitiol (tropolone monoterpene)	*Thujopsis dolabrata, T. plicata* (Cupressaceae)	Carboxypeptidase A (3 μM), Collagenase (24 μM), Thermolysin (61 μM)	[98]
α-Thujaplicin (= 2-Isopropyltropolone) (tropolone monoterpene)	*Thujopsis dolabrata* (Cupressaceae)	Carboxypeptidase A (32 μM)	[99]
γ-Thujaplicin (tropolone monoterpene)	*Thujopsis dolabrata, T. plicata* (Cupressaceae)	Carboxypeptidase A (11 μM), Collagenase (19 μM), Thermolysin (69 μM)	[98]

Non-protein inhibitors of serine proteases

Plants have evolved protein and non-protein defenses directed against serine proteases such as trypsin and chymotrypsin involved in digestion of plant protein by microbial pathogens and herbivores. However such compounds may also be active against other serine proteases. Thus a range of plant-derived phenolics inhibit leucocyte elastase [87, 88], the most potent (IC_{50} values < 3-5 μM) being the flavonols morin and myricetin [87], the anthocyanidin pelargonidin [87] and grape procyanidin tannin [88] (Table 3). The flavan-3-ols epicatechin 3-gallate, epigallocatechin 3-gallate and gallocatechin 3-gallate are potent inhibitors of proteasome chymotrypsin-like activity (IC_{50} values 194, 86 and 187 nM, respectively) [100] (Table 3). The coumarin dicoumarol is an anticoagulant by inhibition of vitamin K-dependent carboxylation of protein glutamate carboxylation that precedes Ca^{2+} binding and the blood clotting serine protease activation cascade [3, 101-103] (Table 3).

A wide range of terpenes variously inhibit serine proteases such as blood clotting factors, C-3 convertase, cathepsin G, chymotrypsin, elastase, furin, leukocyte elastase, plasmin, proprotein convertase, thrombin and trypsin [104-114) (Table 3). A wide range of anti-inflammatory triterpenes are inhibitors of trypsin and chymotrypsin [104-106]. The anti-inflammatory triterpenes oleanolic acid and ursolic acid inhibit leukocyte elastase (K_i values 6 and 4 μM, respectively) [108]. The diterpene andrographolide and related compounds inhibit pro-protein convertases such as furin [107]. The abortefacient dimeric sesquiterpene gossypol inhibits spermatozoal acrosin [110, 111]. The cyclic lactone euphane triterpenes GR133487 and GR133686 are potent inhibitors of a range of serine proteases, having IC_{50} values of 4 nM for thrombin [112] (Table 3).

A range of phenolics are inhibitors of prolyl endopeptidase [115-119], the most potent inhibitors being (-)-epicatechin 3-O-gallate [116], 1,2,3,4,6-penta-O-galloyl-β-D-glucose [117] and 1,2,3,6-tetra-O-galloyl-β-D-glucose [117] (IC_{50} values 52 , 170 and 25 nM, respectively) (Table 3). The monoterpenes rosiridin [117] and sacranoside A [115] and the triterpene glycoside β-sitosterol-3-O-β-D-glucose [118] are relatively weak inhibitors of prolyl endopeptidase (Table 3).

Table 3. Non-protein inhibitors of serine proteases
For details see the legend to Table 1.

Compound (class)	Selected plant sources (Family)	Protease specificity (IC_{50}) [K_d, K_i]	Ref.
Serine proteases - e.g. chymotrypsin (CHY), trypsin (TRY), elastase (ELA), Leucocyte elastase (LELA)			-
Phenolics			
Baicalein (flavone)	*Scutellaria* spp. (Lamiaceae)	LELA (25 μM)	[87]
Catechin 3-gallate (flavan-3-ol)	*Camellia sinensis* (Theaceae)	Proteasome CHY-like activity (124 nM)	[100]
Delphinidin (anthocyanidin)	*Punica granatum* (Punicaceae), *Solanum tuberosum* (Solanaceae)	LELA (25 μM)	[87]
Dicoumarol (= Dicumarol; Dicumol; Dicoumarin; Dufalone; Melitoxin) (coumarin)	*Melilotus* sp. (Fabaceae), *Anthoxanthum* sp. (Poaceae) (from 4-Hydroxycoumarin in decomposing hay)	Anticoagulant by inhibiting Vitamin K-dependent protein glutamate carboxylation & thence	[3, 101-103]

		Ca^{2+}-binding for blood clotting protease activation	
Epicatechin 3-gallate (flavan-3-ol)	*Camellia sinensis* (Theaceae)	Proteasome CHY-like activity (194 nM)	[100]
Epigallocatechin (flavan-3-ol)	*Camellia sinensis* (Theaceae)	Proteasome CHY-like activity (1200 µM)	[100]
(-)-Epigallocatechin 3-gallate (flavan-3-ol)	*Davidsonia pruriens* (Davidsoniaceae), *Hamamelis virginiana* (Hamamelidaceae), *Camellia sinensis* (Theaceae)	Proteasome CHY-like activity (86 nM)	[100]
Fisetin (flavonol)	*Acacia* spp. (Fabaceae)	LELA (16 µM)	[87]
Gallocatechin 3-gallate (flavan-3-ol)	*Davidsonia pruriens* (Davidsoniaceae), *Hamamelis virginiana* (Hamamelidaceae), *Camellia sinensis* (Theaceae)	Proteasome CHY-like activity (187 nM)	[100]
Morin (flavonol)	*Chlorophora tinctoria, Morus alba* (Moraceae)	LELA (5 µM)	[87]
Myricetin (flavonol)	*Soymida febrifuga* (Meliaceae)	LELA (4 µM)	[87]
Pelargonidin (anthocyanidin)	*Fragaria* sp. (Rosaceae), *Pelargonium* spp. (Geraniaceae)	LELA (< 3 µM)	[87]
Procyanidin (condensed tannin)	*Vitis vinifera* (Vitaceae)	LELA (4 µM)	[88]]
Quercetin (flavonol)	*Rhododendron cinnabarinum* (Ericaceae)	LELA (20 µM)	[87]
Vitamin K$_1$ (= Phylloquinone; 3-Phytomenadione) (naphthoquinone);	Widespread; e.g. *Vaccinium corymbosum* (Ericaceae), *Medicago sativa* (Fabaceae), *Castanea* sp. (Fagaceae), *Triticum aestivum* (Poaceae)	Dihydrovitamin K (Koagulations-Vitamin)-dependent protein glutamate carboxylation enables Ca^{2+}-binding for blood clotting protease activation	[3, 101-103]
Terpenes			
Acetyl-11-keto-β-boswellic acid (triterpene)	*Boswellia serrata* (Burseraceae)	LELA (15 µM)	[104]
Amidiol (= Taraxast-20(30)-ene-3β,16β-diol) (triterpene)	*Chrysanthemum mortifolium* (Asteraceae)	CHY (96 µM) [53 µM], TRY (195 µM) [143 µM]	[105]
α-Amyrin (= α-Amyrenol; Viminalol) (ursane triterpene)	*Alstonia boonei* (Apocynaceae), *Balanophora elongata* (Balanophoraceae)	CHY (23 µM) [18 µM], LELA (at 20 µM), TRY (41 µM) [29 µM]	[104, 106]
α-Amyrin linoleate (= α-Amyrin *cis*-9,*cis*-12-octadecadienoic acid acid ester] (ursane triterpene ester)	Semi-synthetic from α-Amyrin	CHY (16 µM) [28 µM], TRY (15 µM) [16 µM]	[106]
α-Amyrin palmitate (= α-Amyrin hexadecanoic acid ester) (ursane triterpene ester)	Semi-synthetic from α-Amyrin	CHY (24 µM) [6 µM]	[106]
Andrographolide (diterpene)	*Andrographis paniculata* (Acanthaceae)	Furin [200 µM]	[107]
β-Boswellic acid (triterpene)	*Boswellia serrata* (Burseraceae)	Leukocyte ELA (at 20 µM)	[104]

Brein (= Urs-12-ene-3β,16β-diol) (triterpene)	*Chrysanthemum mortifolium* (Asteraceae)	CHY (120 μM) [110 μM], TRY (~100 μM)	[105]
Brein 3-*O*-myristate (= Urs-12-ene-3β,16β-diol 3-*O*-myristate) (triterpene)	*Chrysanthemum mortifolium* (Asteraceae)	CHY (78 μM) [114 μM]	[105]
Brein 3-*O*-palmitate (= Urs-12-ene-3β,16β-diol 3-*O*-palmitate) (triterpene)	*Chrysanthemum mortifolium* (Asteraceae)	CHY (42 μM) [110 μM]	[105]
Calenduladiol (= Lup-20(29)-ene-3β,16β-diol) (triterpene)	*Chrysanthemum mortifolium* (Asteraceae)	CHY (120 μM) [57 μM], TRY (~100 μM)	[105]
Cycloartenol (= Cycloart-24-en-3β-ol) (triterpene)	*Taraxacum officinale* (Asteraceae)	CHY (140 μM) [420 μM], TRY (82 μM) [25 μM]	[105]
Dammaradienol (= Dammara-20,24-dien-3β-ol) (triterpene)	*Helianthus annuus* (Asteraceae)	CHY (130 μM) [60 μM]	[105]
Erythrodiol (triterpene)	*Conyza filaginoides, Solidago virga-aurea* (Asteraceae), *Olea europaea* (Oleaceae)	LELA	[108]
Faradiol (= Taraxast-20-ene-3β,16β-diol) (triterpene)	*Chrysanthemum mortifolium* (Asteraceae)	CHY (160 μM) [68 μM], TRY (130 μM) [113 μM]	[105]
Faradiol 3-*O*-myristate (= Taraxast-20-ene-3β,16β-diol 3-*O*-myristate) (triterpene)	*Chrysanthemum mortifolium* (Asteraceae)	CHY (32 μM) [30 μM], TRY (> 100 μM)	[105]
Faradiol 3-*O*-palmitate (= Taraxast-20-ene-3β,16β-diol 3-*O*-palmitate) (triterpene)	*Chrysanthemum mortifolium* (Asteraceae)	CHY (72 μM) [58 μM], TRY (82 μM) [86 μM]	[105]
18-β-Glycyrrhetinic acid (Glycyrrhetic acid; Glycyrrhetin) (triterpene sapogenin)	*Glycyrrhiza glabra* (Fabaceae)	LELA [185 μM]	[108]
Glycyrrhizin (triterpene saponin)	*Glycyrrhiza glabra* (Fabaceae)	Thrombin interactions (but not catalytic activity)	[109]
Gossypol (dimeric sesquiterpene)	*Gossypium hirsutum, Montezuma speciosissima* (Malvaceae)	Acrosin, Azocoll proteinase	[110, 111]
GR133487 (5,5-*trans*-fused cyclic lactone euphane triterpene)	*Lantana camara* (Verbenaceae)	α-Chymotrypsin (10 nM), Factor XIa (1 μM), Plasmin (6 μM), α-Thrombin (4 nM), Trypsin (120 nM)	[112]
GR133686 (5,5-*trans*-fused cyclic lactone euphane triterpene)	*Lantana camara* (Verbenaceae)	Cathepsin G (2 μM), α-Chymotrypsin (70 nM), Factor XIa (0.7 μM), Plasmin (4 μM), α-Thrombin (4 nM), Trypsin (70 nM)	[112]
Hederagenin (triterpene)	*Hedera helix* (Araliaceae), *Spinacia oleraceae*	LELA [62 μM], pancreatic ELA (41 μM)	[108, 113]

	(Chenopodiaceae)		
Heliantriol C (= Taraxast-20-ene-3β,16β,22α-triol) (triterpene)	*Chrysanthemum mortifolium* (Asteraceae)	CHY (> 100 μM)	[105]
Heliantriol C 3-*O*-myristate (= Taraxast-20-ene-3β,16β,22α-triol 3-*O*-myristate) (triterpene)	*Chrysanthemum mortifolium* (Asteraceae)	TRY (34 μM) [40 μM]	[105]
Heliantriol C 3-*O*-palmitate (= Taraxast-20-ene-3β,16β,22α-triol 3-*O*-palmitate) (triterpene)	*Chrysanthemum mortifolium* (Asteraceae)	CHY (> 100 μM)	[105]
Lupeol (= Fagasterol; Monogynol B; β-Viscol) (lupane triterpene)	*Alstonia boonei* (Apocynaceae), Asteraceae, *Phyllanthus emblica* (Euphorbiaceae), *Lupinus luteus* (Fabaceae)	CHY (22 μM) [8 μM], TRY (34 μM) [22 μM]	[106]
Lupeol linoleate (=Lupeol -9,*cis*-12-octadecadienoic acid acid ester) (lupane triterpene FA ester)	Semi-synthetic from Lupeol	CHY (> 50 μM), TRY (10 μM) [7 μM]	[106]
Lupeol palmitate (= Lupeol hexadecanoic acid ester) (lupane triterpene FA ester)	Semi-synthetic from Lupeol	CHY (> 50 μM), TRY (6 μM) [10 μM]	[106]
Maniladiol (= Olean-12-ene-3β,16β-diol) (triterpene)	*Helianthus annuus* (Asteraceae)	CHY (~100 μM)	[105]
Maniladiol 3-*O*-myristate (= Olean-12-ene-3β,16β-diol 3-*O*-myristate) (triterpene)	*Chrysanthemum mortifolium* (Asteraceae)	CHY (78 μM) [26 μM], TRY (73 μM) [267 μM]	[105]
Maniladiol 3-*O*-palmitate (= Olean-12-ene-3β,16β-diol 3-*O*-palmitate) (triterpene)	*Chrysanthemum mortifolium* (Asteraceae)	CHY (84 μM) [120 μM], TRY (97 μM) [190 μM]	[105]
(24*S*)-25-Methoxy-cycloartanediol (= (24*S*)-25-Methoxycycloartane-3β,24-diol) (triterpene)	*Chrysanthemum mortifolium* (Asteraceae)	TRY (110 μM)	[105]
24-Methylenecycloartenol (= 24-Methylcycloart-24 (24′)-en-3β-ol) (triterpene)	*Helianthus annuus* (Asteraceae); Cycloartenol widespread	CHY (~100 μM)	[105]
Neoandrographolide (= Andrographolide *O*-glucoside) (diterpene)	*Andrographis paniculata* (Acanthaceae)	Proprotein convertase (PPC) PPC-1, PPC-7, Furin (a PPC) (54 μM)	[107]
Oleanolic acid (oleanane triterpene)	*Luffa cylindrica* (Cucurbitaceae), *Rosmarinus officinalis* (Lamiaceae), *Olea europaea* (Oleaceae)	C3-convertase (at 200 μM), LELA [6 μM], pancreatic ELA (5 μM)	[108, 113, 114]
Ruscogenin (triterpene)	*Ruscus aculeatus* (Liliaceae)	Pancreatic ELA (120 μM)	[113]

Succinoyl-andrographolide (diterpene)	*Andrographis paniculata* (Acanthaceae)	Furin (a PPC) & PPC-1, PPC-7 [< 30 μM]	[107]
Taraxerol (= Taraxer-14-en-3β-ol) (triterpene)	*Taraxacum officinale* (Asteraceae)	TRY (> 100 μM)	[105]
Δ^7-Tirucallol (= Tirucalla-7,24-dien-3β-ol) (triterpene)	*Chrysanthemum mortifolium* (Asteraceae)	CHY (98 μM) [72 μM], TRY (140 μM) [152 μM]	[105]
Ursolic acid (= Malol; Malolic acid; Micromerol; Prunol; Urson) (triterpene)	Widespread; *Prunella vulgaris*, *Salvia triloba* (Lamiaceae), *Malus*, *Pyrus* (Rosaceae)	LELA [4 μM]	[104, 108]
Uvaol (triterpene)	*Diospyros kaki* (Ebenaceae), *Crataegus pinnatifida* (Rosaceae), *Debregeasia salicifolia* (Urticaceae)	LELA [16 μM]	[108]
Prolyl endopeptidase (PEP)			
Phenolics			
Arbutin (= Hydroquinone-β-D-glucopyranoside) (phenol glucoside)	*Rhodiola sacra* (Crassulaceae), *Origanum majorana* (Lamiaceae), *Pyrus communis* (Rosaceae)	PEP (391 μM)	[115]
(-)-Epicatechin (flavan-3-ol)	Widespread; *Rheum palmatum* (Polygonaceae), *Camellia sinensis* (Theaceae)	PEP (28 μM)	[116]
(-)-Epicatechin 3-*O*-gallate (flavan-3-ol gallic acid ester)	*Cinnamomum* sp. (Lauraceae), *Rheum palmatum* (Polygonaceae), *Camellia sinensis* (Theacaeae)	PEP (52 nM)	[116]
(-)-Epigallocatechin 3-*O*-gallate (flavan-3-ol gallic acid ester)	*Davidsonia pruriens* (Davidsoniaceae), *Hamamelis virginiana* (Hamamelidaceae), *Cinnamomum* sp. (Lauraceae), *Camellia sinensis* (Theacaeae)	PEP (1470 nM)	[113]
Gallic acid (= 3,4,5-Trihydroxybenzoic acid) (phenolic acid)	Widespread; basic constituent of the hydrolysable tannins (gallotannins); *Mangifera indica* (Anacardiaceae)	PEP (487 μM)	[115]
Gallic acid 4-*O*-β-D-(6-*O*-galloyl)glucopyranoside (glucose gallic acid ester)	*Rhodiola sacra* (Crassulaceae), *Rheum palmatum* (Polygonaceae)	PEP (9 μM)	[116]
3-*O*-Galloyl-epigallocatechin-(4β→8)-epigallocatechin-3-*O*-gallate ester (condensed tannin)	*Camellia sinensis* (Theacaeae), *Rhodiola sacra* (Crassulaceae)	PEP (440 nM)	[115]
4-*O*-(β-D-Glucopyranoside)-gallic acid (phenolic glycoside)	*Rhodiola sacra* (Crassulaceae)	PEP (215 μM)	[115]
4(4-Hydroxyphenyl)-2-butanone 4'-*O*-β-D-(2,6-di-*O*-galloyl)glucopyranoside (phenolic glycoside)	*Rheum palmatum* (Polygonaceae)	PEP (11 μM)	[116]

4(4-Hydroxyphenyl)-2-butanone 4'-*O*-β-D-(2-*O*-galloyl-6-*O*-cinnamoyl)-glucoside (phenolic glycoside)	*Rheum palmatum* (Polygonaceae)	PEP (0.7 μM)	[116]
4(4-Hydroxyphenyl)-2-butanone 4'-*O*-β-D-(6-*O*-galloyl-2-*O*-cinnamoyl)-glucoside (phenolic glycoside)	*Rheum palmatum* (Polygonaceae)	PEP (8 μM)	[116]
Licuraside (= Isoliquiritigenin-4-β-D-apiofuranosyl-2'-β-D-Glc; 2',4',4-Trihydroxychalcone-4-β-D-apiofuranosyl-2'-β-D- glucoside) (chalcone glycoside)	*Glycyrrhiza glabra* (Fabaceae)	PEP (27 μM)	[116]
Luteolin (= 5,7,3',4'-Tetrahydroxyflavone) (flavone)	Widespread, Brassicaceae, Lamiaceae, Fabaceae, Scrophulariaceae; *Apium graveolens* (Apiaceae)	PEP (0.6 μM)	[118]
1,2,3,4,6-Penta-*O*-galloyl-β-D-glucose (galloyl ester)	*Rhodiola sachalinensis* (Polygonaceae)	PEP (170 nM)	[117]
Protocatechuic acid (= 3,4-Dihydroxybenzoic acid) (phenolic acid)	Widespread; *Allium cepa* (Liliaceae), *Helianthus* (Asteraceae), *Eucalyptus* (Myrtaceae), *Picea* (Pinaceae), *Olea* (Oleaceae), *Rheum* (Polygonaceae) spp.	PEP (28 μM)	[115]
Purpurogallin (= 2,3,4,6-Tetrahydroxy-5H-benzocyclohepten-5-one) (bicyclic phenolic)	*Dryophanta divisa* gall on *Quercus pedunculata* (Fagaceae)	PEP (16 μM)	[119]
Quercetin (= 3,5,7,3',4'-Pentahydroxyflavone) (flavonol)	Widespread; *Allium cepa* (Liliaceae), *Oenothera biennis* (Onagraceae)	PEP (0.6 μM)	[118]
Rhodionin (flavonol glycoside)	*Rhodiola sachalinensis* (Polygonaceae)	PEP (22 μM)	[117]
Rhodiosin (flavonol glycoside)	*Rhodiola sachalinensis* (Polygonaceae)	PEP (41 μM)	[117]
1,2,3,6-Tetra-*O*-galloyl-β-D-glucose (galloyl ester)	*Rhodiola sachalinensis* (Polygonaceae)	PEP (25 nM)	[117]
1,2,6-Tri-*O*-galloylglucose (glucose gallic acid ester)	*Rheum palmatum* (Polygonaceae)	PEP (0.4 μM)	[116]
cis-3,5,4'-Trihydroxystilbene 4'-*O*-β-D-(6-*O*-galloyl)- glucoside (stilbene glycoside)	*Rheum palmatum* (Polygonaceae)	PEP (3 μM)	[116]
3,5,4'-Trihydroxystilbene 4'-*O*-β-D-(2-*O*-galloyl)-glucoside (stilbene glycoside)	*Rheum palmatum* (Polygonaceae)	PEP (3 μM)	[116]
3,5,4'-Trihydroxystilbene	*Rheum palmatum*	PEP (15 μM)	[116]

4'-O-β-D-(6-O-galloyl)-glucoside (stilbene glycoside)	(Polygonaceae)		
3,5,4'-Trihydroxystilbene 4'-O-β-D- glucoside (stilbene glycoside)	*Rheum palmatum* (Polygonaceae)	PEP (23 μM)	[116]
Terpenes			
Rosiridin (monoterpene glycoside)	*Rhodiola sachalinensis, R. sacra* (Crassulaceae)	PEP (84 μM)	[117]
Sacranoside A (monoterpene glycoside)	*Rhodiola sachalinensis, R. sacra* (Crassulaceae)	PEP (348 μM)	[115]
β-Sitosterol-3-O-β-D-glucoside (phytosterol glycoside)	Widespread; *Caryophyllus flos* (Myrtaceae)	PEP (48 μM)	[118]

PLANT PROTEIN AND PEPTIDE PROTEASE INHIBITORS

While a few very potent non-peptide protease inhibitors (PIs) have been isolated from plants many plant protease inhibitor proteins (PIPs) have evolved to have protease interaction K_i values in the nanomolar and picomolar range. These extraordinary affinities derive from the matching of the PI protein amino sequence about the scissile peptide bond (P1-P1') and evolution of adjacent sequences to fit and interact appropriately within the target protease active site [1, 120, 121]. The structure and function of the different classes of PI proteins from plants are succinctly but comprehensively reviewed below.

Aspartate protease inhibitors

A variety of aspartic protease inhibitor (API) proteins have been resolved from plants [122-135] of which the best characterized at the gene and protein level are those from *Solanum tuberosum* (potato) (Solanaceae) [124-134] (Table 4). The potato aspartic protease inhibitor proteins are typically about 190 residues (about 20 kDa), have 3 disulphide bridges, are homologous to the soybean trypsin inhibitor (Kunitz) family PIs [133] and can also inhibit trypsin [124-134] (Table 4).

Table 4. Plant aspartic protease inhibitor proteins
The number of amino acid residues (aa), cysteines (Cys) and disulphide bonds (S-S) is given together with molecular mass (Da) and the scissile bond amino acid sequence at the the reactive site of the protease inhibitor protein (P1-P1', the one letter amino acid code being employed).

Plant species (other name) (Family)	Protease inhibitor protein	Protease specificity (IC$_{50}$) [K$_d$, K$_i$] (reactive site)	Ref.
Cucurbita sp. (squash)	SQAPI - 2 isoforms (96 aa; 10	Pepsin [2 nM]; *Glomerella*	[122]

(Cucurbitaceae)	kDa monomers; 21 kDa dimer)	*cingulata* secreted aspartic protease [20 nM]	
Lycopersicon esculentum (tomato) (Solanaceae)	Jasmonic acid-inducible potato API homologue	Cathepsin D	[123, 124]
Solanum dulcamara (bittersweet) (Solanaceae)	Jasmonic acid-inducible potato API homologue	Cathepsin D	[124]
Solanum melongena (aubergine) (Solanaceae)	Jasmonic acid-inducible potato API homologue	Cathepsin D	[124]
Solanum tuberosum (potato) (Solanaceae)	PDI, NID, PI-8, PI-13, p749, API-13, clone 4, CathInh, gCDI-A1, pA1 (188-189 aa; 21 kDa; 6 Cys; 3 S-S)	Cathepsin & Trypsin; Cathepsin D [1 nM, NDI], Trypsin (R67-F68, NDI;	[124-134]
Solanum tuberosum (potato) (Solanaceae)	PI-4, PIG, CDI homologue (186-189 aa; 20-21 kDa; 5 Cys)	Cathepsin D, Trypsin (R64-F65, PIG	[128, 130]
Vicia sativa (bean) (Fabaceae)	API	Cathepsin D	[135]

Cysteine protease inhibitor proteins from plants

Specific kinds of cysteine proteases are involved in a variety of processes such as lysosomal proteolysis [1, 17], signalling and cell division [21-26], apoptosis [21-27], seed germination [17], malarial host protein digestion [28] and periodontal disease [136]. Plants like other eukaryotes need to inhibit the activity of exogenous cysteine proteases and also of endogenous cysteine proteases (e.g. in dormant seed) [17]. They do this by elaborating cysteine protease inhibitor proteins referred to as phytocystatins or plant cystatins that are homologous to animal cystatins [17, 20]. Cystatins have been isolated from a wide range of plants as well as various unrelated cysteine inhibitor proteins [137-198] (Table 5).

The phytocystatins are exemplified by the rice oryzacystatins that have a high affinity for cysteine proteases such as cathepsin H and papain. The oryzacystatins have molecular masses of 11-12 kDa, have zero cysteine content and have structures involving an α-helix linked to 5 antiparallel β sheets [168, 171-181] (Table 5). The phytocystatins typically have molecular masses in the range 9-17 kDa but some are much larger such as *Brassica* (Chinese cabbage) BCPI-1 and BCPI-2 (23 kDa) [150], *Helianthus* (sunflower) multicystatin (32 kDa) having 3 phytocystatin domains [166, 167] and the *Solanum* (potato) multicystatin (87 kDa) which has 5 phytocystatin domains [189, 190].

Plant cysteine PIs not structurally related to the phytocystatins include a family of 6 kDa heterodimeric cysteine protease inhibitors from *Ananas cosmosus* (pineapple) (Bromeliaceae) that are homologous to Bowman-

Birk serine PIPs (BBPIPs) [137-141]; *Hordeum* (barley) lipid transfer proteins (LTPs) that inhibit malt cysteine endopeptidases [169]; and *Solanum* (potato) Kunitz PI-type cysteine PIs [185-188] (Table 5).

Table 5. Protein cysteine protease inhibitor (CPI) proteins.
For details see the legend to Table 4.

Plant species (other name) (Family)	Protease inhibitor protein	Protease specificity (IC$_{50}$) [K$_d$, K$_i$] (reactive site)	Ref.
Ananas comosus (pineapple) (Bromeliaceae) [stem]	*Ananas* BI-I, BI-II, BI-III, BI-IV, BI-V, BI-VI (6 kDa; A (41 aa, 7 Cys)-(S-S)$_2$- B (11 aa; 2 Cys); homologous to BBPIPs	Cysteine proteases – Bromelain [0.7 μM], Cathepsin L [0.2 μM], Papain [0.7 μM]; Chymotrypsin (weak), Trypsin (weak)	[137-140]
Arabidopsis thaliana (mouse-ear cress) (Brassicaceae)	Cysteine protease inhibitor (CPI) protein homologues (genes) (11-51 kDa proproteins)	Cysteine protease inhibitor protein homologues	[141-146]
Arabidopsis thaliana (mouse-ear cress) (Brassicaceae)	PRL1 & PRL2 (proprotein 23 kDa)	Cysteine protease inhibitor protein (cystatin) homologues	[147]
Arabidopsis thaliana (mouse-ear cress) (Brassicaceae)	FL3-27 (gene) (proprotein 11 kDa; stress-induced homologue of cowpea cysteine protease inhibitor protein)	Cysteine protease inhibitor protein (cystatin) homologue	[148]
Ambrosia artemisiifolia (short ragweed) (Asteraceae) [pollen]	IPC1/5 (gene) (proprotein 92 aa; 10 kDa; no Cys)	Oryzacystatin I & II homologue	[149]
Brassica campestris (Chinese cabbage) (Brassicaceae) [flower bud]	BCPI-1 (199 aa; 23 kDa; 0 Cys), BCPI-2 (207 aa; 23 kDa; 0 Cys)	Papain	[150]
Carica papaya (papaya, paw-paw) (Caricaceae) [leaf]	Papaya Cst (99 aa; 11 kDa; 1 Cys)	Caricain [1 nM], Chymopapain [0.4 nM], Papaya proteinase IV [3 nM] Papain [0.4 nM]	[151]
Carica papaya (papaya, paw-paw) (Caricaceae) [recombinant]	*Carica* papain pro-region (107 aa; 12 kDa)	Caricain [8 nM], Chymopapain [12 nM], Papain [2 nM], Papaya proteinase IV [3]	[152]
Carica papaya (paw-paw) (Caricaceae) [recombinant]	*Carica* proteinase IV pro-region (106 aa; 12 kDa)	Caricain [34 nM], Chymopapain [15 nM], Papain [20 nM], Papaya proteinase IV [1 μM]	[152]
Castanea sativa (chestnut) (Fagaceae) [seed]	CsC (*C. sativa* cystatin) (102 aa; 11 kDa; 0 Cys)	Cathepsin B [473 nM], Chymopapain [366 nM], Ficin [65], Papain [29 nM], Trypsin [3489 nM]	[153]
Chelidonium majus (celandine) (Papaveraceae) [leaf, stem]	Chelidostatin (90 aa; 10 kDa; 0 Cys; phytocystatin)	Cathepsin H [8 nM], Cathepsin L [56 pM], Papain [110 pM]	[154]
Daucus carota (carrot) (Apiaceae) [cell culture, seed]	Carrot phytocystatin EIP18 (= Extracellular, Insoluble Cystatin of Carrot, EICC) (133 aa; 15 kDa; 0 Cys; phytocystatin)	Chymopapain, Chymotrypsin, Papain	[155]

Dianthus caryophyllus (carnation) (Asteraceae) [flower]	DC-CPIn (98 aa; 11 kDa; 0 Cys)	Carnation petal proteinase (0.2 nM), Papain (4 nM)	[156]
Glycine max (soybean) (Fabaceae) [leaf; L1 constitutive; N2 & R1 induced by wounding & Methyljasmonate]	L1 (scL) (92 aa; 11 kDa; 0 Cys), N2 (scN) (103 aa; 12 kDa; 0 Cys), R1 (100 aa; 11 kDa; 0 Cys),	Papain [19000 nM, L1; 57 nM, N2; 21 nM, R1], Western corn rootworm gut protease (4000 nM, L1; 50 nM, N2; 200 nM, R1)	[157-160]
Glycine max (soybean) (Fabaceae) [seed]	Soybean cystatin (13 kDa)	Cathepsin B [25 nM], Papain [0.1 nM]	[161]
Glycine max (soybean) (Fabaceae) [seed]	Soybean cystatin SC (245 aa; 27 kDa; 1 Cys; homologue of 11 kDa phytocystatins plus N- & C-terminal extensions)	Papain (10-100 nM IC$_{50}$ for SC & SC constructs devoid of N-terminal or C-terminal extension or both extensions)	[162]
Helianthus annuus (sunflower) (Asteraceae)	Sunflower cystatin Sca (82 aa; 9 kDa; 0 Cys; phytocystatin)	Cathepsin H (weak), Ficin [2 µM], Papain [6 nM]	[163, 164]
Helianthus annuus (sunflower) (Asteraceae)	Sunflower cystatin Scb (101 aa; 11 kDa; 0 Cys; phytocystatin)	Cathepsins B, L & H, Ficin [3 µM], Papain [0.2 nM]	[163, 165]
Helianthus annuus (sunflower) (Asteraceae)	Sunflower multicystatin SMC (282 aa; 32 kDa; 0 Cys; 3 phytocystatin domains)	Papain [48 nM]	[166, 167]
Hordeum vulgare (barley) (Poaceae) [seed]	Hv-CPI (107 aa; 12 kDa; 1 Cys; phytocystatin)	Chymopapain [160 nM], Ficin [22 nM], Papain [20 nM]	[168]
Hordeum vulgare (barley) (Poaceae) [seed]	*Hordeum* Lipid Transfer Proteins 1 & 2 (= LTP 1 & 2) (7 kDa proteins; 8 Cys; 4 S-S)	Barley malt cysteine endoproteinases	[169]
Malus domestica (apple) (Rosaceae) [immature fruit]	Apple cystatin (11 kDa; phytocystatin)	Bromelain, Ficin, Papain [0.2 nM]	[170]
Oryza sativa (Poaceae) [seed]	*Oryza* Oryzacystatin-I (OC-I) (102 aa;11 kDa; 0 Cys; phycocystatin; α-helix- 5 antiparallel β sheets)	Cathepsin H [1 µM], Papain [32 nM; 36 nM]	[150, 168, 171-181]
Oryza sativa (Poaceae) [seed]	*Oryza* Oryzacystatin-II (OC-II) (107 aa; 12 kDa; 0 Cys; phycocystatin; α-helix- 5 antiparallel β sheets)	Cathepsin H [25 nM], Papain [830 nM; 1 µM]	[150, 168, 172, 175]
Pennisetum glaucum (pearl millet) (Poaceae) [seed]	*Pennisetum* Cysteine PI (24 kDa; 3 Cys; antifungal protein)	Papain	[182]
Persea americana (avocado) (Lauraceae) [fruit]	Avocado cystatin (100 aa; 11 kDa; 0 Cys; phytocystatin)	Papain	[183]
Saccharum officinarum (sugar cane) (Poaceae)	Sugar cane cystatins Sc1- Sc25 (Phytocystatins)	Cysteine proteases	[184]
Solanum tuberosum (Solanaceae) [tuber]	Potato Cysteine Protease Inhibitors (PCPIs) (20-25 kDa; Kunitz PI homologues – for details see Table 12)	Cysteine proteases (e.g. Papain, Cathepsins B, H & L) – for details see Table 12	[185-188]
Solanum tuberosum (Solanaceae) [tuber]	Potato multicystatin (PMC) (proprotein: 757 aa; 87 kDa; 0 Cys; 8 phytocystatin domains); other homologous genes	Papain (PMC, five 10 kDa 1-domain fragments & 32 kDa 3-domain fragment)	[189, 190]
Sorghum bicolor (sorghum)	Sorghum CPI (proprotein: 130	Papain	[191]

(Poaceae)	aa; 14 kDa; 0 Cys; phytocystatin)		
Triticum aestivum (wheat) (Poaceae) [seed]	Wheat cystatins 1- 4 (WC1-WC4) & gWC2 (processed: 9-13 kDa; 0 Cys; phytocystatins)	Cathepsin L [3 nM, WC1; 7 nM, WC4], Cathepsin H [7 nM, WC1; 11 nM, WC4], Cathepsin B [5 μM, WC1; 7 μM, WC4], WCP-P3 [1 nM, WC1; 0.5 nM, WC4]	[192]
Vigna unguiculata (cowpea) (Fabaceae) [seed]	Vu CPI (97 aa; 11 kDa; 0 Cys; phytocystatin)	Cysteine proteases (inferred)	[193]
Wisteria floribunda (Fabaceae) [seed]	WCPI-3 (17 kDa; phytocystatin)	Papain [6 nM]	[150, 194]
Zea mays (corn) (Poaceae) [seed]	Corn Cystatin I (= CC-I) (proprotein: 135 aa; 15 kDa; 0 Cys; phytocystatin)	Cathepsin B [290 nM; 4 μM], Cathepsin H [5 nM; 6 nM], Cathepsin L [7 nM; 17 nM], Chymopapain [92 nM], Ficin [83 nM], Papain [23 nM, 37 nM], WCP-P3 [2 nM]	[150, 168, 192, 195-197]
Zea mays (corn) (Poaceae) [seed]	Corn Cystatin II (= CC-II) (proprotein: 134 aa; 15 kDa; 0 Cys; phytocystatin)	Cathepsin B [130 nM], Cathepsin H [1 nM], Cathepsin L [0.1 nM], Papain [66 nM]	[196]
Zea mays (corn) (Poaceae) [seed]	Corn Cystatin II' (= CC-II') (proprotein: 134 aa; 15 kDa; 0 Cys; phytocystatin)	Cysteine proteases (inferred)	[197, 198]

Metalloprotease inhibition by plant-derived peptides and proteins

Synthetic peptide inhibitors have been developed for a variety of proteases [199-204]. Peptide inhibitors of the metalloprotease angiotensin I converting enzyme (ACE) are of major importance as hypertensive agents [13, 31]. A variety of peptides derived from protease-catalyzed hydrolysis of corn α-zein [202-203] or of wheat germ protein [199, 204] inhibit ACE (Table 6). The most potent of such plant-derived ACE inhibitory peptides is Ile-Val-Tyr (IVY) (K_i 0.1 μM) [199, 204]. Further plant-derived peptide ACE inhibitors include the tripeptide glutathione [73, 82], the glutathione -related peptide γ-L-glutamyl-(+)-allyl-L-cysteine sulphoxide [73, 82, 200, 201] and the tripeptide His-His-Leu (HHL) from fermented soybean [201] (Table 6).

Table 6. ACE inhibition by plant peptides and peptides derived from plant protein proteolysis
Peptide sequences are given in both the three-letter and one-letter codes. *Triticum aestivum* (wheatgerm) protein was proteolysed by *Bacilllus lichenoformis* alkaline protease [199] and *Zea mays* (corn seed) α-zein was proteolysed by thermolysin [203, 204].

Peptide	Plant source	Protease specificity (IC$_{50}$) [K$_d$, K$_i$]	Ref.
Ala-Phe (= AF)	*Triticum aestivum* (Poaceae)	ACE (15 μM)	[199]

(dipeptide)	[proteolysate]		
Ala-Pro-Gly-Ala-Gly-Val-Tyr (= APGAGVY) (septapeptide)	*Triticum aestivum* (Poaceae) [proteolysate]	ACE (2 µM)	[199]
Asp-Ile-Gly-Tyr-Tyr (= DIGYY) (pentapeptide)	*Triticum aestivum* (Poaceae) [proteolysate]	ACE (3 µM)	[199]
Asp-Tyr-Val-Gly-Asn (= DYVGN) (pentapeptide)	*Triticum aestivum* (Poaceae) [proteolysate]	ACE (0.7 µM)	[199]
γ-L-Glutamyl-(+)-allyl-L-cysteine sulphoxide (dipeptide)	*Allium ursinum* (Alliaceae)	ACE (98 µM)	[73, 82, 200, 201]
Glutathione (tripeptide)	*Allium ursinum* (Alliaceae)	ACE (10 µM)	[73, 82]
Gly-Gly-Val-Ile-Pro-Asn (= GGVIPN) (hexapeptide)	*Triticum aestivum* (Poaceae) [proteolysate]	ACE (0.7 µM)	[199]
His-His-Leu (= HHL) (tripeptide)	*Glycine max* (soybean) (Fabaceae) [fermented soybean paste]	ACE (5 µM)	[201]
Ile-Arg-Ala (= IRA) (tripeptide)	*Zea mays* (Poaceae) [proteolysate]	ACE (6 µM)	[202, 203]
Ile-Arg-Ala-Gln-Gln (= IRAQQ) (pentapeptide)	*Zea mays* (Poaceae) [α-Zein proteolysate]	ACE (160 µM)	[202, 203]
Ile-Val-Tyr (= IVY) (tripeptide)	*Triticum aestivum* (Poaceae) [proteolysate]	ACE (0.5 µM) [0.1 µM]	[199, 204]
Ile-Tyr (= IY) (dipeptide)	*Triticum aestivum* (Poaceae) [proteolysate]	ACE (2 µM)	[199]
Leu-Ala-Ala (= LAA) (tripeptide)	*Zea mays* (corn) (Poaceae) [proteolysate]	ACE (13 µM)	[202]
Leu-Ala-Tyr (= LAY) (tripeptide)	*Zea mays* (Poaceae) [proteolysate]	ACE (4 µM)	[202]
Leu-Arg-Pro (= LRP) (tripeptide)	*Zea mays* (Poaceae) [proteolysate]	ACE (0.3 µM)	[202, 203]
Leu-Asn-Pro (= LNP) (tripeptide)	*Zea mays* (Poaceae) [proteolysate]	ACE (43 µM)	[202, 203]
Leu-Gln-Gln (= LQQ) (tripeptide)	*Zea mays* (Poaceae) [proteolysate]	ACE (100 µM)	[202, 203]
LQP (= Leu-Gln-Pro) (tripeptide)	*Zea mays* (Poaceae) [proteolysate]	ACE (2 µM)	[202, 203]
Leu-Leu-Pro (= LLP) (tripeptide)	*Zea mays* (Poaceae) [proteolysate]	ACE (57 µM)	[202, 203]
Leu-Ser-Pro (= LSP) (tripeptide)	*Zea mays* (Poaceae) [proteolysate]	ACE (2 µM)	[202, 203]
Leu-Tyr (= LY) (dipeptide)	*Triticum aestivum* (Poaceae) [proteolysate]	ACE (6 µM)	[199]
Phe-Tyr (= FY) (dipeptide)	*Zea mays* (Poaceae) [proteolysate]	ACE (25 µM)	[202, 203]
Thr-Ala-Pro-Tyr (= TAPY) (tetrapeptide)	*Triticum aestivum* (Poaceae) [proteolysate]	ACE (14 µM)	[199]
Thr-Phe (= TF) (dipeptide)	*Triticum aestivum* (Poaceae) [proteolysate]	ACE (18 µM)	[199]
Thr-Tyr-Leu-Gly-Ser (=	*Triticum aestivum* (Poaceae)	ACE (0.9 µM)	[199]

TYLGS) (pentapeptide)	[proteolysate]		
Thr-Val-Pro-Tyr (= TVPY) (tetrapeptide)	*Triticum aestivum* (Poaceae) [proteolysate]	ACE (2 μM)	[199]
Thr-Val-Val-Pro-Gly (= TVVPG) (pentapeptide)	*Triticum aestivum* (Poaceae) [proteolysate]	ACE (2 μM)	[199]
Tyr-Leu (= YL) (dipeptide)	*Triticum aestivum* (Poaceae) [proteolysate]	ACE (16 μM)	[199]
Val-Ala-Ala (= VAA) (tripeptide)	*Zea mays* (Poaceae) proteolysate]	ACE (13 μM)	[202, 203]
Val-Ala-Tyr (= VAY) (tripeptide)	*Zea mays* (Poaceae) [proteolysate]	ACE (16 μM)	[202, 203]
Val-Phe (=VF) (dipeptide)	*Triticum aestivum* (Poaceae) [proteolysate]	ACE (18 μM)	[199]
Val-Phe-Pro-Ser (= VFPS) (tetrapeptide)	*Triticum aestivum* (Poaceae) [proteolysate]	ACE (0.5 μM)	[199]
Val-Ser-Pro (= VSP) (tripeptide)	*Zea mays* (Poaceae) [proteolysate]	ACE (10 μM)	[202, 203]
Val-Tyr (= VY) (tripeptide)	*Triticum aestivum* (Poaceae) [proteolysate]	ACE [3 μM]	[204]

Several plant proteins have been isolated that inhibit the metalloprotease carboxypeptidase A [205-217] (Table 7), notably potato carboxypeptidase inhibitor PCI [207-217] (Table 7). PCI is a small, cysteine-rich protein with a compact "knotted" structure determined by 3 disulphide links. The C-terminal region inserts into the active site of the carboxypeptidase. The C-terminal glycine is cleaved and remains trapped in the active site, this representing an example of suicide inactivation [207-216].

Table 7. Carboxypeptidase protease inhibitor proteins from plants
For details see the legend to Table 4.

Plant species (other name) (Family)	Protease inhibitor protein	Protease specificity (IC$_{50}$) [K$_d$, K$_i$] (reactive site)	Ref.
Lycopersicon esculentum (*Solanum lycopersicon*) (tomato) (Solanaceae)	Tomato MCPI (37 aa; 4.1 kDa; 6 Cys; 3 S-S)	Metallo-carboxypeptidase (V37 = C-terminal V)	[205, 206]
Solanum tuberosum (potato) (Solanaceae)	Potato carboxypeptidase inhibitor (PCI) (= PCIA; <EHA-G38) (38 aa; 4 kDa; 6 Cys); PCIB (<EQHA-G39), PCIC (A-G36)	Metallo-carboxypeptidase (G38 = C-terminal G) (PCIA)	[207-216]
Solanum tuberosum (potato) (Solanaceae)	Potato carboxypeptidase inhibitor GM7 (PCI homologue) (proprotein: 87 aa; 6 Cys; 3 S-S)	Metallo-carboxypeptidase (inferred C-terminal G)	[217]

Serine protease inhibitor proteins from plants

As reviewed earlier, serine proteases include a variety of extracellular enzymes involved in digestion, blood clotting, wound and infection responses, inflammation, prohormone processing, angiogenesis and tissue remodelling and intracellular enzymes involved in proprotein processing and other cytosolic proteolysis [1-6]. Plants produce a great variety of serine protease inhibitor proteins that can be grouped into various major classes, namely the Bowman-Birk, cereal bifunctional, Kunitz, potato type I, potato type II, mustard family, serpin and squash family protease inhibitor proteins (PIPs). In addition some other types of plant PIPs have been identified, including the arrowhead, defensin, lipid transfer protein and napin PIPs [1, 120, 121]. These various kinds of plant PIPs are reviewed succinctly below.

Bowman-Birk protease inhibitor proteins

A large number of Bowman-Birk PIPs (BBPIPs) have been isolated and structurally characterized [218-269] (Table 8). This complexity has been variously reviewed [1, 120, 121, 224, 228]. The Fabaceae BBPIPs are typically double headed PIPs having 7 disulphide links and molecular masses of 7-9 kDa. Nearly all the legume BBPIPs have been isolated from seeds (Table 8). The best studied BBPIP is BBI-1 from *Glycine max* (soybean) seeds [226-233]. This 8 kDa protein has two inhibitory loops that inhibit trypsin and chymotrypsin, respectively. The amino acids about the scissile bond (P1-P1′) are Lys16-Ser17 (trypsin inhibition) and Leu43-Ser44 (chymotrypsin inhibition), these reactive site structures reflecting the specificity of trypsin and chymotrypsin in cleaving peptide bonds after basic and hydrophobic amino acids, respectively. Soybean BBPIP has a structure resembling a bow-tie with the trypsin- and chymotrypsin-inhibitory loops separated by two anti-parallel β-strands [227, 229, 231]. An exposed loop lying above the β-strands can be phosphorylated by Ca^{2+}-dependent protein kinase [233, 269] and eukaryote cyclic AMP-dependent protein kinase (PKA) [270], noting that a variety of small, basic, disulphide-rich plant defensive proteins have been shown to be phosphorylated by protein kinases [120, 121, 269, 270]. BBI-1 has chemopreventive anticarcinogenic activity [228, 232].

Table 8. Bowman-Birk serine protease inhibitors from legumes
Legume (Fabaceae) Bowman-Birk protease inhibitor proteins (BBPIPs) are typically double-headed protease inhibitors with 14 cysteines (i.e. 7 S-S links) and molecular masses of 7-9 kDa. All the BBPIPs listed were isolated from seeds except for the *Medicago sativa* (alfalfa) leaf BBPIPs. For other details see the legend to Table 4.

Plant species (common name)	Bowman-Birk protease inhibitor proteins	Protease specificity (IC$_{50}$) [K$_d$, K$_i$] (reactive site)	Ref.
Arachis hypogaea (peanut)	A-I, A-II, B-I, B-II, B-III	Chymotrypsin, Trypsin [2 nM, B-III]	[218, 219]
Canavalia gladiata (white sword bean, Japanese jackbean)	SBI-1, SBI-2, SBI-3	Chymotrypsin (L48-S49), Trypsin (L21-S22)	[220]
Canavalia lineata (jackbean)	CLTI-I & CLTI-II	Chymotrypsin [41 & 57 nM] (L48-S49), Trypsin [3 & 7 nM] (K21-S22)	[221]
Dioclea glabra (*D. leiophylla*)	DgTI	Trypsin (K13-S14) [0.5 nM], (H40-S41)	[222]
Dipteryx alata (Brazilian tree)	DATIa, DATIb, DATIc, DATId, DATIe	Trypsin	[223]
Dolichos biflorus (horsegram)	HGI-III	Chymotrypsin (F51-S52), Trypsin (K24-S25)	[224]
Erythrina variegata (Indian coral tree)	EBI	Chymotrypsin (L38-S39), Trypsin (K12-S13)	[225]
Glycine max (soybean)	BBI-1	Chymotrypsin (L43-S44) (0.5), Trypsin (K16-S17) (0.3)	[226-233]
Glycine max (soybean)	C-II	Chymotrypsin, Elastase, Trypsin	[234, 235]
Glycine max (soybean)	D-II	Trypsin	[234, 236]
Glycine max (soybean)	E-I (= PI-I)	Chymotrypsin, Trypsin	[236-238]
Lonchocarpus capassa (apple leaf seed)	DE-1, DE-2, DE-3, DE-4	Chymotrypsin (F52-S53), Trypsin (R25-S26) (DE-4)	[239]
Macrotyloma axillaris (*Dolichos axillaris*)	DE-3, DE-4	Chymotrypsin (F53-S54, DE-3; L52-S53, DE-4), Trypsin (K24-S25, DE-3; K23-S24, DE-4)	[240]
Medicago scutellata (snail medic)	MsTI	Trypsin (R16-S17) [2 nM]	[241]
Medicago sativa (alfalfa leaves)	ATI (Alfalfa trypsin inhibitor), ATI-18, ATI-21 (wounding-induced leaf BBIPs)	Trypsin	[242, 243]
Phaseolus angularis (*Vigna angularis*) (adzuki bean)	I (ABI-I), I-A, I-B, I-A′, II, II′, I I-A	Chymotrypsin (Y53-S54), Trypsin (K26-P27) (I)	[244-249]
Phaseolus aureus (*Vigna radiata*) (mung bean)	BBPIP	Trypsin (K20-S21; R47-S48)	[250]
Phaseolus lunatus (lima bean)	BBPIP	Chymotrypsin (L53-S54), Trypsin (K26-S27)	[251, 252]
Phaseolus vulgaris (kidney bean)	PFSI	Chymotrypsin, Trypsin	[253]
Phaseolus vulgaris (kidney bean)	PVI-3	Chymotrypsin (F56-S57), Trypsin (K29-S30)	[254]

Phaseolus vulgaris (kidney bean)	PVI-4	Chymotrypsin (F57-S58), Trypsin (K30-S31)	[254]
Phaseolus vulgaris (garden bean)	TI-II, TI-II′	Elastase (A25-S26), Trypsin (R52-S53)	[255]
Pisum sativum (pea)	PsTI-I, PsTI-II, PsTI-III, PsTI-IVA, PsTI-IVB, PsTI-V	Chymotrypsin (Y42-S43), Trypsin (K16-S17) [1 nM] (PsTI-II)	[256-260]
Torresea acreana (Amburana acreana)	TaTI	Chymotrypsin, Plasmin, Trypsin (K15, R42), Factor XIIa	[261]
Torresea cearensis (Amburana acreana)	TcTI	Chymotrypsin [50 nM], Plasmin [36 nM], Trypsin [1 nM] (K15, H42), Factor XIIa [1450 nM]	[262]
Vicia angustifolia (common vetch)	VAI	Chymotrypsin (Y42-S43), Trypsin (R16-S17)	[263]
Vicia faba (broad bean, fava bean)	FBI	Chymotrypsin, Trypsin	[264, 265]
Vigna unguiculata (cowpea)	BTCI	Chymotrypsin, Trypsin	[266, 267]
Vigna unguiculata (cowpea)	CPTIs	Chymotrypsin, Trypsin	[268]

A variety of cereal (Poaceae) BBPIPs have been isolated and characterized [224, 271-284] (Table 9). Whereas the legume (Fabaceae) BBPIPs are typically double-headed, 7-9 kDa proteins with 7 disulphide links (Table 8), many cereal BBPIPs are 7-9 kDa and single-headed with 4-5 disulphides such as the *Coix, Hordeum* Bsi1, *Setaria*, type II *Triticum* and *Zea* BBPIPs (Table 9). However some cereal BBPIPs are of the double-headed kind including particular *Hordeum, Oryza* and *Triticum* BBPIPs (Table 9).

Table 9. Cereal (Poaceae) Bowman-Birk protease inhibitor proteins
For details see the legend to Table 8.

Plant species (common name)	Bowman-Birk protease inhibitor proteins	Protease specificity (IC$_{50}$) [K$_d$, K$_i$] (reactive site)	Ref.
Coix lachryma jobi (Job's tears) (Poaceae) [seed]	*Coix* BBI TI-1; TI-2 (64 aa; 7 kDa; 10 Cys)	Trypsin (R17-S18)	[271]
Hordeum vulgare (barley) (Poaceae) [rootlet]	Barley rootlet BBI (124 aa; 14 kDa; 20 Cys)	Trypsin (R17-S18, R75-S76)	[272]
Hordeum vulgare (barley) (Poaceae) [seed]	Barley seed BBI (120 aa; 14 kDa; 20 Cys)	Trypsin (R14-S15, R73-S74)	[273]
Hordeum vulgare (barley) (Poaceae) [coleoptile]	Bsi1 (fungus-induced) (89 aa; 9 kDa; 10 Cys)	Serine protease (inferred)	[274]
Oryza sativa (rice) (Poaceae) [seed, bran]	RBTI (133 aa; 15 kDa; 18 Cys)	Trypsin (K17-P18, K83-M84)	[275]
Oryza sativa (rice) (Poaceae)	OsBBPI (195 aa; 21 kDa; 18 Cys); OsBBTI1 (136 aa; 15 kDa; 18 Cys); OsBBTI1 (254 aa; 28 kDa; 24 Cys);	Trypsin	[276]

Oryza sativa (rice) (Poaceae) [seed, bran]	RBBI-8 (133 aa; 15 kDa; 16 Cys)	Chymotrypsin (weak), Trypsin (K16-T17, R84-M85)	[277]
Setaria italica (foxtail millet) (Poaceae) [seed]	Setaria FMTI-II (67 aa; 7 kDa; 10 Cys)	Trypsin (K16-S17)	[278]
Setaria italica (foxtail millet) (Poaceae) [seed]	Setaria FMTI-III (67 aa; 7 kDa; 10 Cys)	Trypsin (K16-S17)	[279]
Triticum aestivum (wheat) (Poaceae) [seed]	Triticum BBI Type I (double headed) - I, I-2a, I-2b (126 aa), I-2c (14 kDa; 18 Cys)	Trypsin (R17-S18, I-2b, domain I)	[280]
Triticum aestivum (wheat) (Poaceae) [seed]	Triticum BBI Type II (single-headed) - II-4 (61 aa; 6 kDa; 9 Cys), II-5, II-6a, II-6b, II-7a	Trypsin (K15-S16, II-4)	[280]
Triticum aestivum (wheat) (Poaceae) [seed]	WTI (BBI Type I, single headed) (71 aa; 8 kDa; 12 Cys)	Trypsin (R19-M20)	[281]
Zea mays (Poaceae) [seedling]	WIP1 (gene) (wounding-induced expression) (72 aa; 9 kDa; 10 Cys)	Chymotrypsin (F52-S53, Y82-S83)	[282-284]
Zea mays (Poaceae)	WIP1 variants (genes) (9 kDa; 10 Cys)	Chymotrypsin	[284]
Zea luxurians (Poaceae)	WIP1 variants (genes) (9 kDa; 10 Cys)	Chymotrypsin	[284]

Various non-legume, non-Poaceae Bowman-Birk protease inhibitors have been isolated that have sequence homology to the other BBPIPs [137-140, 285-288] (Table 10). The *Ananas* BBPIPs are disulphide-linked heterodimers [137-140]. The *Helianthus* BBPIP SFTI-1 is a small cyclic peptide (cyclotide) [285, 286] whereas the *Solanum tuberosum* BBPIP is similar to the single-headed Poaceae BBPIPs [287, 288].

Table 10. Non-legume, non-cereal Bowman-Birk protease inhibitor homologues
For details see the legend to Table 8.

Plant species (common name)	Bowman-Birk protease inhibitor proteins	Protease specificity (IC$_{50}$) [K$_d$, K$_i$] (reactive site)	Ref.
Ananas comosus (pineapple) (Bromeliaceae) [stem]	Ananas BI-I, BI-II, BI-III, BI-IV, BI-V, BI-VI (6 kDa; A (41 aa, 7 Cys)-(S-S)$_2$-B (11 aa; 2 Cys); homologous to BBIs	Trypsin (weak), Chymotrypsin (weak), Cysteine proteases (Bromelain, Cathepsin L, Papain)	[137-140]
Helianthus annuum (sunflower) (Helianthus annuus (Asteraceae) [seed]	Helianthus BBI SFTI-1 (14 aa; 1.5 kDa; 2 Cys; cyclotide = cyclic peptide; synthetic acyclic analogue: Trypsin K$_i$ 12 nM)	Cathepsin G [0.2 nM], Chymotrypsin [7 μM], Elastase [105 μM], Thrombin [136 μM], Trypsin [0.1-0.5 nM] (K5-S6) (acyclic SFTI-1 [12 nM])	[285, 286]
Solanum tuberosum (Solanaceae) [tuber]	Potato BBI (PBBI) (74 aa; 8 kDa; 8 Cys)	Serine protease	[287, 288]
Solanum tuberosum (Solanaceae) [tuber]	BBI-homologue (10 kDa; lacks key conserved BBI Cysteines)	Serine protease (inferred)	[289]

Cereal dual function α-amylase/trypsin inhibitor proteins

The cereal dual function α-amylase/trypsin inhibitor proteins are cysteine-rich, disulphide-rich, double-headed, 13-16 kDa, dual function inhibitor proteins that inhibit both of the digestion enzymes α-amylase and trypsin [290-325] (Table 11). Thus the *Zea* (corn) member of this family, corn Hageman factor inhibitor (CHFI), is a double-headed 14 kDa protein that inhibits α-amylase and the serine proteases trypsin and blood clotting Factor XIIa [323-324] (Table 11). The structures of the bifunctional α-amylase/trypsin inhibitor proteins from *Eleusine* (ragi) (RBI) [292-295] and *Zea* (corn) (CHFI) [325] have been determined. These proteins are structurally similar to the lipid transfer proteins, being composed of a bundle of 4 α-helices together with a short β-sheet element connected by loops, the α-amylase- and protease-inhibitory domains being separately located [325].

Table 11. Cereal dual function α-amylase/trypsin inhibitor proteins
For details see the legend to Table 4.

Plant species (other name) (Family)	Protease inhibitor protein	Protease specificity (IC$_{50}$) [K$_d$, K$_i$] (reactive site)	Ref.
Eleusine coracana (ragi, Indian finger millet) (Poaceae) [seed]	RBI (Ragi bifunctional I) = RATI (Ragi α-Amylase and Trypsin Inhibitor) (122 aa; 13 kDa; 10 Cys)	Trypsin (R34-L35) [1 nM], (α-Amylase inhibitor, K$_i$ 11 nM)	[290-297]
Hordeum vulgare (barley) (Poaceae) [seed]	Homologous Barley chloroform-methanol soluble proteins a-e = Barley CMa-e = BTAI-CMa-e, "CMx" (~16 kDa; 10 Cys)	Trypsin (CMc & CMe) [CMa (= BTAI-CMa) inhibits α-Amylase]	[298-303]
Hordeum vulgare (barley) (Poaceae) [seed]	CMc (~13 kDa; 10 Cys)	Trypsin	[298, 300]
Hordeum vulgare (barley) (Poaceae) [seed]	CMe (121 aa; 13 kDa; kDa; 10 Cys)	Trypsin	[298, 302]
Hordeum vulgare (barley) (Poaceae) [seed]	CMa (~13 kDa; 10 Cys)	[CMa (= BTAI-CMa) inhibits α-Amylase]	[298-300]
Hordeum vulgare (barley) (Poaceae) [seed]	CMb, CMd, CMx, BMAI-1, BDAI-1 (13-15 kDa; 10 Cys)	Homologues of CMa, CMc & CMe	[298-299, 303, 304]
Oryza sativa (Poaceae) [seed]	RAP (Rice allergenic protein) = RA17 (135 aa; 15 kDa; 10 Cys)	Cereal αAmylase /Trypsin inhibitor family homologue	[305, 306]
Oryza sativa (Poaceae) [seed]	RA5 (14 kDa; 10 Cys), RA14 (15 kDa; 10 Cys)	Cereal α-Amylase /Trypsin inhibitor homologue	[306]
Sorghum biolor (sorghum) (Poaceae)	SIα4 (13 kDa; 10 Cys), SIα5 (13 kDa; 10 Cys)	Cereal α-Amylase /Trypsin inhibitor family homologue	[307]

		(α-Amylase inhibitors)	
Triticum aestivum (wheat) (Poaceae)	Chloroform-methanol soluble proteins CM1, CM2, CM3, CM16, CM17 (13-16 kDa; 10 Cys; heterotetramers)	Cereal α-Amylase /Trypsin inhibitors	[308-311]
Triticum aestivum (wheat) (Poaceae)	CMX1, CMX2, CMX3 (11-15 kDa; 10 Cys)	Cereal α-Amylase /Trypsin inhibitor homologues	[312]
Triticum aestivum (wheat) (Poaceae)	0.19 inhibitor (14 kDa; 10 Cys; homodimer)	Cereal α-Amylase /Trypsin inhibitor homologue - α-Amylase inhibitor	[313-316]
Triticum aestivum (wheat) (Poaceae)	0.28 inhibitor (= WMAI-1) (13 kDa; 10 Cys; monomer)	Cereal α-Amylase /Trypsin inhibitor homologue - α-Amylase inhibitor	[317, 318]
Triticum aestivum (wheat) (Poaceae)	0.39 inhibitor (13 kDa; 10 Cys; monomer)	Cereal α-Amylase /Trypsin inhibitor homologue - α-Amylase inhibitor	[319]
Triticum aestivum (wheat) (Poaceae)	0.53 inhibitor (13 kDa; 9 Cys; 4 S-S)	Cereal α-Amylase /Trypsin inhibitor homologue - α-Amylase inhibitor	[320, 321]
Triticum aestivum (wheat) (Poaceae)	(12 kDa; monomer)	Cereal α-Amylase /Trypsin inhibitor homologue - α-Amylase inhibitor	[322]
Zea mays (corn) (Poaceae)	Corn Hageman Factor inhibitor (CHFI) (127 aa; 14 kDa; 10 Cys)	Trypsin, Factor XIIa (Hageman Factor) (inhibits α-Amylase)	[323-325]

Plant Kunitz serine protease inhibitor proteins

A large number of Kunitz-type protease inhibitor proteins (PIPs) have been isolated from plants [326-435] (Table 12). These *circa* 20 kDa proteins fall into two major classes which are nevertheless related through sequence homology. The Kunitz PIPs from seeds of the legumes (Fabaceae) *Acacia confusa* [326-329], *Adenanthera pavonina* [330], *Enterolobium contorsiliquum* [347, 348] and *Prosopis juliflora* [395, 396] and some *Solanum tuberosum* Kunitz PIPs [429] are disulphide-linked heterodimers. Other Kunitz PIPs are typically *circa* 20 kDa proteins having 4-6 cysteines (i.e. 2-3 disulphide linkages). Some such *circa* 20 kDa Kunitz PIPs such as those from *Alocasia macrorrhiza* [332, 333] and *Canavalia lineata* [342, 343] can form homodimers, for example by intermolecular disulphide linkage [342, 343]. Miraculin is a 25 kDa glycoprotein from *Richadella dulcifica* that is homologous to the Kunitz PIPs, forms a disulphide-linked homodimer and converts a sour taste to sweet [416-420]. Kunitz PIPs from *Swartzia pickellii* are also glycosylated [430].

While most Kunitz PIPs inhibit serine proteases, some Kunitz PIP homologues inhibit aspartic proteases or cysteine proteases. Thus various

Solanum (potato) Kunitz PIPs inhibit the aspartic protease cathepsin D as well as trypsin [125-134] and potato cysteine protease inhibitor (PCPI) inhibits a variety of cysteine proteases [185-188]. The crystal structures of soybean trypsin inhibitor (STI) [362, 368] and of *Erythrina* trypsin inhibitor (ETI) [350] have been determined. The structure of this type of plant Kunitz serine PIP involves a β-barrel formed by 6 loop-linked anti-parallel β-strands with a "lid" formed by 6 further loop-linked anti-parallel β-strands. The scissile bond is located within a loop that extends out from the surface of the β-barrel [350, 362, 368].

Table 12. Plant Kunitz serine protease inhibitor proteins
For details see the legend to Table 4.

Plant species (other name) (Family)	Protease inhibitor protein	Protease specificity (IC$_{50}$) [K$_d$, K$_i$] (reactive site)	Ref.
Acacia confusa (Fabaceae) [seed]	ACTI-A (ACTI-A (136 aa; 15 kDa)-S-S- ACTI-B (39aa; 4 kDa)	Chymotrypsin [0.6 nM]. Trypsin [0.3 nM] (K64-I65)	[326-329]
Adenanthera pavonina (Brazilian Carolina tree) (Fabaceae) [seed]	DE5 (138 aa; α chain (15 kDa)-S-S-β chain (38 aa; 4 kDa)	Trypsin (R64-I65)	[330]
Albizzia julibrissin (silk tree) (Fabaceae) [seed]	A-II & A-III (22 kDa), B-I & B-II (19 kDa) (α chain-S-S-β chain heterodimeric KPIs)	Trypsin & Chymotrypsin (AII, AIII); Chymotrypsin & Elastase (BI, BII)	[331]
Alocasia macrorrhiza (giant taro) (Araceaee) [tuber]	GTPI (184 aa; 20 kDa; homodimer)	Chymotrypsin (L56-A57), Trypsin	[332, 333]
Arabidopsis thaliana (mouse-ear cress) (Brassicaceae)	Genes encoding KPI-like proteins (22 kDa)	Serine protease (inferred)	[334]
Bauhinia bauhinioides (Fabaceae)	KPI (20 kDa)	Trypsin, Kallikrein	[335]
Bauhinia mollis (Fabaceae)	KPI (20 kDa)	Trypsin	[335]
Bauhinia pentandra (Fabaceae)	KPI (20 kDa)	Trypsin, Kallikrein, Factor XIIa	[335]
Bauhinia variegata (mountain ebony) (Fabaceae)	*B. variegata* lilac (Bvl) BvlTI-1, BvlTI-2 & BvlTI-3; *B. variegata* candida (Bvc) BvcTI-1, BvcTI-2 & BvcTI-3 (167 aa; 19 kDa)	Trypsin [7 nM, BvcTI; 1 nM, BvlTI], R63-I64 (BvcTI-3)	[336]
Brassica napus (rape) (Brassicaceae) [leaf]	BnD22 (drought-induced; 22 kDa)	Serine protease	[337]
Brassica oleraceae (cabbage) (Brassicaceae)	BoKPI (22 kDa)	Serine protease	[338]
Brassica oleraceae (cabbage) (Brassicaceae)	Water-soluble chlorophyll-carrying protein (WSCPI); heat-stress-induced protein (HTI) (24 kDa)	Serine protease	[339-341]
Canavalia lineata (Fabaceae) [seed]	CLSI-II (190 aa; 21 kDa; 5 Cys; 2 S-S; 40 kDa dimer per intermolecular S-S)	Subtilisin BPN' [28 nM], Subtilisin Carlsberg [11 nM], Proteinase K [31 nM]	[342, 343]
Canavalia lineata	CLSI-III (183 aa; 20 kDa; 5	Subtilisin BPN' [16 nM],	[342,

(Fabaceae) [seed]	Cys; 2 S-S; 40 kDa dimer per intermolecular S-S)	Subtilisin Carlsberg [20 nM], Proteinase K [58 nM]	343]
Canavalia lineata (Fabaceae) [seed]	*Canavalia* CLTI-III (21 kDa)	Chymotrypsin [540 nM], Trypsin [4 nM] (R68-G69)	[344]
Carica papaya (Caricaceae) [latex]	*Carica* KPI (183 aa; 20 kDa; 4 Cys; 2 S-S; glycosylated)	Chymotrypsin [700 nM], Trypsin [300 nM]	[345]
Delonix regia (Fabaceae)	DrTI (185 aa; 20 kDa; 4 Cys; 2 S-S)	Kallikrein [5 nM], Trypsin [22 nM]	[346]
Enterolobium contorsiliquum (Fabaceae) [seed]	*Enterolobium* EcTI (174 aa; α (134 aa; 15 kDa), β (40 aa; 4 kDa); 4 Cys; 2 S-S; heterodimeric KPI)	Chymotrypsin [120 nM], Factor XIIa [15 nM], Kallikrein [5 nM], Plasmin [18 nM], Trypsin [2 nM] (R64-I65)	[347, 348]
Erythrina caffra (Fabaceae) [seed]	*Erythrina* DE-3 (172 aa; 19 kDa; 4 Cys; 2 S-S)	Chymotrypsin, Trypsin (R63-S64)	[349, 350]
Erythrina caffra (Fabaceae) [seed]	*Erythrina* DE-1, DE-2, DE-4 (164-166 aa; 18 kDa; 4 Cys; 2 S-S)	DE-1 (Chymotrypsin , Trypsin), DE-2 (Chymotrypsin), DE-4 (Trypsin)	[349]
Erythrina latissima (Fabaceae) [seed]	DE-1 & DE-3 (18-22 kDa)	DE-1 (Chymotrypsin), DE-3 (Trypsin)	[351]
Erythrina variegata (Fabaceae) [seed]	*Erythrina* ETI-a (172 aa; 19 kDa), ETI-b (176 aa; 20 kDa)	ETI-a - Plasminogen activator, Trypsin (R63-S64); ETI-b - Trypsin (R63-S64)	[352, 353]
Erythrina variegata (Fabaceae) [seed]	ECI (179 aa; 20 kDa; 4 Cys; 2 S-S)	Chymotrypsin [98 nM] (L64-S65)	[354-358]
Glycine max (soybean) (Fabaceae) [seed]	Tia (= KTi3), Tib & Tic (181 aa; 20 kDa; 5 Cys)	Trypsin [0.2 nM, 3 pM - Tia; 20 nM - Tib] (R63-I64)	[359-368]
Glycine max (soybean) (Fabaceae) [seed]	KTi1 & KTi2 (2 of ~10 soybean KPI genes) (179 aa; 20 kDa; 5 Cys)	KPI homologues	[364, 369]
Glycine max (soybean) (Fabaceae) [seed]	p20 (182 aa; 20 kDa; 5 Cys)	Trypsin (K64-I65) (also binds GTP)	[370]
Glycine max (soybean) (Fabaceae) [seed]	KTi-S (183 aa; 20 kDa; 5 Cys)	Trypsin (K65-I66)	[371]
Hordeum vulgare (barley) (Poaceae) [seed]	BASI (barley α-Amylase & Subtilisin inhibitor) (177 aa; 20 kDa;)	Subtilisin [0.2 nM] (V67-A68) (also inhibits α-Amylase)	[372-375]
Ipomoea batatas (sweet potato) (Solanaceae)	Sporamin A & Sporamin B (many genes in Sporamin A & B sub-families) (22 kDa; 4 Cys; 2 S-S)	Trypsin	[376-382]
Lycopersicon esculentum (tomato) (Solanaceae)	*Lycospersicon* TI gene (188 aa; 21 kDa; 6 Cys; 3 S-S)	Cathepsin D, Trypsin (R67-F68) (inferred)	[383]
Lycopersicon esculentum (tomato) (Solanaceae)	*Lycopersicon* Miraculin-like protein (LeMir) (181 aa; 20 kDa; 6 Cys; 3 S-S)	STI (Kunitz) & Miraculin homologue (Miraculin converts sour taste to sweet)	[384]
Lycopersicon esculentum (tomato) (Solanaceae)	TPP11 (gene expressed in flower tissue) (20 kDa; 6 Cys; 3 S-S)	STI (Kunitz) homologue	[385]
Nicotiana glauca x N. langsdorffii (tobacco) (Solanaceae)	TobTID91 (gene), TRP (tumour-related protein) (~ 20 kDa; 5 Cys)	STI homologue (plant tumour-related Kunitz homologue)	[386, 387]

Nicotiana glutinosa (tobacco) (Solanaceae)	NgPI (proprotein: 241 aa; 27 kDa; 5 Cys)	Kunitz PI homologue (inferred PI)	[388]
Oryza sativa (rice) (Poaceae) [seed]	Rice α-Amylase-Subtilisin Inhibitor (RASI) (176 aa; 19 kDa; 4 Cys; 2 S-S) (minor variants also)	Subtilisin (A67-A68) (& also inhibits α-Amylase)	[389]
Pisum sativum (pea) (Fabaceae)	α-Fucosidase (20 kDa; 4 Cys)	No protease inhibitor activity found (but KTI homologue & trypsin-resistant)	[390, 391]
Populus balsamifera (poplar) (Salicaceae)	gwin3 (proprotein: 200 aa; 22 kDa; 5 Cys)	Inferred KPI (wounding-induced Kunitz homologue)	[392]
Populus tremuloides (aspen) (Salicaceae)	PtTI1 (4 Cys), PtTI2 & PtTI3 (5 Cys) (proprotein: 24 kDa)	Trypsin (PtTI2)	[393]
Populus trichocarpa x P. deltoides (poplar) (Salicaceae)	5 win3 genes (e.g. win3.3, win3.6, win3.12) (proprotein: 24 kDa; 5 Cys)	Inferred KPIs (wounding induced)	[394]
Prosopis juliflora (Algaroba tree) (Fabaceae) [seed]	PjKTI (α chain (137 aa; 14 kDa)-S-S-β-(38 aa; 4 kDa)	Trypsin [0.6 nM] (R64-I65)	[395, 396]
Psophocarpus tetragonolobus (winged bean) (Fabaceae) [seed]	Pt nodulin (proprotein: 201 aa; 22 kDa; processed: 21 kDa; 4 Cys; 2 S-S)	Trypsin (50 nM) (inferred K56-S57)	[397]
Psophocarpus tetragonolobus (winged bean) (Fabaceae) [seed]	WTI-1 (= WBTI-1) (variants WTI-1A, WTI-1B) (172 aa; 19 kDa; 4 Cys; 2 S-S)	Trypsin [3 nM] (R64-S65)	[398]
Psophocarpus tetragonolobus (winged bean) (Fabaceae) [seed]	WBTI-2 (182 aa; 20 kDa; 4 Cys; 2 S-S)	Trypsin	[399]
Psophocarpus tetragonolobus (winged bean) (Fabaceae) [seed]	WTI-2, WTI-3, WCI-1 (22 kDa), WCI-2, WCI-3, WCI-4, WCI-X, WTCI-1, WBTI-2a, Psophocarpin B1 (20 kDa; 4 Cys; 2 S-S) (& further variants)	Chymotrypsin - WTI-2 [70], WTI-3 [70 nM], WCI-1 [6 nM], WTCI-1 [2 nM]; Trypsin - WTI-2 [4 nM], WTI-3 [0.2 nM], WTCI-1 [0.6 nM]	[400-413]
Psophocarpus tetragonolobus (winged bean) (Fabaceae) [seed]	WBA-1 (175 aa; 19 kDa; 2 Cys; 1 S-S)	No inhibition of Trypsin (Kunitz homologue)	[414, 415]
Richadella dulcifica (miracle fruit) (Sapotaceae) [fruit]	Miraculin (MIR) (25 kDa glycoprotein; 220 aa; 24 kDa; 7 Cys; 3 S-S; 1 interchain S-S)	Soybean KPI homologue (converts sour to sweet taste)	[416-420]
Salix viminalis (Salicaceae)	SWIN1.1 (proprotein: 199 aa; 21 kDa; 7 Cys; 3 S-S)	Trypsin (inferred; R62-I63), putative KPI	[421]
Salix viminalis (Salicaceae)	SVTI (15 kDa)	Trypsin (Kunitz PI homologue)	[421]
Schizolobium parahybum (Fabaceae) [seed]	SPC (20 kDa; 4 Cys; 2 S-S)	Chymotrypsin [59 nM]	[422]
Solanum tuberosum (potato) (Solanaceae)	PDI, NID, PI-8, PI-13, p749, API-13, clone 4, CathInh, gCDI-A1, pA1 (188-189 aa; 21 kDa; 6 Cys; 3 S-S)	Cathepsin & Trypsin; Cathepsin D [1 nM, NDI], Trypsin (R67-F68, NDI)	[125-134]
Solanum tuberosum (potato) (Solanaceae)	PI-4, PIG, CDI homologue (186-189 aa; 20-21 kDa; 5 Cys)	Cathepsin D, Trypsin (R64-F65, PIG	[128, 130]
Solanum tuberosum (potato)	Potato cysteine protease	Actinidin [27 nM],	[185-

(Solanaceae)	inhibitor PCPI 8.3 (= p34021) (180 aa; 20 kDa,; 4 Cys; 2 S-S; Kunitz homologue; PCPI subclass I; 58% β structure)	Ananain [60 pM], Chymopapain [210 nM], Cathepsin B [100-320 nM], Cathepsin L [70-400 pM], Ficin [210 nM], Fruit Bromelain [33 nM], Papain [3 nM], Pinguinain [35 nM], Proteinase III [150 nM], Stem Bromelain [190 nM],	188]
Solanum tuberosum (potato) (Solanaceae)	PCPI 9.4 (22 kDa; Kunitz homologue subclass II)	Cathepsin B [90 nM], Cathepsin H [400 nM], Cathepsin L [60 pM], Papain [100 nM]	[185, 186, 188
Solanum tuberosum (potato) (Solanaceae)	PCPI 6.6 (25 kDa; Kunitz homologue; PCPI subclass III)	Cathepsin B [100 nM], Cathepsin L [2 nM], Papain [50 nM]	[185, 186, 188
Solanum tuberosum (potato) (Solanaceae)	Other PCPIs (Kunitz PI homologues; cf PCPI 8.3) - PCPI 3, PCPI 5, pT1, PKI-2 (PCPI subclass I), PCPI 7, PCPI 8, PCPI 10 (PCPI subclass II) (~ 20-25 kDa)	Cathepsin B [100 nM], Cathepsin L [0.4 nM	[186]
Solanum tuberosum (Solanaceae) [tuber]	g-CDI-B1, pF4 & pKEN14-28 KPI genes (20-24 kDa; 4 Cys; 2 S-S)	Trypsin (inferred)	[423]
Solanum tuberosum (Solanaceae) [tuber]	Kunitz PI homologue AM66 (20 kDa)	Trypsin (inferred)	[217]
Solanum tuberosum (Solanaceae) [tuber]	PKI-1 (20 kDa)	Trypsin	[424, 425]
Solanum tuberosum (Solanaceae) [tuber]	pTI (= PKI-2) (180 aa; 20 kDa; 4 Cys; 2 S-S)	Chymotrypsin, Subtilisin, Trypsin	[424, 425]
Solanum tuberosum (Solanaceae) [tuber]	p340 & p34021 (22 kDa; 4 Cys; 2 S-S)	Trypsin	[423, 426-428]
Solanum tuberosum (Solanaceae) [tuber]	Potato serine protease inhibitors PSPI-21-5.2 (21 kDa, pI 5.2; 150 aa A-S-S-36 aa B; 4 Cys; 2 S-S), PSPI-21-6.3 (21 kDa, pI 6.3; 150 aa A-S-S-37 aa B; 4 Cys; 2 S-S)	Chymotrypsin, Elastase, Trypsin	[429]
Swartzia pickellii (Fabaceae) [seed]	SWTI (aglycone & 2 glycosylated forms) (aglycone: 174 aa; 19 kDa; 2 Cys; 1 S-S),	Chymotrypsin [123 nM], Kallikrein [20 nM], Plasmin [3 nM], Trypsin [47 nM] (Q65-I66)	[430]
Theobroma cacao (cocoa) (Malvaceae) [seed]	CoATi (21 kDa; 7 Cys; 3 S-S)	Chymotrypsin [2300 nM], Trypsin [95 nM]	[431-433]
Triticum aestivum (wheat) (Poaceae) [seed]	WASI (Wheat α-Amylase & Subtilisin inhibitor) (180 aa; 20 kDa Kunitz-related protein; 4 Cys; 2 S-S)	Subtilisin (& α-Amylase)	[434, 435]

Potato Type I serine protease inhibitors

Proteins of the potato type I serine protease inhibitor protein family (Type I PIPs) are 7-9 kDa proteins having either no cysteines or 2 cysteines involved in a disulphide link [436-483] (Table 13). These proteins are important in defence of plants from herbivores and pathogens. Thus Type I PIPs are induced in *Lycopersicon esculentum* (tomato) [460, 461], *Solanum tuberosum* (potato) [476] and *Zea mays* (corn) [481] leaves by wounding. TMV-infection of *Nicotiana tabacum* (tobacco) leaves induces a Type I PIP [468] and a type I PIP is induced in corn leaves by fungal infection [483]. A Type I PIP is induced in tomato fruit by the ripening-promoting growth regulator ethylene [462], indicating a developmental role for Type I PIPs as well as a role in defense of plants against herbivory and pathogens. The crystal structure of *Hordeum vulgare* (barley) Type I PIP CI-2 has been determined [455] and solution structures for CI-2 [456] and *Linum usitatissimum* (flax) Type I PIP (LUTI) [459] have also been determined. These structures involve a series of parallel (2) and anti-parallel (6) β-strands bounded on one side by an α-helix and on the other by a loop carrying the scissile bond of the serine protease binding site [455, 456, 459].

Table 13. Potato Type I serine protease inhibitors
For details see the legend to Table 4.

Plant species (other name) (Family)	Protease inhibitor protein	Protease specificity (IC$_{50}$) [K$_d$, K$_i$] (reactive site)	Ref.
Amaranthus caudatus (amaranth) (Amaranthaceae) [seed]	Amaranth Trypsin & Subtilisin inhibitor (ATSI) (69 aa; 8 kDa; 2 Cys; 1 S-S); minor des-A1R2 form (67 aa)	Cathepsin G [122 nM], Chymotrypsin [0.4 nM], Factor XIIa [440 nM], Plasmin [38 nM], Subtilisin [0.4 nM], Trypsin [0.3 nM] (K45-D46)	[436]
Amaranthus hypochondrionacus (Amaranthaceae) [seed]	Amaranth trypsin inhibitor (AmTI) (8 kDa; 2 Cys; 1 S-S)	Trypsin (K45D46)	[437]
Arabidopsis thaliana (mouse-ear cress) (Brassicaceae)	*Arabidopsis* genes encoding ~ 8 kDa Potato type 1 serine protease inhibitors	Trypsin (inferred)	[438]
Canavalia lineata (Fabaceae) [seed]	*Canavalia* CLSI-1 (65 aa; 7 kDa; 0 Cys)	Subtilisin BPN' [96 nM] (A41-D42), Subtilisin Carlsberg [7 nM], *Streptomyces* alkaline protease [170 nM], *Aspergillus* protease (10 nM), Proteinase K (1 nM),	[439]

		Bacillus protease (1 nM), Pronase (100 nM)	
Cucurbita maxima (squash) (Cucurbitaceae) [seed]	CMTI-V (68 aa; 7 kDa; 2 Cys; 1 S-S)	Activated Hageman Factor (XIIa) [41 nM], Trypsin [16 nM] (K44-D45)	[440, 441]
Cucurbita maxima (squash) (Cucurbitaceae) [seed]	Pumpkin fruit trypsin inhibitor (PFTI) (8 kDa; 0 Cys)	Trypsin	[440, 442]
Cucurbita maxima (squash) (Cucurbitaceae) [seed]	Af4 gene product (67 aa; 8 kDa; 0 Cys)	Trypsin (K43-D44)	[440, 442]
Cucurbita maxima (squash) (Cucurbitaceae) [seed]	Bm7 & Pr10 (67 aa; 8 kDa; 0 Cys)	Chymotrypsin (L43-D44)	[440, 442]
Fagopyrum esculentum (buckwheat) (Polygonaceae) [seed]	BWI-1 (69 aa; 8 kDa; 2 Cys; 1 S-S) (IT1 = BWI-1; IT2, IT4 homologues)	Trypsin [1 nM] (R45-D46), Chymotrypsin [67 nM], Cathepsin G [200 nM]	[443-445]
Fagopyrum esculentum (buckwheat) (Polygonaceae) [seed]	BWI-4a (67 aa; 7 kDa; 2 Cys; 1 S-S) (BWI-4a′ = G40-BWI-4a)	Trypsin (R43-D44) (BWI-4a)	[446, 447]
Fagopyrum esculentum (buckwheat) (Polygonaceae) [seed]	BWI-3c (8 kDa); BWI-4c (6 kDa)	Trypsin, Chymotrypsin, bacterial Subtilisin-like proteases	[448]
Fagopyrum esculentum (buckwheat) (Polygonaceae) [seed]	BTI-1 & BTI-2 (69 aa; 8 kDa; 2 Cys; 1 S-S); BTI-3 (third homologue)	Trypsin (R45-D46) (BTI-1 & BTI-2)	[449]
Hordeum vulgare (barley) (Poaceae)	CI-1a & CI-1b (74 aa; 8 kDa; 0 Cys)	Subtilisin Carlsberg [0.2 & 0.2 nM], Subtilisin BPN′ [1 & 3 nM], Elastase (2 & 2 nM), Chymotrypsin [2300 & 3300 nM] (L55-N56 & L55-D56)	[440, 450, 451]
Hordeum vulgare (barley) (Poaceae)	CI-1c (77 aa; 8 kDa; 0 Cys)	Chymotrypsin (L53-N54)	[440, 450, 452]
Hordeum vulgare (barley) (Poaceae) [seed]	CI-2 (84 aa; 9 kDa; 0 Cys)	Chymotrypsin (M59-E60)	[440, 450-457]
Linum usitatissimum (flax) (Linaceae) [seed]	LUTI (69 aa; 8 kDa; 2 Cys; 1 S-S)	Chymotrypsin [400 nM], Cathepsin G [70 nM], Subtilisin BNP′ [280 nM], Subtilisin Carlsberg [230 nM], Trypsin [0.8 nM] (K45-Q46)	[458, 459]
Lycopersicon esculentum (tomato) (Solanaceae) [leaf]	TI-1 (75 aa; 8 kDa; 2 Cys; 1 S-S)	Chymotrypsin (L45-D46)	[440, 460, 461]
Lycopersicon esculentum (tomato) (Solanaceae) [leaf]	ERI (71 aa; 8 kDa; 2 Cys; 1 S-S)	Glu-specific protease? (E47-D48)	[462]
Lycopersicon peruvianum (tomato) (Solanaceae) [leaf]	DI-1 (71 aa; 8 kDa; 2 Cys; 1 S-S)	Trypsin (K47-D48)	[440, 463, 464]
Momordica charantia (bitter gourd) (Cucurbitaceae) [seed]	BGIA (68 aa; 7 kDa; 2 Cys; 1 S-S)	Streptomyces griseus Glutamate Endopeptidase [70 nM], Subtilisin Carlsberg	[465]
Momordica charantia (bitter gourd)	MCI-3 (62 aa; 7 kDa; 0 Cys)	Trypsin	[466]

(Cucurbitaceae) [seed]			
Nicotiana glauca x *N. langsdorffii* (tobacco) (Solanaceae)	GTI (72 aa; 8 kDa; 2 Cys; 1 S-S)	Serine protease (Q47-D48)	[440, 467]
Nicotiana tabacum (tobacco) (Solanaceae)	PI-Ia (TIMPb), PI-Ib (TIMPa) (71 aa; 8 kDa; 2 Cys; 1 S-S)	Serine protease (E47-D48, PI-Ia & PI-Ib)	[440, 468, 469]
Nicotiana sylvestris (tobacco) (Solanaceae)	NSTI-I (71 aa; 8 kDa; 2 Cys; 1 S-S)	Serine protease (K48-D49)	[440, 470]
Sambucus nigra (European elder) (Adoxaceae)	*Sambucus* SPI (79aa; 9 kDa; 4 Cys; 2 S-S)	Serine protease (T55-D56)	[471]
Solanum tuberosum (potato) (Solanaceae) [tuber]	PIN1 (71 aa; 8 kDa; 2 Cys; 1 S-S)	Serine protease (M47-D48)	[440, 472]
Solanum tuberosum (potato) (Solanaceae) [tuber]	PI-I variants A, B, C & D (71 aa; 8 kDa; 2 Cys; 1 S-S)	Chymotrypsin (M47-D48)	[440, 473-476]
Solanum tuberosum (potato) (Solanaceae) [tuber]	PI-I variant (71 aa; 8 kDa; 2 Cys; 1 S-S)	Chymotrypsin (L47-D48)	[440, 477]
Vicia faba (broad bean) (Fabaceae) [seed]	*Vicia* PI-I (62 aa; 7 kDa; 0 Cys)	Subtilisin (A38-D39)	[478]
Vigna (*Phaseolus*) *angularis* (adzuki bean) (Fabaceae) [seed]	ASI-I (92 aa; 11 kDa; 0 Cys) & processing variant ASI-II (N-terminal 19 aa missing)	Subtilisin (A68-D69)	[479]
Zea mays (corn) (Poaceae) [seed]	Maize protease inhibitor (MPI) (73 aa; 8 kDa; 2 Cys; 1 S-S)	Chymotrypsin, Elastase	[480-483]

Potato Type II serine protease inhibitors

A variety of potato Type II serine protease inhibitor proteins (Type II PIPs) have been resolved from particular Solanaceae sources including *Capsicum annuum* (pepper, paprika), *Lycopersicon esculentum* (tomato), *Nicotiana alata*, *N. glutinosa* and *N. tabacum* (tobacco), *Solanum americanum*, *S. melongena* (eggplant) and *S. tuberosum* (potato) [484-514] (Table 14). These cysteine-rich PIPs derive from precursors having multiple protease inhibitory domains which are proteolytically processed to yield single domain PIPs. These PIPs are exemplified by potato chymotrypsin inhibitor 1 (PCI-1), a 6 kDa serine PIP having 8 cysteines and accordingly 4 disulphide linkages stabilizing the structure [502, 503]. Multidomain potato type II PIP precursors have been resolved which have up to 3 PIP domains [504-514]. The *Nicotiana alata* precursor protein (NaPI II) is a 6 domain, disulphide-rich protein that adopts a circular conformation. Subsequent processing yields monomeric 6 kDa protease inhibitors, namely 4 single chain trypsin inhibitors (T1-T4), a single chain chymotrypsin inhibitor (C1) and a disulphide-linked heterodimer chymotrypsin inhibitor C2 which is formed from the linkage of the processed N- and C-terminal end regions of the precursor [491-496]. The various Solanaceae Type II PIPs have protease interaction K_i values in the nanomolar and picomolar range (Table 14). The monomeric *Nicotiana alata* [495] and *Solanum tuberosum* [503] Type II PIPs have structures stabilized by 4 disulphide linkages and involving 3 anti-parallel β-strands and a large looped region containing the protease-binding site [495, 503]. Solanaceae Type II PIPs can be induced by wounding [512]. Methyl jasmonate (a signalling molecule for systemic plant response to wounding) and abscisic acid (involved in downstream signal transduction in wounding) both induce potato Type II PIP expression [512].

Table 14. Potato Type II serine protease inhibitors
For details see the legend to Table 4.

Plant species (other name) (Family)	Protease inhibitor protein	Protease specificity (IC$_{50}$) [K$_d$, K$_i$] (reactive site)	Ref.
Capsicum annuum (paprika) (Solanaceae) [seed]	PSI-1.1 (55 aa; 6 kDa; 8 Cys; 4 S-S) & PSI-1.2, PSI-1.3	Chymotrypsin [47 nM], Pronase [59 nM], Trypsin [0.5 nM] (R38-Y39)	[484]
Capsicum annuum (paprika) (Solanaceae) [leaf - wounding-inducible]	Pepper leaf PIs – PLPI-34, PLPI-35, PLPI-40, PLPI-41, PLPI-43, PLPI-45, PLPI-46 (6 kDa; 8 Cys; 4 S-S)	Chymotrypsin [80 pM (PLPI-34), 0.3 nM (35), 0.1 nM (40), 0.6 nM (41), 0.7 nM (43), 1 nM (45), 0.3 nM (46)], Trypsin [5 nM (34), 4 nM (35), 10 nM (45), 6 nM (46)]	[485]
Lycopersicon esculentum (tomato) (Solanaceae)	TI-II (123 aa; 14 kDa; 2 x 8 Cys; 2 x 4 S-S; 2 x ~ 60 aa inhibitory domains)	Chymotrypsin (F62-N63), Trypsin (R5-E6)	[486, 487]
Lycopersicon esculentum (tomato) (Solanaceae)	ARPI (= TR8) (proprotein 223 aa; 25 kDa; 3 x 8 Cys; 3 x 4 S-S; 3 x ~ 64 aa repeated inhibitory domains)	Trypsin (K29-E30, repeat 1; K93-E94, repeat 2; K157-E158, repeat 3)	[488, 489]
Lycopersicon esculentum (tomato) (Solanaceae)	CEVI57 (proprotein 201 aa; 21 kDa; 3 x 8 Cys; 3 x 4 S-S; 3 x repeated domains)	Trypsin (K32-E33; K147-E148)	[490]
Nicotiana alata (ornamental tobacco) (Solanaceae) [stigma]	6 domain PI precursor NaProPI (NaPI II) (proprotein: 397 aa; 44 kDa; 29 aa signal peptide; 6 x 8 Cys; 6 x 4 S-S; 6 PI domains yielding 6 kDa PIs C1, C2 (S-S-linked heterodimer), T1-T4)	Chymotrypsin (L34-N35, C1; L92-N93, C2), Trypsin (R143-N144, T1; R208-N209, T2; R266-N267, T3; R317-N318, T4)	[491-495]
Nicotiana alata (ornamental tobacco) (Solanaceae) [leaf, stigma]	4 domain PI-II precursor NaPI IV (proprotein: 291 aa; 32 kDa; 4 x 8 Cys; 4 x 4 S-S; 4 PI domains yielding 6 kDa PIs C1, C2 (S-S-linked heterodimer), T1 (see above) & novel T5)	Chymotrypsin (C1 & C2), Trypsin (T1, T5)	[496]
Nicotiana glutinosa (Solanaceae) [flower, leaf	PI-II precursors NGPI-1 (8 repeated PI domains → PIs I-1	Inferred Chymotrypsin (L-N, NGPI-1-derived I-7 &	[497]

expression]	to I-7), NGPI-2 (6 repeated PI domains → PIs II-1 to II-5) (each inferred monomeric PI: 6 kDa; 8 Cys; 4 S-S)	NGPI-2-derived II-5), Trypsin (R-N, I-1 to I-6 & II-1-II-4)	
Nicotiana tabacum (tobacco) (Solanaceae) [leaf]	TTI-1, TTI-2, TTI-3, TTI-4, TTI-5, TTI-6 (6 kDa) (TTI-1: 53 aa; 6 kDa; 8 Cys; 4 S-S)	Trypsin (R38-N39, TTI-1)	[498]
Nicotiana tabacum (tobacco) (Solanaceae) [leaf, stress/wounding-induced]	PI-II precursor PI2-1(173 aa; 19 kDa; 3 x 8 Cys; 3 x 4 S-S; 3 PI domains)	Trypsin (K5-E6, domain 1)	[499]
Solanum americanum (Solanaceae) [phloem]	SaPIN2a & SaPIN2b (13 kDa; 2 inhibitory domains) (processed SaPIN2a: 121 aa; 13 kDa; 8 Cys; 4 S-S)	Chymotrypsin (F62-E63, putative); Trypsin (R5-E6, putative)	[500]
Solanum melongena (aubergine, eggplant) (Solanaceae) [fruit]	Aubergine PI-II (52 aa; 6 kDa; 8 Cys; 4 S-S)	Trypsin (R38-N39)	[501
Solanum tuberosum (potato) (Solanaceae) [tuber]	PTI (51 aa; 6 kDa; 8 Cys; 4 S-S)	Trypsin [R38-N39]	[502]
Solanum tuberosum (potato) (Solanaceae) [tuber]	PCI-I (52 aa; 6 kDa; 8 Cys; 4 S-S)	Chymotrypsin [L38-N39], *Streptomyces griseus* Proteinase B	[502, 503]
Solanum tuberosum (potato) (Solanaceae) [tuber]	PI-II variants B, C & D (homodimers 21 kDa, monomers 11 kDa);	Chymotrypsin [20 nM, PI-II mixture]	[504]
Solanum tuberosum (potato) (Solanaceae) [tuber]	PI-II K (proprotein 153 aa; processed: 123 aa; 14 kDa; 16 Cys; 8 S-S)	Chymotrypsin (L62-N63), Trypsin (R5-E6)	[505]
Solanum tuberosum (potato) (Solanaceae) [root]	Pin2-R (proprotein 207 aa; processed: 177 aa; 19 kDa; 24 Cys; 12 S-S; PI domains I-III)	Chymotrypsin (L5-E6, I; L120-E121, III), Trypsin (R62-N63, II)	[506]
Solanum tuberosum (potato) (Solanaceae) [leaf]	Pin2-L (proprotein 154 aa; processed: 123 aa; 14 kDa; 16 Cys; 8 S-S; PI domains I & II)	Chymotrypsin (L5-E6, I; L62-N63, II)	[506]
Solanum tuberosum (potato) (Solanaceae) [leaf]	PI-II (proprotein 154 aa; processed: 124 aa; 14 kDa; 16 Cys; 8 S-S; PI domains I [= PCI-I] & II [= PTI homologue])	Chymotrypsin [0.9 nM] (L5-E6, I), Trypsin [0.4 nM] (R62-N63, II)	[507-513]
Solanum tuberosum (potato) (Solanaceae) [leaf]	Other PI-II variants pin2-CM7, cDNA2 (14 kDa; 16 Cys; 8 S-S; PI domains I & II)	Chymotrypsin (L-E, L-N), Trypsin (K-E, R-E, R-N)	[507, 509, 514]

Mustard (Brassicaceae) 7 kDa trypsin inhibitors

The mustard family (Brassicaceae) PIPs are 7 kDa proteins with 8 cysteines in highly conserved positions that form 4 disulphide linkages in a particular pattern of connectivity. The mustard PIPs are variously potent inhibitors of serine proteases such as trypsin, chymotrypsin, thrombin, plasmin and blood clotting factors Xa and XIIa [515, 520] (Table 15).

Table 15. Mustard (Brassicaceae) 7 kDa trypsin inhibitors
The mustard (Brassicaceae) 7 kDa trypsin inhibitors have molecular masses of about 7 kDa and have 8 cysteines (i.e. 4 disulphide links). For other details see the legend to Table 4.

Plant species (Family) [part]	Protease inhibitor name (properties)	Protease specificity (IC$_{50}$) [K$_d$, K$_i$] (reactive site)	Ref.
Arabidopsis thaliana (mouse-ear cress) Brassicaceae)	ATTI-1 (~ 60 aa) (deduced sequence from chromosome1)	Trypsin (inferred) (R20-I21 if DRKC N-terminal sequence; cf. [2])	[515, 516]
Arabidopsis thaliana (mouse-ear cress) Brassicaceae)	ATTI-A (60 aa), ATTI-A' (60 aa), ATTI-B (61 aa), ATTI-C (55 aa), ATTI-D (59 aa) (deduced sequences from chromosome 2 [2])	Trypsin (inferred), ATTI-A (R19-I20), ATTI-A' (R19-I20), ATTI-B (R19-I20), ATTI-C (R19-I20), ATTI-D (K19-F20)	[144, 516]
Brassica napus (Brassicaceae) [seed]	RTI (= RTI-IIIB = D1-P60-RTI-III) (60 aa; 6.7 kDa)	Chymotrypsin (410 nM), Trypsin (300 pM) (R20-I21)	[516, 517]
Brassica napus (Brassicaceae) [seed]	RTI-IIIA (= 5-oxoP-G62-RTI-III) (62 aa, 7 kDa)	Chymotrypsin (420 nM), Trypsin (303 pM) (R21-I22)	[516, 518]
Brassica napus (Brassicaceae) [seed]	RTI-IIIC = S1-P59-RTI-III) (59 aa; 7 kDa)	Trypsin >> Chymotrypsin (R19-I20)	[516, 518]
Brassica oleraceae (cabbage) (Brassicaceae) [seed]	Thrombin inhibitor (~10 kDa; possible member of mustard TI family)	Trypsin (0.2 µM), Thrombin, Xa, XIIa, Plasmin	[519]
Sinapis alba (Brassicaceae) [seed]	MTI-2 (= MTI-2F, D1-Q63) (63 aa; 7 kDa)	Chymotrypsin (500 nM), Trypsin (160 pM) (R20-I21)	[516, 520]
Sinapis alba (Brassicaceae) [seed]	"MTI-2" = mixture of 62-64 aa MTI-2 analogues: MTI-2A, MTI-2B, MTI-2C, MTI-2D, MTI-2E & MTI-2F (MTI-2)	Chymotrypsin (500 nM), Trypsin (159 pM) (MTI-2F R20-I21)	[516]

Plant Serpins

The serpins (serine protease inhibitors) are a major class of relatively high molecular weight eukaryote serine PIPs. The mechanism of action of the serpins involves cleavage of the scissile bond and separation of the N-terminal and C-terminal side amino acids by about 7nm, enabling a serpin strand to insert into a β-sheet [521-523]. The human serpins antitrypsin, antichymotrypsin and plasminogen activator inhibitor I have compact structures involving 2 major adjoining anti-parallel β-sheet domains above a bundle of 4 α-helices [521]. A variety of 42-45 kDa plant serpins have been isolated (or resolved at the gene level) [144, 515] that are homologous to those from other eukaryotes [144, 515, 521-536]. Plant serpins have been resolved from plants of the Brassicaceae, Cucurbitaceae and Poaceae families that inhibit various serine proteases (Table 16).

Table 16. Plant serpins
For details see the legend to Table 4.

Plant species (other name) (Family)	Protease inhibitor protein	Protease specificity (IC$_{50}$) [K$_d$, K$_i$] (reactive site)	Ref.
Arabidopsis thaliana (mouse-ear cress) (Brassicaceae)	20 serpin homologues, notably ~ 45 kDa proteins variously encoded by chromosomes 1 & 2	Serine protease (inferred)	[144, 515]
Cucurbita maxima (squash) (Cucurbitaceae) [phloem]	*Cucurbita maxima* phloem serpin-1 (CmPS-1) (42 kDa)	Elastase (V347-G348; V350-S357)	[524]
Hordeum vulgare (barley) (Poaceae)	BSZ4 (43 kDa)	Cathepsin G (M358-S359)	[525, 527-529]
Hordeum vulgare (barley) (Poaceae)	BSZ7 (44 kDa)	Trypsin (R297-S298)	[530, 531]
Hordeum vulgare (barley) (Poaceae)	BSZx (44 kDa)	APC, Cathepsin G; Chymotrypsin & Cathepsin G (L355-R356), Elastase, Factors VIIa/TF, Xa, XIIa (weak), Kallikrein, Thrombin; Trypsin & Cathepsin G (R356-S357)	[525, 526, 532]
Secale cereale (rye) (Poaceae) [seed]	Rye serpins RSZa, RSZb, RSZc1, RSZc2, RSZd, RSZe & RSZf (43 kDa)	Chymotrypsin (Q-S, RSZb, RSZc2, RSZd, RSZe; Q-M, RSZc1; Y-M, RSZf)	[533]
Triticum aestivum (wheat) (Poaceae) [seed]	WSZ1 (= WSZ1a) (44 kDa)	Chymotrypsin & Cathepsin G (Q357-Q358)	[525, 528, 534-536]
Triticum aestivum (wheat) (Poaceae) [seed]	WSZ1b (44 kDa)	Chymotrypsin & Cathepsin G (Q357-Q358)	[536]
Triticum aestivum (wheat) (Poaceae) [seed]	WSZ1c (44 kDa)	Chymotrypsin & Cathepsin G (L357-Q358)	[536]
Triticum aestivum (wheat) (Poaceae) [seed]	WSZ2a (44 kDa)	Chymotrypsin & Cathepsin G (L357-S358)	[536]
Triticum aestivum (wheat) (Poaceae) [seed]	WSZ2b (44 kDa)	Chymotrypsin & Cathepsin G (L356-R357), Plasmin (R357-Q358)	[536]

Squash family serine protease inhibitors

The squash family (Cucurbitaceae) family of serine PIPs are very small (*circa* 3 kDa), cysteine-rich proteins (6 cysteines being involved in 3 disulphide linkages) [537-561]. These small proteins have extraordinary affinities for the target serine proteases (K$_i$ values in the nanomolar and picomolar range) (Table 17). The serine proteases variously inhibited by squash family PIPs include elastase, trypsin, kallikrein and blood clotting protease factors Xa, XIa and XIIa (Table 17). Of particular note are the squash family PIPs MCoTI-I and MCoTI-II from *Momordica cochinensis*

(Vietnamese squash) which are cyclic peptides (cyclotides) [541]. The small size, exceptional stability and protease inhibitor potency of these cyclic squash family PIPs make them potentially very useful lead compounds and structural scaffolds for pharmaceutical development [541]. The structures of various squash family PIPs have been determined, namely those of *Cucurbita maxima* CMTI-I [544, 545], *Cucurbita pepo* CPTI-II [545], *Momordica charantia* MCTI-A [556, 557] and the cyclotide *Momordica cochinchinensis* McoTI-II [541]. The linear squash PIP structures involve anti-parallel β-strand elements beneath a loop which contains the protease binding site [544, 545, 556, 557].

Table 17.   Squash family serine protease inhibitors
The squash family serine protease inhibitors all have molecular masses of about 3 kDa and have 6 cysteines (i.e. 3 disulphide links). For other details see the legend to Table 4.

Plant species (Family) [part]	Protease inhibitor name (properties)	Protease specificity (IC$_{50}$) [K$_d$, K$_i$] (reactive site)	Ref.
Bryonia dioica (red bryony) (Cucurbitaceae) [seed]	BD-TI-II (29 aa; 3.2 kDa; 6 Cys)	Trypsin (R5-I6)	[537]
Citrullus vulgaris (watermelon) (Cucurbitaceae) [seed]	CVTI-1 (30 aa; 3.4 kDa; 6 Cys)	Trypsin (R5-I6)	[537]
Cucumis melo (oriental pickling melon) (Cucurbitaceae) [seed]	*Cucumis* CMeTI-A (= CMCTI-II) (29 aa; 3.3 kDa; 6 Cys)	Trypsin [160 pM] (K5-I6)	[538]
Cucumis melo (oriental pickling melon) (Cucurbitaceae) [seed]	*Cucumis* CMeTI-B (29 aa; 3.2 kDa; 6 Cys)	Trypsin [470 pM] (R5-I6)	[538]
Cucumis melo (oriental pickling melon) (Cucurbitaceae) [seed]	*Cucumis* CMCTI-I (28 aa; 3.1 kDa; 6 Cys)	Trypsin [127 pM], Lysyl endopeptidase [207 pM] (K4-I5)	[539]
Cucumis melo (oriental pickling melon) (Cucurbitaceae) [seed]	*Cucumis* CMCTI-II (= CMeTI-A) (29 aa; 3.3 kDa; 6 Cys)	Trypsin [118 pM; 160 pM], Lysyl endopeptidase [15 pM] (K5-I6)	[539]
Cucumis melo (oriental pickling melon) (Cucurbitaceae) [seed]	CMCTI-III (30 aa; 3.4 kDa; 6 Cys; N-terminal <E)	Trypsin [78 pM], Lysyl endopeptidase [62 pM] (K6-I7)	[539]
Cucumis melo (Chinese melon, Hami melon) (Cucurbitaceae)	HMTI-I (29 aa; 3 kDa; 6 Cys)	Trypsin (at nM) (R5-I6)	[540, 541]
Cucumis sativus (cucumber) (Cucurbitaceae) [seed]	CSTI-IIb (32 aa; 3.5 kDa; 6 Cys)	Trypsin [1 pM] (K5-I6)	[542]
Cucumis sativus (cucumber) (Cucurbitaceae) [seed]	CSTI-IV (30 aa; 3.4 kDa; 6 Cys)	Trypsin (at nM) (R5-I6)	[542]
Cucurbita maxima (pumpkin) (Cucurbitaceae) [seed]	CMTI-I (= ITD-I) (29 aa; 3.3 kDa; 6 Cys)	Xa [240 µM], XIIa [25 nM], Kallikrein [11 µM], Plasmin [40 nM], Trypsin [3 pM] (R5-I6)	[542-545]
Cucurbita maxima (pumpkin) (Cucurbitaceae) [seed]	CMTI-III (= ITD-III) (29 aa; 3.3 kDa; 6 Cys)	Xa [23], XIIa [70 nM], Kallikrein [130], Trypsin [1 pM] (R5-I6)	[542, 543, 546,

			547]
Cucurbita maxima (pumpkin) (Cucurbitaceae) [seed]	CMTI-IV (32 aa; 3.7 kDa; 6 Cys)	Trypsin [17 pM] (R8-I9)	[542]
Cucurbita pepo (zucchini) (Cucurbitaceae) [seed]	CPGTI-I (= CPTI-II) (29 aa; 3.3 kDa; 6 Cys)	Trypsin [1 pM] (K5-I6)	[542]
Cucurbita pepo (squash) (Cucurbitaceae) [seed]	CPTI-II (29 aa; 3.3 kDa; 6 Cys)	XIIa [0.8 μM], Kallikrein [260 μM], Plasmin [59 nM], Trypsin [1 pM] (K5-I6)	[542, 544, 545]
Cucurbita pepo (squash) (Cucurbitaceae) [seed]	CPTI-III (32 aa; 3.7 kDa; 6 Cys)	Trypsin [8 pM] (K8-I9)	[542]
Ecballium elaterium (squirting cucumber) (Cucurbitaceae) [seed]	*Ecballium* EETI-II (30 aa; 3 kDa; 6 Cys)	Trypsin [1.3 pM] (R4-I5)	[548]
Echinocystis lobata (Cucurbitaceae) [seed]	ELTI-I (33 aa; 3.6 kDa; 6 Cys)	Cathepsin G [79 nM], Trypsin [15 pM] (R9-I10)	[549]
Echinocystis lobata (Cucurbitaceae) [seed]	ELTI-II (29 aa; 3.2 kDa; 6 Cys)	Cathepsin G [91 nM], Trypsin [3 pM] (R5-I6)	[549]
Lagenaria leucantha (bottle gourd) (Cucurbitaceae) [seed]	LLDTI-I (= LLTI-I = <E-LLDTI-II = <E-LLTI-II) (30 aa; 3.4 kDa; 6 Cys; N-terminal <E)	Trypsin [240 pM; 360 pM] (R6-I7)	[550, 551]
Lagenaria leucantha (bottle gourd) (Cucurbitaceae) [seed]	LLDTI-II (= LLTI-II) (29 aa; 3.3 kDa; 6 Cys)	Factor Xa [41 μM], Factor XIIa [1 μM], Kallikrein [27 μM], Trypsin [65 pM; 96 pM] (R5-I6)	[546, 550, 551]
Lagenaria leucantha (bottle gourd) (Cucurbitaceae) [seed]	LLDTI-III (= LLTI-III) (30 aa; 3.4 kDa; 6 Cys)	Factor Xa [19 μM], XIIa [4 μM], Kallikrein [200 μM], Trypsin [30 pM] (R6-I7)	[546, 550, 551]
Luffa acutangula (ridged gourd) (Cucurbitaceae) [seed]	LATI-I (28 aa; 3 kDa; 6 Cys)	Trypsin (R4-I5)	[552]
Luffa acutangula (ridged gourd) (Cucurbitaceae) [seed]	LATI-II (29 aa; 3 kDa; 6 Cys)	Trypsin (R5-I6)	[552]
Luffa cylindrica (sponge gourd, towel gourd) (Cucurbitaceae) [seed]	LCTI-I (29 aa; 3.2 kDa; 6 Cys)	Trypsin (R5-I6)	[553]
Luffa cylindrica (sponge gourd, towel gourd) (Cucurbitaceae) [seed]	LCTI-II (30 aa; 3.3 kDa; 6 Cys)	Xa [780 μM], XIIa [75 nM], Kallikrein [20 μM], Trypsin	[546, 553]
Luffa cylindrica (sponge gourd, towel gourd) (Cucurbitaceae) [seed]	LCTI-III (30 aa; 3.3 kDa; 6 Cys)	Xa [100 μM], XIIa [4 nM], Kallikrein [38 μM], Trypsin (at nM)	[546, 553]
Luffa cyclindrica (towel gourd) (Cucurbitaceae) [seed]	TGTI-I (28 aa; 3.0 kDa; 6 Cys)	Trypsin (R4-I5)	[540]
Luffa cyclindrica (towel gourd) (Cucurbitaceae) [seed]	TGTI-II (29 aa; 3.1 kDa; 6 Cys)	Trypsin (R5-I6)	[540]
Momordica charantia (bitter gourd) (Cucurbitaceae) [seed]	MCEI-I (28 aa; 3.2 kDa; 6 Cys)	Elastase [300 nM; 970 nM] (L5-I6)	[554, 555]
Momordica charantia (bitter gourd)	MCEI-II (29 aa; 3.3 kDa; 6 Cys)	Elastase [9 nM] (L6-I7)	[555]

(Cucurbitaceae) [seed]			
Momordica charantia (bitter gourd) (Cucurbitaceae) [seed]	MCEI-III (30 aa; 3.4 kDa; 6 Cys)	Elastase [4 nM] (L7-I8)	[555]
Momordica charantia (bitter gourd) (Cucurbitaceae) [seed]	MCEI-IV (31 aa; 3.6 kDa; 6 Cys)	Elastase [5 nM] (L8-I9)	[555]
Momordica charantia (bitter gourd) (Cucurbitaceae) [seed]	MCTI-A (28 aa; 3.1 kDa; 6 Cys)	Trypsin [170 pM] (R5-I6)	[556, 557]
Momordica charantia (bitter gourd) (Cucurbitaceae) [seed]	MCTI-I (30 aa; 3.4 kDa; 6 Cys; N-terminal <E)	Xa [100 μM], XIIa [13 nM], Kallikrein [110 μM], Trypsin [67 pM; 12 nM] (R6-I7)	[542, 550, 551]
Momordica charantia (bitter gourd) (Cucurbitaceae) [seed]	MCTI-II (28 aa; 3.2 kDa; 6 Cys)	Xa [1 μM], X1a [18 μM], XIIa [56 nM], Kallikrein [100 μM], Trypsin [25 pM; 0.8 nM] (R5-I6)	[546, 554, 555]
Momordica charantia (bitter gourd) (Cucurbitaceae) [seed]	MCTI-II′ (27 aa; 3 kDa; 6 Cys)	Trypsin (at nM) (R4-I5)	[558]
Momordica charantia (bitter gourd) (Cucurbitaceae) [seed]	MCTI-III (30 aa; 3.3 kDa; 6 Cys; N-terminal <E)	Xa [59 μM], XIIa [2 μM], Kallikrein [140 μM], Trypsin [190 nM] (R6-I7)	[546, 555]
Momordica cochinchinensis (Vietnamese squash) (Cucurbitaceae) [seed]	MCoTI-I (34 aa; 3.5 kDa; 6 Cys; cyclic peptide)	Trypsin (K10-I11)	[541]
Momordica cochinchinensis (Vietnamese squash) (Cucurbitaceae) [seed]	MCoTI-II (34 aa; 3.5 kDa; 6 Cys; cyclic peptide)	Trypsin (K10-I11)	[541]
Momordica cochinchinensis (Vietnamese squash) (Cucurbitaceae) [seed]	MCoTI-III (30 aa; 3.4 kDa; 6 Cys; linear peptide; N-terminal pyroglutamyl)	Trypsin (R6-I7)	[541]
Momordica repens (Cucurbitaceae) [seed]	MRTI-I (29 aa; 3.3 kDa; 6 Cys)	Trypsin (R5-I6)	[559]
Tricosanthes kirilowii (Cucurbitaceae) [seed]	TKTI-I (27 aa; 3 kDa; 6 Cys)	Trypsin (R3-I4)	[541, 560, 561]
Tricosanthes kirilowii (Cucurbitaceae) [seed]	TKTI-II (27 aa; 3 kDa; 6 Cys)	Trypsin (R3-I4)	[541, 560, 561]

Other plant serine protease inhibitor proteins

A variety of other kinds of plant serine PIPs have been isolated [562-573] (Table 18), namely the arrowhead PIPs [562-566] and some representatives of some major classes of plant antifungal proteins, namely some defensins [121, 567, 568], some lipid transfer proteins (LTPs) [121, 169, 569, 570] and some napins [121, 571-573] (Table 18). The *Sagittaria sagittifolia* (arrowhead) protease inhibitors A and B (API-A and API-B) have 150 amino acid residues and 3 disulphide bridges. They are double-headed 16 kDa proteins that inhibit trypsin, chymotrypsin and kallikrein [562-566]. Defensins are antifungal proteins of about 6 kDa having 4 disulphide links and having a structure involving 3 antiparallel β-strands linked to an α-helix [121, 574-578]. LTPs are antifungal proteins interacting with membrane phospholipids and having a cup-like structure composed of α-helices, the central hydrophobic cavity being involved in phospholipid sequestration [121, 577, 579]. The napins are membrane-active, *circa* 14 kDa heterodimeric antifungal proteins composed of disulphide-linked 4 kDa and 10 kDa subunits and having a structure similar to that of the LTPs [121, 577, 580]. The protease inhibitory activity of some defensins, lipid transfer proteins and napins suggests that these particular proteins may have anti-herbivore as well as antifungal roles [121, 567-573]. The lack of protease inhibitory activity in glycosylated forms of a *Phaseolus angularis* seed LTP suggests that the protease-inhibitory activity may be dormant in the seed and that the active aglycone could be generated by deglycosylation during germination or after seed protein ingestion [121].

Table 18. Other serine protease inhibitor proteins
For details see the legend to Table 4.

Plant species (Family) [part]	Protease inhibitor name (properties)	Protease specificity (IC$_{50}$) [K$_d$, K$_i$]	Ref.
Arrowhead PIs			
Sagittaria sagittifolia (arrowhead) (Alismataceae)	Arrowhead PIs API-A & API-B (150 aa; 16 kDa; 6 Cys; 3 S-S; double-headed PIs)	Chymotrypsin, Kallikrein, Trypsin	[562-566]
Defensin (γ-Thionin) PIs			
Cassia fistula (Fabaceae) [seed]	*Cassia* 5467 Da defensin PI (5 kDa; 8 Cys; 4 S-S)	TRY (2 μM) [homologous *C. fistula* Defensin with Y25 instead of K25 is inactive]	[567]
Phaseolus angularis (adzuki bean) (Fabaceae) [seed]	*Phaseolus* 5412 Da Defensin PI (5 kDa; likely 8 Cys, 4 S-S)	TRY (0.5 μM) [*P. angularis* Defensin lacking N-terminal Arg may also be a TI]	[121]
Vigna unguiculata (cowpea) (Fabaceae) [seed]	Cp-thionin (5 kDa; 8 Cys; 4 S-S)	TRY (high affinity, 1:1 complex)	[568]
Lipid transfer protein (LTP) PIs			
Eleusine coracana (ragi, finger millet) (Poaceae) [seed]	*Eleusine* double-headed TRY-α-Amylase inhibitor I-2 (LTP homologue)	Subtilisin (α-Amylase inhibitor) (double headed inhibitor)	[569]
Hordeum vulgare (barley) (Poaceae) [seed]	*Hordeum* Lipid Transfer Proteins 1 & 2 (= LTP 1 & 2) (7 kDa proteins; 8 Cys; 4 S-S)	Barley malt cysteine endoproteinases	[169, 570]

Phaseolus angularis (adzuki bean) (Fabaceae) [seed]	Phaseolus 9 kDa LTP PI (9 kDa)	TRY (0.4 µM) [10 kDa glycosylated forms of this LTP are inactive]	[121]
Napin PIs			
Brassica napus (kohlrabi) (Brassicaceae)	Brassica TIBN (14 kDa; S-S-linked heterodimer)	TRY [50 µM]	[571]
Brassica nigra (Cruciferae)	Brassica BN (16 kDa; S-S-linked heterodimer)	CHY (at 2 µM), Subtilisin (at 2 µM), TRY [20 µM]	[572]
Sinapis arvensis (charlock) (Brassicaceae) [seed]	Sinapis TISA-1, TISA-2 (16 kDa; S-S-linked heterodimer)	CHY (at 2 µM), Subtilisin (at 2 µM), TRY [7 µM]	[573]

CONCLUDING COMMENTS

This review has succinctly summarized what is presently known of non-protein and protein protease inhibitors from plants. The affinities of the non-protein inhibitors for particular proteases are generally much lower than those of plant protease inhibitor proteins (PIPs). Nevertheless the non-protein protease inhibitors may provide structure/activity starting points for development of pharmaceutically useful compounds of much higher affinity. The plant PIP literature has been comprehensively surveyed in this review. However electronic databases such as EMBL and SWISSPROT contain further accessible plant PIP sequences [581]. The array of potent plant PIPs reflects the co-evolution of plant defensive proteins and insect resistance [582]. Potent, stable, protease inhibitor proteins have potential transgenic crop agriculture applications as well as potential chemotherapeutic applications.

ABBREVIATIONS

ACE	=	Angiotensin I converting enzyme
AIDS	=	Acquired immunodeficiency syndrome
API	=	Aspartic protease inhibitor
BBPIP	=	Bowman Birk protease inhibitor protein
CHY	=	Chymotrypsin
CPI	=	Cysteine protease inhibitor
ECE	=	Endothelin-converting enzyme
ELA	=	Elastase
HIV-1	=	Human immunodeficiency virus 1
IC_{50}	=	Concentration for 50% inhibition

K_d	=	Dissociation constant
K_i	=	Enzyme-inhibitor dissociation constant
KPI	=	Kunitz protease inhibitor
LELA	=	Leucocyte elastase
LTP	=	Lipid transfer protein
MMP	=	Matrix metalloprotease
NEP	=	Neutral endopeptidase
PEP	=	Prolyl endopeptidase
PI	=	Protease inhibitor
PIP	=	Protease inhibitor protein
PPC	=	Proprotein convertase
TI	=	Trypsin inhibitor
TRY	=	Trypsin
X, Xa	=	Factor X, Factor Xa

REFERENCES

[1]   Laskowski, M.; Kato, I.; *Annu. Rev. Biochem.*, **1980**, *49*, 593-626.

[2]   Lodish, H.; Berk, A.; Zipursky, S.L.; Matsudaira, P.; Baltimore, D.; Darnell, J., *Molecular Cell Biology*, 4th edn, Freeman: New York, **2000**.

[3]   Matthews, C.K.; van Holde, K.E.; Ahern, K.G., *Biochemistry*, 3rd edn, Addison Wesley Longman: New York, **2000**.

[4]   Nelson, D.L.; Cox, M.M., *Lehninger Principles of Biochemistry*, 3rd edn, Worth: New York, **2000**.

[5]   Berg, J.M.; Tymoczko, J.L.; Stryer, L.; Clarke, N.D., *Biochemistry*, 5th edn, Freeman: New York, **2002**.

[6]   Voet, D.; Voet, J.G.; Pratt, C.W., *Fundamentals of Biochemistry*, John Wiley: New York, **1999**.

[7]   Richter, C.; Tanaka, T.; Yada, R.Y.; *Biochem. J.*, **1998**, *335*, 481-490.

[8]   Hoegl, L.; Korting, H.C.; Klebe, G.; *Pharmazie*, **1999**, *54*, 319-329.

[9]   Khan, A.R.; Khazanovich-Bernstein, N.; Bergmann, E.M.; James, M.N.G.; *Proc. Nat. Acad. Sci. USA*, **1999**, *96*, 10968-10975.

[10]   Hirschowitz, B.I.; *Yale J. Biol. Med.,* **1999**, *72*, 133-143.

[11]   Rochefort, H.; Liaudet-Coopman, E.; *APMIS*, **1999**, *107*, 86-95.

[12]   Howlett, D.R.; Simmons, D.L.; Dingwall, C.; Christie, G.; *Trends Neurosci.*, **2000**, *23*, 565-570.

[13]   Dostal, D.E.; Baker, K.M.; *Circulation Res.*, **1999**, *85*, 643-650.

[14]   Wlodawer, A.; Erickson, J.W.; *Annu. Rev. Biochem.*, **1993**, *62*, 543-585.

[15]   Francis, S.E.; Gluzman, I.Y.; Oksman, A.; Knickerbocker, A.; Mueller, R.; Bryant, M.L.; Sherman, D.R.; Russell, D.G.; Goldberg, D.E.; *EMBO J.*, **1994**, *13*, 306-317.

[16]   Semenov, A.; Olson, J.E.; Rosenthal, P.J.; *Antimicrob. Agents. Chemother.*,

1998, *42*, 2254-2258.

[17]   Turk, B.; Turk, V.; Turk, D.; *Biol. Chem.*, **1997**, *378*, 141-150.

[18]   Berti, P.J.; Storer, A.C.; *J. Mol. Biol.*, **1995**, *246*, 273-283.

[19]   Cummins, P.M.; O'Connor, B.; *Biochim. Biophys. Acta*, **1998**, *1429*, 1-17.

[20]   Brown, W.M.; Dziegielewska, K.M.; *Protein Sci.*, **1997**, *6*, 5-12.

[21]   Sorimachi, H.; Suzuki, K.; *J. Biochem.*, **2001**, *129*, 653-664.

[22]   Huang, Y.; Wang, K.K.; *Trends Mol. Med.*, **2001**, *7*, 355-362.

[23]   Solary, E.; Eymin, B.; Droin, N.; Haugg, M.; *Cell Biol. Toxicol.*, **1998**, *14*, 121-132.

[24]   Yamashima, T.; *Progr. Neurobiol.*, **2000**, *62*, 273-295.

[25]   Wang, K.K.; *Trends Neurosci.*, **2000**, *23*, 20-26.

[26]   Ruiz-Vela, A.; Serano, F.; Gonzalez, M.A.; Abad, J.L.; Bernad, A.; Maki, M.; Martinez, A.C.; *J. Exp. Med.*, **2001**, *194*, 247-254.

[27]   Stennicke, H.R.; Salvesen, G.S.; *Biochim. Biophys. Acta*, **2000**, *1477*, 299-306.

[28]   Rosenthal, P.J.; Sijwali, P.S.; Singh, A.; Shenai, B.R.; *Curr. Pharm. Des.*, **2002**, *8*, 1659-1672.

[29]   Ventura, S.; Gomis-Ruth, F.X.; Puigserver, A.; Aviles, F.; Vendrell, J.; *Biol. Chem.*, **1997**, *378*, 161-165.

[30]   Vendrell, J.; Querol, E.; Aviles, F.X.; *Biochim. Biophys. Acta*, **2000**, *1477*, 284-298.

[31]   Baudin, B.; *Clin. Chem. Lab. Med.*, **2002**, *40*, 256-265.

[32]   Nagase, H.; *Biol. Chem.*, **1997**, *378*, 151-160.

[33]   Brew, K.; Dinakarpandian, D.; Nagase, H.; *Biochim. Biophys. Acta*, **2000**, *1477*, 267-283.

[34]   Polgar L.; *Cell Mol. Life Sci.*, **2002**, *59*, 349-362.

[35]   Li, X-C.; Jacob, M.R.; Pasco, D.S.; El Sohly, H.N.; Nimrod, A.C.; Walker, L.A.; Clark, A.M.; *J. Nat. Prod.*, **2001**, *64*, 1282-1285.

[36]   Zhang, Z.; El Sohly, H.N.; Jacob, M.R.; Pasco, D.S.; Walker, L.A.; Clark, A.M.; *Planta Med.*, **2002**, *68*, 49-54.

[37]   Abuereish, G.M.; *Phytochemistry*, **1998**, *48*, 217-221.

[38]   Kräusslich, H-G.; Wimmer, E.; *Ann. Rev. Biochem.*, **1988**, *57*, 701-754.

[39]   Vaishnav, Y.N.; Wong-Staal, F.; *Annu. Rev. Biochem.*, **1991**, *60*, 577-630.

[40]   Bowman, M.J.; Chmielewski, J.; *Biopolymers*, **2002**, *66*, 126-33.

[41]   Menendez-Arias L.; *Trends Pharmacol. Sci.*, **2002**, *23*, 381-388.

[42]   Xu, H.X.; Wan, M.; Dong, H.; But, P.P.; Foo, L.Y.; *Biol. Pharm. Bull.*, **2000**, *23*, 1072-1076.

[43]   Brinkworth, R.I.; Stoemer, M.J.; Fairlie, D.P.; *Biochem. Biophys. Res. Commun.*, **1992**, *188*, 631-637.

[44]   Hu, C-Q.; Chen, K.; Shi, Q.; Kilkuskie, R.E.; Cheng, Y-C.; Lee, K-H.; *J. Nat. Prod.*, **1994**, *57*, 42-51.

[45]   Wang, H.K.; Xia, Y.; Yang, Z.Y.; Natschke, S.L.; Lee, K.H.; *Adv. Exp. Med. Biol.*, **1998**, *439*, 191-225.

[46]   Ma, C.; Nakamura, N.; Hattori, M.; Kakuda, H.; Qiao, J.; Yu, H.; *J. Nat. Prod.*, **2000**, *63*, 238-242.

[47]   Mahmood, N.; Piacente, S.; Pizza, C.; Burke, A.; Khan, A.I.; Hay, A.J.; *Biochem. Biophys. Res. Commun.*, **1996**, *229*, 73-79.

[48]   Chen, S.X.; Wan, M.; Loh, B.N.; *Planta Med.*, **1996**, *62*, 381-382.

[49]   Vlietinck, A.J.; De Bruyne, T.; Apers, S.; Pieters, L.A.; *Planta Med.*, **1998**, *64*, 97-109.

[50]   Ma, C-M.; Nakamura, N.; Hattori, M.; *Chem. Pharm. Bull.*, **2001**, *49*, 915-917.

[51]   Ma, C.; Nakamura, N.; Hattori, M.; Zhu, S.; Komatsu, K.; *J. Nat. Prod.*, **2000**, *63*, 1626-1629.

[52]   Min, B.S.; Hattori, M.; Lee, H.K.; Kim, Y.H.; *Arch. Pharm. Res.*, **1999**, *22*, 75-77.

[53]   Ma, C.; Nakamura, N.; Miyashiro, H.; Hattori, M.; Shimotohno, K.; *Chem. Pharm. Bull.*, **1999**, *47*, 141-145.

[54]   Paris, A.; Strukelj, B.; Renko, M.; Turk, V.; Pukl, M.; Umek, A.; Korant, B.D.; *J. Nat. Prod.*, **1993**, *56*, 1426-1430.

[55]   Xu, H.X.; Zeng, F.Q.; Wan, M.; Sim, K.Y.; *J. Nat. Prod.*, **1996**, *59*, 643-645.

[56]   Yang, X-W.; Zhao, J.; Cui, Y-X.; Liu, X-H.; Ma, C-M.; Hattori, M.; Zhang, L-H.; *J. Nat. Prod.*, **1999**, *62*, 1510-1513.

[57]   Min, B.S.; Jung, H.J.; Lee, J.S.; Kim, Y.H.; Bok, S.H.; Ma, C.M.; Nakamura, N.; Hattori, M.; Bae, K.H.; *Planta Med.*, **1999**, *65*, 374-375.

[58]   Brill, G.M.; Kati, W.M.; Montgomery, D.; Karwowski, J.P.; Humphrey, P.E.; Jackson, M.; Clement, J.J.; Kadam, S.; Chen, R.H.; McAlpine, J.B.; *J. Antibiot.*, **1996**, *49*, 541-546.

[59]   Leong-Toung, R.; Li, W.; Tam, T.F.; Karimian, K.; *Curr. Med. Chem.*, **2002**, *9*, 979-1002.

[60]   Alur, H.H.; Desai, R.P.; Mitra, A.K.; Johnston, T.P.; *Int. J. Pharm.*, **2001**, *212*, 171-176.

[61]   Polya, G.M., *Biochemical Targets of Plant Bioactive Compounds*, Taylor & Francis: London, **2003**, *in press*.

[62]   Kim, H.E.; Oh, J.H.; Lee, S.K.; Oh, Y.J.; *Life Sci.*, **1999**, *65*, PL33-40.

[63]   Bormann, H.; Melzig, M.F.; *Pharmazie*, **2000**, *55*, 129-132.

[64]   Murphy, A.; Taiz, L.; *Plant Physiol. Biochem.*, **1999**, *37*, 413-430.

[65]   Murphy, A.; Taiz, L.; *Plant Physiol. Biochem.*, **1999**, *37*, 431-443.

[66]   Murphy, A.; Peer, W.A.; Taiz, L.; *Planta*, **2000**, *211*, 315-324.

[67]   Murphy, A.S.; Hoogner, K.R.; Peer, W.A.; Taiz, L.; *Plant Physiol.*, **2002**, *128*, 935-950.

[68]   Melzig, M.F.; Bormann, H.; *Planta Med.*, **1998**, *64*, 655-657.

[69]   Sjolie, A.K.; Chaturvedi, N.; *J. Hum. Hypertens.*, **2002**, *16* Suppl. 3, S42-46.

[70]   Ogino, T.; Sato, T.; Sasaki, H.; Okada, M.; Maruno, M.; *Natural Medicines,* **1998**, *52*, 172-178.

[71]   Kinoshita, E.; Yamakoshi, J.; Kikuchi, M.; *Biosci. Biotechnol. Biochem.*, **1993**, *57*, 1107-1110.

[72]   Shimuzu, E.; Hayashi, A.; Takahashi, R.; Aoyagi, Y.; Murakami, T.; Kimoto, K.; *J. Nutr. Sci. Vitaminol.*, **1999**, *45*, 375-383.

[73]   Wagner, H.; *J. Ethnopharmacol.*, **1993**, *38*, 105-112.

[74]   Inokuchi J, Okabe H, Yamauchi T, Nagamatsu A, Nonaka G, Nishioka I.; *Life Sci.*, **1986**, *38*, 1375-1382.

[75]   Kameda, K.; Takaku, T.; Okuda, H.; Kimura, Y.; Hatano, T.; Agata, I.; Arichi, S.; *J. Nat. Prod.*, **1987**, *50*, 680-683.

622

[76]   Bormann, H.; Melzig, M.F.; *Pharmazie*, **2000**, *55*, 129-132.

[77]   Melzig, M.; Bormann, H.; Heder, G.; Siems, W-E.; Hostetmann, K.; *Pharmazie*, **1998**, *53*, 804-805.

[78]   Lacaille-Dubois, M.A.; Franck, U.; Wagner, H.; *Phytomedicine*, **2001**, *8*, 47-52.

[79]   Wagner, H.; Elbl, G.; *Planta Med.*, **1992**, *58*, 297.

[80]   Sanz, M.J.; Terencio, M.C.; Paya, M.; *Pharmazie*, **1993**, *48*, 152-153.

[81]   Uchida, S.; Ikari, N.; Ohta, H.; Niwa, M.; Nonaka, G.; Nishioka, I.; Ozaki, M.; *Jpn. J. Pharmacol.*, **1987**, *43*, 242-246.

[82]   Wagner, H.; Elbl, G.; Lotter, H.; Guinea, M.; *Pharm. Pharmacol. Lett.*, **1991**, *1*, 15-18.

[83]   Somanadhan, B.; Smitt, U.W.; George, V.; Pushpangadan, P.; Rajasekharan, S.; Duus, J.O.; Nyman, U.; Olsen, C.E.; Jaroszewski, J.W.; *Planta Med.*, **1998**, *64*, 246-250.

[84]   Loffler, B.M.; J. *Cardiovasc. Pharmacol.*, **2000**, *35*, S79-82.

[85]   Patil, A.D.; Freyer, A.J.; Eggleston, D.S.; Haltiwanger, R.C.; Tomcowicz, B.; Breen, A.; Johnson, R.K.; *J. Nat. Prod.*, **1997**, *60*, 306-308.

[86]   Lolis, E.; Petsko, G.A.; *Annu. Rev. Biochem.*, **1990**, *59*, 597-630.

[87]   Sartor, L.; Pezzato, E.; Dell'Aica, I.; Caniato, R.; Biggin, S.; Garbisa, S.; *Biochem. Pharmacol.*, **2002**, *64*, 229-237.

[88]   Maffei Facino, R.; Carini, M.; Aldini, G.; Bombardelli, E.; Morazzoni, P.; Morelli, R.; *Arzneim. Forsch./Drug Res.*, **1994**, *44*, 592-601.

[89]   Kusano, A.; Seyama, Y.; Nagai, M.; Shibano, M.; Kusano, G.; *Biol. Pharm. Bull.*, **2001**, *24*, 1198-1201.

[90]   Demeule, M.; Brossard, M.; Pagé, M.; Gingras, D.; Béliveau, R.; *Biochim. Biophys. Acta*, **2000**, *1478*, 51-60.

[91]   Makimura, M.; Hirasaw, M.; Kobayashi, K.; Indo, J.; Sakanaka, S.; Taguchi, T.; Otake, S.; *J. Periodont.*, **1993**, *64*, 630-636.

[92]   Garbisa, S.; Sartor, L.; Biggin, S.; Salvato, B.; Benelli, R.; Albini, A.; *Cancer*, **2001**, *91*, 822-832.

[93]   Annabi, B.; Lachambre, M-P.; Bousquet-Gagnon, N.; Pagé, M.; Gingras, D.; Béliveau, R.; *Biochim. Biophys. Acta*, **2002**, *1542*, 209-220.

[94]   Sazuka, M.; Imazawa, H.; Shoji, Y.; Mita, T.; Hara, Y.; Isemura, M.; *Biosci. Biotechnol. Biochem.*, **1997**, *61*, 1504-1506.

[95]   Nagase, H.; Ikeda, K.; Sakai, Y.; *Planta Med.*, **2001**, *67*, 705-708.

[96]   Robert, A.M.; Robert, L.; Renard, G.; *J. Fr. Ophtalmol.*, **2002**, *25*, 351-355.

[97]   Kweifio-Okai, G.; De Munk, F.; Rumble, B.A.; Macrides, T.A.; Cropley, M.; *Res. Commun. Mol. Pathol. Pharmacol.*, **1994**, *85*, 45-55.

[98]   Inamori, Y.; Shinohara, S.; Tsujibo, H.; Okabe, T.; Morita, Y.; Sakagami, Y.; Kumeda, Y.; Ishida, N.; *Biol Pharm Bull.*, **1999**, *22*, 990-993.

[99]   Morita, Y.; Matsumura, E.; Tsujibo, H.; Yasuda, M.; Sakagami, Y.; Okabe, T.; Ishida, N.; Inamori, Y.; *Biol. Pharm. Bull.*, **2001**, *24*, 607-611.

[100]   Nam, S.; Smith, D.M.; Dou, Q.P.; *J. Biol. Chem.*, **2001**, *276*, 13322-13330.

[101]   Stirling, Y.; *Blood Coagul. Fibrinolysis*, **1995**, *6*, 361-373.

[102]   Hoult, J.R.; Paya, M.; *Gen. Pharmacol.*, **1996**, *27*, 713-722.

[103]   Harborne, J.B.; Baxter, H., *Phytochemical Dictionary*, Taylor & Francis: London, **1993**.

[104]   Safayhi, H.; Rall, B.; Sailer, E.R.; Ammon, H.P.; *J. Pharmacol. Exp. Ther.*, **1997**, *281*, 460-463.

[105]   Rajic, A.; Akihisa, T.; Ukiya, M.; Yasukawa, K.; Sandeman, R.M.; Chandler, D.S.; Polya, G.M.; *Planta Med.*, **2001**, *67*, 599-604.

[106]   Rajic, A.; Kweifio-Okai, G.; Macrides, T.; Sandeman, R.M.; Chandler, D.S.; Polya, G.M.; *Planta Med.,* **2000**, *66*, 206-210.

[107]   Basak, A.; Cooper, S.; Roberge, A.G.; Banik, U.K.; Chretien, M.; Seidah, N.G.; *Biochem. J.*, **1999**, *338*, 107-113.

[108]   Ying, Q.L.; Rinehart, A.R.; Simon, S.R.; Cheronis, J.C.; *Biochem. J.*, **1991**, *277*, 521-526.

[109]   Francischetti, I.M.; Monteiro, R.Q.; Guimaraes, J.A.; Francischetti, B.; *Biochem. Biophys. Res. Commun.*, **1997**, *235*, 259-263.

[110]   Yuan, Y.Y.; Shi, Q.X.; Srivastava, P.N.; *Mol. Reprod. Dev.*, **1995**, *40*, 228-232.

[111]   Johnsen, O.; Mas Diaz, J.; Eliasson, R.; *Int. J. Androl.*, **1982**, *5*, 636-640.

[112]   Weir, M.P.; Bethell, S.S.; Cleasby, A.; Campbell, C.J.; Dennis, R.J.; Dix, C.J.; Finch, H.; Jhoti, H.; Mooney, C.J.; Patel, S.; Tang, C.M.; Ward, M.; Wonacott, A.J.; Wharton, C.W.; *Biochemistry*, **1998**, *37*, 6645-6657.

[113]   Facino, R.M.; Carini, M.; Stefani, R.; Aldini, G.; Saibene, L.; *Arch. Pharm. (Weinheim)*, **1995**, *328*, 720-724.

[114]   Kapil, H.; Sharma, S.; *J. Pharm. Pharmacol.*, **1994**, *46*, 922-923.

[115]   Fan, W.; Tezuka, Y.; Komatsu, K.; Namba, T.; Kadota, S.; *Biol. Pharm. Bull.*, **1999**, *22*, 157-161.

[116]   Fan, W.; Tezuka, Y.; Kadota, S.; *Chem. Pharm. Bull. (Tokyo)*, **2000**, *48*, 1055-1061.

[117]   Fan, W.; Tezuka, Y.; Ni, K.M.; Kadota, S.; *Chem Pharm Bull (Tokyo)*, **2001**, *49*, 396-401.

[118]   Lee, K.H.; Kwak, J.H.; Lee, K.B.; Song, K.S.; *Arch. Pharm. Res.*, **1998**, *21*, 207-211.

[119]   Inamori, Y.; Muro, C.; Sajima, E.; Katagiri, M.; Okamoto, Y.; Tanaka, H.; Sakagami, Y.; Tsujibo, H.; *Biosci. Biotechnol. Biochem.*, **1997**, *61*, 890-892.

[120]   Polya, G.M.; *Recent Res. Devel. Phytochem.*, **1997**, *1*, 95-110.

[121]   Polya, G.; *Current Topics Peptide Protein Res.*, **2001**, *4*, 37-54.

[122]   Christeller, J.T., Farley, P.C.; Ramsay, R.J.; Sullivan, P.A.; Laing, W.A.; *Eur. J. Biochem.*, **1998**, *254*, 160-167.

[123]   Werner, R.; Guitton, M.C.; Muhlbach, H.P; *Plant Physiol.*, **1993**, *103*, 1473.

[124]   Kreft, S.; Ravnikar, M.; Mesko, P.; Pungercar, J.; Umek, A.; Kregar, I.; Strukelj, B.; *Phytochem.*, **1997**, *44*, 1001-1006.

[125]   Mares, M.; Meloun, B.; Pavlik, M.; Kostka, V.; Baudys, M.; *FEBS Lett.*, **1989**, *251*, 94-98.

[126]   Ritonja, A.; Krizaj, I.; Mesko, P.; Kopitar, M.; Lucovnik, P.; Strukelj, B.; Pungercar, J.; Buttle, D.J.; Barrett, A.J.; Turk, V.; *FEBS Lett.*, **1990**, *267*, 13-15.

[127]   Strukelj, B.; Pungercar, J.; Ritonja, A.; Krizaj, I.; Gubensek, F.; Kregar, I.; Turk, V.; *Nucleic Acids Res.*, **1990**, *18*, 4605.

[128]   Strukelj, B.; Pungercar, J.; Mesko, P.; Barlic-Maganja, D.; Gubensek, F.; Kregar, I.; Turk, V.; *Biol. Chem. Hoppe Seyler*, **1992**, *373*, 477-482.

624

[129]   Hannapel, D.J.; *Plant Physiol.*, **1993**, *101*, 703-704.
[130]   Maganja, D.B.; Strukelj, B.; Pungercar, J.; Gubensek, F.; Turk, V.; Kregar, I.; *Plant Mol. Biol.*, **1992**, *20*, 311-313.
[131]   Strukelj, B.; Ravnikar, M.; Mesko, P.; Poljsak-Prijatelj, M.; Pungercar, J.; Kopitar, G.; Kregar, I.; Turk, V.; *Adv. Exp. Med. Biol.*, **1995**, *362*, 293-298.
[132]   Hildmann, T.; Ebneth, M.; Pena-Cortes, H.; Sanchez-Serrano, J.J.; Willmitzer, L.; Prat, S.; *Plant Cell*, **1992**, *4*, 1157-1170.
[133]   Ishikawa, A.; Ohta, S.; Matsuoka, K.; Hattori, T.; Nakamura, K.; *Plant Cell Physiol.*, **1994**, *35*, 303-312.
[134]   Herbers, K.; Prat, S.; Willmitzer, L.; *Plant Mol. Biol.*, **1994**, *26*, 73-83.
[135]   Roszkowska-Jakimiec, W.; Bankowska, A.; *Rocz. Akad. Med. Bialymst.*, **1998**, *43*, 245-249.
[136]   Kadowski, T.; Nakayama, K.; Okamoto, K.; Abe, N.; Baba, A.; Shi, Y.; Ratnayake, D.B.; Yamamoto, K.; *J. Biochem.*, **2000**, *128*, 153-159.
[137]   Reddy, M.N.; Keim, P.S.; Heinrikson, R.L.; Kezdy, F.J.; *J. Biol. Chem.*, **1975**, *250*, 1741-1750.
[138]   Lenarcic, B.; Ritonja, A.; Turk, B.; Dolenc, I.; Turk, V.; *Biol. Chem. Hoppe-Seyler*, **1992**, *373*, 459-464.
[139]   Hatano, K.; Kojima, M.; Tanokura, M.; Takahashi, K.; *Eur. J. Biochem.*, **1995**, *232*, 335-343.
[140]   Hatano, K.; Kojima, M.; Tanokura, M.; Takahashi, K.; *Biochemistry*, **1996**, *35*, 5379-5384.
[141]   Kaneko, T.; Katoh, T.; Sato, S.; Nakamura, A.; Asamizu, E.; Tabata, S.; *DNA Res.*, **2000**, *7*, 217-221.
[142]   Salanoubat, M.. *et al.* ; *Nature*, **2000**, *408*, 820-822.
[143]   Nakamura, Y.; Sato, S.; Kaneko, T.; Kotani, H.; Asamizu, E.; Miyajima, N.; Tabata, S.; *DNA Res.*, **1997**, *4*, 401-414.
[144]   Lin, X. *et al.*; *Nature*, **1999**, *402*, 761-768.
[145]   Sato, S.; Kotani, H.; Nakamura, Y.; Kaneko, T.; Asamizu, E.; Fukami, M.; Miyajima, N.; Tabata, S.; *DNA Res.*, **1997**, *4*, 215-230.
[146]   Bevan, M. *et al.*; *Nature*, **1998**, *391*, 485-488.
[147]   Nemeth, K.; Salchert, K.; Putnoky, P.; Bhalerao, R.; Koncz-Kalman, Z.; Stankovic-Stangeland, B.; Bako, L.; Mathur, J.; Okresz, L.; Stabel, S.; Geigenberger, P.; Stitt, M.; Redei, G.P.; Schell, J.; Koncz, C.; *Genes Dev.*, **1998**, *12*, 3059-3073.
[148]   Seki, M.; Narusaka, M.; Abe, H.; Kasuga, M.; Yamaguchi-Shinozaki, K.; Carninci, P.; Hayashizaki, Y.; Shinozaki, K.; *Plant Cell*, **2001**, *13*, 61-72.
[149]   Rogers, B.L.; Pollock, J.; Klapper, D.G.; Griffith, I.J.; *Gene*, **1993**, *133*, 219-221.
[150]   Lim, C.O.; Lee, S.I.; Chung, W.S.; Park, S.H.; Hwang, I.; Cho, M.J.; *Plant Mol. Biol.*, **1996**, *30*, 373-379.
[151]   Song, I.; Taylor, M.; Baker, K.; Bateman, R.C.; *Gene*, **1995**, *162*, 221-224.
[152]   Taylor, M.A.; Baker, K.C.; Briggs, G.S.; Connerton, I.F.; Cummings, N.J.; Pratt, K.A.; Revell, D.F.; Freedman, R.B.; Goodenough, P.W.; *Protein Engineering*, **1995**, *8*, 59-62.
[153]   Pernas, M.; Sanchez-Mongas, R.; Gomez, L.; Salcedo, G.; *Plant Mol. Biol.*,

1998, *38*, 1235-1242.

[154] Rogelj, B.; Popovic, T.; Ritonja, A.; Strukelj, B.; Brzin, J.; *Phytochemistry*, **1998**, *49*, 1645-1649.

[155] Ojima, A.; Shiota, H.; Higashi, K.; Kamada, H.; Shimma, Y.; Wada, M.; Satoh, S.; *Plant Mol. Biol.*, **1997**, *34*, 99-109.

[156] Sugawara, H.; Shibuya, K.; Yoshioka, T.; Hashiba, T.; Satoh S.; *J. Exp. Bot.*, **2002**, *53*, 407-413.

[157] Zhao, Y.; Botella, M.A.; Subramanian, L.; Niu, X.; Nielsen, S.S.; Bressan, R.A.; Hasegawa, P.M.; *Plant Physiol.*, **1996**, *111*, 1299-1306.

[158] Botella, M.A.; Xu, Y.; Prabha, T.N.; Zhao, Y.; Narasimhan, M.L.; Wilson, K.A.; Nielsen, S.S.; Bressan, R.A.; Hasegawa, P.M.; *Plant Physiol.*, **1996**, *112*, 1201-1210.

[159] Koiwa, H.; Shade, R.E.; Zhu-Salzman, K.; Subramanian, L.; Murdock, L.L.; Nielsen, S.S.; Bressan, R.A.; Hasegawa, P.M.; *Plant J.*, **1998**, *14*, 371-379.

[160] Koiwa, H.; Shade, R.E.; Zhu-Salzman, K.; D'Urzo, M.P.; Murdock, L.L.; Bressan, R.A.; Hasegawa, P.M.; *FEBS Lett.*, **2000**, *471*, 67-70.

[161] Brzin, J.; Ritonia, A.; Popovic, T.; Turk, V.; *Biol. Chem. Hoppe-Seyler*, **1990**, *371*, 167-170.

[162] Misaka, T.; Kuroda, M.; Iwabuchi, K.; Abe, K.; Arai, S.; *Eur. J. Biochem.*, **1996**, *240*, 609-614.

[163] Kouzuma, Y.; Kawano, K.; Kimura, M.; Yamasaki, N.; Kadowaki, T.; Yamamoto, K.; *J. Biochem. (Tokyo)*, **1996**, *119*, 1106-1113.

[164] Kouzuma, Y.; Tsukigata, K.; Inanaga, H.; Doi-Kawano, K.; Yamasaki, N.; Kimura, M.; *Biosci. Biotechnol. Biochem.*, **2001**, *65*, 969-972.

[165] Doi-Kawano, K.; Kouzuma, Y.; Yamasaki, N.; Kimura, M.; *J. Biochem. (Tokyo)*, **1998**, *124*, 911-916.

[166] Kouzuma, Y.; Inanaga, H.; Doi-Kawano, K.; Yamasaki, N.; Kimura, M.; *J. Biochem. (Tokyo)*, **2000**, *128*, 161-166.

[167] Inanaga, H.; Kobayasi, D.; Kouzuma, Y.; Aoki-Yasunaga, C.; Iiyama, K.; Kimura, M.; *Biosci. Biotechnol. Biochem.*, **2001**, *65*, 2259-2264.

[168] Gaddour, K.; Vicente-Carbajosa, J.; Lara, P.; Isabel-Lamoneda, I.; Diaz, I.; Carbonero, P.; *Plant Mol. Biol.*, **2001**, *45*, 599-608.

[169] Jones, B.L.; Marinac, L.A.; *J. Agric. Food Chem.*, **2000**, *48*, 257-264.

[170] Ryan, S.N.; Laing, W.A.; McManus, M.T.; *Phytochem.*, **1998**, *49*, 957-963.

[171] Abe, K.; Emori, Y.; Kondo, H.; Suzuki, K.; Arai, S.; *J. Biol. Chem.*, **1987**, *262*, 16793-16797.

[172] Kondo, H.; Abe, K.; Nishimura, I.; Watanabe, H.; Emori, Y.; Arai, S.; *J. Biol. Chem.*, **1990**, *265*, 15832-15837.

[173] Chen, M.S.; Johnson, B.; Wen, L.; Muthukrishnan, S.; Kramer, K.J.; Morgan, T.D.; Reeck, G.R.; *Protein Expr. Purif.*, **1992**, *3*, 41-49.

[174] Masoud, S.A.; Johnson, L.B.; White, F.F.; Reeck, G.R.; *Plant Mol. Biol.*, **1993**, *21*, 655-663.

[175] Michaud, D.; Nguyen-Quoc, B.; Yelle, S.; *FEBS Lett.*, **1993**, *331*, 173-176.

[176] Hosoyama, H.; Irie, K.; Abe, K.; Arai, S.; *Biosci. Biotechnol. Biochem.*, **1994**, *58*, 1500-1505.

[177] Leplé, J-C.; Bonadé-Bottino, M.; Augustin, S.; Delplanque, A.; Dumanois Le

Tan, V.; Pilate, G.; Cornu, D.; Jouanin, L.; *J. Mol. Breeding,* **1995**, *1*, 319-326.

[178]   Benchekroun, A.; Michaud, D.; Nguyen-Quoc, B.; Overney, S.; Desjardins, Y.; Yelle, S.; *Plant Cell. Rep.,* **1995**, *14*, 585-588.

[179]   Urwin, P.E.; Atkinson, H.J.; Waller, D.A.; McPherson, M.J.; *Plant J.,* **1995**, *8*, 121-131.

[180]   Urwin, P.E.; Lilley, C.J.; McPherson, M.J.; Atkinson, H.J.; *Plant J.,* **1997**, *12*, 455-461.

[181]   Nagata, K.; Kudo, N.; Abe, K.; Arai, S.; Tanokura, M.; *Biochemistry,* **2000**, *39*, 14753-14760.

[182]   Joshi, B.N.; Sainani, M.N.; Bastawade, K.B.; Gupta, V.S.; Ranjekar, P.K.; *Biochem. Biophys. Res. Commun.,* **1998**, *246*, 382-387.

[183]   Kimura, M.; Ikeda, T.; Fukumoto, D.; Yamasaki, N.; Yonekura, M.; *Biosci. Biotechnol. Biochem.,* **1995**, *59*, 2328-2329.

[184]   Reis, M.E.; Margis, R.; *Genetics & Molecular Biology,* **2001**, *24*, 291-296.

[185]   Brzin, J.; Popovic, T.; Drobnic-Kosorok, M.; Kotnik, M.; Turk, V.; *Biol. Chem. Hoppe-Seyler,* **1988**, *369*, 233-238.

[186]   Rowan, A.D.; Brzin, J.; Buttle, D.J.; Barrett, A.J.; *FEBS Lett.,* **1990**, *269*, 328-330.

[187]   Krizaj, I.; Drobnic-Kosorok, M.; Brzin, J.; Jerala, R.; Turk, V.; *FEBS Lett.,* **1993**, *333*, 15-20.

[188]   Gruden, K.; Strukelj, B.; Ravnikar, M.; Poljsak-Prijatelj, M.; Mavric, I.; Brzin, J.; Pungercar, J.; Kregar, I.; *Plant Mol. Biol.,* **1997**, *34*, 317-323.

[189]   Waldron, C.; Wegrich, L.M.; Merlo, P.A.O.; Walsh, T.A.; *Plant Mol. Biol.,* **1993**, *23*, 801-812.

[190]   Walsh, T.A.; Strickland, J.A.; *Plant Physiol.,* **1993**, *103*, 1227-1234.

[191]   Li, Z.; Sommer, A.; Dingermann, T.; Noe, C.R.; *Mol. Gen. Genet.,* **1996**, *251*, 499-502.

[192]   Kuroda, M.; Kiyosaki, T.; Matsumoto, I.; Misaka, T.; Arai, S.; Abe, K.; *Biosci. Biotechnol. Biochem.,* **2001**, *65*, 22-28.

[193]   Fernandes, K.; Sabelli, P.; Barratt, D.; Richardson, M.; Xavier-Filho, J.; Shewry, P.; *Plant Mol. Biol.,* **1993**, *23*, 215-219.

[194]   Hirashiki, I.; Ogata, F.; Yoshida, N.; Makisumi, S.; Ito, A.; *J. Biochem. (Tokyo),* **1990**, *108*, 604-608.

[195]   Abe, M.; Abe, K.; Kuroda, M.; Arai, S.; *Eur. J. Biochem.,* **1992**, *209*, 933-937.

[196]   Abe, M.; Abe, K.; Domoto, C.; Arai, S.; *Biosci. Biotechnol. Biochem.,* **1995**, *59*, 756-758.

[197]   Irie, K.; Hosoyama, H.; Takeuchi, T.; Iwabuchi, K.; Watanabe, H.; Abe, M.; Abe, K.; Arai, S.; *Plant Mol. Biol.,* **1996**, *30*, 149-157.

[198]   Abe, M.; Domoto, C.; Watanabe, H.; Abe, K.; Arai, S.; *Biosci. Biotechnol. Biochem.,* **1996**, *60*, 1173-1175.

[199]   Matsui,T.; Li, C.H.; Osajima, Y.; *J. Peptide Sci.,* **1999**, *5*, 289-297.

[200]   Sendl, A.; Elbl, G.; Steinke, B.; Redl, K.; Breu, W.; Wagner, H.; *Planta Med.,* **1992**, *58*, 1-7.

[201]   Shin, Z.I.; Yu, R.; Park, S.A.; Chung, D.K.; Ahn, C.W.; Nam, H.S.; Kim, K.S.; Lee, H.J.; *J. Agric. Food Chem.,* **2001**, *49*, 3004-3009.

[202] Miyoshi, S.; Kaneko, T.; Ishikawa, H.; Tanaka, H.; Maruyama, S.; *Ann. N.Y. Acad. Sci.*, **1995**, *750*, 429-431.

[203] Miyoshi, S.; Ishikawa, H.; Kaneko, T.; Fukui, F.; Tanaka, H.; Maruyama, S.; *Agric. Biol. Chem.,* **1991**, *55*, 1313-1318.

[204] Matsui, T.; Li, C.H.; Tanaka, T.; Maki, T.; Osajima, Y.; Matsumoto, K.; *Biol. Pharm. Bull.*, **2000**, *23*, 427-431.

[205] Hass, G.M.; Hermodson, M.A.; *Biochemistry*, **1981**, *20*, 2256-2260.

[206] Martineau, B.; McBride, K.E.; Houck, C.M.; *Mol. Gen. Genet.*, **1991**, *228*, 281-286.

[207] Ryan, C.A.; Hass, G.M.; Kuhn, R.W.; *J. Biol. Chem.*, **1974**, *249*, 5495-5499.

[208] Hass, G.M.; Nau, H.; Biemann, K.; Grahn, D.T.; Ericsson, L.H.; Neurath, H.; *Biochemistry*, **1975**, *14*, 1334-1342.

[209] Nau, H.; Biemann, K.; *Anal. Biochem.*, **1976**, *73*, 175-186.

[210] Hass, G.M.; Ryan, C.A.; *Methods Enzymol.*, **1981**, *80*, 778-791.

[211] Rees, D.C.; Lipscomb, W.N.; *J. Mol. Biol.*, **1982**, *160*, 475-498.

[212] Leary, T.R.; Grahn, D.T.; Neurath, H.; Hass, G.M.; *Biochemistry*, **1979**, *18*, 2252-2256.

[213] Clore, G.M.; Gronenborn, A.M.; Nilges, M.; Ryan, C.A.; *Biochemistry*, **1987**, *26*, 8012-8023.

[214] Villanueva, J.; Canals, F.; Prat, S.; Ludevid, D.; Querol, E.; Aviles, F.X; *FEBS Lett.*, **1998**, *440*, 175-182.

[215] Neumann, G.M.; Thomas, I.; Polya, G.M.; *Plant Science*, **1996**, *114*, 45-51.

[216] Molnar, A.; Lovas, A.; Banfalvi, Z.; Lakatos, L.; Polgar, Z.; Horvath, S.; *Plant Mol. Biol.*, **2001**, *46*, 301-311.

[217] Banfalvi, Z.; Molnar, A.; Molnar, G.; Lakatos, L.; Szabo, L.; *FEBS Lett.*, **1996**, *383*, 159-164.

[218] Norioka, S.; Ikenaka, T.; *J. Biochem. (Tokyo)*, **1983**, *94*, 589-599.

[219] Kuorkawa, T.; Hara, S.; Norioka, S.; Teshima, T.; Ikenaka, T.; *J. Biochem. (Tokyo)*, **1987**, *101*, 723-728.

[220] Park, S.S.; Sumi, T.; Ohba, H.; Nakamura, O.; Kimura, M.; *Biosci. Biotechnol. Biochem.*, **2000**, *64*, 2272-2275.

[221] Terada, S.; Fujimura, S.; Kimoto, E.; *Biosci. Biotechnol. Biochem.,* **1994**, *58*, 376-379.

[222] Bueno, N.R.; Fritz, H.; Auerswald, E.A.; Mentele, R.; Sampaio, M.; Sampaio, C.A.; Oliva, M.L.; *Biochem. Biophys. Res. Commun.*, **1999**, *261*, 838-843.

[223] Kalume, D.E.; Sousa, M.V.; Morhy, L.; *J. Protein Chem.*, **1995**, *14*, 685-693.

[224] Prakash, B.; Selvaraj, S.; Murthy, M.R.; Sreerama, Y.N.; Rao, D.R.; Gowda, L.R.; *J. Mol. Evol.*, **1996**, *42*, 560-569.

[225] Kimura, M.; Kouzuma, Y.; Abe, K.; Yamasaki, N.; *J. Biochem. (Tokyo)*, **1994**, *115*, 369-372.

[226] Odani, S.; Ikenaka, T.; *J. Biochem. (Tokyo)*, **1972**, *71*, 839-848.

[227] Voss, R.H.; Ermler, U.; Essen, L.O.; Wenzl, G.; Kim, Y.M.; Flecker, P.; *Eur. J. Biochem.*, **1996**, *242*, 122-131.

[228] Birk, J.; *Int . J. Peptide Protein Res.*, **1985**, *25*, 113-131.

[229] Werner, M.H.; Wemmer, D.E.; *Biochemistry*, **1992**, *31*, 999-1010.

[230] Odani, S.; Ikenaka, T.; *J. Biochem. (Tokyo)*, **1973**, *74*, 697-715.

628

[231]   Koepke, J.; Ermler, U.; Warkentin, E.; Wenzl, G.; Flecker, P.; *J. Mol. Biol.*, **2000**, *298*, 477-491.

[232]   Kennedy, A.R.; *Pharmacol. Ther.*, **1998**, *78*, 167-209.

[233]   Neumann, G.M.; Condron, R.; Polya, G.M.; *Plant Science*, **1994**, *96*, 69-79.

[234]   Odani, S.; Ikenaka, T.; *J. Biochem. (Tokyo)*, **1976**, *80*, 641-643.

[235]   Odani, S.; Ikenaka, T.; *J. Biochem. (Tokyo)*, **1977**, *82*, 1523-1531.

[236]   Odani, S.; Ikenaka, T.; *J. Biochem. (Tokyo)*, **1978**, *83*, 737-45.

[237]   Chen, P.; Rose, J.; Love, R.; Wei, C.H.; Wang, B.C.; *J. Biol. Chem.,* **1992**, *267*, 1990-1994.

[238]   Hwang, D.L-R.; Davis Lin, K-T.; Yang, W-K.; Foard, D.E.; *Biochim. Biophys. Acta*, **1977**, *495*, 369-382.

[239]   Joubert, F.J.; *Phytochemistry*, **1984**, *23*, 957-961.

[240]   Joubert, F.J.; Kruger, H.; Townshend, G.S.; Botes, D.P.; *Eur. J. Biochem.*, **1979**, *97*, 85-91.

[241]   Ceciliani, F.; Tava, A.; Iori, R.; Mortarino, M.; Odoardi, M.; Ronchi, S.; *Phytochem.*, **1997**, *44*, 393-398.

[242]   McGurl, B.; Mukherjee, S.K.; Kahn, M.L.; Ryan, C.A.; *Plant Mol. Biol.*, **1995**, *27*, 995-1001.

[243]   Brown, W.E.; Takio, K.; Titani, K.; Ryan, C.A.; *Biochemistry*, **1985**, *24*, 2105-2108.

[244]   Ishikawa, C.; Nakamura, S.; Watanabe, K.; Takahashi, K.; *FEBS Lett.*, **1979**, *99*, 97-100.

[245]   Yoshikawa, M.; Kiyohara, T.; Iwasaki, T.; Ishii, Y.; Kimura, N.; *Agric. Biol. Chem.*, **1979**, *43*, 787-796.

[246]   Ishikawa, C.; Watanabe, K.; Sakata, N.; Nakagaki, C.; Nakamura, S.; Takahashi, K.; *J. Biochem. (Tokyo)*, **1985**, *97*, 55-70.

[247]   Tsunogae, Y.; Tanaka, I.; Yamane, T.; Kikkawa, J.; Ashida, T.; Ishikawa, C.; Watanabe, K.; Nakamura, S.; Takahashi, K.; *J. Biochem. (Tokyo)*, **1986**, *100*, 1637-1646.

[248]   Kiyohara, T.; Yokota, K.; Masaki, O.; Iwasaki, T.; Yoshikawa, M.; *J. Biochem. (Tokyo)*, **1981**, *90*, 721-728.

[249]   Zhang, Y.; Luo, S.; Tan, F.; Qi, Z.; Xu, L.; Zhang, A.; *Sci. Sin. B,* **1982**, *25*, 268-277.

[250]   Wilson, K.A.; Chen, J.C.; *Plant Physiol.*, **1983**, *71*, 341-349.

[251]   Tan, C.G.; Stevens, F.C.; *Eur. J. Biochem.*, **1971**, *18*, 515-523.

[252]   Stevens, F.C.; Wuerz, S.; Krahn, J.; In Fritz H., Tschesche H., Greene, L.J., Truscheit E. (eds.), *Proteinase Inhibitors* (Bayer-Symp. V), pp.344-354, Springer-Verlag: Berlin, **1974**.

[253]   De Carvalho, P.G.; Bloch, C.; Morhy, L.; da Silva, M.C.; de Mello, L.V.; Neshich, G.; *J. Protein Chem.*, **1996**, *15*, 591-598.

[254]   Funk, A.; Weder, J.K.; Belitz, H-D.; *Z. Lebensm. Unters. Forsch.*, **1993**, *196*, 343-350.

[255]   Wilson, K.A.; Laskowski, M.; *J. Biol. Chem.*, **1975**, *250*, 4261-4267.

[256]   Domoney, C.; Welham, T.; Sidebottom, C.; Firmin, J.L.; *FEBS Lett.*, **1995**, *360*, 15-20.

[257]   Ferrasson, E.; Quillien, L.; Gueguen, J.; *J. Protein Chem.*, **1995**, *14*, 467-475.

[258] Quillien, L.; Ferrasson, E.; Molle, D.; Gueguen, J.; *J. Protein Chem.*, **1997**, *16*, 195-203.

[259] Pouvreau, L.; Chobert, J.M.; Briand, L.; Quillien, L.; Tran, V.; Gueguen, J.; Haertle, T.; *FEBS Lett.*, **1998**, *423*, 167-172.

[260] Li de la Sierra, I.; Quillien, L.; Flecker, P.; Gueguen, J.; Brunie, S.; *J. Mol. Biol.*, **1999**, *285*, 1195-1207.

[261] Tanaka, A.S.; Sampaio, M.U.; Mentele, R.; Auerswald, E.A.; Sampaio, C.A.; *J. Protein Chem.*, **1996**, *15*, 553-560.

[262] Tanaka, A.S.; Sampaio, M.U.; Marangoni, S.; de Oliveira, B.; Novello, J.C.; Oliva, M.L.; Fink, E.; Sampaio, C.A.; *Biol. Chem.*, **1997**, *378*, 273-281.

[263] Shimokawa, Y.; Kuromizu, K.; Araki, T.; Ohata, J.; Abe, O.; *Eur. J. Biochem.*, **1984**, *143*, 677-684.

[264] Asao, T.; Imai, F.; Tsuji, I.; Tashiro, M.; Iwami, K.; Ibuki, F.; *J. Biochem. (Tokyo)*, **1991**, *110*, 951-955.

[265] Asao, T.; Imai, F.; Tsuji, I.; Tashiro, M.; Iwami, K.; Ibuki, F.; *Agric. Biol. Chem.*, **1991**, *55*, 707-713.

[266] Morhy, L.; Ventura, M.M.; *An. Acad. Bras. Cienc.*, **1987**, *59*, 71-81.

[267] De Freitas, S.M.; de Mello, L.V.; da Silva, M.C.; Vriend, G.; Neshich, G.; Ventura, M.M.; *FEBS Lett.*, **1997**, *409*, 121-127.

[268] Hilder, V.A.; Barker, R.F.; Samour, R.A.; Gatehouse, A.M.; Gatehouse, J.A.; Boulter, D.; *Plant Mol. Biol.*, **1989**, *13*, 701-710.

[269] Polya, G.M.; *Recent Res. Devel. Phytochem.*, **1998**, *2*, 287-302.

[270] Polya, G.M.; *Current Topics Phytochem.*, **1999**, *2*, 51-67.

[271] Ary, M.B.; Shewry, P.R.; Richardson, M.; *FEBS Lett.*, **1988**, *229*, 111-118.

[272] Nagasue, A.; Fukamachi, H.; Ikenaga, H.; Funatsu,G.; *Agric. Biol. Chem.,* **1988**, *52*, 1505-1514.

[273] Song, H.K.; Kim, Y.S.; Yang, J.K.; Moon, J.; Lee, J.Y.; Suh, S.W.; *J. Mol. Biol.*, **1999**, *293*, 1133-1144.

[274] Stevens, C.; Titarenko, E.; Hargreaves, J.A.; Gurr, S.J.; *Plant Mol. Biol.*, **1996**, *31*, 741-749.

[275] Tashiro, M.; Hashino, K.; Shiozaki, M.; Ibuki, F.; Maki, Z.; *J. Biochem. (Tokyo)*, **1987**, *102*, 297-306.

[276] Rakwal, R.; Agrawal, G.K.; Jwa, N-S.; *Gene,* **2001**, *263*, 189-198.

[277] Li, N.; Qu, L-J.; Liu, Y.; Li, Q.; Gu, H.; Chen, Z.; *Protein Expr. Purif.*, **1999**, *15*, 99-104.

[278] Tashiro, M.; Asao, T.; Hirata, C.; Takahashi, K.; Kanamori, M.; *J. Biochem. (Tokyo)*, **1990**, *108*, 669-672.

[279] Tashiro, M.; Asao, T.; Hirata, C; Takahashi, K.; *Agric. Biol. Chem.*, **1991**, *55*, 419-426.

[280] Odani, S.; Koide, T.; Ono, T.; *J. Biochem. (Tokyo)*, **1986**, *100*, 975-983.

[281] Poerio, E.; Caporale, C.; Carrano, L.; Caruso, C.; Vacca, F.; Buonocore, V.; *J. Protein Chem.*, **1994**, *13*, 187-194.

[282] Rohrmeier, T.; Lehle, L.; *Plant Mol. Biol.*, **1993**, *22*, 783-792.

[283] Eckelkamp, C.; Ehmann, B.; Schopfer, P.; *FEBS Lett.*, **1993**, *323*, 73-76.

[284] Tiffin, P.; Gaut, B.S.; *Mol. Biol. Evol.*, **2001**, *18*, 2092-2101.

[285] Luckett, S.; Santiago Garcia, R.; Barker, J.J.; Konarev, A.I.; Shewry, P.R.; Clarke, A.R.; Brady, R.L.; *J. Mol. Biol.*, **1999**, *290*, 525-533.

[286] Korsinczky, M.L.J.; Schirra, H.J.; Rosengren, K.J.; West, J.; Condle, B.A.; Otvos, L.; Anderson, M.A.; Craik, D.J.; *J. Mol. Biol.*, **2001**, *311*, 579-591.

[287] Stiekema, W.J.; Heidekamp, F.; Dirkse, W.G.; Van Beckum, J.; Haan, P.; Ten Bosch, C.; Louwerse, J.D.; *Plant Mol. Biol.*, **1988**, *11*, 255-269.

[288] Hendriks, T.; Vreugdenhil, D.; Stiekema, W.J.; *Plant Mol. Biol.*, **1991**, *17*, 385-394.

[289] Mitsumori, C.; Yamagishi, K.; Fujino, K.; Kikuta, Y.; *Plant Mol. Biol.,* **1994**, *26*, 961-969.

[290] Campos, F.A.P.; Richardson, M.; *FEBS Lett.*, **1983**, *152*, 300-304.

[291] Shivaraj, B.; Pattabiraman, T.N.; *Biochem. J.*, **1981**, *193*, 29-36.

[292] Strobl, S.; Muhlhahn, P.; Bernstein, R.; Wiltscheck, R.; Maskos, K.; Wunderlich, M.; Huber, R.; Glockshuber, R.; Holak, T.A.; *Biochemistry*, **1995**, *34*, 8281-8293.

[293] Strobl, S.; Maskos, K.; Wiegand, G.; Huber, R.; Gomis-Ruth, F.X.; Glockshuber, R.; *Structure*, **1998**, *6*, 911-921.

[294] Gourinath, S.; Srinivasan, A.; Singh, T.P.; *Acta Crystallogr. D Biol. Crystallogr.*, **1999**, *55*, 25-30.

[295] Gourinath, S.; Alam, N.; Srinivasan, A.; Betzel, C.; Singh, T.P.; *Acta Crystallogr. D Biol. Crystallogr.*, **2000**, *56*, 287-293.

[296] Maskos, K.; Huber-Wunderich, M.; Glockshuber, R.; *FEBS Lett.*, **1996**, *397*, 11-16.

[297] Alam, N.; Gourinath, S.; Dey, S.; Srinivasan, A.; Singh, T.P.; *Biochemistry*, **2001**, *40*, 4229-4233.

[298] Barber, D.; Sanchez-Monge, R.; Mendez, E.; Lazaro, A.; Garcia-Olmedo, F.; Salcedo, G.; *Biochim. Biophys. Acta*, **1986**, *869*, 115-118.

[299] Medina, J.; Hueros, G.; Carbonero, P.; *Plant Mol. Biol.*, **1993**, *23*, 535-542.

[300] Rasmussen, S.K.; Johansson, A.; *Plant Mol. Biol.*, **1992**, *18*, 423-427.

[301] Grosset, J.; Alary, R.; Gautier, M.F.; Menossi, M.; Martinez-Izquierdo, J.A.; Joudrier, P.; *Plant Mol. Biol.*, **1997**, *34*, 331-338.

[302] Odani, S.; Koide, T.; Ono, T.; *J. Biol. Chem.*, **1983**, *258*, 7998-8003.

[303] Mena, M.; Sanchez-Monge, R.; Gomez, L.; Salcedo, G.; Carbonero, P.; *Plant Mol. Biol.*, **1992**, *20*, 451-458.

[304] Lazaro, A.; Sanchez-Monge, R.; Salcedo, G.; Paz-Ares, J.; Carbonero, P.; Garcia-Olmedo, F.; *Eur. J. Biochem.*, **1988**, *172*, 129-134.

[305] Izumi, H.; Adachi, T.; Fujii, N.; Matsuda, T.; Nakamura, R.; Tanaka, K.; Urisu, A.; Kurosawa, Y.; *FEBS Lett.*, **1992**, *302*, 213-216.

[306] Adachi, T.; Izumi, H.; Yamada, T.; Tanaka, K.; Takeuchi, S.; Nakamura, R.; Matsuda, T.; *Plant Mol. Biol.*, **1993**, *21*, 239-248.

[307] Bloch, C.; Richardson, M.; *Protein Seq. Data Anal.*, **1992**, *5*, 27-30.

[308] Gautier, M.F.; Alary, R.; Joudrier, P.; *Plant Mol. Biol.*, **1990**, *14*, 313-322.

[309] Garcia-Moroto, F.; Marana, C.; Mena, M.; Garcia-Olmedo, F.; Carbonero, P.; *Plant Mol. Biol.*, **1990**, *14*, 845-853.

[310] Gautier, M.F.; Alary, R.; Lullien, V.; Joudrier, P.; *Plant Mol. Biol.*, **1991**, *16*, 333-334.

[311] Lullien, V.; Alary, R.; Joudrier, P.; Gautier, M.F.; *Plant Mol. Biol.*, **1991**, *16*, 373-374.

[312] Sanchez de la Hoz, P.; Castagnaro, A.; Carbonero, P.; *Plant Mol. Biol.*, **1994**, *26*, 1231-1236.

[313] Maeda, K.; Kakabayashi, S.; Matsubara, H.; *Biochim. Biophys. Acta*, **1985**, *828*, 213-221.

[314] Okuda, M.; Satoh, T.; Sakurai, N.; Shibuya, K.; Kaji, H.; Samejima, T.; *J. Biochem. (Tokyo)*, **1997**, *122*, 918-926.

[315] Oda, Y.; Matsunaga, T.; Fukuyama, K.; Miyazaki, T.; Morimoto, T.; *Biochemistry*, **1997**, *36*, 13502-13511.

[316] Amano, M.; Ogawa, H.; Kojima, K.; Kamidaira, T.; Suetsugu, S.; Yoshihama, M.; Satoh, T.; Samejima, T.; Matsumoto, I.; *Biochem. J.*, **1998**, *330*, 1229-1234.

[317] Poerio, E.; Caporale, C.; Carrano, L.; Pucci, P.; Buonocore, V.; *Eur. J. Biochem.*, **1991**, *199*, 595-600.

[318] Garcia-Maroto, F.; Carbonero, P.; Garcia-Olmedo, F.; *Plant Mol. Biol.*, **1991**, *17*, 1005-1011.

[319] Caporale, C.; Carrano, L.; Nitti, G.; Poerio, E.; Pucci, P.; Buonocore, V.; *Protein Seq. Data Anal.*, **1991**, *4*, 3-8.

[320] Maeda, K.; Hase, T.; Matsubara, H.; *Biochim. Biophys. Acta*, **1983**, *743*, 52-57.

[321] Maeda, K.; Wakabayashi, S.; Matsubara, H.; *J. Biochem. (Tokyo)*, **1983**, *94*, 865-870.

[322] Sanchez-Monge, R.; Gomez, L.; Garcia-Olmedo, F.; Salcedo, G.; *Eur. J. Biochem.*, **1989**, *183*, 37-40.

[323] Mahoney, W.C.; Hermodson, M.A.; Jones, B.; Powers, D.D.; Corfman, R.S.; Reeck, G.R.; *J. Biol. Chem.*, **1984**, *259*, 8412-8416.

[324] Wen, L.; Huang, J.K.; Zen, K.C.; Johnson, B.H.; Muthukrishnan, S.; MacKay, V.; Manney, T.R.; Manney, M.; Reeck, G.R.; *Plant Mol. Biol.*, **1992**, *18*, 813-814.

[325] Behnke, C.A; Yee, V.C.; Trong, I.L.; Pedersen, L.C.; Stenkamp, R.E.; Kim, S.S.; Reeck, G.R.; Teller, D.C.; *Biochemistry*, **1998**, *37*, 15277-15288.

[326] Lin, J.Y; Chu, S.C.; Wu, H.C.; Hsieh, Y.S.; *J. Biochem. (Tokyo)*, **1991**, *110*, 879-883.

[327] Hung, C.H.; Lee, M.C.; Lin, J.Y.; *Biochem. Biophys. Res. Commun.*, **1992**, *184*, 1524-1528.

[328] Wu, H.C.; Lin, J.Y.; *J. Biochem. (Tokyo)*, **1993**, *113*, 258-263.

[329] Hung, C.H.; Lee, M.C.; Lin, J.Y.; *FEBS Lett.*, **1994**, *353*, 312-314.

[330] Richardson, M.; Campos, F.A.P.; Xavier-Filho, J.; Macedo, M.L.R.; Maia, G.M.C.; Yarwood, A.; *Biochim. Biophys. Acta*, **1996**, *872*, 134-140.

[331] Odani, S.; Odani, S.; Ono, T.; Ikenaka, T.; *J. Biochem. (Tokyo)*, **1979**, *86*, 1795-1805.

[332] Argall, M.E.; Bradbury, J.H.; Shaw, D.C.; *Biochim. Biophys. Acta*, **1994**, *1204*, 189-194.

[333] Mathews, A.; Llewellyn, D.J.; Wu, Y.; Dennis, E.S.; *Plant Mol. Biol.*, **1996**, *30*, 1035-1039.

632

[334]   Gosti, F.; Bertauche, N.; Vartanian, N.; Giraudat, J.; *Mol. Gen. Genet.*, **1995**, *246*, 10-18.

[335]   Sampaio, C.A.; Motta, G.; Sampaio, M.U.; Oliva, M.L.; Araujo, M.S.; Stella, R.C.; Tanaka, A.S.; Batista, I.F.; *Agents Actions Suppl.,* **1992**, *36*, 191-199.

[336]   Di Ciero, L.; Oliva, M.L.; Torquato, R.; Kohler, P.; Weder, J.K.; Camillo Novello, J.; Sampaio, C.A.; Oliveira, B.; Marangoni, S.; *J. Protein Chem.*, **1998**, *17*, 827-834.

[337]   Downing, W.L.; Mauxion, F.; Fauvarque, M.O.; Reviron, M.P.; de Vienne, D.; Vartanian, N.; Giraudat, J.; *Plant J.,* **1992**, *2*, 685-693.

[338]   Williams, D.L.; Kain, W.C.; Broadway, R.M.; *Plant Physiol.*, **1997**, *114*, 747.

[339]   Nishio, N.; Satoh, H.; *Plant Physiol.*, **1997**, *115*, 841-846.

[340]   Annamalai, P.; Yanagihara, S.; *J. Plant Physiol.*, **1999**, *155*, 226-233.

[341]   Satoh, H.; Nakayama, K.; Okada, M.; *J. Biol. Chem.*, **1998**, *273*, 30568-30575.

[342]   Terada, S.; Fujimura, S.; Katayama, H.; Nagasawa, M.; Kimoto, E.; *J. Biochem. (Tokyo)*, **1994**, *115*, 392-396.

[343]   Terada, S.; Katayama, H.; Noda, K.; Fujimura, S.; Kimoto, E.; *J. Biochem. (Tokyo)*, **1994**, *115*, 397-404.

[344]   Terada, S.; Fujimura, S.; Kino, S.; Kimoto, E.; *Biosci. Biotechnol. Biochem.*, **1994**, *58*, 371-375.

[345]   Odani, S.; Yokokawa, Y.; Takeda, H.; Abe, S.; Odani, S.; *Eur. J. Biochem.*, **1996**, *241*, 77-82.

[346]   Pando, S.C.; Oliva, M.L.; Sampaio, C.A.; Di Ciero, L.; Novello, J.C.; Marangoni, S.; *Phytochemistry*, **2001**, *57*, 625-631.

[347]   Batista, I.F.; Oliva, M.L.; Araujo, M.S.; Sampaio, M.U.; Richardson, M.; Fritz, H.; Sampaio, C.A.; *Phytochemistry*, **1996**, *41*, 1017-1022.

[348]   Batista, I.F.; Nonato, M.C.; Bonfadini, M.R.; Beltramini, L.M.; Oliva, M.L.; Sampaio, M.U.; Sampaio, C.A.; Garratt, R.C.; *Acta Crystallogr. D Biol. Crystallogr.*, **2001**, *57*, 602-604.

[349]   Joubert, F.J.; *Int. J. Biochem.*, **1982**, *14*, 187-193.

[350]   Onesti, S.; Brick, P.; Blow, D.M.; *J. Mol. Biol.*, **1991**, *217*, 153-76.

[351]   Joubert, F.J.; Carlsson, F.H.; Haylett, T.; *Hoppe-Seyler's Z. Physiol. Chem.*, **1981**, *362*, 531-538.

[352]   Kouzuma, Y.; Suetake, M.; Kimura, M.; Yamasaki, N.; *Biosci. Biotechnol. Biochem.*, **1992**, *56*, 1819-1824.

[353]   Kouzuma, Y.; Yamasaki, N.; Kimura, M.; *J. Biochem. (Tokyo)* , **1997**, *121*, 456-463.

[354]   Kimura, M.; Kouzuma, Y.; Yamasaki, N.; *Biosci. Biotechnol. Biochem.*, **1993**, *57*, 102-106.

[355]   Kimura, M.; Harada, N.; Iwanaga, S.; Yamasaki, N.; *Eur. J. Biochem.*, **1997**, *249*, 870-877.

[356]   Kuramitsu, J.; Iwanaga, S.; Yamasaki, N.; Kimura, M.; *Biosci. Biotechnol. Biochem.*, **1996**, *60*, 1469-1473.

[357]   Iwanaga, S.; Yamasaki, N.; Kimura, M.; *J. Biochem. (Tokyo)*, **1998**, *124*, 663-669.

[358]   Iwanaga, S.; Nagata, R.; Miyamoto, A.; Kouzuma, Y.; Yamasaki, N.; Kimura M.; *J. Biochem. (Tokyo)*, **1999**, *126*, 162-167.

[359]   Ozawa, K.; Laskowski, M.; *J. Biol. Chem.*, **1966**, *241*, 3955-3961.

[360]   Brown, J.R.; Lerman, N.; Bohak, Z.; *Biochem. Biophys. Res. Commun.*, **1966**, *23*, 561-565.

[361]   Koide, T.; Ikenaka, T.; *Eur. J. Biochem.*, **1973**, *32*, 417-431.

[362]   Sweet, R.M.; Wright, H.T.; Janin, J.; Chothia, C.H.; Blow, D.M.; *Biochemistry*, **1974**, *13*, 4212-4228.

[363]   Kim, S.H.; Hara, S.; Hase, S.; Ikenaka, T.; Toda, H.; Kitamura, K.; Kaizuma, N.; *J. Biochem. (Tokyo)*, **1985**, *98*, 435-448.

[364]   Jofuku, K.D.; Schipper, R.D.; Goldberg, R.B.; *Plant Cell*, **1989**, *1*, 427-435.

[365]   Jofuku, K.D.; Goldberg, R.B.; *Plant Cell*, **1989**, *1*, 1079-1093.

[366]   Song, S.I.; Kim, C.H.; Baek, S.J.; Choi, Y.D.; *Plant Physiol.*, **1993**, *101*, 1401-1402.

[367]   De Meester, P.; Brick, P.; Lloyd, L.F.; Blow, D.M.; Onesti, S., *Acta Crystallogr. D Biol. Crystallogr.*, **1998**, *54*, 589-597.

[368]   Song, H.K.; Suh, S.W.; *J. Mol. Biol.*, **1998**, *275*, 347-363.

[369]   Perez-Grau, L.; Goldberg, R.B.; *Plant Cell*, **1989**, *1*, 1095-1109.

[370]   Ashida, Y.; Matsushima, A.; Tsuru, Y.; Hirota, T.; Hirata, T.; *Biosci. Biotechnol. Biochem.*, **2000**, *64*, 1305-1309.

[371]   Gotor, C.; Pintor-Toro, J.A.; Romero, L.C.; *Plant Physiol.*, **1995**, *107*, 1015-1016.

[372]   Yoshikawa, M.; Iwasaki, T.; Fujii, M.; Oogaki, M.; *J. Biochem. (Tokyo)*, **1976**, *79*, 765-773.

[373]   Svendsen, I.; Hejgaard, J.; Mundy, J.; *Carlsberg Res. Commun.*, **1986**, *51*, 43-50.

[374]   Leah, R.; Mundy, J.; *Plant Mol. Biol.*, **1989**, *12*, 673-682.

[375]   Vallee, F.; Kadziola, A.; Bourne, Y.; Juy, M.; Rodenburg, K.W.; Svensson, B.; Haser, R.; *Structure*, **1998**, *6*, 649-659.

[376]   Hattori, T.; Nakagawa, T.; Maeshima, M.; Nakamura, K.; Ashai, T.; *Plant Mol. Biol.*, **1985**, *5*, 313-320..

[377]   Hattori, T.; Nakamura, K.; *Plant Mol. Biol.*, **1988**, *11*, 417-426.

[378]   Hattori, T.; Yoshida, N.; Nakamura, K.; *Plant Mol. Biol.*, **1989**, *13*, 563-572.

[379]   Wang, S.J.; Lin, C.T.; Ho, K.C.; Chen, Y.M.; Yeh, K.W.; *Plant Physiol.*, **1995**, *108*, 829-830.

[380]   Yeh, K.W.; Chen, J.C.; Lin, M.I.; Chen, Y.M.; Lin, C.Y.; *Plant Mol. Biol.*, **1997**, *33*, 565-570.

[381]   Yeh, K.W.; Lin, M.I.; Tuan, S.J.; Chen, Y.M.; Lin, C.Y.; Kao, S.S.; *Plant Cell Reports*, **1997**, *16*, 696-699.

[382]   Yao, P.L.; Hwang, M.J.; Chen, Y.M.; Yeh, K.W.; *FEBS Lett.*, **2001**, *496*, 134-138.

[383]   Werner, R.; Guitton, M.C.; Muhlbach, H.P.; *Plant Physiol.*, **1993**, *103*, 1473.

[384]   Brenner, E.D.; Lambert, K.N.; Kaloshian, I.; Williamson, V.M.; *Plant Physiol.*, **1998**, *118*, 237-247.

[385]   Milligan, S.B.; Gasser, C.S.; *Plant Mol. Biol.*, **1995**, *28*, 691-711.

[386]   Fujita, T.; Kouchi, H.; Ichikawa, T.; Syono, K.; *Plant J.*, **1994**, *5*, 645-654.

[387]   Karrer, E.E.; Beachy, R.N.; Holt, C.A.; *Plant Mol. Biol.*, **1998**, *36*, 681-690.

[388] Park, K.S.; Cheong, J.J.; Lee, S.J.; Suh, M.C.; Choi, D.; *Biochim. Biophys. Acta*, **2000**, *1492*, 509-512.

[389] Ohtsubo, K-I.; Richardson, M.; *FEBS Lett.*, **1992**, *309*, 68-72.

[390] Augur, C.; Benhamou, N.; Darvill, A.; Albersheim, P.; *Plant J.*, **1993**, *3*, 415-426.

[391] Augur, C.; Stiefel, V.; Darvill, A.; Albersheim, P.; Puigdomenech, P.; *J. Biol. Chem.*, **1995**, *270*, 24839-24843.

[392] Bradshaw, H.D.; Hollick, J.B.; Parsons, T.J.; Clarke, H.R.; Gordon, M.P.; *Plant Mol. Biol.*, **1990**, *14*, 51-59.

[393] Haruta, M.; Major, I.T.; Christopher, M.E.; Patton, J.J.; Constabel, C.P.; *Plant Mol. Biol.*, **2001**, *46*, 347-359.

[394] Hollick, J.B.; Gordon, M.P.; *Plant Mol. Biol.*, **1993**, *22*, 561-572.

[395] Negreiros, A.N.M.; Carvalho, M.M.; Filho, J.X.; Blanco-Labra, A.; Shewry, P.R.; Richardson, M.; *Phytochemistry*, **1991**, *30*, 2829-2833.

[396] Oliveira, A.S.; Pereira, R.A.; Lima, L.M.; Morais, A.H.A.; Melo, F.R.; Franco, O.L.; Bloch, C.; Grossi-de-Sà, M.F.; Sales, M.P.; *Pesticide Biochem. Physiol.*, **2002**, *72*, 122-132.

[397] Manen, J.F.; Simon, P.; Van Slooten, J.C.; Osteras, M.; Frutiger, S.; Hughes, G.J.; *Plant Cell*, **1991**, *3*, 259-270.

[398] Yamamoto, M.; Hara, S.; Ikenaka, T.; *J. Biochem. (Tokyo)*, **1983**, *94*, 849-863.

[399] Caldwell, J.B.; Strike, P.M.; Kortt, A.A.; *J. Protein Sci.*, **1990**, *9*, 493-499.

[400] Shibata, H.; Hara, S.; Ikenaka, T.; Abe, J.; *J. Biochem. (Tokyo)*, 99, 1147-1155.

[401] Kortt, A.A.; Burns, J.E.; Strike, P.M.; *Biochem. Int.*, **1990**, *22*, 543-552.

[402] Habu, Y.; Fukushima, H.; Sakata, Y.; Abe, H.; Funada, R.; *Plant Mol. Biol.,* **1996**, *32*, 1209-1213.

[403] Sakata, Y.; Chiba, Y.; Fukushima, H.; Matsubara, N.; Habu, Y.; Naito, S.; Ohno, T.; *Plant Mol. Biol.*, **1997**, *34*, 191-197.

[404] Ghosh, S.; Singh, M.; *Protein Expr. Purif.*, **1997**, *10*, 100-106.

[405] Dattagupta, J.K.; Podder, A.; Chakrabarti, C.; Sen, U.; Mukhopadhyay, D.; Dutta, S.K.; Singh, M.; *Proteins*, **1999**, *35*, 321-331.

[406] Ravichandran, S.; Sen, U.; Chakrabarti, C.; Dattagupta, J.K.; *Acta Crystallogr. D Biol. Crystallogr.*, **1999**, *55*, 1814-1821.

[407] Habu, Y.; Peyachoknagul, S.; Umemoto, K.; Sakata, Y.; Ohno, T.; *J. Biochem. (Tokyo)*, **1992**, *111*, 249-258.

[408] Minami, M.; Morisawa, G.; Hayashi, M.; Sakata, Y.; Habu, Y.; Iwabuchi, M.; Meshi, T.; *Plant Cell Physiol.*, **1999**, *40*, 109-113.

[409] Shibata, H.; Hara, S.; Ikenaka, T.; *J. Biochem. (Tokyo)*, **1988**, *104*, 537-543.

[410] Peyachoknagul, S.; Matsui, T.; Shibata, H.; Hara, S.; Ikenaka, T.; Okada, Y.; Ohno, T.; *Plant Mol. Biol.*, **1989**, *12*, 51-58.

[411] Habu, Y.; Sakata, Y.; Fukasawa, K.; Ohno, T.; *Plant Mol. Biol.*, **1993**, *23*, 1139-1150.

[412] Dattagupta, J.K.; Chakrabarti, C.; Podder, A.; Dutta, S.K.; Singh, M.; *J. Mol. Biol.*, **1990**, *216*, 229-231.

[413] Kortt, A.A.; Burns, J.E,; Caldwell, J.B.; Ferro, T.; Strike, P.M.; *J. Protein Chem.*, **1991**, *10*, 183-188.

[414] Kortt, A.A.; Strike, P.M.; De Jersey, J.; *Eur. J. Biochem.*, **1989**, *181*, 403-408.

[415]   Dayan, S.M.; Van Donkelaar, A.; Kortt, A.A.; *J. Biol. Chem.*, **1987**, *262*, 10287-10289.

[416]   Theerasilp. S.; Kurihara, Y.; *J. Biol. Chem.*, **1988**, *263*, 11536-11539.

[417]   Theerasilp, S.; Hitotsuya, H.; Nakajo, S.; Nakaya, K.; Nakamura, Y.; Kurihara, Y.; *J. Biol. Chem.*, **1989**, *264*, 6655-6659.

[418]   Takahashi, N.; Hitotsuya, H.; Hanzawa, H.; Arata, Y.; Kurihara, Y.; *J. Biol. Chem.,* **1990**, *265*, 7793-7798.

[419]   Igeta, H.; Tamura, Y.; Nakaya, K.; Nakamura, Y.; Kurihara, Y.; *Biochim. Biophys. Acta*, **1991**, *1079*, 303-307.

[420]   Masuda, Y.; Nirasawa, S.; Nakaya, K.; Kurihara, Y.; *Gene*, **1995**, *161*, 175-177.

[421]   Saarikoski, P.; Clapham, D.; von Arnold, S.; *Plant Mol. Biol.*, **1996**, *31*, 465-478.

[422]   Souza, E.M.; Mizuta, K.; Sampaio, M.U.; Sampaio, C.A.; *Phytochemistry*, **1995**, *39*, 521-525;

[423]   Ishikawa, A.; Ohta, S.; Matsuoka, K.; Hattori, T.; Nakamura, K.; *Plant Cell Physiol.*, **1994**, *35*, 303-312.

[424]   Walsh, T.A.; Twitchell, W.P.; *Plant Physiol.*, **1991**, *97*, 15-18.

[425]   Yamagishi, K.; Mitsumori, C.; Kikuta, Y.; *Plant Mol. Biol.*, **1991**, *17*, 287-288.

[426]   Stieckema, W.J.; Heidekamp, F.; Dirkse, W.G.; Van Beckum, J.; De Haan, P.; Ten Bosch, C.; Louwerse, J.D.; *Plant Mol. Biol.*, **1988**, *11*, 255-269.

[427]   Suh, S-G.; Peterson, J.E.; Stieckema, W.J.; Hannapel, D.J.; *Plant Physiol.,* **1990**, *94*, 40-45.

[428]   Suh, S-G.; Stieckema, W.J.; Hannapel, D.J.; *Planta*, **1991**, *184*, 423-430.

[429]   Valueva, T.A.; Revina, T.A.; Mosolov, V.V.; Mentele, R.; *Biol. Chem.*, **2000**, *381*, 1215-1221.

[430]   Do Socorro, M.; Cavalcanti, M.; Oliva, M.L.; Fritz, H.; Jochum, M.; Mentele, R.; Sampaio, M.; Coelho, L.C.; Batista, I.F.; Sampaio, C.A.; *Biochem. Biophys. Res. Commun.*, **2002**, *291*, 635-639.

[431]   Tai, H.; McHenry, L.; Fritz, P.J.; Furtek, D.B.; *Plant Mol. Biol.*, **1991**, *16*, 913-915.

[432]   Spencer, M.E.; Hodge, R.; *Planta*, **1991**, *183*, 528-535.

[433]   Kochhar, S.; Gartenmann, K.; Juillerat, M.A.; *J. Agric. Food Chem.*, **2000**, *48*, 5593-5599.

[434]   Mundy, J.; Hejgaard, J.; Svendsen, I.; *FEBS Lett.*, **1984**, *167*, 210-214.

[435]   Maeda, K.; *Biochim. Biophys. Acta*, **1986**, *871*, 250-256.

[436]   Hejgaard, J.; Dam, J.; Petersen, L.C.; Bjorn, SE.; *Biochim. Biophys. Acta*, **1994**, *1204*, 68-74.

[437]   Valdes-Rodriguez, S.; Blanco-Labra, A.; Gutierrez-Benicio, G.; Boradenenko, A.; Herrera-Estrella, A.; Simpson, J.; *Plant Mol. Biol.*, **1999**, *41*, 15-23.

[438]   Asamizu, E.; Sato, S.; Kaneko, T.; Nakamura, Y.; Kotani, H.; Miyajima, N.; Tabata, S.; *DNA Res.*, **1998**, *5*, 379-391.

[439]   Katayama, H.; Soezima, Y.; Fujimura, S.; Terada, S.; Kimoto, E.; *Biosci. Biotechnol. Biochem.*, **1994**, *58*, 2004-2008.

[440]   Beuning, L.L.; Spriggs, T.W.; Christeller, J.T.; *J. Mol. Evol.*, **1994**, *39*, 644-654.

636

[441]   Krishnamoorthi, R.; Gong, Y.X.; Richardson, M.; *FEBS Lett.*, **1990**, *273*, 163-167.

[442]   Murray, C.; Christeller, J.T.; *Biol. Chem. Hoppe-Seyler*, **1995**, *376*, 281-287.

[443]   Belozersky, M.A.; Dunaevsky, Y.E.; Musolyamov, A.X.; Egorov, T.A.; *Biokhimiia*, **1996**, *61*, 1743-1750.

[444]   Belozersky, M.A.; Dunaevsky, Y.E.; Musolyamov, A.X.; Egorov, T.A.; *FEBS Lett.*, **1995**, *371*, 264-266.

[445]   Dunaevsky, Y.E.; Gladysheva, I.P.; Pavlukova, E.B.; Beliakova, G.A.; Gladyshev, D.P.; Papisova, A.I.; Larionova, N.I.; Belozersky, M.A.; *Physiologia Plantarum*, **1997**, *101*, 483-488.

[446]   Belozersky, M.A.; Dunaevsky, Y.E.; Musolyamov, A.K.; Egorov, T.A.; *IUBMB Life*, **2000**, *49*, 273-276.

[447]   Belozersky, M.A.; Dunaevsky, Y.E.; Musolyamov, A.K.; Egorov, T.A.; *Biochemistry (Mosc)* , **2000**, *65*, 1140-1144.

[448]   Tsybina, T.A.; Dunaevsky, Y.E.; Musolyamov, A.K.; Egorov, T.A.; Belozersky, M.A.; *Biochemistry (Mosc)*, **2001**, *66*, 941-947.

[449]   Pandya, M.J.; Smith, D.A.; Yarwood, A.; Gilroy, J.; Richardson, M.; *Phytochemistry*, **1996**, *43*, 327-331.

[450]   Williamson, M.S.; Forde, J.; Kreis, M.; *Plant Mol. Biol.*, **1988**, *10*, 521-535.

[451]   Greagg, M.A.; Brauer, A.B.; Leatherbarrow, R.J.; *Biochim. Biophys. Acta*, **1994**, *1222*, 179-186.

[452]   Svendsen, I.; Boisen, S.; Hejgaard, J.; *Carlsberg Res. Commun.*, **1982**, *47*, 45-53.

[453]   Svendsen, I.; Hejgaard, J.; Chavan, J.K.; *Carlsberg Res. Commun.*, **1984**, *49*, 493-502.

[454]   Longstaff, C.; Campbell, A.F.; Fersht, A.R.; *Biochemistry*, **1990**, *29*, 7339-7347.

[455]   McPhalen, C.A.; James, M.N.; *Biochemistry*, **1987**, *26*, 261-269.

[456]   Ludvigsen, S.; Shen, H.Y.; Kjaer, M.; Madsen, J.C.; Poulsen, F.M.; *J. Mol. Biol.*, **1991**, *222*, 621-635.

[457]   Neira, J.L.; Davis, B.; Ladurner, A.G.; Buckle, A.M.; Gay, G., Fersht, A.R.; *Fold. Des.*, **1996**, *1*, 189-208.

[458]   Lorenc-Kubis, I.; Kowalska, J.; Pochron, B.; Zuzlo, A.; Wilusz, T.; *Chem. Bio. Chem.*, **2001**, *2*, 45-51.

[459]   Cierpicki, T.; Otlewski, J.; *J. Mol. Biol.*, **2000**, *302*, 1179-1192.

[460]   Graham, J.S.; Pearce, G.; Merryweather, J.; Titani, K.; Ericsson, L.; Ryan, C.A.; *J. Biol. Chem.*, **1985**, *260*, 6555-6560.

[461]   Johnson, R.; Narvaez, J.; An, G.; Ryan, C.; *Proc. Natl. Acad. Sci. U S A*, **1989**, *86*, 9871-9875.

[462]   Margossian, L.J.; Federman, A.D.; Giovannoni, J.J.; Fischer, R.L.; *Proc. Natl. Acad. Sci. U S A*, **1988**, *85*, 8012-8016.

[463]   Wingate, V.P.; Ryan, C.A.; *J. Biol. Chem.*, **1991**, *266*, 5814-5818.

[464]   Wingate, V.P.; Broadway, R.M.; Ryan, C.A.; *J. Biol. Chem.*, **1989**, *264*, 17734-17738.

[465]   Ogata, F.; Miyata, T.; Fujii, N.; Yoshida, N.; Noda, K.; Makisumi, S.; Ito, A.; *J. Biol. Chem.*, **1991**, *266*, 16715-16721.

[466]   Zeng, F-Y.; Qian, R-Q.; Wang, Y.; *FEBS Lett.*, **1988**, *234*, 35-38.

[467]   Fujita, T.; Kouchi, H.; Ichikawa, T.; Syono, K.; *Plant Cell Physiol.*, **1993**, *34*, 137-142.

[468]   Linthorst, H.J.; Brederode, F.T.; van der Does, C.; Bol, J.F.; *Plant Mol. Biol.*, **1993**, *21*, 985-992.

[469]   Heitz, T.; Geoffroy, P.; Stintzi, A.; Fritig, B.; Legrand, M.; *J. Biol. Chem.*, **1993**, *268*, 16987-16992.

[470]   Criqui, M.C.; Plesse, B.; Durr, A.; Marbach, J.; Parmentier, Y; Jamet, E.; Fleck, J.; *Mech. Dev.*, **1992**, *38*, 121-132.

[471]   Coupe, S.A.; Taylor J.E.; Roberts, J.A.; *Plant Cell Environ.*, **1997**, *20*, 1517-1524.

[472]   Beuning, L.L.; Christeller, J.T.; *Plant Physiol.*, **1993**, *102*, 1061.

[473]   Richardson, M.; *Biochem. J.*, **1974**, *137*, 101-112.

[474]   Richardson, M.; Cossins, L.; *FEBS Lett.*, **1974**, *45*, 11-13.

[475]   Richardson, M.; Cossins, L.; *FEBS Lett.*, **1975**, *52*, 161.

[476]   Cleveland, T.E.; Thornburg, R.W.; Ryan, C.A.; *Plant Mol. Biol.*, **1987**, *8*, 199-207.

[477]   Lee, J.S.; Park, J.S.; *Singmul. Hakhoe. Chi.*, **1989**, *32*, 69-78.

[478]   Svendsen, I.; Hejgaard, J.; Chavan, J.K.; *Carlsberg Res. Commun.*, **1984**, *49*, 493-502.

[479]   Nozawa, H.; Yamagata, H.; Aizono, Y.; Yoshikawa, M.; Iwasaki, T.; *J. Biochem. (Tokyo)*, **1989**, *106*, 1003-1008.

[480]   Chevalier, C.; Bourgeois, E.; Pradet, A.; Raymond, P.; *Plant Mol. Biol.*, **1995**, *28*, 473-485.

[481]   Tamayo, M.C.; Rufat, M.; Bravo, J.M.; San Segundo, B.; *Planta*, **2000**, *211*, 62-71.

[482]   Breitler, J.C.; Cordero, M.J.; Royer, M.; Meynard, D.; San Segundo, B.; Guiderdoni, E.; *J. Mol. Breeding*, **2001**, *7*, 259-274.

[483]   Cordero, M..J.; Raventos, D.; San Segundo, B.; *Plant J.*, **1994**, *6*, 141-150.

[484]   Antcheva, N.; Patthy, A.; Athanasiadis, A.; Tchorbanov, B.; Zakhariev, S.; Pongor, S.; *Biochim. Biophys. Acta*, **1996**, *1298*, 95-101.

[485]   Moura, D.S.; Ryan, C.A.; *Plant Physiol.*, **2001**, *126*, 289-298.

[486]   Graham, J.S.; Pearce, G.; Merryweather, J.; Titani, K.; Ericsson, L.H.; Ryan, C.A.; *J. Biol. Chem.*, **1985**, *260*, 6561-6564.

[487]   Johnson, R.; Narvaez, J.; An, G.; Ryan, C.; *Proc. Natl. Acad. Sci. U S A*, **1989**, *86*, 9871-9875.

[488]   Taylor, B.H.; Young, R.J.; Scheuring, C.F.; *Plant Mol. Biol.*, **1993**, *23*, 1005-1014.

[489]   Young, R.J.; Scheuring, C.F.; Harris-Haller, L.; Taylor, B.H.; *Plant Physiol.*, **1994**, *104*, 811-812.

[490]   Gadea, J.; Mayda, M.E.; Conejero, V.; Vera, P.; *Mol. Plant Microbe Interact.*, **1996**, *9*, 409-415.

[491]   Atkinson, A.H.; Heath, R.L.; Simpson, R.J.; Clarke, A.E.; Anderson, M.A.; *Plant Cell*, **1993**, *5*, 203-213.

[492]   Heath, R.L.; Barton, P.A.; Simpson, R.J.; Reid, G.E.; Lim, G.; Anderson, M.A.; *Eur. J. Biochem.*, **1995**, *230*, 250-257.

638

[493]   Nielsen, K.J.; Hill, J.M.; Anderson, M.A.; Craik, D.J.; *Biochemistry*, **1996**, *35*, 369-378.
[494]   Heath, R.; McDonald, G.; Christeller, J.T.; Lee, M.; Bateman, K.; West, J.; *J. Insect Physiol.*, **1997**, *43*, 833-842.
[495]   Lee, M.C.; Scanlon, M.J.; Craik, D.J.; Anderson, M.A.; *Nat. Struct. Biol.,* **1999**, *6*, 526-530.
[496]   Miller, E.A.; Lee, M.C.S.; Atkinson, A.H.O.; Anderson, M.A.; *Plant Mol. Biol.*, **2000**, *42*, 329-333.
[497]   Choi, D.; Park, J.A.; Seo, Y.S.; Chun, Y.J.; Kim, W.T.; *Biochim. Biophys. Acta,* **2000**, *1492*, 211-215.
[498]   Pearce, G.; Johnson, S.; Ryan, C.A.; *Plant Physiol.*, **1993**, *102*, 639-644.
[499]   Balandin, T.; van der Does, C.; Albert, J.M.; Bol, J.F.; Linthorst, H.J.; *Plant Mol. Biol.*, **1995**, *27*, 1197-1204.
[500]   Xu, Z.F.; Qi, W.Q.; Ouyang, X.Z.; Yeung, E.; Chye, M.L.; *Plant Mol. Biol.*, **2001**, *47*, 727-738.
[501]   Richardson, M.; *FEBS Lett.*, **1979**, *104*, 322-326.
[502]   Hass, G.M.; Hermodson, M.A.; Ryan, C.A.; Gentry, L.; *Biochemistry*, **1982**, *21*, 752-756.
[503]   Greenblatt, H.M.; Ryan, C.A.; James, M.N.; *J. Mol. Biol.*, **1989**, *205*, 201-228.
[504]   Bryant, J.; Green, T.R.; Gurusaddaiah, T.; Ryan, C.A.; *Biochemistry*, **1976**, *15*, 3418-3424.
[505]   Thornburg, R.W.; An, G.; Cleveland, T.E.; Johnson, R.; Ryan, C.; *Proc. Natl. Acad. Sci. USA*, **1987**, *84*, 744-748.
[506]   Dammann, C.; Rojo, E.; Sanchez-Serrano, J.J.; *Plant J.*, **1997**, *11*, 773-782.
[507]   Keil, M.; Sanchez-Serrano, J.; Schell, J.; Willmitzer, L.; *Nucleic Acids Res.*, **1986**, *14*, 5641-5650.
[508]   Sanchez-Serrano, J.; Schmidt, R.; Schell, J.; Willmitzer, L.; *Mol. Gen. Genet.*, **1986**, *203*, 15-20.
[509]   Beekwilder, J.; Schipper, B.; Bakker, P.; Bosch, D.; Jongsma, M.; *Eur. J. Biochem.*, **2000**, *267*, 1975-1984.
[510]   Jongsma, M.A.; Bakker, P.L.; Peters, J.; Bosch, D.; Stiekema, W.J.; *Proc. Natl. Acad. Sci. U S A*, **1995**, *92*, 8041-8045.
[511]   Beekwilder, J.; Rakonjac, J.; Jongsma, M.; Bosch, D.; *Gene*, **1999**, *228*, 23-31.
[512]   Xu, D.; McElroy, D.; Thornburg, R.W.; Wu, R.; *Plant Mol. Biol.*, **1993**, *22*, 573-88.
[513]   Johnson, R.; Narvaez, J.; An, G.; Ryan, C.; *Proc. Natl. Acad. Sci. U S A*, **1989**, *86*, 9871-9875.
[514]   Murray, C.; Christeller, J.T.; *Plant Physiol.*, **1994**, *106*, 1681.
[515]   Theologis, A. *et al.*; *Nature*, **2000**, *408*, 816-820.
[516]   Ruoppolo, M.; Amoresano, A.; Pucci, P.; Pascarella, S.; Polticelli, F.; Trovato, M.; Menegatti, E.; Ascenzi, P.; *Eur. J. Biochem.,* **2000**, *267*, 6486-6492.
[517]   Ceciliani, F.; Bortolotti, F.; Menegatti, E.; Ronchi, S.; Ascenzi, P.; Palmieri, S.; *FEBS Lett.*, **1994**, *342*, 221-224.
[518]   Ascenzi, P.; Ruoppolo, M.; Amoresano, A.; Pucci, P.; Consonni, R.; Zetta, L.; Pascarella, S.; Bortolotti, F.; Menegatti, E.; *Eur. J. Biochem.*, **1999**, *261*, 275-284.

[519]   Carter, T.H.; Everson, B.A.; Ratnoff, O.D.; *Blood*, **1990**, *75*, 108-115.

[520]   Menegatti, E.; Tedeschi, G.; Ronchi, S.; Bortolotti, F.; Ascenzi, P.; Thomas, R.M.; Bolognesi, M.; Palmieri, S.; *FEBS Lett.*, **1992**, *301*, 10-14.

[521]   Marshall, C.J.; *Phil. Trans. R. Soc. Lond. B*, **1993**, *342*, 101-119.

[522]   Worrall, D.M.; Blacque, O.E.; Barnes, R.C.; *Biochem. Soc. Trans.*, **1999**, *27*, 746-750.

[523]   Irving, J.A.; Pike, R.N.; Lesk, A.M.; Whisstock, J.C.; *Genome Research*, **2000**, *10*, 1845-1864.

[524]   Yoo, B.C.; Aoki, K.; Xiang, Y.; Campbell, L.R.; Hull, R.J.; Xoconostle-Cazares, B.; Monzer, J.; Lee, J.Y.; Ullman, D.E.; Lucas, W.J.; *J. Biol. Chem.*, **2000**, *275*, 35122-35128.

[525]   Dahl, S.W.; Rasmussen, S.K.; Hejgaard, J.; *J. Biol. Chem.*, **1996**, *271*, 25083-25088.

[526]   Dahl, S.W.; Rasmussen, S.K.; Petersen, L.C.; Hejgaard, J.; *FEBS Lett.*, **1996**, *394*, 165-168.

[527]   Hejgaard, J.; Rasmussen, S.K.; Brandt, A.; Svendsen, I.; *FEBS Lett.*, **1985**, *180*, 89-94.

[528]   Rosenkrands, I.; Hejgaard, J.; Rasmussen, S.K.; Bjørn, S.E.; *FEBS Lett.*, **1994**, *343*, 75-80.

[529]   Brandt, A.; Svendsen, I.; Hejgaard, J.; *Eur. J. Biochem.*, **1990**, *194*, 499-505.

[530]   Rasmussen, S.K.; Klausen, J.; Hejgaard, J.; Svensson, B.; Svendsen, I.; *Biochim. Biophys. Acta*, **1996**, *1297*, 127-130.

[531]   Lundgard, R.; Svensson, B.; *Carlsberg Res. Commun.*, **1989**, *54*, 173-180.

[532]   Rasmussen, S.K.; *Biochim. Biophys. Acta*, **1993**, *1172*, 151-154.

[533]   Hejgaard, J.; *FEBS Lett.*, **2001**, *488*, 149-153.

[534]   Rosenkrands, I.; Hejgaard, J.; Rasmussen, S.K.; Bjorn, S.E.; *FEBS Lett.*, **1994**, *343*, 75-80.

[535]   Rasmussen, S.K.; Dahl, S.W.; Norgard, A.; Hejgard, J.; *Plant Mol. Biol.*, **1996**, *30*, 673-677.

[536]   Østergaard, H.; Rasmussen, S.K.; Roberts, T.H.; Hejgaard, J.; *J. Biol. Chem.*, **2000**, *275*, 33272-33279.

[537]   Otlewski, J.; Whatley, H.; Polanowski, A.; Wilusz, T.; *Biol. Chem. Hoppe Seyler*, **1987**, *368*, 1505-1507.

[538]   Lee, C.F.; Lin, J.Y.; *J. Biochem. (Tokyo)*, **1995**, *118*, 18-22.

[539]   Nishino, J.; Takano, R.; Kamei-Hayashi, K.; Minakata, H.; Nomoto, K.; Hara, S.; *Biosci. Biotechnol. Biochem.*, **1992**, *56*, 1241-1246.

[540]   Ling, M-H.; Qi, H-Y.; Chi, C-W.; *J. Biol. Chem.*, **1993**, *268*, 810-814.

[541]   Hernandez, J.F.; Gagnon, J.; Chiche, L.; Nguyen, T.M.; Andrieu, J.P.; Heitz, A.; Trinh Hong, T.; Pham, T.T.; Le Nguyen, D.; *Biochemistry*, **2000**, *39*, 5722-5730.

[542]   Wieczorek, M.; Otlewski, J.; Cook, J.; Parks, K.; Leluk, J.; Wilimowska-Pelc, A.; Polanowski, A.; Wilusz, T.; Laskowski, M.; *Biochem. Biophys. Res. Commun.*, **1985**, *126*, 646-52.

[543]   Wilusz, T.; Wieczorek, M.; Polanowski, A.; Denton, A.; Cook, J.; Laskowski, M.; *Biol. Chem. Hoppe Seyler*, **1983**, *364*, 93-95.

[544] Grzesiak, A.; Buczek, O.; Petry, I.; Szewczuk, Z.; Otlewski, J.; 1: *Biochim Biophys Acta*, **2000**, *1478*, 318-324.

[545] Helland, R.; Berglund, G.I.; Otlewski, J.; Apostoluk, W.; Andersen, O.A.; Willassen, N.P.; Smalas, A.O.; *Acta Crystallogr. D Biol. Crystallogr.*, **1999**, *55*, 139-148.

[546] Hayashi, K.; Takehisa, T.; Hamato, N.; Takano, R.; Hara, S.; Miyata, T.; Kato, H.; *J. Biochem. (Tokyo)*, **1994**, *116*, 1013-1018.

[547] Liu, J.; Gong, Y.; Prakash, O.; Wen, L.; Lee, I.; Huang, J.K.; Krishnamoorthi, R.; *Protein Sci.*, **1998**, *7*, 132-141.

[548] Favel, A.; Mattras, H.; Coletti-Previero, M.A.; Zwilling, R.; Robinson, E.A.; Castro, B.; *Int. J. Pept. Protein Res.*, **1989**, *33*, 202-208.

[549] Stachowiak, D.; Polanowski, A.; Bieniarz, G.; Wilusz, T.; *Acta Biochim. Pol.*, **1996**, *43*, 507-13.

[550] Matsuo, M.; Hamato, N.; Takano, R.; Kamei-Hayashi, K.; Yasuda-Kamatani, Y.; Nomoto, K.; Hara, S.; *Biochim. Biophys. Acta*, **1992**, *1120*, 187-92.

[551] Hamato, N.; Takano, R.; Kamei-Hayashi, K.; Hara, S.; *Biosci. Biotechnol. Biochem.*, **1992**, *56*, 275-279.

[552] Haldar, U.C.; Saha, S.K.; Beavis, R.C.; Sinha, N.K.; *J. Protein Chem.*, **1996**, *15*, 177-184.

[553] Hatakeyama, T.; Hiraoka, M.; Funatsu, G.; *Agric. Biol. Chem.*, **1991**, *55*, 2641-2642.

[554] Hara, S.; Makino, J.; Ikenaka, T.; *J. Biochem. (Tokyo)*, **1989**, *105*, 88-91.

[555] Hamato, N.; Koshiba, T.; Pham, T.N.; Tatsumi, Y.; Nakamura, D.; Takano, R.; Hayashi, K.; Hong, Y.M.; Hara, S.; *J. Biochem. (Tokyo)*, **1995**, *117*, 432-437.

[556] Huang, Q.; Liu, S.; Tang, Y.; Zeng, F.; Qian, R.; *FEBS Lett.*, **1992**, *297*, 143-146.

[557] Huang, Q.; Liu, S.; Tang, Y.; *J. Mol. Biol.*, **1993**, *229*, 1022-1036.

[558] Miura, S.; Funatsu, G.; *Biosci. Biotechnol. Biochem.*, **1995**, *59*, 469-473.

[559] Joubert, F.J.; *Phytochemistry*, **1984**, *23*, 1410-1410.

[560] Ling, M.H.; Chi, C.W.; Shaw, P.C.; *Acta Biochim. Biophys. Sin.*, **1996**, *28*, 233-239.

[561] Chen, X.M.; Qian, Y.W.; Chi, C.W.; Gan, K.D.; Zhang, M.F.; Chen, C.Q.; *J. Biochem. (Tokyo)*, **1992**, *112*, 45-51.

[562] Chi, C-W.; Zhu, D-X.; Lin, N-Q.; Xu, L-X.; Tan, F-L.; Wang, L-X.; *Biol. Chem. Hoppe Seyler*, **1985**, *366*, 879-885.

[563] Yang, H-L.; Luo, R-S.; Wang, L-X.; Zhu, D-X.; Chi, C-W.; *J. Biochem. (Tokyo)*, **1992**, *111*, 537-545.

[564] Xu, W-F.; Tao, W-K.; Gong, Z-Z.; Chi, C-W.; *J. Biochem. (Tokyo)*, **1993**, *113*, 153-158.

[565] Xie, Z-W.; Luo, M-J.; Xu, W-F.; Chi, C-W.; *Biochemistry*, **1997**, *36*, 5846-5852.

[566] Luo, M-J.; Lu, W-Y.; Chi, C-W.; *J. Biochem. (Tokyo)*, **1997**, *121*, 991-995.

[567] Wijaya, R.; Neumann, G.M.; Condron, R.; Hughes, A.B.; Polya, G.M.; *Plant Science*, **2000**, *159*, 243-255.

[568] Melo, F.R.; Rigden, D.J.; Franco, O.L.; Mello, L.V.; Ary, M.B.; Grossi da Sa, M.F.; Bloch, C.; *Proteins Struct. Function Genet.*, **2002**, *48*, 311-319.

[569]   Campos, F.A.P.; Richardson, M.; *FEBS Lett.*, **1984**, *167*, 221-225.

[570]   Svensson, B.; Asanao, K.; Jonassen, I.B.; Poulsen, F.M.; Mundy, J.; Svendsen, I.B.; *Carlsberg Res. Commun.*, **1986**, *51*, 493-500.

[571]   Svendsen, I.B.; Nicolova, D.; Goshev, I.; Genov, N.; *Carlsberg Res. Commun.*, **1989**, *54*, 231-239.

[572]   Genov, N.; Goshev, I.; Nikolova, D.; Georgieva, D.N.; Filippi, B.; Svendsen, I.; *Biochim. Biophys. Acta*, **1997**, *1341*, 157-164.

[573]   Svendsen, I.B.; Nicolova, D.; Goshev, I.; Genov, N.; *Int. J. Peptide Protein Res.*, **1994**, *43*, 425-430.

[574]   Almeida, M.S.; Cabral, K.M.; Kurtenbach, E.; Almeida, F.C.; Valente, A.P.; *J. Mol. Biol.*, **2002**, *315*, 749-757.

[575]   Thomma, B.P.; Cammue, B.P.; Thevissen, K.; *Planta*, **2002**, *216*, 193-202.

[576]   Broekaert, W.F.; Terras, F.R.G.; Cammue, B.P.A.; Osborn, R.W.; *Plant Physiol.*, **1995**, *108*, 1353-1358.

[577]   Broekaert, W.F.; Cammue, B.P.A.; De Bolle, M.F.C.; Thevissen, K.; De Samblanx, G.W.; Osborn, R.W.; *Crit. Rev. Plant Sci.*, **1997**, *16*, 297-323.

[578]   Bruix, M.; Jiminez, M.A.; Santoro, J.; Gonzalez, C.; Cililla, F.J.; Mendez, E.; Rico, M.; *Biochemistry*, **1993**, *32*, 715-724.

[579]   Kader, J-C.; *Annu. Rev. Plant Physiol. Plant Mol. Biol.*, **1996**, *47*, 627-654.

[580]   Rico, M.; Bruix, M.; Gonzalez, C.; Monsalve, R.I.; Rodriguez, R.; *Biochemistry*, **1996**, *35*, 15672-15682.

[581]   De Leo, F.; Volpicella, M.; Licciulli, F.; Liuni, S.; Gallerani, R.; Ceci, L.R.; *Nucleic Acids Res.*, **2002**, *30*, 347-348.

[582]   Jongsma, M.A.; Bolter, C.; *J. Insect Physiol.*, **1997**, *43*, 885-895.

Atta-ur-Rahman (Ed.) *Studies in Natural Products Chemistry, Vol. 29*

GENERATING AND SCREENING A NATURAL PRODUCT LIBRARY FOR CYCLOOXYGENASE AND LIPOXYGENASE DUAL INHIBITORS

QI JIA

Department of Natural Products Chemistry, Unigen Pharmaceuticals, Inc. 100 Technology Drive, Broomfield, Colorado 80021, USA

ABSTRACT: Natural products are a highly diversified structural pool for the discovery of potential drug leads. Recent technological developments in the fields of high throughput purification, LC/PDA/MS/MS/NMR for structural dereplication, and informatic databases have fundamentally changed the study of bioactive natural products and significantly accelerated the lead discovery process. This article summarizes the current state of parallel fractionation and purification technology for generating natural product fraction and compound libraries. Structural dereplication options using a combination of on-line and off-line LC/PDA/MS/NMR techniques coupled with biological screening processes are reviewed. Adequate usage of structural, spectroscopic, and other databases with related physical and biological properties will also be discussed. The characteristics and advantages of a technology platform - Phytologix™ in generating and screening extract and fraction libraries will be detailed. Finally, this article will review recent publications on natural products with dual cyclooxygenase and lipoxygenase inhibitory activities and present the discovery and development of a novel anti-inflammatory ingredient - Univestin™ by screening a proprietary natural product library.

INTRODUCTION

Natural product chemistry has not only formed the scientific basis for traditional use of medicinal plants, but also plays an important role in drug discovery. Over 50% of the most-prescribed drugs in the US have a natural product origin [1]. Based on a review of drugs approved between 1983 and 1994, 78% of the antibacterial drugs, 75% of platelet aggregation inhibitors, 61% of anticancer drugs, 48% of anti-hypotensive drugs, 47.6% of antiulcer drugs, and 32.5% of the anti-inflammatory drugs have a natural origin [2]. With more than 173,000 known structures, natural products have been demonstrated as a highly diversified structural resource for the discovery of new chemical entities [3]. Based on the analyses of 10,495 natural products and 5757 trade drugs, natural products possess 1748 different ring systems compared to 807 different ring systems found in trade drugs [4]. Approximately 35% of the ring systems found in trade drugs are also found in natural products, however only 17% of the ring systems found in natural products have identical counterparts in trade drugs. Natural products are not only functional structural leads, but also have very similar architecture and pharmacophoric properties as trade drugs. The average calculated molecular weight of natural products is almost identical to that of trade drugs (356 vs. 360), and the average log p values are slightly higher for the natural products (2.9) than for trade drugs (2.5) [5]. Natural products have fewer hydrogen donors per molecule and fewer nitrogens per molecule than trade drugs. They have a much higher number of bridgehead atoms than trade drugs and synthetic drugs and have many more chiral centers per molecule [6]. Both natural products and trade drugs have a similar average number of oxygens per molecule and the same percentage of compounds with at least two "rule-of-5" violations [7].

With the advancement of new technology, such as combinatorial syntheses, computational drug design and super high throughput screening, there is increasing interest in the design of small molecule libraries using natural products as templates [8,9]. Combinatorial libraries can be generated in solution, however, most of the libraries generated to date rely on solid-phase synthesis techniques, including solid-phase extractions, used predominantly in the purification of the targeted synthetic compounds [10]. Unfortunately, there are significant limitations in this synthetic approach in generating libraries from complex natural product templates particularly with compounds containing multiple-rings and multiple chiral center skeletons. An obvious limitation for a semi-synthetic approach is that certain skeletal modifications and crucial functional group positions cannot be diversified. Most published compound libraries were generated using collections of starting materials and a certain reaction or reaction sequence that must be optimized under specific conditions [11]. Additionally, to develop a synthetic library requires a significant amount of information regarding the relationship between structure and activity to define the potential sites on the template that could be modified. Thus, the general approach using combinatorial synthetic chemistry involves the identification of a specific type of compound based upon available pharmacological profiles and the dissection of structures into scaffolds or templates.

The development of high throughput screening technology began in the mid 1980's. Robotic operation coupled with laboratory information management systems in combination with miniaturized signal reading systems enable the screening of literally millions of samples per assay per annum. [12,13]. The use of high-density arrays with 2400-9600 wells per plate were facilitated the screening of the largest compound libraries in as few as several hundred plates within only days or weeks [14]. Natural product libraries have been screened against a variety of biological [15,16], biochemical [17], and genomic targets [18]. Display cloning technology has also been developed for functional identification of natural product receptors using cDNA-phage display [19].

The novel, mechanism based, automatic, and miniaturized high throughput screening (HTS) technology presents both an opportunity and a challenge for natural product chemistry. Offering limited numbers of crude extracts that contain numerous, different components at various concentrations no longer fits with those sensitive and defined HTS models. Natural product samples need to be generated in much bigger numbers in short periods of time with an automatic and standardized methodology that not only offers satisfactory resolution of separation, but also reproducibility and reasonable economy. Accelerated dereplication and hit identification technology has to be developed and integrated into the screening and discovery processes. Finally, the assay directed isolation, purification and elucidation process must be accelerated and automated to improve the efficiency, enhance the quality of the outputs and reduce lead-time from primary screen to the discovery of lead compounds. This review will summarize the current efforts in natural product library generation and structure dereplication. It will use the screening of cyclooxygenase and lipoxygenase dual inhibitors as an example to illustrate the potential of a technology platform – PhytoLogix[TM] with a summary of other known natural product dual inhibitors.

GENERATING NATURAL PRODUCT LIBRARIES BY HIGH THROUGHPUT TECHNOLOGY

The design of focused natural product libraries has its roots in combinatorial synthesis and computational chemistry [20,21]. Efforts have been made to design libraries that target specific types of compounds [22], focus on specific therapeutic targets or ADME as primary criteria [23]. Many different types of natural product templates have been developed and natural product libraries have been successfully generated including alkaloid-like libraries from compounds such as benzylamines [24], quinazolines [25], indoly diketopiperazines [26], mappicine analogues [27], yohimbine analogues [28], oligoheterocycles [29], and flavonoid-like libraries from compounds such as flavone analogues [30] and benzopyrans [31,32].

Synthetic libraries generally contain purified compounds in a quantity of 1-2 mg at approximately 70-80% purity based on HPLC analyses. Due to the co-existence of other chemical reagents resulting from the synthetic processes, without further purification, the screening assays could be significantly impacted by false positives, false negatives and other complications. Designing a combinatorial library demands careful optimization of reaction selectivity and efficiency to avoid low yield, difficulty of purification and loss of chiral centers. It has been demonstrated that a specific, desirable biological property of a natural product can be improved even with rather small libraries by integrating simple functional group modifications [33].

Grabley et al. have published an extensive review on the discovery of drugs from natural product-based libraries [34]. Molecular diversity can be derived from the careful selection of natural sources such as bacteria, fungi, microalgae, plants, invertebrates, from terrestrial and marine sources. Grabley et al. utilized automated, multiple steps of chromatographic solid phase extractions (SPE) to generate a high-quality sample library for high throughput screening (HTS). Several combinations of polystyrene based resins, such as XAD-16, ENV$^+$, and Amberchrom 161 were used in the SPE process. A concept of physiochemical/chemical screening was introduced as an approach to pre-select promising secondary metabolites from natural sources based on physical properties or chemical reactivities. Using this approach, after chromatographic separation of crude extracts, the samples were analyzed by a standardized, reverse phase HPLC coupled to a diode array detector, a mass spectrometer, or a NMR instrument. The characteristics of each sample were determined based on the collected spectroscopic data. The chemical reactivity screening was based on the combination of a high-performance thin-layer chromatography (HPTLC) method and colorization reactions of natural products with selected chemical reagents. Grabley *et al.* also developed methods towards automated solid-phase synthesis in order to modify the functional groups on natural product skeletons on the multi-milligram quantity. In collaboration with other research groups, Grabley *et al.* reported a throughput of 800 compounds per year and a library of 3500 natural compounds, derivatives and analogues in a period of collections over 3 years. Each sample contained 10-20mg of material and was dissolved in DMSO and stored in 96-well microplate format. The library was submitted to screening partners in quantities of 1mg per sample with information about chemical and physical properties, known biological activities, references and suppliers.

A collaborative project, designed to generate a non-redundant pure compound library with a collection of 6,700 chemical entities in a quantity of ≥ 5 mg and a purity of $\geq 80\%$, was also reported [35]. The library generation was designed as a robust, economic, and high throughput process that included intelligent biomaterial selection, predictive profiling, economic scale-up, standardized isolation process, high throughput structure elucidation, and effective data management. The biomaterials consisted of 679 species of plants, 2665 bacterial strains and 1425 fungal strains. Plant selection was based on chemotaxonomy and focused on well-investigated plant families utilizing knowledge of the biosyntheses of diverse chemicals in a significant quantity. For microorganisms, species known for long and difficult cultivation conditions or with scale-up and re-supply problems were excluded. Biomaterials were pre-screened before extraction to meet the criteria of at least 4-5 non-ubiquitous secondary metabolites in sufficient amount. The profiling process utilized a standardized HPLC method with a 30 min gradient coupled with ELSD, DAD and MS detectors. Unambiguous identification of redundant compounds was based on the HPLC retention time less than 0.2 min after calibration with external standards. In this profiling process, only about one in twenty microbial extracts contained the necessary number of compounds at the required concentration. The isolation was then carried out via flash column chromatography to process 1.0 to 3.5 g of each extract. A total of 6700 compounds comprising 4000 distinct secondary metabolites were obtained. Within all isolations, 450 compounds appeared 2-3 times and were characterized as redundant compounds. Around 60 compounds appeared 4-30 times and were viewed as ubiquitous. They found that the output from plants was better than expected with an average of 7.7 different compounds isolated from each plant. Bacterial extracts yielded an average of 5.9 compounds, whereas fungi generated around 7.6 compounds. Significant efforts were utilized in profiling and further eliminating redundant compounds from microbial extracts. Only 10% of bacterial strains (4% of profiled extracts) and 15% of fungal strains (5% of profiled extracts) were selected for scale-up and isolation. The mass minimum of 5 mg per compound proved to be a major hurdle. One milligram per compound was more achievable.

Structural information was collected with unambiguous determinations of molecular weights for 3362 compounds with the average molecular weight of 520 Daltons [35]. Most of the compounds had molecular weights within 200-700 Dalton range. Only 8% of all analyzed compounds had molecular weights higher than 1000 Daltons. The structural dereplication procedures included search referenced retention times and molecular weights based on LC/MS data and comparisons with a commercial database [3]. 2D - NMR and other techniques were utilized to further define substructures and to provide full structural elucidations. The pure compound library generated from the above method was screened against nine different targets and shown to be superior to synthetic libraries with regard to response rates and positive confirmation rates.

Eldridge et al. filed a patent application for a methodology of generation of natural product libraries directly from natural sources suitable for high throughput screening and product discovery [36]. The method is as follows: (a) thorough extraction of a biological source material with alcohol/water or hexane followed by alcohol/water; (b) removal of known bioassay interferences such as polyphenols, tannins, fatty acids, phospholipids, tri-, di-, and mono-glycerides, from the solvent extracts by elution of the extracts through a polyamide or a reverse phase column. (c) solid phase extraction to process the eluent with step gradients and to collect limited fractions (typically 4) by using silica gel, amino, diol, polyamide, C2, C4, C6, C8 and C18 as the potential packing materials; (d) purification of fractions by HPLC to generate the compound library based on detecting hits or bioactive fractions.

The selection methodology was based on the properties of SPE fractions and required "check and continue" and "custom chromatography" features with combination of MS, DAD detection mechanisms, Gilson liquid handling system, and customized computer program for data analysis. The purified compounds were collected with standardized concentrations generated by an automated system. This methodology, however, has several drawbacks. First, using mixtures of alcohol and water as a solvent will not extract all potential biologically active components from the biomass. A good example is polysaccharides, which will not dissolve in alcohol/water and therefore, would not be extracted from the plant biomass. Polysaccharides are a very important class of natural products having known immune regulatory and anti-tumor effects and have been used in the pharmaceutical, nutraceutical and cosmetic industries. Second, polyphenol and tannins are biologically active ingredients that contribute to the efficacies of many popular herbal products [37,38]. EGCG, multiple catechin and phenolic compounds from green teas [39], grape seeds and grape skins [40] are other examples. Removal of these components from plant extracts using the method described by Eldridge et al. will result in a potential loss of bioactive components. Finally, the process is time consuming, expensive, and difficult to automate and standardize.

Stewart et al. have also reported the efforts at Molecular Nature Ltd. to generate a pure natural product library [41]. Compounds for this library were isolated utilizing parallel, normal phase column chromatography followed by C-18 and/or ion exchange chromatography. To be accepted into the library, the compounds must be > 90% pure with structural verification via a combination of HPLC, NMR, MS and GC/MS.

A method to make a secondary metabolite library from a microbial culture broth was reported by Schmid et al [42]. The library was generated using a novel, automated process based on multistep fractionation of a supernatant from broth through an Amberlite XAD-16 column followed by chromatographic column fractionations with a styrene-divinylbenzene resin, reverse phase C-8 and C-18 and other types of solid phase extractions (SPE). These efforts led to higher purity compounds in each fraction based on an automatic procedure with limited manual intervention. However, this method has several limitations. For example, SPE uses step-gradients that lead to limited fraction numbers in large volumes. Therefore, it is not a suitable method to collect fractions in a 96-well format.

Technological developments in genomics, enzymology and bioengineering have resulted in methods for generating natural products utilizing combinatorial biosynthesis [43,44]. For example, multiple genetic modifications of the erythromycin polyketide synthase have produced a novel unnatural, compound library [45]. Combinatorial biosynthetic libraries have been constructed by cloning large fragments of DNA isolated from soil into a Streptomycete host [46] and through glycosyltransferase catalyzed transformation [47].

Recently, a technology platform, referred to as Phytologix[TM], for the discovery and development of novel pharmaceutical, nutraceutical and cosmetic agents has been reported [48]. Phytologix[TM] focuses on documented medicinal plants and other biomaterials. Due to thousands of years of historic use, these medicinal plants and other biomaterials have already been pre-selected and clinically tested for human consumption. Thus, they are most likely to yield safe and efficacious pharmaceutical, nutraceutical and cosmetic products in contrast to a randomized collection of biomaterials. The Phytologix[TM] discovery program includes a relational database and a BioInformatics driven assessment of the medicinal plant library. This database enables the prioritization of plants for screening based upon traditional use. The pre-selection based upon traditional use of plants and other natural biomaterials is unique and critical to ensure a high positive hit rate, a safe product and a short discovery cycle.

The PhytoLogix[TM] Discovery Process relies upon multiple standardized extraction and fractionation protocols that allow the generation of diversified extracts and fractionation libraries in a high throughput format at a reasonable cost. Each biomass was extracted with a medium polarity solvent combination, such as methylene chloride/methanol in a ratio of 1:1 followed by water extractions. The efficiency of the extraction methodology was validated by further extraction of the biomass with methanol after the organic and aqueous extractions. It was found that only a small amount of extractible material was present in methanol extracts with no unique HPLC chromatogram profiles based on two different detection methods, Photo Diode Array and ion trap mass spectrometers.

This method of extraction offers several advantages. First and foremost, the dual extraction strategy provides a more complete and extensive natural product profile from each biomass. Another advantage of the extraction methodology described herein is that the extraction process yields enough material for further fractionation and bioassays. For example, extraction of 30 grams of biomass, generates approximately 1-5 grams of organic and 1-4 grams of aqueous extract. These quantities provide enough material for a number of screens and further fractionations. The crude organic and aqueous extracts can be dissolved in DMSO and deionized (DI) water, respectively, at a concentration of around 50 mg /ml and stored in a 96-deep-well plate with 88 samples per plate. The samples in the extract master plates can be aliquoted and screened with high throughput models.

The uniqueness of PhytoLogix[TM] platform is applying a novel approach to generate a fraction library. Generally, organic extract (400 mg) is dissolved under sonication into a minimum amount of organic solvents and loaded onto a prepacked flash column (2 cm ID x 8.2 cm, 10 g silica gel). The column is dried under vacuum until all solvent is evaporated. Then the column is eluted in parallel using a Hitachi high throughput purification (HTP) system with an unique gradient mobile phase of (A) 50:50 EtOAc:hexane and (B) methanol from 100% A to 100% B in 30 minutes at a flow rate of 5 mL/min. The separation is monitored using a broadband wavelength UV detector and the eluents are collected in 88 fractions in a 96-deep-well plate at 1.9 mL per well using a Gilson fraction collector. The sample plate is dried under low vacuum and centrifugation. DMSO (1.5 mL) is added to each well to dissolve the samples and the 96-deep-well plate is stored at $-70\,^{\circ}$C until analysis.

Figures 1 illustrates the separation efficiency of the high throughput purification fractionation method on an organic extract from the roots of *Pulsatilla chinensis*. Every other HTP fraction was spotted on a silica gel

652

TLC plate and developed with 60% EtOAc in hexane. The TLC plate was spread with the coloration agent, anisaldehyde, in sulfuric acid. The HTP yielded impressive separation of different types of compounds based on polarity. The separated components were distributed in 6-10 cells and in most cases each cell contained either a single compound or at most less than three compounds. Figure 2 depicts the weight distribution of the sample in each well. There were several peaks in the weight distribution profile matched with that of the TLC spots.

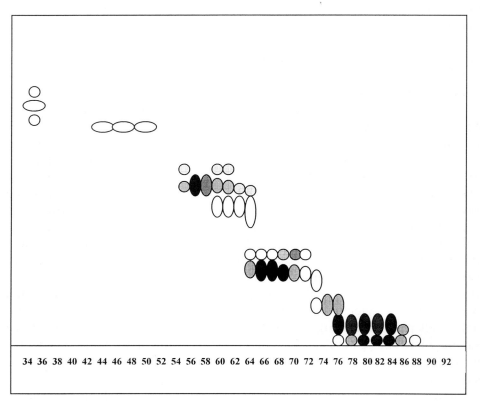

Fig. (1). TLC analysis of HTP fractions from an organic extract of *Pulsatilla chinensis*.

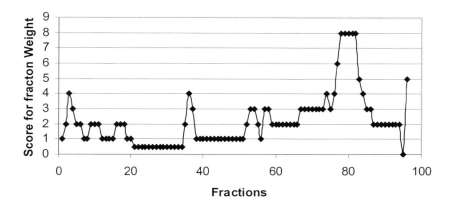

Fig. (2). Weight distribution of HTP fractions from an organic extract of *Pulsatilla chinensis*.

Under a separate fractionation protocol, the aqueous extract (750 mg) is dissolved in deionized (DI) water (5 mL), filtered through a 1 μm syringe filter and transferred to a 4 mL HPLC vial. The solution is then injected by an autosampler onto a prepacked, reverse phase column (C-18, 15 μm particle size, 2.5 cm ID x 10 cm with a pre-column insert). The column is eluted with a gradient mobile phase of (A) water and (B) methanol from 100% A to 100% B in 20 min followed by 100% methanol for 5 minutes at a flow rate of 10 mL/min. The separation is monitored using a broadband wavelength UV detector and the eluent is collected in 88 fractions in a 96-deep-well plate at 1.9 mL/well using a Gilson fraction collector. Figure 3 depicts the HTP/UV chromatogram of a reverse phase fractionation of aqueous extract from the leaves of *Camellia sinensis*.

The reverse phase HTP separations of aqueous extracts are highly reproducible. The aqueous extracts from whole plant of *Ainsliaea henryi* were separated twelve times on 4 parallel C-18 columns and a total of twelve 96-deep well plates generated. The HTP/UV chromatograms from 12 column separations were identical and the samples were combined based on the same well position from the twelve plates (data not shown).

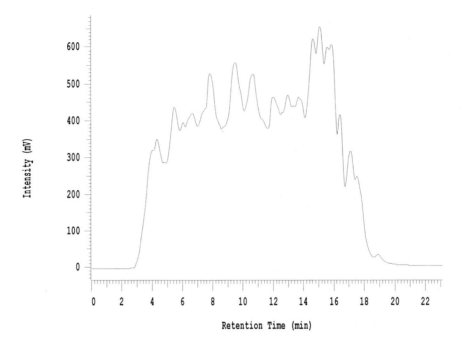

Fig. (3). HTP/UV chromatogram of an aqueous extract from the leaves of *Camellia sinensis*

One of the advantages of the high throughput fractionation methodology is that the HTP column can separate 400 mg crude extract, and each fraction resulting from the high throughput purification will contain milligrams of materials that can be dissolved into a solution at concentrations of 1 mg/ml. The method has not only separated individual component present in the crude extracts, but also significantly enriched minor active components in the plant extracts yielding a much greater chance of being detected in the screening process.

Another unique characteristic of this technology is the potential for online structure information collection. This HTP method utilizes a high-pressure chromatography system in a parallel processing mode, i.e., multiple simultaneous column runs coupled to a robot controlled liquid handling system triggered to deliver chromatographic eluent (containing individual chemical compounds) based upon pre-programmed time, volume, and spectroscopic signals. The HTP system directs the sample to the liquid handling systems. An aliquot of the eluent is then dispensed to an ion trap mass spectrometer with a super sonic ionization chamber where the molecular ion and fragmentation pattern of the compound are determined. From the mass spectrum, it is possible to derive the molecular weight and general structural information regarding the components of the fractions. This information is compared to a structural library [3] by computer analysis to confirm purity and tentative identification. The HTP eluent can also be directed through PDA, NMR, ELSD, and other detectors to obtain spectroscopic, concentration and other physical and chemical information. The online information collection is much more efficient coupled with real time data compilation, rather than analyzing collected fractions in a separate process. Additionally, the method is just as accurate. In the example illustrated in Figures 4-5, the molecular ions of the active compound from COX inhibitory fraction D3, which derived from an aqueous extract of the leaves of *Camellia sinensis*, were obtained from online HTP/MS data collection and from off-line HPLC/MS analysis of the individual well.

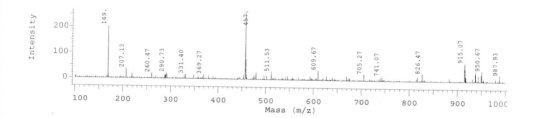

Fig. (4). Online HTP/MS spectrum of a HTP fraction D3 from an aqueous extract of the leaves of *Camellia sinensis*.

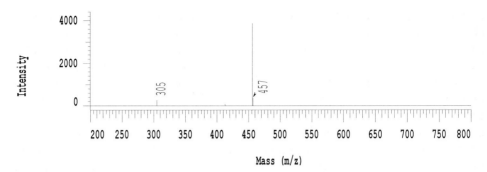

Fig. (5). Offline HPLC/MS spectrum of fraction D3 from an aqueous extract of *Camellia sinensis.*

Finally, a significant advantage of the methodology is the high throughput and low cost of operation. It takes approximately 40 minutes to complete two HTP normal phase column fractionations and 20 minutes to complete two HTP reverse phase column fractionations. Daily HTP throughput for organic extracts is 14 column fractionations to yield 1232 fractions and 32 column fractions with an output of 2618 fractions for aqueous extracts. The normal phase columns can only be used once, but, the reverse phase columns can be reused approximately sixty times with appropriate column washing between each run. The C-18 column performance is closely monitored with a known compound mixture of aloe chromones (data not shown).

To demonstrate the value of an extract library, 774 plant extracts were screened for inhibition of the purified mushroom tyrosinase in the 96 well plate format in an attempt to identify a novel skin whitener for cosmetic usage. The concentrations of L-Dopa substrate and tyrosinase enzyme were scaled down linearly using a modified method of Pomerantz [49]. From the screening, 43 organic extracts were identified with tyrosinase inhibitory activities that were equivalent to a hit rate of 5.6%. It was significantly higher than the 0.78% hit rate for the aqueous extracts. Since the targeted indication is a cosmetic product for use as a skin whitener, the compounds in organic extracts with lower polarity should have better skin penetration. Finally, Six medicinal plants with active fractions (IC$_{50}$<100 µg/mL, 0.78%

confirmed hits) were selected for large-scale isolation and purification and 6 active compounds were identified as the output of the screening project.

Direct screening of the fractionation library has its own value, since each fraction will contain enriched and limited number of compounds that the likelihood of obtaining false positives and false negatives will be eliminated. Figure 6 demonstrates graphically the inhibition of COX-1 and COX-2 enzymes by various HTP fractions from the aqueous extract of the leaves of *Camellia sinensis*. After screening all 88 fractions from the 96-well plate, a total of 8 HTP fractions exhibited greater than 60% COX inhibition as illustrated in Figure 6. Following the dereplication process described in next section, ten individual compounds were identified in active and surrounding fractions. There are many components in the crude aqueous extract that could interfere with the assay or cover up the potency. However, utilizing HTP technology, these components have been separated out into other wells. This process greatly enhances the positive hit rate from a single data point in the crude extract to eight positive fractions from which multiple bioactive compounds have been identified.

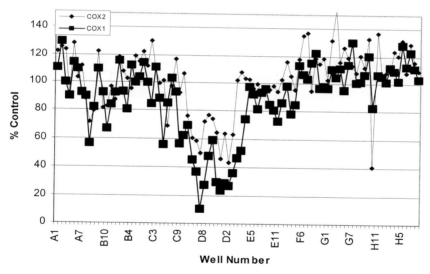

Fig. (6). COX-1 (□) and COX-2 (♦) enzymes inhibition profile of HTP fractions from *Camellia sinensis*.

STRUCTURE DEREPLICATION

"Dereplication" refers to a process of analyzing, without isolation, a natural product, a fraction or an extract for physical, spectroscopic and structural information, comparing the information with internal and commercial databases, reaching a conclusion on the existence of novel and/or known compounds, and determining the strategy of further investigations. The efficiency and the quality of the dereplication program will determine the lead-time from obtaining the primary screen results to the discovery of novel chemical entities. Recent developments in high throughput purification, LC/PDA, LC/MS/MS and LC/NMR for online structure dereplication and creation of informatic databases have fundamentally changed the way in which bioactive natural products are dereplicated.

Hook *et al*. reviewed the general aspect of dereplication technology in modern high-throughput screening of natural products [50]. The dereplication process not only includes hardware selection, protocol development, fraction splitting and data collection, but also requires database implementation, data analysis and system automation. Technology advancement has significantly changed the landscape of hardwares utilized in studies of natural products. Due to the severe limitations of throughput and the quality of information they provide, traditional separation techniques such as thin layer chromatography (TLC) and paper chromatography (PC) have been replaced by gas chromatography (GC) and high performance liquid chromatography (HPLC).

Gas chromatography is a powerful technique to dereplicate low to moderate polarity natural products especially when it is coupled with ESI-MS. The reproducibility of ionization and fragmentation pattern from GC/MS make it one of the most efficient and reliable techniques in structure dereplication. There are different GC/MS libraries commercially available that contain mass spectra of hundred thousands of compounds. Unfortunately, approximately 80% of all known natural compounds are nonvolatile or thermally unstable and therefore incompatible with GC/MS methods. Due to the diversity of HPLC columns and the broad selections of solvent combinations and gradients, HPLC is capable of separating almost any kind of natural products. The development of HPLC detectors and information technology further expands the quality of information obtained about analytes. The coupling of photo diode array and mass chromatographic detectors with HPLC is now more and more prevalent for rapid characterization of natural products.

Strege summarized the technique of high-performance liquid chromatography-electrospray ionization mass spectrometry (HPLC-ESI-MS) in dereplication of natural products. In contrast to earlier electron impact ionization (EI), ESI technique is applicable to virtually any "ion" in solution with a "soft ionization" process. A comparison of ESI with fast atom bombardment (FAB), matrix assisted laser desorption ionization (MALDI), atmospheric pressure chemical ionization (APCI) and other techniques demonstrates its superior sensitivity, compatibility and reliability when coupled with HPLC [51].

Dereplication of active crude extracts using HPLC/UV/MS, coupled with biological activity data obtained on sub-fractions was reported by Cordell and Shin [52]. In this study, active plant extracts were analyzed with an HPLC system using a C-18 column eluted with an acetonitrile/water gradient in 30 min and coupled with a photodiode array detector and a ESI-MS in positive and negative dual modes. The UV absorption properties, molecular weight and ion fragmentation information from extracted ion chromatograms (ELC) of extracts were analyzed against published information about the genus and species by using external databases, such as NAPRALERT and the Dictionary of Natural Products [3]. One of the examples presented by Cordell and Shin was dereplication of cytotoxic chloroform extract from the roots of *Begonia parviflora* L. (Begoniaceae). The bioactive fractions showed two distinct peaks on the HPLC/UV chromatogram with retention times of 9.8 and 12.5 min. The negative ion chromatograms from EI/MS analysis of the bioactive area showed the existence of four compounds with m/z values of 401, 515, 557, and 559. Searching the NAPRALERT database revealed four known compounds that have been isolated from *Begonia* species and matched with the molecular ion profile in the dereplication process. All four compounds, hexanorcucurbitacin D (MW=402), cucurbitacin D (MW=516), cucurbitacin B (MW=558), and dihydrocucurbitacin B (MW=560), were previously reported to possess cytotoxicity and co-injections of authentic samples confirmed the existence of those known compounds in the active extract. Therefore, no further isolation was carried on for this plant.

How to improve the efficiency and accuracy of the dereplication process and how to integrate the bio-active profile with UV and MS chromatograms for active identification are two focused efforts in the PhytoLogix[TM] platform [53]. First, an internal structure and spectroscopic database was developed with more than 300 known compounds that possess representative structural skeletons of common natural products in a quantity of 5-50 mg and a purity of >90% (HPLC). Each compound was dissolved in methanol (1 mg/ml) and then further diluted and concentrated for analysis of individual compounds due to different UV absorption and mass ionization properties. The sample solutions were analyzed by HPLC using a Luna C18 column (2x50 mm, 3 μm) at a flow rate of 0.4 mL/min and a temperature of 35°C. The column was eluted in 8.5 min with a gradient system of ACN and water. The eluent was analyzed with a PDA detector with wavelength from 200-500 nm and an ion trap mass spectrometer with Super Sonic Ionization (SSI) source, positive and negative charge detections. The retention time, UV spectrum, molecular ion and fragmentations were recorded and saved in a searchable library. Table 1. sets forth representative information in the database for flavonoids, alkaloids, caffeic acids, terpenoids, chromones, anthraquinones, iridoids, acetophenones, and coumarins. In the dereplication process, unknown bioactive fractions were analyzed under the same conditions and the HPLC peaks from PDA detection were searched in the UV library for structural skeleton and compound type. The molecular ion of the peak was then used for initiating a molecular weight search using a database like Dictionary of Natural Products, [3] combined with other searchable fields, such as, plant Latin name, type of compound, UV spectrum. Finally the retention time was used to get a general idea about the polarity, log P, solubility, and other physical properties of the compound.

The uniqueness of the Phytologix[TM] dereplication process was demonstrated in a process to evaluate the HTP fractions derived from the aqueous extract of green tea for inhibition of COX peroxidase. A total of 24 HTP fractions surrounding the COX inhibition peaks as shown in Figure 7 were analyzed using standardized HPLC. After obtaining and evaluating retention times, UV and MS data, all of the major components in each of the

24 cells have been dereplicated and identified as known flavan and flavone types of compounds.

Table 1.   **Typical information in the Structure and Spectroscopic Database**

Name	MW	Structure	UV spectra
7,8-Dihydroxy-flavone	254.24	[480-40-0] 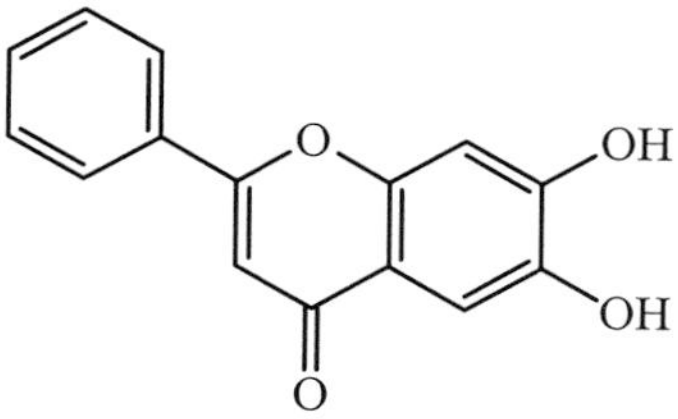	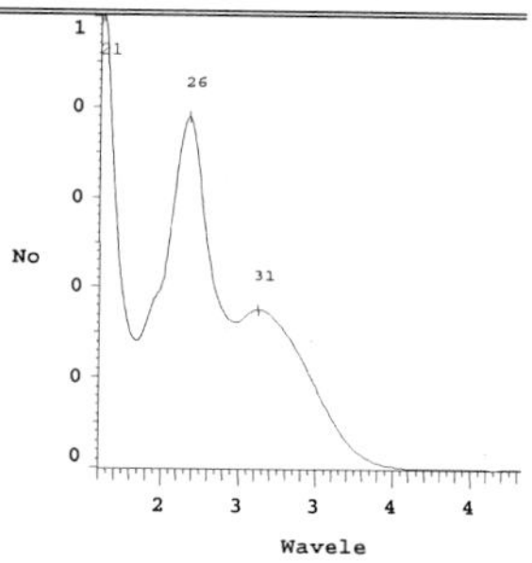
Corynanthine	354.45	[483-10-3] 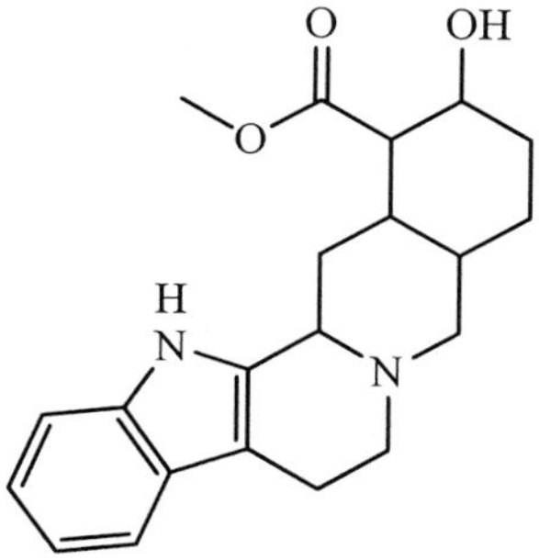	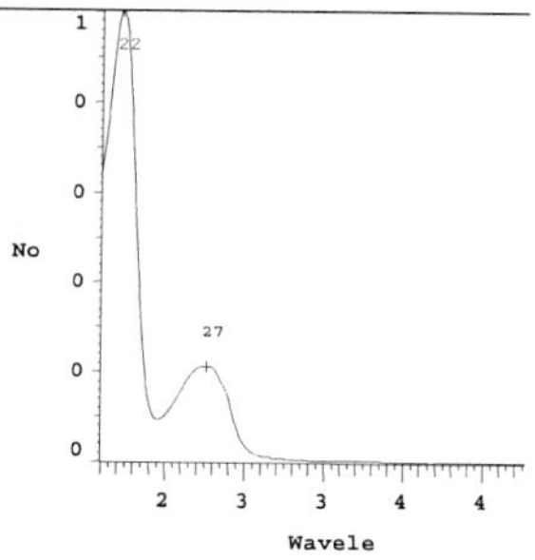
Caffeic acid	180.16	[331-39-5] 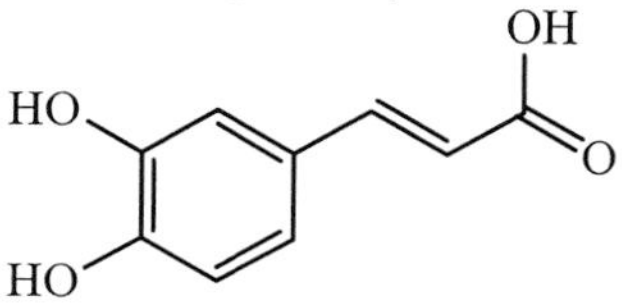	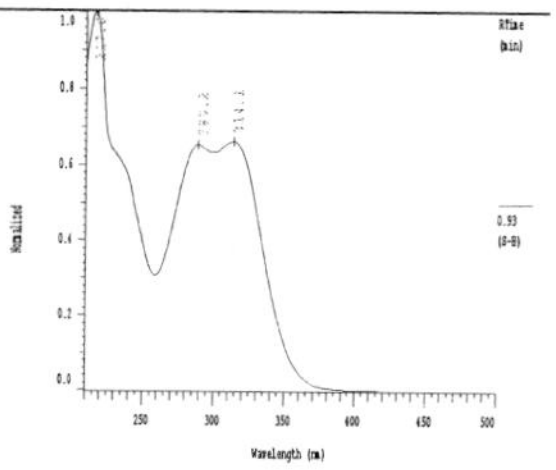

Ursolic acid	456.70	[77-52-1] 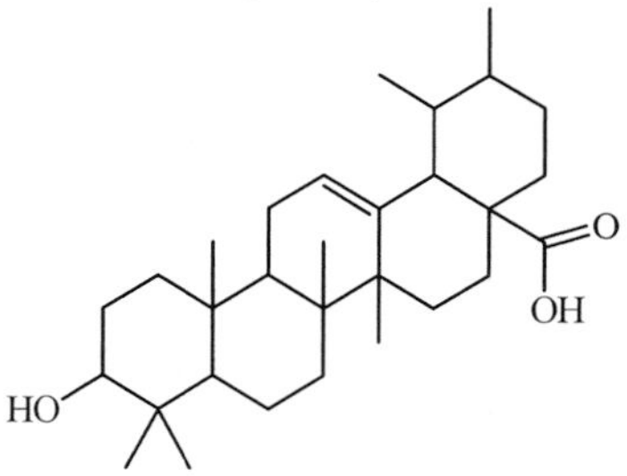	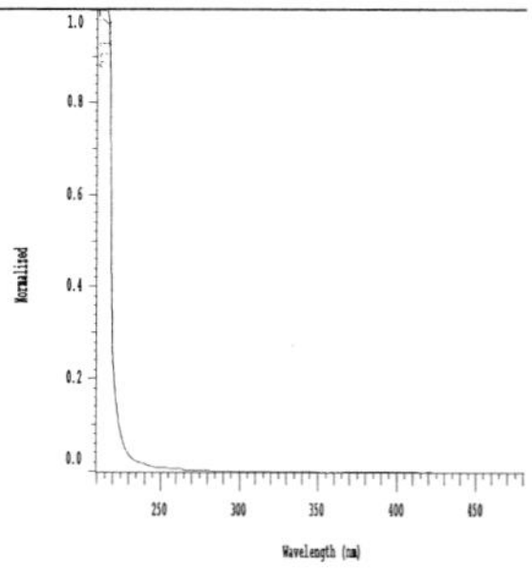
Aloesin	394.38	[30861-27-9] 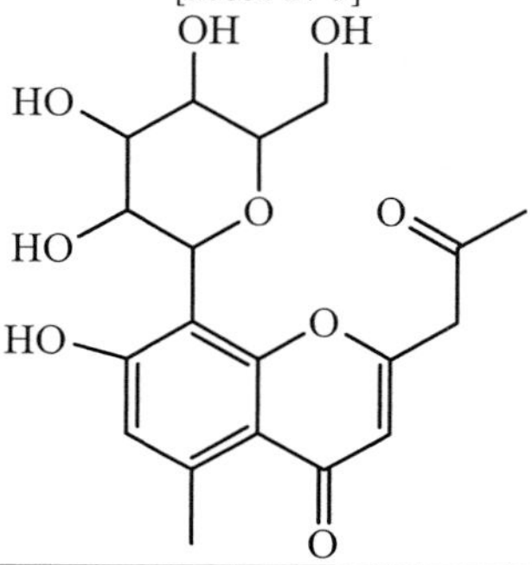	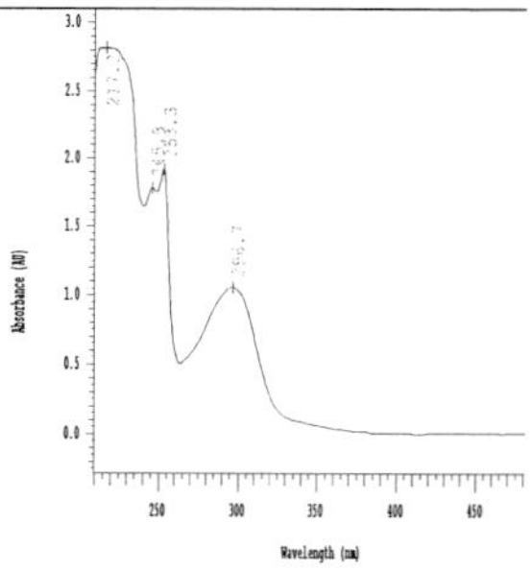
Rhein	284.2	[478-43-3] 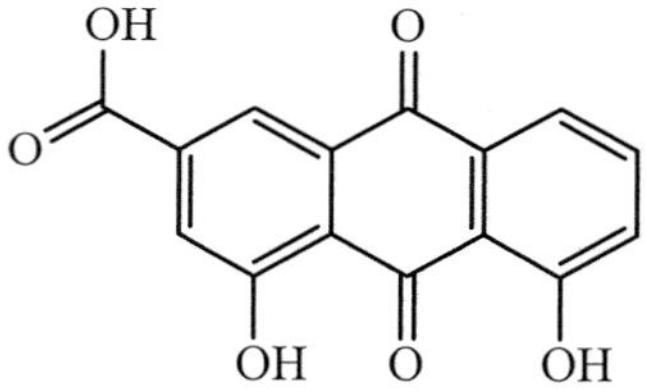	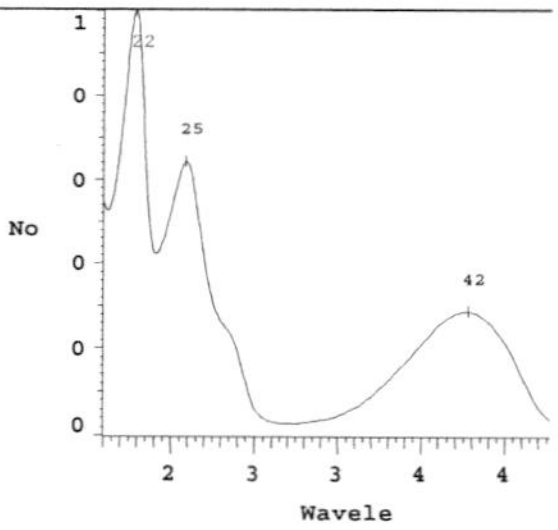

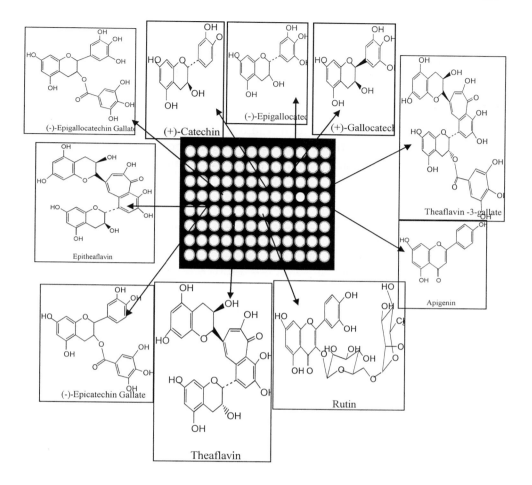

Fig. (7). Structure dereplication results of COX inhibitory HTP fractions from the aqueous extract of green tea.

Each compound was distributed among 3-4 individual cells. Since the COX inhibitory activity of catechins and flavonoids are well known, the conclusion from the dereplication process is that these active fractions are not worth pursuing.

The following example describes the results of the dereplication of an HTP fraction library for inhibition of melanin formation in an *in vitro* assay with a B16 cell line [54]. Briefly, following the inhibition and cell viability assay, the active organic extract from the whole plant of *Mallotus repandus* was fractionated with HTP. All of the HTP fractions were tested for melanin inhibitory activities. As shown in Figure 8, there are three major peaks exhibiting > 50% inhibitions of melanin synthesis and seven minor peaks exhibiting weaker inhibitions. The sharp activity peaks indicate the quality of the separations, which distributed the active components in three to five cells.

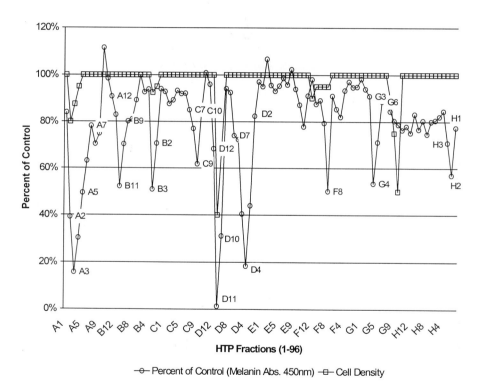

Fig. (8). The melanin production inhibition and cell toxicity of HTP fractions from *Mallotus repandus*.

Since the melanin formation assay was run against a cell viability assay, the activity peak maximum at fraction D11 was most likely due to cytotoxicity. The dereplication of another active peak located from fractions D7 to D2 is illustrated in Figure 9. Every active fraction was analyzed by HPLC/PDA/MS. There was a peak located at Rt=16.33 min in the LC/MS total ion chromatogram of active fractions. This peak showed the same pattern of increasing to decreasing intensity as the trend exhibited by the melanin inhibition activities of those fractions.

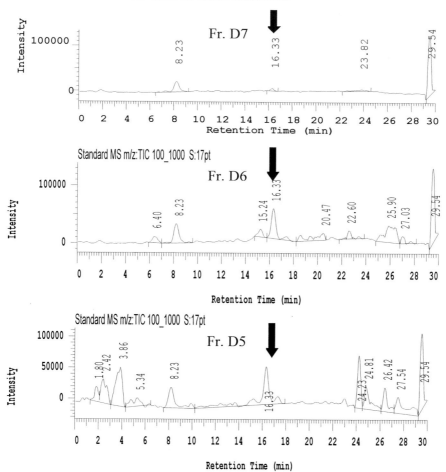

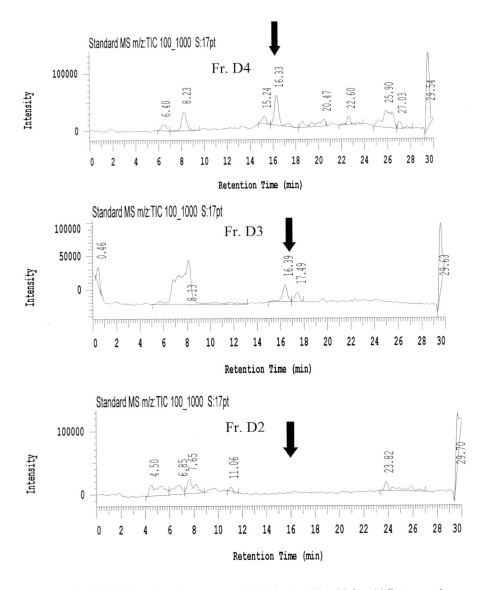

Fig. (9). LC/MS total ion chromatograms of HTP fractions D7 to D2 from *Mallotus repandus*.

Further analysis of the UV spectrum revealed that this compound is a gallic acid derivative. A search of the Dictionary of Natural Products for molecular ion and plant genus name leads to the identification of a known compound – Pterocaryanin B. The compounds with poly-hydroxyl groups in their structures are very well known as tyrosinase inhibitors and are responsible for the inhibition of melanin synthesis. Therefore, no further isolation was necessary.

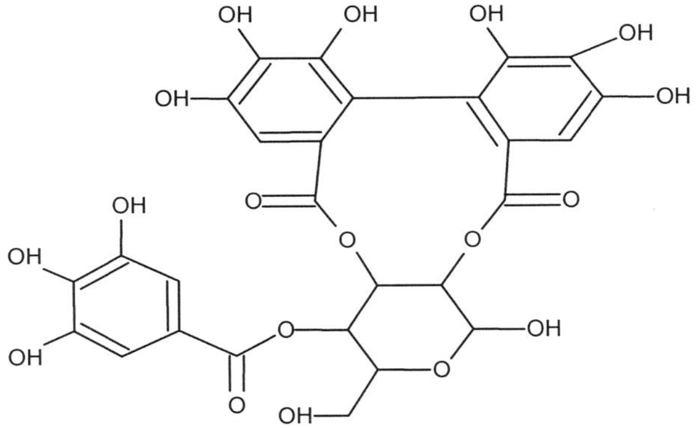

Fig. (10). The chemical structure of Pterocaryanin B

LC-ESI-MS technology was also utilized for quantitatively differentiating crude natural extracts as described by Julian *et al* [55]. Briefly, ethanol/water extracts from fungal cultures were separated on a dual-column HPLC system with C-18 columns in 25 min using an acetonitrile/water/ammonium acetate gradient. A Similarity Index was developed from the HPLC retention time and mass to charge ratio from the positive ion mode of an ESI-MS instrument. Ions with mass between 150 and 2000 amu and retention time between 3 and 25 min were included in the construction of the array. The resulting array was a 200 x 1200 rectangular image with retention time on the x-axis, mass on the y-axis, and pixel intensity representing the sum of the intensity of the ions assigned to the pixel. The custom C programs based computation of the Similarity Index between two samples yielded a quantitative conclusion. This method had a significant advantage to allow effective comparisons of the large data sets and to reach quick and reasonable decisions. This methodology, however, had evident limitations since it require great level control of reproducibility on HPLC separation and MS ionization efficiency and instrument stability. Also Similarity Index contained limited structural information and limited types of compounds had reasonable molecular ion peaks in the positive mode. As demonstrated by Wolfender's report [56], there is no single ionization interface allowing the optimum ionization of all the secondary metabolites within a single crude plant extract. Different ionization techniques, such as ES, APCI, TSP or CF-FAB are required in conjunction with LC/DAD and MS/MS.

Considering the power and successful structure dereplication with electron ionization mass spectra obtained from GC/MS, any attempts to standardize MS/MS fragmentation from LC/MS are worth applauding. To generate a searchable library of MS/MS fragmentation spectra with reproducibility will significantly expand the structural information collected from mass spectrometry. Baumann and others made such efforts by using a quadrupole ion trap mass spectrometer and electrospray or atmospheric pressure chemical ionization [57]. The methodology utilized wideband excitation and normalized collision energy together with the ability of the ion trap to provide daughter ion spectra from parent ions at very low abundance. The samples were flow injected with concentrations of 0.1-10 ng/μL and the MS/MS spectra were searched using NIST algorithms. The MS/MS library contained the mass spectra of over 600 compounds including drugs, drug metabolites, pesticides, herbicides, natural products, and dyes.

Bradshaw *et al.* have disclosed a rapid and facile method for the dereplication of a purified natural product library [58]. The method integrates the molecular weight from low resolution MS data with the exact count of the number of methyl, methylene and methine groups obtained from ^{1}H-^{13}C NMR correlation data. Those structure properties were converted into a searchable text file that could be downloaded into a customized software program with chemical structure information in a specific format – SMILES. In the program, more than 120,000 unique structures were derived from commercial databases, such as Dictionary of Natural Products and Beilstein.

The most common combination, two methyls, zero methylenes and six methines (2-0-6) only occurred 461 times in the program. After combining NMR data with molecular weight, the numbers of possible structures could have been significantly reduced. One of the examples presented in the publication was searching a combination of three methyls, four methylenes and four methines with MW=250, which yielded 40 possible structures. However, further analyses of the number of proton-bearing sp^2 carbons and the number of methylene groups that attached to oxygen based on the NMR data, only four candidates remained. The final and unequivocal identification of the structure as 3β-hydroxycinnamolid was confirmed by comparison with the literature NMR data [58].

Evidentially, the direct coupling of HPLC with NMR spectrometry provides much more structural information and significantly enhances the quality of the conclusions in the dereplication process. There are currently five main options that can be utilized for HPLC/NMR under either isocratic or gradient conditions: continuous-flow, stop-flow, "time-sliced" stop-flow, peak collection into capillary loops for post-chromatographic analysis and automatic peak detection with UV or MS triggered NMR acquisition [59]. This technology requires efficient NMR solvent suppression pulse sequences to detect signals from low concentration analytes in the presence of large signals from HPLC solvents. The current development of new techniques can suppress the solvent NMR resonance in complex solvent combinations and gradients. However, utilizing deuterated solvents, such as D_2O and acetonitrile-d_3, will make multiple solvent suppressions easier. The issues of HPLC flow rates and NMR flow cell sizes have to be addressed to achieve the best field homogeneity without compromising chromatographic resolution. Finally, the NMR instrumentation sensitivity needs to be improved to accommodate the dynamic continuous-flow detection and new software has to be developed to address all related data processing. To couple both NMR and MS into a single chromatographic analysis and to acquire both data simultaneously posts some significant challenges. The HPLC solvent selection has to be compromised among the requirements of optimum ionization efficiency for MS, signal suppression for NMR, and HPLC resolution. Since NMR has relatively low sensitivity, large volumes and high concentrations of analytes are required to achieve the best signal to noise ratio. As a result, generally 4.6-mm HPLC columns are used at flow rates between 0.5-1.0 ml/min. Such high flow rates and concentrations of analytes can easily be accommodated by mass spectrometers. The best way to configure the NMR and mass spectrometers is in parallel. By splitting a minor fraction to MS, there is little effect on sensitivity but greatly enhances the lifetime of the ion source without compromising chromatographic resolution. The applications of HPLC-NMR and HPLC-NMR-MS in natural products and other pharmaceutical fields have been extensively reviewed by Lindon *et al* [59].

The chemical structure of natural products can be identified quickly with a limited amount of materials by utilizing NMR equipment containing cryo probes [60]. More predictable chemical shifts coupled with a reasonable amount of published and internal NMR data [61] will significantly improve the time and accuracy of the structure elucidation process [62,63,64].

Since the primary goal of dereplication is to identify known compounds from active extracts or fractions to avoid unnecessarily isolating these known compounds, the selection of an adequate structure database to evaluate information collected is critical to the dereplication process [65]. Chemical Abstracts Service's Registration File available on the Scientific and Technology Network, STN International, which includes CA, NAPRALERT, REGISTRY, BEILSTEIN, MEDLINE and many other sub-databases and Capman & Hall's "Dictionary of Natural Products", [3] which includes the Bioactive Natural Product Database and DEREP databases, were two of the most comprehensive databases utilized in natural product dereplication. The Registry File contains over 12 million compounds with a process known as file crossover to link together structure, physical properties, and biological information. To overcome the magnificent size of the Registry File, strategies have to be developed by utilizing a combination of formula weight, carbon count, structure fragments, bioactivity and taxonomy. It was estimated that each natural product online dereplication costs an average of $300. However, a saving of $50,000 was incurred in isolation and identification time if the dereplication revealed a known structure for the interested sample. Early dereplication also added the benefit of focusing more resources for the discovery of novel bioactive compounds.

One of the biggest challenges in the dereplication process is how to collect, store, analyze, and report the information collected from internal research and public domains. An issued patent invented by Dietzman offered a comprehensive approach [66]. A computer implemented Natural Products Information System (NAPIS) was capable of processing natural product data, such as chemical structures, geographic location, textual data, taxonomy, genus synonyms, taxonomic classifications as well as natural product images, such as photographs, slides, video, geographic maps, tables, and charts. It has two main features: a) integrates sample collection with chemical and biological activity data; and b) generates an interface to link internal data with existing commercially available data. To communicate with external databases, NAPIS used linkage on genus and species name, Chemical Abstract Registry Number or the National Oceanographic Data Center (NODC) taxonomic code. NAPIS utilized customized computer programs to perform information dissemination, standardized data entry, capture and exchange information, and to link stored specimen photographs with geographical information system (GIS) data. It efficiently addressed two fundamental requirements in natural product research: recollection and dereplication in drug discovery process.

DISCOVERY OF NATURAL COX AND LOX DUAL INHIBITORS

Liberation and metabolism of arachidonic acid (AA) from the cell membrane results in the generation of pro-inflammatory metabolites through the cyclooxygenase (COX), the lipoxygenase and the cytochrome P-450 pathways. The two most important pathways are mediated by the enzymes of 5-lipoxygenase (5-LOX) and cyclooxygenase (COX). These are parallel pathways that result in the generation of leukotrienes and prostaglandins respectively and play important roles in the initiation and progression of the inflammatory response. These vasoactive compounds are chemotaxins that promote infiltration of inflammatory cells into tissues, and their presence will also serve to prolong the inflammatory response [67]. Consequently, the enzymes responsible for generating these mediators of inflammation have become the targets for many new anti-inflammatory agents. Inhibition of the cyclooxygenase is the mechanism of action attributed to most nonsteroidal anti-inflammatory drugs (NSAIDS).

There are two distinct isoforms of the COX enzyme (COX-1 and COX-2) that share approximately 60% sequence homology, but differ in expression profiles and function. COX-1 is a constitutive form of the enzyme that can be detected in most tissues and is typically expressed at constant levels throughout the cell cycle. It is linked to the production of physiologically important prostaglandins that help regulate normal physiological functions such as platelet aggregation, protection of cell integrity in the stomach, maintenance of normal kidney function, and safeguarding of a healthy pregnancy [68]. COX-1 is also found in neuron and throughout the brain and is most abundant in the forebrain where prostaglandins may be involved in complex, integrative functions [69].

The second isoform, COX-2, is undetectable in most mammalian tissues, but its expression can be induced rapidly in cells involved in inflammation such as fibroblasts, monocytes, and vascular endothelia in response to pro-inflammatory cytokines, such as interleukin-1β (IL-1β) interferon gamma, and tumour necrosis factor-alpha (TNF-α) as well as other mitogens [70]. The COX-2 enzyme catalyzes the productions of pro-inflammatory prostaglandins PGE_2 and anti-inflammatory PGD_2 and $PGF_{2\alpha}$ from arachidonic acid (AA). It has been constitutively expressed in brain, testes, tracheal epithelia and kidney. It plays an important role in several bodily functions such as wound healing, maintaining kidney function, and cyclical hormonal production inducing ovulation, and sustaining vascular prostacyclin production [71].

Arachidonate-5-lipoxygenase is particularly found in cells of myeloid origin, such as polymorphonuclear leukocytes, macrophages, eosinophils, mast cells, monocytes, basophils and B lymphocytes that are involved in inflammatory and immune reactions. The leukotrienes generated by the 5-lipoxygenase pathway have been implicated as important mediators of chronic inflammation processes by increasing microvascular permeability, acting as potent chemotactic agents, inducing cell adhesion, modulating pain induced by inflammation reactions, adjusting certain immune responses, and initiating powerful spasmogenic actions in airway smooth muscle and in the vasculature [72]. LTB_4 has the most potent chemotactic and chemokinetic effects [73] and is elevated in the gastrointestinal mucosa of patients with inflammatory bowel disease [74] and within the synovial fluid of patients with rheumatoid arthritis [75,76]. The biological and pathological properties of leukotrienes suggest that a 5-lipoxygenase inhibitors can be utilized in the prevention and treatment of asthma, rheumatoid arthritis, ulcerative colitis and many other allergic and inflammation conditions.

Because the mechanism of action overlap with those of NSAID's, COX inhibitors can be used to treat pain and swelling associated with inflammation in transient conditions and chronic diseases. The major application for COX inhibitors is to alleviate symptoms associated with rheumatoid arthritis and osteoarthritis. While rheumatoid arthritis is largely an auto-immune disease and osteoarthritis is caused by the degradation of cartilage in joints, the up-regulation of COX-2 by cytokines occurs not only in monocytes and macrophages, but also in chondrocytes, osteoblasts, and synovial microvessel endothelial cells [77]. Therefore, reducing the inflammation associated with arthritis provides a significant increase in the quality of life for those suffering from these diseases [78]. Additionally, recent scientific progress has identified correlations between COX-2 expression, general inflammation and the pathogenesis of Alzheimers Disease (AD) [79]. In animal models, transgenic mice that over express the COX-2 enzyme have neurons that are more susceptible to damage. The National Institute on Aging (NIA) has launched clinical trials to determine whether NSAIDs can slow the progression of Alzheimer's disease. Naproxen (a non-selective NSAID) and rofeCOXib (Vioxx, a COX-II specific selective NSAID) will be evaluated.

In addition to inflammation, another potential role for natural COX and LOX dual inhibitors is in the prevention and treatment of cancers [80]. Over expression of COX-2 has been demonstrated in various different human malignancies. COX-2 inhibitors have also been shown to be efficacious in the treatment of animal models of skin, breast and bladder tumors. While the mechanism of action remains to be completely defined, the over expression of COX-2, in excess of production of prostaglandin E_2 and 5-hydroxyeicosatetraenoic acid (5-HETE) have been shown to inhibit apoptosis and increase the invasiveness of tumerogenic cell types [81,82]. It is probable that the enhanced production of PGE_2 and 5-HETE promotes cellular proliferation, and consequently, increases angiogenesis [83].

Inhibitors of both the COX and LOX pathways of arachidonic acid metabolism convert mouse melanoma and human fibrosarcoma cells to a non-invasive state by reducing the production of MMP-2, an enzyme required for the degradation of basement membranes [84]. At least in animal models, both COX and 5-LOX inhibitors significantly reduced tumor growth and final tumor weight associated with lymphocytic infiltration and increased tumor tissue level of PGE_2 [85]. There are many clinical studies that evaluate COX-2 and LOX inhibitors for their potential in preventing and treatment of different type of cancers [86]. Aspirin use (non-specific NSAID) has reduced the incidence of colorectal cancer by 40-50% [87] and mortality by 50% [88]. In 1999, the FDA approved the COX-2 inhibitor CeleCOXib for use in FAP (Familial Adenomatous Polyposis) to reduce colorectal cancer mortality. It is thought that other cancers, with evidence of COX-2 involvement, may be successfully prevented and/or treated with COX-2 inhibitors [89,90].

In order to address classical NSAID toxicity, especially gastrointestinal ulceration and haemorrhage, two strategies have been implemented in the drug discovery process. The first concept is searching selective inhibitors of COX-2 to reduce gastrointestinal side effects by sparing COX-1 protective functions in gastric mucosa [91]. This effort has lead to the successful launch of several commercial drugs, such as Celecoxib and Rofecoxib, which are 375 and >800-fold selective against COX-2. In clinical trials, COX-2 selective inhibitors demonstrated significant potency against pain and other inflammation symptoms with lower incidence of gastrointestinal events. However, the shortcomings of selective COX-2 inhibitors have gradually emerged. They promote allergic and asthmatic attacks, cause acute renal failure, congestive heart failure, exacerbate coronary and cerebrovascular diseases, delay broken bone growth and ulcer healing, suppress the immune system making one susceptible to viral meningitis attack, and promote development of ulcers from patients with gastric erosions or with *Helicobactor pylori* infection [92]. With recent reports that a significant anti-inflammatory effect for some highly selective COX-2 inhibitors was only observed after the dosage level reached to levels inhibition of COX-1 activity [93], together with anti-inflammatory prostanoid generation by COX-2 enzyme at a later phase of the inflammation process [94], further challenged the concept of selective COX-2 inhibitors.

Recent anti-inflammatory efforts have been focused on searching for dual acting agents that inhibit both cyclooxygenase and lipoxygenase [95,96]. The importance of blocking the inflammatory effects of PGE_2, as well as, those of multiple leukotrienes (LT) was based on the recent discovery that the significant drawbacks for selective COX-2 inhibitors are associated with shunting of the arachidonic acid to the lipoxygenase pathway and causing the over production of pro-inflammatory, chemotactic, gastro-damaging, and bronchoconstrictive leukotrienes [97]. NSAID induced gastric inflammation was largely due to metabolites of 5-LOX, particularly LTC_4 and LTB_4 [98]. Leukotrienes contribute to a significant amount of the gastric epithelial injury by stimulating leukocyte infiltration, occluding microvessels, reducing mucosal blood flow and releasing mediators, proteases, and free radicals. The selective 5-LOX inhibitors have demonstrated significant reduction in the severity or prevention of indomethacin-induced ulcer formation [99]. Another example to demonstrate the need for dual inhibitions was aspirin-induced asthma (AIA) syndrome. By inhibiting COX pathways, aspirin and other COX inhibitors divert arachidonic acid metabolites to the LOX pathway causing increased bronchoconstrictive leukotrienes release along with an increase in the levels of cysteinyl leukotrienes that lead to chronic rhinoconjunctivitis, nasal polyps, and asthma akin to a protracted viral respiratory infection. The prevalence of AIA in the asthmatic population is about 10 to 20% and anti-leukotriene drugs have been utilized in the treatment of patients with AIA [100].

Dual inhibitors also demonstrate other therapeutical benefits. They reduced the coronary vasoconstriction in arthritic hearts in a rat model [101], and significantly decreased angiotensin II-induced contractions in human internal mammary artery [102]. Opioid receptor activation can cause a presynaptic inhibition of neurotransmitter release mediated by LOX metabolites of arachidonic acid in midbrain neurons. The efficacy of opioids was enhanced synergistically by treatment of brain neurons with COX and LOX dual inhibitors. This report might lead to development of CNS analgesic medications involving combinations of lowered doses of opioids and COX/LOX dual inhibitors [103]. The COX and 5-LOX dual inhibitors also can prevent lens protein-induced ocular inflammation in both the early and late phases [104].

Dual inhibitors of COX and 5-LOX will not only suppress prostaglandins that contribute to acute inflammation conditions, but also address the accumulation of phagocytic leukotrienes that are directly associated with chronic inflammation symptoms. With the existence of COX-1 inhibition activities, dual inhibitors will also possess the cardio protective functions. Those characteristics suggest that there may be distinct advantages to dual inhibitors of COX and 5-LOX over selective COX-2 inhibitors and NSAIDs. This concept has been proved to be valid on *in vivo* models with synthetic drug candidates [105].

There are two basic types of assays routinely utilized for evaluating the COX or LOX inhibition activities of natural products: purified enzyme based biochemical screening and cell line based *in vitro* assays. The purified COX enzymes were utilized to measure prostaglandin production originally described by White and Glassman [106]. With the discovery of the isoenzyme COX-2 which reveals the importance of COX-2 as an inducible inflammation mediator in physiological and pathological processes, the development of an assay targeting COX-2 inhibition became very important. Noreen et al modified the White and Glassman assay and introduced a radiochemical *in vitro* assay for discovery of natural COX-1 and COX-2 inhibitors [17]. The assay utilized purified microsome COX-1 enzyme prepared from bovine seminal vesicles and COX-2 enzyme from sheep placental cotyledons. The COX enzymes were activated with a cofactor solution that contained l-epinephrine, reduced glutathione and hematin in oxygen-free-Tris-HCl buffer. With/without pre-incubation of enzyme and test solution, $[1-^{14}C]$-arachidonic acid was added and incubated for 15 minutes at $37^{\circ}C$. After termination of the reaction with HCl and carrier solution, the prostaglandin products were separated by column chromatography before counting the radioactivity. The amount of enzyme, concentrations of cofactors, reaction times, requirement of pre-incubation and termination conditions were optimized and the assay was validated with known drugs and natural products with COX-1 and COX-2 inhibition activity and selectivity. The assay was designed for screening of natural products with IC_{50} measurement and selectivity evaluation. The only limitation was sample throughput since a chromatographic process was required for each sample to separate prostaglandin products before quantification.

Recently, a scintillation proximity assay (SPA) has been developed for screening COX-2 inhibitors from plant constituents. The scintillation proximity assay (SPA) uses a radioisotopic assay technique based on immobilization of radio labeled molecules on the surface of scintillant-containing microspheres. Polyvinyl toluene (PVT) beads are precoated with anti-rabbit antibody that captures the primary anti-prostaglandin E_2 (rabbit) antibody. A tritium-labeled PGE_2 tracer is added, which competes with the unlabeled PGE_2 from the enzyme reaction for the binding sites of the PGE_2 antibodies. Only β-particles emitted from immobilized isotopes are in close proximity to the scintillation core and are capable of producing detectible light emissions. Since there is no interference from radioisotopes and tested extracts from solution, further separation and purification are not necessary. The SPA method is rapid, easy to handle, amenable for automation, and suitable for complex mixtures such as plant extracts [107].

Other purified enzyme assays include quantitative measurements of oxygen consumption with a Clark electrode in prostaglandin biosynthesis [108], radioimmunoassay (RIA) [109], enzyme immunoassay (EIA). [110,111], and a pulsed ultrafiltration liquid chromatography-mass spectrometry (LC-MS) binding assay [112].

Using a cell line based assay to evaluate the efficacy of natural products on COX and LOX pathway requires the selection of appropriate cell lines and quantification method for related eicosanoids generated from different but parallel pathways. The most common cell based assay is a lipopolysaccharide (LPS) endotoxin-challenged human whole blood *ex vivo* assay. The separation and quantification method is essential to measure thromboxane B_2 (TXB_2) production for COX-1 pathway and prostaglandin E_2 (PGE_2) release for COX-2 pathway [113]. Human platelet are also widely utilized to evaluate COX-1 and lipoxygenase pathway by measuring thromboxane B_2 (TXB_2) and 12-hydroxy-5,8,10,14-eicosatetraenoic acid (12-HETE) productions [114], respectively. Calcium ionophore A23187 stimulates leukocytes and peripheral blood cells and is widely used for evaluation of 5-LOX pathway by quantifying 5-hydroxyeicosatetraenoic acid (5-HETE) [115]. The other *in vitro* method for 5-LOX pathway was measuring leukotriene C_4 (LTC_4) production from ionophore-stimulated peritoneal macrophages [116]. Recently, Reynaud *et al* improved an HPLC method that could separate and quantify thromboxane B_2 (from COX-1 pathway), 12-hydroxy-heptadecatrienoic acid (HHT) (from thromboxane A_2 synthase pathway), 12-hydroxyeicosatetraenoic acid (from 12-LOX pathway), and arachidonic acid (from phospholipase A_2 pathway) [117]. Those eicosanoids are formed by stimulation of human platelets *in vitro* with collagen. They are then extracted with ethyl acetate, and derivatized with anthryl diazomethane (ADAM) before HPLC analysis. The HPLC separation requires a reverse phase column, an acetonitrile-water gradient and a fluorescence detector. Even though, it is not a high throughput assay, the HPLC method still has significant advantages with the potential of evaluating simultaneously four different inflammation pathways from the cell-based assay.

Since COX enzymes have been purified and screening methods widely utilized for natural products, a high throughput screening of an extract library for COX inhibitory activities was the first step to find the dual inhibitors [118]. The high throughput, *in vitro* assay was modified from a publication that measured inhibition of the peroxidase activity of COX enzymes [119]. Briefly, a known concentration of plant extracts was titrated against a fixed amount of the COX-1 and COX-2 enzymes, respectively. A cleavable, peroxide chromophore was included in the assay to visualize the peroxidase activity of each enzyme in the presence of arachidonic acid as a cofactor. Typically, assays were performed in a 96-well format. After screening 1230 plant extracts from 615 medicinal plants, a total of 15 organic extracts (1.2%) and 7 aqueous extracts (0.6%) were confirmed by measurement of the dose response and IC_{50} values for the concentrations required to inhibit 50% of the enzyme's activity. In this assay, the organic extracts from the roots of *Scutellaria baicalensis* Georgi. (Lamiaceae) and the bark exudates of *Acacia catechu* Willd (Minosaceae) were the most efficacious (IC_{50}= 2-10 µg/ml). Extracts from *Scutellaria spp.* that demonstrated the greatest selectivity against COX-2 were *Scutellaria lateriflora* L. (COX-2 IC_{50} = 30 µg/ml and COX-1 IC_{50} = 80 µg/ml).

In order to identify the active compounds from the plant extracts, a high throughput fractionation process was executed. The active organic and aqueous extracts were fractionated using two different HTP methodologies as illustrated in previous section. The fractions were collected as described into a 96 deep-well plate. Each of the fractions was tested for its ability to inhibit

COX activity. In Figure 11, the most potent inhibitory activity of HTP fractions were derived from the roots of *Scutellaria baicalensis*, and located at two major fractions, E11 and F11. The activity profile of the HTP fractions showed three separate peaks of inhibition suggesting that there are multiple compounds contributing to the observed inhibitory effects of the whole extract.

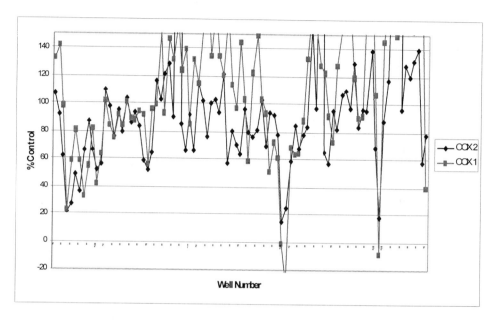

Fig. (11). COX inhibition profile of HTP fractions from organic extract of the roots from *Scutellaria baicaensis*.

Separation, purification and identification have yielded baicalein, baicalin and other Free-B-Ring flavonoids as the major active components in the organic extract from the roots of *Scutellaria baicalensis*, and *S. orthocalyx*. In Table 2, we demonstrated the potencies of five Free-B-Ring flavonoids isolated from three different *Scutellaria* species.

Table 2. Inhibition of COX enzymes by purified Free-B-Ring Flavonoids at a concentration of 20 µg/ml.

Free-B-Ring Flavonoids	Inhibition of COX-1	Inhibition of COX-2
Baicalein	107%	109%
5,6-Dihydroxy-7-methoxyflavone	75%	59%
7,8-Dihydroxyflavone	74%	63%
Baicalin	95%	97%
Wogonin	16%	12%

The IC_{50} values for a purified Free-B-Ring flavonoid, baicalin, from the roots of *Scutellaria baicalensis* were 0.44 µg/ml/unit of COX-1 enzyme and 0.28 µg/ml/unit of COX-2 enzyme. Its aglycone - baicalein showed similar potencies with IC_{50} for COX-1 as 0.18 µg/ml and 0.28 µg/ml for COX-2 enzyme. It has been reported that oroxylin, baicalein and wogonin inhibit 12-LOX activity without affecting cyclooxygenase from the homogenate of human platelets *in vitro*[120].

Acacia is a genus of leguminous trees and shrubs distributed worldwide in tropical and subtropical areas that includes more than 1000 species belonging to the family of Leguminosae and the subfamily of Mimosoideae. Historically, most of the plants and extracts of the *Acacia* genus have been utilized as astringents to treat gastrointestinal disorders, diarrhea, indigestion and to stop bleeding. The organic extract from bark exudates of *Acacia catechu* was a potent inhibitor against human/ovine COX-2 (IC_{50}= 3/6 µg/ml) and ovine COX-1 (IC_{50} = 2.5µg/ml). The *Acacia* extract also directly inhibited 5-LOX activity in a cell free *in vitro* system and demonstrated an IC_{50} of 70 µg/ml. The dual inhibitory active components have been isolated and identified as catechin, epicatechin, robinetinidol-(4-6/8)-catechin and other flavans. The IC_{50} values for purified catechin from *Acacia catechu* were 0.11 µg/ml/unit and 0.42 µg/ml/unit, respectively for COX-1 and COX-2 enzymes, and 1.38 µg/mL/unit of 5-LOX enzyme.

Based on Free-B-Ring flavonoids from the roots of *Scutellaria baicalensis* and flavans from the bark exudates of *Acacia catechu*, a novel product called Univestin™ was developed. This ingredient was a potent COX and LOX dual inhibitor with IC_{50} = 0.18 µg/ml/unit for COX-1 and 0.41 µg/ml/unit for COX-2 inhibition. Univestin™ shows potent inhibition of purified potato 5-LOX with dose response and almost completely inhibited production of LPS-induced LTB_4 in the THP-1 cultures at concentrations between 3 and 10 µg/ml. This result was confirmed by the 48 hr incubation of HT-29 cells with 3 µg/ml Univestin™. However, under the same conditions, ibuprofen only showed a 20% reduction of LTB_4 production over the same time period. Oral administration of Univestin™ at a dosage of 50mg/kg inhibited mouse ear-swelling caused by topical application of arachidonic acid. Then arachidonic acid was injected into the ankle joints of mice, Univestin™ also completely inhibited swelling compared to control groups without Univestin™.

NATURAL PRODUCTS WITH DUAL COX & LOX INHIBITION ACTIVITIES

1. Fatty Acids & Alcohols

It is not surprising that fatty acids inhibit COX and LOX enzymes due to their structural similarities with arachidonic acid. A supercritical fluid extract from the fruits of *Sabal serrulata* (also called *Serenoa repens* Small. Arecaceae) has been utilized for the treatment of benign prostatic hyperplasia (BPH) and non-bacterial prostatitis. The extract was demonstrated as a dual inhibitor of COX and 5-LOX pathways with IC_{50} at 28.1 µg/ml and 18.0 µg/ml, respectively. A further evaluation of the supercritical carbon dioxide extract showed the acidic lipophilic fraction, most likely fatty acids, had the same dual inhibitory activities as the parent extract [121].

Several fatty acids and derivatives were isolated from the dichloromethane extract of *Angelica pubescens* Maxim *f. biserrata* Shan et Yuan (Umbelliferae) by following *in vitro* 5-LOX and COX-1 inhibitory assays. Linoleic acid showed potent inhibition against PGE_2 production from purified COX-1 enzyme. Falcarindiol, which is an acetylenic alcohol, showed higher potency against LOX than COX (Table 3) [122]. Linolenic acid is an omega-3 fatty acid that exists in fixed oil from different species of *Ocimum* (Labiatae) [123]. The content of linolenic acid in the oil showed positive correlation with the ability of inhibiting PGE_2, leukotriene and arachidonic acid induced paw edema by blocking COX and LOX pathways [124].

Table 3.   COX and LOX inhibitory activities of fatty acids [122]

Fatty Acids	IC_{50} (COX-1, μM)	IC_{50} (LOX, μM)
Linoleic acid	13.3	27.9
Oleic acid	58.2	36.6
Palmitic acid	67.3	>100
Falcarindiol	66.0	9.4
11(S),16(R)-dihydroxy-octadeca-9Z,17-diene-12,14-diyn-1-yl acetate	73.3	24.0

2. Acetylenes

Linear acetylenic compounds, such as pentadeca-6,8,10-triynoic acid, isolated from the bark of an Ecuadorian medicinal plant, *Heisteria acuminata* (Humb. & Bonpl.) Engl. (Olacaceae), showed potent COX inhibition with IC_{50} around 13 μM. However, the compounds with double bond, such as cis-hexadec-11-en-7,9-diynoic acid, had selectivity against LOX with 45% inhibitory activity at 5 μM [125].

The acetylene compounds – Panaxynol and falcarindiol, isolated from the roots of *Saposhnikovia divaricata* (Turcz.) Schischk. (Umbelliferae), selectively inhibited 5-LOX ($IC_{50} = 2$ μM), leukocyte-type 12-LOX ($IC_{50} = 1$ μM), platelet-type 12-LOX ($IC_{50} = 67$ μM), and 15-LOX ($IC_{50} = 4$ μM). The same compounds had only marginal effects on COX activities with $IC_{50} >$ 100 μM [126]. *Bidens campylotheca* Schultz Bip. *ssp. campylotheca* (Compositae) is a Hawaiian folk medicine for throat, stomach and asthmatic disorders. The polyacetylenes isolated from the aerial parts of the plants, such as safynol-2-O-isobutyrate, exhibited significant inhibitory effects on both LOX (100% inhibition at 9.6 μg/ml) and COX (IC_{50} = 10.0 μM) [127].

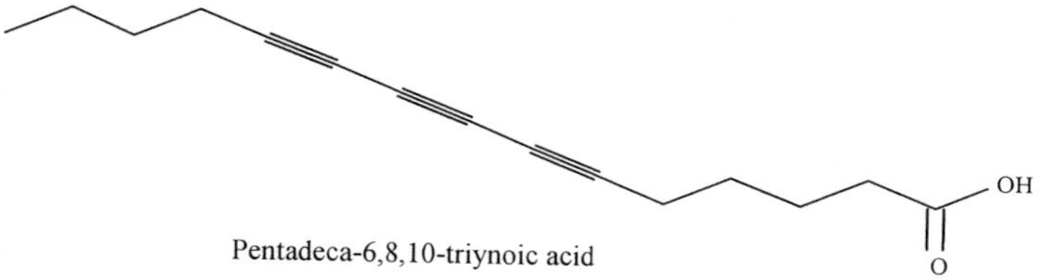

Pentadeca-6,8,10-triynoic acid

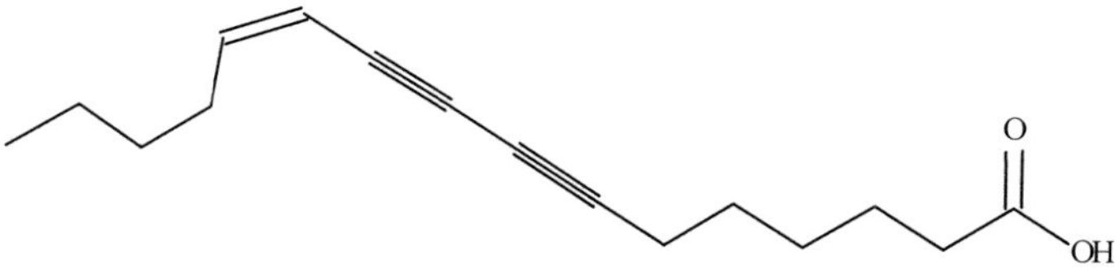

Safynol-2-O-isobutyrate

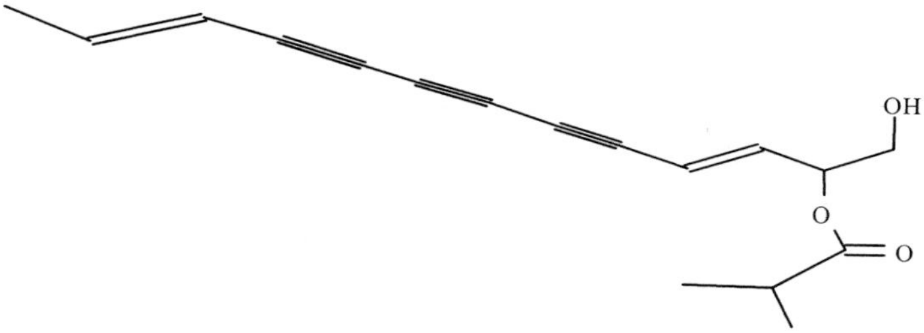

Cis-hexadec-11-en-7,9-diynoic acid

Fig. (12). Chemical structures of dual active acetylenes.

3. Alkamides

Polyunsaturated alkamides have been isolated from *Aaronsohnia*, *Achillea*, *Anacyclus*, and *Echinacea* species. Due to the structural similarity of alkamides to arachidonic acid, it is likely that alkamides act as competitive inhibitors of COX and LOX enzymes. At a concentration of 50 μg/ml, all twenty tested alkamides from *Achillea* and *Echinacea* species showed 21-

75% COX inhibition using purified enzyme from sheep seminal vesicles. However, only seven compounds had 5-LOX inhibition activity. The most active dual inhibitor was dodeca-2E,4E,8Z,10E/Z-tetraenoic acid isobutylamide isolated from the roots of *Echinacea angustifolia DC* (Asteraceae) with 55% inhibition of COX at a concentration of 50 µg/ml and 62% inhibition of 5-LOX at a concentration of 50 µM [128]. Similar isobutylamides with shorter hydrocarbon chains purified from a traditional Chinese herb, *Piper sarmentosum* Roxb. (Piperaceae), only showed weak effects on COX-1 and 5-LOX enzymes even though the crude extract to be a potent dual inhibitor with IC_{50} at 19 µg/ml and 10 µg/ml, respectively [129]. Increasing the carbon chain length, decreasing the unsaturated carbons on the acid part, and using dopamine to replace isobutylamine would significantly improved the potency against 5-LOX [130].

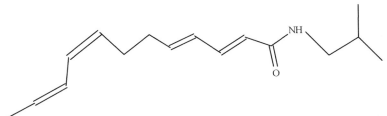

Fig. (13). The chemical structure of dodeca-2E,4E,8Z,10E/Z-tetraenoic acid isobutylamide

4. Aliphatic Sulfides

There are only 140 aliphatic sulfides isolated from natural sources [3]. Garlic (*Allium sativum* L. Liliaceae) and other *Allium* species are the major sources of the natural sulfides. Garlic (*Allium sativum*) and wild garlic (*Allium ursinum*) showed inhibition activity against LOX, COX, thrombocyte aggregation, and angiotensin I-converting enzyme (ACE) [131]. Cepaenes are a class of aliphatic sulfides isolated from *Allium cepa* A. with double bond(s) and a sulfinyl group. Cepaenes isolated from chloroform extract of onion juice inhibited COX and LOX enzyme activities by more than 75% at 10 µM and 1 µM concentrations, respectively. Cepaenes with two double bonds like 1-Propenyl 1-(1-propenyl sulfinyl)propyl disulfide showed higher activity than those, such as methyl 1-propenyl disulfide 5-Oxide, with only one double bond. αβ-Unsaturated thiosulfinates and aromatic thiosulfinates also showed dual inhibitions [132].

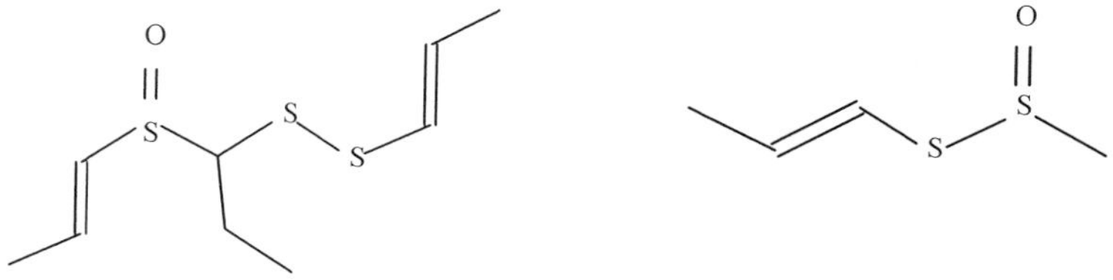

1-Propenyl 1-(1-propenyl sulfinyl)propyl disulfide   Methyl 1-propenyl disulfide 5-Oxide

Fig. (14). Chemical structures of two dual active aliphatic sulfides.

5. Iridoids

Iridoids represent a class of 941 known cyclopentano[c]pyran monoterpenoids [3] and have been reported with hepatoprotective, choleretic, vasoconstriction, anti-viral, anti-microbial, anti-inflammatory and analgesic activities. Seven iridoid glycosides, isolated from *Scrophularia scorodonia* L. (Scrophulariaceae), have been evaluated for effects on PGE_2, and LTC_4 production using calcium ionophore stimulated mouse peritoneal macrophages. TXB_2 release from human platelets was also evaluated. Aucubin was the only iridoid, which showed inhibitions of both LOX pathway (58% inhibition at 100 μM) and COX-1 pathway ($IC_{50} = 12$ μM). However, it was not active against PGE_2 production at 100μM concentration. A similar compound, harpagoside, showed similar potency against TXB_2 production ($IC_{50} = 10$ μM), but weak inhibitory activities on LTC_4 (17% inhibition at 100 μM) and PGE_2 (31% inhibition at 100 μM) releases [133]. Iridoid compounds, oleuropeoside and ligustroside, from the leaves of *Phillyrea latifolia* L. (Oleaceae) had significant inhibitory activity on PGE_2 production from calcium ionophore-stimulated mouse peritoneal macrophages. The IC_{50} values for inhibitions of the COX pathway were 47 μM for oleuropeoside and 48.53 μM for ligustroside. Both compounds were inactive against LTC$_4$ productions [134].

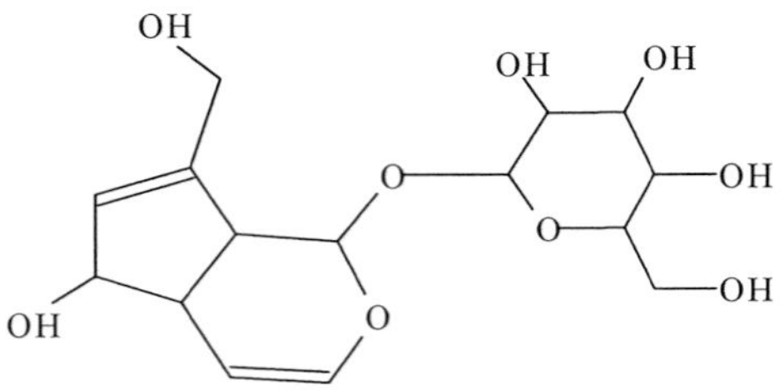

Aucubin 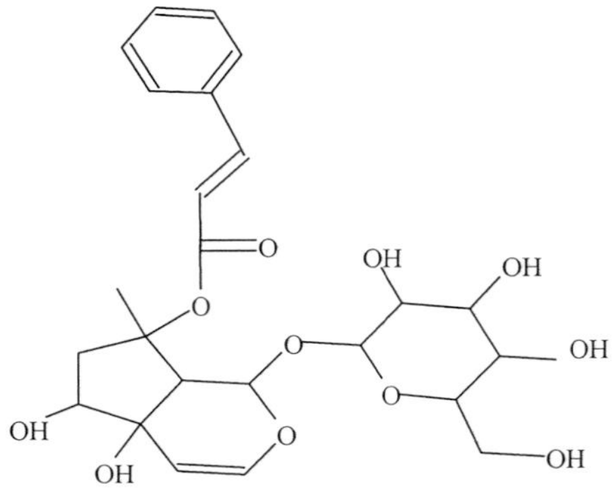   Harpagoside

Fig. (15). Chemical structures of dual active iridoids.

6. Sesquiterpenes

More than 11,000 Sesquiterpenes have been isolated from natural sources [3]. Feverfew (*Tanacetum parthenium* (L.) Sch. Bip., Compositae) is a common herb utilized for fever, arthritis and migraine. The leaves of feverfew contain large amounts of sesquiterpene lactones. The chloroform extracts of fresh leaves and a commercial leaf product show dose-dependent inhibiting the production of throboxane B$_2$ (TXB$_2$) and leukotriene B$_4$ (LTB$_4$) in rat and human leukocytes. Sesquiterpene lactones isolated from the leaves of feverfew, such as parthenolide and tanaparthin-α-peroxide, demonstrate potent dual inhibitions of COX and LOX pathways with IC$_{50}$ for COX at 6 & 17 μg/ml, and for LOX at 12 & 17 μg/ml, respectively [135].

Buddleja globosa Hope (Buddlejaceae) from Chili has been utilized for wound healing and treatment of ulcers. A sesquiterpene compound called buddledin A was isolated from the plant and showed dual inhibitions of COX with IC$_{50}$ = 13.7 μM and of 5-LOX with IC$_{50}$ = 50.4 μM [136].

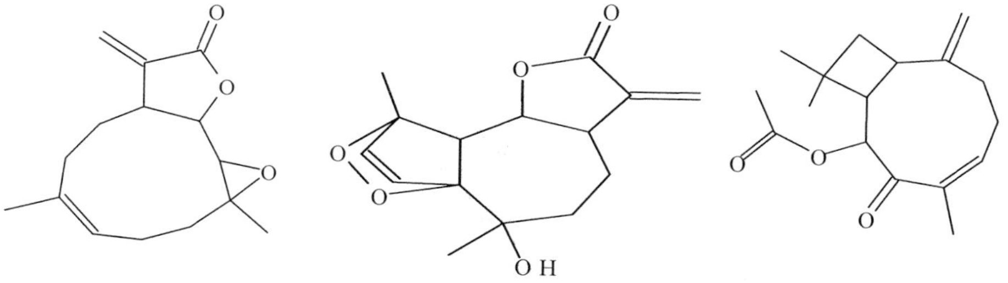

| Parthenolide | Tanaparthin-α-peroxide | Buddledin A |

Fig. (16). Chemical structures of dual active sesquiterpenes.

7. Diterpenes

Diterpenes also represent a group of 10,541 known natural compounds [3]. The biological activities for diterpenes include antiviral, antimicrobial, antiinflammation and anti-tumor activities. Some diterpenes reported as COX and LOX dual inhibitors as listed in Table 3. Clerodane diterpenes, such as teucrin A from *Teucrium* species (Labiatae), are not COX inhibitors, but potent LOX antagonists. The abietane diterpenoids from *Salvia aethiopis* (Labiatae) significantly inhibited PGE$_2$ and LTC$_4$ production from both calcium ionophore A23187 and LPS induced mouse peritoneal macrophages. Lagascatriol isolated from *Sideritis angustifolia* (Labiatae) was selective against COX-1 pathway. However, this compound did not directly inhibit iNOS and COX-2 enzyme activities, but rather it modifies the induction of iNOS and COX-2 enzymes. The compound 11,12-Dihydroxy-6-oxo-8,11,13-abietatriene also affected NO production [137].

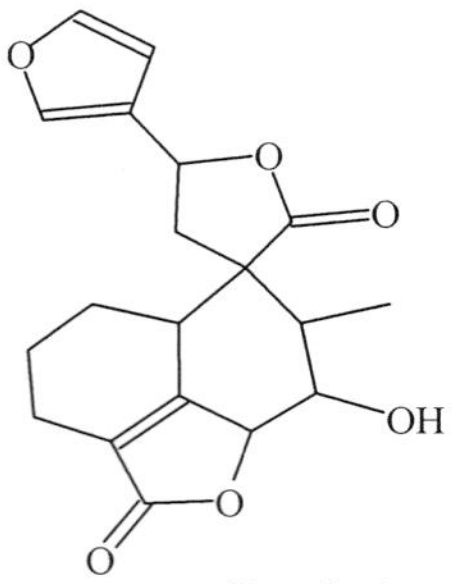

Teucrin A

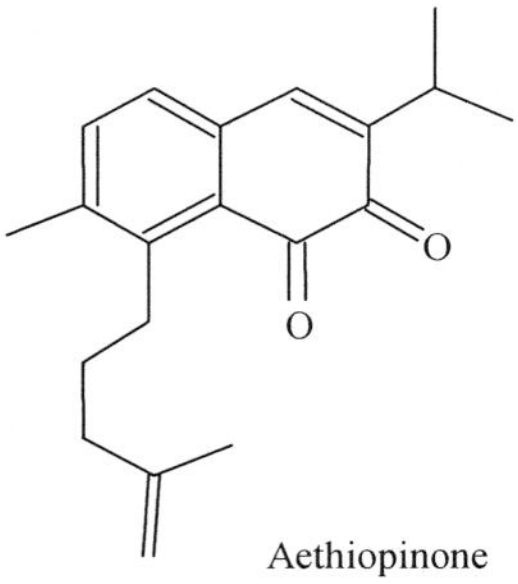

Aethiopinone

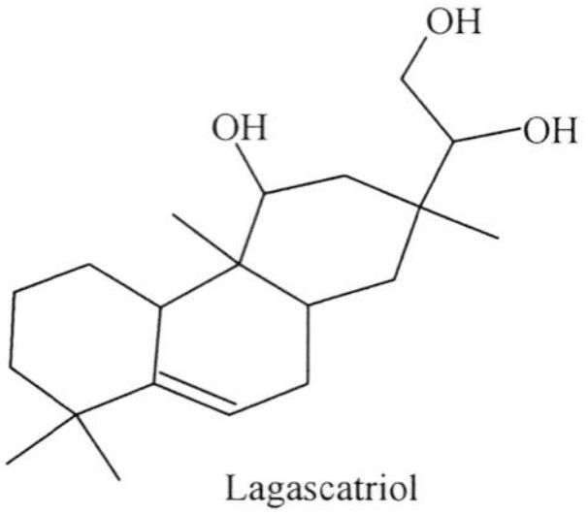

Lagascatriol

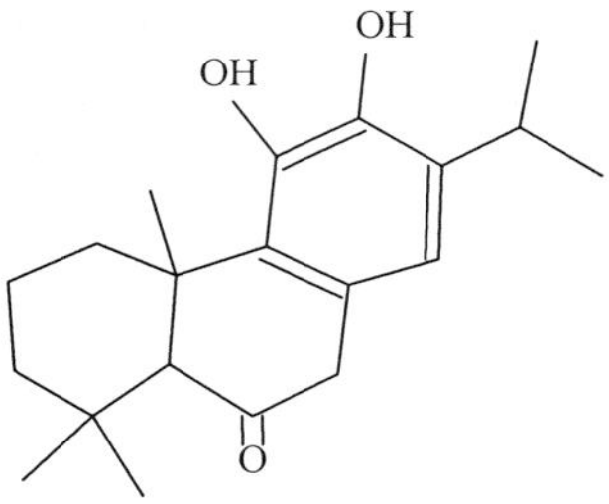

11,12-Dihydroxy-6-oxo-8,11,13-abietatriene

Fig. (17). Chemical structures of dual active diterpenes

Table 3. **COX and LOX dual inhibition activities from diterpenes.**

Diterpenes	IC_{50} (COX-1, μM)	IC_{50} (COX-2, μM)	IC_{50} (LOX, μM)
Teucrin A	not active	not active	4.51
Aethiopinone	6.25	/	4.39
Lagascatriol	3.12	12.5	2.92
11,12-Dihydroxy-6-oxo-8,11,13-abietatriene	6.25	6.25	2.10

O-Methyl-pseudopterosine is a semisynthetic derivative of pseudopterosine A isolated from the soft coral *Pseudopterogorgia elizabethae*. It belongs to the amphilectane diterpenoids with topical anti-inflammatory activity. The mechanism of anti-inflammation has been investigated with enzyme inhibition assays targeting four arachidonic acid metabolic pathways and a complimentary assay determining the potential inhibition of interleukine-1β stimulated prostaglandin E_2 production. It was found that O-Methyl-pseudopterosine inhibited COX-1 (IC_{50} = 80.3 μM) but not COX-2 enzyme activities. It was a weak inhibitor against 5-LOX (IC_{50} = 253 μM) and not at all effective against 15-LOX enzymes. However, IL-1β (1 ng/ml) stimulated PGE_2 production was inhibited about 46% by 22.4 μM of O-Methyl-pseudopterosine [138].

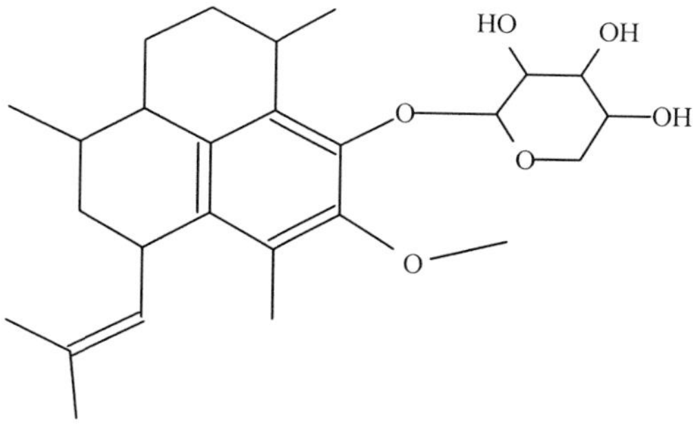

Fig. (18). Chemical structure of O-Methyl-pseudopterosine

E-isolinaridial and *E*-isolinaridial methylketone are two neo-clerodane diterpenoids isolated from the aerial parts of *Linaria saxatilis var. glutinosa* (Schrophulariaceae). Benrezzouk *et al.* evaluated these compounds on multiple inflammation pathways and found that they were not COX-1/COX-2 inhibitors [139]. However, *E*-isolinaridial and *E*-isolinaridial methylketone were dual inhibitors against purified phospholipase A₂ and 5-lipoxygenase. *E*-isolinaridial and *E*-isolinaridial methylketone showed potent inhibition of cell-free 5-LOX enzyme with IC_{50} values of 0.42 μM and 1.41 μM, respectively. These compounds also decreased A23187 induced neutrophils leukotriene B_4 production with IC_{50} values of 0.20 μM and 1.32 μM , respectively. Both compounds inhibited human synovial $sPLA_2$ enzyme in a dose dependent manner with IC_{50} values of 0.20 μM and 0.49 μM, respectively.

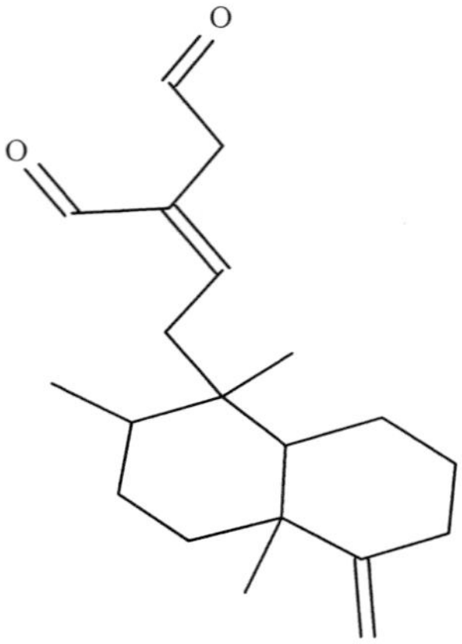

E-Isolinaridial 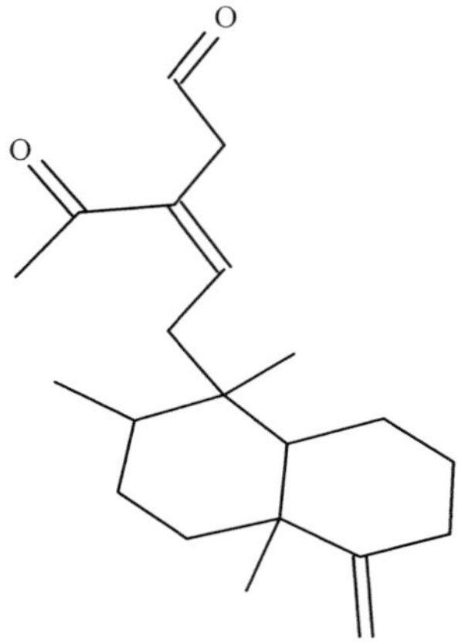   E-Isolinaridial methylketoine

Fig. (19). Neo-clerodane diterpenes from *Linaria saxatilis var. glutinosa.*

Using the same strategy, Benrezzouk *et al.* also reported a naphthoquinone diterpene, aethiopinone, from the roots of *Salvia aethiopis* L. (Labiatae), possessed 5-LOX inhibition activity with IC_{50} value of 0.11μM without affecting COX and PLA_2 pathways [140].

Fuscoside B is a lobane diterpenoid isolated from the Caribbean gorgonian *Eunicea fusca*. It has been evaluated for the inhibitory activities against COX-1, 5-LOX, 12-LOX, PLA_2, and other eicosanoid biosynthesis pathways and found only to irreversibly inhibit the 5-LOX pathway with an IC_{50} value of 10 μM. [141,142]

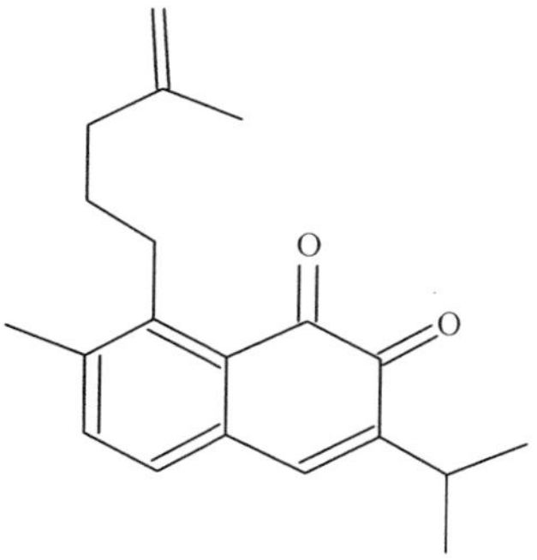

Aethiopinone

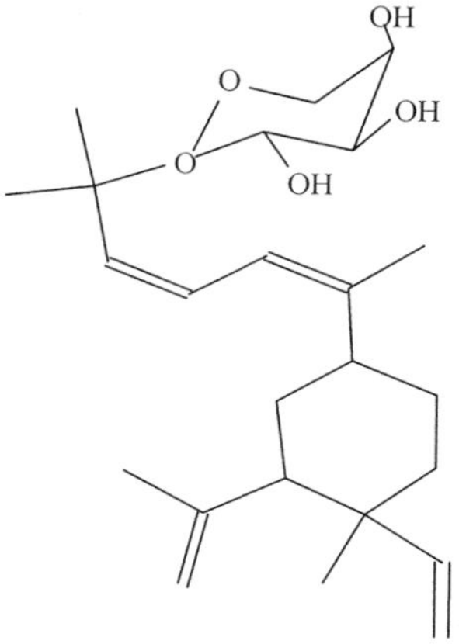

Fuscoside B

Fig. (20). Chemical structures of naphtoquinone and lobane diterpenes

8. Triterpenoids

Triterpenoids represent a class of over 9300 known natural compounds with 30 carbons and a multiple-ring skeleton [3]. A pentacyclic triterpenoid, ursolic acid, was isolated from the flowers of *Calluna vulgaris* (L.) Hull (Ericaceae) by following the 5-LOX enzymatic inhibitory assay. Using the radiochromatography method to quantify the effects of ursolic acid on arachidonic acid metabolite in different blood cells, it was found that ursolic acid was a more potent inhibitor of the 5-LOX pathway than that of 12-LOX and 15-LOX pathways in macrophages. The COX pathway also was inhibited but to a much lower extent (38% inhibition at 1μM). In differenciated HL60 leukemic cells, ursolic acid at a concentration of 1μM decreased 5-HETE production by 86%, 15-HETE by 59% and prostaglandins by 50% [143].

Boswelic acids are pentacyclic triterpenoids isolated from gum resin secreted by trees of *Boswellia serrata* Roxb. (Burseraceae). β-Boswelic acid and acetyl-11-keto-β-boswelic acid showed dual inhibition of 5-LOX and human leukocyte elastase (HLH) but had no impact on the COX pathway [144].

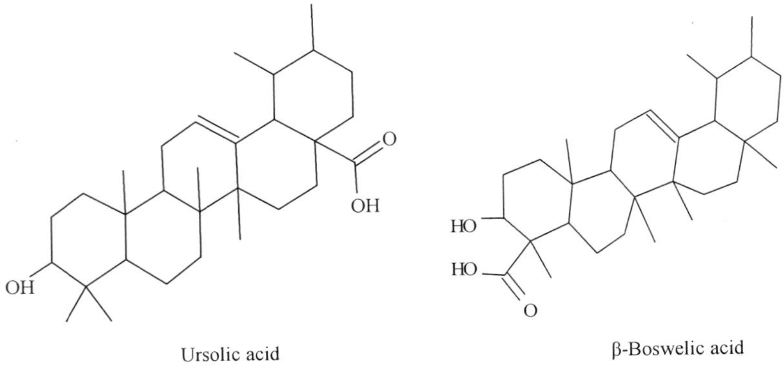

| Ursolic acid | β-Boswelic acid |

Fig. (21). Chemical structures of two dual active triterpenens

9. Alkaloids

The dual inhibitory activities of plant extracts have been reported in many different species of plants. The root bark from *Zanthoxylum chalybeum* Engl. (Rutaceae) has been utilized in Africa to reduce swelling. The dichloromethane extract showed significant inhibition of the 5-LOX pathway (70%) and COX pathway (26%) at a concentration of 25 μg/ml. The benzophenanthridine alkaloids in the extract might be the actives [145]. Most of the published alkaloids with potent COX and LOX dual inhibition activities were derived from quantitative structure-activity relationship studies and total syntheses. It is possible to convert known COX inhibitors like etodolac by replacing hydrogen with methoxyquinoline to yield a potent COX and LOX dual inhibitor, 6-(2-quinolinylmethoxy) etodolac [146]. Dihydropyrrolizine is a novel class of nonantioxidant dual inhibitors of COX and 5-LOX enzymes. (2,3,-Dihydro-1H-pyrrolizin-5-yl)acetic acid with para-substituent of phenyl and 4-methoxyphenyl groups was the most potent dual inhibitor of COX-1 and 5-LOX with IC_{50} values of 0.1 μM and 0.24 μM, respectively [147].

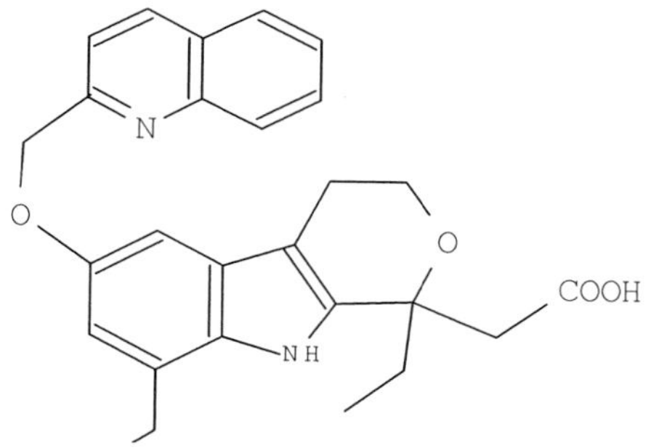

6-(2-quinolinylmethoxy)-etodolac 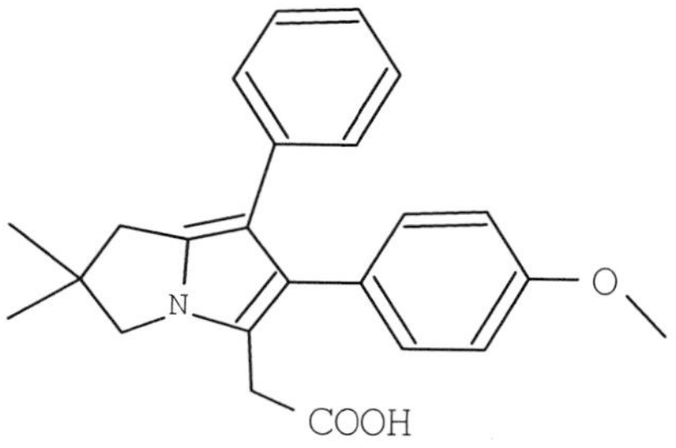   (2,3,-Dihydro-1H-pyrrolizin-5-yl) acetic acid

Fig. (22). Chemical structures of two synthetic dual COX and LOX inhibitors

The synthesized pyrroloquinazoline alkaloid as shown in Fig. 23 selectively inhibited COX-2 (IC$_{50}$ = 1.2 μM) over COX-1 (IC$_{50}$ > 10 μM) activity in human monocyte assays. In a purified ovine enzyme assay, the COX-1 activity was not affected even at a concentration over 50 μM compared to an IC$_{50}$ value of 20.5 μM for COX-2 inhibition. The LOX efficacy was demonstrated by measuring LTB$_4$ production in a mouse air pouch model [148].

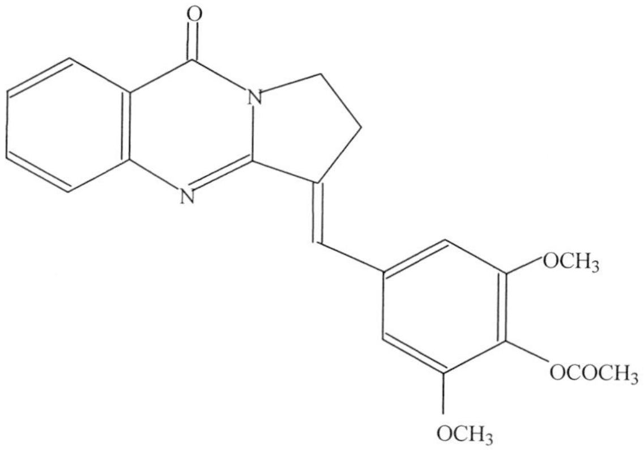

Fig. (23). Structure of 3-(4'-acetoxy-3',5'-dimethyl)benzylidene-1,2-dihydropyrrolo[2,1-b]quinazoline-9-one

10. Quinones

There are more than 300 known quinones isolated from natural sources [3]. With comparatively low molecular weight and polarity, quinones usually exist in oils or lipophilic extracts of plants. The seeds of *Nigella sativa* L. (Ranunculaceae) have been utilized in the Middle East for pain and stiffness in the joints and for the treatment of bronchial asthma and eczema. From the fixed oil of the seeds, a COX and LOX dual inhibitor, thymoquinone, was isolated. It demonstrated potent and dose related reductions of TXB_2 and LTB_4 releasing from rat leukocytes stimulated with calcium ionophore A23187. The IC_{50} values against COX and LOX pathways were 3.5 μg/ml and < 1 μg/ml, respectively. Thymoquinone also showed potent effectiveness (IC_{50} = 0.81 μg/ml) against Fe^{3+}-ascorbate induced peroxidation of phospholipid liposomes in ox brain [149].

The rhizomes of *Atractylodes lancea* (Thunb.) DC. and *Atractylodes chinensis* (Bunge) Koidz. (Compositae) have been utilized in China for treatment of rheumatic diseases, digestive disorders, mild diarrhea and influenza. Bioassay guided separation and purification of active components from the hexane extract of *Atractylodes lancea* yielded two quinone type compounds, atractylochromene and 2-[(2E)-3,7-dimethyl-2,6-octadienyl]-6-methyl-2,5-cyclohexadiene-1,4-dione. As a potent COX-1 and 5-LOX dual inhibitor, atractylochromene had IC_{50} values of 3.3 μM and 0.6 μM, respectively. 2-[(2E)-3,7-dimethyl-2,6-octadienyl]-6-methyl-2,5-cyclohexa-diene-1,4-dione was a more selective inhibitor against 5-LOX with IC_{50} = 0.2 μM, but still maintained weak inhibition of COX-1 pathway (IC_{50} = 64.3 μM) [150]. A series of similar compounds, prenyl-hydroquinones have been synthesized as structural analogs of marine compounds isolated from the sponge *Ircinia spinosula* [151]. For mono- to tetra- prenyl-1,4-hydroquinone, the IC_{50} values of 5-LOX enzyme inhibition were from 0.2 μM to 7.0 μM. Those hydroquinones also inhibited PGE_2, TNFα and nitrite productions stimulated by lipopolysaccharide (LPS) in J774 cells. However, prenyl hydroquinones showed no direct inhibition on COX-1 and COX-2 enzymes [152].

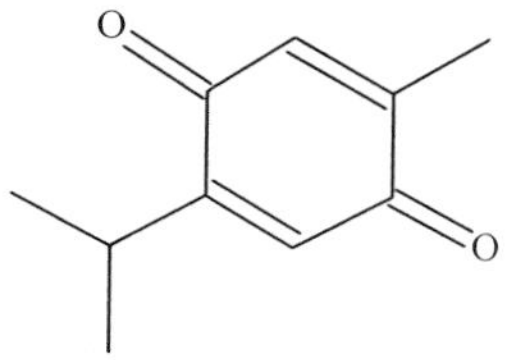

Thymoquinone   Atractylochromene

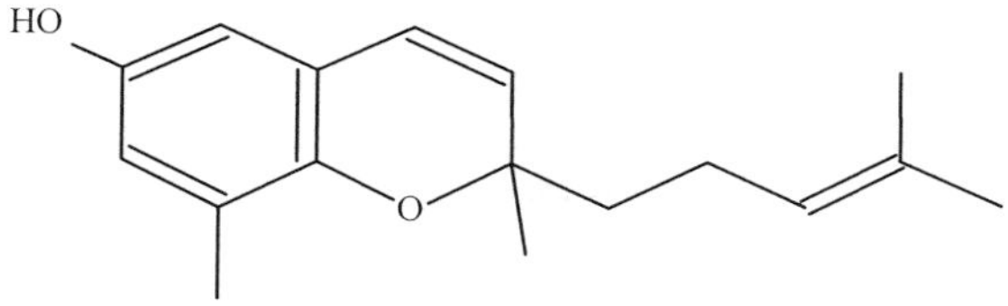

Fig. (24). Chemical structures of two dual active quinones.

11.   Phenolic Compounds

Curcumin, originally isolated from an Indian medicinal plant, *Curcuma longa* L. (Zingiberaceae) is the major yellow pigment in turmeric, curry and mustard. Curcumin is one of the most extensively studied natural phenolic compounds with anti-inflammatory [153], anti-oxidant [154], and chemopreventive activities [155,156,157]. Curcuminoids and other diarylheptanoids were characterized as dual inhibitors of arachidonic acid metabolism by reducing PGE_2 production in the cyclooxygenase pathway and inhibiting 5-HETE production in human neutrophils with same IC_{50} values of 8.0 μM [158].

Based on the structure of curcumin, a series of styrylpyrazoles, styrylisoxazole, and styrylisothiazoles were synthesized and found to be dual inhibitors of COX and 5-LOX tested in rat basophilic leukemia cells. The most potent COX and 5-LOX dual inhibitor had IC_{50} values at 0.9 μM and 1.2 μM, respectively [159].

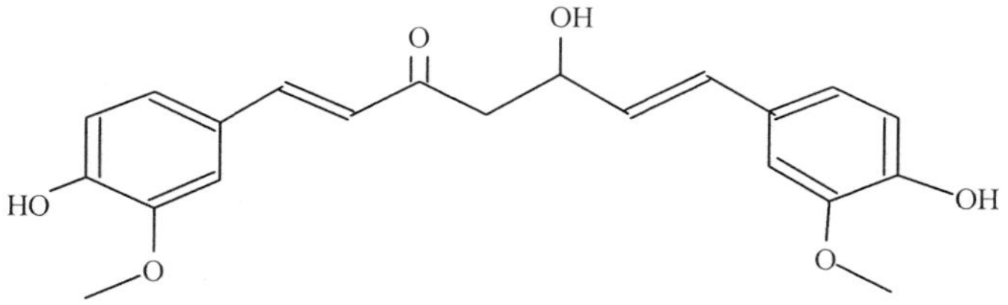

Fig. (25). Chemical structure of Curcumin

Notopterygium incisum T. (Umbelliferae) has been widely used in China for treatment of headaches, common colds, and rheumatism. The hexane extract of the underground parts from the plant showed dual inhibitions against purified COX enzyme and 5-LOX pathway using intact porcine leukocytes. Three active compounds were isolated from the extract. The phenolic compound, phenethyl ferulate, was the most potent dual inhibitor with IC_{50} values against COX and 5-LOX of 4.35 μM and 5.75μM, respectively. The dual inhibitory activities of bornyl ferulate were about two times lower than phenethyl ferulate with IC_{50} values against COX and 5-LOX of 12.0 μM and 10.4 μM, respectively. It was evident that the bulky monoterpene ring in bornyl ferulate decreased enzyme-binding capacity compared to the flat aromatic ring. The third active component was a linear acetylenic compound called falcarindiol with 5-LOX selectivity (IC_{50} = 9.4 μM) over COX activity (IC_{50} = 66 μM) [160].

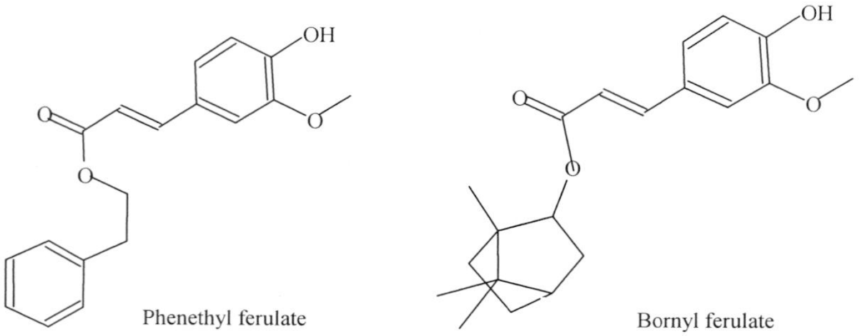

Phenethyl ferulate   Bornyl ferulate

Fig. (26.). Chemical structures of two dual active ferulates.

Macfadyena unguis-cati (L.) A. Gentry (Bignoniaceae) is a variable climbing plant widespread in tropical America and in Western India. This plant has been utilized for snakebite, dysentery, rheumatism, in the treatment of venereal disease and malaria. The COX-1 and 5-LOX dual inhibitons were observed on crude extracts that contained two caffeic esters, chlorogenic acid and ioschlorogenic acid [161]. Similar activities have been observed on phenolic compounds isolated from virgin olive oil. The most potent 5-LOX inhibitor isolated was hydroxytyrosol with IC_{50} at 15 μM [162].

Chlorogenic acid is one of the active components in a herb *Urtica dioica* L. (Urticaceae) and has been utilized for the treatment of rheumatoid arthritis. Chlorogenic acid from the plant demonstrates moderate inhibitions against the 5-LOX pathway with $IC_{50} = 83$ µM and COX pathway with $IC_{50} = 38$ µM [163]. However, caffeic acid has been reported as a weak inhibitors against 5-LOX ($IC_{50} = 200$ µM) and showed no inhibition on COX pathway [164].

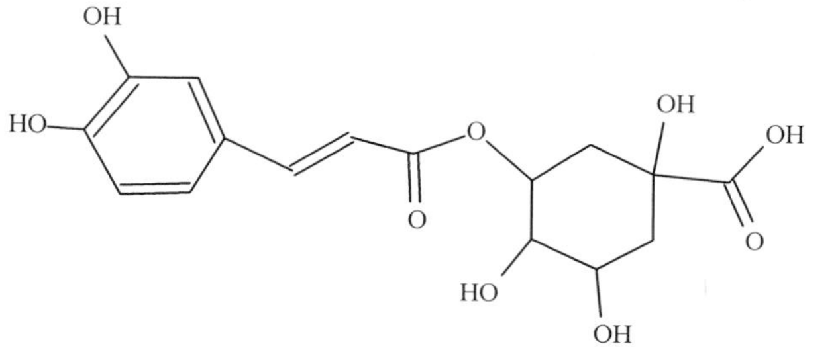

Fig. (27). Chemical structure of Chlorogenic acid

Gossypol has been isolated from the roots, stems, and seeds of the cotton plant and reported as a COX inhibitor based on *in vitro* assay in neutrophils [165] and an inhibitor of LOX and prostaglandin synthase from rat basophilic leukemia cells [166]. Gossypol showed as a potent inhibitor of leukotrienes B_4 and D_4 production and also reduced PTF-induced contraction of guinea-pig lung parenchyma [167].

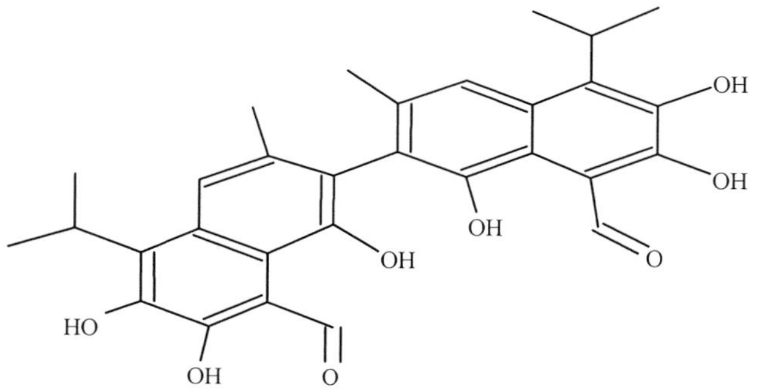

Fig. (28). Chemical structure of Gossypol

Hinokitiol is a tropolone type natural compound isolated from the wood of *Chymacyparis taiwanesis*. The compound has been utilized as a natural anti-microbial agent in hair tonics, toothpastes, cosmetics and food supplements. Hinokitiol was evaluated on five different arachidonic acid metabolic pathways for the mechanism of action of anti-inflammatory effects. It has been found to be a potent inhibitor with IC_{50} values of 0.1 µM against platelet-type 12-LOX and 50 µM against leukocyte-type 12-LOX. It also inhibited soybean 5-LOX enzyme (IC_{50} = 17 µM). However, hinokitiol had almost no effects on COX-1 and COX-2 enzymes. Similar inhibition profiles were also observed on synthetic tropolone derivatives [168].

2-Substituted-1-naphthols are synthetic phenolic compounds with dual inhibitory activities. The most active compound, 2-allyl-1-naphthol, is a very potent 5-LOX inhibitor with an IC_{50} value in nanomolar range and also has moderate COX enzyme inhibition activity. The compound also possesses excellent topical anti-inflammatory potency in the mouse ear edema model [169].

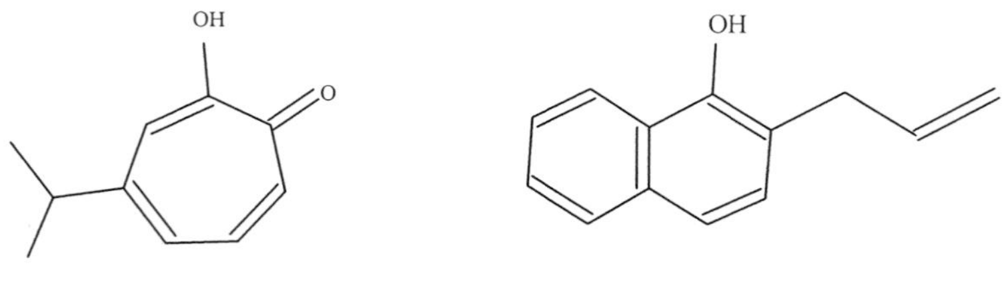

Hinokitiol   2-Allyl-1-naphthol

Fig. (29). Chemical structure of dual active tropolone and naphthol compounds

12. Coumarins

Coumarins are widely distributed in high plants with more than 3,560 compounds isolated [3]. Even though many coumarin-containing plants have been utilized for inflammation-related diseases and conditions, there are not many coumarins act as potent COX & LOX dual inhibitors. Osthol, osthenol from *Angelica pubescens F.* Biserrata (Apiaceae) [170], esculetin, fraxetin and dephnetin have been characterized as dual inhibitors [171]. However, the dihydroxylated coumarins were more selective against 5-LOX and only a few coumarins showed potent inhibition against COX activity. 5,7-Dihydroxy-4-methylcoumarin is one of the few coumarins with potent COX selectivities (IC_{50} = 1 μM). Daphnetin had similar IC_{50} values against both LOX and COX enzymes [172]. 6-(3-Carboxybut-2-enyl)-7-hydroxy-coumarin, which was isolated from the roots of a central European perennial herb, *Peucedanum ostruthium* (L.) Koch (Apiaceae), strongly reduced PGE_2 production based on a rabbit ear-swelling model. The effective concentration for the coumarin was 1.3 μg/ml compared to 1 μg/ml for indomethacin. It also demonstrated inhibitory effect on the 5-LOX pathway with IC_{50} = 0.25 μM [173].

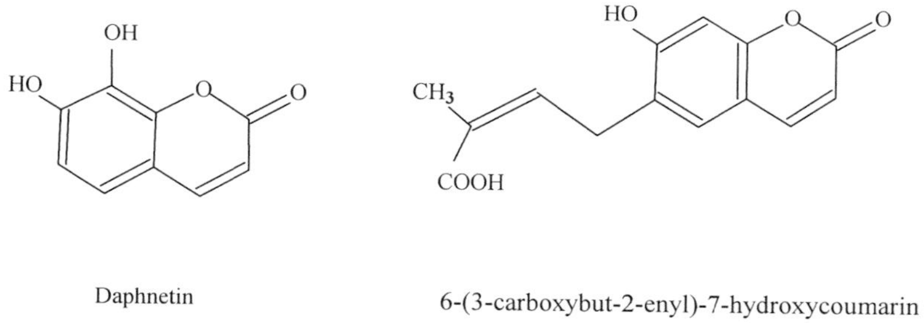

Daphnetin   6-(3-carboxybut-2-enyl)-7-hydroxycoumarin

Fig. (30). Chemical structures of two dual active coumarins.

In Hoult's report, sixteen natural and synthetic coumarins were evaluated in rat peritoneal exudate leukocytes for the impact on production of throboxane B_2 and leukotriene B_4. (Table 4.) Four coumarins exhibited inhibition of the 5-LOX pathway and had a similar structure feature containing ortho-dihydroxyl groups. The most potent 5-LOX inhibitors had 6,7-dihydroxyl groups with and without other substitutions. The coumarins with COX-1 inhibition activities were much less predictable. Surprisingly, 5,7-Dihydroxy-4-methylcoumarin was the most potent COX-1 inhibitor, however without 5-LOX activity. The only dual inhibitor was 7,8-Dihydroxycoumarin (Daphnetin) with moderate and equal potency against both pathways [174].

Table 4. COX-1 and 5-LOX inhibition activities of synthetic coumarins

Compound	IC50 (COX-1, µM)	IC50 (5-LOX, µM)
7-Methoxy-coumarin	53	Inactive at 100
7-Hydroxy-coumarin	Inactive at 100	Inactive at 100
6,7-Dihydroxy-coumarin	Inactive at 100	8.3
7,8-Dihydroxy-coumarin	21	20
6,7-Dihydroxy-4-methyl-coumarin	Inactive at 100	10
5,7-Dihydroxy-4-methyl-coumarin	1	Inactive at 100
7,8-Dihydroxy-4-methyl-coumarin	Inactive at 100	>100
7,8-Dihydroxy-6-methoxy-coumarin	Inactive at 100	75

13. Flavonoids

Flavonoids are very widely distributed in many families of high plants. Around 10,000 individual flavonoids have been isolated [3]. Flavonoids are contained in vegetables, fruits, seeds, nuts, grains, species, and teas. They are one of the most important ingredients in the human diets with an average intake of 100 – 800 mg/daily [175]. Flavonoids also attribute to major actives in many medicinal plants, such as Ginkgo, grape seeds, soybeans, and licorice [176]. Pharmacological studies have shown that flavonoids to be anti-oxidants, anti-inflammation, anti-microbial [177], anti-fungus, and anti-viral agents [178], and insect repellants. Many flavonoids exhibit enzyme inhibition activities, estrogen binding and gene expression regulation [179]. Flavonoids have also been utilized for improving coronary perfusion & peripheral vascular resistance, protecting against venous diseases, hepatic diseases, cancer, heart disease, stroke, and dementia [180].

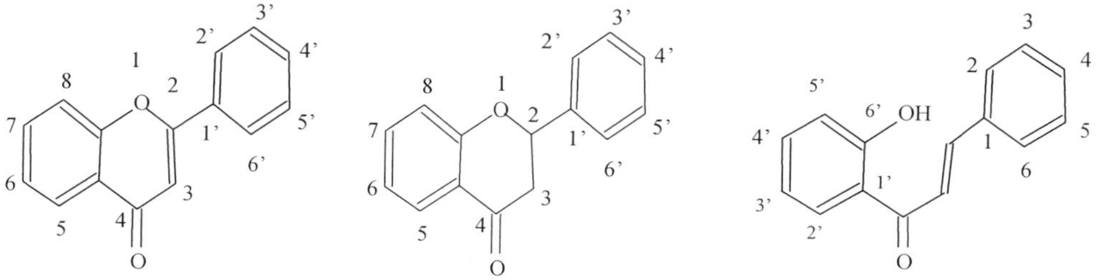

Flavone   Flavanone   Chalcone

Fig. (31). Structural skeletons of three types of flavonoids

The fruits of pomegranate (*Punica granatum*, L. Punicaceae) are popular in the East and far East countries. The dried pericarp has been utilized in folk medicine for colic, colitis, diarrhea, dysentery, leucorrhea, menorrhagia, oxyuriasis, paralysis and rectocele. The flavonoid extract from cold pressed seed oil showed 31-44% inhibition of sheep COX-1 enzyme and 69-81% inhibition of soybean lipoxygenase [181].

Hypolaetin-8-glucoside is a plant-derived flavonoid isolated from *Sideritis mugronensis*, which has been utilized in Spanish folk medicine as an antirheumatic and digestive agent. The compound showed anti-inflammatory and antigastric ulcer efficacy in rats. Moroney *et al.* compared hypolaetin-8-glucoside, hypolaetin, with 14 other flavonoids for inhibition of eicosanoid production via the 5-lipoxygenase and cyclooxygenase pathways [182]. As shown in Table 5, hypolaetin-8-glucoside was a moderate 5-LOX inhibitor without COX activity. Its aglycone, hypolaetin, showed potent 5-LOX inhibition with $IC_{50} = 4.5$ μM and also inhibited the COX pathway with a moderate potency. Three other glycoside/aglycone pairs (rutin/quercetin, gossypin/gossypetin, naringin/naringenin) also demonstrated the same pattern of weaker inhibitions by glycosides (Table, 5, 6, and 7.)

Table 5.   Dual COX and LOX inhibition activities from flavones.

Compound	Substitution Position	IC_{50} (COX, μM)	IC_{50} (LOX, μM)	Ref.
Flavone	None	8	32	[182]
Crysin	5,7-OH	5	18	[182]
Kuwanon C	5,7,2',4'-OH 6,8-diprenyl	62	12	[189]
Hypolaetin	5,7,8,3',4'-OH	70	4.5	[182]
6-Methoxy luteolin	5,7,3',4'-OH 6-methoxy	61	97	[185]
6-Methoxy luteolin-3'-methyl ether	5,7,4'-OH 6,3'-dimethoxy	89	84	[185]
Hypolaetin-8-glucoside	5,7,3',4'-OH 8-sugar,	>1000	56	[182]
Morusin	5,2',4'-OH 3-isoprenyl 7,8-b-dimethylpyran	1.6	2.9	[190]
Artonin E	5,2',4',5'-OH 3-isoprenyl 7,8-b-dimethylpyran	2.5	0.36	[190]

It was evident that the polyhydroxylated flavones and flavonols had the most potent 5-LOX activities and good selectivities for COX inhibitions. Hydroxyl group positions of the B-ring were critical with enhanced 5-LOX potency by 3',4'-di-substitutions, and detrimental effects caused by 2'-substitutions. A-ring substitution groups on the 5- and/or the 7- positions were also important to maintain 5-LOX activities. For COX inhibition activities, few or free substitutions in the B-ring would enhance COX potency and selectivity [182]. The Free-B-Ring flavonoid, Gnaphalin, which was isolated from the Mediterranean herb *Helichrysum picardii* Boiss & Reuter (Asteraceae), showed two-fold more potency against TXB_2 production from COX-1 pathway than the inhibition of the 5-LOX metabolite LTB_4 release [183]. Sobottka et al. evaluated the anti-inflammatory effects of synthetic flavonols and found similar results [184].

Highly methylated, lipophilic flavonoids have been isolated from feverfew plants *Tanacetum parthenium* (L.) Schultz Bip. (Asteraceae) and *Tanacetum vulgare* L. (Compositae). 6-Methoxyleuteolin and 6-hydroxyleuteolin-6,3'-dimethylether showed moderate COX and LOX dual inhibition activities. (Table 5.). Polymethyl ethers of 6-hydroxykaepferol had selective inhibitory COX activities with IC_{50} values around 25 μM whereas both compounds were two to seven times less potent against the 5-LOX pathway (Table 6.) [185]. Two flavonoids, cenaureidin and 5,3'-dihydroxy-4'methoxy-7-carbomethoxyflavonol, were isolated from *Tanacetum microphyllum* DC. (Asteraceae) and showed inhibitory activities against soybean lipoxygenase with IC_{50} values of 20 μM and 29 μM, respectively. Both flavonols showed moderate inhibition of cyclooxygenase activities (Table 5.) [186].

Since both cyclooxygenase and lipoxygenase catalyze stereospecific free radical peroxidations of arachidonic acid at their active sites, it is widely believed that the dual COX and LOX inhibition activities of flavonoids derive from their non-selective free radical scavenging or iron chelating functions rather than specific enzyme inhibition. However, Laughton et al. [187] found that the antioxidant activities of flavonoids as measured in the microsomal system were not significantly correlated ($r = 0.37$, $P > 0.05$) with lipoxygenase inhibition. There was also no correlation between antioxidant activity and ability to inhibit cyclooxygenase ($r = 0.05$, $p > 0.05$). This was consistent with a previous report of non-correlation between inhibiting low-density lipoproteins and their potencies of COX and 5-LOX inhibitions [188]. The influence of flavonoids on iron ion-dependent damage of DNA was measured to compare with inhibitory activities against COX and LOX. Flavonoids, which were potent and selective cyclooxygenase inhibitors, were inactive in the assay and not iron chelators. Flavonoids which were proved to be potent and selective 5-LOX inhibitors, promoted DNA damage initially at low concentrations, but then declined markedly with further increase in concentrations. Therefore, there were strong correlations between 5-LOX selective inhibition of the flavonoids with their abilities to reduce ferricytochrome c at low concentrations and iron chelating capacity at high concentrations.

Table 6. **Dual COX and LOX inhibition activities of flavonols**

Compound	Substitution Position	IC$_{50}$ (COX, μM)	IC$_{50}$ (LOX, μM)	Ref.
3-Hydroxy-flavone	3-OH	1	16	[182]
4'-hydroxy-flavonol	3,4'-OH	7.3	Inactive	[184]
4'-methoxyflavonol	3-OH 4'-methoxy	7.2	20% 10μM	[184]
2'-methoxyflavonol	3-OH 2'-methoxy	3.3	Inactive	[184]
3'-hydroxy-flavonol	3,3'-OH	41% 10μM	Inactive	[184]
Galangin	3,5,7-OH	7	20	[182]
7,4'-Dimethoxy-flavonol	3-OH 7,4'-dimethoxy	25% 10μM	8.7	[184]
Kaempferol	3,5,7,4'-OH	20	20	[182]
Fisetin	3,7,3',4'-OH	80	11	[182]
Gnaphalin	5,7-OH 3,8-dimethoxy	40	80	[183]
Morin	3,5,7,2',4'-OH	180	160	[182]
Quercetin	3,5,7,3',4'-OH	16	3.5	[182]
6-Methoxy-Kaempferol-3,4'-dimethyl ether	5,7-OH 3,6,4'-Trimethoxy	27	58	[185]
Centaureidin	5,7,3'-OH 3,6,4'-trimethoxy	318	20	[186]
5,3'-dihydroxy-4'-methoxy-7-carbomethoxyflavonol	3,5,3'-OH 4'-methoxy, 7-carbomethoxy	60	29	[186]
Gossypetin	3,5,7,8,3',4'-OH	inactive	10	[182]
Myricetin	3,5,7,3',4',5'-OH	56	13	[182]
Quercetagetin 3,6,3'-Trimethyl ether	5,7,4'-OH 3,6,3'-methoxy	22	167	[185]
Rutin	5,7,3',4'-OH 3-sugar	450	45	[182]
Gossypin	3,5,7,3',4'-OH 8-sugar,	630	250	[182]

Prenylated flavonoids contain one or two isoprenyl, geranyl, dimethylallyl, and lavandulyl on the skeleton. They have limited distribution and are mostly isolated from the Moraceae family. Chi *et al.* isolated 19 prenylated flavonoids from six different genera of Chinese medicinal plants and evaluated inhibition activities on eicosanoid metabolisms with multiple cell line models [189]. Two 8-lavandulylated flavanones, kurarinone and sophoraflavanone G, were discovered to be dual inhibitors. They possessed the most potent COX-1 inhibitory activities with IC_{50} less than 1 μM comparable with that of indomethacin. Sophoraflavanone G isolated from the roots of *Sophora flavescense* Art (Fabaceae) also had potent 5-LOX inhibition activity with an IC_{50} below 0.25 μM.

Table 7.　　　　**Dual COX and LOX inhibition activities of flavanones.**

Compound	Substitution Position	IC_{50} (COX, μM)	IC_{50} (LOX, μM)	Ref.
Naringenin	5,7,4'-OH	100	16	[182]
Naringin	7,4'-OH 7-sugar,	320	>500	[182]
Sophoraflavanone G	5,7,2',4'-OH 8-lavandulyl	0.1-0.6	0.09-0.25	[189]
Kurarinone	7,2',4'-OH 5-methoxy, 8-lavandulyl	0.6-1	22	[189]
Sophoraflavanone D	5,7,2',4',6'-OH 6-geranyl	inactive	>100	[189]
Kenusanone C	3,5,7,2',4',6'-OH 6-geranyl, 8-isoprenyl	>100	inactive	[189]
5-Methyl-sophoraflavanone B	7,4'-OH 5-OMe, 8-isoprenyl	>100	>100	[189]

Sanggenon B and sanggenon D are dihydroflavonols isolated from the root bark of *Morus mongolica* Schneider (Moraceae). Both compounds are potent 5-LOX inhibitors with IC_{50} values of 6 μM and 4 μM, respectively. Sanggenon B and sanggenon D inhibited COX-1 enzyme with $IC_{50} = 42$ μM and 59 μM, respectively, and were weak COX-2 inhibitors with $IC_{50} > 100$ μM and 73 μM, respectively [189]. Psoralidin is a coumestan flavonoid obtained from the seeds of *Psoralea corylifolia* L. (Fabaceae). Psoralidin showed potent dual inhibition against COX-1 and 5-LOX with IC_{50} values of 23 μM and 8.8 μM, respectively. [189]

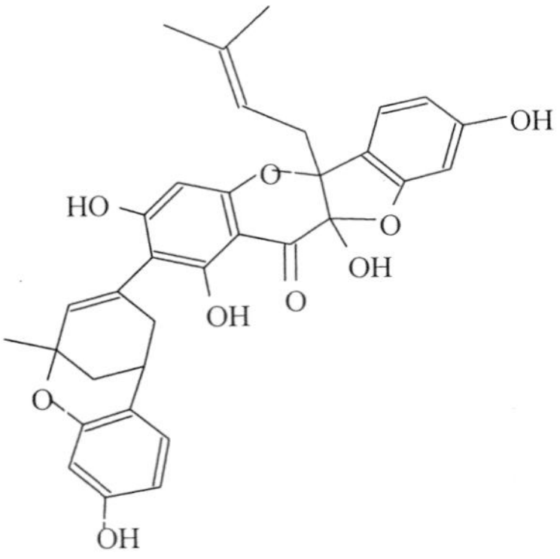

Sanggenon B

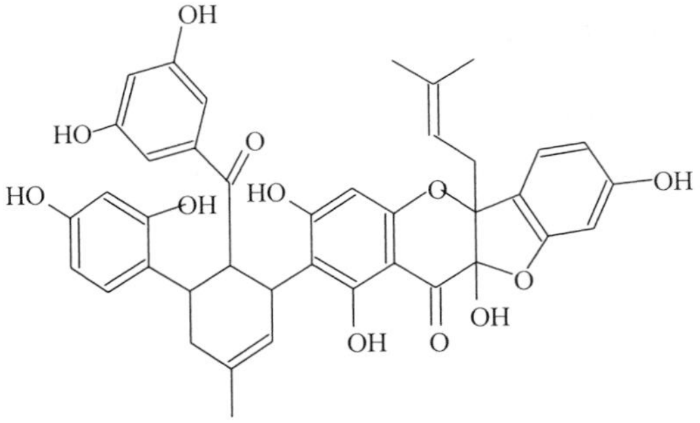

Sanggenon D

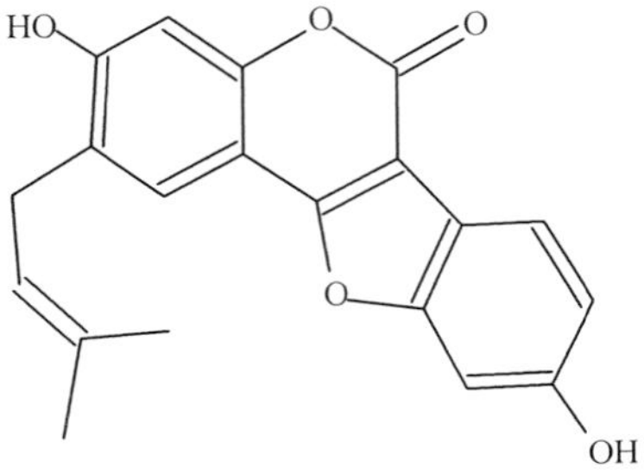

Psoralidin

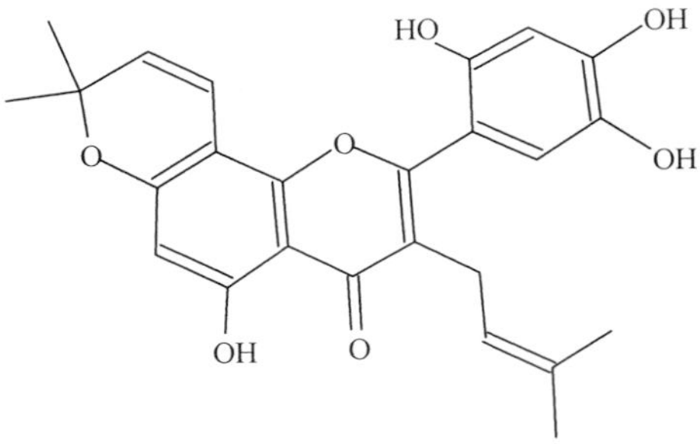

Artonin E

Fig. (32). Chemical structures of four dual active flavonoids.

All eight natural prenyl-flavones evaluated by Reddy *et al.* showed potent inhibitory effects on 5-lipoxygenase with IC_{50} from 0.36 to 4.3 μM. Artonin E isolated from the bark of *Artocarpus communis* Forst. (Moraceae) was the most potent dual LOX and COX inhibitor with IC_{50} values of 0.36 μM and 2.5 μM, respectively. It also inhibited leukocyte and platelet 12-

lipoxygenases (IC_{50} = 2.3 µM and 11 µM), and 15-lipoxygenase (IC_{50} = 2.3 µM) from rabbit reticulocytes [190].

Chalcones with 3,4-dihydroxycinnamoyl structures have been reported to inhibit 12-lipoxygenase and cyclooxygenase in the mouse epidermis [191] and to act as anti-allergic agents [192]. A total of fifty-three 3,4-dihydroxychalcones were synthesized to evaluate inhibitory activities against cyclooxygenase and 5-lipoxygenase. Chalcones with 3,4-dihydroxyl substitutions showed very potent 5-LOX inhibition activities with IC_{50} values in nanomolar range (Table 8.) and moderate COX inhibitions. The 5-LOX inhibition activities were directly associated with inhibition of lipid peroxidation that lead to a perception of the mechanism of action as antioxidation/free radical scavenging. The COX activities were remarkably increased by alkoxyl substitutions at 2'-, or 3'- or 2',5'- positions with 2',5'-dimethoxy-3,4-dihydorxychalcone as one of the most potent dual inhibitors [193]. A prenylated chalcone, kuraridin, from the roots of *Sophora flavescense* Art (Fabaceae) also had potent dual COX and 5-LOX inhibition activities with IC_{50} values of 1 µM and 6.9 µM, respectively [189].

Table 8.   Dual COX and LOX inhibition activities of chalcones.

Compound	Substitution Position	IC_{50} (COX, µM)	IC_{50} (LOX, µM)	Ref.
3,4-Dihydroxy chalcone	3,4-OH	34	0.034	[193]
2',3,4-Trihydroxy chalcone	2',3,4-OH	44	0.023	[193]
3,4-Dihydroxy-2'-methoxy-chalcone	3,4-OH 2'-methoxy	13	0.027	[193]
3,3',4-Trihydroxy chalcone	3,3',4-OH	140	0.0042	[193]
3,4-Dihydroxy-3'-methoxy-chalcone	3,4-OH 3'-methoxy	15	0.0065	[193]
3,4,4'-Trihydroxy-chalcone	3,4,4'-OH	320	0.040	[193]
2',4-Dihydroxy-3-methoxy-chalcone	2',4-OH 3-methoxy	17	63	[193]
3,4-Dihydroxy-2',5'-dimethoxy-chalcone	3,4-OH 2',5'-dimethoxy	9.2	0.0078	[193]
2',3,4-Trihydroxy-5'-methoxychalcone	2',3,4-OH 5'-methoxy	41	0.0041	[193]
Kuraridin	2,4,2'4'-OH 3'-lavandulyl, 6'-OMe	1	6.9	[189]

Catechin and its isomer epicatechin inhibit prostaglandin endoperoxide synthase with an IC_{50} value of 40 µmol/L [194]. Five flavan-3-ol derivatives, including (+)-catechin and gallocatechin, isolated from four plant species: *Atuna racemosa, Syzygium carynocarpum, Syzygium malaccense* and *Vantanea peruviana* exhibit equal to weaker inhibitory activities against COX-2 with IC_{50} values ranging from 3.3 µM to 138 µM [195]. (+)-Catechin, isolated from the bark of *Ceiba pentandra*, inhibits COX-1 with an IC_{50} value of 80 µM [196]. Commercially available pure (+)-catechin inhibits COX-1 with an IC_{50} value of around 183 to 279 µM depending upon the experimental conditions, with no selectivity for COX-2 [17].

Catechin derivatives purified from green tea and black tea, such as epigallocatechin-3-gallate (EGCG), epigallocatechin (EGC), epicatechin-3-gallate (ECG), and theaflavins showed inhibition of COX and LOX dependent metabolism of arachidonic acid in human colon mucosa and colon tumor tissues [197]. Epiafzelechin isolated from the aerial parts of *Celastrus orbiculatus* Thunb. (Celastraceae) exhibited a dose-dependent inhibition of COX-1 activity with an IC_{50} value of 15 µM and also demonstrated anti-inflammatory activity against carragenin-induced mouse paw edema following oral administration at a dosage of 100 mg/kg [198].

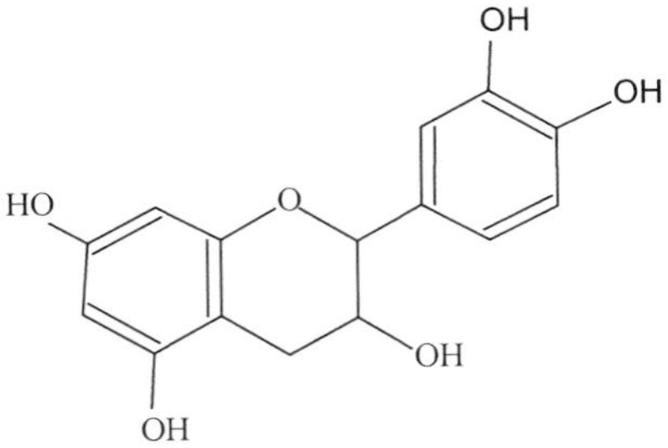

Catechin

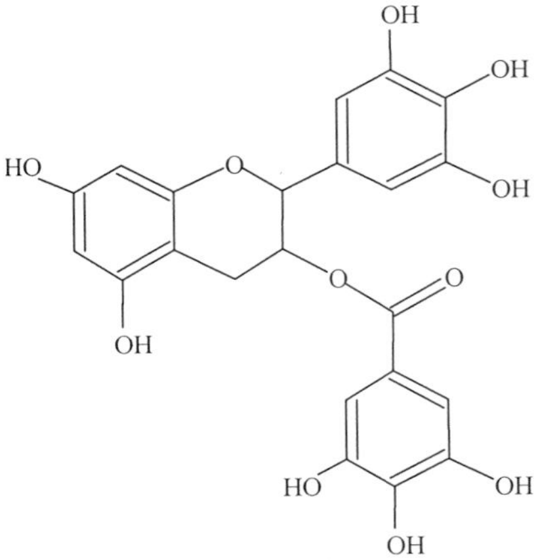

Epigallocatechin-3-gallate (EGCG)

Fig. (33). Chemical structures of two anti-inflammatory flavans.

14. Other natural dual inhibitors

Medicarpin, 6-demethylvignafuran and 6-hydroxy-2-(2-hydroxy-4-methoxy-phenyl)benzofuran were isolated from a COX and 5-LOX dual inhibitory methylene chloride extract of *Dalbergia odorifera* T. Chen (Leguminosae). All three compounds were potent 5-LOX inhibitors with IC_{50} values of 0.5 μM, 0.11 μM, and 0.05 μM, respectively. Both 6-demethylvignafuran and 6-hydroxy-2-(2-hydroxy-4-methoxyphenyl)benzofuran inhibited prostaglandin production in the mouse mastocytoma cells with IC_{50} values of 1.2 μM and 3.9 μM, respectively. However, the COX enzyme inhibitory activity of 6-hydroxy-2-(2-hydroxy-4-methoxy-phenyl)benzofuran was not observed in a sheep microsomal assay [199].

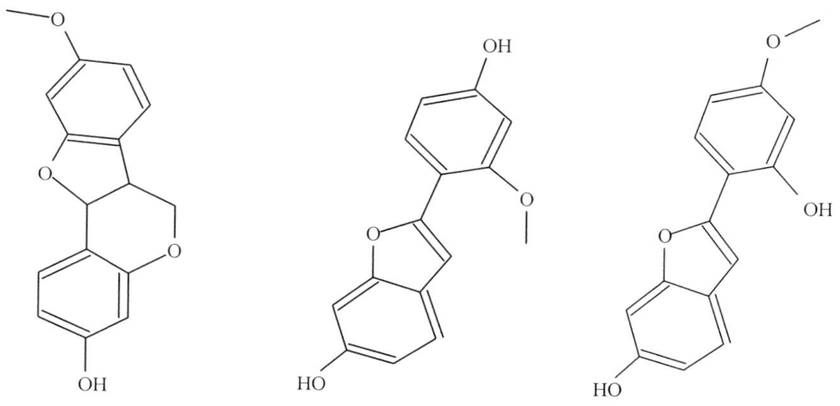

| Medicarpin | 6-demethyl-vignafuran | 6-Hydroxy-2-(2-hydroxy-4-methoxyphenyl) benzofuran |

Fig. (34). Chemical structures of three dual active benzofurans.

The fat soluble extract from rye pollen also showed dual inhibition against COX and 5-LOX pathways with IC_{50} values of 5 $\mu g/ml$ and 80 $\mu g/ml$, respectively. The clinical usage of the pollen extract for benign prostate diseases might due to such dual inhibition activities [200].

CONCLUSION

On November 18, 2002, in an article entitled "New-drug dearth surprisingly troubling", *USA Today* reported that the FDA approved 66 new drugs in 2001, compared to more than 120 per year in 1996-1997. By the end of September 2002, only 46 drugs were brought to USA market. In the term of new chemical entities of the approved drugs, year 2001 only yielded 24, a significant decrease compared to 53 in 1996. The drug discovery is "at the slowest rate in a decade and largely unexpected". The urgent needs of chemo-diversity from the pharmaceutical industry and the demands for scientific data on health-promoting herbs and dietary supplements from world wide consumers will bring a renaissance of natural product research in the near future. The integration of high throughput separation technology, structure dereplication capability, and laboratory information management system in the studies of natural product chemistry will get us prepared to capture this great opportunity.

ACKNOWLEDGEMENTS

The author wishes to thank Dr. Bruce Burnett and Dr. Ted Bartlett for their valuable scientific input, Ms. Mei-Feng Hong for her contributions on PhytoLogix^{TM} technology platform development, Mr. Eric Rhoden, Dr. Timothy Nichols, Mr. Kenneth Jones, Ms. Stacia Silva and Dr. Bruce Burnett for screening supports, Mr. Scott Waite, Dr. Ji-Fu Zhao, Ms. Thu Phan, Ms. Susan Vance and Ms. Miriam Cortes-Guzman for generation of the compound library, Dr. Ruidan Chen, Dr. Chad Orstrander and Dr. Peter Grosshans from Hitachi Instruments, Inc. for their technical assistance. The author gratefully acknowledges Mr. B. William Lee of Namyang Aloe Co. Ltd. for his vision and support on natural product research at Unigen Pharmaceuticals, Inc.

REFERENCES

[1]   Newman D.J.; Cragg, G.M.; Snader, K.M.; Nat. Prod. Rep. **2000**, 17, 215-234.
[2]   Cragg , G.M.; Newman, D.J.; Snader, K.M.; J. Nat. Prod. **1997**, 60, 52-60.
[3]   Dictionary of Natural Products, Chapman and Hall/CRC, **2002**, version 11:1.
[4]   Lee, M.L.; Scheneider, G.; J. Comb. Chem. **2001**, 3, 284-289.
[5]   Bemis, G.W.; Murcko, M.A.; J. Med. Chem. **1996**, 39, 2887-2893.
[6]   Henkel, T.; Brunne, R.M.; Müller, H.; Reichel, F.; Angew. Chem. Int. Ed. **1999**, 38, 643-647.
[7]   Lipinski, C.A.; Lombardo, F.; Dominy, B.W.; Feeney, P.J.; Adv. Drug Delivery Rev. **1997**, 23, 3-25.
[8]   Hall, D.G.; Manku, S.; Wang, F.; J. Comb. Chem. **2001**, 3, 125-150.
[9]   Wang, J.; Ramnarayan, K.; J. Comb. Chem. **1999**, 1, 524-533.
[10]   Desai M.C.; Zuckermann, R.N.; Moos, W.H.; Drug Devel. Res. **1994**, 33, 174-188.
[11]   Weber, L.; Curr. Opinion Chem. Biol. **2000**, 4, 295-302.
[12]   Spencer, R.W.; Bioeng. Comb. Chem. **1998**, 61, 61-67.
[13]   Lin, B.B.; J. Food Drug Anal. **1995**, 3, 233-242.
[14]   Oldenburg, K.R .; Zhang, J.; Chen, T.; Maffia, A.; Blom, K.F.; Combs, A.P.; Chung, T.D.Y.; J. Biomol. Screening, **1998**, 3, 55-62.
[15]   Virador, V.M.; Kobayashi, N.; Matsunaga, J.; Hearing, V.J.; Anal. Biochem. **1999**, 270, 207-219.
[16]   Kingston, D.I.; Abs. Papers Amer. Chem. Soc. **2001**, 221, ORGN 199.
[17]   Noreen, Y.; Ringbom, T.; Perera, P.; Danielson, H.; Bohlin, L.; J. Nat. Prod. **1998**, 61, 2-7.
[18]   Ghai, G.; Boyd, C.; Csiszar, K. Ho, C-T.; Rosen, R.T.; U.S. Patent, **1999**, #5,955,269.
[19]   Sche, P.P.; McKenzie, K.M.; White, J.D.; Austin, D.J.; Chem. Biol. **1999**, 6, 707-716.
[20]   Wessjohann, L.A.; Curr. Opinion Chem. Biol. **2000**, 4, 303-309.
[21]   Kolb, V.M.; Prog. Drug Res. **1998**, 51, 185-217.
[22]   Stahura, F.L.; Xue, L.; Godden, J.W.; Bajorath, J.; J. Mol. Model. **2000**, 6, 550-562.
[23]   Shu, Y-Z.; J. Nat. Prod. **1998**, 61, 1053-1071.
[24]   Green, J.; J. Org. Chem. **1995**, 60, 4287-4290.
[25]   Wang, H.; Ganesan, A.; J. Comb. Chem. **2000**, 2, 186-194.
[26]   Loevezijin, A.V.; Maarseveen, J.H.V.; Stegman, K.; Visser, G.M.; Koomen, G-J.; Tetrahedron Lett. **1998**, 39, 4737-4740.
[27]   Josien, H.; Curran, D.P.; Tetrahedron, **1997**, 53, 8881-8886.
[28]   Ni, Z-J.; Maclean, D.; Holmes, C.P.; Murphy, M.M.; Ruhland, B.; Jacobs, J.W.; Gordon, E.M.; Gallop, M.A.; J. Med. Chem. **1996**, 39, 1601.
[29]   Boger, D.L.; Fink, B.E.; Hedrick, M.P.; J. Am. Chem. Soc. **2000**, 122, 6382-6394.

712

[30]   Marder, M.; Viola, H.; Bacigaluppo, J.A.; Colombo, M.I.; Wasowski, C.; Wolfman, C.; Medina, J.H.; Rúveda, E.A.; Paladini, A.C.; Biochem. Biophys. Res. Commun. **1998**, 249, 481-485.

[31]   Nicolaou, K.C.; Pfefferkorn, J.A.; Roecker, A.J.; Cao, G-Q.; Barluenga, S.; Mitchell, H.J.; J. Am. Chem. Soc. **2000**, 122, 9939-9976.

[32]   Mason, J.S.; Morize, I.; Menard, P.R.; Cheney, D.L.; Hulme, C.; Labaudiniere, R.F.; J. Med. Chem. **1999**, 42, 3251-3264.

[33]   Hall D.G.; Manku, S.; Wang, F.; J. Comb. Chem. **2001**, 3, 125-150.

[34]   Grabley, S.; Thiericke, R.; Sattler, I.; Ernst Schering Res. Found Workshop, **2000**, 32, 217-252.

[35]   Bindseil, K.U.; Jakupovic, J.; Wolf, D.; Lavayre, J.; Leboul, J.; van der Pyl, D;. Drug Discovery Today, **2001**, 6, 840-847.

[36]   Eldridge, G.; Ghanem, M.; Zeng, L.; WO, **2001**, 01/33193, A2.

[37]   Kolodziej, H.; Kayser, O.; Kiderlen, A.F.; Ito, H.; Hatano, T.; Yoshida, T.; Foo, L.Y.; Planta Med. **2001**, 67, 825-832.

[38]   Abe, I.; Kashiwagi, Y.; Noguchi, H.; Tanaka, T.; Ikeshiro, Y.; Kashiwada, Y.; J. Nat. Prod. **2001**, 64, 1010-1014.

[39]   No, J.K.; Soung, D.Y.; Kim, Y.J.; Shim, K.H.; Jun, Y.S.; Rhee, S.H.; Yokozawa, T.; Chung, H.Y.; Life Sci. **1999**, 65, PL241-246.

[40]   Cantos, E.; Espin, J.C.; Tomas-Barberan, F.A.; J. Agric. Food Chem. **2001**, 49, 5052-5058.

[41]   Stewart, M.; Nash, R.J.; Chicarelli-Robinson, M.I.; Saponins, in Food, Feedstuffs and Medicinal Plants, Oleszek and Marston (eds.), **2000**, pp. 73-77.

[42]   Schmid, I.; Sattler, I.; Grabley, S.; Thiericke, R.; J. Biomol. Screening, **1999**, 4, 15-25.

[43]   Khosla, C.; J. Org. Chem. **2000**, 65, 8127-8133.

[44]   Hutchinson, C.R.; Curr. Opinion Micorb. **1998**, 1, 319-329.

[45]   McDaniel, R.; Thamchaipenet, A.; Gustafsson, C.; Fu, H.; Betlach, M.; Ashley, G.; Proc. Natl. Acad. Sci. USA, **1999**, 96, 1846-1851.

[46]   Wang, G.Y.; Graziani, E.; Waters, B.; Pan, W.; Li, X.; McDermott, J.; Meurer, G.; Saxena, G.; Andersen, R.J.; Davies, J.; Org. Lett. **2000**, 2, 2401-2404.

[47]   Thorson, J.S.; Nikolov, D.B.; Lesniak, J.; Jiang, J.; Biggins, J.B.; Barton, W.A.; Abst. Papers Amer. Chem. Soc. **2001**, 221, Carb. 19.

[48]   Jia, Q.; Hong, M.F.; Waite, W.; and Vance, V.; The 42nd Annual Meeting Amer. Soc. Pharmacognosy, **2001**, Oaxaca City, Oaxaca, MÉXICO.

[49]   Pomerantz, S.H.; J. Biol. Chem. **1991**, 241, 161-168.

[50]   Hook, D.J.; Pack, E.J.; Yacobucci, J.J.; Guss, J.; J. Biomol. Screening, **1997**, 2, 145-152.

[51]   Strege, M.A.; J. Chromatogr. B, **1999**, 725, 67-78.

[52]   Cordell, G.A.; Shin, Y.G.; Pure Appl. Chem. **1999**, 71, 1089-1094.

[53]   Jia, Q.; Hong, M.F.; Waite, W.; Rhoden, E.; Vance, V.; Ostrander, C.; Grosshans, P.; 49th ASMS Conf. Mass Spectrom. Allied Topics, **2001**, Chicago, Illinois, USA.

[54]   Jones, K.; Hughes, J.; Hong, M.; Jia, Q.; Orndorff, S.; Pigment Cell Res. **2002**, 15, 335-340.

[55]   Julian, Jr. R.K.; Higgs, R.E.; Gygi, J.D.; Hilton, M.D.; Anal. Chem. **1998**, 70, 3249-3254.

[56]   Wolfender, J-L.; Rodriguez, S.; Hostettmann, K.; Wagner-Redeker, W.; J. Mass Spectr. Repid Commun. In Mass Spectr. **1995**, S35-S46.

[57]   Baumann, C.; Cintora, M.A.; Eichler, M.; Lifante, E.; Cooke, M.; Przyborowska, A.; Halket, J.M.; Rapid Comm. Mass Spectrom. **2000**, 14, 349-356.

[58]   Bradshaw, J.; Butina, D.; Dunn, A.J.; Green, R.H.; Hajek, M.; Jones, M.M.; Lindon, J.C.; Sidebottom, P.J.; J. Nat. Prod. **2001**, 64, 1541-1544.

[59]   Lindon, J.C.; Nicholson, J.K.; Wilson, I.D.; J. Chromatogr. B, **2000**, 748, 233-258.

[60]   Russell, D.J.; Hadden, C.E.; Martin, G.E.; Gibson, A.A.; Zens, A.P.; Carolan, J.L.; J. Nat Prod. **2000**, 63, 1047-1049.

[61]   Smith, S.K.; Cobleigh, J.; Svetnik, V.; J. Chem. Inf. Comput. Sci. **2001**, 41, 1463-1469.

[62]   Patchkovskii, S.; Thiel, W.; J. Computational Chem. **1999**, 20, 1220-1245.

[63]   Grzonka, M.; Davies, A.N.; J. Chem. Inf. Comput. Sci. **1998**, 38, 1096-1101.

[64]   Schütz, V.; Purtuc, V.; Felsinger, S.; Robien, W.; Fresenius J. Anal. Chem. **1997**, 359, 33-41.

[65]   Corley, D.G.; Durley, R.C.; J. Nat. Prod. **1994**, 57, 1484-1490.

[66]   Dietzman, G.R.; US patent, **1999**, #5,978,804.

[67]   Funk, C.D.; Sci. **2001**, 294, 1871-1875.

[68]   Dannhardt, G.; Kiefer, W.; Eur. J. Med. Chem. **2001**, 36, 109-126.

[69]   Yamagata, K.; Andreasson, K.I.; Kaufman, W.E.; Barnes, C.A.; Worley, P.F.; Neuron, **1993**, 11,371-378.

[70]   Herschmann, H.R.; Cancer Metastasis Rev. **1994**, 13, 241-256.

[71]   Smith, W.L.; Garavito, R.M.; De Witt, D.L.; J. Biol. Chem. **1996**, 271, 33157-33160.

[72]   Heller, A.; Kock, T.; Schmeck, J.; Van Ackern, K.; Drugs, **1998**, 55, 487-496.

[73]   Moore *Prostanoids: pharmacological, physiological and clinical relevance,* Cambridge University Press, N.Y. **1985**, pp. 229-30.

[74]   Sharon, P.; Stenson, W.F.; Gastroenterology, **1984**, 86, 453-460.

[75]   Klickstein, L.B.; Shapleigh, C.; Goetzl, E.J.; J. Clin. Invest. **1980**, 66, 1166-1170.

[76]   Rae, S.A.; Davidson, E.M.; Smith, M.J.;.Lancet, **1982**, 1122-1124.

[77]   De Brum-Fernandes, A.J.; Laporte, S.; Heroux, M.; Lora, M.; Patry, C.; Menard, H.A.; Biochem. Biophys. Res. Commun. **1994**, 198, 955-960.

[78]   Weinberg, J.B.; Immunol. Res. **2000**, 22, 319-341

[79]   Ho, L.; Purohit, D.; Haroutunian, V.; Luterman, J.D.; Willis, F.; Naslund, J.; Buxbaum, J.D.; Mohs, R.C.; Aisen, P.S.; Pasinetti, G.M.; Arch Neurol 2001 Mar;58(3):487-92

[80]   McCarty, M.F.; Med. Hypotheses, **2001**, 56, 137-154.

[81]   Dempke, W.; Rie, C.; Grothey, A.; Schmoll, H.J.; J. Can. Res. Clin. Oncol. **2001**, 127, 411-417.

[82]   Moore, B.C.; Simmons, D.L.; Current Med. Chem. **2000**, 7, 1131-1144.

[83]   Fenton, B.M.; Beauchamp, B.K.; Paoni, S.F.; Okunieff, P.; Ding, I.; Am. J. Clin. Oncol. **2001**, 24, 453-457

[84]   Reich, R.; Martin, G.R.; Prostaglandings, **1996**, 51, 1-17.

[85]   Scioscia, K.A.; Synderman, C.H.; Rueger, R.; Reddy, J.; D'Amico, F.; Comsa, S.; Collins, B.; Amer. J. Otolaryngology, **1997**, 18, 1-8.

[86]   Myers, C.; Koki, A.; Pamukcu, R.; Wechter, W.; Padley, R.J.; Urology, **2001**, 57, 73-76.

[87]   Giovannucci, E.; Egan, K.M.; Hunter, D.J.; Stampfer, M.J.; Colditz, G.A.; Willett, W.C.; Speizer, F.E.; N. Engl. J. Med. **1995**, 333, 609-614

[88]   Smalley, W.; Ray, W.A.; Daugherty, J.; Griffin, M.R.; Arch. Intern. Med. **1999**, 159, 161-166,

[89]   Jaeckel, E.C.; Raja, S.; Tan, J.; Das, S.K.; Dey, S.K.; Girod, D.A.; Tsue, T.T.; Sanford, T.R.; Arch. Otolaryngol. Head Neck Surg. **2001**, 127, 1253-1259.

[90]   Kirschenbaum, A.; Liu, X.; Yao, S.; Levine, A.C.; Urology, **2001**, 58 (suppl.), 127-131.

[91]   DeWitt, D.L.; Mol. Pharmac. **1999**, 4, 625-631.

[92]   Rainsford, K.D.; J. Physiol. – Paris, **2001**, 95, 11-19.

[93]   Wallace, J.L.; Chapman, K.; McKnight, W.; Br. J. Pharmac. **1999**, 126, 1200-1204.

[94]   Gilroy, D.W.; Colville-Nash, P.R.; Willis, D.; Chivers, J.; Paul-Clark, M.J.; Willoughby, D.A.; Nature Med. **1999**, 5, 698-701.

[95]   Parente, L.; J. Rheumatol. **2001**, 28, 2375-2382.

[96]   Bertolini, A.; Ottani, A.; Sandrini, M.; Pharmac. Res. **2001**, 44, 437-450.

[97]   Celotti, F.; Laufer, S.; Pharmac. Res.; **2001**, 43, 429-436.

[98]   Kirchner, T.; Aparicio, B.; Argentieri, D.C.; Lau, C.Y.; Ritchie, D.M.; Prostaglandins Leukot. Essent. Fatty Acids, **1997**, 56, 417-423.

[99]   Fosslien, E.; Annals Clin. Lab. Sci. **1998**, 28, 67-81.

[100]   Babu, K.S.; Salvi, S.S.; Chest, **2000**, 118, 1470-1476.

[101]   Gok, S.; Ulker, S.; Huseyinov, A.; Hatip, F.B.; Cinar, M.G.; Evinc, A.; Pharmac. **2000**, 60, 41-46.

[102]   Stanke-Labesque, F.; Devillier, P.; Bedouch, P.; Cracowski, J.L.; Chavanon, O.; Bessard, G.; Cardiovascular Res. **2000**, 47, 376-383.

[103]   Christie, M.J.; Vaughan, C.W.; Ingram, S.L.; Inflamm. Res. **1999**, 48, 1-4.

[104]   Chang, M.S.; Chiou, G.C.Y.; J. Ocular Pharmac. **1989**, 5, 353-360.

[105]   Fiorucci, S.; Meli, R.; Bucci, M.; Cirino, G.; Biochem. Pharmac. **2001**, 62, 1433-1438.

[106]   White, M.J.; Glassman, A.T. Prostaglandins, **1974**, 7,123-129.

[107]   Huss, U.; Ringbom, T.; Perera, P.; Bohlin, L.; Vasänge, M.; J. Nat. Prod. **2002**, 65, 1517-1521.

[108]   Van der Oudearaa, F.J.G.; Buytenhek, M.; Methods Enzymol. **1982**, 86, 60-68.

[109]   Kim, Y.P.; Yamada, M.; Lim, S.S.; Lee, S.H.; Ruy, N.; Shin, K.H.; Ohuchi, K.; Biochim. Biophys. Acta, **1999**, 1438, 399-407.

[110]   Subbaramaiah, K.; Chung, W.J.; Michaluart, P.; Telang, N.; Tanabe, T.; Inoue, H.; Jang, M.; Pezzuto, J.M.; Dannenberg, A.J.; J. Biol. Chem. **1998**, 273, 21875-21882.

[111]   Ueda, N.; Kaneko, S.; Yoshimoto, T.; Yamamoto, S.; J. Biol. Chem. **1986**, 261, 7982-7988.

[112]   Nikolic, D.; Habibi-Goudarzi, S.; Corley, D.G.; Gafner, S.; Pezzuto, J.M.; van Breeman, R.B.; Anal. Chem. **2000**, 72, 3853-3859.

[113]   Brideau, C.; Kargman, S.; Liu, S.; Dallob, A.L.; Ehrich, E.W.; Rodger, I.W.; Chan, C-C.; Inflamm. Res. **1996**, 45, 68-74.

[114]   Nogata, Y.; Yoza, K-I.; Kusumoto, K-I.; Kohyama, N.; Sekiya, K.; Ohta, H.; J. Agric. Food Chem. **1996**, 44, 725-729.

[115]   Kuhl, P.; Shiloh, R.; Jha, H.; Murawski, U.; Zilliken, F.; Prostaglandings, **1984**, 28, 783-805.

[116]   Moroney, M.A.; Alcaraz, M.J.; Forder, R.A.; Carey, F.; Hoult, J.R.S.; J. Pharm. Pharmac. **1988**, 40, 787-792.

[117]   Reynaud, D.; Sun, A.; Demin, P.; Pace-Asciak, C.R.; J. Chromatogr. B, **2001**, 762, 175-180.

[118]   Jia, Q.; Nichols, T.; Rhoden, E.; Waite, S.; Hong, M.; Burnett, B.; The 43rd Ann. Meeting Amer. Soc. Pharm. 3rd Monroe Wall Symp., 2002, New Brunswick, NJ.

[119]   Raz, A.; Needleman, P.; J. Biol. Chem. **1990**, 269, 603-607.

[120]   You, K.M.; Jong, H.G.; Kim, H.P.; Arch. Pharm. Res. **1999**, 22, 18-24

[121]   Breu, W.; Hagenlocher, M.; Redl, K.; Tittel, G.; Stadler, F.; Wagner, H.; Drug Res. **1992**, 42, 547-551.

[122]   Liu, J.H.; Zschocke, S.; Reininger, E.; Bauer, R.; Planta Med. **1998**, 64, 525-529.

[123]   Singh, S.; Majumdar, D.K.; India J. Exp. Biol. **1997**, 35, 380-383.

[124]   Singh, S.; India J. Exp. Biol. **1998**, 36, 1028-1031.

[125]   Kraus, C.M.; Neszmélyi, A.; Holly, S.; Wiedemann, B.; Nenninger, A.; Torssell, K.B.G.; Bohlin, L.; Wagner, H.; J. Nat. Prod. **1998**, 61, 422-427.

[126]   Alanko, J.; Kurahashi, Y.; Yoshimoto, T.; Yamamoto, S.; Baba, K.; Biochem. Pharmac. **1994**, 48, 1979-1981.

[127]   Redl, K.; Breu, W.; Davis, B.; Bauer, R.; Planta Med. **1994**, 60, 58-62.

[128]   Müller-Jakic, B.; Breu, W.; Pröbstle, A.; Redl, K.; Greger, H.; Bauer, R.; Planta Med. **1994**, 60, 37-40.

716

[129] Stöhr, J.R.; Xiao, P-G.; Bauer, R.; Planta Med. **1999**, 65, 175-177.

[130] Tseng, C-F.; Iwakami, S.; Mikajiri, A.; Shibuya, M.; Hanaoka, F.; Ebizuka, Y.; Padmawinata, K.; Sankawa, U.; Chem. Pharm. Bull. **1992**, 40, 396-400.

[131] Sendl, A.; Elbl, G.; Steinke, B.; Redl, K.; Breu, W.; Wagner, H.; Planta Med. **1992**, 58, 1-7.

[132] Wagner, H.; Dorsch, W.; Bayer, Th.; Breu, W.; Willer, F.; Prostaglandins Leukot. Essent. Fatty Acids, **1990**, 39, 59-62.

[133] Benito, P.B.; Lanza, A.M.D.; Sen, A.M.S.; Galindez, J.D.S.; Matellano, L.F.; Gómez, A.S.; Martínez, M.J.A.; Planta Med. **2000**, 66, 324-328.

[134] Diaz, A.M.; Abad, M.J.; Fernandez, L.; Recuero, C.; Villaescusa, L.; Silvan, A.M.; Bermejo, P.; Biol. Pharm. Bull. **2000**, 23, 1307-1313.

[135] Sumner, H.; Salan, U.; Knight, D.W.; Hoult, J.R.S.; Biochem. Pharmac. **1992**, 43, 2313-2320.

[136] Liao, Y-H.; Houghton, P.J.; Hoult, J.R.S.; J. Nat. Prod. **1999**, 62, 1241-1245.

[137] Heras, B. de las; Abad, M.J.; Silván, A.M.; Pascual, R.; Bermejo, P.; Rodríguez, B.; Villar, A.M.; Life Sci. **2001**, 70, 269-278.

[138] Scherl, D.S.; Afflitto, J.; Gaffar, A.; J. Clin. Periodontol. **1999**, 26, 246-251.

[139] Benrezzouk, R.; Terencio, M.C.; Ferrándiz, M.L.; Feliciano, A.S.; Gordaliza, J.M.; Miguel del Corral, J.M.; de la Puente, M.L.; Alcaraz, M.J.; Life Sciences, **1999**, 64, 205-211.

[140] Benrezzouk, R.; Terencio, M.C.; Ferrandiz, M.L.; Hernandez-Perez, M.; Rabanal, R.; Alcaraz, M.J.; Inflamm. Res. **2001**, 50, 96-101.

[141] Jacobson, P.B.; Jacobs, R.S.; J. Pharmac. Exp. Ther. **1992**, 262, 866-873.

[142] Jacobson, P.B.; Jacobs, R.S.; J. Pharmac. Exp. Ther. **1992**, 262, 874-882.

[143] Najid, A.; Simon, A.; Cook, J.; Chable-Rabinovitch, H.; Delage, C.; Chulia, A.J.; Rigaud, M.; FEBS, **1992**, 299, 213-217.

[144] Safayhi, H.; Rall, B.; Sailer, E-R.; Ammon, H.P.T.; Pharmac. **1997**, 281, 460-463.

[145] Müller-Jakic, B.; Müller, M.; Pröbstle, A.; Johns, T.A.; Bauer, R.; Planta Med. **1993**, 59, suppl. A664.

[146] Kreft, A.F.; Failli, A.A.; Musser, J.H.; Kubrak, D.M.; Banker, A.L.; Steffan, R.; Demerson, C.A.; Nelson, J.A.; Shah, U.S.; Gray, W.; Marshall, L.A.; Holloway, D.; Sturm, M.; Carlson, R.P.; Berkenkopf, J.; Grimes, D.; Weichman, B.M.; Chang, J.Y.; Drugs Exptl. Clin. Res. **1991**, 17, 381-387.

[147] Laufer, S.A.; Augustin, J.; Dannhardt, G.; Kiefer, W.; J. Med. Chem. **1994**, 37, 1894-1897.

[148] Rioja, I. Terencio, M.C.; Ubeda, A.; Molina, P.; Tarraga, A.; Gonzalez-Tejero, A.; Alcaraz, M.J.; Eur. J. Pharmac. **2002**, 434, 177-185.

[149] Houghton, P.J.; Zarka, R.; Heras, B.; Hoult, J.R.S.; Planta Med. **1995**, 61, 33-36.

[150] Resch, M.; Steigel, A.; Chen, Z-L.; Bauer, R.; J. Nat. Prod. **1998**, 61, 347-350.

[151] Gil, B.; Sanz, M.J.; Terencio, M.C.; Giulio, A.; De Rosa, S.; Alcaraz, M.J.; Payá, M.A.; Eur. J. Pharmac. **1995**, 285, 281-288.

[152]   Terencio, M.C.; Ferrándiz, M.L.; Posadas, I.; Roig, E.; de Rosa, S.; Giulio, A.D.; Payá, M.; Alcaraz, M.J.; Arch Pharmac. **1998**, 357, 565-572.
[153]   Rao, T.S.; Basu, N.; Siddiqui, H.H.; India J. Med. Res. **1982**, 75, 574-578.
[154]   Sharma, O.P.; Biochem Pharmac. **1976**, 25, 1811-1812.
[155]   Conney, A.H.; Lou, Y-R.; Xie, J-G.; Osawa, T.; Newmark, H.L.; Liu, Y.; Chang, R.L.; Huang, M-T.; Proc. Soc. Exp. Biol. Med. **1997**, 216, 234-245.
[156]   Rao, C.V.; Rivenson, A.; Simi, B. Reddy, B.S.; Cancer Res. **1995**, 55, 259-266.
[157]   Huang, M-T.; Newmark, H.L.; Frenkel, K.J.; Cellular Biochem. Suppl. **1997**, 27, 26-34.
[158]   Flynn D.L.; Rafferty, M. F.; Boctor, A.M.; Prostaglandins Leukot. Med. **1986**, 22, 357-360.
[159]   Flynn, D.L.; Belliotti, T.R.; Boctor, A.M.; Connor, D.T.; Kostlan, C.R.; Nies, D.E.; Ortwine, D.F.; Schrier, D.J.; Sircar, J.C.; J. Med. Chem. **1991**, 34, 518-525.
[160]   Zschocke, S.; Lehner, M.; Bauer, R.; Planta Med. **1997**, 63, 203-206.
[161]   Duarte, D.S. Dolabela, M.F.; Salas, C.E.; Raslan, D.S.; Oliveiras, A.B.; Nenninger, A.; Wiedemann, B.; Wagner, H.; Lombardi, J.; Lopes, M.T.P.; J. Pharm. Pharmac. **2000**, 52, 347-352.
[162]   Puerta, R.de la; Gutierrez, V.R.; Hoult, J.R.S.; Biochem. Pharmac. **1999**, 57, 445-449.
[163]   Obertreis, B.; Giller, K.; Teucher, T.; Behnke, B.; Schmitz, H.; Drug Res. **1996**, 46, 52-56.
[164]   Kohyama, N.; Nagata, T.; Fujimoto, S-I.; Sekiya, K.; Biosci. Biotech. Biochem. **1997**, 61, 347-350.
[165]   Salari, A.; Braquet, P.; Borgeat, P.; Prostagl. Leukot. Med. **1984**, 13, 53-60.
[166]   Hamasaki, Y.; Tai, H.H.; Biochem. Biophys. Acta, **1985**, 834, 37-41.
[167]   Touvay, C.; Vilain, B.; Sirois, P.; Soufir, M.; Braquet, P.; J. Pharm. Pharmac. **1987**, 39, 454-458.
[168]   Suzuki, H.; Ueda, T.; Juránek, I.; Yamamoto, S.; Katoh, T.; Node, M.; Suzuki, T.; Biochem. Biophys. Res. Comm. **2000**, 275, 885-889.
[169]   Batt, D.G.; Maynard, G.D.; Petraitis, J.J.; Shaw, J.E.; Galbraith, W.; Harris, R.R.; J. Med. Chem. **1990**, 33, 360-370.
[170]   Liu, J.H.; Zschocke, S.; Reininger, E.; Bauer, R.; Planta Med. **1998**, 64, 525-529.
[171]   Kimura, Y.; Okuda, H.; Arichi, S.; Baba, K.; Kozawa, M.; Biochim. Biophys. Acta, **1985**, 834, 224-229.
[172]   Hoult, J.R.S.; Payá, M.; Gen. Pharmac. **1996**, 27, 713-722.
[173]   Hiermann, A.; Schantl, D.; Planta Med. **1998**, 64, 400-403.
[174]   Hoult, J.R.S.; Forder, R.A.; Heras, B. de las; Lobo, I.B.; Payá, M.; Agents Actions, **1994**, 42, 44-49.
[175]   Formica, J.V.; Regelson, W.; Fd. Chem. Toxic. **1995**, 33, 1061-1080.
[176]   Craig, W.J.; Am. J. Clin. Nutri. **1999**, 70(suppl.), 491S-499S.

[177]   Ohemeng, K.A.; Schwender, C.F.; Fu, K.P.; Barrett, J.F.; Bioorg. Med. Chem. Lett. **1993**, 3, 225-230.

[178]   Bernard, D.L.; Smee, D.F.; Huffman, J.H.; Meyerson, L.R.; Sidwell, R.W.; Chemotherapy, **1993**, 39, 203-211.

[179]   Wenzel, U.; Kuntz, S.; Brendel, M.D.; Daniel, H.; Cancer Res. **2000**, 60, 3823-3831.

[180]   Commenges, D.; Scotet, V.; Renaud, S.; Jacqmin-Gadda, H.; Barberger-Gateau, P.; Dartigues, J.F.; Eur. J. Epidemiol. **2000**, 16, 357-363.

[181]   Schubert, S.Y.; Lansky, E.P.; Neeman, I.; J. EthnoPharmac. **1999**, 66, 11-17.

[182]   Moroney, M-A.; Alcaraz, M.J.; Forder, R.A.; Carey, F.; Hoult, J.R.S.; J. Pharm. Pharmac. **1988**, 40, 787-792.

[183]   Puerta, R. de la; Forder, R.A.; Hoult, J.R.S.; Planta Med. **1999**, 65, 507-511.

[184]   Sobottka, A.M.; Werner, W.; Blaschke, G.; Kiefer, W.; Nowe, U.; Dannhardt, G.; Schapoval, E.E.S.; Schenkel, E.P.; Scriba, G.K.E.; Arch. Pharm. Pharm. Med. Chem. **2000**, 333, 205-210.

[185]   Williams, C.A.; Harborne, J.B.; Geiger, H.; Hoult, J.R.S.; Phytochem. **1999**, 51; 417-423.

[186]   Abad, M.J.; Bermejo, P.; Villar, A.; Gen. Pharmac. , **1995**, 26, 815-819.

[187]   Laughton, M.J.; Evans, P.J.; Moroney, M.A.; Hoult, J.R.S.; Halliwell, B.; Biochem. Pharmac. **1991**, 42, 1673-1681.

[188]   Whalley, C.V.; Rankin, S.M.; Hoult, J.R.S.; Jessup, W.; Leake, D.; Biochem. Pharmac. **1990**, 39, 1743-1750.

[189]   Chi, Y.S.; Jong, H.G.; Son, K.H.; Chang, H.W.; Kang, S.S.; Kim, H.P.; Biochem. Pharmac. **2001**, 62, 1185-1191.

[190]   Reddy, G.R.; Ueda, N.; Hada, T.; Sackeyfio, A.C.; Yamamoto, S.; Hano, Y.; Aida, M.; Nomura, T.; Biochem. Pharmac. **1991**, 41, 115-118.

[191]   Nakadate, T.; Aizu, E.; Yamamoto, S.; Kato, R.; Prosglandins, **1985**, 30, 357-367.

[192]   Eda, S.; JP. **1986**, #61 76,433.

[193]   Sagawa, S.; Nihro, Y.; Ueda, H.; Izumi, A.; Miki, T.; Matsumoto, H.; Satoh, T.; J. Med. Chem. **1993**, 36, 3904-3909.

[194]   Kalkbrenner, F.; Wurm, G.; von Bruchhausen, F.; Pharmac. **1992**, 44, 1-12.

[195]   Noreen, Y.; Serrano, G.; Perera, P.; Bohlin, L.; Planta Med. **1998**, 64, 520-524.

[196]   Noreen, Y.; el-Seedi, H.; Perera, P.; Bohlin, L.; J. Nat. Prod. **1998**, 61, 8-12.

[197]   Hong, J.; Smith, T.J.; Ho, C-T.; August, D.A.; Yang, C.S.;. Biochem. Pharmac. **2001**, 62, 1175-1183.

[198]   Min, K.R.; Hwang, B.Y.; Lim, H.S.; Kang, B.S.; Oh, G.J.; Lee, J.; Kang, S.H.; Lee, K.S.; Ro, J.S.; Kim, Y.; Planta Med. **1999**, 65, 460-462.

[199]   Miller, D.K.; Sadowski, S.; Han, G.Q.; Joshua, H.; Prostaglandins Leukot. Essent. Fatty Acids, **1989**, 38, 137-143.

[200]   Loschen, G.; Ebeling, L.; Drug Res. **1991**, 41, 162-167.

Atta-ur-Rahman (Ed.) *Studies in Natural Products Chemistry, Vol. 29*

THE CHEMISTRY AND BIOLOGY OF LAPACHOL AND RELATED NATURAL PRODUCTS α AND β-LAPACHONES

ÁNGEL G. RAVELO, ANA ESTÉVEZ-BRAUN, ELISA PÉREZ-SACAU

Instituto Universitario de Bio-Orgánica "Antonio González", Instituto Canario de Investigación del Cáncer. Avda. Astrofísico Francisco Sánchez Nº2, 38206. Universidad de La Laguna. La Laguna, Tenerife, Spain. Fax:34922318571. e-mail: agravelo@ull.es. aestebra@ull.es

ABSTRACT: This review describes a group of compounds related to the prenyl naphthoquinone lapachol. They exhibit interesting biological activities. Various aspects of the identification, biosynthesis, chemistry and biological properties are discussed.

1-INTRODUCTION

Lapachol (**1**) is a beautiful yellow and crystalline pigment (P.F. 140°). It is found in diverse woods like "green heart" of Surinam, woods of taigú, lapacho and bethabarra. Some time ago, these woods were imported from the western coast of Africa to build quality goods, e.g., fishing rods. The substance forms a sodium salt soluble in water which presents a brilliant red colour and it can be isolated by extraction of the powdered wood with a cold solution of sodium carbonate 1% [1]. The structural elucidation and the study of some interesting transformations were carried out by Prof. Samuel C. Hooker [2]. Hooker characterized the pigment as a 2-hydroxy-1,4-naphthoquinone with an isoprenyl chain attached at C-3. The isoprenyl chain is based on γ,γ, dimethyl-allyl radical. The structure was confirmed by synthesis: the condensation of the silver salt of lawsone with dimethyl allyl bromide yielded lapachol [3], Fig.(**1**).

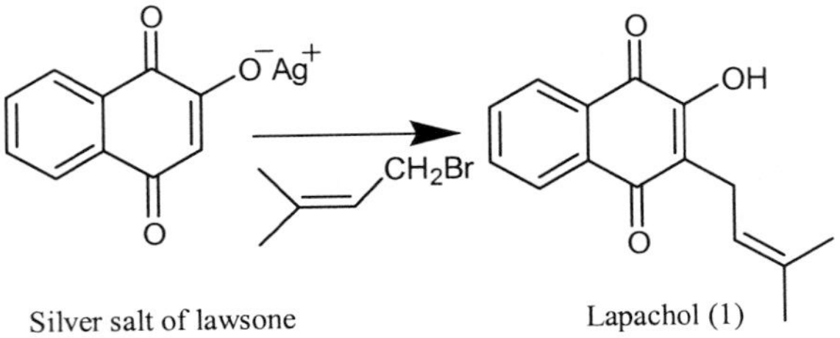

Fig.(1). Preparation of lapachol from lawsone

Lapachol crystallize in two different systems: triclinic LAPA I and monoclinic LAPA II. In both crystalline modifications of lapachol, the naphthoquinone ring system is approximately planar, and the planar unsaturated side chain is twisted about 90° with respect to the ring system. The crystal packings of LAPA I and II show that the molecules, in both cases, form dimers through OH----H hydrogen bonds around centres of symmetry [4]. Lapachol is considered one of the main component of Lapacho [5], an evergreen tree, with rosy coloured flowers, belonging to the Bignoniaceae family. The bark of the tree is used medicinally, specifically, the inner lining of the bark called the phloem. So, the native indians of Brazil, northern Argentina, Paraguay, Bolivia and other South American countries have used lapacho for medicinal purpose for thousands of year. Lapacho is applied externally and internally for the treatment of fevers, infections, colds, flu, syphilis, cancer, respiratory problems, skin ulcerations and boils, dysentery, gastro-intestinal problems of all kinds, debilitating conditions such as arthritis and prostatitis, and circulation disturbances [6]. Other diseases reported to be cured with lapacho are lupus, diabetes, Hodgkins disease, osteomyelitis, Parkinson's disease, and psoriasis [7-10].

Because of such extensive application in folk medicine, it is easy to understand the interest that the lapachol, the related naphthoquinones α-lapachone, β-lapachone and other naphthoquinones derivatives, have aroused as potential drugs. In fact, a broad variety of biological activities have been described for this kind of compounds. Some examples are antitumor promoting activities [11-12], inactivators of human cytomegalovirus protease [13], antiprotozoal activity [14], trypanocidal activity [15], anticancer activity [16], and antibacterial and antifungal activities [17].

2-CYCLIZATION AND INTERCONVERTION OF *ORTHO* AND *PARA*-NAPHTHOQUINONES

Hooker discovered [18] that the course of reaction of lapachol and strong acids apparently depended on the specific acid employed. In a mixture of acetic acid-hydrocloric acid (95%), lapachol was cyclized to the *para* quinone α-lapachone, but in concentrated sulfuric acid to the isomeric *ortho* quinone β-lapachone. Hooker also established a quantitative isomerization of α-lapachone to β-lapachone in sulfuric acid and the reverse in hydrochloric acid, and hence a probable inversion of equilibrium, Fig.(2). Later, Ettlinger [19] measured the ionization constants in strongly acid solutions. It was found that β-lapachone is a base stronger that α-lapachone. For this reason, although α-lapachone is more stable than β-lapachone in conditions of not ionization, the β-oxonium ion is more stable that the corresponding α-oxonium ion. In concentrated sulphuric acid the equilibrium favours the β-oxonium, and the β-lapachone precipitates after dilution with water. The ionization in AcOH-HCl is not strong enough to displace the equilibrium, and the more stable α-lapachone is separated when water is added gradually.

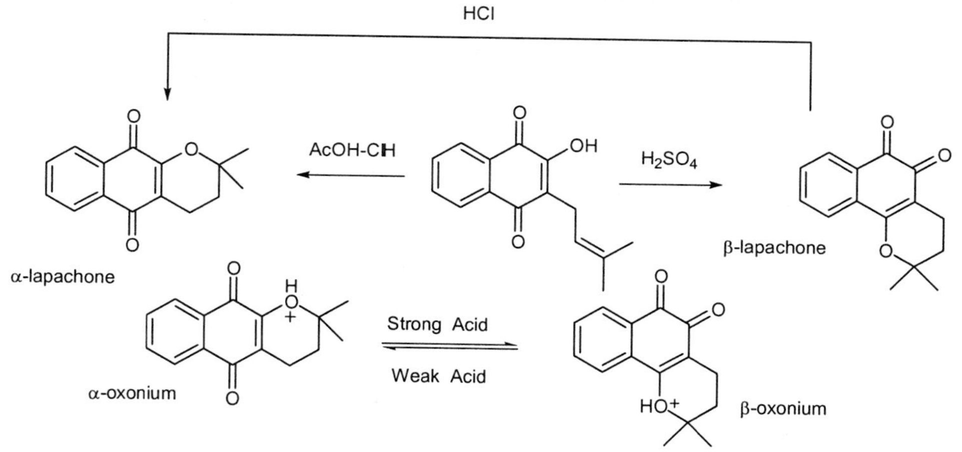

Fig. (2). Cyclisation of lapachol in acid media

The cyclization mechanism is explained by the protonation of the double bond in the side chain, producing the stable terciary carbocation. This carbocation is intramolecularly trapped by the hydroxyl group located on the C-2, or by the oxygen located at C-4, to yield α-lapachone or β-lapachone, respectively, Fig.(3).

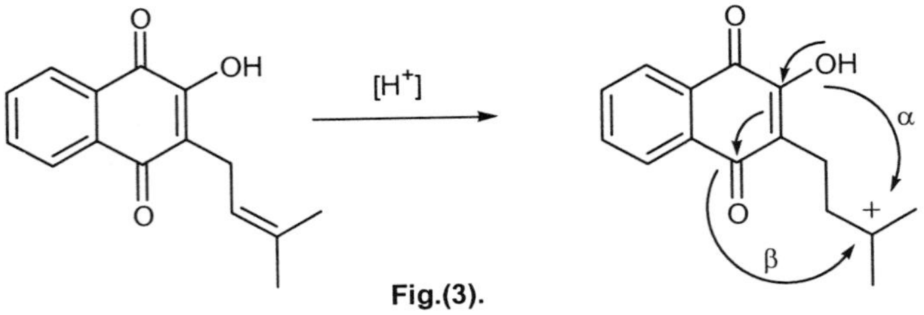

Fig.(3).

3-ISOLATION AND IDENTIFICATION OF LAPACHOL AND LAPACHONES

The vast majority of naphthalene derivatives found in nature are quinones, and the others are mainly related naphthols or naphthyl eters. The apparent distinction between naphthoquinones of plant and animal origin has now disappeared. In the animal kingdom only echinoderms are known to elaborate naphthoquinones, but they are found throughout the plant kingdom [20]. Lapachol is the most abundant quinone in the wood of various Bignoniaceae found mainly in tropical areas all around the world. For example it represents the 7.5% in weight in *Tecoma araliaceae* [21]. Recently it has been reported that the specie *Radermachera xylocarpa* (Bignoniaceae) is a highly efficient source of lapachol: the chloroform extract of the powdered stem bark contains 91.2% of lapachol and 5% of α-lapachone [22]. This and related *para* and *ortho* naphthoquinones have been isolated from Bignoniaceae, Jungladaceae, Lyttraceae, Polypodiaceae, Solanaceae, Rubiaceae, Proteaceae, Leguminosae, Verbenaceae, Schrophulariaceae, Gesneriaceae, Balsaminaceae; from lichens, Arthoniales and Parmeliaceae; and from alga Phaeophyta, *Candsburgia quercifolium*.

Table 1 gives an up-to-date compilation of selected natural products in the lapachol, furan and pyran-naphthoquinones classes, together with their occurrence and the biological activities investigated for each.

Table 1. Selected members of the natural naphthoquinones related to lapachol.

N° Compound	Structure	Occurrence and biological properties
1		Wood and roots of Bignoniaceae [21]: *Bignonia leucoxyton* [23-24]; *B. graciles* [25]; *Tabebuia avellanedae* [26-27]; *T. chrysantha* [28]; *T. guayacan* [29], *T. rosea* [30]; *T. penthaphylla* [31]; *T. palmeri* [32]; *T. chrysotricha* [33]; *T. ochraceae* [34]; *T. serratifolia* [35]; *Radermachera sinica* [36]; *R. xylocarpa* [22]; *Kigelia pinnata* [37]; *Zeyheria montana* [38]; *Newbouldia laevis* [39]; *Catalpa longissima* [40]; *Tecomella undulata* [41]; *Haplophragma adenophyllum* [42]; *Markhamia stipulata* [43]; Stem bark of *Avicennia alba* (Avicenniaceae) [44]; Stem bark of *Bauhinia guaianensis* [45] (Leguminoseae); Roots of *Tectona grandis* (Verbenaceae) [46]; Wood of *Diphysa robinoides* (Leguminoseae)[47]; Roots of *Conospermum teretifolium* (Proteaceae) [48]; Heartwood of *Hibiscus tiliaceus* (Malvaceae) [49]. Biological effects: antitumor (Section 8.1), antifungal and antibacterial (Section 8.2), trypanocidal (Section 8.3.1), leishmanicidal (Section 8.3.2).
2		Brown alga *Landsburgia quercifolia* (Cystoseiraceae) [50]. Biological effects: antifungal and cytotoxic [50].
3		Cell cultures: *Catalpa ovata* (Bignoniaceae) [51-52].
4		Roots of *Galium mollugo* (Rubiaceae) [53].
5		Wood of *Tabebuia guayacan* (Bignoniaceae) [29].
6		Roots and heartwood of *Tectona grandis* (Verbenaceae) [46]. Biological effect: cytotoxic [46].

Nº Compound	Structure	Occurrence and biological properties
7	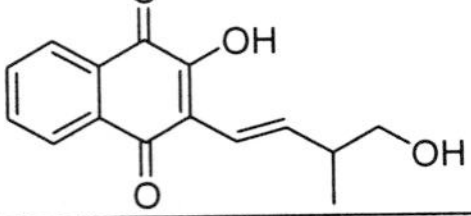	Seeds of *Lomatia arborescens* (Proteaceae) [54].
8	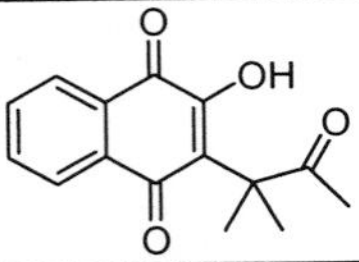	Cell cultures of *Stretocarpus dunnii* (Gesneriaceae) [55].
9	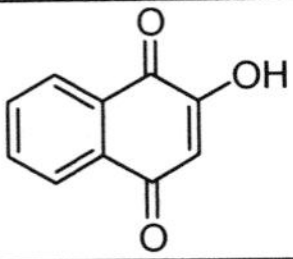	Leaves of *Lomatia ferruginea* (Proteaceae) [56]. Biological effect: trypanocidal [57].
10	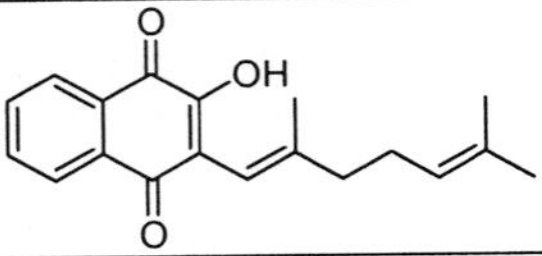	Roots of *Conospermum teretifolium* (Proteaceae) [58].
11	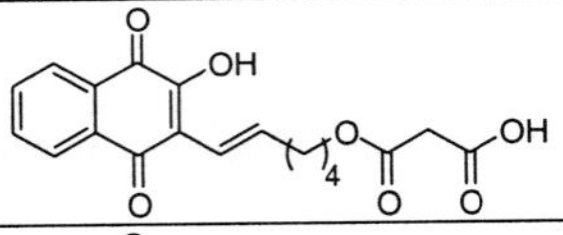	Fruits of *Trichia floriformis, Metatrichia vesparium* (Myxomycetes) [61], *Tectona grandis* (Verbenaceae) [46]. Biological effect: cytotoxic [46].
12	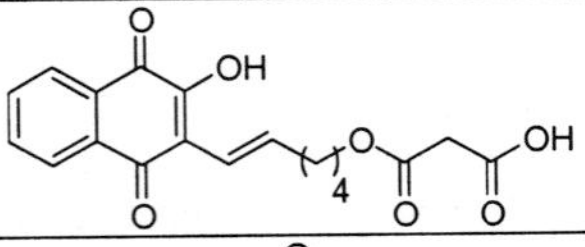	Fruits of *Metatrichia vesparium* (Myxomycetes) [59].
13	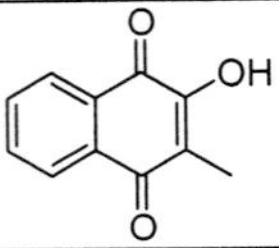	*Asplenium laciniatum, A. indicum* (Polypodiaceae) [60].
14	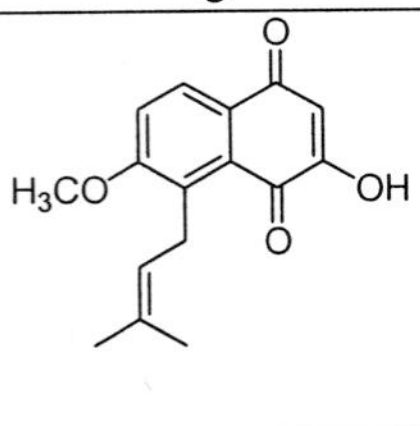	*Conospermun brownii* (Proteaceae) [58].
15	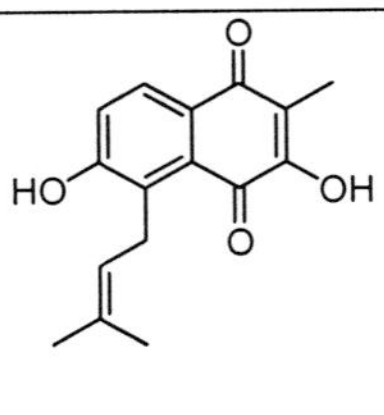	*Conospermun teretifoliumi* (Proteaceae) [58].

N° Compound	Structure	Occurrence and biological properties
16	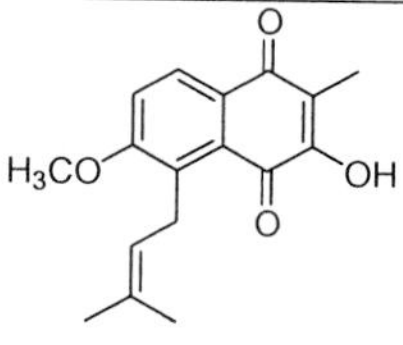	*Conospermun teretifoliumi* (Proteaceae) [58].
17	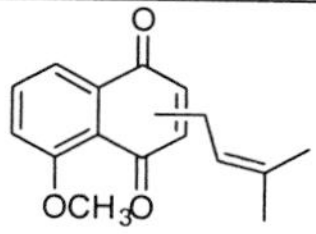	*Rubia cordifolia* (Rubiaceae) [61].
18	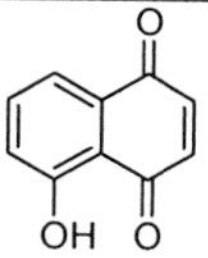	*Juglans nigra* (Jungladaceae) [62], *Pterocarpa frascinifolia* [63].
19	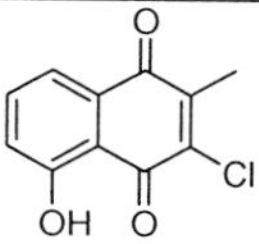	*Plumbago zeylanica* (Plumbaginaceae) [64].
20	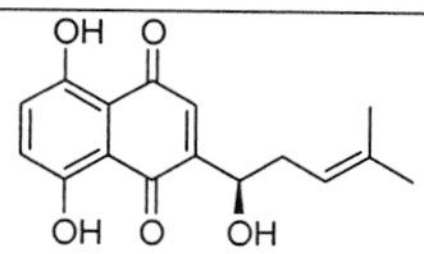	Roots of *Lithospermun ssp* (Boraginaceae) [65]. Biological effects: Antitumour, antiamebic, antipyretic, analgesic, antifungal, chemppreventive, antiinflammatory [65].
21	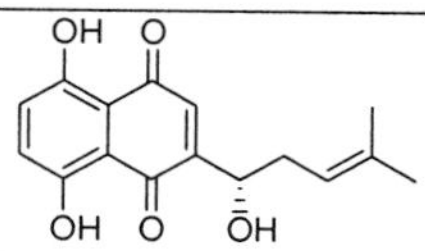	Roots of *Alkanna tinctoria, Arnebia hispidissima, A. nobilis, A. tinctoria, Macrotomia cephalotes, M. euchroma, Onosma echioides, O. paniculata* [65]. Biological effects: wound healing, antiinflammatory, antibacterial, inhibition of topoisomerase-I, antithrombotic, antitumor [65].
22	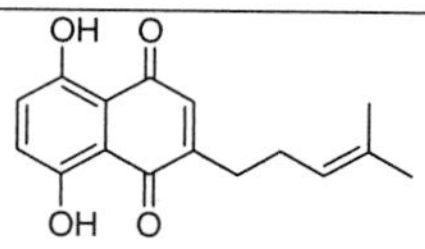	Roots of *Arnebia nobilis* [66], *Alkanna hirsutissima* [67]. Biological effects: antidermatophytic, antibacterial, antitumor [65].
23	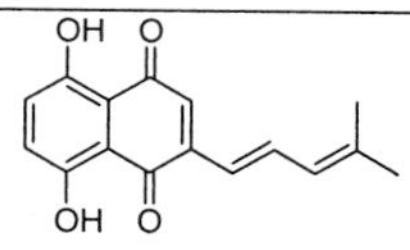	Roots of *Lithospermum erythrorhizon* [68], *Macrotomia euchroma* [69].

N° Com- pound	Structure	Occurrence and biological properties
24		Lichen *Cetraria islandica* (Parmeliaceae) [70].
25		Heartwood of *Tabebuia pentaphylla* (Bignoniaceae) [71].
26		Roots of *Putoria calabrica* (Rubiaceae) [72]; heartwood of *Lippia sidoides* (Verbenaceae)[73].
27		Wood of *Catalpa ovata* (Bignoniaceae) [74].
28		Roots of *Rubia oncotricha* (Rubiaceae) [75].
29		Wood and tissue cultures of *Catalpa ovata* (Bignoniaceae) [76] [77].

N° Com-pound	Structure	Occurrence and biological properties
30	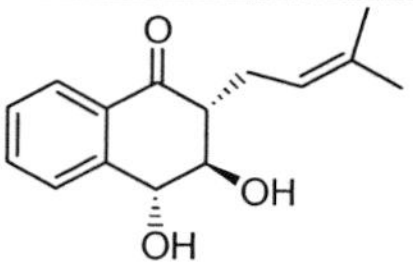	Roots of *Ekmanianthe longiflora* (Bignoniaceae) [78].
31	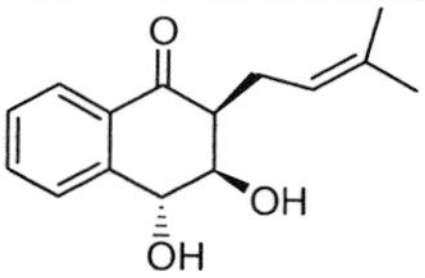	Roots of *Ekmanianthe longiflora* (Bignoniaceae) [78].
32	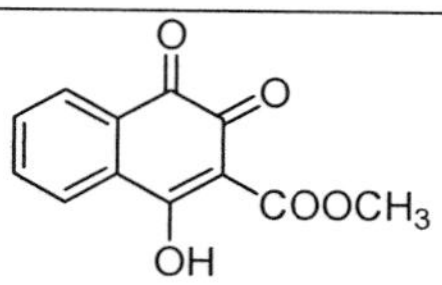	Roots of *Rubia oncotricha* (Rubiaceae) [75].
33	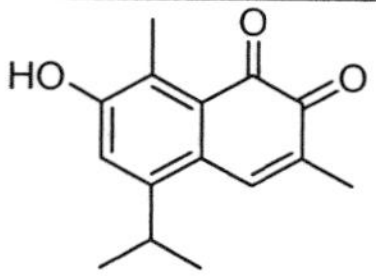	Heartwood of *Azanza garckeana* (Malvaceae) [79].
34	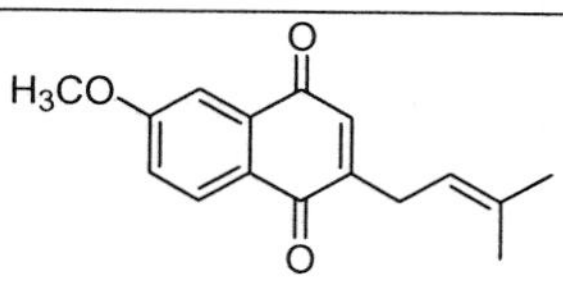	Roots of *Rubia cordifolia* (Rubiaceae) [80].
35	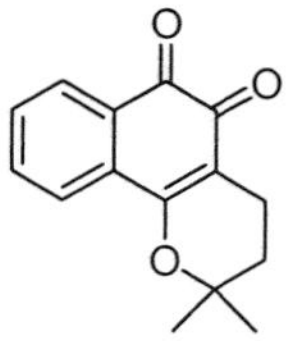	Heartwood and roots of Bignoniaceaes: *Haplophragma adenophyllum* [81]; *Phyllarthron comorense* [82]; *Tabebuia guayacan* [29]; *T. pentaphylla* [83]; Roots of *Tectona grandis* (Verbenaceae) [84]. Biological Effects: Antibacterial and antifungal [17]; inhibition of topoisomerase II [16], induce apoptosis [85].
36	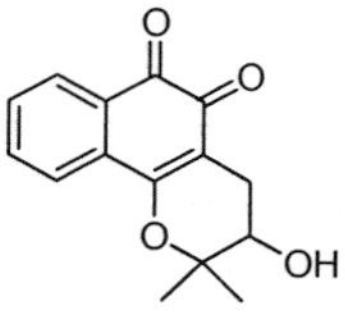	Wood of *Stenocarpus salinus* (Proteaceae) [86].

N° Compound	Structure	Occurrence and biological properties
37	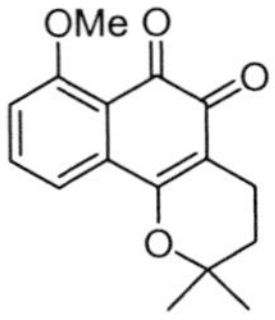	Roots of *Catalpa ovata* (Bignoniaceae) [87]. Biological Effect: Chemopreventive [87].
38	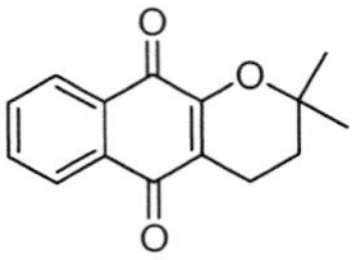	Wood and cell cultures of *Catalpa ovata* [88-89], wood of *Haplophragma adenophyllum* [90], *Tabebuia guayacán* [91]; *T. pentaphylla* [83], *Zeyhera tuberculosa* [92]; Roots of *Ekmanianthe longiflora* (all Bignoniaceae) [78]. Biological Effect: inhibition of topoisomerase II [16].
39	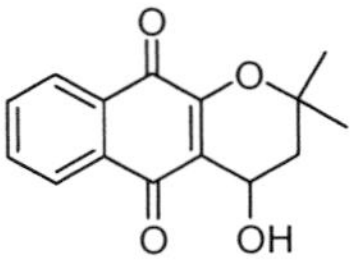	Wood and cell cultures of *Catalpa ovata* [88-89]; *Zeyhera tuberculosa* [92].
40	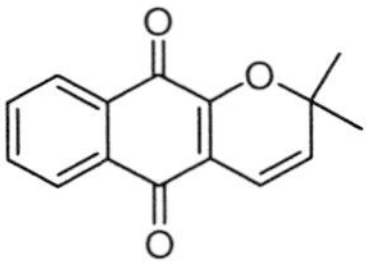	Roots of *Ekmanianthe longiflora* [78]; roots of Zeyhera montana [37]; *Kigelia pinnata* [93]; roots of *Tectona grandis* (Verbenaceae) [46]; roots and leaves of *Newbouldia laevis* [94-95]; wood of *Tabebuia palmeri* [96]; wood of *T. chrysotricha* [97]; wood of *T. serratifolia* [98]; heartwood of *Tecomella undulatum* [99]; heartwood of *Dolichandrone crispa* [100]; heartwood of *T. avellanedae* [101]. Biological effect: cancer chemopreventive [102]; Antibacterial and antifungal [93]
41	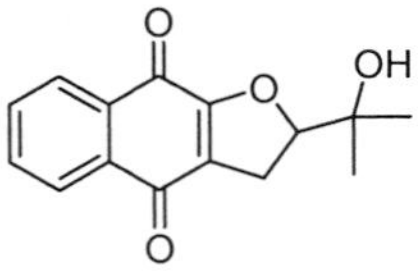	Wood of *Stenocarpus salinus* (Proteaceae) [86]; *Avicennia rumphiana; A. alba* (Avicenniaceae) [103]. Biological effect: cancer chemopreventive [102].
42	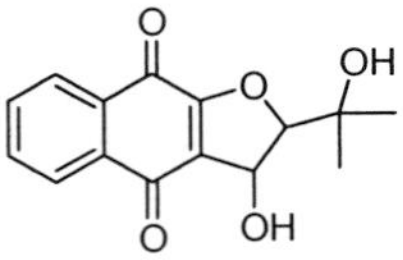	*Avicennia rumphiana; A. alba* (Avicenniaceae) [103]. Biological effect: Cancer chemopreventive [102].
43	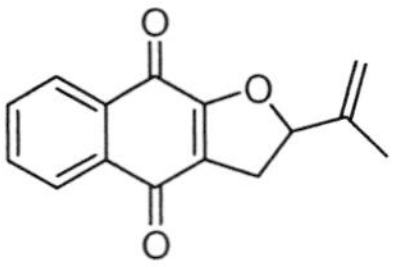	*Radermachera sinica* [104]; wood and tissue cultures of *Catalpa ovata* [105-106] *Cresencia cujete* [107-108], roots of *Ekmanianthe longiflora* [78] (all Bignoniaceae). Biological effect: antitumoral [78]
44	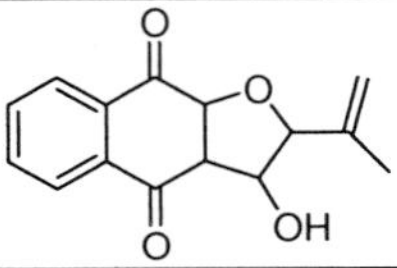	Wood of *Radermachera sinica* [104], tissue culture *Catalpa ovata* [105-106]. Biological effect: cancer chemopreventive [102].

N° Com-pound	Structure	Occurrence and biological properties
45		Wood of *Radermachera sinica* [104] and *Markhamia hildebrantii* [109] (Bignoniaceae); Roots of *Ekmanianthe longiflora* (Bignoniaceae) [78]. Biological effect: antitumoral [78].
46		Stem bark of *Tabebuia cassinoides* [110], wood of *Paratecoma peroba* [111] (Bignoniaceae). Roots of *Ekmanianthe longiflora* (Bignoniaceae) [78]. Biological effect: antitumoral [78].
47		Wood of *Tabebuia impetiginosa* [112], *Kigelia pinnata* [93], *Crescencia cujete* [107-108]). Roots of *Ekmanianthe longiflora* [78] (all Bignoniaceae). Biological effect: antitumoral [78] [107].
48		Roots of *Tabebuia cassinoides* (Bignoniaceae) [110]. Biological effect: antitumoral [78].
49		Leaves and cell cultures of *Streptocarpus dunii* (Gesneriaceae) [113].
50		Leaves and cell cultures of *Streptocarpus dunii* (Gesneriaceae) [113]. Leaves of *Calceolaria integrifolia* (Scrophulariaceae) [114].
51		Leaves and cell cultures of *Streptocarpus dunii* (Gesneriaceae) [113].
52		Leaves and cell cultures of *Streptocarpus dunii* (Gesneriaceae) [113].
53		Leaves and cell cultures of *Streptocarpus dunii* (Gesneriaceae) [113].

The naphthoquinones with acid hydroxyl function like lapachol are easily extracted with bases. They are also extracted in a Soxhlet apparatus with EtOH, and the resulting extract is then exhaustively extracted with AcOEt to yield a residue rich in naphthoquinones [115-116]. The naphthoquinones are mainly isolated from the wood and heartwood but there some examples of isolation from the roots [46, 48, 53, 58, 65, 66, 75, 78]. The study of commercial herbal products sold as Taheebo, Pau d'Arco, Lapacho, and Ipe roxo carried out by Girard *et al* [117], showed important differences in the content of lapachol and other related naphthoquinones present in extracts of *Tabebuia ssp*: *Tabebuia rosea*, *T. cassinoides*, *T. impetiginosa*, and *T. chrisantha*. Similar works using HPLC techniques have been carried out by Awang *et al.* [118].

The most representative data of the UV and IR of selected natural naphthoquinones spectra are shown in Table 2.

Table 2. UV and IR data of representative natural naphthoquinones.

Compound	IR ν_{max} cm^{-1}	UV λ_{max} nm (log ε)
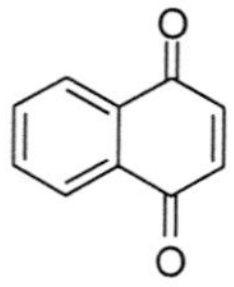	CCl$_4$ 1675 [119]	CHCl$_3$: 245 (4.34), 257 h (4.12), 335 (3.48).
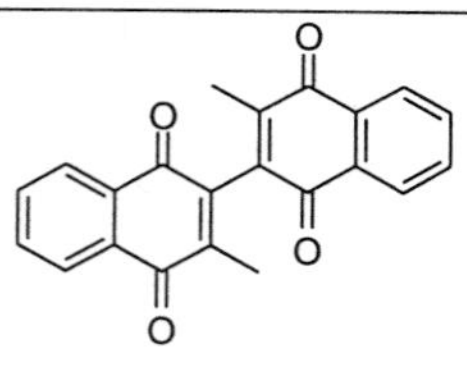	KBr 1668 [120]	CHCl$_3$: 246 (4.58), 263 (4.51), 335 (3.75).
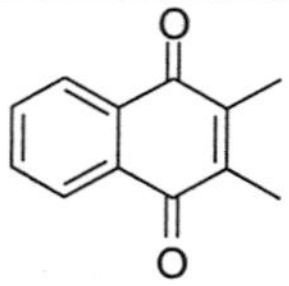	KBr 1660 [121]	CHCl$_3$: 243 (4.26), 249 (4.26), 260 (4.28), 269 (4.28), 330 (3.80)

	IR	UV
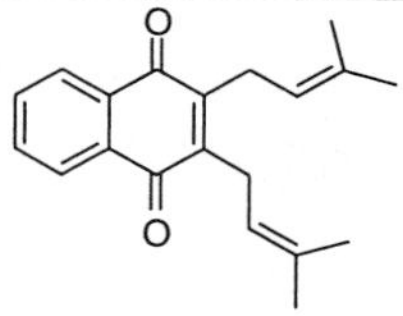 [122]	Nujol: 1660, 1615	EtOH: 245 (4.25), 260 (4.19), 268 (4.20), 327 (3.25).
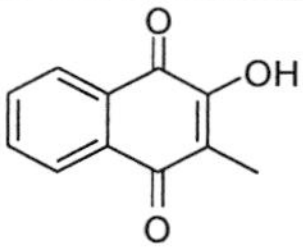 [123]	KBr: 3322, 1660, 1637, 1591	CHCl$_3$: 250, 280, 335
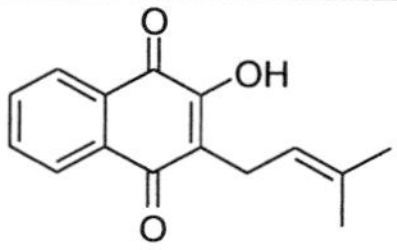 [124]	KBr: 3322, 1660, 1637, 1591	EtOH: 245 (4.46), 249 (4.44), 262, (4.34), 273 (4.31), 331 (3.5).
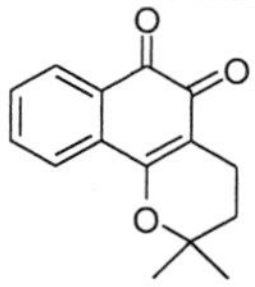 [125]	KBr: 1690, 1640, 1632, 1598	EtOH: 256, 282, 330, 431
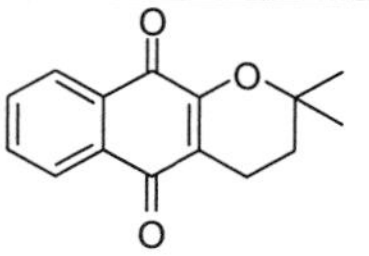 [125]	KBr: 1678, 1640, 1610, 1595	EtOH: 251 (4.45), 282 (4.22), 332 (3.44), 375 (3.15)
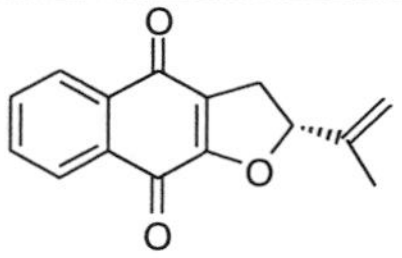 [126]	KBr: 1667	EtOH: 245 (4.46), 249 (4.44), 262 (4.34), 273 (4.31), 331 (3.5)

The work realized by Dauson *et al* [127] is of particular interest for the ^{1}H NMR and ^{13}C NMR assignments of these type of molecules. The isomers α-lapachone and β-lapachone can be easily distinguished, attending to the pattern that present the linear and angular pyranonaphthoquinones in their NMR spectra. In the ^{1}H NMR spectra, the signals corresponding to the aromatic protons appear in two different forms. In the case of α-cycled compounds, the aromatic protons have a symmetric environment and the hydrogens H$_5$ and H$_8$ are equivalent as

732

the hydrogens H_6 and H_7. In the case of β-cycled compounds, the molecule is less symmetric and the four aromatic protons present different chemical shifts, Fig.(4). These features are also observed in the ^{13}C NMR spectra. The signals of the aromatic carbons appear in two groups with very similar chemical shifts in the α-cycled compounds, while the aromatic carbons appear as six signals well differenced for the β-cycled series.

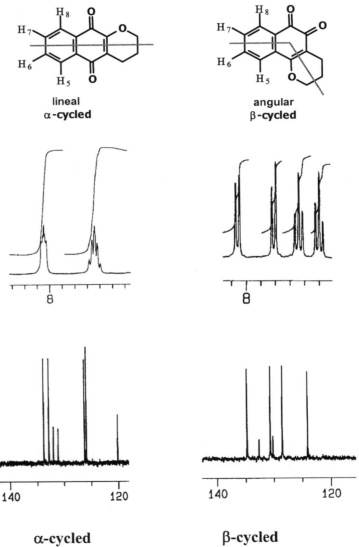

Fig. (4). Spectroscopic differences between α and β-cycled naphthoquinones

All mentioned above is corroborated by HMBC experiments, which also can be used to discern between α and β–cycled isomers. In the α-cycled compounds appear two long range correlations between H-8 and carbonyl C-1, and H-5 and the carbonyl-C-4, while in the β-cycled only one correlation between H-8 and C-1 carbonyl is observed, Fig. (5).

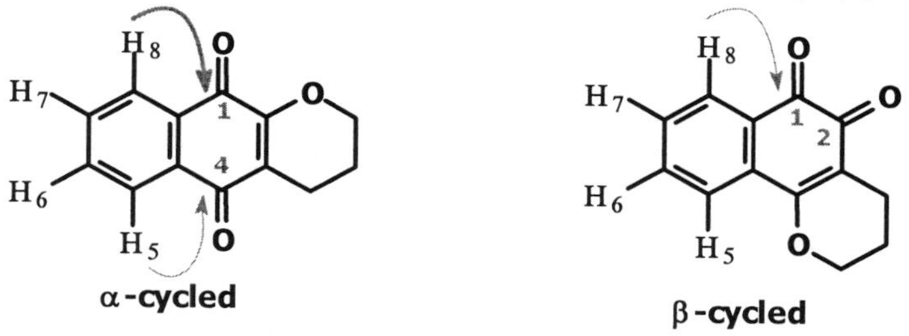

Fig. (5). Clave HMBC correlations to discern between α and β-cycled naphthoquinones

Prof. Burton has recently prepared and characterized (by NMR) benzo[a] phenazines obtained by condensation of β-lapachone with 1,2-phenylenediamines [128].

4-PRODUCTION OF PRENYL NAPHTHOQUINONES FROM CELL TISSUE CULTURES

Dunnione (**50**), α-dunnione (**49**), dehydrodunnione (**51**), streptocarpone (**8**), 7-hydroxydunnione (**52**) and 8-hydroxydunnione (**53**) were isolated from *Steptocarpus dunnii* cell cultures. Cell suspension culture were obtained by subculture of the callus tissues of *S. dunnii* in Linsmaier-Skoog liquid medium supplemented with 10^{-5} M indole-3-acetic acid and 10^{-6} M kinetin [113]. Inouye *et al* examined the naphthoquinones congeners in callus culture of *Catalpa ovata* [51-52]. Mainly using the GC-Ms technique, they have succeeded in identifying seven known constituents previously obtained from the wood of the intact plant, catalpalactone (**54**), catalponol (**55**), catalpone (**27**), and four α-lapachones derivatives including α-lapachone (**38**) itself. They also found six congeners, menaquinone-1 (**3**), 1-hydroxy-2-methylanthraquinone and four dehydro-iso-α-lapachones including 2*R*,3*R*-3-hydroxy-dehydro-iso-α-lapachone (**44**) among the constituents of the callus tissues.

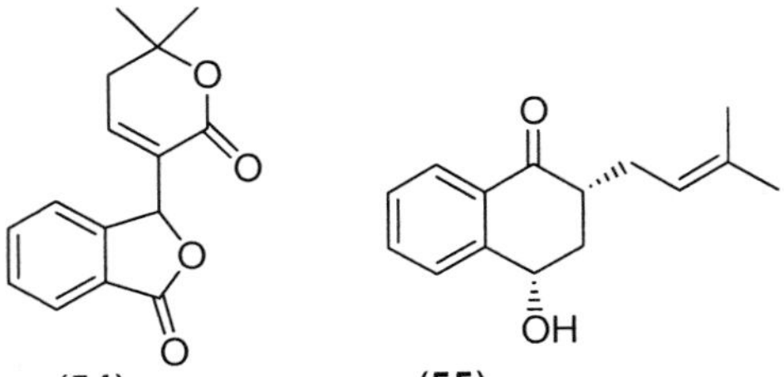

Fig. (6). Structure of catalpalactone (54) and catalponol (55).

Inouye *et al.* demonstrated through the simultaneous administration of (2*R*)-[1-^{14}C] catalponone and (2*S*)-[8-3H] catalponone to the wood of the same plant that the main pathway for the biosynthesis of naphthoquinone congeners, including catalpalactone, catalponol and 4,9-dihydroxy-α-lapachone, proceeds through the 2*R*-enantiomer of catalponone [129]. The induction of naphthoquinones formation in *Impatiens balsamina* cell cultures was achieved by using parent plants with high yielding of 2-methoxy-1,4-naphthoquinone as initiated explants. The cell cultures were capable of producing two naphthoquinones, lawsone (**9**), and an unidentified one, which was more polar than lawsone. The time-course of growth and lawsone production in *I. balsamina* cell culture were also established [130].

5-BIOSYNTHESIS OF LAPACHOL

The extensive research into the production of prenyl naphthoquinones from cell cultures has facilitated the elucidation of several stages in the biosynthesis of these and related natural products [131]. The generally accepted biosynthethic route is shown in Fig. (**7**). These terpenic quinones are derived from chorismic acid via its isomer isochorismic acid. Additional carbons for the naphthoquinone skeleton are provided by 2-oxoglutaric acid, which is incorporated by a mechanism involving the coenzyme thiamine diphosphate (TPP). 2-oxoglutaric acid is decarboxylated in the presence of TPP to give the TPP anion of succinic semialdehyde, which attacks isochorismic acid in a Michael-type reaction. Loss of the thiamine cofactor, elimination of pyruvic acid, and then dehydration yield the intermediate o-succinylbenzoic acid (OSB). This is activated by formation a coenzyme A ester, and a Dieckmann-like condensation allows ring formation. The dihydroxynaphthoic acid is the more favoured aromatic tautomer from the hydrolysis of the coenzyme A ester. This compound is now the substrate for alkylation and methylation. However, the terpenoid fragment is found to replace the carboxyl group, and the decarboxylated analogue is not involved. The transformation of 1,4-dihydroxynaphthoic acid to the isoprenylated naphthoquinone appears to be catalysed by a single enzyme, and can be rationalized as shown in Fig. (8). This involves alkylation, decarboxylation of the resultant β-ketoacid, and finally an oxidation to the *p*-quinone. Replacement of the carboxyl function by an isoprenyl substituent is found to proceed via a disubstituted intermediate in *Catalpa* (Bignoniaceae) [52][53][129], and this can be transformed to deoxylapachol and then menaquinone-1, Fig. (9). Lawsone is formed by an oxidative sequence in which hydroxyl replaces the carboxyl.

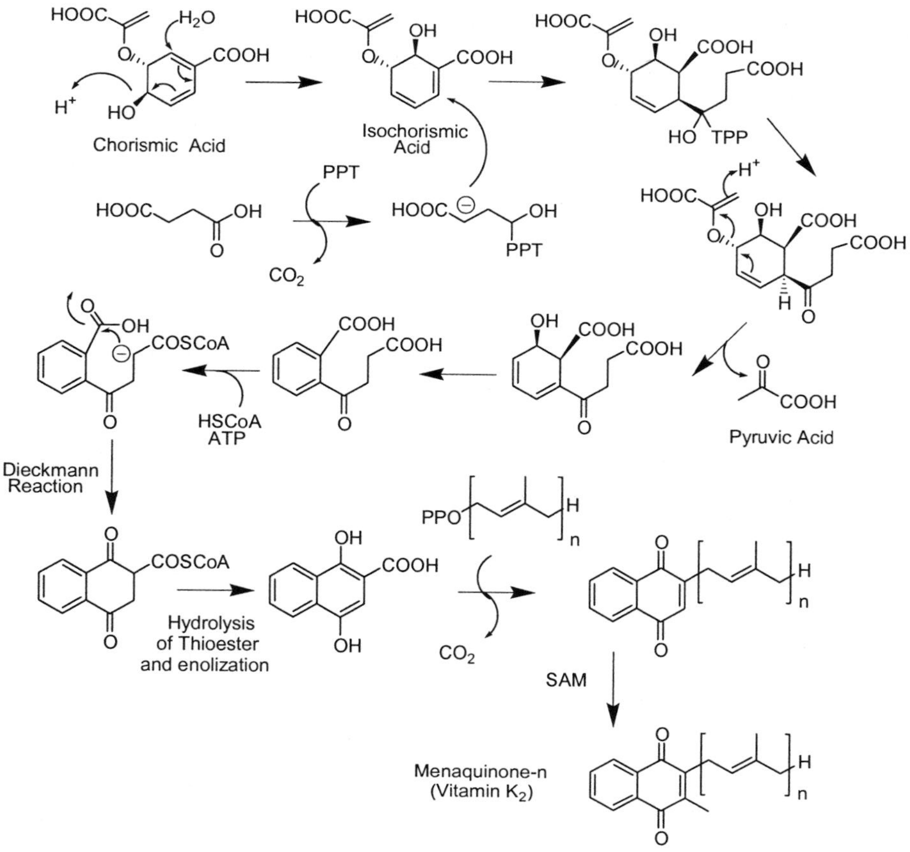

Fig.(7). Biosynthesis of isoprenyl 1,4-naphthoquinones.

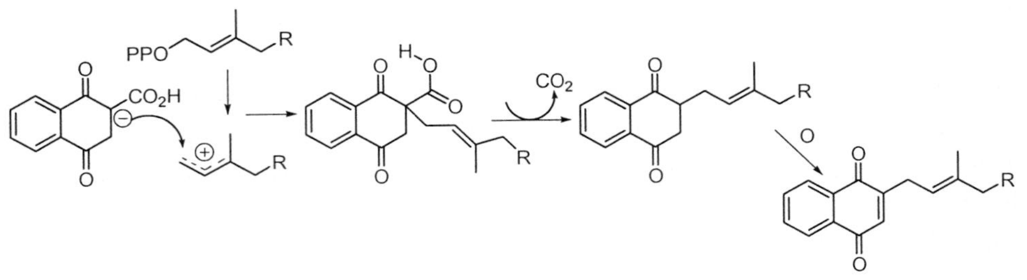

Fig.(8).Transformation of 1,4-dihydroxynaphthoic acid to the isoprenylated naphthoquinone

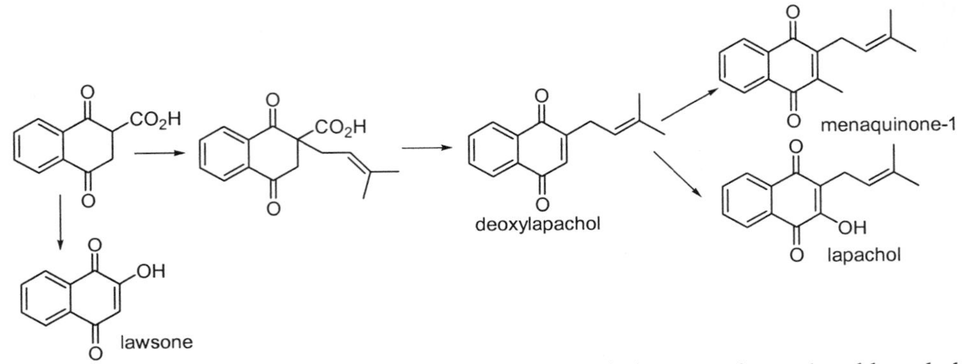

Fig.(9). Formation of lawsone, deoxylapachol, menaquinone-1 and lapachol

6-CHEMISTRY AND REACTIVITY OF LAPACHOL AND LAPACHONE

The dominant feature of the chemistry of lapachol and related compounds is the ease for the side chain to cyclise onto an oxygen function, which gives a variety of pyran and furan naphthoquinones. Some representative examples concerning to the chemistry of lapachol are shown in Fig.(10). Lapachol is oxidised to phthalic acid by nitric acid, and on zinc dust distillation yield naphthalene and 2-methyl-propene [132]. Oxidation of lapachol with lead dioxide in acetic acid gives a neutral orange-yellow compound with molecular formula $C_{30}H_{26}O_6$ ("lapachol peroxide"), evidently a dehydro-dimer (**55**)[133]. Its formation can be regarded as a C-O coupling of the radicals (**56**) and (**57**), Fig (**11**). The dimer compound is remarkably sensitive to base, and reverts to lapachol upon warming in alkaline solution. Recently, Stepanenko *et al* isolated an product identical to lapachol peroxide from the lichen *Cetraria islandica*

[70]. Although it presents diverse stereogenic centers, the authors do not indicate if this compound was found as a racemic mixture. Dehydro-α-lapachone (58) and dehydro-β-lapachone (59) are formed by oxidative cyclisation of lapachol with DDQ [134-135]. The reduction of quinones to hydroquinones is a quick, quantitative, and reversible process, which is common both to *ortho* and *para* quinones [136]. The transformation of lapachol to β-lapachone in acid media was treated previously in section-2.

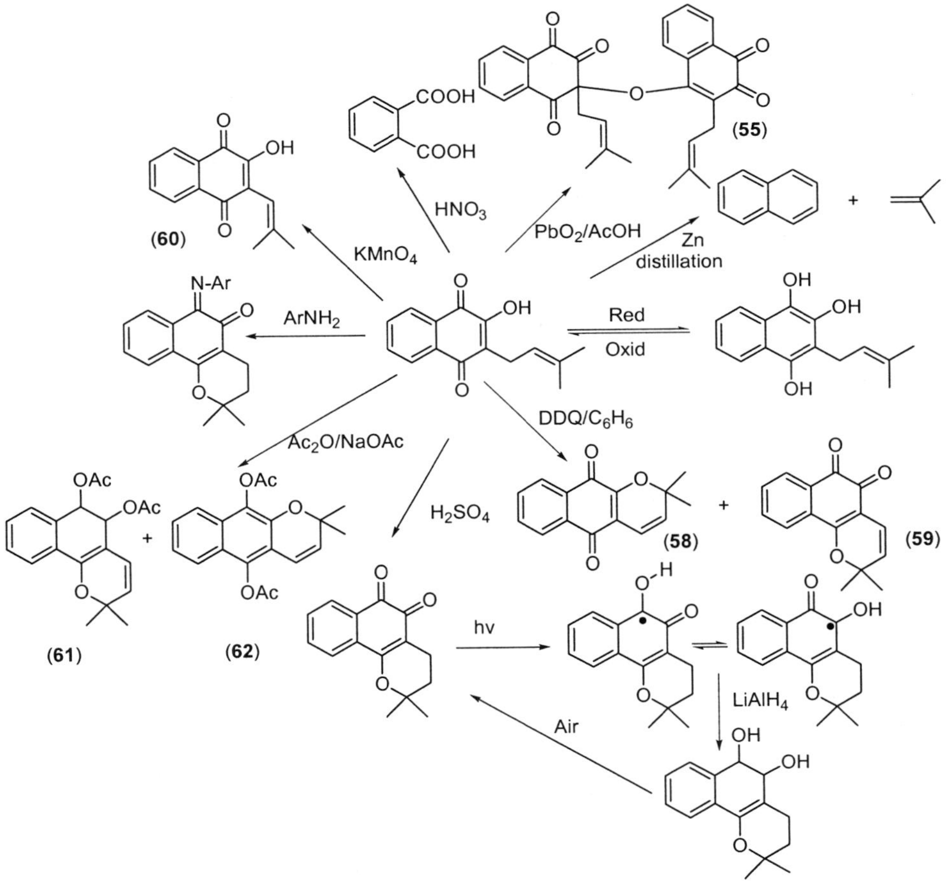

Fig. (10). Chemistry of lapachol.

The β-lapachone absorbing into the visible (λ_{max} 424 nm in benzene) is cleanly and efficiently reduced to the corresponding semiquinone radical upon photolysis in degassed solutions, with alcohols, amines and β-amino alcohols [134]. Lapachol is converted by alkaline permanganate in the cold into the next lower homologue (60), in which the CH_2 has been eliminated and the double bond shifted from the β,γ to the α,β position [137]. This reaction is known as Hooker oxidation and its mechanism was elucidated by Fieser et al.[138].

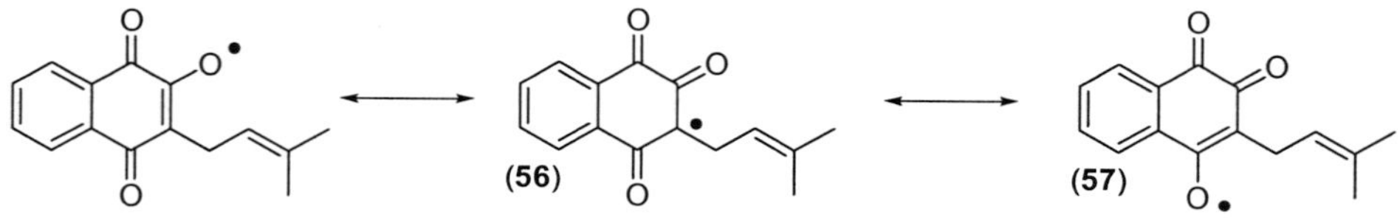

Fig.(11). Radicals implied in the formation of lapachol peroxide (55).

The preparation of the acetates (61) and (62) from lapachol was the first example of the formation of a chromenol by base-catalysed cyclisation of an allylquinone. The reaction proceeds by normal acetylation, apparently giving both isomers (63) and (64), followed by proton abstraction from the activated methylene group and rearrangement of the dienone system, Fig. (12) [139].

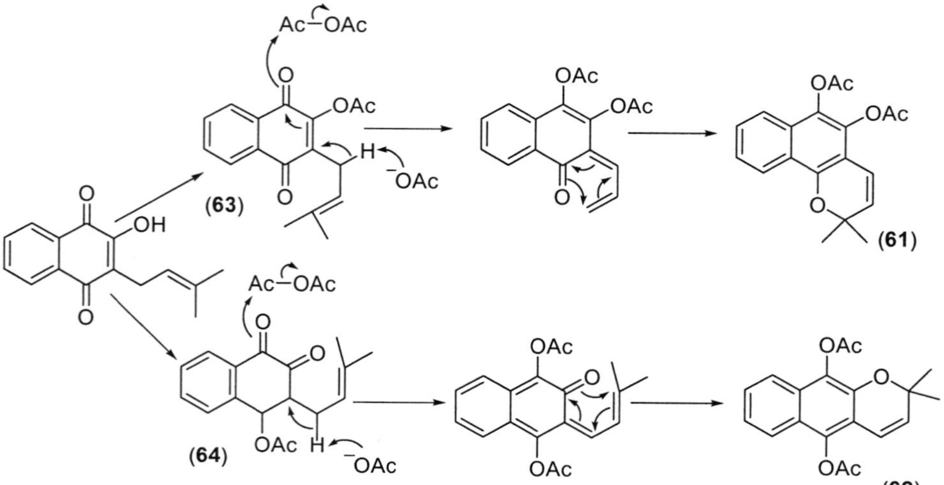

Fig.(12). Plausible formation of diacetates (61) and (62).

A new photochemical conditions with catalytic quantities of β-lapachone have been recently reported [140]. The β-lapachone acts as photosensitizer to promote selective C_{16}-C_{21} bond cleavage of catharanthine **(65)** yielding 21α-cyano-16α-(methoxycarbonyl) cleavamine **(66)** quantitatively, Fig. **(13)**.

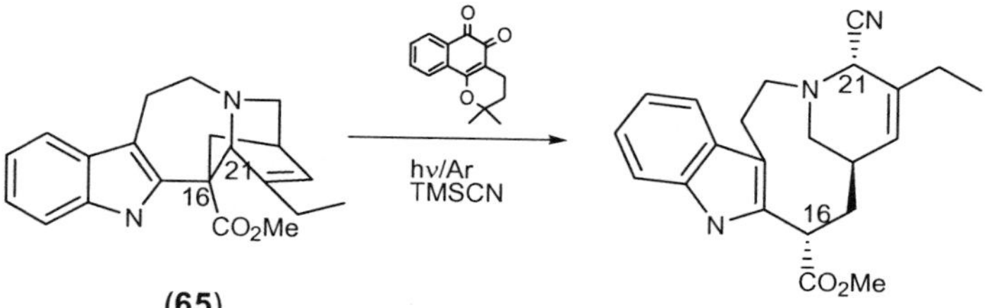

(65)   **(66)**

Fig.(13). Use of β-lapachol as photosensitizer

Complexes of lapachol with diverse metals like copper (II), iron (II), iron (III), chromium (III), aluminium (III), yttrium (III), samarium (III), gadolinium (III), and dysprosium (III) have been investigated [141-144]. Direct electrochemical synthesis of some metal derivatives of lapachol have been carried out [145]. Several β-lapachone hydrazo compounds were synthesized and characterized using spectroscopic techniques including X-ray analyses [146]. Selective aromatic reduction in pyranonaphthoquinones has also been reported [147-148].

7-SYNTHESIS OF LAPACHOL AND LAPACHONES

A synthesis of lapachol using reaction conditions better than those used by Fieser was carried out by Fridman *et al* [149].They used the lithium salt of 2-hydroxy-1,4-naphthoquinone prepared *in situ* instead of the silver salt used for Fieser [150]. The lithium salt was prepared in situ by addition of lithium hydride to the frozen solution of the quinone in dimethyl sulfoxide, Fig. **(14)**. As the solution thawed, the lithium quinone was slowly formed and was then alkylated with 3,3-dimethylallyl bromide. Lapachol was thus obtained in 40% yield.

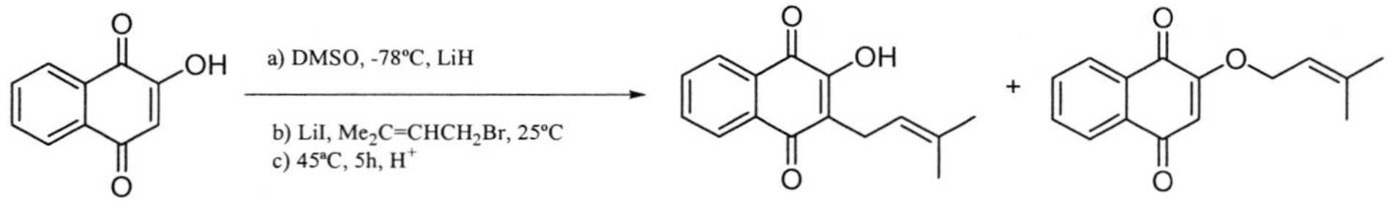

Fig.(14). Improved approach to synthesize lapachol from lawsone

Derivatives of β-lapachol with different substituents in aromatic ring have been obtained from methylindenes adequately substituted [151]. The methylindenes were synthesized from the corresponding 1-indanones and oxidized with dichromate to the phenylacetic acids. Cyclization of the corresponding ethyl esters as shown in Fig. (15) gave the intermediate naphthalenediones. Alkylation of the potassium salts with 1-bromo-3-methyl-2-butene gave the lapachols (67), together with varying amounts of the enol ethers (68), which on treatment with sulfuric acid gave back the starting material. Sulphuric acid treatment of (67) in a manner anologous to the well-known lapachol to β-lapachone conversion gave the desired β-lapachones carrying substituents in the benzene ring.

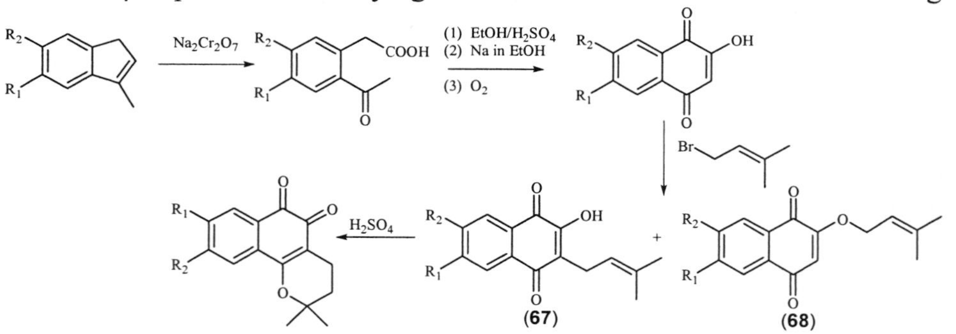

Fig.(15). Approach to lapachol and β-lapachone derivatives.

Syntheses of α- and β- lapachones and their homologues were carried out by photochemical side chain introduction to quinone. Thus, 1,4-naphthoquinone photochemically reacted with 3-methyl-2-butenal to give 2-(3-methyl-2-butenoyl)-1,4-naphthalenediol (69), regiospecifically. The product (69) was successively treated with acid, with dichloroaluminiun hydride, and finally with iron (III) chloride to derive in turn to cromanon (70), dihydropyran (71), and β-lapachone (35). β-lapachone was easily transformed to α-lapachone (38) by treating with acid. Likewise, from other α,β-unsaturated aliphatic aldehydes and 1,4-naphthoquinones α- and β-lapachone analogues were prepared [152].

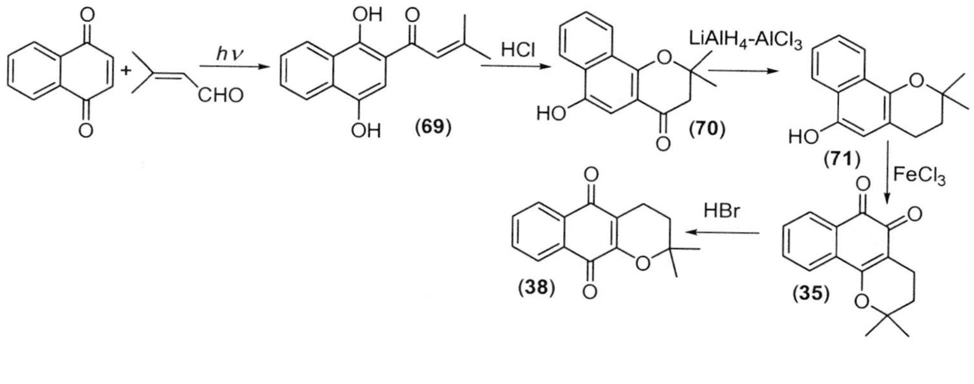

Fig.(16). Formation of β-lapachone and α-lapachone from 1,4-naphthoquinones.

Two short syntheses of β-lapachone from readily available naphthols and 3-methylbut-2-enal via a mild phenyl-boronic acid mediated cyclization to 2H-chromenes have been reported. Catalytic hydrogenation of 2H-chromenes with H$_2$/Pd-C in ethyl acetate afforded the corresponding naturally occurring chromanes (**72**) and (**73**). Oxidation of the adequate chromane with an excess of cerium ammonium nitrate (CAN) furnished β-lapachone in 62% yield [153].

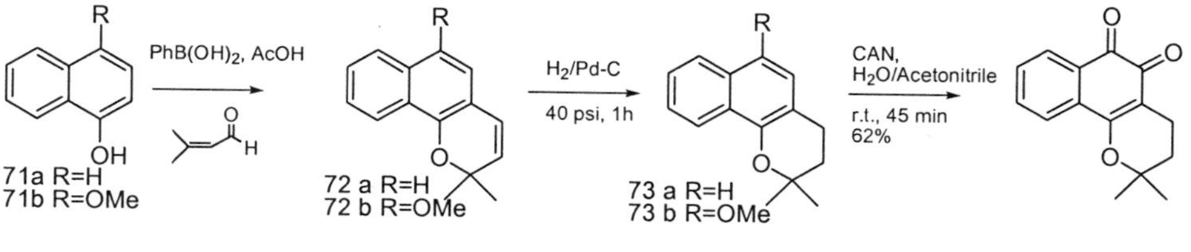

Fig.(17). Synthesis of β-lapachone from naphthols.

β-lapachones carrying various substituents in position 2 were also synthesized starting with lawsone (**9**) and the corresponding allyl bromide adequately substituted, followed of cyclization with concentrated H$_2$SO$_4$ [154].

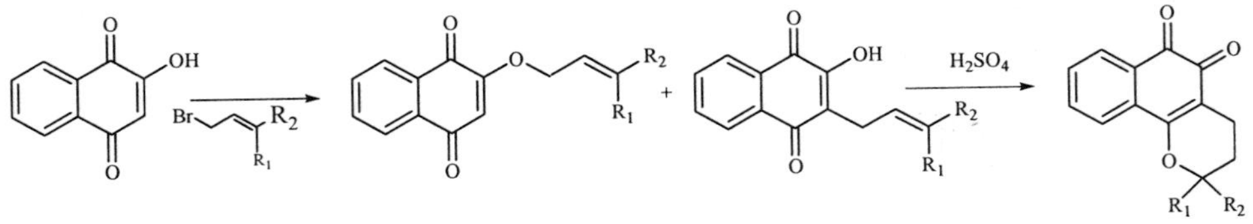

Fig.(18). Synthesis of β-lapachone derivatives

Snieckus *et al* have developed a general combined metalation-cross coupling methodology, and its application to a short synthesis of β-lapachone, Fig. **(19)** [155].

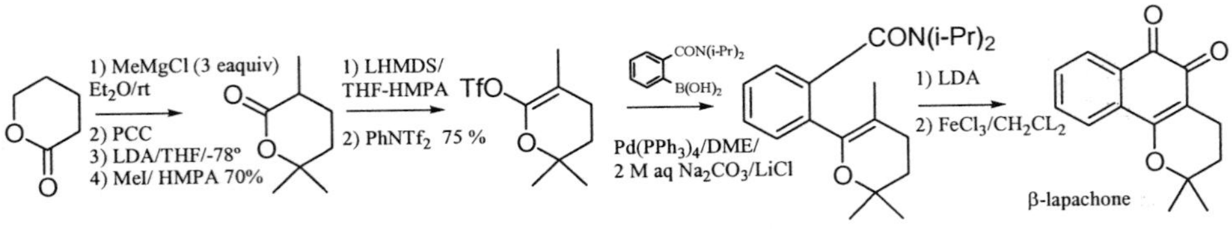

Fig.(19). Snieckus's synthesis of β-lapachone.

Some naturally occurring prenylated juglone derivatives have been synthesized from 2,8-dihydroxy-1,4-naphthoquinone as it is illustrated in Fig. **(20)** [156].

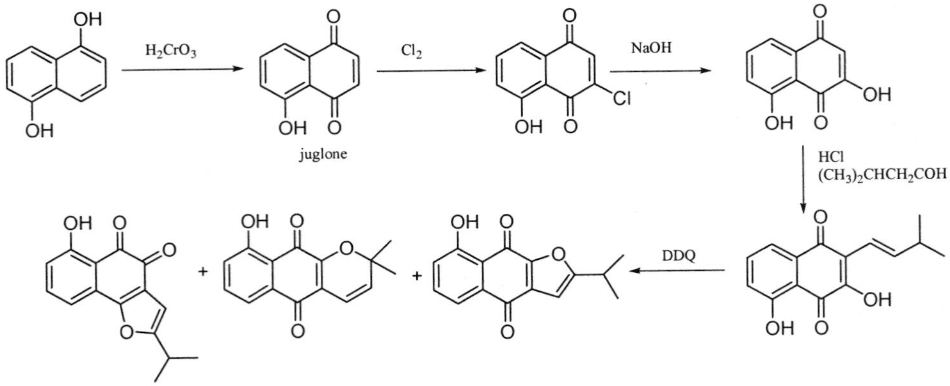

Fig.(20). Obtention of prenylated juglone derivatives

Barnes *et al* have carried out the synthesis of β-lapachone in eight steps with an overall yield of 23% starting from α-naphthol. The yields from the various steps were sufficiently good to assume that various derivatives of β-lapachone could be obtained in reasonable amounts, [157].

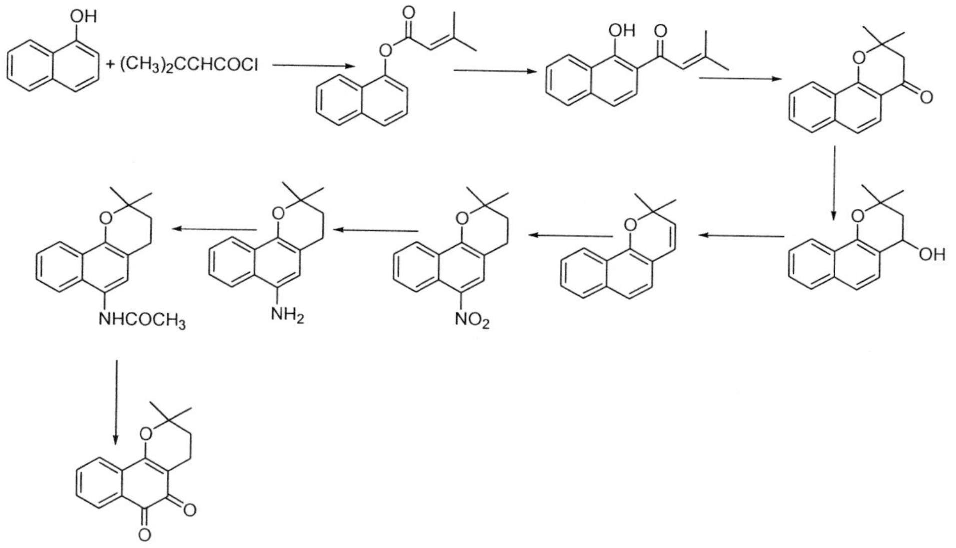

Fig.(21). Barnes's approach

It has been recently reported by Nair *et al* an efficient protocol for the synthesis of α, β-lapachones derivatives based on hetero Diels-Alder trapping of 3-methylene-1,2,4-[3*H*] naphthalenetrione [158].

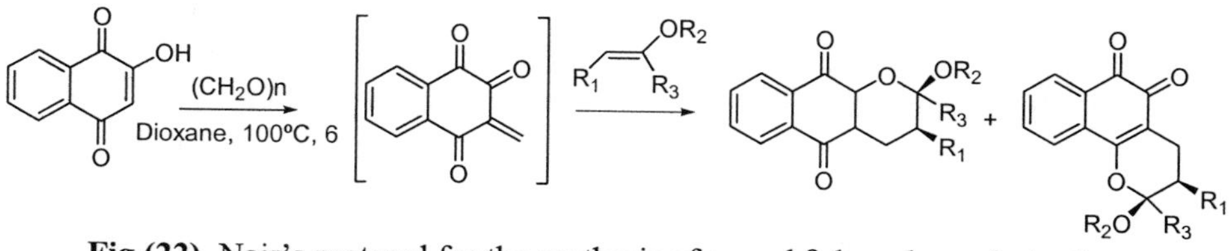

Fig.(22). Nair's protocol for the synthesis of α, and β-lapachone derivatives.

8-BIOLOGICAL ACTIVITY OF LAPACHOL, α-LAPACHONE, β-LAPACHONE AND THEIR DERIVATIVES

Biological investigations over the last years have shown that many of the medicinal properties claimed for Lapacho, Taheebo, Pau D'Arco or Ipe Roxo are attibuted to the active components lapachol, α-lapachone, β-lapachone and their derivatives.

8-1 Antitumor activity

The increasing problem of cancer in the developed world, it now being second only to cardiovascular disease as a cause of death, has motivated the extensive growth of cancer research in recent years. Mass screening programs of natural products and synthetic compounds by the National Cancer Institute of the USA have identified the quinone moiety as a pharmacophore that commonly affords cytotoxic activity [159].

In this sense, several natural naphthoquinones related to lapachol have shown cytotoxic activity. So, for example, the naphthoquinones (45), (46) and (47), isolated from the roots of *Ekmanianthe longifora*, exhibited broad cytotoxic activity in the range 0.2-4.0 μg/mL against a panel of nine human cancer cell lines and one murine cell line. α-lapachone (38) showed marginal activity against the SW626 (ovarian adenocarcinoma) cell line (3.0 mg/mL). Compound (45) was further evaluated in a 25 cell-line Oncology Diverse Cell Assay, representing a diverse group of mouse and human tumors, fibroblast, and normal bovine endothelial cells [78]. Deoxylapachol (2) was isolated by bioactivity-guided assay as the major cytotoxic component of the brown alga *Landsburgia quercifolia* [50] and it had a P-388 IC$_{50}$ of 0.6 μg/mL. 5 hydroxy-lapachol (6) and lapachol (1) isolated from the root heart wood of *Tectona grandis* were both found to be cytotoxic [46]. Bioassay-directed fractionation of the MeCOEt extract of *Crescentia cujete* resulted in the isolation of seven furanonaphthoquinones [107] showing selective activity towards DNA-repair-deficient yeast mutants. Compound (47) was also isolated as cytotoxic component of *Tabebuia impetiginosa* [112]. Table 1 shows other examples of cytotoxic natural related compounds.

Some synthetic lapachol derivatives have also showed cytotoxic activity. Burton *et al* have prepared mono(arylimino) derivatives of β-lapachone. Some of these derivatives had good scores with net cell kills in preliminary in vivo testing hollow fiber assays against a standard panel of 12 human tumor cell lines [160].Twelve substituted 1,4-naphthoquinones were tested against the ascitic form of sarcoma 180 in Swiss mice. Statistical analysis showed that the most important molecular parameter determining their effectiveness in prolonging the life of mice bearing this tumor is their redox potentials [161]. Zalkow *et al* have synthesized a monoimine quinone, namely 2-methyl-(Z)-4-phenylimino)naphth[2,3-d]oxazol-9-one, which in "in vitro" tests showed a selective activity for some solid cancer tumors [162]. Enamine derivatives of lapachol were synthesized and evaluated for cytotoxicity against A549 human breast cancer cells [163]. Seven new 1,4-naphthoquinones structurally related to lapachol were recently synthesized from lawsone and oxygenated arylmercurials. These compounds can also be seen as pterocarpan derivatives where the A-ring was substituted by the 1,4-naphthoquinone nucleus. Some of these compounds exhibited effects against proliferation of the MCF-7 human breast cancer cell line [164]. Two glucoside derivatives of lapachol were prepared and one of them, namely the tetracetyl glucoside, increased the life span of mice with lymphocytic leukemia P-388 by 80% [165].

The cytotoxicity of quinones is often attributed to DNA modification, thiol and amine depletion by alkylation, oxidation of essential protein thiols by activated oxygen species, and/or glutathione disulfide (GSSG) [166]. The variable capacity of quinones to accept electrons is due to the electron-attracting or donating substituents at the quinone moiety [167] which modulate the redox properties responsible for the resulting oxidative stress. The molecular basis of quinone toxicity is the enzyme-catalyzed reductuin to semiquinone radicals which then reduce oxygen to superoxide anion radicals thereby regenerating the quinone. This oxygen activation and redox cycling lead to increased levels of hydrogen peroxide and glutathione disulfide (GSSG). Recently, several studies on different targets for antineoplastic naphthoquinones have been published in an effort to found new specific anticancer compounds. Dual-specifity protein phosphatases are a subclass of protein tyrosine phosphatases which have a central role in the complex regulation of signalling pathways that are involved in cell stress responses, proliferation and death. Diverse natural and synthetic naphthoquinones were found to inactivate Cdc25 phosphatases in an irreversible manner by covalently binding at or near the active site [168-169]. Frydman *et al* have published a biomimetic model of topoisomerase II poisoning by naphthoquinones which is based on the reaction of β-lapachone with 2-mercaptoethanol to yield some Michael-type adducts. The results obtained suggest that the cytotoxic effects of the naphthoquinones derive, at least in part, from their alkylation of exposed thiol residues on the topo II-DNA complex [170-171]. The studies carried out by Oliveira-Brett *et al* agree with the hypothesis mentioned above. They studied the reduction of β-lapachone in the presence of L-cysteine and 2-mercaptoethanol and the results indicated that *β*-lapachone does not interact directly with dsDNA or ssDNA but interact with topoisomerasa and/or DNA-bound topoisomerase [172]. Recently, *Bastow et al* have reported a novel mechanism of DNA topoisomerasa II inhibition by pyranonaphthoquinone derivatives including β-lapachone and α-lapachone. Bastow *et al* studied this mechanism of action and concluded that various naphthoquinones (like β-lapachone and α-lapachone). inhibited topoisomerasa II by inducing religation and dissociation of the enzyme from DNA in the presence of ATP [173]. Nobuhiro *et al* found that the incubation of lapachol with P 450 reductase caused a cleavage of DNA which was reduced in the presence of Cu, Zn-superoxide dismutase, catalase, and hydroxyl radical scavengers. The results indicate that lapachol is bioactivated by P 450 reductase to reactive species, which promote DNA scission through the redox cycling generation of superoxide anion radical [174]. An important process in cancer research is the apoptosis. β-lapachone can induce apoptosis in certain cells, such as in human promyelocytic leukemia cells (HL-60) and human prostate cancer cells (DU-145, PC-3, and LNCaP). It operates before or at early times during, DNA synthesis by causing a block in G_0/G_1 of the cell cycle [175]. The proteasome is a large multisubunit proteolytic complex that participates in the degradation of proteins critical for cell cycle regulation. The proteasome inhibitors can overcome drug resistance. This effect is thought to be mediated through prevention of activation of NF-κB (nuclear transcription factor-κB) by proteasome inhibition. NF-κB is considered as one of the key factors involved in apoptotic pathways [176]. Aggarwal *et al* have described the effect of β-lapachone on tumor necrosis factor (TNF) induced activation of the NF-κB [177].

Cancer chemoprevention is regarded as a promising avenue for cancer control. Most cancer prevention research is based on the concept of multistage carcinogenesis: initiation → promotion → progression. Among these stages, in contrast to both initiation and progression stages, animal studies indicate that the promotion stage takes a long time to occur and may be reversible, at least in its earlier stages. Epstein-Barr virus (EBV) is an ubiquitous human γherpes-virus that is associated with several malignancies and diseases, together with some types of lymphomas in immunocompromised hosts (e.g., AIDS and post-transplant patients) [178]. To search for possible cancer chemopreventive agents, Tokuda *et al* developed a primary *in vitro* assay on Epstein-Barr virus early antigen (EBV-EA) activation induced by 12-O-tetradecanoylphorbol-13-acetate (TPA) as promoter, and *in vivo* assay using two-stage carcinogenesis tests in mouse skin. The inhibitory effects on EBV-EA activation in more than

50 quinones, including several natural occurring naphthoquinones, have been studied. Several natural occurring naphthoquinones and some lapachol analogues showed a great cancer preventive potential [12, 179-184].

8.2-Antibacterial and antifungal activities

The mechanism of action of antimicrobial naphthoquinones from the fungus *Fusarium* was studied by Haraguchi *et al* [185]. They explain that naphthoquinones acted as the electron acceptors for bacterial diaphorase, stimulating the generation of superoxide anion and hydrogen peroxide, and this fact could explain the antibacterial activities of these compounds. There are many examples in the literature of lapachol and lapachol derivatives exhibiting antimicrobial and antifungal activities. For example, several natural ocurring naphthoquinones from *Kigelia pinnata* [94], *Newbouldia laevis* [186] and *Rhinacanthus nasutus* [187]. De Kimpe *et al* [188] and Cameron *et al* have synthesized naphthopyran quinones which exhibit antibiotic activity [189]. Several lapachol derivatives were synthesized from lapachol and tested for their antibacterial activity [190]. 3-hydrazino-naphthoquinones as analogs of lapachol were synthesized and all of them showed antibacterial activity larger than lapachol [191]. Bieber *et al* published the effects of medium composition on the antifungal susceptibility test to lapachol, β-lapachone and synthetic analogues [192]. Guiraud *et al* compared bactericidal and fungicidal activities of lapachol and β-lapachone, concluding that fungi were considerably more sensitive than bacteria, particularly to β-lapachone [193].

8.3-Antiparasitic activities

Fournet and Muñoz published a review on natural products as trypanocidal, antileishmanial and antimalarial drugs. Among the active natural products, they include some naphthoquinones related to lapachol [194].

8.3.1-Antitrypanosomal activity

Chaga's disease, caused by the parasite *Trypanosoma cruzi*, is endemic in Latin America, and represents a very serious public health problem in several countries. At present, the only available therapeutic agent for Chaga's disease is *benznidazole*. This nitroheterocyclic compound is not always effective, and its use is associated with severe side effects. In 1984 W.H.O. recommended the use of *crystal violet* in blood banks in endemic areas in order to prevent the transfusional transmission of Chaga's disease. No serious side effects have been reported following use of this dye, although the temporary bluish colour conferred to blood is not well accepted by the assisted population [194].

Lapachol and β-lapachone exhibit activity against trypomastigotes forms of *T. cruzi*. Ventura-Pinto *et al* synthesized nine heterocyclic naphthooxazole and naphthoimidazole derivatives, and twenty two derivatives of β-lapachone tested against these parasites [195-196]. Among the cyclofunctionalised products, the oxazolic and imidazolic derivatives showed 1.5 to 34.8 times higher activity than crystal violet. Lapachol derivatives were also tested against trypamastigoste bloodstream forms of *T. cruzi* and the triacetoxy derivative of reduced lapachol showed trypanocidal activit, killing 95.7% of the parasites at a concentration of 42 µg/ml [197]. A new phenazine compound obtained from the reaction of β-lapachone with aniline resulted 9 times more active against *T. cruzi* trypomastigotes than crystal violet [198]. Phenazines are a class of heterocycles with a wide spectrum of microbicidal activities. Such substances are also inhibitors of DNA topoisomerasa II, leading to antitumor activity, and are capable of intercalating DNA. Due to its redox character, the trypanocidal effect of the phenazine derivatives may be attributable to the production of reactive intermediates of oxygen. The role of free radicals in the trypanocidal activity has been already established [199]. Goulart *et al* also tested several naphthoquinones against *Trypanosoma cruzi* amastigotes demonstrating the relationship between redox potencial and trypanocidal activity [200]. The higher activity of *ortho*-quinones as compared to their *para*-quinones isomers support the mechanism redox cycling by quinones. A similar effect was observed by Romanha *et al*. They described an increase of activity against *Trypanosoma cruzi* in naphthofuranquinones with methoxy groups which enhance the redox potential of quinonoid ring [201]. Recently, Davioud-Charvet *et al* have studied the correlation between redox cycling activities and *in vitro* antitrypanosomal activities, demonstrating that naphthoquinones are specific and potent inhibitors of TcTR (*Trypanosoma cruzi Trypanothione Reductasa*) [202-203].

8.3.2-Leishmanicidal activity

Leishmaniasis is a parasitic infection transmitted by the bite of an infected female sandfly whose hosts are animals or humans. The parasites can also be transmitted directly from person to person through the sharing of infected needles which is often the case with the Leishmania/HIV co-infection. It is considered to be endemic in 88 countries in tropical and subtropical areas and the south of Europe. Since the fifties, the pentavalent antimony compounds sodium stilbogluconate (Pentostam) and meglumine antimonate (Glucantime) have been the first-line treatment of leishmaniasis. These drugs are proven their efficacy, administrated via intravenous or intramuscular, and have acceptable cost, but they also present disadvantages: parenthal administration, long time therapy; albeit almost always reversible, toxic effect and the necessity of health infrastructure to cope with the long period of patient hospitalization. Five years ago, hexadecylphosphocholine (miltefosine), an antineoplasic agent, was identified as the first effective oral treatment in visceral infection [194]. The in vitro and in vivo leishmanicidal activity of lapachol has been evaluated against intracellular amastigotes of Leishmania braziliensis, then tested in L. braziliensis-infected hamsters, and its efficacy compared with the Pentostam as reference drug [204]. The authors concluded that lapachol tested in vitro exhibited an anti-amastigote effect, whereas in vivo it did not prevent the development of lesions at an oral dose 300 mg/Kg/day for 42 days. This fact suggest that lapachol in vivo might possibly inhibit the microbicidal functionig of macrophages, or it might be transformed into inactive metabolite(s). Kayser et al evaluated the in vitro leishmanicidal activity of monomeric and dimeric naphthoquinones [205]. They used both a direct cytotoxicity assay against extracellular promastigote Leishmania donovani, L. infantum, L. enriettii and L. major and a test against intracellular amastigote L. donovani (residing within murine macrophages). All the tested naphthoquinones showed higher antiparasitic EC_{50} values than the well-established reference reagents which were incorporated in the assays for validating all results. Monomeric derivatives also had significant activity against intracellular L. donovani amastigotes, but showed high and non-selective toxicity for all tested cell cultures. These authors suggest that elements of the mitochondrial respiratory chain are targets for these compounds, and the strong similarity in mammals and amastigote Leishmania might then be a reason for the non-selective toxicity observed. Significant differences in antiprotozoal and citotoxic activities between monomeric and dimeric naphthoquinones may well result from many parameters including the number of units, planarity, and specific nature and pattern of aromatic functional groups [205].

8.3.3-Antimalarial activity.

Malaria is one of the oldest parasitic diseases. The difficulty of malaria control is aggravated by the appearance of strains of *Plasmodium falciparum* resistant to antimalarials, as well as resistance of the vector mosquitoes to DDT and other insecticides. The molecule quinine isolated from the bark of the *Cinchona* sp. tree, represents the model for the synthesis of the majority of drugs currently used for malarial treatment [194]. Davioud-Charvet *et al* [206] describe the screening of a library of 1,4-naphthoquinones (NQ), which led to the identification and characterization of several NQ alcanoic acids as novel glutathione reductase inhibitors. From these lead structures, a series of prodrug esters were designed and synthesized in order to improve the cell penetration properties of the otherwise negatively charged agents. The best results were obtained with the most active prodrugs ester as inhibitors of PfGR (*Plasmodium falciparum* Glutathione Reductase) conjugated with a 4-anilinoquinoline. The double drug was successfully tested *in vitro* against both CQ-resistant and CQ-sensitive *P. falciparum* strains and *in vivo* against *P. berghei* in infected mice .

8.4-Antiviral

Human rhinoviruses (HRVs) are the single most significant causative agents of the common cold. HRV 3C-protease, a cysteine protease, plays a critical role in the replication cycle of HRVs and thus constitutes a potential therapeutic target for the control of HRVs and common cold. Singh *et al* established the structure-activity relationships of various naphthoquinones as HRV 3C-proteasa inhibitors and compared the activities of *ortho-* versus-*para-* quinones. The results indicated that the mode of action of the two classes of compounds is different [207].

Human cytomegalovirus (HCMV) disease is the most common life-threatening opportunistic viral infection in the inmunocompromised. HCMV protease, a serine protease, plays a critical role in capsid assembly and viral maturation and is an attractive target for antiviral chemotherapy. Slater *et al* investigated the interaction of various 1,4-naphthoquinones derivatives with HCMV protease. They identified potent irreversible naphthoquinones inhibitors of HCMV protease which covalently modify

Cys202 [208]. Human immunodeficiency virus (HIV) is the primary cause of acquired immunodeficiency syndrome (AIDS). In an effort to find new drugs preventing the growth of HIV, Masao *et al* developed an *in vitro* assay method of RNase H activity associated with reverse transcriptase (RT) from HIV-1. Some 1,4-naphthoquinones moderately inhibited RNase H activity [209]. Several natural occurring naphtoquinones have showed antiretroviral activity [210-211].

8.5-Other activities

Prokopenko *et al* described that some naphthoquinones normalized or increased the immune response in rats after intensive physical load. The naphthoquinones induced the development of immunostimulating properties in heavy red blood cells by direct action on cell membranes or indirectly through proteolytic enzymes secreted by hepatocytes [212]. Other activities described for several naphthoquinones related to lapachol are: antipsoriatic [213], larvicidal and insecticidal, [214-215] molluscicidal [216], and mutagenicity in salmonella [217]. The potential use of lapachol in cosmetics has been published too [218].

9-TOXICITY AND OTHER PHARMACOLOGICAL ASPECTS

The toxic effects of orally administered lapachol were studied in rodents, dogs, and monkeys [219-220]. The single dose oral LD50 was 0.621 g/Kg in BALB/c mice and >2.4 g/Kg in rats. The male and female mice LD50 values were 0.487 and 0.792 g/Kg. Beagle dogs were given daily oral doses of 0.25, 0.50, 1.0 or 2.0 g/Kg, 6 days a week, for a total of 24 doses without lethal effect. Cynomolgus monkeys were treated on the same schedule and received doses of 0.0625, 0.125, 0.25, 0.5, or 1.0 g/Kg/day. Signs of toxicosis in both dogs and monkeys included moderate to severe anemia, reticulocytosis, and elevated serum alk. phosphatase activity and prothrombin some times occurred in dogs. Leukopenia, thrombocytopenia, and azotemia were noted only in monkeys. The anemia was most pronounced during the first 2 weeks of treatment, and despite continued treatment, recovery was well underway in most animals by the end of the treatment period. Similar pharmacological studies were carried out by other authors [221]. Lapachol was also evaluated for its estrogenicity and antinidational activity in mouses. [222]. The importance of pharmacological studies of tumors caused by methylcholanthrene and lapachol have been emphasized by Sandoval *et al.* [223].

ACKNOWLEDGEMENTS
The authors gratefully acknowledge grants from the Ministerio de Ciencia y Tecnología (PPQ 2000, 1655, C02-01) and Instituto Canario de Investigación del Cancer (ICIC).

REFERENCES

[1]   Thomson, R.H. In *Naturally Occurring Quinones*. Ed. Chapman and Hall. London, **1987**, 139.
[2]   Hooker, S. *J. Chem. Soc.,* **1936**, *69*, 1355-1381.
[3]   Fieser, L. *J. Chem. Soc.* **1928**, *50*, 439-465.
[4]   Larsen, I.K.; Andersen, L.A.; Pedersen, B.F. *Acta Cryst.* **1992**, C48, 2009-2013.
[5]   Joshi, K.C.; Prakash, L.; Singh, P. *Phytochemistry*, **1973**, 12, 469-470.
[6]   Mueller, K.; Sellmer, A.; Wiegrebe, W. *J. Nat. Prod.* **1999**, *62*, 1134-1136.
[7]   de Santana C F; de Lima O g; d' Albuquerque I L; Lacerda A L; Martins D.G.A. Revista Do Instituto de antibioticos, Universidade Federal de Pernambuco, **1968**, *8*, 89-94.
[8]   Kustrak, D. *Farmaceutski Glasnik* , **2001**, *57*, 215-222.
[9]   Shimofuruya, Hiroshi; Fujii, Asami; Suzuki, Ikukatsu; Kunieda, Yoshihiko. *Nippon Kagaku Kaishi* , **2002**, *3*,481-483.
[10]   Garofano, Dona. U.S. Pat. Appl. Publ., **2001**, 5 pp., Cont.-in-part of U.S. Ser. No. 190,932.
[11]   Fujiwara, A.; Mori, T.; Iida, A.; Ueda, S.; Nomura, T.; Tokuda, H.; Nishino, H. *J. Nat. Prod.* **1998**, *61*, 629-632.
[12]   Pérez-Sacau E.; Estévez-Braun, A.; Ravelo, A.G.; Ferro, E.; Tokuda, H.; Mukainaka, T.; Nishino, H. Ferro, E. *Biorg. Med. Chem.* **2003**, *11*, 483-488.
[13]   Ertl, P.; Cooper, D.; Geoffrey A.; Slater, M.J. *Biorg. Med. Chem. Lett.* **1999**, *9*, 2863-2866.
[14]   Hudson, A.T.; Randall, M. *Parasitol.***1985**, *90*, 45-55.
[15]   Goulart, M.O.; Zani, C.I.; Tonholo, J.; Freitas, de Abreu, F.C. , Oliveira, A.B.; raaslan, D.S.; Starling, S. Chiari, E. *Biorg. Med. Chem. Lett.* **1997**, *7*, 2043-2048.
[16]   Krishnan, P.; Bastow, K.F.; *Biochem Pharmacol.* **2000**, *60*, 1367-1379.
[17]   Guiraud, P.; Steiman, R.; Campos-Takaki, G.M., Seigle-Murandi, F.; Simeon de Buochberg, M.S.; *Planta Med.* **1994**, *60*, 373-374
[18]   Hooker, S. *J. Chem. Soc.,* **1892**, *61*, 611-647.
[19]   Ettlinger, M. *J. Chem. Soc.* **1950**, *72*, 3090 .
[20]   Mann, J. In *Secondary metabolism*. Oxford University Press. **1978**, pp. 242-247.
[21]   Thomson, R.H., In *Naturally Occurring Quinones*, Academic Press, London, 1971, pp. 203-209.
[22]   Shetgiri, N.P; Kokitkar, S.V.; Sawant, S.N. *Acta Pol. Pharm.* **2001**, *58*, 133-135.

[23] Oesterle, O.A. *J. Chem. Soc.* **1916**, *112*, 505.

[24] Oesterle, O.A. *Arch. Pharm.* **1916**, *254*, 346-8.

[25] Prakash, L.; Singh, R. *Pharmazie*, **1980**, *35*, 649-50.

[26] de Lima, O. Goncalves; D'Alburquerque, I. L.; de Lima, C.; Dalia, M.H. *Rev. Inst. Antibiot., Univ. Recife* **1962**, *4*, 3-17.

[27] Burnett, A.R.; Thomson, R.H. *J. Chem. Soc.* **1967**, *21*, 2100-4.

[28] Burnett, A.R.; Thomson, R.H. *J. Chem. Soc.* **1968**, *7*, 850-3.

[29] Manners, G.D.;Jurd, L. *Phytochemistry* **1976**, *15*, 225-6.

[30] Krishna, J.C.; Singh, P.; Singh, G. *Indian J. Chem.* **1976**, *14*, 637-8.

[31] Prakash, L.; Singh R.; *Pharmazie*, **1980**, *35*, 813.

[32] Villegas, J.R.; Amato, S.; Castro, I.; Castro, O.; Jacobson, U. *Fitoterapia*, **1995**, *66*, 281-2.

[33] Grazziotin, J.D.; Schapoval, E.E.S.; Chaves, C.G.; Gleye, J. Henriques, A.T. *J. of Ethnopharmacology*, **1992**, *36*, 249-51.

[34] Zani, C.; de Oliveira, A. De Oliveira, G.; *Phytochemistry*, **1991**, *30*, 2379-81.

[35] Vidal-Tessier, A.M.; Delaveau, P.; Champion, B.; Jacquemin, H. *Ann. Pharm Franc.* **1988**, *46*, 55-7.

[36] Kenichiro, I.; Cheng-Chang, C.; Hiroyuki, I.; Kaouru, K. *J. Chem. Soc. Perkin Transactions 1.* **1981**, *11*, 2764-70.

[37] Jacome, R.L.; de Oliveira, A.; Raslan, D.S. Mullerr A.; Wagner, H. *Quimica Nova*, **1999**, *22*, 175-177.

[38] Shehata, I.A. *Bull. Fac. Pharm*, **2000**, *38*, 109-115.

[39] Houghton, P.; Pandey, R.; Hawkes, J. *Phytochemistry*, **1994**, *35*, 1602-3.

[40] Chauhan, A.K.; Dobhal, M.P.; Uniyal, P.N. *Herb. Pol.* **1988**, *34*, 3-5.

[41] Joshi, K.; Sharma, A.; Singh, P. *Planta Med.* **1986**, *1*, 71-2.

[42] Joshi. K.; Singh, P.; Pardasani, R.; Singh, G. *Planta Med.* **1979**, *37*, 60-3.

[43] Joshi, K.C.; Singh, P.; Pardasani, R.; Singh, G. *Planta Med.* **1978**, *34*, 219-21.

[44] Ito, C.; Katsuno, S.; Shinya, K.; Kondo,Y.;Tan, H.; Furukawa, H. *Chem. Pharm. Bull.* **2000**, *48*, 3439-3443.

[45] Viana, E.P.; Santa-Rosa, R.S.; almeida, S.S. Santos, L.S. *Fitoterapia*, **1999**, *70*, 111-112.

[46] Rafiullah M. K., Suleiman M. M. *Phytochemistry*, **1999**, *50*, 439-442.

[47] Sagrero-Nieves, L. *J. Nat. Prod.* **1986**, *49*, 547.

[48] Cannon, J.R.; Joshi, K.R.; McDonald, I. A.; Retallack, R.W.; Sierakowski, A.F.; Wong, L.C.H. *Tetrahedron Lett.* **1975**, *32*, 2795-8.

[49] Ali, S.; Singh, P.; Thomson, R.H.; *J. Chem. Soc. Perkin Trans. I.* **1980**, 257.

[50] Perry, Nigel B.; Blunt, John W.; Munro, Murray H. G. J. Nat. Prod.**1991**, *54*(4), 978-85.

[51] Inoue, K.; Ueda, S.; Shiobara, Y.; Kimura, I.; Inouye, H.. *J. Chem. Soc., Perkin Transactions I*, **1981**, *4* ,1246-58.

[52] Inouye, H.; Ueda, S.; Inoue, K.; Shiobara, Y.; Wada, I. *Tetrahedron Lett.* **1978**, *46*, 4551-4.

[53] Heide, L.; Leistner, E. *J. Chem. Soc. Chem. Comm.* **1981**, *7*, 334-6.

[54] Michael M.; Ronald H. T. *Phytochemistry*, **1973**, *12*, 1351-1353.

[55] Inoue, K; Ueda, S.; Nayeshiro, H.; Inouyet, H. *Phytochemistry*, **1983**, *22*, 737-741.

[56] Mehendale, A.; Thomson, R.H. *Phytochemistry*, **1975**, *14*, 801-802.

754

[57] Goulart, M.; Zani, C.L.; Tonholo, J.; Freitas, L.R. Abreu, F.C.; Oliveira, A.B. Raslan, D.S.; Starling, S.; Chiari, E. *Bioorg. Med. Chem.*1997, *15*, 2043-2048.

[58] Cannon, J. R.; Joshi, K. R.; McDonald, I. A.; Retallack, R. W.; Sierakowski, A. F.; Wong, L. C. H. *Tetrahedron Lett.*1975, *32*, 2795-8.

[59] Lothar, K; Li, G; Steglich, H,. *Liebigs Annalen der Chemie* 1982, *9*, 1722-9.

[60] Gupta, R. B.; Khanna, R. N. Current Science 1979, *48*, 977-9.

[61] Dosseh, C.; Tessier, A. M.; Delaveau, P. *Planta Med.* 1981, *43*, 360-6.

[62] Gupta, R.; Ravindranath, B.; Seshadri, T. R. *Phytochemistry*, 1972, *11*, 2634-2636.

[63] Moore R.E.; Singh, H. Chang, C.W.J. Scheueer, P.J. *J.Org. Chem.* 1966, 31, 3638.

[64] Sidhu, G. S.; Sankaram, A. V. B. *Tetrahedron Letters* 1971, *26*, 2385-8.

[65] Papageorgiou, V.P.; Assimopoulou, A.N.; Couladouros, E.A.; Hepworth, D.; Nicolaou, K.C. *Angew. Chem. Int. Ed.* 1999, *38*, 270-300.

[66] Shukla, Y.N.; Tandon, J.S.; Dhar, M.M. *Indian J. Chem.* 1973, *11*, 528-529.

[67] Afzal, M.; Tofeeq, M. *J. Chem. Soc. Perkin Trans. I* 1975, *14*, 1334-1335.

[68] Bai, G. Jin, X.J. *Chem. Res. Chin. Univ.* 1994, *10*, 263-265 [*Chem. Abstr.* **1995**, *122*, 183154h]

[69] Kyogoku, K; Terayama, H. Tachi, Y.; Suzuki, T.; Komatsu, M. *Shoyakugaku Zasshi* 1973, *27*, 24-30.[*Chem. Abstr.* **1974**, *80*, 112549u]

[70] Stepanenko, L. S.; Krivoshchekova, O. E.; Dmitrenok, P. S.; Maximov, O. B. *Phytochemistry* 1997, *46*, 565-568.

[71] Rohatgi, B. K.; Gupta, R. B.; Roy, D.; Khanna, R. N. *Indian Journal of Chemistry*, 1983, *22B*, 886-9.

[72] Gonzalez Gonzalez, A.; Cardona, R. J.; Medina, J. M.; Rodriquez Luis, F. *Anales de Quimica* 1974, *70*, 858-9.

[73] Lemos, T.L.G.; Costa, S.M.O.; Pessoa, O.D.L.; Braz-Filho, R. *Magn. Reson. Chem.* 1999, *37*, 908-911.

[74] Inouye, H.; Ueda, S.; Inoue, K.; Shiobara, Y. *Phytochemistry* 1981, *20*, 1707-1710.

[75] Itokawa, H.; Qiao, Y.; Takeya, K. *Phytochemistry* 1991, *30*, 637-640.

[76] Simoes, F.; Michavila, B.R.; García-Alvarez, M.C. *Phytochemistry* 1986, *25*, 755-756.

[77] Wenkert, E.; Campello, J.McChesney, J.D.; Watts, D.J. *Phytochemistry* 1974, *13*, 2545-2549.

[78] Peraza-Sánchez, S.; Chávez, D.; Hee-Byung, C.; Young Geun S.; García, R.; Mejía, M.; Fairchild, C. R ; Lane, K. E.; Menendez, A. T. ; Farnsworth, N.R.; Cordell, G.; Pezzuto, J. M.; Kinghorn, A. D. *J. Nat. Prod.* 2000, *63*, 492-495.

[79] Letcher, R.M.; Shirley, I.M. *Phytochemistry* 1992, *31*, 4171-4172.

[80] Dosseh, C.; Tessier, A. M.; Delaveau, P. *Planta Medica* 1981, *43*, 141-7.

[81] Joshi, K.C.; Singh, P.; Pardasani, R.T.; Singh, G. *Planta Med.* 1979, *37*, 60-63.

[82] Joshi, K.C.; Prakash, L.; Singh, P.; *Phytochemistry* 1973, *12*, 469-470.

[83] Rohatgi, B.K.; Gupta, R.B. Khanna, R.N. *Indian J. Chem.* 1983, *22B*, 886-889.

[84] Joshi, K.C.; Singh, P.; Pardasani, R.T. *Planta Med.* 1977, *32*, 71-75.

[85] Watanabe, K.; Kubota, M.; Hamahata, K.; Lin, Y.W. Usami, I. *Biochem. Pharm.* 1999, *57*, 763-774.

[86] Mock, J.; Murphy, S. T.; Ritchie, E.; Taylor, W. C. *Australian J. Chem.* **1973**,

26, 1121-30.

[87] Fujiwara, A.; Mori, T.; Iida, A.; Ueda, S.; Hano, Y.; Nomura, T.; Tokuda, H.; Nishino, H. J. Nat. Prod. **1998**, *61*, 629-632.

[88] Inouye, H.; Okuda, T.; Hayashi, T. *Chem. Pharm. Bull.* **1975**, *23*, 384.

[89] Ueda, S.; Inoue, K.; Shiobara, Y.; Kimura, I.; Inouye, H. *Planta Med.* **1980**, *40*, 168.

[90] Joshi, K.C.; Singh, P.; Pardasani, R.T. *Planta Med.* **1979**, *37*, 60-63

[91] Manners, G.; Jurd, L.; Stevens, K. *J. Chem. Soc. Chem. Comm.* **1974**, 388-389.

[92] Duarte,W.M.; Gottlieb, O.R.; Oliveira, G.C. *Phytochemistry* **1976**, *15*, 570.

[93] Binutu, Oluwatoyin A.; Adesogan, Kayode E.; Okogun, Joseph I.. *Planta Medica* **1996**, 62, 352-353.

[94] Gafner, Stefan; Wolfender, Jean-Luc; Nianga, Malo; Stoeckli-Evans, Helen; Hostetmann, Kurt. *Phytochemistry* **1996**, *42*, 1315-1320.

[95] Houghton, Peter; Pandey, Rita; Hawkes, Jane E. *Phytochemistry* **1994**, *35*, 1602-3.

[96] Villegas, J. R.; Amato, S.; Castro, I.; Castro, O.; Jacobson, U. *Fitoterapia* **1995**, *66*, 281-2.

[97] Grazziotin, J. D.; Schapoval, E. E. S.; Chaves, C. G.; Gleye, J.; Henriques, A. T. *J. of Ethnopharmacology* **1992**, *36*, 249-51.

[98] Vidal-Tessier, A. M.; Delaveau, P.; Champion, B.; Jacquemin, H. *Ann. Pharmaceutiques Francaises* **1988**, *46*, 55-7.

[99] Joshi, K. C.; Sharma, A. K.; Singh, P. *Planta Med.* **1986**, *1*, 71-2.

[100] Prakash, L.; Singh, R. *Pharmazie* **1980**, *35*, 122-3.

[101] Burnett, A. R.; Thomson, R. H. *J. Chem. Soc.* **1967**, *21*, 2100-4.

[102] Itoigawa, M.; Ito, C.; Tan, H.T.W.; Okuda, M.; Tokuda, H.; Nishino, H; *Cancer Letters* **2001**, *174*, 135-139.

[103] Ito, C.; Katsuno, S.; Kondo, Y.; Furukawa, T.H. *Chem. Pharm. Bull.* **2000**, *48*, 339-343.

[104] Inoue, K; Chen, C.; Inouye, H.; Kuriyama, K., *J. Chem. Soc., Perkin Trans. 1* **1981**, *11*, 2764-70.

[105] Inouye, H.; Ueda, S.; Inoue, K.; Shiobara, Y.; Wada, I. *Tetrahedron Lett.***1978**, *46*, 4551-4.

[106] Fujiwara, A; Mori, T.; Iida, A.; Ueda, S.; Hano, Y.; Nomura, T.; Tokuda, H.; Nishino, H. *J. Nat. Prod.* **1998**, *61*, 629-632.

[107] Heltzel, C. E.; Gunatilaka, A. A. L.; Glass, T. E.; Kingston, D. *J. Nat. Prod.* **1993**, *56*, 1500-5.

[108] Heltzel, C. E.; Gunatilaka, A. A. L.; Glass, T. E.; Kingston, David G. I. *Tetrahedron* **1993**, *49*, 6757-62.

[109] Chen, C. C.; Lee, M. H. *Huaxue* **1986**, *44*, 61-4.

[110] Rao, M. M.; Kingston, D.G. *J. Nat. Prod.* **1982**, *45*, 600-4.

[111] Sanderman, W.; Simatupang, M. H.; Wendeborn, W. *Naturwissenschaften* **1968**, *55*, 38.

[112] Fujimoto, Y.; Eguchi, T.; Murasaki, C.; Ohashi, Y.; Kakinuma, K.; Takagaki, H.; Abe, M.; Inazawa, K.; Yamazaki, K. et al. Dep. Chem., Tokyo Inst. *J. Chem. Soc, Perkin Trans. I* **1991**, *10*, 2323-7.

[113] Inoue, K.; Ueda, S.; Nayeshiro, H.; Inouye, H.. *Phytochemistry* **1983**, *22*, 737-41.

[114]   Ruedi, P.; Eugster, C.H. *Helv. Chim. Acta* **1977**, *60*, 945-7.
[115]   Zani, C.L.; de Oliveira, A.B. *Phytochemistry* **1991**, *30*, 2379-2381.
[116]   Itokawa, H.; Quiao, Y.; Takeya, K. *Phytochemistry* **1991**, *30*, 637-640.
[117]   Girard, M.; Kindack, D.; Dauson, B.A.; Ethier, J.-C., Awang, D.V.C. Gentry, A.H. *J. Nat. Prod.* **1988**, *51*, 1023-1024.
[118]   Awang, D.V.C.; Kindack, D.; Dawson D.V.C *J. Chromatogr.* **1986**, *368*, 439.
[119]   Müller, W.U.; Leistner, E. *Phytochemistry* **1978**, *17*, 1739-1742.
[120]   Werbin, H.; Strom, E.T.; *J. Am. Chem. Soc.* **1968**, *90*, 7296-7301.
[121]   Zeeck, A.; Christiansen, P. *Liebigs Ann. Chem.* **1969**, *724*, 172-182.
[122]   Evans, D.A.; Hoffman, J.M. *J. Am. Chem. Soc.* **1976**, *98*, 1983-4.
[123]   Fieser, L.F. *J. Biol. Chem.* **1940**, *133*, 391-396.
[124]   Desai, H.K.; Gawad, D.H.; Govindachari, T.R. *Indian J. Chem.* **1970**, *8*, 851-853.
[125]   Joshi, K.C.; Prakash, L.; Singh, P. *Phytochemistry* **1973**, *12*, 942-943.
[126]   Roberts, J.C.; Thompson, D.J. *J. Chem. Soc.* **1971**, 3488-3492.
[127]   Dauson, B.A.; Girard, M.; Kindack, D.; Fillion, J.; Awang, D.V.C. *Mag. Reson. Chem.* **1989**, *27*, 1176-1177.
[128]   Benedetti-Doctorovich, V.; Escola, N.; Burton, G. *Magn. Reson. in Chemistry* **1998**, *36*, 592-532.
[129]   Inouye, H.; Ueda, S.; Inoue, K.; Shiobara, Y. *Phytochemistry*, **1981**, *20*, 1707-1710.
[130]   Panichayupakaranat, P. *Pharmaceutical Biology*, **2001**, *39*, 1-4.
[131]   Decwick, P.M. In *Medicinal Natural Products*, J. Wiley & Sons, **2001**, 158-164.
[132]   Paterno, E. *Gazz. Chim. Ital.* **1882**, *12*, 337-622.
[133]   Hooker, S.C. *J. Am. Chem. Soc.* **1936**, *58*, 1168-1173.
[134]   Thomson R.H. in *Naturally Occurring Quinones*, Academic Press, London and New York, **1971**, pp. 209-211.
[135]   Dudley, K.; Chiang, R.W. *J.Org.Chem.* **1969**, *34*, 120-126.
[136]   Ball, E.G. *J. Biol. Chem.* **1936**, *114*, 649-655.
[137]   Hooker, S. *J. Chem. Soc.* **1936**, *58*, 1174-1178.
[138]   Fieser, L. Hartwell, J.L.; Seligman, A.M. *J. Chem. Soc.* **1936**, *58*, 1223-1228.
[139]   Fieser, L. *J. Chem. Soc.* **1926**, *48*, 3201-3214.
[140]   Cocquet, G.; Rool, P.; Ferroud, C. *Tetrahedron Lett.* **2001**, *42*, 839-841.
[141]   Sawhney, S.S. Bains, S.S. *Termochimica Acta* **1983**, *71*, 381-6.
[142]   Sawhney, S.S.; Matta, S.D.; Jain, Renu; Kashyap, R.K. *Termochimica Acta* **1983**, *70*, 367-71.
[143]   Sawhney, S.S.; Vohra, Neelam. *Acta Ciencia Indica, Chemistry* **1980**, *6*, 183-7.
[144]   Sawhney, S.S. Bhatia, B.M.L. *Termochimica Acta* **1981**, *43*, 243-7.
[145]   de Oliveira, E.H.; Medeitros, G.E.A.; Peppe, C. Brow, M.A. Tuck, D.G. *Can. J. Chem.* **1997**, *75*, 499-506.
[146]   Carvalho,C.M.; Ferreira, V.F.; Pinto, A.V.; Pinto, M.C.M; Harrison, W. *Dyes and Pigments* **2002**, *52*, 209-214.
[147]   Ferreira, V.F.; Pinto, A.V.; Pinto, M.C.; Silva, M. *Anais da Academia Brasileira de Ciencias* **1987**, *59*, 329-333.
[148]   Ferreira, V.F.; Pinto, A.V.; Pinto, M.C.; Silva, M. *J. Chem. Res. S.* **1987**, *1*, 26.
[149]   Sun, J.S.; Geiser, A.H.; Frydman, B. *Tetrahedron Lett.* **1998**, *39*, 8224-8225.

[150] Fieser, L.F. *J. Am. Chem. Soc.* **1927**, *49*, 857-864.

[151] Schaffner-Sabba, K.; Schmidt-Ruppin, K.H.; Wehrli, W.; Schuerch, A.R.; Wasley, W.F. *J. Med. Chem.* **1984**, *27*, 990-994.

[152] Maruyama, K.; Naruta, Y. *Chemistry Lett.* **1977**, 847-850.

[153] Alves, G.B.C.; Lopes, R.S.C.; Lopes, C.C.; Snieckus, V. Synthesis 1999, *11*, 1875-1877.

[154] Schaffner-Sabba, K.; Schmidt-Ruppin, K.H.; Wehrli, W.; Schuerch, A.R.; Wasley, J.W.F. *J. Med. Chem.* **1984**, *27*, 990-994.

[155] Brandao, M.A.; Braga de Oliveira, A.; Snieckus, V. *Tetrahedron Lett.* **1993**, *34*, 2437-2440.

[156] Kapoor, N.K.; Gupta, R.B.; Khanna, R.N. *Tetrahedron Lett.* **1980**, *21*, 5083-5084

[157] Amaral, A.C.; Barnes, R.A. *J.Het.Chem.* **1992**, 1457-1460.

[158] Nair, V.; Treesa, P.M. *Tetrahedron Lett.* **2001**, *42*, 4549-4551.

[159] Driscoll, J.S.; Hazard, G.F.; Wood, H.B. Goldin, A. *Cancer Chemother. Rep. Part 2*, **1974**, *4* , 1-362.

[160] Chenna, P.H.; Benedetti-Doctorovich, V.; Baggio, R.F.; Garland, M.T.; Burton, G. *J. Med. Chem.* **2001**, *44*, 2486-2489.

[161] Hodnett, E.M.; Wongwiechintana, C.; Dunn III, W.J.; Marrs, P. *J. Med. Chem.* **1983**, *26*, 570-574.

[162] Benedetti-Doctorovich, V.; Burgess, E.M.; Lampropoulos, J.; Lednicer, D.; Derveer, D.; Zalkow, L.H. *J. Med. Chem.* **1994**, *37*, 710-712.

[163] Oliveira, M.F.; Lemos, T.L.G.; De Mattos, M.C.; Segundo, T.A.; Santiago, G.M.P.; Braz-Filho, R. *Anais da Academia Brasileira de Ciencias* **2002**, *74(2)*, 211-221.

[164] da Silva, A.; Buarque, C.; Brito, F.; Aurelian, l.; Macedo, L.; Malkas, L.Hickey, R.; Lopes, D.; Noel, F. *Biorg. Med. Chem.* **2002**, *10*, 2731-2738.

[165] Linardi, M.C.; Oliveira, M.M.; Sampaio, M.R.P. *J. Med. Chem.* **1975**, *18(11)*, 1159-1161.

[166] O'Brien, P.*J. Chem. Biol. Interact.* **1991**, *80*, 1-41.

[167] Zuman, P. In *Substituent effects in organic polarography;* Plenum Press: New York, **1967**.

[168] Lazo, J.S.; Aslan, D.C.; Southwick, E.C.; Cooley, K.A.; Ducruet, A.P.; Joo, B.; Vogt, A.; Wipf, P. *J. Med. Chem.* **2001**, *44*, 4042-4049.

[169] Lyon, M.A.; Ducruet, A.P.; Wipf, P.; Lazo, J.S.; *Nature Reviews.* **2002**, *1*, 961-976.

[170] Neder, K.; Marton, L.J.; Frydman, B. *Cell Mol. Biol.* **1998**, *44*, 465-474.

[171] Frydman, B.; Marton, L.; Sun, J.S.; Neder, K.; Witiak, D.T.; Liu, A.A.; Wang, H-M.; Mao, Y.; Wu, H-Y.; Sanders, M.M.; Liu, L.F. *Cancer Res.* **1997**, *57(4)*, 620-627.

[172] Oliveira-Brett, A.M.; Goulart, M.O.F.; Abreu, F. *Bioelectrochemistry* **2002**, *56*, 53-55.

[173] Krishnan, P.; Bastow, K.F. *Biochem. Pharm.* **2000**, *60*, 1367-1379.

[174] Kumagai, Y.; Tsurutani, Y.; Shinyashiki, M.; Homma-Takeda, S.; Nakai, Y.; Yoshikawa, T.; Shimojo, N. *Environmental Toxicology and Pharmacology.* **1997**, *3(4)*, 245-250.

758

[175]   Planchon, S.M.; Wuerzberger, S.; Frydman, B.; Witiak, D.T.; Hutson, P.;
        Church, D.R.; Wilding, G.; Boothman, D.A. *Cancer Res.* **1995**, *55*, 3706-3711.
[176]   Watanabe, K.; Kubota, M.; Hamahata, K.; Lin, Y-W.; Usami, I. *Biochem.
        Pharm.* **2000**, *60*, 823-830.
[177]   Manna, S.K.; Gad, Y.P.; Mukhopadhyay, A.; Aggarwal, B.B. *Biochem. Pharm.*
        1999, *57*, 763-774.
[178]   Miller, G. *Epstein-Barr virus. In Virology, Second Edition,* B.N. Fields, ed.
        (New york: Raven Press), **1990**, pp.1921-1958.
[179]   Ueda, S.; Umemura, K.; Dohguchi, K.; Matsuzaki, T.; Tokuda, H.; Nishino, H.;
        Iwashima, A. *Phytochemistry.* **1994**, *36*, 323-325.
[180]   Kapadia, G.J.; Balasubramanian, V.; Tokuda, H.; Konoshima, T.; Takasaki, M.;
        Koyama, J.;Tagahaya, K.; Nishino, H. *Cancer Lett.* **1997**, *113*, 47-53.
[181]   Ito, C.; Itoigawa, M.; Furukawa, H.; Ichiishi, E.; Mukainaka, T.; Okuda, M.;
        Ogata, M., Tokuda, H.; Nishino, H. *Cancer Lett.* **1999**, *142*, 49-54.
[182]   Fujiwara, A.; Mori, T.; Iida, A.; Ueda, S.; Hano, Y.; Nomura, T.; Tokuda, H.;
        Nishino H.; Furukawa, H. *J. Nat. Prod.* **1998**, *61*, 629-632.
[183]   Ito, C.; Itoigawa, M.; Tan, H.T.W. Tokuda, H.; Mou, X.Y.; Mukainaka, T.;
        Ishikawa, T.; Nishino H.; Furukawa, H. *Cancer Lett.* **2000**, *152*, 1287-1292.
[184]   Itoigawa, M.; Ito, C.; Tan, H.T.W.; Okuda, M.; Tokuda, H.; Nishino H.;
        Furukawa, H. *Cancer Lett.* **2001**, *174*, 135-139.
[185]   Haraguchi, H.; Yokoyama, K.; Oike, S.; Ito, M.; Nozaki, H. *Archives of
        Microbiology.* **1997**, *167*, 6-10.
[186]   Gafner, S.; Wolfender, J-L.; Nianga, M.; Stoeckli-Evans, H.; Hostetmann, K.
        Phytochemistry **1996**, *42(5)*, 1315-1320.
[187]   Kuwahara, S.; Nemoto, A.; Hiramatsu, A. *Agric. Biol. Chem.* **1991**, *55(11)*,
        2909-2911.
[188]   Kesteleyn, B.; De Kimpe, N. *J. Org. Chem.* **2000**, *65*, 640-644.
[189]   Cameron, D.W.; Crosby, I.T.; Feutrill, G.I. *Tetrahedron Lett.* **1992**, *33*, 2855-
        2856.
[190]   Vasanth, S.; Jayakaran, D.S.; Raj, V.P.; Srinivasan, V. *J. of the Indian Chem.
        Soc.* **2002**, *79(9)*, 765-767.
[191]   Oliveira, C.G.T.; Miranda, F.F.; Ferreira, V.F.; Freitas, C.C.; Rabello, R.F.;
        Carballido, J.M.; Correa, L.C.D. *J. of the Brazilian Chem. Soc.* **2001**, *12(3)*,
        339-345.
[192]   Campos-Takaki, G.M.; Steiman, R.; Seigle-Murandi, F.; Silva, A.A.; Bieber, L.
        Revista de Microbiología **1992**, *23(2)*, 106-111.
[193]   Guiraud, P.; Steiman, R.; Campos-Takaki, G.M.; Seigle-Murandi, F.;
        Buochberg, M.S. *Planta Med.* **1994**, *60*, 373-374.
[194]   Fournet, A.; Muñoz, V. *Curr.Topics in Med. Chem.* **2002**, *2*, 1215-1237.
[195]   Pinto, A.V.; Pinto, C.N.; Do Carmo, M.; Pinto, M.C.F.R; Santa Rita, R.;
        Pezzella, C.A.C.; Castro, S.L. *Arzneimittel-Forschung.* **1997**, 47(1), 74-79.
[196]   Pinto, C.N.; Dantas, A.P.; De Moura, K.C.G.; Emery, F.S.; Polequevitch, P.F.;
        Pinto, M.C.F.R.; Castro, S.L.; Pinto, A.V. *Arzneimittel-Forschung.* **2000**,
        50(12), 1120-1128.
[197]   Dos Santos, A.F.; Ferraz, P.A.L.; Caxico de Abreu, F.; Chiari, E.; Goulart,
        M.O.F. ; Sant' A. *Planta Med.* **2001**, *67*, 92-93.

759

[198] Neves-Pinto, C.; Malta, V.R.S.; Pinto, M.C.F.R.; Santos, R.H.A.; Castro, S.L.; Pinto, A.V. *J. Med. Chem.* **2002**, *45*, 2112-2115.
[199] Morello, A. *Comp. Biochem. Physiol.* **1988**, *90C*, 1-12.
[200] Goulart, M.O.F.; Zani, C.L.; Tonholo, J.; Freitas, L.R.; Abreu, F.C.; Oliveira, A.B.; Raslan, D.S.; Starling, S.; Chiari, E. *Bioorganic Med. Chem. Lett.* **1997**, *7*, 2043-2048.
[201] Ribeiro-Rodrigues, R.; Santos, W.G.; Oliveira, A.B.; Snieckus, V.; Zani, C.L.; Romanha, A.J. *Bioorg. Med. Chem. Lett.* **1995**, *5*, 1509-1512.
[202] Salmon-Chemin, L.; Lemaire, A.; De Freitas, S.; Deprez, B.; Sergheraert, C.; Davioud-Charvet, E. *Bioorg. Med. Chem. Lett.* **2000**, *10*, 631-635.
[203] Salmon-Chemin, L.; Buisine, E.; Yardley, V.; Kohler, S.; Debreu, M-A.; Landry, V.; Sergheraert, C.; Croft, L.; Krauth-Siegel, R.L.; Davioud-Charvet, E. *J. Med. Chem.* **2001**, *44*, 548-565.
[204] Teixeira, M.J.; Almeida, Y.M.; Viana, J.R.; Holanda, J.G.F.; Rodrigues, T.P.; Romulo, J.; Prata, C.; Coelho, I.C.B.; Rao, V.S.; Pompeu, M.M.L. *Phytother. Res.* **2001**, *15*, 44-48.
[205] Kayser, O.; Kiderlen, A.F.; Laatsch, H.; Croft, S.L. *Acta Tropica.* **2000**, *77*, 307-314.
[206] Davioud-Charvet, E.; Delarue, S.; Biot, C.; Schwöbel, B.; Boehme, C.C.; Müssigbrodt, A.; Maes, L.; Sergheraert, C.; Grellier, P.; Schirmer, R.H.; Becker, K. *J. Med. Chem.* **2001**, *44*, 4268-4276.
[207] Singh, S.B.; Graham, P.L.; Reamer, R.A.; Cordingley, M.G. *Bioorganic Med. Chem. Lett.* **2001**, *11*, 3143-3146.
[208] Ertl, P.; Cooper, D.; Allen, G.; Slater, M. *Bioorganic Med. Chem. Lett.* **1999**, *9*, 2863-2866.
[209] Byung-Sun, M.; Miyashiro, H.; Hattori, M. *Phytotherapy Research.* **2002**, *16(2)*, 57-62.
[210] Kraus, G.A.; Melekhov, A.; Carpenter, S.; Wannemuhler, Y.; Petrich, J. *Bioorganic Med. Chem.* **2000**, *10*, 9-11.
[211] Vlietinck, A.J.; Bruyne, T.; Apers, S.; Pieters, L.A. *Planta Med.* **1998**, *64*, 97-109.
[212] Rybnikov, V.N.; Laskova, I.L.; Prokopenko, L.G. *Antibiotics and Chemotherapy.* **1997**, *42*, 6-10.
[213] Mueller, K.; Sellmer, A.; Wiegrebe, W. *J. Nat. Prod.* **1999**, *62(8)*, 1134-1136.
[214] Oliveira, M.F.; Lemos, T.L.G.; De Mattos, M.C.; Segundo, T.A.; Santiago, G.M.P.; Braz-Filho, R. *Anais da Academia Brasileira de Ciencias.* **2002**, *74*, 211-221.
[215] Jacobsen, N.; Pedersen, L.E.K.; Cheminova, A-S.; Lemvig, D. *Pesticide Science.* **1986**, *17(5)*, 511-516.
[216] Feitosa, S.A.; Ferraz, P.A.L.; Pinto, A.V.; Pinto, M.C.F.R.; Goulart, M.O.F.; Santa'Ana, A.E.G. *International J. for Parasitology.* **2000**, *30(11)*, 1199-1202.
[217] Tikkanen, L.; Matsushima, T.; Natori, S.; Yoshira, K. *Mutation Research.* **1983**, *124(1)*, 24-34.
[218] Franco, E.P.M.; De Gouvea, M.C.B.L.F.; Barreto, D.W. *Aerosol & Cosmetics.* **1986**, *8(48)*, 6-13.
[219] Oleson, J.J.; Morrison, R. K.; Brown, D.E.; Timmens, E.K.; Tassini, R.A. U.S. Gov. Res. Develop. Rep. **1967**, *67*, 55-6.

[220]   Morrison, R.K.; Brown, D.E.; Oleson, J.; Cooney, D.A.; Smith, J.L. *Toxicol. Appl. Pharm.* **1970**, 17, 1-11.

[221]   Magalhaes, J.R.; Ramos, P.R.; *Anais Acad. Bras. Ciencias* **1970**, *42*, 257-261.

[222]   Vinecta, S.; Sumita, Y.; Anita, N. *Phytotherapy Res*. **1995**, *9*, 139-141.

[223]   Sandoval, N.A.; Rodriguez, C.P.; Martinez, N.R. *Acta Fisiol. Farmacol. Terap. Latinoamericana* **1996**, 46, 257-264.

Atta-ur-Rahman (Ed.) *Studies in Natural Products Chemistry, Vol. 29*
© 2003 Elsevier B.V. All rights reserved

THE CHEMISTRY AND PHARMACOLOGY OF THE GENUS *DORSTENIA* (MORACEAE)

BONAVENTURE T. NGADJUI[a] BERHANU M. ABEGAZ[b]

[a]*Department of Organic Chemistry, Faculty of Science ,University of Yaounde 1 BP 812 Yaounde, Cameroon*
[b]*Department of Chemistry , Faculty of Science , University of Botswana, Private Bag, 0022, Gaborone, Botswana*

ABSTRACT: The genus *Dorstenia* contains many plants that are used as anti-infection, anti-snakebite and anti-rheumatic remedies in the medicinal plant therapy of several countries in Africa, Central and South America. The genus is now recognized as a rich source of prenyl and geranyl-substituted coumarins, chalcones, flavanones, flavones and flavonols. Three naturally occurring styrenes were reported from *D. barnimiana*. The geranyl substituted furocoumarin, O-[3-(2,2-dimethyl-3-oxo-2H-furan-5-yl)butyl]bergaptol, has been found in many species. *D. poinsettifolia* furnished the unusual substituted 4-phenyldihydrofurocoumarin. Prenylated and geranylated flavonoids have so far been reported only from African *Dorstenia* species. 5, 3'-Digeranyl-3,4,2',4'-tetrahydroxychalcone isolated from *D. proropens*, is the only example of a *bis*-geranylated chalcone in the literature. A part from 7,8- (2,2-dimethylpyrano-4'-methoxyflavanone (dorspoinsettifolin) from *D. poinsettifolia* and 3',4'- (2,2-dimethylpyrano-6-prenyl-7-hydroxyflavanone (dinklagin A) from *D. dinklagei*, all flavanone derivatives named dorsmanins A - J described here are from *D. mannii*. These dorsmanins appear to be derived from 6,8-diprenyl-5,7,3',4'-tetrahydroxyflavanone and a possible biosynthetic scheme for them is presented. *D. Zenkeri* yield a bichalcone derivative which is believed to arise from dorstenone another prenylated bichalcone isolated from *D. barteri*. The latter is probably formed via an enzymatic Diels-Alder reaction of 3'-prenyl-4,2',4'-trihydroxychalcone (isobavachalcone) and its dehydroderivative. *D. psilurus* provided all the triprenylated flavonoids found so far in *Dorstenia* and, of these, dorsilurin E is unique having ring B of the flavonoïd structure modified to a dienone. The pharmacological data on this genus are scanty. Extracts of *D. multiradiata* show antileishmanial activity. Extracts and /or compounds from other species show anti-inflammatory, analgesic, anti-oxidant and citotoxic activities.

1- INTRODUCTION

The genus *Dorstenia* (Moraceae) consisting of approximately 170 mostly tropical species [1] is indigenous to many countries in Africa, Central and South America. *Dorstenia* is a large genus occurring in the tropics around the world. There are succulent and non-succulent species. Most of the succulent species come from Africa. It belongs to the fig and mulberry family, and has also an unusual flower arrangement. The flowers are grouped in a structure called hypanthodium, and many in this genus have a common name of "shield flowers". When the seeds are ripe, they are expelled at distances of several feet. *Dorstenia* species are used as anti-snakebite, anti-infection and anti-rheumatic remedies in the medicinal plant therapy. *D. psilurus* in Cameroon , *D. brasiliensis* in Brazil and *D. contrajerva* in Panama and Mexico are used for such purposes. In Addis-Ababa, Ethiopia, *D. barnimiana* is sold in Merkato for the treatment of a variety of diseases but most significantly as remedy for goots. In Cameroon these plants are used generally to treat infections and wounds while the rhizomes of *D. psilurus* are reputed spices for the ethnodietary preparation *na'a poh* [2]. "Carapia" is a *Dorstenia* based drug formulation in the cultural medicine of Brazil used for the treatment of skin diseases.

The first report on a plant belonging to this genus was that of Casagrande et al in 1974 describing the isolation and characterization of syriogenin , a cardenolide steroid [3]. In 1988 Woldu et al reported the presence of unusual styrenes and benzofuran derivatives in *D. barnimiana* [4]. Interest in this genus has been increased over the last decade with close to 45 papers published on over more than 25 *Dorstenia* species. These reports appear to be based on studies conducted on plants originating from Brazil, Mexico, Panama, Ethiopia, Tanzania and Cameroon. The chemistry of the genus *Dorstenia* has recently been reviewed [5]. Besides the usual sterols and fatty acids this genus is recognised as rich source of styrenes, benzofuran derivatives, prenylated and geranylated coumarins, C-prenylated flavonoids and terpenoids,. Many of these secondary metabolites are referred to in the literature only by their botanically derived trivial names. The first section of this review deals with benzofuran and coumarin derivatives. The ethnomedical information of some species is described at the first mention of the plant and this information is concentrated in this first section. In the remaining sections flavonoids and terpenoids are discussed. Flavonoids represent the most impressive structural diversity observed for a single genus. The chalcones are grouped into one section; this is then followed by C-prenylated flavanones, flavones and flavonols grouped according to the degree of prenylation and by the geranylated ones. The review ends with a discussion of the scanty data available on the pharmacology of crude extracts and pure compounds isolated from the genus *Dorstenia*.

2- BENZOFURAN AND COUMARIN DERIVATIVES

Coumarins have been recognised as widely distributed plant products. The vast majority carry an oxygen substituent at C-7; consequently 7-hydroxycoumarin (umbelliferone) (1) is often regarded as the parent , in a structural sense and also biogenetically, of a large number of the structurally more complex coumarins. A number of coumarins and benzofuran derivatives have been isolated from nine *Dorstenia* species most of which are originated from Central and South America. These species are *D. brasiliensis* from Brazil [6], *D. excentrica*, *D. drakena* and *D. lindeniana* from Mexico [7], *D. contrajerva* from Panama [8] and Mexico and *D. gigas* [11] collected at Socotra Island, Yemen; the African ones being *D. barnimiana* from Ethiopia [4], *D. poinsettifolia* [9], *D. psilurus* [10] and *D. elliptica* [Ngadjui, B.T., Fotso, S., Dongo, E., Abegaz, B.M., unpublished results] from Cameroon. In Addis Ababa *D. barnimiana* is sold for the treatment of gout and also for various skin diseases. Samples of *D. barnimiana* bought from Merkato in Addis Ababa as well as tubers of the plant collected from a site just south of the city led to the isolation of the first examples of natural styrenes (2), and (3), which contain a furan moiety. These, as well as 4, were found to be devoid of any significant antimicrobial activity when tested against a variety of organisms [4]. In Cameroon, a decoction of the leaves of *D. psilurus* is used to treat rheumatism, snakebites, headache and stomach disorders. Examination of the rhizomes of *D. psilurus* led to the isolation of psoralen (5) [10] in addition to the benzofuran derivative (6). This methyl ester 6 and the corresponding ethyl ester have been synthesized from psoralen [12] in two steps by partial hydrogenation and subsequent treatment with a 1:1 mixture of ethanol and methanol. *D. excentrica* yielded umbelliferone (1), psoralen (5) and a diastereoisomer prandiol (7), with its acetate 8 [7]. *D. cayapiaa* yielded the unusual furanocoumarin (9) [13] which was found in other *Dorstenia* species [6-8, 14]. *D. brasiliensis* is used in Brazil for treating malaria, typhoid fever and against snakebites. Kuster et al [6] have identified from this species bergapten (10) psoralen (5), O-[3-(2,2-dimethyl-3-oxo-2H-furan-5-yl)butyl]bergaptol (9) and traces of a compound which from mass spectroscopic data was concluded to be a dihydroderivative of 9.

1

2 R=H
3 R=OMe

4

5 R=H
10 R=OMe
16 R=OH
17 R=O-αL-Rha-(1→6)-βD-Glc

6

9

8

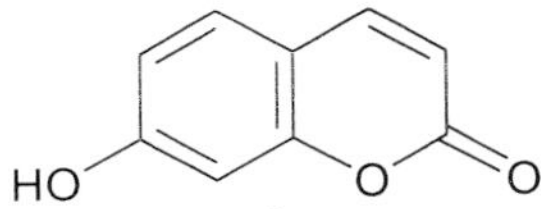

In a search for biologically active substances from South American medicinal plants [15, 16] the roots of this plant (*D. brasiliensis*) were recently been reinvestigated [17]: (1"-hydroxy-1"-methylethyl)-psoralen (**11**), (2'S, 3'R)-3'-hydroxymarmesin (**12**), (2'S, 3'R)-3'-hydroxymarmesine 4'-O-β-D-glucopyranoside (**13**) and (2'S)-marmesin 4'-O-α-L-rhamnopyranosyl(1→6)-O-β-D-glucopyranoside (**14**) Fig.(**1**) were isolated and characterized. **11** was initially isolated from the genus *Phebalium* [18] and synthesized by Zubia et al [19] and Stanjek et al [20]. **12** was isolated also from *Brosimum gaudichaudii* [21]. The monoglucoside (**13**) was initially obtained from *Angelica archangelica* and *A. silvestris* [22]. **14** was also identified in the roots of *Murraya koenigii* [23].

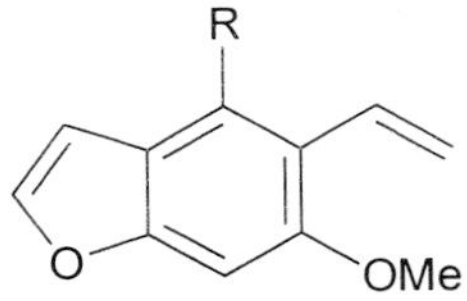

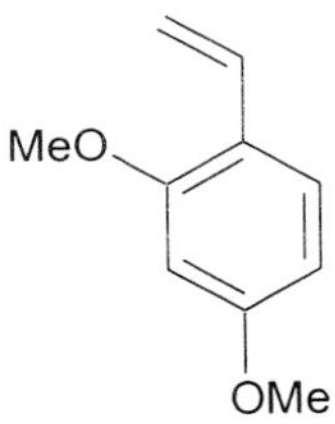

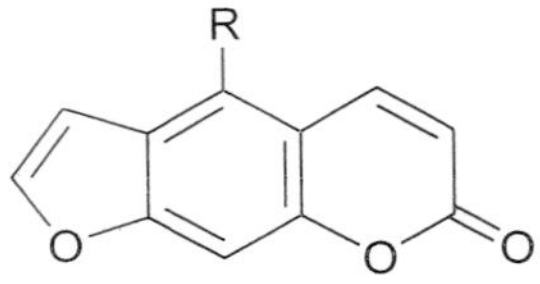

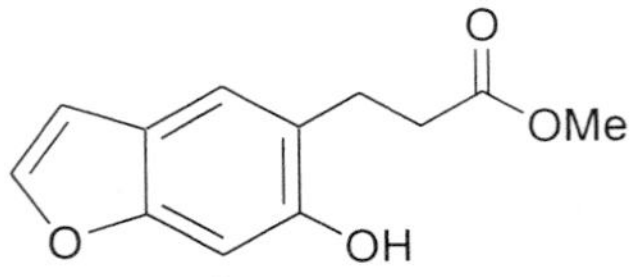

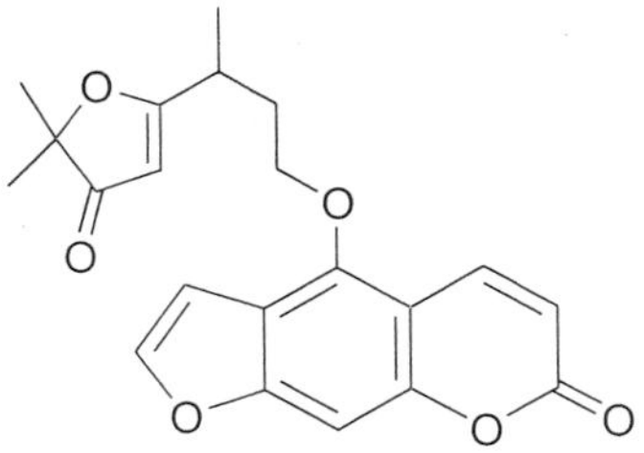

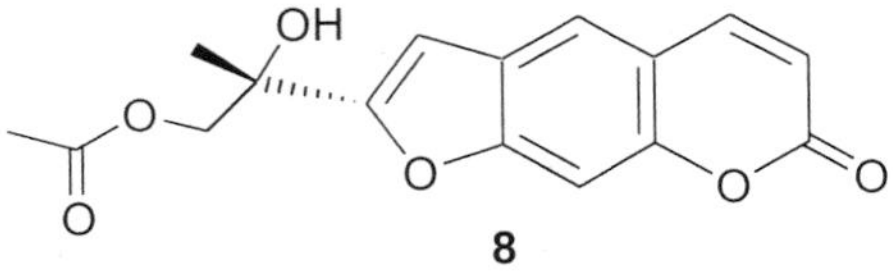

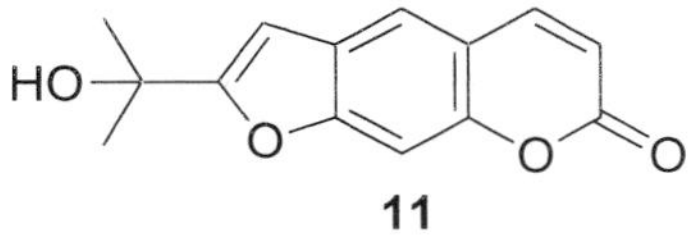

11

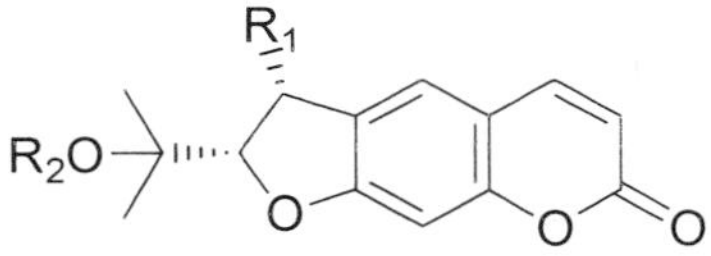

12 R₁=OH R₂=H
13 R₁=OH R₂=β-D-Glc
14 R₁=H   R₂=αL-Rha(1→6)-β-D-Glc

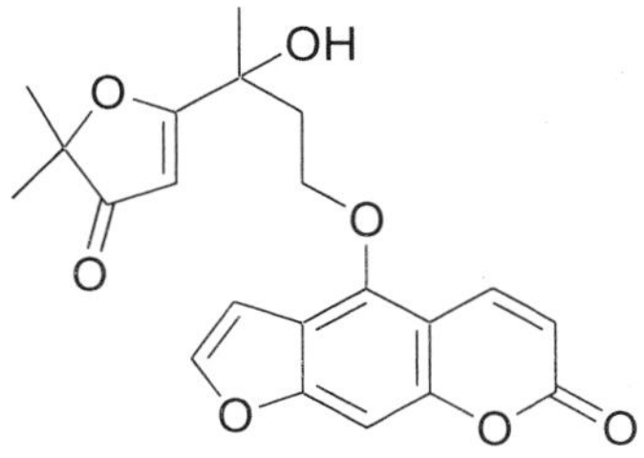

15

Fig. (1). Dihydrofurocoumarins of *D. brasiliensis*

12, 13 and **14** are three dihydrofuranocoumarins. *D. contrajerva* is a medicinal plant used by the Kuna Indians of Panama as a cold remedy and for the treatment of snakebites and muscle aches and for the treatment of rattlesnake bites in South-eastern Mexico. Swain et al [14] examined this plant and isolated bergapten and a derivative of the same furanocoumarin whose structure was subsequently revised as **9** by Terreaux et al [7] and confirmed by X-ray crystallography. This same plant also yielded a dihydrofurocoumarin, dorsteniol (prandiol) (**7**). The structure and relative stereochemistry of dorsteniol was confirmed by total synthesis and X-ray data on the acetate (**8**) [8]. Cameroonian *D. elliptica* also gave two geranylated furanocoumarins **9** and **15** in addition to bergapten [Ngadjui, B.T., Fotso, S., Dongo, E., Abegaz, B.M., unpublished results] .The spectroscopic data of Compounds **15** and **9** are too close; the difference is the additional 16 amu in **15**. Recently Caceres et al [23a] have isolated bergaptol (**16**) and a new glycosylated furanocoumarin, O-α-L-rhamnopyranosyl-(1→6)-β-D-glucopyranosyl-bergaptol (**17**) from the aerial parts of *D. contrajerva*. Rojas-Lima et al [24] have recently reported known and novel furocoumarins from the roots of *D. excentrica*, *D. drakena* and *D. lindeniana*. These authors reveal that the roots of several *Dorstenia* species are used in Mexican folk medicine against skin diseases. *D. drakena* yielded bergapten and **9**. A study of light-activated toxins of the Moraceae revealed the occurrence of phototoxins in all investigated species. The phototoxicity was most likely due to the presence of psoralen and bergapten [25]. *D. poinsettifolia* is used in Cameroon for the treatment of yaws and infected wounds. The twigs of this plant yielded Poinsettifolactone (**18**), a novel 4-phenyldihydrocoumarin [9]. The structure of **18** was established on the basis of spectral analysis of the HMBC, DEPT, HMQC and COSY. It belongs to a small group of dihydrocoumarins isolated earlier from *Pityrogramma calomelanos* [26, 27] and *Calophyllum thwaitesii* [28].

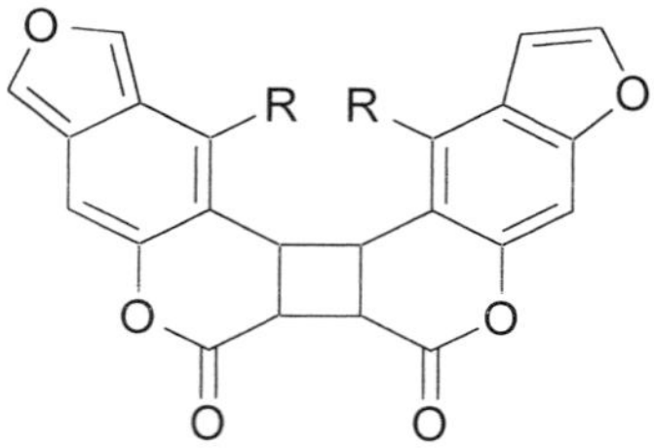

19 R=H
20 R=OMe

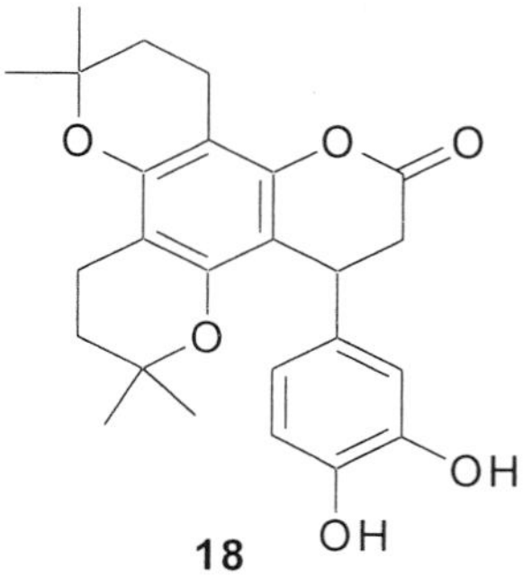

18

D. lindeniana furnished small quantities of psoralen and the dimeric compounds **19** and **20**. The structure of the psoralen dimer **19** was established by X-ray diffraction analysis, while those of the known compounds were identified by comparison of their spectral data with those reported for the same compounds in the literature. Compound **19** was also synthesized by irradiation of psoralen (**5**) using a medium pressure mercury vapor lamp. This led to a mixture of isomers with the head to head, *cis-syn* dimer being the major product which by spectral

comparison was identical to the natural dimer isolated from the roots of
D. lindeniana. These authors argued that while the [2+2]
cyclodimerization can lead to eight possible configurational isomers, the
isolation of only a single isomer in the crude extracts together with the
absence of photo-oxidation products were considered as confirmation of
the enzymatic nature of the dimerization process.

Recently a series of linear and angular prenylated furanocoumarins and
benzofuran derivatives were isolated from leaves and twigs of *D. gigas* a
plant occurring endemically on Socotra Island (Yemen) [11]. *D. gigas* is a
succulant shrub growing endemically on steep limestone and granite cliffs
at Socotra Island (Yemen) [29]. Chomatographic separation of the *n*-
heptane extract afforded the known linear furanocoumarins:
dimethoxychalepensin (**21**) isoimperatorin (**22**), furopinnarin (**23**),
swietenocoumarin B (**24**), 5-methoxy-8-geranyloxypsoralen (**25**), cnidilin
(**26**), and additionally the new linear furocoumarins 27 and **28** Fig.(2)
and the angular one **36** Fig.(3). Their structures were elucidated on the
basis of spectral data and some chemical transformations as well as by
comparison with data reported in the literature. The only hitherto known
natural source of dimethoxychalepinsin (**21**) were callus cultures of *Ruta
graveolens* [30], its synthesis was reported by Hernàndez-Galàn et al [31].
A new benzofuran derivative whose structure was determined as 2-(*p*-
hydroxybenzyl)-6-methoxybenzofuran (**29**) was also isolated from this *n*-
heptane extract. The structure of **29** was determined on the basis of NOE
difference, HMBC and COSY long-range experiments. The presence of
one hydroxy group in this compound was confirmed by the formation of a
trimethylsilyl ether exhibiting the molecular ion at m/z 326 obtained by
GC-MS.

29

The angular furanocoumarin dorstegin (**41**) was also isolated from the
n- heptane extract. NOE irradiation of H-4 in **41** caused an enhancement
of H-4' suggesting the pyran ring annelation at position 5 and 6. Ten

more polar furanocoumarins were identified in the ethyl acetate extract: the known compounds oxypeucedaninhydrate (**30**), byakangelicin (**31**) swietenocoumarin F (**32**), and seven new furocoumarins **33, 34, 35** Fig.(**2**) and **37- 40** Fig.(**3**) including three structural isomers of swietenocoumarin F. Compound **32** was previously isolated as a racemate from the back of *chloroxylon swietenia* belonging to the Rutaceae [32]. However, switenocoumarin F isolated from *D. gigas* exhibited optical activity ($[\alpha]_D$ -17°). Compounds **37, 38, 39** and **40** showed the typical UV spectrum of angular furanocoumarins. Compound **37**, the main component of the ethyl acetate extract of this plant, represented a structural isomer of switenocoumarin F (**32**). ^{1}H NMR and EIMS fragmentation of **32** and **37** revealed similar results indicating the presence of a 3-methyl-2, 3-dihydroxybutyl residue. The vicinal hydroxy groups were confirmed by formation of the corresponding methylboronate which gave the molecular ion at m / z 342. Long-range correlation of H-1" A/B with C-4a, C-5, C-6 and C-2" as well as NOE correlation between H-4 and H-1" A/B indicated the attachment of this side chain at C-5. The OMe-protons exhibited HMBC correlations to C-6 and NOE correlations with H-1" A/B, H-2"and H-2'. The position of the glucosyl moiety in **38** was deduced from HMBC and NOESY experiments. The structure of **40** was supported by the MS fragmentation pattern showing the degradation of the side chain m/z 269 $[M-CH_2OH]^+$, m/z 229 $[M-C_4H_7O]^+$, m/z 216 $[M-C_5H_8O]^+$. After methylsilylation no molecular ion was observed, but a prominent fragment at m/z 282 resulting from the loss of TMSiOH. Previously the isolation of a coumarin bearing the 4-hydroxy-3-methyl-but-2-enyl residue at position C-8 (arnottinin) was reported from the xylem of *Xanthoxylum arnottianum* [33]. The furanocoumarins found in *D. gigas* display a complex structural diversity not comparable to the compounds identified in previously investigated *Dorstenia* species. A special feature of the *D. gigas* furanocoumarins seems the incorporation of prenyl units either as 1, 1-dimethylallyl or 3-methylbut-2-enyl groups in oxidized or non oxidized form. Prenylation at C-8 lead to the formation of the observed angular structures. An additional heterocyclic ring was formed in compound **41** by the interaction of a prenyl substituent at C-5 and oxygen at C-6. The variety of furanocoumarins detected in others species of the genus *Dorstenia* is restricted to linear and angular structures with methoxy substituents. The furanocoumarin pattern of *D. gigas* exhibit similarities to that of some Rutaceae and Alpiaceae species.Prenylated furanocoumarins are widely distributed in these two families; however , oxidation of the side chain is more frequent in Rutaceae.1, 1-dimethylallyl substitution at C-3 as found in **21** is rare and has been detected only in Rutaceae [34].

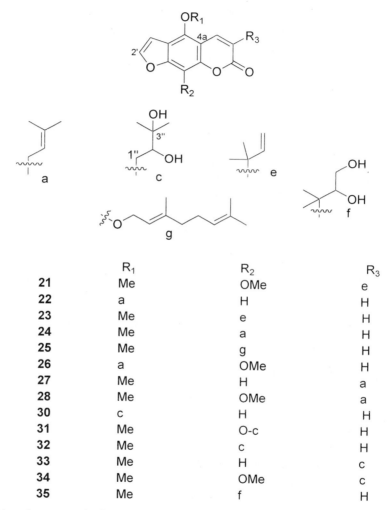

	R_1	R_2	R_3
21	Me	OMe	e
22	a	H	H
23	Me	e	H
24	Me	a	H
25	Me	g	H
26	a	OMe	H
27	Me	H	a
28	Me	OMe	a
30	c	H	H
31	Me	O-c	H
32	Me	c	H
33	Me	H	c
34	Me	OMe	c
35	Me	f	H

Fig.(2). Linear furanocoumarins from *D. gigas*

770

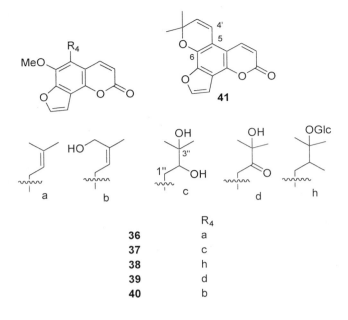

Fig. (3). Angular furanocoumarins of Dorstenia gigas

3- PRENYLATED AND GERANYLATED CHALCONES

A range of simple as well as complex prenylated and geranylated flavonoid derivatives has been isolated and characterized from a number of *Dorstenia* species. It is interesting to note that all the flavonoids so far reported from this genus appear to be only from African *Dorstenia* species; but more data would have to be obtained to confirm, if indeed this is truly so and if this observation has any chemotaxonomic significance. Monochalcones as well as bichalcones are reported from this genus.

3-1 Prenylated monochalcones

3-1-1 Monoprenylated monochalcones

A part from 4, 2', 4'- trihydroxychalcone (**42**) isolated from the twigs of *D. zenkeri* all chalcones obtained from this genus carry at least one prenyl

(3-methylbut-2-enyl) group in oxidised or non oxidised form. This species also yielded isobavachalcone (**43**), dorsmanin A (**44**) and the hydroxydimethyldihydropyrano derivative (**45**) [35]. The NMR spectra of compound **43** were found to be identical to those reported for 4-hydroxyisocordoin isolated from *Cordoa piaca* (*Lonchocarpus sp.*, Leguminosae) [36], and for isobavachalcone isolated from callus culture of *Glycyrrhiza uralensis* (Leguminosae) [37]. An isomeric compound of dorsmanine A, called crotmadine (**46**) in which the 2'-hydroxyl function is involved in the formation of the dimethyldihydropyrano ring has been reported from *Crotalaria madurensis* [38]. These two compounds **44** and **46** show a difference in the downfield ^{1}H-NMR region. The sharp peak at δ13.97 in **44** was absent in **46**; also the bathochromic shift induced by the addition of aluminium chloride observed in **44** was not seen in **46**. Decoction of the leaves of *D. mannii* is used for the treatment of rheumatism and stomach disorders. The twigs of this plant yielded also **44**, 4-hydroxylonchocarpin (**47**) and isobavachalcone (**43**) [39]. **42** and 4-methoxylonchocarpin (**48**) were obtained from *D. poinsettifolia* [9]. **48** was previously reported from *Milletia pachycarpa* and *Pongamia glabra* [40]. *D. kameruniana* [41] and *D. ciliata* [42] yielded also isobavachalcone. The chroman derivative **44** is undoubtedly formed by cyclization of isobavachalcone (**43**). What is more uncertain is whether this cyclization occurs during the isolation process or takes place enzymatically in the plant. However, the co-occurrence of dorsmanine A (**44**) and 4-hydroxylonchocarpin (**47**) in *D. mannii* [39] suggests that such cyclization reaction in *Dorstenia* may very well be enzymatic. Thus cyclization of isobavachalcone (**43**) would yield dorsmanine A (**44**). The latter would then undergo subsequent hydroxylation in the dihydropyrano ring at 3"-, or 4"-position followed by dehydration to give the dehydro-derivative, 4-hydroxylonchocarpin (**47**). It is noteworthy that the 3"-hydroxyderivative (**45**) has been recently isolated from *D. zenkeri*. Initially it was not clear whether the hydroxyl group was located at the 3" or the 4"-position. But this was readily established from HMBC studies where the C-H proton signal of the oxymethine group was found to have a long range correlation with C-2", C-4" (^{2}J) and C-2"(Me)$_2$, C-3' (^{3}J). furthermore the two protons signals of the C-4" showed correlations with the two oxygenated carbons at C-2" and C-4'.

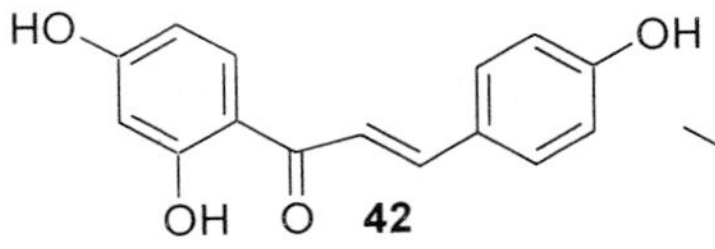

3-1-2 Di- and triprenylated monochalcones

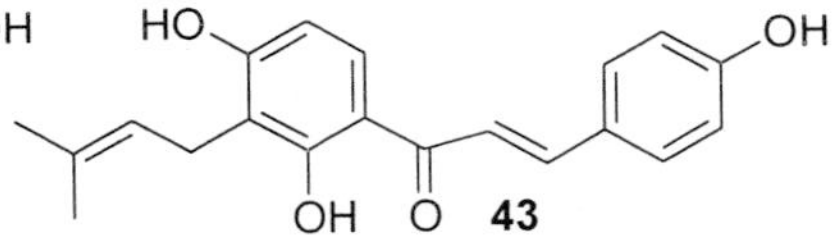

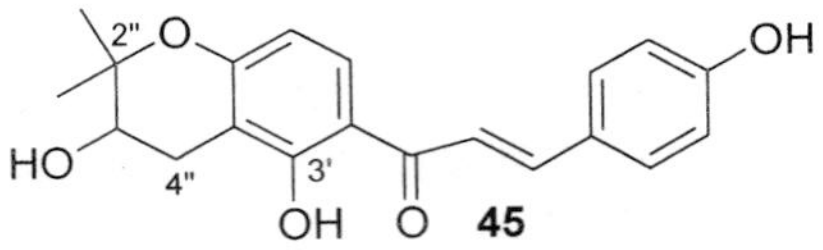

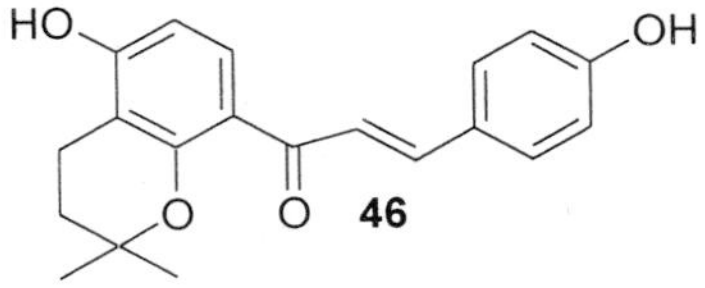

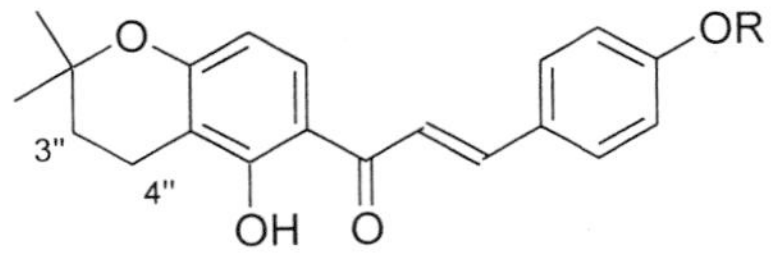

The diprenylated chalcone stipulin (**49**) was isolated from *D. kameruniana* [41], *D. ciliata* [42], and *D. dinklagei* [43]. It has also been reported earlier from the legume *Dalbergia stipulacea* [44]. Comparison of the measured chemical shifts with those reported in reference [44], were identical. But these authors reported a chemical shift value of 25.7 for C-1" and C-1'", and values of 29.5 and 28.4 ppm for C-4" and C-4'" or (E)-Me, respectively. Our analysis of the DEPT spectrum led us to conclude that these values have to be revised. The correct assignments are 26.0 for C-1" and C-1'" and values of 29.5 and 28.8 for C-4" and C-4'". Stipulin has been tested against growth profiles and viability of HL-60 promyelocytic leukaemia cells. It was not toxic at low concentration but was extremely toxic above 100 μM. *D. kameruniana* also yielded the novel *bis*-dihydropyran derivative (**50**) [41] which is should be formed as we said earlier by cyclization of **49**. The twigs of *D. barteri* var. subtriangularis yielded recently [Ngameni, B., Ngadjui, B.T., Watchueng, J., Abegaz, B.M. unpublished results] stipulin and three new diprenylated

chalcones: bartericin A (**51**), bartericin B (**52**) and Bartericin C (**53**).
Paratocarpin D in which the non oxidized prenyl (3, 3-dimethylallyl)
group is at C-5' and 2-hydroxy-3-methylbut-3-enyl moiety at C-3, isomer
of bartericin A (**51**) was isolated from *Paratocarpus venenosa* [45]. **52**
and **53** are two dihydrofuranochalcones bearing a 1-hydroxy-1-
methylethyl group in the furan ring. The structures of these secondary
metabolites were determined on the basis of spectroscopic analysis,
especially , NMR spectra in conjunction with 2D-NMR experiments,
COSY, HMQC and HMBC. Bartericin C (**53**) is formed by cyclization of
52 which can generate bartericin A (**51**) by dihydrofuran ring opening and
dehydration . The methanol extract of *D. barteri* var. multiradiata were

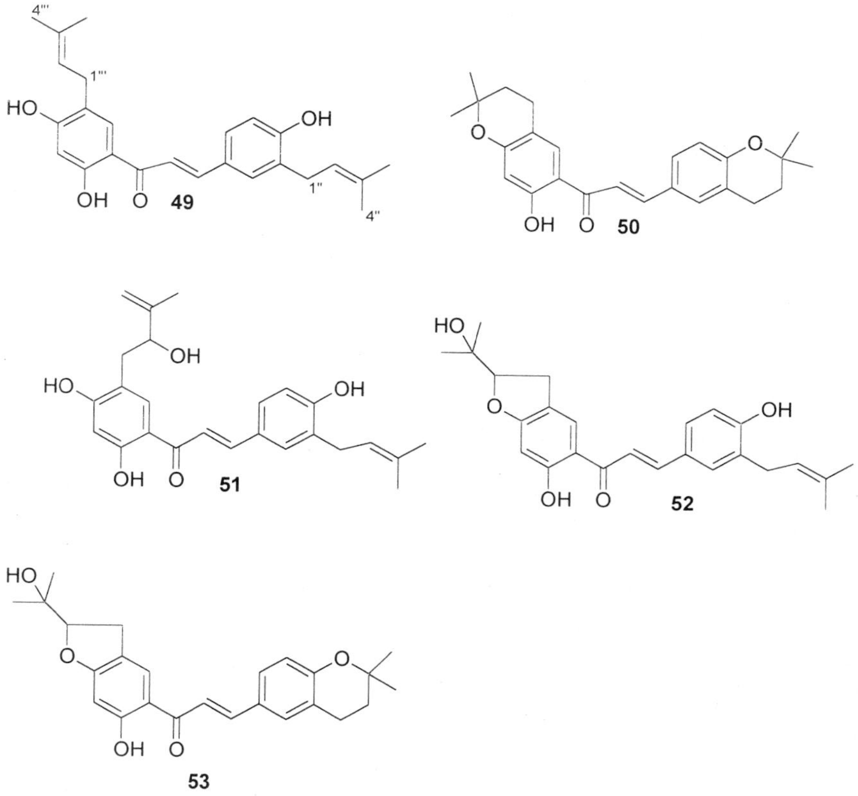

Fig. (4). Diprenylated chalcones of D. barteri and D. kameruniana

found to be active against the amastigote stages of *Leishmania*. The antileishmanial activity was later traced to chalcones [46].

We have so far isolated only one triprenylated chalcone, sophoradin (**54**) from the organic extract of the aerial parts of *D. involuta*. This compound was previously reported in the roots of *Sophora subprostrata* [47].

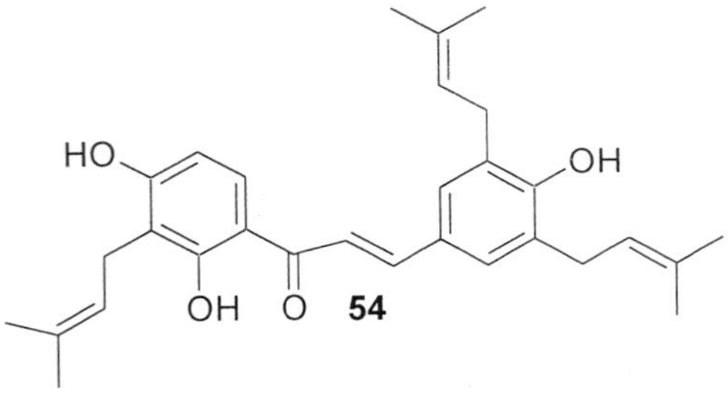

3-1-3 Geranylated monochalcones

Poinsettifolin B (**55**) and proropensin (**56**) two geranylated monochalcones have recently been isolated from *D. poinsettifolia* [48] and *D. proropens* [35], respectively. Poinsettifolin B (**55**) obtained as yellow needles from *D. poinsettifolia* [48] is a novel prenylated and modified geranylated derivative. Its structure was established on the basis of ^{1}H and ^{13}C-NMR and other spectral features. The location of the various substituent groups was determined with the help of ^{1}H-^{1}H COSY, HMQC and HMBC correlations. In poinsettifolin B, the 2-geranyl moiety has undergone cyclization with the 3-hydroxy group. The two geranyl groups in proropensin (**56**) are not oxidised. Two *trans*- arrangements are possible in this compound. The first is linear structure as shown in **56**. The second is angular conformation which would orient the β-proton close to H-6'. Selective NOESY experiments were performed by irradiating both the α- and β-proton signals and it was found that only irradiation of the former resulted in the nuclear Overhauser enhancement of the H-6' signal. This result is consistent with the conformation of the chalcone as shown in **56** in the NMR solvent used. Survey of the chemical literature reveals that bis-geranylated chalcones have not been reported so far. However, over then chalcones containing one geranylated substituent have been reported in the literature. These include 4'-

geranyloxy-4, 2'-dihydroxychalcone from *Mellettia ferruginea* [49], poinsettifolin B from *D. poinsettifolia* [48], flemingins A (**57**), B (**58**), C (**59**), D (**60**), E (**61**) and F (**62**) [50] from *Flemingia rhodocarpa* and *F. congesta*, flemiwallichins A (**63**), B (**64**), C (**65**), D (**66**), E (**67**), and F (**68**) [50] from *Flemingia wallichii*, boesenbergins A (**69**) and B (**70**) [50] from *Boesenbergia pandura* and xanthoangelols A (**71**) and B (**72**) [50, 51] from *Angelica keiskei*. 5, 3'-(3,7-Dimethyl-2,6-octadienyl)-3,4,2',4'-tetrahydroxychalcone, proropensin (**56**), is the first member of digeranylated monochalcone so far identified from the literature.

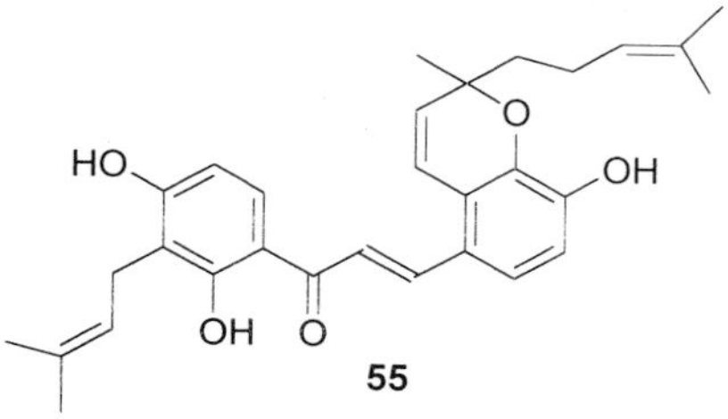

55

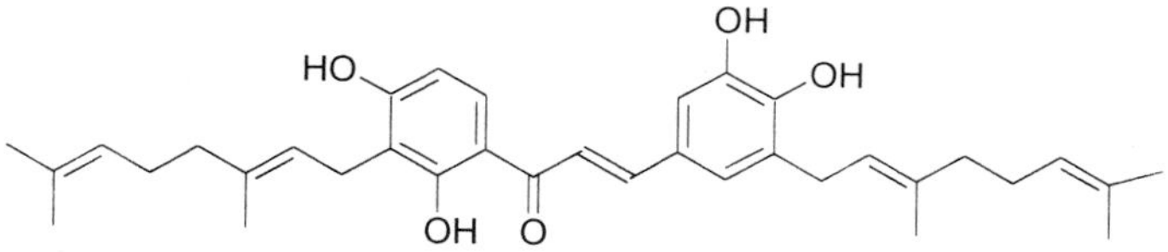

56

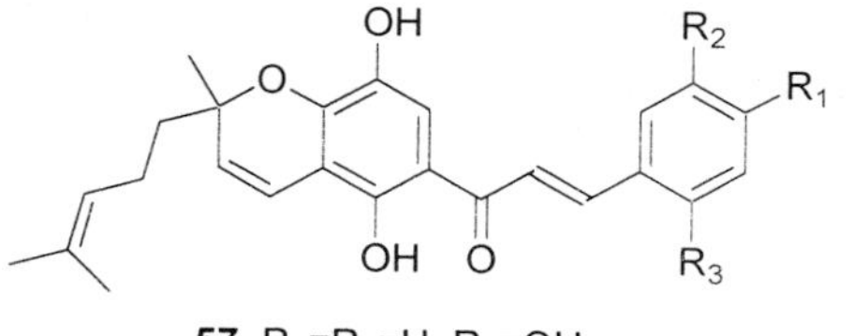

57 $R_1=R_2=H$ $R_3=OH$
60 $R_1=OH$ $R_2=R_3=H$
59 $R_1=H$ $R_2=R_3=OH$

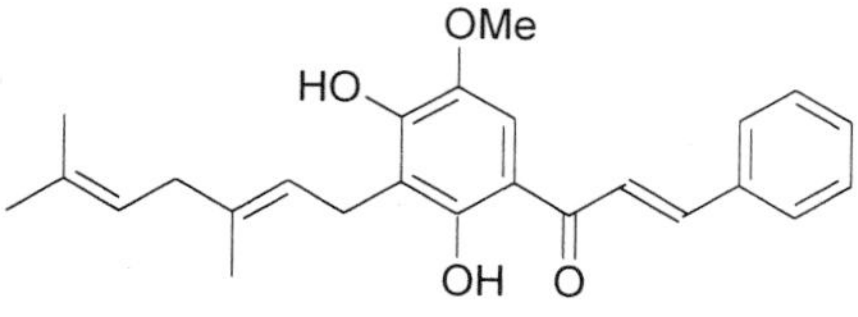

58

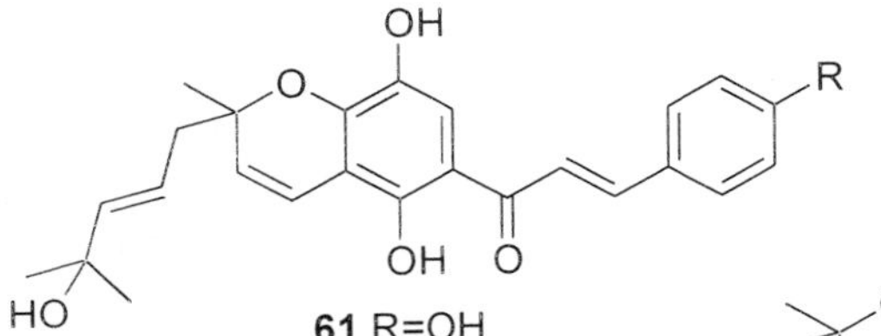

61 R=OH
68 R=H

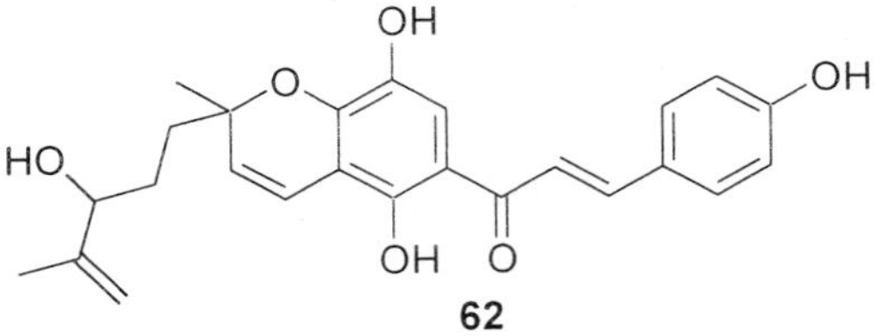

62

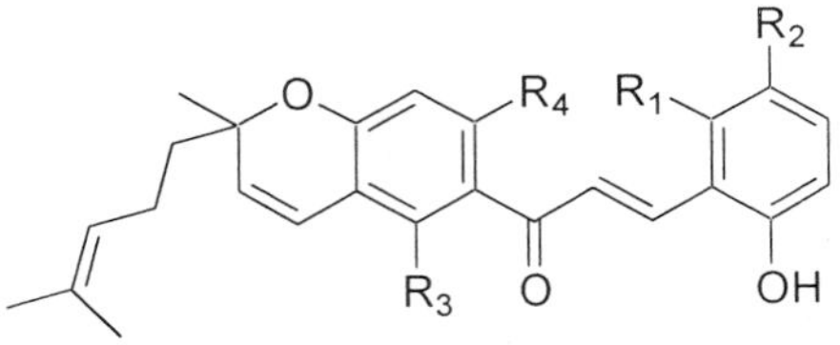

63 R$_1$=OH   R$_2$=R$_3$=R$_4$=H
64 R$_1$=H   R$_2$=R$_3$=R$_4$=OH
66 R$_1$=R$_4$=H R$_2$=R$_3$=OH

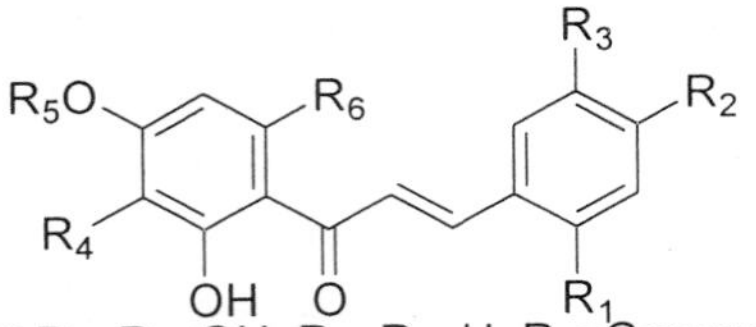

65 R$_1$=R$_3$=OH R$_2$=R$_5$=H R$_4$=Geranyl R$_6$=OMe
67 R$_1$=R$_3$=R$_4$=R$_6$=H R$_2$=OH R$_5$=Geranyl

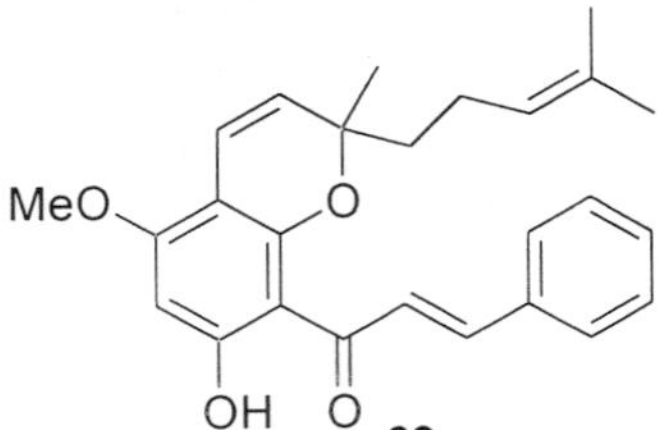

69

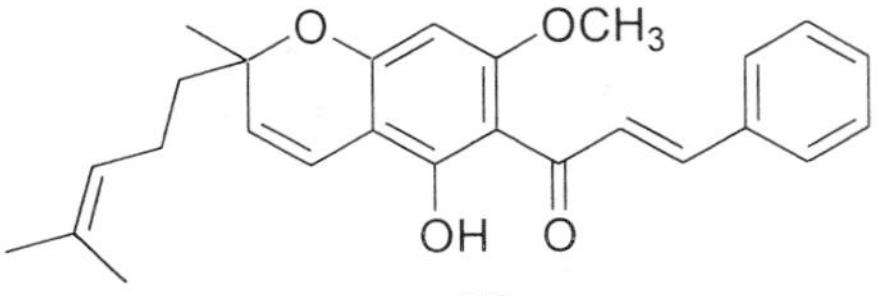

70

(Fig. 5) contd....

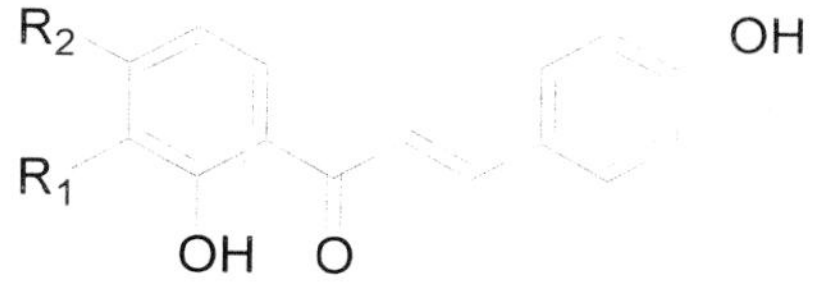

71 R$_2$=OH R$_1$=Geranyl
72 R$_2$=OMe R$_1$=(E)-6-hydroxy-3,7-dimethyl-2,7-octadienyl

Fig. (5). Natural geranylated chalcones (55-72) of *Dorstenia* and other plant species

3-2 Prenylated bichalcones

Dorstenone (**73**) a new and possibly enzymatic, Diels-Alder adduct was recently been reported from the Cameroonian species *D. barteri* [52]. The relative configuration of dorstenone has been established by comparison of the ^{1}H and ^{13}C NMR data of its methylcyclohexene ring with those of the corresponding ring in similar compound like brosimone B [53] and kuwanon V [54], all of known stereochemistry. The configuration of **73** was further confirmed by the NOESY experiment. This compound is believed to arise by intermolecular [4+2] cycloaddition reaction of the α, β- unsaturated double bond of isobavachalcone (also called 4-hydroxyisocordoin, isolated from *Cordoa piaca*) (**43**) [55] as the dienophile and the isoprenyl portion of its dehydroderivative (**74**) as the diene Fig. (**6**). The observed large optical rotation [α] value of −371° was attributed to the enzymatic nature of the reaction. Comparison of the optical rotations of **73** ([α]$_D$−371°) with that of kuwanon V ([α]$_D$ +145°) confirmed the diastereoisomeric relationship of the two compounds and further supported structure **73** for dorstenone. The isolation of **43** in the same plant added more support to the Diels-Alder type reaction that was envisaged to take place enzymatically. It was observed from the ^{1}H NMR spectrum that dorstenone (**73**) undergoes epimerisation in solution. This behavior coupled with the high negative optical rotation is typical of these complex flavonoid type adducts [56-59]. *D. zenkeri* from Cameroon also yielded a related compound **75** which may very well be derived from dorstenone (**73**) by interaction of the vinyl double bond of the cyclohexene moiety with the phenolic oxygen as shown in Fig. (**6**).

75, however, showed only a weak rotation indicating that racemization may have occurred at some stage. The structures of **73** and **75** were determined by using extensive ^{1}D and ^{2}D NMR and especially through gradient-enhanced HMQC, HSQC, HMBC, and NOESY techniques. TOCSY experiments were utilized in **75** to identify the various mutually coupled protons around the bicyclic system. A one-dimensional experiment using a Bruker pulse program selmlgp and a spin locking time of 250 ms provided a partial spectrum showing transfer of magnetization on the proton in the bicyclic ring. Diels-Alder adducts arising from two chalcones or a chalcone and a diene-substituted flavone are not entirely uncommon [60]. Although over forty such compounds are reported from Moraceous plants only two are so far known from *Dorstenia* genus; Its possible to speculate that the dehydroderivative of isobavachalcone (**74**) is probably formed *via* hydroxylation of the prenyl side chain followed by dehydration to form the diene. Although there is no evidence to link **45** as an intermediate to the formation of the diene, the introduction of the hydroxyl group at position 3" of the dihydropyran ring is interesting. Dehydration of **45** will give 4-hydroxylonchocarpin (**47**) which can be considered as starting material in the synthesis of these bichalcones **73** and **75**. The diene **74** can also be formed from rearrangement reaction (pyran ring opening) of **47**, Fig. (**6**). The cycloaddition of **74** and **43** could, in principle, lead to any of the two isomers **75** and **76**, but the relative stereochemistry was deduced from extensive ^{2}D NMR analysis which also enabled us to reject an alternative mode of cycloaddition which would give rise to **76**, by reaction of the same diene with the dienophile in the reverse orientation. Of particular significance in support of the latter conclusion was the echo-anti-echo ROESY experimental evidence, which clearly showed enhancement of the H-20 as result of interaction with the H-5.

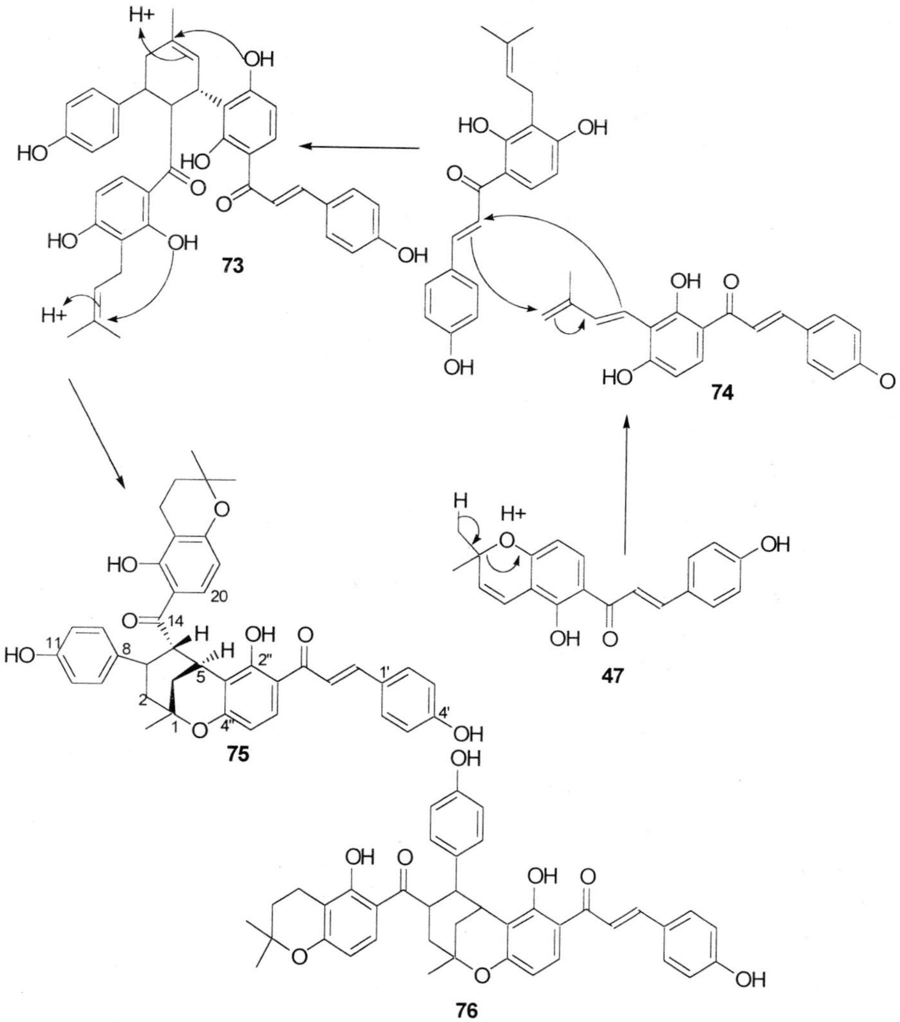

Fig.(6). Modified Diels Alder adduct **75** arising from a diene and dienophile of two chalcones **74** and **43**

Recently we have carried out thermal reaction of **47** [Ngadjui, B.T., Ngameni, B. Abegaz, B.M., unpublished results]. this compound was heated in the sealed tube in the oil bath at 200°C for 3hrs. The thermolysis of **47** yielded a mixture of compounds. Chromatographic separation of this mixture gave the dihydrochalcone (**77**), the flavanone (**78**) and traces of a compound which from mass spectroscopic data was concluded to be a dimer of **47**. It is also of interest to note that such Diels-Alder adducts possess hypotensive properties [60].

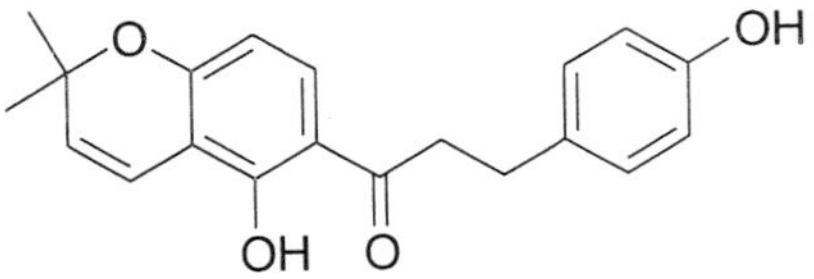

77

4- ISOPRENOID FLAVANONES

From the biosynthetic relationship among different classes of flavonoids, flavanones derived from the cyclization of chalcones. Although there is a quite number of biflavanones in the literature, no biflavanone has so far been identified from the *Dorstenia* genus.

4-1 Monoprenylated flavanones

Only two monoprenylated flavanones have so far been identified from *Dorstenia* species. As we have mentioned earlier **78** was obtained from the thermal reaction of 4-hydroxylonchocarpin (**47**). Twigs of *D. poinsettifolia* yielded dorspoinsettifolin (**79**). Cyclization of **48** in formic acid and methylation of **78** with diazomethane both readily gave **79**. An isomeric compound, 6-methoxyisolonchocarpin, has been reported from *Lonchocarpus subglauscescens* [61].

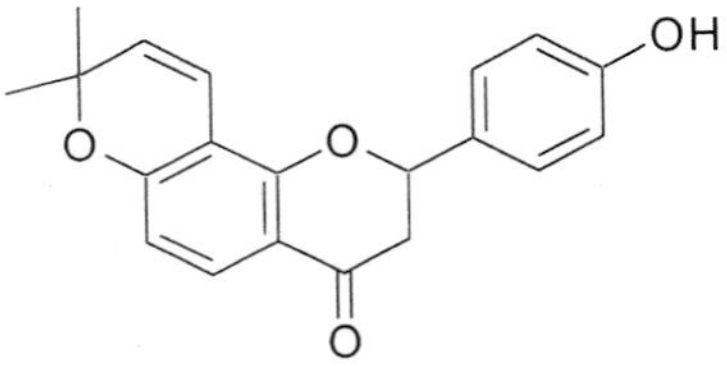

78

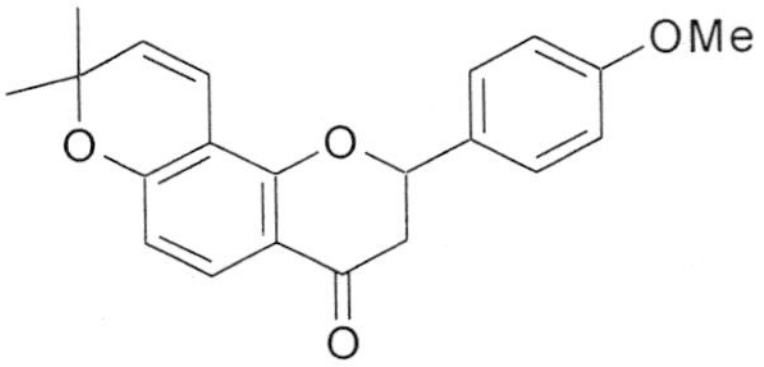

79

4-2 Diprenylated flavanones

Eleven simple as well as complex diprenylated flavanones have been identified from *D. dinklagei* [43] and *D. mannii* [39,62,63]. Almost 80% of the 11 diprenylated flavanones carry out one or two dimethylpyrano or dimethyldihydropyrano or furano side attachments, most often at positions 7, 8 ; 6, 7; 3', 4'; as other prenylations involve positions 8 and 6. A part from dinklagin A (**80**) isolated recently from *D. dinklagei* [43] all these diprenylated flavanones are from *D. mannii*, Fig.(**7**). Two possibilities were considered for dinklagin A one with the pyran in ring B and the prenyl group at C-6 (**80**) or an alternative structure with the pyran in ring A and the prenyl group at C-3'. This issue was settled by HMBC data which indicated a cross peak for the methylene carbon of the prenyl group (δ29.4) , and the downfield proton signal at δ 7.71 assigned to H-5. Thus the prenyl group is without doubt located at C-6. Euchrenone a_5 (**81**) and abyssinone III (**82**) isomers of **80** were isolated , respectively, from *Euchresta formosana*[64] and *Erythrina abyssinica* [65].

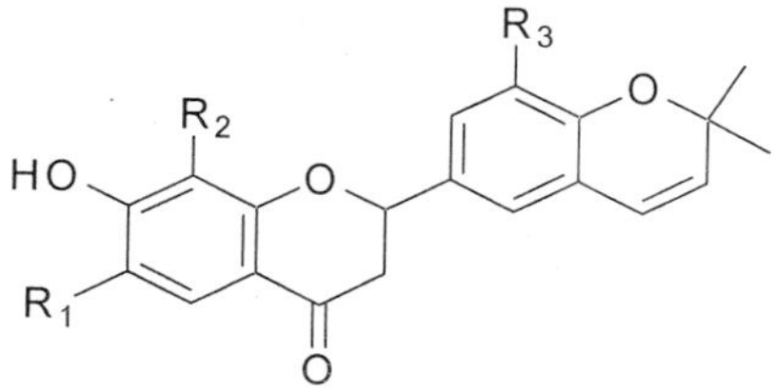

80 R_1=Prenyl R_2=R_3=H
81 R_1=R_3=H R_2=Prenyl
82 R_1=R_2=H R_3=Prenyl

6, 8-Diprenyleriodictyol (**83**), the simplest diprenylated flavanone, is a citotoxic compound first reported by Harborne et al in 1993 from *Vellozia coronata* and *V. nanuzae* [66]. The structure of dorsmanine E (**84**) was established from analysis of the spectroscopic data for the presumed natural product as well by synthesizing it from 6, 8-diprenyleriodictyol (**83**) upon treatment of the latter with methanolic HCl. Dorsmanin B (**85**) [39] as dorspoinsettifolin (**79**) and **78** lack the 5-hydroxyl functionality that is common in most of the flavonoids isolated so far from *Dorstenia* spp and the range of compounds isolated from *D.*

mannii is a good illustration of the diversity of structures that this taxon is capable of producing. The 5-hydroxyl derivative of **85** is in fact a known compound and has been found in *Paratocarpus venenosa*, a plant belonging to the family Moraceae [45].

Both dorsmanin F and epidorsmanin F (**86/ 87**) occur as epimeric mixtures (*vide infra*), as do dorsmanin G and epidorsmanin G (**88 / 89**) [62,63] which could not be separated by usual chromatographic procedures. **86/ 87** and **88 / 89** are recognized as the 3'-hydroxyl derivatives of lonchocarpols C (**90 / 91**) and D (**92/ 93**) respectively, reported from the legumes *Lupinus luteus* [67] and *Lonchocarpus minimiflorus* [68]. The clarifications of the structures of dorsmanins F and G with respect to which of the two possible prenyl groups at position 6, and 8 is involved in the cyclization with the 7-OH group is of interest and is based on NMR spectroscopic evidence which is described here. Dorsmanin F (**86 / 87**) has an angular 2, 3-dihydrofuran and a prenyl substituent at C-6. Tahara et al [68,69] have shown that the chemical shift of the 5-OH signal is predictably influenced by the mode of cyclization to form the dihydrofuran ring, i.e. cyclization of the 8-prenyl (8→7 [O]) shifts the signal of the 5-OH to the lower, whereas 6-prenyl cyclization (6→7 [O]) shifts it to higher field. In 6, 8- diprenyleriodictyol (**83**), (uncyclized), the 5-OH is reported to occur at *ca* δ 12.30. In dorsmanine F the corresponding signal is at δ 12.74 consistent with the (8→7 [O]) cyclization and the attachment of the other prenyl group at C-6. The structure of dorsmanine F was determined to be 7, 8-[2''-(1-hydroxy-1-methylethyl)-dihydrofurano]-6-prenyl-5, 3', 4'-trihydroxyflavanone (**86/ 87**). The ^{1}H NMR of dorsmanins F (**86 / 87**) and G (**88 / 89**) showed duplicate resonances for the chelated 5-OH. This is also observed in the ^{13}C NMR spectra. Detail NMR analysis indicated that both consist of two pairs of diastereoisomers resulting from epimerisation of the asymmetric centre in the side chain attachment. Efforts to separate each set of diastereoisomers to dorsmanin F, epi-dorsmanin F (**86 / 87**) and dorsmaninG, epidorsmanin G (**88 / 89**) were unsuccessful. Each pair of diastereoisomers has been characterized as a mixture.The EIMS of each diastereoisomer showed mass fragments at m / z 352 and 59 [Me$_2$C=OH]+ typical for flavonoids possessing a 2-(1-hydroxy-1-methylethyl)dihydrofurano group [68,70,71]. Oxidation of the 8-prenyl group in **83** yielded dorsmanine H (**94**) meanwhile cyclization of the 6-prenyl lead to dorsmanins I (**95**) and J (**96**) [63]. The NMR spectroscopic data of **95** and **96** are too similars. The NMR similarities of **95** and **96** together with two additional amu in the EIMS of **96**, led us to determine the structure of dorsmanine J as 8-prenyl-6,7-(2, 2-dimethyldihydropyrano)-5, 3', 4'-trihydroxyflavanone. Confirmation of the location of the prenyl group at position 8 in **95** and **96** was obtained from HMBC data which showed correlations of the benzylic proton signals of the prenyl substituent with two of the oxygenated aromatic carbon signals assigned to C-7 and C-9 but not to the C-5 signal at δ 157.3. The assignment of the latter signal to C-5 was established by observing the long range methine (CH) correlation to the only chelated – OH signal at δ 12.46. Likewise , the vinyl proton signal of the pyran ring at C-4" showed correlation to the C-5 signal. From the foregoing spectroscopic data, the structures of dorsmanine I and J were deduced to be **95** and **96** respectively. The ^{13}C NMR signals of dorsmanin I (**95**) were fully assigned using DEPT spectra and by comparison of the measured values with those reported for lupinifolin (**97**) isolated from *Tephrosia lupinifolia* (Leguminosae) [71]. A 3'-methylether derivative (**98**), of **95** was isolated and characterized by Lin et al from the roots of *Derris laxiflora* (Leguminosae) [72].

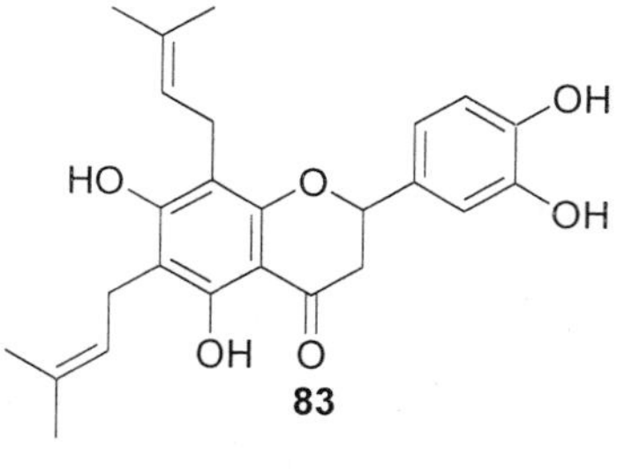

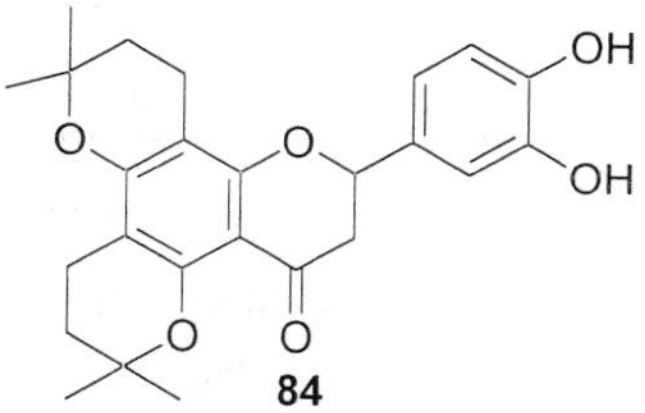

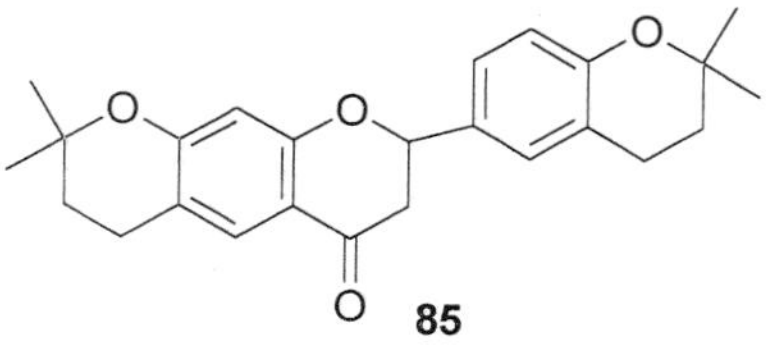

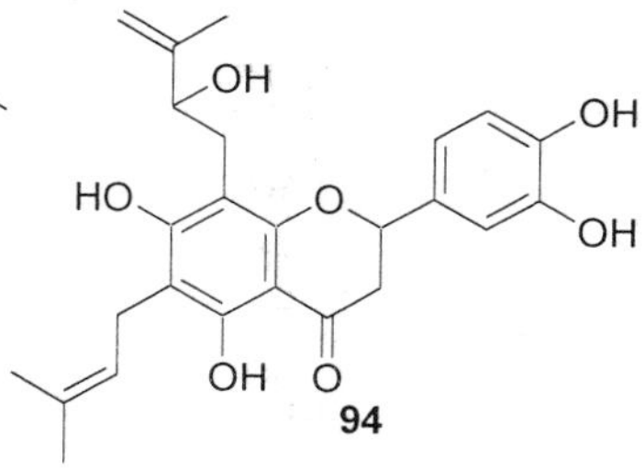

(Fig. 7) contd....

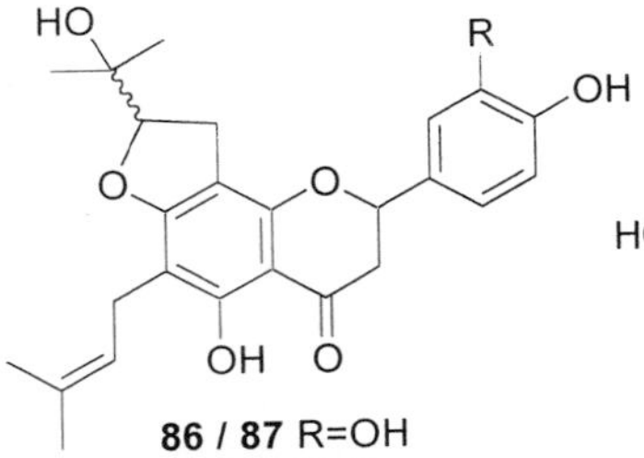

86 / 87 R=OH
90 / 91 R=H

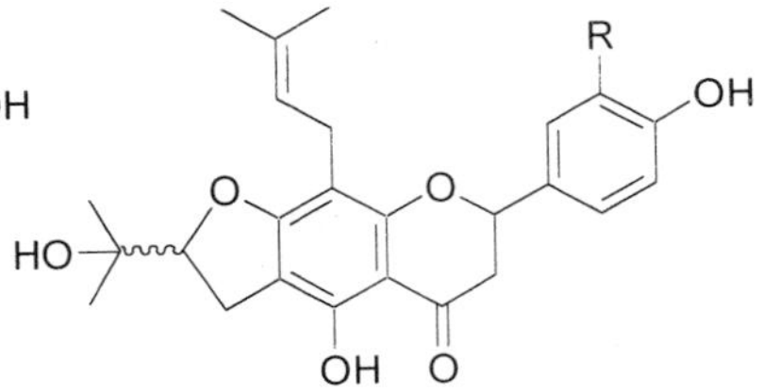

88 / 89 R=OH
92 / 93 R=H

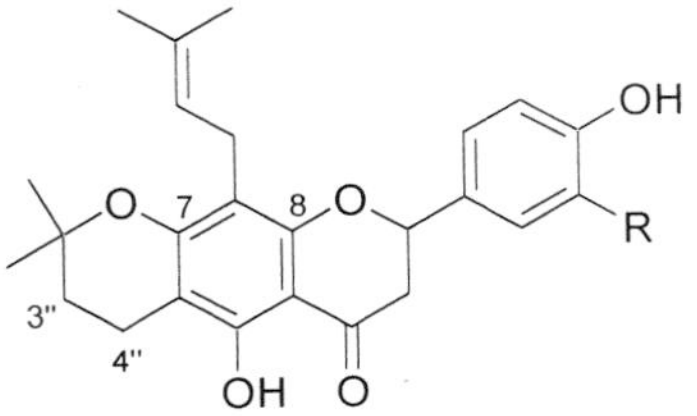

96 R=OH
95 R=OH $\Delta^{3'',4''}$
97 R=H $\Delta^{3'',4''}$
98 R=OMe $\Delta^{3'',4''}$

Fig.(7). Diprenylated flavanones of *D. mannii* **83-89**, **94-96** and other species

The results from our investigation of *D. mannii* indicated that this plant is a rich source of diprenylated flavanones. It is possible to speculate the biosynthesis of most of the dorsmanins **94**, **86-96** from 6, 8-diprenyleriodictyol (**83**) which is in fact present in the plant in rather significant amount. Fig.(**8**) shows various epoxide intermediates arising from **83**, which undergo subsequent opening by phenolic hydroxyls to yield the various diprenylated dorsmanins.

5- PRENYLATED FLAVONES

Fourteen flavone derivatives have so far been identified from a number of *Dorstenia* species: *D. ciliata* [42], *D. dinklagei* [43], *D. elliptica* [Ngadjui et al unpublished results], *D. kameruniana* [41], *D. mannii* [39],

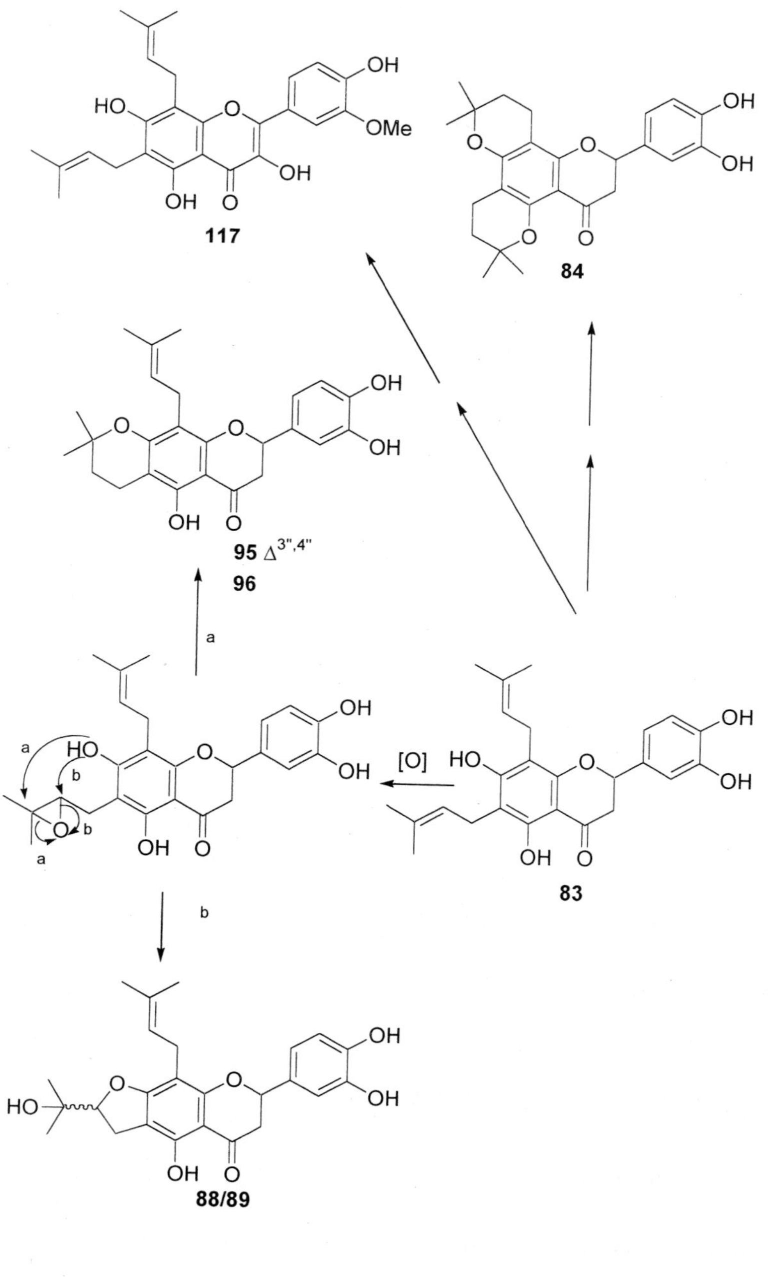

117

84

95 $\Delta^{3'',4''}$
96

a

[O]

83

b

88/89

(Fig. 8) contd….

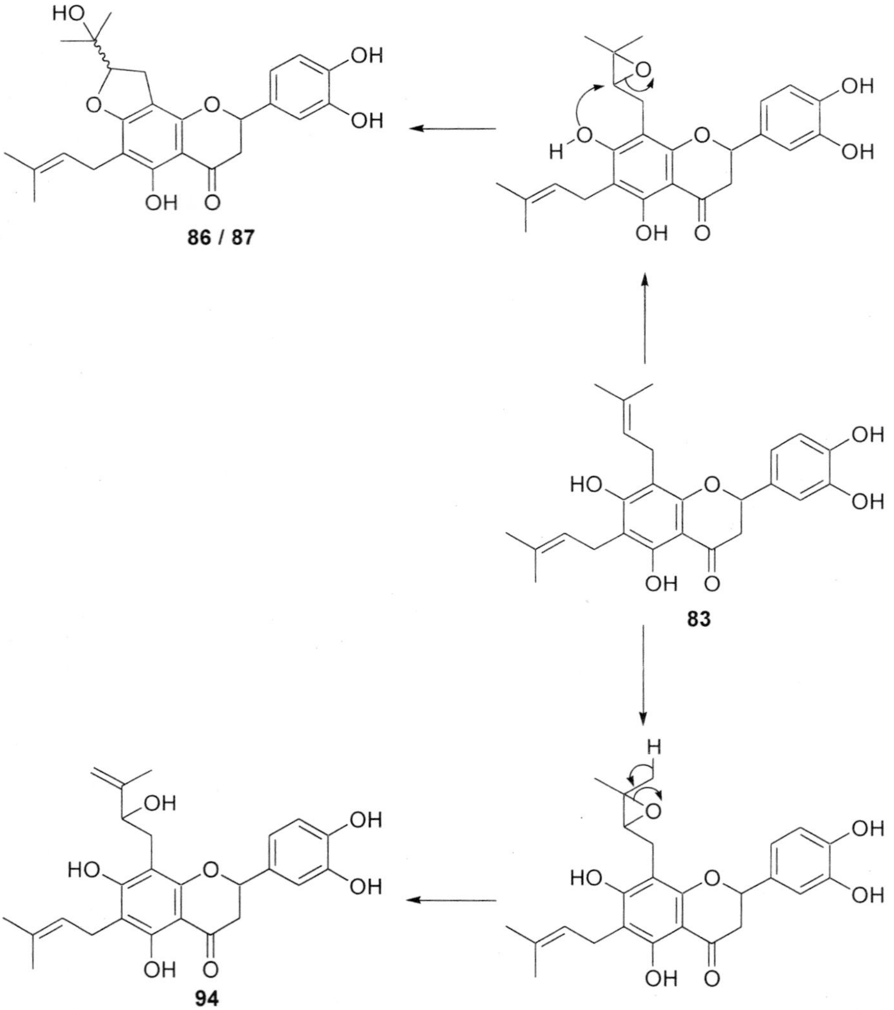

Fig. (8). Biogenetic transformation of 6,8-diprenyleriodictyol **83** to various dorsmanins (84, 88/89, 95,96, 117, 94, 86/87)

D. poinsettifolia [9], *D. psilurus* [10,73] and *D. zenkeri* [35]. These represent the second most abundant class of prenylated flavonoids with nine mono- and four tri-prenyl, as well as the dimethylpyrano, hydroxydimethyldihydropyrano and dihydrofurano derivatives. Notable among the prenylated flavones is the frequent substitution at position 6 mostly with 3,3-dimethylallyl. A part from 5, 7, 3', 4'-tetrahydroxyflavone (99) isolated from *D. zenkeri* [35] all these flavones carry one or three prenyl groups either in ring A, B or the heterocyclic ring C. Diprenylated flavone has not been so far isolated from *Dorstenia* species meanwhile eleven diprenylated flavanone derivatives were isolated from this genus.

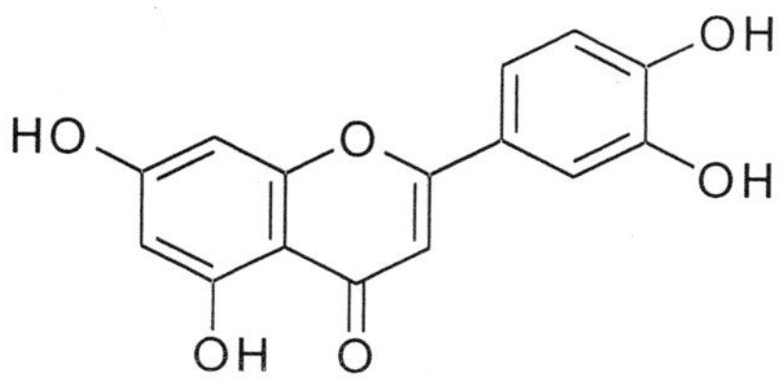

99

5-1 Monoprenylated flavones

Nine monoprenylated flavones with simple 3, 3-dimethylallyl, 2-hydroxy-3-methyl-3-butenyl, dimethyldihydropyrano or furano substitutions have so far been identified from the genus *Dorstenia*. The incorporation of the prenyl unit can take place in any ring of the flavone skeleton. It is interesting to note that all the monoprenylated flavones reported so far appear to carry the prenyl group either at C-6 or C-8. 8-Prenyl, 5-hydroxyflavone (100) isolated from the twigs of *D. elliptica* [Ngadjui, B.T., Fotso, S., Dongo, E., Abegaz, B.M., unpublished results] is the only example which lacks the 4'-hydroxyl function that is common in most of flavonoids obtained so far in the genus *Dorstenia*. The simple *Dorstenia* flavones include 6-prenyl apigenin (101), isolated from *D. cilata* [42], *D. dinklagei* [43] and *D. kameruniana* [41] and licoflavone C (102), the 8-prenylated isomer from *D. poinsettifolia*. 6-Prenyl apigenin was isolated earlier from *Maclura pomifera* by Delle Monache et al [74]. The simple dimethyldihydropyran (103) has been reported from *D. kameruniana* [41] and *D. dinklagei* [43]. From the spectroscopic analysis two possibilities were considered for the structure of this compound, one with a linear dihydropyran (103) or an alternative structure with an angular 7,8-chroman ring. The ^{13}C chemical shift at δ95.5 ppm as a doublet strongly favours the linear 6,7-chroman junction [75]. Dinklagin B (104) the 3''-

hydroxyderivative of **103** was recently isolated from the twigs of *D. dinklagei* together with dinklagin C (**105**). The structure of dinkagin B was supported both by the ^{13}C NMR spectrum and CIMS. The CIMS showed a fragment ion at *m/z* 283 diagnostic for the hydroxydimethyldihydropyran degradation [76]; another important ion fragment at *m/z* 337 due to the loss of a molecule of water was also observed. The ^{13}C NMR signals were fully assigned using DEPT spectra and by comparison of measured values with those reported for ficuisoflavone (**106**) isolated from the bark of *Ficus microcarpa* [77]. The hydroxymethylbutenyl group in dinklagin C can be substituted either in position 6 (**105**) or in position 8. It was unambiguously fixed at C-6 because of the chemical shift of the carbon doublet at δ95.8 characteristic of unsubstituted C-8 5-hydroxyflavone [75]. The location of this group was readily established from HMBC studies where the two protons signals of the C-1" showed correlations with the two oxygenated sp^2 carbon signals at C-5 and C-7. Ephedroidin (**107**) and **108** isomers of dinklagin C were isolated, respectively , from *Genista ephedroides* [78] and *Vancouveria hexandra* [79]. Compound **105** can derived from dinklagin B (**104**) through enzymatic opening of the 3-hydroxy-2, 2-dimethyldihydropyran ring coupled with the loss of a proton from one of the methyl groups.

D. ciliata yielded Ciliatins A (**109**) and B (**110**) and **111**. Two possibilities were also considered regarding the orientation of the isopropenyldihydrofuran group in ciliatin A, located in ring A: one with the linear furan (**109**) or an alternative structure with angular furan group. This question is also related to the designation of whether the singlet signal at δ_H 6.48 is appropriate for H-6 or H-8. The signal of H-3 was undoubtedly assigned using HMBC correlations observed between this proton and the carbonyl and C-1' signals. The chemical shift of δ_C 93.4, which was correlated by HMQC experiments to this aromatic proton signal 6.48 is consistent with its location at C-8 [75]. The structure of ciliatin B as **110** was unequivocally established by its hemisynthesis from **111**; refluxing this latter compound in the mixture of HCl and MeOH readily gave **110** after the workup. Compound **106** has been reported under other names, for example, 6-prenylchrysoeriol [80] canniflavone [81], as well as cannflavin B [82]. No di-prenylated flavone has so far been identified from this genus.

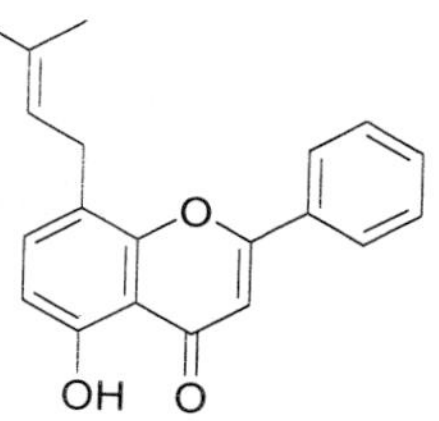

100

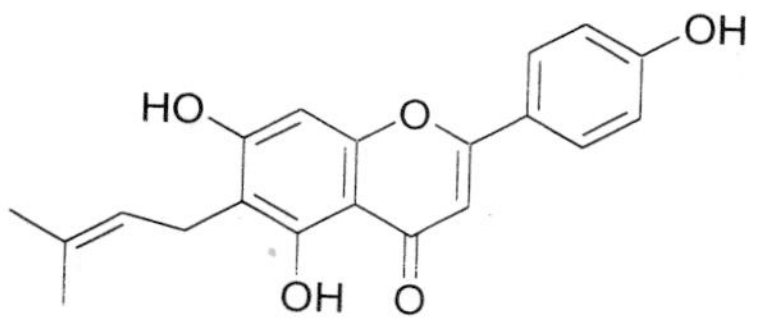

101

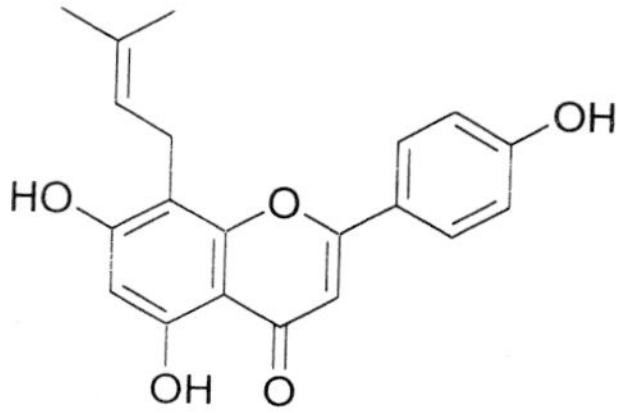

102

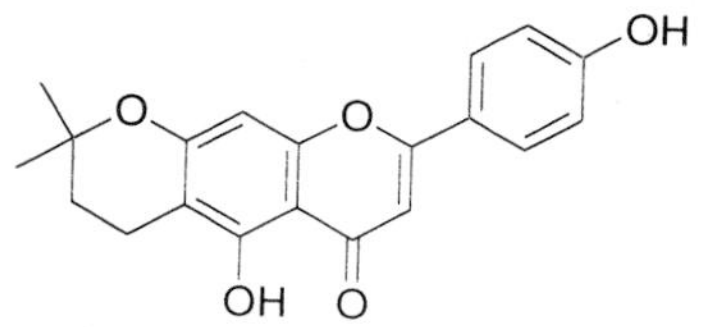

103

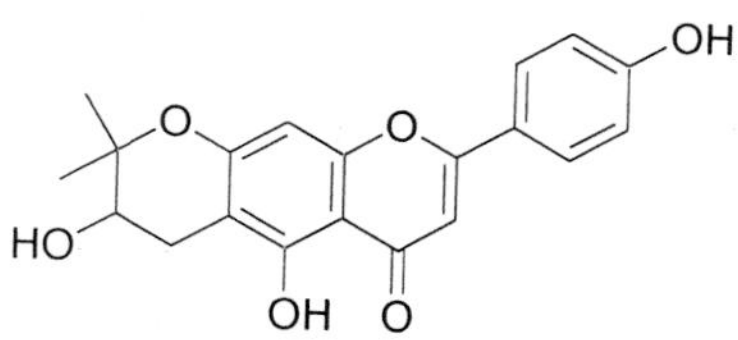

104

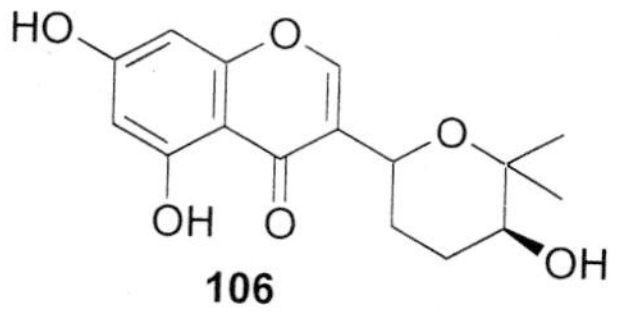

106

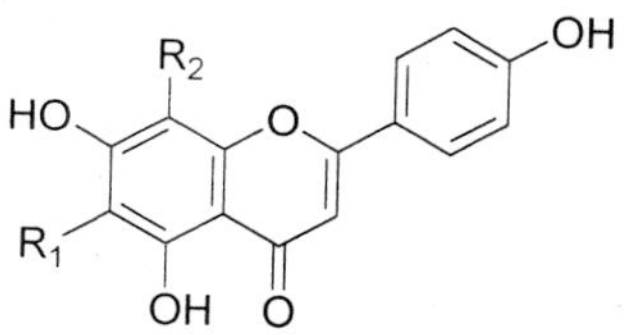

105 R₂=H,   R₁=

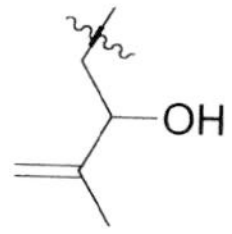

107 R₁=H   R₂=

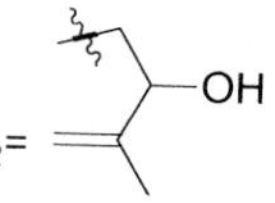

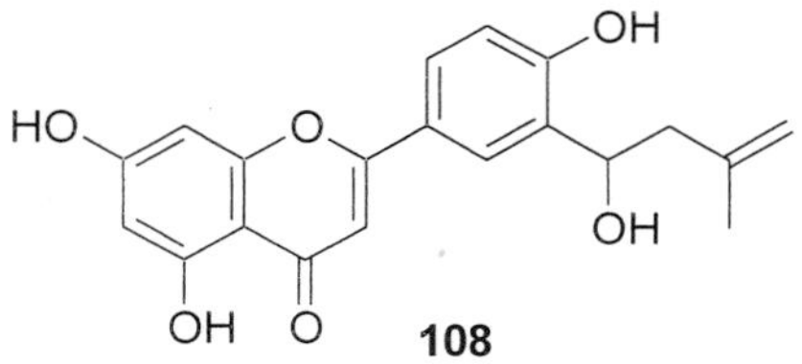

Fig. (9). Monoprenylated flavones (**100-111**) of genus *Dorstenia* with their isomers from other plants species

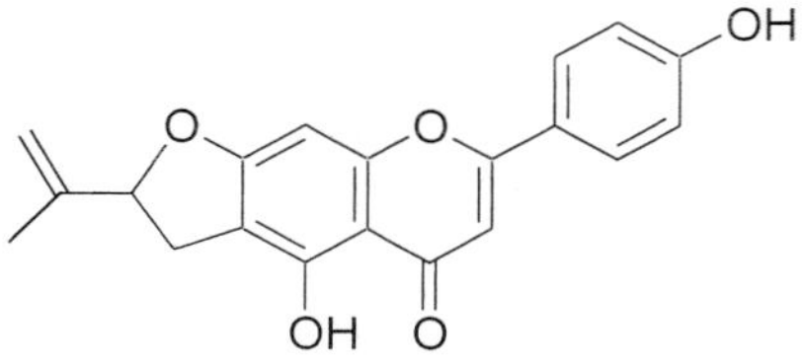

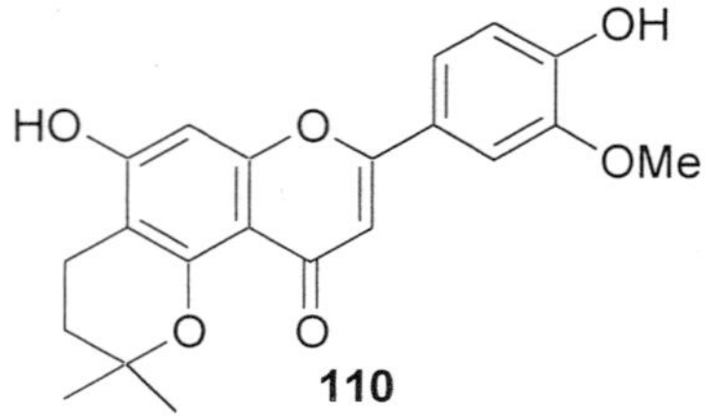

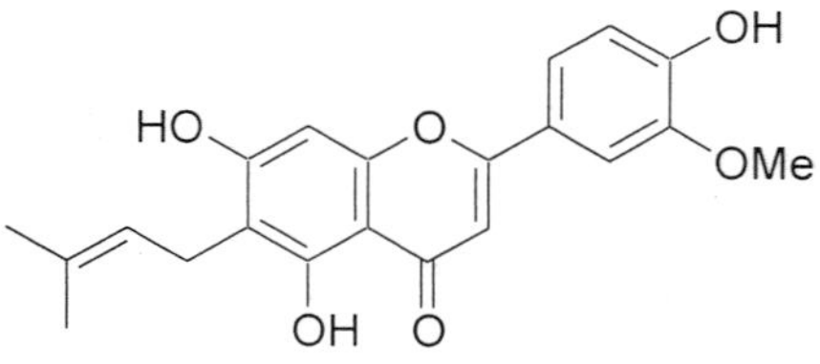

5-2 Triprenylated flavones

Five triprenylated flavones have been reported and all of them are from the roots of *D. psilurus*. [10, 73] Fig. (**10**). Two of the prenyl group are located at C-6 and C-8 in ring A. The third prenyl group is located either at C-3 or C-3'. The prenyl units are incorporated as 3, 3-dimethylallyl (3-methylbut-2-enyl), 2-hydroxy-3-methylbut-3-enyl and 2, 2-dimethyl-dihydropyrano groups. The three prenyl units in Dorsilurin A (**112**) are simple incorporation of 3, 3-dimethylallyl groups at positions 6,8 and 3'. Dorsilurin A is a labile compound. A chloroform solution which was kept in a NMR tube overnight, upon re-examination revealed that a number of decomposition products has formed; many chelated OH signals were observed in the downfield NMR region. This decomposition may be due probably to the cyclization of the 3-methylbut-2-enyl groups with the phenolic hydroxyl groups at C-5, C-7 and C-4'. An isomer of Dorsilurin A, with the third 3,3-dimethylallyl group at position 3 instead of 3', artelasticin (**113**) has been reported from *Artocarpus elasticus* [83]. Dorsilurin B (**114**) carry two intact prenyl group at positions 3 and 6 but the third one has undergone modification to hydroxydimethyldihydropyran at positions 7, 8. The structure of this compound was established on the basis of spectroscopic analysis. Like dorsilurin A , **114** is also quite labile and a chloroform solution of it when freed of solvent gave an orange-red resinous material. Dorsilurin D (**115**) isolated as an oil carry also two intact prenyl group at positions 3 and 6; the third one oxidised as 2-hydroxy-3-methylbut-3-enyl was located at C-8. The ^{13}C NMR of dorsilurin D gave additional support to the attachment of a carbon residue and not an oxygen function at C-3 since the chemical shift of the latter was observed at δ120.7. An oxygen substituent at this position would have led to a signal at ca. δ139 which is not observed [75]. Irradiation of one of the two methylene proton signals of the prenyl groups gave a 2 % NOE enhancement of the doublet δ7.16. Regarding the position of the 2-hydroxy-3-methylbut-3-enyl side chain, and the second prenyl group, two possibilities were considered: one with the 2-hydroxy-3-methylbut-3-enyl side chain located at C-8 and the 3, 3-dimethylallyl group at C-6 (**115**) or the alternative structure with the prenyl substituent at C-8. The co-occurrence of dorsilurins B (**114**) and D (**115**) and the chemical shift of 112.8 favors the attachment of the non oxidised prenyl group at C-6 [75]. An attractive biogenetic speculation would be that dorsilurin D (**115**) could arise from **114** through enzymatic opening of the 3-hydroxy-2,2-dimethyldihydropyran ring coupled with the loss of a proton from one of the methyl groups. It was evident from the spectroscopic analysis that dorsilurin E (**116**) carry three prenyl groups at C-3, C-6 and C-8 with the former two cyclized (8→7[O] and 6→5[O] to form two dihydropyrano rings. The prenyl group at C-3 was also cyclized but it was not clear at this stage if it had formed an eight member ring with an oxygen functionality at C-2' of the ring B. Close investigation of the HMBC and HSQC spectra showed that the third 2, 2-dimethyldihydropyran ring was formed between the prenyl group at C-3 and the oxygen at C-4 and that ring B was present as a β-hydroxy-dienone moiety. An isomeric dienone having the keto-function at 2' and an hydroxyl at 4' would be an alternative structure which would be fully consistent with the observed spectroscopic data. The carbon chemical shift values of **116** were assigned by analysis of the ^{2}D NMR experiments HSQC and HMBC. The question of E-Z stereochemistry with respect to the double bond connecting the two rings (B and C) was investigated. Preliminary investigations employing ^{2}D NOESY experiments revealed small but noticeable enhancements of the signals due to the methylene proton signal at δ 2.74 upon irradiation of the aromatic proton signal at δ 7.49. This suggests a Z configuration for the double bond as shown in **116**. It is interesting that a triprenylated derivative like artelasticin (**113**) is a very probable biogenetic precursor for this highly modified flavone (**116**).

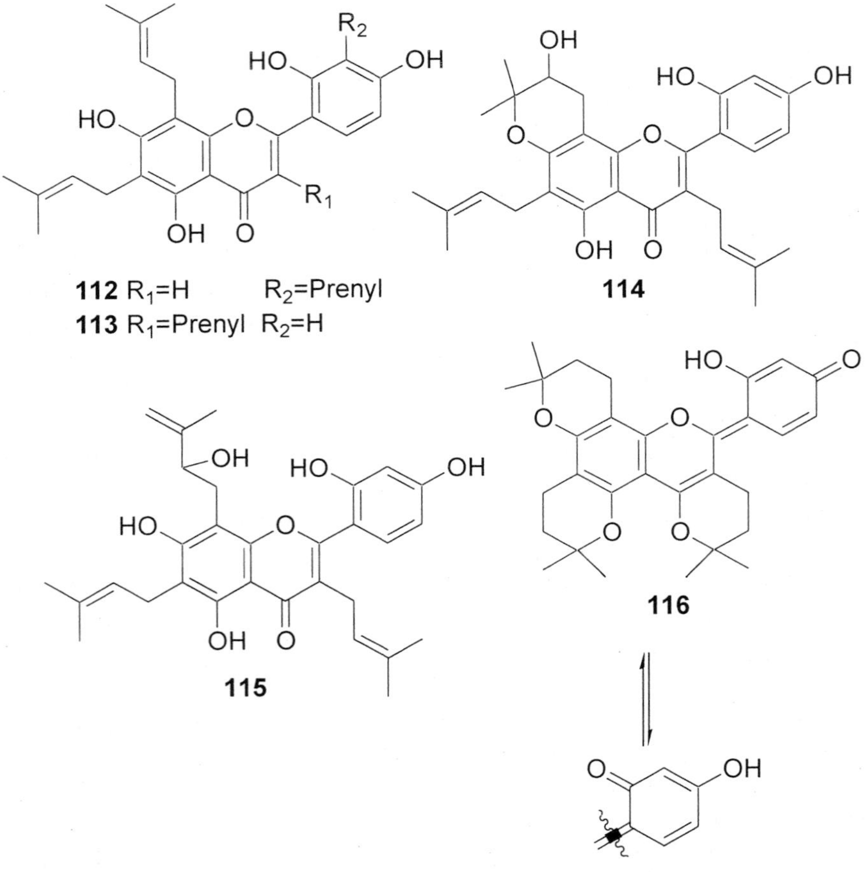

112 R$_1$=H   R$_2$=Prenyl
113 R$_1$=Prenyl R$_2$=H

114

115

116

Fig.(10). Triprenylated flavones **112-116** of *D. psilurus*

6- PRENYLATED AND GERANYLATED FLAVONOLS

Only five prenylated flavonol derivatives have so far been isolated and characterized from the genus *Dorstenia*. Two of them dorsmanin D (**117**) and **118**, from D. mannii [39] and D. psilurus [73], respectively, carry a simple prenyl group incorporated as 3,3-dimethylallyl units at C-6 and C-8 in ring B. The methoxyl group in **117** upon irradiation caused a 2 % enhancement of the *meta* coupled doublet at δ 7.73. This finding together with the observed chemical shift of the OMe group at δ_C 55.5 suggested that the methoxyl function was located at C-3'. The structure of dorsmanin D (**117**) was determined as 6, 8-bis-(3,3-dimethylallyl)-5, 7, 4' trihydroxy-3'-methoxyflavonol. The proposed structure was further confirmed by the ^{13}C NMR data which was fully assigned using DEPT spectra and by comparison of measured values with those reported for its isomer broussoflavonol B (**119**) isolated from *Broussonetia papyrifera*, Moraceae [84]. Dorsilurin C (**120**) is a triprenylated flavonol with two intact prenyl groups located at C-6 and C-8 as in **117** and **118**. The third prenyl group is incorporated as 2, 2-dimethylpyran substituent in ring B, although prenylation in ring B is not common. The observed chemical shifts of the three aromatic proton signals of the ring B are consistent with a 3'[O], 4'[C] substitution instead of the more common 3'[C], 4'[O] pattern. However , 3, 5, 7, 3'-tetraoxygenated flavonoids are not entirely uncommon [85].The proposed structure of dorsilurin C was confirmed by the ^{13}C NMR data and EIMS spectra.

Besides the mono- and bis-geranylated chalcones , **55** and **56**, respectively, only two examples of such compounds have been reported so far from *Dorstenia*. One of these is dorsmanine C (**121**), a compound isolated from D. *mannii* [39] and D. *tayloriana* from Cameroon and Tanzania, respectively. The structure of dorsmanine C (**121**) was determined to be 7,8-(2,2-dimethylpyrano)-6-geranyl-5, 3', 4'-trihydroxyflavonol by UV, IR, EIHRMS ^{1}D and ^{2}D NMR analysis. The second compound is poinsettifolin A (**122**) and was isolated from a Cameroonian herb D. *poinsettifolia* [48]. Dorsmanin C (**121**) and poinsettifolin A (**122**) are interesting in several aspects. First, to our knowledge, they are the only flavonol with a   complex substitution pattern obtained so far from the family Moraceae, secondly poinsettifolins A (**122**) and B (**55**)   belong to the small group of C-geranylated flavonoids in which the 8- or 2-geranyl group has undergone cyclization with the 7- or 3- hydroxyl group. Moreover both geranylated compounds carry each one prenyl unit and are isomeric differing only in the substitution pattern in ring A. In poinsettifolin A the geranyl substituent located at position 8 has undergone a cyclization with the 7-hydroxyl group, meanwhile dorsmanine C carry the geranyl group intact at C-6 and the prenyl unit as 2, 2-dimethylpyrano group at 7 and 8 positions.

117

118 R=H
119 R=Me

120

121

122

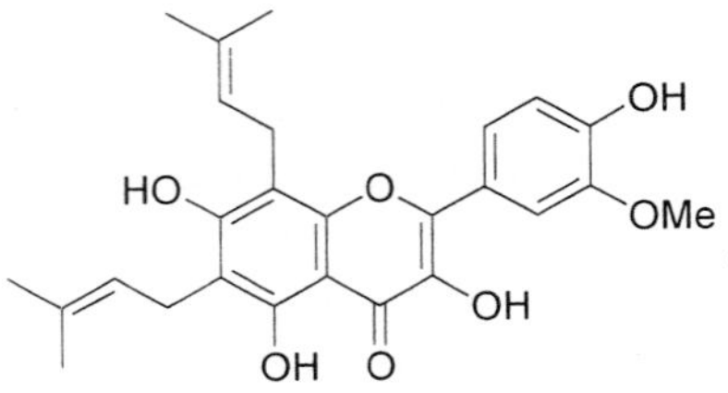

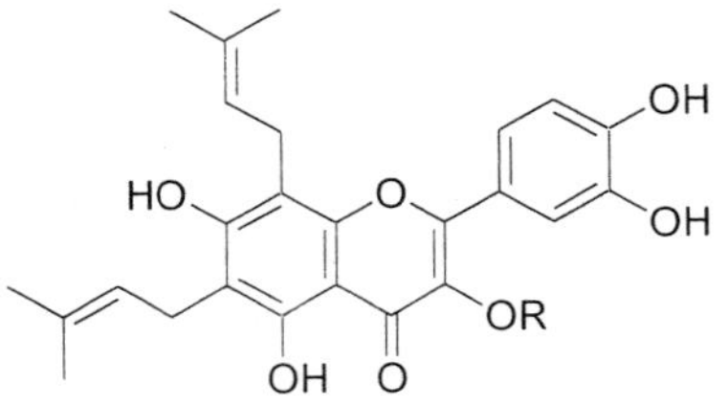

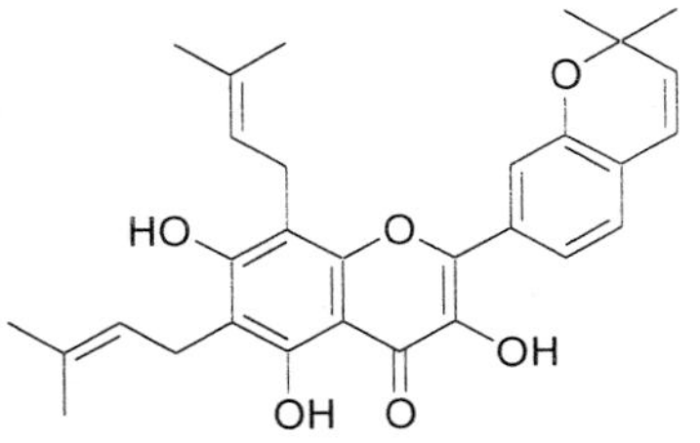

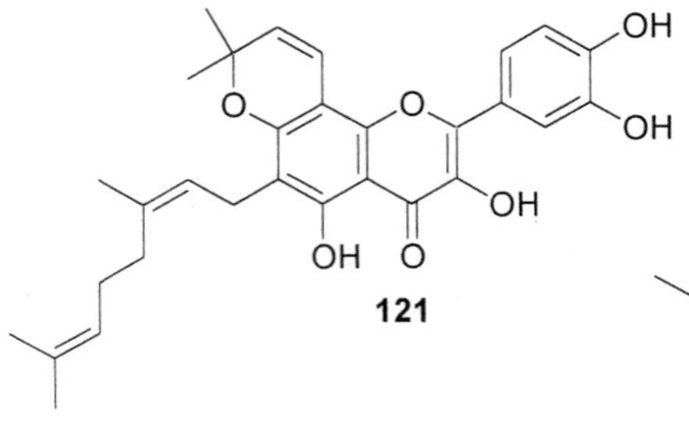

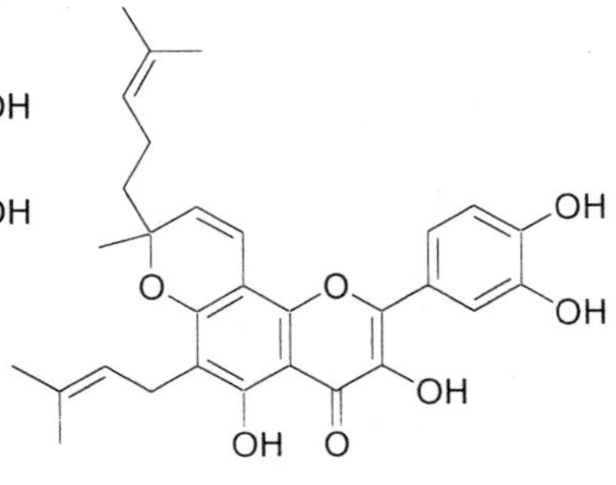

Fig.(11). Prenylated **117-120** and geranylated **121**, **122** flavonols

7- TERPENOIDS

The genus *Dorstenia* is not rich in terpenoids. Nine of them Fig.(**12**) are so far been isolated and characterised from *D. barnimiana* [4], *D. bryioniaefolia* [86], *D. poinsettifolia* [48] and *D. brasiliensis* [17]. The rhizomes of *D. bryioniaefolia* yielded the common triterpenoids α-amyrine (**123**) and β-amyrin (**124**) with its acetate (**125**) [86]. The triterpene lupeol acetate (**126**) was obtained from the Ethiopian marketed plant *D. barnimiana* [4]. Butyrospermol acetate (**127**) was isolated from the Cameroonian herb *D. poinsettifolia* [48] initially published by Galbraith et al [87]. Recently the roots of *D. brasiliensis* yielded 14 α-hydroxy-7, 15-isopimaradien-18-oic acid (**128**) reported for the first time by Bruno et al and two *seco*-adianane-type triterpenoids, dorstenics acid A (**129**) and B (**130**) [88].The ^{1}H-^{1}H COSY and HMQC spectra of **129** revealed the presence of isopropyl and isopropenyl proups together with a propionic acid moiety. The connectivities of these groups were investigated by HMBC experiments and the relative stereochemistry of **129** was determined by the NOESY spectrum analysis. Thus the structure of dorstenic acid A was confirmed as the 3, 4-*seco*-adianane-type triterpenoid **129** with a carboxyl group at C-2. The ^{1}H and ^{13}C NMR spectra of dorstenic B (**130**) were very similar to those of **129** except for the presence of a carbonyl group (δ 212.0) and the disappearance of an isopropenyl group. The HMBC spectrum of **130** showed cross peaks due to the long-range correlations betweenH-10 and C-2, C-5 and C-6. The structure of dorstenic B (**130**) was elucidated to be a deisopropenyl derivative of **129**. Dorstenics acid A and B showed moderate cytotoxicity against leukaemia cells (L-1210): IC_{50} = 5 and 40 μM, and HL-60: IC_{50} = 10 and 40 μM), respectively.

8- PHARMACOLOGICAL ACTIVITIES

In view of the importance of the *Dorstenia* based crude drug preparation, "carapia" in Brazil, capillary gas chromatographic procedures for the quantification of furocoumarins and other constituents from rhizomes and aerial parts of *D. tubinica*, *D. asaroides* and *D.vitifolia*, and a high resolution gc/ms analytical method for the non polar constituents of

D.bahiensis, D. bryoniifolia, D. carautae, D.cayapiaa and *D. heringerii* have bee described [89,90].

The majority of the plants described in this report are established items of trade and are sold in African traditional markets. The occupy important

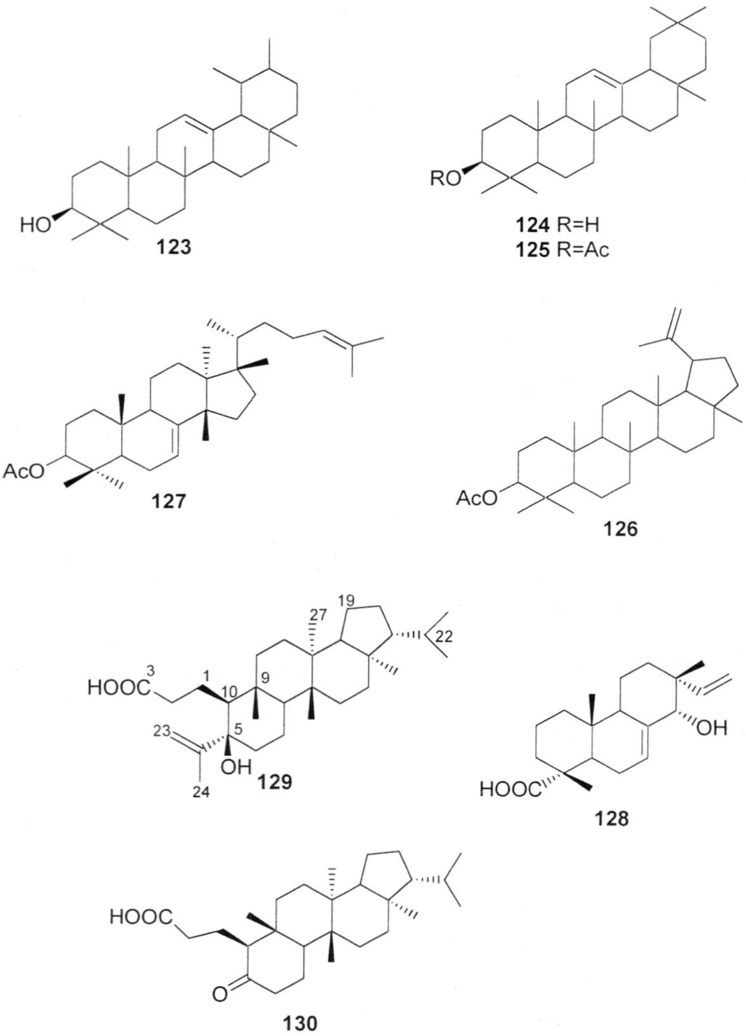

Fig.(12). Terpenoids 123-130 of Dorstenia spp

positions in the material medical of the culturally developed medical practices of many communities. The observation that these items are acquired by consumer directly from the market, often without prescriptions from traditional herbalists, may justify the classification of these plant drugs as a group that has broader acceptability among the indigenous population.

A study of light-activated toxins of the Moraceae revealed the occurrence of phototoxins in all investigated *Dorstenia* species [11]. The phototoxic properties were most likely due to the presence of furanocoumarins especially psoralen (**5**) and bergapten (**10**) [11,91]. Bergapten is recognised as : antiapertif, anticonvulsant, antihistaminic, anti-inflammatory, hypotensive, insecticide and molluscicide meanwhile psoralen is cytotoxic (ED_{50}=2.30 µg/ml), anti-tumor promoter (IC13 = 50 µg/ml CPB38), artemicide (LC_{50} = 5.93 µg/ml), bacteriophagicide, antimutagenic, antileukodermic M11 and viricide [96].

The roles that flavonoids play as anticancer, radical-scavenging, anti-inflammatory and hepatoprotective agents have also been documented [92]. A number of biflavonoids show HIV-1 reverse transcriptase activity [93] and still others are found to inhibit the interleukin-1β-induced procoagulant activity of adherent human monocytes [94]. Recently luteolin was shown to possess adenosine A_1 receptor binding activity [95]. There is now intense activity centered around investigations of the possible roles of simple flavonoids such as apigenin, quercetin, luteolin, etc. in the prevention of the cancer [96]. There is no scientific evidence to relate the empirically determined beneficial uses of *Dorstenia*-based plant preparations to the flavonoids present and especially to those and the others that are discussed in this chapter. There are, however, few reports dealing with the pharmacological effects of *Dorstenia* extracts and compounds.

Iwu et al [97] have reports based on the evaluation of plant-extract of *D. multiradiata* for antileishmanial activity using a mechanism-based radiorespirometric micro-technique. Extracts were found to be active at concentrations of 50 µg/ml or less against a visceral leishmania isolate. A number of *Dorstenia* species used traditionally as anti-snake venom were subjected to a pharmacological screening process and were found to possess analgesic and anti-inflammatory activities [98]. Many of the flavonoids isolated from African *Dorstenia* show moderate to good anti-oxidant activities [Croft, unpublished results]. The cytotoxic properties of

6,8-diprenyleriodictyol (**83**) (ED_{50} values < 4μg/ml in several cell lines) have been reported [99,100]. Stipulin (**49**) and 6-prenylapigenin (**101**) have been tested against growth profiles and viability of HL-60 promyelocytic leukaemia cells. Compound (**49**) was toxic to these cells and an ED_{50} of 50 μM was determined. 6-prenylapigenyl (**101**) was not as toxic at low concentrations but was extremely toxic at 100 μM [41]. The roots of *D. brasiliensis* [96,101] are worldwide used (see Table 1 below).

Table 1. Ethnobotany worldwide uses of *D. brasiliensis* [99-101]

	Description
Properties/ Actions:	Antacid, Analgesic, Antifatigue, Anti-inflammatory, Aromatic, Bitter, Diaphoretic, Digestic, Diuretic, Emmemogogue, Febrifuge, Purgative, Tonic, Stimulant, Stomachic, Tonic
Countries	Ethnobotany uses
Argentina	Diaphoretic, Diuretic, Emetic, Emmenagogue, Fever, Tonic
Brazil	Anemia, Bite (snake), Constipation, Cystitis, Diuretic, Diaphoretic, Diarrhea, Dysentery, Emetic, EMMenagogue, Febrifuge, Gastritis, Purgative, Stimulant, Stomachic, Tonic, Ulterine
Elsewhere	Alexipharmic, Cough, Diaphoretic, Gastritis, Stimulant

Prenylated flavonoids isolated from *D. mannii* have antioxidant activity [102] against LDL oxidation. The antioxidant activities of three of these flavonoids, namely: 6,8-diprenyleriodictyol (**83**) dorsmanin C (**121**) and dorsmanin F/ epidorsmanin F (**86/87**), were compare to the common, non prenylated flavonoid, quercetin (5,7,3',4'-tetrahydroxyflavonol). All these prenylated flavonoids tested were found to be potent scavengers of the free radical DPPH (1,1-diphenyl-2-picrylhydrazyl), although only 6,8-diprenyleriodictyol (**83**) dorsmanin C (**121**) showed significant binding to copper ions. They are more potent scavengers than BHT (butylated hydroxytoluene), a common antioxidant used as a food additive. The prenylated flavonoids also inhibited Cu^{++} – mediated oxidation of the human low density lipoprotein (LDL). Dose response studies indicated that the prenylated flavonoids were effective inhibitors of lipoprotein oxidation with IC_{50} values < 1μM. They had similar inhibitory potency to quercetin and inhibitory activity was not directly related to either Cu binding. The prenylated flavonoids, unlike quercetin, did not show any pro-oxydant activity at high doses in Cu^{++}-mediated lipoprotein oxidation system. The medicinal action of *D. mannii* may be related to the high concentration of potent antioxidant prenylated flavonoids in this species. There has been a considerable amount of work examining the struture – antioxidant activity relationships for the flavonoids [103-105] but very little information is available on the effect of prenylation on flavonoid antioxidant activity. Ko et al. [106] examined the radical scavenging and antioxidant properties of several prenylflavones isolated from *Artocarpus heterophyllus*, a medicinal plant from South East Asia. Several of the compounds tested inhibited iron-induced lipid peroxidation and scavenged DPPH, particularly those with free hydroxyl groups on the A and B ring. No comparison was made in that study with non prenylated flavonoids, so that no conclusions could be made on the effect of prenylation on antioxidant activity. The effect of prenylation on lipid solubility may also make these compounds more readily absorbed or associated with lipoprotein particles in the blood. Conclusions on the actual absorption and bioavailability of these compounds must await detailed in vivo studies.

9- CONCLUSION

Interest in the studies in the genus *Dorstenia* has been increasing over the last decade with close to 45 papers published on over 25 species Table (**2**). These reports appear to be based on the studies conducted from plants originating in Brazil, Panama, Mexico, Cameroon, Ethiopia and Tanzania. It shows that the genus *Dorstenia*, like its sister genera, *Ficus*, *Atrocarpus* and *Morus*, is a rich source of diverse flavonoids (chalcones, flavanones, flavones and flavonols) with even greater structural diversity in the prenyl and geranyl C-constituents. It is interesting to note that all the flavonoids so far reported from this genus appear to be only from African *Dorstenia* species; but more data would have to be obtained to confirm, if indeed this is truly so and if this observation has any chemotaxonomic significance. In addition this genus also contains coumarin, benzofuran and styrene derivatives. *D. mannii* may be considered as a potential source of 6,8-prenyleriodictyol as multigram quantities of the substance can be obtained from ca 1kg of plant material. Several bioassay guided searches for bioactive substances from higher plants have led to the realization that flavonoids are responsible for many kinds of pharmacological effects. As a result, one is observing renewed interest in what otherwise would have been considered as well-known and well-studied class of metabolites. The diversity of structures presented in this report and the potential to discover many more novel structures will make this genus the subject of much research in the future.

Table 2. Chemical screening of some investigated *Dorstenia* species

Plant species	Benzofu rans	Couma rins	Chalco nes	Bichal cones	Flavan ones	Flavon es	Flavon ols
D.africana	+	+	-	-	-	-	-
D.asaroides	+	+	-	-	-	-	-
D.bahiensis	+	+	-	-	-	-	-
D.barteri	-	-	+	-	-	-	-
D.brasiliensis	+	+	-	-	-	-	-
D.bryoniifolia	+	+	-	-	-	-	-
D.cayapiaa	+	+	--	-	-	--	-
D.ciliata	-	-	+	-	-	+	+
D.contrajerva	+	+	-	-	-	-	+
D.dinklagei	-	-	+	-	+	+	-
D.drakena	+	+	-	-	-	-	-
D.elliptica	+	+	+	-	-	+	-
D.excentrica	+	+	-	-	-	-	-
D.gigas	+	+	-	-	-	-	-
D.heringeri	+	+	-	-	-	-	-
D.involuta	-	-	+	-	-	-	-
D.kameruniana	-	-	+	-	-	+	-
D.lindeniana	+	+	-	-	-	-	-
D.mannii	-	-	+	-	+	+	+
D.multiradiata	+	+	-	-	-	-	-
D.poinsettifolia	+	+	+	-	+	+	+
D.proropens	+	+	+	-	-	-	-
D.psilurus	+	+	+	-	-	+	+
D.zenkeri	-	-	+	+	-	-	-

+ : Presence of the class compound in the plant species
- : Absence of the class compound or it has not yet be identified.

ACKNOWLEDGEMENTS

The authors are grateful to the International Programs in Chemical Sciences (IPICS) for fostering intra-African cooperation under the auspices of the Network for Analytical and Bioassay Services in Africa (NABSA). B.M.A. acknowledges financial support from the University of Botswana administered by the Faculty Research and Publication Committee.

REFERENCES

[1]   Mabberley, D.J., *The plant book*, Cambridge University press, **1987**.

[2]   Thomas, D.W., Thomas, J.M., Bromely, W.A. and Mbenkum, F.T., *Korup Etnobotany Survey*, WWF (Surrey), A31, **1989**.

[3]   Casagrande, C.; Ronchetti, F. ; Russo, G. ; *Tetrahedron.*, **1974**, 30, 3587-3589.

[4]   Woldu, Y.; Abegaz, B.; Botta, B.; Delle Monache, G.; Delle Monache, F.; *Phytochemistry*, **1988**, 27, 1227-1228.

[5]   Abegaz, B.M.; Ngadjui, B.T.; Dongo, E.; Bezabih, M.-T.; *Current Organic Chemistry*, **2000**, 4, 1079-1090.

[6]   Kuster, R.M.; Bernardo, R.R.; Da Silva, J.R.; Parente, J.P.; Mors, W.B.; *Phytochemistry*, **1994**, 36, 221-223.

[7]   Terreaux, C.; Maillard, M.; Stoeckli-Evans, H.; Gupta, M.P.; Downum, K.R.; Quirkes, J.M.E.; Hostettmann, K.; *Phytochemistry*, **1995**, 39, 645-647.

[8]   Tovar-Miranda, R.; Cortés-Garcia, R. ; Santos-Sanchez, N.F. ; Joseph-Nathan, P. ; *J.Nat.Prod.*, **1998**, 61, 1216-1220.

[9]   Ngadjui, B.T. ; Kapche, G.W.F. ; Tamboue, H. ; Abegaz, B.M. ; Connolly, J.D. ; *Phytochemistry*, **1999**, 51, 119-123.

[10] Ngadjui, B.T. ; Dongo, E. ; Happi, E.N. ; Bezabih, M.-T. ; Abegaz, B.M. ; *Phytochemistry*, **1998**, 48, 733-737.

[11] Kanke, K. ; Porzel, A. ; Masaoud, M. ; Adam, G. ; Schmidt, J. ; *Phytochemistry*, **2001**, 56, 611-621.

[12] Lin, Y.-L. ; Kuo, Y.-H.; *Heterocycles*, **1992**, 34, 1555-1560.

[13] Llabres, G.; Baiwir, M.; Vilegas, W.; Pozetti, G.L.; Vilegas, J.H.Y.; *Spectrochim. Acta*, **1992**, 48, 1347-1353.

[14] Swain, L.A. ; Quirke, J.M.E., ; Winkle, S.A. ; Downum, K.R. ; *Phytochemistry*, **1991**, 30, 4196-4198.

[15] Satoh, M.; Satoh, Y.; Fujimoto, Y.; *Natural Medecines*, **2000**, 54, 97-100.

[16] Ohkoshi, E.; Makino, M.; Fujimoto, Y.; *Chemical and Pharmaceutical Bulletin*, **2000**, 48, 1774-1775.

[17] Uchiyama, T.; Hara, S.; Makino, M.; Fujimoto, Y.; *Phytochemistry*, **2002**, 60, 761-764.

[18] Quader, M.A.; El-Turbi, J.A.; Armstrong, J.A.; Gray, A.I.; Waterman, P. G.; *Phytochemistry*, **1992**, 31, 3083-3089.

[19]   Zubia, E.; Luis, F.R.; Massanet, G.M.; Collado, I.G.; *Tetrahedron*, **1992**, 48, 4239-4246.

[20]   Stanjek, V.; Miksch, M.; Boland, W.; *Tetrahedron*, **1997**, 52, 17699-17710.

[21]   Vilegas, W.; Pozetti, G.L.; *J. Nat. Prod.*, **1993**, 56, 416-417.

[22]   Lemmich, J.; Havelund, S.; Thastrup, O.; *Phytochemistry*, **1983**, 22, 553-555.

[23]   Srivastava, S.; Srivastava, S.; *Journal of Indian Chemical Sociaty*, **1993**, 70, 655-659

[23a]   Caceres, A.; Rastrelli, L.; De Simone, F.; De Martino, D.; Saturnino, C.; Saturnino, P.; Aquino, R. ; *Fitoterapia*, **2001**, 72, 376-381.

[24]   Rojas-Lima, S. ; Santillan, R.L.; Dominguez, M-A. ; Gutiérrez, A. ; *Phytochemistry*, **1999**, 50, 863-865.

[25]   Swain, L. A.; Downum, K.R.; *Biochemical Systematics and Ecology*, **1990**, 18, 153-156

[26]   Asai, F.; Iinuma, M.; Tanaka, T.; Mizuno, M.; *Phytochemistry*, **1991**, 30, 3091-3094.

[27]   Asai, F.; Linuma, M.; Tanaka, T. ; Mizuno, M. ; *Heterocycles*, **1992**, 33, 329-335.

[28]   Samaraweera, U. ; Sotheeswaran, S. ; Sultanbawa, M.U.S. ; *J.Chem.Soc., Perkin* 1, **1983**, 703-706

[29]   Miller, A.G.; Cope, T.A.; *Flora of the Arabian Peninsula and Socota*, 1. University Press, Edinburgh, **1996**.

[30]   Reinhard, E.; Corduan, G.; Brocke, W.; *Herba Hungarica* **1971**, 10, 9-25.

[31]   Hernàndez-Galàn, R.; Salvà, J.; Massanet, G.M.; Collado, J.G.; *Tetrahedron*, **1993**, 49, 1701-1710.

[32]   Bhide, K. S.; Mujumbar, R.B.; Rama Rao, A.V.; *Indian Journal of Chemistry*, **1977**, 15B, 440-444.

[33]   Igutchi, S.; Goromaru, T.; Noda, A.; *Chem. Pharm.Bull*, **1975**, 23, 934-936.

[34]   Murray, R.D.H.; Méndez, J.; Brown, S.A.; John Wiley & Sons Ltd, Bristol, **1982**.

[35]   Abegaz, B.M.; Ngadjui, B.T.; Dongo, E.; Ngameni, B.; Nindi, M.N.; Bezabih, M.-T.; *Phytochemistry*, **2002**, 59, 877-883.

[36]   Delle Monache, G.; De Mello, J.F.; Delle Monache, F.; Marini-Bettolo, G.B.; De Lima, G.D.; Coelho, J.S.D.B.; *Gazz. Chimica Ital.*, **1974**, 104, 861-865.

[37]   Kobayashi, M.; Noguchi, H.; Sankawa,U.; *Chem. Pharm. Bull.*, **1985**, 33, 3811-3814.

[38]   Bhakuni, D.S.; Chaturvedi, R.; *J. Nat Prod.*, **1984**, 47, 585-588.

[39]   Ngadjui, B.T.; Abegaz, B.M.; Dongo, E.; Tamboue, H.; Kouam F.; *Phytochemistry*, **1998**, 48, 349-354.

[40]   Singhal, A.K.; Barua, N.C.; Sharma, R.P.; Baruah, J.N.; *Phytochemistry*, **1983**, 22, 1095-1097.

[41]   Abegaz, B.M.; Ngadjui, B.T.; Dongo, E.; Tamboue, H.; *Phytochemistry*, **1998**, 49, 1147-1150.

[42]   Ngadjui, B.T.; Ngameni, B.; Kouam, S.F.; Abegaz, B.M.; *Bull. Chem. Soc. Ethiop.* **2002**, in press.

[43]   Ngadjui, B.T.; Dongo, E.; Abegaz, B.M.; Fotso, S.; Tamboue, H.; *Phytochemistry*, **2002**, 61, 99-104.

[44]   Bhatt, P.; Dayal, R.; *Phytochemistry*, **1992**, 31, 719-723.

[45]   Hano,Y.; Itoh, N.; Hanoka, A.; Nomura, T.; *Heterocycles*, **1995**, 41, 2313-2326.

[46]   Ayafor, J.F.; Tsopmo, A.; Tene, M.; Kamnaing, P.; Ngnokam, D.; Sterner, O.; Iwu, M.; *Proceeding of the 7th NAPRECA Symposium on Natural Products Dar es Salaam, Tanzania, August 17-22*, **1997**, 99-106.

[47]   Komatsu, M.; Tomimori, T.; Hatayama, K.; Makiguchi,Y.; Mikuriya, N.; *Chem. Pharm. Bull.*, **1969**, 17, 1299-1305.

[48]   Tsopmo, A.; Tene, M.; Kamnaing, P.; Ngnokam, D.; Ayafor, J.F.; Sterner, O.; *Phytochemistry*, **1998**, 48, 345-348.

[49]   Dagne, E.; Bekele, A.; Noguchi, H.; Shibuya, M.; Sankawa, U.; *Phytochemistry*, **1990**, 29, 2671-2673.

[50]   Barron, D.; Ibrahim, R.K.; *Phytochemistry*, **1996**, 43, 921-982.

[51]   Hano, Y.; Suzuki, S.; Nomura, T.; Ueda, S.; *Heterocycles*, **1989**, 807-813.

[52]   Tsopmo, A.; Tene, M.; Kamnaing, P.; Ayafor, J.F.; Sterner, O.; *J. Nat. Prod.*, **1999**, 62, 1432-1434.

[53]   Messana, I.; Ferrari, F. ; De Mello, J.F. ; De Araujo, M.C.M. ; *Heterocycles*, **1989**, 29, 683-690.

[54]   Ikuta (née Matsumoto), J. ; Fukai, T. ; Nomura, T. ; Ueda, S. ; *Chem. Pharm. Bull.* **1986**, 34, 2471-2478.

[55]   Kobayashi, M.; Noguchu, H.; Sankawa, U.; *Chem. Pharm. Bull.*, **1985**, 33, 3811-3816.

[56]   Hano, Y. ; Itoh, M. ; Koyama, N. ; Nomura, T. ; *Heterocycles*, **1984**, 22, 1791-1800.

[57]   Hano, Y. ; Aida, M. ; Nomura, T. ; *J. Nat. Prod.*, **1990**, 53, 391-395.

[58]   Nomura, T. ; Fukai, T. ; *Heterocycles*, **1981**, 15, 1531-1567.

[59]   Nomura, T. ; Fukai, T. ; Matsumoto, J. ; Imashimizu, A. ; Tereda, S. ; Hama, M. : *Planta Med.*, **1982**, 46, 167- 174.

[60]   Nomura, T. ; Hano, Y. ; Aida, M. ; *Heterocycles*, **1999**, 47, 1179-1185.

[61]   Magalhaes, A.F. ; Tozzi, A.M.G.A. ; Saes, B.H.L.N. ; Magalhaes, E.G. : *Phytochemistry*, **1996**, 42, 1459-1462.

[62]   Ngadjui, B.T. ; Dongo, E. ; Tamboue, H. ; Kouam, F. ; Abegaz, B.M. : *Phytochemistry*, **1999**, 50, 1401-1406.

[63]   Ngadjui, B.T. ; Kouam, S.F. ; Dongo, E. ; Kapche, G.W.F. ; Abegaz, B.M. : *Phytochemistry*, **2000**, 55, 915-919.

[64]   Mizuno, M. ; Tamura, K.I. ; Tanaka, T. ; Iinuma, M. ; *Phytochemistry*, **1989**, 28, 2811-2812.

[65]   Kamat, V.S. ; Chuo, F.Y. ; Kubo, I. ; Nakanishi, K. ; *Heterocycles* , **1981**, 15, 1163-1170.

[66]   Harborne, J.F. ; Greenham, J. ; Williams, C.A. ; Eagles, J. ; Markham, K.R. ; *Phytochemistry*, **1993**, 34, 219-224.

[67]   Roussis, V. ; Ampofo, S.A. ; Wiemer, D.F. ; *Phytochemistry*, **1987**, 26, 2371-2376.

[68]   Tahara, S. ; Katagiri, Y. ; Ingham, J.L. ; Mizutani, J. : *Phytochemistry*, **1994**, 36, 1261-1266.

[69]   Tahara, S. Ingham, J.L. ; Hanawa, F. ; Mizutani, J. ; *Phytochemistry*, **1991**, 30, 1683-1687.

[70]   Nakahara, S. ; Tahara, S. ; Mizutani, J. ; Ingham, J.L. ; *Agric. Biol. Chem.*, **1986**, 50, 863-869.

804

[71]   Smaberger, T.M. ; Vleggaar, R. ; Weber, J.C. ; *Tetrahedrons*, **1974**, 30, 3927-3931.

[72]   Lin, Y.L., Chen, Y.L. ; Kuo, Y.H. ; *Chem. Pharm. Bull.* **1991**, 39, 3132-3135.

[73]   Ngadjui, B.T. ; Tabopda, T. ; Dongo, E. ; Kapche, G.W.F. ; Sandor, P. ; Abegaz, B.M. ; *Phytochemistry*, **1999**, 52, 731-735.

[74]   Delle Monache, G. ; Scurria, R. ; Vitalis, A. ; Botta, B. ; Monacelli, B. ; Pasqua, G. ; Palocci, C. ; Cernia, E. ; *Phytochemistry*, **1994**, 37, 893-895.

[75]   Agrawal, P.K., carbon-13 NMR of flavonoids, Elsevier, Amsterdam, **1989**.

[76]   Drewes, S. I Budzikiewicz ed. Progress in Mass Spectrometry Verlag Chemie **1973** vol 2. pp. 3-18.

[77]   Li, Y.C.; Kuo, Y.-H.; *J. Na. Prod.*, **1997**, 60, 292-293.

[78]   Pistelli, L.; Beroli, A.; Giachi, I.; Manunta, A.; *J. Nat.Prod*, **1998**, 61, 1404-1406.

[79]   Iinuma, M.; Kanie, Y.; Tanaka, T.; Mizuno, M.; Lang, F.A.; *Heterocycles*, **1993**, 35, 407-413.

[80]   Crombie, L.; Crombie, W. M. L. ; Jamieson, S. V. ; *Tetra. Lett.* **1980**, 21, 3607-3609.

[81]   Crombie, L.; Crombie, W. ; *J. Chem.Soc.,Perkin Trans*, **1982**, 1455-1458.

[82]   Barrett, M.L., Scutt, A.M., Evans, F.J., *Experintia*, **1986**, 42, 452-

[83]   Kijjoa, A.; Cidade, H.M.; Pinto, M. M. M.; Gonzalez, M. J. T. G.; Anantachoke, C.; Gedris, T.; Herz, W.; *Phytochemistry*, **1996**, 43, 691-697.

[84]   Matsumoto, J.; Fujimoto, T.; Takino, C.; Saitoh, M.; Hano, Y.;   Fukai T.; Nomura, T.; *Chem. Pharm. Bull.*, **1985**, 33, 3250-3254.

[85]   Lin, M.; Li, J. B.; Li, S. Z.; Yu, D. Q.; Liang, X. T.; *Phytochemistry*, **1992**, 31, 633-636.

[86]   Vilegaz, J. H. Y.; lancas, F. N.; Vilegaz, W.; Pozetti, G. L.; *Phytochemistry*, **1993**, 5, 230- 232.

[87]   Galbraith, M. N.; Miller, C. J.; Ramson, J. W. L.; Ritchie, E.; Shannon, J. S., Taylor, W. C.; *Aust. J. Chem.*, **1965**, 18, 226-229.

[88]   Bruno, M. ; Savona, G. ; Gadea, F. F. ; Rodriguez, B. ; *Phytochemistry*, **1986**, 25, 475-477.

[89]   Cardoso, C.A.L.; Vilegas, W.; Honda, N.K.; *Chromatographia*, **1999**, 50, 11-18.

[90]   Vilegas, J.H.Y.; Lancas, F.M.; Vilegas, W.; Pozetti, G.L.; *Journal of the Brazilian Chemical Sociaty*, **1997**, 8, 529- 534.

[91]   Swain, L. A.; Downum, K.R.; *Biochemical Systematics and Ecology*, **1990**, 18, 153-156.

[92]   Harborne, J.B.; Baxter, H. *The Handbook of Natural Flavonoids*, John Wiley & Sons, Chichester, **1999**.

[93]   Lin, Y.-M.; Derson, J.; Flavin, M.T.; Pai, Y.-H.S.; Mata-Greewood, E.; Penguspart, T.; Pezzuto, J.M.; Schinazi, R.F.; Huges, S.H.; Chen, F.-C.; *J. Nat. Prod.*, **1997**, 60, 884-889.

[94]   Lale, A. ; Herbert, J.M. ; Augereau, J.M. ; Billon, M. ; Leconte, M. ; Gleye, J. ; *J. Nat. Prod.*, **1996**, 59, 273-275.

[95]   Ingkaninan, K.; Ijzerman, A.P.; Verpoorte, R.; *J. Nat. Prod.*, **2000**, 63, 315-318.

[96]   Duke, J.A.; *Handbook of Biologically Active Phytochemicals and their Activities*, CRC Press, Boca Raton, London, **1992**.

[97]   Iwu, M.M.; Jackson, J.E.; Tally, J.D.; Klayman, D.L.; *Planta Medica*, **1992**, 58, 436-441.

[98]   Ruppelt, B.M.; Pereira, E.F.R.; Goncalves, L.C.; Pereira, N.A.; *Memoria do Instituto Oswaldo Cruz*, **1991**, 86, 203-207.

[99]   Mahidol, C.; Prawat, H.; Ruchirawat, S.; Lihkitwitayawuid, K.; Lin, L.; Ze, Cordell, G.A.; *Phytochemistry*, **1997**, 45, 825-827.

[100]   Seo, E.K.; Silva, G.L.; Chai, H.B.; Chagwedera, T.E.; Farnsworth, N.R.; Cordell, G.A.; Pezzuto, J.M.; Kinghorn, A.D.; *Phytochymistry*, **1997**, 45, 509-512.

[101]   Ruppelt, B.M.M.; Pereira, E.F.; Goncalves, L.,C.; Pereira, N.A.; *Mem Inst. Cruz, Rio de janiero*, **1991**, 86:Supp II, 203-205.

[102]   DufallK.G. ; Ngadjui, B.T. ; Abegaz, B.M. ; Croft, K.D. ; Kouam,F.S. ; (in preparation)

[103]   Rice-Evans, C.A. ; Millar, N.J. ; *Free Radic. Biol.*, **1996**, 20, 933-956.

[104]   Cao,G.; Sofic,E.; Prior, R.L; *Free Radic. Biol.*, **1997**, 22, 749-760.

[105]   Chen,Z.Y.; Chan, P.T., Ho, K.Y.; Fung, K.P.; Wang, J.; *Chem. Phys. Lipids*, **1996**, 76, 157-163.

[106]   Ko, F.N.; Cheng, Z.J., Lin, C.N.; Teng, C.M.; *Free Radic. Biol.Med*, **1998**, 25, 160-168.